# Vegetable Genetic Resources
## Principles and Management

## The Author

**Dr D R Bhardwaj** obtained post-graduate and Ph. D. degrees (Horticulture with specialization in Vegetable Science) from Institute of Agricultural Sciences, B.H.U. Varanasi (U.P.). Dr. Bhardwaj has an experience of 27 years in vegetable crops improvement (research and development) under Indian Council of Agricultural Research, New Delhi. Presently he is working as a Principal Scientist (Vegetable Science) after joining of scientist (Vegetable Science) through ARS in 1991. He focused on augmentation of genetic resources and genetic improvement of vegetable crops since 1992 and continuing till date. He has identified 7 novel/unique genetic stocks which were reregistered/patented with ICAR-NBPGR, New Delhi. He is the first person who developed and reported gynoecism in bitter gourd, seedless pointed gourd (parthenocarpic line), andromonoecism in sponge gourd, cluster bearing sponge gourd, and cms system in tropical carrots; which are being utilized in development of varieties/hybrids in the country. He has developed 9 varieties/hybrids in different vegeWtable crops which are known for high yield and quality attributes and all are very popular among the vegetable growers. He has published more than 95 research papers in reputed national and international journal, 31 papers in proceedings, 32 popular articles, 125 abstracts and 36 book chapters on vegetable crops. He is in credit of 3 books, 3 technical bulletins, 1 information bulletin (on Biodiversity of vegetable crops), 5 training manuals and 6 extension leaflets. He is recipient of Dr. Rjendra Prasad Award in 2013 and 2014 conferred by ICAR, New Delhi. He is awardee of Dr. B. R. Ambedkar Young Fellow Award (2009); ISVS Fellow award (2013); ISPGR Award (2013); Fellow Award of Horticultural Sciences (2014) and Young Scientist Associate Award (2014).

# Vegetable Genetic Resources
## Principles and Management

*Author*

## D R Bhardwaj

*Principal Scientist (Vegetable Science)*
*Division of Crop Improvement*
*ICAR-IIVR, Varanasi - 221 305*
*(Uttar Pradesh)*

**2018**

# Daya Publishing House®
*A Division of*

# Astral International Pvt. Ltd.
New Delhi – 110 002

*Published by*          :   **Daya Publishing House®**
                            A Division of
                            **Astral International Pvt. Ltd.**
                            – ISO 9001:2008 Certified Company –
                            4760-61/23, Ansari Road, Darya Ganj
                            New Delhi-110 002
                            Ph. 011-43549197, 23278134
                            E-mail: info@astralint.com
                            Website: www.astralint.com

*Digitally Printed at*   :   **Replika Press Pvt. Ltd.**

# Preface

Nearly 42.0 per cent of the world population depends on agriculture directly and indirectly, which is more so in the developing countries. In India, 17 per cent human and 11 per cent animal populations of world are depending on agriculture. To meet the demand of increasing population, it is essential to produce more diversified food commodities for food and nutritional securities. The food security of people does not depend merely on what is cultivated but it needs realization that what is available in a particular situation. Three major physical resources in the world comprise land, water and the biological diversity principally governed by various environmental conditions for existence. Environmental conduciveness and stresses have a profound effect on the distribution of natural vegetation and evolution. Plant biodiversity in general and agro-biodiversity in particular not only maintains the diverse ecosystem on the planet but also acts as vital links of all the food chains. Biological resources of the earth are vital to humanity too because they depends upon them for clean environment, food security, health care, social needs, sources of livelihood, trade, industrial growth and economic development. Due to the ever-increasing human and domestic animal populations and demand for food, feed and fiber, introduction, collection, conservation and sustainable use of the existing biodiversity is one of the most serious challenges of the mankind. Since the beginning of agriculture, farmers have attempted to identify and use the available variants to their need and advantage. The genetic diversity of crop plants and their relatives are the raw materials in development of improved cultivars. Over the past 20 years, a series of technical innovations have opened opportunities to improve the efficiency and effectiveness of plant breeding programme. Commercial seed companies in developed countries and in some developing countries have adopted these innovations as trade articles. In many regions of the world, potential arable land is not being cultivated because of insufficient water, temperature extremes, excessive salinity or physically unfavorable soil conditions, which is proving a major challenge for survival of living organisms. Despite our reliance on plants, flash point seems to have reached as between 60000 and 100000 plant species are reportedly threatened worldwide. In this context it essential to focus on sustainable use for development of new crop varieties, benefit-sharing and capacity building assume greater importance. Plant genetic resources (PGR) are the basic and essential materials for development of improved crop varieties/hybrids designed to combine high yield potential with superior quality, resistance to diseases and pests, and also better adaptation to abiotic stress environments. Today, the sustainable growth in agricultural productivity is becoming a matter of prime importance in most of the countries. There is an urgent need for strategic action plant and innovative frame work for vegetable germplasm enhancement actions at global, national, regional, and local levels. Good management and ecological sustainable use of vegetable genetic resources is the key to development and prosperity of any nation. India has rich vegetable genetic resources with important implication for research and conservation. The strategies has to be supported by a wide range of organizations and institutions namely conservation and research organizations such as protected area management boards, botanic gardens, gene banks, universities, research institutes, network of NGOs and the private sectors. India, being the signatory to several international agreements and treaties related to PGR including CBD, TRIPS, Cartagena Protocol on Biosafety, ITPGRFA and Nagoya Protocol, is prepared to fulfill all the international commitments. This book entitled "**Vegetable Genetic Resources: Principles and Management**" will prove to be a critical input in strengthening Plant Genetic Resources (PGR) including the areas of farmers' rights, benefit sharing, harmonization of provisions of different conventions and implementation of the convention on biological diversity at regional level. Apart from these, germplasm introduction, exploration, collection, evaluation, trait specific identification, conservation mechanisms *etc.* are well highlighted. The document may prove handy material for students, researchers, the farmers and organizations who have interest in conservation and utilization of vegetable genetic diversity. The ultimate goal is that India should become one of the leaders in quality vegetable genetic resource holdings that form the basis of future vegetable production and base industries.

*D.R. Bhardwaj*

# Contents

## 9. Source Availability and Utilization of Vegetable Genetic Resources     293

## 10. Genetic Erosion and Conservation of Vegetable Genetic Resources     345

## 11. Biotechnological Approaches in Vegetable Genetic Resource Management and Utilization     399

*of Solanaceous Crops – In vitro Culture and Plant Regeneration of Cucurbitaceous Crops – In vitro Regeneration of Okra – In vitro Culture and Plant Regeneration of Cole Crops – In vitro Culture and Plant Regeneration of Bulb Crops – In vitro Culture and Plant Regeneration of Rhizomatous Crops – Photo Autotrophic Propagation (PAP) – In vitro Culture and Plant Regeneration of Bamboo – Elimination of Viruses – In vitro Culture of Bamboos – In vitro Culture and Plant Regeneration of Root and Tuber Crops – In vitro Culture and Plant Regeneration of Cassava – In vitro Culture and Plant Regeneration of Sweet Potato – Yams – In vitro Culture and Plant Regeneration of Aroids – In vitro Culture and Plant Regeneration of Sesbania grandiflora – In vitro Culture and Plant Regeneration of Aquatic Vegetables – Guidelines for Germplasm Acquisition – Haploid Development – References*

*Protection to a New Plant Variety – Procedure of Registration of Plant Varieties – Grouping and Categorization of Proposals – Who Can Apply for Variety Protection? – Form of Application – Acceptance and Advertisement of Registration Application – Opposition to the Acceptance – Duration, Terms and Effect of Registration – Compulsory Licensing – Exceptions – Fee Structure – Issuance of Registration Certificate – Maintenance of Registered Varieties – DUS in Vegetable Crops – Statement of Distinctness of Candidate Variety – National Gene Fund (NGF) – Seed Bill 2004 – Objective of the Seed Bill 2004 – Implementable Aspects – Salient Features of Seed Bill Act, 2004 – References*

# 1

# Vegetable Genetic Resources: Centre of Origin, Diversity and Distribution

## Introduction

Human beings may have existed as species for about 2 million years yet it was only about 10, 000 years ago that they started growing food, rather than merely gathering it from wild state and hunting wildlife.The resource base in agricultural production system includes land, air, water, enery and biogenetic diversity. With the origin of settled agriculture, the interest in environmental assets began. Wherever man has gone, his plants have gone with him because of its utility. Land and water are the most important factors in agricultural production and biodiversity perpetuation. When the transition from 'gathering' to 'growing' food took place, there was immediate interest in the collection and utilization of useful plants and animals from native flora and fauna. Most of the important economic plants and animals were domesticated thousand of years ago.

## Availability of Nutritious Vegetable Genetic Resources in the Indian Gene Centre and World

Under vegetables several crops are widely grown all over the country throughout the year for their edible, leaves, stems, flower buds and open flowers, fruits, seeds, roots *etc*. Availability of nutritious vegetable genetic resources in the Indian gene centre and world are given in Table 1.1.

**Table 1.1: Vegetable Crops and Availability of Genetic Resources in Indian and World Gene Centre**

| Crop | Availability of Genetic Resources |
|---|---|
| *Solanum* species (Tomato) | Tomato is a warm season crop, sensitive to frost and typically cultivated in sub-tropical and cold climate. It is a native of South America and Mexico, however, domestication took place in Mexico (Baswell, 1949) and now tomato is grown all over the world.The greatest variability occurs in Mexico, Central America and Coastal Peru.The wild species of *Solanum* (*Lycopersicon*) have shown to be a valuable source of resistance to various disease and pest, which is vital requirement in any breeding programme. *S. pimpinellifolium* (*Lycopersicon pimpinellifolium*) is resistance to *Fusarium* wilt, early blight and leaf curl disease and is a rich source of ascorbic acid; *S. cheesmanii* (*L. cheesmanii*) is resistance to salinity and a source of high total soluble solids (TSS); *S. chilense* (*L. chilense*) is tolerant to leaf curl disease; *S.chemielewskii* (*L. chemielewskii*) has high content of TSS; *S. habrochaites* (*L. hirsutum*) is an excellent source of resistance to several pests and to leaf curl disease; *S. arcanum* (*L. peruvianum*) is resistant to root knot nematode, tobacco mosaic virus (TMV), leaf curl and has high ascorbic acid content and *L. chilense* is good source of drought resistance. |
| *Solanum* species (Brinjal) | Brinjal or eggplant being the crop of Indian origin has developed some secondary variability in China (Vavilov, 1926). A large number of land races/traditional grown cultivars have been developed in different agro-ecological zones of India and these land races possess valuable genes for resistance to biotic and abiotic stresses and adaptation to various environments. Maximum diversity and distribution of *Solanum* species was noted in southern India, foothills of Himalaya and North Eastern Region (NEH).The widely distributed species in the region includes *S. torvum, S. indicum, S. insanum, S. surattense, S. pubescens* and *S. khasianum*. |

| Crop | Availability of Genetic Resources |
|---|---|
| *Capsicum* species (Chillies and Bell pepper) | *Capsicum* species (Chillies or Pungent pepper and Bell pepper) or *Simla mirch*; *Capsicum annuum* L. are usually grown in warm to hot and humid climate throughout the country. The large fruited bell pepper originated from tiny pungent and pointed fruited wild species *C. annuum*. It is considered to be the cool season crop and grown in sub-tropical and temperate regions. There are four other species viz. *C. frutescens*, *C. chinense*, *C. baccatum* and *C. pubescens*, which are commonly grown in southern and Central America except *C. frutescens*, which is also grown in some parts of USA. Due to continuous selection, long history of cultivation and popularity of the crop, sufficient genetic variability has been generated and now India is treated as secondary centre of diversity for chillies. In hot chillies (*Capsicum annuum* L.), rich variability in plant and fruit morphological traits occurs throughout the country particularly in south peninsular region, north eastern region, in foothills of Himalaya and Gangetic plains. The variability includes in plant type, fruit size/shape (long, short, pointed, smooth, wrinkled), bearing habit (fruit facing upwards, horizontal, downwards) and pungency in addition to annual and perennial types. However, in bell pepper less variability has been observed. Mostly introduced lines are grown particularly in mid hill of Himalaya and parts of hilly regions of Central India including Karnataka and Maharashtra. |
| *Cucurbita* species | The genus *Cucurbita* comprised of about 27 species, both wild and cultivated, mostly concentrated in the tropical and sub-tropical regions of Central and South America. In this genus, few species are of commercial importance viz. field pumpkin (*C. moschata* Duch. Ex. Poir.), Winter squash (*C. maxima* Duch.), Summer squash or vegetable marrow (*C. pepo* L.), Winter squash pumpkin (*C. mixta* Pang. Syn. *C. angyrosperma*) and Malabar gourd or fig leaf gourd (*C. ficifolia* Boucha). The vegetable marrow (*C. pepo* L.) is introduced crop which withstands in cooler climate. *C. texa* Gray grows wild in Texas (Chadha and Lal, 1993). |
| *Cucumis* species | The genus Cucumis comprises of about 26 species. The major crops of economic importance are cucumber (*C. sativus*), muskmelon (*C. melo*), snapmelon (*C. melo* var. *momordica*), and longmelon (*C. melo* var. *utilissimus*). The Indian sub-continent is considered to be the centre of origin for *C. sativus* and centre of diversity for *C. melo* (Zeven and DeWet, 1982). The wild species, *C. hardwickii* is found growing in natural habitats in the foothills of Himalayas. The free hybridization with cultivated *C. sativus*, with no reduction in fertility in $F_2$ generation suggested that *C. hardwickii* is likely progenitor of cultivated cucumber. The cucumber, *C. sativus* var. *sativus* has 3-5 lobed leaves, ovary usually 3 placentiferous with fruits oblong, obscurely trigonus or cylindric. However, *C. sativus* var. *sikkimensis* has 7-9 lobed leaves, ovary often 5 placentiferous, fruits ovoid-oblong, adapted in temperature humid climate. Muskmelon (*C. melo*) based on the distribution of diversity can be grouped into following six sub-sets (Munger and Robinson, 1991). |
| -*Cucumis melo* var. *agrestis* (Kachri) | It is wild type with slender vines and small inedible fruits, probably synonym of *Cucumis melo* var. *callosus* (*Cucumis callosus*). It is easily available in Indian subcontinent. |
| -*Cucumis melo* var. *cantaloupensis* Naud. (Muskmelon or cantaloupe) | Medium sized fruits with netted, warty or scaly surface, flesh usually orange but sometimes green, aromatic or musky flavor. Fruits dehiscent at maturity, usually andromonoecious. It is easily available in Indian subcontinent. |
| -*Cucumis melo* var. *flexuosus* Naud. (*Cucumis melo* var. *utilissimum* Duthie and Fuller) | Snakemelon, snake cucumber, Tar-Kakari, fruits long and slender consumed at immature stage, monoecious. It is easily available in Indian subcontinent. |
| -*Cucumis melo* var. *momordica* | Snapmelon or 'Phut' is a monoecious crop grown in India and Asian countries. It has white to pale orange, less sweet pulp. The smooth surface of the fruit starts cracking at the time of maturity. It is easily available in Indian subcontinent. |
| *Cucumis melo* var. *conomon* (Sweet or pickling melon) | It is generally andromonoecious in nature and bears small fruits with green skin, white flesh. |
| -*Cucumis melo* var. *inodorous* Naud. (winter melon) | The fruit is smooth or wrinkled surface with white or green flesh, and lacking musky odor. It is also andromonoecious in nature and usually requires more time for maturity. |
| *Luffa* species | The Indian gene centre has diversity in genetic resources of *Luffa* species. The genus comprised of 9 species in the world, of which 7 species (*L. acutangula*, *L. cylindrica*, *L. echinata*, *L. graveolens*, *L. hermaphrodita*, *L. tuerosa* and *L. umbellata*) is native to India. There is ambiguity with regards to *L. tuberosa* and *L. umbellata* because they are considered synonym to species of *Momordica* and *Cucurbita*, respectively (Chadha and Lal, 1993). The sponge gourd (*L. cylindrica*) and ridge gourd (*L. acutangula*) has rich diversity in vine and fruit morphological characteristics throughout India particularly in North Eastern Region (NEH) including Sikkim, West Bengal, Western, Central and Southern India. *L. hermaphrodita* considered to be originated from *L. graveolens* is another potential species distributed in some parts of North-Central India. *L. acutangula* var. *amara* grows in peninsular India is a wild relative of *L. cylindrica* and *L. echinata* in natural habitats is in Western Himalayas, Central India and Gangetic plains. |
| *Lagenaria* species | Under genus *Lagenaria*, 6 species are reported, of which *L. abyssinica*, *L. siceraria*, and *L. leucantha* are common. Among them *L. siceraria* is generally cultivated in all tropical parts of the world, especially in India and few African countries. Remaining species are wild, perennial and dioecious in nature. Two wild species, *L. abyssinica* and *L. bravifolia* are perennial in nature. There are suggestions that *Lagenaria* occurs in wild form in South America and in India (De Candolle, 1982). |
| *Citrullus* species | The genus *Citrullus* has two species of economic importance i.e., *C. lanatus* (Thunb.)-called watermelon and *C. vulgaris* Schrad var. *fistulosus*-called roundmelon or round gourd. Based on the evidence (De Candolle, 1982) and linguistic data (Filov and Vilenskaya, 1972), *C. lanatus* is considered to be originated in Africa and reached India in pre-historic times. Cultivation of large watermelon *C. lanatus* var. *citroides*, is comparatively recent and Soviet varieties grown today still have shape of their African ancestors (*C. lanatus* var. *caffer*). Shimotsuma (1963) reported that *C. vulgaris*, *C. colocynthis*, *C. ecirrhosus* and *C. naudinianus* are related and cross compatible with each other. It is considered that *C. colocynthoides* is a probable ancestor of watermelon. |

| Crop | Availability of Genetic Resources |
|---|---|
| *Trichosanthes* species | *Trichosanthes* has 22 species and is reported to be of Indian origin particularly Indo-Malayan region. Among these, *T. anguina* L. (Snake gourd) and *T. dioica* Roxb. (Pointed gourd) are cultivated throughout the country. The major zone for distribution of diversity for *T. dioica* in north-central and north-eastern India including West Bengal whereas for *T. anguina* is distributed throughout the country. However, the rich diversity in *T. anguina* has been observed in north-eastern states, West Bengal, Malabar Coast and in Eastern Ghats in low and mid hills. |
| *Momordica* species | The genus *Momordica* has 60 species of which 7 species have been reported to occur in India. The cultivated species is *M. charantia* L. (bitter gourd) is grown all over the country in tropical and subtropical climates. *M. dioica* (Spine gourd or Kartoli) grows all over West Bengal, Asom, parts of Bihar and adjoining area; *M. cochinchinensis* (Sweet gourd or Kakrol) is most popular particularly in Tripura, Asom and West Bengal. Occurrence of other species like *M. cymbalaria* is confined mainly in Western Ghats, Maharashtra, southwards and eastern peninsular tract, *M. denudata* in peninsular tract, *M. macrophylla* in northeastern region and *M. subangulata* in North Eastern Hills (NEH), Eastern Ghats and Deccan peninsula. |
| *Benincasa* species | *Benincasa hispida* Cogn. known as ash gourd, white gourd or wax gourd is reported to be a native of Java and Japan. It was domesticated in India during pre-historic time. It is widely grown all over the country in tropical and sub-tropical climates and possesses variability in fruit morphological characteristics and quality. |
| *Coccinia* species | *Coccinia* has 35 species distributed in tropical Africa and Asia. Out of these, only one species i.e. *C. grandis* is under cultivation. The related species are *C. histella* and *C. sessilifolia*, which are widely cultivated in several countries like Africa, Central America, China, Malaysia, Australia and other tropical Asian countries (Pier, 2001). |
| *Sechium* species | Genus *Sechium* is said to be originated in mountainous region of America and Mexico. It includes *Microsechium compositum, Microsechium gintonii, S. compositum, S. edule, S. hintoni* and *S. jamaicense*. Principally it was confined to tropical and sub-tropical climates and particularly in mid hill conditions. Maximum variability occurs in Sikkim. In Meghalaya and Mizoram it is grown on commercial scale and is popular in Darjeelimg Hills, also. |
| *Beta* species | *Beta* species includes *Beta adanensis, B. altissima, B. atriplicifolia, B. bengalensis, B. bourgaei, B. brasiliensis, B. campanulata, B. chilensis, B. cicla, B. corolliflora, B. lomatogona, B. macrocarpa, B. macrorhiza, B. maritime, B. maritima* subsp. *danica, B. maritima* var. *atriplicifolia, B. maritima* var. *erecta, B. maritima* var. *prostrata, B. nana, B. orientalis, B. palonga, B.a patellaris, B. patula, B. perennis, B. procumbens, B. trigyna, B. trojana, B. vulgaris, B. vulgaris* cv. *conditiva, B. vulgaris* cv. *saccharifera, B. vulgaris* f. *rhodopleura, B. hybrida, B. vulgaris* sub sp. *adanensis, B. vulgaris* subsp. *cicla, B. vulgaris* subsp. *flavescens, B. vulgaris* subsp. *lomatogonoides, B. vulgaris* var. *glabra, B. vulgaris* var. *cicla, B. vulgaris* var. *crassa, B. vulgaris* var. *atriplicifolia, B. vulgaris* subsp. *macrocarpa, B. vulgaris* var. *altissima, B. vulgaris* subsp. *vulgaris, B. vulgaris* subsp. *provulgaris, B. vulgaris* subsp. *patula, B. vulgaris* subsp. *maritime, B. vulgaris* subsp. *orientalis, B. vulgaris* var. *erecta, B. vulgaris* var. *flavescens, B. vulgaris* var. *foliosa, B. vulgaris* var. *grisea, B. vulgaris* var. *macrocarpa, B. vulgaris* var. *maritime, B. vulgaris* var. *orientalis, B. vulgaris* var. *perennis B. vulgaris* var. *pilosa, B. vulgaris* var. *prostrata, B. webbiana, B. intermedia* and *Beta vulgaris* var. *trojana*. The spinach beet is also known as Indian Spinach beet (*B. vulgaris* var. *bengalensis* Hort.) and considered as a close relative of beet root. It is grown in the plains and hills of India from Punjab to North Eastern Region (NEH) including West Bengal, Odisha, Madhya Pradesh, Rajasthan, Gujarat and Maharashtra. |
| *Amaranthus* species | The origin of various species of cultivated amaranth is complicated because the wild ancestors are pan tropical cosmopolitan weeds (Mohideen and Irulappan, 1993). *Amaranthus spinosus, A. hybridus* and *A. dubius* are tropical types. *Amaranthus retroflexus, A. viridis, A. lividus* and *A. graecizans* are more temperate hot season weeds. In all probably; it might have originated in India and have spread to neighboring countries by traders, buddhist monks, and others. In India, a number of domesticated forms are available all over the country especially in northeastern states, West Bengal, Odisha, Tamil Nadu, Andhra Pradesh, Karnataka and Kerala. Distribution pattern of *A. blitum* (weed throughout India), *A. caudatus* (wild throughout India), *A. gangeticus* (warmer parts of India), and *A. spinosus* (weed throughout India) are reported. Wide ranges of variability in plant morphological characteristics are available in different parts of the country. In Odisha and northeastern states a special type is grown where stem is consumed after full growth and different genotypes are grown in different parts of the year. |
| *Spinacia* species | Genus *Spinacia* belong to the family Chenopodiaceae and said to originate in southwest Asia. There is well description about four types of plants with reference to sex expression, which are: |

1. Extreme males-Plants are small and fast emergence of flower takes place (Roja, 1925).

2. Vegetative males- Plants are comparatively taller.

3. Monoecious-In this group basically male, female or in early stage completely female and some time latter male flower or male and female in equal quantity emerged.

4. Female- Plants are comparatively taller and remain longer time in vegetative stage. Late emergence of flower takes place but all the flowers appeared at the same time.

The monoecious plants may predominantly staminate, pistillate or pur pistillate early but with some staminate flower late or almost equally staminate in the plains and in foothills. Variability occurs in plant morphological characteristics, seed type and foliage yield attributing component.

| | |
|---|---|
| *Chenopodium* species | Genus *Chenopodium* belong to the family Chenopodiaceae. *Chenopodium album* is found as weed and *C. murale* in hills upto 3000 m. In both the species maximum variabiliorty can be observed in Upper Gangetic plains, extending to northern hills and elsewhere only sporadic distribution. |
| *Basella* species | Climbing spinach, Malabar spinach or 'Poi' is grown all over the country particularly in the plains and mid-hills during hot humis season. The genus *Basella* includes *B. alba, B. cordifolia, B. rubra, B. tuberosa* and *B. vesicaria*. Rich variability is found in plant morphological characteristics including stem fleshiness, leaf shape and size in South India, northeastern region including Asom, West Bengal and Odisha. |

| Crop | Availability of Genetic Resources |
|---|---|
| *Trigonella* species | The genus *Trigonella* is having 22 species, which are having different chromosome numbers. The 15 species having 2n=16 chromosomes includes *T. anguina, T. arabica, T. bolansae, T. calliceras, T. coerulea, T. corniculata, T. cretica, T. foenum-graecum, T. gladiata, T. glomerate, T. gordej, T. hamosa, T. lypokyi, T. monspeliach, T. radiata:* three species have 2n=18 (*T. stellata, T. striata, T. ornithopodioides*): two species have 2n=44, (*T. geminiflora, T. grandiflora*) (Fedorov, 1974) and another species have 2n= 44 (*T. polycerata*).. Two species of *Trigonella* are of economic importance. These are *T. foenum-graecum* L., the common Methi and the other is *T. corniculata* Kasuri or Champa methi. It is found growing wild in North Western parts of India (Bailey, 1950). |
| *Portulaca* species | Parslane (*Portulaca oleracea* L.) is said to be indigenous to Himalaya. It occurs in several forms during hot weather, in the cultivated and wild state. The cultivated forms have broad leaves whereas in wild; both broad and small leaves types are seen growing. It is tiny creeper with small, succulent fleshy leaves and tender stems, which is cooked like spinach. |
| *Brassica* species | Chinese cabbage (*Brassica pekinensis* and *Brassica chinensis*) is indigenous to China and eastern Asia where it has been in cultivation since 5th century. There are two types of Chinese cabbage- one that heading type (*Brassica pekinensis*) and other open leaf type (*Brassica chinensis*). The former forms an erect, moderately compact, usually cylindrical heads, while the later develops clusters of succulent leaves. Rich diversity in plant type occurs in foot hills of Himalaya and northeastern states including Sikkim. |
| *Allium* species | The genus *Allium* includes onion, garlic and their relatives like leeks, shallots and chives that are considered prized vegetables due to food, medicine or religious purposes. The genus *Allium* is represented by about 750 species which are widely distributed over the warm temperate and temperate zones of the northern hemisphere. Regions of high species diversity occur in Turkey, Iran, North Iraq, Afghanistan, Soviet, middle Asia and West Pakistan and a second less pronounced centre of species diversity occurs in Western North America. About 40 species of *Allium* are reported in India of which 7 species are cultivated, the main being onion and garlic. Much diversity has built up in India in these crops as they have been cultivated from ancient times and selections have been made in different agro-ecological zones. |
| *Raphanus* species | Genus *Raphanus* includes *R. caudatus, R. landra, R. lyratus, R. pterocarpus, R. raphanistrum, R. raphanistrum* subsp. *landra, R. sativus, R. sativus* var. *caudatus, R. sativus* var. *longipinnatus, R. sativus* var. *niger, R. sativus* var. *oleiferus, R. sativus* var. *oleiformis, R. sativus* var. *sativus, R. violaceus* and *XBrassicoraphanus* species. *R. sativus* var. *niger,* seems to be progenitor of common radish. Decreasing in crop variability has been observed to China and from there to Japan. |
| *Daucus* species | Genus *Daucus* includes 16l species world wide of which only two (*D. carota* and *D. sativus* DC) are being cultivated. The cultivated types were evolved as selection from interspecific cross between *D. carota* and *D. sativus*. Some prominant species are: *D. aureus, D. bicolor, D. bocconei, D. brachiatus, D. broteri, D. capillifolius, D. carota, D. carota* cv. *atrorubens, D. carota* subsp. *carota, D. carota* subsp. *commutatus, D. carota* subsp. *drepanensis, D. carota* subsp. *gadecaei, D. carota* subsp. *gummifer, D. carota* subsp. *hispanicus, D. carota* subsp. *hispidus, D. carota* subsp. *maritimus, D. carota* subsp. *maximus, D. carota* var. *atrorubens, D. carota* var. *boissieri, D. carota* var. *commutatus, D. carota* var. *sativus, D. crinitus, D. durieua, D. gingidium, etc.* In these genus two types i.e. Asiatic and European types are more popular. |
| *Abelmoschus* species | The family Malvaceae has 82 genera and 1,400 species. In India, about 111 species belonging to more than 20 genera are reported. The centre of diversity of *Abelmoschus* includes West Africa (Benin, Togo, and Guinea), India and Southeast Asian countries i.e. Burma, Indochina, Indonesia and Thailand (Chevalier 1940; Van Borssum Waalkes 1966; Siemonsma 1982a; Hamon 1988; Hamon and Hamon 1991; Bisht *et al.* 1995). *A. esculentus* (L.) Moench) is the commercially cultivated species of okra. However *A. caillei* is also cultivated for its fresh fruit in West Africa. The geographical distribution of *A. caillei* is from northern Guinea extending to southeast Cameroon. *A. esculentus* cultivated mostly in humid parts of Guinea, mostly in an intercropped system with *A. caillei* (Hamon and Charier 1983; Hamon 1988). At present okra is an important crop of India, Afghanistan, Burma, Japan, Malaysia, Brazil, Pakistan, Iran, Western Africa, Ghana, Ethiopia, Cyprus, Middle East, Greece, Yugoslavia, Bangladesh, Turkey, the Caribbean, Southern United States and the USA. |
| | India is the largest producer of okra in the world. In India it is commercially grown in Uttar Pradesh, Odisha, Bihar, Maharashtra, Haryana, Chhattisgarh, Asom, West Bengal, and Punjab during spring season and in Gujarat, Andhra Pradesh, Karnataka and Tamil Nadu during winter season. Other okra growing states are Jharkhand, Chhattishgarh, and Asom. Based on different floras, floristic accounts and other published evidence expressed main areas of concentration of wild species are as follows: |

☆ Semi-arid tracts of north and northwestern India are for *A. ficulneus* and *A. tuberculatus*.

☆ *Tarai* range and the foothills of the Himalayas are for *A. crinitus, A. manihot* and *A. tetraphyllus* or *A. tetraphyllus* var. *pungens*.

☆ Western and Eastern Ghats and also peninsular tracts are for *A. manihot* (including *tetraphyllus* type), *A. angulosus* mainly confined to hilly tracts up to 2000 m in south India with only meager distribution elsewhere.

☆ Northeastern regions are for *A. crinitis* and *A. manihot*, mostly var. *pungens*.

☆ In some of the wild taxa intra-specific variation exists and has been taxonomically identified as in the *A. manihot* var. *tetraphyllus* and *pungens*, in *A. angulosus* var. *purpureus* and *grandiflorus*, in *A. moschatus* var. *biakensis* and *rugosus*, and in *A. tuberculatus* var. *deltoidefolius*.

The pioneer work of de Candole (1986) and N.I. Vavilov (1926) led to the wide spread recognition that the diversity of crops was centred in certain geographical regions which differ from aquatic to mountains. India has about 46, 042 species of plants (more than 11 per cent of the world's known flora). It is true that natural resources can be managed effectively, efficiently and optimally in the hills as well as in plains and niches as they will positively influence land and water resources. Since the hills and mountains are spread over a great range of elevation from 200 to 7,500 m or so and climatically take latitude from tropical to temperate, there is a great diversity in its natural vegetation. In addition, variation in precipitation, associated with the alignment and altitude of hill and mountain ranges determines the altitudinal adaption, growth and variety of vegetation. The knowledge of genetic diversity in crop plants and their wild relatives; the pattern of their distribution in different phytogeographical regions and the areas of concentration of diversity are of special significance to all concerned with genetic resources, in general and to the explorer, in particular, for undertakimng plant explorations as well as for *in situ* conservation of biodiversity.

## Cultural and Biological Diversity

Cultural and biological diversity are closely interwined. Biodiversity is at the centre of many religions and cultures, while worldwide influence biodiversity through cultural taboos and norms which influence how resources are used and managed. As a result of many people biodiversity and cultutre cannot be considered independently of one another. This is particularly true for the more than 400 million indigenous and local community members for whom the Earth's biodiversity is not only a source of well being but also the foundation of their cultural and spiritual indentities. The close association between biodiversity and culture is particularly apparent in sacred sites, those areas which are held to be of importance because of their religious or spiritual significance. Through the application of traditional knowledge and customs unique and important biodiversity has been protected and maintained in many of these areas over time.

Ecosystem services can be divided into four categories:

**Table 1.2: Division of Ecosystem Services**

| Sl.No. | Services | Descriptions |
|---|---|---|
| 1. | Provisioning services | It covers the supply of goods of direct benefit to people, and often with a clear monetary value, such as timber from forests, medicinal plants, and fish from the oceans, rivers and lakes. |
| 2. | Regulating services | It covers the range of vital functions carried out by ecosystems which are rarely given a monetary value in conventional markets. They include regulation of climate through the storing of carbon and control of local rainfall, the removal of pollutants by filtering the air and water, and protection from disasters such as landslides and coastal storms. |
| 3. | Cultural services | Cultural services not providing direct material benefits, but contributing to wider needs and desires of society, and therefore to people's willingness to pay for conservation. They include the spiritual value attached to particular ecosystems such as sacred groves, and the aesthetic beauty of landscapes or coastal formations that attract tourists. |
| 4. | Supporting services | Supporting services not of direct benefit to people but essential to the functioning of ecosystems and therefore indirectly responsible for all other services. Examples are the formation of soils and the processes of plant growth. |

## Why Biodiversity Matters?

Biodiversity is the variation that exists not just between the species of plants, animals, microorganisms and other forms of life on the planet but also within species, in the form of genetic diversity and at the level of ecosystem in which species interact with one another and with the physical environment. No doubt today crop diversity is the foundation of our food. The diversity is of vital importance to people, because it underpins a wide range of ecosystem services on which human societies have always dependent, although their importance has often been greatly undervalued or ignored. When elements of biodiversity are lost, ecosystem become less resilient and their services threatened. More homogeneous, less varied landscapes or aquatic environments are often more vulnerable to sudden external pressures such as diseases and climatic extremes. For millennia, food plants have been domesticated, selected, exchanged, and improved by farmers in traditional ways, within traditional production systems (Plucknett *et al.*, 2014.).

## Biodiversity vis-à-vis Swing in Food Habits

Global dietary patterns transformed radically in the past 50 years, presenting both a boom and a threat to the health and well being of populations everywhere. Worldwide 30 per cent or more people are now obese than those who are underfed. It is revealed that more than one billion of the world's population lack access to food or are chronically malnourished. On the flip side, a 2006 World Health Reported that upto 2015 there were 2.3 billion overweight adults and more than 700 million obese. Similarly 1.4 billion adults were overweight and 65 per cent of the world's population in countries where overweight and obesity kills more people than underweight (WHO, 2012). This double burden suggests that half (47 per cent) of the global population is suffering from some form of nutritional disorder. Today hunger and malnutrition reaches almost 1000 million people. As a consequence, 15 million people die every year, it means that more than 41,000 people every day, the majority of whom are children. Satisfying the growing demand for animal products while simultaneously sustaining productive assets of natural resources is one of the major challenges for agriculture. FAO projected that by 2050; global average per capita food consumption could rise to 3130 kcal per day. Worldwide, meat consumption per capita per year will increase from 41 kg in 2005 to 52 kg in 2050, reaching an average of 44 kg in developing countries and 95 kg in developed countries (Bruinsma, 2011). Coupled with urbanization and increasing desk bound lifestyles, there is an unprecedented rise in obesity and subsequently, non concommunicable diseases, such as cardiovascular disease, diabetes, and hypertension. From a biological conversion point of view, animal population systems consume more energy in feed than they generate in animal products.

Of a total of 300,000 plant species, 10,000 have been used for human food since the origin of agriculture. Out of these, only 150-200 species have been commercially cultivated and of which only four species *viz.* rice, wheat, maize and potatoes supply 50 per cent of the world's energy needs, while 30 crops provide 90 per cent of the world's caloric intake. Three crops- maize, wheat and rice-make up an estimated 87 per cent of all food grain production and occupy approximately 40 per cent of the global agricultural landscape. Today's model of food production in which we rely on only around 100 crops species for about 90 per cent of per capita supplies of food from plants. Healthy ecosystems provide a diverse range of food sources and support entire agricultural systems. A healthy diet includes multiple food groups, made of diverse foods. A diverse diet is the basis of food pyramids and nutrition guidelines around the world. Nutritional security, where dietary diversity plays an important role is a vital component of food security. Food security greatly depends on the conservation, exchange and wise use of agricultural biodiversity and the genetic resources that constitute such diversity. The main challenge to increase food security is not just food production, but access to food. In addition, it is not simply a matter of delivering more calories to more people. It should be noted that most hungry people in the world (over 70 per cent) live in rural areas. The heavy reliance on a narrow diversity of food crops puts future food and nutrition security at risk. During the last two centuries, as a consequence of agricultural and industrial development and the progressive amalgamation of cultural and eating habits, accentuated of late, due to the globalization and interdependence process, the number of crops and the diversity within them has been progressively reduced.

## Agrobiodiversity: Some Potential Benefits

Anthropogenic activities have profoundly re-shaped the earth's land, oceans, air and biodiversity to such an extent that geologists have proposed a new epoch called the 'Anthropocene' marking the end of 'Holocene' (Since 12,000 years). This new epoch is being said to begin from 1950, when radioactive elements from nuclear testing were likely spread all over the globe, characterized by mass extinctions, plastic pollution, and spike in carbon emmissions in the atmosphere (Waters *et al.*, 2016). Consequently, biological diversity got reduced; the earth became warmer with greater incidence of natural catastrophic events. A recent scientific finding shows that about 58 per cent of the world's land surface, and 9 out of 14 of the world's terrestrial biomes, have fallen below 'safe threshold' of biodiversity, impacting a wide range of services provided by biodiversity, including crop pollination, waste decomposition, regulation of the global carbon cycle, and cultural services that are critical to human well-being (Newbold *et al.*, 2016).

☆ Agricultural biodiversity can play a role in sustaining soil health, food and habitat for important pollinators and natural pest predators that are fundamental to agricultural production.

☆ Diverse crops and land use attracts and sustains a variety of pollinators that contribute to the production of over 75 per cent of the world's most important crops (Klein *et al.*, 2007).

☆ Crop biodiversity is vital for functioning of ecosystems and provision of ecosystem services (Naeem *et al.*, 1994) and also provides dietary needs and services that consumers demand as economies change (Smale, 2006).

☆ Genetic resources provide the building blocks that permit conventional plant breeders and biotechnologists to develop new commercial varieties and other biological products.

☆ Agricultural biodiversity can provide a cost-effective way for farmers to manage pests and diseases, since an estimated 10-16 per cent of global harvests are lost due to plant disease each year (Strange and Scott, 2005; Oerke, 2006).

☆ It also provides conditions for natural pest predators that help farmers save on insecticide costs. For example, lady beetles that eat cotton aphids-valued at US$4.96 saved for every year additional lady beetle per 100 cotton plants according to a study in China (Zhou Ke, 2013).

☆ Using diversity allows farmers to limit the spread of pests and diseases without investing in high chemical inputs.

☆ Use of diversity within varieties and across locations, serves both conservation and crop improvement purposes. It offers the maximum opportunity for the continuation of co-evolutionary processes, without compromising either diversity or productivity.

☆ Agricultural biodiversity plays an important part in maintaining cultural identity and traditional knowledge; whether it involves passing on knowledge about local medicinal plants or traditional recipes.

☆ Agricultural biodiversity promotes healthy functioning of ecosystems and helps ensure the resilience of agriculture as it intensifies in meeting growing demands.

☆ Agricultural biodiversity can contribute to health and nutrition.

☆ Genetic improvement of crop plants is a great benefit to humanity. It is one of the least costly and most effective ways to increase production of food, fibre and plant based products, to resist pests and diseases, to meet new market opportunities and to address the challenges of abiotic stresses such as drought, temperature and climate change. Production of locally adapted crop species will diversify overall cropping systems and reduce the risk of food insecurity.

## Historical Aspects of Agro-biodiversity and Vegetable Crops

Human needs for food, clothing, shelter and other basic amenities may be ideally met from the abundant biodiversity available on the earth planet on a sustainable basis. Unprecedented three global challenges *i.e.* hunger, environment degradation and population growth are greatly being observed today. No doubt, agriculture sector is the only option which can address all three challenges efficiently in both developed and developing country. Agricultural biodiversity refers to the *'biological assortment exhibited among crops, animals and other organisms used for food and agriculture, as well as the web of relationships that bind these forms of life at ecosystem, species and genetic levels'*. Biological diversity in short can be termed as 'biodiversity' has been the backbone of food, health and livelihood for human ever since civilization started'. Agriculture biodiversity is a subcomponent of biodiversity and is important for food and agriculture and plays a vital role in productivity and livelihoods of all farmers, regardless of resource endowment or geographical location. It is the variety and variability of living organisms (plants, animals, microorganisms) that are involved in food and agriculture. It is chief source of livelihood for rural communities in developing countries. According to the Food and Agriculture Organization of the United Nations, ever since the beginning of agricultural history, human race has already used more than 10,000 plant species for feeding. Interaction across the living world and human societies led to the domestication of a wide range of plants. In terms of existence of species diversity on earth, current estimates vary drastically between 8 and 125 million species. The Convention on Biological Diversity (CBD) endorses a possible occurrence of around 14 million species. Out of these, only less than 1.75 million species have been described and formally named so far. This includes 0.27 million plant species and 1.32 million animal species (Table 1.3).

## Recent Trends in Biodiversity

Across the range of biodiversity measures, current rates of change and loss exceed those of the historical past by several orders of magnitude and show no indication of slowing. At large scales, across biogeographic realms and ecosystems (biomass), declines in biodiversity are recorded in all parts of the world. Among well studied groups of species, extinction rates of organisms are high and increasing (medium certainty), and at local levels both populations and habitats are most commonly found to be in decline.

☆ Habitat conversion to agricultural use has affected all biogeographical realms.

☆ The majority of biomass has been greatly modified. Between 20 per cent and 50 per cent of 9 out of 14 global biomass have been transformed to crop lands.

☆ Between 12 per cent and 52 per cent of species within well-studied higher taxa are threatened with extinction, according to the IUCN Red List.

☆ Among a range of higher taxa, the majority of species are currently in decline.

☆ Genetic diversity has declined globally, particularly among domesticated species.

☆ Globally, the net rate of conversion of some ecosystems has begun to slow, and in some regions; ecosystems are returning to more natural states largely due to reductions in the rate of expansion of cultivated land, though in some instances such trends reflect the fact that little habitat remains for further conversion.

☆ Translating biodiversity loss between different measures is not simple: rates of change in one biodiversity measures may underestimate or overestimate rates of change in another.

☆ Biotic homogenization, defined as the process whereby species assemblages become

### Table 1.3: Estimates of Global Biological Diversity

| Group | Described Species (Number) | Estimated Total Species (Number) |
| --- | --- | --- |
| Bacteria | 4,000 | 1,000,000 |
| Protoctists (algae, protozoa *etc.*) | 80,000 | 600,000 |
| Animals | 1,320,000 | 10,600,000 |
| Fungi | 70,000 | 1,500,000 |
| Plants | 270,000 | 400,000 |
| **Total** | **1,744,000** | **Approx. 14,000,000** |

*Source*: Convention on Biological Diversity. On Line Status and Trends of Global Biodiversity <http://www.cbd.int/gbo1/chap-01.shtml>

increasingly dominated by a small number of widespread species, represents further losses in biodiversity that are often missed when only considering changes in absolute number of species.

☆ While biodiversity loss has been a natural part of the history of Earth's biota, it has always been countered by origination and, except for rare events, has occurred at extremely slow rates. Currently, however, loss far exceeds origination and rates are orders of magnitude higher than average rates in the past.

☆ The patterns of threat and extinction are not evenly distributed among species but tend to be concentrated in particular ecological or taxonomic groups.

## Agrobiodiversity: The Scenario in India

India is one of the 17 megadiverse countries, possesses 12 megagene biodiversity centres and is a Vavilovian Centres of Origin of crop plants. About 29 per cent of the flowering plants occurring in India are endemic and three of the 25 hot-spots of biodiversity exist here. India is one of them. India is geographically (extreme variations in altitude), agro-ecologically (climate and soils), sociologically, economically and culturally (long history of cultivation and its ancient heritage) diverse. Not only this, India has highly diverse habitats, such as coastal, plateau, desert, plains, hills and mountain systems. All these factors helped in the build up of useful diversity/variability, particularly for wide adaptability, stress tolerance and quality traits in the useful crop plants. Worldwide this sub-continent is considered extremely rich in its plant genetic wealth. A unique and interesting feature of this region is that about 60 per cent of its floristic wealth is endemic. Two out of the 25 global hotspots of biodiversity, *viz.* the Indo-Buirma and Western Ghats are located in India. As per history of human civilization, around 2,50,000 plant species have been described so for and a much larger number still remains to be duly recognized yet only about 3,000 out of them were picked up from wild by human beings for their use and grown on varying scales since the beginnings of agriculture, considered to be nearly 10,000 years ago. Endemic Indian biodiversity is also significant: 4,950 species of flowering plants, 16,214 insects, 110 amphibians, 214 reptiles, 69 birds and 38 mammals are endemic to the country. More than 49,000 plant species (including fungi and lower plants), 17,500 species of higher plants and nearly 87,000 animal species have been described in the country, which respectively constitute 11.80 per cent and 7.25 per cent of the global biodiversity. It possesses about 12 per cent of world flora with 5725 endemic species of higher plants belonging to about 141 prevelent genera and over 47 families. About 166 species of crops including 25 major and minor crops have originated and/or developed diversity in India. It

is also considered to be the centre of origin or diversity of our crops. Currently, 479 species of both indigenous and exotic plants are cultivated in different agro-ecological regions of India.

**Table 1.4.: Biological Diversity and Number of described Species**

| Taxonomic Group | Number of Species | | Per cent (India/ World) |
|---|---|---|---|
| | India | World | |
| Angiosperms | 17,500 | 250,000 | 7.00 |
| Gymnosperms | 48 | 650 | 7.40 |
| Pteridophytes | 1,200 | 10,000 | 12.00 |
| Bryophytes | 2,850 | 14,500 | 19.70 |
| Lichens | 2,075 | 13,500 | 15.00 |
| Fungi | 14,500 | 70,000 | 20.70 |
| Algae | 6,500 | 40,000 | 16.30 |
| Virus/Bacteria | 850 | 8,050 | 10.60 |
| Protistas (Protozoans) | 2,577 | 31,250 | 8.24 |
| Molluscs | 5,070 | 6,6535 | 7.62 |
| Arthropods | 68,389 | 987,949 | 6.90 |
| Chordates | 4,994 | 48,451 | 10.40 |
| Protochordates | 119 | 2,105 | 5.65 |
| Pisces | 2,546 | 21,723 | 11.72 |
| Amphibians | 240 | 5,150 | 4.66 |
| Reptilians | 460 | 5,817 | 7.91 |
| Aves | 1,232 | 9,026 | 13.66 |
| Mammalians | 397 | 4,629 | 8.58 |
| **Total Planta** | **45,523** | **40,6700** | **11.80** |
| **Total Animalia** | **86,905** | **1,196,903** | **7.25** |

*Source*: Third National Report (India) to CBD.

<http://www.cbd.int/doc/world/in-nr-03-en.doc>

## Loss of Agro-biodiversity: Some Terrible Realities

In 2002, the world's leaders agreed to achieve a significant reduction in the rate of biodiversity loss by 2010. Having reviewed all available evidence, including national reports submitted by Conferences of Parties (CoPs), the *Global Biodiversity Outlook* concludes that the target has not been met. Moreover, the *Outlook* warns, the principal pressures leading to biodiversity loss has not just constant but are, in some cases, intensifying. Biodiversity underpins the functioning of the ecosystems on which we depend for food and fresh water, health and recreation, and protection from natural disasters. Its loss also affects us culturally and spiritually.

☆ Around 940 species of cultivated plants are threatened globally (Khoshbakht and Hammer, 2007).

☆ Over 47, 677 species may soon disappear, as per the assessments made by the International Union for the Conservation of Nature.

☆ Nearly 12 per cent of birds, 21 per cent of mammals, 30 per cent of amphibians, 27 per cent of coral reefs and 35 per cent of conifers and cycads are currently facing extinction.

☆ More than 70 per cent of global grasslands, 50 per cent of savannas, 45 per cent of temperate deciduous forests and 27 per cent of tropical forests are lost due to agricultural expansion.

☆ Advent of the green revolution heralded huge loss of farmer's varieties, which were replaced by a few high-yielding varieties and hybrids.

☆ When a species or the diversity within a species is lost, the genes that could be important for improving resistance in crops to pests and diseases or adapting to the effects of climate change are also lost.

☆ Introduction of genetic led technologies leads to replacement of a large number of local varieties with a few more uniform, high-yielding strains and deletion of indigenous species (Swaminathan, 2000).

☆ Current extinction rates are 1,000-10,000 times greater than background extinction rates (Rockstrom *et al.*, 2009); a disaster that has far greater consequences than economic collapse or nuclear war (E.O. Wilson, 1994).

## Concept of Centres of Origin and Crop Diversity

The study of origin of agriculture and its spread provides clues to the geographical distribution of centres of domestication. These centres were found to be located in areas of maximum diversity. In the forefront of contributors to the global dispersal of crop plants and their wild relatives has been well assessed by Vavilov (1927); who believed that time has been the only factor that influenced the dispersal of species and their increase in variation. He strongly suggested that centres of origin of species coincide with the areas where greater diversity occurs in species. A well known hypothesis for the origin of agriculture is the 'Rubbish-Heap hypothesis' which says that early humans gathered roots and seeds for their food. Such plants actively colonized the base around their dwellings, which were rich with the discarded rubbish. Indian gene centre holds a prominent position among the eight Vavilovian centres of origin of crop plants. Of the 2, 40, 000 economic plant species distributed over 12 megagene centres (Zeven and de Wet, 1982), about 160 species of economic importance are reported in this region. The Centres of Origin *sensu* Vavilov of Culivated Plants also list following cultivated plants to have Indian origin: *Phaseolus aconitifolius, Phaseolus calcaratus* (Rice bean); *Dolichos biflorus, Vigna sinensis* (Asparagus bean); *Solanum melongena* (eggplant); *Raphanus caudatus* (Rat's

tail radish); *Colocasia antiquorum* (Taro yam); *Cucumis sativus* (cucumber); *Corchorus olitorius* (Jute). In addition, the list of 'Food Crops of the World', published by the World Conservation Monitoring Centre also includes the following crops as having Indian origin: *Colocasia esculenta* (Taro); *Brassica juncea* (mustard seed); *Cucumis sativus* (cucumber); *Solanum melongena* (Eggplant) *etc.*

**Table 1.5: Special Components and Vegetable Crop**

| Special Components | Crop |
|---|---|
| Crops of Indian origin | Eggplant, cucumber, *Luffa* (ridge gourd and sponge gourd), pointed gourd, colocasia, elephant foot yam |
| Secondary centare of diversity | Okra, melons, chillies and *Cucurbita* species |
| Rich diversity in introduced crops | French bean, cowpea, tomato and several cucurbits |

## Centres of Origin

Centres of origin play a significant role in overall improvement of vegetable crops.

*"Area where cultivated plant species show much greater variation than any other place of the world"* is called as centres of origin.

Each of the world's basic vegetable crops originated in relatively confined geographic regions. Vavilov recognized the areas as primary locales or centre of origin where the plants exhibited most diverse forms and were domesticated for the first time. Origin of species and evolution of diversity in it are two different aspects in terms of ecological requirements. Later is more often observed to be correlated with ecological diversity. It is generally accepted that cultivated plants were not distributed uniformly throughout the world. Certain areas exhibited much diversity of a species or crop though the plants were not domesticated first there. Even today, certain areas show far greater diversity than others for certain cultivated crops and their wild relatives. These areas possessing rich species diversity as a result of migration of species from primary centre were term as secondary centres. Tropical areas of the world were the major holders of genetic diversity of cultivated plants. About 400 different species constitute the global diversity in vegetable crops, representing the centre of origin and/or diversity. The region possessing maximum diversity at global levels is tropical America, tropical Asia and the Mediterranean lists the diversity in major vegetable securing in different regions. In the tropical Asian region, both India and China hold maximum diversity. Knowledge regarding gene centres of cultivated plants and their possible wild ancestors (progenitors) serve as guidelines to collect main sources of genetic resistance to parasitic and non-parasitic diseases, insect pests and nematodes.

## Criteria for Assigning Geographical Region

The criteria for assigning a particular geographical region to specific vegetable crops are:

(i) The availability of considerable crop variability

(ii) The presence of wild forms

(iii) Existence of source of resistance to many of the diseases

(iv) Historical records

(v) Fossil records

(vi) Archaeological findings

The regions overlap for a number of crops. But 9 major and 3 minor centres in the old world (Agrica, Asia and Europe) and new world (America) respectively have been identified as the areas for the origin and diversity of the vast majority of vegetables.

Genetic wealth includes both primitive/native cultivars and their wild and weedy species is distributed to these 12 primary centres of diversity (Zeven and Zhukovsky, 1975, Zeven and de Wet, 1982). They have dealt elaborately in the Dictionary of Cultivated Plants and their Centres of Diversity, listing species for different mega centres and the range and extent of distribution of genetic/varietal/specific diversity *etc.* This includes centres of origin and distribution of 218 vegetable crop species in different regions of diversity *viz.* Chinese-Japanese (56), Indo-Chinese-Indonesian (31), Australian (1), Hindustani (11), Near Eastern (4), Mediterranean (24), African (36), European- Siberian (29), South American (18), Central American and Mexican (6), and North American (2) (Arora, 1985).

(i) **Primary centre of origin**: In 1926, he proposed that crop plants/cultivated plant evolved from wild species in the areas showing great diversity and termed them as primary centre of origin.

(ii) **Secondary centre of origin**: Due course of time, the crops moved to other areas primarily due to human activities; these areas generally lack the richness in variation found in the primary

centres of origin. But in some areas, certain crop species show considerable diversity of forms although they did not originate there; such areas are known as secondary centre of origin.

## Vavilo's Criteria in Determining the Centre of Origin

The concept of centres of origin was given by Vavilov based on his studies of a vast collection of plants at the Institute of Plant Industry, Leningrad. He was Director of this institute from 1916 till 1936. Vavilov's centres of origin generally concide with the ancient centres of civilization also. This is understandable because as the early man left his nomadic habits of hunting and food gathering and gradually adopted a sedentary habit, he initially chose river valley basins for his earliest settlements. He found that his needs steadily began to increase, and the food materials that he was able to gather from naturally occurring plants were inadequate to meet his full needs. He was thus compelled to take up cultivation gradually. For cultivation, he selected types having superior attributes as higher yield, better quality, better adaptability to the prevailing environment, and some resistance to the most serious diseases and pest affecting them. As the community's requirements increased steadily, he began to explore newer areas in search of more land for cultivation, and for newer species of plants to meet his increasingly varried requirements. Almost simultaneously, a process of commerce and barter trade began to develop. With this, trade and caravan routes also began to establish gradually.

On the basis of geographical survey, a Russian plant explorer and geneticist, Nikolai I. Vavilov, one of the first scientists to recognize the importance of plant genetic diversity, summarized that the cultivated plants have originated from the eight geographical centres of origin and two sub-centres (Figure 1.1).

Vavilov also postulated the "Law of Homologous Series in Variation", which states that characters found in one species also occur in other related species. For

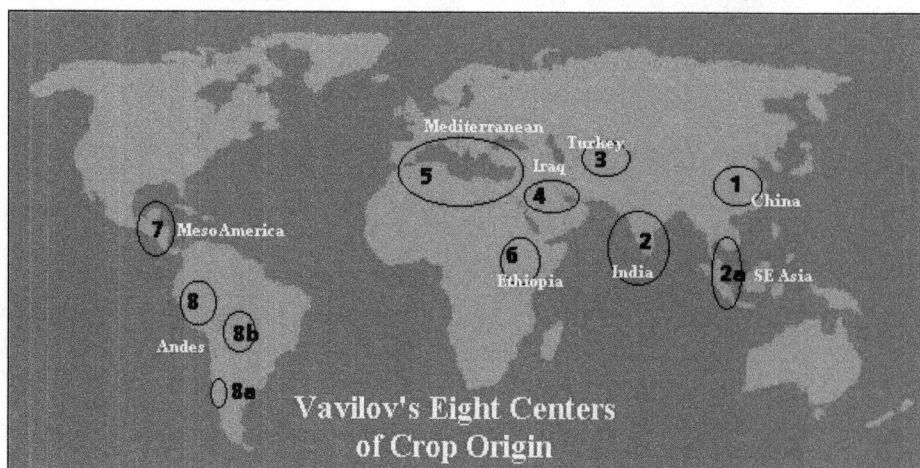

**Figure 1.1: Vavilo's Centres of Crop Origin.**

**Table 1.6: Centre of Origin as Proposed by N.I. Vavilov (1926, 1935) for Vegetable Crops**

| Sl.No. | Centre of Origin | Primary Centre of Origin | Secondary Centre of Origin |
|---|---|---|---|
| 1. | Abyssinian Centre | Onion, okra, pea, lablab bean | Broad bean (*Vicia faba*) |
| 2. | Asia Minor Centre (Syn. Near East or Persian Centre) | Carrot, cabbage, lettuce, | *Brassica capestris*, turnip |
| 3. | Central American Centre (Syn.) Mexican Centre) | Rajma (*P. vulgaris*), lima beans, melons, pumpkin, sweet potato, arrowroot, chillies | - |
| 4. | Central Asia Centre (Syn. Afghanistan Centre) | Mung, radish, muskmelon, carrot, onion, garlic, spinach | Rajma (*P. vulgaris*), cowpea, turnip |
| 5. | China Centre | Soybean, radish, bunda (*Colocasia* sp.), brinjal | - |
| 6. | Hindustan Centre [Divided into: (1) Indo-Burma and (2) Siam-Malaya-Java Centres] Mediterranean Centre | Cowpea, mung bean, brinjal, cucumber, Indian radish, broad bean, *Brassica* sp., onion, garlic, beets, lettuce, asparagus | - |
| 7. | South American Centre [Divided into: (1) Peru (2) Chile, and (3) Brazil-Paraguay Centres. | Potato, lima bean, pumpkin, tomato, cassava | - |
| 8. | U.S.A. Centre | Jerusalem artichoke | - |

example, diploid (2x), tetraploid (4x) and hexaploid (6x) wheat show a series of identical contrasting characters. Thus a character absent in a species but found in a related species, is likely to be found in the collections of that species made from the centre of its origin. Vavilov also recognized secondary centres of origin and pointed out that valuable forms are found far removed from primary area of origin. A good example is the Washington Navel Orange discovered in Brazil wheras the basic centre of diversity of citrus in south eastern Asia.

These centres were found to be characterized by the accumulation of dominant gene in the middle portion and recessive gene in the periphery. This possiobility led to the high concentration of genetic diversity in the secondary centre rather than the primary centre of origin. The centres of diversity located by Vavilov have proved to be fertile promising collecting areas for future explorations. Hawkes (1983) recorded the undermentioned criteria used by Vavilov in determining the centre of origin of crop plants:

1. Differentiate plants into specific and intraspecific taxa on morphological and genetical basis.

2. Determine the areas of distribution of such species and groups of species.

3. Establish the distribution of genetic diversity and determine the geographical centres where this is at maximum, especially those centres with endemic forms/characters.

4. Determine the centre where diversity of genetically allied species is concentrated.

5. Correlate the above centres with the areas of concentration of nearest wild relatives.

6. Compare centres of origin of group of cultivated plants with certain specialized parasite(s).

7. Support the above with the evidences from archaeology, linguistics and history.

## Law of Homologous Series

It was noted that similar variation in two or more related crops could be observed. Vavilov expressed this concept in genetic terms as "Law of Homologous Series in Variation" and used this principal as a clue to the characters remaining to be discovered. Naked grain form of barley, Oats and millets, were found in China. Marked differences, observed in wheat, barley, peas, lentils and chickpea collected from Mediterranean region and southern Asiatic region. The former are found to be large fruited/seeded while the later were small fruited/seeded.

## Transformation of Concept of Origin

There has been considerable change in the concept of origin ever since Vavilov (1951) propounded them. It was considered, later that the presence of wild relatives was essential for designating the centre of origin. Subsequently several researchers revised the boundaries supported with the concepts developed by them. Schiemann (1951) criticized the Vavilovian gene centre theory especially in relation to Ethiopean and central Asian centres which were considered by Vavilov as the centre of origin of Durum wheat (*Triticum durum*) and barley (*Hordeum vulgare*) and that of hexaploid wheat (*T. aestivum*) on the other. The criticism is primarily based on the fact that none of wild progenitors were presented in these centres, hence termed them as "Accumulation centres". Zohary (1970) also criticized the Vavilovian concept of centres of origin. Similarly, Kuckuck (1963) criticized the Vavilovian concept as he did not find the dominant alleles in the centres of origin and recessives in the periphery, *e.g.*, bread wheat and sorghum as claimed by Vavilov.

## Classification of Gene Centres as per Zhukovsky (1965)

(i) **Megagene centres:** Zhukovsky (1965) a close associate of Vavilov proposed 12 megagene centres of crop plant diversity. The new areas added to vavilov's eight centres were Australia, Africa and Siberia followed by revision of the boundaries to make 12 centres.

(ii) **Microgene centres:** Microgene centres of wild growing species related to crop plants where the cultigens first originated were also indicated.

(iii) **Primary microgene centres:** Zhukovsky (1965) also distinguished between the 'primary microgene centres' for the restricted areas where the cultigen actually first originated and the secondary megagene centres which are the areas into which they have now spread.

Zeven and Zhukovsky (1975) in "Dictionary of cultivated plants and their centres of diversity" have listed the species for different megagene centres. Zeven and de Wet (1982) prefer the term region to centre. By and large, these 12 regions have wider coverage and more acceptability.

## Centres and Non-centres

Harlan (1971) recognized only 3 main centres each with more or less connected but large and diffuse non-centres (Table 1.7).

**Table 1.7: Proposed Centres as per Harlan (1971)**

| Centres | Non-centres |
|---|---|
| North Chinese-B1 | South East asian and South Pacific non-centre-B2 |
| North East centre-A1 | African non-centre- A2 |
| Meso American Centre-C1 | South American non-centre- C2 |

**Micro centres**-Harlan also recognized smaller areas/pockets of varietal and/or racial diversity within Vavilovian Centre and termed these as micro centres. Darlington (1973) has increased the number of centres of origin to sixteen. Hawkes (1983) further identified small 'micro' centres for several crops.

**Table 1.8: Crop Species in 'Micro' Centre**

| Sl.No. | Micro centres | Crop |
|---|---|---|
| 1. | New Guinea | Sugarcane (*Saccharum officinarrum*) |
| 2. | Solamon Islands and Fiji | *Musa* species |
| 3. | North Western Europe | *Avena strigosa, Secale cereale* |
| 4. | United States/China | *Helianthus annuus, H. tuberosus* |
| 5. | Southern Chile | *Bromus, Mangifera* species |
| 6. | Brazil | *Manihot esculenta, Ananas comosus* |

## Nuclear Centres and Regions of Diversity

Hawkes (1983) advocated that agriculture began not once but several times, more or less simultaneously and in different regions of the world. He envisaged centre of agricultural origin from which farming spread into one or more regions for which he proposed the name Nuclear Centres and Regions of Diversiy. He linked up his nuclear centres with the archaeological evidences to provide proofs of the agricultural origins. He recognized the following Nuclear Centres and regions of diversity.

**Table 1.9: Nuclear Centres and Regions of Diversity**

| | Nuclear Centres | | Regions of Diversity |
|---|---|---|---|
| A. | Northern China | I | China |
| | | II | India |
| | | III | South East Asia |
| B. | The Near East | IV | Central Asia |
| | | V | The Near East |
| | | VI | The Mediterranean |
| | | VII | Ethiopia |
| | | VIII | West Africa |
| C | Southern Mexico | IX | Meso America |
| D | Central to Southern Peru | X | Northern Andes (Venezuela to Bolivia) |

There are several such regions where crop actually did not originate. This argument is based on the wild progenitors, absence of archaeological remains to suggest antiquity of a crop species. These are regions into which the crops perhaps spread from the nuclear centres in the past in which due to spatial isolation in time and intensive human selection played a pre-dominant role in the increase of genetic diversity.

The vegetable crops in different centre of origins are given below:

**Table 1.10: Centre of Origin and Geographical Area Covered**

| Sl.No. | Centre of origin | Country/area includes |
|---|---|---|
| 1. | The China centre of origin | Mountaineous regions of central and western China and neighbouring low lands. |
| 2. | The Himalayan centre of origin (a) Indo-Burma (b) Siam-Malaya-Java | Burma, Asom, Malaya, Java, Sumatra, Philippines but excludes North-West India, Punjab and North-Western frontier |
| 3. | The Central Asia centre of origin | Afghanistan, North-West India, North-West Frontier Province and Kashmir. Tadjikistan, Uzbekistan. |
| 4. | The Mediterranean centre of origin | Mediterranean region, European countries |
| 5. | The Asia minor centre of origin | Enterior of Asia minor, Transcaucasia, Iran, high lands of Turkmenistan |
| 6. | The Abyssinian centre of origin | Ethiopia and hill country |

| Sl.No. | Centre of origin | Country/area includes |
|--------|-----------------|----------------------|
| 7. | The Central American centre of origin | South Mexico, Central America |
| 8. | The South American centre of origin | Peru, Bolivia, Ecuador, Colombia, parts of Chile, Brazil, Paraguay |
|  | (a) Peru |  |
|  | (b) Chile |  |
|  | (c) Brazil-Paraguay |  |
| 9. | USA centre of origin | US, Canada *etc.* |

In these centres of origin, the vegetable crops have long been exposed to the selective pressure of local pathogens and insect pests, and have developed inherent basic resistance. Leppik (1970) reviewed the role of gene centres of plants as source of diseases and pest resistance. He opined that systematic exploration of primary and secondary gene centres of a particular vegetable may provide additional gene pools for resistance breeding.

Eight main centres of origin were originally proposed by Vavilov in 1936 and these centres are:

1. China Centre of Origin
2. Hindustan Centre of Origin
3. Central Asia Centre of Origin
4. Asia Minor Centre of Origin
5. Mediterranean Centre of Origin
6. Abyssinia Centre of Origin
7. South Mexican and Central Centre of Origin
8. South America Centre of Origin

## (1) The China Centre of Origin

This centre consists of the mountainous regions of central and western China and the neighboring lowlands. It is the largest and the oldest independent centre of origin. Territorial jurisdiction of the centre includes central and western China alongwith adjoining low lands. This centre is extremely rich in plant genetic resources. Vavilov named origin of 136 species of cultivated plants in this centre. The crop that originated in this area (primary centre of origin) is soybean (*Glycine max*), radish (*Raphanus sativus*), bunda (*Colocasia antiquorum*), and other several species of *Brassica* and *Allium*, brinjal (*Solanum melongena*), some species of *Cucumis* and *Cucurbita*. In addition, it is secondary centre of origin for several crop plants *e.g.* rajma (*P. vulgaris*), cowpea (*Vigna unguiculata*), and turnip (*Brassica rapa*).

## (2) The Hindustan Centre of Origin

This centre includes Burma, Asom, Malaya Archipelago, Java, Borneo, Sumetra and Philippines, but excludes Nort-West India, Punjab, and North-Western Frontier provinces. In the primary centre of origin of cowpea (*Vigna unguiculata*), mung bean (*Vigna radiata*), brinjal (*Solanum melongena*), cucumber (*Cucumis sativus*),

Indian lettuce (*Lactuca indica*), certain species of yams (*Dioscorea* spp.) Indian radish (*Raphanus indicus*) and turmeric (*Curcuma domestica*).

## (3) The Central Asia Centre of Origin

It includes Indian sub-continent *i.e.*, North-Western India (Punjab, Jammu and Kashmir), including Myanmar but excludes North-West Frontier Provinces, all of Afghanistan, the Soviet Republics of Tadjikistan and Uzbekistan and Tian-Shan. It is also known as the Afghanistan centre of origin. One hundred seventeen species of cultivated plants have originated in this centre. The vegetable crops that originated in this centre (primary centre of origin) are peas (*Pisum sativum*), broad bean (*Vicia faba*), mung bean (*Vigna radiata*), radish (*Raphanus indicus*) muskmelon (*Cucumis melo*), pumpkin (*Cucurbita moschata*), carrot (*Daucus carota*), onion (*Allium cepa*), garlic (*Allium sativum*), and spinach (*Spinacea oleracea*). It is secondary centre of origin of rye (*Secale cereale*).

## (4) The Asia Minor Centre of Origin

This is also known as the Near East or the Persian centre of origin. It includes the interior of Asia Minor, the whole of Transcaucasia, Iran and highlands of Turkmenistan. The vegetable crop species that originated in this region (primary centre of origin) include carrot (*Daucus carota*), cabbage (*Brassica oleracea* var. *capitata*), species of *Allium*, and lettuce (*Lactuca indica*). It is the secondary centre of origin of rape (*Brassica campestris*), black mustard (*Brassica nigra*), leaf mustard (*Brassica japonica*), and turnip (*B. compestris* L. var. *rapifera*).

## (5) The Mediterranean Centre of Origin

The centre provides adequate biodiversity at least in 84 cultivated plants species. Large seeded variants have evolved in response to Mediterranean environments in in several field crops. Many valuable cereals and legumes originated in this area. Some of the vegetable include (primary centre of origin) pea (*Pisum sativum*), broad bean (*Vicia faba*), several species of *Brassica i.e.* rape (*Brassica campestris*), black mustard (*Brassica nigra*), cabbage (*Brassica oleracea* var. *capitata*), and turnip (*Brassica rapa*), onion (*Allium cepa*), garlic (*Allium sativum*), beets, lettuce (*Lactuca indica*) artichoke, and asparagus (*Asparagus officinalis*).

## (6) The Abyssinian Centre of Origin

The Abyssinian Centre of Origin includes Ethiopia and hill country of Eritrea (adjoining areas situated in Africa). This centre owes to the origin of 38 cultivated species. It is the primary centre of origin for certain vegetables like lablab bean (*Dolichos lablab*), pea (*Pisum sativum*), onion (*Allium cepa*), and okra (*Abelmoschus esculentus*). It is the secondary centre of origin for broad bean (*Vicia faba*) as suggested by Guzhov (1984).

## (7) The South Mexican and Central American Centre of Origin

This includes the region of South Mexico and Central America. It is also referred to as the Mexican centre of origin. This centre owes to the origin of 49 cultivated species. The plants that originated (primary centre of origin) here are, Rajma (*P. vulgaris*), lima bean (*P. lunatus*), muskmelon (*Cucumis melo*), pumpkin (*Cucurbita moschata*), sweet potato (*Ipomoea batatas*), arrowroot (*Cana edulis*), and chillies (*Capsicum annuum*).

## (8) The South American Centre of Origin

This centre includes the high mountainous regions of Peru, Bolivia, Ecuador, Columbia, parts of Chile and Brazil, and whole of Paraguay. Forty five crop sopecies are native to this centre. The crop that originated in this centre (primary centre of origin) are many species of potatoes (*Solanum* species), lima bean, pumpkin (*Cucurbita moschata*), tomato (*Solanum* species), and cassava (*Manihot esculenta* L.). Later Vavilov identified two sub-centres in this centre namely:

**(a) Chile sub-centre:** Chile sub-centre includes four crops namely, common potato, tarweed, brome grass and strawberry.

**(b) Brazilian-Paraguayan sub-centre:** In this region thirteen crop species were found originated in this sub-centre. Important among them are rubber tree, peanut, cacao, and pineapple.

### Table 1.11: Origin of World Vegetable Crops and their Chromosome Number

| Sl.No. | Centre of Origin | Vegetables | Chromosome Number (2n=) |
|---|---|---|---|
| 1 | Ethiopia | Bottle gourd | 22 |
| | | Celery | 22 |
| | | Cucumber | 14, 28 |
| | | Cowpea | 22, 24 |
| | | Fenugreek | 16 |
| | | Garden pea | 14 |
| | | Okra | 130 |
| | | Onion | 16, 32 |
| | | Watermelon | 22 |
| | | West Indian gherkin | 24 |
| 2 | Mediterranean | Asparagus | 20 |
| | | Cabbage | 18 |
| | | Beets | 18, 19, 20, 27 |
| | | Cauliflower | 18 |
| | | Celery | 22 |
| | | Garlic | 16-secondary |
| | | Great headed garlic | 16 |
| | | Kale | 38 |
| | | Knol-khol | 18 |

| Sl.No. | Centre of Origin | Vegetables | Chromosome Number (2n=) |
|---|---|---|---|
| | | Leek | 32 |
| | | Lettuce | 18, 36+B |
| | | Onion | 16, 32-secondary |
| | | Swede | 38 |
| | | Swiss chard | 18, 36 |
| | | Turnip | 20 |
| 3 | Asia Minor | Beans | 22-secondary |
| | | Cabbage | 18 |
| | | Carrot | 18 |
| | | Chive | 16, 24, 32 |
| | | Kale | 38 |
| | | Kurrat | 16, 24, 32 |
| | | Lettuce | 18, 36+B |
| | | Mustard | 36 |
| | | Onion | 16, 32-secondary |
| | | Pea | 14 |
| | | Rat tail radish | 18 |
| | | Brussels sprout | 18 |
| 4 | Central Asiatic (Afghanistan, Turkestan) | Bottle gourd | 22 |
| | | Broad bean | 12, 14, 24 |
| | | Carrot | 18 |
| | | Garlic | 16 |
| | | Mustard | 36 |
| | | Onion | 16, 32 |
| | | Pea | 14 |
| | | Spinach | 12, 24 |
| | | Turnip | 20 |
| | | Welsh onion | 16 |
| 5 | Indo-Burma | Amaranths | 32, 34 |
| | | Ash gourd | 24 |
| | | Ceylon spinach or Poi (*Basella*) | 44, 60 |
| | | Bread fruit | 54, 56 |
| | | Brinjal | 24 |
| | | Bitter gourd | 22 |
| | | Chekkurmanis | Not Known |
| | | Cluster bean | 14 |
| | | Coleus | 32 |
| | | Colocasia | 28, 36, 48 |
| | | Cowpea | 22 |
| | | Cucumber | 14, 28 |
| | | Curry leaf | 18 |
| | | Dolichos bean | 22, 24, 44 |
| | | Drumstick | 22 |
| | | Elephant foot yam | 26, 28 |
| | | Indian spinach | 44, 60 |
| | | Longmelon | 24, 48 |
| | | Mango ginger | 42 |
| | | Muskmelon | 24, 48-secondary |

| Sl.No. | Centre of Origin | Vegetables | Chromosome Number (2n=) |
|---|---|---|---|
| | | Pointed gourd | 22 |
| | | Ridge gourd | 26 |
| | | Rozelle | 36, 72 |
| | | Roundmelon (Tinda) | 22 |
| | | Smooth gourd | 26 |
| | | Snake gourd | 22 |
| | | Snapmelon | 24, 48 |
| | | Spinach beet | 18, 24 |
| | | Sweet basil | 48 |
| | | Winged bean | 26 |
| | | Yam | 40 |
| 6 | Siam, Malaya, Java | Ash gourd | 24 |
| | | Bread fruit | 54, 56 |
| | | Colocasia | 28 |
| | | Curry leaf | 18 |
| | | Elephant foot yam | 26, 28 |
| | | Ginger | 22 |
| | | Turmeric | 32, 62, 64 |
| | | Yam | 40 |
| 7 | China | Adzuki bean | 22 |
| | | Coleus | 32 |
| | | Cowpea | 22, 24-secondary |
| | | French bean | 22-secondary |
| | | Leaf mustard | 24-secondary |
| | | Muskmelon | 24, 48-secondary |
| | | Radish | 18 |
| | | Rakkyo | 16, 32 |

| Sl.No. | Centre of Origin | Vegetables | Chromosome Number (2n=) |
|---|---|---|---|
| | | Soybean | 40, 80 |
| | | Turnip | 20 |
| 8 | Mexico-Guatemala | Amaranth | 32, 34 |
| | | Chilli | 24 |
| | | Chowchow | 24 |
| | | French bean | 22 |
| | | Jackbean | 22, 44 |
| | | Lima bean | 22 |
| | | Runner bean | 22, 24 |
| | | Squash | 24, 40 |
| | | Sweet potato | 90 |
| | | Tomato | 24 |
| 9 | Peru, Ecquador, Bolivia | Chilli | 24 |
| | | Baby corn | 20 |
| | | Edible canna | 18, 27 |
| | | Pumpkin | 24 |
| | | Runner bean | 22, 24 |
| | | Squash | 24, 40 |
| | | Tapery bean | 22 |
| | | Tomato | 24 |
| | | *Xanthosoma* | 12, 13 |
| 10 | Southern Chile | Potato | 48 |
| 11 | Brazil, Paraguay | Tapioca | 36 |
| 12 | United States | Globe artichoke | 34 |
| | | Jerusalem artichoke | 102 |

**Table 1.12: Bulb, Tuber as Vegetatively Propagated Vegetable Crops and Specific Area of Availability**

| Sl.No. | Botanical Name | Common Name | Specific Area of Availability |
|---|---|---|---|
| | | ***Allium*** | |
| 1. | *Allium ampeloprasum* var. *ampeloprasum* | Great headed garlic | Mediterranean, Western Himalayas |
| 2. | *A. ampeloprasum* var. *porrum* | Leek | Mediterranean, Western Himalayas |
| 3. | *A. cepa* var. *cepa* | Common onion | Central Asia/Throughout India |
| 4. | *A. cepa* var. *aggregatum* | Multiplier onion, Shallot | Central Asia |
| 5. | *A. chinense* | Rakkyo | China, North Eastern Region |
| 6. | *A. fistulosum* | Welsh onion | China, N. Eastern Region, Western Himalaya |
| 7. | *A. sativum* | Garlic | Central Asia, Throughout India |
| 8. | *A. schoenoprasum* | Chives | Europe Western Himalaya |
| 9. | A. tuberosum | Chinese onion | North Eastern Region |
| | | ***Dioscorea*** | |
| 10. | *Dioscorea alata* | Greater Yam | India, Western and Eastern Ghats, N. E. Hills |
| 11. | *D. esculenta* | Lesser Yam | Asom-Burma Western and Eastern Ghats, N.E. Hills |
| 12. | *D. deltoidea* | - | India, Himalayan foothills |
| 13. | *D. prazeri* | | India, Himalayan foothills |
| 14. | *D. rotundata* | | Africa, Kerala in India |

| Sl.No. | Botanical Name | Common Name | Specific Area of Availability |
|---|---|---|---|
| | | **Colocasia** | |
| 15. | *Colocasia esculenta* var. *esculenta* | Taro-Dasheen | India, North Eastern Region, Eastern Ghats |
| 16. | *C. esculenta* var. *antiquorum* | Taro-eddoe | Probably in China, Japan, Eastern Ghats and Deccan in India |
| | | **Alocasia** | |
| 17. | *Alocasia macororrhiza* | Giant Alocasia | Asom-Bengal region, North Eastern Region |
| | | **Amorphophallus** | |
| 18. | *Amorphophallus campanulatus* | Elephant foot yam | South East Asia, Deccan peninsula, wild in N.E. Region |
| | | **Xanthosoma** | |
| 19. | *Xanthosoma sagittifolium* | Cocoyam | Tropical America, Peninsular Island regions |
| 20. | *X. violaceum* | Cocoyam | Tropical America, Peninsular Island regions |
| | | **Manihot** | |
| 21. | *Manihot esculenta* | Cassava | Tropical America, Western Ghats, sporadic in eastern region, N.E. Hills |
| | | **Ipomoea** | |
| 22. | *Ipomoea batatas* | Sweet potato | South America, Bihar and adjoining areas |
| | | **Zingiber** | |
| 23. | *Zingiber officinale* | Ginger | South East Asia, Western Eastern Himalaya, N.E. Region, Eastern Western Peninsular regions |
| | | **Curcuma** | |
| 24. | *Curcuma amada* | Mango ginger | India, Eastern Peninsular, East Ghats, Deccan |
| 25. | *C. angustifolia* | Indian arrowroot | India, Central parts, Himalayan foot hills |
| 26. | *C. aromatica* | Yellow zedoary | Throughout India |
| 27. | *C. caesia* | Black zedoary | India particularly Eastern India |
| 28. | *C. domestica* | Turmeric | Throughout India |
| 29. | *C. zedoaria* | Zedoary | India particularly N. E. India, Eastern Ghats |
| | | **Maranta** | |
| 30. | *Maranta arundinacea* | Arrowroot | Caribbean, Western Ghats, N.E. Hills |
| | | **Canna** | |
| 31. | *Canna edulis* | Edible canna | South America |
| | | **Helianthus** | |
| 32. | *Helianthus tuberosu* | Jerusalem artichoke | North America, Eastern and N.E. region of India |
| | | **Cynara** | |
| 33. | *Cynara scolymus* | Globe artichoke | Mediterranean and Northern India |
| | | **Coleus** | |
| 34. | *Coleus parviflorus* | - | Africa and Kerala in India |

## Centre of Diversity of Vegetable Crops

Diversity in vegetable species occurs in the form of land races, traditional cultivars, wild edible forms and related non-edible wild and weedy species. Origin of a species usually occurs in a specific set of ecological conditions while diversification demands variable ecosystem and the same situation may be available at the place of origin or at places of subsequent introduction often species diversity overlapped areas of several centres. Hence now more appropriately centre of origins have been re-designated as centres of diversity. Species diversity is mainly distributed in seven-eight geographical regions, which represent the centres of origin and/or diversity as well. Native genetic diversity of Indian gene centre exhibit variable form/landraces/primitive types in different crops and their wild relatives.

## Characteristics of Centre of Diversity

The main characteristics of centre of diversity are as under (Sharma, 2009):

(i) All centres described lie in tropical or sub-tropical regions of the world.

(ii) The regions are show greater altitude variation.

(iii) The areas are marked by greater temperature fluctuation wider humanity and soil variation.

(iv) More impregnation of atmosphere with ultra violet and cosmic radiation in these regions

(v) Availability of varied ecological niche at short distance.

(vi) More seasonal and diurnal variations.

(vii) As one goes away from the centre of diversity, the plant population of species becomes more homogenous.

(viii) The areas are mountainous with considerable topographical variation.

(ix) Centres are often isolated by high mount peaks, rivers or deserts.

(x) Diversity patterns of species shows overlapping often more them one centres of diversity.

## Centres of Diversity at Global Levels

The concept that centres of diversity represent centres of origin has been seriously questioned. Plants of a species growing in different environments are likely to be different *i.e.* diverse. Thus a plant species is likely to show a greater variation in a region with varied climatic and other ecological conditions. Area with mountains and valleys show considerable variation in the prevalent environment; therefore, plant species would show a greater variation in such areas. Interestingly, the centres of origin are situated in such mountain-valley areas. Further, the centres of diversity of many species have shifted with time due to a shift in the area of the greatest cultivation and introduction of the crop species into an area with greater ecological diversity than where they existed before. These processes have given rise to secondary centre of origin of plants. Consequently, several species have two or more centres of diversity, and it is often difficult to determine as to which of them is the real centre of origin (Singh, 2012). The significance of variation within a species is less widely appreciated but is critical, particularly for agriculture.

Thus the centres of origin may be more appropriate called centres of diversity. Zhukovsky, in 1965, recognized 12 mega gene centres of crop plant diversity (Table 1.13) and a number of micro-gene centres of wild relatives of crop species. The cultivated forms are believed to have first originated in these micro-gene centres. These centres may not be the centres of origin of the species concerned, but they are the areas of the maximum diversity of these species. This serves as an extremely useful guide to plant explorers but indicating the region where they should search for variation in a given species. Within the large centres of diversity, small areas may exhibit a much greater diversity than the centre as a whole; these areas are known as micro-centres. Evolution appears to proceed at a more rapid rate in such micro-centres and they are important for plant collection as well as for an experimental study of plant evolution. Global diversity of vegetables is recognized in the form 400 plant species which is distributed in 9 major and 3 minor centres covering both old and new world. The regions possessing maximum diversity at global levels rich in diversity are tropical America, Africa, tropical Asia and Mediterranean. In the tropical Asian region, both Indian and Chinese regions show maximum diversity. Since long, Indian subcontinent is well known as an important centre of origin and diversity of a large number of vegetable crops. It has two 'hotspots' of biodiversity in the Eastern Himalayas and Western Ghats. At least 147 agri-horticultural crop species are believed to have originated in India. For the underutilized, neglected and less known domesticated/cultivated vegetables, this diversity is mainly confined to seven regions, the Chinese-Japanese region, Indo-chinese-Indonesian region, Hindustan region, Mediterranean region, European-Siberian region, African, and South American regions where approximately 495 species of the 540 species occurred worldwide (Arora, 1985).

### Table 1.13: The Regions of Diversity of Vegetable Crops

| Sl.No. | Centre of Diversity | Vegetable Crops |
|---|---|---|
| 1. | Chinese-Japanese region | Leafy mustard, bamboo shoot |
| 2. | Indo-Chinese-Indonesian region | Winged bean, cucurbits (ash gourd, bitter gourd, bottle gourd *etc.*), bamboo shoot, taros, yams |
| 3. | Australian region | - |
| 4. | Hindustan region | Lablab bean, cowpea, eggplant, okra, cucumber, taros, yams, ginger, turmeric |
| 5. | Central region | *Allium* (onion and garlic), spinach, peas, beetroot, faba bean, melons |
| 6. | Near-Eastern region | Faba bean, French bean, pea, *B. oleracea*, *Allium*, melon |
| 7. | Mediterranean region | *B. oleracea*, lettuce, beetroot, faba bean, radish, |
| 8. | African region | Cowpea, bottle gourd, okra, cucumber, yams, melons |
| 9. | European-Siberian region | Black mustard, lettuce |
| 10. | South American region | Potato, sweet potato, lima bean, amaranth, *Chenopodium* species, *Cucurbita* species, tomato, cassava |
| 11. | Central American and Mexican region | French bean, potato, *Cucurbita* species, chilli, amaranth, *Chenopodium* spp. |
| 12. | North American region | Jerusalem artichoke |

*Source*: Based on Zeven and Zhukovsky, 1975 and Zeven and de Wet, 1982.

## Diversity of Crop Plants

Vavilov revealed that out of the wide range of plant diversity in the tropical and subtropical regions of the world, the major food crops have come mainly from high mountain valleys, isolated from each other to a large extent and with a great habitat range. There people made useful selection in several cereal and vegetable crops, which were eventually domesticated and cultivated. Several weedy species were never or only temporarily domesticated, remaining as weeds but often hybridizing by chance with the cultivated ones and thus, enhancing the diversity in cultivated plants.

(1) **Primary centre of diversity or primary gene centre:** Primary gene centre represents the centre of origin. It is considered that the species first evolved here and subsequently variation was brought here by various evolutionary factors. Diversity of a vegetable crop here is characterized by wild edible, wild non-edible species with varying degree of crossability, several land races and ecotypes differing for morphological, physiological, biochemical and genetical parameters. Primary centre of diversity includes brinjal (*Solanum melongena*), cucumber (*Cucumis sativus*), dolichos bean (*Lablab purpureus*), drumstick (*Moringa oleifera*), Indian lettuce (*Lactuca sativa*), Indian spinach (*Basella rubra*), ivy gourd (*Coccina cordifolia*), pointed gourd (*Trichosanthes dioica, T. cucumerina*), ridge gourd (*Luffa acutangula*), smooth gourd (*Luffa cylindrica*), sword bean (*Canavalia gladiate*), winged bean (*Psophocarpus tetragonolobus*), *Alocasia esculenta*, *Alocasia macrohiza, Amorphophallus campanulatus, Citrullus lanatus var. fistulosus, Coccinia cordifolia, Colocasia esculenta, Cucumis sativus, Dioscorea species, Lagenaria siceraria, Luffa acutangula, L. aegyptiaca, L. hermaphrodita, Rumex vesicarius etc.*

(2) **Secondary centre of diversity or secondary gene centre:** Secondary centre of diversity or secondary gene centres have evolved in response to dispersal or dissemination of species to new ecological either due to natural shift or due to introduction of the species to an area with much ecological diversity by human being. Several species often overlap two or more centres of diversity, creating ambiguity about their real centre of origin. However, secondary centres may differ from centre of origin in terms of presence and magnitude of diversity among wild relatives or their primitive forms. Secondary centre of diversity includes okra (*Abelmoschus esculentus*), tomato (*Solanum esculentum*), pumpkin (*Cucurbita* species), chayote (*Sechium edule*), chilli (*Capsicum* species), cucumber (*Cucumis sativus*), bitter gourd (*Momordica charantia*), bottle gourd (*Lagenaria siceraria*), snake gourd (*Trichosanthes anguina*), brassicas (*Barassica* species), cowpea (*Vigna sinensis*), pumpkin (*Cucurbita moschata*) and radish (*Raphanus sativus*)

Distributions of major vegetable species along with primary and secondary centre of diversity are as under shown in Table 1.14.

### Table 1.14: The Centre of Diversity (Whole world) of Major Vegetable Crops

| Sl.No. | Gene Centre | Primary Centre of Diversity | Secondary Centre of Diversity |
|---|---|---|---|
| 1. | Chinese-Japanese region | Eggplant, wax gourd, Chinese cabbage, Kangkong, welsh onion | Watermelon, amaranth |
| 2. | Indo-Chinese-Indonesian region | Wax gourd, sponge gourd, ridge gourd, bitter gourd, sword bean, winged bean, taro, chayote, cucumber, bottle gourd, yam | Chinese cabbage, bottle gourd, cucumber, yam, bean, amaranth, yard long bean, Kangkong |
| 3. | Hindustani centre | Eggplant, wax gourd, cucumber, ridge gourd, bitter gourd, sponge gourd, hyacinth bean, drumstick, okra, Kangkong | Watermelon, melon, Roselle, bottle gourd, amaranth |
| 4. | Central Asia | Onion, garlic, carrot, spinach | Eggplant, watermelon, melon, cauliflower |
| 5. | Near East | Onion, garlic, leek, beet | Okra |
| 6. | Mediterranean | White cabbage, cauliflower, watermelon, broccoli, radish | Sweet pepper, garlic, okra |
| 7. | African | Eggplant, watermelon, melon, bottle gourd, cowpea, okra, Roselle, locust bean | Onion, shallot, lima bean, white cabbage, amaranth |
| 8. | European-Siberian | Lettuce | Onion, white cabbage, common bean, cauliflower, spinach, carrot |
| 9. | Central America and Mexican region | Tomato, hot pepper, pumpkin, squash, yam, bean, sweet potato, common bean | - |
| 10. | South American region | Tomato, hot pepper, cassava, pumpkin, lima bean, chayote, amaranth, sweet potato | Common bean |
| 11. | North American | - | Tomato, Eggplant, melon, watermelon, pepper, squashes, onion, lettuce, lima bean, okra, pumpkin |

*Source*: Gupta *et al.*, 1995.

**Table 1.15: Agroecological Regions of India**

| Sl.No. | Region | Temperature | | Rainfall (mm) | Soil |
|---|---|---|---|---|---|
| | | Month | °C | | |
| 1. | Humid Western Himalayan Region | January | 8°-4°C | 1500 | Forest and Hill, mountain and meadow soils |
| | | July | 5°-30°**C** | | |
| 2. | Humid Bengal-Asom Basin | January | 10°-25°C | 2000 | Alluvial, red sandy, and laterite soils |
| 3. | Humid Eastern Himalayan Region and Bay Islands | January | 11°-24°C | 2000 | Riverine, alluvial, red and laterite soils |
| 4. | Sub- humid Sutlej Ganga Alluvial Plains | January | 6°-23°C | 1000 | Riverine, alluvial soils with saline and alkaline pathches |
| | | July | 25°-41°C | 1500 | |
| 5. | Sub-Humid to Humid Eastern and South Eastern Uplands | January | 16°-28°C | 750 | Red and yellow, red sandy, black and coastal alluvial soils |
| | | July | 27°-35°C | 1500 | |
| 6. | Arid Western Plains | January | 10°-22°C | 250 | Alluvial, sandy desert and black soils |
| | | July | 28°-45°C | 750 | |
| 7. | Semi-Arid Lava Plateaus and Central Highlands | January | 13°-29°C | 500 | Predominantly black soils, red sandy and red loamy soils |
| | | July | 26°-42°C | 1000 | |
| 8. | Humid to Semi-Arid Western Ghats and Karnataka Plateaus | January | 20°-29°C | 750 | Predominantly red sandy soils black, coastal alluvial and laterite soils |
| | | July | 28°-38°C | 2000 | |

## Diversity in Agroecological Regions/ Systems of India

India has a rich diversity of the agroecosystems as similar to richness in terms of between and within species diversity. Eight agro-ecological regions have been recognized in India, on the basis of physiography, climate soil and associated factors (Murthy and Pandey, 1978) wherein the entire native diversity in plant genetic resources is distributed.

## Agroecological Systems of India

ICAR has clearly mentioned about 20 distinct agroecological regions in the country, which belong to six broad ecological systems on the basis of climates, souil characteristics and average length of growing period of annual crops in their agroecosystems.

**Table 1.16: Agroecological Systems of India**

| Sl.No. | Agroecological Regions/Systems | Description |
|---|---|---|
| 1. | Arid ecosystem | Average growing length of period less than 90 days: |
| | | (i)  Western Himalaya-cold arid ecoregion with shallow skeletal soils |
| | | (ii)  Western Plain, Kachchh and part of Kathiawar Peninsula- hot arid ecoregion with desert and saline soils |
| | | (iii)  Deccan Plateau- hot arid ecoregion with red and black soils |
| 2. | Semi- arid ecosystem | Average length of growing period ranging between 90 and 150 days: |
| | | (iv)  Northern Plain and Central Highlands includiong Aravallis-hot semi arid ecoregion with alluviul derived soils |
| | | (v)  Central (Malwa) highlands, Gujarat Plains and Kathiawar Peninsula- hot semi arid ecoregion with medum and deep black soils |
| | | (vi)  Deccan Plateau hot semi arid ecoregion with shallow and medium (with inclusion of deep) black soils |
| | | (vii)  Deccan (Telangana) Plateau and Eastern ghats- hot semi arid ecoregion with red and black soils |
| | | (viii)  Eastern Ghats, Tamil Nadu uplands and Deccan (Karnataka Plateau)- hot semi arid ecoregion with red loamy soils |
| 3. | Sub-humid ecosystem | Average length of growing period ranging between 150 and 180 (to 210)+ days: |
| | | (ix)  Northern Plain-hot subhumid (dry) ecoregion with alluvium-derived soils |
| | | (x)  Central Highlands (Malwa, Bundelkhand and Satpura)- hot subhumid ecoregion with black and red soils |
| | | (xi)  Eastern Plateau (Chattishgarh)- hot subhumid ecoregion with red and yellow soils |
| | | (xii)  Eastern (Chotanagpur) Plateau and Eastern Ghats- hot subhumid ecoregion with red and lateritic soils |
| | | (xiii)  Eastern Plain- hot subhumid (moist) ecoregion with alluvium-derived soils |
| | | (xiv)  Western Himalaya- warm subhumid to humid with inclusion of perhumid ecoregion with brown forest and podzolic soils |

| Sl.No. | Agroecological Regions/Systems | Description |
|---|---|---|
| 4. | Humid perhumid ecosystem | Average length of growing period ranging above 210 days: |
| | | (xv) Bengal and Asom Plain- hot sub-humid (moist) to humid (inclusion of perhumid) ecoregion with alluvium-derived soils |
| | | (xvi) Eastern Himalaya-warm perhumid ecoregion with brown and red hill soils |
| | | (xvii) Northeastern Hills (Purvanchal)- warm perhumid ecoregion with red and lateritic soils |
| 5. | Coastal ecosystem | (xviii) Eastern Coastal Plain- hot subhumid to semi arid ecoregion with coastal alluvium-derived soils and average growing period ranging between 90 and 210+ days. |
| | | (xix) Western Ghats and Coastal Plain- hot humid perhumid ecoregion with red, lateritic and alluvium-derived soils and average growing period ranging above 210 days |
| 6. | Island ecosystem | (xx) Ilands of Andaman-Nicobar and Lakshadweep-hot humid to perhumid island ecoregion with red and sandy soils and average growing period ranging above 210 days. |

*Source*: Handbook of Agriculture, ICAR, New Delhi, 2009.

## Indian Gene Centre and Diversity

The Indian gene centre is one of the 12 mega-biodiversity centre (One of the Vavilovian Centres of Origin) of crop plant diversity with two major hot spots (Western Ghats and North Eastern Hill Region) which holds rich floristics wealth of over 17, 500 species of higher plants and about one-third of endemic flora including several vegetable crops (Zeven and de Wet, 1982). India's share of crops is 44 per cent as compared to the world average of 11 per cent. The Indian sub-continent is well known as an important centre of origin for large number of agri-horticultural crops. Over 15, 000 species of flowering plants are native to Indian centre of origin. High priority is accorded to nearly 500 species of cultivated plants where 166 crop species (25 major and minor crops) plants have originated in Indian Sub-Continent and over 320 species of wild ancestral/forms (relatives) of crop plants are reported. In Indian gene centre is also recognized for its native wealth of PGR with over 800 species of ethno-botanical importance and 1,200 species are known to possess mecicinal and aromatic value. Indian gene centre has been contributed useful plant species such as yam, taro, eggplant, edible banana, cucumber, pepper and many other plant species of economic value.

## Agro-climatic Zones and Vegetable Diversity in India

India is located between 8° N to 38° N latitude and 68° E to 97.5° E-longitude and exhibits extreme variations in edapho-climatic situations, agro-climatic regions and floristic diversity. The altitudinal variations are observed from below sea level to above vegetational limits in the Himalayas *i.e.* more than 3600 m above mean sea level; the climates change from monsoon to temperate/alpine in the northern/Himalayan zone. Occurrence of 20 agro-climatic regions in India based on physiographic, climatic and cultural features has been reported. It has highly fertile Indo-gangetic plains. Climatic variation of the country includes cold and hot arids, hot semi-arids, hot sub-humids, and hot humids. The cultural ethnic diversity includes over 550 tribal communities of 227 ethnic groups spread over 5,000 forested villages. In short, it supports all kinds of tropical, sub-tropical and temperate vegetation or agro-biodiversity. Indian subcontinent is well known since long as an important centre of origin and diversity of a large number of crops including vegetable crops. Its floristic diversity is very rich with about 30 per cent of 49,000 plant species being endemic. Moreover its agro-biodiversity is extremely rich with about 30,000 to 50,000 landraces of different

**Table 1.17: The Prominent Plant Families and Species Diversity in India**

| Crop Group | Prominent Families | Total spp. |
|---|---|---|
| Cereals and millets | Poaceae | 51 |
| Legumes | Fabaceae | 31 |
| Oilseeds | Cruciferae, Pedaliaceae | 12 |
| Fiber crops | Tiliaceae, Malvaceae | 24 |
| Spices and condiments | Liliaceae, Zingiberaceae, Pipraceae | 27 |
| Plantation crops | Palmae, Theaceae | 16 |
| Vegetables | Araceae, Dioscoreaceae, Compositeae, Malvaceae, Solanaceae, Cucurbitaceae | 54 |
| Fruits | Rutaceae, Rosaceae, Musaceae, Anacardiaceae, Rhamnaaceae, Mystaraceae, Moraceae | 109 |
| **Total** | | **324** |

**Table 1.18: Vegetable Crop Diversity in India**

| Species | Origin | Diversity/Domestication in India |
|---|---|---|
| *Luffa acutangula* | India | W.C and India |
| *Luffa cylindrica* | Japan, Brazil, India | Concentration in eastern peninsular India |
| *Lagenaria siceraria* | Latin America, Africa | Throughout India |
| *Cucurbita pepo* | North America, Mexico | Secondary centre of diversity |
| *Cucurbita maxima* | North and Central America | Secondary centre of diversity, Concentration in NEH regions |
| *Cucurbita moschata* | Mexoco and Central America | Tropical and sub-tropical regions |
| *Cucumis melo* | South Africa | Domestication in India-North West and Indo-Gangetic plains |
| *Cucumis sativus* | India (Himalaya) | Throughout India |
| *Trichosanthes anguina* | South Asia, India | Malabar coast, Eastern Ghats, NEH regions |
| *Trichosanthes dioica* | India (Karnataka and West Bengal) | Uttar Pradesh, Bihar, West Bengal and Asom |
| *Benincasa hispida* | South East Asia, Indo-China | Tropical and sub-tropical regions |
| *Praecitrullus fistulosus* | India | North West and Indo-Gangetic plains |
| *Citrullus lanatus* | West Africa | Uttar Pradesh and Maharashtra |
| *Solanum melongena* | India | Throughout India |

crops. Around 80 species of major and minor vegetables, apart from several wild/gathered kinds occurs in Indian condition. Bestowed with, immensely rich landrace diversity in major agri-horticultural crops (166 species) and enormously rich in wild relatives (324 species). Thus makes the Indian sub-continent unique and interesting to many in the world.

The Indian region is covered under two centres of origin:

**(a)** Indo-Chinese

**(b)** Hindustani

About 22 per cent diversity of vegetable crops is found in the region. It is primary centre of variability of brinjal (*Solanum melongena*), ridge gourd (*Luffa acutangula*), sponge gourd (*Luffa cylindrica*), snake gourd (*Trichosanthes anguina*), Amaranth (*Amaranthus tricolor*) and Dolichos bean (*Lablab purpureus*). Indian region is also secondary centres of diversity for several vegetables, namely okra (*Abelmoschus esculentus*), onion (*Allium cepa*), garden pea (*Pisum sativum* var *hortense*), cowpea (*Vigna unguiculata*), chillies (*Capsicum annuum*), watermelon (*Citrullus lanatus*), cluster bean (*Cyamopsis tetragonoloba*), muskmelon (*Cucumis melo*), bottle gourd (*Lagenaria siceraria*), ivy gourd (*Coccinia cordifolia*) and several *Brassica* leafy vegetables. Diversification is well recognized in species originating here. Cucumber shows variation in range of adaption to diverse climatic conditions, fruit shape, size, colour and flavor. Vegetable diversity is mainly distributed in eight eco-geographical regions or phyto-geographical zones (Chatterjee, 1939) of India. By and large, tropical vegetable crops are grown throughout the country as suitable agro-climates are available; only temperate vegetables have more specific requirements and are grown mainly in the Himalayan ranges and to some extent in higher ranges of the Western Ghats; also these are well adapted as *rabi* (cold season) in Northern/North Western Plains

and in the tarai foot hills. The species of wild okra like *Abelmoschus ficulneus, A. manihot, A. tuberrosus etc.* are fairly distributed in Western and Eastern Ghats, Gangetic plains and north Western/North Eastern India. *A. tuberculatus*, a tall type with tuberculate fruits has provided resistance for yellow-vein mosaic virus in breeding of Pusa Sawani, a widely grown cultigen for so long but today it is highly susceptible to YVMV. The wild forms of eggplant, *Solanum incanum, S. melongena* var. *insanum, S. khasianum, S. sisymbrifolium* and *S. surratense etc.* having useful genes for diseases and pests resistance, occur more frequently in tarai region and the eastern peninsular region. The other wild germplasm include species of *Momordica, Trichosanthes* and *Cucumis*, the former two are predominant in the north east and peninsular region and the latter in foot hills and north eastern region in Himalayas. Similarly several species of wild ginger and turmeric (*Zingiber* and *Curcuma* species) and in root crops such as wild yams (*Dioscorea*) and taros (*Alocasia* and *Colocasia*) occur particularly in the humid tropical habitats in the Western and the Eastern Ghats.

## 1. North-Western and Eastern Himalayan Region

Under Western and Eastern Himalayan region, only temperate vegetables are predominantly found. Enormous diversity occurs in *Allium* species- leek, shallot and other introductions of *Allium sativum* and *Allium cepa*. The cold and drought tolerant types in *Allium* and *Trigonella* occur in disturbed cold habitats of North West Himalayas. Sporadic diversity also occurs in asparagus, spinach, chenopods, amaranths, and *Beta vulgaris*. Other vegetables like Brassicas, squash, *Cucurbita* spp., *Cucumis*, chillies, bell pepper, peas, faba bean, cowpea, horse radish, artichoke, potato, *Colocasia*, tomato, parsley, coriander, ginger, *Sechium edule* and *Cyclanthera pedata* do expresses a lot of variability.

**Table 1.19: Endemic Wild Edibles of Indian Himalaya**

| Himalayan Region | Wild Edible Species |
|---|---|
| Trans North – West Himalaya | Allium consanguinium |
| | Caralluma tuberculata |
| | Campanula cashmiriana |
| | Sedum tibeticum |
| | Ribes nigrum |
| | Linaria incana |
| West Himalaya | Allium strachey |
| | Cordia vestita |
| | Ribes uva-crispa var. sativum |
| Central Himalaya | Mahonia sikimensis |
| | Robus treutleri |
| | R. wardii |
| | Sterculia roxburghii |
| East Himalaya | Calamus erectus |
| | Caryota obtuse |
| | Livistonia jenkisiana |
| | Begonia rubrovenia |
| | B. episcopalis |
| | Streptolirion volubile |
| | Stixix suaveolens |
| | Gaultheria discolor |
| | Baliospermum calycinum |
| | Garcinia pedunculata |
| | G. sopsopia |
| | G. cowa |
| | G. stipulate |
| | Illicium griffithii |
| | Musa sanguine |
| | Phoenix rupicola |
| | Rubus insignis |
| | Spiradiclis bifida |
| Trans/North–West/West | Berberis zabeliana |
| North–West/West | Lonicera parvifolia |
| West/Central | Angelica glauca |
| West/Central/East | Decaisnea insignis |
| Central/East | Elaeagnus pyriformis |
| | Elaeocarpus sikkimensis |
| | E. floribundus |
| | Saurauia punduana |

*Source*: Samant and Dhar, 1997.

## 2. North-Eastern Region including Asom

The north east as a whole is recognized as a primary centre of origin (Vavilov, 1926) of 152 species of cultivated plants, where several groups of tribals are still practicing a life style which enables them to utilize plants as vegetables, medicinal and other economic uses without causing disturbances to the valuable species. In this region, most of the temperate vegetable crops predominate. North Eastern Region including Asom has maximum diversity of leafy vegetable crops like *Amaranthus* and *Brassica* species. Other vegetables like chillies, tomato, brinjal, okra, taros, yams and cucurbits are also grown in this tract. Several kinds of beans like winged bean, Jack bean, French bean, sword bean and lima bean are specialized due to edaphic and climatic factors. In the lower tract, rich diversity occurs for cucumber, pointed gourd chayote, bitter gourd, spine gourd (*Momordica dioica* Roxb.) and meetha karela (*Cyclanthera pedata*). Among the wild species, *Abelmoschus manihot* (pungens form), *Alocasia macrorrhiza, Amorphaphallus bulbifera, Colocasia esculenta, Cucumis hystrix, Cucumis trigonus, Dioscorea alata, Luffa graveolens, Moghania vestita, Momordica cochinchinensis, Momordica macrophylla, Momordica subangulata, Trichosanthes cucumerina, Trichosanthes dioica, Trichosanthes dicaelosperma, Trichosanthes khasiana, Trichosanthes ovata, Trichosanthes truncata* and *Solanum indicum* are prevalent.

## 3. The Northern Plains/Gangetic Plains including Tarai Region

The region is one of the richest pockets of diversity for major vegetable crops. Due to year round cultivation, more diversity emerges and it is well fitted into the cropping patterns. Rich diversity can be observed in *Cucumis* species, *Luffa* species, *Cucurbita* species, *Benincasa hispida, Lagenaria siceraria, Momordica* species, *Trichosanthes* species, *Solanum* species, *Capsicum* species, *Abelmoschus* species, *Brassicas* and tuber crops.

## 4. North Western/Indus Plains

In this region, variability exists for *Cucumis* species, *Momordica* species, *Citrullus* species, *Solanum* species, *Amaranthus, Chenopodium, Abelmoschus* species, *Capsicum* species and *Allium* species. There are certain sporadic pockets, which are rich in indigenous germplasm like Cucurbits, okra, eggplant and garlic. Specific adoptable diversity of *Caralluma* species may be spotted in this region. The wild species of *Momordica balsamina, Citrullus colocynthis* and *Cucumis prophetarum* are prevalent.

## 5. The Central Region/Plateau

More diversity occurs for *Cucurbita* species, ash gourd, round gourd, bitter gourd, pointed gourd, ridge gourd, okra, eggplant, chillies, tomato, root and bulbous crops and sporadically in leafy vegetables. As indigenous vegetables, more diversity exists in cucurbits, eggplant, okra and chillies. The wild species like *Cucumis setosus, Cucumis trigonus, Luffa acutangula var. amara, Momordica cymbalaria* etc. are more commonly prevail in this region.

## 6. The Western and Eastern Peninsular Region

This region is extremely rich for cucurbits (cucumber, bitter gourd, bottle gourd and squashes) diversity, eggplant, okra and chillies (both annual and perennial types). Diverse landrace of snake gourd occurs in the western and for *Luffa* species and eggplant in eastern region. Sporadic diversity can also be spotted for leafy vegetables like *Amaranthus, Brassicas, Chenopodum, Spinach, Beta vulgaris, Basella rubra* and *Basella alba*. Several wild species like *Abelmoschus angulosus, A. moschatus, A.manihot, A. ficuleneous, Amorphophallus campanulatus, Colocasia antiquorum, Cucumis hystrix, C. setosus, C. trigonus, Luffa acutangula* var. *amara, Luffa graveolens, Luffa umberrata, Momordica cymbalaria, M. denticulata, M. dioica, M. cochinchinensis, M. subangulata, Solanum indicum, S. melongena, Trichosanthes anamalaiensis, T. bracteata, T. cordata, T. cuspidata, T. horsfieldii, T. piniana, T. perottitiana, T. nerifolia, T. himalensis, T. multiloba and T. villosa* are quite prevalent in this region.

Rich genetic diversity occurs in several crop plants (about 166 species) and their wild relatives (320 species) antiquity of agriculture and ethnic diversity in the sub-continent have played a major role in the diversification of crop resources in this region. In addition, there has been a continuous stream of introductions of new crops and their cultivars by man since the ancient times.However the widely grown prominent species in this region are hardly 20-25 (Harlan, 1971; Hawkes, 1983). Vegetable crops like garlic, potato, sweet potato, tomato, chillies, French bean *etc.* were brought by the

Mughals, Spaniards, Portugese and the British. Thus, both indigenous and well adapted exotic set of materials constitute a well balanced matrix of crop diversity in Indian sub-continent.

Some hills of the country with respect to vegetaional diversity are mentioned in Table 1.21.

## Vegetable Scenario with Respect to their Origin

About 400 species constitute the global diversity in vegetable crops. Ever growing population and increasing demand on agriculture, particularly during the 20th century, led to destruction of habitats rich in diversity and consequently narrowing down the versatility of food baskets. Food production would fail to keep pace with increase in the demand for food by growing population in many developing countries. This food gap would be doubled making some of world's poorest people even more vulnerable to hunger and possible famine. However, today's food is based on 150 species only and hardly about 12 species provide more than 80 per cent of food calories consumed by humans. Infact, only four species *viz.* maize, wheat, rice and potato provide more than half of the required calories.

**Table 1.20: Distribution of Major Vegetable Crops Variability in Agro-ecological Regions of India**

| Sl.No. | Agro ecological Regions | Geographical Ranges | Variability in vegetable crops |
|---|---|---|---|
| 1. | Humid Western Himalayan Region | Jammu and Kashmir, Himachal Pradesh and parts of U.P. | Cucurbits (*Sechium edule*), radish, carrot, turnip, peas, cowpea, chillies, brinjal, potato, okra, baby corn, spinach, fenugreek, amaranth, *Solanum khasianum, Solanum hirsutum, Basella rubra* |
| 2. | Humid Bengal/Asom Basin | West Bengal and Asom | Cucurbits (*Cucumis sativus* var. *sikkimensis, Edgeria dargelingensis, Melothria Asomica, Momordica cochinchinensis, Sechium edule, Tuladiantha cordifolia*), radish, cowpea, chillies, brinjal, okra, spinach, beet, *Abelmoschus manihot* ssp. *manihot, Amaranthus, Solanum indicum, Solanum khasianum, Solanum surattense, Basella rubra* |
| 3. | Humid Eastern Himalayan Region and Bay Lands | Arunchal Pradesh, Nagaland, Manipur, Mizoram, Tripura, Meghalaya, Andaman and Nicobar Islands | Cucurbits (*Cucumis hystrix, Luffa echinata, Sechium edule*), radish, cowpea, pea, chillies, brinjal, okra, spinach amaranth, *Abelmoschus manihot* ssp. *tetraphyllus, Solanum khasianum, Solanum torvum, Solanum sisymbrifolium, Solanum ferox, Solanum verbasifolium* |
| 4. | Sub-humid Sutlej Ganga Alluvial Plains | Punjab, U.P. and Bihar | Cucurbits (*Cucumis hardwickii, Cucumis trigonus*), radish, peas, brinjal, okra, spinach beet, fenugreek, onion, garlic, *Abelmoschus manihot* ssp. *tetraphyllus* var. *pungens, Abelmoschus tuberculatus, Solanum indicum, Solanum khasianum, Solanum torvum, Solanum surattense, Solanum hispidum* |
| 5. | Humid Eastern and South Eastern Uplands | East M.P., Odisha and A.P. | Cucurbits, radish, carrot, cowpea, chillies, brinjal, okra spinach, amaranth, garlic, *Abelmoschus manihot* ssp. *manihot, Solanum surattense, Solanum torvum* |
| 6. | Arid Western Plains | Haryana, Rajasthan and Gujarat | Cucurbits (*Citrullus colocynthes*), cauliflower, radish, carrot, peas, cowpea, chillies, brinjal, okra, spinach beet, fenugreek, onion, garlic, amaranth, *Abelmoschus tuberculatus, Abelmoschus ficuleneus, Abelmoschus manihot* ssp. *tetraphyllus, Solanum torvum, Solanum nigrum* |
| 7. | Semi-Arid Lava Plateau and Central Highlands | Maharashtra and West M.P. | Cucurbits (*Cucumis setosus, Luffa acutangula* var. *acutangula*), cauliflower, radish, carrot, cowpea, chillies, brinjal, okra, spinach, fenugreek, amaranth, onion, *Solanum surattense, Solanum torvum, Solanum nigrum, Solanum khasianum* |
| 8. | Humid to Semi-Arid Western Ghats and Karnataka Plateau | Karnataka, Tamil Nadu, Kerala and Lakshadweep Islands | Cucurbits (*Luffa acutangula* var. *acutangula, Melothria angulata*), chillies, brinjal, okra (*Abelmoschus crinitus, Abelmoschus angulosus, Abelmoschus ficuleneus, Abelmoschus moschatus, Abelmoschus manihot* var. *tetraphyllus, Solanum trilobatum, Solanum indicum, Solanum incanum, Solanum pubescens, Solanum surattense, Solanum torvum,, Basella rubra* |

*Source*: Gupta *et al.*, 1995.

**Table 1.21: Hills of the Country with Respect to Vegetaional Diversity**

| State | Area Description and Characterization of Agro-Eco-Zones | |
|---|---|---|
| Andhra Pradesh | High altitude and Tribal Zone | Northern borders of the state in the district of Srikakulam, Vijanagaram, Vishakhapatnam, East Godavari and Khammam |
| Arunachal Pradesh | | Dibang Valley, Changlang, East Siang, West Siang, West Kameng, Lower Subansiri, Upper Subansiri, and Lohit |
| Asom | | Karbi Anglong, and North Cachar |
| Bihar | | Hills parts of Gaya and Aurangabad |
| Chhattishgarh | | North Hill Zone of Chhattishgarh, Bastar Plateau Zone, Kymore Plateau (Durg, Raipur, Bilaspur, Sarangarh and Gharghoda tehsils of Raigarh and Kankar tehsils of Bastar) |
| Gujarat | South Gujarat Heavy Rainfall Zone | Hilly areas of Dangs and Valsad district; North Gujarat Zone; 500 to 1,090 m a.m.s.l |
| Haryana | | Shiwalik Hills, Dissected plains in the foot-hills and southern tract with Aravalli hills |
| Himachal Pradesh | Sub-Montane and Low Hills Sub Tropical Zone | Uplands of Chamba, Kangra, Solan, Hamirpur, Sirmaur and Bilaspur |
| | Mid Hills Sub-Humid Zone | Parts of Chamba, Kangra, Mandi, Solan, and Shimla |
| | High Hills Temperate Wet | Kullu and Parts of Chamba, Kangra, Mandi, Sirmaur and Shimla |
| | High Hills Temperate Dry Zone | Kinnaur, Lahaul-Spiti and parts of Chamba |
| Jharkhand | Western Plateau | Palamu, Lohardaga, Gumla and Ranchi |
| | Central and North-eastern Plateau | Hazaribagh, Giridih, Devghar, Godda, Sahibganj, Santhal Pargana, Dhanbad |
| | South eastern Plateau | Paschim and Purba Singhbhum |
| Karnataka | Hilly zone | North Kannada, Chikmangalur, Dharwad, Coorg and Hassan |
| Kerala | High Range | The mountainous land (750 to 2,500 m m.s. l.)-Wynad, and Idukki districts and Eastern part of other districts bordering the Western Ghats |
| | High Land | Hilly tracts (750 to 750 m m.s. l.) on western side of the Western Ghats covered with forests |
| Madhya Pradesh | Satpura Hill Zone | Jabalpur, Panna, Satna, Rewa, Seoni and Gopandbana districts and Deosar tehsil of Sidhi |
| | Hill Zone | Jhabua and Dhar |
| Maharashtra | Sub-Montane Zone | Spread over Nasik, Pune, Satara, Kohlapur and Sangli |
| Manipur | | Senapati, Xhandel, Tamenglang, Ukhrul and Churachandpur |
| Meghalaya | | Jaintia Hills, East Khasi Hills, East and West Garo Hills |
| Mizoram | | Aizawal, Lunglei and Chimtuipui |
| Nagaland | | Kohima, Phek, Zonheboto, Wokha and Mokokchong, Tuensang and Mon |
| Odisha | Some parts of North-Eastern Ghat Zone (300 to 800 m, above mean sea level) | Phulbani |
| | Eastern Ghat Highland Zone 150 to 1,000 | Koraput, Nawaranpur and Jeypore |
| Punjab | Sub-Montane Undulating Zone | Gurdaspur |
| | Undulating Plain Zone | Hoshiarpur and Ropar and parts of Gurdaspur |
| Rajasthan | Aravali Hill | Sirohi, Udaipur, Bhilwara and Chittorgarh |
| Sikkim | | East Sikkim, North Sikkim, West Sikkim and South Sikkim |
| Tamil Nadu | High Altitude and Hilly Zone | Nilgiris, Shevroys, Elagiri-Javachi, Kollimalai, Pachdi Malai, Annamalais, Palnis and Podhezai Malai |
| Tripura | | Tripura |
| Uttarakhand | Hill Zone | Uttarkashi, chamoli, Pauri Garhwal, Tehri Garhwal, Dehra Dun, Rudraprayag, Bageshwar, champawat, Pithoragarh, Almora and Nainital. The Zone comprise of 4 sub-Zones: |
| | a. Sub-tropical | 250-1,200 m.a. m.s. l. |
| | b. Sub-temperate | 1,200 to |
| | c. Temperate | 1, 700 to 3,500 m.a. m.s. l. |
| | d. Alpine Zone | 3,500 m.a. m.s. l. and above |

| State | | Area Description and Characterization of Agro-Eco-Zones |
|---|---|---|
| Jammu and Kashmir | a. Low altitude sub-tropical zone | Kathua, footlands of Jasrata, Samba and Jambu |
| | b. Mid to High Altitude Intermediate Zone | Poonch, Rajauri and Doda. Altitude range 1,500-3,333 m. a. m.s. l. |
| | c. Mid to High Altitude Temperate Zone | Ananatnag, Pulwama, Srinagar, Badgam, Baramula and Kupwara. Plain valleys have an altitude of about 1, 700 m which rises to 2, 160 m and to 2, 660 m to 3,333 m on the upper belts. Altitude further rises to 4, 666 m in snow bound areas |
| | d. Cold Arid Zones | Leh and Kargil districts of Ladakh from 2,660 m to 9,300 m or more. Ladakh one of the loftiest inhabited regions of the world is located in the zone |
| West Bengal | Hill Zone | Darjeeling district except Siliguri Sub-Division. Altitudes varies above 60-90 M in plains up to 4,000 m. classified in 3 altitudinal ranges |
| | a. High Altitude | 1,650-2,150 m. a. m.s. l. |
| | b. Medium Altitudinal range | 1,250-1,650 m. a. m.s. l. |
| | c. Foot hills | Below 1,250 m. |

*Source:* Handbook of Agriculture, ICAR, New Delhi.

Agricultural biodiversity is the first link in the food chain developed and safeguard by indigenous people throughout the world and it makes an essential contribution to feeding the world (Nakhauka, 2009). Agrio-biodiversity has ecological, genetic, economic, scientific, educational and cultural values (Wale, 2004). It can provide smallholder farmers with more crop options and help cushion the effects of intense events such as droughts or floods. India demonstrates that the conservation and use of agro-biodiversity are the two key components of the livelihood development of farming communities. There is no country that considers it to be self-sufficient for all the genetic diversity in all of its crop plants for all time. Each country benefits from having access to plant genetic resources kept in other countries. Overall archaeological records depict an ancient occurrence of *Dolichos*; *Cucurbita*; *Cucumis* (melon), *Phaseolus*; *Pisum etc.* in the country.

**Table 1.22: Centre of Origin and Historical Records of Vegetable Crops**

| Vegetable Crops | Historical Aspects |
|---|---|
| Artichoke | Artichokes are native to the Mediterranean. They were eaten by the Greeks and Romans and later by the Arabs. However after the fall of Rome artichokes were rare in Europe until the 15th century when they were grown in Italy. From there artichokes spread to the rest of Europe. |
| Asparagus | Asparagus is native to the Eastern Mediterranean region. Asparagus was grown by the Greeks and Romans and it became a popular vegetable in Europe in the 16th century. |
| Aubergine | Aubergines or eggplants are native to India. Later they spread to China and by the 15th century they were being grown in southern Europe. |
| Beet root | Beetroot is descended from wild sea beet, which grew around Europe and Asia. Eating beetroot only really became popular in England in the 18th century. |
| Broad Bean | Broad beans are native to the Middle East and South Asia. They were known to the Ancient Greeks and they have been eaten in Europe ever since. |

| Vegetable Crops | Historical Aspects |
|---|---|
| Broccoli | It is not known for certain when broccoli was first eaten. The Romans ate a vegetable that may have been broccoli. It was certainly eaten in France and Italy in the 16th century. Broccoli was introduced into in England in the 18th century. It first became popular in the USA in the 1920s. |
| Brussel sprout | Brussels sprouts became popular in most Europe in the 16th century. They became popular in England in the 17th century. Brussels sprouts were grown in the USA from the 19th century. |
| Butter bean | Butter beans are native to Central America. They were first recorded in Europe in 1591. |
| Cabbage | Cabbages are native to southern Europe. They were grown by the Greeks and the Romans and in Europe they have been a popular vegetable ever since. Cabbages were brought to North America in the 16th century. |
| Carrot | Carrots are native to Asia and spread to the Mediterranean area. Carrots were grown in Europe in the Middle Ages they and have been popular ever since. |
| Cauliflower | Cauliflower is believed to come from Asia Minor. In Europe they were first eaten in Italy. However in the 16th century the cauliflower spread throughout Europe. Cauliflower was first grown in North America in the late 17th century. |
| Celery | Celery is native to the Mediterranean. Wild celery was known to the Greeks and Romans. However cultivation of celery only began in Europe in the 17th century. |
| Chickpea | Chickpeas are native to the Middle East. They were popular with the Romans and they have been eaten in Europe ever since. |
| Chilli | Chilies are from Central America where they have been grown for thousands of years. The Aztecs were fond of chilies and the Spanish brought them back to Europe. Chilies came to England in 1548. |
| Cucumber | Cucumbers are native to south Asia. They were grown by the Greeks and Romans. Cucumbers were also grown in England in the Middle Ages. The Spaniards introduced cucumbers into the New World in 1494. |

| Vegetable Crops | Historical Aspects |
|---|---|
| Kidney bean | Kidney Beans are native to South America. They were common in England by the mid-16th Century. |
| Leek | Leeks are believed to be native to central Asia. They were grown by the Egyptians. The Greeks and Romans also grew leeks and the Romans are believed to have introduced them to Britain. The leek is the symbol of Wales. According to legend Welsh soldiers wore a leek in their caps to distinguish themselves from their Saxon enemies during a battle. |
| Lentil | Lentils are very ancient. They have been eaten since prehistoric times. Lentils are native to Asia and they were eaten the Egyptians, the Greeks and the Romans. They were also eaten in India. |
| Lettuce | Lettuce is an ancient vegetable. It is native to the Mediterranean area. The Egyptians, the Greeks and the Romans ate lettuce. The Spaniards took lettuce to the New World. |
| Olive | Olives are native to the Eastern Mediterranean and people have grown them since prehistoric times. Olives were very important to the Egyptians, the Greeks and the Romans. |
| Onion | It is not known for certain where onions come from but it was probably Asia. Onions were one of the first vegetables grown by people. They were eaten by the Egyptians, the Greeks and the Romans. During the Middle Ages onions were one of the staple foods of people in Europe. |
| Parsnip | Parsnips are thought to be native to the Mediterranean region. The Romans grew them and they were a popular vegetable in the Middle Ages. However in England parsnips became less popular once potatoes became common in the 18th century. |
| Peas | Peas are native to Asia and they were one of the earliest vegetables grown by human beings. The Greeks and Romans grew peas and during the Middle Ages peas were an important part of the diet of ordinary people in Europe. |
| Potato | Potatoes are native to South America and they were grown by the native people for thousands of years before Europeans discovered them. The Spaniards took potatoes to Europe in the 16th century and they were first introduced to England in 1586. However at first potatoes were regarded as a strange vegetable and they were not commonly grown in Europe until the 18th century. In the 1840s potatoes in Ireland were afflicted by potato blight and the result was a terrible famine as the people had come to rely on potatoes for their staple food. |
| Pumpkin | Pumpkins are native to Central America. The Native Americans used them as a staple food. Pumpkins were adopted as a food by European colonists. Meanwhile Christopher Columbus brought pumpkin seeds to Europe. |
| Radish | Radishes are native to Asia. They were grown by the Egyptians, the Greeks and the Romans. Radishes were taken to the New World in the 16th century. The word radish comes from the Latin word radix, meaning root. |

| Vegetable Crops | Historical Aspects |
|---|---|
| Runner bean | Runner beans are native to central America and were grown there long before they were discovered by Europeans in the 16th century. Runner beans were first grown in England in the 17th century. |
| Spinach | Spinach is native to Asia. However it was unknown to the Greeks and Romans. It was first grown in Persia. Later it was grown by both the Arabs and the Chinese. The Arabs introduced spinach to southern Europe and by the 14th century it was eaten in England. |
| Tomato | Tomatoes are native to South America. The Spaniards came across them in the 16th century. However tomatoes were unknown in England until the end of the 16th century. |
| Turnip | Turnips are native to northern Europe. They were grown by the Romans and during the Middle Ages turnips were a staple food of poor people in Europe. In the 18th century Charles 'Turnip' Townshend pioneered growing turnips to feed cattle. |

**Table 1.23: Some Important Crop Species Introduced in India**

| Crop | Period | Introduced from | Remarks |
|---|---|---|---|
| Potato | Early 10th century | South America | - |
| Tomato | Early 10th century | South America | - |
| Cabbage | 18th century | Mediterranean | By East India Company |
| Cauliflower | 18th century | Mediterranean | By East India Company |

The wide gap in the number of species estimated to occur and those taxonomically identified is attributed to *inter alia*. The second National Report (India) to CBD in 2001 had indicated that around 350 taxonomists were available in the country. Out of these, 130 taxonomists covered seed plants, 85 insects, and 76 vertibrates. Only around 70 per cent of the country's land area has been surveyed so far to describe the species diversity therein. According to estimates, more than 500 years would be still required to discover all biodiversity on earth with the current rate of identification of new taxa and strength of scientific infrastructure. Biodiversity plays a central role in the sustainability of productivity and other agro-ecosystems services like nutrient cycling, carbon sequestration, pest regulation and pollination. Ever since agriculture and settlement came into existence some 10-12 thousand years ago, the human intervention in nature has given rise to a set of biodiversity, which became recurrently dependent for further evolution and development on the human activities rather than competition. This may be termed as agricultural/farm biodiversity, including the biodiversity of domesticated plants and animals. Such evolution in agriculture-related species has led to the emergence of distinct manifestations and types as well as their co-adaptations. This further gave rise to diverse agroecological systems and complexes.

## What is Biological Diversity?

*'Biological Diversity' means the variability among living organism from all sources including, inter alia, terrestrial, marine and other aquatic ecosystem and the ecological complexes of which they are part (UNEP,2002), and this includes diversity within species (genetic diversity), among species and that of ecosystems (Article 2 of the Convention on Biological Diversity).*

In brief, biodiversity can be defined as *"the quality, range or extent of differences among the biological entities in a given situation".*

## What is Germplasm?

Germplasm or plant genetic resources may be defined as:

*"The sum total of hereditary material, i.e. all the alleles of various genes, present in a crop species and its wild relatives".*

It refers to genetic material of plant origin of actual or potential value in the form of seed, vegetative propagule, tissue, cell, pollen, DNA molecule *etc.* containing the functional unit of heredity that can be utilized in crop improvement. This is generally referred to as a germplasm and includes varieties, landraces, and wilds/weedy relatives of economically important plant species available in India. Plant genetic resources (PGR) related information needs to be collected, studied, documented and conserved for posterity. Exotic vegetable crops that occupy a prominent place in Indian farming systems are potato, tomato, chilly, French bean, cowpea, vegetable soybean *etc.*

## Types of Plant Genetic Resources

Availability of complete gene pool in a particular species forms its genetic resources. It also consists of distant related or unrelated genera as in case of transgenic technology that makes gene transfers possible across the genera, family or even kingdom. The genetic resources constitutes of the gene pool that is available and amenable to transfers within the 'pool' through natural or artificial breeding methodologies. Genetic resources are broadly grouped into two major types:

☆ Wild genetic resources (wild species/forms and wild relatives)

☆ Cultivated genetic resources

## (A) Wild Genetic Resources (Wild Species/ Forms and Wild Relatives)

*"Gene pools representing populations of wild species occurring in nature are called wild genetic resources".*

The Indian sub-continent endowed with diversity in vegetable crops including eggplant (*Solanum melongena*), cucumber (*Cucumis sativus*), ridge gourd (*Luffa acutangula*) and sponge gourd (*Luffa cylindrica*). Approximately 80 species of major and minor vegetables occur, several of which exist only in the wild form and are generated for edible use (Seshadri, 1987). Mostly cultivated crop species are directly derived from wild species. The wild relatives include all other species, which are related to the crop species descending during their evolution. Wild relatives are generally difficult to hybridize with their cultivated counterparts. Both, wild forms and wild relatives are sources of valuable genes for insect and disease resistance, tolerance to abiotic stresses like drought, cold, salinity *etc.*, and even for quality traits and yield. Wild species and putative ancestral forms of vegetable crops contain valuable genes that are of immense use in vegetable breeding programme using conventional methods or modern tools of biotechnology. These genetic resources can be utilized for the development of new cultivars, strains and hybrids and also in restructuring the existing ones that lack one or the other attribute. In nature, several species are flourishing and evolving in their natural habitat. In the beginning of civilization, the first step in the development of cultivated plant was their domestication, which is a process of bringing wild species under human management. Knowingly or unknowingly, the man started selection according to their needs. Under domestication, crop species have changed considerable as compared to that of wild species. It has been observed that many wild species have been domesticated and have become crop weeds like *Coccinia* species. As and when they reached into the farmer's field they thrive well due to availability of nutrients applied by farmers and compete with agricultural crop plants. Similarly, several species are on the verge of extinction because of habitat reduction, fragmentation of holding, destruction of natural landscape, deforestation and encroachment of human upon the habitat of wild species by increasing area under agriculture.The domestication of wild species is still an ongoing process and is likely to continue till life exists on the earth. This is because the human needs are likely to change with time. Consequently, the wild species of little importance today may assume great significance tomorrow. With the development in science new breeding methods were invented to increase productivity and production to feed fast developing world. The distributional pattern of the wild plant genetic resources in different botanical/phytogeographical regions and the areas of their concentration where rich diversity of wild species still continues to perpetuate, are of special significance for undertaking programmes on collection as well as for *in situ* conservation of biodiversity. The distribution pattern of wild species in different phytogeographical regions reveals seven zones as follows:

**Table 1.24: Occurrence of Wild Genetic Resources in different Phytogeographical Regions**

| Sl.No. | Phytogeographical Regions | Genera and Species | Number of Wild Species | Common species |
|---|---|---|---|---|
| 1. | Western Himalayas | 5 genera and 8 species | 125 | *Abelmoschus manihot* (*tetraphyllus* forms), *Cucumis hardwickii, C. trigonus, Luffa echinata, Luffa graveolens, Solanum incanum, Trichosanthes multiloba, T. himalensis, Cucurbita* species, *Allium* species, *Zingiber* species, *Brassica* species. |
| 2. | Eastern Himalayas | 4 genera and 4 species | 82 | *Abelmoschus manihot, Cucumis trigonus, L. graveolens, Neoluffa sikkimensis, Cucurbita* species, *Allium* species, *Zingiber* species, *Brassica* species. |
| 3. | North-Eastern Regions | 10 genera and 17 species | 132 | *Abelmoschus manihot* (*pungens* forms), *Alocasia macrorrhiza, Amorphophallus bulbifera, Cucumis sativus, Cucumis hystrix, Dioscorea alata, Colocasia* species, *Luffa graveolens, Moghania vestita, Momordica cochinchinensis, M. macrophylla, M subangulata, Benincasa hispida, Sechium edule, Trichosanthes cucumerina, T. dioica, T. dicaelosperma, T. khasiana, T. lobata, T. truncata, Solanum indicum, Capsicum* species. |
| 4. | Gangetic plains | 4 genera and 8 species | 66 | *Abelmoschus tuberculatus, Abelmoschus manihot* (*tetraphyllus* forms), *Luffa echinata, Momordica cymbalaria, M. dioica, M. cochinchinensis, Solanum incanum, S. indicum, Cucumis* species. |
| 5. | Indus plains, North-Western plains | 3 genera and 3 species | 45 | *Momordica balsamina, Citrullus colocynthis, Cucumis prophetarum, Abelmoschus tuberculatus, Abelmoschus manihot* (*tetraphyllus* forms). |
| 6. | Western peninsular region and Malabarcoast | 7 genera and 7 species | 145 | *Abelmoschus angulosus, A. moschatus, A. manihot* (*pungens* forms), *A. ficulneus, Amorphophallus campanulatus, Cucumis setosus, C. trigonus, Luffa graveolens, Momordica cochinchinensis, Solanum indicum, Trichosanthes anamalaienisis, T. bracteata, T. enticula, T. horsfildii, T. perottitiana, T. nervifolia, T. villosa,.* |
| 7. | Eastern peninsular region/Deccan plateau | 8 genera and 21 species | 91 | *Amorphophallus campanulatus, Abelmoschus manihot, A. moschatus, Colocasia antiquorum, Cucumis hystrix, C. setosus, Luffa acutangula* var. *amara, L. graveolens, L. enticula, Momordica cymbalaria, M. enticulata, M. dioica, M. cochinchinensis, M. subangulata, Solanum indicum, S. melongena* (*insanum* types), *Trichosanthes bracteata, T. cordata, T. lepiniana, T. himalensis, T. multiloba.* |
| 8. | Island regions | 3 genera and 15 species | 23 | *Capsicum* species, *Colocasia* species, *Alocasia* species, *Xanthosoma* species |

*Source*: Rana *et al.*, 1995 and Singh (2010).

Indiscriminate extraction of biological material from the wild threatens their survival, even to the extent of pushing some species into extinction. It is hence important to develop sustainable harvesting regimes that do not undermine the capability of a plant species to survive in the wild.

## (B) Cultivated Genetic Resources

The presently cultivated plants in the country have been derived from two sources:

(i) Indigenous and

(ii) Introduced

### The New Crops/Introduced Crops

The introduced crops have come from various continents and countries, namely Western Asia, Africa, China, Southeast Asia and Pacific Islands, the New World and the Europe. The new crops introduced in the country over the past few centuries by Moughals and the Spanish, Portuguese and British voyagers include maize, potato, sweet potato, tomato, beans, onion, garlic, and chillies. In due course, the introduced crop species naturalized and adapted to Indian conditions and farmer's interventions led to generation of further diversity and traditional knowledge in these crops. Consequently these crops have become integral part of the Indian agriculture. The Arabs possibly brought with them clove, coriander, cumin and fennel. The introduced and naturalized species of several crops on being isolated climatically and spatially from their original homes over time and space. Gene pools that have been used or are being used from the cultivated source constitute cultivated genetic resources. Nearly 97.6 per cent cultivated varieties are being used in breeding programmes. They form a major part of a working collection. They are good sources of genes for yield, quality, biotic and abiotic stresses, *etc.* On the basis of their economic importance cultivated genetic resources can be grouped as:

☆ Field crops *i.e.,* Cereals, pulses, oilseeds *etc.*

☆ Commercial crops *i.e.,* sugarcane, cotton, jute *etc.*

☆ Vegetable crops *i.e.,* Cucurbitaceous, Solanaceous, Leafy vegetables *etc.*

☆ Medicinal and aromatic plants *i.e.,* Sufed Musali

☆ Fodder crops *i.e.,* Lucerne, Barseem

☆ Ornamental and flowering plants *i.e.,* Cactus, Marigold *etc.*

☆ Forestry and fuel plants *i.e.*, Popular plant.

☆ Condiments and spices *i.e.*, Black Pepper, Cardamom, Chillies.

☆ Bio-diesel yielding plants *i.e.*, Jatropha.

Alternatively, they may be termed based on their place of origin as:

(i) Indigenous (from the country in question) or

(ii) Exotic (from another country)

Cultivated genetic resources also include materials which are generated through different breeding programme from any organization, institutes, university *etc*. These genetic resources may be:

## 1. New Varieties

Several new varieties/hybrids are being developed through various breeding methods like introduction, selection, hybridization, mutations, polyploidy *etc*. As per the requirement, desirable lines are being selected and identified as a new variety on the basis of genotypic superiority and adoption. Genetic resources of released varieties of vegetable in India have grown to number above 385 by 2007 (Sharma, 2009). Since these varieties have been evolved through planned breeding, are ordinarily highly homogenous and genetic variability is ordinarily present only among varieties (Singh, 2012).

## 2. Old or Obsolete Varieties

Old or obsolete varieties can be defined as *"The varieties which were developed by systematic breeding efforts, and were once commercially cultivated but are no more grown"*. Improved varieties of recent past are known as obsolete cultivars. It is true that when new varieties/hybrids are released and become popular in a particular area, old varieties or hybrids are not grown by farmers due to change in production system, climate change, outbreak of a pest or pathogen, change in consumption preferences, or demand of processing industries. These obsolete varieties/hybrids serve as a main source of genes for adoption and quality traits which can be utilized for future breeding programmes. For example, in bottle gourd varieties Pusa Naveen, Pusa Summer Prolific Long, Pusa Summer Prolific Round *etc*. are out of cultivation but have been good source for earliness, fruit shape, quality, and better yield. Similarly, okra varieties Parbhani, Karanti, Pusa Sawani *etc*. were most popular traditional varieties before introduction of high yielding okra varieties. Now there are no more cultivated.

## 3. Landraces

A review of definitions and classification of land races has been presented by Zeven (1998). During some 90 years various researchers have coined definitions of a landrace. Some are too concise whereas others are merely long descriptions. The term landrace is not mentioned in the International Code of Nomenclature for cultivated Plants-ICNCP (Trehane, 1995). The item landrace cannot be included in the term cultivar as the cultivar is described as a taxon that had been selected for a particular attribute or combination of attributes, and that is clearly distinct, uniform and stable in its characteristics and that, when propagated by appropriate means, retains those characteristics. No or only limited human selection is carried out to maintain a landrace, it may clearly be distinct from other landraces, but repeated cultivation specially under other circumstances, often results in a different appearance of the landrace. Hence, a land race is not uniform and stable, and thus is different from a cultivar. The first definition was given by K. Von Ruemker in 1908 who stated for cereal varieties, that landraces are varieties, which in the region, of which they carry the name, were grown since time immemorial. In fact Von Ruemker says in his definition that a landrace is a landrace, because it carries the name of the region where it has been grown for a long time. He further mentioned that landraces are adapted to their growing conditions and that no human selection is carried out. If grown outside its native region it will continue to preserve its original characters and characteristics. Further, he added in his elucidation that a landrace is also typified by its own characters. U.J. Mansholt in 1909 pointed out that landraces have a high stability of their characteristics and great resistance capacity to tolerate adverse influences. Their production capacity, however, is less than that of cultivars and when grown outside their home region their genetic composition will change (Zeven, 1998). In 1912, H. Kiessling defined a landrace (of a particular region) as a mixture of forms (phenotypes) with certain external uniformity and with a composition specific for that region and great adaptability to the natural and technical-ecoomical conditions of that home region. Natural selection determines the frequencies (of the phenotype), including those of mutants and segregants. The best adapted phenotypes and therefore, genotypes will increase in frequency. In the same year, E. Tschermak discussed Kiessling's definition. He stated that a landrace was introduced from one region into another, and he added that a landrace as a variety, may be given the name of particular region, where it has ben grown since time immemorial. Furthermore, its genetic composition is influenced by its composition at the time of introduction into the present home region and by later changes. J Schindler in 1998 stated that a landrace should not be compared with a cultivar, because yield stability is the major characteristics of a land race, wheras a high yielding capacity under optimal conditions characterizes a cultivar. C. Fruwirth and Th. Roemer in 1921 noted that landraces have been cultivated for many generations under adverse conditions in a particular region without conscious selection and have become adapted to those adverse conditions. In 1930 Fruwirth redefined landraces. In his new definitions, he included the capacity of a landrace to adapt itself to a new region by changing its genetic composition. Selection by man is mostly done unconsciously, nature selects for characters such as frost tolerance, drought tolerance and low temperature adaptation (Zeven, 1998). Banga in 1944 defined a landrace as a population which naturally develops in a certain region under the influence of the regionally prevailing conditions of climate, soil and management without or with only little mass selection (Zeven, 1998). He realized the complexity of a landrace. Therefore, he did not define a landrace, but described it as follows "While landrace populations are variable, diversity is far from random. They consist of mixture of genotypes all of which are reasonably adapted to the region in which they evolved but which differ in details as to specific adaptations to particular conditions within the environment. They differ in reaction to diseases and pests, some lines being resistant or tolerant to certain races of pathogens and some to other races. Some components of the population are susceptible to prevalent pathogenic races, but not all, and no particular race of pathogen is likely to buildup to epiphytotic proportions because there are always resistant plants in the populations. "Landraces tend to be rather low yielding but dependable. They are adapted to rather crude land preparation, seeding, weeding, and harvesting procedures of traditional agricultutre. They are also adapted to low soil fertility, they are not very demanding, partly because they do not produce very much". According to Harlan (1975), "Landraces have a certain genetic integrity. They are recognizable morphologically; farmers have names for them and different landraces are understood to differ in adaption to soil type, time of seeding, date of maturity, height, nutritive value, uses and other properties". Most important, they are genetically diverse. Such balanced populations-variable, in equilibrium with environment and pathogens and genetically dynamic are our heritage from past generations of cultivars. They are the result of millennia of natural and artificial selections and are the basic resources upon which future plant breeding must depend.

Brown (1978) also described landraces as geographically or ecologically distinctive populations which are conspicuously diverse in their genetic composition both between populations and within them. They differ from their wild relatives because they have evolved under cultivation upon which most of them have come to rely for their survival. They differ from the cultivars developed by modern scientific plant breeding in that they have not been deliberately intensively selected to a predetermined reduced level of genetic heterogeneity.

Hawkes (1983) refers to landraces as 'highly diverse populations and mixtures of genotypes'. He further says that 'genetic resources (the total genetic diversity of cultivated species and their wild relatives) can be classified into various types of materials'. Two of them:

☆ **Old landraces:** are obtained from remote areas or small garden plots where the new highly bred

cultivars have not been bred as cultivars, but under natural and artificial selection (notably largly of an unconscious nature), have become adapted to the conditions under which they are cultivated.

☆ **Primitive landraces:** Primitive form of crop plants are collected from the Old Vavilovian centres of origin and diversity. These are primitive varieties which have evolved over centuries or millennia through both natural and artificial selection without systematic and sustained plant breeding effort, have been acquired by farmers according to his/her preference. These are genetically heterogenous but morphologically homogenous and contain many genes as source for resistance/tolerance to many stresses.

Martin and Adams (1987) defined a landrace of a common bean (*Phaseolus vulgaris* L.) as a genotypic mixture of a predominantly inbreeding species that is grown by a subsistence farmer at a particular farm site. Similarly, Rieger *et al.* (1991) gave a definition of a landrace for the first time in the 5th edition of their 'Glossary of Genetics'. They defined a landrace as 'any of the geographically distinct populations which evolve under cultivation and are conspicuously diverse in their genetic composition both within and between populations. They differ from varieties developed by modern breeding in that they have not been deliberately selected to a predetermined level of heterogeneity and performance'.

Astley (1991) stated that "The conscious and unconscious selections of variant within crop populations by primitive agricultuists led to localized diversification within population, now termed landraces". This definition would have been clear if Astely had not used plurals for population and landrace. Further he described that selection results in 'localized diversification', meaning that each farmer may develop his own distinct landrace. Hence, the diversification also refers to 'between populations'. He did not explain term 'primitive'.

Voss (1992) defined a landrace as a variety that is more or less in a state of homeostasis. He described farmer's mixtures of bean (*Phaseolus vulgaris* L.) in Central Africa, as consisting of several landraces; in his definition each bean seed colour phenotype is a separate landrace.

Prosperi *et al.* (1994) defined a landrace as an assemblage of genotypes of the same species on which the grower in a certain region and with certain growing methods carried out more or less directed mass selection during several generations.

A new word was introduced by Cleveland *et al.* (1994). They used a term folk varieties. A folk variety of a crop may be grown by various farmers; material collected from one farm (household) was described as a farmer's population. They equated the folk variety to terms found in literature; landraces, traditional varieties and primitive varieties. The term folk variety was used by Soleri and Smith (1995) when describing Hop maize landraces. An advantage of the term folk variety is that it can be applied to both landrace and garden-race.

Louette *et al.* (1997) described a landrace as farmer's variety which has not been improved by formal breeding programme. Teshome *et al.* (1997) defined, or better described, landraces as variable plant populations adapted to local agroclimatic conditions which are named, selected and maintained by the traditional farmers to meet their social, economic, cultural and ecological needs. They continue that in the absence of farmer's manipulations, landraces may not exist in the ecologically dynamic forms that are known today. Thus landraces and farmers are interdependent, in need of each other for their survival. As landraces have a rather complex nature it is not possible to give an all-embracing definition as it could result in description. May be Mansholt's (1909) amended definition is still the best: an autochthonous landrace is a variety with a high capcity to tolerate biotic and abiotic stress resulting in a high yield stability and intermediate yield level under a low input agricultural system (Zeven, 1998). In fact all these land races exhibit valuable genetic variability and are well adapted to the local soil type, climatic conditions *etc.* which can survive under unfavorable conditions (stresses) but low in yield. There is now preponderance of variable landrace forms/primitive types belonging to different crop groups like vegetables, legume *etc.* grown in the diverse phytogeographical and agroecological regions of the country. Rich diversity occurs in okra, eggplant, cucumber, melons, *Luffa*, pointed gourd, snake gourd yams *etc.* There are several local land races of vegetable crops which are still popular among the growers.

Rich crop diversity is available in India in terms of both between and within species variability. Landraces, traditional cultivars and farmer's varieties in several agri-horticultural plant species are abundant. Crop in which rich diversity occurs in India include cluster bean, okra, eggplant, cucumber, melons, ginger, turmeric *etc.* Among tuberous crops, rich variability exists in sweet potato, *taros*, and yams. Native resouirces are also available in *Coleus* species, sword bean, velvet bean *etc.*

## 4. Breeding Lines with Specific Genes

In modern breeding programme, breeders are developing gene/trait specific lines, which are not only required for better growth and development but also to suit in a particular environment.

*"Breeding lines are lines/populations evolved in different breeding programmes".*

**Table 1.25: Popular Local Land Races of Vegetable Crops**

| Crop | Popular Land Races | Place |
|------|--------------------|-------|
| Brinjal | 'Ram Nagar Giant' (Giant fruit size with less seed and soft pulp) | Ram Nagar, Varanasi (U.P.) |
| | 'Cuttak Pindi' (very small, tasty and soft fruit) | Cuttak (Odisha) |
| | 'Bundelkhand Deshi' (drought tolerance) | Bundelkhand (U.P.) |
| | 'Katchpachia Baigan' (small size cluster bearing, suitable for distance transportation) | |
| | 'Dudhiya Baigan' (cluster bearing with milky white round to oval fruits) | |
| Chilli | 'Bhoot Jolokia'/'Naga Jolokia' (having highest pungency) | Asom and Manipur |
| | 'Dabi Badgy' | Dharwad (Karnataka) |
| | 'Stuff Type' (Bharua Mircha) cultivated land race | Ghazipur, Azamgarh (Eastern U.P.) |
| Radish | 'Jaunpuri Mooli' (extra-large sized root) | Jaunpur (U.P.) |
| Muskmelon | 'Mau Naspati' | Mau (U.P.) |
| Cucumber | 'Naga Khira' | Nagaland |
| Bitter gourd | 'Jaunipuri Baramasi', 'Faizabadi', 'Faizabad Jhalari' | Jaunpur, Faizabad (Eastern U.P.) |
| Bottle gourd | 'Faizabadi long', and 'Tumba type' | Faizabad (Eastern U.P.) and Bihar |
| Watermelon | 'Rajasthan Local' | Rajasthan |
| | 'Ganga Tirahawa' | Eastern U.P. |
| Muskmelon | 'Lucknow Safed' | Lucknow (U.P.) |
| | 'Ganga Tirahawa' | Eastern U.P. |
| | 'Naspati' | Mau (U.P.) |
| Peas | 'Batri' | Western U.P. |
| | 'Kinnouri' | Himachal Pradesh |
| | 'Shillong Local' | Shillong (Meghalaya) |
| French bean | 'Solan Local' | Himachal Pradesh |
| | 'Meghalaya Local' | Meghalaya |
| | 'Manipur Local' | Manipur |
| | 'Sikkim Local' | Sikkim |
| | 'Arunachal Local' | Arunachal Pradesh |
| Cowpea | 'Kala Jhabra' | Eastern U.P. |
| | 'Chitkabra' | Eastern U.P., S. India |
| Cauliflower | 'Hazipur Local' (early and mid-season cultivars) | Bihar |
| Pointed gourd | 'Deshi' | Eastern U.P. |
| | 'Chitkabara, Goloua' | Bihar |
| | 'Safeda' | W. Bengal, Asom |
| Ivy gourd | Local land races | M.P., Jharkhand |
| Longmelon | Local land races | Rajasthan |

These lines possess narrow genetic base and often contain valuable gene combinations. It includes nearly homozygous lines, mutant lines, and lines derived from biotechnological means.

## 5. Transgressive Lines

After hybridization in cross pollinated vegetable crops, several segregating populations are advanced according to expression of trait, are called transgressive segregants and are selected to develop new varieties/breeding stocks.

## 6. Transgenic Lines

A transgenic line is one which carries foreign gene(s) that is stably integrated and expressed in the genome. Transgenic technology has been successfully used in 120 species of 35 families including major cucurbits like cucumber, muskmelon and other crops like tomato, brinjal, chilli *etc.*

## 7. Special Genetic Stock

Genetic stock often associated with distinct morphological physiological and biochemical changes, which regulate functioning of chromosome or a gene. It includes lines carrying gene mutations, chromosomal aberrations, markers genes, *etc.* During the process of primary, secondary and tertiary trisomics have been generated in several crops which enhance the number and usage of genetic resources. A plant species collection may be not good enough to be utilized directly as a new crop or new variety but it may possess desirable genes for other economic traits such as quality, resistance to disease, pest and abiotic stresses. For instance wild species of tomato *S. pimpinellium* is genetic stock for bacterial wilt, *Fusarium* wilt, ascorbic acid and β-carotene content and *Solanum cheesmani,* is source for salinity tolerance. Similarly wild species of potato (*Solanum demissum, S. antipoviczii*) confer resistance to late blight and *Solanum saltense* confer resistance to early blight. Similarly, *Solanum stoloniferum* is source for virus X, frost, Colorado beetle and nematodes. The genetic stocks serve as trait reservoirs and are used in hybridization program to transfer desirable characters to cultivated or popular varieties through backcross method of breeding.

## 8. Mutants

Variations may also be created through induced mutation with the help of mutagens but most of the mutants are recesive and not stable. There are several crops where available variability is very less, which is a constraint for breeders. Now these problems can be solved by mutation breeding (gene mutation or point mutations). Over a hundred years ago after the rediscovery of the Mendel's law of inheritance, Hugo de Vries in 1901 and 1903 presented an integrated concept concerning the occurrence of sudden, shock-like changes (leaps) of existing traits, which lead to the origin of new species and variation. In his experiments with evening primrose, de Vries observed many aberrant types, which he called mutants. De Vries in 1904 had referred to the new species and variation and also had referred to the new types of radiations (X-rays and gamma rays, discovered earlier by scientists like Rontgen, Beckrel and Curie) and suggested that they might be applied to induce mutations artificially. In 1927 Hermann J. Muller induced genetic mutations in fruit fly Drosophila with X-rays and gamma rays in barley and maize, which became the foundation stone for mutation breeding in crop plants (Van Harten, 1998). Crop improvement programmes through mutation breeding were initiated in 1930s in Sweden, USSR and Germany. The Swedish programme covered several crop species and generated a lot of information on mutation breeding. Later several countries including China, India, the Netherlands, USA and Japan took up the task of crop improvement through mutation breeding approaches. The discovery of radio-mimetic properties of a large number of chemicals further strengthened the arsenals of mutation breeders. Chemical mutagens came into existence in 1946 when Auerbach and Robson demonstrated the mutagenicity of nitrogen mustard in *Drosophila*. This followed the discovery of an array of chemicals that could induce mutation in living organisms. The most commonly used and widely studied chemical mutagens in plants belong to alkylating group, Ethylmethanesulphonate (EMS) is the one used most widely in crop plants (Gaul *et al.,* 1966). A coordinated programme on mutation breeding in rice was initiated in South East Asia in 1964 by the International Atomic Energy Agency (IAEA). By the end of the 20th century, nearly 2252 mutant varieties of crop plants including cereals, oilseeds, pulses, vegetables, fruit, fibers and ornamentals have been developed (Maluszynski *et al,* 2000).

Mutation breeding experiments for the improvement of vegetable crops was initiated in nineteen sixties in Japan, USSR, Belgium and India, and later in other countries including China, Bulgaria, Italy and USA. By the end of the 20th century, nearly 67 mutant varieties of vegetable crops belonging to about twenty-four different species have been released for cultivation worldwide. Among the countries, China occupies the first place followed by India in resect of developing the maximum number of mutant varieties. Solanaceous vegetables have the largest share (48 per cent) of the total mutant varieties of the vegetable crops. Physical mutagens such as gamma rays, X-rays and fast neutron were used to develop about 50 per cent of the mutant varieties. Gamma rays doses ranging from 80 to 630 Gy were used in different experiments. Increased yield, earliness and disease resistance were the major characters improved through induced mutations. In India, nearly eleven varieties of vegetable crops have been developed through mutation breeding approaches, majority of them by using gamma radiation. Mutation breeding is envisaged as a potential tool for overall improvement of vegetable crops. Vegetable crops account for 67 mutant varieties, that is 2.9 per cent of the total of 2252 mutant varieties released for cultivation worldwide. These varieties belong to about twenty-four.

**Table 1.26: Number of Mutant Varieties for Cultivation in different Vegetable Crops**

| Botanical Name | Common Name | Number of Varieties Released |
|---|---|---|
| *Solanum lycopersicum* | Tomato | 4 |
| *Solanum tuberosum* | Potato | 4 |
| *Solanum melongena* | Brinjal | 10 |
| *Capsicum annuum* | Green Pepper | 4 |
| *Allium cepa* | Onion | 4 |
| *Brassica pekinensis* | Chineses cabbage | 1 |
| *Brassica oleracea* var. *acephala* | Kale | 1 |
| *Brassica campestris* | Turnip | 2 |
| *Beta vulgaris* | Beet root | 1 |
| *Spinach oleracea* | Spinach | 6 |
| *Lactuca sativa* | Lettuce | 1 |
| *Raphanus sativus* | Radish | 1 |
| *Abelmoschus esculentus* | Okra | 1 |
| *Allium macrostemon* | Chinese Garlic | 1 |
| *Amaranthus* sp. | Amaranth | 1 |
| *Colocasia esculenta* | Taro | 1 |
| *Cucumis sativus* | Cucumber | 2 |
| *Luffa acutangula* | Ridge gourd | 1 |
| *Momordica charantia* | Bitter gourd | 1 |
| *Phaseolus coccineus* | Scarlet Runner bean | 1 |
| *Ipomea batatas* | Sweet potato | 4 |
| *Trichosanthes anguina* | Snakegourd | 1 |
| *Brassica oleracea* var. *acephala* | Kale | 1 |
| *Dolichos lablab* | Hyacinth bean | 1 |

For example in tomato alone a number of characters were improved through different mutagenes (Table 1.27).

### 9. Wild Forms/Wild Relatives

In general wild forms are wild species from which crop species were derived. They freely cross with the concerned crop species. Wild relatives are difficult to hybridize with crops than are the wild forms. Wild forms/wild relatives are valuable gene source for insect and disease resistance, tolerance to abiotic stresses, and even for quality traits and yield. It is true that wild species, which are closely related to the cultivated species, have been extensively used in resistance breeding. In some cases wherein wild species were collected and used in experimental taxonomy and other studies, wild species and their hybrids have often been dissipated after the conclusion of such studies. Collection and conservation of wild relatives of most of the cultivated species has been very meager. Therefore, collection and conservation of wild species should be takenup on priority basis. Conventional interspecific hybridization via sexual process is always useful to the plant breeders. Based on the chromosome behaviors of wide hybrids and the resulting chromosome constitutions in their progenies; chromosome manipulation of wide hybrids for crop improvement is classified into three main categories:

i. Incorporation of singe-chromosome or chromosome fragment from a wild species into an existing crop in order to enhance crop genetic diversity (Qi *et al.*, 2007).

ii. Incorporation of all the alien chromosomes by chromosome doubling in order to produce amphidiploid and partial amphidiploid.

iii. Elimination of all alien chromosomes in order to induce crop haploid which can be doubled to enhance breeding efficiency or facilitate genetic analysis (Pratap *et al.*, 2010)

The utilization of wild relatives depends on the degree of relationship between the wild relatives and cultivated. For the successful introgression requires some degree of homology between the cultivated genome and the wild donor genome. The greater degree of homology, the likely it's the chromosomal conjugation will occur in the hybrid ($F_1$).

### 10. Weed Races

In areas, where the cultigens and its closely related weed-wild taxa occur together, one often finds races of weedy taxa. Such crop-weed complex occurs in many cultivated plants.

**Table 1.27: Mutant Varieties of Tomato Released World Cide for Cultivation**

| Crop/variety | Year of Release | Country | Mutagenic Cose/Parent Used | Character Improved |
|---|---|---|---|---|
| Luch 1 | 1965 | USSR | Gamma rays [Pushkinsky] | Earliness and yield |
| S-12 | 1969 | India | Gamma rays | Reduced plant height |
| Pusa lal | 1972 | India | Gamma rays | Fruit ripening |
| Kyoryokureikou | 1974 | Japan | Gamma rays, Pollen of L. peruvianum | Resistance to TMV and wilt |
| PKM-1 | 1980 | India | Gamma rays, 25 kr [Annanj] | Yield and plant type |
| CO-3 | 1981 | India | EMS-0.1 per cent, seeds of Co-1 | Plant type and vitamin C content |
| Rannii Nush | 1983 | - | - | - |

## Types of Germplasm

### (a) Introduction of Exotic Germplasm

Basically come from various continents and countries, namely Western Asia, Africa, China, Southeast Asia and Pacific Islands, the New World and the Europe. There are certain crops likeTomato, peas, French bean, scarlet bean, Lima bean, winged bean, Faba bean, Brassicae (cauliflower, cabbage, knol-khol, turnip, leafy mustard *etc.*), radish, carrot, *Beta*, capsicum, Chayote, potato, *Cyclanthera pedata*, *Allium* spp. (onion, leek, shallot, garlic), asparagus, artichoke and parsley where so many valuable introductions have been made and these exotic germplasm has contributed a lot in Indian conditions. Some of the introduced genotypes are given in Table 1.28.

### Table 1.28: Vegetable Varieties Established in India through introduction

| Crop | Botanical Name | Introduced Varieties | Country |
|------|----------------|----------------------|---------|
| Tomato | *Solanum lyopersicum* | Roma | South America |
| | | Marvel | |
| | | Sioux | U.S. |
| | | Ponderosa | |
| | | Marglobe | U.S. |
| | | Best of All | |
| | | Fireball | Canada |
| | | La Bonita | U.S. |
| | | Dwarf Money Maker | Isreal |
| | | Devlin's Choice | |
| | | Red Cloud | |
| | | Balkan | Bulgaria |
| Brinjal | *Solanum melongena* | Black Beauty | |
| | | Wynad Giant | |
| Chilli (Hot pepper) | *Capsicum annuum* | Hot Portugal | |
| Sweet pepper/ Bell pepper | *Capsicum annuum* var. *frutesense* | Chinese Giant | |
| | | Hungarian | |
| | | Bull Nose | |
| | | California Wonder | U.S.A. |
| | | Yolo Wonder | California |
| | | World Beater | |
| | | R-449 | Russia |
| Okra | *Abelmoschus esculentus* L. Monch. | Perkins Long Green | U.S. |
| | | Ghana Red | Ghana (Africa) |

| Crop | Botanical Name | Introduced Varieties | Country |
|------|----------------|----------------------|---------|
| Peas | *Pisum sativum* | Arkel | England |
| | | Early Badger | U.S. |
| | | Meteor | England |
| | | Little Marvel | England |
| | | Alderman | |
| | | New Line Perfection | |
| | | Bonneville | U.S. |
| | | Multi-Freezer | |
| | | Thomas Laxton | |
| | | Sylvia | Sweden |
| | | Dwelich Commando | Wisconsin |
| French bean | *Phaseolus vulgaris* | Giant Stringless | Sweden |
| | | Jampa | Mexico |
| | | Contender | U.S. |
| | | Kentucky Wonder | U.S.A. |
| | | Bountyful | U.S.A. |
| | | Top Crop | U.S.A. |
| | | Wates | U.S.A. |
| | | Premier | Sweden |
| | | BKW-74 | Sweden |
| Cowpea | *Vigna unguiculata* | Philippines Early | Philippines |
| Cauliflower | | D-96 | |
| | | Improved Japanese | Isreal |
| | | Snowball-16 | Holland |
| Cabbage | | Copenhagen Market | United Kingdom |
| | | Golden Acre | |
| | | Jersey Wake Gield | |
| | | Chieftain | |
| | | September | German Democratic Repblic |
| | | Pride of India | United Kingdom |
| Radish | *Raphanus sativus* | Japanese White | |
| | | White Icicle | |
| | | Chinese Pink | |
| | | Rapid Red White Tipped | |
| | | Scarlet Globe | |
| | | Pusa Chetki | Denmark |
| Knol-Khol | | White Vienna | |
| | | Purple Vienna | |
| | | King of North | |

| Crop | Botanical Name | Introduced Varieties | Country |
|------|----------------|----------------------|---------|
| Carrot | | Nantes | |
| | | Nantes Half Long | |
| | | Chantay | |
| | | Danverse | |
| | | Zeno | German Democratic Republic |
| Turnip | | Purple Top White Globe | |
| | | Golden Ball | |
| | | Snow Ball | |
| | | Early Milan | |
| | | Red Top | |
| | | Early Grano | |
| | | Brown Spanish | |
| Watermelon | *Citrullus lanatus* | New Hampshire Midget | U.S.A. |
| | | Charleston Grey | |
| | | Improved Shipper | |
| | | Asahi Yamato | Japan |
| | | Sugar baby | U.S.A. |
| | | Shipper | U.S.A. |
| Cucumber | *Cucumis sativus* | China | |
| | | Straight Eight | |
| | | Poinsett | U.S.A. |
| Potato | | Magnum Bonnum | |
| | | Craigs Defiance | |
| | | Up-To-Date | |
| Summer Squash | *Cucurbita pepo* | Australian Green | Australia |
| | | Pattypan | U.S.A. |
| Onion | *Allium cepa* | Early Grano | U.S.A. |
| | | Bermuda Yellow | Philippines |
| | | Red Granno | America |
| Garden beet | | Detroit Dark Red | |
| | | Crimson Globe | |
| | | Crosby Egyptian | |
| | | Early Wonder | |
| Sweet potato | *Ipomoea batatas* | FA 17 white | China |

## (b) Indigenous Germplasm

Eggplant, sponge gourd, ridge gourd, snake gourd, round gourd, *Cucumis melo*, 'Phoot', Kakri, cucumber, leafy vegetables *viz. Basella, Chenopodium, Amaranthus, Brassica*, Spinach, Rumex/Sorrel; Taro, Yam, Elephant Foot Yam, Ginger, *Allium* spp., Bitter gourd *etc.*

## (c) Germplasm of Exotic Origin for which India is a Centre of Diversity

Okra, pumpkin, chilli, cowpea, Brassicae, Chyote and coriander.

## (d) Underutilized/Under Exploited Vegetables

In tribal areas of India many minor vegetables are grown in the field as well as in kitchen garden. Such crops are consumed locally. The genetic resources of such minor vegetable crops include over 50 species (Arora, 1955). Some important species are *Allium rubellum, Amaranthus* spp., *Apium graveolens, Asparagus officinalis, Basella alba, Basella rubra, Bamboo* spp., *Fagopyrum cymosum, Hibiscus sabdariffa, Ipomoea aquatica, Lactuca sativa, L. indica, Malva* spp., *Moringa oleifera, Ocimum* spp., *Parkia roxburghii, Polygonum* spp., *Pilea* spp., *Portulaca oleracea, Sesbania grandiflora, Solanum torvum, Trichosanthus* spp., *Trigonella* spp., *Vigna* spp., *Zizania latifilia*, several aroids and yams.

## (e) Wild Relatives and Related Species of Vegetable Crops

The wild relatives and related species of vegetable crops belongs to the categories, legumes-31 spp., vegetables-54 spp., spices and condiments-21 sp. (Arora and Nayar, 1984) and these are distributed in the western and eastern Himalayas, North Eastern Region, Gangetic Plains, Indus Plains and in the western and eastern peninsular regions. Botanically these can be classified as follows:

**Table 1.29: Family and Species**

| Family | Species |
|--------|---------|
| Brassicaceae (Cruciferae) | *Brassica, Lepidium* |
| Malvaceae | *Abelmoschus, Hibiscus, Malva* |
| Fabaceae | *Canavalia, Cicer, Dolichos, Moghania, Trigonella, Vigna* |
| Cucurbitaceae | *Citrullus, Coccinia, Luffa, Momordica, Neoluffa, Trichosanthes* |
| Umbelliferae | *Carum* |
| Compositae | *Chicorium* |
| Solanaceae | *Solanum* |
| Amranthaceae | *Amaranthus* |
| Chenopodiaceae | *Chenopodium* |
| Polygonaceae | *Fagopyrum, Polygonum, Rumex* |
| Zingiberaceae | *Alpinia, Curcuma, Zingiber* |
| Dioscoreaceae | *Dioscorea* |
| Liliaceae/Amaryllidaceae | *Allium* |
| Araceae | *Alocasia, Amorphophallus, Colocasia* |

## The Biogeographic Classification of India (Rodgers and Pawar, 1990)

The Indian gene centre holds a prominent position among the eight Vavilovian Centres of the crop plant origin. Rich genetic diversity occurs in about 166 crop plants including vegetable species and there wild relatives (about 320 species) in the different regions. At one time we can find out almost all the vegetable in some or other part of the country. More than 100 cultivated, lesser known and wild texa of different vegetables are available in our country (Table 1.30).

## Identification of Genetic Stocks/Germplasm for Registration

The distribution of germplasm in a particular geographical area with ecological conditions, agricultural systems and cultural patterns that make possible the survibility and use of the biodiversity in that area in relation to specific genetic stock. Through the concentration of variability, could have ethical and cultural origin besides biological evolution. The farmers family/concerned personnel/scientist is able to identify genetic variations (specific genetic stock/varieties) based on phenotypic elements growing period, soil adaptation, colour and shape, flavor and quality and time is a variable factor to define.

## Gene Pool Concept and Classification

Gene pool concept was first proposed by Harlan and de Wet (1971). It is a proposition of some guidelines for classification or grouping of genetic diversity based on cross-compatibility relationship that can simplify use of genetic diversity.

*Gene pool consists of all the genes and their alleles present in all such individuals, which hybridize or can hybridize with each other* (Singh, 2012).

Harlen (1992) developed *'gene pool'* concept by assigining the constituent taxa to primary, secondary

**Table 1.30: Distribution and Diversity of Cultivated/Lesser Known/Wild Taxa of Indigenous Vegetables**

| Crops | Group | Cultivated/Lesser Known/Wild Taxa |
|---|---|---|
| Solanaceous vegetable | Cultivated | *Solanum lycopersicum* (tomato), *Capsicum annum* (sweet pepper ant hot pepper), *Capsicum frutescens* (bird chillies or tobacco pepper), *Solanum melongena* (Brinjal). |
| | Lesser Known | *Cyphomandra betacea* (tree tomato), *Physalis exocarpa* (tomatillo), *Physalis pruinosa* (husk tomato), *Solanum nigrum* (garden husk berry). |
| | Wild | *Solanum microcarpum, S. gilo, S. xanthocarpum, S. insanum, S. surattense, S. torvum, S. forex, S. varbasyfolium, S. hirsutum, S. hispidum, S. pubescens* |
| Leguminous vegetable | Cultivated | *Pisum sativum* (pea), *Dolichos lablab var. typicus* (dolichos bean), *Vigna unguiculata* (cowpea), *Vigna sinensis* (cowpea), *Phaseolus vulgaris* (French bean), *Vicia faba* (broad bean), *Cyamopsis tetragonoloba* (cluster bean) |
| | Lesser Known | *Dolichos lablab var. lignosus* (bush type dolichos), *Vigna cylindrica* (fodder type cowpea), *Vigna sesquipedalis* (asparagus bean), *Phaseolus lunatus* (lima bean), *Phaseolus acontifolius* (moth bean), *Canavalia gladiata* (sword bean), *Moringa oleifera* (drumstick, horse radish tree, west Indian bean), *Ipomoea muricata* (clove bean) |
| | Wild | *Vigna dekindtiana, V. mensenesis* |
| Leafy Vegetables | Cultivated | *Amaranthus tricolor* (badi chaulai), *Amaranthus blitum* (chhoti chaulai), *Beta vulgaris* (beat leaf), *Beta vulgaris var. bengalensis* (spinach) |
| | Lesser Known | *Basella alba* (poi), *Basella rubra* (poi-coloured), *Ipomea aquatica* (water leaf) *Murraya koenigi* (curry leaf) *Spinacea oleracea* (spinach), *Hibiscus altissima* (sorrel), *Chenopodium album, C. murale* (chenopods) |
| Root crops | Cultivated | *Raphanus sativus var. radicula* (European type radish), *Raphanus sativus var. niger* (tropical type radish), *Brassica rapa, Daucus carota* (carrot) |
| | Lesser Known | *Raphanus sativus var. mongari* (rat tail or mongri radish), *Beta vulgaris* (beet root) |
| Cucurbits | Wild | *Luffa echinata, L. graveolens, L. tuberose, L. umbellata, Cucurbita ficifolia, C. texana, C. ecuadorensis, C. pendantifolia, Trichosanthes multiloba, T. himalensis, T. khasianum, T. ovata, T. ovigera, T. villosa, T. neviflora, T. truncata, T. palmata, T. nerifolia, T. caspidata, Cucumis hardwickii, C. hystrix, C. zeheri, Momordica balsamia, M. cymbalria, M. denundata* |
| | Cultivated | *Luffa acutangula* (ridge gourd), *Luffa cylindrica* (sponge gourd), *Cucurbita moschata, C. maxima, C. pepo, Trichosanthes dioica* (pointed gourd), *T. anguina* (sanke gourd), *Coccinia cordifolia* (ivy gourd), *Cucumis sativus* (cucumber), *C. melo* (muskmelon), *Momordica charantia* (bitter gourd), *Lagenaria siceraria* (bottle gourd), *Benincasa hispida* (ash gourd), *Citullus lanatus* (watermelon) |
| | Lesser Known | *Luffa hermaphrodita* (satputia), *Cucurbita mixta, Cucumis melo var. momordica* (snapmelon), *C. callosus* (kachari), *Momordica cochinchinensis* (kakrol), *M. dioica* (kartoli) |
| Other vegetables | Wild | *Abelmoschus manihot, A. moschatus, A. tuberculantus, A. tetraphyllus var. tetraphyllus, A. tetraphyllus var. pungens, A. crinitus, A. ficulneus, A. angulosus* |
| | Cultivated | *Abelmoschus esculentus* (lady's finger), *Brassica oleracea var. botrytis* (Indian cauliflower), *Allium cepa* (onion), *A. sativum* (garlic) |
| | Lesser Known | *Allium cepa* (multiplier onion), *Eleocharis dulcis* (water chestnut) |

and tertiary gene pools. At the intra-specific level, cultivars are grouped into races and sub-races.

The gene pool of a crop is classifieds into three groups:

  (i)  Primary gene pool
 (ii)  Secondary gene pool
(iii)  Tertiary gene pool

## (i) Primary Gene Pool (GP1)

Primary gene pool includes all the strains (wild and/or weedy as well as cultivated races) of the concerned crop species. Crossing (inter mating) between members of primary gene pool (species/germplasm) is quite easy and resulting fertile hybrids are vigorous, express normal meiotic chromosome pairing and recombination, normal gene segregation and seed fertility. *"Members of primary gene pool are the most commonly used in breeding programmes, and most breeders rely primarily or entirely on this material"* (Singh, 2012).

It can be divided into two subspecies:

  i.  Subspecies A: Include the cultivated races
 ii.  Subspecies B: Include the spontaneous races

## (ii) Secondary Gene Pool (GP2)

Members of secondary gene pool are all those species that hybridize with those of the primary gene pool with some to considerable difficulty and the hybrids are at least partially fertile. The difficulty observed in hybridization is due to difference in ploidy, chromosome alterations or genetic barriers. Gene transfer from secondary gene pool to primary gene pool is possible but usually difficult (poorly or not at all). *"Members of secondary gene pool are often used in breeding programme"* (Singh, 2012). This gene pool is available for use; however, the plant breeder or geneticist will have to put an extra effort to overcome the cross-ability barriers with application of various possible cyto-genetic manipulations to establish a fertile hybrid.

## (iii) Tertiary Gene Pool (GP3)

Members of tertiary gene pool represent the extreme outer limit of the potential germplasm (wild species)/genetic reach. They cross with the members of primary gene pool with considerable to great difficulty, and hybrids, if produced are anomalous, lethal or completely sterile. Gene transfer from tertiary gene pool to secondary gene pool is relatively easier. "Tertiary gene pool is used only occasionally in breeding programmes, and that to by a group of researchers having the competence and the patience for tackling the associated problems" (Singh, 2012). Gene transfer is either not possible with known techniques, requires embryo culture or grafting to obtain hybrids, doubling chromosome number or using bridging species or biotechnological techniques. It is rather ill defined.

# Survey and Characterization of Genetic Materials

We have to make the survey and to array the data and then analyze it under historical, geographical, agro-ecological, cultural, social and economic points of view. Variable such as habitat, number and cultivated area in each zone, zone importance, crop and wild species concentration, uses, customs, accessibility and relation between hot spot are the prime factors (Verma *et al.,* 2008a, and Verma *et al.,* 2008b).The problems was very well foreseen by Russian botanist and plant breeder N.I. Vavilov, who planned and successfully embarked on research of plant genetic resources. Vavilov and his associates organized a total of 180 trips in the Soviet Union and 40 trips abroad in 64 countries and huge volume of collections were generated, comprising 250 thousands different varieties and specimens. Vavilov studied the world's germplasm collections thoroughly and made the following observations:

  (i)  Plant species are unevenly distributed over the planet earth.

 (ii)  Some parts of the world are characterized by extra-ordinary diversity of species, namely South-Eastern China, Indo-China, India, the Malay, Archipelago, South East Asia, tropical Africa, Ethiopia, Central and South America, the Mediterranean basin, the Near East *etc.*

(iii)  World's regions with sparse species composition are Siberia, all of Central and Northern Europe and North America.

 (iv)  Striking differential plant species concentration at certain places. For instance, Costa Rica and Salvador in Central America, each having area hundred times smaller than US have as many species as all of North America (US and Canada combined).

From the above observations his conclusions were:

  (i)  The diversity of forms, varieties and species of a cultivated plant distributed in a specific pattern indicating speciation process were geographical localized.

 (ii)  The whole independent cultivated flora was associated with a particular region and was their centre of origin.

(iii)  Vavilov hypothesized that maximum concentration of interaspecific and interspecific diversity of cultivated plants was likely to occur in area of their centre of origin.

 (iv)  Most cultivated plant species had never crossed the boundaries of the main centres of origin.

  (v)  Centres of origin of cultivated plants were separated from one another by mountain chains, deserts or expanses of water giving rise to independent agricultural civilization.

(vi) Some cultivated crops overlapped to more than one centre and were associated with two or more centres of diversity.

## Important Genera and Species of Vegetable Crops in the Indian Gene Centre

The distribution patter of the wild species in different botanical/phyto geographical regions and the area of their concentration where rich diversity of wild species still continues to perpetuate, are of special significance for undertaking programmes on collection as well as *in situ* conservation of biodiversity. Important genera with number of species occurring in India (given in parenthesis) are as follows:

### Table 1.31: Important Genera with Number of Species in Vegetable Crops

| Family | Genera | Number of Species |
|---|---|---|
| Amaranthaceae | Amaranthus | 40 |
| Chenopodiaceae | Chenopodium | 8 |
| Leguminosae | Canavalia | 4 |
| | Dolichos | 6 |
| | Lallab | 1 |
| | Trigonella | 12 |
| | Vigna | 10 |
| | Cicer | 1 |
| Malvaceae | Abelmoschus | 6 |
| | Malava | 5 |
| Umbelliferae | Carrum | 3 |
| Compositae | Cichorium | 3 |
| Cucurbitaceae | Citrullus | 2 |
| | Coccinia | 1 |
| | Cucumis | 5 |
| | Luffa | 4 |
| | Neoluffa | 1 |
| | Momordica | 4 |
| Dioscoreaceae | Dioscorea | >40 |
| Others | Colocasia | 1 |
| | Fagopyrum | 4 |
| | Moghania | 20 |
| | Polygonum | 80 |
| Alliaceae | Allium | 30 |
| | Alpinia | 15 |
| | Zingiber | 4 |
| | Curcuma | 18 |
| **Total** | | 407 |

India is 'primary centre of origin' for crops like eggplant, smooth gourd, ridge gourd, cucumber *etc.* and secondary centre for cowpea, okra, chilli, pumpkin and several cole crops. India is a homeland of 167 cultivated species and 329 wild relatives of crop plants. Wild plants have played an important role in human diets due to their higher nutritional value than cultivated species. Wild species and putative ancestral forms of vegetable crops contain genes that are of immense value in vegetable breeding programme using conventional methods or modern biotechnology. These genetic resources can be utilized in the development of new cultivars, strains and hybrids and also in restructuring of the existing ones that lack one or more attributes. The information on distributional pattern of the wild plant genetic resources in different phytogeographical regions and the areas of rich diversity of wild species still continues to perpetuate and are of special significance for undertaking programmes on collection as well as for *in situ* conservation of biodiversity. The important families possessing wild genetic diversity are Brassicaceae (*Brassica*), Malvaceae (*Abelmoschus*), Leguminoceae (*Canavalia, Lablab, Trigonella, Vigna*), Cucurbitaceae (*Citrullus, Coccinia, Luffa, Momordica, Neoluffa, Trichosanthes*), Solanaceae (*Solanum*), Amaranthaceae (*Amaranthus*), Dioscoreaceae (*Dioscorea*), Amaryllidaceae (*Allium*), Araceae (*Alocasia, Amorphophallus, Colocasia*), Zingiberaceae (*Alpinia galangal*), Coronaceae (*Cornus capitata* and *Cornus controversa*) and Mulluginaceae (*Gisekia pharnaceoide*) are widely consumed during food shortage but in the South and West (Deccan region) the leaves are used as greens, as are the leaves of *Glinus trianthemoides*. The species of wild okra like *Abelmoschus ficulneus, A. manihot, A. tuberosus etc.* are fairly distributed in western and Eastern Ghats, Gangetic plains and north western/north eastern India. *A. tuberculatus*, a tall type with tuberculate fruits has provided resistance for yellow vein mosaic virus in breeding of okra of Pusa Sawani, widely grown cultigens. The wild forms of eggplant *i.e., Solanum incanum, S. melongena* var. *insanum, S. khasianum, S. sisymbrifolium* and *S. surrateense etc.*, having useful genes for diseases and pest resistance, occur more frequently in tarai region and the eastern peninsular region. The other wild germplasm include species of *Momordica, Trichosanthes* and *Cucumis*, the former two are predominant in the north east and peninsular region and the latter in foot hills and northeastern region in Himalayas. Several species of wild ginger (*Zingiber* species) and turmeric (*Curcuma* species) and in root crops such as wild yams (*Dioscorea* species) and taros (*Alocasia* species and *Colocasia* species) occur particularly in the humid tropical habitats in the western and the Eastern Ghats.

Extant of genetic diversity comprising native species and landraces occurs more in Western Ghats and North-eastern Himalayas. The richness of plant diversity is largely due to ecological diversity superimposed with tribal and ethnic diversification, plant usages and religious rituals. Crops in which rich diversity still occurs in India include cowpea (*Vigna unguiculata*), common bean (*Lablab purpureus*), cole crops (*Brassica* species), okra (*Abelmoschus esculentus* and related species), brinjal (*Solanum melongena* and related species), sweet potato

(*Ipomoea batatas*), taros (*Colocasia* and *Alocasia*), yams (*Dioscorea esculenta, Dioscorea alata, Dioscorea deltoidea*), sword bean (*Canavalia* species), velvet bean (*Mucuna* species), and elephants foot yam (*Amorphophalus* species). *Solanum* spp. are widely distributed in the north eastern region; yams in western ghats and northeastern states, chives, leeks and other wild *Allium* spp. in Kumaon and Garhwal Himalayas; cluster bean in Western arid zone; Lablab bean in Deccan plateau; cucurbits in Rajasthan and Madhya Pradesh and leafy vegetables like *Amaranthus*, and *Fagopyrum* spp. in Western Himalayan region.

## Specific Adaptability of Vegetable Crop Species in different Area/Region

Tribal dominated region inhabited by different ethnic group provide another interesting situation for gene pool sampling of primitive types of vegetable germplasm. Such areas often possess rich diversity of various vegetable crops. The tribes hold primitive cultivars of different vegetable crops and usually do not exchange seeds between the ethnic groups. In tribal areas particularly in Northeastern parts of Madhya Pradesh, Chhattisgarh, West Bengal, Bihar and Odisha, village markets/haats are most desirable place to collect diversity in a number of vegetable crops. The areas that remain to be covered include remote and less accessible areas, hilly terrain and tribal dominated belt. There are several vegetable crops in the country which require specific environment to adaptation. Specific adaptability of various vegetable crop species in different area/region is given in Table 1.32.

## Availability of Useful Genera and Species in Arid Environments

More than 80 per cent of total land in South Asia suffers from some kind of stress. Out of this, drought stress condition is experienced in more than half of the land (Dent, 1980). Out of the total 594 plant species reported from the Thar, 6 per cent have edible foliage. Grasses account for 13 per cent of the flora, 0.14 per cent of which produce seeds 'akin' to the minor millets. Edible fruit bearing trees constitute 1.7 per cent of the vegetation (Bhandari, 1974).

**Table 1.32: Specific Adaptability of Vegetable Species in different Regions of Country**

| Crop | Area/Region | Remarks |
|---|---|---|
| Common bean | High altitudes of Himachal Pradesh and Western Arunachal Pradesh | Large variability |
| Lablab bean | Tripura | Large variability with respect to pod colour |
| Soybean | Sikkim | Pods used as vegetable and cultivated on bunds of paddy fields |
| Tree bean | Mizoram, Manipur, Nagaland | Local preference |
| Horse-radish | Deccan Plateau | Local preference |
| Potato | Lahaul Valley | High yield (Average Tuber weight 1.0 kg) |
| *Solanum gilo* | North Eastern Region | Introduced from Africa and naturalized |
| Brinjal | Bundelkhand (U.P.) and Tripura | Primitive types available (Bholanath and Sheonath types) |
| Tomato | NEH (Meghalaya) | Cultivated in paddy lands from February to April |
| Bell pepper | Mizoram | Variability for fruit colour (white, red and black) |
| Sweet Potato | Tripura | Large variability and Cross-4 highly successful |
| Radish | Meghalaya and Arunachal Pradesh | Radish Sel.1 highly successful |
| Red pepper | Asom | Baghi chillies, said to have high capsaicin content (only medicinal use) |
| Chow-chow | Mizoram, Karnataka, Maharashtra | Highly naturalized |
| Pointed gourd | Bihar | Abundant |
| Bitter gourd | Tamil Nadu, Kerala, U.P. (around Amroha town) Bihar, West Bengal, Maharashtra | Highly specific adaptability |
| Cucumber | Bihar (Hajipur) | Grown as winter crop in February |
| Sweet gourd/Kakrol | Mizoram, Tripura, West Bengal, Bihar, Vindhya Hills of U.P. | A specialty favorites with natives |
| Spine gourd | Maharashtra, U.P., M.P., Chhattisgarh, Bihar, Jharkhand | Very nutritive vegetable |
| Watermelon | Rajasthan, Punjab, Haryana, Western U.P., Karnataka, M.P. | Near river beds only, develop more sweetness in arid zones |
| Muskmelon | Rajasthan, Eastern U.P. (Mau, Deoria, Azamgarh), Punjab | Develop more sweetness in arid zones |
| Buffalo gourd | Rajasthan | Accumulate more starch in roots and provide foliage, watery fruit (flesh contain more water) for animals |
| Buck wheat | Himachal Pradesh (Kinnaur) and Sikkim | Used as leafy vegetable in early stage |
| Ferns | Himachal Pradesh (Kinnaur) and Sikkim | Juvenile fronds used as vegetable |

**Table 1.33: Useful Vegetable Crops of the Arid Environments**

| Leafy | *Amaranthus gracilis, A. spinosa, A.hybridus, Achyranthus aspera, Acalypha indica, Portulaca oleracea, P. quadrifida, Chenopodium album, C. murale, Ipomea aquatica* |
|---|---|
| Sprouted buds | *Bambusa arundinacea, Moringa oleifera* |
| Seeds | *Citrullus lanatus, C. colocynthis, Vigna trilobata,* |
| Fruits | *Cordia myxa, Moringa oleifera, Cucumis callosus, Citrullus lanatus, Momordica dioica, Coccinia grandis* |
| Underground parts | *Asparagus racemosus* |
| Medicinal | *C. colocynthis* |

*Source*: Shankarnaraya and Saxena, 1988.

## Gaps and Opportunities

Management of landscapes for both production of food and conservation of ecosystem services and wild biodiversity forms the basis for "Ecoagriculture" (Mc Neely and Scherr, 2003). Agricultural land scapes must become net producer of ecosystem services rather than consumer services. Scarcity of economic resources in world countries is not only an obstacle to the protection of wild species, but also a major cause of genetic erosion, as people search for fuel-wood or convert virgin areas into farmland. It calls on close collaboration with landscape planners, political leaders, farmers and community groups and a broad range of professionals from ecology, agronomy, and economics amongst other disciplines within mixed-use landscape. Ecosystem services blend the domains of ecology, economic and social sciences; eco-nutrition brings together the science of nutrition, agronomy and ecology. Globalization of the food system has contributed to environmental degradation and biodiversity loss, while lowering prices for diets high in energy but low in variety and vital nutrients. Currently, the global food system is estimated to contribute to 30 per cent of global greenhouse gas emissions. Time-tested and almost carbon neutral ecological systems of agricultural production are particularly relevant for our present era of the threat of a climate chaos. Owing to the importance of the agricultural biodiversity of a particular region for world agriculture, loss of agrobiodiversity shoud not be simply considered a problems of that region but a serious world problem (FAO, 2008). Therefore, it should be considered a global responsibility of all the countries to sustain these threatened agro-ecosystems and conserve endangered plant genetic resources. The Indian sub-continent is immensely rich in agro-biodiversity and genetic resources of vegetable crop species. More important crops for which native diversity would still need more emphasis for collection include *Cucumis* and other *Cucumis* species, pointed gourd, yam, taro, sweet potato, chillies muskmelon, lablab bean, *Trigonella* and

winged bean. Priority areas for cucurbits include Indo-Gangetic plains, foothills of Himalayas, peninsular tract, Kachchh areas in Gujarat, eastern Ghats, Kodur and adjoining areas in Andhra Pradesh and particularly for *Cucumis melo* Barabanki, Unnao, Lucknow, Baghpat, Tarai, western Uttar Pradesh, Chhotanagpur, West Bengal, Asom, Karnataka and Andhra Pradesh, *Cucumis melo* var. *utlissimus* (tar kakri) diversity is available throughout India, parval (*Trichosanthes dioica* Roxb.) in Bengal plains and Asom valley; ivy gourd (*Coccinia* species) in North-eastern hill region; *Luffa* and *Lagenaria* from Indo-Gangetic Plains, NW plains; snake gourd (*Trichosanthes anguina*) is southern peninsular tract and Kerala, Karnataka and adjoining pockets; lablab bean in eastern peninsular tract, Odisha, Goa and southwards; cassava in Kerala, Asom, Tripura, Andhra Pradesh, Tamil Nadu and Madhya Pradesh (tribal tract of Bastar); sweet potato in Bihar, Odisha, Uttar Pradesh and West Bengal; *Dioscorea* species from Andaman and Nicobar islands, Bihar, North-eastern hill region; *Amorphophallus* species from Andaman and Nicobar islands, Andhra Pradesh and West Bengal; carrot (purple deshi types) need to be collected from South-East of Gujarat and adjoining Rajasthan. Rice bean is to be collected from parts of orissa, Santhal Pargana (Bihar) North-Eastern hills; winged bean from parts of Kerala, Anjangaon Surji near Akot (Akola), broad bean (*Phaseolus multiflorus*) from Himachal Pradesh and Uttar Pradesh hills, Jammu and Kashmir and Nilgiri hills and *Lepidium sativum* (Halim)- used as leafy vegetables from Palampur area in Gujarat. The species that need specific hunting are *Chenopodium album* from Gangetic plain extending up to northern hills; *Citrullus colocynths* from North-western plains and Gujarat; *Cucumis hardwickii* from foothills of Himalayas, Dehradun/Mussoorie; *Cucumis hystrix* from eastern plains to North-Eastern hills, Tura range in Meghalaya and Mishmi hills, *Cucumis setosus* from eastern India and upper Gangetic plains, *Cucumis prophetarum* from Sirohi and Abu areas in Rajasthan. Similarly, *Momordica cochinchinensis* from eastern Bihar, West Bengal and central peninsular tract, *Trichosanthes bracteata* from Himalayan ranges, Andaman and Nicobar Ilands, peninsuar region; *Trichosanthes cucumerina* (wild relative of *Trichosanthes anguina*) sporadically from all over India, *Amorphophallus bulbifera* from Khasi hills, eastern Himalayas in Sikkim and *Amorphophallus campanulatus* from Deccan plateau, *Dioscorea alata* from western and North-eastern Himalayas. India is, and will remain, a world model in between the gene rich and the technology – rich divide.

In particular, the following strategies may be adopted for future genetic resource management for sustainable agriculture:

☆ Accelerated genetic resource inventrization, technology development, assessment and refinement to maintain sustainability and competitive advantage.

☆ Organizing and strengthening genetic resource database and providing access to information through electronic networking.

☆ Harnessing the eco-regional potentials, particularly in raifed coastal and hill eco-region and hot spots.

☆ Diversification and enlargement of food basket and nutrition parlour for enhanced sustainability, increased profitability and improved human health.

☆ New initiatives in the areas of processing, product development and value-addition of suitable crops and varieties by appropriate exploitation of their genetic resource.

☆ Developing ability, appropriate institutional arrangements and policy framework for handling intellectual property rights related issues.

☆ Evolving long-term human resource development strategy, taking advantage of modern human resource management principles and ensuring continued competence and skills through national and international training.

☆ Developing Centres of Excellence in basic and strategic research on plant genetic resource, including biotechnology and DNA fingerprinting, to innovate/improve cutting edge technologies and address other important issues of national, regional or global importance.

☆ Enhance public and private investment in research on genetic resource handling and use and confidence building to increase share of private sector in research and development, exchange and collaborative use of genetic resources for food and agriculture.

☆ Encourage networking for exchange, evaluation and sustainable utilization of plant genetic resources.

# References

1. Anonymous. 2005. AVRDC Medium Term Plan. Available online at http://www.avrdc.org.

2. Anonymous. 1993. General information of protection of new varieties of plants under UPOV Convention. UPOV Publication No. 408(E), p. 51.

3. Arora, R.K. 1985. Genetic resources of less known cultivated food plants. *ICAR-NBPGR Sco. Monogr.* No., **9: 125**.

4. Astley, D. 1991. Exploration: methods and problems of exploration and field collection, p. 11-22. In J.G. Hawkes (ed.). Genetic conservation of world crop plants. Academic Press. Also issued as *Biol. Journal of Linnean Soc.*, **42(1): 87**.

5. Bailey, L.H. 1950. The Standard Encyclopedia of Horticulture. The MacMillan Co., New York.

6. Bhatt, K.C., Srivastava, U and Duhoon, S.S. 1998. Ethnobotany in relation to plant genetic resources. In: *Germplasm Collecting: Principles and Procedures* (eds. Gautam, P.L., Dabas, B.S., Srivastava, U. and Duhoon, S.S.), ICAR-NBPGR, New Delhi, p. 115-122.

7. Brown, A.D.H. 1978. Isozymes plant population genetics structure and genetic conservation. *Theor. App. Genet.*, **52: 145-157**.

8. Chadha, M.L. and Lal, T. 1993. Improvement of cucurbits. In: *Advances in Horticulture*, Vol. 5 Vegetable crops. (Eds.) Chadha, K.L. and G. Kalloo. Malhotra Publ. House, New Delhi.

9. Cleveland, D.A., Soleri, D. and Smith, S.E. 1994. Folk Crop Varieties: do they have a role in sustainable agriculture. *Bioscience*, **44: 740-751**.

10. de Candolle, A. 1982. Origine des plantes cultivers. Paris (English Translation, 1886, Kogan Paul).

11. de Candolle, A. 1986. Origin of cultivated plants. Haffner, New York.

12. Dalington, C.D. 1973. *Chromosome, Botany and the Origin of Cultivated Plants*, Allen and Unwin. London, p. 237.

13. Dent, F.J. 1980. Priorities for alleviating soil related constraints of food production in the tropics. IRRI. p. 79-107.

14. Gupta, P.N., Rai, M. and Rana, R.S. 1995. Centres of origin and genetic variability of vegetable crops. In: Genetic Resources of Vegetable Crops, p. 52-62, ICAR-NBPGR, New Delhi.

15. Guzhov, Y. 1989. Genetics and Plant Breeding for Agriculture. Mir Publishers, Moscow. p. 13-33.

16. Harlan, J.R. 1971. Agricultural origins: centres and non-centres. *Science*, **174: 468-474**.

17. Harlan, J.R. 1971. Crops and man. American Society of Agronomy, Madison, Wisconsin, p. 295.

18. Harlan, J.R. 1975. Our vanishing genetic resources. *Science*, **188: 618-621**.

19. Hawkes, J.G. 1983. The diversity of crop plants. Combridge, London. p. 184.

20. Kuckuck, H. 1963. Present views on Vavilov's gene centre theory. *Plant Genet. Resources Newsletter*, **12: 8-10**.

21. Louette, D., Charrier, A. and Berthaut, J. 1997. *In situ* conservation of maize in Mexico: genetic diversity and maize seed management in a traditional community. *Econ. Bot.*, **51: 20-38**.

22. Martin, G. and Adams, M.W. 1987. Land races of *Phaseolus vulgaris* in variation. *Econ. Bot.*, **41: 190-203**.

23. Mohideen, M.K. and Irulappan, I. 1993. Improvement of *Amaranthus*. In: *Advances in Horticulture*. Vol. 5 Vegetable crops. (Eds.) Chadha, K.L. and G. Kalloo. Malhotra Publ. House, New Delhi, p. 305-323.

24. Munger, H.M. and Robinson, R.W. 1991. Nomenclature of *Cucumis melo* L. *Cucurbit Genet. Coop. Rep.*, **14:** 33.

25. Murthy, S.R. and Pandey, S. 1978. Delineation of agro-ecological regions of India. 11th Congress, International Soc. of Soil Sci. Edmonton, Canada, 17-27, June.

26. Paroda, R.S. and Arora, R.K. 1991. Plant Genetic Resources: Conservation and Management, IBPGR, ROS and SEA, New Delhi.

27. Pier. 2001. "Invasive plant species: *Coocina grandis*" Pacific Island Ecosystem at risk. United States Department of Agriculture, Natural Resources Conservation Services. The Maui Weekly Wrap. *Maui News*, p. 10-18.

28. Plucknett, D., Smith, N., Williams, J.T. and Anishetty, N.M. 2014. *Genebanks and the World's Food*. Princeton University Press.

29. Pratap, A., Choudhary, A.K. and Kumar, J. 2010. *In vitro* techniques towards genetic enhancement of food legumes-a review. *J. Food Legumes*, **23:** 169-185.

30. Prosperi, J.M., Demarquet, F., Angevain, M. and Mansat, P. 1994. Evaluation agronomique devarieties de pays de sainfoin (*Onobrachis sativa* L.) originaties du sud-est de la France. *Agronomie*, **14:** 285-298.

31. Qi, L.L., Frieb, B., Peng, Z. and Gill, B.S. 2007. Homeologus recombination, chromosome engineering and crop improvement. *Chromosome Research*, **15(1):** 3-19.

32. Rana, R.S. 1995. Conservation and management of plant genetic resources in India. In: Gentic resources of vegetable crops: Management, conservation and utilization.(Eds.) Gupta, P.N., Rai, M. and Kochher, S. Published by ICAR-NBPGR, New Delhi, p. 4.

33. Rieger, R., Michaelis, A., Green, M.M. 1991. Glossary of Genetics. Classical and Molecular. 5th edition. Springer-Verlag. p. 553.

34. Roberts, H.F. 1929. Plant hybridization before Mendel. Princeton University Press, Princeton. p. 374.

35. Samant, S.S. and Dhar, U. 1997. Diversity, endemism and economic potential of wild edible plants of Indian Himalaya. *Int. J. Sustain. Dev. World Ecol.*, **4:** 179-191.

36. Schiemann, E. 1951. New results on the history of cultivated cereals. *Heredity*, **5:** 305-320.

37. Seshadri, V.S. 1987. Genetic resources and their utilization in vegetable crops. In: Plant Genetic Resources: Indian perspective, p. 335-343.

38. Shankarnaraya, K.A. and Saxena, S.K. 1988. Life supporting arid zone plants in famine food. In: Life Support Plant Species Diversity and Conservation, Paroda, R.S., Kapoor, Promilla, Arora, R.K. and Bhag Mal (Eds.). MBPGR, New Delhi, p. 55-59.

39. Singh, B.D. 2012. Plant Breeding: Principal and Methods, Kalyani Publishers, p. 18.

40. Singh, H. P. 2010. Managing genetic resources of horticultural crops. *Indian Horticulture* (May-June, p. 3-16)

41. Soleri, D. and Smith, S.E. 1995. Morphological and phonological comparisons of two Hopi maize varieties conserved *in situ* and *ex situ*. *Econ. Bot.*, **49:** 56-77.

42. Stoskopf, N.C. 1999. Plant Breeding: Theory and Practices. Scientific Publishers, Jodhpur. p. 1-26.

43. Teshome, A., Baum, B.R., Fahrig, L., Torrance, J.K., Arnason, T.J. and Lambert, J.D. 1997. Sorghum (*Sorghum bicolor* L. Moench) landrace variation and classification in north Shewa and South Welo, Ethiopia. *Euphytica*, **97:** 255-263.

44. Trehane, P. 1995. International Code of Nomenclature for Cultivated Plants. p. 174.

45. Vavilov, N.I. 1926. Studies on the origin of cultivated plants. *Bull. Appl. Bot.*, **26 (2):** 248.

46. Vavilov, N.I. 1951. Phytogeographical basis of plant breeding. The origin, variation, immunity and breeding of cultivated plants (K.J. Choster, translator). *Chronica Botanica*, **13:** 366.

47. Voss, J. 1992. Conserving an increasing on-farm genetic diversity: farm management of varietal bean mixtures in Central Africa, pp 34-51. In J.L. mock and R.E. Rhodes (eds.) diversity, farmer's knowledge and sustainability. Cornell Univ. Press, p. 278.

48. Waters, C.N., Zalasiewicz, J., Summerhayes, C., Barnosky, A.D., Poirier, C., Galuszka, A., Cearreta, A., Edgeworth, M., Ellis, E.C., Ellis, M. and Jeandel, C. 2016. The Anthropocene is functionally and stratigraphically distinct from the Holocene. *Science*, **351(6269):** p.aad2622

49. Weather Wax, P. 1954. Indian corn in old America. Macmillan Company. New York. p. 253.

50. Zeven, A.C. 1998. Landraces: A review of definitions and classifications. *Euphytica*, **104:** 127-139.

51. Zeven, A.C. and J.M.J. de Wet. 1982. Dictionary of cultivated plants and regions of diversity. Wageningen, p. 259.

52. Zeven, A.C. and Zhukovsky, P.M. 1975. Dictionary of cultivated plants and their centres of diversity, PUDOC, Wageningen, p. 259.

53. Zeven, A.C. and Zhukovsky, P.M. 1975. Dictionary of cultivated plants and their centres of diversity. Centres for Agricultural Publishing and Documentation, Wageningen, p. 219.

54. Zohary, D. 1970. Centres of diversity and centres of origin. p. 33-42. In: O.H. Frankel and E. Bennett (eds.) Genetic Resources in Plants Oxford, Blackwell.

55. Zhukovsky, P.M. 1965. Genetic and botanical irregularities in the evolution of cultivated plants. *Genetika Mosc.*, **1**: 41-49 (In Russian; English summary).

56. Zirkle, C. 1932. Some forgotten records of hybridization and sex in plants. *J. Hered.*, **23**:443-448.

# 2

# Evolution and Domestication of Vegetable Genetic Resources

## Introduction

Agriculture is fairly recent invention, only about 10,000 years old. Since agriculture began, generation after generation of farmers have artificially bred crops to select for more desirable traits, like size and taste. Needless to say, favourite vegetables have change drastically over the centuries. Kennedy explains that while ancient "Wild watermelon" weighed no more than 80 g, modern watermelons can range from 2-8 kg in market. And the Guinness World Record for the heaviest watermelon recorded exceeded 121 kg in the year 2000. Thousands of year of human induced evolution have worked miracles on several vegetable crops. Due to change in climatic conditions and cultural traditions only a part of the world vegetable diversity is cultivated. Major evolutions have affected most of the cultivated species at several levels:

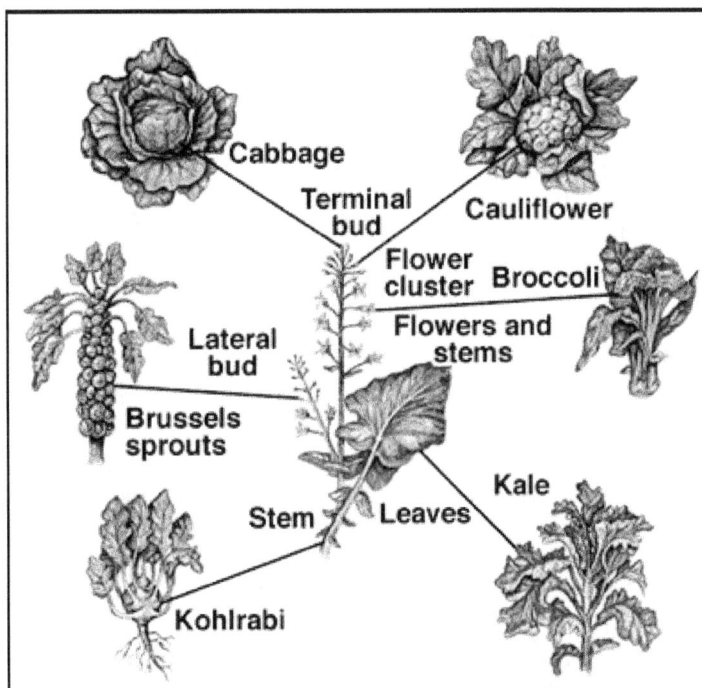

**Evolution and Selection in *Brassica.***

(i) An enlargement of the number of cultivated species, with the introduction for instance of *Brassica*.

(ii) New culti-groups or botanical varieties have been introduced or created introducing market segmentation for instance, parthenocarpic cucumber.

(iii) A tremendous increase of the number of cultivable cultivars with dense 20000 cultivars.

## Evolution of Crop Plants

The agro-biodiversity is well demonstrated in the traditional agricultural systems where it essentially maintains biodiversity in terms of mixture of crops as well as the variability of other farm related biological organisms and farming systems. It contributed to the system health and sustainability in several ways. For example, it ensures a sustainable harvest from mixture of crops that may show differential survival under adverse situations or calamities. It provides endemicity and equilibrium for crop health vis-à-vis production. In short, evolution acting through natural selection represents an ongoing interaction between a species genome and its environment over the course of multiple generations. It is apparent that selection by nature and humans has been responsible for the evolution of crop plants. However, selection is effective in altering a species only when genetic variability exists in the populations of that species. Origin and evolution of genetic resources has proceeded variedly amid changing continental, regional and global environments. Further, the evolutionary courses were greatly affected by domestication, introduction and selection activities undertaken by man.

Interventions by human being had both positive and negative impacts on evolution of crop plant genetic resources.

### Positive Impacts

Positive impacts in the sense that some species when introduced to new locations, expressed huge amount of genetic variation and rapidly evolved so many new ecotypes.

### Negative Impacts

Negative aspect of introductions was noticed in the form of encroachment of habitat of indigenous species by exotic ones.

In fact, domestication, introduction and selection are very basic to crop improvement. They will continue to play pivotal role even in the future.

## Evolutionary Mechanisms of Variation

The patterns of evolution of various crops are therefore, considered here according to the mode of origin of genetic variation crucial for evolution of that species. Many vegetable crops are found to have evolved through either intergeneric, interspecific or inter-varietal crosses under natural aided circumstances. Whether it is natural, traditional or through advanced gene manipulation techniques 'distant hybridization' may be interspecific or intergeneric. The variation generated by various aforesaid means is subjected to repeated selection. Gametic and somatic sieves are continuously operating in an individual to eliminate undesirable or unfavorable genetic changes. Only those genetic variations are permitted to pass to progenies, which add to the survival of the organisms, thus evolving to the newer types. Vegetatively propagated roots and tubers crop landraces are genetically fixed once created. This facilitated the study of their conservation dynamics on farm (McKey *et al.*, 2012), even though the lack of time series data, geographical benchmarking, and other factors still pose challenges for field-based crop evolutionary studies.

There are three major mechanisms in which genetic variability has arisen in various crop species are as under (Singh, 2012):

(1) Mendelian variation (generated mainly by gene mutation)

(2) Inter-specific hybridization

(3) Intergeneric hybridization

(4) Polyploidy

### (1) Mendelian Variation (Generated mainly by gene mutation)

In nature several crop have been evolved through variation generated by gene mutation, and natural hybridization between different genotypes within the same species, followed by segregation and recombination. As result variability present in any species has originated from gene mutation occurred in nature. It has been observed that most of the gene mutations occurred in nature are harmful and are eventually eliminated but some mutations are beneficial and are retained in the population. However, beneficial mutations which are low in frequency are retained and contribute to evolution of species. On the basis of phenotypic effect, mutations may be grouped into two broad categories:

(a) Macro-mutation

(b Micro-mutation

### (a) Macro-Mutation

A macro-mutation produces a large and distinct morphological effect, and often affects several characters of the plant. A single macro mutation has led to the differentiation of modern maize (*Zea mays*) like sweet corn, baby corn, pop corn *etc.* Macro-mutation has affected the positions of male and female inflorescences, habit of the plant and several other characters. Similarly, cabbage (*Brassica oleracea* var. *capitata*), cauliflower (*B. oleracea* var. *botrytis*), broccoli (*B. oleracea* var. *italica*), and Brussels' sprouts *B. oleracea* var. *gemmifera*) have all originated from a common wild species/wild cabbage called coleworts (*B. sylvestris*) and they differ from each other with respect to a few major genes.

### (b) Micro-Mutation

The greater part of genetic variation, however, has resulted from mutations with small and less drastic effects, *i.e.*, micro-mutations. It is true that micro-mutations have only small effects; they tend to be accumulated in a population as a consequence of natural selection. Human being selected desirable plant types leading to the differentiation of domesticated species from the wild ones.

Several important vegetable crops have evolved through differentiation of domesticated species from the wild ones adopting the mendelian variation *e.g.*, beans (*Phaseolus* sp.), peas (*Pisum sativum*), tomato (*Solanum lycopersicum*) and many other crops.

## (2) Inter-Specific Hybridization

A system shows tendency to be uniform or self-centreed while the nature tends to generate variability in a system. In the biological world too, nature has endowed various mechanisms to promote variability.

Inter-specific hybridization may be defined as *"Crossing of two different species of same genus/plants usually takes place by sexual fusion"*.

It is usually practiced to transfer desirable genes from wild species to cultivated species of plants. Interspecific crosses may be fully fertile like tomato (*Solanum lycopersicum* x *S. peruvianum*), partially fertile or sterile like wheat (6X x 4X). In the cross combination, $F_1$ is generally more vigorous than the parents, and segregation and recombination in $F_2$ and later generations produce a vast range of genotypes since the parental species are likely to differ from each other for a large number of genes. Segregating generations may be weak and un-desirable. In general Inter-specific hybrids are highly sterile and do not set seed. There is little evidence to suggest that inter-specific hybridization contributed to any great extent in the evolution of crop species. The inter-specific hybrids may have repeatedly back crossed to one of the parental species. As a result, most of the genotype of that parental species, to which the hybrid had repeatedly backcrossed, would be recovered along with few or several genes from the other parental species. This process is known as introgressive hybridization.

*"Introgressive hybridization leads to the transfer of some genes from one species into another"*.

The modern maize is isolated to have developed through introgressive hybridization between the primitive maize and a wild grass. Fertility level of interspecific crosses depends on the homology of chromosomes in the parental species. In the case of sterile crosses, amphidiploidy is induced with colchicine and the fertility is restored.

## (3) Inter-Generic Hybridization

Term inter-generic hybridization refers to *'cross between two different genera of the same family'*. In general such crosses are not commonly used in crop improvement but such crosses become desirable in a number of situations. Inter-generic crosses sometimes become essential when the desirable genes are not present in the same genus, but they are present in allied genera. $F_1$ hybrids of this type of crosses are always sterile. However, they can be made fertile by chromosome doubling through colchicine treatment. Inter-generic hybridization has been used successfully in the development of the synthetic cereal, for example-triticale.

### Problems associated with the Wide Cross

☆ Cross incompatibility

☆ Hybrid inviability

☆ Hybrid sterility

**Table 2.1: Problems Associated with the Wide Cross**

| | |
|---|---|
| Cross incompatibility | Inability of the pollen grains of one species or genus to affect fertilization in another species or genus. This problem can be overcome through reciprocal crosses, bridge crosses, using pollen mixture, pistil manipulations or use of growth regulators *etc*. |
| Hybrid inviability | Inability of the hybrid zygote or embryo. In some cases, zygote formation occurs but further development of the zygote is arrested and similarly in other cases after the completion of the initial stages of development, the embryo gets aborted due to either unfavorable interactions between the chromosomes of the two species or unfavorable interaction of the endosperm with the embryo or reciprocal crosses (Mukherjee *et al.*, 1991; Mukherjee, 1999), application of growth hormones and embryo rescue techniques this problem can be overcome. |
| Hybrid sterility | Inability of a hybrid to produce viable offspring. This is more prominent in the case of intergeneric crosses. The reason for hybrid sterility is the lack of structural homology between the chromosomes of the two species. This may lead to meiotic abnormalities like chromosome scattering, chromosome extension, lagging of chromosome in the anaphase, formation of anaphase bridge, development of chromosome rings, chains, and irregular as well as unequal anaphase separations. These irregularities may lead to aberrations in chromosome structure. Sterility base structural differences can be overcome by amphidiploidization. Hence techniques like alien addition and alien substitution through manipulation of chromosomes can be effective tool for wide cross. |

## Techniques to make Wide Crosses Successful

Below mentioned techniques can be used to made wide cross successful.

**Table 2.2: Tips for Getting Success in Wide Cross**

| | |
|---|---|
| Selection of plants | The most compatible parents available should be selected for the crosses. |
| Reciprocal crosses | Reciprocal crosses may be attempted when one parental combination fails. |
| Manipulation of ploidy | Diplodization of solitary genomes to make them paired will be helpful to make the cross fertile. |
| Bridge crosses | When two parents are incompatible, a third parent that is compatible with both the parents can be used for bridge crosses and thus it becomes possible to perform cross between the original parents. |
| Use of pollen mixture | Unfavorable interaction between pollen and pistil in the case of wide crosses can be overcome to some extent by using pollen mixture. |
| Manipulation of pistil | Decapitation of the style will sometimes prove helpful in overcoming incompatibility. |
| Application of growth hormones | Pollen tube growth can be accelerated by using growth hormones like IAA, NAA, 2,4-D and Gibberelic acid. |
| Chromosomal manipulation | Chromosomal manipulation can be used in wide or distant hybridization. The chromosome behaviors in $F_1$ hybrids provide the essential genetic basis for chromosome manipulation. The induction of homologous pairing in $F_1$ hybrid plants followed by the incorporation of a single chromosome fragment from an alien or a wild species into an existing crop species by translocating chromosomes has been used in the production of translocation lines. Beneficial traits from wild plants can be transfer into crops to bridge the species gap via alien chromosome translocation lines. Chromosome doubling in somatic cells or gametes of $F_1$ hybrids followed by the incorporation of all alien chromosomes has been used in the production of amphidiploids. Amphidiploidy can be used as a bridge to move a single chromosome from one species to another or for the development of new crops. |

## (4) Polyploidy

Polyploidy has evolved in response to complete sterility of generally, autopolyploidy leads to increased vigour, larger flowers and fruits *etc.* over the diploid forms.

### (a) Autotriploid (3x)

The commercial banana (*Musa sapientum*) is an autotriploid (3x); it has larger and seedless fruits in comparison to the diploid banana. Triploid varieties are known in watermelons (*Citrullus vulgaris*), sugarbeets (*Beta vulgaris*) and some other crops.

### (b) Autotetraploid

The commonly grown potato (*S. tuberosum*) may be regarded as an autotetraploid, although inter-specific hybridization may also be involved. *S. tuberosum* has 3x, 2x and 4x types. Some of the 2x progeny obtained from the 4x potato are fully fertile and as vigorous as the 4x types, indicating that it is largely an autotetraploid.

Some other crop species like sweet potato (6x) is also considered to be autopolyploid. Thus autopolyploidy has played a limited role in crop evolution.

### (c) Allopolyploid

Allopolyploidy, in contrast, has been considerably more important in crop evolution than autopolyploid. Allopolyploidy results from chromosome doubling of Inter-specific $F_1$ hybrids. About 50 per cent of the crop plants are allopolyploids.

**Some of the important allopolyploid crop plants are:**

## Domestication

The term 'domestication' refers to: "*A plant genotype which has been adopted by human being activity*".

Domestication may be defined as:

"*Bringing wild species under human management*".

When natural phenomenon plays their own role in existence, a lot of variation or diversity in individuals can occurred.

"*Biological diversity or simply biodiversity, is the sum of life on Earth (plants, animals and microbes) encompassing all levels of biological organization from genomics to species to ecosystems*'.

Approximately 1.8 million species are known as a result of 300 years of the biological exploration of the planet. Astonishingly, an estimated 15050 million species await discovery and basic description. India figures among the biodiversity rich countries in the world, more than 8 per cent of the identified higher plant and animal species on nearly 2.5 per cent of the global land mass. In terms of plant species about 17, 500 species of higher plants are estimated to occur in the country in its 16 major vegetation types. About 33 per cent of the species are endemic. Indians have domesticated more than two dozen crop plant species. There are more than 380 cultivated/semi-cultivated crop species. The monumental work of N.I. Vavilov on the "Centres of Origin of Cultivated Plants" created awareness about the prevalence of genetic diversity in certain phyto-geographical regions and this subsequently resulted in organized to these regions for collecting plant germplasm. In the history of agriculture, however, human beings have selectively domesticated several species occurring in nature and provided them with alternative environment under agriculture for succession. Man's arrival on the earth is estimated to be around 2 million years ago which should have been associated with collection of wild plants and fruits for food. Though evidences exist that farming practices were adopted in Neolithic era for better harvest. On archeological grounds it can be estimated that such efforts first occurred in Southeast Asia. Consequently many crop plants of economic importance had been domesticated by 13,000 B.C. As per the historical background, adequate knowledge of cultivation existed for rice (*Oryza sativa*), beans (*Phaseolus, Vigna, Vicia* and *Glycine* species). Probably temperate genera of horticultural crops were domesticated in this region (Stoskopf, 1999; Roberts, 1929; Weather Wax, 1954 and Zirkle, 1932). Plants domesticated primarily for aesthetic enjoyment in and around the home are usually called house plants or ornamentals, while those domesticated for large-scale food production are generally called crops. A distinction can be made between those domesticated plants that have been deliberately altered or selected for special desirable characteristics, termed cultigens, and those plants that are used for human benefit, but are essentially no different from the wild populations of the species. Domestication is a sustained multi-generational relationship in which one group of organisms assumes a significant degree of influence over the reproduction and care of another group to secure a more predictable supply of resources from that second group (Zeder, 2015). Charles Darwin (1868) recognized the small number of traits that made domestic species different from their wild ancestors. He was also the first to recognize the difference between conscious selective breeding in which humans directly select for desirable traits, and unconscious selection where traits evolve as a by-product of natural selection or from selection on other traits (Diamond, 1997; Larson *et al.*, 2014; Olsen and Wendel, 2013. There is a genetic difference between domestic and wild populations. There is also such a difference between the domestication traits that researchers believe to have been essential at the early stages of domestication, and the improvement traits that have appeared since the split between wild and domestic populations (Olsen and Wendel, 2013; Doust *et al.*, 2014; Larson, 2014). Domesticated traits are generally fixed within all domesticates, and were selected during the initial episode of domestication of that animal or plant, whereas improvement traits are present only in a proportion of domesticates, though they may be fixed in individual breeds or regional populations (Doust *et al.*, 2014; Larson, 2014; Meyer *et al.*, 2013). Domestications interfered in evolutionary process of species by encouraging selective disseminations of a species or its variety.

## Historical Background of Domestication

Domestication began over 11,000 years ago when humans began agriculture. The earliest human attempts at plant domestication occurred in South-Western Asia. There is early evidence for conscious cultivation and trait selection of plants by pre-Neolithic groups in Syria: grains of rye with domestic traits have been recovered from Epi-Palaeolithic (c. 11,050 BCE) contexts at Abu Hureyra in Syria, but this appears to be a localised phenomenon resulting from cultivation of stands of wild rye, rather than a definitive step towards domestication (Hillman *et al.*, 2001). According to Braice Wood (1960) first successful attempt to cultivate crops for food production took place in a large area falling in between the rivers Tigris and Euphrates, comprising present day Iraq, parts of Iran, Turkey, Lebanon, Israel and Jordan. Imprints found on clay vessels suggest that wheat, barley, rye, oat and millets were produced between 9000 and 7000 BC. Similarly, domestication activities also occurred independently in south central Mexico between 6700-5000 B.C. where squash, avocados and maize were cultivated by human being for food. The materials were eventually brought to their respective regions and evaluated. As per the historical facts, the spread of such knowledge created interest among plant breeding in the 20th century. Nearly 300,000 plant species are documented in the world but around 2, 50,000 plant species have been described till date and a much larger number still remains to be duly recognized. It is known that there are 75,000 edible plants on the planet but out of these populations 5,000 in use indicates domestication. Only about 3,000 out of them were picked up from the wild our ancient ancestors for their use and grown on varying scales since the beginnings of agriculture. The earliest human attempts at plant domestication likely occurred in southwestern Asia and the Middle East about 10,000 years ago. The first domesticated vegetable crops were generally annuals with large seeds, or fruits including legumes such as peas, and not herbaceous vegetable crops (Zohary and Hopf, 1988), which may be due to their suitability and needs. Here it can be presumed that their selection may be different from wild species commonly occurring in nature.

As per archaeological evidence by 10,000 BCE the bottle gourd (*Lagenaria siceraria*) plant was one of the first crops to be domesticated in Africa (Decker-Walter *et al.*, 2001; Whitaker, 1971), used as a container before the advent of ceramic technology, appears to have been domesticated. Bottle gourd might have been dispersed to the New World by ocean currents or by human migration in pre-historic times (Erickson, *et al.*, 2005; Morimoto *et al.*, 2005). The domesticated bottle gourd reached the America from Asia by 8000 BCE, most likely due to the migration of peoples from Asia to America (Erickson *et al.*, 2005). Vegetative plants like sweet potato were likely collected in the tropics, semi-tropical regions of the America, Southeast Asia and Africa. When digging wild roots and tubers with sharp sticks, some of the roots undoubtedly remained in the ground to regrow. Some of the roots and tubers may have dropped or been discarded near encampments, only to regrow in areas with adequate rainfall. In the Americas cultivation of squash, corn, potato and beans was adapted. The Middle East was especially suited to these species; the dry-summer climate was conducive to the evolution of large-seeded annual plants, and the variety of elevations led to a great variety of species. As domestication took place humans began to move from a hunter-gatherer society to a settled agricultural society. This change would eventually lead, some 4000 to 5000 years later, to the first city states and eventually the rise of civilization itself. Continued domestication was gradual, a process of trial and error that occurred intermittently. Over time perennials and small trees began to be domesticated including apples and olives. Some plants were not domesticated until recently such as the macadamia nut and the pecan. In other parts of the world very different species were domesticated. In the Americas squash, maize, beans, and perhaps manioc (also known as cassava) formed the core of the diet. In East Asia millet, rice, and soy were the most important crops. Some areas of the world such as Southern Africa, Australia, California and southern South America never saw local species domesticated. Domesticated plants often differ from their wild relatives in the way they spread to a more diverse environment and have a wider geographic range. They may also have a different ecological preference; flower and fruit simultaneously; may lack shattering or scattering of seeds, and may have lost their dispersal mechanisms completely; have larger fruits and seeds, and so lower efficiency of dispersal; may have been converted from a perennial to annual; have lost seed dormancy and photoperiodic controls; lack normal pollinating organs; may have a different breeding system; may lack defensive adaptations such as hairs, spines and thorns, protective coverings and sturdiness; may have better palatability and chemical composition, rendering them more likely to be eaten by animals; may be more susceptible to diseases and pests;

### Table 2.3: Historical Steps in Domestication and Plant Breeding

| Period/Year | Researchers | Description |
| --- | --- | --- |
| 700 B.C | - | Babylonians and Assyrians pollinated date palm artificially for increasing the fruit set. |
| 1694 | Camerarius | Describe sexes in plant and studied fertilization. |
| 1717 | Thomas Fairchild | Crossing of two tobacco varieties. |
| 1760-1766 | Joseph Kolreuter | Systematic hybridization in tobacco |
| 1707-1778 | Carolus Linnaeus | Basis of plant classification (classes, families and genera). |
| 1759-1835 | Thomas Andrew Knight | First man to use artificial hybridization for practical plant improvement. |
| 1840 | John LeCouteur and Patrick Shirroff | Individual plant selection and progeny test to develop self pollinated crops. |
| 1856 | Vilmorin | Refined individual plant selection and used it in improving sugar content of sugar beet called "Vilmorin isolation principle". |
| 1890 | H. Nilson | Progeny test by careful observation and maintaining systematic record and established 'entire single plant and not single spike or single seed'. |
| 1809-1882 | Charles Darwin | Origin of species by his 'theory of organic evolution' in 1876 |
| 1876 | Charles Darwin | Darwin's book 'Cross fertilization in vegetable kingdom' |
| 1833 | Robert Brown | Prominent body within the cell termed as 'Nucleus' |
| 1822-1884 | Gregor John Mendel | Hybridization in garden pea and established 'law of inheritance' in 1866. |
| 1888 | Waldeyer | Thread like structure termed as 'Chromosome' |
| 1899 | Hopkins | Concepts of 'ear to row selection' in cross pollinated crops. |
| 1900 | de Vries, vonTshermack and Correns | Confirmation of Gregor John Mendel work through their hybridization experiments. The work led to the foundation of gene (character) transmission from generation to generation and has been interpreted in the form of Mendel's laws of heredity. |
| 1900 | Weissmann | Proposed 'Germplasm' theory. |
| 1901 | Hugo deVries | Mutation theory (sudden heritable changes in the organisms). |
| 1904 | Freeman | Reported to wheat rust heritable controlled by single dominant gene. |
| 1908 | East | Concept of 'hybrid vigor and inbreeding depression'. Hypothesis of over dominance to explain basis of heterosis. |
| 1909 | Shull | Concept of 'hybrid vigor and inbreeding depression'. Hypothesis of overdominance to explain basis of heterosis. |
| 1918 | Jones | Developed double cross concept in maize |
| 1922 | Harlan and Pope | Developed backcross method for transfer of disease resistance gene. |
| 1927 | Muller | Discovered mutagenic action of x-ray on Drosophila. |
| 1928 | Davis | Concept of top crossing for early screening in maize |
| 1929 | Stadler | Mutagenic effects of x-ray on 'barley and maize' |
| 1943 | Jones and Clark | Used 'male sterility' in onion for exploitation of hybrid vigour. |
| 1945 | Hull | Developed and used recurrent selection method for 'specific combining ability' in maize. |

may develop seedless/parthenocarpic fruits; may have undergone selection for double flowers, which may involve conversion of stamens into petals; may have become sexually sterile and therefore only reproduce vegetatively (Wikipedia.org, 2016).

It is also true that domestication process of wild species likely to be continue in the future due to change with human needs. It may be little importance today may be great significance tomorrow. In the Indian history, Aryans have been referred at several instances as crop cultivars. Aryan Rishi strongly advocated about vegetarian diet for long and healthy life. As a pattern of science the initial domestication of animals impacted most on the genes that controlled their behavior, but the initial domestication of plants impacted most on the genes that controlled their morphology (seed size, plant architecture, dispersal mechanisms) and their physiology (timing of germination or ripening) (Zeder, 2012; Zeder, 2006).

The history of plant domestication for example in New Guinea is complex and yet to be fully untraveled for most staples. Those are the product of intra and inter-specific hybridization with Southeast Asian (SE) cultivars *viz.* bananas, sugarcane, taro and some yams (Lebot, 1999). Carol Lentifer (University of Queensland) outlined a multi-locus and multi-phased domestication process for two section of *Musa* banana. Domestication of AA diploid varieties of 'Musa section' eventually gave rise to the most important groups of cultivated bananas in the world. At present 200 species are domesticated as food crops and 150 have entered the world commerce. In general < 20 provide most of the World Food (Barthlott, *et al.*, 1999, WRI, 2006). Among the 20 only three (rice, wheat and maize) are the major components. This reflects very narrow dietary base of a highly developed organism (Human being) on the earth. There are 49 major and minor crops endemic to the country (M.S. Randhawa, *History of Agriculture in India*, ICAR, New Delhi), including rice, 4 minor millets, 4 pulses, Indian mustard, 9 vegetables, 5 tuber crops, 11 fruits, 5 spices, 5 sugarcane, 7 fibre crops, and *Cannabis* (medicinal plant). In development of cultivated plants initial step was undertaken as domestication of wild species.

Conscious cultivation and trait selection of plants may have occurred in what is today Syria as early as 11,050 BC, but this appears to have been a localized phenomenon rather than a definitive step towards domestication (Hillman *et al.*, 2001). In India, domestication and breeding works started on several getable crops (cucurbits, leafy, tuberous) which originated in Indian sub-continent. However, some historical journeys covered are as under shown in Table 2.4.

## Genotypic Modification under Domestication

The agro-environment is expected to have changed their equilibrium under domestication/farming over

**Table 2.4: Historical Journey of Domestication**

| Period/Year | Description |
|---|---|
| 1900 | Establishment of 'Department of Agriculture' by Lord Curzon, Ciceroy and Governor General of India. The 'Department of Agriculture' introduced and domesticated several temperate vegetables such as cauliflower, cabbage, tomato, peas, carrot, lettuce, celery *etc.* |
| 1940-1946 | Seed production of temperate vegetable at Quetta (now in Pakistan) |
| 1948-1949 | Success in seed production of exotic vegetables (cauliflower, cabbage, knol-khol, garden beet, carrot, radish, turnip and capsicum) at Srinagar (J&K) and Katrain (H.P.). |
| 1968 | Establishment of ICAR-'Indian Institute of Horticultural Research, Bengaluru (Karnataka). |
| 1970 | Start of 'All India Coordinated Vegetable Improvement Project (AICVIP) with headquarter at IARI, New Delhi |
| 1971 | First hybrid ($F_1$) 'Pusa Meghdoot' of bottle gourd developed at IARI, Regional Station, Katrain (H.P.) released for commercial cultivation. |
| 1987 | Elevation of 'All India Coordinated Vegetable Improvement Project (AICVIP) to the level of Project Directorate which was shifted to Varanasi (U.P.) in 1992. |
| 1994 | Promoting 'Breeder Seed Production' (BSP) of vegetables by funding through National Seed Project (NSP). |
| 1995 | Promotion of "Hybrid Research in Vegetables" through a-hoc project. |
| 2004 | Launch of 'Horticulture Technology Mission Project' with the aim to gearing of machinery for transfer of technologies and enhancing production of planting material in horticultural crops. |

the millennia, giving rise to new diversity that is well adapted under these conditions. Precise sequence of events during the evolution of crop plants under domestication is not known. Presumably, in the initial stages, considerable genetic variability existed in each domesticated species. This variability was action upon by both natural and artificial selections. It may be expected that humans always tried to pick out the plant types, which better suited their needs. For example, they would obviously have selected for larger fruits and seeds. As per record of planned and systematic selection, goes only as far back as middle of the 19th century. Before this period, selection efforts would obviously have been primitive, unrecognized and lacking focus.

## Domestication Syndrome

Domestication syndrome is a term often used to describe the suite of phenotypic traits arising during domestication that distinguish crops from their wild ancestors (Olsen and Wendel, 2013; Hammer, 1984). Domestication of crop is believed to have occurred independently in the following at least six region: (i) Mesoamerica (ii) the Southern Andes, (iii) the Near East (iv) Africa (probably the Sahel and the Ethiopian highlands) (v) South East Asia and (vi) China, (v) South

East Asia, and (vi) China. In spite of the geographical diversity of these centres, a remarkably similar set of traits seems to have been selected in widely different crops; these traits are called domestication syndrome traits (Table 2.5).

## Zones of Domestication of Major Crops

Domestication of crop is believed to have occurred independently in the following at least six regions (Singh, 2012):

(i) Mesoamerica-Corn, common bean, squash, sweet potato, tomato, upland cotton

(ii) The Southern Andes-Potato, common bean, peppers, cassava, pineapple, pima cotton

(iii) The Near east-Wheat, barley, pea, lentil, chickpea, grape

(iv) Africa (probably the Sahel and the Ethiopian highlands)-Sorghum, African rice, coffee, okra, melon, cowpea, white yam

(v) South East Asia and Eastern India-Mungbean, cucumber, banana, plantain

(vi) Southern China-Asian rice, soybean, citrus, tea, cabbage

Due to domestication certain useful changes occurred are as under (Table 2.6):

Almost all the characteristics of plant species have changed under domestication due to accumulation of spontaneous mutations. The characters that show more distinct change are those that have been objects of selection and are still plant breeding objectives in many cultivated species.

The term 'cultivar' represents a particular line or bred of a domesticated type. Wild types are associated with wild niche, on the other hand domesticates and cultivars are associated with the managed habitats. However vice versa is also possible.

## Land Races

Relative reproductive isolation and differential agro climatic conditions created numerous landraces of our crops. Landraces do not have a fixed complement of genes (alleles) because subsistence farmers are always trading seeds and the conditions change over time span of hundreds of years.

## Domestication of Land Races and Speciation

All the traditional starch-rich staples (major and minor), most vegetables are vegetatively propagated. Over time, vegetative propagation is thought to have selected for parthenocarpic varieties in several species, including bananas and yam cultivars. Vegetative propagation, whether of parthenocarpic and sterile varieties or not, is a way of genetically isolating a cultivated or adapted plant from surrounding populations. It is also a mechanism through which plants can be moved beyond the natural range in which sexual reproroduction is viable. Due to climate, absence of pollinators or low density in the landscape may all inhibit sexual reproduction in a plant.

## Selection under Domestication

When different genotypes present in a population reproduce at different rates, it is called selection. A population may be defined as *"the group of individuals, which mate or can mate freely with each other"*. Thus a population consists of individuals of a single species growing in the same locality. Selection process can be grouped in to two depending on the agency responsible (Singh, 2012):

(i) Natural Selection

(ii) Artificial Selection

### (i) Natural Selection

Selection due to natural forces like climate, soil, biological factors, *e.g.*, diseases, insect pests, *etc.*, and

**Table 2.5: The different Traits Comprising Domestication Syndrome**

| Selection at Growth Stage | Selection Trait | | |
|---|---|---|---|
| | General Feature | Specific Trait | Example Crop(s) |
| Seedling | Increased seedling vigour | Loss of seed dormancy | |
| Reproductive | Increased rate of selfing | - | |
| | Adoption of vegetative reproduction | - | Sugarcane, cassava |
| | Increased in seed yield | Loss of seed dispersal | Legumes |
| | | More compact growth habit | Legumes |
| | | Increased number or size of inflorescence | Maize, Wheat |
| | | Increased number of grains/ inflorescence | Maize |
| | | Changed photoperiod sensitivity | Legumes, rice |
| | | Colour, size, taste, texture | Many crops |
| | | Reproduction in toxic substances | Cassava, lima bean, cucurbits |

*Source*: Singh, 2012.

**Table 2.6: Changes Occurred under Domestication**

| Sl.No. | Factors | Beneficial Changes |
|---|---|---|
| 1. | Reduction in pod shattering | Cultivated species of legume vegetables |
| 2. | Elimination of dormancy | Mungbean, cowpea, green peas and French bean |
| 3. | Decrease in toxins | Erucic acid lowering in *Brassica* and the bitterness in Cucurbitaceous vegetables |
| 4. | Plant type (indeterminate growth habit) | Indeterminate growth habit has been observed in tomato, pea *etc.* |
| 5. | Plant canopy modification | Plant height, tillering habit, branching and leaf characters show alterations in domesticated types like bushiness in cowpea, lablab bean |
| 6 | Shorter life cycle | Cucurbits |
| 7. | Increase of fruit size and weight | Fruit size in cultivated tomato is 10-20 times higher than wild types. Fruit size and weight increased in potato, onion, and cucurbits. In beans and green peas, grain size increased. |
| 8. | Promotion of asexual reproduction | Potato (*Solanum tuberosum*), Sweet potato (*Ipomoea batatas*) |
| 9. | Promotion of sexual reproduction | Potato (*Solanum tuberosum*) through true potato seeds (TPS) |
| 10. | Preference for polyploidy | Potato (*Solanum tuberosum*), sweet potato (*Ipomoea batatas*) |
| 11. | Shift from cross pollination to self pollination behavior | Wild species of tomato, chillies and brinjal are cross pollinated species but their cultivated types are self-pollinated. |
| 12. | Increase in economic yield | Almost in all vegetable crops |
| 13. | Shift in sex form | Gynoecism in bitter gourd, Andromonoecism in bottle gourd |
| 14. | Increase in economic traits | Seed less (parthenocarpiness) pointed gourd, TSS in muskmelon and watermelon |
| 15. | Drastic reduction in genetic variability | Drastic reduction in genetic variability of species has been observed under domestication. |
| 16. | Exploitation of hybrid vigour | Hybrid vigour is being exploited by developing hybrids ($F_{1s}$) in tomato, potato, brinjal, okra, cauliflower knolkhol (kohlrabi) chilli, capsicum *etc.* |

other factors of the environment is called natural selection. It occurs in natural populations *i.e.*, wild forms and wild species, and determines the course of their evolution. Generally, all the genotypes of the population reproduce; the population becomes more adapted to the prevailing environment, and considerable genetic variability is retained. In 1962, Nicholson proposed that natural selection may be seen to operate through two mechanisms:

### (a) Environmental Selection

Environmental selection acts against all such genotypes that are unable to cope with the environmental stresses. As a result, the population consists, ultimately, of only those genotypes that are capable of surviving the prevalent environmental stresses and are also able to reproduce.

### (b) Competition

Natural selection through competition occurs in crop populations where a plant takes up more water, nutrients or light than another, and at the expense of the other.

Therefore, the more successful is a plant in exploiting resources, the greater will be its potential to be represented in the succeeding generations and it will be selected through competition.

### (ii) Artificial Selection

In contrast, artificial selection is carried out by humans, and is confined to domesticated species. It allows only the selected plants to reproduce, ordinarily makes plants more useful to humans, and generally leads to a marked decline in genetic variability in the selected progenies/populations. Usually, plants become less adapted to the natural environment and they have to be grown under carefully managed conditions. Our present day crops are the products of continued artificial selection.

## Types of Selection

Selection is grouped into the following three types depending mainly on the type of phenotypic class favored by it:

(i) Directional selection

(ii) Stabilizing selection

(iii) Disruptive selection

In nature, however, selection would be either directional selection, stabilizing selection or disruptive selection depending on the state of evolution of the population. While a population is adapting to new environmental conditions (either in a new area where it has been introduced or in the same region after a change in the environment) there will be directional selection to increase its fitness. But once the population has become adapted, directional selection will be succeeded by stabilizing and disruptive selections.

### (i) Directional Selection

In plant breeding situations, selection is almost always directional aiming to achieve the maximal expression of targeted characters.

*"When individuals having the extreme phenotype for a trait or a group of traits are selected for, it is called directional selection".*

Directional selection usually selects for such gene combinations that produce a fully balanced phenotype; such a phenotype results in the maximum yield under artificial selection, and in the maximum fitness under natural selection. Once such gene combinations are established, they are protected from further changes by genetic linkages and sometimes, by a change in the mode of reproduction, *e.g.*, from cross pollination to self-pollination. In cross pollinated populations, directional selection will favour alleles showing dominance in the appropriate direction, and genes showing 'desirable' epistatic interactions will also be selected for. As a result, characters subjected to prolonged directional selection will show high directional dominance and/or epistasis.

### (ii) Stabilizing Selection

*When selection favours the intermediate phenotype and acts against the extreme phenotypes, it is termed stabilizing selection.* In nature, it follows directional selection, and strives to maintain the co-adapted gene complexes generated by the latter. It may be pointed out that for such characters that directly affect fitness, *e.g.*, viability and fertility, selection will always be directional. Therefore, stabilizing selection occurs only for those traits that do not affect fitness directly and it faours those genotypes whose phenotypic expression clusters around the population mean. In such characters, therefore, the 'maximal' expression is not the 'optimal expression'. Stabilizing selection disfavour dominance; if dominance is present, it is bidirectional, *i.e.*, some alleles show dominance in one direction, while some others show dominance in the opposite direction. Similarly, epistasis is also selected against. Thus stabilizing selection accumulates alleles showing additive gene action.

### (iii) Disruptive Selection

Disruptive selection succeeds directional selection in such natural population that are subjected to distinct ecological niches that may be spatial temporal (seasonal or long-term cycles) or functional (*e.g.*, males and females of species) in nature. In such 'ecological niche' a different 'phenotype optima' is selected for so that the population ultimately consists of two or more recognizable forms; such a selection is called disruptive selection. The consequences of such a selection depend mainly on the following two factors:

(a) Whether the different optimal phenotypes are independent of or dependent on each other for their maintenance or function, and

(b) The rate of gene flow between them. For example, the male and female forms of a single species are completely interdependent in function, *i.e.*, reproduction, and show 100 per cent gene exchange. At the other extreme, a species may occupy a habitat that is fragmented into two or more independent niches, and in each niche a different phenotypic optima is selected for. In such a case, if the selection pressure is high enough and continued long enough, genetic barriers to crossing may be arise leading to genetic separation of these forms, and eventually to their evolution as distinct species.

Disruptive selection maintains polymorphism in a population. Further, it shows such features as frequency-dependence (*e.g.*, less frequent alleles being more favoured), density-dependence, cyclical nature, *etc.* Since disruptive selection is 'directional' in nature within each 'ecological niche' of the habitat, it favours dominance and epistasis. In addition, it often leads to the establishment of integrated 'supergenes', *e.g.*, in case of male and female forms of a species.

"A supergene is a set of closely linked genes that together lead to the development of a specific optimal phenotype, *e.g.* male or female form.

Human being as per their need selected similar set of traits (domestication syndrome traits) in widely different crops.

## Ways of Selecting the Materials

The Division of Germplasm Exchange of ICAR-NBPGR, New Delhi is following two ways of selecting the material for introduction into the country:

(a) One is by scanning the current world literature available with respect to promising varieties developed all over the world and making selections according to their suitability to Indian conditions. Scientists in the Division have been assigned a group of crops and each one makes this a regular exercise to scan the literature, and select, suitable accessions which would be introduced.

(b) The second way is where a research worker requests Bureau for his specific needs of material to be introduced based on research programmes.

## Pattern of Distribution

An examination of the data shows certain interesting trends in pattern of plant genetic resource distribution:

i. The movement of plants might have been a two-way process, even though this was being carried out by traders, conquerors and

**Table 2.7: Distribution of Wild and Weedy Species of Cultivated Plants in South West Asia and South Asia**

| Sl.No. | Cultivated Plants Originated in south West Asian Centre | Species Present in both the Centres | Species Present in only South West Asian Centre | Species Present in only South Asian Centre | Remarks |
|---|---|---|---|---|---|
| 1. | *Cucumis melo* L.<br><br>*C. flexuosus* L.<br>*C. sativus* spp. *antasiticus* Gabarev | Two spp.,<br>*C. callosus*<br>*C. prophetarum* | One sp. *C. chali* | Two spp.,<br>*C. hystrix*<br>*C. sitosus* | *ca.* 70 spp., warmer parts of the world, especially, Africa |
| 2. | *Cucurbita pepo* L. | Two spp.<br>*C. maxima*<br>*C. moschata* | Nil | One sp., *C. ficifolia* | *ca.* 30 spp., tropics and sub-tropics |
| 3. | *Lepidium sativum* L. | Two spp.<br>*L. latifolium*<br>*L. ruderale* | 12 spp.<br>*L. ancheri*<br>*L. caespilosum*<br>*L.campestre*<br>*L.cartilagineum*<br>*L.chalepense*<br>*L.hirtum*<br>*L.lyratum*<br>*L.nebrodense*<br>*L.perfoliatum*<br>*L.spinosum*<br>*L.vesicartum*<br>*L.graminifolium* | Three spp.,<br>*L. apetalim*<br>*L. capitalum* | *ca.* 150 spp. |
| 4. | *Brassica campestris* L. subsp. *oleifera* Metz.<br><br>*B. juncea* Czern. var. *sarektana* Sinsk.,<br>*B. nigra* L. var. *pseudocampestris* Sinsk. | Nil | Four spp.,<br>*B. cretica*<br>*B. deflexa*<br>*B. elongata*<br><br>*B. tournefortii* | Two spp.,<br>*B. napus*<br>*B. rapa* | More than 200 spp., world-wide, but more in north temperate zone specially in Mediterranean region |
| 5. | *Beta vulgaris* L. | Nil | Six spp.,<br>*B. adanensis*<br>*B. corolliflora*<br>*B. lomatogana*<br>*B. macrorhiza*<br>*B. maritima*<br>*B. trigyna* | One sp.,<br>*B. Palonga* | *ca.* 6 spp., Europe and Mediterranean region |
| 6. | *Daucus carota* L. | Nil | Six spp.,<br>*D. aureus*<br>*D. broteri*<br>*D. conchitae*<br>*D. guttatus*<br>*D. involucratus*<br>*D. littoralis* | | |
| 7. | *Eruca sativa* Mill. var. *orientalis* Sinsk. | Nil | Nil | One sp.,<br>*E. vesicaria* | *ca.* 6 spp., Mediterranean and north-east Asia |
| 8. | *Allium cepa* L.<br>*Allium porrum* L. | Two spp. | | | |

| Sl.No. | Cultivated Plants Originated in south West Asian Centre | Species Present in both the Centres | Species Present in only South West Asian Centre | Species Present in only South Asian Centre | Remarks |
|---|---|---|---|---|---|
| 9. | *Petroselinum crispum* (Miller) A.W. Hill<br><br>*P. hortense* Hoff. | Nil | Nil | Nil | *ca.* 5 spp., Europe and Mediterranean region |
| 10. | *Lactuca sativa* L. | four spp.,<br>*L. crambifolia*<br>*L. orientalis*<br>*L. serriola*<br>*L. undulata* | Eighteen spp. | Twenty one spp. | *ca.* 100 spp., temperate Eurasia with an extension to tropical south Africa |
| 11. | *Portulaca oleracea* L. | Nil | One sp.,<br>*P. sativa* | Three spp.,<br>*P. pilosa*<br>*P. qudrifida*<br>*p. wightiana* | *ca.* 200 spp., tropics and sub-tropics |

*Source*: Nayar *et al.*, 1994.

settlers. Prehistoric evidences show that greater movement took place from south-west Asia to south Asia.

ii. Some of the plants which had their origin in the Middle East have not got established in south Asia. This may have happened for various reasons such as climatic differences, personal preferences, human habits, *etc.*

iii. Several of the plants of the Middle East may have been introduced into south Asia from very ancient times as could be inferred from the prevalence of genetic differences. Few of them may have undergone considerable differentiation and even further speciation in their new habitat in south Asia, and fourthly, some of genera show wide distribution and variation in south Asian region.

# References

1. Darwin, Charles. *1868*. The Variation of Animals and Plants under Domestication. *London: John Murray*. OCLC 156100686.

2. Decker-Walters, D., Staub, J., Lopez-Sese, A. and Nakata, E. 2001. *Diversity in landraces and cultivars of* bottle gourd [*Lagenaria siceraria* (Mol.) Standl., Cucurbitaceae] as assessed by random amplified polymorphic DNA. *Genet. Resour. Crop Evol.*, **48**: 369-380.

3. Doust, A. N.; Lukens, L.; Olsen, K. M.; Mauro-Herrera, M.; Meyer, A.; Rogers, K. 2014. "Beyond the single gene: How epistasis and gene-by-environment effects influence crop domestication". *Proceedings of the National Academy of Sciences*, **111** (17): 6178–6183.

4. Diamond, J. 1997. Guns, Germs, and Steel. *Chatto and Windus*, London. ISBN 978-0-09-930278-0.

5. Erickson, D.L., Smith, B.D., Clarke, A.C., Sandweiss, D.H., Tuross, N., Smith, Clarke., Sand Weiss, Tuross. 2005. "An Asian origin for a 10,000-year-old domesticated plant in the Americas". *Proc. Natl. Acad. Sci. U.S.A.*, **102 (51)**: 18315–20.

6. Hammer, K (1984). "Das Domestications syndrom". *Kulturpflanze*, **32**: 11–34.

7. Hillman, G., Hedges, R., Moore, A., Colledge, S., Pettitt, P., Hedges, M. and Colledge; P. (2001). "New evidence of late glacial cereal cultivation at Abu Hureyra on the Euphrates". *Holocene*, **11 (4)**: 383–393.

8. Larson, G. 2014. "The Evolution of Animal Domestication" *(PDF). Annual Review of Ecology, Evolution and Systematics*, **45**: 115–36.

9. Larson, G.; Piperno, D. R.; Allaby, R. G.; Purugganan, M. D.; Andersson, L.; Arroyo-Kalin, M.; Barton, L.; Climer Vigueira, C.; Denham, T.; Dobney, K.; Doust, A. N.; Gepts, P.; Gilbert, M. T. P.; Gremillion, K. J.; Lucas, L.; Lukens, L.; Marshall, F. B.; Olsen, K. M.; Pires, J. C.; Richerson, P. J.; Rubio De Casas, R.; Sanjur, O. I.; Thomas, M. G.; Fuller, D. Q. 2014. "Current perspectives and the future of domestication studies". Proceed. National Academy of Sciences, **111 (17)**: 6139–6146.

10. Lebot, V. 1999. Biomolecular evidence for plant domestication in Sahul. *Genetic Resources and Crop Evolution*, **46**: 619-628.

11. McKey, M., Celis-Gamboa, C., Chumbiauca, S., Salas, A. and Visser, R.G.F. 2012. Hybridization between wild and cultivated potato species in the Peruvian Andes and biosafety implications for development of GM potatoes. *Euphytica*, **164**: 881-892.

12. Meyer, Rachel S.; Purugganan, Michael D. 2013. "Evolution of crop species: Genetics of domestication and diversification". *Nature Reviews Genetics*, **14 (12)**: 840–52.

13. Morimoto, Y., Maundu, P., Fujimaku, H. and Morishima, H. 2005. Diversity of landraces of the white-flowered gourd [*Lagenaria siceraria* (Mol.) Standl.] and its wild relatives in Kenya: fruit and seed morphology. *Genet. Resour. Crop Evol.*, **52**: 737-747.

14. Mukherjee, A. 1999. Tuber Crops: In: Biotechnology and its application in horticulture. Edited by Ghosh, S.P., Narosa Publishing House, New Delhi, p. 267-294.

15. Mukherjee, A., Unnikrishnan, M. and Nair, N.G. 1991. Growth and morphogenesis of immature embryos of sweet potato (*Ipomoea batatas* L.) *in vitro*. *Plant Cell Tissue and Organ Culture*, **26**: 97-99.

16. Nayar, N.M., Ahmedullah, M. and Singh, R. 1994. Spread of south-east Asian cultivated plants and their wild and wedy relatives to south Asia and beyond. *Indian J. Plant Genet. Resour.*, **7(1)**: 1-12.

17. Olsen, K.M., and Wendel, J.F. 2013. A bountiful harvest: genomic insights into crop domestication phenotypes. *Annu. Rev. Plant Biol.*, **64**: 47–70.

18. Whitaker, T.W. 1971. Endemism and pre-Columbian migration of bottle gourd [*Lagenaria siceraria* (Mol.) Standl.]. In: Man across the sea. Edited by: Riley, C.L., Kelly, J.C., Pennington, C.W. and Rounds, R.L. University of Texas Press, Austin; **78**: 218.

19. Zeder, M.A. 2006. Archaeological approaches to documenting animal domestication. In Documenting Domestication: New Genetic and Archaeological Paradigms, (eds.) Zeder, M.A., Bradley, D.G., Emshwiller, E., Smith, B.D., p. 209–27. Berkeley: Univ. Calif. Press.

20. Zeder, M.A. 2012. "The domestication of animals". *Journal of Anthropological Research*, **68 (2)**: 161–190.

21. Zeder, M. A. 2015. "Core questions in domestication Research". *Proceedings of the National Academy of Sciences of the United States of America*, **112 (11)**: 3191–8.

# 3

# Introduction and Augmentation of Vegetable Genetic Resources

## Plant Introduction

In order to have proper understanding of past, present and future of plant introduction, it is desirable to analyse various definitions of plant introduction proposed appropriately in relation to the periods through which plant introduction has passed and is expected to pass in future.

Most crop specialists, the term plant introduction means:

*"The transfer of a living genetic entity from a local where the plant usually has survived through several generations to a new location".*

It may range from the wild and primitive forms to the more highly developed economic varieties, or to cultivars which are the result of man's manipulation of desirable characters.

Or

According to Frankel (1957) plant introduction is *"Transposition of a genetic entity from an environment to the one in which it is untried".*

Or

According to Allard (1960) plant introduction is *"The acquisition of superior varieties by importing them from other areas.*

Or

*"Plant introduction comprises shifting of a genotype or a group of genotypes (species or its variety) of plants into a new area or region where they were not being grown before".*

Or

*"Plant introduction it its broader sense could be defined as domestication of wild plants and their relatives which can be used to develop as new agricultural plants mainly for industrial purpose to meet new and more exacting demands".*

Thus, it encompasses the entire gamut of modern techniques of plant breeding (such as transfer, addition or substitution of genes, chromosome segments or entire chromosome brought about through near or distant hybridization by employing wide array of techniques such as back crossing, embryo culture, polyploidy, mutations *etc.*), the ultimate goal of which is the introduction of improved varieties of plants into cultivation, anywhere in the world. Plant introduction includes new varieties of a crop already grown in an area, wild relatives of the crop species or a totally new crop species for the area. Often plant materials are introduced from other countries but movement of crop varieties from one area into another within the same country also constitute introduction (Singh, 2012). Usually the migration of crops or their varieties accompanied the man's migration. Due to population growth, changing climate, expansion of trade, reduced vegetation, discovery of new lands/area *etc.* are the major cause of peoples and tribes migration. Introductions have been the most important activity in vegetable improvement globally. With the discovery of new world, exchange of germplasm was accelerated. A wide range of germplasm exchange was noticed throughout the world for cereals, pulses, oilseeds, fruit plants and vegetables. According to recent studies, average crop genetic resources dependency among various regions of the world is more than 50 per cent and for regions it may be as high as 100 per cent for the most important crops. Crop genetic resources dependency is defined as:

*"The degree of resistance of plant breeding/crop production of a country/region on exotic germplasm".*

Plant introductions can be categorized depending upon political, geographical and climatic boundaries such as (Sharma, 2009):

(i) **Inter-continental introduction**: When the plant introduction is between two continents.

(ii) **Inter-regional introduction**: Introduction between two regions of a continent.

(iii) **Inter-state Introduction**: Introduction between two states.

(iv) **Inter-district introduction**: Introduction from one district to another.

(v) **Inter- locality introduction**: Introduction between two localities of a district.

(vi) **Inter-climatic introduction**: Production perspective of plant species have further increased with their introduction from one climatic zone to other.

It may be realized in two ways:

(a) Introduction of a species from temperate to sub-tropical or *vice-versa.*

(b) Introduction of a species from tropical to sub-tropical and *vice-versa.*

## Historical Background of Plant Introduction

For many thousands of years, plants have travelled around the world alongwith people journeying to differtent countries for *e.g.* tomato, maize, pepper (chilli) (South America), banana (South East Asia), carrots (Afghanistan), potato (Andean region, South America), Onion (Centrl Asia) and pumpkins (tropical America). In historic times, the transfer of plants from one region to another must have been very slow, but the movement of plants originating in parts of the world was possible among the old world countries much earlier because of geographic contact but the exchange between the new world and old world was possible only after the discovery of America by Colombus in 1942, and the European colonization, soon after. As per the historical evidences many crop plant travelled into new areas from their place of origin due to migration of human being. The Portuguese, English, French and Dutch during 16th century have introduced many plants to different lands. The last five centuries have witnessed much give and take among the economically useful flora of different regions in the world. This exchange in the beginning was mostly at the species level, and was made in an empirical way; still it has to be appreciated because of the achievements being very significant. It is true that many varieties of vegetable crops of our country are introduction from other countries. For several centuries the agencies of plant introductions were invaders, settlers, traders, travelers, explorers, pilgrims and naturalists. During last 150 years, plant introduction work has further accelerated mainly for the introduction of new species. As a horticultural crops Muslim invaders introduced in India cherries and grapes from Afghanistan by 1300 A.D. but as a vegetable crops, in the 16th century A.D., Portuguese introduced chillies, and sweet potato but certain crops like chilli and potato were known. East India Company brought cabbage, cauliflower and other vegetables from Mediterranean region. During 19th century, a number of Botanic Gardens played an important role in plant introduction. The Kolkata Botanic Gardens was established in 1781. During and after the last part of 19th century, various agricultural and horticultural research stations were established in the country and make a lot of efforts to introduced horticultural and agricultural plants.

**Table 3.1: Some Important Vegetable Crop Species Introduced in India**

| Vegetable Crops | Period | Source of Introduction | Remarks |
|---|---|---|---|
| Potato | Early 10th century | South America | - |
| Tomato | Early 10th century | South America | - |
| Cabbage | 18th century | Mediterranean region | By East India Company |
| Cauliflower | 18th century | Mediterranean region | By East India Company |

In recent year, several reputed institutions and organisations have entered in this field of activities. Today, the plant breeders and other plant scientists have realized that a stage has reached where the introduction of altogether new economic species has limited possibility but there is a way much wider scope in improving agronomic potential of the presently grown economic species by genetic improvement through a more judicious use of extensive germplasm. Therefore, greater emphasis has been laid on location and assemblage of varietal wealth through correspondence or through joint explorations. This is the real junction where the significance of exchange of genetic resources and its introduction was realised.

The present activities in plant introduction aim at (a) exploring the possibility of utilizing proven exotic stocks in hybridization or (b) providing populations for selection.

## Types of Plant Introduction

The cultivation zones of field crops and vegetable have increased with inter-climatic zone introductions. Depending on the utilization of material, introductions may be classified into two categories:

(1) Primary introduction

(2) Secondary introduction

## (1) Primary Plant Introduction

To make a distinction of convenience, Bennet (1965) proposed that the introduction of wild plant into cultivation and the successful transfer of cultivars, with their genotypes unaltered, to new environment could together be designated as "primary plant introduction'.

Primary plant introduction is concerned as *"The introduced variety is well adapted to the new environment; it is released for commercial cultivation without any alteration in its genotype"*.

Or

*"The introduction of wild plants into cultivation and successful transfer of cultivars with their genotypes unaltered to new environments could together be designated as primary plant introduction"*.

It may be a new crop or a new variety of the crop already in cultivation. For instance in the 16th century Portuguese introduced potato, sweet potato, chillies, tomato and several other crops in India. In 18th century East India Company introduced cabbage, cauliflower, *etc.* in India.

**Table 3.2: Some Vegetable Introductions Used Directly in Cultivation (in India)**

| Vegetable Crop | Variety | Country |
|---|---|---|
| Tomato | Sioux, Fireball, Marglobe, Best of All, Roma, Dwarf, La Bonita | USA |
| | Money Maker | Israel |
| Watermelon | Asahi Yamato, Shim Yamato | Japan |
| | Sugar Baby, Hampshire Midget | USA |
| Cucumber | Straight Eight | USA |
| Summer squash | Australian Green | Australia |
| Pea | Bonneville | U.S.A., |
| | Arkel | France |
| French bean | Contender | USA |
| Cabbage | Golden Acre | Denmark |
| Cauliflower | Snowball 16 | Holland |
| | Improved Japanese, 96D | Israel |
| | Early Snowball | Japan |
| Knol-Khol | White Vienna and Purple Vienna | Vietnam |
| Carrot | Nantes (temperate type) | USA |
| Radish | Japanese White | Japan |
| | China Red | China |
| Capsicum | California Wonder Bell Pepper, Yolo Wonder and Bull Nose | USA |
| Hot pepper | Tabasco | Africa |
| Onion | Early Grano | USA |
| | Yellow Skinned, Bermuda Yellow | Philippines |
| Okra | Ghana Red | Ghana |

*Source*: Gautam *et al.*, 2001.

Primary introduction is less common, particularly in countries having well organized crop improvement programmes. There are many examples of readymade varieties being adapted in new lands which have held the field for long as direct introductions. Several primary introductions, in the past, proved useful as popular varieties and continued even recently and many of them are still grown on large scale under Indian conditions. During the past, introduction of new vegetable varieties in every crop are becoming very popular among farmers or being used in selections. Example can be cited in every crop in vegetable group where more than half the varieties grown in India are either direct introductions or selections thereof. Some useful collections release as cultivars after desired improvement

Some of the notable introductions during recent years are bacterial wilt and *Fusarium* wilt resistant lines of tomato from Taiwan, nematode resistant varieties 'Quinte' and 'Early Rouge' from Canada. Carrot variety Beta III has high carotene content. This has been used by many human nutrition programmes in India to cope up Vitamin A deficiencies. In sweet chilly, varieties like Bell Pepper, California Wonder, Yolo Wonder and Bull Nose have been found promising. Yolo Wonder is large fruited, 3-4 lobed, medium thick flesh and tolerant to tomato leaf curl virus. California Wonder is medium tall, late maturing, square shaped, smooth, thick walled and having green fruits. In hot pepper, variety 'Tabasco' a West African introduction has also been found promising under Indian conditions. In cabbage Drum Head, Danish Ball Head and Late Flat Dutch are the promising exotic introductions which were reported widely under cultivation. In radish Rapid Red, Scarlet Globe, French Breakfast and Rapid Red White Tipped are the exotic promising introductions. In turnip Purple Top White Globe, Snow ball, Golden Ball are some of the useful introductions in India. In okra Perkins Long green and Velvet Green were recommended for direct cultivation in India. In carrot 'Nantes Half Long' is also promising exotic introduction of economic importance. Similarly, in beet root Crimson Globe, Detroit Dark Red and Crosby's Egyptian are found suitable introduction for direct cultivation. In summer squash 'Zucchini Long' and 'Patty Pan' are promising introduction which are still being used as cultivar in India. Similarly, in pumpkin Large Cheese, Kentucky Field, Butternut and African Bell are promising exotic introductions. In lettuce three exotic introductions *viz.*, Grand Rapids, Slobat and Chinese Yellow have been found promising.

In watermelon high yielding varieties tolerant to *Fusarium* wilt have been introduced from USA. In muskmelon a multiple disease resistant line AC70-54 has resistance to gummy stem blight, powdery and downy mildew. In cucumber also, gynoecious line Gy-4 has been introduced which is being used for breeding high yielding multiple disease resistance including mildews and *Fusarium* wilt. In Chinese cabbage, heat tolerant and black rot resistant lines have been introduced from Taiwan. In cabbage, lines resistant to club rot, black rot and heat tolerant have also come from Taiwan. In capsicum, TMV tolerant, nematode tolerant and high yielding varieties World Beater and Yolo Wonder have all proved useful introductions. In chillies, hot peppers lines from AVRDC, Taiwan have been very successful at several locations in India.

A number of accessions of vegetable crops and their wild relatives were introduced from different countries. This includes *Citrullus lunatus* (8, all from USA), *Cucumis melo* (15, from France, Japan), *Cucumis sativus* (15 from Japan, USA), *Cucumis heptadactylus* (1 USA), *Cucumis metuliferus* (16 USA), *Cucumis anguria* (94 USA), *Cucurbita pepo* (2 from USA), *Cucurbita species* (4 from Algeria, Japan), *Luffa cylindrica* (1 Japan) and *Lagenaria siceraria* (1 Japan,). A number of vegetable germplasm were introduced from other countries are given in Table 3.3.

### (2) Secondary Introduction

Other than primary plant introduction is called secondary plant introduction.

Both these types of introduction activities have run parallel since the beginning of agri-horticulture, but progressively the emphasis has shifted from 'primary' to 'secondary'. Further, according to M.W. Hardas (personal communication), plant introduction is an experimental dispersal of plants, arranged by man under phyto-sanitary conditions.

## Principles of Plant Introduction

☆ The study of geography of plants has shown that the distribution of plant species on the earth is not uniform. Vavilov (1951) proposed "Gene centre concept of cultivated plants and their wild progenitors".

☆ Agro-climate' concept (climatically homologous regions): introduction of plant genetic resources from region having similarity in climate is expected to be more fruitful for its success.

☆ Adaptability introduced crop will have a greater chance of success, if it has a considerable amount of adaptability permitting it to fit into wide variety of environment.

☆ A new crop plant should be introduced with maximum range of variation adequate to the objectives and depending on the breeding systems.

## Sources of Introductions

1. Introduction from centres of diversity: Primary or Secondary centre.

2. National organisations: USDA (USA), IPI (Russia), CSIRO (Australia), CENARGEN (Brazil) and others.

3. International organisations: CGIAR institutions *viz.* CIP (Peru), CIAT (Colombia), World Vegetable Centre (earlier AVRDC) Taiwan, IITA (Nigeria).

**Table 3.3: Introduction of Vegetable Genetic Resources**

| Vegetable Crops/Genera | No. of Accessions Obtained | Source Countries |
|---|---|---|
| *Lycopersicum* | 3489 | Australia, Bulgaria, Canada, Cuba, Denmark, France, Italy, Japan, the Netherlands, the Philippines, Poland, Taiwan, USA, USSR |
| Capsicum | 2241 | Australia, Bulgaria, Czechoslovakia, Hungary, Italy, the Netherlands, Nigeria, the Philippines, Taiwan, USA, USSR, Yemen |
| Potato | 1250 | Australia, Bangladesh, Canada, China, Germany, Hungary, Indonesia, Japan, Kenya, Mexico, Nepal, the Netherlands, New Zealand, Peru, Sri Lanka, Sweden, Switzerland, Taiwan, UK, USA |
| Brinjal | 431 | Australia, Bangladesh, Denmark, France, Hungary, Japan, the Philippines, Sri Lanka, USA, USSR |
| *Allium* | 710 | Australia, Brazil, Germany, Hungary, Indonesia, Iran, Nigeria, Bulgaria, Canada, Cuba, Denmark, France, Italy, Japan, the Netherlands, the Philippines, Poland, Syria, Taiwan, USA, USSR |
| *Brassica* | 1740 | Canada, France, Italy, Japan, Korea, the Netherlands, UK, USA, USSR |
| Cucurbits | 1939 | France, Hungary, Korea, Mexico, the Netherlands, Nigeria, the Philippines, Turkey, UK, USA, USSR, Zambia |
| Leafy vegetables | 139 | Italy, the Netherlands, Poland, USA, USSR |
| Garden pea | 274 | Australia, Bangladesh, Bulgaria, Czechoslovakia, Italy, New Zealand, USA, USSR |
| French bean | 524 | Australia, Bangladesh, Bulgaria, Czechoslovakia, Italy, New Zealand, USA, USSR |
| Carrot, radish | 646 | Australia, Brazil, Denmark, Egypt, France, Hungary, Italy, Japan, Korea, the Netherlands, Poland, Taiwan, Turkey, UK, USA, USSR, Zambia |
| Okra | 997 | Bangladesh, Brazil, Ivory Coast, Nigeria, the Philippines, Singapore, Sri Lanka, Sudan, Turkey, UK, USSR |

*Source*: ICAR-NBPGR, New Delhi Data Base, 1976-1990.

4. FAO/IBPGR
5. Gene Banks
6. National arboreta/botanical gardens/private nurseries *etc.*

## Introduction from Abroad

ICAR-NBPGR, New Delhi is the major institution in India which has the mandate of introduction of plant genetic resources of diverse agri-horticultural and agri-silvicultural plants. The Division of Germplasm Exchange of ICAR-NBPGR owes its birth to the erstwhile Plant Introduction Division of IARI, which has remained engaged since 1940 in the important task of introducing elite strains, varieties, promising genetic stocks and improved cultivars of various crops from different parts of the world into India under strict quarantine control. During the past, introduction of new vegetable varieties in every crop are becoming very popular among farmers or being used in selections. Example can be cited in every crop in vegetable group where more than half the varieties grown in India are either direct introductions or selections thereof (Table 3.4).

Some of the notable introductions during recent years are bacterial wilt and *Fusarium* wilt resistant lines of tomato from World Vegetable Center, Taiwan and nematode resistant varieties Quinte and Early Rouge from Canada. Carrot variety Beta III has high carotene content. This has been used by many human nutrition programmes in India to cope up Vitamin A deficiencies. In watermelon high yielding varieties tolerant to *Fusarium* wilt have been introduced from USA. In muskmelon a multiple disease resistant line AC70-54 has resistance to gummy stem blight, powdery and downy mildew. In cucumber also, gynoecious line Gy-4 has been introduced which is being used for breeding high yielding multiple disease resistance including mildews and *Fusarium* wilt. In Chinese cabbage, heat tolerant and black rot resistant lines have been introduced from Taiwan. In cabbage, lines resistant to club rot, black rot and heat tolerant have also come from World Vegetable Center, Taiwan. In capsicum, TMV tolerant, nematode tolerant and high yielding varieties World Beater and Yolo Wonder have all proved useful introductions. In chillies, hot peppers lines from World Vegetable Center, (AVRDC), Taiwan have been very successful at several locations in India.

### Table 3.4: Some Promising Introductions (Exotic Collections)

| Crop | Lines | Attributes | Country |
|---|---|---|---|
| Tomato | Patriot (EC-179883) | Resistant to *Fusarium* wilt and Root Knot Nematode | USA |
| | Karboreta | Resistant to TMV and tracking, rich in β-carotene (3-4 mg/100g) and ascorbic acid (20-30 mg/100 g) | Bulgaria |
| | CL-1131-0-013-0-6, CL1131-0-0-43-8-1, CL-5915-154, CL-5915-153D41-D, CL 143-0-4-B-1 | Resistant to TMV and heat | Taiwan |
| | EC202430 | Wilt resistant line | Canada |
| | EC202431 | Early maturity | Canada |
| | EC-173859 | Resistant to TMV (race-1) | Taiwan |
| | EC-182761 to 874 | Tolerant to bacterial wilt and leaf mould; resistant to blight, powdery mildew | Taiwan |
| | EC 243066 to 69, EC 253374 to 98 EC 321425 and 26, EC 321898 | Heat leaf spot and Root Knot Nematode resistant | Taiwan |
| | EC312339 | Resistant to leaf curl virus | Taiwan |
| | EC198416 | Heat tolerant | Taiwan |
| | EC 316299 to 301 | Gene for slow ripening | USA |
| | EC204194 to EC204198 | Resistant to *Verticillium* wilt, TMV, *Alternaria*, *Fusarium oxysporum* race-1 and race-2, tolerant to nematodes | USA |
| | EC310299-EC310301 | Have genes for slow ripening, suitable for long distance transport and long shelf life. | USA |
| | EC129571, Geralton | Good paste type collection | Australia |
| | EC287750, Red Lander | Resistant to bacterial wilt, *Verticillium* wilt with determinate growth habit | Australia |
| | EC130044 Lycoperea XVF | Good paste type collection | New Zealand |
| Brinjal | EC99488, SM70540, Arangnees | Resistant to *Anthracnose* | USA |
| | EC104107, Florida Market | Resistant to *Phomopsis* fruit rot | USA |
| | EC163147, Wianmanolo Long | Very promising, high fruit yielding type | USA |

| Crop | Lines | Attributes | Country |
|------|-------|-----------|---------|
| Chilli/*Capsicum* | EC 175959 to EC 175966 (Sweet pepper) | White, sweet, erect with 'L' gene Resistant to red spider, blocky types | Hungary |
| | EC 203581-EC203603(Sweet pepper) | Conical shape with CMV tolerance | Hungary |
| | Nigold (EC-264584) | Yellow fruited, more branching, resistant to wind damage | Hungary |
| | Marold Giant (EC 304963) | Fruits with thick wall, spreading habit, strong upright, tolerant to TMV | USA |
| | EC-371245 and 46 | High level of resistant to Root Knot Nematode | USA |
| | *C. chinense* (EC 260643) | Resistant gene (*L-3*) for some TMV races | Netherlands |
| | *C. frutescens* (EC 260642) | Resistant gene (*L-2*) for all Dutch races of TMV | Netherlands |
| | *C. chacoense* (EC 260644) | Resistant gene (*L-4*) for all Dutch races of TMV | Netherlands |
| | EC-381007 to 01 | Suitable for table purpose; cherry shaped fruits and bear in clusters | Taiwan |
| | EC-332333 | Early and resistant to TMV | Taiwan |
| | EC-354883 | Heat tolerant and resistant to CMV | Taiwan |
| | EC-347246 to 48 and EC 347259 | White fruited, tolerant to CMV | Taiwan |
| | EC-347284 | High level of resistant to *Phytophthora* | Taiwan |
| | EC312342-EC312349 | Resistant to viruses PVY, PUMV, CYMV, and CVMV | Taiwan |
| Onion | EC196631-EC196639 | Flat top, long day with tolerance to *Fusarium oxysporum* | Italy |
| Onion | EC187221-EC187225 | Yellow colosured bulb, good for storage | Germany |
| Garlic | EC280378 | Very pungent with large white clove | France |
| Cabbage | EC187228-EC187230 | Club rot resistant, male fertile cultivars | Canada |
| Cabbage | EC287707-EC287708 | Resistant to yellow (caused by *Fusarium oxysporum* f. sp. *conglutinens*) and tolerant to black rot *Xanthomonas campestris* | Japan |
| Cauliflower | EC205372-EC205373 | Smooth, medium maturity in 60-70 days and disease tolerant | USA |
| Cauliflower | EC175800-EC175806 | Multiple disease-resistant lines | USA |
| Watermelon | EC217073-EC217074 | Moderately to highly resistant to Anthracnose, Fusarium wilt and gummy stem blight | USA |
| Muskmelon | EC178496 | Early strain, creamy yellow fruits, bushy plant vigorous and very productive; aphid resistant, free from blight, resistant to WMV-1 and WMV-2, sulphur tolerant, powdery mildew resistant. | USA |
| Muskmelon | Green Ice | Resistant to race-3 of powdery mildew, *Sphaerotheca fuligin*ea, high in vitamin C content | USA |
| Muskmelon | WI998 | Resistant to *Fusarium* wilt (race 0, 1, 2) and melon aphid | USA |
| Cucumber | EC178495 (*Cucumis sativus*) | Scab and mosaic resistant | USA |
| Cucumber | EC222198-EC222201(*C. zagiltatus*) | Resistant to fruit fly | Holland |
| Garden pea | EC292160 | Resistant to *Ascochyta* blight and fungal root rot | USSR |
| French bean | EC297176-EC297180 | Bushy, shelling type, rich in protein | USSR |
| French bean | Sanilac | Resistant to alpha, beta and gamma races of *Colletotrichum lundemuthianum* and races of BCMV | USA |
| French bean | BC-6 | Resistant to alpha, beta and gamma races of *Colletotrichum lundemuthianum* and races of BCMV | USA |
| Lettuce | EC174981 | Mosaic free | USA |
| Lettuce | EC216893-EC216894 | Summer adaptable | UK |
| Carrot | EC178385 | High vitamin A content | Italy |
| Carrot | EC187207, Beta-III | Good flavor eating quality high in vitamin A content | USA |
| Carrot | EC277678 | Blunt and short root type, good flavor high cerotine content and resistant to *Alternaria* blight. | USA |
| Radish | EC170668; Summer Top | A very good introduction, attractive shape and colour | Taiwan |
| Okra | EC187251; Stock No. 510171 | Very attractive fruit, smooth and green | The Philippines |
| Potato | EC170542 | Highly tolerant to brown rot | Madagascar |
| Potato | EC175394 | Resistant to several diseases | USA |

In tomato brinjal and chillies some priority species for specific traits are identified which are given in Table 3.5.

**Table 3.5: List of Priority Species for Specific Traits in Solanaceous Vegetables**

| Crop | Genus | Priority species | Resistant against |
|---|---|---|---|
| Tomato | *Solanum* | *S. pimpinellifolium, S. esculentum* var. *cerasiforme, S. peruvianum* | Late blight |
| | | *S. hirsutum* f. *glabratum, S. pimpinellifolium* | Early blight |
| | | *S. pimpinellifolium, S. hirsutum* | Fusarium wilt |
| | | *S. pimpinellifolium, S. hirsutum* f. *glabratum* | Leaf curl virus |
| | | *S. pimpinellifolium* | Bacterial wilt |
| Brinjal | *Solanum* | *S. khasianum* | Shoot and fruit borer |
| | | *S. gilo* | *Phomopsis* rot |
| | | *S. integrifolium* | *Phomopsis* rot |
| | | *S. sisymbrifolium, S. macrocarpon* | Shoot and fruit borer, aphids, root knot nematodes |
| | | *S. macrosperma* | Drought |
| Chilli/Sweet pepper | *Cacpsicum* | *C. baccatum* | *Phytophthora* rot and cucumber mosaic virus |
| | | *C. chinense* | Leaf curl virus, tomato mosaic virus and *Verticillium* wilt |
| | | *C. frutescens* | Leaf curl virus |

**Table 3.6: Some Promising and Useful Introductions (Exotic Collections)**

| Crop | Accessions | Country | Promising and Useful Attributes |
|---|---|---|---|
| *Cucumis sativus* | EC-399914 -37 | UK | Resistant to CMV, downy mildew |
| | EC-398030 | China | Early maturing, determinate, fruits in cluster |
| | EC-398966-67 | USA | Resistant to angular leaf spot |
| | EC- 398968-70 | USA | Resistant to Anthracnose |
| | EC- 398971 73 | USA | Resistant to fruit rot |
| | EC- 398974-90 | USA | Resistant to downy mildew |
| | EC- 398991-9007 | USA | Resistant to leaf spot |
| | EC-382737-39 | USA | Gynoecious lines |
| | EC565796 | Netherlands | Early type |
| | EC565750 | Netherlands | Large size with green skin |
| | EC565767 | Netherlands | Female line |
| | EC-497645, EC-497646, EC-497647, EC106285, EC118292, EC163683, EC173856, EC170662, EC110578, EC129081, EC129353, EC168084, EC122063, EC251636, EC315481, EC315486, EC339059, EC337827, EC339074, EC357829, EC362940, EC357840, EC362954, EC362957, EC361424, EC367856, EC367857, EC257463, EC315489, EC237288, EC114147, EC163602, EC362915, EC362946, EC357839, EC357832, EC339072 | | High yield |
| | EC041824, EC007764, EC164855, EC169308, EC164653, EC122063, EC037311, EC031515, EC126955, EC035220, EC163709, EC164332, EC154660, EC164677, EC164674, EC162515, EC154560, EC162598, EC125754, EC315479, EC315484, EC321425, EC320584 | | Heat tolerant lines |
| | EC315487, EC320578 | | Moderate heat tolerant lines |
| | EC382737-39, EC329300 | USA | Gynoecious lines |
| | EC738814-739038 | Netherlands | Recombinant lines |
| *Cucumis melo* | EC-399866-212 | UK | Resistant to downy mildew, powdery mildew and CMV |
| | EC- 382726-36 | France | Germplasm with male sterility genes MS1 to MS5, white fruit (gene), *Fusarium* wilt resistant (genes form-1, form-3) |
| | EC589374 and EC612133 | USA and Vietnam | Resistant to powdery and downy mildew |

| Crop | Accessions | Country | Promising and Useful Attributes |
|---|---|---|---|
| | EC 612132 | USA and Vietnam | Tolerant to *Fusarium oxysporum* strains 0 and 1 |
| | EC612134 | USA and Vietnam | Yellow skin, cream flesh |
| | EC-468986, EC-477977, EC-477978, EC-477979, EC-477980, EC-477981 | USA and Vietnam | High yield |
| | EC348140-43 | USA | $F_1$ hybrids |
| | EC382726-36 | France | Male sterile genes |
| *Cucumis* species | EC-382500-69 | USA | Different *Cucumis* species with nematode resistance |
| *Citrullus lanatus* | EC-393240-43 | USA | Small medium round fruits and red fleshed for breeding purpose |
| *Citrullus lanatus* | EC-402549 | USA | Multiple disease resistant to race-2, anthracnose, *Fusarium* wilt and gummy stem blight with early maturity |
| *Cucurbita pepo* | EC-380995 | USA | Unique, small, oblong early maturing, the flesh comes out in strings, can be bakes or used in breads |
| | EC- 380996 | USA | Hull-less seeded pumpkin; seeds can be popped and eaten like snack, high in nutrition and good source of Zinc. |
| *Citrullus lunatus* | EC-380989-91 | USA | Yellow fleshed watermelon, early maturing, sweet crisp and fleshy |
| | EC-382753 | USA | Breeding line SSDL resistant to *Fusarium* wilt and anthracnose |
| | EC-378523-23 | USA | *Fusarium* wilt resistant |
| | EC572745-47 | Taiwan | Globe shaped, firm fleshed, small seeded, sugar content 11-12 per cent and diseases resistant |
| | EC678820 | Taiwan | Heat tolerance, strong resistance to BW, blight and rainfall |
| | EC678821 | Taiwan | Heat tolerance, resistance to BW and blight |
| | EC678822 | Taiwan | Tolerance to heat, rainfall, BW and blight |
| Bitter gourd | EC-399808 | - | - |
| Sponge gourd | EC-305586 | - | - |
| Ridge gourd | EC-284347 | - | - |
| Bottle gourd | EC-305378 | - | - |
| Spinach | EC-284349 | - | - |
| Tomato | Sioux, Fireball, Marglobe, Best of all, Roma, Money Maker | - | - |
| | EC14997-99, EC580988-020, EC637360-65, EC612852-57 | Taiwan, Canada, USA, France, Sri Lanka and Cuba | Resistant to TY, BW, TMV |
| | EC664585-88 | -do- | Early maturing lines |
| | EC664589-599 | -do- | Cold resistant lines |
| | EC659266-71, EC572692-708 | -do- | Beta carotene content |
| | EC657640, EC654678-86, EC606703-4, EC580001, EC588212-230, EC586975, EC570024-28, EC571821-829, EC635529-33, EC 612867-69, EC562072-74, EC606704, EC610636, EC568938-44, EC600051, EC 610627 | -do- | Tomato mosaic virus and *Fusarium* wilt, gray leaf spot pathogen and early blight |
| | EC654694-99, EC581525, EC560340, EC611883-91, EC581035-1043, EC592058-59, EC570019, EC 610636 | -do- | Heat tolerant and plum shaped tomato |
| | EC-671595-96, EC675836 | Taiwan | TMV resistance |
| | EC-671597-98 | Taiwan | TMV, FW2 resistance |
| | EC675830 | Taiwan | BW, TMV resistance |
| | EC675831 and EC675833 | Taiwan | BW, FW-1 TMV resistance |

| Crop | Accessions | Country | Promising and Useful Attributes |
|---|---|---|---|
| | EC675832 | Taiwan | GLS, BW, FW-1 TMV resistance |
| | EC675834 | Taiwan | TMV, LB, FW1, FW-2, ST resistance |
| | EC675835 | Taiwan | GLS, BW, FW1 TMV, Gemini virus (Ty2) resistance |
| | EC678814 and EC678815 | Taiwan | Resistance to BW, Virus and blight |
| | EC678817 | Taiwan | Resistance to low and high temperature |
| | EC678818 | Taiwan | Resistance to virus, leaf mould blight, RKN, long shelf life |
| | EC687094 | Taiwan | Resistance to TYLCV, determinate, shape oblong, size medium to medium large, green colour stem |
| | EC687095 | Taiwan | Resistance to TYLCV, square round, medium large, green colour stem, semi-determinate |
| | EC687096 | Taiwan | Resistance to TYLCV, square round, medium size, green stem, determinate |
| | EC687097 | Taiwan | Resistance to TYLCV, determinate |
| | EC687098 | Taiwan | Resistance to TYLCV, shape round, size medium, purple stem, determinate |
| | EC687099-103 | Taiwan | red colour fruit, oblong shape, resistance to TYLCV, Gemini virus (Ty2) TMV and FW1 |
| | EC687104-05 | Taiwan | Indeterminate, red colour fruit, oblong shape, resistance to TYLCV, Gemini virus (Ty2) TMV and FW1 |
| | EC687106 | Taiwan | Semi-determinate, glove fruit shape, resistance to TMV, BW and FW1 |
| | EC687107 and EC687108 | Taiwan | Indeterminate, orange colour fruit, oblong shape, resistance to TMV, BW and FW1 |
| | EC690981 | Taiwan | Uniform shoulder, square round shape, semi determinate type, homozygous for resistance for Gemini virus (Ty1, Ty2 and Ty3), FW2 and TMV |
| | EC690982 | Taiwan | Uniform shoulder, square round plum shape, determinate type, homozygous, resistance for Gemini virus (Ty1, Ty2 and Ty3), FW2 and TMV |
| | EC690983 | Taiwan | Green shoulder, homozygous for resistance for Gemini virus (Ty1, Ty2 and Ty3), homozygous allele susceptible to FW2 |
| | EC690984 | Taiwan | Green shoulder, oblong, semi determinate homozygous allele resistance for Gemini virus (Ty1, Ty2 and Ty3), FW2 and TMV |
| | EC690985 | Taiwan | Green shoulder, oblong, determinate homozygous allele resistance for Gemini virus (Ty1, Ty2 and Ty3), FW2 and TMV |
| | EC690986 | Taiwan | Uniform shoulder, square shape, semi determinate type, homozygous allele resistance for Gemini virus (Ty1, Ty2 and Ty3), FW2 and TMV |
| | EC690990 | Taiwan | Uniform shoulder, plum square oblong shape, determinate type, homozygous allele resistance for Gemini virus (Ty2 and Ty3), and TMV |
| | EC692274 | Taiwan | Resistant to gray leaf spot, BW, FW1, TMV and begomovirus, heat tolerance |
| | EC692275 | Taiwan | Resistant to BW, FW1, TMV and heat tolerance |
| | EC692276 | Taiwan | Resistant to BW, FW1, TMV and heat tolerance |
| | | Taiwan | Heat tolerant, fruit large size and globe shaped |
| | EC692277 | Taiwan | Resistant to FW2, gray leaf spot and early blight |
| | EC565215 | | Rich in β-carotene |
| | EC570015-16 | -do- | Globe shaped large fruited type |
| | EC 604747 | -do- | Long shelf life, indeterminate type, elongated fruit |
| | EC 611208-09 | -do- | Tolerant to salt and drought |
| | EC347359-68 | Taiwan | $F_1$ hybrid |
| | EC343391-95 | Taiwan | $F_1$ hybrid (Cherry tomato) |

| Crop | Accessions | Country | Promising and Useful Attributes |
|------|-----------|---------|--------------------------------|
| | EC346011 | USA | Stamenless line |
| | EC346013 | USA | Corolla and stamenless line |
| | EC346014 | USA | Green pistillate type |
| | EC737661-62 | Taiwan | Resistant to gray leaf spot, race 1 of FW, TMV, susceptible to late blight, FW race 2, TyLCV type 1, 2 and 3. |
| | EC753215 - 32 | Taiwan | Resistant to TYLCD, late blight, race 2 of the FW, gray leaf spot resistance, fruit shape round blocky and semi- determinate type. |
| | EC751801-13 | Taiwan | Resistant to TYLCD, late blight, FW race 2, gray leaf spot, fruit shape round blocky |
| C. pepo | EC-516625 to EC-516665 and EC-473294 to EC-473296 | | High yield |
| Brinjal | EC304992 | | Long purple colour |
| | EC304975, EC329327 | | Purple long |
| | EC305049 | | Deep purple oval |
| | EC316263 | | Green Oval |
| | EC305096 | | Large round type |
| | EC304973 | | Thin and long fruit type |
| Bell pepper | California Wonder, Yolo Wonder, Oshkosh, Ruby king, King of North, Early Giant, Chinese Giant, World Beater | | High yield |
| Chillies | EC656670–687 | Taiwan and | Restorer lines |
| | EC596949-60 | USA | Restorer lines |
| | EC637341-42 | -do- | Male sterile lines |
| | EC611331-64 | -do- | Tolerant to aphids |
| | EC612322 | -do- | Bird Chilli |
| | EC596931-36 and EC559416-9426 | -do- | CVMV resistant lines |
| | EC 631661-90 | -do- | Potato virus Y and BW |
| | EC596937-48, EC571257-266 | -do- | CMS A and B lines |
| | EC572236-39 | -do- | Chilli Veinal Mosaic Virus and Poty virus Y resistant lines |
| | EC559491-496 | -do- | Resistant to Anthracnose and Capsicum mosaic virus |
| | EC559424-9508, EC570005-12, EC568925-29 | -do- | Potyvirus mosaic virus Y and bacterial wilt |
| | EC596749-50 | -do- | Heat tolerant lines |
| | EC582593 | -do- | drought tolerant and excellent keeping quality, suitable for ornamental and culinary applications |
| | EC668798, EC668801, EC668803, EC668805, EC668806, EC668808, EC668812, EC668814, EC692278, EC692280, | Taiwan | A lines |
| | EC668799, EC668800, EC668802, EC668804, EC668807, EC668813, EC668815, EC692279, EC692281, | Taiwan | B lines |
| | EC668809 | Taiwan | R line |
| | EC668810, EC668811 and EC668816 | Taiwan | Restorer lines |
| | EC678809 | China | Fresh use, green, pungent, resistant to heat and humidity |
| | EC673074 | Taiwan | Cayenne type |
| | EC678805 | China | Dark green, strong pungent, heat and humidity resistance |
| | EC678806 | China | Fresh use, plant height 80-90 cm, light green turn red and smooth |
| | EC678807 | China | Fresh and dry use, green, high pungency |
| | EC678808 | China | Fresh use, pungency, heat and humidity resistance |
| Onion | Texas Early Grano, EC-4144 | USA | Mild flavored, large sized, yellow, suitable for fresh use |
| | Bermuda Yellow, EC-109123 | Philippines | Non-bolting salad onion |

| Crop | Accessions | Country | Promising and Useful Attributes |
|------|-----------|---------|--------------------------------|
| | Giza-6, Giza-20, EC-200936, EC-200937 | Sudan | High yielding, large sized, dark red pungent types |
| | Nu Mex Br-1, EC-169372 | USA | Short day, resistant to bolting and pink root disease |
| | Sweet Sandwich, EC-169373 | USA | Dehydrator/salad type |
| Garlic | EC-244949, EC-244858 | Taiwan | Large bulbs, purple cloves |
| | EC-210991 | Egypt | Compact white bulbs, bold cloves |
| | EC-158250 | Taiwan | Large light pink bulbs, bold cloves Bold white cloves, field tolerant to purple blotch |
| Sweet potato | EC-332805 | Puerto Rico | Red skin and yellow flesh |
| | C-12 | Peru | Deep orange, semi erect, resistant to leaf scales, drought tolerant |
| Okra | Ghana Red | Ghana (Africa) | |
| Radish | Japanese White | Japan | |
| | Red Tail Radish | | |
| | China Red | China | |
| Turnip | Snow Ball, Purple Top, White Globe | | |
| Peas | Early Badger, Bonneville, Sylvia, Asauji, Arkel | | |
| Cauliflower | Snowball-16, Improved Japanese | Japan | |
| | EC678824 and EC678825 | China | Heat tolerance |
| Cabbage | EC678826 | China | Heat tolerance |
| | EC304718-25 | USA | Cytoplasmic male sterile and restorer line |
| Carrot | EC-178385 | Italy | High Vitamin A |
| | EC-187201 | USA | High Vitamin A |
| | EC-27883 to EC-274886 | Netherlands | Male sterile and restorer line, resistant to club rot |
| Indian bean | EC305789 (270.38) | | More pods/plant |

## Introduction of Trait Specific Germplasm for Hybrid Seed Production

Hybrids have been recognized world over for their superiority in different quantitative and qualitative traits. The development of superior hybrids takes 5-12 years depending on pollination mechanism involving in the crop concerned. Once a superior hybrid has been developed, it can be released in other places having similar agro-climatic condition. Thus, a lot of cost and energy can be saved. A large number of $F_1$ hybrids have been directly introduced to India by private and public sectors through ICAR-NBPGR. Additionally, germplasm possessing monoecy/dioecy, self-incompatibility and male sterility, have been introduced to facilitate hybrid seed production.

## Introduction of Transgenic

Modern biotechnology involving the use of rDNA technology/genetic engineering has emerged as a powerful tool with many potential for improving the quantity and quality of food supply. Food derived from genetically modified crops, commonly referred to as genetically modified food ingredients have already become available worldwide with aim of enhancing

**Table 3.7: Introduced Hybrids and Germplasm to Facilitate Hybrid Seed Production in India**

| Vegetable crop | Exotic Collection (EC) | Desirable Features | Country |
|----------------|------------------------|--------------------|---------| 
| Tomato | EC347359-68 | $F_1$ hybrids | Taiwan |
| | EC343391-95 | $F_1$ hybrids (Cherry tomato) | Taiwan |
| | EC346011 | Stamenless line | USA |
| | EC346013 | Corolla and stamenless line | USA |
| | EC346014 | Green pistillate type | USA |
| Cucumber | EC382737-39 | Gynoecious line | USA |
| | EC329300 | Gynoecious line | USA |
| Muskmelon | EC348140-43 | $F_1$ hybrids | USA |
| | EC382726-36 | Male sterile genes | France |
| Cabbage | EC304718-25 | Cytoplasmic male sterile and restorer lines | USA |

productivity, decreasing the use of certain agricultural chemicals, modifying the inherent properties of crops, improving the nutritional value or even increasing shelf life. The government has certain rules on GM research like Roule-1989 (The Ministry of Environment and Forests), Recombinant DNA Guidelines 1990 (Department of Biotechnology), Guidelines for Research in Transgenic Plants, 1989 (DBT brought out separate guidelines for carrying out research in transgenic plants), Seed Policy, 2002 (The seed policy 2002 issued by Ministry of Agriculture – a separate section (No. 6) on transgenic plant varieties.

Seeds of transgenic plant varieties for research purposes will be imported only through the ICAR-NBPGR, New Delhi as per EPA, 1986. Transgenic crops/varieties will be tested to determine their agronomic value for atleast two seasons under the All India Coordinated Project Trials of ICAR, in coordination with the tests for environment and bio-safety clearance as per the EPA before any varieties is commercially released in the market. After the transgenic plant variety is commercially released, its seed will be registered and marketed in the country as per the provisions of the Seed Act. After commercial release of a transgenic plant variety, its performance in the field, will be monitored for at least 3 to 5 years by the Ministry of Agriculture and State Departments of Agriculture.

## Status of Transgenic Crops in India

The trials on vegetable crops are being conducted by both public and private sectors institutions and the target traits include insect resistance, herbicide tolerance, viral and fungal disease resistances and stress tolerance.

**Table 3.8a: Transgenic Crops under Development and Field Trials**

| Crop | Organization | Gene |
|------|--------------|------|
| Brinjal | ICAR, IARI, New delhi | Cry1AB |
| | MAHYCO, Mumbai | Cry1AC |
| Cauliflower | MAHYCO, Mumbai | Cry1AC |
| | Sungrow Seed Ltd., New Delhi | Cry1AC |
| Cabbage | Sungrow Seed Ltd., New Delhi | Cry1AC |
| Okra | MAHYCO, Mumbai | Cry1AC |
| Potato | ICAR-CPRI, Shimla | Cry1AB |
| | ICAR-NCPGR, New Delhi | Ama-1 |
| Tomato | MAHYCO, Mumbai | Cry1AC |
| | ICAR-NCPGR, New Delhi | OXDC |

*Source*: Department of Biotechnology, Government of India.

Several transgenic seed material have been introduced to develop transgenic varieties (Table 3.8b).

## Procedural Steps in Plant Introduction

Introduction of germplasm consist of the following steps:

(1) Procurement

(2) Quarantine

(3) Cataloguing

(4) Evaluation

(5) Multiplication

(6) Distribution

### (1) Procurement

Any individual or institution can introduce germplasm in India, but all the introductions must be routed through ICAR-NBPGR, New Delhi. The individual or institution concerned may follow any one of the following two routes for this purpose.

(i) He/It may take direct request to an individual or institution abroad, who has the desired germplasm, to send it to him/it through ICAR-NBPGR, New Delhi. It is mandatory that a copy of such requests be sent to ICAR-NBPGR, New Delhi in order to facilitate a speedy distribution of germplasm when they arrive. The germplasm should reach the bureau at least 2 months in advance of the date of sowing, and the packet should contain the address of actual user. This route is preferable to the second one since the concerned individual or institution can tap his/its contacts abroad and thereby, ensure availability of the desired germplasm.

(ii) Alternatively, whenever the concerned individual or institution is unable to procure the desired germplasm on his/its own, a request or import of the required germplasm is submitted to ICAR-NBPGR, New Delhi. It is important to give as much detailed information about the germplasm required and their proposed utilization afterwards. It is of great help if the indenter indicates possible source (s) of the germplasm but this not essential; the bureau makes efforts to obtain the request germplasm even if their source is not indicated by the indenter, but this may lead to delays and failures. It may be emphasized that scientists and

**Table 3.8b: Transgenic Seed Material Introduced in India**

| Name of Crop | Botanical Name | Lines | Traits | Source | Indenter |
|--------------|----------------|-------|--------|--------|----------|
| Tomato | *Solanum lycopersicum* | EC676413 | Insect resistant gene ARg | USA | Bejo Sheetal, Jalna |
| Cabbage | *Brassica oleracea* L. var. *capitata* | EC753911-15 | HO4 and Cry 1B genes | USA | ICAR-NBPGR, New Delhi |

institutions should import only such germplasm that are useful and essential to them since their procurements, quarantine and subsequent handling require funds, labour and considerable time of the scientists involved in the process.

## Gift or Purchase

The germplasm obtained from other countries may be:

(1) As gifts from individuals/institutions

(2) In exchange of germplasm provided (in the past, present or future) by the ICAR-NBPGR, New Delhi to the concerned individual or institution

(3) Purchased or

(4) Collected through an exploration.

Gift, exchange and purchase are arranged from scientists, research institutes, botanic gardens, agencies responsible for germplasm conservation, private nurseries and research organizations, international crop research institutes, *etc.* of other countries. ICAR-NBPGR, New Delhi maintains good exchange and working relationship with similar agencies of 80 different countries, and participates in the activities of International Plant Genetic Resources Institute (IPGRI). IPGRI aims at free exchange of plant genetic resources among the different countries, and is often helpful in arranging the supply of needed germplasm. In addition, many scientists abroad have contacts with the scientists of ICAR-NBPGR and/or other scientists in the country; this is often helpful in the procurement of germplasm.

## Exchange of Germplasm

The ICAR-NBPGR has brought out a brochure 'Guidelines for the exchange of seed/planting materials' which has been widely circulated amongst scientist in India. The guidelines for import of seed/planting material for research purpose have been revised (Anon, 1989) in view of the enactment of new Seed Development Policy by govt. of India, which has also been circulated amongst scientists. The government has made the issuance of import permit mandatory for import of all seed/planting material for research purposes.

### (a) Import of Germplasm

As per the New Seed Development Policy of the Government of India, every case of import of seed/planting material into India has been made obligatory for all plant breeders, researchers intending to import seed/planting material, to fulfill the following two mandatory requirements of the plant, fruits and seeds (Regulation of Import into India) order of 1984.

(i) An import permit before import of any material and phytosanitory certificate from the country of origin

(ii) These two documents must accompany every seed/planting material consignment that is imported from abroad.

The Director, ICAR-NBPR, New Delhi has been authorized to issue import permit and receive imported seed/planting material for its quarantine inspection and clearance in respect of researchers ICAR Institutes/Centres, Central Agricultural Universities, State Agricultural Universities, and ICRISAT, Hyderabad.

All request for indenting germplasm from abroad are to be made to the ICAR-NBPGR, New Delhi in prescribed application proforma giving specific details of the required material stating the sources/country as well as address of the organization at least 15 days in advance after signing a declaration form, so that the 'Import permit' is issued and sent to concerned scientist (s) well in time. After obtaining import permit the indentor should send it to the organization/scientist abroad that has agreed to supply the seed/planting material. He may further be requested to send the above import permit alongwith seed/planting material and also a 'phytosanitary certificate' that is to be issued by the authorized agency of the exporting country.

Concerned scientist/organization abroad is advised to take into consideration the following requirements for mailing the materials to India:

☆ Only healthy, viable and clean seed materials (free from soil, pests, pathogens and weeds) are to be forwarded without any seed treatment. The Quarantine Division, ICAR-NBPGR prefers to receive seed material tha has not been coated/treated heavily with chemicals so as to facilitate proper quarantine examination/inspection of imported samples. It may, however, be fumigated, if considered necessary.

☆ The material is required to be accompanied with the 'Import Permit' (which is to be sent to them along with our request letter) and phytosanitary certificate with additional declaration, if any, based on crop inspection certifying that the material is free from particular pathogen(s)/insects (s).

☆ It is to be further ensured that the package of seed/planting material must be addressed to the Director, ICAR-NBPGR New Delhi who has to take delivery of the seed/planting material and conduct quarantine examination.

☆ The seed material may preferably be sent by first class airmail, while perishable propagules (scion woods, bud woods, rhizomes, suckers or rooted plants) may preferably be sent by air freight through any commercial airline, operating between source country and the Indira Gandhi International Airport, New Delhi, so as to avoid delay in receipt and clearance. If unavoidable, the material can be sent on charge collect basis. An intimation regarding the dispatch of such perishable material to the Director, ICAR-NBPGR New Delhi through telegram or telex, email, phone call, will facilitate in prompt receipt/clearance of material, soon after its arrival. This will also help to avoid payment of demurrage charges.

☆ Full particulars of seed/planting material as well as the address of concerned scientist in India, to whom, the material is to made available after its quarantine clearance by the Bureau.

☆ The germplasm should be obtained in small quantities not exceeding 3000 to 4000 seeds each, while in case of plant propagules *viz.* scion woods, bud woods, suckers *etc.* it should be as minimum as possible, but not exceeding six in each case. In case of rooted plants, it should not be more than two in each collection.

The material is either received as air parcel, if in small quantity, or through air cargo, if it is in bulk. Usually the air mail parcels are delivered by the Department of posts in the Bureau whereas for bulk consignments, intimation is received from concerned airlines about their landing at the airport. The staff of the Division then initiates the process of clearance of these consignments from customs and finally from airport. On receipt of consignments, each lot is designated an import quarantine number and sent to Plant Quarantine Division where it is unopened for checking of possible infestation of pests/diseases. The quarantine scientists do detailed studies and if found to be free from diseases, return the lot to Germplasm Exchange Division for its further processing. Here, each collection is systematically documented and assigned an exotic collection number. the material is then properly repacked and is dispatched to user scientists/institutes on whose request the material has been introduced. However, for the sake of appropriate management consideration and precaution, a small part of it is sent to National Gene Bank (NGB) for reference sample under medium/long term seed storage. In cases of material being infected, it is tried to salvage it to maximum possible extent and release disease free material. However, in case the seed treatments are done to material supplied, such material is grown in post entry quarantine nursery for observance of possible hidden infestations. After prescribed period of growth *i.e.* seeding or whole plant quarantine, if there appears to be no infestation, the material is released to indenting scientists/institutes, as in former case.

### (b) Export of Germplasm

Exchange of germplasm involves not only introductions but also the supply of seed and other materials to collaborating scientist/organizations abroad. Export of vegetable germplasm is also made on the basis of request received by Bureau/ICAR Institutes/Agricultural Universities in India under various protocols/work plans/memorandum of understanding with different countries/CGIAR institutions. The quarantine regulation for export of seeds/planting material are based on the International Plant Protection Convention (1951) and or modified from time to time according to the specific requirement of the importing country.

ICAR-NBPGR, New Delhi is exporting a large number of vegetable germplasm to a number of countries. Highest export of vegetable germplasm numbering 2365 accessions of 16 genera and 23 species were exported to as many as 9 countries during the year 1992-93. The export of vegetable tuber crops involving 180 accessions was highest in 1992-93. Important germplasm of *Abelmoschus esculentum, Amaranthus caudatus, Allium cepa, Allium sativum, Brassica oleracea* var. *botrytis, Brassica oleracea* var. *capitata, Brassica chuk, Brassica rapa, Capsicum annuum, Capsicum frutescence, Cymposis tetragonoloba, Daucus carota, Lablab perpureus, Solanum lycopersicum, Phaseolus vulgaris, Pisum sativum, Raphanus sativum, Solanum incanum, Solanum integrifolia, Solanum melongena, Solanum torvum,* and *Spinacea oleracea* and some cucurbits like *Cucumis melo* (1), *Cucurbita pepo* (4), *Cucumis sativus* (1), *Luffa cylindrica* (1) to Italy; *Momordica charantia* (1) to Taiwan; *Sechium edule* (2) and *Triochosanthes dioica* (1) were exported during 1986-1994 to different countries. During the year 2001-2002 few cucurbitaceous germplasm like cucumber (1), bitter gourd (1) Ridge gourd (1) and sponge gourd (1) were again exported under phytosanitary certificate to Iraq.

**Table 3.9: Export of Germplasm**

| Crop | Number of Samples | Country |
|---|---|---|
| *Cucumis melo* | 6 | DPR, Korea |
| *Citrullus lanatus* | 1 | DPR, Korea |
| *Cucumis sativus* | 1 | DPR, Korea |
| *Momordica charantia* | 1 | DPR, Korea |
| *Citrullus lanatus* | 1 | Bangladesh |
| *Cucumis sativus* | 1 | Iraq |
| *Lagenaria siceraria* | 1 | Bangladesh, Iraq, Taiwan |

Since 1986 the Bureau has exported vegetable and tuber germplasm to 53 and 16 countries respectively as shown in Table 3.10.

The following guidelines are to be observed while responding to such requests:

1. Request for seed/planting materials received from concerned organizations/agencies abroad are to be forwarded to the Director, ICAR-NBPGR with relevant information (Import Permit) so that prompt action on the supply of desired material could be taken.

**Table 3.10: Countries Involved in Exports of Vegetable Crops Germplasm**

| Crops | Countries |
|---|---|
| Vegetable crops | Afghanistan, Angola, Argentina, Bangladesh, Brazil, Bulgaria, Canada, China, Coasta Rica, Czechoslovakia, DPR Korea, Egypt, Ethiopia, Fiji, France, Greece, Hungary, Indonesia, Iran, Iraqw, Italy, Japan, Kuwait, Liberia, Malaysia, Mexico, Mongolia, The Netherlands, Nepal, New Guinea, Nicaragua, The Philippines, Poland, Republic of Yemen, Saudi Arab, Senegal, Singapore, Solomon, Sri Lanka, Surinam, Taiwan, Tanzania, Thailand, Tunisia, Turkey, U.K., USA, USSR, Vietnam, Yeman, Zimbabwe, West Germany and West Indies. |
| Tuber vegetable crops | Australia, Bangladesh, Belgium, Bulgaria, Egypt, France, Mongolia, Nepal, Pakistan, Peru, The Philippines, Sri Lanka, Surinam, U.K. and USA. |

2. No seed/planting materials should be sent to the Director, ICAR-NBPGR, New Delhi, unless asked for.

3. The dispatch of the seed/planting material is to be channelized through the Bureau so that prompt inspection of the material could be done from quarantine angle and phytosanitary certificate be issued.

4. Only the healthy, viable seeds/planting materials (free from debris, weeds, pests and diseases *etc.*) should be sent to the Bureau, in small quantities along with full details of the material and the name and address of recipient in foreign country. The quarantine inspection/clearance and dispatch normally takes 7 to 10 days.

5. No seed dressing with insecticides or fungicides be given, while dispatching the seed/planting materials to the Bureau.

### Restrictions on Export and Introduction of Plant Materials

Introduction of certain plant species in India is restricted. For example cotton, berseem, linseed and sugarcane seeds cannot be obtained through letter post.

### Procedure for Exchange of Seed/Planting Materials

The exchange of plant material on worldwide basis has been carried out, with well-defined procedures, by countries which have well established plant introduction organisations, like USA, Russia, Australia, Canada, Brazil, Japan *etc.* for channelizing import and export of plant material under quarantine control.it is mandatory to plant introduction agencies; to follow a set of procedures for introduction of germplasm, either through plant exploration in the respective country or collection through contacts or by correspondence'. The ICAR-NBPGR, New Delhi had brought out a brochure, 'Guidelines for the exchange of seed/planting materials'

(NBPGR, 1986) in view of the enactment of New Seed Development Policy, by the Government of India.

## (2) Import Regulations/Plant Quarantine in India

Plant quarantine can be defined as:

*"Legal enforcement of measures aimed to prevent pests and pathogens from spreading, or to prevent them from multiplying further in case they have already found entry and have established in a new area"*. Though Plant quarantine measures may not guarantee an everlasting protection against the entry of exotic species but will certainly check or delay the introduction of these unwanted organisms and their subsequent establishment in hitherto clean areas. In more simple way quarantine means to keep materials in isolation to prevent the spread of diseases *etc.* present in them to the other materials. In the case of plant introduction, all the introduced plant propagules are thoroughly inspected for contamination with weeds, diseases and insect pests. If procured materials are suspected to be contaminated are fumigated or are given other treatments to get rid of the contamination. If necessary, the materials are grown in isolation for observation of diseases, insect pests and weeds. This entire process is known as quarantine and the rules prescribing them are known as quarantine rules. In India import and export of seeds, plant products and planting materials are regulated by the rules and regulations framed under the Destructive Insects and Pest (DIP) Act 1914, extended by the Directorate of Plant Protection, Quarantine and Storage, Ministry of Agriculture and Irrigation (1976) and subsequently revised several times. According to the DIP Act 1914 all plant produce imported in India must be free from diseases, insect pests and weeds. The main objectives of this act is to prevent the introduction into the country, and the transport from one state to another, of any insect, fungus or other pests, which may be destructive to crops. The seed was not originally included in the DIP Act, but in 1984, the government of India passed the 'Plants, Fruits and Seeds (Regulation of Imports into India) order of 1984 (Ministry of Agriculture and Cooperation, 1985) which came into import for 17 important crops. The main features of the order are as follows:

1. Seeds have been brought under the purview of the DIP Act.

2. No consignment can be imported into India without an official 'Import Permit'.

3. No consignment can be imported into India without an official, 'Phytosanitary certificate' issued by the official Plant Quarantine agency of the exporting country.

4. Post entry isolation of specific crops at approved locations is stipulated.

The latest regulation enacted under the DIP Act is the 'Plants, Fruits and Seeds Order, 1989 (PFS). This was necessitated to cater to the needs of 'The New Policy on Seed Development' (NPSD) of Govt. of India which came into force on 1st October, 1988 with the objective to make available to the Indian farmers the best genetic materials in the world to increase our agricultural productivity and to encourage the private sector seed industry in India not only to fulfill domestic requirements but also to develop export potential. While liberalizing import, care has been taken that there is absolutely no compromise on plant quarantine requirements.

Although there are several requirements under the FPS Order, 1989, but for our purpose the most important requirements are that:

(a) No consignment shall be imported into India without a valid import permit issued by the competent authority. For importing germplasm, Director, ICAR-NBPGR, New Delhi has been authorized by the Govt. of India to issue import permits, both for government institutions as well as private seed companies. For bulk consignments the import permits is issued by the Plant Protection Adviser to the Govt. of India.

(b) No consignment shall be imported, unless accompanied by phytosanitary certificate by an official of the exporting country.

(c) Seeds/planting materials requiring isolation growing under detention shall be grown in an approved post-entry quarantine facility.

Import of soil, earth, sand, compost, and plant debris accompanying seeds/planting materials shall not be permitted.

(d) Hay, straw or any other material of plant origins shall not be used as packing material.

The various plant produces may be grouped into two categories on the basis of their intended use:

(1) Those for consumption and

(2) Those for cultivation and research.

The quarantine of these two groups of plant produces is carried out by separate agencies.

(1) **Plant produce imported for consumption-** The quarantine of this category of plant produces is the responsibility of the Directorate of Plant Protection, Quarantine and Storage. The directorate has quarantine and fumigation centres at 8 seaports, 7 airports and at the entry points of 7 land routes. These centres examine the plant produces imported for consumption and carry out their fumigation, if necessary.

(2) **Plant propagules imported for cultivation and research-** The quarantine of plant propagules is done by the following three agencies depending on the nature of the concerned plant species.

(I) ICAR-NBPGR, New Delhi is responsible for the quarantine of all the propagules of agricultural and horticultural species.

(II) The quarantine of propagules of forest trees is carried out by the Forest Research Institute, Dehradun (Uttarakhand).

(III) The propagules of remaining plant species are handled by the Botanical Survey of India, Kolkata.

The quarantine of germplasm pertaining to agricultural and horticultural crops is done by entomologists, plant pathologists and nematologists of the Quarantine Section of ICAR-NBPGR, New Delhi at the headquarters of the Bureau. According to the quarantine laws, only those propagules that are free from diseases, insect pests and weeds can be allowed to enter the country (beyond the point of quarantine). The quarantine laws cover not only the propagules but also their packing materials, especially of plant origin, and other materials accompanying them.

## Quarantine Procedure

The following generalized description relates to the quarantine activity of the ICAR-NBPGR, New Delhi; it has been designed to prevent the entry of weeds, diseases and pests.

1. It is essential that the propagules must be clean, healthy and free from weeds and insect pests. The sender should not treat propagules with any fungicides or insecticides. If necessary, the sender may only fumigate the seeds/propagules before sending them, and indicate the same in the phytosanitary certificate.

2. Each imported entry or sample must be accompanied by a 'phytosanitary certifrcate' from the scientist/institution sending the sample/entry. In this certificate the sender certifies that the seeds/propagules being sent are free from weeds, diseases and pests. All the entries not accompanied by an authentic phytosanitary certificate are either returned to the sender or -destroyed by the Bureau.

3. The entries accompanied by an authentic phytosanitary certificate are examined closely with the help of a magnifying glass/microscope and screened with X-rays. X-ray examination is helpful in the detection of insects *etc.* present within the propagules.

The healthy entries free from diseases, insect pests and weeds are identified and sent to the recipient scientists/institutions. Contaminated entries are detained by the Bureau and attempts are made to free them from the contaminating weeds, insects and pathogens.

4. Contaminated entries may be fumigated if it is considered that such a treatment would rid them of the contaminating insect or pathogen.

5. If needed or considered desirable, the contaminated entries may be grown in isolation in an effort to isolate some healthy plants. Seeds/propagules from such healthy plants may be collected and sent to the indentor. Entries suspected to be contaminated by pathogens, particularly viral pathogens, are also grown to monitor disease symptoms and the presence of pathogens. In advanced countries, sophisticated techniques like immunological assays *etc.* are used to detect the presence of pathogens, especially viral pathogens, in germplasm introductions.

6. Samples/entries that are heavily contaminated are destroyed by the Bureau.

The process of quarantine (except growing in isolation) at the ICAR-NBPGR, New Delhi takes at least three weeks. The quarantine of short-lived propagules is done at top priority. The elimination of these contaminated samples prevented the entry of more than 12 new pathogens and insects in this country. In addition, this process of quarantine has prevented the introduction of many new biotypes of several pathogens and insect pests.

## Purpose and Achievements of Plant Introduction

The main purpose of plant introduction is to improve the plant wealth of the country and to make available to breeders the germplasm required for their breeding programmes. The chief objectives and achievements of plant introduction are:

1. **Entirely new crops:** Through plant introductions entirely new crop species can be obtained. Almost all the countries in the world have obtained some entirely new crops through introduction. Many of our important vegetable crops, *e.g.* potato, tomato *etc.* are introductions. Some recently introduced crops are soybean, gobhi sarson (*Brassica napus*), Karan sarson/ Ethiopian mustard (*Brassica carinata*) and sugar beet (1960).

2. **Direct release as varieties and selection from introduction:** Sometimes introductions are directly released as superior commercial varieties for cultivation. Several introduction of tomato (Sioux), vegetable cowpea, cauliflower, onion, lettuce, watermelon, *etc.* have also been directly released as varieties.

From exotic introductions, several selections/ varieties could be directly released after their acclimatization in different locations and initial performance or evaluation. Also the exotic materials possessed several promising traits, which could be incorporated in to indigenous verities through breeding. Thus several varieties selections could be developed possessing better yield, adaptability to biotic and abiotic stresses, and fitted under different agro-climates and cropping patterns: such as in tomato, pungent/ sweet pepper, eggplant, cucurbits, peas, French bean, cowpea, pumpkin, okra, bitter gourd, cauliflower, radish, turnip, other Brassicae, carrot and onion. As example 'Pusa Lal' and 'Pusa Sunehari' sweet potato, 'Pusa Barsati' vegetable cowpea and 'Japanese White' and '40 Days' radish, *etc.* A list of vegetable crops varieties introduced and released directly or used for breeding new varieties of vegetables by ICAR-IARI/ICAR-NBPGR is being given in the Table 3.11.

**Table 3.11: Vegetable Varieties Introduced and Released Directly for Cultivation**

| Crop | Introduction/Selection |
|---|---|
| Tomato | Sioux, Marglobe, Best of All, Fire Ball, La Bonita, Balkan |
| Brinjal | Pusa Purple Long, Pusa Purple Round |
| Hot Chilli | NP 46A, Pusa Red |
| Guar | Pusa Sadabahar, Pusa Mansuri, Pusa Shadabahar |
| Cauliflower | Snowball, Pusa Katki, Improved Japanese |
| Lablab bean | IC-16862, Pusa Bunch |
| Sponge gourd | Pusa Chikni |
| Ridge gourd | Pusa Nasdar |
| Okra | Pusa Makhmali, Ghana Red |
| Peas | Early Badger, Bonneville, Sylvia, Asauji, Arkel, Harbhajan |
| Cowpea | Pusa Barsati, Pusa Phalguni |
| Garlic | IC-49373, IC-49382, T86/Sel-1 |
| Onion | Early Grano, Pusa Red, Pusa Ratnar, Ratnar Selection, Brown Selection |
| Carrot | Pusa Keasr |
| Amaranth | Bari Chauli |
| Bottle gourd | Pusa Summer Prolific Long, Pusa Summer Prolific Round |
| Fenugreek | Kasuri Methi, Pusa Early Bunching |

3. **Utilization in crop improvement programme:** Often the introduced genetic resources are used for hybridization with local varieties to develop improved varieties. For example 'Pusa Ruby" tomato was derived from a cross between 'Meeruty and Sioux', an introduction from USA. Similarly, 'Pusa Early Dwarf' tomato derived from the cross 'Meeruty' and 'Red Cloud'. Other examples are 'Pusa Kesar' of carrot, and 'Pusa Kanchan' of turnip.

4. **Saving a crop from a diseases or pest:** Sometimes a crop is introduced into a new area to protect it from a divesting diseases or pest.

5. **Utilization in scientific studies:** The main purpose of germplasm collection is to use in studies like biosystematics, evolution and origin of plant species, *etc*. First concept in this regard was developed by N.I. Vavilov and put forth the theory of homologous series in variation from the study of a vast collection of plant type.

6. **Used for Aesthetic value:** Ornamentals, shrubs and lawn grasses are introduced to satisfy the finer sensibilities of humans. These plants were used for decoration and are of great value in social life. Similarly, several introductions were made in pumpkin for variegated fruit colour.

## Merits of Plant Introduction

☆ Major source of entirely new crop plants

☆ Source of superior varieties either directly, after selection or incorporation through hybridization programme.

☆ Feasible means of germplasm collection.

☆ Helps a lot in crop improvement programme *i.e.* direct release as varieties, selection or hybridization and selection.

☆ Helps in identification of new disease free area to protect the damage through introduction.

## Demerits of Plant Introduction

All the time introductions will be beneficial, it is not true. There are certain disadvantages of plant introduction:

☆ **Introduction of weeds:** Through plant introductions several obnoxious weeds like *Argemone Mexicana, Eichornia crassipes* and *Phylaris minor etc.,* entered in India from other countries

☆ **Introduction of diseases:** The burning example of introduction of disease is 'late blight of potato' which was introduced in India from Europe in 1883. Similarly 'bunching top of banana' arrived in India from Ceylon in 1940. Not only this, some other diseases were also introduced in India along with plant materials.

☆ **Introduction of insect pests:** The burning example of introduction of insect pest is 'potato tube moth' which was introduced in India from Italy in 1900.

☆ **Ornamentals-Turned-Weeds:** Some introduced ornamental species may become noxious weeds in their new habitat. For example water hyacinth and *Lantana camara* were both introduced in India as ornamental plants, but they are now noxious weeds. Similarly, in certain parts of world ivy gourd (*Coccina* species) also behaving as a noxious weed in certain areas.

☆ **Threat to Ecological balance:** It is true that some introduced species sometimes disturb the ecological balance in their new home, and may cause serious damage to the ecosystem. For example Eucalyptus species introduced from Australia, causes rapid depletion of subsoil water reserves.

## Plant Introduction Agencies in India

The government of India has approved three National institute/organisations, to act as official introduction and quarantine agencies for exchange (import/export) of seeds/planting materials for different group of crop plants, especially for research purpose:

1. The ICAR-National Bureau of Plant Genetic Resources, New Delhi for agri-horticultural and agri-silvicultural plants under the control of Department of Agricultural Research and Education.

2. The Forest Research Institute (FRI), Dehradun (Uttarakhand) forestry plants under the control of Department of Environment and Forests.

3. The Botanical Survey of India (BSI), Kolkata, for plants of botanical interest under the control of Department of Environment and Forests.

4. The import/export of bulk quantities of plant materials for commercial purposes is handled by the Directorate of Plant Protection, Quarantine and Storage, Ministry of Agriculture, Government of India, through its 27 Plant Quarantine and Fumigation Stations at different seaports, airports and land frontiers.

## References

1. Allard, R.W. 1960. Principles of Plant Breeding. John Wiley and Sons, London.

2. Allard, R.W. 1970. Population structure and sampling methods. In: Frankel, O.H. and Bennett, E. (eds.), Genetic Resources in Plants-Their Exploration and Conservation. Blackwell, Oxford and Edinburgh, p. 97-107.

3. Bennet, E. 1965. Plant Introduction and genetic conservation. pp. 27-113. In: Genecological aspect of an urgent world problem. Scottish Plant Breed. Res. Sta.

4. Gautam, P.L., Karihaloo, J.L., Kumar, A. Kochhar, S. and Singh, B.M. 2001. Vegetable germplasm management: Development, issues and strategies. In: Emerging scenario in vegetable research and development (G. Kalloo and Singh, K., eds.). Research Periodicals and Book Publishing House, New Delhi, India, p. 45-57.

5. Frankel, O.H. 1957. The biological system of plant introduction. *Jr. Aust. Inst. Agric. Sci.*, **20:** 302-307.

6. Frankel, O.H. and Bennett, E. (eds.). 1970. Genetic Resources in Plants: Their Exploration and Conservation. Blackwell, Oxford, p. 538.

7. Frankel, O.H. and Hawkes, J.G. (eds.). 1975. Crop Genetic Resources for Today and Tomorrow. Cambridge Univ. Press, London, p. 492.

8. Singh, B.D. 2012. Plant Introduction. In: Plant Breeding: Principles and Methods, Kalyani Publishers, p.54.

9. Sharma, J.P. 2009. Vegetable Genetic Resources. In: Principles of Vegetable Breeding. Kalyani Publishers. p. 20-38.

# 4

# Explorations and Collections of Vegetable Genetic Resources

## Introduction

Indian subcontinent is one of the centres of origin of vegetable crops (Zeven and Zhukovsky, 1975) and about 199 wild relatives have been reported from diverse ecological habitats and geographical regions of India (Arora and Nayar, 1984). Nayar *et al.* (2003) recorded 92 species of vegetable and edible tubers under cultivation in India. About 521 species are used as leafy vegetables and 145 as roots/tubers from wild edible sources, which contribute to the richness in diversity (Arora and Pandey, 1996). Germplasm is the basic tool and had played most vital role in the improvement of cultivated plants including vegetables. It is universally agreed that a catastrophic loss of crop diversity has been taking place during last few decades, and that this process of genetic erosion is likely to continue at an even greater speed in future. It is also true that plant breeders need genetic diversity as a basis for development of new high yielding, better adapted and more resistant varieties to biotic and abiotic stresses. Concerted efforts are being made to collect, characterize, preserve and exploit genetic diversity before it disappears from the nature. Plant breeders need, not only genetic resources of major crops but also minor crops, including living genetic resources of their related wild species for improvement. Carefully designed plans for collecting crop diversity with established regional and crop priorities are currently undertaken. Plant explorations are conducted from within the country and also from abroad in the areas of diversity distribution of the crop plants and their wild relatives. These surveys are made mainly for collection of propagules, (seed and other living parts), agri-horticultural importance, and thus it differs from general floristic surveys. Thus it includes collection of local landraces of crop plants and their wild relatives. It is imperative that planning for such survey is to be made with utmost care. While planning, it is essential to consider the appropriate time of collection *i.e.* maturity period of the crop(s), vulnerability of the area *etc.* In addition, knowledge of agro-ecology of the area is essential. Our knowledge of the taxonomy, population genetics and breeding objectives of cultivated plants shows very clearly that exploration for genetic resources purpose is a discipline in its own right. It is also beneficial to have knowledge on ethno botany of the area such as local inhabitants, the crops grown by them *etc.* which are very fruitful in survey and collection. The long history of cultivation and continuous selection has witnessed several landraces/primitive types suited to different climatic and soil conditions in different vegetable crops including the introduced ones. Collecting this diversity is an important and foremost activity for the sustainable management of our rich vegetables wealth. The objectives are to assemble the desired genetic material in viable, healthy and sufficient quantity along with adequate documentation, following national legislative protocols. Contributions of the following are worth mentioning.

In the Indian context, Emperor Akbar (1542-1605) established a mango orchard (*Lakhi Bagh*) in Darbhanga

(Bihar) during his regime. Emperor Ashoka (304-232 BC) patronized the establishment of fruit and shade trees in his kingdom. Some historical events are given in Table 4.1.

**Table 4.1: Historical Events for Explorations/Collections**

| Persons Involved | Area of Work |
|---|---|
| Queen Hatsheput (Egypt, 3500 years ago) | Collected resin of myrrh plant and frankincense trees near Somalia |
| Christopher (Italy-in 1492) | Potatoes, sweet potatoes, maize, tomatoes, peanut, cassava, cacao, peppers, tobacco, beans and squashes |
| Thomas Jefferson (3rd President of USA) | Vanilla, tea, olives |
| Nicolai Vavilov (Russia) | Centres of origin |
| Carlos Ochoa (Peru) | Wild and endangered species of potato |
| John George Jack (Canada) | Tree genetic resources |
| Jack Harlan (USA) | Centres of diversity |
| Gregory, W.C. | Peanut germplasm |
| Krapovickas, A. | Peanut germplasm |
| Brown, W.L. | Maize germplasm |
| Hawkes, J.G. (UK) | PGR science |
| Otto Frankel (Australia) | PGR science |
| Zeven, A.C. (The Netherlands) and Zhukovsky, P.M. | Centres of origin |

## Hotspots of Biodiversity

The unique gift of nature of diverse climate and distinct seasons in the country, make it possible to grow an array of vegetables as commercial crop or diversity *per se* in nature. Hotspot of Biodiversity is relatively a recent concept related to the richness in variety and variability of species of plants, animals and micro-organism in a definite region which is also facing a serious threat of destruction. The British biologist 'Norman Mayers' coined this term in 1988 as a biogeographic region characterized both by exceptional levels of plant endemism *i.e.* plants regularly found in a particular locality and by serious levels of habitat loss in that region. In 1990, Mayers identified eight hotspots facing serious threat from human activities. Later, a total of 20 different hotspots of biological diversity all over the world were identified with about 49, 550 endemic species of higher plants *i.e.* 20 per cent of the world's total. Out of those 20 hotspots, 12 are in tropical rainforest region (Hawaii, Colombian Choco, West Ecuador, West Amazonia, forest area of Brazil, Eastern Madagascar, Peninsular Malaysia, East Himalayas, North Borneo, Philippines, Queensland of Australia and New Caledonia). The remaining eight are California Floristic Province, Central Chile, Ivory Coast, Cape Floristic Province, East Arc Forests of Tanzania, Western Ghats, Sri Lanka and South West Australia in other climatic ecosystem. In 1989, the ecosystems

Conservation International (C.I.) organization adopted Mayer's hotspots as its institutional blueprint and in 1996 it was decided to undertake a reassessment of the hotspot concept. Accordingly after three years of extensive global review, the criteria were laid down to identify and declare a region as hotspot of biodiversity. They are:

**(a)** The place must contain at least 1500 species of vascular plants (>0.5 per cent) of the world's total) as endemics and

**(b)** It must have lost at least 70 per cent of its original habitat.

In 1999, C.I. identified 25 biodiversity hotspots present in different countries of the world. The total area covering the above 25 hotspots constituted only 11.8 per cent of the total land area of the earth but held about 44 per cent of the world's plants and 35 per cent of the terrestrial vertebrates as endemic species. But due to the rapid loss of the habitat this land area had been reduced by 87.8 per cent of its original extent such that the wealth of biodiversity was restricted to only 1.4 per cent of earth's total land surface. In 2005, C.I. has revised and updated the list of hotspots (Table 4.2).

**Table 4.2: Hotspots of Agro-biodiversity in different Continents**

| Asia Pacific | East Melanesian Islands |
|---|---|
| | Himalaya |
| | Indo-Burma |
| | Japan |
| | Mountains of South West China |
| | New Caledonia |
| | New Zealand |
| | The Philippines |
| | Polynesia-Micronesia |
| | Southwest Australia |
| | Sundaland |
| | Wallacea |
| | Western Ghats and Sri Lanka |
| Europe and Central Asia | Caucasus |
| | Irano-Anatolian |
| | Mediterranean Basin |
| | Mountains of Central Asia |
| North and Central America | California Floristic Province |
| | Caribbean Islands |
| | Madrean Pine-Oak Woodlands |
| | Mesoamerica |
| South America | Atlantic Forest |
| | Cerrado |
| | Chilean Winter Rainfall-Valdivian Forest |
| | Tumbes-Chaco-Magdalena |
| | Tropical Andes |

| Asia Pacific | East Melanesian Islands |
|---|---|
| Africa | Cape Floristic Region |
| | Coastal Forests of Eastern Africa |
| | Eastern Afromontane |
| | Guinean Forests of West Africa |
| | Home of Africa |
| | Madagascar and the Indian Ocean Islands |
| | Maputaland-Pondoland-Albany |
| | Succulent Karro |

*Source*: Mohanty and Tripathy, 2011.

This collection contains the description of 34 regions in different corners of the earth with just 1.4 per cent of the total land but support about 60 per cent of all plant species. There are two recognized hotspots of India and some more awaits their inclusion in the list. The unique flora and fauna of all these hotspots are on verge of extinction due to rapid destruction of their habitats. Thus alarming situation warrants immediate attention in the International Year of Biodiversity, 2010, to save those organisms from extinction.

## Collections and Augmentation of Genetic Diversity

Augmentation of germplasm through exploration and collection particularly in developing countries is still limited despite the wide recognition (Gill, 1984). Collection refers to tapping of genetic diversity from various sources and assembling the same at one place. The exploration and collection is a highly scientific process. Endemic Indian plant wealth supplanted with new species and forms that have transgressed national boundaries and enriched our flora which got diversified on being isolated climatically and spatially. The past linkages with Indo-Chinese-Indonesian, Chinese-Japanese and the Central Asian region helped considerably in augmenting our crop plant resources. The influx of genetic material in distant past from Mediterranean and African regions has also resulted in the accumulation and diversification of enormous genetic variability. The ancient travellers, invaders and religious missionaries have also contributed significantly towards enriching the Indian gene centre. A germplasm collection of a crop species consists of a large number of lines, varieties and related wild species of the crop. Introduction in earlier times were unsystematic and were carried with sole aim to meet personal requirements. The situation led to evolution of systematic procedures in Augmentation/Introduction of genetic resources. Now, every country is realizing the importance of plant genetic resources. International programme related to plant genetic resources are being planned, executed and monitored by International Board for Plant Genetic Resources (IBPGR), Rome while at national level every country developed its body to regulate activities of germplasm management and exchange.

*"The process of obtaining germplasm accessions (for a germplasm collection) is known as collection of germplasm".*

## Reasons for Collecting Germplasm

*"Germplasm collections aim at minimizing the detrimental effects of genetic erosion by collecting and preserving the variability in crops and their related species".*

Collection of germplasm is essential for livelihood but it is difficult to predict the demand of germplasm. One does not know what tomorrow's need may be and what germplasm may be able to fulfill them. The more diversity is conserved and made available for future use, the better the chance of fulfilling future's demand. But in practice, some prioritization is necessary both at species level and geographic regions. The main reason that can be put forward for collecting germplasm of assigned gene pool in given area are that (Srivastava, 1998):

**(i)** It is in danger of extinction or even erosion;

**(ii)** A clear need exists for it, as it was expressed by user both at national and international levels;

**(iii)** The diversity it represents is missing from or insufficiently represented in, existing *ex-situ* germplasm collections;

**(iv) Rescue collecting:** Rescue collections are conducted when genetic diversity is imminently threatened in an area and *in-situ* conservation methods are not feasible but germplasm collecting may be warranted *e.g.* the emergency programme construction of Sardar Sarover Dam in Gujarat by the Government of India. The catchment area of this dam will engulf 320 villages. All vegetation including those with endemic distribution will be at risk and in such circumstances; high priorities are assigned for rescue collection. To capture the diversity from the areas under the influence of natural calamities, four rescue missions were planned and executed. These rescue missions were undertaken in: super cyclones affected parts of Odisha (1999), earthquake hit areas of Gujarat (2001), drought affected areas of Gujarat, Rajasthan, Haryana, Odisha, Madhya Pradesh, Uttar Pradesh and Vindhyachal hills region (Uttar Pradesh) and natural calamity affected areas of Uttarakhand (2013).Total 2413 accessions including cereals/pseudo-cereals (697), millets (147), oilseeds (100), legumes (460), vegetables (384), spices/condiments (67), medicinal and aromatic plants (130) and other agri-horticultural crops (428) were collected. Wild relatives *viz. Abelmoschus crinitus, A. manihot, Atylosia scarabaeoides, Citrullus colocynthis, Coccinia indica, Crotolaria burhia, C. medicagenea, Curcuma amada,*

*Solanum pimpinellifolium, S. sisymbrifolium, Musa acuminata, Musa* (seeded type), *Oryza granulata, O. nivara, O. rufipogon, Solanum nigrum, Solanum torvum,* and *S. viarum* were collected. Wild ginger (*Zingiber* sp.) having 1.5 meter long rhizome was also collected from Odisha (Panwar *et al.,* 2016). The devastating earthquake that hit Nepal on 25th April 2015 was most severe in rural farm households particularly in remote and risk-prone mountainous regions where farmer's dependence on food security was high from self-saved and locally exchanged seeds of traditional crops and varieties. Aftermath of the earthquake, various national government and international relief agencies made efforts to rescue human beings, livestock and valuable assets but no immediate initiatives were made to rescue endangered seeds and native crop varieties. Considering the critical role of local crop diversity in the livelihood of smallholder farmers in remote and risk-prone mountainous regions, Bioversity International and National Gene Bank of Nepal jointly initiated a study on rescue seed collection, conservation and repatriation of local crop genetic resources that are endangered from disasters. The main objective of the study was to assess and measure loss of on-farm diversity of traditional crops in earthquake affected areas. Rescue seed collection missions and information collection on seed losses were carried out in most severely affected villages of 10 earthquake affected districts. The collected information was analysed for diversity assessment, regeneration and processing for their safe storage in national gene bank. The most endangered and valuable local diversity based on farmers demand are planned for repatriation to same communities and community seed banks for on-farm conservation and strengthening local seed system. This rescue strategy will be useful to promote both *ex situ* and *in situ* conservation and help to safeguard native crop diversity for future generation in disaster affected areas (Gauchan *et al.,* 2016).

(v) **Collecting for immediate use:** Local communities especially engaged in traditional medicine are continuously collecting germplasm for immediate use.

The main purpose of the exploration may be to collect wild species and primitive cultivars. These are, in general, not particularly well represented in some of the larger collections; even some so-called "world Collection" may not include more than a few samples of related wild and weedy species.

## Prioritization of Cultivated Species

There is an increasing need to broaden the range of plant species for the diversification of agriculture.

A massive untapped potential of species exists for meeting the needs for food, fodder, feed, fibre, energy and industrial products. The prospects of these species to become agricultural crops vastly differ due to several factors. The cost of domesticating and bringing a new species into agriculture, horticulture or forestry is very high and the time required is many years. The change of environmental conditions poses problems and, therefore, an environment similar to the natural one should be selected for initial evaluation to understand the potential of the crop. The new crop has to find a place in the existing or modified cropping system or a place in an area not presently used. It should be sufficiently commercially attractive by justifying a change. Quality of the produce, marketability, profitability and people's preference are also important consideration. Keeping in view the various factors, a few plants with a good promise for commercial exploitation have been selected for undertaking intensified research efforts under the All India Coordinated Research Project on Underutilized Plants. These species requiring priority attention are listed in Table 4.3.

During phase I, mainly multi-crop explorations were carried out for collection of seed material of field crops including vegetables with emphasis on conservation. The priority was to collect the prevailing genetic diversity of primary genepool with focus on landraces and primitive/local cultivars from farmers' fields and disturbed habitats. Under NATP (Plant Diversity), the multi-crop and region specific explorations were continued but in hitherto un-explored areas, which resulted in the enrichment of genetic as well as species diversity. With the involvement of various partners, the priority shifted to crop-specific missions with emphasis of horticultural crops, vegetatively propagated species and wild relatives. The priority for above crop categories continued in the present era too, however, utilization of collected diversity for the collection of trait-specific germplasm from distinct ecological habitats, hot spots, *etc.* Besides, need-based re-visits are also planned to earlier explored areas for collecting germplasm for specific purpose/use.

## Prioritization of Crop Wild Relatives

Crop wild relatives (CWR) are wild species closely related to landraces and commercial crops. CWR can be a source of genes for food crops, however, researchers have neglected these species and they are becoming more threatened in the wild. Prioritization of CWR process was conducted by Maluleke *et al.* (2016) in Africa using criteria such as potential use in crop improvement, socio-economic value, threat and relative distribution with 292 species recorded. They surveys three provinces (Mpumalanga, Limpopo and Kwazulu-Natal). A feasible number of 31 species were identified in terms of conservation status, using the field survey criteria on selected species that occurs and does not occur in protected areas, including species on oldest

**Table 4.3: Crops/Species, Usage and Adaptation/Areas of Cultivation**

| Crops/Species | Usage | Adaptation/Areas of Cultivation |
|---|---|---|
| **A. First Priority Crops** | | |
| Amaranth (*Amaranthus* spp.) | Grains and leafy vegetable | Northern hills, north and south Indian Plains |
| Buckwheat (*Fagopyrum* spp.) | Leafy vegetable | Temperate Himalayan region, southern India hills |
| Rice bean (*Vigna umbellata*) | Pulse, fodder | Hot humid climates, sub-tropical hilly regions |
| Winged bean (*Psophocarpus tetragonolobus*) | Pulse, vegetable, fodder, edible roots | Sub humid tropical parts of north-eastern region, central/eastern peninsular region |
| Faba bean (*Vicia faba*) | Pulse, vegetable, fodder | Eastern, northern and peninsular India |
| Tumba (*Citrullus colocynthis*) | Seed oil for industrial use; pulp of roots used in medicine | Warm arid and sandy tracts of north-west, central and southern India |
| **B. Second Priority Crops** | | |
| Chenopod (*Chenopodium* spp.) | Grains, leafy vegetable | Himalayan region |
| Adzuki bean (*Vigna angularis*) | Vegetable | Northern hilly region |
| Kankora (*Momordia dioica*) | Vegetable | Central India. |
| Saltbush (*Atriplex* spp.) | Fodder | Arid situations, salt affected lands |
| Bamboo (*Bamboosa* spp.) | Vegetable | North-eastern region, Western Ghats |
| **C. Third Priority Crops** | | |
| Bambara groundnut (*Vigna subterranea*) | Grains rich source of carbohydrate and lysine | Dry arid tracts |
| Vegetable wild rice (*Zizania coduciflora*) | Fodder | Marshy lands and swampy waterlogged areas |
| Euphorbia (*Euphorbia* spp.) | Source of hydrocarbons | Sandy and rocky soils in arid region. |

*Source*: Bhag Mal. 1992.

records (above 30 years). Parameters such as collection date, collector, species name, genus, family, land use, associated species, plant height, locality, co-ordinates, population size and specimen for herbarium were collected using collection form. Out of these collected species, families with the higher number of species in all province were Solanaceae (7 taxa) and on Poaceae (2 taxa). Species were found in unprotected and protected areas and recorded. Although most of these species occurred in all three provinces across the country, *Oryza longistaminata* only occurred in one province (Limpopo) in wetland in Nylsvley Nature Reserve. Kwazulu-Natal had significantly high number of 34 localities with Miscanthusjunceus found in 5 of these localities on private land (plantations), compared to Mpumalanga with 18 and Limpopo with 12 localities. Among the 31 species, 20 were collected along roadsides, especially in disturbed areas, whereby they are exposed to threats that need sufficient attention for conservation. Similarly, Ng'uni' *et al.* (2016) also admitted that national actions towards conservation of plant genetic resources do not adequately cover *ex situ* CWR occurring in the country and there is no active *in situ* conservation of CWR. Through the three years EU-ACP supported SADC project, Zambia prioritized a total of 30 CWR taxa from the generated partial checklist consisting of 459 taxa. Occurrence data of priority CWR taxa were compiled from several sources including national herbaria, the national plant genetic resources collections database and other online sources. The populated occurrence data of priority CWR taxa were subjected to spatial analysis employing DIVA-GIS version 7.5 and CAPFITOGEN tools to establish species distribution, species richness, gaps in *in situ* and genebank collections to objectively identify priority sites for *in situ* conservation and *ex situ* collecting. The results generated from these analyses have provided information on priority CWR species richness indicating the spatial differences in richness of the priority CWR in the country, distribution of priority CWR of cowpea (*Vigna* spp.), *Dioscorea* spp. and *Solanum* spp. Gap analysis has indicated priority CWR taxa not actively conserved through both *in situ* and *ex situ* strategies. Consistent with the national development of agenda, results of the spatial analyses of occurrence data of priority CWR taxa objectively provide the basis for decision making on targeted sites for the conservation of CWR taxa in Zambia. In India, a total of 588 species have been shortlisted as CWR of 168 crops belonging to 14 crop-groups (Pradheep *et al.*, 2016). Collection of database indicates that out of 588 species only 243 amounting to 10,529 accessions were collected so far; accounting to about 30 per cent of total wild germplasm collected. However, this excludes those accessions involving wild/weedy/semi-wild populations of 142 crop taxa as well as those identified only up to genus level. Significant collections were made in some CWR under crop genera of *Abelmoschus, Cucumis, Momordica,* and *Solanum.*

## Characters to be Collected

Collection of genetic diversity requires an understanding of what is a character and how does it vary in a population. Plant character refers to form, structure, behavior or function of a plant which can be considered separately from the whole plant for the purpose of interpretation or comparison. Characters are qualitative or quantitative and can be measured, counted or otherwise assessed. Those used in identification, characterization and delimitation called analytic and others synthetic, individual or between populations of individuals. Variations of individual characters and of a gap of characters- pattern variation- can be studied in the field populations and/or in the experimental garden.

## Issues Related to Collection

FAO Commission on Plant Genetic Resources is presently in preparation of voluntary code of conduct for the responsibilities of collectors, financial donors and research organizations dealing with the conservation and use of genetic resources. A Biodiversity Convention has been proposed and signed by several countries. The convention aims at preservation and use of genetic diversity for sustainable development and equitable sharing of the benefits arising out of the utilization of genetic resources, including both accesses to genetic resources as well as appropriate transfer of technologies. In the context of Farmer's Right, it is collector's responsibility to properly record the name and complete address of farmers or tribal people who permitted the collector to collect a representative sample from the target site. It would help in eventually rewarding the informal innovators for selecting and preserving the genetic diversity of specific target species. Biodiversity Convention (Earth Summit) also recognizes the rights of informal innovators. The Indian Patent Act (1960) was revised accordingly.

## Ways of Augmenting Germplasm

This process takes into account six important items, viz.,

(i) Sources of collection

(ii) Priority of collection

(iii) Agencies of collection

(iv) Methods of collection

(v) Methods of sampling and

(vi) Sample size.

Germplasm collection can be done by two ways:

(1) Exploration and collection of genetic resources

(2) Procurement of genetic resources

## (1) Exploration and Collection of Genetic Resources

Explorations are trips for collection of cultivated forms like land races, open-pollinated varieties, *etc.*, wild forms and wild relatives of crop plants.

*"Explorations are the primary source of all the germplasm present in various germplasm collections"*. "Exploration refers to collection trips".

In the activities of Plant Genetic Resources (PGR), exploration is one of the most important issues. A number of cultivated vegetables crops and their wild relatives possess enormous variability and genes resistant to biotic and abiotic stresses. It has rich array of vegetable wealth both of indigenous and introduced types distributed in diverse ecological habitats and geographical regions. Due to long history of cultivation and continuous selection, several land races/primitive types suitable to different climatic and soil conditions have been developed in various vegetable crops. For example amaranths offer a great deal of diversity in terms of various traits of economic importance for collection and utilization (Joshi and Mehra, 1983). In order to preserve this diversity of amaranths several region specific explorations were undertaken in the last four decades by the ICAR-NBPGR, Regional Station, Shimla (Himachal Pradesh). During these explorations, diverse ecological regions comprising Shimla, Solan, Bilaspur, Kullu, Mandi, Lahual and Spiti, Sirmour, Chamba, Kangra and Kinnaur of Himachal Pradesh; Uttar Kasi, Pauri, Almora, Pithoragarh and Nainital of Uttarakhand; J&K; Sikkim, Drjeeling and North East Hill Regions were covered.

### Exploration Missions

Exploration, by and large, is something of a personal art. The humane element within you would play as much role as would your scientific thinking and understanding. Never forget that crop plants and man are intricately knit together.

Broadly two kinds of explorations are formulated based on:

(i) Priority of crops and

(ii) The area/region

    (a) Crop specific

    (b) Region specific/multi crop collection

Frequently the past collections:

(i) May have been collected by breeders for a limited purpose,

(ii) May not have been population sample,

(iii) May not have been conserved properly for long term conservation,

(iv) May have suffered from genetic erosion or drift since they were collected,

(v) May have been collected from easily accessible areas, ignoring remote or even fairly accessible sites

(vi) May have been lost by neglect, fungal and pest attacks *etc.*

## Merits of Plant Exploration

Exploration has its own merit in germplasm augmentation (Singh, 2012):

(i) It is the source of virtually all genetic diversity stored in gene bank

(ii) It is the only means of collecting and conserving the threatened genetic diversity.

(iii) It often provides access to material of special interest *e.g.* new genes (=alleles), new species, *etc.*

## Limitation of Plant Exploration

Exploration also has its own limitation in germplasm augmentation (Singh, 2012):

(i) It is tedious, time taking and expensive.

(ii) It poses various hardships to the collectors, *e.g.,* in boarding, transportation, *etc.*, especially in remote areas.

## Objective of Exploration

Explorations are planned to fulfill mainly the two major objectives:

**(i) Collection of germplasm needed by breeders to develop new variety:** This is most important objective to collect the germplasm from various sources. The germplasm accessions collected for this purpose possess the specific traits that are required by breeders either in the immediate future or in the foreseeable future.

**(ii) Collection of the variability for conservation:** In nature ample variability exists in the crop plants and their relatives. Therefore, it is important to collect the existing variability. For this purpose, germplasm samples are collected without any references to the presence of specific traits; the only consideration for collection is that as many diverse types are collected as possible.

## Types of Exploration

The planning of collecting expeditions on a national basis should incorporate the regulations set out by the particular country involved. It is sometimes necessary to collect only one crop on an expedition, as for instance when there is an intermediate threat to indigenous land races of a crop because of their replacement by improved cultivars. Very often, however, several related species

are all grown together or groups of even unrelated species occur in the same rotation. When this is so, an attempt should be made to collect this wider range of crops since it will probably be very expensive and not very cost effective to send several expeditions to the same region, each for its own particular crops. In some instances, multi crop collecting may not be practical.

Depending on the purpose, explorations are of following types:

**(i)** Multi-crop exploration

**(ii)** Single crop specific

**(iii)** Target specific

**(iv)** Area specific

### (i) Multi-crop exploration

The purpose of this type of exploration is to collect genetic variability available in the cultivated as well as wild relatives in general of a given region (referred as region-specific exploration). Here the target is seasonal crops such as rabi crops, *kharif* crops or *Zaid* crops. Sometimes the objective may be group specific as *rabi* cereals, *kharif* pulses or *rabi* oilseeds *etc.* In these, exploration missions, one should have extensive knowledge of the plant genetic resources. Usually these are planned when no systematic collecting in the area has been conducted before or/when the area is difficult to reach and future visits are therefore unlikely. Emergency rescue collecting also fall under the multi crop exploration. For example Sardar Sarover catchment area in Gujarat, Maharashtra and Madhya Pradesh areas, imminently threatened to genetic erosion, need multi-species (crops) collecting. In some instances, multiple crop collecting may not be practical.

### (ii) Single Crop Specific Exploration

Crop specific exploration include only specific crop with the mission to assemble as many genetic diverse types as possible from its genepool. In the year 2010-12, ICAR-NBPGR, New Delhi has taken the lead to collect the specific crop (*i.e.,* Brinjal, Chillies, *Trichosanthes etc.*) base exploration and collection. Species specific (or genepool) mission is relatively with high reliability and less complicated. At the same time its eco-geographic distribution should be known in more detail. The samples should preferably be representative of the local diversity that exists in a given area.

### (iii) Target Specific Exploration

Target specific exploration is organized by crop base institute with a well-defined purpose for instance searching for genetic variability against prevalent diseases or pests of a crop. Similar explorations conducted with objectives of collecting germplasm have resistance to environmental stresses or having some better quality traits are target specific exploration. This warrants the involvement of breeders, crop physiologists and biotechnologists in collection of trait-specific germplasm for maximum likelihood through the use of advanced techniques and tools. Populations in marginal areas or distinctive isolated habitats have a higher likelihood of representing unique traits (Bothmer and Seberg, 1995). As consideration, these are needed in tomato for long shelf life, biotic and abiotic resistance, and in onion for higher TSS, in garlic for bulb and clove size, in chillies for higher pungency lines *etc.*

### (iv) Area Specific Exploration

Target areas to be identified after studying the ecogeographic and pest information. For abiotic stress tolerance, it is preferable to collect from areas where species has been exposed to stress factor for a considerable time. Such explorations are based on the information of crop genetic diversity available in a region. For validating area specific exploration, Vavilov (1926) designated certain areas as centre of origin of cultivated crop plants on the basis of genetic diversity found in those areas for cultivated crops. An area specific exploration depends usually on two facts:

(a) Agro climatic specific and

(b) Species specific

Gautam *et al.* (2001) and Ram and Srivastava (1999) surveys certain areas and on the basis of adaptation, suggested that each and every niche has specific climatic conditions which promote flowering, fruiting and perpetuation of specific crop. For example explorations for cucumber in Indo-Gangetic plains, Sub-Himalayan tract, Western Ghats and Eastern Peninsular Region, for *Cucumis* sp. in Jammu and Kashmir, Himachal Pradesh, Uttar Pradesh, Rajasthan, Andhra Pradesh, and

### Table 4.4: Priority Crops/Specific Areas for Collection

| Sl.No. | Crop | Genera and Species | Type of Collection | Region to be Explored |
|---|---|---|---|---|
| 1. | Cucumber | *Cucumis sativus* | Local landraces | Indo-Gengetic plains, Sub-himalayan tract, Western Ghats, Eastern peninsular tract |
| 2. | | *Cucumis hardwickii* | - | Western Himalayan foot-hills |
| 3. | | *Cucumis propheterum* | - | North West plains |
| 4. | | *Cucumis setosus* | - | Eastern India and upper Gangetic plains |
| 5. | Muskmelon | *Cucumis melo* | Local landraces | Uttar Pradesh, Rajasthan, Andhra Pradesh, Karnataka, Jammu and Kashmir |
| 6. | Other *Cucumis* species | *Cucumis* | Local landraces | Uttar Pradesh, Rajasthan, Andhra Pradesh, Karnataka, Jammu and Kashmir |
| 7. | Pointed gourd | *Trichosanthes dioica* | Local landraces | Bihar, Bengal (plains), Asom (valley), Uttar Pradesh (Gangetic Plains), Odisha |
| 8. | | *Trichosanthes bracteata* | Wild | Himalayan ranges, Eastern India, Andaman and Nicobar Islands |
| 9. | Ivy gourd | *Coccinia* species | Local landraces | Eastern M.P., West Bengal, Uttar Pradesh, Bihar and NEH region |
| 10. | *Luffa* | *Luffa* species | Wild and cultivated | Indo-Gengetic plains, *Tarai* region, north eastern plains |
| 11. | Bottle gourd | *Lagenaria siceraria* | Wild and cultivated | Indo-Gengetic plains, *Tarai* region, north eastern plains |
| 12. | Snake gourd | *Trichosanthes cucumerina* | Local landraces | Southern peninsular tract, and Kerala |
| 13. | Bitter gourd | *Momordica* species | Wild and cultivated | Uttar Pradesh, Kerala, Karnataka, Andhra Pradesh, Bihar and Jharkhand |
| 14. | Sweet gourd | *M. cochinchinensis* | Local landraces | Penisular region, West Bengal and NEH region |
| 15. | Spine gourd | *M. dioica* | Local landraces | Central peninsula tract, Andhra Pradesh, Telangana |
| 16. | Pumpkin | *Cucurbita* species | Wild and cultivated | NEH region |
| 17. | Ash gourd | *Benincasa hispida* | Wild and cultivated | NEH region |
| 18. | Cho-Cho | *Sechium edule* | Local landraces | NEH region |
| 19. | Lablab bean | *Lablab purpureus* | Local landraces | Eastern peninsular region, Odisha, Bihar, Gujarat, Maharashtra, Goa and Southward |
| 20. | Sweet potato | *Ipomoea batatas* | Local landraces | Eastern peninsular region, Odisha, Bihar, Gujarat, Maharashtra, Goa and Southward |
| 21. | Colocasia | *Colocasia* | Local landraces | Kerala, Andhra Pradesh, Bihar, Uttar Pradesh, West Bengal and NEH region, Nilgiri and Annamalai hills |
| 22. | Dioscorea | *Dioscorea* | Local landraces | Andaman and Nicobar Islands, Bihar and NEH region |
| 23. | Chenopod | *Chenopodium* | Wild and Local landraces | Upper Gangetic Plains extending to northern hills, Himachal Pradesh, Haryana, Punjab, Jammu and Kashmir |
| 24. | Tumba | *Citrullus colocynthis* | Wild and Local landraces | North West plains, Rajasthan and Gujarat |
| 25. | Elephant Foot Yam | *Amorphophallus* species | - | Khasi Hills, Sikim, West Bengal |

Karnataka and for pointed gourd in Bihar, Bengal and Asom. North Eastern Region has scope for exploration of variability in edible banana and bamboo shoots. Ram *et al.* (1999), found specific variability in specific area for certain vegetable crops in Eastern Uttar Pradesh. For spine gourd, exploration can be done in Bundelkhand and adjoining areas of Uttar Pradesh (Rathi *et al.,* 2006).

Similarly, exploration of vegetable germplasm by BCKV, Kalyani (West Bengal) are given in Table 4.5.

## Who is Responsible for Vegetable Genetic Resource Collection in India?

ICAR-National Bureau of Plant Genetic Resources (NBPGR), New Delhi is the mandatory National Active Germplasm Site (NAGS) for vegetable germplasm collection (base collection) in the Indian National Plant Genetic Resources System (IN-PGRS). Vegetable genetic resource collection can also be done by designated NAGS centres of ICAR-NBPGR. Initially, IPGRI, Rome has assigned crop responsibilities for 5 vegetables *viz.* okra, eggplant, lablab bean (all global), chillies and radish (Asian) to ICAR-NBPGR, New Delhi.

## What to be Collected?

In vegetable crops collection may be done for seed grown crops, vegetatively propagated crops and fruit trees which are used as vegetables. As matters of facts, a good number of vegetables are of Indian/Asian origins are available in nature and a good amount of diversity exists in various parts of the country. These includes eggplant, lablab bean, cluster bean, radish, few cucurbits (cucumber, roundmelon, sponge gourd, ridge gourd, snake gourd, pointed gourd, muskmelon, snapmelon, longmelon and Indian squash), leafy vegetables (Indian spinach and climbing spinach) and tuber vegetables

(*Colocasia* and Elephant foot yam). It is true that some of the vegetables are introduced and have acclimatized/adapted throughout the country leading to good amount of variability *viz.,* cauliflower, cabbage, French bean, tomato, pumpkin, ash gourd, bottle gourd, bitter gourd, garlic and potato. However, these are being constantly threatened due to high yielding varieties/hybrids and also due to shrinkage of natural habitat.

### (i) Collection of Seed

In most of the cases seeds are collected. In a given eco-geographic region, it is desired to sample as many as individual plants as possible at a particular site rather than the frequent sampling from different populations. Also, seeds from several fruits should be gathered from each plant sampled to increase the diversity of the genes in the sample.

### (a) Field Collection

Field collection may be cultivated or wild materials. The overall sampling strategy depends on the kind of species and especially its breeding behavior and gene flow between populations *etc.* however, this is often not known in advance. Therefore, one should try to cover the whole region by taking random population samples at wide intervals. The size of these intervals depends on environmental diversity. Thus, if an area seems to be fairly uniform in climate, soil type, vegetation, farming practices, crop cultivars and altitude the intervals can be quite large. However, if these factors are changing quickly, then frequently sampling should be made. More intensive sampling in specific areas depending on the records of the evaluation of material collected previously. This two stage collecting is essential for searching of defined genotypes *e.g.* drought or

**Table 4.5: Exploration and Collection of Vegetable Germplasm by BCKV, Kalyani (West Bengal)**

| Sl.No. | Crop | Place of Exploration | No. of Accession Collected | Programme Undertaken |
|---|---|---|---|---|
| 1. | Brinjal | Amdanga, Bongaon, Barasat, 24 Parganas (N), Ranaghat-I and II, Nadia, Jangipur, Murshidabad, Dhandinguri, Patlakhau, Cooch Behar | 40 | BCKV, UBKV and ICAR-NBPGR, New Delhi |
| 2. | Chilli | Basanti, Nimpith, Kakdwip, 24 Parganas (S), Kusmundi, Dianjpur (S), Beldanga, Jiaganj, Jangipur, Murshidabad, Balarampur, Pundibari, Haldibari, Cooch Behar, Dhupgiri. Moinaguri, Jalpaiguri | 60 | BCKV, UBKV and ICAR-NBPGR, New Delhi |
| 3. | Pointed gourd | Katwa, Burdwan, Manikchak, Malda, Balurghat, Jangipur, Murshidabad, Khagrabari and Dinhata, Cooch Behar, Kakdwip, 24 Parganas (S) | 21 | IIVR, BCKV and UBKV |
| 4. | Spine gourd | Ranaghat-I and II, Nadia, Chakdah, Aismali, Nadia, Asom, Tripura | 11 | BCKV, Kalyani |
| 5. | Bottle gourd | Gadamara, 24 Parganas (N), Sheorafully, Hooghly, Amdanga, Chakdah, Nadia, Baharampur, Murshidabad, | 15 | BCKV and ICAR-NBPGR, New Delhi |
| 6. | Pumpkin | Gadamara, 24 Parganas (N), Baidyabati, Sheorafully, Hooghly | 24 | BCKV and ICAR-NBPGR |
| 7. | Dolichos bean | Gadamara, 24 Parganas (N), Santiniketan, Birbhum, Haripal, Hooghly, Garbeta, Midnapur, Sitalkuchi, Cooch Behar | 15 | BCKV and UBKV |
| 8. | Okra | Sheorafully, Tarakeswar, Hooghly, Kanthi, Midnapur, Chakdah, Nadia, Basirhat and Amdanga, 24 Parganas (N), | 18 | BCKV, Kalyani |
| 9. | Cucumber | Balagarh, Hooghly, Chakdah, Nadia, Amdanga, Barasat, 24 Parganas (N), | 8 | BCKV, Kalyani |

*Source*: Chattopadhyay, 2011.

disease resistance *etc.* Similarly, special sampling of disjunct populations, peripheral populations and those occupying geographical remote and often peculiar or distinct ecological niches. Where ever possible the wild populations should be sampled at least twice in different years, because climate changes from year to year may change the frequency of certain biotypes in the population. An explorer has to be careful in collecting from the field having more than one variety or its wild forms/progenitor/crossable wild relatives occurring adjacently. It requires a strategic approach by the collector in ensuring proper sampling of the germplasm.

### (b) Collection from Farmer's Stores, Markets, Shops etc.

It may not be always be possible to take field samples over a whole region adequately, even using a coarse grid through lack of time. In any case it is often advisable to investigate and collect from farmers store bins, from local shops, and markets, or even the good offices of the local officials. Much useful information may be gathered in this way, the collections may be made more easily. In region oriented multiple crop expeditions, the farmers stores are very important. When sampling from shops and markets is being done it is essential to ascertain how much mixing has occurred especially if the seed is being sold for consumption rather than as seed for sowing. In general, many market seed stalls offer mixtures matched to consumer demand, *e.g.* grain legumes with a particular cotyledon colour when split, and this may represent selected seed lots with mixtures of genotypes different from those when sown by the farmer.

In the above collection programme random sampling is strongly recommended. This is often spoken of as non-selective sampling. Sampling error is minimized when a large sample is taken but if the variation in each population is high large samples may be needed. For a wild species this must be determined in relation to the variation observed in the population, the size of the colony and environmental factors. Some collectors advice that widespread populations whether cultivated or wild should be divided into sub-populations, each of which will then require a different collection number and set of locally and habitat records. In some cases these data may require for individual plants, as with very sparsely occurring wild species,. However, other advice against this, since then the natural plant community with its characteristics feature is lost.

### (ii) Collection of Vegetative Materials

There are several vegetable crops which are propagated through their vegetative parts like pointed gourd (stem cuttings and sprouted roots), ivy gourd (stem cuttings), sweet gourd (tuberlets), spine gourd (tubers) *etc.* Many root and tuber crops such as potatoes and particularly the related wild species reproduced by seed as well as vegetatively, again for seed collection techniques in these crops may be followed. Some of the difficulties in collecting root crops are listed below:

(a) They are slower to collect because they need to be dug up.

(b) They need to be collected at right stage of maturity.

(c) It is bulky and perishable in nature.

(d) Collections are difficult to store.

(e) The clonal population in wild species may be widespread and consequently attempts at random sampling may merely result in collecting identified clonal material rather than populations.

### Collection of Wild Materials

Wild relatives are an invaluable source for vegetable improvement to tackle both biotic and abiotic stress. Since ancient times they have served as the source for crop domestication and improvement. Wild species as exist as populations in the normal way, but each genotype may be propagated itself vegetatively over quite a large area.

(i) Taxonomically unique, rare and narrowly endemic species deserve high priority in collection programme.

(ii) In case of cosmopolitan/polymorphic species, there is more probability of getting variants. Often they have sub-species/forms/ecotypes, hybrid swarms and crossable wild relatives in the nearby vicinity, which may possess the target trait.

(iii) Need repeat explorations for marking herbarium specimens/collecting germplasm, *etc.*, due to differences in maturity/phenology. Moreover, it may not be always possible to collect both crop and related wild species at the same time.

Careful observations should be made in the target area of apparently similar phenotype, and sample should not be made too close together. The following strategies are to be adopted:

(a) Collect bulk sample of 10-15 individuals from each propagule

(b) The area of target collecting site for this population sample would be about 100 x 100 m or less if population is smaller.

(c) Avoid duplicates.

(d) Sample as many sites as possible in reference to collect a large number of individuals from few sites. Choose sampling over as broad an environmental range as possible.

(e) Make voucher sample.

Today, crop wild relatives that are threatened in the wild and are only partially conserved in gene banks have been rediscovered as an essential resource for crop improvement programs aiming to make major crops more resilient to climate change.

### Collection of Cultivated Materials

(a) Try to complete seeds or fruits (*Sechium edule*) where ever possible. If not, take bud wood cuttings, suckers (curry leaf) *etc*.

(b) If material is grown from seeds, treat the whole village as collecting site and make a random population sample from 10-15 individuals.

(c) If it is clonally propagated from selected variety sample distinct variety.

(d) Sample as many sites as possible

(e) Take care of collected seeds, bud wood related to their viability

These are vegetatively propagated clones. They are not populations, but small parts of what were once populations that have been very strongly selected by the farmers. They must be sampled selectively (non-randomly) in complete contrast to seed collections.

The agreed methods for these are as follows:

**(a)** Collect distinct varieties from market or village

**(b)** Repeat this process at 10-50 km intervals over the area.

**(c)** Collect a complete range of morphotypes at every collecting site, no matter whether they seem to be same as those collected previously.

**(d)** Duplicates can be eliminated during characterization.

**(e)** Supplement with seed collection where possible and give the same accession number.

### (iii) Collection of Fruit Crops Utilized as Vegetables

Similarly there are many tree plants like: drumstick, curry leaf, plantain, jackfruit, bread fruit, tree bean *etc.* are useful as vegetables. Collecting such beneficial fruit trees is complicated because:

(a) Some trees possess seeds known as recalcitrant and storage life is small. Hence, seeds should be sown immediately after collection.

(b) Collection of woody cutting and bud-wood is generally preferred but handling is difficult particularly bud sticks in case of drumsticks.

### Collection of Wild Relatives

Wild relatives in vegetable crops assume great importance in crop improvement as a source of disease and pest resistance, stress tolerance *etc*. It provides basic information on species relationship and clue to crop evolution pattern. *Cucumis sativus* (cucumber), closely related to *C. hardwickii* (now in rare occurrence)

has resistance to cucumber green mottle mosaic virus, downy/powdery mildew; *C. hystrix* (resistant to root knot nematode, powdery mildew); *C. callosus, C. melo* var. *momordica*, *C. melo* var. *agrestis* (all drought tolerant)

(a) Collect seeds from up to 10-15 individuals in some 10 ha and put all together as single sample.

(b) Takes as many seeds as possible per sample.

(c) Repeat at intervals, depending on climate, altitude or soil differences.

(d) Do not sample more than one tree in each clump or group.

(e) Make arrangement for quick dispatch of material.

## Steps in Germplasm Exploration and Collection

There are various steps in collection of vegetable genetic resources like planning, making contacts with local research organization, gathering equipment and preparation, meeting with local researchers/government, sorting out of collected samples, reporting to the Headquarters, preparation and publication of reports and delivering/distributing collected samples. In general, some important steps are involved in executing systematic collection missions are given below:

☆ Planning

☆ Gathering required information and preparation

☆ Making contacts with local research organizations

☆ Meeting with local researchers and other officials including NGSs/NGOs

☆ Collecting the germplasm and herbarium specimens

☆ Recording passport data and relevant information

☆ Sorting out and packing collected germplasm and its identification, if required

☆ Reporting and preparation of report and

☆ Processing of collected germplasm.

Some details are as under:

### (1) Survey and Exploration

In vegetables a large group of crops are annuals and propagated by seeds, while some are perennials and propagation is done by vegetative means. The principles for both groups of crops differ to some extent. Initially the survey is conducted for a particular crop in their region of diversity known as coarse grid survey (conducted in unexplored areas to capture the overall variability) and further intensive survey is undertaken from the source locality for a particular genotype having desirable gene, which is known as fine grid survey (build-up collections for specific trait(s) known to exist in identified pockets in previously explored areas). In coarse grid survey a series of collection is made at a wide interval over the whole area. Further the collection from the area, where interesting and intensive variation appear, the second expedition and intensive survey would then provide fine grid sampling by concentrating upon the gene pool in this region. Hawkes (1976) is of the view that the above sampling strategy may be attempted wherever possible depending upon the availability of time and resources.

### Pre-requisite for Collecting

Before proceeding on expedition for gene pool sampling in vegetable crops a curator should have sound knowledge and understanding about the following:

1. Knowledge of area
2. Knowledge of people (cultural communities, ethnic groups)
3. Socio-religious customs
4. Eco-edaphic conditions
5. The vegetable crop grown including their primitive/landraces/local types
6. Their wild relatives
7. Distribution of diversity
8. Keen observation of the variation in plants and environment
9. Biotic and abiotic stress and their identifying key characteristics
10. Understanding of the gene pool concept
11. Breeding system and population structure
12. Coarse and fine grid sampling
13. Propagation/seed multiplication techniques

Information which would guide the collection includes.

### (2) Exploration Planning

Germplasm collecting should be carried out judiciously as it entails high expenditure, labour and meticulous planning. Study of floras, visit to herbaria and study of online databases are recommended to acquire information on the range of distribution, locality of availability, diversity pattern and period of collecting, particularly for crop wild relatives. Broadly two kinds of explorations are formulated based on:

☆ The priority of crops, and
☆ The area/region

Both, crop specific and region specific/multi crop collection are advocated depending on the situation. If the collecting is planned for crops where already a substantial collection exists with the collecting organization, a gap analysis should be carried out based on information from database of genebanks (*e.g.* National Genebank of ICAR-NBPGR), catalogues and other publications. The crop specific explorations are undertaken as per the needs of the breeders. However the multi crop specific explorations are undertaken to cover probity areas, which hold rich crop genetic diversity in various vegetable crops, and maturity period, coincided at the same time. The exploration is carried out by explorer, crop specialist/researcher, extension worker, and by botanist. Collecting of crop wild relatives (CWR), threatened, medicinal and endemic species may entail collecting missions in difficult-to-access terrains. The duration of exploration vary according to the mission; within the country it is about 20 to 45 days whereas in the foreign country larger duration is required, as otherwise it will be more expensive and less remunerative. In all expedition planning to other countries it is of the greatest importance to enlist the enthusiastic support of the host country and to ensure that government officials, agricultural scientists and extension officers are fully aware of the importance of the exploration, conservation and efficient utilization of their national plant genetic resources. While planning, it would be appropriate to give a thought on immediate inclusion or exclusion of species for their collection in case where:

☆ Difficult to establish, multiply, characterize/ evaluate and conserve;

☆ Species with recent history of cultivation using introduced germplasm in a particular area

☆ Species propagated exclusively through vegetative means or where germplasm needs to be collected in the form of bud-woods from remote inaccessible areas with difficulty in transport/shipment/packing of grmplasm;

☆ Areas already explored or species collected by other organizations and the germplasm readily available for use;

☆ Species represented by single/a few plants in given area.

### (3) Selection of Site

Planning of route is essential prior to go to expedition. The tentative route map may be prepared. Collection 'site' indicates the place of sample collection. This includes cultivated fields, natural habitats (mountains, valley, river sites, sea shore, forest *etc.*), local village market, threshing floors and farmers seed store *etc.* In case of annual vegetable crops, the collection site is individual field or farm (if farmers use different seed stock) or a group of field or farms (if farmers use common seed stock). Further the collection site may be reduced if crop is self-pollinated and increased if the crop is cross-pollinated and showing high genetic diversity. In case of wild and weedy or weedy forms relatives the selection of appropriate sampling site to collect diversity is more difficult. First, such species are often continuously distributed in nature or they follow the island model or the stepping stone model (Jain, 1975). In such cases, there are no artificial boundaries defining discrete, relatively homogenous populations as in cultivated crops. Thus Marshal and Brown (1975) suggested that the plant explorer must decide where, in the continuum he/she will collect and the total size of the target area to be sampled (*e.g.* 1000, 10,000 or 100,00 m²).

Gene flow through pollen and seed dispersal is extremely limited in wild vegetable species. Natural population of wild species often shows marked genetic differences over small distances (Marshal and Brown, 1975). Thus, two points need careful attention as outlined below (Koppar, 1998):

(i) The most important population factories the size of the breeding unit which is a function of the number and density of plant per site, the mating system, and level of pollen and seed dispersal.

(ii) The most important ecological factor is the degree of environmental heterogeneity for such variables as soil types, aspects, slope, moisture, regime and associated vegetation.

### (4) Area of Collection

The knowledge of the crop diversity in a particular area is also essential before undertaking the exploration trip. If the collecting mission is targeted to unexplored areas, prior information about the flora and genetic diversity prevailing in the region should be gathered. The area to be covered by explorations is usually the centres of origin of the concerned crops. In addition, collections should also be made from the peripheral regions of the crop distribution, and even in areas where it was introduced in comparatively recent times. It may be expected that these areas the crops, would be exposed to environmental stresses and special mutations may have been selected for. Tribal dominated region inhabited by different ethnic group provide another interesting situation for gene pool sampling of primitive types of vegetable germplasm. Such areas often possess rich diversity of various vegetable crops. The tribal hold primitive cultivars of different vegetable crops and usually do not exchange seeds between the ethnic groups. In tribal area particularly in Northeastern parts of Madhya Pradesh, Chhattisgarh, West Bengal, Bihar and Odisha, village market/haat are most desirable place to collect diversity in a number of vegetable crops. The areas that remain to be covered include remote and less accessible areas, hilly terrain and tribal dominated belt. The most effective area of exploration should be determined on the basis of knowledge accumulated so far and that obtained in the previous surveys. Information on environment of the target area must be synthesized with information on target taxa. This would include data on distribution, phenology, genetic diversity, reproductive biology and ethno-botany. Information on diseases/pests and cultural practices are also needed. Although primary concern of the explorers, is the collection of the variants available, but they should also know the social circumstances in the area. These skills can be developed through contact or correspondence with the local persons or botanists. Government permission is often required to enter in certain areas for which prior permission or intimation may be developed before undertaking the exploration of the area. The ICAR-NBPGR Regional Station Jodhpur (Rajasthan), established in 1965, has collected and conserved the germplasm of arid legumes namely, moth bean (1839 accessions), cowpea (2166 accessions), mungbean (2911 accessions), and cluster bean (7753 accessions) in its genebank. The germplasm includes land races, obsolete varieties, wild forms and wild relatives, breeding lines, varieties in cultivation and genetic stocks and has primarily been collected from the hot arid and semi-arid regions of Rajasthan, Gujarat, Haryana and Maharashtra provinces of India (Singh and Singh, 2016).

### (5) Season/Time of Collection

The maturing stage of cultivated crops germplasm is the most favorable season for collecting seeds and for the checking characteristics of the crop. The farmers are working in the field, one can get some important and useful information about crops and cultivation practices. Since the seeds of wild species are easily shattered after maturity, collection should be done before the maturing stage of these plants. For vegetative propagules, collecting time may be quite different. Then the decision will depend on:

(i) Breeding system of the species: Autogamous species can be collected as seeds.

(ii) Whether a population structure exists: This will not be the case for traditional varieties clonally propagated for centuries but may still obtain for sexually fertile, out-crossing wild species.

(iii) Each seed is potentially a new variety whereas each vegetative sample will represent on establishment variety. Collecting storage organ which are immature or which are already sprouting may result in significant loss of viability during transport. Information is needed at the planning stage in deciding on the best time to collect.

Correct timing will make possible the followings (Koppar and Rai, 1994):

(a) Largest amount of genetic diversity

(b) To be in areas at the right time

(c) To look for weedy forms around the field boarders.

(d) To look for related wild species.

### (6) Sampling Sites (Route maps) and Number

The planning of the route requires detailed maps, such as those on the scale of 1 to 5, 00,000. It is essential to have the maps beforehand. It is difficult to obtain detailed map of the foreign countries. Therefore it is essential to obtain detailed maps before entering the local area for collection. The location of sampling sites within the collection area should be decided with care and in advance. This decision should be based on the changes in ecological, agricultural and social conditions in the area, and also on soil patterns and changes in agricultural practices. If there is considerable variation for these factors, the sampling sites should be closer; otherwise they should be relatively farther apart. Generally, sites for cultivated materials would be scattered over the entire area, while those for wild species would occur in clusters. A preliminary survey of the area may be done before the actual collection. Number of sites depends on the availability of variability and pollination nature (self and cross pollinated) of crop. However, each site provides opportunity for sampling different set of alleles, if no information on the distribution of variation is available. The number of sites often depends on the length of the season, relative abundance of the target species and roughness of the terrain *etc.* Changes in echo- habitats/niches and farming system should also be taken into consideration for deciding number of sites. The optimum number of sites for sampling for a given species could be fifty. The presence of indicator species may help in pin-pointing the sites for collecting germplasm for abiotic stress tolerant materials (Bhandari and Pradheep, 2010).

☆ For cultivated species, sampling sites in order of preference should be farmer's field, backyard/kitchen garden, threshing yard, farm store, local village market, *etc.* Inaccessible areas of valleys, isolated hills, villages at the edge of deserts, forests, mountains and isolated coastal belts may hold rich genetic diversity, potential/trait-specific germplasm and crop wild relative (CWR).

☆ Collecting mission should be started first from drier tracts (*vs.* humid), un-irrigated areas (*vs.* irrigated), valleys (*vs.* hills) to capture maximum available diversity, since crops grown in those areas mature first.

☆ The crop often varies depending on the ethnic diversity and different array of material may be collected even from contiguous belts occupied by different tribes.

☆ Sites having abiotic stress situations *viz.* saline habitat, un-irrigated/drought conditions, desert (cold and hot), flood-prone areas should be target areas for sampling of promising genotypes.

☆ To collect biotic stress-tolerant material, hot-spots of pest incidence should be visited to collect healthy plants in field where severe pest damage is evident.

The frequency of sampling (number of samples per site) should be decided based on variability observed on-the-spot. In general, more sites per target area are preferred to sample the targeted species rather than sampling from a few sites.

## (7) Optimum Sample Size

The main objective of exploration and collection is to assemble maximum genetic diversity/variation by selecting limited number of plants/seeds from a population or several population. It is true that diversity within populations upon the number and frequency of all alleles across all the loci as well as the population structure. It indicates that sample size required for accumulation of gene depends on genetic variation present in the population(s). The sampling theory propounded for germplasm collection and regeneration for conservation is almost four decades old. In addition to a population sample, a selective sample can be made if the expression of the targeted trait is obvious.

### Determining Factors for Sample Size

There are following factors which determine the theoretical sample size:

(i) Genetic diversity and sample size

(ii) Aim of exploration

(iii) Population structure/distribution

    (a) Alleles and allelic frequencies

    (b) Number of loci

    (c) Degree of association/mating system

    (d) Ploidy level

    (e) Degree of certainty

(iv) Cost considerations/allocation of resources or efforts

### (i) Genetic Diversity and Sample Size

Genetic diversity is a crucial factor in deciding the sample size. It is possible only when diversity is measure in a particular location/region. Marshall and Brown (1975) suggested that the most important measure of genetic diversity is the average number of alleles per locus and the measure of genetic diversity (variance) based on quantitative characters may be unreliable indicators of diversity in a population at the level of individual gene. They defined the optimum sample size per site as the number of plants required to obtain, with 95 per cent certainty, all the alleles at a random locus occurring in the target population with frequency greater than 0.05. They also proposed a diallelic single locus diploid model and estimated 59 random unrelated gametes or 30 random genotypes/plants for out breeding sexual or apomictic species and 59 random individuals for self-fertilizing species.

### (ii) Aim of Exploration

The main purpose of the exploration may be to collect wild species and primitive cultivars. These are, in general, not particular well represented in some of the larger collections. Sample size may vary depending upon as to whether the team is interested in collecting the allelic diversity, genotypic diversity or the phenotypic diversity. The criteria based on capturing allelic diversity has been found to be the optimum requiring low number of individuals to be sampled for a given probability of conservation and allelic frequencies. Again sample size governed by the criteria of inclusion of these alleles *i.e.*

☆ Capturing at least a single or multiple copies of rare allele(s) or capturing at least a single or multiple copies of all the alleles,

☆ Representative sample

In case of genetic conservation, the property of allelic richness is considerably more important than the property of allelic evenness (Marshall and Brown, 1975). They clearly defined the aim of exploration as *"The collection of at least one copy of each variant occurring in the target populations with frequency greater than 0.05."*

They are likely to be greater interest to plant breeders than rare variants. Rare variants represent newly arising mutants or recombination, or deleterious genes or genes combination maintained in the population by a balance between mutation, migration or recombination and selection.

### (iii) Population Structure/Distribution

Allard (1970) first discussed population structure and sampling methods and identified clearly the critical problem facing the plant explorers and proposed a sample size of 200 plants/10 seeds for wild oat population. He stressed that most plant species contain remarkable stores of genetic variation and consist of millions of different genotypes and a plant explorer can hope to sample only a fraction of the variation that occurs in nature. Sampling strategy depends upon the kind of species and especially its breeding system, amount of gene flow between populations *etc.* (Hawkes, 1980). Thus in a particular species, the sample size will primarily depend upon the number of alleles, their frequencies, number of loci and ploidy level as well as the mating behavior *etc.* Lawrence *et al.* (1995) suggested a sample of about 172 plants for conserving all or very nearly all of the polymorphic genes with high probability provided their frequency is not less than 0.05, irrespective of whether the individuals of the species set all of their seed by self or by cross fertilization or a mixture of both.

### (a) Alleles and Allelic Frequencies

Both the number of alleles and allelic frequency affect the sample size. It is the rare allele and not the common allele which affects the sample size severely. Sample size increases at an exponential rate as the allele is progressively rarer. The corresponding sample sizes for rare allele frequencies 0.05, 0.03, 0.01 and 0.001 in a diploid with two alleles and single locus population are 59, 99, 299 and 2999. How many alleles the sample will include? The expected number of selectivity neutral alleles (k) in a sample of S gametes, from equilibrium population of size N based on the neutral allele theory of

Kimura and Crow (1964) and later developed by Brown and Brigs (1991) are approximately:

$$k = \theta \log_e [S+\theta)/\theta] +0.6$$

Where $\theta=4N\mu > 0.1$ and $D > 10$ and $\mu$ is the mutation rate.

This formula shows that the number of alleles in a sample increases in proportion to the logarithm of sample size.

Gregorius (1980) calculated the probability of losing an allele while sampling for genotypes and suggested that when allelic frequency decrease from 0.05 to 0.03 or from 0.02 to 0.01, the required sample size approximately doubles.

### (b) Number of Loci

The sample size increases with the number of loci. Crossa *et al.* (1993) while discussing optimal sample size for regeneration, suggested sample sizes of 160-210 plants for capturing alleles at frequencies of 0.05 or higher in each of 150 loci, with 90-95 per cent probability. When allele frequencies are unknown, an equation for estimating an optimal sample size for capturing all the alleles, (k-1) rare alleles having an identical frequency of $p_o$ and 'kth' allele occurring at a frequency of $1-(k-1)$ $p_o$ at a number of independent loci was also developed by the said authors. Assuming that k-1 alleles occur at an identical low frequency of $p_o$ and that the 'kth' allele occur at a frequency of $1-(k-1)p_o$, at all the 1 independent loci with the same number of alleles, Crossa *et al.* (1993) determined the number of individuals required to capture at least a copy of all the alleles as:

$$n> \{\log[1-(P)^{1/\lambda}] - \log(k-1)\}/\log (1-p_o)$$

Where, P is the probability of conservation or degree of certainty.

Thus, for a diallelic population with a rare allele having a frequency ($p_o$) of 5 per cent, the approximate sample size is $-20\log [1-(P)^{1/dikudd}]$. If 15 alleles are considered at all the loci, then the sample size becomes almost double when the number of loci, increase from 1 to 150 for capturing at least a copy of all the alleles at all the loci with a 90 per cent probability of conservation. Qualset (1975) suggested a sample size as large as 500 plants based on the probability model that at least one plant of very rare genotype (<0.01) is included in the sample. On a sample size of 90 individuals was suggested by Namkoong (1988) for an average loss of one allele at a frequency of 0.05 at any of the 100 loci.

### (c) Degree of Association/Mating System

The sample size also depends upon the degree of association between genes within individuals (Crossa *et al.*, 1993). If there is a perfect association between genes within individuals at any locus, then the required sample size is exactly double than that of a population when there is no association. Thus, the sample size is almost double under exclusive selfing and it decreases on deviation from selfing and almost becomes half when the individuals mate at random.

### (d) Ploidy Level

Polysomic inheritance is more conservative of genetic variability (Bray, 1983).Number of individuals to be sampled from a tetraploid population are almost half as compared to that of sampling from diploid individuals. However, under exclusive selfing, the sample size is the same irrespective of the ploidy level when other parameters held constant (Sapra *et al.*, 1997). They also proposed a 2k-polyploidy model for multi-allelic and multiplied situation to tally with those already by some of the earlier workers. The model also establishes the conservative characteristics of genetic variability of polysomic inheritance under chromosomal segregation.

### (e) Degree of Certainty

The higher the degree of certainty, the more is the sample size. The number of individuals to be sampled at 95 per cent certainty for capturing two alleles at a single locus and the rare allele having a frequency of 5 per cent is 59 and becomes 180 when 99.00 per cent certainty is required. Thus, there is a drastic increase in sample size with marginal gains. In most of the practical situations, 95 per cent certainty is optimum.

### (iv) Cost Considerations/Allocation of Resources or Efforts

Ultimate aim of exploration is to collect maximum diversity with a given amount of resources. Since the resources are limited, therefore, explorer should optimize the sampling strategy accordingly. In this situation, collectors have to make a decision whether to visit many sites spending a relatively small amount of effort to individual sites, or in contrast, do an intensive collection for a few sites. The sample size may be depending upon this optimization process.

### (8) Collection Priorities

The species to be collected are determined by the needs of the breeders, and by the level of threat to the concerned species. The need-based priorities will depend on the country in question, while the threat-based priorities have been developed by FAO; emergency situations are indicated by 'E' while lower level priorities are denoted by I, II and III

### (9) Sampling Procedure

Sampling methods differ depending on reproductive behavior, whether; one is to collect annual seed crops, vegetatively propagated crop or tuber crop. The objective of sampling is to capture the maximum amount of genetic diversity with the minimum number and size of samples. Bogyo *et al.* (1980) suggested a phenotypic model and sampling strategy to be based on conscientious selection for visible phenotypic traits

rather than on random sampling, however phenotypic model has been disagreed by many. The collection of plant materials from sampling sites may be:

### (i) Random Sampling

Random sampling is usually practiced to collect maximum genetic variability of species, irrespective of relative frequency or rarity of any genes or linked genetic complexes. In practice, random sampling is usually employed. For this, at each collection site, single spike or few pods/fruits are randomly collected at every second or third place along with a number of transacts through the crop at least at 50 places scattered over an area or sample are taken in the clustered form. Hawkes (1976) suggested that bulked seed samples from up to 50 individuals and certainly not more than 100 should be collected randomly as multiple samples at one site. Sample should be clustered and cluster should be spread as evenly as possible throughout the collecting area. The sampling technique may be appropriate in case of wild and weedy species.

### (ii) Biased Sampling

The random sampling is not feasible in many of the vegetables like eggplant and most of the cucurbits because farmers keep few selected fruits only for seed purpose and many times ripened fruit is not available in the field. Under this situation the explorer stick to biased sampling. Bennett (1970) argued for biased sampling for capturing rare variants. Harlan (1975) strongly pleaded about subjective, biased sampling. For biased sampling, crop specialist category of explorer is desirable because he has to be aware of the morphological and physiological characters of the crop to be sampled. These characters include pattern, rate and habit of growth (Fordham, 1971) and seed source in relation to quality and germination requirement (Flint, 1970). Sampling of specific genotype forms a known source; also form important part in biased sampling.

### (iii) Concept of Coarse and Fine Grid Sampling

Harlan (1975) pointed out that grain and field crops, in general, tend to be somewhat mobile being exchanged from village to village to the point that a rather coarse grid is likely to pick up but the very rare items. Jain (1975) recommends for the collection of few economically important field crops, if possible at least two collecting trips to the same area. In the first, a coarse grid series of collection is made at wide intervals over the whole area. The variation in samples from this is later studied in the field. In collections from areas, where interesting or intensive variation appears, the second expedition would then provide *fine grid* collecting by concentrating the gene pool sampling. Hawkes (1976) is of the view that the above sampling strategy may be attempted wherever possible depending up on the availability of time and resources.

### (10) How Long?

Exploration within the country are of shorter duration, about 3-4 weeks would be deal. When exploration is organized in foreign countries, longer duration is required, as otherwise they are more expensive and less remuneration. If larger ecological range is involved (altitudinal/latitudinal variation), collecting season is prolonged to 3-4 months to cover diverse terrain.

### (11) Field Records

Adequate field records must be maintained during collection. For this purpose, ICAR-NBPGR, New Delhi has developed "Passport Data Book" (Minimum Data Sheet) which can be modified as per need.

### Sampling from Several Populations (Multi Sites)

Model for capturing variability from a target population spread over several sites was first proposed by Oka (1969, 1975) suggested:

☆ Wild species-20 populations (sites) with 5-30 plants

☆ Semi wild species-30-40 populations with 20-40 plants

There were several drawbacks in Oka's model. One of the major discrepancies in Oka's model was that it didn't consider the allocation of sampling resources. The issue concerning sampling from several populations was also discussed by Marshall and Brown (1975) by recognizing four types of alleles:

1. Common, widely distributed
2. Common, locally distributed
3. Rare, widely distributed
4. Rare, locally distributed

The explorer may aim at collecting alleles which are locally common. Marshall and Brown (1975, 1983) recommended that where little or no information is available on the distribution of variation in nature, the optimal strategy is:

(i) To collect 50-100 plants per site

(ii) To sample as many sites as possible within the time available,

(iii) To ensure that sampling sites represent as broad a range of environments as possible.

Brown and Marshall (1995) suggested about 50 individuals from each of the ecogeographical area and up to a total of about 50 populations for the mission.

Theoretical and practical issues proposed by Brown and Marshall (1995)

Brown and Marshall (1995) also discussed in detail, several Theoretical and practical issues related to basic sampling strategy *viz.*:

(i) Measurement of variation

(ii) Expected number of alleles in the sample

(iii) Number and location of sampling sites

(iv) Number of individual plants sampled at a site

(v) Random vs non random sampling

(vi) Number and type of propagules per plant

(vii) Modifications to basic strategy for different species

(viii) Modifications to basic strategy when sampling for specific goals

(ix) Modifications to basic strategy when re-sampling of a region.

In his study Yonezawa *et al.* (1985, 1989) incorporated accomplishment factor (R) in his model when sampling from multi-site target population and pointed out that overall efficiency of the expectation depends upon the number of sites visited rather than the number of plants. The strategy suggested fewer plants from more populations for collection of allelic diversity and about 50 plants for genotypic multiplicity. Lawrence *et al.* (1995) recommended a sample size of 172 divided by number of sites visited. Thus, if, say ten populations/sites are visited, we need only draw a random sample of 17 or 18 plants from each in order to achieve a very high probability of conserving at least one copy of each allele at a large number of loci.

### Sampling of Seeds/Plant

Brown and Marshall (1995) in his basic sampling strategy suggested to sample sufficient seeds or vegetative material per plant to ensure representation of each original plant in all duplicates. Hawkes (1980) in his field collection manual suggested a sample of 100 plants x 50 seeds for a highly heterogenous population and 50 plants x 50 seeds for a fairly homogeneous population. However, he did not give any mathematical justification for the said sample sizes. Yonezawa and Ichihashi (1989) were the first to consider the genotypic multiplicity of seed embryos on a single plant sample. They pointed out that the sizes suggested by Hawkes (1980) are excessive in most of the cases.

On the gene basis Weir (1990) defined the genetic diversity at a single locus as one minus the sum of squares of allelic frequencies.

### Seed Samples

Depending on the plant species and objective of the collecting mission, seed should be collected.

☆ The sample size increases with the increase in diversity level. Hawkes (1976) suggested that bulked seed samples from up to 50 per cent plants (minimum) should be collected from one site. However, this becomes difficult particularly in the crops like brinjal and cucurbits where availability of this much number of fruits is very rare at a site.

☆ In case of cross pollinated vegetable each sample size should be of 10,000 to 12,000 seeds and in case of self-fertilized crop sample size should be of nearly 8000 seeds besides meeting the requirement of characterization, evaluation and related studies. In case of species with extremely small-sized seeds, low seed-set, asynchronous maturity and low seed viability, extra care should be taken to collect sample of adequate size.

☆ Random sampling should be done by collecting single spike/panicle or fruit/berry/pod from at least 50 plants along with a number of transects throughout the field to obtain a representative and adequate sample.

☆ In case of population that appears to be highly variable, larger seed sample (seeds from >50 plants) should be collected through bulking. Non-random (selective) sample should be done in case any distinct variant is observed and that should be given a different collection number.

☆ When a population of wild species with few individuals is encountered at a site, seeds should be collected from multiple plants to capture maximu7m representative samples. In case of certain wild and semi-domesticated species occurring with scattered populations (treated as sampling site), the seed should be bulked. However, it should be ensured that such collecting does not deplete the populations of farmers' planting stocks or wild species, or totally remove significant genetic variation. Limit collecting of seed/propagules/cuttings from wild species so that the survival of population in its natural habitat is assured.

☆ Period of exploration should be of at least 10 working days (excluding journey period) in India and more than a month when organized in foreign countries.

☆ For seed producing crops and species, collecting should be undertaken when seeds are physiologically mature. In case of species that undergo seed shattering at maturity (*e.g.* wild species) mission should be executed earlier (7-10 days before physiological maturity). Further, longer duration mission (2-3 weeks) and repeat visits are suggested for collecting adequate sample size of wild species.

### *Propagules Sample and Period Suitable for Collecting*

Depending on the plant species and objective of the collecting mission, vegetatively propagated material, *in vitro* material and pollen should be collected.

☆ For vegetatively-propagated crops/species, the targeted areas should be surveyed first for identification and marking of elite types at the time of flowering/fruiting and subsequently the collections should be made at the appropriate time.

☆ Period of exploration should be of at least 10 working days (excluding journey period) in India and more than a month when organized in foreign countries.

☆ Sampling and transporting of vegetative propagules should be done with care, making sure that they reach the planting sites at the earliest, as they tend to undergo rapid deterioration after collecting.

☆ For large tubers, only a portion, *e.g.* head or proximal ends in yams, crown or tuber in taro and other aroids (*Amorphophallus*, *Dioscorea*, *etc.*) should be collected.

☆ In case of scion collection for budding and grafting, a minimum of 10 rootstocks per sample should be collected. In case of cuttings and rooted suckers 15-20 cuttings may be sufficient.

### *Pollen Collection*

Pollen should be collected during peak flowering period from flowers with freshly dehisced anthers on bright sunny days, normally between 8-10 a.m.

### *DNA Collection*

Healthy leaves, subjected to proper drying (ensuring infection-free status0 should be collected for DNA extraction (molecular analysis studies).

### *Herbarium Collection*

Herbarium specimens, especially of the distinct types and wild relatives should be collected for proper identification/authentication. Wherever possible, efforts should be made to collect economic products of the collected germplasm, as supportive material.

### ICAR-NBPGR Sampling Model

ICAR-NBPGR recently has developed a general expression based on the Yonezawa and Ichihashi (1989) model for estimating the number of plants and number of seeds per plant required to capture (k-1) number of rare alleles having identical low frequency ($p_o$) at all the $\lambda$ independent loci when the number of seeds are taken as zero the formula gives the same results as reported by Brown and Marshall (1975), Chapman (1983) and Crossa *et al.* (1993). The developed formula estimates the number of plants accurately when numbers of seeds (n) are taken as more than five. However, we get an under estimate of number of plants when n is less than five. The general expression is as follow:

$$m > (A-B)/(nC+D)$$

*where,*

m = number of plants

n = number of seeds/plants

A = log $\{1-(P)^{1/\lambda}\}$

B = log (k-1)

C = log $\{s+(1-s)(1-p_o)\}$

D = log $\{(1- p_o)(1-f) + (1-p_o) f\}$

k = number of alleles

$\lambda$ = number of independent loci

f = s/(2-s); f is the inbreeding coefficient and s is the selfing rate

P = probability of conservation

$p_o$= frequency of rare alleles

## Logistic Preparation before Departure for Exploration and Implementation

☆ The explorers/collectors should be well-versed with the nature and extent of diversity and have information on reproductive biology, mating systems and seed physiology of the target crop/species to be collected. Collectors should also have information on previous collecting missions carried out in the region where the current mission is being planned.

☆ Collaborator(s), if required, should be identified and contacted. A ream consisting of two or three members should be formed preferably with a botanist/breeder as the team leader.

☆ Planning should be done well in advance to facilitate preparations for the proposed missions except those to be carried out under special situations like rescue collecting.

☆ Information on topography, climatic conditions and vegetation needs to be gathered to finalize the itinerary of collecting mission. Knowledge about population and reproductive biology, phenology, taxonomy and ethnobotany should be acquired in advance. As far as possible, explorers should established local contacts especially at grass-root level to seek the social, cultural, ethnic and other required information of use.

☆ Prior permission should be obtained from the concerned authorities for collecting in protected (biosphere reserves, sanctuaries, national parks) and restricted areas (border areas, Ladakh, some states in North Eastern Hill Region).

☆ Information on major pests in area of collecting should be gathered, so that necessary precautions are taken for pest-free germplasm collecting and transportation. Phytosanitary regulations should be followed in case of transportation of material from foreign country.

☆ The explorers/collectors should be suitably equipped for each exploration and collecting mission.

Before departure for exploration and collection, it essential to make sure on the following points:

### (a) Making Contact with Administrative and Research Organization

Before departure for exploration and collection curator/explorer should list out the local agencies (NGOs) and contact with appropriate official in the areas of exploration/survey. It is also important to established administrative (State Agricultural Departments, Block Development Offices and their branches) and scientific contacts (regional/National Institutes, Central Agricultural University, State Agricultural University, Krishi Vigyan Kendra National Training Organizations *etc.*). The responsible organization or institute/laboratory in the target area and the possibility of an exploration will be very helpful in making trip successful. If necessary, a final official letter from the "Director of the Institute" should be provided to the approaching organization for the activities and movement. One may contact the curator (a responsible explorer and researcher of specific crop) and international/national research institutes responsible for the germplasm collection of target crops, if necessary. Information on the area proposed to be visited, routs to be followed, camping sites, distances between places enroute jeepable or to be covered on foot, location of petrol pumps, crops and their harvesting time *etc.*, should be discussed with local people. Also correspond with scientists engaged in crop specific research programmes to know about the nature of variability available in local landraces and their distribution.

### (b) Acquisition of Minimum Information

For acquisition of minimum information, it is essential to survey local conditions in the target area, especially the distribution of crop species and range of their diversity. Information about social and religious customs in the area is also very important. Office hours and national holydays differ considerably from one country to another. If curator enters into the area without such basic information will be forced to change his/he itinerary later.

### (c) Studying Existing Germplasm

A visit to national/regional herbaria is also recommended before departure of the curator. This will be helpful to curator in investigating the existing collections of seeds or herbaria of the target crop species. If the explorer is not familiar with the species, he should visit gene bank or botanical gardens; or seed museums to see the seed samples and or living material with his own eyes. Based on the mission needs, the explorer should get familiar with the plant species to develop a visual expression of the taxa. This should facilitate in hunting the desired material in the areas of diversity or distribution. This is essentially importance while collecting wild relatives and related taxa of crop plants. It is advisable that the provenance data on the herbarium sheets may be studied critically and notes may be taken on flowering/fruiting/habitat/altitude *etc.* For specific species explorations, this information is very useful especially for locating materials of taxa with a narrow range of distribution.

### (d) Team Composition and Selection of Members

The team composition should include a researcher from the local institute, who is well informed about local condition besides his expertise in the target crop. I general, collection work can be carried out by botanists or plant explorers, crop researchers or extension workers. Researcher who is a team member should have adequate knowledge of morpho-agronomic characters of the varieties/cultivars, environmental, edaphic conditions, major diseases and pests *etc.* and prevailing cultural practices in the areas to be surveyed. Larger teams usually prove cumbersome. A two member team is considered ideal. A local scientist must be a member of such a party. One man explorations have been effective in the past with local guides employed from the area of survey. Since the primitive germplasm occurs in areas usually dominated by natives, larger teams create suspicion, while smaller teams are welcomed and can build up confidence quickly through the local officer/ guide. The farmers of underdeveloped tracts holding native landraces do not like/permit a visit to their fields, and the smaller the party, the less the interference and

the more the possibility to survey such sites. When the crops have been harvested and are in the threshing yard or lying near the hutments of the dwellers, larger team moving in the village creates a stir, interferes in the privacy of the natives, while one member with a local helper becomes a guest. He, not only collects the desired diversity, but also gets a share of the native's hospitality. But in some specific exploration missions, larger teams are unavoidable. Here, depending on the need, the team should have a crop scientist, plant pathologist *etc.* (it is an additional advantage if one such member knows driving also). By and large, small teams are more effective and mobile. When a team has more members, the team leader must coordinate collecting activities right from the initial planning. Sometimes, team members are from different centres/organisations and an effective contact is to be maintained with them by the team leader. Before the collection programme starts, the team leader must thoroughly brief the party members.

### (e) Equipments

For effective exploration and collection programme, team should choose necessary equipments/tools appropriately to the purpose and methods of collection. Exploration kit and passport data book are the first and the foremost items needed for any exploration. This will vary according to the material to be collected, climate, local conditions, mode of travel *etc.* The list of necessary equipments are given in the below mentioned table (Srivastava, 1998) but the items constitute an optimal list for an ideal collecting mission, if certain items are not available good collections can even be made.

**Table 4.6: Equipments Needed during Exploration and Collection**

| Activities | Articles Required |
| --- | --- |
| Reference material | Regional/national flora, digital herbarium, lap-top and accessories, list of local names of plants, roadmap, vegetation/ climate map, list of rest houses/lodges, hotels, resting/stay places and list of local contacts (phones, fax, email). |
| Essential collection items | Samples bags (cloth, net, polyethylene and others), Haversack/backpack, seed envelopes, tag labels, drying sheets, vasculum, pencil, ballpoint pen, rubber bands, secateurs/dissecting scissors, knives, road maps, towel, digger, stapler, markers, hand gloves, cloth sheets, seed envelops, polythene bags, reports on regional flora, passport data book, and field note book for recording other relevant information, mossgrass, rubber bands, packing tape, *sutli*-cord (thick and thin), digger, torch light, permanent marker, |
| Scientific equipments | Global positioning system (GPS), Small portable altimeter, Field compass, Digital Camera with additional memory card/Film rolls (Slides and Prints)/cameras (35 mm SLR/high resolution), Large piece of cloth for photography in the fields, scalpels, dissecting needles, a pair of binoculars, digital vernier caliper, portable balance, pH kit/soil kit, and pocket lenses/magnifying glasses (10X), campass, measuring tape, aluminium and tag labels, presses in which voucher specimens can be dried, absorbent paper for pressing specimens, and old newspaper for individual specimens, |
| Camping gear | Tents (mats, mosquito nets), sleeping bags, flash lights (with extra batteries), candles, water bottles, candle, match box, other camping and cooking items depending on needs. |
| Transportation and camping equipment | A four wheel drive vehicle, packing material, and transporting boxes. It is also essential to have spare parts and fuels in the field or remote places to make proper survey of flora. In the case of expedition of very long duration *i.e.* more than 30 days and in remote or hilly places, it is essential to carry camping equipments besides carrying cooking and food articles. |
| Clothing | Jackets, shirts, trousers, hiking boots/hunter shoes, hand gloves, rain suit, rucksacks, sun glasses *etc.*. |
| Medicines (First aid box) | All the medicines as prescribed for first aid-box and other few common medicines like Antibiotic, cough drops, anti-malaria pills, Paracetemol/Combiflam, Dependal-M, Unienzyme, Disprin, Enteroquinol, Vicks, antiseptic cream, liquid/ lotion, cotton packs, band aids, first aid box, Furacin powder, dressing gauge, and water purifying tablets. |
| Others | Water filter, steel boxes, printed slips with institutes, address, plastic jars, formaldehyde, alcohol, tarpaulins, and ropes. |

## (f) Duration of the Visit

Duration of the visit vary according to the mission. Explorations within the country are of shorter duration; about 45 days would be ideal, though one month is the general limit. When organizing in foreign countries, larger duration (two to three months may be needed) is required, as otherwise they are more expensive and less remunerative. This would depend much on the area of visit, crops involved, mission needs and the harvesting time *etc*. If larger ecological range is involved (altitudinal/latitudinal variation), collecting season is prolonged and 3-4 months may be needed to cover diverse terrains. Variation in sowing dates is related to weather conditions, topography *etc*. for every 1000 m. in hills, the temperatures falls by 5°C slowing vegetative growth and delaying maturity by 3-4 weeks. Since resurveying in foreign land may not be feasible, long duration explorations are considered more practical.

## (g) Transporting Collection

If the mission is to send plants/collections (for example: drumstick sticks cutting and collection) back to the institute for regeneration quickly, halting in between at a bigger town may be necessary for dispatch of materials. Incidentally, this halt can also be utilized for reequipping provisions, jeep servicing/checkup, and purchase of several other articles needed by the team. In explorations of long duration, interaction between transportation and sampling would be needed.

## (h) After Expedition Care

Follow up would be needed so that all collections are safe, indexing/accessioning *etc.*, done and collections passed on for quarantine. As an explorer, be sure that your notes are complete for writing the final report.

## Preparation of Exploration Report

It is considered that an exploration is unaccomplished unless the report of the exploration is finalized. The manuscript should contain following items (Srivastava, 1998):

  (i) Title of the exploration programme

 (ii) Author and Organization (s) involved

(iii) Dates, itinerary, major explorations areas

(iv) Purpose, data on collection

 (v) Physiography and agro-climatic of the areas surveyed

(vi) Sampling strategies and areas surveyed

(vii) Useful information and observations

(viii) Discussion on extent of diversity, variability pattern *etc*. on collected material

(ix) Impressions/overviews

 (x) Future plans for managing collected samples

Additional following tables and figures should also be attached in final report:

  (i) Itinerary

 (ii) Collection lists

(iii) Route map of exploration area

(iv) List of collaborating/cooperating organization (s)

As per the ICAR-NBPGRNew Delhi 'Guidelines for Management of plant Genetic Resources in India' 2016 following contents is also necessary:

☆ Name of the organizations(s)

☆ Name of the scientist(s)/person(s) involved

☆ Collaborating organization(s)

☆ Objectives of the collecting mission

☆ A description of the environment of the target area

☆ An account of the logistics and scientific planning

☆ Details of the execution of the mission (timing, itinerary, sampling strategy and collecting techniques)

☆ A summary of the results (areas surveyed along with route maps, germplasm and herbarium specimens collected, ethnobotanical/traditional/indigenous knowledge documented and extent and magnitude of diversity collected with elite germplasm, if any).

☆ Role of women in conservation of diversity

☆ Details of sharing germplasm and information

☆ Photographs of material depicting diversity/variability

☆ An account on loss of germplasm, if any

☆ Difficulties encountered during collecting mission

☆ Recommendations for follow-up actions(s)

☆ Acknowledgements

## Other Guidelines for Germplasm Exploration Activities

There are no set rules for germplasm exploration activities but certain guidelines are given which are as under (Srivastava, 1998):

## (a) Working Hours

Germplasm exploration and collection activities start daily on sunrise after breakfast. It is advisable to lunch early and finish germplasm exploration and collection activities and travelling 2 hours before sunset. To save the time, it desirable to have packed lunches.

## (b) Range and Distance of Activities

If germplasm exploration and collection activities are well chalked out and team members are travelling by car/jeep, it is possible to cover distance of 100-150 km. per day, whereas on foot 10 km. is standard. No hard and fast line can be set for distances to be covered daily; depending on diversity in land forms, topography, and vegetation. The distance may also vary with the availability of crop diversity and other local factors, it may vary. In daily routine, when halting at a place, 50-100 km. is enough from camp-site and much less when more sites are surveyed on foot. Much time will be needed if wild relatives are to be collected.

### (c) Herbarium Collection and Photography

A visit to a regional/national herbarium is recommended. Based on mission needs, the explorer should get conversant with the plant species to develop a visual impression of the taxa. This would facilitate work in hunting for the desired material in areas of diversity/distribution. This is of particular importance while collecting wild relatives and related taxa of crop plants. The data on the herbarium sheets may be studied critically and notes may be taken on the flowering/fruiting/habitat/altitude *etc.* for species specific explorations; this information is very helpful especially for locating materials of taxa with a narrow range of distribution. Regional herbaria provide a key reference to such locations or collecting sites, *e.g. Vigna grandis* (northern strip of Western Ghats), *Citrus indica* (Garo hills, Meghalaya), *Musa sikkimensis* (Sikkim and bordering areas). During germplasm exploration and collection, herbarium must be collected on route and sufficient number of photograph should be taken for record and study.

### (d) Base Camp Activities

After reaching at base camp or other lodging places, the following jobs/activities should be carried out immediately:

(i) Drying of sample when seed is wet.

(ii) To prepare herbarium by pressing them on sheets, this will be necessary for taxonomical studies in future.

(iii) The records of the collected samples should be completed (if not done earlier while in the field) by the end of the day in passport data book (sample data sheet developed by the Exploration Division of the ICAR-NBPGR, New Delhi appended).

(iv) Sorting of the collected samples should be done accurately each day.

(v) Before going to bed, update to your diary. Record activities of each day, map references of collection sites, the distance travelled and any other relevant information.

### (e) Planning Itinerary

Information assembled as above would help in the planning of proper itinerary. A provisional route and time scheduled can be worked out based on the harvesting time, agro-ecology of the terrain and the distribution of crop diversity.

### Suggestions

Sometime it happen that explorer or curators' approaching in unforeseen situations like drought prone tract and crop failure has occurred (in the exploration sites). The growers may get angry with you, and refuse to part with his material. Listen to him patiently, place your request humbly, make him realize the importance of your work and his importance in turn. Quite likely, you may bring him around and he agrees to allow you to sample you to collect, do not lose your patience keep your temper always cool. When permitted, do not be over greedy to take more material than what has been agreed to, and samples carefully do not spoil/disturb his field/threshing yard, courtyard, *etc.* Before leaving, never forget to convey to him your gratitude.

### Handling of Collected Germplasm during Explorations

It is extremely important to handle germplasm material efficiently during the explorations. Procedures of handling different kinds of germplasm materials are mentioned below:

### (a) Threshing, Cleaning and Processing of Dry Pods per Fruits

The dry seeds from individual fruits can be threshed manually after harvesting by beating with sticks or rubbing and splitting by hand. The trash is the removed by winnowing. Seeds while collecting should be fully mature. It has been observed that an explorer when collects physiologically pods/fruits and then spread without opening cotton bags/muslin cloth bags and spread in the room along with labels and other information. It is essential to drying of seeds, otherwise there can be fungal infection to the seed and the material will be damaged. In Brassica vegetable crops seeds should be dried properly otherwise oil present in the seed will develop rancidity and the viability will be lost.

### (b) Extracting, Cleaning and Processing of Seeds of Fleshy Fruits

In tomato, seeds of fully ripe fruits can be extracted right after harvest (same day in the evening). The fruits are crushed in plastic bucket/container. Add sufficient quantity of water so that seeds should sinks to the bottom of the container. Remove the water and other inert materials from the bucket/container. Collect the seed and again washed. After proper cleaning and drying the seeds, keep in paper bag. In Cucurbits (cucumber, snapmelon, longmelon *etc.*) mature fruits are cut into halves lengthwise and seeds with the attached core are scooped into a plastic container to ferment or put on sieve and are cleaned under water pressure and dried in shade. In eggplant, fruit is cut or macerated and the seeds are separated from the pulp by washing. Watermelon, muskmelon pumpkin and ash gourd seeds are split into halves, usually longitudinally and the seeds are separated manually or washed off from the pulp.

### (c) Handling Tree Vegetable Crops

There are several tree plants like: drumstick, curry leaf, plantain *etc.* are useful as vegetables. In perennial drumsticks the propagating sticks (1.5-2.00 m in length) are collected from the mother tree by cutting and planted in the field and within 2 month sticks are established as a plant.

### Transportation of Germplasm

Sometime, long duration exploration and collection programme is organized to places which are far away from Headquarters or material is vegetatively propagated like pointed gourd (*Trichosanthes dioica* Roxb.), ivy gourd (*Coccinia cordifolia* Cogn.), drumstick (*Moringa oleifera* Lam.) *etc.*, are recruited and they accompany the team. Courier services connecting the destination from the different important cities help in carrying the important material/perishable material to the Headquarters quickly. The runners/carriers, depending on the perishability of the material, or at times for limited viability, bring back the material quickly to the Headquarters from the collecting sits. In addition, the exploration team, should make use of the facilities available in routs *i.e.* Post office, airports, transport agencies to properly pack the material and send these to the Headquarters

☆ Seeds samples/physiological mature fruits should be placed in muslin cloth bags.

☆ Care should be taken at the time of labeling, as most often this turns out to be critical at the time of post collecting handling.

☆ To avoid deterioration of species with recalcitrant seeds, vegetative propagules, leaves and pollen, prior arrangements should be preferably made for the en-route transportation of collected material to the place of its establishment/ maintenance. In case of bud wood collection/ cuttings, the material should be wrapped in moss grass.

### The Ingredients of Success

The success of germplasm collecting programme depends on following basic points:

(i) Plan well ahead

(ii) Involve local people

(iii) Be prepared to be flexible

(iv) Develop a search image

(v) Choose collecting and processing techniques with care

(vi) Document the collection scrupulously

### (2) Procurement

The germplasm can be augmented through procurement from scientists and other agencies concerned with germplasm conservation, from research institutions, individuals or companies through ICAR-National Bureau of Plant Genetic Resources (NBPGR), New Delhi.

### Passport Data Recording during Collection

Data gathering is very important aspects during collection of vegetable genetic resources. When a collector decides to collect a sample, he has to record data on geographic location, habitat, socio-economic and genetic factors associated with the creation of the variation and evolution of the plant populations.

*"Pin pointing the important observations on samples and their collection sites are called the passport data and are recorded in the passport data book".*

ICAR-NBPGR, New Delhi has developed suitable passport data book with the consideration of IPGRI/ USDA/other organizations. Not only this ICAR-NBPGR has developed a comprehensive passport data book with due emphasis on the collection sites and plant characteristics' to better suit our local requirements and amicability for computerization. Material is passed on to the germplasm curator/National Gene Bank along with the purposeful passport data and subsequently supplemented with evaluation data before its final storage in the National Gene Bank. The availability of such data could help in planning of the exploration route in future collection/recollection programmes and to decide ecological regions and analogous agro-climatic conditions where such materials are to be evaluated. This way, passport data are an asset to understand nature, cause and even consequences of variation and evolution of plant population both within and between sites. It is very useful to the curators/breeders and other users to draw valid conclusions about the utility of the material. With the implementation of Convention on Biodiversity and patenting of genetic resources, the passport data have assumed supreme significance. An inverse relationship exists between the time spent in collecting the samples and recording passport data information. A collector can neither afford to loose genetic resources available nor to record passport data, so he has to make a purposeful compromise between the two depending upon the feasibility and relative importance of these two aspects.

Bennett (1970), Bunting and Kuckuck (1970), Harlan (1974) and Hawkes (1976) suggested to record data on the broad based agro-ecological studies supplemented with the information on specific locality of collection. Hawkes (1976) suggested that the passport data to be recorded on pre numbered data sheets (see passport data sheet as developed by ICAR-NBPGR-appended). Data sheet is of convenient size with tear out labels. These tear out labels should bear collector number as recorded on data sheet for identification of the collected samples, herbarium specimen, photographs *etc.* of the same sample.

## Accessioning

*"The systematic numbering of genetic resources with a specific number which helps in easy identification of each germplasm is called accessioning".*

### (i) Accessioning for Indigenous Materials

The indigenously collected material is received in the Exploration Division along with a list of germplasm indicating collector number and important passport data such as botanical name, common name, local name, data of collection, site (village, district and state), type of material and specific characteristics of a particular accessions. Each germplasm is assigned Indigenous Collection (IC) number and same number is entered into the IC Register. The germplasm is sent to the active collection sites under IC number for evaluation. After initial establishment, evaluation and successful seed increase, the material is conserved in seed genebank. Accessioning was started in India in 1940 and 1st number (EC1) was assigned to *Cynodon plectostachya* procured from UK. The last accession number is EC407404 assigned to 'Tarapuri' variety of eggplant (*Solanum melongena* L.) from Bangladesh.

### (ii) Accessioning for Exotic Materials

All Plant Introductions (PI) from abroad have to pass through quarantine inspection. After quarantine clearance they are assigned Exotic Collection (EC) number in the National Register of Exotic Collection. The other information on botanical name/varietal name, common name, donor's, institute address, source country, taxonomic status, pedigree longitude/latitude, specific characters and recipients address are also recorded in EC Register.

### Post Exploration Handling of Vegetable Genetic Resources

The passport data books, field notes and the post-harvest observations on the collected vegetable genetic resources are required for writing the final report which will be used for conservation and utilization. After care and follow up is essential to ensure that the collected germplasm should be suitable for sowing, evaluation and conservation into National Gene Bank. In post explorations handling the following operations are important:

(1) Drying of samples
(2) Threshing and cleaning
(3) Observations on post-harvest characters
(4) Processing and
(5) Disposal

### (1) Drying of Samples

The samples collected from the field are generally moist and thus immediate drying is essential to avoid rotting and fungal development. Drying of seeds/pods/ fruits *etc.* can be done either in electric driers or in shade as per the available facilities. Electric drying is useful during rainy/cloudy weather when sun drying is not feasible but during summer sun drying is best option.

### (2) Threshing and Cleaning

After drying, the collected samples are threshed either mechanically or manually and cleaned. Sufficient care has to be taken to avoid any damage to the seeds during threshing and admixture of the seeds. The seeds must be cleaned thoroughly to remove the trash content and other seeds.

### (3) Observations on Post-harvest Characters

It is difficult to take all the in the field while collecting the material. The observations on important post-harvest characters like seed weight, length and breadth of seed/pods/fruits *etc.*, quality characters, disease and pest damage *etc.*, are taken in the laboratory before the material is processed further.

### (4) Processing

The clean and dried samples are divided into two parts and put into envelops specifying common name, botanical name, collector number, IC/EC number and date of collection. Two separate sets of all the accessions are prepared and three copies of list of germplasm with passport data, one each for collector, gene bank manager and crop curator, are developed.

### (5) Disposal

One complete set of the collected material with passport data is handed over to the gene bank manager for conserving temporarily as a reference material and other set along with passport data to the crop curator for evaluation and characterization.

### Initial Establishment and Multiplication

The materials collected during exploration may not be in sufficient quantities and may loose viability during transportation. Therefore, special care is needed for the initial establishment of the collected genetic resources to avoid the risk of loosing a particular accession due to poor adaption, diseases and pests complex of the new environment, admixture through contamination or error and change in the genetic composition through conscious or unconscious selection. This is true in case of wild relative as their germination and survival under cultivated condition is often very difficult and poor. If the seed dormancy is the problem, necessary measure like seed treatment/scrapping of hard seed coat *etc.* are to be followed to ensure germination. The use of glass house/green house may be required for photo/ thermo-sensitive types for initial establishment of the material. It is essential to increase seed stocks sufficiently in one cycle so that the harvested seeds can be used for evaluation, differentiation and storage. It is always wiser to save a portion of the seeds. It will serve as a reference sample and in case the first effort fails, this portion can be used to raise another crop. The multiplication should be done at the site where the climate and soil conditions are analogous, as nearly as possible, to that of its collection site. During initial establishment and multiplication, the data on the morphological and other traits of major importance are recorded. The identification of duplicates and promising accessions is also done at this stage. The plant quarantine clearance can also be initiated at this stage.

In some cases, some rare genotypes unique morphotypes, wild relatives with shattering habit/shy seed setting and other such material of specific interest to the breeders, where collection of sufficient quantities of material is not possible for the collector, the initial establishment and multiplication becomes indispensible step between collection and evaluation. For the initial establishment of vegetatively propagated material, it is essential that the sufficient numbers of rootstocks are raised well in advance of collection so that the collected parts are immediately grafted without any loss of viability.

## Efforts in Exploration/Collection of Genetic Resources

### International Efforts in Plant Germplasm Collection

"N.I. Vavilov All Union Scientific Research Institute of Plant Industry (VIR)" is the most important institute on germplasm exploration/collection. Vavilov first time started systematic germplasm collection programme and collected 50,000 samples of crop plants from over 50 countries in the 1920s and 1930s. After this event, American, German, Swedish, British and Latin American collectors made several collecting missions. Recognizing the importance, developing countries also started integrating with global plant genetic resources system in the year 1960, where crop diversity was in abundance. In 1965, a panel of experts on crop germplasm exploration was set up with Sir Otto Frankel as Chair. International Biological Survey (IBP) came in vogue after discussions and various meetings held on germplasm. Frankel published a document "FAO/IBP Survey of Crop Genetic Resources in their centres of origin" in 1973. In a series manner, FAO organized technical conference on plant genetic resources in 1967, 1973 and 1981 wherein Sir Otto Frankel, Erna Bennett, R.O. Whyte, Jack R. Harlan, T.T. Chang, Jack G. Hawkes and others played an important role in preparation of roadmap for germplasm collection, conservation and utilization. In 1972, the UN Conference on Human Environment gave responsibility to FAO to establish an international forum and in 1974, "International Board for Plant Genetic Resources (IBPGR)" was established under the aegis of CGIAR to promote and coordinate plant genetic resources worldwide. As a main activity, IBPGR initiated germplasm collection in collaboration with national plant genetic resources programmes. Many International Agricultural Research Centres (IARCs) came forward for this novel cause and contribute significantly in several exploration expeditions with IBPGR. Over the years, the ICAR-NBPGR had sponsored some 650 exploration mission in over 130 countries and a total of 2, 00,000 samples collected. "IBPGR" now International Plant Genetic Resources Institute continued to be actively involved in germplasm collecting apart from strengthening collection and other conservation activities in national plant genetic resources system in developing countries.

International Plant Genetic Resources Institute's (IPGRI) South Asia Office is located within the ICAR-NBPGR; NAAS campus, in New Delhi. This office coordinates plant genetic resources activities in 6 countries *viz.* Bangladesh, Bhutan, India, Maldives, Nepal and Sri Lanka. International Plant Genetic Resources Institute's South Asia office functions under the MoU with ICAR/DARE, Government of India. Within India, the plant genetic resources progrmmes are developed with ICAR-NBPGR mainly as per national priorities, based on biennial work plans with emphasis to assist the national PGR programme (Table 4.7).

Explorations outside India under international collaborations are also done time-to-time. Explorations made outside India with international collaboration are given in Table 4.8.

#### Table 4.7: Collaboration of ICAR-NBPGR in Germplasm Exploration and Collection at International Level

| Sl.No. | PGR Activity | Year | Collaboration | Crops |
|--------|-------------|------|--------------|-------|
| 1 | Exploration and germplasm collection | 1990-1992 | ICAR-NBPGR-BARI, BARC, Bangladesh | Eggplant, okra |
| | | | ICAR-NBPGR-NARC, Nepal | Eggplant, okra |
| | | | ICAR-NBPGR- PGRC, Sri Lanka | Eggplant, okra |

#### Table 4.8: Explorations of Germplasm Outside India with International Collaboration

| Country | Crops Collected |
|---------|----------------|
| Central Asian Republics of the former USSR | Cassava, okra, winged bean, velvet bean, cucurbits, *Capsicum*, *Hibiscus* and tomato. |
| East Africa-Malawi and Zambia | Tuber crops, legumes, *Amaranthus* and other vegetables |
| Kenya, Sudan and Ethiopia, Bangladesh | Eggplant |

Extensive Germplasm collections are maintained by the USDA Plant Germplasm System at the Plant Introduction Station in Iowa (Clark *et al.,* 1991) and by the Vavilov Institute in Leningrad, USSR (Robinson, 1989). Another major germplasm repository is maintained by the Peoples' Republic of China, Nigeria, Costa Rica, and the Philippines (Esquinas-Alcazar and Gulic, 1983). However, for foreign explorations, guidelines of the ICAR/Government of India need to be followed to develop and finalize the mission. For such explorations, the international code of conduct for plant germplasm collecting and transfer may also be followed (http://www.fao.org/docrep/x5586e/x5586e0k.htm).

## IPGRI

The International Plant Genetic Resources Institute (IPGRI) has played significant role in germplasm collection and conservation by offering expertise, training and funding for research. In India IPGRI office for South Asia is located in the NAAS campus and there is an active collaboration between them based on biennial work plan. In addition to supporting joint

**Table 4.9: Scope of Germplasm Collection**

| Crop and Species Covered | | Scope of Collection | | Institute |
|---|---|---|---|---|
| Crop | Species | Global | Regional | |
| Faba bean | | | | CNR, Bari Italy |
| *Phaesolus* spp. | Wild spp. | | | JBNB, Bruxelles Belgium |
| | Cultivated spp. | | | CIAT, Colombia |
| | | | | NPGS, United States |
| | | | European | FAL. Braunschweig, Germany West |
| Soybean | | | | NAIR, Tsukuba, Japan |
| | | | | NPGS, United States |
| | Wild spp. | | | CSIRO, Canberra, Australlia |
| *Vigna* spp | Wild spp. | | | JBNB, Bruxelles Belgium |
| | *V. mungo* | | | ICAR-NBPGR, New Delhi, India |
| | *V. radiata* | | | IPB, Los Banos, Philippines |
| | | | | AVRDC, Taiwan |
| | *V. umbellata* | | | ICAR-NBPGR, New Delhi, India |
| | *V. unguiculata* | | | IITA, Nigeria |
| | | | | NPGS, United States |
| Winged bean | | | | IPB, Los Banos, Philippines |
| | | | | TISTR, Bangkok |
| **Root crop** | | | | |
| Carrot | | | | IHR, Welles Bourne, United Kingdom |
| Cassava seed | | | | CIAT, Colombia |
| *Solanum* spp. | | | | CIP, Peru |
| Sweet potato | | | | NPGS, United States |
| | | | Asian | AVRDC, Taiwan |
| | | | | NAIR, Tsukuba, Japan |
| *Allium* | | | | CGN, Wageningen, Netherlands |
| | | | | IHR, Welles Bourne, United Kingdom |
| | | | | NPGS, United States |
| | | | South and South East European | RCA, Tapioszele, Hungary |
| | | | Asian | NAIR, Tsukuba, Japan |
| *Amaranthus* | | | | NPGS, United States |
| | | | Asian | ICAR-NBPGR, New Delhi, India |
| *Capsicum* | | | | CATIE, Turriabla, Costa Rica |
| | | | | AVRDC, Taiwan |
| | | | Asian | ICAR-NBPGR, New Delhi, India |
| Crucifer | *Brassica corinata* | | | FAL. Braunschweig, Germany West |
| | | | | PGRC/E, Addis Ababa, Ethiopia |

| Crop and Species Covered | | Scope of Collection | | Institute |
|---|---|---|---|---|
| Crop | Species | Global | Regional | |
| | *Brassica oleracea* | | | CASS, Beijing, People's Republic of China |
| | | | | IHR, Welles Bourne, United Kingdom |
| | | | | CGN, Wageningen, Netherlands |
| | *Raphanus* spp. | | | CASS, Beijing, People's Republic of China |
| | | | | IHR, Welles Bourne, United Kingdom |
| | | | Asian | ICAR-NBPGR, New Delhi, India |
| | Wild spp. | | | Universidad Politecnica Madrid, Spain |
| | | | | Tohoku University Sendai, Japan |

exploration and collection programme in this region, the IPGRI helps in funding projects related to plant genetic resources. Under the MoU for scientific and technical cooperation between ICAR and IPGRI, under International Network for Improvement of Banana and Plantain (INIBAP)-a programme of IPGRI, a work plan of cooperation had been completed.

### Table 4.10: Estimates of Germplasm Holdings in the Five Largest National Plant Germplasm Systems and Major International Systems

| Category/Centre | Category/Concerned | Total |
|---|---|---|
| United States | All crops | 557,000 |
| China | All crops | 400,000 |
| Former Soviet Union | All crops | 325,000 |
| IRRI | Rice | 86,000 |
| ICRISAT | Sorghum, millet, chickpea, peanut, pigeon pea | 86,000 |
| ICARDA | Cereals, legumes, forages | 77,000 |
| India | All crops | 76,000 |
| CIMMYT | Wheat, maize | 75,000 |
| CIAT | Common bean, cassava, forages | 66,000 |
| Japan | All crops | 60,000 |
| IITA | Cowpea, rice, root crops | 40,000 |
| World Vegetable Centre (AVRDC), Taiwan | Alliums (Onion, garlic, shallot), Chinese cabbage, common cabbage, eggplant, mungbean, pepper, soybean, tomato, other vegetables of regional importance | 38,500 |
| CIP | Potato, sweet potato | |

*Source*: Managing Global Genetic Resources: Agricultural Crop Issues and Policies.

In China, Institute of Vegetable and Flowers (IVF) has coordinated exploration tour into Yunnan, Tibet and Shennongjia region of Hubei province with the mission to collect dwarf squash, naked seed squash, wild bitter melon, wild Chinese chive, wild green onion and wild carrot have been discovered. Meanwhile IVF coordinated with 1,000 scientific staff from 32 provincial institutes to request 40, 000 accessions of samples out of which 21,000 were recorded and preserved both in the long term National Gene Bank and midterm Exchange Banks with related data base. These materials belong

to 27 families, 67 genera and 132 species (50 originated from Chin). Cambridge, U.K., Wisconsin, and U.S.A. collected several genotypes of potato in their region to provide solid foundation for breeding projects.

### Table 4.11: Vegetable Germplasm at World Vegetable Centre (formerly AVRDC) Taiwan

| Crop | Number of Accessions |
|---|---|
| **Globally important crops** | |
| Soybean | 14195 |
| Pepper | 7498 |
| Tomato | 7217 |
| Mungbean | 5646 |
| Eggplant | 2613 |
| *Brassica* | 1688 |
| *Allium* | 1079 |
| **Sub total** | **39936** |
| **Regionally important crops** | |
| *Vigna unguiculata* | 1390 |
| *Phaseolus* bean | 613 |
| *Luffa* | 618 |
| *Vigna mungo* | 481 |
| *Cucumis* | 458 |
| *Amaranthus* | 434 |
| *Vigna unguiculata* spp. *sesquipedalis* | 341 |
| *Abelmoschus* | 334 |
| *Lablab* | 253 |
| *Pisum* | 216 |
| *Vigna unguiculata* spp. *unguiculata* | 81 |
| Others | 2948 |
| **Sub total** | **8533** |
| **Total** | **48469** |

### Germplasm Collection at World Vegetable Centre (AVRDC), Taiwan

At present World Vegetable Centre (AVRDC), Taiwan houses the world's largest collection of valuable germplasm of vegetable crops in the public sectors. The AVRDC gene bank has assembled nearly 60,000 accessions of vegetables including specie covering 170 genera and 436 species from 156 countries. Many of these accessions are relatives of the cultivated forms and represent a unique and invaluable resource for plant breeders worldwide.

### National Efforts in Plant Germplasm Collection

Systematic work on plant exploration and germplasm collection was initiated in 1946 with emphasis on major field crops. Late Dr. Harbhajan Singh, botanist was possibly the most distinguished plant explorer of India and he is remembered today as "Indian Vavilov". He has given the new concept and identity to the discipline of germplasm exploration, collection, and maintenance. Systematic plant exploration and collection work was initiated in India with the establishment of central agency for this purpose in 1946 in the Division of Botany, ICAR-IARI, New Delhi under his guidance. He initiated germplasm collection programme considering both mission-oriented on specific crops or multi-crop collecting/non-specific in India and in the neighbouring countries. In the subsequent period, emphasis was laid to collect both genetic and species diversity from diverse agro-ecological regions across the country. It is true that initially sporadic collections of indigenous germplasm of various crops were made but during this period no attention was given on vegetable crops (Table 4.12).

**Table 4.12: Multi Crop Collection**

| Scientist Involved | Period | Crop |
|---|---|---|
| Howard, A. and Howard, G.L.C. | 1910 | Wheat |
| B.P. Pal | 1910 | Wheat |
| Burkil, I.H. and Finlow, R.S. | 1907 | Jute |
| Bezbaruah, H.P. | 1968 | Tea |
| ThuljaramRao, J. and T.N. Krishnamurthy | 1968 | Sugarcane |
| Shaw *et al.* | 1931-1933 | Several legumes |
| Govidaswamy | 1955-60 | Rice |
| R.H. Richharia | 1965, 1977-79 | Rice |
| S.D. Sharma and S.V.S. Sastry | 1967-1972 | Rice |

Efforts have made to collect various vegetables through exploration from different parts of the country in collaboration with ICAR-NBPGR, State Agricultural Universities and ICAR Institutes. A number of crop specific explorations were undertaken to collect major and minor vegetables. This includes tomato, brinjal, chillies, okra, pea, cowpea, tropical cauliflower, kale, radish, Asiatic carrot, amaranths, spinach, bitter gourd, pointed gourd, cucumber, snapmelon, kakrol (Sweet gourd), kartoli (Spinet gourd), pumpkin, sponge gourd, ridge gourd *etc.* Mega explorations programme were initiated under National Agricultural Technology Programme on Plant Biodiversity (NATP on Plant Biodiversity) in mission mode manner and a large number of collections were made in different crops. A special drive for germplasm collection was launched by the Bureau during 2011-2016. A total of 3,103 germplasm samples (190 taxa) were collected

in exploration trips in all eight states *viz.* cereals and millets (992), pseudocereals (104), pulses (265), oilseeds (107), fibres (61), vegetables (964), fruits (158), spices and Medicinal and Aromatic Plants (73), crop wild relatives (322) and minor economic species (57). In vegetables, rich diversity was collected in leafy mustard, chilli, brinjal, garlic, cowpea and cucurbits. During the exploration two new crops-*Plukenetia corniculata* (leafy vegetable) and *Cucumis metuliferus* (salad vegetable) were augmented some rare species specific to NEH region collected for the first time are *Vigna nepalensis*, *Cucumis javanicus*, *Trichosanthes himalensis* and *T. cordata*.

**Table 4.13: International Collaboration in Plant Genetic Resources Collection from India**

| Indian Scientist (s) | Scientist of other Countries | Area Visited and Crop | Year |
|---|---|---|---|
| Dr. Harbhajan Singh | Howard S. Gentry of USDA | Punjab and Shimla hills | 1953 |
| Dr. Harbhajan Singh | Prof. Von Soest of Netherlands | Kashmir and Himachal Pradesh | 1959 |
| Dr. Mehra and Dr. Arora | Dr. J.R. Witcombe of the University of North Wales | North-West Himalayas (wheat, maize, barley, rice and few more crops) | 1978 |
| Dr. Mehra and Dr. Arora | Dr. J.R. Witcombe of the University of North Wales | Ladakh (*Triticum compactum* and wild barley) | 1978 |

Scientist from outside countries had undertaken several exploration and collection missions in India *i.e.* German scientist for collection of tropical forages; Royal Horticultural Society, U.K for ornamental Orchids; Simmonds for *Musa* species; Australian scientist for legumes and grasses; US scientist for cucumber and melons; Japanese botanists for collecting clones of *Thea*, *Brassica* species, different minor millets and Sesame; Canadian team for collection of wild *Brassica* species and ICAR-IARI scientists for collection of wild rice. Under several Indo-International progrmmes, plant exploration and collection of specific crops and groups of crops were undertaken by ICAR-IARI (erstwhile Plant Introduction Division) and ICAR-NBPGR in collaboration/financial support from certain international organizations (Table 4.14).

The involvement of large number of stakeholders under National Agricultural Technology Project (NATP) during 1999-2005 resulted in collection of significant diversity especially from remote, inaccessible and land-locked areas. The assemblage of wider gene pool, its safe conservation and access, provides a greater choice to the breeders and molecular biologists in developing the new varieties. However, the scientists are now increasingly looking for trait-specific germplasm with respect to quality, and tolerance/resistance to abiotic and biotic stresses.

**Table 4.14: Collection Vegetable Crops (within India) with International Collaboration**

| Year | Collaboration/Scheme | Areas Surveyed | Crops Collected | No. of Collections |
|---|---|---|---|---|
| 1966-70 | ICAR-IARI/PL-480 scheme | Different parts of India | Cluster bean (vegetable and gum types) | 942 |
| 1966-70 | ICAR-IARI/PL-480 scheme | Different parts of India | Legumes | 8,926 |
| 1977-78 | Indo-Soviet protocol | Russia | Legumes, okra, cassava, winged bean | 7,300 |
| 1979-80 | Indo-Soviet protocol | Mali, Niger | Legumes | 962 |
| 1980 | FAO/IBPGR | Malawi and Zambia | Legumes, vegetables, tuber crops | 3, 922 |
| 1984 | ICAR-NBPGR/Australia | Wild herbaceous grasses and legumes | Central, Deccan, Peninsular India | 1000 |
| 1989-93 | ICAR-NBPGR/IPGRI | India, Nepal, Sri Lanka, Bangladesh *i.e.* (i) North-eastern region (ii) Parts of India, Nepal and Sri Lanka (iii) South Asia | Okra and eggplant and their wild relatives | 4, 665 |
| 1992 | ICAR-NBPGR/Canada (Indo-Canadian) | Kumaun hills of India (Himalayas) | *Brassica* spp. | 18 |
| 1992 | Indo-USAID collaborative explorations | Parts of Rajasthan, Madhya Pradesh and Uttarakhand | Cucumber and Melons | 194 |
| 1992 | | | Snapmelon | 236 |
| 1992 | | | *Cucumis callosus* | 156 |
| 1992 | | | Snake cucumber/kakari (*C. melo* var. *utilissimus*) | 24 |
| 1992 | | | Other cucurbits | 48 |
| | IPGRI collaborative explorations | | Okra, eggplant | 4665 |

## Status of Vegetable Genetic Resources at ICAR-NBPGR, New Delhi

Under augmentation of vegetable genetic resources ICAR-NBPGR, New Delhi made several trips to collect important agri-horticultural crops and their wild relatives in different agro-ecological zones/habitats.

**Table 4.15: Indigenous Germplasm Collections made Prior to the Creation of the ICAR-NBPGR**

| Period | Number of Collections | Crops |
|---|---|---|
| 1946-195 | 625 | Cereals, millets, legumes, oil seeds, Vegetables, fiber yielding crops and other economic plants including wild relatives |
| 1951-1955 | 3,363 | |
| 1956-1960 | 4,245 | |
| 1961-1965 | 4,470 | |
| 1966-1970 | 1,902 | |
| 1971-1975 | 10,737 | |
| 1976 up to July | 5,603 | |

Till 2010, out of 2663 explorations conducted, vegetable germplasm were collected in 1525 explorations including 154 vegetable-specific ones. A total of 51,229 accessions of vegetable germplasm were assembled which contribute to about 20 per cent of total collections. The assembled germplasm contributed to both genetic and species diversity of 47 genera, mainly *Abelmoschus, Allium, Capsicum, Colocasia, Cucumis, Cucurbita, Lagenaria, Luffa, Momordica, Pisum* and *Solanum*. Details

**Table 4.16: Exploration Undertaken and Germplasm Collected by ICAR-NBPGR, New Delhi**

| Year | Exploration Undertaken | Germplasm Samples Collected | | |
|---|---|---|---|---|
| | | Cultivated | Wild | Total |
| 1976 | 4 | 1,987 | 138 | 2,125 |
| 1977 | 6 | 5,099 | 24 | 5,123 |
| 1978 | 7 | 938 | 11 | 949 |
| 1979 | 11 | 4,256 | 54 | 4,310 |
| 1980 | 7 | 4,559 | 22 | 4,581 |
| 1981 | 16 | 6,031 | 271 | 6,302 |
| 1982 | 9 | 3,575 | - | 3,575 |
| 1983 | 9 | 3,000 | - | 3,000 |
| 1984 | 10 | 3,525 | - | 3,525 |
| 198 | 25 | 8,008 | 55 | 8,063 |
| 1986 | 43 | 8,391 | 97 | 8,488 |
| 1987 | 44 | 7,115 | 178 | 7,293 |
| 1988 | 52 | 7,646 | 154 | 7,800 |
| 1989 | 59 | 12,503 | 934 | 13,437 |
| 1990 | 49 | 6,122 | 899 | 7,021 |
| 1991 | 57 | 5,854 | 996 | 6,850 |
| 1992-93 | 57 | 5,969 | 272 | 6,241 |
| 1993-94 | 21 | 1,580 | 175 | 1,755 |
| 1994-95 | 31 | 2,497 | 291 | 2,788 |
| 1995-96 | 16 | 997 | 511 | 1,508 |
| 1996-97 | 34 | 1,793 | 297 | 2,090 |
| **Total** | **567** | **1,01,445** | **5,379** | **1,06,824** |

of collections in important genera having rich diversity are depicted in Table 4.17.

**Table 4.17: Vegetable Genetic Resources Collected during 1946-2010**

| Sl.No. | Family | Genus | Number of Collection |
|--------|--------|-------|----------------------|
| 1. | Alliaceae | Allium | 2107 |
| 2. | Amaranthaceae | Amaranthus | 1209 |
| 3. | Araceae | Colocasia | 1770 |
| | | Amorphophallus | 985 |
| 4. | Cucurbitaceae | Benincasa | 722 |
| | | Cucumis | 4820 |
| | | Cucurbita | 2510 |
| | | Lagenaria | 2039 |
| | | Luffa | 3105 |
| | | Momordica | 2024 |
| | | Trichosanthes | 849 |
| 5. | Dioscoraceae | Dioscorea | 538 |
| 6. | Fabaceae | Cyamopsis | 5171* |
| | | Lablab | 630 |
| | | Pisum | 1743* |
| | | Vicia | 534 |
| 7. | Malvaceae | Abelmoschus | 3519 |
| 8. | Solanaceae | Capsicum | 5416 |
| | | Lycopersicon | 787 |
| | | Solanum | 7105 |

*= Includes grain types.

Due to prominence of multi-crop explorations in the earlier phase, on an average, large number of collections in each exploration (~180) were assembled which narrowed down (~65) during the ongoing phase as the emphasis is laid on crop(s)/trait-specific explorations. The share of horticultural crops including vegetables in collected germplasm doubled from 29 to 58 per cent in last two phases. A progressive increased trend was noticed in the share of wild accessions (including wild relatives of crops) from about 6 to over 35 per cent in the collected germplasm.

## Status of Vegetable Genetic Resources at PCPGR GBPUA&T, Pantnagar (Uttarakhand)

The elite lines of various vegetable crops imported as follows:

**Table 4.18: Elite Collection of Vegetable Crops**

| Crop | Number of Accessions |
|------|----------------------|
| Tomato | 211 |
| Grain/vegetable soybean | 153 |
| Sweet pepper | 93 |
| Chilli pepper | 50 |
| Eggplant | 5 |

| Crop | Number of Accessions |
|------|----------------------|
| Amaranth | 16 |
| **Total** | **531** |

## Role of All India Coordinated Research Programme (Vegetable Crops) in Germplasm Augmentation

All India Coordinated Research Project (Vegetable Crops) or AICRP (VC) has made efforts to collect all possible types of germplasm lines available in the country, especially of indigenous vegetables. Though AICRP (VC) coordinating centres have collected a lot of germplasm lines of nearly vegetable crops (Table 4.19).

**Table 4.19: Crop-wise Germplasm Status in India under AICRP (VC)**

| Group of Vegetables | Crop | Total Collection |
|---------------------|------|------------------|
| Solanaceous group | Tomato | 4813 |
| | Brinjal | 2372 |
| | Chilli | 2909 |
| | Capsicum | 61 |
| | Paprika | 44 |
| | **Total** | **10199** |
| Brassica group | Cauliflower | 603 |
| | Cabbage | 120 |
| | Broccoli | 53 |
| | Kale | 7 |
| | **Total** | **783** |
| Legumes | Garden bean | 1427 |
| | French bean | 1547 |
| | Indian beans | 1006 |
| | Sword bean | 2 |
| | Cowpea | 890 |
| | Cluster bean | 43 |
| | **Total** | **4915** |
| | Okra | 2431 |
| Bulb crop | Onion | 511 |
| | Garlic | 485 |
| | **Total** | **996** |
| Root crops | Radish | 28 |
| | Carrot | 366 |
| | Turnip | 4 |
| | **Total** | **398** |
| Cucurbits | Summer squash | 15 |
| | Ivy gourd | 87 |
| | Bitter gourd | 454 |
| | Bottle gourd | 751 |
| | Pointed gourd | 162 |
| | Sponge gourd | 525 |
| | Ridge gourd | 556 |
| | Snake gourd | 24 |
| | Ash gourd | 255 |

| Group of Vegetables | Crop | Total Collection |
|---|---|---|
| | Spine gourd | 57 |
| | Pumpkin | 412 |
| | Cucumber | 607 |
| | **Total** | **3905** |
| Melon | Longmelon | 112 |
| | Roundmelon | 34 |
| | Watermelon | 280 |
| | Muskmelon | 1174 |
| | Snapmelon | 203 |
| | **Total** | **1803** |
| Leafy vegetable | Amaranth | 249 |
| | Spinach | 6 |
| | *Chenopodium* | 17 |
| | Lettuce | 55 |
| | Fenugreek | 67 |
| | Lai Patta | 19 |
| | Drumstick | 68 |
| | **Total** | **481** |
| | Grand Total | 25911 |

However, there are many areas of the country which are still unexplored and further it needs to make strategy to collect germplasm from these in phases. On the other side, attempts have made to conserve the vegetable germplasm lines at ICAR-NBPGR, New Delhi in national gene bank for long term conservation.

## Precautions during Germplasm Collection and Movement

Seeing the natural balance it is true that both, crop plants and the pests, have evolved together and through long and continuous association, have developed a sort of live and let live relationship. Hence the centres of origin and diversity in plants are also the centres of origin and diversity in pests. It is of paramount importance, therefore, that before embarking on any exploration mission, the germplasm collection scientists should orient themselves with the pests and diseases of the specific crops and their symptoms. This would help them in collection of pest free germplasm. Further, while collecting the material they should ensure that no soil and infected plant debris should accompany the collected germplasm material. During the transportation of the collected material, care should be taken to avoid spoilage or damage to the germplasm materials. They should be transported in safe and sealed containers.

**Table 4.20: Status of Cultivated and Wild Germplasm of Important Vegetable Crops Conserved in NGB**

| Crop Name | Botanical Name | Total |
|---|---|---|
| Tomato | *Solanum lycopersicum* | 2337 |
| | *S.esculentum var. cerasiforme, S. hirsutum, S. peruvianum, S. Pimpinellifolium, S. chilense* | 162 |
| Brinjal | *Solanum melongena* | 3485 |
| | *S. aculeatissimum, S. aethiopicum, S. albicans, S. viarum, S. albicaule, S. anguivi, S. aviculare, S. giganteum, S. gilo, S. khasianum, S. incanum, S. indicum, S. insanum, S. nigrum, S. macrocarpon, S. pubescens, S. verbascifolium, S.torvum, S. seaforthianum, S. Setosissimum, S. torvum, S. lasiocarpum* | 763 |
| Chilli | *Capsicum annuum, C. baccatum, C. chinense, C. frutescens* | 2568 |
| Okra | *Abelmoschus esculentus* | 2416 |
| | *Abelmoschus angulosus, A. crinitus, A. ficulneus, A. manihot subsp manihot, A. pungens, A. manihot subsp. tetraphyllus, A. moschatus, A. tuberculatus, A. caillei* | 716 |
| Onion | *Allium cepa* | 834 |
| | *Allium albidum, A. altaicum, A. auriculatum, A. stracheyi, A. fistulosum, A. griffithianum, A. tuberosum, A. senescens, A. ledebouranum, A. oreoprasum, A. oschaninii, A. pskemense, A. ramosum* | 117 |
| Ash gourd | *Benincasa hispida* | 222 |
| Bottle gourd | *Lagenaria siceraria* | 722 |
| Bitter gourd | *Momordica charantia* | 369 |
| | *M. dioica, M. cochinchinensis, M. balsamina, M. subangulata* | 159 |
| Ridge gourd | *Luffa acutangula, L. hermaphrodita, L. pentandra, L. echinata, L. tuberosa* | 294 |
| Sponge gourd | *Luffa cylindrica* | 383 |
| Round gourd | *Praecitrullus fistulosus* | 108 |
| Ivy gourd | *Coccinia grandis, C. indica, C. cordifolia* | 23 |
| Snake gourd | *Trichosanthes anguina, T. bracteata, T. cucumerina, T. cuspidata, T. Lobata, T. lepiniana* | 218 |
| Cucumber | *Cucumis sativus* | 343 |
| Cucumis species | *C. melo var. utilissimus, C. prophetarum, C. argrestis, C. callosus, C. hardwickii, C. trigonus* | 544 |
| *Cucurbita* spp | *Cucurbita maxima, C. pepo, C. moschata* | 353 |
| Muskmelon | *Cucumis melo* | 793 |
| Watermelon | *Citrullus lanatus* | 257 |
| Pea | *Pisum sativum* | 260 |

| Crop Name | Botanical Name | Total |
|---|---|---|
| Leafy vegetable | *Trigonella corniculata, Spinacia oleracea, Amaranth cruentus* | 723 |
| Cabbage | *B. oleracea* var. *capitata* | 86 |
| Cauliflower | *B. oleracea* var. *botrytis* | 149 |
| Beet root | *Beta vulgaris* | 46 |
| Carrot | *Daucus carota* | 74 |
| Radish | *Raphanus sativus* | 253 |
| Lablab bean | *Lablab purpureus* | 293 |
| Cowpea | *Vigna unguiculata* | 225 |
| French bean | *Phaseolus vulgaris* | 118 |
| **Total** | | **20188** |

All the collected germplas material must always be subjected to quarantine examination and clearance before it is grown in the field away from its place of collection. By not subjecting the material to quarantine check and clearance, the germplasm collector may risk his entire collection if even one serious pest gets through. Similarly healthy materials should be put in cold store facilities for medium or long term conservation since under cold temperature conditions; many pests are known to survive for long periods and in some cases may even outlive the seed. In fact, low temperature storage of infected seeds has been suggested for conservation of diversity in plant pathogens.

### (3) Cataloguing

Each germplasm accession is given an accession number. This number is prefixed, in India, with IC (indigenous collection), EC (exotic collection) or IW (indigenous wild). Information on the species and variety names, place of origin, adaptation and on its various features or descriptors is also recorded. The usefulness of an accession, in fact, the plant breeders only know the entire germplasm collection, when the information about the features of the accessions becomes available to them. Therefore, catalogues of the germplasm collections for various crops are publishes by the gene banks every year. The amount of data recorded during evaluation is huge. Its compilation, storage and retrieval is now done using special computer programmes.

### (4) Evaluation

Evaluation consists of assessment of the germplasm accessions for their various features or traits of some known or potential use in breeding programmes. Generally, germplasm accessions are evaluated for morphological, physiological, biochemical, plant pathological (*i.e.,* disease resistance), entomological (*i.e.,* insect resistance) and other features. The characters assessed must be related to the need of the breeders since they are the ones who are going to utilize the germplasm. Characterization, preliminary evaluation and further advance evaluation of germplasm are prerequisite for utilization of plant genetic resources. Characterization and preliminary evaluation consists of recording a limited number of additional traits, thought desirable by the consensus of specialists/users of the particular crops/species, which help in identifying useful germplasm. This is very important for identification and help in designating core collection, identifying duplicates and planning future exploration. Advanced evaluation consists of recording a number of additional descriptors though to be useful in crop improvement. These include important agronomic traits, stress tolerance, disease and pest resistance and quality characters *etc.*

### (5) Multiplication

The germplasm accessions requested by breeders/researchers are multiplied and supplied to them, usually without cost. Ordinarily, active collections are used for this purpose. This is a very important activity of gene banks since it is the very purpose for which they are established. Generally, only a limited quantity of seed is provided to each worker. It is expected that each breeder/research worker will report back to the gene bank his assessment of the important characters of the accessions used by him.

### (6) Distribution

It is very essential to distribute the collected genetic resources among the researchers with the proper identity and traits. So, that plant breeders can use the reported germplasm as quickly as possible.

### Gaps, Opportunities and Future Thrust in Exploration and Collection of Germplasm

In future, the exploration and germplasm collection should be intensified considering crop/trait specific collaborative programme:

☆ Priority crops of national importance

☆ Trait specific exploration/collection should be intensified.

☆ Priority areas would contain significant genetic diversity of vegetable crops and their wild relatives, areas relatively un-explored/under explored; areas where change/development in agriculture or industrialization/deforestation necessitate immediate collection.

☆ Endangered economic species

☆ Genes for biotic and abiotic stress and other specific traits in hot-spot areas of diversity.

☆ Crop collecting would lay emphasis on areas of crop variability/distribution with reference to native types; also specific collection for particular trait.

☆ Re-survey/collection in those areas from where useful genotypes have already been identified based on evaluation studies and biochemical/molecular linked desirable traits.

☆ Gaps in collection being conserved under Indian plant genetic resource (PGR) system.

More important crops for which native diversity would still need more emphasis for collection include *Cucumis* and other *Cucumis* species, pointed gourd, yam, taro, sweet potato, chillies muskmelon, lablab bean, *Trigonella* and winged bean. Priority areas for cucurbits include Indo-Gangetic plains, foothills of Himalayas, peninsular tract, Kachchh areas in Gujarat, eastern Ghats, Kodur and adjoining areas in Andhra Pradesh and particularly for *Cucumis melo* Barabanki, Unnao, Lucknow, Baghpat, Tarai, western Uttar Pradesh, Chhotanagpur, West Bengal, Asom, Karnataka and Andhra Pradesh, *Cucumis melo* var. *utilissimus* (tar kakri) diversity is available throughout India, parval

(*Trichosanthes dioica* Roxb.) in Bengal plains and Asom valley; ivy gourd (*Coccinia* species)in North-eastern hill region; *Luffa* and *Lagenaria* from Indo-Gangetic Plains, NW plains; snake gourd (*Trichosanthes anguina*) is southern peninsular tract and Kerala, Karnataka and adjoining pockets; lablab bean in eastern peninsular tract, Odisha, Goa and southwards; cassava in Kerala, Asom, Tripura, Andhra Pradesh, Tamil Nadu and Madhya Pradesh (tribal tract of Bastar); sweet potato in Bihar, Odisha, Uttar Pradesh and West Bengal; *Dioscorea* species from Andaman and Nicobar islands, Bihar, North-eastern hill region; *Amorphophallus* species from Andaman and Nicobar islands, Andhra Pradesh and West Bengal; carrot (purple *deshi* types) need to be collected from South-East of Gujarat and adjoining Rajasthan. Systematic efforts will be made to analyse gaps in collection and cover the areas recommended by Crop Advisory Committees/Coordinated crop workshops/symposia *etc.* To accomplish collection and conservation of fast eroding genetic resources, the ICAR-NBPGR, New Delhi will involve effective collaboration with ICAR crop based institutes/NRCs/Crop Coordinating units/SAUs/State Agriculture Departments/other national/international agencies and governmental organizations, research foundations, NGOs including defense personnel/organizations.

## NATIONAL BUREAU OF PLANT GENETIC RESOURCES, NEW DELHI-110012

### PASSPORT DATA SHEET

Date_____ Collector's No._____ Accession No._____

Species Name_____ Common Name (English)_____

Cultivar/Vernacular name _____ Regions Explored_____

Village/Block_____ District_____ State_____

Latitude_____ N/S Longitude_____ E/W Altitude_____ m

Temp_____ Rainfall_____

| SOURCE | : | 1. Natural wild 2. Disturbed wild 3. Farmer's field 4. Threshing yard 5. Fallow |
| | | 6. Farm store 7. Market 8. Garden 9. Institute 10. _____ |
| STATUS | : | 1. Wild 2. Weed 3. Landrace 4. Primitive cultivar 5. Breeders line |
| FREQUENCY | : | 1. Abundant 2. Frequent 3. Occasional 4. Rare |
| MATERIAL | : | 1. Seeds 2. Fruit 3. Inflorescence 4. Roots 5. Tubers 6. Rhizomes 7. Suckers |
| | | 8. Live plants 9. Herbarium 10. _____ |
| SAMPLE TYPE | : | 1. Population 2. Pure line 3. Individual plant |
| SAMPLE METHOD | : | 1. Bulk 2. Random 3. Selective (non-random) |
| HABITAT | : | 1. Cultivated 2. Disturbed 3. Partly disturbed 4. _____ |
| DISEASE SYMPTOMS | : | 1. Susceptible 2. Mildly susceptible 3. Tolerant 4. Resistant 5. Immune |
| INSECT/PEST/ | : | 1. Mild 2. Moderate 3. High |

NEMATODE INFECTION

| | | |
|---|---|---|
| CULTURAL PRACTICES | : | 1. Irrigated 2. Rainfed 3. Arid 4. Wet 5. _____ |
| SEASON | : | 1. Kharif 2. Rabi 3. Spring-summer |

Approx. Sowing Date_____ Approx. Harvesting Date_____

| | | |
|---|---|---|
| ASSOCIATED CROP | : | 1. Sole 2. Mixed with_____ |
| SOIL COLOUR | : | 1. Black 2. Yellow 3. Red 4. Brown 5. _____ |
| SOIL TEXTURE | : | 1. Sandy 2. Sandy loam 3. Loam 4. Silt loam 5. Clay 6. Silt |
| STONINESS | : | 1. Stony 2. Pulverized 3. _____ |
| LAND ASPECT | : | 1. Level 2. Crest 3. Escarpment 4. Rounded summit 5. Upper summit 6. _____ |
| SLOPE | : | 1. Terrace 2. Lower slope 3. Open depression 4. Closed depression |
| TOPOGRAPHY | : | 1. Swamp 2. Flood plain 3. Level 4. Undulating 5. Hilly dissected |

6. Steeply dissected 7. Mountainous 8. Valley

| | | |
|---|---|---|
| AGRONOMIC SCOPE | : | 1. Very poor 2. Poor 3. Average 4. Good 5. Very good |

*ETHNOBOTANICAL USES*

| | | |
|---|---|---|
| PART(S)KIND | : | 1. Stem 2. Leaf 3. Root 4. Fruit 5. Flower 6. Whole plant 7. Seed 8. Others |

1. Food 2. Medicine 3. Fibre 4. Timber 5. Fodder 6. Fuel 7. Insecticide/Pesticide

8. Others

| | | |
|---|---|---|
| INFORMANT(S) | : | 1. Local Vaidya 2. House-wife 3. Old folk 4. Grazier/Shepherd 5. Others |

FARMER'S/DONOR'S NAME:_____ ETHNIC GROUP_____

Collector No.        Collector No.        Collector No.        Collector No.

PLANT CHARACTERISTICS/USES ADDL. NOTES:_____

_____

_____

_____

_____

_____

_____

_____

_____

*Source*: ICAR-NBPGR, New Delhi.

## IC Number Performa for Collected Germplasm

### PASSPORT INFORMATION DATA

*Collector/Developer's Name*: Ex. Dr D.R. Bhardwaj, Principal Scientist (Vegetable Science)

*Address*: Ex. ICAR-Indian Institute of Vegetable Research, Post Office-Jakhini (Shahanshahpur) Varanasi-221305 (U.P.)

| Sl.No. | Collector No. | IC No. | Crop Name | Botanical Name | Vernacular Name | Cultivar/ Hybrid/ Wild | Types of Material | Date of Collection/ Develoment |
|---|---|---|---|---|---|---|---|---|
| | | | | | | | | |

| Sl.No. | Source | Frequency | Sample Type | Habitat | Village | District | State |
|---|---|---|---|---|---|---|---|
| | | | | | | | |

# References

1. Allard, R.W. 1960. Principles of Plant Breeding. John Wiley and Sons, London.

2. Allard, R.W. 1970. Population structure and sampling methods. In: Frankel, O.H. and Bennett, E. (eds.), Genetic Resources in Plants-Their Exploration and Conservation. Blackwell, Oxford and Edinburgh, p. 97-107.

3. Arora, R.K. and Nayar, E.R. 1984. Wild Relatives of Crop Plants in India. ICAR-NBPGR Scientific Monograph No.7, National Bureau of Plant Genetic Resources, New Delhi. p. 90.

4. Arora, R.K. and Pandey, A. 1996. Wild Edible Plants of India. ICAR-National Bureau of Plant Genetic Resources, New Delhi, p. 294.

5. Bennett, E. 1965. Plant Introduction and genetic conservation. p. 27-113. In: Genecological aspect of an urgent world problem. *Scottish Plant Breed. Res Sta.*

6. Bennett, E. 1970. Tactics of plant exploration. In: Genetic Resources in Plants: Their exploration and conservation. Frankel, O.H. and Bennett, E. (eds.), Blackwell, Oxford. p. 157-179.

7. Bezbaruah, H.P. 1968. Genetic improvement of tea in North-east India-its problems and possibilities. *Indian J. Genet.*, **28A:** 126-134.

8. Bhag, M. 1992. Biodiversity utilization and conservation in underutilized plants: Indian perspective.

9. Bhandari, D.C and Pradeep K. 2010. Trait specific germplasm collection: What ecology can offer? In: Training manual on molecular markers for efficient management and enhancing utilization of plant genetic resources. ICAR-NBPGR, New Delhi. p. 35-41.

10. Bogyo, T.P., Poreceddu, E. and Perrino, P. 1980. Analysis of sampling strategies for collecting genetic material. *Econ. Bot.*, **34:** 160-174.

11. Bothmer, R. Von and and Seberg, O. 1995. Strategies for the collecting of wild species. In: L. Guarino, V. Ramanatha Rao and R. Reid (Eds.) Collecting Plant Genetic Diversity. Technical Guidelines. CABI International, U.K., p. 93-111.

12. Bray, R.A. 1983. Strategies for gene maintenance. In: Mclvor, J.G. and Bray, R.A. (eds.). Genetic resources of forage plants. CSIRO, Melbourne.

13. Brown, A.H.D. and Briggs, J.D. 1991. Sampling strategies for genetic variation in *ex situ* collections endangered plant species. In: Falk, D.A. and Holsinger, K.E. (eds.). Genetics and conservation of rare plants. Oxford Univ. Press, New York, p. 99-119.

14. Brown, A.H.D. and Marshall, D.R. 1995. A basic sampling strategy: Theory and practice. In: Collecting Plant Genetic Diversity. Guarino, L., Ramanath Rao, V. and Reid, R. (eds.). CAB International. Wallingford, Oxon OX108DE, UK, p. 727.

15. Bunting, A.H. and Kuckuck, H. 1970. Ecological and agronomic studies related to plant exploration. In: Genetic Resources in Plants: Their exploration and conservation. Frankel, O.H. and Bennett, E. (eds.), Blackwell, Oxford. p. 181-188.

16. Burkil, I.H. and Finlow, R.S. 1907. Races of jute. *Agric. Ledger*, **14**: 41-137.

17. Chattopadhyay, A., Dutta, S., Hazra, P. and Sarkar, G. 2011. Conservation and Utilization of Vegetable Germplasm at Bidhan Chandra Krishi Viswavidyalya. Nat. Symp. on 'Vegetable Biodiversity' Held at JNKVV, Jabalpur (M.P.), p. 131-140.

18. Clark, R.L., Widrlechner, M.P., Reitsma, K.R. and Block, C.C. 1991. Cucurbit Germplasm at the North Central Regional Plant Introduction Station, Ames, Iowa. *HortSci.*, **26**: 326 and 450-451.

19. Crossa, J., Harnandez, C.M., Bretting, P., Eberhart, S.A. and Taba, S. 1993. Statistical genetic considerations for maintaining germplasm collections. *Theor. Appl. Genet.*, **86**: 673-678.

20. Esquinas-Alcazar, J.T. and Gulick, P.J. 1983. Genetic resources of Cucurbitaceae. Int. Board for Plant Genet. Resources, Rome.

21. Fordham, R.A. 1971. Field population of deermice with supplemental food. *Ecology*, 52: 138-146.

22. Frankel, O.H. 1957. The biological system of plant introduction. *Jr. Aust. Inst. Agric. Sci.*, **20**: 302-307.

23. Frankel, O.H. and Bennett, E. (eds.). 1970. Genetic Resources in Plants: Their Exploration and Conservation. Blackwell, Oxford, p. 538.

24. Frankel, O.H. and Hawkes, J.G. (eds.). 1975. Crop Genetic Resources for Today and Tomorrow. Cambridge Univ. Press, London, p.492.

25. Gauchan, D., Joshi, B.K., Ghimire, K.H. and Sthapit, B. 2016. Assessment of loss of on-farm diversity of traditional crops in earthquake affected mountain regions of Nepal. 1st International Agrobiodiversty Congress (Session: III. *In-situ* and On-farm conservation), held at New Delhi, 6-9 November 2016; p. 810.

26. Gauchan, D., Joshi, B.K., Sthapit, S., Ghimire, K., Gautam, S., Poudel, K., Sapkota, S., Neupane, S., Sthapit, B. and Vernoy, R. 2016. Post-disaster revival of the local seed system and climate change adaptation: A case study of earthquake affected mountain region of Nepal. *Indian J. Plant Genet. Resour.*, 29(3): 119.

27. Gautam, P.L., Karihaloo, J.L., Kumar, A. Kochhar, S. and Singh, B.M. 2001. Vegetable germplasm management: Development, issues and strategies. In: Emerging scenario in vegetable research and development (G. Kalloo and Singh, K., eds.). Research periodicals and Book Publishing House, New Delhi, India, p. 45-57.

28. Gill, K.S. 1984. Research imperatives beyond the green revolution in the third world. In: Human Fertility, Health and Food Impact on Molecular Biology and Technology. p. 195-231. D. Phelt (ed.). United Nations Fund for Population Activities, New York.

29. Gregorious, H.R. 1980. The probability of losing an allele when diploid genotypes are sampled. *Biometrics*, **36**: 643-652.

30. Harlan, J.R. 1974. Cereals-Sorghum. In: Handbook of Plants Introduction in Tropical Crops. *FAO Agricultural Studies*, **93**: 12-16.

31. Harlan, J.R. 1975. Geographic patterns of variation in some cultivated plants. *J. Heredity*, 66: 182-191.

32. Harlan, J.R. 1975. New uses of old herbals. *Non Solus*, 2: 12-20.

33. Harlan, J.R. 1975. Our vanishing genetic resources. *Science*, 188: 618-621.

34. Hawkes, J.G. 1976. Manual for field collectors-seed crops. FAO, Rome, Italy, p.33.

35. Hawkes, J.G. 1976. Sampling gene pools. Proc. NATO Conf. Conservation of threatened plants. Sec. I. *Ecology*. Plenum London.

36. Hawkes, J.G. 1980. Crop genetic resources field collection manual. International Board for Plant Genetic Resources and European Association for Research on Plant Breeding (Eucarpia), IBPGR, Secretariat, Rome. p. 37.

37. Howard, A. and Howard, G.L.C. 1910. Wheat in India, its production, varieties and improvement. Thacker Spinn and Co. Kolkata.

38. Jain, S.K. 1975. Population structure and the effects of breeding system. In: Frankel, O.H and Hawkes, J.G. (eds.) Crop Genetic Resources for Today and Tomorrow. p. 15-36. Cambridge University Press, Cambridge.

39. Joshi, B.D. and Mehra, K.L. 1983. Genetic variability and distribution of amaranths genetic resources in the Himalayas. XVth International Congress of Genetics, New Delhi, Dec., 12-21. Abs. Nop. 983, V.L. Chopra *et al.* (eds.), Oxford-IBH. New Delhi.

40. Kimura, M. and Crow, J.F. 1964. The number of alleles that can be maintained in a finite population. *Genetics*, **49**: 725-738.

41. Koppar, M.N. 1998. Strategies in plant genetic resources: Exploration and collection. In: Plant Germplasm Collecting: Principles and Procedures Gautam, P.L.; Dabas, B.S.; Srivastava, U. and Duhoon, S.S. (eds.), ICAR-NBPGR, New Delhi.

42. Koppar, M.N. and Rai, M. 1994. Guidelines for germplasm collection expedition. In: Plant Genetic Resources: Exploration, Evaluation and Maintenance. Pub. ICAR-NBPGR, New Delhi. p. 61-69.

43. Lawrence, M.J., Marshall, D.F. and Davies, P. 1995. Genetics of genetic conservation. II. Sample size when collecting seed of cross pollinated species and the information that can be obtained from the evaluation of material held in gene banks. *Euphytica*, 84: 101-107.

44. Maluleke, N., Mokoena, M., Raimondo, D.C., Dulloo, M.E. and Magosbrehm, J. 2016. Identification and collection of priority crops wild relatives in three province of South Africa. 1st International Agro-biodiversity Congress (Abs. Session: Crop Wild Relatives: Back to the Wild to Save the Future), held at New Delhi; 6-9 November 2016, p. 320.

45. Marshall, D.R. and Brown, A.H.D. 1975. Optimum sampling strategies in genetic conservation. In: Frankel, O.H. and Hawkes, J.G. (eds.) Crop Genetic Resources: Today and Tomorrow, p. 53-80, Cambridge University Press, Cambridge.

46. Marshall, D.R. and Brown, A.H.D. 1983. Theory of forage plant collection. In: McIvor J.G. and Bray, R.A. (eds.) *Genetic Resources of Forage Plants*, p. 135-148.

47. Mohanty, R.B. and Tripathy, B.K. 2011. Hotspots of biodiversity. *Agrobios News Letter*, 10(7): 69-70.

48. Namkoong, G. 1988. Sampling for germplasm collections. *HortSci.*, 23: 79-81.

49. Ng'uni', D., Munkombwe, G., Mwila, G., Brehm, J.M., Maxted, N., Thormann, I., Gaisberger, H. and Dulloo, M.E. 2016. Spatial analysis of occurrence data of CWR taxa as tools for selection of sites for conservation of priority CWR in Zambia. 1st International Agro-biodiversity Congress (Abs. Session: Crop Wild Relatives: Back to the Wild to Save the Future), held at New Delhi; 6-9 November 2016, p. 321.

50. Oka, H.I. 1969. A note on the design of germplasm preservation work on grain crops. *SABRAO Newsl.*, 1: 127-137.

51. Oka, H.I. 1975. Consideration on the population size necessary for conservation of crop germplasm. Pages 57-63 in T. Matsau (ed.). JIBP Synthesis S., University of Tokyo Press, Tokyo.

52. Panwar, N.S., Rathi, R.S. and Bhatt, K.C. 2016. Agro-biodiversity captured in rescue missions executed in natural calamities affected areas of India. 1st International Agro-biodiversity Congress (Abs.), held at New Delhi; 6-9 November, 2016' p. 13.

53. Pradheep, K., Ahlawat, S.P., Bhatt, K.C. and Pandey, A. 2016. Prioritization of vis-à-vis Germplasm collection status of crop wild relatives in India. 1st International Agro-biodiversity Congress (Abs. Session: Crop Wild Relatives: Back to the Wild to Save the Future), held at New Delhi; 6-9 November 2016, p. 322.

54. Qualset, C.O. 1975. Sampling germplasm in a centre of diversity: An example of disease resistance in Ethiopian Barley. In: Frankel, O.H. and Bennett, E. (eds.), Crop genetic resources today and tomorrow. Cambridge Univ. Press, Cambridge.

55. Ram, D. and Srivastava, U. 1999. Some lesser known minor cucurbitaceous vegetables: Their distribution, diversity and uses. *Indian J. Pl. Genet. Reour.*, 12(3): 307-316.

56. Ram, D., Gupta, V.K., G. KAlloo, Tomar, J.B. and Srivastava, U. 1999. Green hot spot: Eastern Uttar Pradesh-A core site for intensive vegetable cultivation and production. *Indian J. Pl. Genet. Reour.*, 12(2): 247-253.

57. Rathi, R.S., Ram, D., Phogat, B.S. and Raiger, H.L. 2006. Collection of spine gourd (*Momordica dioica* Roxb. Ex Willd) from Bundelkhand and adjoining areas. *The Indian Forester*, 132(6): 757-762.

58. Robinson, R.W. 1989. Genetic resources of the Cucurbitaceae, p. 85a-85j. In: Thomas, C.E. (ed.). Proc. Cucurbitaceae 89: Evaluation and enhancement of cucurbit germplasm. USDA Vegetable Lab.

59. Sapra, R.L., Narain, P. and Chauhan, S.V.S. 1997. A general model for sampling size determination for collecting germplasm. *J. Biosci.*, 23: 647-652.

60. Sidhu, A. S., Yadav, S.K., Islam, S., Singh, S.P. and Singh, S. 2008. Germplasm Introduction in Vegetable Crops: Achievements and Opportunities. In: Training Manual on Germplasm Exchange: Policies and Procedures in India, Arjun Lal, Deep Chand, VandanaTyagi, NidhiVerma, S.K. Yadav, Vandana Joshi, S.P. Singh and Surender Singh (eds). p. 256 – 267.

61. Singh, B.D. 2012. Plant Introduction. In: Plant Breeding: Principles and Methods, Kalyani Publishers, p.54.

62. Singh, O.V. and Singh, A.K. 2016. Plant genetic resources of arid legumes are the sources of useful genes for biotic stress resistance and high yield. 1st International Agrobiodiversity Congress (Abs. Session: Trait Discovery and Enhanced Use of PGR), 6-9 November, held at New Delhi, p. 197.

63. Srivastava, U. 1998. Exploration and germplasm collection: planning and logistics. In: Plant Germplasm Collecting: Principles and Procedures (Eds. Gautam, P.L.; Dabas, B.S.; Srivastava, U. and Duhoon, S.S.), ICAR-NBPGR, New Delhi.

64. Thuljaram Rao, J. and T.N. Krishnamurthy. 1968. Accelerating genetic improvement in sugarcane *Indian J. Genet.*, **28A:** 88-96.

65. Unnikrishnan, M.S., Edison, S. and Sheela, M.N. 2005. Tuber Crops and their wild relatives. In: *Tami Nadu Biodiversity Strategy and Action Plan* (ed.) R. Annamalai, Published by Tamil Nadu Forest Department Chennai, p. 159-168.

66. Vavilov, N.I. Phyto-geographical basis of plant breeding. The origin, variation, immunity and breeding of cultivated plants (K.J. Choster, translator). *Chronica Botanica*, **13:** 366.

67. Weir, B.S. 1990. Genetic data analysis. Sinauer Associates, Sunderland.

68. Yonezawa, K. 1985. A definition of the optimal allocation of effort in conservation of plant genetic resources with application to sample size determination for field collection. *Euphytica*, 34(2): 345-354.

69. Yonezawa, K. and Ichihashi, H. 1989. Sample size for collecting germplasm from natural plant population in view of genotypic multiplicity of seed embryos borne on a single plant. *Euphytica*, **41:** 91-97.

70. Zeven, A.C. and Zhukovsky, P.M. 1975. Dictionary of Cultivated Plants and Their Centres of Diversity. PUDOC, Wagemningen, the Netherlands. p. 21.

# 5

# Germplasm Evaluation, Characterization and Development of Core Set

## Introduction

Genetic resources as raw materials enable the breeders to develop productive combinations of traits in elite cultivars to fulfill the needs in various agro-systems. Huge collections, introductions and exchanges of germplasm across the globe posed serious management problems in terms of time, money and land required for raising or regeneration of crops on one side and planning for characterization and evaluation, adaptation of suitable lists of characters for data recording and documentation and networking, on the other. The environment invariably influences the expression of most valued traits in crop improvement, such as yield, agronomic performance, insect and disease resistance, quality evaluation (oil, protein content, secondary metabolites *etc*.) and response to abiotic stresses (moisture-deficient or excess; temperature-heat or cold) and salinity. In this situation characterization and evaluation of germplam (among wide variety of genetic resources) is a prerequisite and essential component for utilization in crop improvement. Unless it is properly evaluated, it has no practical utility and hence evaluation forms most important aspect in germplasm resources activity. The detailed evaluation in a particular environment determines the potentiality of an accession for specific purpose in crop improvement. Breeders are interested to know the characteristics of assembled germplasm. The value of any germplasm collection is directly correlated to the extent of information available for each accession. Data on morphological, agronomic and nutritional traits and reaction to biotic and abiotic stresses increases the importance of the germplasm. Assemblage of data on the important characteristics of a germplasm collection not only distinguishes accessions within a species, but also enables grouping of accessions development of core collection, identification of gaps and retrieval of valuable germplasm for breeding programmes. The plant genetic resources handling systems are evolving through several phases of scientific refinement and technical precision-making. The ICAR-NBPGR, New Delhi is essentially users driven national organization devoted to conduct, promote and coordinate activities pertaining to the plant exploration and collection, germplasm exchange, quarantine, multiplication, distribution, characterization, evaluation, documentation and conservation of plant genetic resources. A distinction may be made, for practical purpose, between evaluation and characterization as described in Table 5.1.

## Evaluation of Plant Genetic Resources

Evaluation consists of:

*'Assessment of the germplasm accessions for their various features or traits of some known or potential use'*.

Or

*'Germplasm evaluation is the description of the material in a collection'*.

Above definition comprises of various steps involving the activities beginning from seed increase, characterization, preliminary evaluation and also detailed evaluation and documentation. Generally,

**Table 5.1: Guidelines for Characterization and Evaluation**

| Characterization Base | Evaluation Base |
|---|---|
| **i. Pre-requisites** | **i. Pre-requisites** |
| Characterization of germplasm should be carried out during the initial stage of acquisition. As per international standards, 60 per cent of accessions held in a genebank should be characterized within five to seven years of acquisition or during the first regeneration cycle (FAO, 2013). | Evaluation should be undertaken in germplasm accessions which are already characterized and where enough quantity of planting material is available. They should conform to standardized and calibrated measuring formats. |
| Characterization should be based on standardized and calibrated measuring formats and should follow internationally/nationally agreed descriptor list, such as those published by IBPGR/IPGRI/Bioversity International, UPOV, USDA and ICAR-NBPGR. | Prior to evaluation, planning and specific strategies for evaluation should be developed considering the breeding behavior and biological status of germplasm. |
| Data recording should be preferably be carried out by trained staff. All data should be validated by respective crop curators and documentation officers before its entry in publically accessed publications/databases. | Germplasm, evaluation being a multi-disciplinary activity, should involve a germplasm curator, plant breeder, physiologist, pathologist, entomologist, biochemist, *etc.*, for a meaningful evaluation. A network/coordinated approach at multiple locations under different agro-climatic zones in collaboration with crop-based institutes, Project Directors/Coordinators and All India Coordinated Research Project (AICRP) centres is recommended. |
| The germplasm should have an assigned national identity, along with complete passport information. The latter is essential for selection of appropriate characterization strategy. For example, accessions should preferably be grouped as per their growth habit and maturity duration to avoid the effect competition among accessions. | Dissimilarity in the incidences of pests, severity of abiotic stresses and the fluctuations in environmental and climatic factors in the field impact the accuracy of data and should be mitigated through reasonably replicated, multilocational, multi-season and multi-year evaluations. |
| Biological status of the germplasm should also be known in advance to determine the characterization strategy. For example, in wild species and inbreds which may have few seeds and/or be difficult to germinate, special treatments need to be given for their proper germination and less number of plants may be considered for recording observations. | Measurement of quality traits like oil, protein, secondary metabolites, *etc.* requires specialized equipment which needs a lot of investment. For recording such parameters, strong linkages should be developed with referral laboratories. |
| In addition to agro-morphological traits, genomics including molecular marker technologies may also be utilized for characterization of germplasm, if facilities are available. | |
| **ii. Site for Characterization** | **ii. Site for evaluation** |
| Characterization should be carried out in a suitable environment, preferably at the site or near the site of germplasm collection, for proper expression of traits and also to avoid genetic shift. | Accessions should be evaluated in the area of their adaptation or under similar environmental conditions. |
| Characterization should be performed in a single environment. | The target traits must be prioritized and number of locations should be selected keeping in view the similar environmental matching with the major agro-ecological zones of the crop. |
| | Evaluation trials should be carried out in at least three diverse environments to minimize Genotype x Environment (G x E) interaction. |
| **iii. Field plot layout** | **iii. Experimental details** |
| The field experiment should be conducted with proper experimental design depending upon the quantity of material available and the number of accessions. The blocks should be laid out across the soil fertility gradient. | Experiments should be conducted with proper experimental design, depending upon the number of accessions to be evaluated. The blocks should be laid out across the soil fertility gradient of the experimental plot. |
| Statistically sound replicated design should be selected. Augmented block design (ABD) should be used for large number of accessions (more than 50) with less quantity of planting material. However, for less number of accessions randomized block design (RBD) may be followed. | For large number of accessions and with high soil heterogeneity level, Augmented block design (ABD) should be adopted. In ABD, the checks should be replicated in each block after separate randomization of checks within a block. ABD generates four critical differences (CD) to facilitate comparison between accessions within a block or between the blocks or with the checks. |
| The checks should be replicated in each block after separate randomization of checks within a block in ABD, whereas in RBD, the checks should be randomized together with the accessions in each replication. | For few or promising accessions and less soil heterogeneity evaluation should be done in randomized block design (RBD) wherein the checks should be randomized along with the accessions in each replication. |
| The number of checks should be at least three or more and should represent the type of germplasm. Preferably, at least one recent national check as well as one locally adapted check should be used for meaningful comparison. | The number of checks will depend upon the crop, nature of Germplasm and the parameters to be recorded. At least three or more checks should be used for orthogonal comparison. However, one local check should be used for comparative assessment of germplasm. |

| Characterization Base | Evaluation Base |
|---|---|
| Plot size depends on the crop and the amount of available seed/planting material. In general, single row plot of 3-5 m length with sufficient number of plants (at least 10) for observation should be used for characterization. | The experimental plot should have at least three rows of 3-5 m length for each accession with sufficient gap (as recommended for that crop) in between the accessions for ease of recording observations and to maintain identity of the accession. The observations should be recorded on the plants from the middle row to minimize the border effects. The number and row length should be more for cross-pollinated species than those for self-pollinated ones. |
| For clonally propagated crops, fully grown plants available in the field genebank should be used for characterization. | After initial evaluation, the promising accessions should be further evaluated in a replicated design and under controlled/artificial conditions. |
| | Evaluation experiments conducted for nutrient use efficiency (NUE) should take into account soil heterogeneity within the plot and its associated effects on germplasm. |
| | In fruit crops (less in vegetable crops), the rootstocks germplasm should be evaluated properly for desirable traits and categorized accordingly. Appropriate rootstocks should be used. |

| iv. Raising of crop | iv. Raising of crop |
|---|---|
| Standard agronomic practices as prescribed for raising a good crop should be followed. These include proper row and plant spacing, fertilizer application, weeding, irrigation, plant protection measures, *etc.* | Standard and recommended agronomic practices as prescribed for raising a good crop including proper row and plant spacing, fertilizer application, weeding, irrigation, plant protection measures, *etc.* need to be followed. |
| All plants should be properly labeled with suitable tags, immediately after sowing/planting. | The harvesting schedule should be as per the requirement of the crop. |
| During the period of growth and flowering, attention should be given to minimize natural cross pollination and physical/mechanical contamination. | For evaluation of biotic stresses, plant protection measures should not be applied so that optimum pest population is maintained to record the reaction of the germplasm to the stress. |
| The harvesting schedule should be followed as per the requirement of the crop. For example, in case of horticultural crops and asynchronously maturing crops. Multiple picking is practiced. | |

| v. Population and sample size | v. Population and sample size |
|---|---|
| The appropriate population size will depend upon the breeding behavior of the crop as well as the biological status of the material. | The population size for evaluation depends upon the breeding behavior of the crop as well as the biological status. Generally, population size should be large (>50 plants) for cross-pollinated species as compared to self-pollinated species. |
| If characterization is combined with regeneration, the population size should be large enough in cross-and often cross-pollinated crops to avoid genetic drift. In order to minimize the loss of heterozygosity, inbreeding depression and change in allelic frequency, a minimum number of 50-100 plants are required in a population. | Since evaluation is usually done for quantitative traits, the sample size should be optimum to take observations from randomly selected plants. In case of self-pollinated and vegetatively propagated crops, 10-15 plants are sufficient whereas for cross-pollinated or often cross-pollinated crops, there should be at least 20 plants. |
| The sample size for recording observations will vary depending upon the type of descriptors. For qualitative descriptors, the observation should be visually recorded on a few plants and scored once, while for quantitative descriptors, there should be at least 10-20 plants per accession. | In tree crops (some tree vegetable crops), where generally number of plants required for observation is few, single plant or even individual branch may be considered as sample replicate for recording observations, depending upon the availability of material. |

| vi. Descriptors and descriptor states | vi. Evaluation for agronomic traits |
|---|---|
| The globally accepted descriptors and descriptor states should be used to record the observation for characterization of germplasm. The descriptors developed by Bioversity International (formerly known as IPGRI/IBPGR) or ICAR-NBPGR, minimal descriptors for characterization and preliminary evaluation should be used. | For evaluation of agronomic traits, in large number of accessions with diverse adaption, unreplicated trial may be followed. In case of unreplicated experiments, the number of environments may be increased for greater precision of the experiment. |
| Whenever descriptors and descriptor states are not available, these should be developed by Germplasm curators in consultation with crop experts for a particular crop species. | The stage of recording of descriptors and descriptor states should be strictly followed as defined in the descriptors' list. |
| The descriptor states which are not mentioned in the descriptors' list but observed during characterization should be included in 'others' category. | In fruit tree crops (some tree vegetable crops), only untrained, non-juvenile trees/plants should be used for recording vegetative growth parameters. |
| The descriptor should be recorded by well-trained staff using calibrated and standardized descriptors and descriptor states, as indicated in the internationally/nationally agreed and published descriptors lists. | **vii. Evaluation for Biotic stress** |
| Characterization descriptors should be recorded using standard instruments/items like refractometer for total soluble solids (TSS), RHS Colour Chart for colour, *etc.* | Evaluation for biotic stress should be carried out in specific environments, especially at hot spots. Such screening should be followed by rigorous testing under artificial conditions. |

| Characterization Base | Evaluation Base |
|---|---|
| Notes on occurrence of important insect/pests on the germplasm should also be the part of descriptors and descriptor states of characterization. | Since, there are a number of biotic stresses that are of economic importance, prioritization should be done. It should be noted that the evaluation of Germplasm against biotic stresses means evaluation against the pest and not against the disease because similar disease symptoms may be reported by different pests. |
| Reproducible molecular markers like Amplified Fragment Length Polymorphism (AFLPs), Simple Sequence Repeats (SSRs) and Single Nucleotide Polymorphisms (SNPs) should be used for molecular characterization. | The pests and their race/strain/isolate/biotype in a particular location should be clearly identified employing morphological or molecular tools. |

The identified pest and its race/strain/isolate/biotype should be mass cultured for inoculation.

In case of viruses, infectious viral construct needs to be prepared for challenge inoculation.

Preliminary screening should be done for one year with large number of accessions to narrow down the numbers to a manageable extent. For such a purpose, only per cent disease incidence or pest infestation should be recorded and field tolerant accessions (<20 per cent incidence) to be selected.

For advance screening (at least two years) of the selected field-tolerant accessions, following guidelines need to be followed:

i. The advance screening plot should be different from the germplasm maintenance plot as in the former no plant protection measures are adopted and the germplasm is exposed to the maximum load of inoculum of pest.

ii. The experiments should be done in replicated trials with susceptible check as an infector row. Preferably, one line of infector row is planted after every three rows of germplasm. There should be border rows of susceptible check on all sides to facilitate spread of the pest infestation.

iii. Observations should be recorded at regular intervals. For spreading of the observations on fungal, bacterial and viral diseases, standard evaluation system scale should be followed.

iv. For recording of the data on defoliators, per cent infestation in each accession should be recorded. For sucking insect-pest, the number of insects per unit plant/leaves/inflorescence should be counted.

v. For soil-borne pathogens, germplasm should be grown in sick plots which should be maintained with optimum inoculum load. The susceptible check should show at least 80 per cent infection in sick plot to corroborate the optimum inoculum load.

vi. For air-borne pathogens, spores suspension should be sprayed at regular intervals.

vii. For virus and other vector-borne pathogens advance screening under natural field condition is carried out for two years followed by challenged inoculation of the promising accessions under controlled conditions.

ix. The screening of germplasm for nematode should be done under optimum load of nematode populations in pot culture.

**viii. Evaluation for Biotic stress**

Evaluation of germplasm for tolerance to different abiotic stresses should be carried out under well-defined controlled conditions so that the optimum stress can be imposed at the desired stage.

Standard checks identified for specific abiotic stress should be used for proper comparison of the germplasm.

Descriptors for evaluation like canopy temperature, reflectance, *etc.* at different phenophases should be used for high temperature and drought stress.

Preliminary screening for abiotic stress should be done for one year with large number of accessions under field conditions identified for a particular abiotic stress, to narrow down the numbers to a manageable extent. Further evaluation for abiotic stresses should be performed under controlled conditions as follows:

| Characterization Base | Evaluation Base |
|---|---|
| | i. Evaluation of germplasm for drought tolerance should be undertaken in drought plots with rain-out shelters and with well-defined moisture conditions. |
| | ii. Plots with different water depths should be used to evaluate germplasm for submergence tolerance. |
| | iii. Micro-plots with well-defined electrical conductivity including specific salt/ions should be used for evaluating germplasm under saline/alkaline/acidic conditions. |
| | iv. Germplasm evaluation for light and temperature stresses should be carried out under phytotron facility. |
| | **ix. Recording quality traits** |
| | Germplasm selected for quality evaluation should be properly labeled, accompanied with passport information and with complete experimental details. |
| | Adequate quantity of pure, clean and preferably dry material, free from any infestation and harvested at appropriate stage, should be used for analysis of quality traits. |
| | In aromatic crops, material should be analyzed immediately or transported under controlled conditions to minimize loss of weight and low volatiles. |
| | Material should be stored under ambient conditions to retain the quality compounds. |
| | In medicinal plants, shade-dried material should be analyzed for estimation of active constituents. |
| | Quality data should be expressed either on fresh weight basis (FBW) or dry weight basis (DWB) depending upon the nature of plant material/ germplasm. |
| | Quality parameters should be analyzed using standard protocols in a well-equipped laboratory. |
| | **x. Molecular characterization** |
| | The most common and recent molecular markers in germplasm characterization and evaluation should be used. |
| | Advances in next generation sequencing and the accompanying reduction in costs have led to increased use of sequencing based assays such as the sequencing of coding and non-coding regions, and genotyping-by-sequencing (GBS) in germplasm evaluation. |
| **vii. Documentation** | **xi. Documentation** |
| The characterization data should be documented for collation, analysis and retrieval after through verification. | The recorded data should be standardized to a uniform format and then entered in MS-Excel sheet for integrating into the main database at the earliest to avoid any loss of data. Wherever data loggers are used, these should be downloaded immediately into the computer. |
| The qualitative data may be analyzed for estimating allelic richness and evenness. | The quantitative data should be statistically analyzed following the standard experimental designs. |
| Catalogues may be developed using the qualitative and quantitative data. | |
| **viii. Core and mini core sets** | |
| A core collection contains a subset of accessions from the entire collection that captures, with a minimum of repetitiveness, the genetic diversity of a crop species and its wild relatives. | |
| Whenever the number of accessions is large (>1,000), characterization data is generally utilized to develop core set comprising~10 per cent of the entire collection representing the total variability in the germplasm to bring them to a manageable level. | |
| However, if the core set itself is too large to handle, a mini core set comprising~10 of the core set representing the total variability in the core set is developed for facilitating germplasm management in crop improvement. | |
| Several methodologies may be employed to arrive at the desired core and mini-core sets from the entire germplasm and core collections, respectively. The same should be validated through various statistical tools and techniques. | |

*Source*: ICAR-NBPGR, Guidelines for Management of Plant Genetic Resources in India, 2016.

germplasm accessions are evaluated for morphological, physiological, biochemical, plant pathological (*i.e.,* disease resistance), entomological (*i.e.,* insect resistance) and other features. The characters assessed must be related to the need of the breeders since they are the ones who are going to utilize the germplasm. Evaluation is the most critical step determining the utilization of a collection, and a poorly assessed germplasm collection is unlikely to be of any use to anyone. However, evaluation of germplasm is very difficult and time consuming so that the actual diversity present in many germplasm collections is yet to be assessed. Evaluation requires a team of specialists from the disciplines of plant breeding, physiology, biochemistry, pathology and entomology. Characterization, preliminary evaluation and further advance evaluation of germplasm are pre-requisite for utilization of plant genetic resources. Characterization and preliminary evaluation consists of recording a limited number of additional traits, thought desirable by the consensus of specialists/users of the particular crops/species, which help in identifying useful germplasm. This is very important for identification and help in designating core collection, identifying duplicates and planning future exploration. Advanced evaluation consists of recording a number of additional descriptors though to be useful in crop improvement. These include important agronomic traits, stress tolerance, disease and pest resistance and quality characters *etc.* Therefore, the need of the hour is to assess the worth of the introduced germplasm, which will be helpful in targeting future import requirements of the country (Tiwari, 1998; Sidhu *et al.* 2008). A large number of varieties of self-pollinated and cross pollinated vegetable crops have been released as primary introductions directly or through acclimatization. The introduced germplasm played important role in the development of new high yielding varieties when compared with local germplasm or another exotic germplasm (indicates evaluation) ultimately resulted selection.

'*Characterization is a part of preliminary evaluation only and this includes recording of highly heritable phenotypic characters*'.

## Nature of Existing Collections

The germplasm collection of any crop consists of diverse type of collection such as those derived from:

(i) Centres of diversity (primitive, natural hybrids; wild relatives and related species and genera.

(ii) Areas of cultivation (commercial type, obsolete varieties, primitive varieties), and

(iii) Breeding programmes (pure line from farmer's stock, elite varieties/hybrids, breeding lines mutants, polyploids, intergeneric hybrids).

## Characterization and Evaluation of Germplasm

In broad sense, the characterization and preliminary evaluation in the context of materials in collection, will be the responsibility of crop curators, while further characterization and evaluation should normally be carried out by the user (plant breeder). However, the data from further evaluation should be fed back to the curator/coordinator to record and update the catalogues for further use. Germplasm characterization and evaluation is the mandate of ICAR-NBPGR, New Delhi. Evaluation trials on specific crops as per allocation of crop responsibilities are done on regular basis at various centres. Depending upon the suitability of agro-climatic conditions *etc.* different crops may be evaluated at one or, more locations simultaneously; *e.g.* tomato is evaluated at ICAR-IARI, New Delhi and ICAR-IIVR, Varanasi and ICAR-IIHR, Bengaluru. It is desirable to have at least two years evaluation for documentation and preparation of catalogues. It is important to check-list the steps taken at various stages of germplasm characterization and evaluation for effective documentation. Germplasm collections are grown in uniform nurseries, at one or more locations, to prepare catalogues, using standard set of descriptors. Characterization refers to description of highly heritable traits that are uniformly expressed in all environments. These range from morphological, physiological and agronomical features to molecular markers.

## Guidelines for Preliminary Characterization and Evaluation

A few guidelines for planning/operations/follow-up in relation to characterization and evaluation are presented as under (Singh and Kochhar, 1994):

### 1. Criteria for Selecting Germplasm Lines for Characterization

The following categories of germplasm may be included:

(i) New collections through explorations

(ii) New exotic introductions

(iii) New accessions generated from parasexual methods/vegetative propagules/tissue culture raised propagules *etc.*

(iv) Samples redrawn from gene bank after long intervals to monitor the changes in expression of stable (characterization) traits may also be included, and

(v) Samples procured from other gene banks as duplicate sets to monitor the changes due to the location effect in character expression.

### 2. Characterization and Initial Seed Increase can be Simultaneously done

This would minimize the probability of changes in the genetic equilibrium particularly in outbred crops and thus ensure full expression of basic characterization traits.

### 3. Criteria for Selecting Germplasm Lines for Evaluation

'Evaluation' and 'Further Evaluation' are continuous activities of a germplasm resources programme. The set of germplasm accessions selected for evaluation may be non-overlapping, partially or completely overlapping with the previous year's set. The most optimum choice is evaluation of a set of germplasm for at least two years at one location. A blended choice of materials may be kept in mind for evaluation as follows:

(i) Materials available after initial seed increase/ characterization.

(ii) New accessions procured in sufficient seed quantity from indigenous explorations.

(iii) New exotic introductions obtained after quarantine in sufficient seed quantity.

(iv) Accessions from active collections which were rejected by the gene bank due to low germination or attack by storage pests.

(v) Old accessions from the active collections, selected at random or which were somehow escapes and are not available in long term storage.

### 4. Active Liaison of Evaluation Unit with other PGR Divisions

Germplasm evaluation unit should interact with germplasm Exploration and Collection unit to record the passport data of the accessions under evaluation. Similarly the passport data of exotic accessions should be obtained from Plant Introduction Reporters or Germplasm Exchange Unit. The feed-back from Germplasm Conservation Unit is essential for appropriate decision making for definite inclusion of the entries which were rejected during first year of evaluation.

### 5. Choice of Experimental Plot should be done well in Advance

Taking into consideration plant canopy of the crop under evaluation, adjoining crops, previous crop in rotation in the same plot, fertility and *Rhizobial* status of the experimental plot and irrigation facilities *etc.* A proper time gap between the harvesting operation of the previous crop and sowing of current evaluation experiment may be ensured, so that 2-3 field operations *viz.* preparatory tillage, *paleva* and fine bed preparation *etc.* could be done before planting.

### 6. Seed Packets should be Prepared well in Advance

Seed packet should be properly labeled for Crop Name, Accession Identifier Number (EC/IC/NIC/ Collector No., or Variety name), and Plot Number. For the convenience during sowing operations the packets should be serially arranged as per plot numbers and bundles of 10-15 packets plus one check variety should be made.

### 7. Mechanization Sowing may be done where Feasible

Tractor driven 'HEGE' machine can also be used for planting large number of accessions in order to save manual labour. At least three persons are needed for mechanical operations:

(i) One driver

(ii) One seed operator

(iii) One changer

### 8. Use of Design

The design of experiment would invariably be Augmented Block Design (when size of germplasm is large for evaluation) unless otherwise specified. Number of checks may be 3-5 which should be placed after suitable gap of 10-15 accessions, depending upon the size of the experiment, so as to ensure that each check is replicated 4-5 times and the error degrees of freedom is not less than the minimum required *i.e.* 12.

### 9. Labeling

It should be neatly done by the time the crop is 2 weeks old or 10-15 cm high. A field label should only contain the information relating to: Plot Number (Right hand upper corner), Date of Planting (Right hand lower corner), Crop Name (Upper middle portion) and Accession Identifier Number (Middle portion). Mention of replication number, if applicable, should also be there, in the right hand lower corner.

### 10. Packages of Agronomical Practices

Optimal package of practices for the crop suitable under local conditions should be followed including a judicious balance between pre-emergence weed control and weeding operations.

### 11. Disease/Pest Control

There should be an optimum balance between the screening (data recording) for natural incidence of disease or pests and the more essential part of saving the crop germplasm. Thus fungicides/pesticides should be timely arranged to save the germplasm once the disease incidence increases beyond an optimum level, on visual judgment basis, in the evaluation of experiment.

### 12. Use of Descriptors and their Descriptor States

Descriptor should be selected from the standard IBPGR descriptors, past experience on the crop or specific biotic or abiotic stress situations in such a way so that there is an optimum balance between the work load and the nature of information collected (data recorded).

**Table 5.2: List of Descriptors Published by IBPGR, Rome**

| Sl.No. | Descriptor/Crop | Year of Publication |
|---|---|---|
| 1. | Beet Descriptor | 1980 |
| 2. | *Colocasia* Descriptor | 1980 |
| 3. | Yam Descriptor | 1980 |
| 4. | Cruciferous crop genetic resources | 1981 |
| 5. | Genetic resources of *Amaranthus* | 1981 |
| 6. | *Vigna* species genetic resources | 1982 |
| 7. | *Phaseolus vulgaris* Descriptor | 1982 |
| 8. | Cowpea Descriptor | 1983 |
| 9. | Genetic resources of Cucurbitaceae | 1983 |
| 10. | *Abelmoschus* genetic resources | 1984 |
| 11. | Eggplant Descriptor | 1988 |
| 12. | Winged bean Descriptor | 1989 |
| 13. | *Xenthosoma* Descriptor | 1989 |
| 14. | *Raphanus* Descriptor | 1990 |
| 15. | Descriptor for *Beta* | 1991 |

The ICAR-NBPGR, New Delhi has developed suitable lists of descriptors and their descriptor states for a number of vegetable crops suited to Indian conditions, which are advocated for uniform documentation in the National Plant Genetic Resource System. Similarly, ICAR-IIVR, Varanasi has also developed minimal descriptor lists for vegetable crops.

### 13. Data Recording

Data recording should be judiciously planned so that it is recorded over a period of crop growth without crowding in only at harvest. Recording for characters like initial stand, vigour *etc.* should be completed in the first fortnight. Morphological characters like leaf colour, size, shape *etc.* should be recorded in the third/fourth week. Data recording on flowering/50 per cent flowering/pod setting/maturity *etc.*, type of phenological characters should be well spread over longer period of crop growth due to the fundamental variation in the set of germplasm in the field. Canopy, pod and ovule characteristics can be recorded well ahead of maturity. Most of the morpho-agronomic quantitative traits, height, branching, cluster of pod *etc.*, should be recorded on 5-plant basis at the time of harvest, whereas the seed characteristics can be recorded at post-harvest care stage.

### 14. Post-harvest Operation

Post-harvest operation should be quick, thereby ensuring minimum time-loss for storage in ambient transit conditions. Seed should be dried, if required, in shade, cleaned, labeled properly with duplicate labels (one inside the bag and one with the tying thread). The accessions would be suitably divided into smaller lots for supply to gene-bank, active collections store, duplicate set (if applicable) and sample of laboratory data on seed characteristics *etc.*

### 15. Up-to Dating the Germplasm Accessioning

Allocation of formal Accession Identifier Numbers *i.e.* "IC" Numbers, for the indigenous collections, should be got done for the new accessions, holding 'NIC' or 'Collector's Numbers, in which initial establishment or seed increase has been attained. This would avoid confusion at a later stage such as in cases when the accession is sent to gene bank without allocation of formal 'IC' Numbers.

### 16. Documentation

Computation of 5-plant data means, tabulation of data and their entry in the computer data registers from the manual data sheets should be done along with log-in of passport data for further processing and preparations of catalogues.

## Traits of Characterization/Evaluation

Observations are recorded on qualitative and quantitative traits.

### (i) Quantitative Traits

Quantitative traits subject to environmental factors and are responsible for adaptation and productivity. These include productivity, quality components, resistance to disease and pest as well as tolerance to adverse conditions or stress. For example a quantitative categorization of seed storage proteins profiles of 13 genotypes of *Solanum lycopersicum* L. was performed by Sodium Dodecyl Sulphate Polyacrylamide Gel Electrophoresis (SDS-PAGE). The banding patterns were characterized by 3 clear distinct zones *viz.* A, B and C. The unweighed pair group method using arithmetic average (UPGMA) analysis of 13 tomato genotypes was done and two major clusters obtained through seed proteins analysis expressed better grouping of genotypes. The dendrogram showed that the genotype EC519724 was most dissimilar from other genotypes (Kathayat *et al.*, 2016).

### (ii) Qualitative Traits

Qualitative traits include morphological, physiological and biochemical characters related to survival and scored using a number of checks to be determining variation within and among the traits.

Much emphasis is presently given on multi-locational and multi-displinary approach of germplasm evaluation. For preliminary evaluation locally adapted cultivars should be used as check and screening for specific diseases under controlled conditions or at hot spots should be carried out. Augmented design is invariably used due to the large number of germplasm holdings under evaluation. Care should be taken to minimize natural cross-pollinated contamination and erroneous labeling. While, regenerating care should be taken to preserve original structure and productivity of accession. Main dangers are due to inappropriate handling of materials during sowing/planting, harvesting, threshing, cleaning, sub-sampling, packing and labeling. Frequent regeneration should be also be avoided by producing sufficient seeds during initial seed increase and conservation in medium term storage in case of working collection and long term gene bank as base collection. IPGRI, Rome has developed model lists of descriptors (= characters) for which germplasm accessions of various crops should be evaluated. It should be kept in mind that even such accessions, which do not possess a trait of some value to the breeders, should be retained in the collection. This is because what seems to be of no value today may become a highly valuable feature tomorrow.

At BCKV, Kalyani (West Bengal) a total of 118 lines of tomato, 40 lines of brinjal, 60 lines of chilli, 24 lines of pumpkin, 15 lines of bottle gourd, 8 lines of cucumber, 21 lines of pointed gourd, 11 lines of spine gourd, 15 lines of dolchis bean, 18 lines of okra and 8 lines of cucumber have been maintained both through *in situ* and *ex situ* methods. A brief account of special characteristics of germplasm is presented in Table 5.3.

### Steps of Evaluation

Characterization, preliminary evaluation and further evaluation of germplasm are prerequisite for utilization of plant genetic resources. Until characterization is not done, and attributes/traits are not known, the germplasm has little practical utility. The evaluation of genetic resources is a multidimensional endeavor; involving various disciplines such as cytology, agronomy, genetics and biochemistry *etc.* the multidisciplinary participation to generate wide

spectrum information of the germplasm collection would lead to its meaningful utilization. Various steps in plant genetic resource handling beginning from augmentation to seed increase, characterization, preliminary evaluation, detailed evaluation, utilization, regeneration and documentation need descriptions for practical purpose, as given below:

1. Attributes consideration during characterization and preliminary evaluation
2. Advanced evaluation

### (1) Attributes Consideration during Characterization and Preliminary Evaluation

Characterization and preliminary evaluation consists of recording a limited number of additional traits, thought desirable by the consensus of specialists/users of the particular crops/species, which help in identifying useful germplasm. Evolution data refer to environmentally influenced characters. The important ones include site data; data on plant, leaf, flower, fruits, seeds and reaction to pest and diseases. Passport data-Data collected by curator including origin of the sample or it's known history. This is very important for identification and helps in designating core collection, identifying duplicates and planning future exploration. Preliminary evaluation consists of recording data on a limited number of agronomic traits thought desirable by a consensus of research workers of the particular crop. The traits, in general, have quantitative inheritance and, are influenced environmentally. Thus, to better assess their performance, the germplasm should be evaluated in the agro-ecological region from where the accessions have been collected or in a similar environment. There

### Table 5.3: Characterization of Germplasm At BCKV, Kalyani (West Bengal)

| Crop | Collections | Important Characters |
|------|-------------|----------------------|
| Tomato | 118 | Anthocyanin absent, potato leaf, joint less pedicel, sporogenous male sterility, functional male sterility (non-dehiscent anther, exerted stigma), yellow fruit, deep green fruit, high pigment fruit, slow ripening, resistant to tomato leaf curl virus, tolerant to high temperature, very high bearing. |
| Brinjal | 40 | Pigmentation and non-pigmentation, non-prickly, different flowering and fruiting patter (solitary, cluster and mixed) very big fruit, less susceptibility to shoot and fruit borer and bacterial wilt, less sensitive to high temperature, function male sterility (non-dehiscence anther) *etc.* |
| Chilli | 60 | Purple and yellow corolla, bluish yellow and purple anther, very long oval, conical and small fruit, erect fruit position, different blossom end characters, tolerant to leaf curl complex, vitamin C content *etc.* |
| Bottle gourd | 15 | Globular, oval, pyriform, and elongated fruit shape |
| Pumpkin | 24 | Different fruit shape (globular, round, elongated fruit, oval, flattish *etc.*) very high carotene and sugar content in the pericarp *etc.* |
| Pointed gourd | 21 | Female clones showing different fruit shape (spindle, oval and cylindrical) and size, striation |
| Spine gourd | 11 | Female clones showing different fruit shape (oval and cylindrical) size, and soft or tough rind *etc.* |
| Dolichos bean | 15 | Pole and bush type growth habit, pigmented and non-pigmented vine, different pod colour (green, white, red), very high bearing, tolerant to mosaic *etc.* |
| Okra | 18 | Different plant structure, fruit length and colour and number of ridges present on the fruit and tolerant to YVMV. |
| Cucumber | 8 | Different fruit shape (round, elongated, oval, long *etc.*), tolerant to downy mildew. |

*Source*: Chattopadhyay, *et al.*, 2011.

**Table 5.4: Traits Specific Observations in Vegetable Crops**

| Crops | Characters to be Recorded |
|---|---|
| | **Solanaceous vegetables** |
| Tomato (*Solanum lycopersicum*) | 1. Plant height (cm): To be recorded at peak fruiting stage. |
| | 2. Number of primary branches per plant: To be recorded at peak fruiting stage of at least 5 plants. |
| | 3. Days to 50 per cent flowering: To be recorded as number of days from sowing to the date when at least 50 per cent of the plants show flowers. |
| | 4. Fruit size: To be recorded at near maturity stage: very small (<=20g), Small (>20-30g), Medium (>30-80g), Medium large (>80-100g), Large (>100-175g), Very large (>175g). |
| | 5. Fruit shape: To be recorded at near maturity stage: Flat round, Slightly flattered, Round, Oval, Heart shaped, Lengthened cylindrical (banana type), Plum shaped. |
| | 6. Presence of green shoulder on the fruit: To be recorded on fully matured fruits stage: Absent (uniform ripening), Present (fruit shoulder-upper part of fruit around calyx is green while pistil area of fruits are red). |
| | 7. Fruit colour: To be recorded at near maturity stage: Yellow, Green, Orange, Crimson, Pink, Tangarine, Yellow and red, Tangarine and red, Yellow. |
| | 8. Number of fruits per plant: To be recorded as average of at least 5 randomly selected plants near maturity stage. |
| | 9. Fruit weight (g): To be recorded as average weight of at least 5 fruits near maturity stage. |
| | 10. Number of locules per fruit: To be recorded as average of at least 5 randomly selected fruits near maturity stage. |
| | 11. Pericarp thickness (mm): To be recorded from an equatorial section of the fruits near maturity stage as average of at least 5 fruits by using vernier calipers. |
| | 12. Total soluble solids (per cent): To be recorded at full maturity stage by using hand refractometer. |
| | 13. Titratable Acidity (per cent): To be recorded at full maturity stage. |
| | 14. Yield per plant (kg): To be recorded as average of cumulative yield of all pickings of at least 5 selected plants near maturity stage. |
| | 15. Biotic information about per cent disease incidence and insect pest infestation should be recorded. |
| Brinjal (*Slanum melongena*) | 1. Plant height (cm): To be recorded at peak fruiting stage. : Small (<=50 cm), Medium (50-100cm), Tall (>100 cm). |
| | 2. Number of primary branches per plant: To be recorded at peak fruiting stage of at least 5 plants. |
| | 3. Plant spread: To be recorded at peak fruiting stage of at least 5 plants: Very narrow (<=30 cm), Narrow (>30-40vm), Broad (>60-90cm), Very broad (>90cm). |
| | 4. Days to 50 per cent flowering: To be recorded as number of days from transplanting to date when at least 50 per cent plants show anthesis in first flower. |
| | 5. Fruit shape : To be recorded on marketable stage : Long, Round, Oblong, Oval. |
| | 6. Fruit colour: Recorded at marketable stage of at least 10 fruit: Milky white, Green, Purple, Purple Black, Light purple, Black, variegated *etc*. |
| | 7. Number of fruit per plant: To be recorded as average of at least 5 fruits at marketable stage. |
| | 8. Fruit length and width (cm): To be recorded as average of at least 5 fruits at marketable stage. |
| | 9. Fruit weight (g): To be calculated on the basis of total yield and number of fruits of at least 10 plants at marketable stage. |
| | 10. Yield per plant (kg): To be recorded as average of cumulative yield of all picking of at least 5 selected plants near maturity stage. |
| | 11. Biotic stress susceptibility: General information about per cent disease incidence and insect pest infestation should be recorded. |
| Chilli (*Capsicum annum*) | 1. Plant height (cm): To be recorded as average of 5 randomly selected plants when the fruit in 50 per cent of the plants began to ripe. |
| | 2. Branching habit: To be recorded when plants have ceased its growth: Sparse, Intermediate, Dense. |
| | 3. Days to 50 per cent fruiting: To be recorded as number of days from transplanting to the date when at least 50 per cent plants bear marketable fruit. |
| | 4. Marketable (green) fruit stage: To be recorded at mature (green) fruit stage: White, Yellow, Green, Light green, Orange, Deep purple, Black, Olive green, Chocklet. |
| | 5. Ripe fruit colour : To be recorded on fully ripe fruit stage : White, Lemon yellow, Pale orange yellow, Orange yellow, Pale orange, Orange, Light red, Red, Dark red, Brown, Purple. |
| | 6. Capsaicin content (per cent): Colouring matter (ASTA units). |
| | 7. Fruit shape: To be recorded at mature (green) fruit stage: Long, Very long, Tapering, Conical, Oval. |
| | 8. Number of fruits per plant: To be recorded as average of at least 5 randomly selected plant. |
| | 9. Fruit length and width (cm): To be recorded as average of at least 5 randomly selected fruits. |
| | 10. Fruit weight (g): To be calculated on the basis of fruit yield and number of fruits per plant of at least 5 plants. |

| Crops | Characters to be Recorded |
|---|---|
| | 11. Seed colour: To be recorded at dry seed stage; Light yellow, Deep yellow, Brown, Black. |
| | 12. Capsaicin content (per cent): To be recorded at red ripe fruit stage. |
| | 13. Pungency: To be tested as organoleptic test at marketable green stage: Mild pungent, intermediate, pungent, Highly pungent (scoville scale). |
| | 14. Yield (green fruit) per plant (kg): To be recorded as average of cumulative yield of all pickings at mature green fruit stage of at least 10 plants. |
| | 15. Yield (red fruit) per plant (kg): To be recorded as average of cumulative yield of all pickings at mature green fruit stage of at least 10 plants. |
| Sweet Pepper (*Capsicum annum*) | 1. Plant height (cm): To be recorded as average of 5 randomly selected plants when the fruits in 50 per cent of the plants began to ripe. |
| | 2. Branching habit: To be recorded when plants have ceased its growth: Sparse, Intermediate, Dense. |
| | 3. Days to 50 per cent fruiting: To be recorded as number of days from transplanting to the date when 50 per cent of plants bear marketable fruits. |
| | 4. Fruit colour intermediate stage: To be recorded (fruits) just before the marketable stage: White, Yellow green, Orange, Purple, Deep purple. |
| | 5. Fruit colour at mature fruit stage: To be recorded at near marketable stage of fruits : White, Lemon-yellow, Pale orange-yellow, Pale orange, Orange, Light red, red, dark red, Purple, Brown, Black. |
| | 6. Fruit shape: To be recorded at near marketable stage of fruit: Elongate, Almost round, Triangular, Companulate, and Blocky. |
| | 7. Blossom end fruit shape: To be recorded at neat maturity stage : Pointed, Blunt, Sunken, Sunken and pointed. |
| | 8. Number of fruits per plant: To be recorded as average of at least 5 plants fruits counted at all picking. |
| | 9. Fruit length (cm): To be recorded as average of at least 5 randomly selected at second harvesting stage. |
| | 10. Fruit weight (g): To be calculated on the basis of fruit yield per plant and number of fruits per plant of at least 5 plants. |
| | 11. Yield per plant (kg): To be recorded as average of cumulative yield of all pickings at mature green fruit stage of at least 5 plants. |
| | 12. Biotic stress susceptibility: General information about per cent disease incidence and insect pest infestation should be recorded. |
| Paprika (*Capsium annuum*) | 1. Plant height (cm): To be recorded as average of 5 randomly selected plants when the fruit in 50 per cent of the plants began to ripe. |
| | 2. Branching habit: To be recorded when plants have ceased its growth: Sparse, Intermediate, Dense. |
| | 3. Days to 50 per cent fruiting: To be recorded as number of days from transplanting to the date when at least 50 per cent plants bear marketable fruit. |
| | 4. Marketable (green) fruit stage: To be recorded at mature (green) fruit stage: White, Yellow, Green, Light green, Orange, Deep purple, Black, Olive green, Chocklet. |
| | 5. Ripe fruit colour: To be recorded on fully ripe fruit stage: White, Lemon yellow, Pale orange yellow, Orange yellow, Pale orange, Orange, Light red, Red, Dark red, Brown, Purple. |
| | 6. Capsaicin content (per cent): Colouring matter (ASTA units). |
| | 7. Fruit shape: To be recorded at mature (green) fruit stage: Long, Very long, Tapering, Conical, Oval. |
| | 8. Number of fruits per plant: To be recorded as average of at least 5 randomly selected plant. |
| | 9. Fruit length and width (cm): To be recorded as average of at least 5 randomly selected fruits. |
| | 10. Fruit weight (g): To be calculated on the basis of fruit yield and number of fruits per plant of at least 5 plants. |
| | 11. Seed colour: To be recorded at dry seed stage; Light yellow, Deep yellow, Brown, Black. |
| | 12. Capsaicin content (per cent): To be recorded at red ripe fruit stage. |
| | 13. Pungency: To be tested as organoleptic test at marketable green stage: Mild pungent, intermediate, pungent, Highly pungent. |
| | 14. Yield (green fruit) per plant (kg): To be recorded as average of cumulative yield of all pickings at mature green fruit stage of at least 10 plants. |
| | 15. Yield (red fruit) per plant (kg): To be recorded as average of cumulative yield of all pickings at mature green fruit stage of at least 10 plants. |
| **Cole crops** | |
| Cabbage (*Brassica oleracea* L. var. *capitata*) | 1. Plant spread: To be recorded as average of the distance between two outer leaves of 5 plants. |
| | 2. Leaf colour: To be recorded at marketable stage: Green, Blue green, Dark green, Red, Purple. |
| | 3. Number of non-wrapping leaves: To be recorded as average of 5 randomly selected plants at marketable stage. |
| | 4. Days to 50 per cent head maturity: To be recorded as number of days from transplanting to date when at least 50 per cent plant show marketable heads. |

| Crops | Characters to be Recorded |
|---|---|
| | 5. Duration of head maturity: To be calculated on the basis of days to 50 per cent head formation and days to 50 per cent head maturity. |
| | 6. Head compactness: To be recorded at marketable stage: very compact, compact, medium compact, loose. |
| | 7. Head shape: To be recorded at marketable stage: Flat (drum head), globe (conical), round, oval. |
| | 8. Stalk length (cm): To be recorded at marketable stage as average of 10 plants from ground level to the first non-wrapping leaves. |
| | 9. Head length and width (cm): To be recorded at marketable stage as average length and width of 5 randomly selected heads. |
| | 10. Net plant weight (g) (excluding root): To be recorded at marketable stage as average of at least 5 heads along with the non-wrapping leaf and stalk. |
| | 11. Biotic stress susceptibility: General information about per cent disease incidence and insect pest infestation should be recorded. |
| Cauliflower (*Brassica oleracea* L. var. *botrytis*) | 1. Plant growth habit: To be recorded when heads attains marketable stage: Spreading, Semi spreadling, Semi right, Upright. |
| | 2. Number of leaves: To be recorded as average of at least 5 randomly selected plants at marketable head stage. |
| | 3. Leaf length and width (cm): To be recorded as average of middle leaf of 5 randomly selected plants at marketable stage. |
| | 4. Day to 50 per cent curd maturity: To be recorded as number of days from date of transplanting to date when at least 50 per cent plants show curd initiation. |
| | 5. Curd maturity duration: To be calculated on the basis of days to 50 per cent curd initiation and 50 per cent curd maturity. |
| | 6. Curd shape: To be recorded at marketable stage: Flat, Round, Pointed. |
| | 7. Curd colour: Snow white, White, Cream, Yellow. |
| | 8. Curd compactness: To be recorded at marketable stage : Loose, Medium compact, Very compact. |
| | 9. Stalk depth (cm) : To be recorded as average of 5 plants from ground to the base of the curd at marketable stage. |
| | 10. Curd width (cm): To be recorded as average of 5 randomly selected curds at widest part of the curd at marketable stage. |
| | 11. Gross plant weight (g): To be recorded as average weight of 5 curds along with leaves and stalks at marketable stage. |
| | 12. Marketable curd weight (g): To be recorded as average weight of curd and leaves (pruned to curd level) in 5 randomly selected curds at marketable stage. |
| | 13. Net curd weight: To be recorded as average weight of curd only (exclusive leaves and stalk) at marketable stage. |
| | 14. Physiological disorder: To be recorded at marketable stage: Rinciness, Fuzziness, Leafiness. |
| | 15. Maturity group: To be recorded at flowering stage: Early, Mid, Late. |
| | 16. Biotic stress susceptibility: General information about per cent disease incidence and insect pest infestation should be recorded. |

**Cucurbitaceous vegetables**

| Crops | Characters to be Recorded |
|---|---|
| Bitter gourd (*Momordica charantia* L.) | 1. Vine growth habit: To be recorded on fully grown stage: Short viny, Medium viny, Long viny. |
| | 2. Fruit shape : Tapering/spindle shaped, Elliptical, Oblong, Long cylindrical, Top shaped, Globular. |
| | 3. Fruit surface: To be recorded in cross-section at marketable stage: Smooth, Light tubercle, Deep tubercle. |
| | 4. Fruit skin colour: To be recorded at marketable stage: White, Milky white, Light green, Green, dark green. |
| | 5. Days to first fruit harvest: To be recorded as number of days from date of sowing/transplanting to the date of first marketable fruit harvest. |
| | 6. Number of fruits per plant: To be recorded as average of cumulative number of fruits in all pickings of at least 5 plants at marketable stage. |
| | 7. Fruit weight (g): To be calculated on the basis of fruit yield and number of fruits per plant. |
| | 8. Fruit length and width (cm): To be recorded as average of at least 5 randomly selected fruits at marketable stage. |
| | 9. Yield per plant (kg): To be recorded as average of cumulative yield of all pickings of at least 5 plants. |
| | 10. Biotic stress susceptibility: General information about per cent diseases incidence and insect pest infestation should be recorded. |
| Bottle gourd (*Lagenaria siceraria*) | 1. Plant growth habit : To be recorded on fully grown plant : Short viny <3, Medium viny 3-6 m, long viny >6 m. |
| | 2. Node number at which first female flower appears: To be recorded as the node at which first 5 randomly selected plant of female flower appeared. |
| | 3. Fruit shape: To be recorded at marketable stage: Elliptical, Elongate, Pyriform, Oblong, Club shaped, globular, dumbbell shaped, lengthened cylindrical. |
| | 4. Fruit skin colour: To be recorded at marketable stage: Light green, Green, Dark green, Patchy green. |

| Crops | Characters to be Recorded |
|---|---|
| | 5. Days to first fruit harvest: To be recorded as number of days from date of sowing/transplanting to the date of first marketable fruit harvest. |
| | 6. Number of fruits per plant: To be recorded as average of at least 10 randomly selected fruits at marketable stage. |
| | 7. Fruit weight (kg): To be calculated on the basis of fruit yield and number of fruits per plant of at least 5 plants. |
| | 8. Yield per plant (kg): To be recorded as average of cumulative yield of all pickings in 10 plants. |
| | 9. Biotic stress susceptibility: General information about per cent disease incidence and insect pest infestation should be recorded. |
| Pumpkin (*Cucurbita moschata, C. maxima, C. pepo*). | 1. Plant growth habit : To be recorded on fully grown plant : Short viny <3, Medium viny 3-6 m, Long viny >6 m. |
| | 2. Node number at which first female flower appears: To be recorded as average of node at which first 5 randomly selected plants of female flower appeared. |
| | 3. Mature fruit skin colour : To be recorded at mature stage: Creamy, Yellowish, Green, Red. |
| | 4. Fruit ridge (rib)shape : To be recorded in cross section at marketable stage : Superficial, Rounded/grooved, Deep grooved, Narrowly winged. |
| | 5. Number of ribs per fruit : To be recorded in cross section at full maturity stage of at least 5 fruits. |
| | 6. Days to first fruit harvest: To be recorded as number of days from date of sowing/transplanting to the date of first marketable fruit harvest. |
| | 7. Mature flesh colour : To be recorded of marketable stage of mature fruit : Yellow, Deep yellow, Orange. |
| | 8. Fruit length and width (cm): To be recorded as average of at least 5 randomly selected fruits at marketable stage. |
| | 9. Number of fruits per plant : To be recorded as average total number of fruits in each picking of at least 5 plants. |
| | 10. Total carotenoids (mg/100g of edible portion): To be redorded of mature fruit stage. |
| | 11. Yield at mature stage per plant (kg): To be recorded as average of cumulative yield of all pickings of at least five plants. |
| | 12. Biotic stress susceptibility : General information about per cent disease incidence and insect pest infestation should be recorded. |
| Ash gourd (*Benincasa hispida*) | 1. Plant growth habit : To be recorded on fully grown plant stage: Short viny <3, Medium viny 3-6 m, Long viny >6 m. |
| | 2. Node number at which first female flower appears: To be recorded as average of the node at which first female flower appeared. |
| | 3. Fruit shape (g) : To be recorded at marketable stage : Oblong, Round, Round oval, others. |
| | 4. Fruit weight (g): To be recorded as average of at least 5 randomly selected fruits of mature stage. |
| | 5. Seediness: To be recorded at marketable stage: Low, Medium High. |
| | 6. Yield per plant (kg): To be recorded as average of cumulative yield of all pickings of at least 5 plants and averaged. |
| | 7. Biotic stress susceptibility: General information about per cent disease incidence and insect pest infestation should be recorded. |
| Sponge gourd [*Luffa cylindria* (Syn. *L. aegyptiaca*)] | 1. Plant growth habit : To be recorded on fully grown plant stage : Short viny <3, Medium viny 3-6 m, Long viny >6 m. |
| | 2. Node numbers at which first female flower appears: To be recorded as the node at which first female flower appeared. |
| | 3. Fruit shape : to be recorded at marketable stage : Cylindrical, Club shaped, Spindle type, Elliptical, Elongate. |
| | 4. Fruit skin colour: To be recorded at marketable stage : Light green, Green, Dark green. |
| | 5. Days to first fruit harvest: To be recorded as number of days from date of sowing/transplanting to the date of first marketable fruit harvest. |
| | 6. Number of stripes per fruit: To be counted at marketable stage. |
| | 7. Number of fruits per plant: To be recorded as total number of fruits in each picking of at least 5 plants and average. |
| | 8. Fruit length and width (cm): To be recorded as average of at least 5 fruits from selected plant at marketable stage. |
| | 9. Fruit weight (g): To be recorded as average of at least 5 fruits at marketable stage. |
| | 10. Yield per plant (kg): To be recorded as average of cumulative yield of all pickings of at least 5 plants |
| | 11. Biotic stress susceptibility: General information about per cent disease incidence and insect past infestations should be recorded. |
| Ridge gourd (*Luffa acutangula*) | 1. Plant growth habit : To be recorded on fully grown plant stage : Short viny <3, Medium viny 3-6 m, Long viny >6 m. |
| | 2. Node number at which first female flower appears: To be recorded as the node at which first female flower appeared. |
| | 3. Fruit skin colour : To be recorded at marketable stage : Light green, Green, Dark green. |
| | 4. Number of ridges per fruit: To be recorded as average of at least10 fruit's weight (g). |
| | 5. Fruit shape: To be recorded at full fruiting stage: cylindrical, club shaped, spindle shape, elliptical, globular, Elongate. |
| | 6. Days to first fruit harvest : To be recorded as number of days from date of sowing/transplanting to the date of first marketable fruit harvest. |
| | 7. Number of fruits per plant: To be recorded as total number of fruits in all picking and averaged. |

| Crops | Characters to be Recorded |
|---|---|
| | 8. Fruit weight (g): To be calculated on the basis of number of fruits and fruit yield per plant. |
| | 9. Fruit length and width (cm): To be recorded as average of at least 5 randomly of selected fruits at marketable stage. |
| | 10. Yield per plant (kg): To be recorded as average of cumulative yield of all pickings of at least 5 selected plants. |
| | 11. Biotic stress susceptibility: General information about per cent disease incidence and insect pest infestation should be recorded. |
| Pointed Gourd (*Trichosanthes dioica*) | 1. Node number at which first female flower appeared. |
| | 2. Days to first fruit harvest: To be recorded as number of days from transplanting to the date of first marketable fruit harvest. |
| | 3. Fruit shape: To be recorded at marketable stage: Club shaped, Cylindrical, Oval, Spindle type, Tapering. |
| | 4. Fruit skin Texture: Striped, non-striped, and spotted. |
| | 5. Presence of seed at marketable stage: To be recorded at marketable stage: Inconspicuous, Conspicuous, Others. |
| | 6. Number of fruits per plant: To be recorded as average of total number of fruits in all picking of at least 5 selected plants and averaged. |
| | 7. Fruit length and width (cm): To be recorded as average at least 5 randomly selected fruits at marketable stage. |
| | 8. Fruits weight (g): To be calculated on the basis of fruit yield and numbers of fruits per plant. |
| | 9. Seediness: To be recorded at marketable stage: Low, Medium High. |
| | 10. Yield per plant (kg): To be recorded as average of cumulative yield of all pickings of at least 5 selected plants. |
| | 11. Biotic stress susceptibility: General information about per cent diseases incidence and insect pest infestation should be recorded. |
| Cucumber (*Cucumis sativus*) | 1. Plant growth habit: To be recorded on fully grown plant stage: Short viny <3, Medium viny 3-6 m, Long viny >6 m. |
| | 2. Number of primary branches: To be recorded at the end of flowering stage as average of at least 5 plants (The branch that arises from the main vine is known as primary branch). |
| | 3. Node number at which first female flower appears: To be recorded as the node at which first female flower appeared. |
| | 4. Fruit skin colour: To be recorded at marketable stage: Cream, Yellow, Light green, Green, Dark green, Orange, Pink, Brown. |
| | 5. Fruit strips colour: To be recorded on marketable stage: Absent, White, Green, and Yellow. |
| | 6. Fruit shape: To be recorded on marketable stage: Elliptical elongate, Oblong ellipsoid, Globular (round), Stem-end tapered, Blossom-end tapered. |
| | 7. Fruit bitterness intensity: To be recorded on marketable stage: Absent, Low, Medium, and High. |
| | 8. Days to first fruit harvest: To be recorded as number of days from sowing/transplanting to the date of first marketable fruit harvest. |
| | 9. Fruit length and width (cm): To be recorded as average of at least 5 randomly selected fruits at marketable stage. |
| | 10. Fruit weight (g): To be calculated on the basis of fruit yield and number of fruits per plant. |
| | 11. Number of fruits per plant: To be recorded as average of total number of fruits of 5 randomly selected plants. |
| | 12. Presence of placental cavity (Hollowness): To be recorded at marketable stage: Absent, Present. |
| | 13. Yield per plant (Kg): To be recorded as average of cumulative yield of all pickings of at least 5 selected plants. |
| | 14. Biotic stress susceptibility: General information about per cent disease incidence and insect pest infestation should be recorded. |
| Muskmelon (*Cucumis melo*) | 1. Plant growth habit: To be recorded on fully grown plant stage: Short viny < 2.0 m, Medium viny 2.0-3.0m, Long viny >3.0 m. |
| | 2. Number of primary branches: To be recorded at the end of flowering stage as average of at least 5 plants. |
| | 3. Node number at which first female flower appears: To be recorded as the node number at which first female flowers appeared. |
| | 4. Sex type: To be recorded at flowering/full blossom stage: Monoecious (male and female flowers on same plant), Gynomonoecious (female and hermaphrodite flowers on the same plant), Andromonoecious (male and hermaphrodite flowers on the same plant). |
| | 5. Days to first fruit harvest: To be recorded as number of days from sowing/transplanting to the date of first marketable fruit harvest. |
| | 6. Rind colour: To be recorded at marketable stage. |
| | 7. Fruit skin texture: To be recorded at marketable stage: Smooth, Grainy, wrinkled, Shallow wavy, Netted, Warty, Spiny. |
| | 8. Fruit length and width (cm): To be recorded on the basis of widest position of fruit as average of at least 5 randomly selected fruits at marketable stage. |
| | 9. Number of fruits per plant: To be recorded as total number of fruits of 5 randomly selected plants and averaged. |
| | 10. Fruit weight (g): To be calculated on the basis of total fruit yield and number of fruits per plant. |
| | 11. Flesh colour: To be recorded at marketable stage: White, Green, Yellow, Orange, Salmon orange. |

| Crops | Characters to be Recorded |
|---|---|
| | 12. Flesh flavor: To be recorded at marketable stage: Absent, Mild, Moderate, Strong. |
| | 13. Total soluble solids (kg): To be measured by refractometer just after harvest. |
| | 14. Yield per plant (kg): To be recorded as average of cumulative yield of all pickings of at least 5 selected plants. |
| | 15. Biotic stress susceptibility: General information about per cent disease incidence and insect pest infestation should be recorded. |
| Watermelon (*Citrullus lanatus*) | 1. Plant growth habit : To be recorded on fully grown plant : Short viny <3, Medium viny 3-6 m, Long viny >6 m. |
| | 2. Number of primary branches: To be recorded at the end of flowering stage as average of at least 5 plants. |
| | 3. Node number at which first female flower appears: To be recorded as the node number at first appearance of female flowers. |
| | 4. Days to first fruit harvest: To be recorded as number of days from sowing/transplanting to the date of first marketable fruit harvest. |
| | 5. Fruit length and width (cm): To be recorded on the widest portion of fruit as average of at least 5 randomly selected fruits at marketable stage. |
| | 6. Number of fruits per plant: To be recorded as total number of fruits of 5 randomly selected plants and average. |
| | 7. Fruit weight (g): To be calculated on the basis of total fruit weight and number of fruits per plant. |
| | 8. Flesh colour: To be recorded at marketable stage: Yellow, Light red, Deep red. |
| | 9. Total soluble solids (per cent): To be measured by refractometer at ripe stage. |
| | 10. Yield per plant (kg): To be recorded as average of cumulative yield of all pickings in 5 selected plants. |
| | 11. Biotic stress susceptibility: general information about per cent disease incidence and insect pest infestation should be recorded. |
| Chow Chow (*Sechium edule*) | 1. Plant growth habit: To be recorded on fully grown plant: Short viny, Medium viny, Long viny. |
| | 2. Fruit colour: Light green, Green, Dark green. |
| | 3. Fruit skin texture : Smooth, Spiny. |
| | 4. Fruit shape: To be recorded on marketable stage: Elliptical, Tapering, Pyriform, Elongated, Club shaped, Top shaped, Oblong, Globular, Cylindrical. |
| | 5. Fruit length and width (cm): To be recorded as average at least 5 randomly selected fruits at marketable stage. |
| | 6. Number of fruits per plant: To be recorded as total number of fruits of 5 randomly selected plants and average. |
| | 7. Fruit weight (g): To be calculated on the basis of fruit yield and number of fruits per plant. |
| | 8. Yield of marketable fruits per plant (kg): To be recorded as average of cumulative yield of all picklings in 10 selected plants. |
| | 9. Biotic stress susceptibility: General information about per cent disease incidence and insect pest infestation should be recorded. |
| | 10. Plant growth habit: To be recorded on fully grown plant: Short viny, Medium viny, Long viny. |
| | 11. Fruit shape: To be recorded at marketable stage: Cylindrical, Cylindrical with stem end tapering, club shaped, spindle shape, elliptical, globular, elongate. |
| | 12. Days to first fruit harvest: To be recorded as number of days from sowing to the date of first marketable fruit harvest. |
| | 13. Number of fruits per plant: To be recorded as total number of fruits of 5 randomly selected plants and average. |
| | 14. Fruit weight (g): To be calculated on the basis of fruit yield and number of fruits per plant. |
| | 15. Yield of per plant (kg): To be recorded as average of cumulative yield of all pickings in 5 selected plants. |
| | 16. Biotic stress susceptibility: General information about per cent disease incidence and insect pest infestation should be recorded. |

**Legume vegetable**

| | |
|---|---|
| Pea (*Pisum sativum*) | 1. Pod colour: To be recorded at marketable stage: Green, Light green, Dark green. |
| | 2. Pod shape: To be recorded at marketable stage: Straight, Slightly curved. |
| | 3. Pod length (cm) : To be recorded as average of 10 randomly by selected marketable pods. |
| | 4. Pod width (cm) : To be recorded as average of 10 randomly by selected marketable pods. |
| | 5. Green pod yield per plant(g) : To be recorded as average of cumulative yield of all pickings in 10 randomly selected plants. |
| | 6. Number of pods per plant : To be recorded as average of 10 randomly selected plants. |
| | 7. Pod weight (g) : To be calculated on the basis of pod yield per plant and number of pods per plant. |
| | 8. Days to 50 per cent maturity: To be recorded as number of days from date of sowing to date when at least 50 per cent plants show marketable pods. |
| | 9. Number of seeds per pods: To be counted as average of 10 randomly selected plants pods. |
| | 10. 100 seed weight (g-seed size): To be recorded as weight of 100 randomly selected green and dry seeds to observe the boldness of seed. |

| Crops | Characters to be Recorded |
|---|---|
| | 11. Seed yield per plant : To be recorded as average of cumulative yield of all pickings in 5 selected plants and averaged. |
| | 12. Biotic stress susceptibility: General information about per cent disease incidence and insect pest infestation should be recorded. |
| Cowpea (*Vigna unguiculata*) | 1. Days to 50 per cent maturity: To be recorded as number of days from date of slowing to date when at least 50 per cent plants show marketable pods. |
| | 2. Pod colour : To be recorded at marketable stage: Green, Light green, Dark green. |
| | 3. Pod shape : To be recorded at marketable stage: Straight, Slightly curved, Round flat, Flat. |
| | 4. Pod and width (cm): To be recorded as average of 10 randomly selected marketable pods. |
| | 5. Number of pods per plant : To be recorded as average of at least 5 plant. |
| | 6. Pod yield per plant (Kg) : To be recorded as average of cumulative yield of all pickings in 5 randomly selected plants. |
| | 7. Biotic stress susceptibility: General information about per cent disease incidence and insect pest and infestation should be recorded. |
| French Bean (*Phaseolus vulgaris*) | 1. Plant growth habit : To be recorded at completion of vegetative stage : Bush Semi pole, Pole, Others. |
| | 2. Days to 50 per cent maturity: To be recorded as number of days from sowing to date when at least 50 per cent plants have marketable pods. |
| | 3. Pod length and width (cm) : To be recorded as average of at least 10 randomly selected marketable pods. |
| | 4. Number of pods per plant : To be recorded as total number of pod pickings during crop duration. |
| | 5. Pod weight (g): To be calculated on the basis of total pod yield per plant and number of pods per plant. |
| | 6. Pod colour : To be recorded as base colour of seed: White, Cream, Yellow, Brown, Red, Maroon, Purple, Black, Others. |
| | 7. Seed length and width (mm) : To be recorded as average of 10 randomly selected mature and dried seeds. |
| | 8. Number of seeds per pods : To be recorded as average of 10 pod seeds selected randomly. |
| | 9. 100 seed weight (g-seed size) : To be recorded as weight of 100 randomly selected dry seeds. |
| | 10. Green pod yield per plant (kg): To be recorded as average of cumulative yield of all pickings in 10 randomly selected plants. |
| | 11. Biotic stress susceptibility : General information about per cent disease incidence and insect pest infestation should recorded. |
| Lablab Bean (*Lablab purpureus*) | 1. Plant growth habit: To be recorded at flowering stage : Bush, Semi pole, Pole, Others. |
| | 2. Pod Colour: To be recorded at marketable stage : White, Creamy, Light green, Greenish, Purple, Others. |
| | 3. Pod shape: To be recorded at marketable stage: Straight, curved, intermediate, flat, round, others. |
| | 4. Days to first pod harvest : To be recorded as number of days from sowing to the date of first marketable pod picking. |
| | 5. Pod length and width (cm): To be recorded as average of 10 randomly selected pods at marketable stage. |
| | 6. Number of pods per plant: To be recorded as average of at least of randomly selected plants at marketable stage. |
| | 7. Pod weight (g) : To be calculated on the basis of total pod yield per plant and number of pods per plant. |
| | 8. Number of seeds per pod: To be recorded as average of seeds in 10 randomly selected matured and dry pods. |
| | 9. Pod yield per plant (kg): To be recorded as average of cumulative yield of all pickings in randomly selected 10 plants at marketable stage. |
| | 10. Biotic stress susceptibility : General information about per cent disease incidence and insect pest infestation should be recorded. |
| **Root crops** | |
| Radish (*Raphanus sativus*) | 1. Shoot habit : To be recorded at marketable root harvest stage : erect, semi erect, spreading, others. |
| | 2. Number of leaves per plant : To be recorded as average of at least 5 randomly selected plants of marketable root harvest stage. |
| | 3. Days to 50 per cent root harvest : To be recorded number of days from sowing to the date when at least 50 per cent plants show marketable root. |
| | 4. Root weight (g) : To be recorded as average weight of 10 randomly selected roots at marketable root harvest stage. |
| | 5. Root length and diameter (cm) : To be recorded as average of the widest part of 10 roots selected for root weight observation. |
| | 6. Root Branching : To be recorded at marketable root harvest stage: Absent, Present. |
| | 7. Root skin colour: To be recorded at marketable root harvest stage : White, Creamy white, Pink, red, purple, other. |
| | 8. Root shape: To be recorded at marketable root harvest stage: non swollen tape root, triangular, cylindrical, elliptic, spherical, transverse elliptic, inverse triangle, apically bulbous, horn shaped, branched. |
| | 9. Root pithiness : To be recorded at marketable root harvest stage |
| | 10. Biotic stress susceptibility: General information about per cent disease incidence and insect pest infestation should be recorded. |

| Crops | Characters to be Recorded |
|---|---|
| Carrot (*Daucus carota*) | 1. Shoot habit: Erect, Semi erect, Spreading, Others. |
| | 2. No. per leaves per lant: To be recorded at marketable root harvest stage. |
| | 3. Core colour: To be recorded at marketable root harvest stage:Self colour, White, Creamy white, Green, Light red, Pink, Red, Dark red. |
| | 4. Days to first root harvest: To be recorded as number of days from sowing date to date of first marketable-root harvest stage. |
| | 5. Root colour: To be recorded at marketable root harvest stage: Creamy white, Yellow, Orange, Light red, Dark red, Deep purple. |
| | 6. Root shape : To be recorded at marketable root harvest stage: Conical, Cylindrical, Tapering |
| | 7. Root length and diameter (cm): To be recorded as average of at least 5 randomly selected roots at marketable root harvest stage. |
| | 8. Root forking: To be recorded at marketable root harvest stage: Absent, Present. |
| | 9. Root cracking: To be recorded at marketable root harvest stage: Absent, Horizontal, Vertical. |
| | 10. Root weight (g): To be recorded as average of at least 5 randomly selected roots at marketable root harvest stage. |
| | 11. Biotic stress susceptibility: General information about per cent disease incidence and insect pest infestation should be recorded. |

**Bulbous crops**

| Crops | Characters to be Recorded |
|---|---|
| Onion (*Allium cepa*) | 1. Plant height (cm): To be recorded as average of least 5 plants randomly selected plant just before maturity. |
| | 2. Number of leaves: To be recorded as average of 10 plants at maturity stage. |
| | 3. Maturity of bulb: Flattening of leaves to be recorded from 5th leaves onwards. |
| | 4. Bulb colour: Red, White, Yellow, Others. |
| | 5. Bolting: To be recorded on population basis. |
| | 6. Bulb neck thickness (mm): To be recorded as average of 10 randomly selected bulb just before maturity. |
| | 7. Diameter of bulb (cm): To be recorded as average of 10 randomly selected bulbs. |
| | 8. Doubling of bulb (cm): To be recorded as average 20 bulbs just after harvesting. |
| | 9. Average bulb weight (g): To be recorded as average of 10 randomly selected bulbs. |
| | 10. Total soluble solids (per cent): To be recorded with the help of hand refractometer of at least 5 bulbs at maturity stage. |
| | 11. Dry matter (per cent): To be recorded in 10 randomly selected bulbs at maturity stage. |
| | 12. Biotic stress susceptibility: Per cent incidence of disease *i.e. Stemphyllium* blight, and pest infestation *i.e.* Thrips *etc.* should be recorded as and when appeared. |
| Garlic (*Allium sativum*) | 1. Plant height (cm): To be recorded as average of at least 5 plants just before maturity. |
| | 2. Number of leaves: To be recorded as average of 10 plants at maturity stage. |
| | 3. Bolting (per cent): To be recorded on population basis. |
| | 4. Bulb diameter: To be recorded at harvesting as average of 10 bulbs measured at widest portion of Bulb. |
| | 5. Number of clove bulb: To be recorded at harvesting as average of 10 clones. |
| | 6. Clove length and girth (mm): To be recorded as average of 10 cloves. |
| | 7. Average clove weight: To be recorded as average of 20 cloves randomly selected from different bulb. |
| | 8. Total soluble solids: To be recorded with the help of hand refractometer in 10 bulbs at maturity stage. |
| | 9. Dry matter (per cent): To be recorded at maturity stage. |
| | 10. Biotic stress susceptibility: Per cent disease incidence and insect pest infestation should be recorded as and when appeared. |

**Leafy vegetable**

| Crops | Characters to be Recorded |
|---|---|
| Amaranth (*Amaranthus* spp.) | 1. Plant weight (g): To be recorded as average of 10 randomly selected plants |
| | 2. Leaf weight (g): To be calculated on the basis of leaf weight and under of leaf per plant. |
| | 3. Stem weight (g): To be recorded as average of at least 10 randomly selected plants at marketable stage with leaves. |
| | 4. Stem length (cm): To be recorded as average of 5 randomly selected plant at marketable stage. |
| | 5. Leaf colour: To be recorded. |
| | 6. Stem girth (cm): To be measured as average of 10 randomly selected plant at marketable stage. |
| | 7. Leaf length and breadth (cm): To be calculated as average of 20 leaves of 10 plants. |
| | 8. Yield per plant: To be recorded as average of cumulative yield of all harvesting at marketable stage. |
| | 9. Biotic stress susceptibility: General information about per cent disease incidence and insect pest infestation should be recorded. |

| Crops | Characters to be Recorded |
|---|---|
| | **Others** |
| Okra (*Abrlmoschus esculentus*) | 1. Plant height (cm): To be recorded as average of at least 5 plants selected randomly after final harvest. |
| | 2. Inter-nodal length: To be recorded as average of at least 5 plants selected randomly after final harvest from 6th to 10th internode. |
| | 3. Number of branches per plant: To be counted and averaged at least 5 plants after final harvest. |
| | 4. Number of flowering nodes on main stem: To be recorded the node on which first flower appeared of at least 5 plants selected randomly and averaged. |
| | 5. Number of fruits per plant: To be recorded as average of at least 5 plants. |
| | 6. Fruit colour: To be recorded at marketable stage. |
| | 7. Fruit length and diameter (cm): To be recorded as average of 10 randomly selected fruits at marketable stage. |
| | 8. Number of ridge per fruit: To be counted as average of 10 randomly selected fruits at marketable stage. |
| | 9. Average fruit weight (g): To be recorded as average of 10 randomly selected fruits at marketable stage. |
| | 10. Marketable yield per plant (kg): To be recorded as average yield of at least 5 randomly selected plants. |
| | 11. Biotic stress susceptibility: To be recorded as and when appeared. |
| | 12. Yellow vein mosaic virus: Low susceptibility (LS)-Resistant/Tolerant (1-15 per cent), Medium susceptibility (MS)-16 per cent to 25 per cent, High susceptibility (HS)-26-50 per cent. |
| | 13. Fruit and stem borer: Low susceptibility (LS)-Less than 15 per cent, Medium susceptibility (MS)-16-40 per cent, High susceptibility (HS)-41 per cent and above. |

is no prescribed limit for the number of accessions in a trial planted for preliminary evaluation. Some trials may have many hundreds accessions, each in small and perhaps single line plots. To minimize the inter-genotype competition, accessions should be arranged according to their height, growth habit, maturity and other such traits. Large number of accessions in a trial usually needs a compromise between statistical and practical consideration. If these are ever in conflict, then practical consideration should take precedence over the statistical ones. In general, the preliminary evaluation trials undertaken by the curators have two features in common that set them apart from other field experiments (IPGRI, 2001).

At ICAR-NBPGR Regional Station Jodhpur (Rajasthan) primarily evaluated 470 accessions of cowpea (*Vigna unguiculata* (L.) Walp.), the 510 accessions of moth bean (*Vigna acontifolia* (Jacq.)) and the 435 accessions of mung bean (*Vigna radiata* (L.)) and 1103 accessions of cluster bean (*Cyamopsis tetragonoloba* (L.)) during summer and *kharif* seasons for various agronomical traits (Singh and Singh, 2016).

## (2) Advanced Evaluation

Consists of recording a number of additional descriptors though to be useful in crop improvement. These include important agronomic traits, stress tolerance, disease and pest resistance and quality characters *etc*. Observations are recorded on qualitative and quantitative traits. Qualitative traits include morphological, physiological and biochemical characters related to survival and scored using a number of checks to determine the variation within and among the traits.

Quantitative traits subject to environmental factors and are responsible for adaptation and productivity. These include productivity, quality components, resistance to disease and pest as well as tolerance to adverse conditions or stress. Much emphasis is presently given on multi-locational and multidisplinary approach of germplasm evaluation. For preliminary evaluation locally adapted cultivars should be used as check and screening for specific diseases under controlled conditions or at hot spots should be carried out. Augmented design is invariably used due to the large number of germplasm holdings under evaluation. Care should be taken to minimize natural cross-pollinated contamination and erroneous labeling. While, regenerating care should be taken to preserve original structure and productivity of accession. Main dangers are due to inappropriate handling of materials during sowing/planting, harvesting, threshing, cleaning, sub-sampling, packing and labeling. Frequent regeneration should be also be avoided by producing sufficient seeds during initial seed increase and conservation in medium term storage in case of working collection and long term gene bank as base collection. A mega evaluation executed by ICAR-NBPGR, New Delhi in collaboration with other ICAR institutes and SAUs for several accessions of vegetable crops.

## Main Issues Related to Germplasm Evaluation

The main issues related to germplasm evaluation are discussed below:

☆ **Number of accessions and crops:** A curator needs to evaluate a large number of accessions of various crops. So it is a huge task to evaluate all the accessions, it requires, besides large and diverse technology base, huge land area, resources and times.

☆ *Soil heterogeneity:* Soil heterogeneity; its associated effects and the consequent interactions with germplasm accessions are major factors that curators must accommodate.

☆ *Evaluation environment:* To have reasonable information about the genetic potential of accessions, these are to be evaluated in the area of their adaptation or a similar environment.

☆ *Target environment:* A breeder works for more or less a defined area, though; in that process many widely adapted cultivars have been developed. In contrast, a curator's mandate is to generate information to be accessed by all breeders at least at the national level.

☆ *Target traits:* A curator is concerned with whole array of traits including those, which may gain importance in future. But breeder focuses on the biotic and abiotic stresses in the target area for which he/she is working whereas a curator needs to generate data on all stresses prevalent in the country.

☆ *Number of environments:* The germplasm need to be evaluated over environments (locations, years *etc.*) to have precise information about the genetic potential.

☆ *Large diversity in germplasm:* Germplasm evaluation trials, as a rule includes materials of diverse nature, particularly for adaptation and plant type. The evaluation of such materials results in high genotype-environment interaction. To have realistic value of accessions, it is necessary to have an idea of the effect of genotype-environment interaction on the performance so as to have pragmatic guide of genetic potential.

☆ *Number of plants-*The curator deals more often with heterogeneous raw materials than a plant breeder. Further, he/she is interested not only in the mean performance but the range of expression also. Thus, data need to be recorded on a larger number of plants.

☆ *Inter-plot interference:* When diverse germplasms are grown side by side, large inter-plot competition affecting the performance of the neighbouring accessions, may be expected. To minimize it the accessions should be grouped on the basis of adaptation, plant type and other relevant information.

☆ *Perennial crops:* The evaluation of germplasm of perennial crops is very difficult in terms of time. The recording of data on the same individual over seasons and years results in autocorrelation.

The analysis of such data necessitates application of special statistical techniques.

☆ *Uniformity in evaluation:* Plant breeder often makes visual observations and selects without recording the relevant data. Further, voluminous data remains in plant breeder's book and often goes unreported. Although, it is requested that the evaluation data may be sent to curator, but it is not mandatory and often not complied with. Therefore, it is important to record data in a uniform pattern using common reference/standard checks (s) so that these are of value to a number of scientists.

## Statistical Design for Germplasm Evaluation

The issues that need careful consideration for proper germplasm evaluation are:

(i) The number of accessions,

(ii) Choice of experimental design

(iii) The number of replicates and,

(iv) The number and choice of trial environment.

Bos (1983a,b) showed theoretically that with fixed resources, the expected gain may be greater for a selection based on a preliminary screening of many genotypes, rather than on a more detailed evaluation of a smaller number. Further, in multi-environment testing it is desirable to commit larger resources to environments rather than replications within an environment.

### a) Un-replicated Design

Here the new test accessions are planted in long, narrow plots and check cultivars/genotypes are planted at regular intervals. The indices, as function of check performance, are developed for comparing the performance of the test accessions. The four commonly used indices are:

(i) The performance of nearest check,

(ii) The performance of the two nearest checks

(iii) The weighted mean of these two checks, where weights are inversely related to the distance from the test plot and,

(iv) The mean of all the checks in the range (Kempton and Fox, 1997). One possible way of performing objective comparisons is based on a measure of experimental error computed from the variation among the observations made on the performance of the checks. The least significant difference is then obtained to compare the performance of the test accessions with the mean of the checks under the assumption that the experimental error of the test accessions is similar to that of the checks.

## b) Replicated Design

once promising accessions are identified; these are further evaluated in a replicated design. Generally, the number of accessions selected for the evaluation is too large to be evaluated in common designs, namely randomized complete block design (RCBD), latin square design *etc*. The efficient design for comparing relatively a large number of test accessions with that of check treatments have been developed, namely reinforced incomplete block designs, supplemented balanced block designs, inter-and intra-group balanced block designs with varying replications, balanced bipartite block designs and balanced treatment incomplete block designs.

## c) Augmented Design

In augmented design; the check treatments are replicated in any standard design (complete block, incomplete block, row-column design *etc*.) and are augmented with test treatments, which are un-replicated. The experimenter generally has a large number of accessions and has to opt for their un-replicated evaluation, as he cannot afford to have replications during preliminary screening. However, the check treatments are replicated to get an estimate of experimental error. It appears paradoxically with single replication of the test accessions that one may not be able to make all the possible paired comparisons involving the test treatments, the check treatments and the test and check treatments. Augmented designs are available for 0-way, 1-way, 2-way *etc*. heterogeneity settings. Those eliminating heterogeneity in one direction are called augmented block design and those eliminating heterogeneity in two directions are called augmented row-column designs. Federer and Raghavarao (1975) obtained augmented designs using RCB design and linked block designs for 1-way heterogeneity setting and outlined a general theory of augmented designs. They also developed a method of construction of augmented row-column design using a Youden Square design and provided their analysis including the computation of standard errors of estimable treatment contrasts. Federer *et al*. (1975) described systematic methods of construction of augmented row-column designs and the procedure for their analysis. Lin and Poushinsky (1983) modified the augmented row-column design to adjust the effects of test entries for environmental heterogeneity. The modified design has a split-plot structure.

## d) Augmented Randomized Complete Block Design (ACRB)

In most of the experiments number of accessions (total test treatments) are to be compared with 'c' check treatments using 'n' experimental units (plots) arranged in 'b' blocks such that the size of a block is greater than the number of checks (the blocks can be of unequal size). In an augmented randomized complete block design, check treatments are replicated 'b' times and occur once in every block, and test accessions are not repeated and they occur only once in any one of the blocks. Therefore, in any blocks, number of test accessions is equal to the number of experimental plots minus number of checks.

## e) Incomplete Block Design

There is certain *situations* where the experimenter is interested in all possible paired comparisons rather than comparing test vs. check treatment in a replicated design. Moreover, the effective control of variation through blocking usually requires fairly small block size, while most evaluation experiments include relatively large numbers of accessions and, hence, have large blocks. In these situations, the incomplete block designs like balanced incomplete block (BIB), partially balanced incomplete block (PBIB) designs *etc*. are available.

## f) Generalized Lattice Design

When a large number of germplasm accessions are available but limited land and other resources bound to have a limited number of replicates. For such a situation the generalized lattice design and alpha designs provide a wide range of efficient experimental designs. Suppose we have 'r' replicates of 't' total treatments (test treatments and standard checks) and 't'=bk; there being 'b' blocks each having 'k' plots within each replicate. Thus, there is a two- stage structure for the control of field variation. First, replicates can allow an adjustment for large-scale variation and then within each replicate the blocks provide a second level of adjustment. An incomplete block design, in which blocks can be grouped into replicates of the accessions, is called a resolvable design. There are non- resolvable designs also but these designs do not allow a two-stage removal of field heterogeneity and, thus, are less useful for field experiments. The overall structure of an 'r' replicate resolvable design for 't' treatments with 'b' blocks of size 'k' within each replicate is an example of a generalized lattice design. Special cases of such designs are the square lattice design ($t=b^2$, k=b), the rectangular lattice design [t=b (b-1), k=b-1] cubic lattice design ($t=b^3$, k=3b) circular lattice design ($t=2b^2$, k=2b) *etc*. the lattices provide simple designs for situations where there are many accessions and blocks are reasonably small compared with the size of the experiment (IPGRI, 2001).

## Characterization

  i. Morphological characterization
  ii. Molecular characterization

Stable and unique morphological traits should be effectively used for assessing the degree of genetic variation in the initial as well as regenerated *ex situ* conserved samples. Morphological characterization of *ex situ* collections should be based on standardized format and the data follow internationally agreed descriptors (Breese, 1989; Engels and Rao, 1995).

## Evaluation at International Level

Several genotypes are collected jointly or received personally with full details from other countries are again evaluated with care and distributed to research organizations.

## Descriptors and Descriptor States for Characterization

For meaningful characterization and evaluation adoption of descriptor list is important. A descriptor can be defined as:

**Table 5.5: Evaluation of Indigenous Wild and Landrace Species (Vegetables) of Nepal**

| Botanical Name | Nature | Family | Elevation in Nepal (m) |
|---|---|---|---|
| *Daucus carota* L. | Vegetable | Apiaceae | 1350 |
| *Alocasia navicularis* Koch and Bouche | Vegetable | Araceae | 450 |
| *Amorphophallus bulbifera* (Schott) Blume | Ornamental/Wild | Araceae | 300-900 |
| *Colocasia affinis* Schott | Wild | Araceae | 2000 |
| *Colocasia esculenta* (L.) Schott | Vegetable | Araceae | 300-1200 |
| *Colocasia fallax* Schott | Wild | Araceae | 400-2000 |
| *Bombax ceiba* L. | Ornamental/Timber | Bombaceae | 200-900 |
| *Brassica juncea* (L.) Czern. | Vegetable | Brassicaceae | 640-1340 |
| *Brassica rapa* L. | Vegetable | Brassicaceae | 1340-3200 |
| *Lepidium sativum* L. | Vegetable/Spice | Brassicaceae | 200-3000 |
| *Raphanus sativus* L. | Vegetable | Brassicaceae | 600-3500 |
| *Canna* L. (2 species) | Ornamental/Wild | Cannaceae | 900-1400 |
| *Ipomoea* L. (5 species) | Ornamental/Wild | Convolvulaceae | 200-1200 |
| *Ipomoea eriocarpa* R. Br. | Vegetable | Convolvulaceae | 610-760 |
| *Cucumis callosus* (Rottb.) Cogn | Wild | Cucurbitaceae | - |
| *Cucumis melo* L. | Vegetable | Cucurbitaceae | 200-800 |
| *Cucumis sativus* L. (includes var. *hardwickii* (Royle) Kitam., and var. *sikimensis* Hook.f.) | Vegetable | Cucurbitaceae | 1300-1800 |
| *Cucurbita pepo* L. | Vegetable | Cucurbitaceae | 100-1500 |
| *Lagenaria siceraria* (Molina) Standl. | Vegetable | Cucurbitaceae | 200-2300 |
| *Luffa cylindrica* (L.) Roem | Vegetable | Cucurbitaceae | 1200-1700 |
| *Momordica balsamina* L. | Wild | Cucurbitaceae | 600-1200 |
| *Momordica charantia* L. | Vegetable | Cucurbitaceae | 300-1200 |
| *Momordica dioica* Willd. | Wild | Cucurbitaceae | 1100 |
| *Trichosanthes cordata* Roxb. | Wild | Cucurbitaceae | 500 |
| *Trichosanthes cucumerina* L. | Wild | Cucurbitaceae | 400-1200 |
| *Trichosanthes dioica* Roxb. | Vegetable | Cucurbitaceae | 600 |
| *Trichosanthes ovigera* Blume | Wild | Cucurbitaceae | 200-1700 |
| *Trichosanthes tricuspidata* Lour. | Wild | Cucurbitaceae | 1200-2300 |
| *Trichosanthes wallichiana* (Ser.) Wight | Wild | Cucurbitaceae | 600-2700 |
| *Dioscorea alata* L. | Vegetable/Ornamental | Dioscoraceae | 600-1200 |
| *Dioscorea belophylla* (Prain) Haines | Ornamental/Wild | Dioscoraceae | 200 |
| *Dioscorea bulbifera* L. | Ornamental/Wild | Dioscoraceae | 150-2100 |
| *Dioscorea deltoidea* Griseb. | Medicinal/Ornamental/Wild | Dioscoraceae | |
| *Dioscorea esculenta* (Lour.) Burkill | Ornamental/Wild | Dioscoraceae | 100-1500 |
| *Dioscorea glabra* Roxb. | Ornamental/Wild | Dioscoraceae | 900-2200 |
| *Dioscorea hamiltonii* Hook.f. | Ornamental/Wild | Dioscoraceae | 1200-1700 |
| *Dioscorea hispida* Dennst. | Ornamental/Wild | Dioscoraceae | 600 |
| *Dioscorea kamoonensis* Kunth | Ornamental/Wild | Dioscoraceae | 1800-2200 |
| *Dioscorea melanophyma* Prain and Burkill | Ornamental/Wild | Dioscoraceae | 2000-2500 |
| *Dioscorea pentaphylla* L. | Ornamental/Wild | Dioscoraceae | 600-1500 |
| *Dioscorea prazeri* Prain and Burkill | Ornamental/Wild | Dioscoraceae | 910 |
| *Dioscorea pubera* Blume | Ornamental/Wild | Dioscoraceae | - |

| Botanical Name | Nature | Family | Elevation in Nepal (m) |
|---|---|---|---|
| *Bauhinia purpurea* L. | Food | Fabaceae | 300-600 |
| *Bauhinia variegata* L. | Food/Ornamental/Vegetable | Fabaceae | 150-2000 |
| *Lablab purpureus* (L.) Sweet | Pulse/Vegetable | Fabaceae | 1000-2500 |
| *Pisum sativum* L. | Pulse/Vegetable | Fabaceae | 1200-1400 |
| *Trigonella foenum-graceum* L. | Spice/Vegetable | Fabaceae | 1400 |
| *Vicia faba* L. | Pulse/Vegetable | Fabaceae | 1400 |
| *Vigna angularis* (Willd.) Ohwi and H. Ohashi | Pulse/Vegetable | Fabaceae | 300-700 |
| *Vigna mungo* (L.) Hepper | Pulse/Vegetable | Fabaceae | 450-2100 |
| *Vigna radiata* (L.) R. Wilczek | Pulse/Vegetable | Fabaceae | 100-300 |
| *Vigna umbellata* (Thunb.) Ohwi and H. Ohashi | Pulse/Vegetable | Fabaceae | 450-2100 |
| *Vigna unguiculata* (L.) Walp. | Pulse/Vegetable | Fabaceae | 181 |
| *Allium carolinianum* DC. | Ornamental/Wild | Liliaceae | 4800-5100 |
| *Allium cepa* L. | Ornamental/Vegetable | Liliaceae | 1350 |
| *Allium fasciculatum* Rendle | Ornamental/Wild | Liliaceae | 2800-4500 |
| *Allium hypsistum* Steearn | Ornamental/Wild | Liliaceae | 5500 |
| *Allium prattii* C.H. Wright | Ornamental/Wild | Liliaceae | 2400-4500 |
| *Allium przewalskianum* Regel. | Ornamental/Wild | Liliaceae | 3900-4200 |
| *Allium sativum* L. | Ornamental/Vegetable | Liliaceae | 1350 |
| *Allium sikkimense* Baker | Ornamental/Wild | Liliaceae | 3000-4800 |
| *Allium tuberosum* Spreng. | Ornamental/Wild | Liliaceae | 2300-2600 |
| *Allium wallichii* Kunth | Medicinal/Ornamental/Wild | Liliaceae | 2400-4650 |
| *Asparagus* L. (4 species) | Ornamental/Wild | Liliaceae | 600-2900 |
| *Asparagus racemosus* willd. | Medicinal/Ornamental/Vegetable | Liliaceae | 600-2100 |
| *Abelmoschus manihot* (L.) Medic | Vegetable/Fiber | Malvaceae | 1200 |
| *Hibiscus esculentus* L. | Vegetable | Malvaceae | 1400 |
| *Bambusa nutans* Wall. | Vegetable/Timber | Poaceae | 700-1700 |
| *Dendrocalamus hamiltonii* Nees and Arn. | Vegetable/Timber | Poaceae | 1000-2000 |
| *Solanum melongena* L. | Vegetable | Solanaceae | 1200-1500 |
| *Curcuma angustifolia* Roxb. | - | Zingiberaceae | 1500 |
| *Curcuma aromatica* Salisb. | Spices | Zingiberaceae | 700-1100 |

"A descriptor is a character that is considered important and/or useful in the description of a population/an accession".

It is better to use internationally approved list for uniformity in data processing and storage. IPGRI has developed more than 80 descriptors for different Agri-Horticultural crops, including 15 for vegetable crops. In India, efforts are also being made by IBPGR to evolve suitable descriptors and their descriptor states, as a guideline to document effectively the characterization and evaluation of data, which have been modified by various germplasm handling or user agencies so as to suit their need. Descriptors and their descriptor states are integrated component of a plant genetic resources (PGR) system which engraves a base line for the systematic studies on documentation, indexing and retrieval of accessions. In developmental stages IBPGR provided descriptor lists on a standard format following the advice on descriptors and their descriptor states from respective crop experts throughout the world (Anon., 1985).

## IBPGR Role in Development of Descriptors and Descriptor States

Several descriptors in various crops/groups of crops have been published by IBPGR in collaboration with expert groups from different national and international programs, CGIAR centres. Such lists are organized in two parts *viz.*

(i) Descriptors for characterization and preliminary evaluation and

(ii) For detailed evaluation

Genetic resources organizations/gene bank often deal with catalogues based on characterization and preliminary evaluation, while plant breeders often undertake detailed evaluation. Such catalogues have been prepared for germplasm holdings of several crop species in ICAR-NBPGR, New Delhi and its regional stations. Their scheme encouraged the collection of data uniformly on a minimum of descriptors in the 'Accession', 'Collection' and 'Characterization' and 'Preliminary Evaluation' sectors that should be ideally available for any one accession. The descriptor lists provide an international format and thereby produce a universally understood language for the PGR data (CYMMIT/IBPGR, 1991) in maize. It also provides a rapid, reliable and efficient means for information storage, retrieval and communication for encoding, transformation and interpretation of data throughout the international network on a uniform code. Thus germplasm utilization is also greatly assisted by adopting uniform descriptors lists. The proposal has been flexible in terms of adding up data in other sectors like 'further characterization and evaluation' and 'management' which could serve as examples for creating additional descriptors by any user. Although the IBPGR suggested descriptors and their states were hardly used as such by any gene bank, curator or user yet the format given therein was found useful by the scientists and organization concerned with PGR activities and promoted *mutatis mutandis* the world over.

## Choice of Descriptors

As per the existing scheme(s) data recording on primary sectors of descriptors *viz.*, 'Accession', 'Collection' and 'Characterization' and 'Preliminary Evaluation' is managed directly by the institute related to PGR activities whereas the 'further characterization and evaluation' is expected to be recorded by the users. Nevertheless, the latter is expected to be passed on to the PGR related institute as a feedback to the database management system for the long term use of an accession. A germplasm user would be more interested in its applied aspects *i.e.* breeding progrmmes rather than basic characterization and evaluation for further use. An extended format of descriptors and their descriptor states, in relation to multi-disciplinary PGR management sectors was proposed by Rana *et al.* (1993). The authors proposed this set as an ideal system, which included Quarantine, Storage-management and Germplasm Utilization sectors in addition to Acquision/ Passport and characterization and evaluation parts. It was also observed that to achieve a working optimum under the extended format would remain as flexible as before yet database management system in relation to long term conservation of PGRs should evolve gradually under the concerted and integrated efforts of the explorers, curators, gene bank managers and users alike.

Descriptors differ according to species, and the value of a descriptor depends primarily on the user concerned:

1. Plant breeders tend to choose descriptors of agronomic importance which is useful in crop improvement that are, usually, under polygenic control.

2. Botanists choose morphological descriptors irrespective of their genetic control.

3. Geneticists, on the other hand, favour qualitative traits with monogenic inheritance.

Descriptor list is most important for proper characterization and evaluation (Rana *et al.*, 1994). It is better to use internationally approved descriptor format, which also provides universally understood language for all plant genetic resources data. International Plant Genetic Resources Institute (IPGRI) has developed over 65 descriptors for different Agri-Horticultural crops including about 15 for vegetable crops. The IPGRI format will produce a rapid, reliable and efficient means for information storage, retrieval and documentation. This will greatly assist the utilization of germplasm throughout the international plant genetic resources work. If required the list can be modified depending upon need/situation and breeders' requirement for specific case. Development of descriptor and descriptor states at national level keeping IPGRI descriptor as model may be generate more information to specific situation during characterization and evaluation.

## Role of Germplasm Advisory Committees (GACs) on Descriptor and Descriptor States

ICAR-NBPGR has set up Germplasm Advisory Committees (GACs) on various crops with the objectives to get advice from the renowned and eminent scientists, in the respective crops, to improve capability, efficiency and effectiveness of the services related to crop plants' germplasm. One of the mandates for GACs is, to review descriptors and descriptors states used for evaluation of germplasm collections. Submitted descriptors lists would be critically examined by each member and discussed in a GAC meeting for finalization. Such finalized lists would be circulated to the related crop research stations, throughout the country for uniform adoption and, thereby, documentation of germplasm (Singh and Kochhar, 1994).

## Scoring or Coding of Descriptors States

The following internationally accepted norms for scoring or coding of descriptors states should be followed as indicated below (Dadlani *et al* 1981):

☆ Measurements are made according to the SI system. The unit to be applied is given in square brackets following the descriptor.

☆ Many descriptors, which are continuously variable, are recorded on a 1-9 scale. Sometimes, a section of the stages *e.g.* 3, 5 and 7 are also used, where this has occurred, the full range of codes is available for use by extension of the codes given or by interpolation between them *e.g.* 1= extremely low susceptible and 9= from extremely high susceptible.

☆ Presence/absence of characters are scored as + (present) and 0 (absent).

For descriptors, which are not generally, uniform throughout the accession (*e.g.* mixed collection, genetic segregation) mean and standard deviation could be reported where the descriptor is about continuous or mean where the descriptor is discontinuous.

When the descriptor is in applicable, '0' is used as the descriptor value; *e.g.* if an accession does not form flower, 0 would be scored for the flowering descriptors

## Flower Colour

1. White
2. Yellow
3. Red
4. Purple

☆ Blanks are used for information not yet available.

☆ Standard colour chart, *e.g.* Royal Horticulture Society Colour Chart, Menthuem Handbook of colour, Munsell colour chart for plant tissues are strongly recommended for all ungraded colour characters

☆ Dates to be expressed numerically in the format DD MM YYYY

☆ DD 2 digit to represent the day

☆ MM 2 digits to represent the month

☆ YYYY 4 to represent the year.

Further, the descriptor lists are sub grouped for characterization, evaluation and 'further evaluation' purposes as listed below:

1. Accession data
2. Collection data

## Characterization and Preliminary Evaluation

3. Site data
4. Plant data

## Further Characterization and Evaluation

5. Site data
6. Plant data
7. Stress susceptibility
8. Pest disease susceptibility
9. Alloenzyme composition and zymo type
10. Cytological characters and identifier genes
11. Notes

In another set the multi-crop Descriptor is as under

## 1. Institute Code (INSTCODE)

Code of the institute where the accession is maintained. The codes consist of the 3-letter ISO 3166 country code of the country where the institute is located plus a number. The current set of Institute Codes is available from the FAO website (http://apps3.fao.org/wiews/).

## 2. Accession Number (ACCENUMB)

This number serves as a unique identifier for accessions within a gene bank collection, and is assigned when a sample is entered into the gene bank collection.

## 3. Collecting Number (COLLNUMB)

Original number assigned by the collector(s) of the sample, normally composed of the name or initials of the collector(s) followed by a number. This number is essential for identifying duplicates held in different collections.

## 4. Collecting Institute Code (COLLCODE)

Code of the Institute collecting the sample. If the holding institute has collected the material, the collecting institute code (COLLCODE) should be the same as the holding institute code (INSTCODE). Follows INSTCODE standard.

## 5. Genus (GENUS)

Genus name for taxon. Initial uppercase letter required.

## 6. Species (SPECIES)

Specific epithet portion of the scientific name in lowercase letters. Following abbreviation is allowed: 'sp.'

## 7. Species Authority (SPAUTHOR)

Provide the authority for the species name.

## 8. Subtaxa (SUBTAXA)

Subtaxa can be used to store any additional taxonomic identifier. Following abbreviations are allowed: 'subsp.' (for subspecies); 'convar.' (for con variety); 'var.' (for variety); 'f.' (for form).

## 9. Subtaxa Authority (SUBTAUTHOR)

Provide the subtaxa authority at the most detailed taxonomic level.

## 10. Common Crop Name (CROPNAME)

Name of the crop in colloquial language, preferably English (*i.e.* 'malting barley', 'cauliflower', or 'white cabbage')

## 11. Accession Name (ACCENAME)

Either a registered or other formal designation should be given to the accession. First letter should be uppercase. Multiple names separated with semicolon without space. For example: Rheinische Vorgebirgstrauben; Emma; Avlon.

## 12. Acquisition Date [YYYYMMDD] (ACQDATE)

Date, on which the accession entered the collection where YYYY is the year, MM is the month and DD is the day. Missing data (MM or DD) should be indicated with hyphens. Leading zeros are required.

### 13. Country of Origin (ORIGCTY)

Code of the country in which the sample was originally collected. Use the 3-letter ISO 3166-1 extended country codes.

### 14. Location of Collecting Site (COLLSITE)

Location information below the country level that describes where the accession was collected. This might include the distance in kilometers and direction from the nearest town, village or map grid reference point, (*e.g.* 7 km south of Curitiba in the state of Parana).

### 15. Latitude of Collecting Site 1 (LATITUDE)

Degree (2 digits) minutes (2 digits), and seconds (2 digits) followed by N (North) or S (South) (*e.g.* 103020S). Every missing digit (minutes or seconds) should be indicated with a hyphen. Leading zeros are required (*e.g.* 10——S; 011530N; 4531–S).

### 16. Longitude of collecting site1 (LONGITUDE)

Degree (3 digits), minutes (2 digits), and seconds (2 digits) followed by E (East) or W (West) (*e.g.* 0762510W). Every missing digit (minutes or seconds) should be indicated with a hyphen. Leading zeros are required (*e.g.* 076——W).

### 17. Elevation of Collecting Site [msl] (ELEVATION)

Elevation of collecting site expressed in meters above sea level (msl). Negative values are allowed.

### 18. Collecting Date of Sample [YYYYMMDD] (COLLDATE)

Collecting date of the sample where YYYY is the year, MM is the month and DD is the day. Missing data (MM or DD) should be indicated with hyphens. Leading zeros are required. 1 To convert from longitude and latitude in degrees (°), minutes ('), seconds (''), and a hemisphere (North or South and East or West) to decimal degrees, the following formula should be used: $d°\ m'\ s''=h\ *(d+m/60+s/3600)$ where $h=1$ for the Northern and Eastern hemispheres and $-1$ for the Southern and Western hemispheres *i.e.* $30°30'0''\ S= -30.5$ and $30°15'55''\ N=30.265$.

### 19. Breeding Institute Code (BREDCODE)

Institute code of the institute that has bred the material. If the holding institute has bred the material, the breeding institute code (BREDCODE) should be the same as the holding institute code (INSTCODE). Follows INSTCODE standard.

### 20. Biological Status of Accession (SAMPSTAT)

The coding scheme proposed can be used at 3 different levels of detail: either by using the general codes (in boldface) such as 100, 200, 300, 400 or by using the more specific codes such as 110, 120 *etc.*

100) Wild
110) Natural
120) Semi-natural/wild
200) Weedy
300) Traditional cultivar/landrace
400) Breeding/research material
410) Breeder's line
411) Synthetic population
412) Hybrid
413) Founder stock/base population
414) Inbred line (parent of hybrid cultivar)
415) Segregating population
420) Mutant/genetic stock
500) Advanced/improved cultivar
999) Other (Elaborate in REMARKS field)

### 21. Ancestral Data (ANCEST)

Information about either pedigree or other description of ancestral information (*i.e.* parent variety in case of mutant or selection). For example a pedigree 'Hanna/7*Atlas//Turk/8*Atlas' or a description 'mutation found in Hanna', 'selection from Irene' or 'cross involving amongst others Hanna and Irene'.

### 22. Collecting/Acquisition Source (COLLSRC)

The coding scheme proposed can be used at 2 different levels of detail: either by using the general codes (in boldface) such as 10, 20, 30, 40 or by using the more specific codes such as 11, 12 *etc.*

10) Wild habitat
11) Forest/woodland
12) Shrub land
13) Grassland
14) Desert/tundra
15) Aquatic habitat
20) Farm or cultivated habitat
21) Field
22) Orchard
23) Backyard, kitchen or home garden (urban, peri-urban or rural)
24) Fallow land
25) Pasture
26) Farm store
27) Threshing floor
28) Park
30) Market or shop
40) Institute, Experimental station, Research organization, Gene bank
50) Seed company
60) Weedy, disturbed or ruderal habitat

61) Roadside

62) Field margin

99) Other (Elaborate in REMARKS field)

### 23. Donor Institute Code (DONORCODE)

Code for the donor institute. Follows INSTCODE standard.

### 24. Donor Accession Number (DONORNUMB)

Number assigned to an accession by the donor. Follows ACCENUMB standard.

### 25. Other Identification (Numbers) Associated with the Accession (OTHERNUMB)

Any other identification (numbers) known to exist in other collections for this accession. Use the following system: INSTCODE: ACCENUMB; INSTCODE: ACCENUMB; INSTCODE and ACCENUMB follow the standard described above and are separated by a colon. Pairs of INSTCODE and ACCENUMB are separated by a semicolon without space. When the institute is not known, the number should be preceded by a colon.

### 26. Location of Safety Duplicates (DUPLSITE)

Code of the institute where a safety duplicates of the accession is maintained. Follows INSTCODE standard.

### 27. Type of Germplasm Storage (STORAGE)

If germplasm is maintained under different types of storage, multiple choices are allowed, separated by a semicolon (*e.g.* 20; 30). (Refer to FAO/IPGRI Genebank Standards 1994 for details on storage type.)

10) Seed collection

11) Short term

12) Medium term

13) Long term

20) Field collection

30) *In vitro* collection (Slow growth)

40) Cryopreserved collection

99) Other (elaborate in REMARKS field)

### 28. Remarks (REMARKS)

The remarks field is used to add notes or to elaborate on descriptors with value 99 or 999 (=other).

---

### Table 5.6: Descriptor and their Descriptor States in PGR Evaluation and Documentation: Guidelines

**1. Characterization and evaluation (Curators)** - Based on standard format of preliminary characterization and evaluation data. However, the scheme should be better defined under the following sub-heads.

| | | |
|---|---|---|
| 1.1 | Taxonomical descriptors | Covering identity of the accession at species, subspecies, botanical types *etc.* levels |
| 1.2 | Phenological descriptors | Covering growth stages of the accession *viz.* flowering, maturity, senescence, synchrony *etc.* including seasons of planting |
| 1.3 | Morpho-agronomical descriptors | Covering a minimal list of traits which determine plant and pod characters |
| 1.4 | Seed traits descriptors | A set of minimal traits which would express characteristics of seed |
| 1.5 | Field resistance/tolerance descriptors | Covering reactions to natural incidence of diseases/viruses/nematodes and other pests |
| 1.6 | Any other descriptors | As per standard format or specific requirement of the species |

**2. Characterization and evaluation (Enhancement programmes and/or users)**-Based on standard format of further characterization and evaluation data. The descriptors and their descriptor states which supplement a suitable information retrieval, for particular accession, under any one of the sub-heads 1.1 to 1.5 should be included in the corresponding sub-heads 2.1 to 2.5 in addition; the following points may also be included.

| | | |
|---|---|---|
| 2.6 | **Resistance/tolerance** | Under artificial stress conditions |
| 2.6.1 | Disease resistance | Under epiphytotic conditions |
| 2.6.2 | Pest resistance | Under artificial heavy insect populations |
| 2.6.3 | Drought tolerance | Under artificial water-stress conditions |
| 2.6.4 | Tolerance to alkalinity/salinity | Under artificial testing |
| 2.7 | Adaptability descriptors | |
| 2.7.1 | Multi-location testing | 1. Widely adapted |
| | | 2. Adapted under favourable conditions |
| | | 3. Adapted under stress conditions |
| 2.7.2 | Specific observations | Notes, including GXE interactions |
| 2.8 | Combining ability descriptors | |
| 2.8.1 | General combining ability | 1. High |
| | | 2. Average |
| | | 3. Low |

| 2.8.2 | Specific combining ability | 1. High |
|---|---|---|
| | | 2. Average |
| | | 3.Low |

**3. Specific trait descriptors**

| 3.1 | Quality traits | As per related specifications |
|---|---|---|
| 3.2 | Industrial value traits | As per related specifications |

**4. Utilization descriptors-** Based on feedback from users the database management system may include this sector for future use of the accession.

| 4.1 | The accession was used for | 1. Single trait (specify) |
|---|---|---|
| | | 2. Multiple trait (specify) |
| 4.2 | The accession was used as/after | 1. *per se* selection |
| | | 2. Mass selection |
| | | 3. Pure line selection |
| | | 4. Parent in reciprocal recurrent selection |
| | | 5. Parent in single cross/es |
| | | 6. Parent in double/multiple crosses |
| | | 7. Any other (specify) |

**5. Notes (Descriptive)-** In order to cover any other information not dealt with above

*Source*: Singh and Rana, 1993.

Prefix remarks with the field name they refer to and a colon (*e.g.* COLLSRC: roadside). Separate remarks referring to different fields are separated by semicolons without space.

## Characterization and of Evaluation Vegetable Genetic Resources

### (A) Morphological Characterization and of Evaluation Vegetable Genetic Resources

Morphological characterization and of evaluation vegetable genetic resources for various traits are prime need of the day. Without trait identification, germplasm has no value for breeder or other research workers associated with improvement programme of the crop.

### (a) Evaluation of Indigenous Collection

In India concentrated efforts were made to evaluate collected vegetable germplasm of different part of the country and identify promising accessions in each vegetable crop for meaningful use.

### Crop-wise Evaluation and Characterization of Vegetable Genetic Resources

Crop wise evaluation and characterization is given below:

### Tomato

Seventy one genotypes of tomato were evaluated by Chaurasia and Singh (1998) during 1993-94 and 1994-95 and results revealed significant difference for all the characters studied. Based on Mahalanobis $D^2$ values, genotypes were grouped into 9 clusters I 1993-94 and 10 clusters in 1994-95. Fruit weight showed maximum contribution to the genetic diversity in both years followed by plant height. Considering cluster differences and cluster means of the genotypes KS-7, SVRT-2, Antey and PS-1 were recommended the best parents for hybridization.

### Brinjal

BCKV, Kalyani (West Bengal) has evaluated several genotypes of brinjal and promising accessions are given in Table 5.8.

### Chillies

Verma *et al.* (1998) evaluated 119 accessions of chillies at ICAR-NBPGR, Regional Station, Bhowali, Nainital (Uttar Pradesh)and noted wide range of variability in plant height (50.0-86.0 cm), growth habit (prostrate, compact, erect), leaf and stem pubescence (glabrous, sparse, intermediate, abundant), seedling, leaf and stem colour (green, purple), density of branches, plant canopy (104.0-6348.0 sq cm), leaf length (4.2-15.9 cm), leaf width (1.2-6.8 cm), flower colour, days to 50 per cent flowering (142-172 days), pedicel position, number of pedicel per axil, days to 50 per cent fruiting (161-220 days), fruit set, calyx margin and shape, annular construction at junction of calyx and peduncle, fruit position, fruit colour at immature stage (green, yellow, orange, red, purple, brown/black), fruit shape (elongate, oblate, round, conical, companulate, bell/blocky), fruit shape at peduncle attachment (truncate, cordate, lobate, acute, obtuse), fruit shape at blossom end (pointed, blunt, sunken), presence and absence of neck at base of fruit, fruit cross section (smooth, intermediate, corrugated), fruit pungency, number of fruits per plant (2.1-120.0), fruit length (1.6-12.4 cm), fruit width (0.2-2.6 cm), fruit green weight/10 fruits (7.0-55.0 g), fruit dry weight/10 fruit (1.0-7.5 g), seed colour and pest and disease of varying intensity).

**Table 5.7: Grouping of Tomato Genotypes in different Clusters**

| Cluster No. | No. of Entries | Group of Genotypes (1993-94) | No. of Entries | Group of Genotypes (1994-95) |
|---|---|---|---|---|
| 1 | 60 | NDT-1, PDVR-4, TC-104, H-36, BT-17, NDT-6, Ageta-1, DVRT-2, Sel-10, Ageta-2, Marglobe, ATV-2, Sel. No.4, PDVR-5, Punjab Kesari, Desi Local, ACE, Sel. No.6, IP-11, JS-2, Kalyani Eunish, Sioux, H-24, Roma, BT-3, BDT-11, NDT-96, Dhanshree, Bhagyashree, BT-(12-2), NDT-4, Arka Saurabh, Pant-T-4, Pusa Hyb.-2, IP-10, KT-15, NDT-120, IP-6, Arka Vikas, Sel-2, PDVR-1, Anand T-1, Arka Vishal, IP-8, TC-1, Sel-16, Pusa Sel-4, IP-2, Sel-32, IP-4, Pusa Ruby, Field King, PDVR-3, Pant Bahar, KS-17, Solan Gola, Sel-4, DVRT-1 and KT-10 | 57 | NDT-1, PDVR-4, IP-11, Sel. No.6, H-36, Punjab Kesari, Marglobe, NDT-6, BT-(12-2), BR-17, Pusa Sel-4, Kalyani Eunish, KT-15, PDVR-3, Ageta-1, TC-1, Pant Bahar, DVRT-1, Sel-17, Bhagyashree, Sel-14, Arka Saurabh, TC-104, NDT-11, IP-10, H-24, IP-4, Dhanshree, Ageta-2, Arka Vikas, Sioux, Pusa Ruby, Desi Local, PDVR-5, KS-17, ACE, KT-15, Sel. No.4, NDT-120, Sel- 2, DVRT-1, Sel-16, Sel-32, Pant T-1, NDT-96, UP-6, Sel-18, Sel-4, Deogiri, NDT-4, IP-2, Solan Gola, Anand T-1, Field King, ATV-2, KS-2, BT-3 |
| 2 | 3 | Sel-18, Rutger, Floradale | 4 | KS-7, Punjab Chhuhara, KT-10, Pusa Hyb.-2 |
| 3 | 2 | KS-2, Punjab Chhuhara | 1 | Roma |
| 4 | 1 | PDVR-2 | 3 | ATV-2, Rutger, Arka Vishal, |
| 5 | 1 | Antey | 1 | Pant T-4 |
| 6 | 1 | PS-1 | 1 | DVRT-2 |
| 7 | 1 | Pusa Hyb.-2 | 1 | IP-8 |
| 8 | 1 | Pant T-1 | 1 | Antey |
| 9 | 1 | Deogiri | 1 | PS-1 |
| 10 | - | - | 1 | PDVR-1 |

**Table 5.8: Promising Accessions of Brinjal and their Salient Features**

| Crop | Character | Range | Promising Accessions |
|---|---|---|---|
| Brinjal | Days to 50 per cent flower | 47.00-79.00 | BCB-30, BCB-18 |
| | Fruit length (cm) | 8.70-23.90 | BCB-11, BCB-17 |
| | Fruit girth (cm) | 2.80-10.30 | BCB-30, BCB-6 |
| | Average fruit weight (g) | 52.00-320.00 | BCB-11, BCB-30 |
| | Number of marketable fruit/plant | 5.00-14.00 | BCB-11, BCB-30 |
| | Marketable fruit yield/plant (kg) | 0.39-1.70 | BCB-11, BCB-30, BCB-1 |

*Source*: Chattopadhyay *et al.,* 2011.

**Table 5.9: Promising Accessions Identified in Chillies**

| Sl.No. | Characters | Promising Accessions |
|---|---|---|
| 1. | Days to 50 per cent germination >80 days | EC362911, EC362931, |
| 2. | leaf length >12 cm | EC362918, EC362919, EC362922, P-1939, P-2072, N-1149, N-1382 |
| 3. | leaf width >5 cm | EC362919, EC362925, EC362930, EC362938, P-1639, P-1713, P-1718 |
| 4. | Plant height >60 cm | P-2072, N-1160, N-1382 |
| 5. | Days to 50 per cent flowering <150 days | P-32, P-64, P-173, P-290, P-241, P-272, P-249, EC362905 |
| 6. | Days to 50 per cent fruiting <165 days | P-32, P-64, P-173, P-297, P-641, EC362905 |
| 7. | Number of fruits/plant >60 | P-1399, EC362903, EC362916, EC362930 |
| 8. | Fruit length >9 cm | P-1311, P-2072 |
| 9. | Fruit width >2 cm | P-840, P-190, P-1258, P-2072, N-349, KN-613 |
| 10. | Fruit (10) green weight >60 g | P-64, P-100, P-1258, P-2072, N-349, KN-613 |
| 11. | Fruit (10) dry weight >30 g | P-64, P-190, P-1258, P-1639, N-349, N-368, N-867, N-1149 |
| 12. | Disease free accessions from leaf spot powdery mildew and fruit rot | IC99910, IC99912, IC92150, P-1718, P-1939, N-1015, EC362901, EC362910, EC362913, EC362925 |

*Source*: Verma *et al.,* 1998.

BCKV, Kalyani (West Bengal) has evaluated several genotypes of chillies are as shown in Table 5.10.

### Sweet Pepper

Pandey *et al.* (2016) evaluated 12 genotypes (Feroz, SP-24, DARL-70, DARL-71, EC579997, SP-701, SP-19, SP-6, SP-7, DARK green, Capsicum Violet and California Wonder) found high heritability along with high GCV and high genetic gain was observed in fruit weight, fruit yield per plant, number of fruits per plant, total chlorophyll and fruit width. Total 12 genotypes were evaluated for different horticultural traits at DIBER, Field Station, Pithoragarh (Uttarakhand) and significant difference observed (Pandey *et al.*, 2013).

### Ridge Gourd

In ridge gourd 51 germplasm were evaluated and frequency distribution (per cent) of morphological traits were accessed (Varalakshmi *et al.*, 2016).

Varalakshmi *et al.* (2016 also given details values of traits are as shown in Table 5.14.

### Satputiya (*Luffa hermaphrodita*)

Ram *et al.* (2006) evaluated and characterized 24 accessions of locally grown 'Satputiya' (*Luffa hermaphrodita*) derived form of ridge gourd (*L. acutangula* L.). Out of these genotypes SP-59 recorded highest edible fruit yield/ha (107.9 q) while DR/Y-1 expressed earliness with respect to days to first hermaphrodite flower anthesis (40.6) and maximum mean value for number of fruits per plant (185.3).

### Pointed Gourd

BCKV, Kalyani (West Bengal) has evaluated several genotypes of pointed gourd and promising accessions are given in Table 5.18.

**Table 5.10: Evaluation of Chillies**

| Crop | Character | Range | Promising Accessions |
|------|-----------|-------|----------------------|
| Chilli | Days to 50 per cent flower | 30-90 | BCC-46, BCC-30, BCC-31 (very early) |
| | Fruit length (cm) | 2.93-12.00 | BCC-49, BCC-54, BCC-25, BCC-11, BCC-2 |
| | Number of fruits/plant | 22-156 | BCCH SI-4, BCC-41, BCC-24, BCC-1 |
| | Average fruit weight (g) | 1.53-23.00 | BCC-18, BCCH- Sel.-4, BCC-24 |
| | Green fruit yield/plant (g) | 50.40-890.00 | BCCH SI-4, BCC-24, BCC-1, BCC-23, BCC-18 |
| | Ascorbic acid content (mg/100 g) | 21.00-278.00 | BCC-1, BCC-32, BCC-31, BCC-30, BCC-41 |
| | Capsaicin content (per cent) | 0.10-0.31 | BCCH- Sel.-4, BCC-62, BCC-1 |
| | Oleoresin content (per cent) | 8.89-31.17 | BCCH SI-4, AC-575 |

*Source*: Chattopadhyay *et al.*, 2011.

**Table 5.11: Mean Performance of Genotypes for Economic Traits in Sweet Pepper**

| Genotypes | Fruit Length (cm) | Fruit Width (cm) | Fruit Weight (g) | Fruits/Plant | Yield/Plant (kg) | Ascorbic Acid (mg/100 g) | Total Chlorophyll (kg/100 g) |
|-----------|-------------------|------------------|------------------|--------------|------------------|--------------------------|------------------------------|
| Feroz | 7.6 | 17.1 | 53.0 | 20.0 | 0.950 | 56.60 | 74.82 |
| SP-24 | 8.2 | 23.0 | 86.7 | 24. | 1.580 | 444.94 | 54.28 |
| DARL-70 | 9.6 | 20.6 | 71.7 | 18.0 | 1.470 | 99.87 | 23.73 |
| DARL-71 | 9.6 | 25.2 | 103.0 | 14.0 | 1.220 | 83.63 | 40.03 |
| EC579997 | 8.8 | 13.7 | 35.0 | 39.0 | 1.170 | 128.91 | 44.65 |
| SP-701 | 9.1 | 16.6 | 56.6 | 33.0 | 1.380 | 77.38 | 36.95 |
| SP-19 | 9.6 | 17.0 | 68.3 | 33.0 | 1.520 | 113.87 | 32.57 |
| SP-6 | 10.1 | 21.1 | 88.0 | 20.0 | 1.150 | 104.71 | 52.14 |
| SP-7 | 8.3 | 15.8 | 50.0 | 38.0 | 1.640 | 103.89 | 66.49 |
| Dark Green | 7.2 | 16.3 | 43.3 | 42.0 | 1.570 | 81.16 | 54.14 |
| Capsicum Violet | 5.0 | 15.3 | 30.0 | 28.0 | 0.790 | 58.81 | 50.06 |
| California Wonder | 8.6 | 17.1 | 60.0 | 32.0 | 1.62. | 126.60 | 129.23 |
| **SEM** | **0.105** | **0.649** | **4.947** | **1.076** | **0.098** | **7.644** | **0.585** |
| **CD at 5 per cent** | **0.309** | **1.905** | **14.509** | **3.156** | **0.287** | **22.418** | **0.172** |
| **CD at 1 per cent** | **0.421** | **2.589** | **19.720** | **4.290** | **0.3904** | **30.409** | **0.233** |

*Source*: Pandey *et al.*, 2013.

**Table 5.12: Estimates of Variance and Genetic Parameters for Economic Traits in Sweet Pepper**

| Characters | Genotypic Variance | Phenotypic Variance | GCV (per cent) | PCV (per cent) | Heritability (h2 per cent) | Genetic Advance (per cent) | Genetic Gain (per cent) |
|---|---|---|---|---|---|---|---|
| Fruit length | 1.44 | 1.47 | 14.12 | 14.26 | 97.95 | 2.44 | 28.60 |
| Fruit width | 34.25 | 35.52 | 32.10 | 32.69 | 96.42 | 11.83 | 64.89 |
| Fruit weight | 1417.24 | 1490.65 | 60.77 | 62.33 | 95.07 | 75.61 | 122.06 |
| Number of fruits/plant | 246.85 | 250.33 | 32.06 | 32.69 | 98.60 | 26.36 | 92.07 |
| Fruit yield/plant | 0.208 | 0.236 | 34.08 | 36.30 | 88.14 | 0.883 | 65.99 |
| Fruit yield/plot | 11.66 | 13.59 | 16.63 | 27.41 | 85.79 | 4.274 | 20.85 |
| Ascorbic acid | 376.96 | 552.24 | 20.54 | 24.85 | 68.26 | 33.04 | 34.93 |
| Total chlorophyll | 747.78 | 747.79 | 49.39 | 49.48 | 99.99 | 56.32 | 101.90 |

*Source*: Pandey *et al.*, 2013.

**Table 5.13: Frequency Distribution (per cent) of Morphological Traits of Ridge gourd Germplasm**

| Sl.No. | Morphological Traits | Stage of Observation Recorded | Class | Germplasm (No.) | Frequency |
|---|---|---|---|---|---|
| 1. | Early plant vigour | After 30 days of sowing | Poor | 5 | 9.8 |
| | | | Good | 17 | 33.3 |
| | | | Very good | 29 | 56.9 |
| 2. | Plant growth habit | At fully grown plant | Short viny | 1 | 2.0 |
| | | | Medium viny | 16 | 31.4 |
| | | | Long viny | 34 | 66.7 |
| 3. | Leaf size | At full foliage stage | Small | 1 | 2.0 |
| | | | Medium | 30 | 58.8 |
| | | | Large | 20 | 39.2 |
| 4. | Leaf pubscence | At full foliage stage | Soft | 0 | 0.0 |
| | | | Intermediate | 21 | 41.2 |
| | | | Hard | 30 | 58.8 |
| 5. | Fruit skin colour | At marketable stage | Light green | 3 | 5.9 |
| | | | Green | 33 | 64.7 |
| | | | Dark green | 15 | 29.4 |
| 6. | Fruit skin luster | At marketable stage | Matt | 23 | 45.1 |
| | | | Intermediate | 21 | 41.2 |
| | | | Glossy | 7 | 13.7 |
| 7. | Fruit ridge (rib) shape | In cross-section at marketable stage | Superficial | 9 | 17.6 |
| | | | Rounded/grooved | 0 | 0.0 |
| | | | Intermediate | 17 | 33.3 |
| | | | Deep grooved | 25 | 49.0 |
| 8. | Seediness | At marketable stage | Low | 20 | 39.2 |
| | | | Medium | 19 | 37.3 |
| | | | High | 12 | 23.5 |
| 9. | Fruit taste | At marketable stage | Normal | 29 | 56.9 |
| | | | Sweet | 18 | 35.3 |
| | | | Bitter | 4 | 7.8 |

*Source*: Varalakshmi *et al.* (2016).

### Table 5.14: Morphological Performance of Genotypes

| Sl.No. | Accession | NFF | DFF | Vine Length (cm) | Branch Number | Peduncle Length (cm) | Fruit Length (cm) | Fruit Girth (cm) | Fruit Weight (g) | Fruit Number/ Vine | Fruit Yield/ Vine (kg) | Fruit Yield/ ha (t) |
|---|---|---|---|---|---|---|---|---|---|---|---|---|
| 1. | IC92700 | 8.2 | 47.5 | 4.3 | 5.4 | 7.0 | 22.3 | 10.0 | 142.7 | 22.9 | 1.7 | 23.2 |
| 2. | IC20404 | 5.3 | 41.2 | 3.8 | 5.0 | 5.4 | 14.1 | 10.6 | 101.1 | 18.0 | 1.0 | 14.1 |
| 3. | IC92624 | 7.9 | 49.9 | 2.6 | 2.9 | 7.4 | 20.0 | 10.0 | 120.4 | 18.5 | 1.5 | 18.6 |
| 4. | IC105554 | 8.8 | 44.1 | 3.6 | 5.9 | 6.9 | 15.3 | 11.2 | 120.5 | 19.0 | 2.0 | 28.4 |
| 5. | IC23255 | 8.4 | 42.3 | 2.7 | 4.8 | 7.3 | 20.4 | 10.6 | 141.7 | 8.9 | 1.3 | 17.0 |
| 6. | IC339224 | 7.7 | 44.8 | 3.4 | 4.2 | 6.4 | 18.1 | 10.0 | 151.8 | 12.9 | 1.7 | 22.0 |
| 7. | IC23259 | 6.0 | 41.2 | 3.5 | 5.4 | 7.3 | 17.1 | 10.6 | 123.7 | 15.6 | 1.5 | 21.8 |
| 8. | IC92637 | 7.7 | 39.2 | 3.4 | 5.8 | 7.1 | 18.0 | 10.4 | 130.1 | 21.6 | 1.7 | 26.9 |
| 9. | IC92689 | 6.2 | 46.0 | 3.0 | 4.2 | 6.5 | 18.4 | 10.7 | 153.01 | 7.8 | 0.6 | 8.6 |
| 10. | IC92618 | 7.1 | 42.9 | 2.8 | 3.9 | 6.4 | 16.6 | 11.4 | 112.2 | 16.7 | 1.9 | 23.7 |
| 11. | IC93393 | 13.9 | 48.7 | 2.9 | 3.9 | 9.3 | 21.9 | 11.9 | 163.2 | 9.9 | 1.7 | 20.9 |
| 12. | IC392334 | 7.4 | 47.9 | 3.6 | 4.8 | 8.0 | 16.1 | 10.8 | 143.9 | 16.7 | 1.8 | 25.1 |
| 13. | IC308562 | 6.9 | 44.8 | 3.2 | 4.9 | 6.2 | 13.2 | 9.7 | 100.1 | 24.1 | 1.8 | 26.2 |
| 14. | IC385911 | 9.2 | 45.2 | 3.8 | 9.0 | 8.7 | 24.4 | 10.8 | 175.3 | 15.4 | 2.3 | 31.5 |
| 15. | IC385912 | 9.1 | 45.5 | 3.5 | 4.8 | 7.8 | 12.2 | 12.0 | 118.8 | 18.8 | 1.8 | 23.4 |
| 16. | IC92622 | 7.0 | 37.0 | 3.5 | 4.5 | 7.2 | 15.8 | 15.0 | 117.9 | 12.4 | 2.5 | 19.6 |
| 17. | IC105579 | 7.3 | 48.1 | 3.6 | 6.4 | 7.6 | 18.4 | 14.3 | 151.7 | 15.9 | 1.7 | 23.6 |
| 18. | IC110893 | 10.8 | 40.9 | 3.1 | 3.3 | 4.9 | 10.5 | 12.9 | 80.1 | 33.8 | 1.8 | 23.9 |
| 19. | IC146606 | 8.7 | 46.6 | 3.0 | 13.3 | 5.7 | 16.7 | 14.2 | 116.4 | 14.6 | 1.3 | 19.9 |
| 20. | IC201145 | 6.8 | 46.8 | 3.1 | 8.4 | 7.8 | 17.2 | 14.6 | 149.5 | 14.4 | 1.6 | 25.4 |
| 21. | IC308561 | 7.5 | 41.2 | 2.2 | 3.0 | 5.8 | 11.4 | 16.9 | 100.6 | 18.1 | 1.3 | 16.4 |
| 22. | IC344652 | 8.1 | 49.5 | 3.8 | 7.9 | 8.8 | 22.0 | 15.5 | 168.7 | 9.8 | 1.6 | 20.3 |
| 23. | IC146589 | 8.3 | 49.4 | 3.5 | 6.0 | 7.1 | 16.1 | 13.0 | 114.2 | 17.5 | 1.8 | 23.7 |
| 24. | IC369441 | 8.7 | 46.7 | 2.9 | 4.2 | 7.0 | 13.4 | 12.7 | 100.2 | 21.3 | 1.6 | 21.7 |
| 25. | IC395846 | 9.8 | 56.0 | 2.8 | 6.5 | 6.6 | 13.0 | 13.5 | 131.1 | 12.3 | 1.5 | 20.0 |
| 26. | IIHR-1 | 9.9 | 47.8 | 4.2 | 9.3 | 7.4 | 23.6 | 10.5 | 185.7 | 16.5 | 2.2 | 27.0 |
| 27. | IIHR-2 | 10.0 | 46.4 | 5.5 | 11.0 | 10.7 | 32.2 | 10.7 | 235.6 | 14.3 | 1.9 | 25.6 |
| 28. | IIHR-6 | 11.5 | 48.8 | 5.6 | 9.4 | 10.8 | 31.9 | 10.1 | 228.7 | 11.7 | 2.6 | 34.8 |
| 29. | IIHR-7 | 17.8 | 53.7 | 5.2 | 13.0 | 8.3 | 23.7 | 10.9 | 214.3 | 5.1 | 0.9 | 12.1 |
| 30. | IIHR-10 | 10.1 | 49.4 | 3.8 | 9.8 | 7.6 | 25.7 | 10.2 | 181.9 | 11.3 | 1.7 | 21.7 |
| 31. | IIHR-12 | 10.1 | 46.6 | 3.7 | 6.0 | 7.9 | 26.6 | 10.8 | 188.9 | 9.3 | 1.8 | 23.1 |
| 32. | IIHR-13 | 10.7 | 52.8 | 3.3 | 5.0 | 9.8 | 25.2 | 9.7 | 189.3 | 12.6 | 2.0 | 29.0 |
| 33. | IIHR-14 | 15.3 | 59.1 | 5.6 | 6.0 | 7.8 | 22.1 | 8.1 | 214.8 | 7.3 | 1.2 | 15.7 |
| 34. | IIHR-15 | 11.4 | 49.7 | 3.1 | 7.1 | 8.1 | 27.7 | 10.7 | 174.1 | 9.3 | 1.7 | 23.3 |
| 35. | IIHR-16 | 9.6 | 49.3 | 3.3 | 7.2 | 8.3 | 25.8 | 10.4 | 172.1 | 13.3 | 1.8 | 25.2 |
| 36. | IIHR-21 | 8.6 | 48.7 | 4.4 | 6.4 | 10.3 | 28.8 | 10.6 | 219.4 | 14.1 | 2.8 | 37.9 |
| 37. | IIHR-22 | 8.7 | 50.6 | 3.9 | 11.4 | 8.6 | 26.3 | 10.0 | 210.3 | 10.8 | 2.1 | 28.2 |
| 38. | IIHR-63 | 9.5 | 46.0 | 4.5 | 9.9 | 10.3 | 33.6 | 14.6 | 244.4 | 8.4 | 2.0 | 26.4 |
| 39. | Pusa Nasdar | 12.6 | 51.1 | 5.4 | 10.5 | 10.8 | 29.5 | 10.6 | 224.4 | 6.3 | 1.4 | 19.8 |
| 40. | Arka Sumeet | 15.6 | 55.2 | 5.0 | 11.3 | 9.6 | 41.3 | 11.0 | 300.8 | 5.7 | 1.5 | 20.6 |
| 41. | Arka Sujat | 14.9 | 54.7 | 4.4 | 9.7 | 9.5 | 28.5 | 10.9 | 234.6 | 7.7 | 1.5 | 19.6 |
| 42. | CO-1 | 12.4 | 60.6 | 5.0 | 10.6 | 12.7 | 37.1 | 15.0 | 260.8 | 4.4 | 1.2 | 15.3 |
| 43. | Rekha | 12.6 | 54.0 | 4.5 | 6.9 | 12.2 | 40.9 | 10.3 | 241.5 | 9.7 | 2.6 | 31.7 |
| 44. | Mallika | 8.5 | 41.8 | 5.0 | 6.2 | 11.7 | 33.5 | 11.2 | 367.2 | 12.4 | 2.8 | 36.6 |
| 45. | NS-3 | 9.5 | 42.3 | 4.3 | 6.4 | 11.4 | 32.7 | 11.7 | 258.9 | 13.4 | 2.8 | 37.0 |
| 46. | Naga | 9.5 | 48.3 | 5.0 | 10.8 | 8.1 | 12.4 | 13.4 | 102.8 | 18.1 | 2.1 | 27.1 |
| 47. | Malav-11 | 12.2 | 44.5 | 4.1 | 9.3 | 7.3 | 24.6 | 13.9 | 174.5 | 8.7 | 1.4 | 17.6 |
| 48. | IIHR-17 | 20.0 | 66.4 | 6.5 | 8.9 | 8.9 | 25.5 | 9.9 | 237.5 | 5.3 | 1.3 | 14.4 |

| Sl.No. | Accession | NFF | DFF | Vine Length (cm) | Branch Number | Peduncle Length (cm) | Fruit Length (cm) | Fruit Girth (cm) | Fruit Weight (g) | Fruit Number/ Vine | Fruit Yield/ Vine (kg) | Fruit Yield/ ha (t) |
|---|---|---|---|---|---|---|---|---|---|---|---|---|
| 49. | IC92625 | 8.3 | 49.0 | 3.4 | 4.2 | 7.6 | 15.2 | 15.1 | 154.8 | 10.0 | 1.5 | 17.9 |
| 50. | IC92685 | 6.4 | 42.2 | 2.9 | 3.0 | 5.5 | 13.8 | 13.1 | 79.8 | 17.7 | 1.5 | 19.4 |
| 51. | IIHR-51 | 10.3 | 38.7 | 5.9 | 5.9 | 11.1 | 36.4 | 13.9 | 234.1 | 11.7 | 2.8 | 37.1 |
| | Mean | 9.8 | 47.7 | 3.9 | 6.9 | 8.2 | 22.5 | 11.8 | 169.8 | 13.6 | 1.8 | 23.3 |
| | Sig. | ** | ** | ** | ** | ** | ** | ** | ** | ** | ** | ** |
| | CD at 0.01 | 2.0 | 2.0 | 6.1 | 0.9 | 1.5 | 1.7 | 4.6 | 2.8 | 37.5 | 4.3 | 0.5 |
| | 6.1CV per cent | 17.6 | 11.2 | 20.0 | 19.0 | 18.3 | 18.1 | 20.9 | 19.4 | 28.2 | 22.8 | 22.9 |

** Significant at P=0.01; NFF: Node number for first female flower appearance; DFF: Days taken for first female flower appearance.

### Table 5.15: Mean Performance of 24 Satputiya (*Luffa hermaphrodita*) Genotypes

| Genotype | Days to First Hermaph-rodite Flower Anthesis | Days to First Harvest | Vine Length (cm) | Length of Full Grown Fruit (cm) | Diameter of Full Grown Fruit (cm) | Individual Fruit Weight at Edible Maturity (g) | Days to Edible Maturity from Fruit Set | Number of Fruits/ Plant | Number of Seeds/ Fruit | Seed Weight/ Fruit (g) | 100 Seed Weight (g) | Edible Fruit Yield/ ha (q) |
|---|---|---|---|---|---|---|---|---|---|---|---|---|
| SP-48 | 50.33 | 62.33 | 5.37 | 6.73 | 2.33 | 33.37 | 9.33 | 110.00 | 40.00 | 3.04 | 7.60 | 59.36 |
| SP-5 | 55.67 | 67.33 | 6.20 | 9.20 | 4.23 | 39.50 | 8.33 | 119.33 | 74.33 | 6.86 | 9.23 | 75.43 |
| SP-13 | 51.00 | 69.33 | 5.00 | 9.00 | 2.37 | 30.40 | 11.67 | 113.00 | 29.67 | 2.54 | 9.57 | 54.97 |
| SP-16 | 51.67 | 60.33 | 5.10 | 9.07 | 2.37 | 28.33 | 8.67 | 127.33 | 67.00 | 6.07 | 9.07 | 54.44 |
| SP-193 | 47.33 | 60.33 | 5.23 | 7.43 | 2.73 | 41.17 | 11.67 | 99.67 | 48.67 | 4.56 | 9.37 | 65.60 |
| SP-180 | 49.67 | 59.33 | 6.23 | 8.33 | 3.20 | 2017 | 8.33 | 96.33 | 49.00 | 4.70 | 9.60 | 34.40 |
| RAU LA-2 | 41.67 | 52.33 | 4.13 | 10.67 | 2.37 | 26.57 | 8.67 | 123.33 | 113.00 | 9.13 | 8.07 | 52.54 |
| SP-59 | 45.67 | 60.67 | 7.03 | 8.43 | 2.67 | 36.63 | 13.33 | 171.00 | 73.67 | 7.04 | 9.60 | 107.97 |
| SP-140 | 50.33 | 60.33 | 4.23 | 7.10 | 2.73 | 41.30 | 10.33 | 100.00 | 93.33 | 8.33 | 8.93 | 66.07 |
| SP-6 | 40.67 | 50.00 | 4.43 | 5.47 | 3.27 | 19.33 | 10.67 | 125.67 | 89.67 | 8.18 | 9.13 | 40.05 |
| DR/Y-1 | 40.67 | 50.67 | 4.17 | 9.73 | 2.87 | 29.53 | 7.33 | 185.33 | 140.33 | 12.06 | 8.60 | 87.47 |
| DR/Y-2 | 50.33 | 50.33 | 4.50 | 8.37 | 2.73 | 26.33 | 10.67 | 107.33 | 94.00 | 7.14 | 7.60 | 45.19 |
| DR/Y-3 | 43.67 | 63.37 | 5.30 | 8.50 | 3.20 | 34.10 | 12.00 | 112.00 | 47.67 | 3.58 | 7.53 | 61.03 |
| DR/Y-4 | 41.33 | 59.00 | 5.50 | 7.17 | 2.53 | 35.23 | 13.33 | 158.67 | 21.00 | 1.80 | 8.60 | 89.43 |
| DR/Y-7 | 50.33 | 59.33 | 6.20 | 9.07 | 2.50 | 17.37 | 12.00 | 133.33 | 42.33 | 4.37 | 10.33 | 37.07 |
| DR/Y-8 | 50.67 | 61.67 | 7.20 | 5.83 | 2.63 | 30.10 | 10.00 | 117.67 | 107.33 | 7.31 | 6.83 | 56.76 |
| DR/Y-9 | 45.33 | 60.33 | 6.23 | 8.83 | 3.30 | 20.13 | 11.00 | 145.00 | 68.00 | 6.34 | 9.30 | 46.70 |
| DR/Y-10 | 50.67 | 63.33 | 6.50 | 7.33 | 2.63 | 27.17 | 12.33 | 122.00 | 47.67 | 4.61 | 9.67 | 52.98 |
| DR/Y-11 | 50.67 | 68.67 | 7.20 | 7.50 | 2.50 | 31.00 | 13.00 | 144.33 | 93.67 | 8.12 | 8.67 | 61.91 |
| DR/Y-12 | 46.67 | 67.67 | 4.23 | 7.50 | 3.17 | 28.17 | 17.33 | 110.00 | 82.67 | 7.82 | 9.47 | 49.58 |
| DR/Y-13 | 50.33 | 67.33 | 4.50 | 6.00 | 2.80 | 38.97 | 14.33 | 113.33 | 98.33 | 10.14 | 10.33 | 70.66 |
| DR/Y-14 | 51.33 | 65.67 | 6.23 | 7.00 | 2.97 | 34.93 | 13.00 | 155.00 | 57.67 | 5.19 | 9.00 | 86.63 |
| DR/Y-15 | 48.00 | 64.00 | 6.23 | 10.50 | 2.57 | 27.17 | 13.00 | 108.33 | 135.00 | 10.84 | 8.13 | 47.08 |
| DR/Y-17 | 50.33 | 64.67 | 5.43 | 8.33 | 2.87 | 36.33 | 14.67 | 168.33 | 37.33 | 2.76 | 7.40 | 92.51 |
| SEM+ | 0.8871 | 1.307 | 0.305 | 0.345 | 0.2445 | 1.405 | 0.8764 | 7.054 | 4.786 | 0.4738 | 0.5391 | 3.528 |
| CD at 5 per cent | 1.836 | 2.705 | 0.632 | 0.714 | 0.5061 | 2.9083 | 1.814 | 14.601 | 9.907 | 0.980 | 1.115 | 7.302 |
| CV (per cent) | 2.259 | 2.606 | 6.784 | 5.253 | 10.642 | 5.626 | 9.367 | 6.762 | 8.033 | 9.130 | 7.488 | 6.933 |

*Source*: Ram *et al.*, 2006.

**Table 5.16: Mean, Range, Coefficient of Variability, Heritability and Genetic Advance in *Luffa hermaphrodita***

| Char Actors | Range | Mean | PCV | GCV | Heritability $(h^2)$ | Genetic Advance (GA) | GA as per cent of Mean |
|---|---|---|---|---|---|---|---|
| Days to first hermaphrodite flower anthesis | 40.67-55.67 | 48.09 | 8.59 | 8.29 | 93.1 | 7.92 | 16.47 |
| Days to first harvesting | 50.00-69.33 | 61.44 | 9.54 | 9.18 | 92.5 | 11.18 | 18.20 |
| Vine length (m) | 4.13-7.20 | 5.51 | 18.88 | 17.61 | 87.1 | 1.87 | 33.94 |
| Length of full grown fruit (cm) | 5.47-10.67 | 8.04 | 17.53 | 16.72 | 91.0 | 2.64 | 32.84 |
| Diameter of full grown fruit (cm) | 2.33-4.23 | 2.81 | 17.47 | 13.85 | 62.9 | 0.64 | 22.78 |
| Individual fruit weight (g) at edible maturity | 17.37-41.30 | 30.59 | 22.92 | 22.21 | 94.0 | 13.57 | 44.36 |
| Days to edible maturity from fruit set | 7.33-17.33 | 11.45 | 22.13 | 20.05 | 82.1 | 4.29 | 37.47 |
| Number of fruit per plant | 96.33-185.33 | 127.76 | 20.09 | 18.92 | 88.7 | 46.88 | 36.69 |
| Number of seed per fruit | 21.00-140.33 | 72.97 | 44.83 | 44.10 | 96.8 | 65.22 | 89.38 |
| Seed weight per fruit (g) | 1.80-12.06 | 6.35 | 43.50 | 42.53 | 95.6 | 5.45 | 85.83 |
| 100-seed weight (g) | 6.83-10.33 | 8.81 | 12.25 | 9.69 | 62.6 | 1.39 | 15.78 |
| Edible fruit yield per hectare (q) | 34.40-107.97 | 62.32 | 31.19 | 30.41 | 95.1 | 38.07 | 61.09 |

*Source*: Ram *et al.*, 2006.

**Table 5.17: Phenotypic and Genotypic Correlation Coefficients for the 12 Characters in *Luffa hermaphrodita***

| Chara-cter | Days to Herma-phrodite Flower Anthesis | Days to First Harvest | Vine Length (m) | Length of Fruit (cm) | Diameter of Fruit (cm) | Fruit Weight (g) | Days to Edible Maturity from Fruit Set | Number of Fruit/ Plant | No. of Seed/ Fruit | Seed Weight/ Fruit (g) | 100-Seed Weight (g) | Edible Fruit Yield/ha (q) |
|---|---|---|---|---|---|---|---|---|---|---|---|---|
| | (1) | (2) | (3) | (4) | (5) | (6) | (7) | (8) | (9) | (10) | (11) | (12) |
| (1) rp | | 0.586* | 0.336* | -0.075 | 0.041 | 0.223* | 0.019 | -0.319* | -0.199 | -0.189 | 0.099 | -0.085 |
| rg | | 0.637* | 0.399* | -0.094 | 0.102 | 0.234 | 0.008 | -0.346 | -0.214 | -0.208 | 0.124 | -0.083 |
| (2) rp | | | 0.390* | -0.073 | 0.045 | 0.290* | 0.461* | -0.110 | -0.354* | -0.303* | 0.199 | 0.094 |
| rg | | | 0.456* | -0.093 | 0.060 | 0.325 | 0.563 | -0.139 | -0.374 | -0.310 | 0.324 | 0.110 |
| (3) rp | | | | -0.058 | 0.054 | -0.075 | 0.109 | 0.174 | -0.181 | -0.212* | 0.011 | 0.102 |
| rg | | | | -0.047 | 0.017 | -0.073 | 0.073 | 0.198 | -0.198 | -0.242 | -0.013 | 0.104 |
| (4) rp | | | | | -0.018 | -0.211* | -0.252* | 0.127 | 0.185 | 0.155 | -0.014 | -0.054 |
| rg | | | | | -0.043 | -0.227 | -0.290 | 0.153 | 0.202 | 0.167 | -0.032 | -0.050 |
| (5) rp | | | | | | 0.05 | -0.061 | -0.025 | 0.046 | 0.097 | 0.122 | 0.031 |
| rg | | | | | | 0.051 | -0.101 | -0.055 | 0.021 | 0.098 | 0.164 | 0.027 |
| (6) rp | | | | | | | 0.163 | 0.015 | -0.076 | -0.063 | -0.101 | 0.722** |
| rg | | | | | | | 0.197 | 0.047 | -0.093 | -0.081 | -0.144 | 0.720** |
| (7) rp | | | | | | | | 0.074 | -0.189 | -0.119 | 0.182 | 0.196 |
| rg | | | | | | | | 0.116 | -0.226 | -0.167 | 0.179 | 0.208 |
| (8) rp | | | | | | | | | 0.008 | 0.025 | -0.021 | 0635* |
| rg | | | | | | | | | 0.031 | 0.047 | -0.040 | 0.713* |
| (9) rp | | | | | | | | | | 0.954** | -0.186 | -0.066 |
| rg | | | | | | | | | | 0.972** | -0.201 | -0.077 |
| (10) rp | | | | | | | | | | | 0.090 | -0.044 |
| rg | | | | | | | | | | | 0.012 | -0.058 |
| (11) rp | | | | | | | | | | | | -0.069 |
| rg | | | | | | | | | | | | -0.104 |

*Source*: Ram *et al.*, 2006.

**Table 5.18: Morphological Performance of Pointed gourd**

| Crop | Character | Range | Promising Accessions |
|---|---|---|---|
| Pointed gourd | Node number at which 1st female flowers appears | 9.0-22.0 | BCPG-3, BCPG-9 |
| | Fruit length (cm) | 5.8-9.20 | BCPG-6, BCPG-18 |
| | Average fruit weight (g) | 22-55 | BCPG-6, BCPG-8 |
| | Number of fruits/plant | 35-102 | BCPG-16, BCPG-3 |
| | Fruit yield (q/ha) | 156.60-643.20 | BCPG-3, BCPG-4, BCPG-5, BCPG-16 |
| | Moisture (per cent) | 91.32-64.86 | BCPG-2, BCPG-5, BCPG-13 |
| | Dry matter (per cent) | 5.14-8.68 | BCPG-2, BCPG-5, BCPG-13 |
| | Soluble protein (per cent) | 1.45-3.03 | BCPG-10, BCPG-8 |
| | Ascorbic acid content (mg/100 g) | 9.00-29.70 | BCPG-2, BCPG-13 |
| | Dietary fiber (g/100 g) | 2.35-3.20 | BCPG-3, BCPG-4, BCPG-16, BCPG-21 |
| | $\beta$-carotene (mg/100 g) | 0.16-4.20 | BCPG-4, BCPG-5, BCPG-8 |
| | Storage life (days) | 2-7 | BCPG-4, BCPG-5, |

*Source*: Chattopadhyay *et al.*, 2011.

### Spine Gourd

BCKV, Kalyani (West Bengal) has evaluated several genotypes of spine gourd are given in Table 5.19.

### Pea

Sardana *et al.* (1998) worked out on genetic variation of 69 accessions of pea germplasm of exotic and indigenous origin.

Genotypic coefficient of variation revealed high to moderate variability for all the characters studied, except for days to 50 per cent flowering and days to maturity, which showed low variability. Seed yield/plant, plant height, number of primary branches per plant and seed weight expressed high heritability (broad sense) coupled with high genetic advance; number of clusters per plant, number of pods/plant, pod length and number of seeds/pod showed moderate to high

**Table 5.19: Morphological Performance of Spine gourd Genotypes**

| Crop | Character | Range | Promising Accessions |
|---|---|---|---|
| Spine gourd | Node number at which 1st female flowers appears | 60-85 | BCSG-1, BCSG-4 |
| | Fruit length (cm) | 6.10-9.00 | BCSG-11, BCSG-8 |
| | Average fruit weight (g) | 49.0-72.0 | BCSG-2, BCSG-1, BCSG-8 |
| | Number of fruits/plant | 12-30 | BCSG-1, BCSG-2, BCSG-8 |
| | Fruit yield/plant (kg) | 0.74-2.04 | BCSG-1, BCSG-2, BCSG-8 |

*Source*: Chattopadhyay *et al.*, 2011.

**Table 5.20: List of Indian and Exotic Germplasm of Pea**

| Sl.No. | Place of Collection | Accession Number |
|---|---|---|
| 1. | Bihar, India | IC240779 |
| 2. | Madhya Pradesh, India | IC240778, IC209136, IC209137 |
| 3. | Uttar Pradesh, India | IC209117, IC209119, IC209120, IC209122 IC209127, IC209131, IC209132, IC219020, IC21906, IC240777, IC240780, IC240781 |
| 4. | Rajasthan, India | IC209092, IC209093, IC209095, IC209096, IC209102, IC209103, IC209107, IC209108, IC209109, IC209110, IC209111, IC209112, IC209113, IC219005, IC240773, IC240774, IC240775, IC240776 |
| 5. | USA | EC8257, EC8495, EC8762, EC8763, EC341765, EC341766, EC341771, EC341787, EC341793, EC341986, EC341988, EC341998, EC342007, EC381953, EC381854, EC381855, EC381856, EC381857, EC381858, EC381859, EC381860, EC381861, EC381862, EC381864, EC381865, EC381866 |
| 6. | UK | EC6620, EC385244, EC385245, EC385246 |
| 7. | France | EC269322, EC269538, EC269540, EC269581 |
| 8. | The Netherlands | EC9123 |

*Source*: Sardana *et al.*, 1998.

**Table 5.21: Mean, Range, Mean Sum of Squares and Standard Error of Mean**

| Sl.No. | Characters | Mean Square | | Range | | Mean ± SEM |
|---|---|---|---|---|---|---|
| | | Treatment | Error | | | |
| | (df) | 72 | 144 | | | |
| 1. | Seed yield/plant | 173.57** | 3.82 | 2.92 (IC240781) | 37.39 (EC6620) | 7.55±1.59 |
| 2. | Number of primary branches/plant | 18.87** | 1.45 | 2.0 (IC240774) | 13.70 (IC240781) | 6.09±0.98 |
| 3. | Number of pods/plant | 2013.03** | 364.33 | 19.30 (EC381862) | 120.00 (IC209026) | 63.21±15.58 |
| 4. | Number of clusters per plant | 1623.27** | 192.81 | 17.00 (EC385242) | 106.00 (EC8257) | 54.18±11.33 |
| 5. | Pod length (cm) | 3.66** | 0.17 | 4.40 (IC240781) | 10.60 (EC342007) | 5.97±0.34 |
| 6. | Number of seeds/pod | 4.37** | 0.81 | 2.60 (IC240775) | 7.80 (IC209102) | 5.38±0.73 |
| 7. | 100-seed weight (g) | 58.91** | 0.01 | 4.35 (IC240781) | 26.50 (IC209122) | 17.52±0.09 |
| 8. | Plant height (cm) | 3171.58** | 171.98 | 40.00 (EC341793) | 158.00 (IC209122) | 84.42±10.70 |
| 9. | Days to 50 per cent flow | 119.48** | 16.37 | 77.00 (EC381861) | 109.00 (IC240779) | 98.65±3.30 |
| 10. | Days to maturity | 51.40** | 0.63 | 137.0 (EC381862) | 156.00 (EC8763) | 150.38±0.65 |

** Significant at 1 per cent level.

*Source*: Sardana *et al.*, 1998

**Table 5.22: Estimates of Genetic Parameters for some Agro-morphological Traits in Pea Germplasm**

| Sl.No. | Characters | Coefficient of Variation | | | Heritability in Broad Sense (per cent) | Genetic Advance as per cent of Mean |
|---|---|---|---|---|---|---|
| | | Phenotypic | Genotypic | Environmental | | |
| 1. | Seed yield/plant | 102.88 | 99.57 | 25.86 | 93.68 | 198.54 |
| 2. | Number of primary branches/plant | 44.19 | 39.54 | 19.74 | 80.05 | 72.87 |
| 3. | Number of pods/plant | 47.83 | 37.09 | 30.19 | 60.13 | 59.24 |
| 4. | Number of clusters/plant | 47.78 | 40.30 | 25.63 | 71.21 | 70.06 |
| 5. | Pod length (cm) | 20.20 | 18.94 | 7.01 | 87.95 | 36.59 |
| 6. | Number of seeds/pod | 26.27 | 20.26 | 16.75 | 59.48 | 32.19 |
| 7. | 100-seed weight (g) | 25.29 | 25.28 | 0.65 | 99.93 | 52.17 |
| 8. | Plant height (cm) | 40.55 | 37.46 | 15.53 | 85.32 | 71.27 |
| 9. | Days to 50 per cent flow | 7.22 | 5.94 | 4.10 | 67.73 | 10.07 |
| 10. | Days to maturity | 2.79 | 2.74 | 0.53 | 96.43 | 5.53 |

values whereas days to 50 per cent flowering and days to maturity had low estimates. The wide genetic variability observed in the accessions for yield and yield related traits could be useful for breeding superior genotypes. WEC6620, EC8257, IC240775, IC240779 and IC209107 were identified as superior accessions for utilization.

Seed yield per plant had highest coefficient of variation, both at the phenotypic and genotypic levels and the magnitude was considerably higher than he corresponding values at environment level.

### Indian Bean (Dolichos bean)

BCKV, Kalyani (West Bengal) has evaluated several genotypes of Indian bean and selected few promising accessions are given in Table 5.23.

Islam *et al.* (2016) evaluated and characterized 150 accessions received from Bangladesh and 16 other countries of Asia, Africa and Europe. High Shannon-Weaver Diversity Index (H'≤0.75) was observed in colour of hypocotyl, epicotyl, stem, leaf vein, flower as

**Table 5.23: Promising Accessions and their Salient Features**

| Crop | Character | Range | Promising Accessions |
|---|---|---|---|
| Indian (Dolichos) bean | Days to 50 per cent flower | 57-115 | Gomchi green, RCMDL-1 |
| | Length of pod (cm) | 7.15-15.05 | SEMVAR-8, KDB-413 |
| | Breadth of pod (cm) | 1.44-3.11 | BCDB-1, SEMVAR-8 |
| | Weight of pod (g) | 2.92-8.92 | SEMVAR-8, BCDB-1 |
| | 100 seed weight (g) | 22.2-45.5 | KDB-413, HADB-3 |
| | Pods/plant | 768-1897 | BCDB-1, KDB-413 |
| | Protein content of pod (per cent) | 0.70-5.45 | Swarna Uthkrisht, HADB-3 |
| | Yield/plant (kg) | 5.07-12.21 | BCDB-1, KDB-413, JIB(V)-16 |

*Source*: Chattopadhyay *et al.*, 2011.

well as the shape and size of seed. The accessions were exhibited 4.92 to 16.74 cm edible pod length, 1.35 to 4.36 cm pod width, 2.10 to 17.07 g pod weight and 84 to 537 pods/plant. The genotypic coefficient of variation was ranged from 3.59 to 40.08 per cent from leaf ratio (L:W) and rachis length, respectively among the 19 quantitative characters. Accessions were grouped into ten clusters using $D^2$ values. Number of accessions in each cluster was ranged from 8 to 25. Accessions collected from the same districts of Bangladesh or abroad were distributed into different clusters. No clear relationship was found among the accessions from different geographical region. Crosses between accessions belonging to maximum divergent clusters of CPI 106548 (India), ILRI 14437 (Zimbabwe) and TOT 7905 (Uzbekistan) from cluster IX with accessions of BD 122 (Bangladesh) and BD 8785 (Bangladesh) from cluster I and BD 8770 (Bangladesh) from cluster VI could be done for obtaining better variability to the subsequent generation.

### Cabbage

Rai and Asati (2005) evaluated 8 hybrids of cabbage and found significant genotypic correlation coefficients for all the traits *i.e.* core length, gross and net head weights, number of non-wrapper leaves, leaf length and breadth, days to edible head maturity, longitudinal and equatorial head lengths except stalk length. The yield components mostly exhibited significant and positive interrelationship among themselves, which indicated the need of their simultaneous selection. The leaf breadth had the highest positive direct effect on yield followed by longitudinal head length and leaf length.

The direct effect of longitudinal head length (0.881) on yield was quite close to its significant correlation with yield (0.771), which indicated that this correlation explained the true relationship and direct selection of hybrids through this trait would be effective while formulating selection indices for improvement of yield in cabbage.

### Amaranths

Amaranths germplasm were evaluated for 29 descriptors and a detailed catalogue of 800 accessions was published (Joshi, 1981a and b). The germplasm of 2250 new accessions including 315 exotic accessions has also been evaluated for 40 descriptors as recommended

**Table 5.24: Estimate of Genotypic (G) and Phenotypic (P) Correlation Coefficients among Various Characters in Cabbage**

| Traits | G and P | Stalk Length (cm) | Gross Head Weight (g) | Net Head Weight (g) | Number of Non-wrapper Leaves | Leaf Length (cm) | Leaf Breadth (cm) | Days to Maturity | Longitudinal Head Length (cm) | Equatorial Head Length (cm) | Yield (q/ha) |
|---|---|---|---|---|---|---|---|---|---|---|---|
| Core length (cm) | G | -0.149 | 0.126 | 0.290** | -0.349 | 0.347** | 0.149** | 0.237** | 0.462** | 0.158 | 0.294** |
| | P | -0.075 | 0.089 | 0.122 | 0.164 | 0.236** | 0.187 | 0.182 | 0.357** | 0.126 | 0.208** |
| Stalk length (cm) | G | | 0.805** | 0.605** | 0.756** | 0.392** | 0.478** | 0.132 | 0.097 | 0.227** | -0.005 |
| | P | | 0.574** | 0.439** | 0.515** | 0.241** | 0.349** | 0.11 | 0.086 | 0.120 | -0.005 |
| Gross head weight (g) | G | | | 0.932** | 0.525** | 0.706** | 0.933** | 0.328** | 0.349** | 0.527** | 0.517** |
| | P | | | 0.658** | 0.352** | 0.537** | 0.711** | 0.240** | 0.280** | 0.407** | 0.379** |
| Number of non-wrapper leaves | G | | | | 0.476** | 0.851** | 0.964** | 0.318** | 0.625** | 0.773** | 0.698** |
| | P | | | | 0.319** | 0.598** | 0.699** | 0.272** | 0.500** | 0.597** | 0.517** |
| Number of non-wrapper leaves | G | | | | | 0.460** | 0.491** | 0.107 | 0.394** | 0.236** | 0.458** |
| | P | | | | | 0.323** | 0.368** | 0.081 | 0.221** | 0.155** | 0.340** |
| Leaf length (cm) | G | | | | | | 0.879 | 0.655** | 0.609** | 0.603** | 0.816** |
| | P | | | | | | 0.748 | 0.515** | 0.430** | 0.424** | 0.663** |
| Leaf breadth (cm) | G | | | | | | | 0.279** | 0.509** | 0.788** | 0.709** |
| | P | | | | | | | 0.233** | 0.379** | 0.566** | 0.591** |
| Days to maturity | G | | | | | | | | 0.414** | -0.216 | 0.528** |
| | P | | | | | | | | 0.233** | -0.172 | 0.448** |
| Longitudinal head length (cm) | G | | | | | | | | | 0.436** | 0.711** |
| | P | | | | | | | | | 0.363** | 0.545** |
| Equatorial head length (cm) | G | | | | | | | | | | 0.533** |
| | P | | | | | | | | | | 0.423** |

** Significant at 5 per cent.

*Source*: Rai and Asati (2005).

**Table 5.25: Genotypic Path Coefficient Showing Direct (Diagonals) and Indirect Effects of different Characters on Yield of Cabbage**

| Traits | Core Length (cm) | Stalk Length (cm) | Gross Head Weight (g) | Net Head Weight (g) | Number of Non-wrapper Leaves | Leaf Length (cm) | Leaf Breadth (cm) | Days to Maturity | Longit-udinal Head Length (cm) | Equatorial Head Length (cm) |
|---|---|---|---|---|---|---|---|---|---|---|
| Core length (cm) | -0.179 | 0.027 | -0.023 | -0.052 | 0.063 | -0.062 | -0.035 | -0.042 | -0.083 | -0.028 |
| Stalk length (cm) | -0.017 | 0.113 | 0.091 | 0.068 | 0.085 | 0.044 | 0.054 | 0.015 | 0.011 | 0.025 |
| Gross head wt. (g) | -0.013 | -0.086 | -0.107 | -0.099 | -0.056 | -0.075 | -0.100 | -0.035 | -0.037 | -0.056 |
| Net head weight (g) | -0.390 | -0.814 | -1.254 | -1.346 | -0.641 | -1.145 | -0.298 | -0.428 | -0.842 | -0.040 |
| Number of non-wrapper leaves | 0.086 | -0.187 | -0.130 | -0.118 | -0.247 | -0.114 | -0.122 | -0.026 | -0.097 | -0.058 |
| Leaf length (cm) | 0.142 | 0.161 | 0.290 | 0.349 | 0.189 | 0.410 | 0.361 | 0.269 | 0.250 | 0.247 |
| Leaf breadth (cm) | 0.304 | 0.749 | 1.464 | 1.153 | 0.771 | 1.378 | 1.569 | 0.438 | 0.799 | 1.236 |
| Days to maturity | -0.016 | -0.009 | -0.022 | -0.021 | -0.007 | -0.043 | -0.018 | -0.066 | -0.027 | 0.014 |
| Longitudinal head length (cm) | 0.407 | 0.085 | 0.307 | 0.551 | 0.347 | 0.537 | 0.448 | 0.364 | 0.881 | 0.384 |
| Equatorial head length (cm) | -0.030 | -0.043 | -0.100 | -0.147 | -0.045 | -0.115 | -0.150 | 0.041 | -0.083 | -0.190 |
| **Genotypic correlation with yield** | **0.294** | **-0.005** | **0.517** | **0.698** | **0.458** | **0.816** | **0.709** | **0.528** | 0.771 | **0.533** |

*Source*: Rai and Asati (2005).

by IBPGR Crops Advisory Committee (Grubben and Stolen, 1981) for more than two years. Amaranth genetic resources collected from Shivalik, Kumaon and Garhwal Himalayas exhibited a wide range of variability for various quantitative and qualitative traits. On comparing the coefficient of variability of different regions the material collected from Himachal Pradesh showed more variability for more number of traits than that of Uttar Pradesh and Uttarakhand hills. Seed weight, protein percentage, seed yield, plant height and inflorescence length has shown high coefficient of variation, high heritability and genetic advance indicating their importance in the breeding programs. The correlation and path analysis studies showed significant positive association of grain yield with plant height, leaf length, leaf breadth, inflorescence length and spiklets/spike. Leaf length showed maximum direct effect on grain yield followed by number of leaves per plant, height and test weight (Joshi and Rana, 1995). Genetic divergence studies of 20 genotypes of different eco-geographical origins revealed commendable diversity. Genetic diversity showed no relationship with geographical diversity. Popping size, protein content and grain yield contributed maximum to the genetic diversity. The accession numbers like IC38280 and IC42254-2 were found to be most promising for their utilization in hybridization (Joshi and Rana, 1995). Seventy five core collections were identified and selected from the germplasm collection representing the entire range of known variability in the germplasm holdings with regard to 40 descriptors. These have been multiplied and maintained for distribution to

the crop improvement researchers. Based on 3 years multilocation testing of amaranths 16 elite lines, selection IC42258-1 named 'Annapurna' giving 2.8 t//ha average grain yield in 1986 (Joshi, 1985) has been released by the Central Varietal Release Committee and is currently used as national check both in hills and plains under AICRP on under-utilized crops.

### Chenopod

The entire germplasm of chenopods were evaluated for 10 descriptors developed at ICAR-NBPGR, Regional Station, Shimla (Himachal Pradesh). A wide range of variation was observed for plant height, branching pattern, leaf shape, size and colour, seed shape, size and colour, inflorescence shape, size and colour (Joshi, 1992). Nine promising lines in *C. album* and 6 in *C. quinoa* were selected and a replicated trial was conducted at Shimla and Sangla in Kinnaur district of Himachal Pradesh. NC58613 out-yielded all the collections by giving 1.0 t/ha grain yield followed by IC15022 (0.902 t/ha) and NC58229 (0.9 t/ha), respectively. In total germplasm pool IC15022 was found to be early maturing taking 96 days to mature. None of the exotic collections could surpass the indigenous collection in their yield potential. The black seeded types were more promising in the indigenous collections than that of exotics.

### Cluster Bean

Very less evaluation works related to vegetable type of cluster bean has been done. In seed type total 44 accessions including the two wild species namely *C. senegalensis* and *C. serrata* were evaluated in different

**Table 5.26: Estimates of Coefficient of Variations and other Genetic Parameters for different Seed Traits in French Bean Pooled over Two Years**

| Traits | Range | Mean | Coefficient of Variation | | | Heritability | Genetic Advance | Genetic Advance (per cent) of Mean |
|--------|-------|------|-----------|-----------|--------------|--------------|-----------------|------------------------------------|
| | | | Phenotypic | Genotypic | Environ-mental | | | |
| Days to 50 per cent flowering | 40.00-46.67 | 42.52 | 4.88 | 3.71 | 3.17 | 57.74 | 2.46 | 5.81 |
| Average pod weight (g) | 4.52-8.23 | 5.96 | 15.63 | 11.66 | 10.40 | 55.69 | 1.07 | 17.93 |
| Branches/plant | 2.68-4.27 | 3.29 | 17.60 | 13.12 | 11.73 | 55.57 | 0.66 | 20.15 |
| Pods/plant | 9.13-26.7 | 14.96 | 31.05 | 28.46 | 12.41 | 84.01 | 8.04 | 57.72 |
| Fresh pod yield/plant (g) | 43.65-141.50 | 87.93 | 29.15 | 28.58 | 5.74 | 96.12 | 57.72 | 53.74 |
| Days to seed maturity | 78.50-84.50 | 81.44 | 2.29 | 1.92 | 1.24 | 70.40 | 2.70 | 3.32 |
| Pod length (cm) | 10.55-16.10 | 13.82 | 13.38 | 11.89 | 6.14 | 78.91 | 3.00 | 21.75 |
| Seeds/pod | 4.03-6.33 | 5.59 | 10.33 | 7.87 | 6.69 | 58.00 | 0.69 | 12.34 |
| Biological yield/plant (g) | 14.95-50.67 | 29.23 | 27.49 | 26.40 | 7.69 | 92.18 | 15.26 | 52.21 |
| Plant height (cm) | 29.37-39.57 | 35.47 | 9.27 | 6.83 | 6.27 | 54.28 | 3.68 | 10.36 |
| 100 seed weight (g) | 14.88-48.95 | 33.00 | 25.48 | 24.18 | 8.04 | 90.06 | 15.60 | 47.27 |
| Seed yield/plant (g) | 12.58-38.57 | 22.26 | 40.50 | 35.88 | 18.78 | 78.49 | 14.57 | 65.48 |

*Source*: Sharma *et al.*, 2016.

years. The range of variation is generally conspicuous for all the qualitative as well as quantitative descriptors particularly but more so in case of 100 seed weight (2.5 to 5.3 g), endosperm percentage (35.7-49.8) and gum content per cent (15.9 to 31.8). The seed colour is light pink, pink, purple, light grey, grey and dark grey. Hymowitz (1961) believed that seed coat colour is environmentally controlled but the stocks under observation have shown remarkable stability to retain their pink colour (from Kutch area) when grown elsewhere. Seed size and shape is rather more important than colour. The seed size (100 seed weight) of the introductions varied from 2.3 to 3.9 g and contains 22.4 to 26.4 per cent of gum. The varieties with 100 seed weight of 3.5 g and slightly above with oval shape are more preferred by the guar gum industry to suit the size of the sieve used in processing.

### French Bean

Sharma *et al.* (2016) evaluated 33 accessions (Arka Suvidha, Arka Anoop, DPDFB-1, DPDFB (M), DPDFB-2 (M), DWDFB-I, DWDFB-53, DWDFB-57, HAFB-1, HAFB-2, HAFB-3, HAFB-4, IVRFB-1, IVFB-1, IVFB-2, IVFB-3, JFB-97-1, KPV-2, MFB-1, MFB-2, MFB-3, MFB-4, MFB-5, VLB-8, VLB-9, VLB-2003, VLFB-130, Aparna, Chandini, Falguni, Surya, Arka Komal and Contender) in Randomized Complete Block Design (RCBD) with three replications for two seasons during summer 2008 and 2009 at Vegetable Farm, H.P. Agricultural University, Palampur to explore the genetic variability and G x E interaction for yield and yield related traits.

### Rice Bean

Singh *et al.* (1998) identified stable genotypes as RCRB-1-301 and EC 18585 for days to flowering, RBL-

2, RBL-35 and RCRB-1-301 for days to maturity and IC 16807, S-10 and EC 18585 for plant height, and Dobhal and Gautam (1994) identified RB-49 and RB-40 as the most stable and highest yielding of 11 genotypes. In Nagaland condition, Kumar *et al.* (2014) reported that RBS-53 produced significantly higher growth, grain yield, straw yield, biological yield, harvest index, protein content with maximum gross return, net return, B:C ratio, production efficiency, and economic efficiency.

**Table 5.27: Rice Bean Cultivars and its Economics**

| Cultivar | Gross Return (Rs/ha) | Net Return (Rs/ha) | B:C Ratio | Production Efficiency (kg/ha/day) | Economic Efficiency (Rs./ha/day) |
|----------|---------------------|--------------------|-----------|-----------------------------------|----------------------------------|
| RBS-16 | 29,774 | 19,713 | 2.02 | 8.34 | 231.05 |
| RBS-53 | 33,639 | 23,869 | 2.36 | 9.50 | 271.69 |
| PRR-2 | 27,690 | 17,920 | 1.86 | 7.74 | 204.26 |
| RCRB-4 | 31,548 | 21, 778 | 2.25 | 8.88 | 251.75 |
| LSD (O=0.05) | 1302 | 901 | 0.10 | 0.37 | **10.55** |

*Source*: Kumar *et al.* (2014).

Further, Sarkar *et al.* (2015) evaluated the performance of few indigenous landraces collected from Nagaland and Mizoram and reported that highest yield was recorded in 'Rhusie' cultivar (2.1 t/ha) which was significantly higher than the all other cultivars like 'Rhudi', 'Rhuka', 'Rhuwho-1', 'Rhuwho-2', 'Rhujo', 'Champhal' and 'Kolasib'.

**Table 5.28: Yield Contributing Characters and Yield of Rice Bean**

| Land Races | Plant Height (cm, 65 DAS) | No. of Pods/ Plant | No. of Seeds/ Pod | Test Weight (g) | Yield (t/ha) |
|---|---|---|---|---|---|
| Rhudi | 61 | 6 | 5 | 246 | 1.1 |
| Rhuka | 91 | 7 | 5 | 154 | 0.3 |
| Rhuwho-1 | 67 | 11 | 4 | 8 | 0.1 |
| Rhuwho-2 | 70 | 6 | 5 | 204 | 1.9 |
| Rhusie | 82 | 18 | 6 | 19 | 2.1 |
| Rhujo | 78 | 8 | 6 | 11 | 0.2 |
| Champhal | 64 | 33 | 5 | 104 | 0.2 |
| Kolasib | 76 | 7 | 3 | 148 | 1.8 |
| SEm± | 16.11 | 1.60 | 0.44 | 19.20 | 5.64 |
| CD (O=0.05) | NS | 2.80 | 0.77 | 33.63 | 9.89 |

*Source*: Sarkar *et al.* (2015).

## Jack Bean

Thirty eight accessions of jack bean were augmented from Peninsular India and characterized during 2010-14 and considerable variation was found among the genotypes for morphological characters *viz.*, growth habit (pole and bush), leaf density (sparse, intermediate and dense), stem colour (light green, purple and dark purple), flower colour (purple white), pod shape (straight, intermediate and curved), pod colour (light green and green), pod beak length (short, medium and long), pod curvature (straight, curved and highly curved), pod surface (smooth and wrinkled) and seed colour (brown, reddish-purple, white, grayish-yellow and grayish-orange). While the protein content ranged between 16-21 per cent, the oil content ranged between 0.4-2.3 per cent with fatty acids oleic, linoleic and palmitic being predominant. The plant height, pod weight, number of pods per plant, and test weight were observed with high genetic variability and direct selection would be rewarding for jack bean improvement in India (Sivaraj *et al.*, 2016).

## Garlic

Dhall and Brar (2013) worked out genetic variability, correlation and path coefficient in garlic and reported that values of genotypic correlation coefficients were higher than those of their respective phenotypic correlation coefficients in majority of cases suggesting that genotypic correlation were stronger, reliable and free from environmental factors. Bulb weight was positively and significantly correlated with bulb equatorial diameter, clove weight, clove length, bulb polar diameter, plant height at both genotypic and phenotypic level. Therefore, bulb yield can be improved by selecting genotypes having high value of equatorial diameter, clove length, bulb polar diameter and plant height.

Path coefficient studies indicated that plant height, number of leaves per plant, bulb polar diameter, bulb equatorial diameter and clove diameter had greater direct influence on bulb weight.

**Table 5.29: Genotypic and Phenotypic Correlation Coefficient of 9 Quantitative Characters in Garlic**

| Trait | Plant Height (cm) | No. of Leaves/ Plant | Bulb Weight (g) | Bulb Equatorial Dia. (mm) | Clove Weight (g) | Clove Length (mm) | Clove Diameter (mm) | No. of Cloves/ Bulb |
|---|---|---|---|---|---|---|---|---|
| No. of leaves/plant | -0.1357 | | | | | | | |
|  | -0.1186 | | | | | | | |
| Bulb weight (g) | 0.2784* | -0.1218 | | | | | | |
|  | 0.2773* | -0.0924 | | | | | | |
| Bulb equatorial diameter (mm) | 0.3122** | -0.0737 | 0.8417** | | | | | |
|  | 0.2844* | -0.0819 | 0.7782** | | | | | |
| Clove weight (g) | 0.1018 | -0.1695 | 0.6030** | 0.3615** | | | | |
|  | 0.1017 | -0.1437 | 0.5925** | 0.3478** | | | | |
| Clove length (cm) | -0.4183** | 0.0952 | -0.0463 | -0.1574 | 0.2929* | | | |
|  | -0.4009** | 0.0797 | -0.0425 | -0.1369 | 0.2881* | | | |
| Clove diameter (mm) | 0.0171 | 0.0492 | 0.4263** | 0.1742 | 0.9107** | 0.4350** | | |
|  | 0.0219 | 0.0718 | 0.3862** | 0.1601 | 0.8210** | 0.4061** | | |
| No. of cloves/bulb | 0.2885* | 0.1735 | 0.0476 | 0.3640** | -0.6295** | -0.4275** | -0.7380** | |
|  | 0.2838* | 0.1403 | 0.0457 | 0.3346** | -0.6229** | -0.4242** | -0.6516** | |
| Bulb polar diameter (mm) | 0.2707* | 0.0427 | 0.04327** | 0.4327** | 0.1795 | 0.1296 | 0.1148 | 0.3019** |
|  | 0.2634* | 0.0357 | 0.4429** | 0.3934** | 0.1637 | 0.1240 | 0.0706 | 0.02704* |

*, ** Significant at 5 and 1 per cent levels of significance, respectively.

**Table 5.30: Path Coefficient Analysis Showing Direct and Indirect Effects of Component Characters on Bulb**

| Trait | Plant Height (cm) | No. of Leaves/ Plant | Bulb Polar Diameter (mm) | Bulb Equatorial Dia. (mm) | No. of Cloves/ Bulb | Clove Weight (g) | Clove Length (mm) | Clove Diameter (mm) |
|---|---|---|---|---|---|---|---|---|
| Plant height (cm) | 0.274 | 0.738 | -0.712 | 0.782 | 0.323 | -0.446 | -0.348 | 0.737 |
| No. of leaves/plant | 0.265 | 0.763 | -0.999 | -0.109 | 0.441 | -0.338 | -0.285 | 0.579 |
| Bulb polar diameter (mm) | -0.296 | 0.115 | 0.658 | 0.119 | -0.106 | 0.111 | 0.174 | -0.277 |
| Bulb equatorial dia. (mm) | 0.947 | -0.367 | 0.347 | 0.226 | -0.131 | 0.814 | 0.990 | -0.163 |
| No. of cloves/bulb | -0.576 | -0.219 | 0.457 | 0.193 | -0.153 | 0.100 | 0.131 | -0.208 |
| Clove weight (g) | 0.968 | 0.204 | -0.578 | -0.145 | 0.122 | -0.126 | -0.101 | -0.220 |
| Clove length (mm) | 0.886 | 0.204 | -0.106 | -0.208 | 0.188 | -0.119 | -0.107 | 0.220 |
| Clove diameter (mm) | 0.903 | 0.197 | -0.817 | -0.165 | 0.142 | -0.124 | -0.106 | 0.224 |

## Promising Indigenous Collection in Vegetable Crops

**Table 5.31: Promising Indigenous Collection and Evaluation for Traits**

| Crop | IC, NIC and other Number | Traits |
|---|---|---|
| Brinjal | IC249349, IC090982, IC526796 | Resistance to bacterial wilt |
| | IC296759 | Intermediate, tall, thornless |
| | IC126879, NIC5938, NIC6048 | Long deep purple |
| | IC90764, IC112312, IC090975, IC126869, IC144038, IC144084, IC144166, IC111025, IC111083 | |
| | IC74209A, IC136142, IC144083, IC144061, IC144085, NIC4573 | Green long |
| | IC014405, NIC013009 | Deep purple oblong |
| | IC090957, IC127239, IC113802, IC112821, IC090754, IC090810, IC126884, IC137771, IC144053, IC089910, NIC006910, NIC005803 | Purple |
| | IC144033 | Green oblong |
| | IC144034, IC144174, IC144780 | Purple green oblong |
| | IC137748, IC144078, IC136328 | Creamy white oblong |
| | IC127163 | White with purple stripe oblong |
| | IC90956, NIC003603,, NIC004064, | Deep purple oval |
| | IC112343, IC127018, NIC004283, NIC006878, NIC006885, NIC006886, NIC006960, NIC009430 | Purple oval |
| | IC090950, IC144002, IC144153, | Green oval |
| | IC144146 | Purple green oval |
| | IC144025 | Purple with stripe oval |
| | IC126955, IC137674, IC143995, IC144045, NIC005765 | Purple round |
| | IC144092 | Green round |
| | NIC010459, NIC006067 | Long fruit (>30) |
| | NIC006393, NIC006908 | Large purple type |
| | NIC003222, NIC003228 | Thin and long fruit type |
| | NIC006896, NIC009310, IC090115, IC099649, IC099701, IC099702 | Early type |
| | IC099747, IC111428, NIC003654, NIC005874, NIC005877, NIC005904 | Cluster bearing type |
| Tomato | IC528034 | High carotene |
| | IC564448 | High TSS 6.0°B) |
| | IC296468, IC565013-14 | Resistant to root knot nematodes |
| | IC395328 | Resistant to tomato leaf curl virus and bacterial wilt |
| Chilli | IC119676, IC119639, PSR-1775, JBT12/83 (at Amravati Centre), MNCH-64, BD-119, H-2107, BDS-2573 | High Yield |
| Dioscorea | IC087336, IC04255 (*Dioscorea alata*) | High Yield |

| Crop | IC, NIC and other Number | Traits |
|------|--------------------------|--------|
| Coleus | IC008564, IC085683, IC85686 (*Coleus parviflorus*) | High Yield |
| Okra | IC086781, IC086759, IC086768 (*Manihot esculenta*) | High Yield |
| Elephant Foot Yam | IC070551, IC087260, M-4, M-15 (*Amorphophallyus campanulatus*) | Resistant to Cassava Mosaic disease high yielding |
| Zinger | IC085052,Z90/65, Z90/16, J4905, J5045 (*Zingiber officinale*) | Tolerant to soft rot and high yielding |
| Onion | IC048025, IC048129, IC048145, IC48361, IC48404, IC48411 | High yielding, large sized bulbs |
| | IC038775, IC47958, IC48320, IC048968 | High tillering |
| | IC034444, IC042834, IC047548, IC048025, IC048051 | Early maturing, high yielding |
| | IC047954, IC048145, IC048544, IC48764, IC048807, IC048916 | High TSS (>15 Brix), attractive red bulbs |
| | IC034444, IC48041, IC048045, IC048059, IC048513 | Narrow root zones, compact necks, storage potential |
| | IC042900, IC47972, IC047998, IC048575, IC048710, IC049812 | Field tolerant to purple blotch and *Setemphylium* blight |
| Garlic | NIC0079, NIC0087, NIC002369 | Bold white cloves, good storage potential |
| | C1525, C735 | Compact white bulbs, bold cloves |
| | IC049345, IC049373, IC049382, IC100507 | High yielding large size bulbs, purple cloves |
| | IC035286, IC043398, IC048157, IC049373, IC049382, IC100507, IC100530, IC100530 | Field tolerant to purple blotch and *Setemphylium* blight |
| | IC100530, IC100582 | Field tolerant to *Setemphylium* blight |
| | IC049345, IC100551, G-4 | Tolerant to Garlic Mosaic Virus |

## *Drumstick*

### Table 5.32: Fruit and Seed Characters of *Moringa oleifera* Germplasm

| Sl.No.* | Fruit Length (cm) | Fruit Girth (cm) | Seeds per Fruit | 100-Seed Weight (g) | Per cent Kernel | Kernel Length (mm) | Kernel Dia-meter (mm) | Fruit Colour (dry) | Kernel Shape | Wing Promi-nence | Seed-coat Detach-ability | Remarks |
|---------|-------------------|------------------|-----------------|---------------------|-----------------|--------------------|-----------------------|--------------------|--------------|------------------|--------------------------|---------|
| 1 | 33.34 | 3.44 | 15.40 | 14.70 | 69.68 | 10.32 | 6.41 | LB | Ovate | ++ | † | - |
| 2 | 40.27 | 2.85 | 18.70 | 12.08 | 76.45 | 10.19 | 7.26 | RB | Ovate | ++ | † | Gregarious bearing |
| 3 | 44.38 | 3.06 | 15.30 | 11.24 | 72.95 | 9.18 | 6.11 | DB | Ovate | ++ | † | Gregarious bearing |
| 4 | 28.31 | 2.53 | 16.60 | 10.22 | 79.35 | 8.91 | 5.88 | LB | Ovate | +++ | † | Gregarious bearing |
| 5 | 35.48 | 2.92 | 14.10 | 8.64 | 67.81 | 7.41 | 3.36 | RB | Ovate | + | †† | Big-sized leaflet |
| 6 | 22.67 | 2.34 | 13.50 | 10.32 | 82.21 | 6.76 | 5.05 | RB | Ovate to round | ++ | † | Big-sized leaflet |
| 7 | 37.40 | 2.83 | 14.20 | 9.38 | 76.57 | 8.47 | 5.07 | LB | Ovate | ++ | †† | - |
| 8 | 44.55 | 2.79 | 14.30 | 9.04 | 76.96 | 9.23 | 6.72 | RB | Ovate | +++ | † | Gregarious bearing |
| 9 | 45.19 | 2.97 | 16.10 | 10.04 | 73.71 | 8.04 | 4.18 | RB | Ovate | +++ | † | Gregarious bearing |
| 10 | 35.34 | 2.47 | 15.30 | 9.98 | 81.17 | 10.28 | 6.85 | LB | Ovate | +++ | † | Less bitter fruits; Big-sized leaflet |
| 11 | 47.18 | 2.96 | 14.20 | 13.78 | 76.92 | 10.39 | 8.52 | LB | Ovate | ++ | †† | - |
| 12 | 25.34 | 2.20 | 13.20 | 12.29 | 68.93 | 10.95 | 5.88 | LB | Ovate | + | †† | - |
| 13 | 50.64 | 2.66 | 13.60 | 7.42 | 72.78 | 6.33 | 4.65 | LB | Ovate | ++ | † | - |
| 14 | 40.32 | 2.46 | 13.70 | 11.68 | 80.90 | 10.12 | 6.52 | LB | Ovate | ++ | † | - |
| 15 | 45.21 | 2.75 | 14.90 | 12.22 | 77.33 | 9.55 | 6.84 | DB | Ovate | +++ | † | Fruit bitter free |
| 16 | 47.89 | 2.57 | 13.90 | 7.89 | 79.05 | 8.78 | 5.54 | LB | Ovate | +++ | †† | - |
| 17 | 30.77 | 2.75 | 12.30 | 6.40 | 72.88 | 6.27 | 4.02 | LB | Ovate to round | +++ | † | - |
| 18 | 37.38 | 2.81 | 12.10 | 10.24 | 70.31 | 8.86 | 5.13 | LB | Ovate | +++ | † | Big-sized leaflet |
| 19 | 57.34 | 2.86 | 17.80 | 8.37 | 71.88 | 9.31 | 5.76 | RB | Ovate | +++ | †† | - |
| 20 | 60.44 | 3.13 | 16.40 | 8.44 | 72.04 | 10.11 | 6.07 | LB | Spindle | ++ | † | - |

| Sl.No.* | Fruit Length (cm) | Fruit Girth (cm) | Seeds per Fruit | 100-Seed Weight (g) | Per cent Kernel | Kernel Length (mm) | Kernel Diameter (mm) | Fruit Colour (dry) | Kernel Shape | Wing Prominence | Seed-coat Detach-ability | Remarks |
|---|---|---|---|---|---|---|---|---|---|---|---|---|
| 21 | 42.53 | 2.68 | 13.50 | 11.34 | 67.80 | 9.50 | 5.33 | LB | Spindle | ++ | † | - |
| 22 | 30.56 | 5.62 | 15.30 | 16.78 | 76.98 | 10.27 | 16.46 | DB | Round | +++ | ††† | Fruits fleshy |
| Mean ± S.E. | 40.57 ± 2.11 | 2.76 ± 0.06 | 14.72 ± 0.37 | 10.27 ± 0.46 | 74.65 ± 0.98 | 8.97 ± 0.30 | 5.79 ± 0.26 | - | - | | | |
| Range | 22.67-60.44 | 2.20-3.44 | 12.10-18.70 | 6.40-14.70 | 67.80-82.21 | 6.27-10.95 | 3.36-8.52 | - | - | | | |
| CV (per cent) | 23.88 | 10.23 | 11.45 | 20.43 | 5.99 | 15.20 | 20.62 | - | - | | | |

*Serial No. 1-22 indicates collector Nos.KP/PKS.RC-10-1 to KP/PKS.RC-10-22, respectively, last one is the cultivated type which is not considered for mean, range and CV; LB-Light brown, RB-Reddish brown, DB-Dark brown; †-Easy, ††-Slightly difficult, †††-Difficult

*Source*: Pradheep *et al.*, 2011.

### Sugar Beet

More than 400 sugar beet germplasm comprising *Beta vulgaris*, fodder beets, cms lines, inbred lines and diploid, anisoploid and hybrid varieties from public research institutes and private seed companies from different countries were evaluated for root crop performance at ICAR-IISR Farm, Lucknow (U.P.). Over 25 lines have been shown potential for high sugar, while, 14 varieties showed adaptability to high temperature tolerance and 'IISR Comp-1', 'Pant Comp-3', 'Ramonskaya', 'CIR-PbII/79', 'M.R. Poly', 'Virtus' and S'olid' for salinity tolerance. Genetic improvement efforts and germplasm maintenance activities is going on at Lucknow (U.P.) and Mukteswar. The seed production of Russian diploid variety 'R-06' is possible at altitude above 5000'. Other varieties *viz.*, IISR Comp-1 and LS-6 developed by Lucknow (U.P.) and Pant S-10 by GBPUA&T, Pantnagar (Uttarakhand) were also tried in a limited way. Indigenous diploid varieties recommended for cultivation in India with high temperature (40-45 °C) tolerance varieties maintained (Mall *et al.*, 2016).

### Potato

Potato is one of the most important non-grain food crops of the world but productivity in India is quite low that stands at ~22 metric tonnes per hectare as compared to the European countries that ranges between 30-40 tons per hectare. Potato production in *Terai* region of Uttarakhand is lower than other potato growing belts, because of the low yield. This fluctuation is may be due to narrow genetic base of the germplasm used for development of cultivars. Therefore, considering the above facts potato 28 germplasm and 4 checks were evaluated in Augmented Block Design at G.B.P.U.A. and T., Pantnagar (Uttarakhand). Correlation studied indicated that tuber yield was highly significant and positive correlation with average weight of tuber (0.966), tuber weight per plant (0.650) and leaf length (0.484). These may be key criteria for selecting high tuber yield lines. Path coefficient analysis revealed that average weight of tuber showed positive direct effect (0.854) towards tuber yield followed by tuber weight per plant (0.463), number of tuber per plant (0.348) however, protein (-0.577) showed negative effect. Genotypes 97-P-27 (46.452 t/ha), TPSK-05-06-007 (45.497 t/ha) and J/96-48 (45.404 t/ha) were promising for yield (Shubha *et al.*, 2016).

### Uniformity in Evaluation

Plant breeder often makes visual observations and selects without recording the relevant data. Further, voluminous data remains in plant breeder's book and often goes unreported. Although, it is requested that the evaluation data may be sent to curator, but it is not mandatory and often not complied with. Therefore, it is important to record data in a uniform pattern using common reference/standard checks (s) so that these are of value to a number of scientists.

At G.B. Pant University of Agriculture and Technology, Pantnagar (Uttarakhand) a total of 1529 genotypes of different vegetables were evaluated. The information is summarized in Tables 5.35 and 5.36.

### (b) Evaluation of Exotic Vegetable Genetic Resources for Various Traits

**Chenopod-**Among the exotic chenopods, 'EC180010' was found to be most promising. Most of the exotic collections were found to be more susceptible to leaf spot disease. However, EC201678, EC201682 and EC201687 expressed field resistance to leaf spot.

### Evaluation of Vegetable Genetic Resources at ICAR-NBPGR, New Delhi

A lot of germplasm of vegetable crops are being introduced every year by ICAR-NBPGR, New Delhi. Some of the promising accessions with specific traits are given in Table 5.38

**Table 5.33: Evaluation and Characterization of Vegetable Genetic Resources at different Centres**

| Crop | Evaluating Centre | Traits | Promising Accession(s) |
|------|-------------------|--------|------------------------|
| Bitter gourd | ICAR-NBPGR, New Delhi | High yield | UB-22, UB-66, IC-74158, C-1621, K-3374 |
| | ICAR-IIVR, Varanasi | High yield | IC-44428B, NIC-72285, IC-85604A, IC-85608B, IC-85611, IC-85636, V-89/0-103C, VRBT-29, VRBT-50 |
| | | Maximum number of fruits/plant | Ic-45364A, VRBT-78, Ic-44426D, IC-44428B, IC-45350, NIC-72285, IC-85611 |
| | | CMV Resistance | IC-44428B, IC-72285, IC-85608B, V-89/0-103C, IC-85611, IC-85636, VRBT-50 |
| | | Fruit fly Resistance | IC-85641 and V-89/0-103C |
| Bottle gourd | ICAR-NBPGR, New Delhi | High yield (Summer) | U-10-315, U-11-10, U-8-217, U-8-46, U-8-198, BDJ-864 |
| | | High yield (*Kharif*) | U-11-216A, U-11-50, BDJ-53, BDJ-783, BDJ-896, EC-187247, BDJ-616 |
| | ICAR-IIVR, Varanasi | High yield | U-10-316, IC-92363, VRBG-2, VRBG-4 |
| | | Resistance to Red pumpkin beetle | U-10-316, NIC-1225 |
| | | Resistance to Leaf miner | U-10-316, IC-92363 IC-92418 |
| | | Resistance to downy mildew | U-10-316, IC-92362 |
| Ivy gourd | IGKVV, Raipur | High yield (3.20-21.21kg/plant) | Ac-59, 35, 54, 16, 6, 48, 10, 11, 14, 34, 52, 24, 17, 36, 2, 4, 31, 29 |
| | | Number of fruit/plant (209.33-1292.7) | Ac-35, 59, 11, 5, 16, 54, 24, 17, 10, 51, 61, 22, 58, 60, 53, 32, 31, 63, 37, 23 |
| | | Length of internodes (7.79-11.43cm) | Ac-17, 61, 5,14, 51, 6, 11, 15, 37, 1, 4, 29, 32, 36, 22, 2, 9, 55 |
| | | Diameter of stem (3.30-4.96 cm) | Ac-4, 51, 63, 15, 36, 6, 11, 59, 61, 24, 62, 17, 16, 7, 32, 52, 58, 32, 2, 54, 5, 23, 4, 22, 9, 60, 29, |
| | | Length of fruit (3.97-7.60cm) | Ac-6, 1, 62, 11, 33, 54, 61, 31, 59 |
| | | Diameter of fruit (1.63-3.20 cm) | Ac-59, 52, 31, 14, 35, 37 |
| | ICAR-IIVR, Varanasi | High yield | VRK-05, VRK-10, VRK-20, VRK-35 |
| | | Diameter of stem | U-325/DA/DR/20, 325/DA/DR/36 |
| | | Length of internodes | U-35/DA/DR/48 |
| | | Resistance to mosaic virus | VRK-05, VRK-10, VRK-20, VRK-35 |
| | | Resistance to leaf miner | VRK-01, VRK-06, VRK-22, VRK-04, VRK-31, VRK-33 and VRK-55 |
| Pointed gourd | ICAR-IIVR, Varanasi | High yield | VRPG-13, VRPG-44, VRPG-70, VRPG-72, VRPG-93 |
| | | Number of fruits/plant | VRPG-13, VRPG-44, VRPG-70, VRPG-72, VRPG-88, VRPG-93, VRPG-113 |
| | | Heat tolerance | VRPG-72, VRPG-99, VRPG-19, U-34/BP/DR/07, VRPG-110, VRPG-110A, VRPG-110B |
| | | Resistance to nematode | U-34/BP/DR/44, VRPG-72, VRPG-18, VRPG-96 |
| Ridge gourd | ICAR-NBPGR, New Delhi | | Promising- IC-93399, U-324-138, U-20-202A, U-24-134, Nic-10216 |
| Sponge gourd | ICAR-NBPGR, New Delhi | | Promising- IC-92779, IC-92745 (very early), D-24-163, Nic-10237 |
| Satputia | ICAR-NBPGR, New Delhi | | Promising- DCB-1363 |
| Cucumber | GBPUA&T, Pantnagar | High yielding | EC237658 |
| Brinjal | ICAR-NBPGR, New Delhi | High yielding | IC-126879, NIC-5938, EC-304992 (Long deep purple); IC-90764, IC-112312, IC-90975, IC-126869, IC-144084 (Purple Long); IC- 74209A, IC- 136142, IC- 144083, NIC- 4573 (Green Long), IC-14405, IC-13009 (Deep Purple Oblong); IC-90957, IC-127239, IC-113802, IC-112821 (Purple oblong); IC-127163 (white with purple stripe oblong), IC- 137748, IC- 144078 IC- 136328 (Creamy white oblong) |
| Chillies | ICAR-NBPGR, New Delhi | High yield | IC-119676, IC-119639, PSR-1775, JBT-12/83 MNCH-64, BD-119, BDS-2573 |

**Table 5.34: Promising Accessions for different Traits in Indigenous Collection**

| Crop | Promising Accessions |
|---|---|
| **Brinjal** | |
| Early flowering (<60 days) | IC89953, IC90111A, IC99676, IC137770 |
| Early fruiting (<80 days) | IC89953, IC90111A, IC99676 |
| Dwarf (< 40 cm) | IC89964, IC89990 |
| Tall (> 130 cm) | IC99633, IC111302, IC112662, IC112960 |
| Small fruit (> 30 g/fruit) | IC89947, IC89949, IC90040, IC90105, IC90109, IC90949, IC90893 |
| Large fruit (> 550 g/fruit) | IC90943 |
| High yielding (> 3 kg/plant) | IC89870, IC89913, IC90842, IC90875, IC74235, IC112341, IC112733, IC112735, IC12736, IC111068, IC112666, Pusa Purple Cluster |
| Long fruit (> 35 cm) | IC90102, IC90126, IC90152, IC90157, IC90158, IC111384, IC111410, IC90821, IC111045 |
| High yielding (> 3 kg/plant) and purple round fruit | IC90875, IC90938, IC90958, IC112328 |
| High fruit yield (>3 kg/plant) and purple long fruit | IC74235, IC90890, IC112356, IC11823, Pusa Purple Cluster |
| High fruit yield (>3 kg/plant) | IC90842, IC90821, IC112813, and Green Long Fruit |
| *Phomopsis* blight resistance | IC89883, IC90013, IC90105, IC99571, IC111457, IC111395, IC136571, IC99648, IC128680, IC136326, IC136256, IC90865, NIC13021, NIC13337 |
| Chillies (*Capsicum* spp.) | IC147637, IC147719, IC18882, IC119556 |
| High yielding | NIC19991, NIC92153, NIC92156, NIC92250, NIC20901, NIC99912, P2072, BDJ2254 |
| Long fruits (> 130 cm) | IC136061, IC136252, IC136266 |
| PBNV/Mosaic resistant | IC119611 |
| Drought tolerance | IC23837, NIC23797 |
| Root-knot nematode | IC090869 (Khan *et al.*, 2016) |
| **Bottle gourd** | |
| Summer type | IC1612, 010-315, 08-217, U10-315, BOJ217, BOJ628, BOJ864 |
| *Kharif* type | IC90904, IC92904, BOJ535, BOJ1733, U11-21C-A |
| Early fruiting | IC110389, IC29404, IC 02429, U11-21C-A |
| High yielding | IC144389, IC94404, IC29404, IC92429, IC110389, NIC13182, NIC131188 |
| **Pumpkin** | |
| High yielding | U8-246, K3228, NIC13188, IC92533, IC92571, IC92499, NIC10170 |
| **Snapmelon** | |
| High yielding | IC92305, NIC2518,, 022-107, NIC10174, U22-107 |
| **Ridge gourd** | |
| Early fruiting | NIC20402 |
| High yielding | IC93399, IC92779, IC12136, NIC957, NIC10216, NIC10216, NIC10224, NIC20213 |
| Cluster bearing | NIC10222, NIC10232, NIC10288, NIC10213, NIC10215 |
| **Sponge gourd** | |
| Early fruiting | IC92779, IC9227797, IC92604, IC93445, NIC10237, NIC13236, NIC110235 |
| High yielding | NIC23288, NIC23291, NIC23292 |
| **Bitter gourd** | |
| Long fruiting duration (> 150days) | IC256226, IC256229 |
| **Watermelon** | |
| High yielding | U8-246, K3228 |
| **Okra** | |
| Early fruiting | IC117097, IC117263, IC117276 |
| Short fruit | IC89972, IC89982, IC90097, IC90168, IC111343, IC305717 |
| Multi fruit | IC90075, IC90077, IC90186, IC90202, IC90290 |
| High fruit yield | IC116970, IC116983, IC117002, IC117231, IC89977, IC90072, IC112446 |
| Pod borer resistance | IC7472, IC23624, IC27821, IC27831 |
| Yellow vein mosaic virus resistance | IC7194, IC10262, IC13999, IC43181, IC46018 |
| Root-knot nematode | IC255758, IC306737 (Khan *et al.*, 2016) |

| Crop | Promising Accessions |
|---|---|
| **Onion (seed crop)** | |
| Early maturity | IC48045, IC48025, IC3444, IC80551, IC47691 |
| High tillering | IC989968, IC49207, IC48320, IC48336 |
| **Onion (bulb crop)** | |
| High pungency | IC42911, IC42992, IC49130, IC47959, |
| High TSS | IC35549, IC35851, IC42854M, IC42911, IC47967, IC48051, IC98183, IC48304, IC48996, IC49072, IC49113, IC49127, IC216197, IC35758, IC48573, IC487451, IC316197, IC35758 |
| **Garlic** | |
| Purple blotch resistance | IC48988, IC3222 |
| **Radish** | |
| High yielding | IC143928 |
| **Coriander** | |
| Early maturity | NIC18187, NIC18192, NIC18205 |
| High foliage | IC155513, IC67145, IC363976, IC363981 |
| Late maturing | NIC18237, BDJ81-21 |
| Bold seed | IC143583, IC143584 |
| **Fenugreek** | |
| High foliage | IC143885, IC143967, IC143890, IC144225 |
| High yielding | IC143855, IC143967, IC143890, IC144225 |
| **Spinach** | |
| Foliage type | NIC14420, BDJ516 |
| **French bean** | |
| High yielding | PLB10-1, PLB14-1, NIC22600, PI226868, PLB170 |
| High pod | NIC22620, NIC22608, NIC8809, IC42381, IC147698 |
| Early maturity | NIC8814, IC16544, NIC22611, PI220668, PLB420 |
| High seed number | IC18162, N847, NIC14405, PLB420 |
| **Pea** | |
| Root-knot nematode | IC208367, IC219002 (Khan *et al.*, 2016) |

### Table 5.35: Vegetable Germplasm Evaluation at G.B.P.U.A. and T., Pantnagar, (Uttarakhand)

| Sl.No. | Crop | Total Accessions | Promising Accessions |
|---|---|---|---|
| 1. | Pea | 173 | Pant Uphar, PM-2, PM-3, PMR-17 |
| 2. | Chilli | 686 | Pant C-1, Pant C-3, IC119276, IC119285, IC119290, IC119296, IC119432, IC119525, IC119591, IC119630, IC119640, IC119755 |
| 3. | Okra | 90 | Pant Okra-1, KAC-10, KAC-171 |
| 4. | Brinjal | 34 | Pant Samrat, Pant Rituraj, Pant Baigan-4 |
| 5. | Tomato | 90 | Pant Bahar, Pant T-3, Pant T-4, Pant T-5 |
| 6. | Cauliflower | 25 | Pant Shubhra, Pant Gobhi-2, Pant Gobhi-3, Pant Gobhi-4 |
| 7. | Bitter gourd | 23 | PBIG-1, PBIG-2, PBIG-3, PBIG-56 |
| 8. | French bean | 117 | Pant Anupama, Pant bean-2 |
| 9. | Ridge gourd | 12 | PRG-6, PRG-7 |
| 10 | Sponge gourd | 16 | PSM-35 |
| 11. | Bottle gourd | 30 | PBOG-22, PBOG-40, PBOG-61, PBOG-62 |
| 12. | Ash gourd | 22 | AG-96-1, AG-96-12 |
| 13. | Longmelon | 11 | PLM-1 |
| 14. | Watermelon | 25 | PWM-15, PWM-61, PWM-71, PWM-72 |
| 15. | Pumpkin | 25 | PP-39, PP-46 |
| 16 | Cucumber | 49 | PCUC-8, PCUC-15, PCUC-28 |
| 17 | Muskmelon | 39 | PMM-96-20, PMM-97-10 |

| Sl.No. | Crop | Total Accessions | Promising Accessions |
|---|---|---|---|
| 18. | Fennel | 6 | PFS-1, PFS-2, PFS-3 |
| 19. | Garlic | 21 | PGS-14, PGS-16, PGS-2, PGS-4, PGS-7 |
| 20. | Fenugreek | 11 | PFGS-1 |
| 21. | Coriander | 4 | Pant Haritima |
| 22. | Turmeric | 18 | PTS-20, PTS-4, PTS-11, PTS-27, PTS-3 |
| 23. | Black cumin | 1 | PBC-1 |
| 24 | Ajowain | 1 | PA-1 |

**Table 5.36: Promising Germplasm Evaluated and Characterized at different Organizations**

| Crop | Centre | No. of Collection | Traits | Promising Germplasm |
|---|---|---|---|---|
| Tomato | Pantnagar | 21 | Yield | CLN-2870°, IAC-599 |
| | | | Earliness (days to harvest) | S-108 (114), EC519802 (114) |
| | | | Fruit/Cluster | S-108 (6) and EC519802 (6) |
| | | | Fruits/plant | AC599 (75) |
| | Coimbatore | 75 | Yield | LE231, LE1150 |
| | | | Earliness (50 per cent flowering) | LE18 (54.2) and LE477 (54.2) |
| | | | Fruits/plant | LE231 and LE1150 each of 68.6 |
| | | | Bushiness | LE996 (45.4 cm) |
| | Solan | 5 | Yield | B102 |
| | | | TSS (Brix) | Hawaii-7998 and 101E both 5.0 brix |
| | | | Number of fruits/plant | Hawaii-7998 (25.1) |
| | | | Pericarp thickness (mm) | AC-402 (6.0) |
| | Ludhiana | 2 | Introduced from World Vegetabe Centre (AVRDC), Taiwan | LBR-9-Plants having determinate growth habit, fruits round, very firm, dark red and having 70 g average fruit weight. The line is resistant to late blight. |
| | | | Introduced from World Vegetabe Centre (AVRDC), Taiwan | LBR-9-Plants having determinate growth habit, fruits oval, firm, deep red and having 60 g average fruit weight. The line is resistant to late blight. |
| Brinjal | Bhubaneshwar | 56 | Yield | BBSR-202, BBSR195-1 and SM6-6 |
| | | | Days to 50 per cent flowering | SM6-6 (77.0) |
| | | | Days to edible maturity | SM6-6 (93.0) |
| | | | Bushiness/plant height | IG (69.8 cm) |
| | Raipur | 76 | Round big | IGB-9, IGB-14, IGB-37 |
| | | | Oblong | IGB-1, IGB-7, IGB-19, IGB-20 |
| | ICAR-IIVR | 285 | Max. Fruits/plant | EC316283 (29.39) |
| | | | Yield | EC316283 |
| | | | Days to 50 per cent flowering | EC316283 (54.38) |
| | Kalyani | 46 | Yield | BCB-11, PB-67 |
| | | | Fruits/plant | BCB-11, DBL-02, HBL-25 (all,11.0) |
| | | | Earliness 50 per cent flowering | IBWL2001-1 (50 days) |
| | Vellanikkara | 17 | High yield and bacterial wilt resistant | VKBr-21 (4.200 kg/plant and resistant) |
| | | | | 2-VKBr-2145 (4.140 kg/plant and Resistant) |
| | | | Earliness (days to flowering) | VKBr—26 (29 days) |
| Chillies | ICAR-IIVR | 410 | Resistance to Anthracnose and PepLCV | Bhut Jolokia |
| | | | Pungency | Bhut Jolokia-world hottest chilli 8,85000 SHU |
| | | | Resistant to LCV | C-11-32 Resistant to LCV |
| | | | GMS lines | 1-CMS (9 set, A and B) |
| | | | | 2-GMS (MS-12 and MS-3) |

| Crop | Centre | No. of Collection | Traits | Promising Germplasm |
|------|--------|-------------------|--------|---------------------|
| | | | Ornamental chilli | Numex Twilight |
| | Lam | 346 | Oleoresin (per cent) | GP89 (14.51), GP82 (14.31) |
| | | | Capsanthin (EOA color value) | GP299 (53375), GP132 (47672) and GP155 (45140) |
| | | | Capsacin (per cent) | GP276 (0.581), GP148 (0.571) |
| | Coimbatore | 111 | Yield | CA 197, CA 166, CA 25 |
| | | | Fruits/plant | CA 141 (140.8), CA 60 (400.93), CA 52 (99.6) |
| | Portblair | 26 | Fruit Length (cm) | Small type: CCW (1.2), CCO (1.2), SPG-7 (1.3) |
| | | | | Large type: M-2 (7.5), H-1 and CCG (6.9) |
| | | | Max. plant height (cm) | H-1 (72.0), CARI-1 and H-2 (71.0) |
| | Lam | 346 | Dry chilli yield | Max.-IC397334 and Mini.- IC430010 |
| | | | Bush type | IC255594 (55.0 cm) |
| | | | Tall type | IC26223 (134.0 cm) |
| | | | Plant spread | Max. IC381090 (215.6 cm) and Mini. IC263388 (105.6 cm) |
| | | | Fruits/plant | IC391461 (263.1) |
| | | | Fruits length (cm) | Max. IC255594 (10.4 cm) and Mini. IC260948 (4.0 cm) |
| | | | Seeds/fruit | Max. IC319187 (92.0) and Mini. IC266342 (18.6) |
| | | | Oleoresin content | Max. IC278336 (23.0) and IC383136 (4.4) |
| | SKUAS and T (Srinagar) | 4 | Yield ripe fruit | SH-KC-71, SH-KC-70 |
| | | | Fruits/plant | SH-KC-69 (86), SH-KC-70 (72) |
| | | | Fruit Length (cm) | SH-KC-70(10.32), SH-KC-71(10.31) |
| Capsicum | SKUAS and T (Srinagar) | 4 | Fruit weight (g) | SH-SP-50, SH-SP-52 |
| | | | Fruits/plant | SH-SP-50 (13), SH-SP-52 (12) |
| Paprica | ICAR-IIVR | 12 | Leaf Curl Virus and Anthracnose resistance (under field condition) | BS-35, Bhut Jolokia, MC334 |
| | | | Oleoresin- non pungent | VR-535 |
| | SKUAS and T (Srinagar) | | Red Ripe fruit weight | SH-P-48, SH-P-49 |
| | | | Fruits/plant | SH-P-48 (19), SH-P-46 and SH-P-49(18) |
| | | | Fruit length (cm) | :SH-P-48(11.02), SH-P-49 (10.56) |
| Peas | Ludhiana | **5** | Earliness (50 per cent Flowering) | PM-65 (25) and PM-69 (30 days), Angoori (54 days) and Vasundhara (55 days) |
| | | | Pods/plant | Pea winter (60-65) and Arka Karthik (50-55), ARYAVEER (41.20) |
| | | | Seeds/pod | Arka Karthik (8-9) |
| | | | Pod colour | 1-Dark Green- PM-65 |
| | | | | 2-Green: Pea Winter |
| | ICAR-IIVR | 455 | Early maturing genotypes (days) | VRP-227 (34), VRP-300 (34), VRP-359 (36), VRPE-25-2 (39), VRPE-168 (32), VRP-13 (39), VRPE-90 (40), VRPE-316 (37), VRPE-466 (36), VRPE-119 35) VRP9414(40), EC93810 (45) |
| | | | Mid maturity genotypes | No. 18, VRP-135, VRP-176, VRp-168, VN-53-5, VRP 176, VRp-190, VRP-288, No.19, VRP-298, VRP-12-1, Kashi Shakti |
| | | | Yield | EC-269396 and VRP290 |
| | | | Pods/plant | EC269396 (15) and VRP290(13) |
| French Bean | Dharwad | 9 | | |
| | Rahuri | 48 | Earliness (days to 50 per cent Flowering) | RHRFB-39 (42), RHRFB-16 and RHRFB-2 (43) |

| Crop | Centre | No. of Collection | Traits | Promising Germplasm |
|------|--------|-------------------|--------|---------------------|
| | | | Pods per plant | RHRFB-25 (74.9), RHRFB-11 (70.7) |
| | | | Fruit length (cm) | RHRFB-44 (13.36), RHRFB-47 (13.34) |
| | | | Fruit shape | Flat- RHRFB-25, RHRFB-11, RHRFB-43 |
| | | | | 2-Round-RHRFB-47, RHRFB-48, RHRFB-3 |
| | | | | 3- Round sickle- RHRFB-1, RHRFB-12, 43, 45 |
| | ICAR-IIVR | Bush type (32) | | |
| | | Pole type (56) | | |
| | ICAR-IIHR | 20 | Yield | IC-525251, IC-525241, IC-525228 |
| | | | Pod length (cm) | IC-525251 and IC-525263 (16.75) |
| | | | Pod maturity (days) | IC-525285(44.5), IC-525243(44.5) |
| Lablab bean | ICAR-IIVR | 300 | Green pod yield | VRSEM-501, VRSEM-6 |
| | | | Earliness (1st flowering) | VRSEM-752 (45), VRSEM-201 (54) |
| | | | Pod Length (cm) | VRSEM-934 (15. 57), VRSEM-946 (14.69) |
| | Raipur | 63 | Days to 1st flowering | IS-32 (43), IS-21 (46) |
| | | | First harvest | IS-32 (58 days), IS-21 (62 days) |
| | | | Green pod yield | IS-02 IS-11 |
| | Jabalpur | | Days to 1st flowering | JDL-13(44.8), JDL-14(53.4) |
| | | | Pod yield | JDL7, JDL3 |
| | ICAR-HARP | 33 | Yield | HADB105, HADB107, HADB106 |
| | | | Days to 1st pod harvest | HADB113 (66), HADB119(68), HADB-120 and HADB-122(73) |
| | | | Pod length (cm) | IC249534 (12.73), IC20033 (14.33) |
| | | | Pods/plant | HADB-113 (258), IC249538 (206) |
| | ICAR-IARI | | | IC-016862 |
| Cowpea | Raipur | 38 | Pole type | ICP-1, ICP-3, ICP-4, ICP-6, ICP-7, ICP-8, ICP-9, ICP-11 |
| | | | Semipole type | ICP-2, ICP-10, IC-P12 |
| | | | Bush type | ICP-5 |
| | | | Pod color (maroon type) | ICP2, ICP4, ICP36, ICP38 |
| | | | Pod color (dark green Type) | ICP-6, ICP10 |
| | | | Pod color (dark tan type) | ICP-17 |
| | Periyakulum | 24 | Pod yield | Acc.No. 6, Acc.No. |
| | | | Pods/plant | Acc.No.19 (35), Acc.No.17 (31 |
| | | | Pod weight (g) | Acc.No.6 (23.87), Acc.No.7 (13.69) |
| | ICAR-IIVR | Bush type (294) | Earliness (per cent flowering, days) | VRC-431 (45.0), VRC-425 (45.2), VRC-429 (45.5) |
| | | | Longest (cm) and heaviest pod (g) | VRC-435 (45.2, 11.3), VRC-431 (43.5, 11.0) |
| | | | Yield | KPC-11, KPC-8, VRC-431 |
| | | | Golden bean mosaic resistance | VRC-426, RGC-4, KPC-10 |
| | | Pole type (180) | | |
| | ICAR-IIHR | 26 | Yield | IIHR-114 and IIHR-115 |
| | | | Pods/plant | IIHR-79 (63), IIHR-91 (56) |
| | | | Earliness (50 per cent flowering in days) | IIHR-114 (50) and IIHR-95 (50) |
| | | | Fruits/cluster | IIHR-80 (20), IIHR-91 (14), IIHR-79 (14) |

| Crop | Centre | No. of Collection | Traits | Promising Germplasm |
|---|---|---|---|---|
| Cucumber | ICAR-IIVR | 93 | Yield | VRC-11-2 |
| | | | Fruit wt. (g) | SPP-87 (176 g) |
| | Pantnagar | 8 | Yield | PCUC-104 PCUC-133 |
| | | | Earliness | PCUC-104 (44 days) |
| | | | Fruits/plant | PCUC-104 (11.2) |
| | Rahuri | 34 | Earliness | RHR-29 (37.67 days), RHR-21(41 days) |
| | | | Fruit length (Max.) | RHR-6 (37.22 cm) |
| | | | Fruit length (Mini.) | RHR-34 (4.44 cm) |
| | Solan | 8 | Yield | LC-33, LC-29 |
| Muskmelon | ICAR-IIVR | 68 | Yield | BS-56, VRM-44 |
| | | | Fruits/plant | VRM-7(9.2) |
| | | | TSS (per cent brix) | VRM-26(14.6) |
| | Durgapura | 41 | | |
| | Ludiana | 8 | Max. fruit wt. (g) | MC-2008-15 (1500), Kajri (1000) |
| | | | Mini. Fruit wt. (g) | MC2010-5 (500) |
| | | | High TSS (per cent brix) | Kajri (14), MC2010-15 (10) |
| | | | Netted type | MC2008-8 and Kajri |
| | Rahuri | 37 | Yield | RHRMM-26-1 |
| | | | High TSS (per cent brix) | RHRMM-1 (11.0) and RHRMM-36 (11.0) |
| | | | Fruits/plant | RHRMM-28 (4.50) |
| | | | Flesh colour (light green) | RHRMM-1, RHRMM-4, RHRMM-6, RHRMM-9, RHRMM-15 |
| | | | Flesh colour (orange) | RHRMM-2, RHRMM-3, RHRMM-5, RHRMM-8, RHRMM-10, RHRMM-12 |
| | | | Flesh colour (white) | RHRMM-14 |
| | | | Netted type | RHRMM-2, RHRMM-3, RHRMM-4, RHRMM-5 and RHRMM-7 |
| Watermelon | Durgapura | 3 | | |
| | Ludhiana | 2 | Yellow rind | MC-1- No Strips, Red Flesh, TSS 10 per cent, Fruit wt. (2.225kg) |
| | | | Gray rind | MC-2- red flesh, TSS 10 per cent, Fruit Wt. 2.0 kg |
| Pumpkin | ICAR-IIVR | 152 | Yield | VRPH 15-1 and, VRPH-15 |
| | | | Fruits/plant | VRPK 5-1 (3.14) and VRPK-67 (3.0) |
| | | | Earliness First female flower emergence | VRPK-5-1 (36 days) |
| | | | Fruit weight (kg) | VRPK-207 (7.5) and VRPK-15-1(6.2) |
| | | | Flesh thickness (cm) | VRPK-401(5.2) and VRPK-67 (5.0) |
| | NDUA&T, Faizabad | 20 | Tolerant to water stress | NDPK-5063 |
| | Coimbatore | 15 | Yield | ACC 5, ACC 8 |
| | | | Flesh thickness | ACC 3 (3.65 cm) |
| | | | Less seed | ACC-1 (135.3) |
| | | | Earliness (1st female flower) | ACC 13 (43.9 days) |
| | | | Fruits/plant | ACC 15 (4.75), ACC 8 (5.00) |
| | Hyderabad | 21 | Yield | CM-115, CM-27 |
| | | | Flesh thickness (cm) | CM-31 (4.3), CM-115 (4.1) |
| | | | Av. fruit weight (kg) | CM-115 (6.7) and CM-83 (4.6) |
| | | | Fruits/vine | CM-27 (5.5) |
| Bitter gourd | ICAR-IIVR | 65 | Yield | VR-145 and VR-175 |
| | Vellanikkara | 23 | Earliness (50 per cent flowering) | VKB-182 (72 days, Green), VKB-185(74 days, G), VKB-186 (75 day) |
| | | | Yield | VKB-176 |

| Crop | Centre | No. of Collection | Traits | Promising Germplasm |
|---|---|---|---|---|
| Bottle gourd | ICAR-IIVR | 66 | Yield | KL-4, KL-3-40 –Round |
| | | | | VR-152 Long |
| | | | | Winter type- KD-1 and KD-216.52 |
| | Faizabad | 51 | Yield | RHRBG-18 (9.97) |
| | Rahuri | 33 | Earliness (50 per cent flowering) | RHRBG-6 (67), RHRBG-18 (68) |
| | | | Cylindrical fruit | RHRBG-18, RHRBG-17, RHRBG-10 |
| | | | Fruit length (32 cm) | RHRBG-5 and RHRBG-8 |
| Pointed gourd | ICAR-IIVR | 65 | Yield | VR-112, VR-3, VR-111 |
| | Bhubaneswar | 9 | Yield | BPG-4, BPGAN-02, BPG 1 |
| | | | Max. fruit length (cm) | BPGAN-02, BPG-4 |
| | | | Max. fruit girth (cm) | BPGAN -02 (11.5), BPG-4 (11.2) |
| | Kalyani | 22 | Earliness (days to 1st flowering | BCPG-9 (82), BCPG-15 (83) BCPG-5 (85) |
| | | | Yield | BCPG-8, BCPG-4 |
| | | | Fruits/plant | BCPG-16 (96.80), BCPG-3 (90.10) |
| | | | Downy mildew (Mini. Intensity per cent) | BCPG-8 (14.6), BCPG-4 (18.0) |
| | | | Vine and fruit rot (Mini. Intensity per cent) | BCPG-8 (11.80), BCPG-4 (12.0) |
| | Faizabad | 16 | Yield | NP-801, NP-260, NP-504 |
| | Sabour | 18 | Yield | Rajendra Parwal-1, Rajendra Parwal-2 |
| | | | Fruit length (cm) | Rajendra Parwal-1 (9.90); SPG-2000-01 (8.15) |
| | | | Average fruit wt. (g) | Rajendra Parwal-1 (36), SPG-08-1 (28) |
| | Jorhat | 5 | | - |
| | Navsari | 20 | Yield | NPG-10, NPG-1, NPG-9 |
| Ivy gourd | ICAR-IIVR | 13 | Yield | Long-VRSIG-3 and Oval-VRSIG-9 |
| | Raipur | 32 | Fruit weight (g) | Acc. 48 (25.0) and Acc.-52 (23.0) |
| | | | Earliness (first harvest) | Acc. 05, Acc. 06, Acc. 10, Acc.15, Acc.23, Acc. 34, Acc.35, Acc.36 (73 days) |
| | ICAR-Port Blair | 1 | - | - |
| | Navsari | 19 | Fruit yield | NLG-16, NLG-18, NLG-8 |
| | Vellanikkara | 17 | Earliness (first harvest | CG-27 (45 days), CG-82 (45 days) |
| | | | Fruit yield | CG-23, CG-27 |
| | | | Resistance to mosaic | CG-84 (Highly bitter fruit) |
| | | | Individual fruit weight (g) | CG-23 (16.80), CG-82 (15.0) |
| Sponge gourd | ICAR-IIVR | 68 | Andromonoecious | VRSG 103 (IC590826) |
| Spine gourd | ICAR-IIVR | 14 | | |
| | Navsari | 52 | Yield | NSG-8, NSG-42, NSG-6 |
| | | | Length of fruit (cm) | NSG-52 (5.72), NSG-1 (4.44) |
| | Kalyani | 11 | Fruit yield | BCSG-1, BCSG-2 |
| | | | Fruits/plant | BCSG-1 (30), BCSG-2 (25) |
| | | | Earliness (days to first flowering) | BCSG-4 (64) |
| | ICAR-Port Blair | 1 | | |
| Cho-Cho | ICAR, Research Complex, Barapani | 5 | Max. fruit weight (g) | Coll. No. 1 (475) |
| | | | Fruit colour | Coll. No. 3 (Greenish yellow) |
| | | | | Coll. No. 2, 4, 5 (Light green) |
| | | | | Coll. No. 1 (Dark green) |

| Crop | Centre | No. of Collection | Traits | Promising Germplasm |
|---|---|---|---|---|
| Amaranth | ICAR-IIHR | 6 | Yield | IIHR -181 |
| | | | Leaf number | IIHR -141(50.3) |
| | Coimbatore | 87 | Yield | A155 (Green), A156, A59, A100, (Pink), |
| | | | Plant height (cm) | A146 (305.8) and A145 (303.8) |
| | | | Leaf length (cm) | A162(35.3), A161 (32.3) |
| | | | Leaf width (max.) (cm) | A146(26.5)A102(22.3) |
| | | | Leaf both side red | A69, A83, A88, A89, A102 |
| | | | Leaf both side pink | A21, A100, A149, A160 |
| | | | Green leaf | A1, A2, A3, A6, A7, A11, A17 |
| | Hyderabad | 32 | Leaf yield | IC526829 |
| | | | Stem yield | IC526828 |
| | Jorhat | 25 | Yield | JAS-8, Am-1 and Am-2 |
| | HARP (Ranchi) | 12 | Yield | HAAMTH-48, Red) and HAAMTH, Red) |
| | ICAR-IARI | | Yield | Pusa Lal Chaulai, P. Kirti and P. Kiran |
| | Vellanikkara | 15 | Yield | VKA-6 with 45 cm plant height |
| Cauliflower (Early) | ICAR-IIHR | 30 | Curd weight (g) | Max. IIHR-426 (402.9), IIHR-437 (395.0) |
| | | | | Mini. IIHR-409 (175.0), IIHR-469 (175.0) |
| | | | Non ricey | IIHR-409, IIHR-414, IIHR-415 |
| | | | Copact curd | IIHR-414, IIHR-439, IIHR468 |
| | | | Errect | IIHR-414, IIHR-415, IIHR-416 |
| | Sabour | Early (4) | Curd weight (g) | 93-2(388), 2006-2(360), Samastipur RECF-2(426) |
| | | Mid (8) | Net curd weight (g) | 96-5M (595), 2008-1M (590) |
| | | | Color of curd (white) | 94-2M, 96-5M, 99-1M, 2007-5M, 2008-1M |
| | | | Color of curd (creamy white) | 97-1M, 2001-5M |
| | Katrain | 52 (Late) | Yield | Mukutmani-1 |
| | | | First harvest | Hemantika (107) |
| Kunwari | ICAR-IIVR | 63 | Curd wt. (g) | Amrit Special (Max.)-(750.0) and (Min.)- IIVR EC-2 (420.0) |
| Kataki | ICAR-IIVR | 69 | Curd wt. (g) | HS-Kataki (750.0) and (Min.)- BSB-Kataki-35 (420) |
| Agahani (Mid Season) | ICAR-IIVR | 63 | Curd wt. (g) | BSBV (1900 g) and BSBV64 (1780) |
| Cabbage | Solan | | Yield | C No. 8 |
| | | | Net head weight (g) | C.No. 8 (760) |
| Carrot (Tropical) | Hisar | 20 (Red) | Root length (cm) | HC-251-4(30.3), HC-4-3 (28.3) and HC-251-1-2(27.5), HC-2-2 (21.2), HC-100-2 (21.3) |
| | | | Root dia. (cm) | HC-251-1-2, HC-251-1 and HC251(3.8) |
| | | | Root wt. (g) | HC-1 (130.0) |
| | | 7 (Purple) | Root length (cm) | Max.- HCP-1 (29.2), HCP-1-2(28.3) |
| | | | Root dia. (cm) | HCP and HCP-1-1(3.2) |
| | | | Root weight (g) | HCP-1-1 (89.5) |
| | | 6 (Yellow) | Root length (cm) | HCY-1-3 (30.2), HCY-10 (25.3), HCY-1-2 (25.7) |
| | | | Root dia. (cm) | HCY-1-2 (3.4), HCY-3 (3.5) |
| | | | Root dia. (cm) Mini. | HCy-self-Core (2.7) |
| | | 14 (Black) | Root length (cm) | HCB-160 (40.2), HCB-4 (28.6), HCB-5 (26.2) |
| | | | Root dia. | HCB-1-2, HCB-1,HCB-5, HCB-D |
| | | 6 (Orange) | Root length (cm) | Max.-HCO-A(30.5), HCO5(27.0) |
| | | | Root diameter | HCO-3 and HCO5 (3.1) |
| | | | Root diameter | HCO-4 (2.8)and HCO-1 (2.9) |

| Crop | Centre | No. of Collection | Traits | Promising Germplasm |
|---|---|---|---|---|
| | ICAR-IIVR | 86 (Red), Asian type | Root length (cm) | Pusa Kesar Golden (23.20), Cr-28 (23.0) |
| | | | | L-18264 and Yellow carrot (13.0) |
| | | | Root Dia. (cm) | L-18245 (5.16), L-18242 (4.8) |
| | | | Root weight (g) | Super Red (173.0) |
| | Ludhiana | 2 | Root colour (dark red) | TC-10, C-2 Root Length(26)- Free from hairs |
| | | | Root (red colour) | TC-10, C-1 Root Length(30)- Small Core |
| Carrot (Temperate) | ICAR-Katrain | 36 | | |
| | Srinagar | 4 | Yield | SH-C-155, SH-C-157 |
| | | | Root length (cm) | SH-C-157 |
| | | | Root length | SH-C-156 and SH-C-158 |
| | ICAR-IIVR | 310 | Fruits/plant | IC304450 (18.2), IC111480 (17) |
| | | | Yield | IC304456,VROB |
| | | | YVMV and insect resistance) | IC140986, IC141056, IC140985 |
| | | | Jassid tolerance | VROB-181 (5.3 jassid/leaf), VROB-110 (5.5) |
| | Bhubaneswar | 53 | Days to first harvest | BBSR-09-1 (45), BO-2 (46), BO-13 (46) |
| | | | Yield | BBSR-18, BBSR-48 |
| | Rahuri | 217 | Fruit length (cm) | Max. EC169362 (16.56), EC282282 (15.73) |
| | | | | Mini. IC33953 (6.05), IC902620 (6.76) |
| | | | Bushiness (cm) | IC128889 (60.0) and IC31398-A (65) |
| | ICAR-IIHR | 11 | Yield | IIHR299 |
| | | | Fruits/plant | IIHR299 (30.66), IIHR287(26.73) |
| | | | YVMV Free genotypes | IIHR296, IIHR294, IIHR299 |
| | | | Smallest fruit length (cm) | IIHR286(11.83) |
| | Periyakulum | | Number of pods/plant | Acc. No.3(145), Acc. No.1(132) |
| | | | Pods Length (cm) | Acc.No.-7(75), Acc No.1(71.65) |
| | | | Pod Weight (g) | Acc.No.2(110), Acc. No.7(99.30) |

**Table 5.37: Evaluation of Exotic Vegetable Genetic Resources**

| Crop | Centre | Economic Traits | Promising Accessions |
|---|---|---|---|
| Tomato | ICAR-NBPGR, New Delhi | High yield | EC-106285, EC-118292, EC-163683, EC-170662, EC-110578 |
| | | Heat tolerant | EC-41824, EC-164855, EC-169308, EC-122063, EC-126955 |
| | | Moderate heat tolerant lines | EC-315487, EC-320578 |
| | | Suitable for fresh market | EC-260639, EC-232429, EC-267729, EC-267726 |
| | | Suitable for transportation | EC-161252, EC-100845, EC-110268 |
| | | Paste type | EC-54628, EC-52062 |
| | | Early maturing type | EC-357833, EC-357827, EC-362956, EC-362940, EC-362944, EC-362947 |
| Pumpkin | | High yielding | EC90512 |
| Snapmelon | | High yielding | EC303049 |
| Cucumber | | High yielding | EC237658 |
| Watermelon | | High yielding | EC28435A |
| Muskmelon | | High yielding | EC303044 |

## Evaluation of Vegetable Genetic Resources at GBPUA&T, Pantnagar (Uttarakhand)

GBPUA&T, Pantnagar (Uttarakhand) received a large number of germplasm from World Vegetable Centre, (earlier AVRDC) Shanhua, Taiwan and evaluated them for different horticultural traits. The soybean consignment includes lines suitable as vegetables where 100 seed weight is about 30 g as compared to 10-12 g of grain type soybeans. Vegetable type soybean lines AGS-334, AGS-339 and AGS-352 have done well in preliminary trials at Pantnagar and

**Table 5.38: Promising Accessions of Brinjal Selected at ICAR-NBPGR, New Delhi**

| Trait | Mean | Range | Best Check (Value) | Superior Accessions |
|---|---|---|---|---|
| Plant height (cm) | 84.4 | 34.6-126.0 | Pusa Kranti (90.6) | IC090137, EC169079, IC090932, IC099691, IC090051 (>110) |
| Plant spread (cm) | 120.5 | 62.1-196.6 | Punjab Sadabahar (131.2) | EC169079, IC264470, IC261808, IC261803 (>175) |
| Primary branches | 7.4 | 4.4-11.8 | Pusa Upkar (8.1) | IC347779, IC354525, IC111085, IC353212 (>10) |
| Fruit length (cm) | 11.1 | 4.9-28.9 | Pusa Kranti (16.7) | IC383195, IC398820, EC384619, IC264470, IC265246, EC169079, IC336472 (>20) |
| Fruit Breadth (cm) | 5.1 | 1.8-9.1 | Pusa Uttam (6.3 cm) | IC099676, IC374942 EC316284, IC112905, IC249343, IC112779 (>8) |
| Fruit weight (g) | 120.7 | 19.2-324.8 | Pusa Kranti (183.4) | IC099676, EC316284, IC112738, IC374942, IC361838 (>270) |
| No of fruits/plant | 11.1 | 1.3- 35.3 | P. Shyamala (30.2) | EC490062 (58.6) |
| fruit wt./plant (kg) | 1.0 | 0.08- 3.5 | Pusa Shyamala (1.59) | IC112723, IC112779, EC490062, IC112341 (>2) |
| Per cent Fruit borer infestations (no. basis) | 22.4 | 58.3 | Pusa Shyamala (11.9) | EC315014, EC305163, EC490062, EC316275, IC099731, IC099691 (<8) |

are expected to be released. A set of 25 genotypes of World Vegetable Centre, Shanhua, Taiwan, tomato lines including a cherry tomato CH-154 and additional check cultivar Pant Tomato-3 was evaluated during summer 2003 at the Regional Research Station, Majhera located in the Uttarakhand hills at the height of about 3000 feet. The highest yielding lines were CLN-2237A (835.0 q/ha), CLN-915-93D4-1-0-3 (813.0 q/ha), CL-11-31-0-38-4-0 (730.0 q/ha) and CLN-2123E (615.0 q/ha) in comparison to 252 q/ha which was recorded in the check cultivar Pant Tomato-3. The yield of cherry tomato CH-154 was 91.0 q/ha. The average fruit weight of these lines was 43, 52, 28, 60, 36 and 6 g respectively. Thus, it is clear that top 4 yielding World Vegetable Centre lines were large fruited except CL-11-31-0-38-4-0 which was small fruited. The rest of the germplasm lines were under evaluation. The chief features of the breeding/advance lines received from World Vegetable Centre are summarized in Table 5.39.

**Source of Virus Resistance**

See the Table 5.40.

**Source for Fungal Resistance**

See the Table 5.41.

**Source for Bacterial Wilt**

See the Table 5.42.

**Source for Insect Resistant**

See the Table 5.43.

**Source for Nematode Resistant**

See the Table 5.44.

**Source for Quality Attributes**

See the Table 5.45.

**Table 5.39: Chief Features of Advance Breeding Lines of Tomato Received from World Vegetable Centre**

| Name | Plant Habit | Market Type | Resistance/Tolerance | Fruit Weight (g) |
|---|---|---|---|---|
| CLN-2026D | Determinate | FM/PT | BW, TMV, F-1, F-2, St | 65 |
| CLN-1466P | Determinate | FM | BW, TMV, F-1, St | 110 |
| PT-4719A | Determinate | PT | TMV, F-1 | 70 |
| PT4664B | Determinate | PT | BW, TMV, F-1 | 100 |
| CLN-2037B | Determinate | FM | LB, TMV, BW | 40 |
| CLN-2123C | Determinate | PT | TYLCV, BW, TMV, F-1, St | 60 |
| CLN-2443A | Determinate | FM | TYLCV, BW, TMV, F-1, St | 60 |
| CLN-1462A | Indeterminate | FM | BW, TMV, F-1, St | 100 |
| CL5915-206 | Indeterminate | FM | BW, TMV | 100 |
| CLN-155A | Indeterminate | CH/FM | BW, TMV, F-1 | 40 |
| CH154 | Semi-determinate | CHF-1 | - | 10 |
| CLN-1314G | Indeterminate | FM/beta | TMV, BW | 180 |
| CLN-2070B | Indeterminate | CH/beta | BW, TMV, F-1 | 50 |

FM=Fresh Market; PT= Processing; CH=Cherry tomato, high beta carotene, orange fruited; BW= Bacterial wilt tolerance; TMV= Resistance to tomato mosaic virus (tm2a allele); F-1=Resistance to race1 of *Fusarium oxysporum* f. *lycopersici*; F-2= Resistance to race2 of *Fusarium oxysporum* f. *lycopersici*; St= Resistance to *Stemphyllium solani* (cause of gray leaf spot); TYLCV= Tolerance to yellow leaf curl virus; LB=Late blight resistance.

**Table 5.40: Trait Specific Accessions in Various Vegetable Crops**

| Crop | Evaluating Centre | Traits | Promising Accessions |
|------|-------------------|--------|----------------------|
| Bitter gourd | ICAR-IIVR | CMV resistant | IC-44428B, IC-72285, IC-85608B, V-89/0-103C, IC-85611, IC-85636, VRBT-50 |
| Ivy gourd | ICAR-IIVR | Mosaic virus resistance | VRK-05, VRK-10, VRK-20, VRK-35 |
| | Vellanikkara | Resistance to mosaic | CG-84 (Highly bitter fruit) |
| Brinjal | GBPUA&T, Pantnagar | PBNV/Mosaic resistant | IC119611 |
| Chillies | ICAR-IIVR | PepLCV | Bhut Jolokia |
| | | Resistant to LCV | C-11-32 |
| Paprika | ICAR-IIVR | Leaf Curl Virus | BS-35, Bhut Jolokia, MC334 |
| Okra | GBPUA&T, Pantnagar | YVMV resistant | IC7194, IC10262, IC13999, IC43181, IC46018 |
| Carrot (temperate type) | ICAR-IIVR | YVMV resistant | IC140986, IC141056, IC140985 |
| | ICAR-IIHR | YVMV Free lines | IIHR296, IIHR294, IIHR299 |

**Table 5.41: Identification of Resistance Source**

| **Bottle gourd** | | |
|---|---|---|
| GBPUA&T, Pantnagar | Resistance to downy mildew | U-10-316, IC-92362 |
| **Brinjal** | | |
| | Phomopsis blight resistance | IC89883, IC90013, IC90105, IC99571, IC111457, IC111395, IC136571, IC99648, IC128680, IC136326, IC136256, IC90865, NIC13021, NIC13337 |
| **Garlic** | | |
| GBPUA&T, Pantnagar | Purple blotch resistance | IC48988, IC3222 |
| **Tomato** | | |
| PAU, Ludhiana | Resistant to late blight. | LBR-9 |
| **Paprika** | | |
| ICAR-IIVR, Varanasi | Anthracnose resistance | BS-35, Bhut Jolokia, MC334 |
| **Pointed gourd** | | |
| BCKV, Kalyani | Downy mildew (Mini. intensity per cent) | BCPG-8 (14.6), BCPG-4 (18.0) |

**Table 5.42: Potential Resistant Source against Bacterial Wilt**

| **Brinjal** | | |
|---|---|---|
| Vellanikkara | Bacterial wilt resistant | VKBr-21, VKBr-2145 |

**Table 5.43: Potential Resistant Source against Insects**

| **Bitter gourd** | | |
|---|---|---|
| ICAR-IIVR | Fruit fly resistant | IC-85641 and V-89/0-103C |
| **Bottle gourd** | | |
| | Resistance to red pumpkin beetle | U-10-316, NIC-1225 |
| | Resistance to leaf miner | U-10-316, IC-92363 IC-92418 |
| **Ivy gourd** | | |
| ICAR-IIVR | Resistance to leaf miner | VRK-01, VRK-06, VRK-22, VRK-04, VRK-31, VRK-33 and VRK-55 |
| **Okra** | | |
| GBPUA&T, Pantnagar | Pod borer resistance | IC7472, IC23624, IC27821, IC27831 |
| **Carrot (temperate type)** | | |
| ICAR-IIVR | Insect resistance) | IC140986, IC141056, IC140985 |
| | Jassid tolerance | VROB-181 (5.3 jassids/leaf), VROB-110 (5.5) |

**Table 5.44: Potential Resistant Source against Nematode**

| Pointed gourd | | |
|---|---|---|
| ICAR-IIVR | Resistance to nematode | U-34/BP/DR/44, VRPG-72, VRPG-18, VRPG-96 |

**Table 5.45: Potential Accessions for Processing Attributes**

| Tomato | | |
|---|---|---|
| ICAR-NBPGR | Suitable for fresh market | EC-260639, EC-232429, EC-267729, EC-267726 |
| | Suitable for transportation | EC-161252, EC-100845, EC-110268 |
| | Paste type | EC-54628, EC-52062 |
| Solan | TSS (Brix) | Hawaii-7998 and 101E both 5.0 brix |
| **Onion** | | |
| GBPUA&T, Pantnagar | High pungency | IC42911, IC42992, IC49130, IC47959, |
| | High TSS | IC35549, IC35851, IC42854M, IC42911, IC47967, IC48051, IC98183, IC48304, IC48996, IC49072, IC49113, IC49127, IC216197, IC35758, IC48573, IC487451, IC316197, IC35758 |
| **Chillies** | | |
| ICAR-IIVR | Pungency | Bhut Jolokia-world hottest chilli 8,85000 SHU |
| Lam | Capsanthin (EOA color value) | GP299 (53375), GP132 (47672) and GP155 (45140) |
| | Capsacin (per cent) | GP276 (0.581), GP148 (0.571) |
| | Oleoresin (per cent) | GP89 (14.51), GP82 (14.31), Max. IC278336 (23.0) and IC383136 (4.4) |
| Paprika | | |
| ICAR-IIVR | Oleoresin- non pungent | VR-535 |
| **Muskmelon** | | |
| Rahuri | High TSS (per cent brix) | RHRMM-1 (11.0) and RHRMM-36 (11.0) |
| **Pumpkin** | | |
| NDUA&T, Faizabad | Tolerant to water stress | NDPK-5063 |
| **Cauliflower** | | |
| ICAR-IIHR | Non ricey | IIHR-409, IIHR-414, IIHR-415 |

## Characterization of Tuber Crops

### (i) Cassava

At Central Tuber Crops Research Institute, Thiruvananthapuram (Kerala), in cassava, Santha *et al.* (1996) characterized 1632 accessions for 11 morphological traits under the aegis of the Cassava Germplasm. Network established by CIAT and IPGRI created a computerized 'Database'. The germplasm were screened for the incidence of Cassava Mosaic Disease (the most serious disease of the crop) and about 100 accessions were observed to be resistant. These elite germplasm were further evaluated for yield and yield components and the accessions having superior rating were identified as promising materials for utilization (Santha *et al.,* 2000). Rajendran *et al.* (1993) characterized 820 collections for 26 characters and catalogue was published. The ICAR-NBPGR Regional Station, Thrissur (Kerala) characterized and evaluated 131 accessions collected from southern region for 45 morphological traits biotic stresses under the epiphytotic conditions and a catalogue has been prepared. The results indicate that 56 morphotypes exist in cassava collections from southern region. The accessions IC086765, IC086816, IC086761, IC086826-1, IC86807, IC136816, IC086759

and IC214017 were high yielding and accessions IC086731, IC086734, IC086759, IC086784A, IC086772, IC086769, IC136781, IC136799, IC214027 and IC214028 were tolerant to Cassava Mosaic Disease under the field epiphytotic condition; whereas, all the accessions were susceptible to *Cercospora* leaf spot disease (Edision *et al.,* 2005; Sheela *et al.,* 2007). Nedunchezhiyan and Mohanty (2005) evaluated three short duration (H-65, Sree Parakash and Sree Jaya) and four long duration (M-4, H-97, H-226 and CO-2) varieties during 2001-2002 and four short duration (H-165, Sree Praksh, Sree Jaya and Sree Vijaya) and six long duration (Sree Vishakam, Meg-36, M-4, H-97, H-226 and CO-2) during 2002-2003 at Regional Centre of Central Tuber Crops Research Institute, Dumuduma, Bhubaneswar (Odisha). The yield variation due to varieties was significant during both the years of study. Significantly higher yield was recorded with short duration varieties like H-165, Sree Jaya, Sree Vijaya and Sree Prakash than long duration varieties. However, Meg-36 harvested 10 months after planting recorded higher yield during 2002-03. Long post monsoon dry period during maturity phase may be affecting the yield of long duration varieties. Among the short duration varieties H-165 recorded higher yield during both the years of investigation. The tuber yield of 'Sree Prakash' was very low during the year 2002-03 compared to the year 2001-02. This showed that 'Sree Prakash' is highly vulnerable to climatic fluctuations.

### (ii) Sweet Potato

At Central Tuber Crops Research Institute, Thiruvananthapuram (Kerala), 764 accessions of sweet potato have been characterized and evaluated for 39 morphological descriptions, for which a computerized database is being maintained (Rajendran *et al.* 1992). In addition, 869 accessions have been evaluated for important morphological traits for assessing the spectrum of variability and association of traits such as plant type, vine growth rate, petiole length, number of storage roots and tubers yield per plant. In general, accessions possessing semi-compact plant type, intermediate vine growth rate, long petioles and high number of storage roots, were better yielders (Amma *et al.*, 1996). The accessions were also evaluated for reaction to disease and pests including sweet potato weevil (*Cylas formicarius*), which is the most important pest. Evaluation of 913 accessions at Central Tuber Crops Research Institute, Thiruvananthapuram (Kerala) showed that though all the accessions were infested by the weevil, about 10 per cent were relatively less susceptible having tuber damage from 2-21 per cent. Out of 822 accessions were evaluated under field conditions over years, only 29 were free from all diseases. Among the virus diseases, chlorotic leaf spot was observed in 498 accessions, ring spot in 354 and leaf roll in 73 accessions. Further, the incidence of two fungal diseases, namely, chlorotic leaf distortion and brown leaf spot were observed in 400 and 207 accessions, respectively (Sheela *et al.*, 2007).

### (iii) Yams

Central Tuber Crops Research Institute, Thiruvananthapuram (Kerala), accessions of *Dioscorea alata* were characterized for 46 traits and a catalogue was published (Abraham, *et al.*, 1987). Further using 99 descriptors of IPGRI (IPGRI, 1997), 275 germplasm accessions have been characterized and information stored in computerized database. A core collection was also formed based on 13 key traits Da 7, Da 8, Da 10, Da 13, Da 18, Da22, Da 41, Da 60, Da74, Da 105, Da 123, Da124, Da142, Da144, Da145, Da152, Da170, Da173, Da189, Da194, and Da 212 with excellent taste. Three accessions namely, Da 15, Da 74, and Da 184 were observed to be immune against root-knot nematode (*Meloidogyne incognita*).

In *Dioscorea esculenta*, 109 accessions were evaluated for 36 traits and a catalogue was published (Sheela *et al.*, 2000). Germplasm recorded high coefficient of variation for yield (55.4 per cent) and De 96, De 16 and De 38 were found to be promising accessions. Wide variation was observed for tuber length, leaf size and petiole length; and 24 accessions had long tubers (>10.0 cm). The accessions De 22, De 24, v 29, De 31, De 39, De41,

De55, De 120, De 121 and De123 had excellent cooking quality. Two accessions, namely, De 25 and De 85 were observed to be immune to nematode galls. The variation among accessions with respect to other morphological and agronomic traits is very less (Sheela *et al.,* 2007).

Among *Dioscorea rotundata* accessions, Dr 12, Dr 29, Dr 36, Dr 43, Dr 113, v 147, Dr 157, Dr 164, Dr 171, Dr 180, Dr 233, Dr 267, and Dr 318 were found to be high yielding (>5.0 kg/plant). Two accessions, namely, Dr 318 and Dr 327 had good tuber shape, which is an important attribute.

### (iv) Aroids

The germplasm of *Colocasia* at Central Tuber Crops Research Institute, Thiruvananthapuram (Kerala) was characterized and evaluated for 40 descriptors and two catalogues having information on 304, and 492 accessions have been published (Unnikrishnan *et al.,* 1987; Thankamma Pillai *et al.,* 1993). A third catalogue containing data on 40 descriptors of 415 accessions has also been prepared.

### (B) Characterization of Germplasm using Molecular Markers

Molecular characterization of germplasm accessions is a useful tool for better management and to study genetic diversity and integrity of conserve germplasm. Molecular markers can be used for DNA finger printing, classification and clustering of germplasm collections. The classification of plant germplasm collections to different taxa was largely done on the basis of morphological traits, using subsequently some biochemical markers. Previous studies using molecular tools have been performed on the genetic integrity of genebank accessions of some crop species during regeneration. The availability of DNA based molecular markers has revolutionized classification, characterization and construction of phylogenetic trees. There are two categories of probes or markers which are visualized and scored the following:

(i) Labeling and Southern hybridization

(ii) Primer synthesis and PCR amplification

Restricted fragment length polymorphism (RFLP) from genomic libraries and expressed sequence from cDNA libraries were the first series of molecular markers used for construction of high density molecular maps in tomato. Simple sequence repeats (SSR) popularly known as microsatellite markers were subsequently developed and used as new series of co-dominant markers. The utilization of these markers was cumbersome, time consuming and required radioactive or non-radioactive labeling, hybridization, autoradiography, *etc.* Randomly amplified polymorphic DNA (RAPDs) involving single synthetic primers each of 10 nucleotide length were the first series of PCR base markers. Amplified length polymorphism (AFLP) is one of the latest series of PCR based dominant markers providing high degree of genome-wide polymorphism. Sequence characterized amplification region (SCAR) and sequence tagged microsatellite sites (STMS) are the other of co-dominant sequenced based markers visualized by PCR. The anchored markers and maps are being extensively used for characterization and utilization of germplasm collections. DNA fingerprinting of agri-horticultural crops, conservation and use of genomic resources and GM detection are also important activities at ICAR-NBPGR, New Delhi. About 2300 varieties of >35 crops have been fingerprinted to safeguard against biopiracy. Soengas *et al.* (2009) investigated the effects of regeneration on the genetic integrity of *Brassica oleracea* accessions based on simple sequence repeats (SSRs) and found that there were significant changes in the population structure and the allelic frequency at individual loci due to the action of genetic drift, directional selection, and possibly assortative mating. The new molecular marker system, known as single nucleotide polymorphism (SNP), is widely used in different crops. Fingerprinting of genebank accessions can help manage genetic integrity of the germplasm accessions as well as the molecular diversity.

## Evaluation of Vegetable Genetic Resources for Abiotic Stresses

### Table 5.46: Resistant to Abiotic Stresses

| Crop | Evaluating Centre | Traits | Promising Accessions |
|---|---|---|---|
| Pointed gourd | ICAR-IIVR | Heat tolerance | VRPG-72, VRPG-99, VRPG-19, U-34/BP/DR/07, VRPG-110, VRPG-110A, VRPG-110B |
| Tomato | ICAR-NBPGR | Heat tolerant | EC-41824, EC-164855, EC-169308, EC-122063, EC-126955 |
| | | Moderate heat tolerant lines | EC-315487, EC-320578 |
| Brinjal | GBPUA&T, Pantnagar | Drought tolerance | IC23837, NIC23797 |

Genetic diversity was evaluated among five different species of Asparagus of Asparagaceae (formerly Liliaceae) collected from Central Gujarat (India) using RAPD primers. The RAPD primers resulted in the amplification of 273 bands, among which 258 were polymorphic (94.50 per cent) while only 15 were monomorphic (5.50 per cent). So, high level of genetic diversity and low level of genetic similarity (Cluster analysis based on Dice coefficient) were observed among these five species. A dendogram was constructed using UPGMA method. These investigations provide important baseline data for conservation and inter breeding programs (Lal *et al.*, 2011).

## Network of Vegetable Germplasm Evaluation in India

In India vegetable germplasm collection evaluation and conservations are being done in a network mode at various agro-ecological zones. There is separate AICRP for potato; tuber crops, onion and garlic are established.

### Table 5.47: Centre for AICRP (Vegetable Crops)

| Sl.No. | Centres |
|---|---|
| | **Main Centre** |
| 1 | ICAR-IIVR, Varanasi (Uttar Pradesh) |
| 2 | TNAU, Coimbatore (Tamil Nadu) |
| 3 | ICAR-IIHR, Hessaraghatta, Bengaluru (Karntaka) |
| 4 | ICAR-IARI, Regional Station Katrain (Himchal Pradesh) |
| 5 | PAU, Ludhiana (Punjab) |
| 6 | ICAR-IARI, New Delhi |
| 7 | MPKV, Rahuri (Maharashtra) |
| 8 | Bihar Agricultural University, Sabour (Bihar) |
| | **Co-opting Centres** |
| 10 | OUA&T, Bhubaneswar (Odisha) |
| 12 | Dharwad (Karnataka) |
| 13 | RARS, Durgapura (Rajasthan) |
| 14 | NDUA&T, Faizabad (Uttar Pradesh) |
| 15 | CCS, HAU, Hisar (Haryana) |
| 16 | ANGR, University Hyderabad (A.P.) |
| 17 | JNKVV, Jabalpur (Madhya Pradesh) |
| 18 | AAU, Jorhat (Asom) |
| 20 | Chandrasekhar Azad University of Agriculture and Technology, Kalyanpur (Uttar Pradesh) |
| 21 | BCKV, Kalyani (West Bengal) |
| 22 | Lam (Andhra Pradesh) |
| 23 | GB Pant Univ. of Agril and Technology, Pantnagar (Uttarakhand) |
| 24 | IGKVV, Raipur (Chhatishgarh) |
| 25 | Solan (Himachal Pradesh) |
| 26 | SKUAS T (K), Srinagar |
| 27 | KAU, Vellanikkara (Kerala) |
| 29 | Pusa, Samastipur (Bihar) |
| 30 | Anand (Gujarat) |
| 31 | ICAR, Research Complex, Barapani (Meghalaya) |
| 32 | KKV, Dapoli (Maharashtra) |
| 33 | SKUAS T (J), Jammu Srinagar |
| 34 | ICAR-NBPGR, New Delhi |
| 35 | Navsari |
| 36 | Palampur |
| 37 | Periyakulum |
| 38 | Port Blair |
| 39 | Ranchi (HARP) |
| 40 | Tripura |
| 41 | UAS, Bengaluru |
| 42 | ICAR-VPKAS, Almora (Uttarakhand) |

**Table 5.48: Crop-wise Germplasm Evaluating Centres under AICRP (Vegetable Crops)**

| Sl.No. | Crop | Allotted centre |
|---|---|---|
| 1. | Amaranths | TNAU, Coimbatore (Tamil Nadu), KAU. Vellanikkara, (Kerala), AAU, Jorhat (Asom) |
| 2. | Bitter gourd | TNAU, Coimbatore (Tamil Nadu), KAU. Vellanikkara, (Kerala), AAU, Jorhat (Asom) |
| 3. | Bottle gourd | NDUA&T., Faizabad (U.P.), MPKV, Rahuri (Maharashtra) |
| 4. | Cucumber | MPKV, Rahuri (Maharashtra), Y.S. Parmar Univ. Horti. and Forestry, Solan (Himachal Pradesh); ICAR-Indian Institute of Vegetable Research, Varanasi (U.P.), ICAR-HAFRP, Ranchi (Jharkhand) |
| 5. | Pointed gourd | ICAR-Indian Institute of Vegetable Research, Varanasi (U.P.), ICAR-HAFRP, Ranchi (Jharkhand), NDUA&T., Faizabad (U.P.), BCKV, Kalyani (West Bengal), BAU-Sabour (Bihar) |
| 6. | Ivy gourd | KAU. Vellanikkara, (Kerala), ICAR-Indian Institute of Vegetable Research, Varanasi (U.P.) |
| 7. | Muskmelon | PAU, Ludhiana (Punjab), RAU, Durgapura (Rajasthan) |
| 8. | Pumpkin | ICAR-Indian Institute of Vegetable Research, Varanasi (U.P.), NDUA&T., Faizabad (U.P.), ANGARU-Hyderabad |
| 9. | Tinda | CCS-HAU Hisar (Haryana) |
| 10. | Watermelon | RAU, Durgapura (Rajasthan) |
| 11. | Cho-Cho | ICAR Research Complex, Umiam, Barapani (Meghalaya) |
| 12. | Tomato | ICAR-NBPGR, New Delhi, ICAR-Indian Institute of Vegetable Research, Varanasi (U.P.), ICAR-HAFRP, Ranchi (Jharkhand), PAU, Ludhiana (Punjab), Y.S. Parmar Univ. Horti. and Forestry, Solan (Himachal Pradesh), TNAU, Coimbatore (Tamil Nadu), Lam |
| 13. | Brinjal | ICAR-NBPGR, New Delhi, HAFRP, Ranchi (Jharkhand), OUA&T, Bhubaneshwar (Odisha), ICAR-IARI, New Delhi |
| 14. | Chilli | Lam, TNAU, Coimbatore (Tamil Nadu), AAU, Jorhat (Asom), UAS, Dharwad (Karnataka), BCKV, Kalyani (West Bengal), KAU. Vellanikkara, (Kerala), ICAR-Indian Institute of Vegetable Research, Varanasi (U.P.) |
| 15. | Capsicum | ICAR-Regional Research Station, Kullu Valley Katrain, Y.S. Parmar Univ. Horti. and Forestry, Solan (Himachal Pradesh) |
| 16. | Pea | ICAR-Indian Institute of Vegetable Research, Varanasi (U.P.), CSAU A and T, Kalyanpur, Kanpur (U.P.), Palampur, PAU, Ludhiana (Punjab) |
| 17. | French bean | GBPUA&T, Pantnagar, Udham Singh Nagar (Uttarakhand) |
| 18. | Indian bean | UAS-Dharwad (Karnataka), ICAR-Indian Institute of Vegetable Research, Varanasi (U.P.), ICAR-HAFRP, Ranchi (Jharkhand) |
| 19 | Okra | ICAR-NBPGR, New Delhi, ICAR-Indian Institute of Vegetable Research, Varanasi (U.P.), OUA&T, Bhubaneshwar (Odisha) |
| 20. | Garlic | NHRDF-Nasik (Maharashtra), ICAR-NRC-O and G, Pune (Maharashtra), MPKV, Rahuri (Maharashtra) |
| 21. | Cauliflower (early group) | ICAR-Indian Institute of Vegetable Research, Varanasi (U.P.), BAU-Sabour (Bihar) |
| 22. | Cauliflower (late group) | Y.S. Parmar Univ. Horti. and Forestry, Solan (Himachal Pradesh), ICAR-Regional Research Station, Kullu Valley Katrain, |

## Network of Germplasm Evaluation in Potato

**Table 5.49: Details of Centres under the AICRP (Potato) for Germplasm Evaluation**

| | Location/Institute/University | Latitude/Longitude | Established during |
|---|---|---|---|
| **Headquarter** | | | |
| Shimla | Project Coordinated Unit, CPRI, Shimla (H. P.) | 31.06N 77.13E | 1971(IV Plan) |
| **SAU based AICRP centres** | | | |
| Bhubaneshwar | OUAT, Bhubaneshwar (Odisha) | 20.15N 85.52E | 1971 (IV Plan) |
| Chhindwara | JNKVV, Jabalpur (Madhya Pradesh) | 22.03N 78.59E | 1971 (IV Plan) |
| Deesa | SDAU, SK Nagar | 24.14N 72.13E | 1971 (IV Plan) |
| Jorhat | AAU, Jorhat (Asom) | 26.46N 94.16E | 1971 (IV Plan) |
| Kota | MPUA and T, Udaipur | 25.10N 75.52E | 1971(IV Plan) |
| Hassan | UAS, Bengaluru (Karnataka) | 1301N 76.10E | 1972 (IV Plan) |
| Srinagar | SKUAST (K), Srinagar | 34.15N 74.25E | 1972 (IV Plan) |
| Hisar | CCSHAU, Hisar (Haryana) | 29.10N 75.46E | 1975 (V Plan) |
| Kalyani | BCKV, Kalyani (West Bengal) | 22.34N 88.24E | 1975 (V Plan) |
| Pantnagar | GBPUA&T, Pantnagar (Uttarakhand) | 24.48N 79.09E | 1975 (V Plan) |
| Fiazabad | NDUAT, Fiazabad (Uttar Pradesh) | 26.74N 82.12E | 1987 (VII Plan) |

| | Location/Institute/University | Latitude/Longitude | Established during |
|---|---|---|---|
| Dharwad | UAS, Dharwad (Karnataka) | 15.27N 75.05E | 1993 (VIII Plan) |
| Dholi | RAU, Pusa (Bihar) | 25.55N 85.50E | 1994 (VIII plan) |
| Raipur | IGKV, Raipur (Chhattisgarh) | 21.15N 81.41E | 1999 (IX Plan) |
| Kanpur | CSAUAT, Kanpur (Uttar Pradesh) | 26.28N 80.24E | 2009 (XI plan) |
| Pasighat | CAU, Imphal (Manipur) | 28.40N 95.19E | 2009 (XI plan) |
| Pune | MPKV, Rahuri (Maharashtra) | 18.31N 73.55E | 2009 (XI plan) |
| **CPRI based co-operating centres** | | | |
| Shillong | ICAR-CPRI, Regional Station (Meghalaya) | 25N 91E | 1971 (IV Plan) |
| Patna | ICAR-CPRI, Regional Station (Bihar) | 25.37N 85.13E | 1971 (IV Plan) |
| Kufri | ICAR-CPRI, Regional Station (SPU)* | 31.06N 77.13E | 1976 (V Plan) |
| Modipuram | ICAR-CPRI, Regional Campus (Uttar Pradesh) | 29.01N 77.45E | 1976 (V Plan) |
| Ootacamund | ICAR-CPRI, Regional Station (Tamil Nadu) | 11.24N 76.44E | 1987(VII Plan) |
| Jalandhar | ICAR-CPRI, Regional Station (Punjab) | 31.19N 75.18E | 1993 (VIII Plan) |
| Gwalior | ICAR-CPRI, Regional Station (SPU)* (Madhya Pradesh) | 26.14N 78.10E | 2006 (X Plan) |
| **SAU based voluntary centre** | | | |
| Ranichauri | GBPUA&T, Pantnagar (Uttarakhand) | 29.01N 77.45E | 1984 (VI Plan) |

*SPU: Seed Preparatory Unit.

## Development of Core Collection

The word 'core' means the most central, the innermost part or the heart. The term 'core' has been frequently used in scientific literature and fiction in different contexts, but its proposal in Plant Genetic Resources (PGR) was a radical departure Frankel (1984). Until then, efforts were concentrated on an open ended task, irrespective of continuing cost and use, for collecting as many as possible and securing them in gene banks. Huge germplasm collections representing diverse genetic variability of crop species (cultivated) and their wild relatives have been accumulated by several International Agricultural Research Centres (IARCs) and many national programmes through continuing explorations and exchanges. The volume of these collections has outgrown the management resources. It is true that germplasm collections are reservoirs of biological diversity but search for desirable breeding stocks from these collections often takes very long time. Therefore, scientific community interested in assessing potential of the existing genetic diversity contained in these collections enabling promotion and use of plant genetic resources (PGR). Frankel (1984) first proposed the 'core collection' concept to denote a set which would contain germplasm accessions with minimum repetitiveness (the genetic diversity of crop species and its wild relatives in the whole collections) retaining at the same time most of the genetic diversity of the principal collection. It does not replace the existing collection or material from which it is obtained. Core collection have become accepted as efficient tools for improving conservation and use of collection. Later on this concept was further developed by Frankel and Brown (1984) and Brown (1989a, b, 1992). A pre-requisite towards the development of a core collection is that adequate information regarding proper understanding of genetic diversity among primary constituents of germplasm collection and their proper documentation is available for appropriate analysis and decision making. The Global Plan of Action for the Conservation and Sustainable Utilization of Plant Genetic Resources for Food and Agriculture (FAO, 1996) recommends core collection development as one of the activities needed to improve use of plant genetic resources.

## Concept of Core Collection

### Genesis

Browa (1989a) opined that core collection was most important component of the Plant Genetic Resources (PGR) dogma. Collection, characterization of collections and evaluation data describing the extent of variation in the germplasm and a better understanding of distribution of the genetic diversity, in the base collection it is required for effective management and utilization of plant genetic resources (Hodgkin, 1991; Crossa *et al.*, 1992; Gawley, 1992; Jaradat, 1992; Kresovich *et al.*, 1992; Rana, 1992b). Elimination of duplicates became essential to reduce the maintenance costs involved in huge size of the collections. It was also kept in mind that instead of developing a single core the germplasm would be categorized into 'usable' sets of collection which adequately represented specific traits like disease and pest resistance, tolerance to stress conditions and adaptability to specific ecogeographic range. Alteration of the core sets. It is normally used for a set of designated accessions within an existing collection and is more suitable for situations where population size is no consideration and entries of the core need not be physically separable from the entire collection (Brown, 1992).

## Practical Difficulties towards Development of Core Collection

Difficulties encountered in decision making whether new additions in germplasm would add new information or they would be redundant, selecting a particular set of accessions for evaluation, testing for viability or regeneration or responding to breeders requests are of great concern (Brown, 1989a).

1. Though for theoretical formulation of the core set, much discussion revolves round the conservation of alleles, but information about individual alleles is rarely available in any case alleles don't occur independently.

2. Passport data for most of the accessions may be missing, thus making the data set a non-orthogonal set. Such data make the statistical analysis quite cumbersome.

3. Unexplored material from locations with broad diversity may be lead to under-representative of diversity in the core set.

4. Duplicate set of accessions unnecessarily increase the size of the base collection.

5. Variable migration rates of germplasm material in the historical past and/or well-structured populations in defined regions might lead to uneven differentiation among groups of accessions. Such irregular phenomenon make it difficult to select representative and yet non-redundant set.

6. Lack of proper statistical procedure for classifying the germplasm material into homogeneous groups with respect to qualitative and quantitative descriptors.

## Development of Core Set and Core Subset

With the buildup of huge germplasm assemblages representing variability in crop species and their wild relatives through explorations and introductions, there is an increasing awareness among the scientific community for their management. Plant genetic resources (PGR) management and utilization is of great concern following the gravitation of large collections of crop plants and their relatives in gene banks. Various approaches have been put forward from time to time for the development of representative core sets in certain crops. But no approach is fool proof because of various practical difficulties faced by workers during process of development of core collection. Suggestions for ways of managing large, diverse germplasm collections to facilitate their efficient use despite their size have been proposed (Chang, 1991). The concept of core collection and the back-up reserve collection was proposed as a radical departure from the alternative conservation approaches. Conventionally, the accessions were short listed after characterization and evaluation only by few morpho-agronomic attributes, identifying as the

promising accession(s) for single or multiple traits, for promoting their utilization. Further, the idea of developing core collections has been proposed and debated for a long time leading to suggestions whether breeders need one core or many of them depending upon their requirements. Therefore, breeders depend heavily on limited strains evolved in various breeding programmes and tested in multilocational evaluation trials under different ecosystems (Mahajan, et al., 1996).

Ways and means for improving management or utilization of germplasm collections, the reservoir of candidate genes, through the application of powerful tools of biotechnology, which deserve sharp focus due to recent advances in techniques for rapid screening and gene manipulation, can be studied efficiently on a representative set of core collections than on a random lot out of base collection (Spagnoletti, Zeuli, 1992; Vaughan and Jackson, 1992; Tohme et al., 1992).

Recently, AVRDC-World Vegetable Centre has created a core collection of its *S. pimpinellifolium* collection comprising 322 accessions to make it more easily available to breeders. A subset of this collection was evaluated to assess the effects of salt stress on physiological traits as well as yield related traits with the aim of identifying potential *S. pimpinellifolium* accessions that could be used for salt tolerance breeding in tomato.

## (A) Developing Core Sets

### Definition

**Core sets:** A core collection may be suitably defined as:

*"A subset of base collection sampled in such a manner so as to represent the large collection is called as core set."*

Alternatively, the core is a sample drawn out from the base population, which according to Frankel (1984) would represent with minimum of repetitiveness the genetic diversity of a crop species and its relatives. The definition was again modified by Frankel and Brown (1984), Brown (1989a, b,). This way it is defined as: 'a core collection is a diverse, representative yet severely limited sample (Peacock, 1989; Palmer, 1989). It was elaborated by CENARGEN (1992) to stating that the core collection is a part of of the existing collection and does not replace it. The alternative part or 'reserve' is the one unavailable for general distribution because of restriction by plant breeder's rights, narcotic regulations, *etc.* this usage, however, has neither precedence nor semantics in its favour. Core was also defined by IBPGR, as the freely available part of the germplasm without restriction in contrast with the reserve collections restricted for distribution because of propriety or regulations (IBPGR, 1985). The usage of the core in the latter way is controversial and now redundant. Accessions not included in the core would not be jettisoned, according to the proposal, but retained as the backup 'reserve collection' (Mahajan, et al., 1996). A set of well documented core collections will greatly assist in indexing new accessions into various categories and also provide a guide for conservation and exploration activities (Strauss *et al.,* 1988).

**Core subsets**: Instead of developing a single core from a germplasm population some situation specific subsets may be developed for special traits at several eco-sites assigned under the components of active collections which are ultimately linked to a single base collection. The core subset should include as much as possible of its genetic diversity.

## Why Delineate Core?

The 'core' concept revolves round the investigations on the extent and distribution of genetic diversity in a cultivated species and its relatives followed by the delineation of accessions that optimally represent this available genetic diversity. There are several points in favor of setting up a core germplasm instead of selecting accessions at random or bulking of similar accessions. These may be enlisted as follows:

1. It provides efficient access to the base collection for germplasm utilization for a particular trait and gives further idea, based on passport data of the selected accessions, to screen other germplasm of the same geographical origin, for the trait under question, from reserve collections.

2. It facilitates prioritization of representative diversity in base population for further evaluation and study of molecular or cytological variations.

3. It also facilitates conducting studies in relation to conservation aspects and regeneration criteria on the selected set of germplasm on priority.

4. The PGR management is facilitated in responding to generalized seed requests as more quantity of seed of a systematic representative sample is kept on hand by curators.

5. It is helpful for further research on germplasm management and utilization under limited budgetary provisions of developing countries and also it is a fund-saver for the lavish international germplasm programme aimed at the unique-ness of their product and its enhanced quantum of production across the globe.

Designation of a core is needed for testing general combining ability of diverse representative germplasm with locally adapted strains for yield contributing traits (Frankel and Brown, 1984). Mating of a constant parent with core entries in a test cross can also provide fair estimates of combining ability with delineated set of germplam (Spagnoletti Zeuli, 1992). The combining ability of the standby subsets or reserve collections may be tasted by 'Line x Tester' crossing with known lines out of the designated core subset. The in-depth studies carried out on various basic and applied aspects on

accessions in the core would streamline their effective utilization.

## Core vs Mode of Reproduction of Species

Core collections are more relevant to the:

**(I) Seed propagated crops-** In self-pollinated uniform lines in contrast to the development of synthetic or composite bulks for efficient handling of open pollinated crops genetic resources.

**(II) Clonally propagated crop-** In clonally propagated crop core may be set up by eliminating duplicates or genetic redundant.

## Defining Base Population for Core Delineation

It is true that for a particular species any gene bank cannot represent complete diversity. Therefore modality should be fixed as to work independently or linked to each other through defined protocols and MOU'S for better understanding. The conceptual core would, therefore, become a relative term once applied extensively across the gene banks. It is worthwhile to define base population for delineation of a particular core in terms of:

(i) Its array of germplasm accessions.

(ii) Stratum in terms of geographical or taxonomical diversity.

(iii) Availability of duplicate sets from other gene banks or vice versa.

(iv) Proportion of probable duplicates.

(v) Number and size of stratified sub-groups in case core is not set up at random.

## Assumptions for Core Delineation from a Base Collection

For setting up of a core collection, a few assumptions should be fulfilled such as the base collection includes:

1. The base collection is represented by accessions available from whole agro-ecological range (entire range of adaptability) suitable to the species.

2. All the congeners are included in the population/ to ensure representative inclusion of alleles.

3. A large collection of above 3000 individuals is available which ensures retention of over 70 per cent of these selectively neutral alleles in the whole collection.

4. Wild relatives are included in the base population whenever applicable.

5. Passport data of accessions are available which hint appropriately at the geographic origin, status of germplasm (landrace, old or obsolete cultivar, breeders stock *etc.*) and homo or heterogeneous nature of an entry (core entries).

6. Adequate seed is available for the selected accessions for multiplicational testing and supply to users.

For delineation of core sets in structured populations, sampling procedures from reserve collections were investigated on the principal of maintenance of largest genetic multiplicity within a given amount of resources (time, labour, facilities *etc.*). In this regard Empirical studies indicate that the P-strategy would have the widest range of application over C or L, although strategy G is expected to be superior to others *in situ*ations with known degree of genetic multiplicity within groups whereas the strategies H or M were most useful *in situ*ation where genetic diversity among loci within accessions was positively correlated (Yonezawa and Nomura, 1992). Brown (1989b), however, demonstrated that strategy L was better placed for sampling accessions in *Glycine tomentella* and USDA's barley core collections (Mahajan *et al.*, 1996).

On germplasm delineation; accelerated ageing (AA) test at 45°C and 100 per cent relative humidity for 4 days was conducted by Singh *et al.* (2016) on *Phaseolus vulgaris* germplasm and seed quality was assessed using the physiological parameters such as per cent germination, speed of germination, seedling vigour (root and shoot vigour) and seedling vigour index. Seeds subjected to accelerated aging for a period of 2 days maintained vigour and viability closer to the initial but increase in ageing duration reduced the viability and vigour progressively and significantly. Ageing beyond 2 days resulted in loss of seed vigour, as evidenced by decreased number of vigorous seedlings and several fold increase in the abnormal seedlings. The accelerated ageing test could thus delineate 17 genotypes as distinctly different from others in their storability and could be ranked as good storers. These were IC328372, IC258365, IC361547, IC415517, IC258395, EC500577, IC047655, BASPA, EC500275, EC500336, EC500528, IC037138, EC398536, EC100098, EC500522, EC398563 and EC500250.

## Criteria for Setting up Core set

Aspects like size of the base population, size of the 'core' sample, precision of data including diverse representation of collection, choice of suitable statistical technique are a few important considerations for delineation of a core collection. Such a designated core would strengthen if it covers the maximum amount of diversity present in the germplasm material.

Various statistical approaches have been advocated for core identification based alternatively on the

(i) Population genetic theory (genotype)

(ii) Statistical theory (phenotype)

### (i) Population Genetic Theory

The germplasm base population may carry four conceptual classes of alleles based upon preponderance or rarity of different alleles into their genetic build up (Marshall and Brown, 1975, 1983). The frequency and prevalence as follows:

**Table 5.50: Frequency and Prevalence Population**

| Class | Type of allele in the Genetic Background |
| --- | --- |
| Class I | Widespread and common allele |
| Class II | Widespread and rare allele |
| Class III | Localized and common allele |
| Class IV | Localized and rare allele |

The priority/ies for inclusion of these classes may vary accordingly. Least priority is given to deliberate inclusion of class I as the widespread and common allele is likely to be (high probability) included any one of the indirect selection criteria. Class II variables may be safely represented by taking care of appropriate sample size; localized common allele (Class III) may be aptly included by giving due coverage to geographical and ecological representatives whereas, the localized and real allele has to be deliberately included in the core wherever observed. The most optimal sampling strategies may be chosen accordingly as:

1. Random
2. Stratified-random
3. Proportional to geographical frequencies and
4. Based on statistical analysis of evaluation data (Marshall, 1990) to ensure inclusion of alleles from the above four classes.

Brown (1989a) used the neutral allele model of Kimura and Crow (1964), and its sampling theory (Ewens, 1972; Nei *et al.*, 1975). They assumed random sampling in order to meet the baseline expectations. However, stratified random sampling based on ecological factors can strengthen the inclusion of widespread and rare alleles (class II) in core collections. Localized and rare allele in class IV is highly desirable for a 'core' due to the fear of its being lost during regeneration in routine. The localized rare alleles may also be new recurrent mutations which could be useful for mutation breeding programmes. This class has many a times been put forward in debates as a counter example to proposed strategy (Bogyo *et al.*, 1980). However, the accessions carrying localized rare alleles can be included in the flexible core accomplished on their subsequent detection during exhaustive statistical analysis of evaluation data followed by verification from rapid screening techniques (Mahajan, *et al.*, 1996).

Brown (1989a) showed that in large populations irrespective of different levels of polymorphism, a 10 per cent sample (at least 3,000 individuals) of the entire population retained at least 70 per cent of the 'rare-widespread' group of alleles with a frequency greater than 0.0001 and with a confidence limit of 95 per cent.

Schoen and Brown (1993) discussed assemblage of core collection of wild crop relatives in terms of maximization of allelic diversity. Such issues are helpful for theoretical core collections but information about individual alleles in wild relatives is difficult to obtain due to limited availability of base primers to detect alternate alleles and also because alleles don't occurs independently (Gawley, 1992). The representativeness of core collections in terms of available genetic diversity in the base population can be validated at different levels *viz.* allelic representation in terms of allozymes, RFLP analysis *etc.* (brown, 1989b), phenotypic representation in terms of clustering diversity for quantitative traits (Mackay, 1992; Crossa *et al.*, 1992) and then delineating core entries. Yet another level of representativeness *i.e.* the genetic level which would be most useful for efficient PR utilization and which may be expressed in terms of elaborate multilocational evaluation traits, combining ability analysis and even inheritance studies is difficult to attain due to extremely high level of input resources required to meet such objectives. The definition of the base populations need to be viewed at two different planes depending upon existing strategies of core collections delineation (Mackay, 1992). Accordingly, 'international gene pool based core collection' in terms of efficient management under networking of gene banks and situation specific subsets (Rana, 1992b) in terms as 'specific attributes based subset' (Mackay, 199) of efficient utilization of PGR's strongly fulfill the criteria for core setup under population genetic theory. Inclusion of common-localized alleles poses certain problems for selecting neutral alleles (Mahajan, *et al.*, 1996).

## (ii) Statistical Theory

Drawing of a representative sample from the population is bound to result in some loss of information in terms of adequate representation and remainders. An important task in the formulation of core collection is:

(a) How to identify non-diverse subgroups of accessions

(b) How to carry out sampling based on field data comprising of measurements of several attributes in various accessions across environments.

## Major Sets of Multivariate Techniques

Complexity of the data may be overcome by adopting multivariate statistical techniques which will facilitate the description and interpretation of accessions across environments.

    I. Classification methods

    II. Ordination methods

## I. Classification Methods

The objective of classification is to classify accessions into clusters with more or less similar value for all the descriptors. In this method two well-known classification methods are:

### A. Cluster Analysis

Clustering is the technique of partitioning a set of accessions into clusters such that accessions within a cluster are homogeneous and between clusters they are heterogeneous. The main aim of any numerical cluster analysis is to find discontinuity in the data set by means of a cluster strategy that groups the items. Clustering methods may be hierarchical or non- hierarchical

    **i. Hierarchical clustering methods**: All agglomerative methods start with a dissimilarity matrix and fuse the two individuals which have the smallest dissimilarity between them to form a group with two members. Next, the group individual dissimilarity between this new group and all the remaining individuals is calculated. This set of dissimilarity is added to the matrix of dissimilarities among the remaining individuals to form a new dissimilarity matrix. The procedure is repeated and another fusion is made. The procedure terminates when all the individuals are in one group. The method for computing group-individual and group-group dissimilarity is called the clustering strategy.

    **ii. Non-hierarchical clustering methods**: All non- hierarchical clustering methods operate by guessing a group and then employing some method or algorithm to improve the classification. The procedure is repeated until no further improvement in the classification occurs. The algorithm requires both the criteria for evaluating the improvement in the grouping when individuals are transferred from one group to another and a procedure for determining how individuals should be reallocated to groups to improve the classification.

### B. Discriminate Analysis

Linear discriminate analysis is the most widely used method of classification. The formal purpose of such an analysis is to assign distinct set of items to one or several groups' base on a set of measurements. The method allocates new items to previously defined groups by determining mathematical discriminate functions (Linear combinations of original variables) that minimize the chance of misclassification.

## II. Ordination Methods

Ordination methods summarize multidimensional data by production a low dimensional space in which similar accessions are close together and dissimilar ones are far apart. Spatial representation in two or three dimensions will reflect the relationship in higher dimensions with minimum distortion. Ordination techniques/methods may be used to form core collection in order to:

    a. Study diversity of traits among accessions

b. Complement and confirm grouping obtained through classification

c. Establish whether further classification is needed based on variability shown in lower dimensions, and

d. Choose accessions having the combined traits of interest.

The most widely used Ordination method is the principal component analysis.

## Principal Component Analysis

Principal component analysis is usually applied in order to overcome the multi-collinearity between different descriptors. It consists in dividing the total variability present in the data set into different components (equal to the number of descriptors) such that each component is a linear combination of different descriptors and is independent of the preceding ones. This is achieved by applying orthogonal transformation on the original set of descriptors. Then the first few components explaining more than 75 per cent variability are selected. Thus, a multidimensional data set is reduced to a space of low dimension.

## Sampling Theory

Sampling of a representative sample from the population is always bound to result in some loss of information. The stratified random sampling is found adequately efficient for grouping of germplasm collections into distinct groups (Brown, 1989b) and a better choice over simple random sampling, a default option.

## Grouping of Accessions

The accessions may be grouped into different strata based on the available information relating to geographical regions. Sample selection of accessions within each group may be guided by different strategies. Out of these strategies:

i. **Constant strategy (C):** Selection of equal sample size from each group

ii. **Proportional strategy (P):** Number of accessions being proportional to the number of accessions in a particular group

iii. **Logarithmic strategies (L):** Sample size being proportional to logarithm of population size in each group have been widely acclaimed.

In addition, a few more sampling strategy have been proposed by several researchers *viz.*,

i. **Genetic multicity dependent strategy (G):** Sample size proportional to the amount of genetic multicity available in groups

ii. **Nei's genetic diversity index or H strategy:** This takes into account Nei's genetic diversity index (H) or panmictic heterozygosity to estimate

variation among geographical regions using marker locus data in effective population size (Ne) or indices correlated with Ne.

iii. **M strategy:** M strategy uses non-linear programming methods to pinpoint accessions for maximum diversity within each geographical region and

iv. **R strategy:** R strategy which involves random sampling of accessions irrespective of stratification. Thus a situation specific selection of a particular sampling strategy would make the core collection more meaningful.

## Sampling Strategies for the Core

Once decided up on the optimal sample size for the core, which may be 10 per cent entries of total accessions with a minimum of 3000 per species, the sampling strategies may be drawn out as follows:

i. Simple random sampling should be done as a default option when there is no clear cut evidence to base any grouping in the accessions.

ii. The collection should be first-divided into non-overlapping groups and then random samples be drawn out of each group in case of stratified random sampling. The stratification may be based on statistical and or biological bases.

iii. The statistical analysis primarily includes classification or ordinate multivariate methods. The former includes cluster analysis and discriminant analysis whereas the latter may be carried out as Principal Component Analysis (PCA). It may be clearly understood that the results from classification and analysis are confirmed with those obtained from ordinate analysis.

iv. The passport and evaluation data of the base collections may be extensively used for undergoing hierarchical cluster analysis. The hierarchy of groups may vary according to importance of differentiating traits, *e.g.* taxonomic (species, cytotype), category of materials (wild, landraces, obsolete, old or recent cultivar, breeder's line), country, region, morphological traits *etc.*, in terms of weightage for their exploitation. At least one sample can be drawn from each sub-group for delineating core germplasm.

v. The grouping of characterization and evaluation traits can be done by grades for qualitative descriptors and multivariate analysis for quantitative traits. Representative samples may be drawn from the grades/clusters showing similar character expression. In case of uneven data preference may be given to accessions with more extensive data.

vi. The number of entries in the core may be drawn as per constant strategy (equal number of accessions per group), proportional strategy (a fixed proportion of accessions from each group) or logarithmic strategy (number of accessions included in the core proportional to the logarithm of the size of group). Of these logarithmic and proportional strategies are better than constant strategy but are at per with each other (Brown, 1989b).

vii. Isozyme analysis of accessions likely to be included in the core would provide further resolution about the mono/polymorphic nature of the allele as well as occurrence of polymorphism. More recent techniques like RFLP, RAPD *etc.* will provide information about the strongly linked block of quantitative genes of varying background linked to the marker genes (*e.g.* Quantitative Trait Loci or QTL). These studies help in the inclusion of diverse specific entries into the core germplasm.

## Constraints for Setting up Core

Following constraints have been observed during setting up core:

(i) Base collection is invariably a non-random and sometimes incomplete sample the genetic resources of a crop.

(ii) Making effect by major genes for the character expression due to quantitative genes poses difficulty in selection for most of the entries, based on specific traits, which otherwise may be having a desirable genome.

(iii) Difficulties in determining source country vs country of origin

(iv) Unavariability of appropriate characterization and evaluation data for grouping of accessions.

(v) Non-utility of wild related species directly in breeding programmes unless pre-breeding is done poses difficulty in stratification too.

(vi) Probability of inclusion of localized rare allele is very remote.

## Uses of Core Germplasm

Two fold use of core germplasm is facilitated once a core is set up using any one of the above mentioned strategies. Sampling of accessions using these kinds of operations can be done for management (by curators) as well utilization (by various categories of users) aspects.

(a) Management aspects

(b) Utilization aspects

### (a) Management Aspects

(i) Prompt response to generalized seed requests which may safely cover a representative set of the array of base collection.

(ii) Basic studies on germplasm conservation and regeneration aspects like-temperature and humidity ranges under storage regimes, seed moisture content for longevity under storage, pre-information on seed multiplication or synchrony of wild materials for crossing programmes *etc.*

(iii) Indexing of new germplasm collections by direct comparison with the identified core and similar indexing of reserve collections into germplasm pools (sub-sets).

(iv) Decision making for further collection of a particular category of germplasm accessions from a specific niche based on passport data.

### (b) Utilization Aspects

(i) Basic studies like combining ability and inheritance of complex characters can be judiciously done on a defined sample than on a random lot out of a huge population.

(ii) To validate different grouping under stratified sampling using molecular markers and cytological variations in the representative geographical samples.

(iii) Decisions for pre-breeding of the distant relatives or accessions showing crossability barriers in the core germplasm.

(iv) Covering medium to long term objectives of testing combining ability (general combining ability and specific combining ability) and donor potentials of reserve collections.

## Crop Networking in Relation to Core

Diversity is also restricted due to the closeness among cultivars originated from single breeding programme, co-adaptiveness of the landraces, or co-evolution of primitive cultivars *etc.* various plant genetic resource programmes have duly recognized the need of conserving duplicate sets of germplasm accessions across the gene banks. There is a need of international collaboration to devise appropriate strategies likely to be of importance in maximizing the benefits of investigations on establishing core collections. The setup of international networks for effective exchange of database along with duplicate sets would help in appropriate statistical analysis, reduce the chance of availability of missing data for representative geographical sections of accessions and increase the probability of inclusion of fullest range of diversity of the crop species in its core. The researcher's interest lies in exploiting maximum diversity in the gene banks for the core set up and, if the need be, to go in for further explorations in specific regions either unaccounted for sparsely represented in the core. It is thus desirable to have collaboration between gene banks, users and researchers in order to meet the objectives of maximum availability as well as diversity among the accessions. In other words, the development of crop-networks is closely linked to an effective core collection. The basic idea of establishing crop-networks in order to strengthen core collections could be viewed differently at the international and national levels. Various international agencies like CGIAR, IBPGR, Bioversity International and IARCs share the global responsibilities for crop based PGR programmes. On the other side, the national programmes like USDA, ICAR-NBPGR would tend to fulfill the requirements for farming systems need based agro-ecological situations specific subsets for their respective countries and also contribute to global reserves through collaborative explorations, exchange and duplicate sets *etc.* (Mahajan *et al.*, 1996).

As a suggestion, the region specific smaller PGR units within the national set ups constitute significant and integrated part within the national PGR network (Rana, 1992b). Besides other factors, such as a network within a national PGR programme would be helpful in retrieving further evaluation data (FED) from the multilocationnal tests at National Active Collection Sites (Rana *et al.*, 1993) which would ultimately provide desirable database for constituting core collections. The need based, situation specific subsets, of the developing countries, may be developed by following either of the population genetic approach under stratified sampling using C, P or L strategy as in soybean (Brown, 1989a,b), in case the passport data is complete, or biometrical approach using statistical analysis of the evaluation data on quantitative characters as done in okra (Hamon and van Sloten, 1989).

## Establishment of Core Collections in Vegetable Crops

Till date the number of core collections with published information is very limited. At the same time, the absence of a well-defined approach in the published literature on core germplasm does not substantiate any defined criteria for future guidelines. A survey was initiated at the end of 1989 by IBPGR, Rome to look into the ongoing work on core collections by scientific community throughout the world based on the overall interest shown by various national and international organizations, over 20 projects involving cereals, legumes, forage crop species, fruits and vegetables crops were identified (T. Hodgkins, Pers. Comm.). Core collection for *Glycine* species (Brown *et al.*, 1987), okra (Hamon and van Stolen, 1989), common bean (Tohme

*et al.*, 1992) *etc.* are developed and established. Not only this, information with regards to their methodologies is available.

A few representative projects related to core collections are reviewed hereunder:

### Okra

Okra (Hamon and van Stolen, 1989) established core collection on okra based the variability in passport, characterization and evaluation data from single trial. A core comprising 189 accessions out of 2,283 accessions was developed with specific inclusion of rare types. This core collection on core was set up with a slightly varied approach in order to have a manageable collection scaled down to the needs of the breeders and other users. It suffered major limitations in terms of missing passport data to the tune of 20 per cent for collector's number, 30 per cent for town/province, 33 per cent for latitude/longitude, 60 per cent for vernacular names and 99 per cent for altitude thereby making it difficult to follow a stratified random sampling based on population genetics approach. In order to include widest possible range of variability it was constituted primarily on the basis of rigorous statistical analysis of the available data on agro-morphological characters.

The statistical analysis resorted to were:

(i) **Univariate analysis** *viz.*, for quantitative descriptors and some basic statistical like mean, range, standard deviation, coefficient of variation and frequency distribution for taxonomical or qualitative morphological descriptors,

(ii) **Bivariate analysis** *viz.*, correlations among quantitative traits,

(iii) **Multivariate analysis** *viz.*, factor analysis, principal component analysis and hierarchical clustering methods elucidating the Euclidean distances among various quantitative characters.

Three distinct groups were earmarked at the Euclidean distance level of 0.80 and a choice at level 0.30, which authors availed in this situation, allowed elimination of four descriptors.

Some of the interesting observations for setting up core in okra included using the cultivation system as an important source of information wherein the graphical representation of variability by means of factorial analysis corresponded closely to the farming systems used, long tradition of okra cultivation in a particular country/agro-ecological zone and the choice of morphological descriptors. Generalization of such an indirect approach is, however, difficult to make due to the non-availability of descriptors like farming system in the routine data recorded by collectors. Other significant features emerged from the core collections were that the reduction in number of descriptors obtained through multivariate analysis emerged as another significant feature in the okra core collection. A stepwise approach consisted of running a principal component/factor analysis projecting the variability in limited dimensions. Then a clustering analysis was done which could be later tested by a discriminant analysis. In such an approach the choice made by the curator/database manager leads to a deliberate but controlled loss of information. Nonetheless, the base collection was effectively reduced in size to a manageable representative set for further use. ICAR-NBPGR, New Delhi has also established suitable statistical methodology for core collection in okra.

### Common Bean

A core collection of 1400 accessions in common bean (*Phaselus vulgaris* L.) from 24,000 accessions stored at CIAT using evolutionary, agro-morphology and agro-ecological data was established (Tohme *et al.*, 1992). The corset comprised of 1200 accessions from primary centres, 200 from secondary centres and about 25 a priori selected bean lines as reference. A weighted stratified model was used. First an arbitrary weight was assigned to geographical areas delineated by similar subjective ecological classification. In corollary to the length of adoption of landraces in different farming systems in case of okra, the germplasm from the primary centres, in common bean was assigned a higher grade than from area of recent introduction. The second level of stratification was based on morphological data.

### Soybean

A collection of 111 accessions of perennial soybean (*Glycine* spp.) was developed in Canberra, Australia (Brown *et al.*, 1987) from a collection of around 1400 accessions of 12 species. The criteria in choosing the accessions for developing a dynamic core were:

i. More than one accession per *Glycine* spp. selected to provide replication for generalization of results from any basic or applied research carried out from these accessions at species level,

ii. Geographical coverage in Australia

iii. Inclusion of morphological, cytological and isozymes groups with known intraspecific variation.

iv. Priority inclusion of accessions used for research in the past and the first hand collections.

The core thus constituted was much different from a random set and achieved an overall target of about 10 per cent sampling proportion from each category *i.e.* species, geographical region or political state *etc.* varied drastically (Brown, 1989a).

Further, Brown (1989b) intensively demonstrated the effect of sampling strategy on specific inclusion of accessions in the core *Glycine tomentella*, a species collected extensively from all across its natural range in Far East Asia and Oceania and which comprised 5 diploid (2n=38, 40) and tetraploid (2n=78, 80) groups. In this example the choice of alternate strategies (C, P or L) showed marked effects on the chance of including rare types. For the diploid strategy P gave the best recovery whereas for the tetraploid strategy C was the best option. The strategy L, however, emerged as a good compromise and was thus favoured by Brown (1989b) in cases where the representative set of germplasm showed skewed distribution for rarest variants, *e.g.* in diploid, occurring in the largest group and tetraploids, available in the smallest group, in *Glycine tomentella*.

### Potato

At present the germplasm repository of potato at ICAR-Central Potato Research Institute, Shimla (H.P.) has a total collection of 1400 *S. tuberosum* accessions. Creation of core set is one approach of hastening and prioritization of the germplasm characterization and ultimately its utilization for different traits specific breeding activities. Considering above, a core set of 140 accessions of *S. tuberosum* was extracted from 1400 accessions based on 19 morphological descriptors *viz.*, vegetative, tuber and sprout parameters. The validity of the core set by means and variances comparision and Shannon-Weaver diversity index and phenotypic correlations between different traits revealed that actual diversity of base collection have been conserved. The established core set forms the basis for wider exploitation of potato germplasm for effective utilization in potato improvement programme (Dalamu *et al.*, 2016).

### (B) Developing Core Subsets

Developing core subsets is essential feature aimed at developing a minimal core subset representing the diversity of base collections. There are several considerations which go in favour of developing situation specific core subsets rather than single core collection.

### Situations for Developing Core Subsets

Situations for developing core subsets may be enlisted as:

(i) Logical

(ii) Ethno-botanical

(iii) Agro-climatical

(iv) Biometrical genetical

(v) Managerial and

(vi) Germplasm usage

Genetic variation is contained in the base collection, as the network shared among landraces, cultivars, strains and the related species, which may not be suitable interpreted in a single representative sample. The above mentioned considerations provide critical resolution factors for isolation or pooling of variable types. It appears more scientific to subgroup specific types for adaptability to agro-climatic, geographical, seasonal *etc.* situations besides resistance to specific diseases or stresses or agro-morphological or quality traits for effective management as well as utilization, rather than making a single core (Mahajan, *et al.*, 1996) and it is possible by using appropriate multivariate statistical tools. In the next step, within group selection is carried out for accessions accounting for maximum variability in qualitative and quantitative descriptors. A group may invariably be split into sub-groups equaling total number of combinations for a few important qualitative descriptors at this stage, showing a fairly spread distribution order for accessions explaining maximum diversity within each subgroup through some appropriate multivariate technique (Mahajan, *et al.*, 1996). At the same time, Shannon Diversity Index (SDI) for these accessions is computed, retained up to a stage when the SDI of accessions becomes approximately equal to greater than the corresponding value for the entire group/subgroup. The accessions thus selected from different sub-groups within a group would constitute a core subset. Situation specific subsets may be similarly obtained from other groups.

### Salient Features of Developing Core Subsets

Salient features of developing a core subset delineation programme would be as follows:

(i) Components of active collections would be maintained at diverse agro-ecosites each holding specific responsibility. Yet these components would be linked to single base collection at the National Gene Bank (NGB).

(ii) Upgrading of germplasm from centres and integrating of data files for locations, for networking.

(iii) Generating further evaluation data (FED) on the upgraded germplasm.

(iv) Listing of important landraces, old or obsolete varieties, or unique germplasm.

(v) Rationalization of sampling strategies and development of software.

(vi) Characterization of selected accessions using biological markers.

A sample core set in okra has been worked out with limited size of base collection originated from series of interrelated exploration programmes and covering the variability from South East Asia, including some South African Taxons (SAT). There is an ample variability for a species/taxa and the within species accessions in this act. However, this limited set was exercised at the initial stage to suit the availability softwares of non-hierarchical cluster analysis which could accommodate 500 accessions and 10 quantitative descriptors and also to confirm the validity of the above approach. The 260 okra accessions produced 8 homogeneous unequal groups on the basis of quantitative characters (Bisht *et al.*, 1995). In later assessment of the database, these groups were ascribed to plant types. Each group was split into subgroups depending upon its size, frequency distribution of accessions on qualitative characters and variability observed within groups. Accessions were then selected from within each group/subgroup to account for maximum variability in quantitative characters, using Principal components Technique, and qualitative characters using Shannons Diversity Index (SDI). The selected accessions from various subgroups within a group consisted of a sample core set. This core set comprised of 50 accessions to which 3 rare types were deliberately incorporated making it a final set of 53 accessions.

### References

1. Abraham, K., Nair, S.G. Kumari, S., Unnikrishnan, M., Babu, M.L. and Palaniswami, M.S. 1987. *Catalogue on Genetic Resources of Greater Yam.* Central Tuber Crops Research Institute, Thiruvananthapuram (Kerala).

2. Amma, E.C.S., Sheela, M.N., and Rajendran, P.G. 1996. Evaluation of variability and association of characters in sweet potato germplasm. In: *Tropical Root and Tuber Crops in Food Security and Nutrition* (eds.) Balagopalan, C., Nair, T.V.R., Sundaresan, S., Premkumar, T. and Lakshmi, K.R. p. 144-149.

3. Anonymous 1985. Descriptors for *Vigna aconitifolia* and *A. trilobata* (based upon a list prepared by T.A. Thomas of the ICAR-NBPGR, India). IBPGR Secretariat, Rome.

4. Anonymous, 1985. Report of the III meeting of IBPGR Advisory Committee on Seed Storage. IBPGR, Rome.

5. Bisht, I.S., Mahajan, R.K. and Rana, R.S. 1995. Studies on South Asian Okra (*Abelmoschus esculentus*) germplasm collection I. Measurement and Classification of Genetic Diversity. *Annals of Applied Biology*, 539-550.

6. Bogyo, T.P., Porceddu, E. and Perrino, P. 1980. Analyses of sampling strategies for collecting genetic material. *Economic Botany*, 34: 160-174.

7. Bos, I. 1983a. The optimum number of replications when testing lines or families on a fixed number of plots. *Euphytica*, 32: 311-318.

8. Bos, I. 1983b. About the efficiency of grid selection. *Euphytica*, 32: 885-893.

9. Breese, E.L. 1989. Regeneration and multiplication of germplasm resources in seed genebanks: the scientific background. IBPGR, Rome.

10. Brown, A.H.D. 1989a. The case of core collections. In: The use of Plant Genetic Resources. Brown, A.H.D. *et al.* (eds.), Cambridge University Press, Cambridge. p. 136-156.

11. Brown, A.H.D. 1989b. Core collection: A practical approach to genetic resources management. *Genome*, **31**: 818-824.

12. Brown, A.H.D. 1992. The core collection at the crossroads. Paper presented in International Workshop on Core Collections of Plant Germplasm. Held at CENARGEN, Brasilia, Brazil. Sep., 1992.

13. Brown, A.H.D., Grace, J.P. and Speer, S.S. 1987. Designation of a "core" collection of perennial *Glycine. Soybean Genet. Newsl.*, **14:** 59-70.

14. Chang, T.T. 1991. Guidelines on developing core collections of rice cultigens. In: Rice Germplasm: Collecting, Preservation and Use. IRRI, p. 105-107.

15. Chattopadhyay, A., Dutta, S., Hazra, P. and Sarkar, G. 2011. Conservation and utilization of vegetable Germplasm at Bidhan Chandra Krishi Viswavidyalya. Nat. Symp. On 'Vegetable Biodiversity' Held at JNKVV, Jabalpur, p. 131-140.

16. Chaurasia, S.N.S. and Singh, M. 1998. Genetic divergence in tomato. *Indian J. Plant Genet. Resour.*, **11(2):** 203-206.

17. Crossa, j., Taba, S. and Dalcy, I. 1992. The use of multivariate methods in the process of forming core subsets. Paper presented in International Workshop on Core Collections of Plant Germplasm. Held at CENARGEN, Brasilia Brazil. Sept. 92.

18. CYMMIT/IBPGR. 1991. Descriptors for Maize. CYMMIT/IBPGR, Rome.

19. Dabas, B.S. and Phogat, B.S. 1994. Collection, evaluation and prospects of arid legume. In: Plant Genetic Resources: Exploration, Evaluation and Maintenance. R.S. Rana, Bhag Singh, M.N. Koppar, Mathura Rai, S. Kochhar and S.S. Duhoon (eds.) Pub. By ICAR-NBPGR, New Delhi. p. 184-190.

20. Dadlani, S.A., Singh, B.P. and Singh, R.V. 1981. System of national and international exchange of germplasm and methods of recording followed at ICAR-NBPGR. *Sci. Monogr.* No. 5, ICAR-NBPGR, New Delhi, p. 72-87.

21. Dalamu., Kumar, V., Bhardwaj, V., Kumar, R., Kaur, R.P., Upadhyaya, H.D. and Singh, M. 2016. Development of core set of Indian collection of cultivated potatoes (*Solanum tuberosum* ssp. *tuberosum*). 1st International Agrobiodiversity Congress; (Session: Trait Discovery and Enhanced use of PGR) held at New Delhi; 6-9 November 2016; p. 147.

22. Dhall, R.K. and Brar, P.S. 2013. Genetic variability, correlation and path coefficient in garlic (*Allium sativum* L.). *Veg. Sci.*, **40(1):** 102-104.

23. Edison, S., Velayudhan, K.C., Amma, E.C.S., Santha, V.P., Mandal, B.B., Sheela, M.N., Vimla, B., Unnikrishnan, M. and Zakir, H. 2005. In: *Plant Genetic Resources: Horticultural Crops, Tropical Root and tubers* (eds.) Dhillon, B.S., Tyagi, R.K., Saxena, S. and Randhawa, G.J., Narosa Publishing House, New Delhi, p. 228-250.

24. Engels, J.M.M. and Rao, V.R.1995. (eds.) Regeneration of seed crops and their wild relatives, p. 140-143. Proceedings of a consultation meeting, 4-7 December, 1995. ICRISAT, Hyderabad (India) and IPGRI, Rome.

25. Ewens, W.J. 1972. The sample theory of selectively material alleles. *Theoret. Pop. Biol.,* 3: 87-112.

26. Federer, W.T. and Raghavarao, D. 1975. On augmented design. *Biometrics.* 31: 29-35.

27. Federer, W.T., Nair, R.C. and Raghavarao, D. 1975. Some augmented row column design. *Biometrics.* 31: 361-373.

28. Frankel, O.H. 1984. Genetic perspective of germplasm conservation. In: Genetic Manipulation: Impact on Man and Society. In: W.K. Arber, Limensee, K., Peacock, W.J., Starlinger, P. (eds.), Cambridge University Press, Cambridge, p. 161-170.

29. Frankel, O.H. 1986. Genetic resources: Museum or utility. In: Plant Breeding Symposium. DSIR; 1986. Williams, T.A. and G.S. Wratt (eds.), *Agron. Soc. Newzealand*, Christchurch, p. 3-7.

30. Frankel, O.H. and Brown, A.H.D. 1984. Current plant genetic resources: A critical appraisal. In: Genetics: New Frontiers Vol.4, Oxford and IBH Publishing Co., New Delhi, p. 1-11.

31. Frankel, O.H. and Brown, A.H.D. 1984. Plant genetic resources today: A critical appraisal. p. 249-257. In: Crop Genetic Resources: Conservation and Evaluation (J.H.W. Hoden and J.T. Williams eds.). George Allen and Unwin Ltd., London.

32. Gawley, N.W. 1992. Verifying and validating the representative of a core collection. Paper presented in International Workshop on Core Collections of Plant Germplasm. Held at CENARGEN, Brasilia Brazil in Sept. 92.

33. Grubben, G.J.H. and Stolen, D.H. 1981. Genetic resources of *Amaranthus.* AGP: IBPGR/60/2, FAO, Rome, Italy.

34. Hamon, S. and van Sloten, D.H. 1989.Characterization and evaluation of okra. In: A.H.D. Brown, Frankel, O.H., Marshall, D.R., Wilson, J.T. (eds.). The use of plant genetic resources. Cambridge University Press, Cambridge. p. 173-196.

35. Hodgin, T. 1991. Improving utilization of plant genetic resources through core collections. In: Rice Germplasm-Collecting, Preservation and Use. IRRI, Philippines. p. 99-104.

36. IBPGR, 1985. Report of the II meeting of IBPGR Advisory Committee on Seed Storage, IBPGR, Rome.

37. IPGRI. 1997. Descriptors of Yam (*Dioscorea* spp.). International Plant Genetic Resources Institute, Rome, Italy. p.61.

38. Islam, M.T., Haque, M.S., Islam, M.M. and Haque, M.M. 2016. Genetic diversity of Hayacinth bean collections in Bangladesh. 1st International Agrobiodiversity Congress; (Session: Trait Discovery and Enhanced Use of PGR), held at New Delhi; 6-9 November 2016; p. 187.

39. Jaradat, A.A. 1992. The dynamics of core collections. Paper presented in International Workshop on Core Collections of Plant Germplasm. Held at CENARGEN, Brasilia Brazil. Sept. 92.

40. Joshi, B.D. 1981a. Exploration for amaranths in North West India. Plant Genetic Resources News Letter, p. 41-51. AGP: PGR/48. IBPGR, FAO, Rome, Italy.

41. Joshi, B.D. 1981b. Catalogue on amaranth Germplasm. ICAR-NBPGR, Regional Station, Shimla, p. 42.

42. Joshi, B.D. 1985. "Annapurna" a new variety of grain amaranth. *Indian Farm.*, **25(8):** 29-31.

43. Joshi, B.D. 199. Catalogue on Amaranth Germplasm, ICAR-NBPGR, Regional Station Shimla (H.P.)

44. Joshi, B.D. and Rana, R.S. 1995. Genetic divergence in grain amaranth (*Amaranthus hypochondriacus*). *Indian J. Agric. Sci.*, **65:** 605-607.

45. Kathayat, K., Singh, A. and Singh, A.K. 2016. Seed protein profiling of tomato (*Solanum lycopersicum* L.) genotypes using SDs-PAGE. *Indian J. Plant Genet. Resour.*, **29(2):** 151-155.

46. Kempton, R.A. and Fox, P.N. 1997. Statistical methods for plant variety evaluation. London: Chapman and Hall.

47. Khan, Z, Gautam, N.K., gangopadhyay, K.K., Gawade, B.H. and Dubey, S.C. 2016. Screening of Germplasm for the source of resistance to root-knot nematode (*Meloidogyne incognita*). 1st International Agribiodiversity Congress; (Session: Trait Discovery and Enhanced use of PGR) held at New Delhi; 6-9 November 2016; p. 149.

48. Kimura, M. and Crow, J.F. 1964. The number of alleles that can be maintained in a finite population. *Genetics*, **49:** 725-738.

49. Kresovich, S., Lamboy, W.F., McFerson, J.R. and Forsline, P.L. 1992. Utilizing different types of genetic information to develop useful and representative collections. Paper presented in International Workshop on Core Collections of Plant Germplasm. Held at CENARGEN, Brasilia Brazil in Sept. 92.

50. Kumar, R. Chatterjee, D., Kumawat, N., Pandey, N., Roy, A. and Kumar, M. 2014. Productivity, quality and soil health as influenced by lime in rice bean cultivars in foothills of northeastern India. *The Crop Journal*, **2:** 338-334.

51. Lal, S., Mistry, K.N., Vaidya, P.B., Shah, S.D. and Thaker, R.A. 2011. Genetic diversity among five economically important species of Asaparagus collected from central Gujarat (India) utilizing RAPD markers. *International Journal of Advanced Biotechnology and Research*, **2(4):** 414-421

52. Mackay, M. 1992. One core or many. Presented at the International Workshop on Core Collection of Plant Germplasm. CENARGEN, Brasilia, Brazil, Sep. 1992.

53. Mahajan, R.K., Bisht, I.S., Agrawal, R.C. and Rana, R.S. 1996. Studies on South Asian Okra collection: Methodology for establishing core collection using characterization data. *Gene. Resour. and Crop Evol.*, **43:**249-255.

54. Mahajan, R.K., Kochhar, S., Agrawal, R.C. and Chandel, K.P.S. 1996. Developing a core set or subsets of large germplasm collections: An overview. *Indian J. Plant Genet. Resour.*, **9(1):** 11-30.

55. Mall, A.K., Pathak, A.D. and Kapur, R. 2016. Sugarbeet-germplasm evaluation, seed production and agro-techniques for Sub Tropical India. 1st International Agrobiodiversity Congress (Common Session), held at New Delhi; 6-9 November 2016; p. 118.

56. Marshall, D.R. 1990. Crop Genetic Resources: Current and emerging issues. In: Plant population genetics, breeding and genetic resources. Brown, A.H.D. *et al.* (eds.), Sin Auer Associates Inc., Sunderland, M.A. p. 384-385.

57. Marshall, D.R. and Brown, A.H.D. 1975. Optimum sampling strategies in genetic conservation. In: Crop genetic resources for today and tomorrow. O.H. Frankel and J.G. Hawkes (eds.). Cambridge University Press, Cambridge. p. 53-80.

58. Marshall, D.R. and Brown, A.H.D. 1983. Theory of forage collection. In: Melver, J.C., Bray, R.A. (eds.). Genetic Resources in Forage plants, CSIRO, Melbourne. p. 135-148.

59. NBPGR. 2016. Guidelines for Management of Plant Genetic Resources in India. ICAR-NBPGR, New Delhi, p. 48-53.

60. Nedunchezhiyan, M.and Mohanty, A.K. 2005. Economic evaluation of cassava (*Manihot esculenta*) varieties under rainfed conditions in Odisha. *The Odisha J. Hort.*, **33(1):** 108-109.

61. Nei, M., Maruyama, T. and Chakraborty. 1975. Bottleneck effect and genetic variability in populations. *Evolution*, **29:** 1-10.

62. Peacock, W.J. 1989. Molecular biology and genetic resources. p. 363-376. In: The Use of Plant Genetic Resources, (Eds.) A.D.H. Brown, O.H. Frankel, D.R. Marshall and J.J. Williams. Cambridge University Press, New York.

63. Pradheep, K., Singh, P.K., Pandey, A. and Bhandari, D.C. 2011. Collecting genetic resources of wild *Moringa oleifera* Lam. from Western Himalaya. *Indian J. Plant Genet. Resour.*, **24(3):** 292-298.

64. Rai, N. and Asati, B.S. 2005. Correlation and path coefficient analysis for yield and its traits in cabbage. *The Odisha J. Hort.*, **33(1):** 3134.

65. Rajendran, P.G., Amma, C.S.E. and Lakshmi, K.R. 1992. *Description, Documentation and Evaluation of Sweet Potato Gerrmplasm*, Central Tuber Crops Research Institute, Sreekariyam, Thiruvananthapuram (Kerala).

66. Rajendran, P.G., Sreekumari, M.T., Nair, R.B., Pillai, K.S., Nair, N.G. and Moorthy, S.N. 1993. *Catalogue of Genetic Resources in Cassava*. Tech. Bull. No. 7. Central Tuber Crops Research Institute, Thiruvananthapuram (Kerala).

67. Ram, D., Rai, M., Verma, A. and Singh, Y. 2006. Genetic variability and association analysis in *Luffa* sp. *Indian J. Hort.*, **63(3):** 294-297.

68. Rana, R.S. 1992a. Indian National Plant Genetic Resources Systems. In: Rana, R.S., Sapra, R.L., Agrawal, R.C. and Gambhir, R. (eds.). Plant Genetic Resources-Documentation and information management. ICAR-NBPGR, New Delhi.

69. Rana, R.S. 1992b. Core collections and priorities of national programmes-Indian perspective. Paper presented in International Workshop on Core Collections of Plant Germplasm. Held at CENARGEN, Brasilia Brazil. Sept. 92.

70. Rana, R.S., Saxena, R.K. and Kochhar, S. 1993. Descriptor States in relation to PGR conservation: Fresh Perspective. p. 192-201. In: R.S. Rana *et al.* (eds.) Conservation and Management of Plant Genetic Resources. ICAR-NBPGR, New Delhi.

71. Santha, V.P., Nair, S.G., Lakshmi, K.R. and Sheela, M.N. 1996. Characterization of cassava germplasm by morphological and Isozyme markers. In: *Tropical Root and Tuber Crops in Food Security and Nutrition* (eds.) Balagopalan, C., Nair, T.V.R., Sundaresan, S., Premkumar, T. and Lakshmi, K.R. p. 112-117.

72. Santha, V.P., Thankappan, M. and Misra, R.S. 1993. Breeding for leaf blight resistance in Taro: Present Status and Future Strategies. Proc. National Symposium on Plant Breeding; *Strategies for 2000 A.D. and beyond*. MAU, Aurangabad. ISGPB, Publications.

73. Sardana, S., Kumar, D., Suneja, P. and Kochhar, S. 1998. Variability in Germplasm for some agromorphological traits. *Indian J. Plant Genet. Resour.*, **11(2):** 197-202.

74. Sarkar, N.C., Ghosh, B., Maiti, R., Thorie, M., Rualthankkhuma, C. and Teja, K.C. 2015. Preliminary evaluation of indigenous rice bean landraces under Red Lateritic Belt of West Bengal, India. *International Journal of Bioresources and Stress Management*, **6(1):** 167-169.

75. Schoen, D.J. and Brown, A.H.D. 1993. Conservation of allelic richness in wild crop relatives in aided by assessment of genetic markers. *Proc. Natl. Acad. Sci.*, USA. 90: 10623-10627.

76. Sharma, A., Devi, J. and Katoch, V. (2016). Genetic variability and genotype x environment interaction for seed yield and related traits in French bean germplasm. *Indian J. Plant Genet Resour.*, **29(2):** 156-162.

77. Sheela, M.N., Abraham, K., V.P., Nair and Mohandas, S.G. 2000. *Catalogue on Genetic Resources of Lesser Yam (Dioscorea esculenta* (Lour. Burk). Tech. Bull. No. 29, Central Tuber Crops Research Institute, Thiruvananthapuram (Kerala).

78. Sheela, M.N., Edison, S., Santha, V.P., Amma, C.S.E. Unnikrishnan, M. and Sreekumari, M.T. 2007. Agrobiodiversity Conservation and its Utilization of Tropical Tuber Crops. In: *Agrobiodiversity* (eds). Kannaiyan, S. and Gopalan, A. Published by Associated Publishing Co., New Delhi, p. 186-192.

79. Singh, B. and Kochhar, S. 1994. Descriptors and their descriptor states for evaluation of pulse crops. In: Plant Genetic Resources: Exploration, Evaluation and Maintenance. Rana, R.S., Singh, B., Koppar, M.N., Rai, M., Kochhar, S.and Duhoon, S.S. (eds.), Pub. By ICAR-NBPGR, New Delhi, p. 83-89.

80. Singh, B. and Kochhar, S. 1994. Guidelines for characterization and evaluation of plant genetic resources. In: Plant Genetic Resources: Exploration, Evaluation and Maintenance (Pub. By ICAR-NBPGR, New Delhi). p. 51-59.

81. Singh, B. and Rana, R.S. 1993. Descriptor of Pulse Crops. ICAR-NBPGR, New Delhi, p. 29.

82. Singh, C., Srinivasan, K and Gupta, V. 2016. Physiological manifestation during accelerated ageing in *Phaseolus vulgaris* germplasm. 1st International Agrobiodiversity Congress; (Session: Seed Genebanks Resources) held at New Delhi; 6-9 November 2016; p. 66.

83. Singh, O.V. and Singh, A.K. 2016. Plant genetic resources of arid legumes are the sources of useful genes for biotic stress resistance and high yield. 1st International Agrobiodiversity Congress (Abs. Session: Trait Discovery and Enhanced Use of PGR), 6-9 November, held at New Delhi, p. 197.

84. Sivaraj, N., Pandravada, S.R., Kamala, V. and Babu, B.S. 2016. Characterization and evaluation of jack bean [(*Canavalia ensiformis* L. (DC)], an underutilized wild legume collections from Peninsular India. 1st International Agrobiodiversity Congress, held at New Delhi, 6-9 Nov., 2016.

85. Soengas, P., Cartea, M.E., Lema, M. and Velasco, P. 2009. Effect of regeneration procedures on the genetic integrity of *Brassica oleracea* accessions. *Molecular Breeding*, **23(3):** 389-395.

86. Spagnoletti Zeuli, P.L. 1992. The 'core' collection and the plant breeder. Paper presented in International Workshop on Core Collections of Plant Germplasm. Held at CENARGEN, Brasilia Brazil.

87. Strauss, M.S., Pinoja, J.A. and Cohen, J.I. 1988. Quantification of diversity in *ex situ* plant collections. *Diversity*, **16:** 30-32.

88. Subha, K., Singh, D. and Mukherjee, A. 2016. Morphological and biochemical characterization of potato (*Solanum tuberosum* L.) germplasm in *Terai* region of Uttarakhand. 1st International Agrobiodiversity Congress (Abs. Session: Trait Discovery and Enhances use of PGR), 6-9 November, 2016, held at New Delhi, p. 166.

89. Thankamma Pillai, P.K. and Unnikrishnan, M. 1993. Genetic resources of Taro. Vol. II; Catalogue Series-4. Published by Central Tuber Crops Research Institute, Thiruvananthapuram (Kerala).

90. Tohme, J., Jones, P., Bees, S. and Iwanaga, M. 1992. The CIAT *Phaseolus vulgaris* core collection established by the combined use of evolutionary, characterization and agro-ecological data. Paper presented in International Workshop on Core Collection of Plant Germplasm. Held at CENARGEN, Brasilia, Brazil.

91. Unnikrishnan, M., Edison, S., Sheela, M.N., Abraham, K., Sreeja, T., Sreerekha, V.R. and Asha, V. 2004. Problem and Prospects in conservation of Biodiversity of Tropical tuber crops and allied species. Proc. of *XIV Annual Conference of Indian Association of Angiosperm Taxonomy and National Seminar on Frontiers in Plant Taxonomy and Biodiversity Conservation*, Dec. 29-31, TBGRI and IAAT, Thiruvananthapuram, 2004, Abstract 27(O). p. 44-45.

92. Unnikrishnan, M., Nair, G.G., Pillai, P.K.T., Vasudevan, K., Thankappan, M., Lakshmi, K.R. and Venkateswaralu, T. 1987. 'Sree Rashmi' and 'Sree Pallavi': Two promising varieties of *Colocasia. J. Root Crops.*, **13:** 111-116.

93. Varalakshmi, V., Suchita, Y. and Sanna Manjunath, K.S. 2016. Characterization and evaluation of ridge gourd [*Luffa acutangula* (Roxb.) L.] germplasm. *Indian J. Plant Genet. Resour.*, **29(1):** 66-70.

94. Vaughan, A.D. and Jackson, M.T. 1992. The relationship between the whole collection and its core. Paper presented in International Workshop on Core Collections of Plant Germplasm. Held at CENARGEN, Brasilia Brazil in Sept. 92.

95. Verma, S.K., Pant, K.C. and Muneem, K.C. 1998. Evaluation of chillies germplasm. *Indian J. Plant Genet. Resour.*, **11(2):** 237-239.

96. Yonezawa, K. and Nomura, T. 1992. Sampling strategies for using structured germplasm collections. Paper presented in International Workshop on Core Collections of Plant Germplasm. Held at CENARGEN, Brasilia, Brazil in September 1992.

# 6

# Multiplication, Exchange and Safe Movement of Vegetable Genetic Resources

## Introduction

As is increasingly realized, it is necessary to create the supporting infrastructure and a congenial environment to capitalize on the potential of the new biological technologies (Brenner, 1995). The realization of usefulness of genetic resources and its utilization has created a trade system worldwide. While trade is often regarded as one of the main causes for the loss of biodiversity due to its role in the global movement of pests and diseases. It is also a crucial driver for the preservation of biodiversity as it compels governments and economic operators to implement measures to protect agricultural productive capacity and to consider biodiversity as an economic asset and working capital. Numerous endemic disease agents wreak havoc on such staples as cassava, maize *etc.* and inflict substantial losses on pastures, threatening the livelihoods of vulnerable farmers and severely undermining the calorie intake and the nutritional status of millions in developing countries. It has also been acknowledged that the release of 'Living Modified Organisms' (LMOs) in a contained or open environment could pose risks which would have various direct and indirect impacts. If the benefits of breeding/biotechnology are to be optimized without affecting the environment, effective biosafety regulations must be developed based on sound scientific principles (Persley, *et al.*, 1992; Walsh, 1993; Krattiger and Lesser, 1994). These damages are exacerbated by climate change that increase pest and diseases pressures due alterations in the host-pathogen-vector interaction thereby leading to more frequent

outbreaks and upsurge; and the continuous appearance of new pathogen that are resistant to agrochemicals. No doubt, increased application of pesticide leads to loss of biodiversity due to their devastating impacts on natural enemies, soil fauna and pollinators. These threats are exacerbated by man-caused introductions of invasive alien species that severely undermine the sustainability of farming systems leaving producers helpless towards these unfamiliar invaders against which no traditional management practices work. The WTO Agreement on the application of Sanitary and Phytosanitary (SPS) measures ensures the safe transboundry movement of goods, including foodstuffs. It aims to limit the risk of introduction, through the trade pathway, of pathogen that could otherwise severely damage plant, animal and human health or the environment. Importing plants, animals and their products requires a science-based risk assessment to design adequate risk management measure that guarantee the freedom of imported products from quarantine pests and diseases (Kenza, 2016). Biodiversity mainstreaming is the process of embedding biodiversity consideration into policies, strategies and practices of key public and private actors that impact or rely on biodiversity, so that biodiversity is conserved, and sustained used, both locally and globally. In Nagoya, Japan, in 2010, countries made a commitment to future generations and adopted the strategic plan for Bioversity 2011-2020 and 20 Aichi Biodiversity Targets. The CoP-10 to the CBD has urged parties to develop national regional targets, using the Strategic Plan and its targets. Accordingly, India had revised it National

Biodiversity Action Plan (NBAP) prepared during 2008 by developing National Biodiversity Targets (NBTs), keeping in view of the Aichi Biodiversity Targets as a frame work. The NBT 5 says that, by 2020 measures are adopted for sustainable management of agriculture and NBT 7 advocates that genetic diversity of cultivated plants, farm livestock and their wild relatives including other socio-economically as well as culturally valuable species is maintained, and strategies have been developed and implemented for minimizing genetic erosion and safeguarding their genetic diversity. The concept of "main streaming" biodiversity also include the integration of biodiversity considerations into "cross-sectoral plans, programmes and policies" (Jacob, 2016).

## Multiplication/Seed Increase

The first stage in plant genetic resource activity is seed increase and it is essential to increase sufficient seed in one cycle after keeping part of it as reference sample. The site of initial seed increase should be as close as to the original site to avoid the risk of losing a particular accession due to poor adaption, disease and pest damage. In vegetable crops this aspect is most essential because it consists of large group of crops including annual, biennial as well as requiring cold and dry climate for seed production. In general seed increase followed as in:

(1) **Self-pollinated crops-** One to three rows depending on the quality of seeds and nature of the species should be grown for seed increase.

(2) **Cross pollinated crops-** For cross pollinated crops, sib-mating need to be done to avoid genetic drift during seed increase.

During the multiplication and regeneration of collected or received genetic resources, losses of accessions do occur quite frequently. Sometimes this loss may go up to 50 per cent of the original collections while the genetic fidelity of its sizable underhand is questioned (Singh and Williams, 1984). This fact poses dual problems of frequencies and methods of germplasm collection and regeneration. It is also essential to standardize the strategies and tactics of regeneration because the varying genetic resources encompass quite considerable range of life cycles, mode of reproduction, breeding systems and ploidy levels which interplay in the making up of genetic constitution of an array of populations. Inherently shy seed producers with low multiplication factor/ratio also influence the regeneration strategies.

## Breeding Systems and Regeneration/ Multiplication Methods

It is true that regeneration/multiplication and evaluation activities cannot be considered mutually exclusive of each other as the high genetic integrity achieved through random mating conflict with the expression of recessive alleles which are uncovered,

through selfing/sib-mating, which are otherwise concealed in the heterozygous state. The inherent breeding system is the principal determination in consideration of regeneration procedures. There is preponderance of out-cross among the land races, primitive cultivars and wild relatives of even otherwise inbreeding crop species which show substantial degree of out breeding and heterozygosity. Even low rates of out crossing can generate considerable heterozygosity and heterogeneity therefore it became imperative to the curator for special attention while handling the germplasm materials. Through the finding of several experimentations Burton (1985) suggested that selfing, particularly of those plants that are used to describe an accession, should be exercised to obtain sufficient amount of seed of an accession. The idea behind this was:

(i) Requirement of less labour and time

(ii) Less likely to produce outcross

(iii) Retains maximum alleles responsible for classification and

(iv) Uncovers recessive alleles enabling field screening along with the dominant.

For utilizing the genetic resources for a particular purpose, the gene pool similar and originated within a narrow geographic region should be worked out. The general principals laid out are modified in accordance with the degree of inbreeding/outbreeding of the germplasm within the species in consideration.

## Breeding Systems for Out Breeding Species

The greatest problems are offered by predominantly out-breeding species and hence this group should be considered first in regeneration and multiplication. The significant features of this group of species with regard to the genetic system include:

(1) Outbreeding mechanisms (including self-incompatibility) encouraging random mating.

(2) The loss of heterozygous balance through inbreeding which leads to a decrease in expression of those characters showing directional dominance and hybrid vigour.

(3) The highly heterozygous and heterogeneous nature of the populations rendering them particularly vulnerable to changes in gene and genotype structure through selection.

## Minimizing the Changes of Out Breeding Sample

Vegetable genetic resources collected from natural populations are generally found in unstable equilibrium by the operating forces of natural selection. These forces cannot be emulated in *ex situ* situations. The requirement is, therefore, to minimize selection forces as for as possible. However, in theory, this cannot be achieved by ensuring complete random mating between constituent populations. The principle involved in minimizing changes of the sampled out breeding population can be simply stated as:

(a) Avoiding contamination by foreign pollen and or/mechanical mixtures of seed through proper isolation and seed handling techniques.

(b) Minimizing genetic drift (random loss of alleles) and genetic shift (selective loss of alleles) by ensuring sufficient population size and reducing opportunities for natural selection.

(c) Securing effective random mating through appropriate pollination techniques.

(d) Employing appropriate harvesting methods and post-harvest handling procedures.

All these aspects are interrelated in practice and thus have to be considered as per the requirements of curators handling the large number of accessions. However, their main concern should always be to increase seed as per requirements with population sizes as small as possible (Singh, 1994).

## Principal Hazards and Control Measures

### (1) Mutation

Mutation rates are not of high order ($10^{-5}$ to $10^{-6}$) per generation having no measurable effect. However, it has been shown that mutation rates are very significantly increased as seeds age and/or loss their viability. The minor order mutations causing deterioration can largely be obviated by adopting viability standards of 85 per cent or above.

### (2) Avoiding Pollen Contamination through Isolation

Selection for or against alleles is not allowed to be exercised in accession multiplication. Even little rate of pollen contamination could cause significant genetic changes. Therefore, ensuring adequate isolation from pollen exchange between accessions becomes the single most important step in maintaining genetic integrity during germplasm regeneration. Isolations are obtained by:

**(i)** Spatial or temporal separation

**(ii)** Natural or artificial barriers

**(iii)** Hand crossing accompanied by bagging inflorescences or pair of plants

Keeping isolation distance is very good approach to obtain pure seed but it is difficult to get proper isolation in each crop. Other methods being used to gets pure seeds of accessions are linen bags and purpose built pollen proof glass houses which are quite effective but expensive. Therefore, the crop curators develop their methods for promoting random mating within isolations. However, there is need for developing cost effectiveness of methods for eliminating pollen contamination in maintaining the genetic integrity of the accessions.

### (3) Population Size and Genetic Drift

Of alleles do occur quite frequently consequent upon reduction in population size for various reasons. Random losses can be more powerful force of genetic erosion than natural selection (shift). Initial loss of heterozygosity during generation increase is not of great importance for quite small populations since quite modest decline of 1-2 per cent occur with effective population size in between 25-50. An estimated about 70 per cent of the alleles with frequencies of 0.05 in the source population would still be present after five generations with a constant effective population size of 50. For longer and continued regeneration (5+ generations) minimum population size of 50 or more would be desirable. If population number falls too low in a single generation, then most of the efforts expanded further in raising large number of plants in the proceeding generation are wasted.

### (4) Maintaining Effective Population Size

All plants does not always contribute equal number of gametes to the next generation due to various reasons. In this situation the effective population size (No.) of the population is decreased. Therefore, the curators should try to ensure equal genetic contribution if minimum population sizes are grown or the number of plants for the accession must be increased. Differences in reproductive output can arise from:

(i) Chance

(ii) Uncontrolled variation in the environment

(iii) Genetic differences between plants with regard to mating.

## Pair Crossing Schemes with Maximum Control on Regeneration

A number of options (mating designs or pollination systems) are open ranging from controlled (hand) pair crossing to poly crossing by natural vectors, combined with methods to equalize gamete output (pollen pool) over the cycles of regenerations.

**(a)** Pair cross families kept distinct

**(b)** Mixing equal amount of seed from each pair of crossing

**(c)** Controlled poly cross

   **(i)** Hand crossing random females with mixed pollen

   **(ii)** Maternal lines held distinct as subset populations

   **(iii)** Equal amount of seed from maternal lines maintained in a common seed container.

**(d)** Polycross with natural pollination-equal of seed pooled from each genotype.

**(e)** Bulk harvest from pollinated polycross.

## Avoiding genetic shift due to natural selection

The male and female contribution may vary independently which is subjected to natural selection. Natural selection operates on population in their habitats. Germplasm regeneration under *ex situ* conditions is effect differently by these selection pressures and thereby leading to accelerating the loss of adaptive characteristics. Natural selection operates in two ways:

(i) Through differential survival and

(ii) Through differential fecundity

Both these aspects may result in gene loss or a change of gene frequency resulting in consequent changes in mean performances of the population. Firstly, the efforts should be made to choose environments and cultural practices which reduce the selection pressures. All attempts are made that all plants flower and set seeds (Singh, 1994). However, natural selection can operate at any stage of the life cycle of the plant *i.e.* seed to seed as shown below:

| Viability Selection | : | Fertility Selection |
|---|---|---|
| **Seed** | : | **Seed** |
| Seed sown, seed germinate, seedling emerge, seedling and adult plant interactions with environment *i.e.* climate and soil and disease and insect pests | : | Number of flowers fecundity and number of seeds produced |

## Factors Influencing Genetic Shift

The most pertinent factor concerning shift is that even when the population are grown in apparently suitable areas the differences from the original habitat will still be sufficient to promote differences in the relative seed production capacities of the constituent genotypes. Such genotype environment interaction will also lead to shift of the population. Factors influencing genetic shift and their relation to the regeneration of heterogeneous population are given in Table 6.1.

## Seed Dormancy, Seed Ageing and Seed Germination

Seed dormancy is a particular problem for a number of vegetable species. There can also be considerable variations for dormancy conditions even within accessions. Genetical constitution of the accessions may be altered due to differential ageing and selectively diminished germination. Roos (1984) studied the effect of accelerated ageing on different bean cultivars (genotype) grown in mixtures. It was shown that differential effects of the ageing treatments significantly altered their competitive ability in mixed stands which would bring about genetic shift sufficient to ultimately result in a loss of 75 per cent of the original genotypes after repeated regeneration. However, selection pressure of this kind be minimized by ensuring that percentage viability does not fall too low before the accession is regenerated. This is particularly important for genetically heterogeneous populations (Singh, 1994).

## Pollination, Seed Set and Harvesting

All attempts are made to achieve an effective random mating and securing equal maternal contributions.

### Table 6.1: Regeneration of Heterogeneous Populations: Some Factors Influencing Genetic Shift

| Stage | Factors | Minimize by |
|---|---|---|
| Stored seed | | |
| Germination | Differential genotypic | (i) Regenerate before generation falls to <85 per cent |
| | (i) Longevities | (ii) Artificially break dormancy |
| | (ii) Dormancy | |
| Seedling and vegetative stage | Differential genotypic survival due to: | (i) Regenerate in regions stage as close as possible to that of adaption or under controlled conditions |
| | (i) Interaction with climate and soil factors | (ii) Protected by fungicides, pesticides, *etc.* |
| | (ii) Susceptibility to diseases and pests | (iii) Grown at low densities (*i.e.* spaced plants) |
| | (iii) Competition | |
| Reproductive phase | Differential production of flowers, pollen and seed due to factors listed above | (i) Maximize production from individual genotypes (seed particularly (iii) above) |
| | | (ii) Equalize inflorescences before pollen shed |
| | | (iii) Store equal quantities of seed from maternal parents |
| Harvesting, threshing, drying and packing | Differential maturities and seed shattering | Harvest (bag) heads individually at appropriate stage |
| Storage of high quality seed | Differential maturities may influence storage potential (longevity) | See above |

*Source*: Singh, 1994.

Optimum cultural conditions are also essential for the development of high quality seed which is a pre-requisite of viability and longevity during storage. In heterogeneous populations the problems of harvesting of all plants with uniform ripeness can only be effectively met by harvesting individual plants separately. Variation to seed shedding in wild and weedy relatives of crop plants may pose special problems and lead to selective shift. A possible solution is to raise sufficient plants to permit the selection of a large enough number that flower simultaneously.

The effect of genetic drift and shift may be counterbalanced by saving equal number of seed for long term storage and regeneration and bulk the remainder as a multiplication sample for distribution, evaluation and immediate use.

## Standard Procedures of Multiplication and Regeneration

The relevant theoretical principles and major practical issues while assessing the merits of current multiplication and regeneration procedures or devising new ones for other crops or species. Some technical matters relating to necessary isolation distances and methods, population size and inter-pollination techniques, cultural conditions, natural selection *etc.* are very essential. Some methods of regeneration have been recommended for a number of crop species which differ in their breeding systems, floral morphology, pollination mechanisms as well as growing requirements and multiplication rates. The standards vary between crops and even within crops. Hand pollination has proved prohibitively different and costly. Some of the standard procedures for vegetable crops are as under shown in Table 6.2.

Special problems are attached in in avoiding unconscious selection during regeneration in wild population and species *e.g.* high variability in flowering time and seed production, seed shattering and seed dormancy. In the case of perennials, variation in seed production is negatively correlated with the perennially and selective shift to short lived populations.

## Exchange of Vegetable Genetic Resources

Germplasm enhancement is an essential link between collection/evaluation of germplasm and its utilization to benefit agriculture but in safe mode. Historically, invaders introduced several vegetable crops like tomato, potato, cauliflower and cabbage, as they constituted their staple food. It is also true that exotic resources are a vital input in crop improvement programme as they are the valuable sources of genes for higher yield, desirable quality traits and resistance or tolerance to various biotic and abiotic stresses. National climate change adoption policies on broadening genetic diversity for resilient cropping systems and the prevention of biodiversity loss are also spurring germplasm exchanges. Landraces and wild

### Table 6.2: Study of Standard Procedures for Vegetable Crops

| Sl.No. | Crops/species | Population Size | Isolation | Inter-pollination | Harvesting |
|---|---|---|---|---|---|
| 1. | (a) *Brassica* species (Self-incompatible: up to 100 per cent) | 62 | Hand pollination | Plants randomly paired and bi-parentally mated (bud pollination to avoid in compatibility system) | Total 110 seeds from each of 31 mating pairs which are kept distinct as sub-lines |
| | (b) *Brassica* species (Self-incompatible: up to 100 per cent) | 507 | Insect cage | Honey bees | Bulk |
| 2. | (a) Maize (Monoecious >80 per cent) | 200 | Bag ear and tassels | Bulk sib-mated (mixed pollen from all tassels shedding pollen) by hand | 100 good sibbed ears-bulked |
| | | 20-100 | Bag ear and tassels | Bulk sib-mated (mixed pollen from all tassels shedding pollen) by hand | Equal number of seed from each ear (100) |
| 3. | Cucurbits [Monoecious: High] | 24 | Male and female flowers tied off | Hand pollination | Bulk |
| 4. | Beet [Some self-incompatible: high] | 100 | Isolation tents | Natural | incompatible: |
| 5. | *Cucumis* [Diclinous: high] | 24 | Insect cages | Honey bees (nucleus) | Bulk |
| 6. | Tomato species (Pollination 4-80 per cent) | 20-80 | Spatial barrier crops barrier | Insect pollination (honey bees) | Equal maternal contribution |
| | (a) Tomato species [Autogamous] Nil or Low | Enough for desired seed quantity | Nil | Automatic selfing | Bulk |
| | (b) Tomato species (Facultative incompatible and self-incompatible) [Moderate to high] | As many as possible | Isolate in green house | Hand cross with pollen pool | Bulk |

*Source*: Singh, 1994.

germplasm are mainly exchanged by the genebanks; improved varieties and research material (training sets, international evaluation trials, pre-breeding populations, *etc.*) are mainly exchanged by the breeding programmes, and both the private and public sectors are involved in the distribution of plant propagative materials for commercial and non-commercial uses. As per the historical facts, the spread of such knowledge created interest among plant breeding in the 20th century. Interest also developed in the exchange of elite materials bred by plant breeders. All these movement had good and ill effects. This movement made it possible for the nations of the world to have a rich variety of plants for food and fodder as also plants which led to the establishment of flourishing plant based industries. Exchange of genetic material is, therefore, imperative for a country to have adequate safeguards while introducing planting material and other agricultural products. Plant quarantine provides such safe-guards.

## International Exchange of Genetic Resources

International exchange of genetic resources as botanic seeds, plants or plant parts suitable for propagation has played a crucial role in the growth of international and national genebanks; development of improved high-yielding nutrient-rich crop varieties resilient to pests, diseases, and abiotic stresses; international trade in planting materials; and even international research collaborations. Now exchanges of vegetable genetic resources are quite common among the countries and regions. Studies on the origin and use of food crops indicated that 68.7 per cent of national food supplies are derived from crops with a foreign origin and emphasize the importance of the inter- and intra-continental exchange of germplasm (Khoury *et al.,* 2016). The frequency of international exchange of germplasm is on rise due to the growing demands of global food systems, international research partnerships, and trade linkages. The exchange of plant material on a worldwide basis had been mostly carried out without well-defined procedures, but several countries have established plant introduction organizations for channelizing import and export (exchange of plant materials). India recognized the importance of International cooperation for plant germplasm exchange a long time ago. Some of these earlier vegetable crops introductions like 'Sioux' in tomato, 'Early Grano' in onion, 'Bonneville' and 'Arkel' in table peas from USA and 'Asahi Yamato' in watermelon from Japan were widely accepted by Indian farmers.

## Mechanisms for Germplasm Exchange

Various mechanisms are in place to facilitate germplasm exchange between countries. Exchange includes both import and export of genetic resources and supply to users within the country. Phytosanitary cleanliness is the key criterion for a decision on the export or import of germplasm. The details of exchange are given below as per statutory framework:

The ICAR-NBPGR, New Delhi has brought out brochure 'Guidelines for the Exchange of Seed/Planting Material' which has been widely circulated amongst scientists in India.

### (a) Import of Germplasm

☆ To import plant germplasm including transgenics/genetically modified organisms (GMOs)/living modified organisms (LMOs) for research, experimental, and/or breeding purpose by public and private sector institutions in India, it is mandatory to fulfill the following two requirements, as per Section 6(1) and (2) of the PQ Order, 2003 (www.plantquarantineindia.nic.in):

☆ Obtain an 'Import Permit' (IP) from the Director, ICAR-National Bureau of Plant Genetic Resources, Pusa Campus, New Delhi-110012, India, before import of any material.

☆ Ensure that a Phytosanitary Certificate (PC) is issued from the country supplying the material.

All indent for importing germplasm from abroad are to be made to the ICAR-NBPGR in prescribed application proforma giving specific details of the required material stating the sources/country as well as address of the organization, so that the 'Import Permit' is issued and sent to concerned scientists for sending the same to the ICAR-NBPGR, New Delhi. Concerned scientist/organizations abroad are advised to take into consideration the following requirements for mailing the materials to India:

Only healthy, viable and clean seed material (free from soil, pests, pathogens and weeds) is to be forwarded without any seed treatment so as to facilitate proper quarantine examination. It may, however, be fumigated, if considered necessary.

The material is required to be accompanied with the

(i) 'Import permit' (which is to be sent to them along with our request letter) and

(ii) Phytosanitary Certificate with additional declaration, if any, based on crop inspection certifying that the material is free from particular pathogen(s)/insects (s).

It is to be further ensured that the package of seed/planting material must be addressed to the Director, ICAR-NBPGR New Delhi who has to take delivery of the seed/planting material and conduct quarantine examination.

☆ Import of large consignments of seeds of coarse cereals, pulses, oil seeds and fodder seeds and seeds/stocks material of fruit plant species for propagation for commercial purpose is only permitted based on the recommendations of EXIP Committee of DAC. However, Director, ICAR-National Bureau of Plant Genetic Resources, Pusa Campus, New Delhi-110012, can issue import permit (IP) to procure specified quantities of seeds of the above-mentioned crops for trial purpose or for conservation in the genebank, as per quantities given in Schedule XII of the Plant Quarantine (PQ) Order 2003 [clause 3(4)].

☆ Private organizations/seed companies requesting for IP must be certified R&D organizations. They should provide a copy of valid R&D recognition certified issued from Department of Scientific and Industrial Research (DSIR), Ministry of Science and technology, Government of India.

☆ Certain seeds/planting material from specified countries are prohibited for import due to incidence of pests of quarantine significance to India (Schedule IV of the PQ Order, 2003)

Schedule V- list of plant materials restricted for import, permissible only with the recommendations of authorized institutions with additional declarations and special conditions [under Clause 3(3)(6)(7) and 10 and 11(3)of PQ Order 2003]-here only vegetable crops are being given:

**Table 6.3: Authorized Centre for Receiving Augmented Crop Species**

| Sl.No. | Plant Species/Variety | Authorized to Import |
|--------|----------------------|---------------------|
| 1. | Banana, plantain and Abaca (*Musa* spp.) | Director, ICAR-National Research Centre for Banana, Thongmalai Main Road, Thyanur Post, Tiruchirapalli (Tamil Nadu) |
| 2. | Cassava or tapioca (*Manihot esculenta*) | Director, ICAR-Central Tuber Crops Research Institute, Sreekariyam, Thiruvnanathapuram (Kerala) |
| 3. | Potato (*Solanum tuberosum*) and other tuber-bearing species of Solanaceae | Director, ICAR-Central Potato Research Institute, Shimla (Himachal Pradesh) |
| 4. | Sweet potato (*Ipomoea* spp.) | Director, ICAR-Central Tuber Crops Research Institute, Sreekariyam, Thiruvnanathapuram (Kerala) |
| 5. | Yam (*Dioscorea* spp.) | Director, ICAR-Central Tuber Crops Research Institute, Sreekariyam, Thiruvnanathapuram (Kerala) |

☆ The request for IP from ICAR-National Bureau of Plant Genetic Resources, Pusa Campus, New Delhi-110012, for certain seeds/planting material from specified countries as per Schedule V of the PQ Order 2003, should be submitted along with recommendation of respective authorized institutions. Such germplasm should be procured with a PC with additional declarations.

☆ Every consignment of import must be accompanied by an original copy of the phytosanitary certificate (PC) with additional declaration about freedom from specific pests as specified in the IP or that the pests specified do not occur in the country or state of origin as supported by documentary evidence, thereof. Specific pests for the crops are mentioned in Schedule V and VI of PQ Order 2003.

☆ The IP issuance for germplasm of transgenics/GMOs/LMOs by ICAR-NBPGR, New Delhi is subject to prior approval of the Review Committee on Genetic Manipulation (RCGM) of the Department of Biotechnology (DBT), Ministry of Science and Technology, Government of India (http://dbtindia.nic.in). The RCGM examines the proposal for import of transgenics/GMOs/LMOs for biosafety issues, under the provisions of EPA, 1986. In accordance with this act, all transgenics are regulated items. The RCGM issues a permit letter (valid for one year after date of issue) for authorization to import GMOs/LMOs and their products for R&D purpose with following instructions for use (conditions are subject to revision from time to time).

☆ No transgenics/GMOs/LMOs are allowed for experimentation for commercial production/manufacturing of the product without prior authorization from the Competent Authority.

☆ ICAR-National Bureau of Plant Genetic Resources, Pusa Campus, New Delhi-110012, shall retain up to 5 per cent of the transgenics seeds/planting material in its facility (National Genebank) and keep them under safe custody. ICAR-NBPGR, New Delhi shall further issue a recipt to that effect to the applicant. The seeds retained by the ICAR-NBPGR, shall be used by the Government of India for future reference, if required.

☆ All rDNA materials are to be destroyed and disposed off in accordance with the Recombinant DNA Safety Guidelines 1990, of the DBT, Government of India, after conclusion of experiments.

☆ All experiments carried out should be documented.

The details of wild species introduced (imported) during 2005-2010 by ICAR-NBPGR are depicted in Figure 6.1.

## Issuance of Import Permit (IP)

☆ Any person desirous of importing germplasm into India for research, breeding and/or conservation, should apply to the Director, ICAR-NBPGR, New Delhi, on a prescribed application form (PQ08), for issuance of IP. The application form can be downloaded from ICAR-NBPGR website (www.nbpgr.ernet.in).

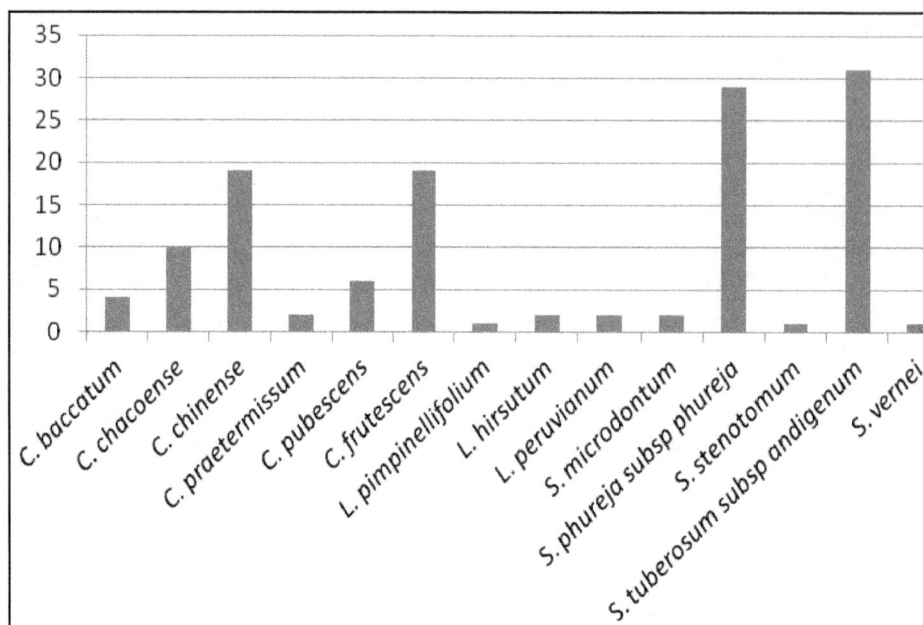

**Figure 6.1**

# PQ Form 08

**ICAR-National Bureau of Plant Genetic Resources**

**Indian Council of Agricultural Research**

**New Delhi-110012, India**

*Application for Permit to Import Germplasm/Transgenics/Genetically Modified Organisms (GMOs)*

*(For Research Purpose)*

*The Director*

*ICAR-National Bureau of Plant Genetic Resources*

*Pusa Campus New Delhi-110012, India*

I hereby apply for a permit in accordance with provisions of clause 6(2) of the Plant Quarantine (Regulation of Import into India) Order, 2003 issued under the Sub-section (1)of section (3) of the Destructive Insects and Pests Act, 1914 (2 of 1914), authorizing the import of plants/planting material for research purpose as per details given below:

1. Name and address of the applicant
2. Research and Development (R&D) status/affiliation of the organization [Please attach relevant documents]
3. Exact description of Seeds/Planting Material to be imported:
    a. Common and botanical name
    b. Germplasm/variety/hybrid/composite/synthetic/provenance/clone/others
    c. Form of material required (seed/rooted plants/scions/tubers/cuttings/bulbs *in vitro* cultures)
    d. Parentage, if known
4. Place of collection/origin of material to be imported (country/state)
5. Whether transgenic/GMO or not?
    [If yes, attach the approval letter issued by RCGM/(DBT) in original]
6. Name and address of the organization/institution producing the material.
7. Number of samples to be imported.
8. Quantity to be imported
    (Separately for each accession/variety/hybrid/transgenic/GMO)
9. Suggested source of availability of material including published reference, if known.

    (a) Whether the aforesaid germplasm/variety/hybrid was imported by you earlier? If so, details thereof (year, quantity, source, *etc.*)

    (b) Was the material shared with other scientists/National Genebank at ICAR-NBPGR?

10. Expected date and arrival in India

11. Mode of shipment (Airmail/Air freight/accompanied baggage)

12. Place where imported seeds/planting material will be grown and scientists under whose supervision the seeds/planting material will be grown.

## Declaration

1. I hereby declare that the germplasm under import has no commercial value/exclusive ownership and may be shared freely for research purpose.

2. The germplasm does not contain any terminator genes.

3. I undertake that the material is exclusively for research purpose.

Place _____

Date _____

*Signature of the Applicant and Address*

(**Source:** *ICAR-NBPGR 'Guidelines for Management of Plant Genetic Resources in India, 2016*)

☆ Along with the application form PQ 08, the applicable processing fee for issuance of an IP should be sent, in the form of a demand draft, in favour of Director ICAR-National Bureau of Plant Genetic Resources Pusa Campus New Delhi-110012, India. the fee is non-refundable and subject to revision

### Fee and Service Charge

(**Source:** *ICAR-NBPGR 'Guidelines for Management of Plant Genetic Resources in India, 2016*)

  i. Fee for Issuance of Import Permit

| Import Permit | Type of Import Permit | Fee Amount (Rs.) | Service Charges @ 14* | Total Fee for Each Permit (Rs.) |
|---|---|---|---|---|
| Public | Non-transgenic | 150.00 | 21.00 | 171.00 |
| | Transgenic | 250.00 | 35.00 | 285.00 |
| Revalidation of existing Import permit | Transgenic | 25.00 | - | 25.00 |
| | Non-transgenic | 50.00 | - | 50.00 |
| Private | Non-transgenic | 300.00 | 42.00 | 342.00 |
| | Transgenic | 550.00 | 77.00 | 627.00 |
| Revalidation of existing Import permit | Non-transgenic | 50.00 | - | 50.00 |
| | Transgenic | 100.00 | - | 100.00 |

*Service and other charges as per prevailing guidelines of Government of India

  ii. Handling and Cargo Clearance Charges for Consignments Received at the New Delhi Airport

| Public Sector | Private Sector |
|---|---|
| Rs. 1500 per consignment | Rs. 3000 per consignment |

  iii. Quarantine Processing fee*

| Item | Public Sector | | Private Sector | |
|---|---|---|---|---|
| | Non-transgenic | Transgenic | Non-transgenic | Transgenic |
| Seeds (per sample) | Rs. 100.00 | Rs. 500.00 | Rs. 200.00 | Rs. 1000.00 |

**Service Charges @ 14 will be applicable**

*The fee above will not be applicable to nurseries of CGIAR Institutions

iv. Qurantine Fee for Vegetative Propagules (VP)/Tissue Culture Tubes (TC)

| Item | Public Sector | | Private Sector | |
|---|---|---|---|---|
| | Non-transgenic | Transgenic | Non-transgenic | Transgenic |
| One sample up to 10VP/10TC tubes | Rs. 100.00 | Rs. 250.00 | Rs. 200.00 | Rs. 500.00 |

*Fee applicable is subject to change with the approval of ICAR, and service charges as per the Government of India notification from time to time

(***Source***: *ICAR-NBPGR 'Guidelines for Management of Plant Genetic Resources in India, 2016*)

☆ ICAR-National Bureau of Plant Genetic Resources Pusa Campus New Delhi shall issue the IP in form of PQ 09 in triplicate, with a yellow/green tag for germplasm and a red/white tag for transgenics/GMOs/LMOs, after validating all information provided in the application.

## ICAR-National Bureau of Plant Genetic Resources
### Indian Council of Agricultural Research
### New Delhi-110012, India

*Permit for Import of Germplasm/Transgenics/Genetically modified organisms for Research Purpose*

Permit No._____     Date of issue_____

Valid up to_____

In accordance with the provision of clause 6(2)of the Plant Quarantine (Regulation of Import into India) Order 2003 issued under Sub-section (1) of section 3 of the Destructive Insects and Pests Act, 1914, I hereby grant permission to import of germplasm/transgenic/genetically modified organisms herein specified

1. Name and address of importer

2. Name and address of exporter

3. Country of origin                                                      Point of Entry

4. Description of Germplasm/Transgenics/Genetically modified organism

5. Variety to be imported

6. Quantity (Weight/Nos.)

7. No. of Packages

8. Mode of Packing

9. The above permission is granted subjected to follow conditions:

    (1) The consignment of germplasm/transgenic shall be free from soil, weed species and plant debris

    (2) (i) The consignment shall be accompanied by a Phytosanitary Certificate/Phytosanitary Certificate (re-export issued by an authorized officer in the country of origin/country of re-export) as the case may be with additional declaration for the freedom from:

    (a)

    (b)

    or that the above specified pests do not occur in the country or state of origin.

    (ii) Certified that the germplasm/transgenic as described above obtained from mother crop/stock which were inspected on regular intervals by an appropriate authority in the country of origin and found free from:

(3) The consignment shall be grown in an approved post-entry quarantine facility established by the importer at_____(name of location of PEQ facility) under the supervision of_____for a period of (days/months)_____(Name and address of Inspection Authority)

(4) The permit is not transferable and valid for one-time import. The permit number shall be quoted on the Phytosanitary Certificate issued at the country of origin or re-export as the case may be.

Place: New Delhi   Seal      Name:_____

Date:_____         Signature:_____

                   Director

                   ICAR-National Bureau of Plant Genetic Resources Pusa Campus New Delhi

*Source*: ICAR-NBPGR 'Guidelines for Management of Plant Genetic Resources in India, 2016

☆ The prescribed quantity of seeds and the planting material permitted for imports is given below:

# Quantity of Germplasm

The maximum permissible limit of sample size of plant germplasm for import and export for research purpose

(*Source*: ICAR-NBPGR 'Guidelines for Management of Plant Genetic Resources in India, 2016)

Small quantity of seed/planting material can be imported for research purpose. The quantity of seed or planting material just sufficient for the establishment of plant is considered optimum and safe.

| **(i) Seed Quantity** | | |
|---|---|---|
| (a) | Large seeded crop species *viz.*, -*Zea mays*, *Helianthus* spp., *Phaseolus* spp., *Arachis* spp., *Dolichos* spp., *Mucuna* spp., *Pisum sativum*, *Cicer* spp., *Vicia* spp., *Canavalia* spp., *Cyamopsis* spp. and others. | 1.00 kg |
| (b) | Small seeded crop species *viz.*, *Allium* spp., *Brassica* spp., *Capsicum* spp., *Lycopersicon* spp., *Solanum melongena* and others. | 200 g |
| (c) | All other plant species | 500 g |
| **(ii) Vegetative Propagules** | | |
| (a) | Number of rooted cuttings/plants/sample | 25 No. (max.) |
| (b) | Number of other vegetative propagules/samples | 50 No. (max.) |

☆ For specific crops the requests for Import Permit need to be routed through Directors of respective ICAR crop-based institutes for non-transgenic as well as for transgenic crops (as per schedule V, Clause 3(3) and 10 and 11 (3) of the PQ Order 2003). **As mentioned above.**

☆ The Import Permit is non-transferable and has a validity of six months from the date of issue and is valid for successive shipment provided the exporter and importer, bill of entry, country of origin and PC are the same for the entire consignment.

☆ Validity of Import Permit may be extended up to one year on request, before the expiry of permit and payment of prescribed fee, if suitable reasons are provided in writing and are justified.

## Shipment and Consignment

(*Source*: ICAR-NBPGR 'Guidelines for Management of Plant Genetic Resources in India, 2016)

☆ Two copies of Import Permit should be sent by the indentor/applicant to the concerned official/supplier (abroad) who has agreed to supply the required germplasm for use in research.

☆ The supplier should be instructed to send back both the copies of Import Permit with the consignment (one pasted outside the parcel and other kept inside), along with other relevant documents, such as phytosanitary certificate (PC).

☆ Instructions should be given to the supplier that seeds/planting material meant for export to India should not be treated with any chemical.

☆ All the parcels should be addressed to The Director, ICAR-National Bureau of Plant Genetic Resources Pusa Campus New Delhi-110012 (India). The port of entry of imported plant germplasm (including transgenic) is New Delhi Airport, only.

☆ Original Import Permit (in duplicate) issued from ICAR-NBPGR, phytosanitary certificate (PC) from source country, proforma invoice and a packing list are mandatory requirements for custom clearance of the consignment.

### Import of Germplasm under ITPGRFA

(ICAR-NBPGR 'Guidelines for Management of Plant Genetic Resources in India, 2016)

☆ For importing PGRFA under MLS of the ITPGRFA, any natural or legal pertson can make a request to any Contracting Party.

☆ Besides, Import Permit issued by Director, ICAR-National Bureau of Plant Genetic Resources Pusa Campus New Delhi-110012, India as per the PQ Order, 2003 and phytosanitary certificate (PC) from the exporting country, an Standard Material Transfer Agreement (SMTA) (Article of the ITPGRFA), signed by both the supplier and recipient of the germplasm, is also required prior to exchange of material.

☆ The material, received from MLS of the ITPGRFA under the conditions of the SMTA, can be transferred (in the form received) to any third party, after entering into a fresh SMTA with the third party (as per Article 12.4 of the ITPGRFA).

## (b) Export of Germplasm

Exchange of germplasm involves not only introductions but also the supply of seed and other materials to collaborating scientist/organizations abroad. Export of vegetable germplasm is also made on the basis of request received by ICAR-NBPGR/ICAR Institutes/Agricultural Universities in India under various protocols/work plans/memorandum of understanding with different countries/CGIAR Institutions. ICAR-NBPGR, New Delhi is exporting a large number of vegetable germplasm to a number of countries.

### Statutory Framework

Despatch of the seed/planting material is also to be channelised through the Bureau so that prompt inspection of the material could be done from quarantine angle and phytosanitary certificate be issued. No seed dressing with insecticides or fungicides be given while dispatching the seed to the Bureau.

(*Source:* ICAR-NBPGR 'Guidelines for Management of Plant Genetic Resources in India, 2016)

☆ Access to all biological resources occurring in or obtained from India, is governed by the Biological Diversity Act (BDA) 2002, and the Biological Diversity Rules, 2004, notified by the Ministry of Environment, Forest and Climate Change (MoEF and CC), Government of India.

☆ As per the provisions of Section 3 of the BDA, all non-Indians (persons or institutes) defined in Section 3(2) of BDA need prior approval of National Biodiversity Authority (NBA) for access to any biological resources occurring in India or its associated knowledge for research, commercial utilization, bio-survey and/or bio-utilization.

☆ Section 3(2) defines non-Indians as:

(a) A person who is not a citizen of India

(b) A citizen of India, who is a non-resident (as defined in clause (30) of section 2 of the Income Tax Act, 1961).

(c) A body corporate, association or organization:

(i) Not incorporated or registered in India; or

(ii) Incorporated or registered in India under any law for the time being in force which has any non-Indian participation in its share capital or management.

☆ All such persons or organizations as defined in section 3(2) are required to seek prior approval of NBA for access to biological resources, in a prescribed form and with payment of applicable fee as displayed in the website of National Biodiversity Authority (NBA) (www.nbaindia.org).

☆ As per section 5 of the Biological Diversity Act (BDA) 2002, exchange of germplasm for collaborative research projects under the bilateral/multilateral agreements, which conform to the Central Government Guidelines made in this regard and approved by the concerned Ministry/Department of the Central/State Governments are exempted from seeking permission of NBA as required in Section 3. This includes government sponsored institutions of India and such institutions in other countries, engaged in collaborative research projects.

### Guidelines for International Collaboration Research Projects Involving Transfer or Exchange of Biological Resources or Information Relating thereto between Institutions, Government-Sponsored Institutions and such Institutions in other Countries

(*Source:* ICAR-NBPGR 'Guidelines for Management of Plant Genetic Resources in India, 2016)

These guidelines came into effect from 8th November, 2006 [Clause (a) of sub-section (3) of section 5 of the Biological Diversity Act (BDA) 2002 (18 of 2003)]

In view of the fact that collaborative research projects have been exempted from obtaining approval of the NBA and that the need for transfer and exchange of biological resources cannot be ruled out in such projects, sponsored under the bilateral and multilateral agreement, Memorandum of Understanding and work plant *etc.* under the International Collaboration Research Projects, the collaborative research project shall clearly state in the proposal:

1. (a) The key investigator(s)in each of the collaborating institution, who shall be responsible for all compliances, and in case of any contravention, this person will be held responsible. Changes in the identity of the key investigator should be intimated to the concerned Department/Ministry of the Central Government.

   (b) Details of biological resources occurring in India and knowledge associated thereto, intended to be exchanged or transferred under the project, *viz.*, biological name, quantity, purpose, source, place of collection and such other activities.

   (c) Value addition, if any to the biological resource and associated knowledge.

   (d) In case the biological resource referred to in (b) above has any special status under any law in force in India or any international agreement, the detail of same may be provided, including necessary clearances from competent authority.

2. The collaborators shall abide by the provisions of existing national laws, regulatory mechanisms and international agreements of treaties.

3. The biological resource(s) and associated knowledge intended to be exchanged or transferred under such projects shall be used only for the research purpose specified in clause (b) of sub paragraph (1) above.

4. The quantity of biological resource(s) to be transferred or exchanged shall be limited to the quantity necessary for experimental purpose, as specified in the proposal and as per the access and material transfer guidelines developed by National Biodiversity Authority.

5. In case the results of research from this project subsequently lead to any Intellectual Property Rights, the collaborating partners shall enter into a fresh agreement with National Biodiversity Authority (established under section 8 of the Act) to ensure sharing of benefit in accordance with provisions of section 6 of the Act, prior to filing of the application for Intellectual Property Rights(s).

6. The voucher specimen of the biological resource occurring in India, transferred or exchanged under the project, shall be sent to the designated repository in accordance with section 39 of the Act.

7. In case the collaborative research projects involve exchange and transfer of dead or preserved specime(s) and/or herbarium(s) of India on loan or on any other terms, for taxonomic studies as required by bonafide scientists/professors of recognized universities and government institute of India who are engaged in pure classical taxonomic studies, this shall be done with the approval of concerned Department/ministries of the Government of India.

8. Collaborators shall not communicate or transfer research results of the collaborative project to any third party in any manner without entering into an agreement with the National Biodiversity Authority for this purpose.

9. Publication of Research [paper(s), book(s), bulletin(s), registered accession(s) and ouput(s)] based on the results of the research of such projects, shall not be done without the prior approval of the Indian collaborator.

10. During the course of the implementation of the project, any knowledge associated with exchanged or transferred biological resource from India, shall be reported to National Biodiversity Authority for facilitating documentation of such knowledge.

11. Any publication(s) relating to knowledge associated with biological resource exchanged and/or transferred from India under the collaborative project, shall acknowledge the knowledge holder.

12. Any new taxon, breed(s), genetic stock(s), culture(s), strain(s), or line(s) discovered or developed through the project shall be reported to the National Biodiversity Authority and a voucher specimen shall be deposited with designed repository in accordance with the Biological Diversity Act, 2002.

13. The collaborative research project shall have to be approved by the concerned Ministry/Department of the State or the Central Government.

14. A copy of the approval along with all relevant details shall be sent to the National Biodiversity Authority.

15. For more details about Biological Diversity Act, 2002, National Biodiversity Authority and related issues, log on to www.nbindia.org

    ☆ Export of seeds/vegetative parts of plants, normally traded as commodities, are exempted from seeking permission from NBA (as per Section 40 of BDA 2002.

    ☆ India signed and ratified the ITPGRFA in June 2002. The DAC and FW, MoA and FW, is the Nodal Department for the implementation of the ITPGRFA and the Joint Secretary (Seeds) is the National

Focal Point (NFP) (as per Article 12.3c of the Treaty). The ICAR-NBPGR, New Delhi is to serve as the single-window system for exchange of germplasm under the ITPGRFA.

☆ Export of PGRFA of crops of the ITPGRFA to member countries of ITPGRFA shall be carried out under a Standard Material Transfer Agreement (SMTA) adopted by the Governing Body (GB) of the ITPGRFA, and as per the notified guidelines of Government of India (Gazette notification of Annex I crops. NBA shall be kept informed of the granting of such exports approved by ICAR-NBPGR/DAC).

A. Food crops (35)-Include vegetables like Bread fruit (*Artocarpus*), Asparagus (*Asparagus*), Beet (*Beta*), Brassica (Genera included are: Brassica, *Armoracia, Barbarea, Camelina, Crambe, Diplotaxis, Eruca, Isatis, Raphanobrassica, Raphanus, Rorippa,* and *Sinapis*).This comprises oilseed and vegetable crops such as cabbage, rapeseed, mustard, cress, rocket, radish and turnip. The species *Lepidium meyenii* (*maca*) is excluded. Carrot, yams, sweet potat, cassava, beans, pea, potato, eggplant, faba bean, cowpea.

B. Forage crops (29)

C. Grass Forages

D. Other Forages

**Table 6.5: Categories of Germplasm Export**

| Category | Remarks |
|---|---|
| Category 1 | Export of germplasm not covered under any collaborative research project with research institutes/counter parts, public-private transfer, private entities (Indian Citizen/Non-Indian), as per Section 3(2) of the BDA, 2002. |
| Category 2 | Export of germplasm under collaborative research projects/work plans, under Section 5 of the BDA, 2002. |
| Category 3 | Export of Annex I crops under ITPGRFA and FAO designed accessions of CG Centres located in India. |
| Category 4 | Indian research/Government institution to carry/send germplasm for non-commercial research/or emergency purpose other than collaborative research. |

## Export of Germplasm under Category 1 (Not covered under collaborative research project)

For export of germplasm not covered under any collaborative research project, applicants are required to seek written approval from the NBA, in the prescribed format (Form 1 under rule 14 of the Biological Diversity Rules, 2004) along with the prescribed fee (http://nbaindia.org/).

☆ After written approval from NBA, applicant should seek an IP from the appropriate authority, in the country of germplasm export.

☆ After obtaining the IP, Plant Quarantine scientists should inspect the germplasm meant for export and issue a PC to the applicant.

☆ The consignment for export should be sent through suitable postal service along with IP, PC and other relevant documents, either by the supplier of germplasm or through ICAR-NBPGR.

## Export of Germplasm under Category 2 (Collaborative research project)

☆ Prior to export of germplasm under collaborative research projects, following documents need to be assembled by the indentor.

☆ A request letter from foreign institution submitted to ICAR/research institution/researcher in India for access to germplasm.

☆ A copy of the Collaborative research agreement, bilateral/multilateral agreement, Memorandum of Understanding, work plans, *etc.* approved by the concerned Ministry/Department of Central/State Governments.

☆ Details of the seed/planting material for export along with the list of material.

☆ IP of the recipient country

☆ Signed copy of ICAR/DARE approved Material Transfer Agreement (MTA) or Standard Material Transfer Agreement (SMTA), as responsible.

☆ Export of germplasm under Category 2 shall be as per the extant relevant national legislations-the provisions contained in the BDA 2002, the Biological Diversity Rules, 2014, the MoEF and CC Notification S.01911(E) dated November 8, 2006 and GSR 8267 dated November 21, 2014 and any further guidelines notified by the Central Government in this regard.

☆ After approval of DARE, the consignment for export should be submitted to Director, ICAR-NBPGR, New Delhi where it should be registered and all details (crop name, no. of package and the receipt country) should be documented.

☆ The seed/planting material for export should be then submitted to Division of Plant Quarantine, ICAR-NBPGR, New Delhi for quarantine inspection as per the IP instruction of the importing country, for issuance of PC.

☆ The seed/planting material along with IP and PC should be dispatched to the receiving country either by the indentor ot by ICAR-NBPGR, New Delhi.

## Export of Germplasm under Category 3 (Crop as mentioned above under ITPGRFA and FAO/Designated Accessions of CG Centres located in India)

☆ Export of germplasm under Category 3 shall be as per the MoEF and CC Notification 5.03232(E), 17-12-2014 and DAC O.M. No. 13-5/2013-SD-V dated 16-2-2015.

☆ The DAC and FW, Government of India is the Nodal Department and Joint Secretary (Seeds) is the Nodal Focal Point (NFP) for access to PGRFA of designated accessions in accordance with the provisions of the ITPGRFA to which India is a party, and relevant extant national legislations.

☆ All requests for access to PGRFA listed in Annex 1 of the ITPGRFA (as per article 11 of ITPGRFA) should be addressed to NFP, either directly or through ICAR-NBPGR, New Delhi. Access to PGRFA requested shall be granted only with the approval of NFP.

☆ On receipt of request for export, the Germplasm Export Facilitation committee (GEFC) constituted for the purpose shall examine on case-to-case basis and submit its recommendations to NFP.

☆ After the approval NFP, ICAR-NBPGR, New Delhi shall coordinate the supply of PGRFA as per established norms.

☆ The access shal be provided to the natural and legal persons from the Contracting Parties (country signatory to ITPGRFA) only.

☆ Facilitated access shall be provided only for the PGRFA, designated by Government of India under the MLs of the ITPGRFA.

☆ In addition to the above-mentioned PGRFA, the germplasm of mandate crops available with International Agricultural Research Centres (IARCs) collected before 1993 (also known as 'FAO designated' accessions or 'in trust'

material), shall also be facilitated for access (as per article 15 of ITPGRFA).

☆ Facilitated access to PGRFA as notified by Government of India under the MLs of the Treaty should be provided only for the purpose of utilization and conservation for research, breeding and training for food and agriculture, provided that such purpose does not include chemical, pharmaceutical and/or other non-food/feed industrial uses (as per article 12.3a of ITPGRFA).

☆ All requests for other purposes including non-food/feed use of PGRFA shall be in accordance with the provisions of extant relevant national legislations.

☆ Access by other countries, which are not Contracting Parties, will be in accordance with the extant relevant national legislations, on a bilateral basis.

☆ Voucher sample along with passport data of all PGRFA access/supplied, shall be conserved and documented by ICAR-NBPGR, New Delhi.

☆ All germplasm shall be exported to the requesting country only under the provisions of SMTA adopted by Governing Body of the ITPGRFA. Specific conditions may be added in SMTA in case of any 'material under development' as per provisions of the ITPGRFA.

☆ The SMTA shall be signed prior to transfer of PGRFA only as hard copies and the Director, ICAR-NBPGR, New Delhi shall be the authorized signatory.

☆ For FAO/designated material being exported by the IARCs, the SMTA shall be signed by the authorized signatory of the provider IARC.

☆ For export of PGRFA, the IP from the importing country shall be required. The transfer of material remains subject to normal phytosantary inspection and clearance.

## Export of Germplasm for Non-Commercial Research/or Research for Emergency Purposes Other than Collaborative Research

☆ Any Indian researcher/Government institution who intends to carry/send the biological resources outside India to undertake basic research other than collaborative research as referred to, in section 5 of the BDA, 2002, shall apply to the NBA in Form'B'.

### Form'B'

(See regulation 13 of MoEF and CC Notification C S R No. 827 dated 21st November, 2014)

Conducting of Non-Commercial Research or Research for Emergency Purpose outside India by Indian Researchers/Government Institution using the Biological Resources

1. Name of the applicant (Indian Researchers/Government Institution)

2. Complete address*
   a. Permanent
   b. Present

3. Name and address of Institution in India

4. Name of the Supervisor or Head of Institution at the place of work in India
5. Name and contact details of the Institution or organization who shall guide the proposed research/receiving the Biological Resources
6. Detail of the Supervisor or Head of Institution or organization who guides the proposed research or recipient of the Biological Resources
7. Name of the funding agency supporting the proposed research
8. Brief description of the research
9. Detail of Biological Resources proposed to be carried along or sent for the research
    i. Name of the Biological Resources (scientific/common name)
    ii. Location of collection (Village/Taluka/Distt./State)
    iii. Quantity required
    iv. Duration of the research
10. If it is for emergency purpose, specify details

*Attach self-attested Address/ID proof such as Aadhar Card/PAN Card/Passport, *etc.*

## Undertaking

I, _____Son/Daughter/Wife/Husband of_____aged_____
residing at_____in_____holding a permanent. I.D. No._____
(PAN Card/Aadhar Card/Passport, *etc.*) hereby declare that all the information provided above is correct and true. I hereby affirm that the Biological Resources shall be used only for the purpose as stated in the application. I shall not share/provide/part/leave behind any Biological Resource at my collaborator's facility/laboratory without approval of the NBA. I, along with my supervisor and collaborator, individually and severally declare that we shall not put to commercial utilization, nor shall seek any IPR claim on the Biological Resources and associated traditional knowledge used in this research collaboration. In case such a situation arises we shall apply to National Biodiversity Authority to seek prior approval. Results, process(es), products or other outcomes arising out of this activity shall be shared with the NBA during and/or upon completion of research intended along with the copy of relevant documents and publications.

Date:_____

Place:_____                                                                 Signature

## Declaration by the Supervisor/Head of Institution

I, _____working as_____in (Name of institution) confirm that details provided by Mr./Dr/Mrs/Ms_____are true and correct

Date:_____

Place:_____                                                                 Signature

                                                                                              Designation

                                                                                              Official Seal

## Declaration by the Recipient/Collaborator

I, _____working as_____in (Name of institution/Organization) hereby affirm that I or my institution/organization shall use the Biological Resources for the purpose as stated in the application and which were sent by_____(Name of institution) or being brought by Mr./Dr/Mrs/Ms_____. The said Biological Resources shall be destroyed in full after the completion of the studies/partnership or upon completion of the studies, the Biological Resources shall be sent back to the institution from where the Biological Resources were received as the case may be. I or the institution I am associated with shall not claim any ownership under instant application nor shall claim any IP Rights over the Biological Resources, derivatives or other such components without prior approval of the applicant, institution affiliated and the National Biodiversity Authority.

                                                                                              Signature

                                                                                              Designation

                                                                                              Official Seal

☆ Any government institution which intends to send biological resources to carry out certain urgent studies to avert emergencies like epidemics, *etc.*, shall apply in 'Form B'.

☆ The NBA shall, on being satisfied with the application, accord its approval within a period of 45 days from the date of receipt of the application.

☆ On receipt of approval of the NBA, the applicant shall deposit voucher specimens in the designated national repositories before carrying/sending the biological resources outside India and a copy of proof of such deposits shall be endorsed to NBA.

The export of vegetable tuber crops involving 180 accessions was highest in 1992-93. Important germplasm of *Abelmoschus esculentum, Amaranthus caudatus, A. cepa, Allium sativum, Brassica oleracea* var. *botrytis, B. oleracea* var. *capitata, Brassica chuk, Brassica rapa, Capsicum annuum, C. frutescence, Cymposis tetragonoloba, Daucus carota, Lablab perpureus, Solanum lycopersicum, Phaseolus vulgaris, Pisum sativum, Raphanus sativum, Solanum incanum, S. integrifolia, S. melongena, S. torvum,* and *Spinacea oleracea* and some cucurbits like *Cucumis melo* (1), *Cucurbita pepo* (4), *Cucumis sativus* (1), *Luffa cylindrica* (1) to Italy; *Momordica charantia* (1) to Taiwan; *Sechium edule* (2) and *Trichosanthes dioica* (1) were exported during 1986-1994 to different countries. During the year 2001-2002 few cucurbitaceous germplasm like cucumber (1), bitter gourd (1) ridge gourd (1) and sponge gourd (1) were again exported under phyto-sanitary certificate to Iraq.

#### Table 6.6: Export of Vegetable Germplasm

| Crop | Number of Samples | Country |
| --- | --- | --- |
| *Cucumis melo* | 6 | DPR Korea |
| *Citrullus lanatus* | 1 | DPR Korea |
| *Cucumis sativus* | 1 | DPR Korea |
| *Momordica charantia* | 1 | DPR Korea |
| *Citrullus lanatus* | 1 | Bangladesh |
| *Cucumis sativus* | 1 | Iraq |
| *Lagenaria siceraria* | 1 | Bangladesh, Iraq, Taiwan |

Since 1986, the bureau has exported vegetable and tuber germplasm to 53 and 16 countries, respectively and their details are given in Table 6.7.

### Procedure for Plant Introduction/Exchange

Introduction or exchange of germplasm consists of the following steps:

(1) Procurement

(2) Quarantine

(3) Cataloguing

#### Table 6.7: Countries Involved in Exports of Vegetable Crops Germplasm

| Crops | Countries |
| --- | --- |
| Vegetable crops | Afghanistan, Angola, Argentina, Bangladesh, Brazil, Bulgaria, Canada, China, Coasta Rica, Czechosolovkia, DPR Korea, Egypt, Ethiopia, Fiji, France, Greece, Hungary, Indonesia, Iran, Iraqw, Italy, Japan, Kuwait, Liberia, Malaysia, Mexico, Mongolia, The Netherlands, Nepal, New Guinea, Nicaragua, The Philipines, Poland, Republic of Yeman, Saudi Arab, Senegal, Singapore, Soloman, Sri Lanka, Surinam, Taiwan, Tanzania, Thiland, Tunesia, Turkey, U.K., USA, USSR, Vietnam, Yeman, Zimbabwe, West Germany, West Indies. |
| Tuber vegetable crops | Australia, Bangladesh, Belgium, Bulgaria, Egypt, France, Mongolia, Nepal, Pakistan, Peru, The Philipines, Sri Lanka, Surinam, and U.K., USA. |

(4) Evaluation

(5) Multiplication and

(6) Distribution

### Movement of Genetic Resources within Country

The germplasm is supplied by ICAR-NBPGR, New Delhi from its active germplasm collections at Headquarters and Regional Stations or through the National Active Germplasm Site (NAGS) on different crops located all over the country. This activity is performed by Germplasm Exchange Division, ICAR-NBPGR, New Delhi.

### Why Safe Movement of Vegetable Germplasm is Essential?/Risk in Germplasm Exchange

Centre of origin and diversity of plants are also the centre of origin of diversity of pests/pathogens. Today, introduction of germplasm of superior crop varieties is an essential activity to improve agricultural productivity for resistance against plant or pathogens, early maturity, high yield, and quality traits. At that time effects of introductions/unregulated germplasm exchange on natural resources and environment were not considered but today it is concrete fact that several obnoxious weeds, insects, nematodes and pathogens (exotic to the country) are consequences of plant introductions; which sometimes may proves harmful to Indian agriculture. These pests have shown more prolific and aggressive behavior in introduced area than the home environment. With the result they have gained economic significance at new location. It is difficult to predict how exactly an introduced pest if established in new location/environment would behave. However, there is a distinct possibility that freed from its natural enemies and competitors and with possibly more susceptible hosts available, it may have more serious effect than in its place of origin. Exchange of germplasm carries simultaneous risk of moving harmful pests and

### Table 6.8: Movement of Vegetable Genetic Resources for Research Purpose

| Crop | Sample | Supplied to | Received from |
|---|---|---|---|
| Ash gourd | 9 | TNAU, Tamil Nadu | Germplasm Exchange Division, ICAR-NBPGR, New Delhi |
| Bitter gourd | 103 | JNKVV, Jabalpur; CHES, Bhubaneswar, KAU, Kerala; APHU, Hyderabad; Kolkata University; KRC Horticulture College Arabhavi | ICAR-IARI, New Delhi, ICAR-IIHR, Bengaluru; ICAR-NBPGR, RS Thrissur |
| Brinjal | 1497 | ICAR-IARI, New Delhi; YSPUHF Solan; KAU, Kerala; APHU, Tadepalligudem; Jindal Crop Sciences Pvt. Ltd., Jalna; PDKV Akola; MPUAT, Udaipur; GBPUA&T, Pantnagar, Dr. PDKV Horticulture College and Res. Inst., Akola; Horticulture Res. Station, Hugeri, Bihar; TNAU, Coimbatore; GKVK, Bengaluru | ICAR-IARI, New Delhi, ICAR-IIHR, Bengaluru, ICAR-NBPGR, New Delhi, ICAR-NBPGR, RS Hyderabad, ICAR-NBPGR, RS Thrissur |
| Bottle gourd | 364 | ICAR-IIHR, Bengaluru; JNKV, Jabalpur; ICAR-IARI, New Delhi; PAJANCOA, Karaikal, Pondicherry; PDKV, Akola; Krishi College, Vellayani; APHU, Hyderabad; KRC Horticulture College, Arabhavi | ICAR-IARI, New Delhi, ICAR-IIHR, Bengaluru, ICAR-NBPGR, New Delhi, ICAR-NBPGR, RS Jodhpur, ICAR-NBPGR, RS Thrissur and ICAR-NBPGR, RS Hyderabad |
| Chilli | 212 | ICAR-IARI, New Delhi; Jindal Crop Sciences Pvt. Ltd Jalna; APHU, HRS, Guntur; PAJANCOA, Karaikal, Pondicherry; Ag. College and Res. Inst., Madurai, ICAR-CIAH Bikaner; Vibha Agrotech; Marathwada Krishi Vishwa Vidyalaya, Rahuri; KRC Horticulture College, Arabhavi; GKVK, Bengaluru | ICAR-IIVR, Varanasi, ICAR-NBPGR, RS Hyderabad, ICAR-NBPGR, RS Bhowali |
| Okra | 711 | JNKV, Jabalpur; Agrl. College, WB; Jindal Crop Sciences Pvt. Ltd. Jalna, ICAR-IIHR, Bengaluru; ITC, Hyderabad; APHU, Hyderabad; IARI, New Delhi; GBPUA&T, Uttarakhand; Bioseed Research India Pvt. Ltd., Hyderabad; PDKV, Akola; Biology School, Hyderabad; Agriculture College Guntur; Annamalai University, TN; Horticulture College, Hyderabad,; JNU, Delhi; IARI, New Delhi, Neelkant Govt. Sanskrit University, Pattambi | ICAR-IARI, New Delhi, IIHR, Bengaluru, ICAR-NBPGR, New Delhi, ICAR-NBPGR, RS Akola, ICAR-NBPGR, RS Thrissur |
| Pumpkin | 2 | JNKV, Jabalpur | ICAR-IARI, New Delhi, ICAR-IIHR, Bengaluru |
| Ridge gourd | 139 | ICAR-IIHR, Bengaluru, Swami Keshwanand RAU, Bikaner, Kar. Hort. Univ., Karnataka | ICAR-NBPGR, New Delhi |
| Roundmelon | 1 | NRC DNA F, ICAR-NBPGR | CIAH, Bikaner |
| Sponge gourd | 96 | ICAR-IARI, New Delhi | ICAR-NBPGR, New Delhi |
| Tomato | 349 | JNKVV, Jabalpur (M.P.); IARI, New Delhi; Karnataka Hort. Univ., Karnataka; NDUAT, Faizabad (U.P.); Inst. Agril Sci. BHU, Varanasi (U.P.), JNU, New Delhi; Krishi Station, Bihar; Horticulture college, Bihar; GKVK Bengaluru; JNKV, Jabalpur (M.P.) | ICAR-IARI, New Delhi, ICAR-IIHR, Bengaluru, ICAR-NBPGR, New Delhi, ICAR-IIVR, Varanasi |
| Tumba | 1 | ICAR-NRC DNA F, ICAR-NBPGR, New Delhi | ICAR-CIAH, Bikaner |
| Watermelon | 70 | ICAR-NRC DNAF, ICAR-NBPGR; Dr. PDKV Akola, KAU, Vellanikkara (Kerala) | ICAR-IIVR, CIAH, ICAR-NBPGR, RS Jodhpur |
| Onion | 8 | Kolkata University, Kolkata (W.B.) | ICAR-NBPGR, RS Bhowali |
| Cucumber | 94 | CSKHPKV, Palampur; DRDL, Pithoragarh, ICAR-NRC DNA F | ICAR-NBPGR, RS Bhowali, ICAR-NBPGR, RS Jodhpur |
| *Trichosanthes cucumerina* var. *anguina* | 51 | IARI, New Delhi | ICAR-NBPGR, New Delhi |

pathogens between geographies and introducing them in new territories where they were not known to exist (*e.g.*, intercontinental spread of *Banana bunchy top virus*). High risk of biotic agents spreading through plant and plant-propagation materials was recognized in the 19th century as a consequence of damaging outbreaks caused by pests and pathogens introduced with germplasm (Klinkowski, 1970). Irish famine due to the epidemic of potato late blight (*Phytophthora infestans*) in the 1845s introduced was a notable example of catastrophic damage by an introduced pathogen from Central-South America. Similarly powdery mildew (*Uncinula nacator*), root eating aphid (*Phylloxera vitifolia*) and downy mildew (*Plasmopara viticola*) of grapes into France in quick succession in mid-19th century from America; coffee rust

into Sri Lanka in 1875 and its subsequent introduction into India in 1876 are prominent examples that clearly demonstrate that introduction and establishment of quarantine pests into new areas can severely damage the crop production and economy of a region/country. Sensing similar risks, quarantine measure were enacted in 1873 for the first time in Germany and the UK as a legal measure for the inspection of potato for Colorado potato tuber beetle imported from the USA. Quarantine measures were subsequently adapted by several countries and eventually by the mid-twentieth century they had been established as a global norm.

In India, laxity in application of proper quarantine measures has resulted in the introduction of several serious exotic pathogens and pests along with imported planting material causing serious losses from time to time. Introduction of certain plant, India restricts, more often prohibits, the export of propagules of black pepper among horticultural crops. There are glaring example of losses caused by the insect-pests and diseases introduced with import of planting materials like seed, propagules, fruits, plant *etc.* As per the historical records the introduction of exotic noxious weeds such as *Lantana camara*, in 1809 (in the early 19th century) from Central America, *Parthenium hysterophorus* from Central and South America and *Phalaris minor* from Mexico in mid-20th century into India have become endemic sources of threat to our crop production and environment. Of the several introduced exotic insect pests the spiraling whitefly, *Aleurodicus disperses* is quite recent, which has established successfully in Karnataka and has been found infesting 253 plant species. Not only these, black rot of crucifers (*Xanthomonas campestris*), smoke disease of banana (*Pseudomonas solanacearum*), bunchy top of banana introduced from Sri Lanka in 1943; potato wart (*Synchytrium endobioticum*) and golden nematode (*Globodera rostochiensis* and *Globodera pallida*) infesting potatoes introduced in 1960s from UK has also established in several parts of the country particularly entire Nilgiri hills. Recently, the introduction of tomato pin worm *Tuta absoluta* in 2014, Jackbeardsley mealy bug (*Pseudococus jackbeardsley*) in 2012, Papaya mealy bug (*Paracoccus marginatus*) in 2007, fluted scale on citrus introduced from Sri Lanka in 1928; San Jose scale in apple in 1930s; These examples clearly indicate that there is continuing threat from exotic pests and pathogen in germplasm exchange and highlights the need for adoption of comprehensive and effective quarantine measures.

## Issues in International Exchange of PGR

### Agencies Responsible for Genetic Resource Exchange

Different nations organized separate plant exchange agencies with adequate facilities for proper inspection and handling of plant materials under exchange. These countries also formulate the quarantine laws, to keep out the risk of the introduction and spread of weeds, pests and diseases with the creation of such facilities, scientists interested in obtaining seed/plant material from abroad for their research were requested to channelize their introductions through such agencies. Exporting agencies were also advised to forward only healthy materials duly accompanied with an official phytosanitary certificate.

### ICAR-NBPGR, New Delhi (India)

Germplasm Exchange Division, ICAR-NBPGR, New Delhi is mainly concerned with the introduction of genetic resources and also look after the reciprocal plant introduction services *i.e.*, export of Indian genetic stocks to scientists working abroad. Scientists of the organizations have studied the working system of the three major and well organized plant introduction agencies of the world located in USA, Australia and Russia. Based on the knowledge and experience, ICAR-NBPGR, New Delhi has established its own working systems which are suitable to our political and economic conditions. The ICAR-NBPGR, New Delhi is recognized by the Government of India as the official agency for the international exchange of agri-horticultural plants and allowed essential relaxations of trade an import control, to facilitate its smooth working.

### International Institute of Tropical Agriculture (IITA)

International Institute of Tropical Agriculture (IITA) in Nigeria is one of the CGIAR centres with an international genebank conserving germplasm of its six mandate crops: banana and plantain (*Musa* spp.), cassava, cowpea and wild *Vigna* species, maize, soybean, and yam. Since its establishment in 1967 the centre has collected and conserved over 33,000 accessions of landraces and wild relatives representing global collections (cowpea and yam) as well as regional resources (cassava, maize, *Musa* and soybean). The centre also has a vibrant international breeding programme developing improved varieties of its six mandate crops. Both the genebank and breeding programmes frequently exchange germplasm for research and other uses. Generally landraces and wild relatives are exchanged from the genebank; breeding programmes often exchanged parental lines, pre-breeding populations, inter-specific hybrids, and inbred lines, and share international multilocational trials; seed system programmes acquire pest-free material for the bulk propagation of planting materials; germplasm or plant products are exchanged for research use.

## Process of Exchange of Plant Genetic Resources

Process of exchange of plant genetic resources includes:

(i) Literature survey

(ii) Assemblage of germplasm

(iii) Administrative requirements of international exchange

(iv) Phytosanitary screening and certification, and

(v) Transfer of materials to the users

### (i) Literature Survey

Only literature provides sufficient information about the germplasm and their specific traits. Therefore, literature survey for desired information on new genetic stocks of agri-horticultural material, as also for other useful plants is regularly carried out with a view to spot useful cultigens and record their outstanding features and sources of availability. Latest issues of publication including periodicals, newsletters, seed list and catalogues *etc.* are screened/searched and desired information documented, country wise lists prepared and requests for procurement of seed/plant materials forwarded to the likely sources.

### (ii) Assemblage of Germplasm

Availability of a large base collection with wide variability among accessions is of utmost importance for any crop improvement programme. This assemblage of desirable plant genetic resources is mainly done in the following three ways:

(a) Collection through exploration within the country

(b) Collection through organized cooperative explorations in other countries particularly in the areas noted as centres of diversity and

(c) Procurement of materials of interest through correspondence with cooperating plant introduction agencies, individual scientists, growers, private seed companies *etc.*, in different countries.

### (iii) Administrative Requirements of International Exchange

Administrative requirements connected with the international movement of plant materials are strictly observed by maintaining links with Indian missions abroad, Ministries of External Affairs, Finance, Commerce, Agriculture *etc.* of the Government of India, customs and postal authorities, Reserve Bank of India, and various surface transport agencies and airlines.

### (iv) Phytosanitary Screening and Certification

It is mandatory that all incoming and outgoing material must be subjected to requisite phytosanitary screening and certification as per international regulations Close links are maintained between the Germplasm Exchange Division and Quarantine Division of ICAR-NBPGR, New Delhi. The ICAR-NBPGR, New Delhi also maintain a close liaison with the Directorate of Plant Protection, Quarantine and Storage, Government of India, Faridabad which is responsible for the examination and release of the seed/plant consignment at specified points of entry into India.

### (iv) Transfer of Materials to the Users

All assembled healthy plant materials are regularly transmitted to various scientific agencies in the country. ICAR-NBPGR, New Delhi also handle sizeable quantity of materials for International Crop Research Institute for Semi-Arid Tropics (ICRISAT), located at Hyderabad (Telangana). Seed materials are also being transferred for conducting international trials/nurseries in major crop plants in cooperation with international institute *e.g.*, World Vegetable Centre (old name-AVRDC), Taiwan, CIP (Peru), Bioversity International, NVRS (UK), USVL (USA), VRS (USA), NIVTS (Japan), IVF (China), SAARC Agriculture Centre, CIAT (Colombia) *etc.*

## Guidelines and Procedures for the Exchange of Seed/Planting Materials

The ICAR-NBPGR has brought out a brochure 'Guidelines for the exchange of seed/planting materials' which has been widely circulated amongst scientist in India. The guidelines for import of seed/planting material for research purpose have been revised (Anon, 1989). The government has made the issuance of import permit mandatory for import of all seed/planting material for research purposes.

The following guidelines are to be observed while responding to such requests:

a) Request for seed/planting materials received from concerned organizations agencies are to be forwarded to the ICAR-NBPGR, New Delhi with relevant information so that prompt action on the supply of desired material could be taken.

b) Dispatch of the seed/planting material is to be channelized through the ICAR-NBPGR, New Delhi so that prompt inspection of the material could be done from quarantine angle and phytosanitary certificate be issued.

c) No seed dressing with insecticides or fungicides is given while dispatching the seed to the bureau.

The salient features of the procedures to be followed both for import as well as export of germplasm are enumerated below:

(A) Import of plant genetic resources from abroad

(B) Export of plant genetic resources

(C) Inland supplies of plant genetic resources

(D) Documentation and dissemination of information on assembled plant genetic resources

### (A) Import of Plant Genetic Resources from Abroad

Research institutes of public and private sectors are interested in importing plants or planting material from various countries.

Procurement of plant genetic resources is generally of three categories:

(i) Latest released varieties, genetic stocks and known wild relatives as reported in literature,

(ii) Request of various Indian scientists for specific requirements for their experiments

(iii) Materials received in response to direct requests of other scientists or for international trials/nurseries to be conducted in India

The enactment of new "Seed Development Policy" by Government of India, which has been circulated widely among plant breeders, scientists, researchers *etc.*, has made it obligatory to fulfill the two requirements for getting the plants, fruits and seed (regulation of import into India) from abroad:

(i) Import permit before import of any material,

(ii) Phytosanitary certificate from the country of origin

As mentioned above both the documents must be accompanied with every seed/planting material consignment that is to be imported from abroad. Concerned organization should request to The Director, ICAR-NBPGR New Delhi who has been authorized to issue Import Permit (IP) and receive imported seed/planting material for its quarantine inspection and clearance in respect of ICAR institutes/centres, state agricultural universities and ICRISAT, Hyderabad. Import Permit is not transferable. The scientists desirous of importing seed/planting material has to apply to the Director, ICAR-NBPGR, New Delhi at least 15 days in advance on a prescribed application form and sign a declaration from so that import permit (IP) could be issued and sent to him/her well in time. After obtaining import permit, the indenter should send it to the scientist abroad/organization that have agreed to supply the seed/planting material. He/she may further be requested to send the above import permit along with seed/planting materials and also a phytosanitray certificate that is to be issued by the authorized agency of the exporting country is another statuary requirement. It proves that the consignment has been examined according to the requirements of the importing country and found to be free from the quarantine pests. The package of seed/planting material must be addressed to the Director, ICAR-NBPGR, New Delhi, who has to take delivery of the consignment and conduct quarantine examinations. In case material is found to be infected/infested with pests, all efforts are made to salvage the material. Only in rare cases, when the material cannot be salvaged it is incinerated. In case of post-entry quarantine (PEQ) examination of the imported material is required, it is done at PEQ greenhouse facilities, at ICAR-NBPGR New Delhi, its Regional Station, Hyderabad and ICRISAT and also at the indentor's PEQ growing facility.

## Requirement for Mailing the Materials

The concerned organization/scientist abroad, need to be advised further to take into consideration the following requirements for mailing the material to the Director, ICAR-NBPGR, New Delhi:

(i) Only healthy viable and clean seeds (free from soil, weeds, pests and pathogens *etc.*) are to be forwarded without any seed treatment. The Quarantine Division prefers to receive seed material that has not been coated/treated heavily with chemicals so as to facilitate proper laboratory inspection of imported samples.

(ii) The seed material may be sent by first class airmail, while perishable propagules (scion woods, bud woods, rhizomes, suckers or rooted plants) may

preferably be sent by air freight through any commercial airline operating between source country and Indira Gandhi International (IGI) airport, New Delhi, so as to avoid delay in receipt and clearance. An intimation regarding the dispatch of such perishable material to the Director, ICAR-NBPGR, New Delhi, through telex, email or fax, will facilitate prompt receipt/clearance of material soon after its arrival. This will also help to avoid payment to demurrage charges.

(iii) Full particulars of seed/planting material as well as the address of concerned scientist in India, to whom the material are to be made available after its quarantine clearance by ICAR-NBPGR, New Delhi.

(iv) The germplasm should be obtained in small quantities not exceeding 3,000 to 4,000 per accession, while in case of plant propagules, *viz.*, scion woods, bud woods, rhizomes, suckers *etc.*, should be as minimum as possible not exceeding four in each. In case of rooted plants, it should not be more than two in each case.

## Recording of Information of Introduced Materials

On arrival of seed/planting materials, it should be arranged taxonomically as genus, species, common name and cultivar name for accessioning in the national record (Exotic Register). Each accession/introduction is assigned with EC (Exotic collection) number which remains unchanged. Other information (Passport Data) like name and address of suppliers (donors), characteristics (if any) of the germplasm, relevant literature/references, date of arrival of seed/planting material consignment, condition of the materials on arrival and name and address of user scientists are also recorded. Indigenous collection (IC) also required similar information.

### Table 6.9: Countries Involved in Imports of Vegetable Crops Germplasm

| Sl.No. | Group of Vegetables | Countries Involved |
|---|---|---|
| 1. | Vegetable crops | Australia, Bangladesh, Bulgaria, Canada, China, Colombia, Costa Rica, Cuba, Denmark, Egypt, France, Germany, Hungary, Israel, Italy, Japan, Korea, Nepal, Netherlands, New Zealand, Philippines, Russia, Singapore, South Africa, Spain, Sri Lanka, Sudan, Syria, Taiwan, Tanzania, Thailand, Turkey, United Kingdom, USSR, Venezuela |
| 2. | Vegetable tuber crops | Argentina, Australia, Bangladesh, Bhutan, Brazil, Canada, Colombia, Egypt, France, Hungary, Japan, Nepal, Netherlands, New Zealand, Peru, Philippines, Poland, Republic of Yemen, Singapore, Sri Lanka, Sweden, Switzerland, Taiwan, United Kingdom, USSR, Vietnam. |

## Quarantine Processing of Imported PGR

In case of PGR including transgenics, the sample size is generally small and the entire sample is examined for presence of pests and transgenics are also tested for ensuring the absence of terminator gene technology (embryogenesis deactivator gene) which is mandatory requirement. During quarantine processing, seeds and planting materials are examined for the presence of unwanted weed seeds, plant debris, soil clods, insect and mite pests, plant parasitic nematodes and pathogens including fungi, bacteria, viruses, *etc*. The external feeders and other incidental insect pest infesting the planting materials are easily detected visually either by the naked eye or with the help of magnifying glass or stereoscopic binocular microscope. Presence of nematode is indicated during visual examination by the observation of galls or swellings on roots, tubers and rhizomes; white, yellow or brown pinhead sized round bodies adhering to roots; swollen or malformed leaf, stem or other tissues or root lesions or unusual root proliferation. Fungal infection is indicated by the presence of sclerotia, smut balls, malformed seeds and fungal fructifications on seed surface. Presence of yellow discolouration around the hilum is suggested of bacterial infection. The discoloured, deformed and shriveled seeds are removed during dry seed examination as these seeds may carry seed transmitted pathogens. The seeds of quarantine weeds for India also need to be detected in imports. Specialized tests used for detection of different groups of pests and all efforts are made to salvage the infected/infested material before release.

## (B) Export of Plant Genetic Resources

The exchange of germplasm involves not only the import (procurement) of plant genetic resources but also the supply of seed/planting materials collaborating organization/scientists abroad (export). The export of Indian stocks fall under two categories, *viz*.

(i) Request for Indian genetic stocks received by the CAR-NBPGR, New Delhi and

(ii) Request received directly by various scientists and scientific institutions in the country.

These are processed in Germplasm Exchange Division (GEX), CAR-NBPGR, New Delhi. The quarantine regulation for export of seed/planting materials are based on the "International Plant Protection Convention (1951)" and are modified from time to time according to the specific requests to supply of seed/planting material abroad, scientists are advised to take care of following points:

(1) The request for seed/planting materials received from organization/agencies abroad should be forwarded to the Director, ICAR-NBPGR New Delhi along with import permit (if received) as well as relevant information, so that decision could be taken regarding its supply.

(2) No seed/planting material should be sent to the Director, ICAR-NBPGR New Delhi unless asked for.

(3) The dispatch of seed/planting material is to be channelized through the ICAR-NBPGR New Delhi, so that required inspection of the material could be done from quarantine angle and phytosanitary certificate issued.

There are several vegetable crops which were exported to different countries.

**Table 6.10: Vegetable Species**

| Sl.No. | Group of Vegetables | Name of Crops |
|---|---|---|
| 1. | Vegetable crops | *Abelmoschus esculentum, Amaranth caudatus, Brassica oleracea* var. *botrytis, Brassica oleracea* var. *capitata, Brassica chuk, Brassica rapa, Capsicum annuum, Capsicum frutescence, Cymposis tetragonoloba, Daucus carota, Lablab perpureus, Solanum lycopersicum, Phaseolus vulgaris, Pisum sativum, Raphanus sativum, Solanum incanum, Solanum integrifolia, Solanum melongena, Solanum torvum,* and *Spinacea oleracea* and some cucurbits like *Cucumis melo* (1), *Cucurbita pepo* (4), *Cucumis sativus* (1), *Luffa cylindrica* (1) to Italy; *Momordica charantia* (1) to Taiwan; *Sechium edule* (2) and *Triochosanthes dioica* (1) |

During the year 2001-2002 few cucurbitaceous germplasm like cucumber (1), bitter gourd (1), Ridge gourd (1) and sponge gourd (1) were again exported under phytosanitary certificate to Iraq.

## Collection of Materials for Export

The request for the supply of seed/planting materials received in ICAR-NBPGR, are entered in register and serially numbered in the Germplasm Exchange Division, ICAR-NBPGR, New Delhi. Requests are then forwarded to various known Indian sources for sparing small samples of the desired materials. Germplasm Exchange Division maintains up to date inventories of indigenous genetic stocks and their sources and remains in touch with these sources till such time the material are received and phytosanitary certificate forwarded to the requesting agencies.

## Preparation of Materials for Export and Mode of Transmission

As per the need of desired foreign indenter, materials are assembled and entered in the Export Quarantine (EQ) register and sent to the Plant Quarantine Division along with EQ form for clearance/examinations. Division certifies export materials in conformity with the quarantine regulations of the importing country. The materials are then suitably packed and forwarded to the indenter along with the phytosanitary certificate and import permits (Tag) if received. The seeds are usually forwarded by the air mail while perishable plant materials are dispatched through commercial airlines. In case of requests received directly by other scientists/ICAR institutes/universities *etc.*, they are advised to forward the materials to this Bureau along with a copy of their original record of all exports is maintained in a dispatch register known as foreign supply record (FS record).

## (C) National/Inland (Domestic) Supply of Plant Genetic Resources

India is a large country and has a large and complex network of agri-horticultural research stations, resulting in a large number of requests from Indian scientists for material already available in the country. If the desired materials are available in the Bureau, these are collected and forwarded to the indenters, or else the requests are forwarded to various sources in India, for direct compliance.

☆ Requests for samples of plant germplasm available with/maintained by ICAR-NBPGR/NAGS should be sent to the Director, ICAR-NBPGR, Pusa Campus, New Delhi-110012, in the prescribed form GEX 01 along with a signed MTA by the indentor of germplasm. The GEX 01 and MTA can also be downloaded from ICAR-NBPGR website (www.nbpgr.ernet.in).

## GEX 01

*Requisition for Supply of Seed/Planting Material for Research from/through ICAR-NBPGR, Pusa Campus, New Delhi-110012*

*The Director*

*ICAR-National Bureau Plant Genetic Resources*

*Pusa Campus, New Delhi-110012*

*Ph:011-25843697, 25841129*

*Fax: 011-258442495*

1. Details of seed/planting material required for research

| Sl.No. | Botanical Name | Crop Name | No. of Accessions (IC/EC) | Seed quantity (per accession/ per sample) | Purpose (Screening/breeding/evaluation/augmentation/multiplication) |
|--------|----------------|-----------|---------------------------|-------------------------------------------|---------------------------------------------------------------------|
|        |                |           |                           |                                           |                                                                     |

2. Thesis/Project title for which request is made (name of the funding agency

3. Objective/Activity for utilization of indented material: (Please attach sheet if required): ————————————

4. Material Transfer Agreement (enclosed): Yes/No

5. Feedback report submitted on germplasm received earlier (if applicable): Yes/No

6. Have you or your Institute developed any variety based on germplasm supplied by ICAR-NBPGR? Yes/No (If yes, please let us know the details)

7. If required by ICAR-NBPGR, will you be able to send viable multiplied seed material of the above seed in sufficient quantity for conservation: Yes/No

8. Signature of the indentor:

   Name:_____ Designation

   Address of the Institute_____ Phone (with STD code_____
   M_____ Fax_____ E-mail)

9. Signature of the Competent Authority with Seal:_____

   (PI/Head of the Department/Director of the Institute)

✩ Acknowledgement of receipt of germplasm from ICAR-NBPGR should be provided by the indentor immediately after receipt of material.

✩ Indantor should retain the unique identity no. assigned by ICAR-NBPGR (IC/EC/ET) while using the germplasm.

✩ Feedback information (in the prescribed format) on the performance or utilization of material should be sent to the Director, ICAR-NBPGR, Pusa Campus, New Delhi-110012.

✩ Whenever requested, sufficient quantity of multiplied seed must be sent back to Director, ICAR-NBPGR at the earliest.

**Material Transfer Agreement**

**Indian Council of Agricultural Research**

**Krishi Bhawan, New Delhi – 110001. INDIA**

Agreed between

**National Bureau of Plant Genetic Resources (NBPGR), New Delhi-110012**

of the Indian Council of Agricultural Research, Krishi Bhawan, New Delhi – 110001, the apex agricultural research organization of India, being the first Party (Provider of the Material) and

1. Being the Second Party (Recipient of the Material)
2. For the Supply/Exchange/Transfer of Genetic Resources for Food and Agriculture/Germplasm/Genetic Material/Genetic Components for Research[2]

   ❏ Within India, not covering persons as described in Section 3(2) of the Biological Diversity Act, 2002 (18 of 2003) (BDA).

   ❏ Within India, wholly or partly covering persons as described in Sec. 3(2) of BDA.

   ❏ Outside India, with Members of the International Treaty for Food and Agriculture (ITPGRFA), and wholly or partly covering persons as described in Sec. 3(2) of BDA.

   ❏ Outside India, with Non-Members of ITPGRFA, and wholly or partly covering persons as described in Sec. 3(2) of BDA.

AS follows:

Recipient Name

Recipient Institution/Organization/Agency/Centre

Recipient Full Address with PIN Code

Phone number

Fax

E-mail

Nature of activities

| Germplasm material (specify)[3] | Crop and varieties |
|---|---|
| Supply made through | NBPGR |
| For Official Use of Supplier | 1. Germplasm Identity (Species name, common name, *etc.*) |
| | 2. Accession Number |
| | 3. Short Description of the Material |
| | 4. Passport Data |

---

1 Mention Name and address of the Second Party

2 Tick mark the appropriate box

3 Specify the type of material involved for supply/transfer *e.g.* seed, tissue culture, DNA *etc.*

I/We agree to abide by the following terms of the MTA and certify that:

i) The germplasm MATERIAL (S) transferred herein as above shall be used only for the purpose of research under my/our direct/close supervision and will not be used for commercial purposes or profit making whatsoever, without prior written approval of the NBA[4]/MoEF[5]/DARE[6]/ICAR[7], Government of India as the case may be. The importer/recipient (Second party) agrees to provide a concept note of research project in which the MATERIAL (S) will be used, including the manner in which to be used. The importer/recipient (Second party) agrees to cease any use of the material in case of suspension of research project at the instance of either party or due to factors beyond the control of either party. Upon such suspension of further research work, both parties will mutually agree for adopting a suitable provision for their preservation. In case of failure of the parties to arrive at an agreement, the materials including derivatives will be destroyed upon 90 days' notice from ICAR-NBPGR, New Delhi.

ii) All information and material supplied by ICAR-NBPGR shall be deemed to have been disclosed or provided to the recipient in confidence. The recipient agrees to preserve the confidential status of the material and information.

iii) The germplasm MATERIAL (S) or its (their) part(s), components or derivatives (including live or dead tissue/ DNA) that can be used to retrieve whole DNA/fragment or sequence or any other genetic information shall not be distributed or transferred to any third country/party, except those directly engaged in research under direct supervision of the recipient (second party), without prior written approval of the NBA/MoEF/ICAR/ DARE, Government of India as the case may be.

iv) Any development of commercial product based on research on gene manipulation/selective breeding programme for genetic improvement shall not be undertaken without written consent of NBA/MoEF/ICAR/ DARE, Government of India as the case may be. Modalities of undertaking any such work will be worked out before its conduct.

v) If any third country/party is to be associated with any commercial development arising out of the germplasm accessed, permission from NBA shall be sought.

vi) The recipient agrees to acknowledge explicitly the name, original identity and source of the material, if used directly or indirectly, in all research publication(s) or other publications, such as, monographs, bulletins, books, *etc.* and shall send a copy of each of the publications to the ICAR-NBPGR.

ix) The intellectual property protection or benefit sharing in respect of derivatives of the material(s) received/ accessed, where applicable, shall be as per the Indian IPR/Biodiversity laws.

x) The recipient agrees to hold the entire responsibility for the quarantine/SPS clearance of the material accessed as specified herein above. The recipient shall abide by the biosafety guidelines of _____ (Name of the importing country/organisation) and shall not hold ICAR-NBPGR/ICAR/DARE, Government of India responsible for any identity/quality/viability/purity/quarantine/biosafety related or any other related matter/hazard that may be attributable to the release of genetic material/resource accessed as specified in this Agreement. The recipient agrees to hold entire responsibility for the importer/indenting country's biosafety and other related hazards due to release of genetic material. The recipient agrees waive all claims against ICAR-NBPGR/ICAR/DARE, Government of India and to defend and indemnify them from all claims and damages/recoveries arising from the use, storage or handling of the material.

xi) The recipient also agrees that the material is for experimental use and is being supplied without any warranties, whatsoever.

xii) The MTA is non-assignable. The recipient agrees to abide by any other conditions that may be set in and conveyed to them from ICAR-NBPGR in respect of this germplasm access/exchange or any Law, Rules, Regulations, *etc.* enacted by Government of India from time to time.

xiii) In case of any dispute between the parties to this MTA, the dispute shall be referred to the Sole Arbitrator to be appointed by the Secretary, DARE, Government of India. The Decision of the Sole Arbitrator shall be final and binding on the Parties. The Arbitration proceedings shall be governed by the Arbitration and Conciliation Act, 1996. The Arbitration proceedings shall be in New Delhi.

---

4 National Biodiversity Authority

5 Ministry of Environment and Forests

6 Department of Agricultural Research and Education

7 Indian Council of Agricultural Research

| AGREED RECIPIENT | PROVIDER |
|---|---|
| Authorised Officer's | Authorised Officer's |
| Name: | Name: |
| Designation: | Designation: |
| Organization/Institute/University Address: | Organization/Institute/University Address: |
| Signature: | Signature: |
| Date: | Date: |
| Recipient Scientist/Person's | Provider Scientist/Person's |
| Name: | Name: |
| Designation: | Designation: |
| Organization/Institute/University Address: | Organization/Institute/University Address: |
| Signature: | Signature: |
| Date: | Date: |

**Definitions**

Extract from Section 3(2) of BDA-2002-

a) a person who is not a citizen of India;

b) a citizen of India, who is a non-resident as defined in clause (30) of Section 2 of the Income-Tax Act, 1961 (43 of 1961);

c) A body corporate, association or organisation:

(i) Not incorporated or registered in India; or

(ii) Incorporated or registered in India under any law for the time being in force which has any non-Indian participation in its share capital or management.

**Proforma for Feedback Information**

| EC/IC Nos. | Crop/Variety Name | Source/Country | Characteristics Observed | Status of Material Maintained/not Maintained | Seed Supplied to NGB (Y/N) (Qty) | Details of the Germplasm Utilized in Crop Improvement |
|---|---|---|---|---|---|---|
| | | | | | | |

## (D) Documentation and Dissemination of Information on Assembled Plant Genetic Resources

The available information on genetic stocks both exotic and indigenous, including the wild relatives of crop plants is assembled and documented in various record of Bureau. These include the national registers of exotic and indigenous collections. This information is compiled in the form of Plant Introduction Reporter (issued quarterly) and Crop Inventories in specific crops. These are distributed to the concerned scientists in the country for their references. The crop breeders/ users can select germplasm materials of their choice for further utilization in crop improvements from these two publications. They can approach directly to the crop recipients of the material indicated in the Plant Introduction Reporter (PIR) as well crop inventories.

### (i) Plant Introduction Reporter (PIR)

Proper documentation of all introduced materials is essential for record and ready reference in the form of Plant Introduction Reporter (PIR) which is freely available to all research organization in the country. The information on Plant Introduction are compiled and brought out regularly for wider circulation among the user agencies within India. The first PIR was brought out in 1949. However the regular publication of PIR started from 1965 onwards.

### (ii) Crop Inventory

Crop inventories on different crops have been prepared enlisting the materials introduced along with relevant information and distribution to the concerned scientists in the country for their future reference.

### Exchange of *in vitro* Generated Genetic Resources

Propagules developed through micro-propagation like *in vitro* slow growth cultures, encapsulated somatic embryos and nodal explants as well as micro-tubers have been used for international exchange of germplasm safely over long distances. The live cultures or other propagules are properly packed in tight containers and sent by air along with phyto-sanitary certificate ascertaining freedom from infections. At the receiving end, the package is opened and cultures stored in light (3,000Lux) in a sterile room for a week at 25°C to remove extra moisture and effect of continuous darkness. Then all the cultures are checked for contamination. Contaminated cultures are destroyed in an incinerator to prevent chance introduction of any pathogen. Plantlets are then either micro-propagated and multiplied or subjected to hardening in soil inside green house or mist house and later transferred to field. The international advisory body on crop germplasm (IPGRI) recommends micro-propagated material to be safe for exchange of germplasm of root and tuber crops across national borders.

### Plant Quarantine (PQ)

Developed and developing countries both emphasizes the need for sustainable efforts in crop improvement strategies, a wide genetic diversity in various crop is backbone to such efforts. To build such diversity, introduction of Plant Genetic Resources (PGR) from different countries of the world is of paramount importance. However, such introductions always carry an element of inadvertent introduction of exotic pests and pathogens into new areas. Therefore it is essential to inspect and quarantine the introduced materials. In general plant quarantine is a government endeavor, enforced through legislative instruments, to prevent the introduction and/or spread of quarantine pests or to ensure their official control, when any planting material, plant products, soil or living organisms are introduced into a new geographical area.

Plant quarantine can be defined as:

*"Legal enforcement of measures aimed to prevent pests and pathogens from spreading, or to prevent them from multiplying further in case they have already found entry and have established in a new area".*

or

In simple manner *"Quarantine means to keep materials in isolation to prevent the spread of diseases, etc. present in them to other materials".*

Plant quarantine can also be defined as:

*"Rules and regulations promulgated by governments to regulate the introduction of plants, planting materials, plant products, soil, living organism etc. with a view to prevent the inadvertent introduction of weeds and pests harmful to the agriculture or the environment of a country or region and if introduced, to prevent their establishment and further spread".*

### Importance of Plant Quarantine

Plant quarantine services facilitate safe introduction of germplasm samples from other countries. Almost all countries regulate the import of plants/plant products, including germplasm, because of the pest risk posed by such imports into its ecosystems, especially agricultural ecosystem. There are many examples in the history of development of world agriculture which show how the introduced pests have been responsible for crop devastation and even created famine conditions in different parts of the world. The well-known Ireland famine of 1845 was the result of almost total failure of potato crop due to the epiphytotic caused by the introduced late blight pathogen (*Phytophthora infestans*) from Central America which resulted in mass migration of Irish people to four corners of the world. In our country also, several serious pests have got introduced from time to time some of which like late blight of potato, and bunchy top of banana have since become widespread. Some others like golden nematode of potato and downy mildew of onion are still localized in certain parts of the country. These examples only highlight the importance of plant quarantine in the present day agriculture of various countries. Plant quarantine required safeguard to the agriculture of a country by excluding the exotic pests, pathogens and weeds and allowing healthy plant material to flow in, and also prohibit or regulate the movement of material within the country. While realizing the importance of such exchange, it is not only necessary but imperative to view critically the quarantine aspects associated with such exchange, as plant quarantine being the first line of defense against introduction of exotic pests, diseases and weeds. Though plant quarantine measures may not guarantee an everlasting protection against the entry of exotic species but will certainly check or delay the introduction of these unwanted organisms and their subsequent establishment in hitherto clean areas. In more simple way quarantine means to keep materials in isolation to prevent the spread of diseases *etc.* present in them to the other materials. Hence, plant quarantine will be justified when it has been taken into consideration the pest introduction risk and adequate safeguard. It is also necessary that while formulating plant quarantine rules and regulations, each country must take associated risk factor into consideration and if required these regulations may be amended from time to time as per the existing situation. It is essential and mandatory that during collection mission the scientists/ collectors should know and get ideas/knowledge about quarantine regulations of the country from where the collections made/to be made and where it is to be utilized. This will help them in collection of pests/ pathogens free germplasm. Such collection(s) must be routed through official plant quarantine channels. Even materials collected within the country also must passes through quarantine laboratories/centres. Bypassing quarantine is danger for country's agriculture. It is to be noted that while conserving the collected material (under low temperature) care should be taken that pests free material is only stored. In India ICAR-National Bureau of Plant Genetic Resources (ICAR-NBPGR), New Delhi is now actively engaged in gearing up its activities and network systems to face future challenges. In recent years, they quarantine of seed and other planting propagules imported annually for research purpose or for commercial purpose, has increased manifold, and in the near future, it will increase further. Seeing the challenges towards import and export of genetic resources, it requires more infra-structural facilities, equipments and trained personnels. It is envisaged that in near future there will be a significant change in the plant quarantine systems. Then the crop loss due to exotic pests/pathogens will be reduced considerably. In the case of plant introduction, all the introduced plant propagules are thoroughly inspected for contamination with weeds, diseases and insect pests. If procured materials are suspected to be contaminated are fumigated or are given other treatments to get rid of the contamination. If necessary, the materials are grown in isolation for observation of diseases, insect pests and weeds. This entire process is known as quarantine and the rules prescribing them are known as quarantine rules.

## Plant Quarantine Rule and Regulations Framed by Government of India for Safe Movement of PGR

International and national regulatory/legislative measures are adopted to prevent introduction of exotic pests. Most countries of the world have promulgated plant quarantine rules and regulations with a view to prevent introduction of exotic pests. Plant quarantine will be justified only when the pest in question has no natural means of spread and when it is based on biological considerations only *i.e.* pest introduction risk and available safeguards. Generally speaking, pest introduction risks are more with vegetative propagules than with true seed. In the case of true seed, risks are more with the deep seated infections than with surface borne contaminations. In the case of vegetative propagules, rooted plants, bulbs, rhizomes, corms carry more risk than the above ground parts like bud wood, unrooted cuttings *etc.* Further, risk are more with pests like viruses, viroids, smuts, downy mildews and certain bacterial carried inside the seed with no external symptoms. In any case, bulk introductions are always highly risky as their thoroughly examination and treatment is very difficult and the plantings area is far too large for proper surveillance.

Quarantine procedures continue to evolve to prevent every possible biotic risk from consignments of propagative and non-propagative plant products. Despite comprehensive measures, pests and pathogens find their way into new territories (Waage *et al.*, 2006) resulting in serious economic damage.

Quarantine measures adopted in India for introduction of plant germplasm are detailed hereunder:

### *Destructive Insects and Pests Act (DIP Act, 1914)*

Rule and regulations framed by Government of India for safe movement of genetic resources long back in 1914 as 'Destructive Insects and Pests Act' (Act was passed on Feb. 3, 1914) by the British government ruling India which was retained revisiting it as per requirements over the years through various amendments. According to the Destructive Insects and Pests Act (DIP), 1914, all plant produce imported in India must be free from diseases, insect pests and weeds. Some main feature of Destructive Insects and Pests Act (DIP), 1914 of India are:

1. It authorizes the central government to prohibit or to regulate the import of any article or class of articles into India or any zone/area of her territory.
2. It authorizes the central government to prohibit or to regulate the export from any state or movement from one state to another within the country (Domestic Quarantine) of plant or planting material.

**Table 6.11: Introduction of some Pest and Pathogenes**

| Year | Crop | Pest and Pathogens | Scientific Name |
|---|---|---|---|
| 1977 | Onion | Onion downy mildew | *Peronospora destrutor* |
| 1961 | Potato | Potato golden nematode | *Globodera rostochinensis* |
| 1953 | Potato | Potato wart | *Synchytrium endobioticum* |
| 1910 | Apple | San Jose Scale | *Quadraspidiotus perniciosus* |
| 1906 | Grapes | Downy mildew of grapes | *Plasmopara viticola* |
| 1883 | Potato | Late blight of potato | *Phytophthora infestans* |
| 1879 | Coffee | Coffee rust | *Hemileia vastatrix* |

So far eight Domestic Quarantine regulations have been promulgated under the provisions of DIP Act, 1914. These include fluted scale (*Icerya purchasi*), San Jose scale (*Quadraspidiotus perniciosus*), apple scab (*Venturia inaequalis*), colding moth (*Cydia pomonella*) and golden nematode (*Globodera rostochiensis* and *Globodera pallida*).

3. It authorizes the state government to make rules for detection, inspection, disinfection or destruction of any article or class of articles in respect of which Central Government has issued notifications.

4. It provides for penalties for person(s) who knowingly contravene the rules and regulations issued under the Act.

5. It also protects the person(s) from any suit or prosecution or other legal proceedings for anything done in good faith under the Act.

Till to date DIP Act has been modified and enacted several times.

### *International Plant Protection Convention (IPPC, 1952)*

India is signatory to the International Plant Protection Convention (IPPC, 1952) of the Food and Agriculture Organization (FAO) since 1952, which requires each country to establish a National Plant Protection Organization (NPPO) to discharge quarantine functions.

### Plants, Fruits and Seeds Order (PFS, 1989)

The latest regulation enacted under the DIP Act is the 'Plants, Fruits and Seeds Order, 1984 (PFS) and was enforced on 24th June, 1985. This was necessitated to cater to the needs of 'The New Policy on Seed Development' (NPSD) of Govt. of India (Gazette notification on 16 Sept.1988) which came into force on 1st October, 1988 with the objective to make available to the Indian farmers the best genetic materials in the world (liberalized the existing rules with regard to import of plant/planting materials) to increase our agricultural productivity, farm incomes, and to encourage the private sector seed industry in India not only to fulfill domestic requirements but also to develop export potential. Order of 1989, APFS amended Order 1991 notified Gazette of India dated 20th Jan.1992. It has been stated in the policy that while importing quality planting materials care has to be taken to ensure that there is absolutely no compromise on the requirement of plant quarantine procedures to prevent entry into the country the exotic pests and weeds detrimental to Indian agriculture and horticulture.

As per the new order:

☆ A number of consignment of (specific crops) seed/planting material can be imported into India without a valid permit;

☆ All of consignment must be accompanied by an official phytosanitary certificate issued by the quarantine authority of the exporting country;

☆ Import of soil in any form is prohibited;

☆ The import of plants/seeds/fruits for consumption is allowed only through authorized entry points;

☆ Use of packing material of plant origin is prohibited;

☆ Post entry isolation growing has been prescribed for certain crops.

While liberalizing import, care has been taken that there is absolutely no compromise on plant quarantine requirements. Although there are several requirements under the Plants, Fruits and Seeds Order (FPS) Order, 1989, but for our purpose the most important requirements are that:

(a) No consignment shall be imported into India without a valid import permit issued by the competent authority. For importing germplasm, Director, ICAR-NBPGR, New Delhi has been authorized by the government of India to issue import permits, both for government institutions as well as private seed companies. For bulk consignments the import permits is issued by the 'Plant Protection Adviser' to the Govt. of India.

(b) No consignment shall be imported, unless accompanied by phytosanitary certificate by an official of the exporting country.

(c) Seeds/planting materials requiring isolation growing under detention shall be grown in an approved post-entry quarantine facility.

Import of soil, earth, sand, compost, and plant debris accompanying seeds/planting materials shall not be permitted.

(d) Hay, straw or any other material of plant origins shall not be used as packing material.

### National Plant Quarantine Organizations

The various plant produces may be grouped into two categories on the basis of their intended use:

(1) Those for consumption and

(2) Those for cultivation and research.

The quarantine of these two groups of plant produces is carried out by separate agencies.

## (1) Plant Produce Imported for Consumption

In India, quarantine regulations are enacted and enforces under the provisions of the "Destructive Insects and Pests Act-1914" (DIP Act, 1914) included a provision under section 4, empowering the officer of the 'Custom Department' to discharge this function till 1946. DIP Act, 1914 have been revised several times, the latest one being the "Plants, Fruits and Seeds (regulation of import into India) Order, 1989". The quarantine of this category of plant produces is the responsibility of the Directorate of Plant Protection, Quarantine and Storage (PPQS). Plant Protection Advisor (PPA) to Government of India was established to handle commercial/bulk materials for consumption or for planting.

### Main Features of "Plants, Fruits and Seeds (regulation of import into India) Order, 1989"

(i) No consignment shall be imported into India without a valid import permit to be issued by a competent authority.

(ii) No consignment shall be imported into India unless the same is accompanied by a phytosanitary certificate issued by an authorized official of the plant quarantine service of the source country.

(iii) All consignments shall be imported only through prescribed ports of entry where these shall be inspected and if necessary, fumigated, disinfected or disinfected by authorized plant quarantine officials.

(iv) Seed/planting materials requiring isolation growing in detention, shall be grown in post-entry quarantine facility approved by the competent authority.

(v) Hay, straw or any other material of plant origin shall not be used as packing material.

(vi) Import of soil, earth, sand, compost, plant debris accompanying seed/planting materials shall not be permitted except for research purpose under special permit issued by the Plant Protection Advisor to the Government of India.

### Directorate of Plant Protection, Quarantine and Storage (DPPQS)

In India, the Directorate of Plant Protection Quarantine and Storage (DPPQS), Faridabad, under the Ministry of Agriculture and Farmers Welfare (MoA and FW), New Delhi is the NPPO for implementation of plant quarantine regulations as per the IPPC, through the Destructive Insects and Pests Act (DIP Act, 1914), amended from time to time. The quarantine of bulk material imported for planting or for consumption is undertaken by the DPPQS, Faridabad, and the network of quarantine stations under it are located all over the country. Directorate of Plant Protection, Quarantine and Storage is headed by the Plant Protection Advisor to the

Government of India. The Directorate has quarantine and fumigation centres at 17 seaports, 11 airports and at the entry points of 7 land routes/frontiers. These centres examine the plant produces imported for consumption and carry out their fumigation, if necessary. To meet its quarantine obligations, this Directorate has established a chain of Plant Quarantine and Fumigation Stations at all the international airports, seaports and land customs stations. Elaborated facilities including post-entry quarantine green-houses are being developed at Delhi, Mumbai, Kolkata and Chennai stations to handle seed and planting materials to be imported in bulk by private grower's seed industry under the New Policy on Seed Development announced in 1988. Under the regime of liberalized trade in agriculture under the World Trade Organization (WTO), to which India is also a signatory, the Plant Quarantine (Regulation of Import into India) Order, 2003 henceforth referred as the PQ Order, 2003, came into force on January 1, 2004 except sub-clause (22) of clause which entered into force on April1, 2004.

## (2) Plant Propagules Imported for Cultivation and Research

The quarantine of plant propagules is done by the following three agencies depending on the nature of the concerned plant species.

(1) ICAR-NBPGR, New Delhi is responsible for the quarantine of all the propagules of agricultural and horticultural species required for research purpose both in public and private sectors..

(2) The quarantine of propagules of forest trees is carried out by the Forest Research Institute, Dehradun (Uttarakhand).

(3) The propagules of remaining plant species are handled by the Botanical Survey of India, Kolkata.

According to the quarantine laws, only those propagules that are free from diseases, insect pests and weeds can be allowed to enter the country (beyond the point of quarantine). The quarantine laws cover not only the propagules but also their packing materials, especially of plant origin, and other materials accompanying them.

### Authorization by Government of India

As detail accountability, germplasm exchange/plant introduction activities were initiated in the mid-1940s and the work was more systematized with the creation of the Division of Plant Introduction at ICAR-IARI, in the 1961 and its subsequent elevation in 1976 into a full-fledged institute of ICAR-National Bureau of Plant Genetic Resources (NBPGR), New Delhi. The ICAR-NBPGR has been authorized by the Government of India to take care of the quarantine, processing and clearance of the materials. To meet the obligation, an independent Division of Plant Quarantine comprising discipline of Entomology, nematology and Plant Pathology/virology with well-equipped modern laboratory facilities and trained staff has been established at the headquarters in ICAR-NBPGR, New Delhi and at Regional Station, Telangana, which mainly deals with the quarantine processing of PGR meant for Southern India including State Agricultural Universities, ICAR institutes, private industry and international institutes *viz.,* International crop Research Institute for Semi-arid Tropics (ICRISAT), CIMMYT, World Vegetable Centre (AVRDC), Taiwan and CIP, Peru. The quarantine of germplasm pertaining to agricultural and horticultural crops is done by entomologists, plant pathologists and nematologists of the Quarantine Section of ICAR-NBPGR, New Delhi at the headquarters of the Bureau. Post entry quarantine (PEQ) nurseries) facilities have also been developed at Bhowali in Uttarakhand and Kanpur in U.P. The ICAR-NBPGR has been instrumental in providing the plant genetic resources exotic as well as indigenous collections to the ICAR crop based institutes, various All India Coordinated Crop Improvement Projects, agriculture as well as other conventional universities, state agricultural departments, private seed companies and other research organizations in the country. The Government of India has also authorized various officers of the ICAR; SAU's, the central and the State Government for inspection, issue of import permit and phytosanitary certificates. A total of 41 Designated Inspection Authority (DIAS) has been authorized till today. ICAR-NBPGR, New Delhi is well equipped with most modern quarantine facilities including a Containment Facility of Level 4 (CL-4) for quarantine processing ofv transgenic germplasm in a risk-free manner.

## Quarantine Risk and Safeguard

Any germplasm imported into India, must be accompanied with an Import Permit (IP) and a Phytosanitary Certificate (PC) (c.f. Chapter 3, Sections 3.1.1 and 3.1.2). as per Section 6 (1) and (2) of the PQ Order, 2003, only ICAR-NBPGR is authorized to issue IP for import of germplasm including transgenics, genetically modified organisms (GMOs) or living modified organisms (LMOs) (true seeds as well as vegetative propagules) meant for research or experimentation. Further, as per Section 6(3) of the PQ Order, 2003, imported consignments of plant germplasm/transgenics/GMOs/LMOs meant for research purpose, are not be opened at the point of entry, but are required to be forwarded to the Director, ICAR-NBPGR, New Delhi, for its quarantine.

### Pest Risk

Analysis of pest risk at the time of introduction of planting material is essential to determine the potential of a pest to cause damage. Successful colonization of exotic pests associated with infected or infested material requires few pre-conditions which includes:

(i) Availability of susceptible host

(ii) Availability of favourable environmental conditions

(iii) Ability to multiply and spread within a short time, and

(iv) Absence of natural enemies.

Risks are more with the introduction of vegetative propagules than true seed. While in case of true seed, risks are more with deep seated infection than surface contaminated seed. Further, risks are more with peats like viruses, viroids, downy mildews and certain fastidious bacteria. It is true that bulk imports create more quarantine problems.

### Safeguard

It is essential to prevent the spread of exotic pests, the plant quarantine system (categorize the quarantine objects) regulates the plant introduction activities in a country (of a country) as follows:

A. **(I) Complete Prohibition-**When the pest risk is very high and safeguard against it is not available in the country-import should be prohibited.

**(II) Post-entry quarantine (High risk-adequate)-** The pest risk is very high but adequate safeguards in the form of post entry isolation facilities, growing facilities or salvaging techniques are available. Imported material is to be grown in post-entry-quarantine facility only.

**(III) Restricted-**Pest risk is not very high and import permit is required stipulating/indicating conditions for entry, inspection and treatment of the material to be imported.

B. The seed health testing (from quarantine angle) procedures must be popularized at strategic point of entries.

C. Each country should be familiar with important pests and diseases prevalent in the region. Exotic pests known to be potentially dangerous to important crops of each count should be listed.

D. The recent development in the field of biotechnology, especially in tissue culture and genetic engineering has opened new avenues in regard to safe exchange of vegetatively propagated crops, specially.

## Quarantine Procedure

For importing germplasm including transgenics/ GMOs/LMOs not listed under Schedules IV, V, VI, VII of the PQ Order, 2003, a pest risk analysis (PRA) needs to be carried out prior to issue of IP. Export inspection and phytosanitary certification of plants and plant products is carried out in accordance with Articles IV of IPPC to meet the legal obligation of the member countries.

PCs issued in the model formats set out under Article V of the IPPC in harmony with prevailing quarantine regulations of the importing country. The Head, Division of Plant Quarantine, ICAR-NBPGR, New Delhi and Officer-in-Charge, ICAR-NBPGR, Regional Station, Telangana, are the designated official from ICAR-NBPGR, to inspect, fumigate or disinfect and to grant PC for plant germplasm for export to foreign countries (as per notification No. 8-97/91-PPI, dated 26-11-1993 of the Ministry of Agriculture, now MoA and FW, GOI) (http://plantquarantineindia.nic.in/PQISPub/Docfiles/notify10.htm).

Quarantine examination is carried out for all germplasm at ICAR-NBPGR, New Delhi, except germplasm indented from southern states of India (Andhra Pradesh, Goa, Karnataka, Kerala and Tamil Nadu), which are inspected at ICAR-NBPGR, Regional Station, Hyderabad. The elimination of these contaminated samples prevented the entry of more than 12 new pathogens and insects in this country. In addition, this process of quarantine has prevented the introduction of many new biotypes of several pathogens and insect pests. The seed material of different vegetables crops were distributed to various institutes, coordinated projects, agricultural universities and other researchers in India.

**Pests Interception Categories**

The pests intercepted can be categorized as

(i) Many that are not known to occur in India;

(ii) Have different races/biotypes/strains not known to occur in India;

(iii) Are present on a new host or are from a country from where they were never reported before;

(iv) An entirely new pest species hitherto unreported in science or

(v) Are reported to be present in India but with a wide host range.

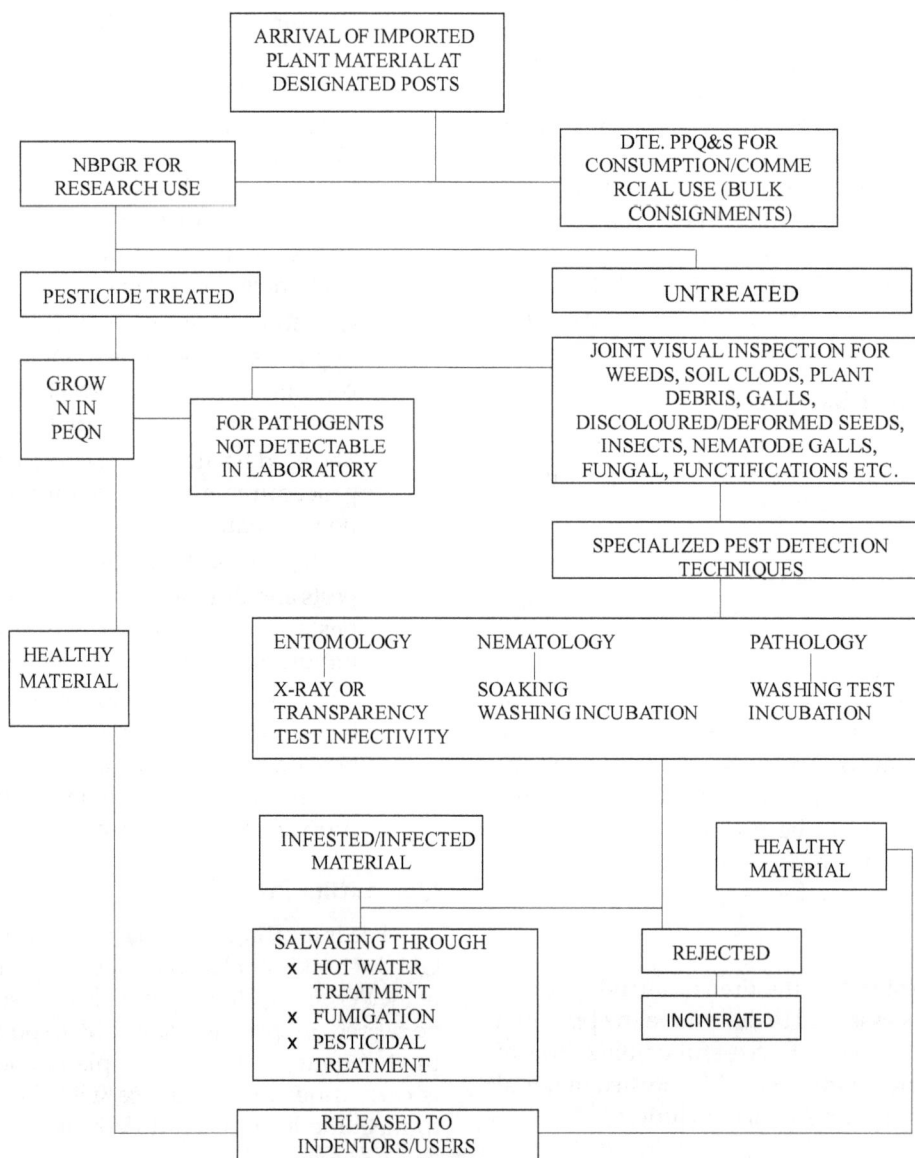

**Figure 6.2: Quarantine Processing of Imported Plant Germplasm.**

Interceptions, especially of pests and their variability not yet reported from India signify the importance of quarantine in preventing the introduction of destructive exotic pests. The third and forth category of pests are not expected in the sample as per the risk analysis which is literature-based and since no records are available on the pest/host their presence is unexpected and important from quarantine view point. The last category-pests with a wide host range are critical and could become invasive in case they find suitable biotic and abiotic environment. Such interceptions signify the success of quarantine; otherwise, these pests would have entered the country and played havoc with the plant biodiversity and agriculture.

## 1. General Safeguard

The following generalized description relates to the quarantine activity of the ICAR-NBPGR, New Delhi; it has been designed to prevent the entry of weeds, diseases and pests. The import of seed/planting material is regulated as per the provisions of the "Plants, Fruits and Seed Order, 1989" (PFS Order): regulation of import into India, issued under the DIP Act, 1914.

1. It is essential that the propagules must be clean, healthy and free from weeds and insect pests. The sender should not treat propagules with any fungicides or insecticides. If necessary, the sender may only fumigate the seeds/propagules before sending them, and indicate the same in the phytosanitary certificate.

2. Each imported entry or sample must be accompanied by a 'phytosanitary certificate' from the scientist/institution sending the sample/entry. In this certificate the sender certifies that the seeds/propagules being sent are free from weeds, diseases and pests. All the entries not accompanied by an authentic phytosanitary certificate are either returned to the sender or -destroyed by the bureau.

3. The entries accompanied by an authentic phytosanitary certificate are examined closely with the help of a magnifying glass/microscope and screened with X-rays. X-ray examination is helpful in the detection of insects, pathogens, weeds *etc.* present within the propagules.

   The healthy entries free from diseases, insect pests and weeds are identified and sent to the recipient scientists/institutions. Contaminated entries are detained by the bureau and attempts are made to free them from the contaminating weeds, insects and pathogens.

4. Contaminated entries may be fumigated if it is considered that such a treatment would get rid them of the contaminating insect or pathogen.

5. If needed or considered desirable, the contaminated entries may be grown in isolation in an effort to isolate some healthy plants. Seeds/

propagules from such healthy plants may be collected and sent to the indenter.

Entries suspected to be contaminated by pathogens, particularly viral pathogens, are also grown to monitor disease symptoms and the presence of pathogens. In advanced countries, sophisticated techniques like immunological assays *etc.* are used to detect the presence of pathogens, especially viral pathogens, in germplasm introductions.

6. Samples/entries that are heavily contaminated are destroyed by the bureau.

The pre-entry requirements for planting material under NPSD are as under:

| | Requirement | | FVS | BT | CS |
|---|---|---|---|---|---|
| i. | Import Permit | | R | R | R |
| ii. | Phytosanitary certificates (PSC:E) | | R | R | R |
| iii. | Additional Declaration | (a) Freedom from soil (E) | R | R | R |
| | | (b) Freedom from weeds | R | NR | NR |
| | | (c) PQ objects | R | R | R |
| iv. | Approval of PQ facilities | | NR | NR | R |

FVS: Flower and Vegetable Seeds; BT: Bulbs and Tubers; CS: Cuttings and Saplings, E: Existing; R: Required; NR: Not Required.

## 2. Inspection

Inspection is undertaken at the point of origin, transit and destination and is decided that the consignment is to be released, treated, detained, returned or destroyed. The process of quarantine (except growing in isolation) at the ICAR-NBPGR, New Delhi takes at least three weeks. The quarantine of short-lived propagules is done at top priority.

| | Requirement | | FVS | BT | CS |
|---|---|---|---|---|---|
| i. | Sampling | | As per ISTA | Minimum 10 no. | or 0.1 per cent |
| ii. | V.I. | | R | R | R |
| iii. | L.I. | | R | NR | NR |
| iv. | | (a) X-ray radiography | | | |
| | | (b) Washing test | | | |
| | | (c) Nematode detection | | | |
| | | (d) Incubation (Insects/ diseases) | | | |
| v. | G.T. | | NR | NR | R by DIA |
| vi. | Total duration (days) | | 45 | 50 | Min:25, Max:50 |

R: Required; NR: Not Required; DIA: Designated Inspection Authority, V.I.: Visual Inspection; L.I.: Laboratory Inspection; G.T.: Grow Out Test; PEQ: Post-Entry Quarantine.

Inspection requirement of Flower and Vegetable Seeds (FVS), Bulbs and Tubers (BT), and cuttings and saplings (CS) is regulated as per provisions of PFS Order, 1989.

### 3. Techniques/Methods for Pests Detection

Success or failure of plant quarantine measures would depend to a very extent on the ability of quarantine of officials to detect various pests that may accompany the introduced materials. Since, for quarantine purposes tolerance levels are zero, detection techniques/procedures have to be sensitive enough to detect/reveal even trace infection/infestation. Pest detection techniques have been developed by the ICAR-NBPGR, New Delhi or by available standardized techniques (Neergaard, 1979) salvaging of infected/ infested planting materials. This techniques have been used for the interceptions of different pests can move more rapidly in vegetable propagules like rooted plants, cuttings, bulbs, rhizomes and tubers. Therefore, all these planting materials are being examined by all the available techniques immediately after their arrival.

### Group of Detection Techniques

Detection techniques may be broadly put in two groups:

    **(i) Generalized tests**

    **(ii) Specialized tests**

### I. Generalized Tests

A very widely used method is the inspection of dry seed with unaided eye or under the low power of a stereoscopic microscope. Generalized tests reveal the presence of a wide range of insects, mites, weeds, soil clods, infected plant debris, discoloured/deformed seeds and fungal fructifications on or mixed with the seed.

    **(a)** Examination of dry seed under UV or NUV light may reveal infections by certain fungi and bacteria through emission of fluorescence of different colours. Examination of seed washings would reveal surface contamination by rusts, smuts, downy mildews and a large number of other fungi.

    **(b)** Examination of vegetative propagules like rooted plants, cuttings, bulbs, corms, rhizomes *etc.* can similarly be examined unaided eye or under low power of microscope for presence of insects, mites, fungal fructifications and symptoms of infestation/infection.

### Examination of Dry Seed

Examination of dry seed includes following methods for:

    (a) Insects and mites

    (b) Nematode

    (c) Fungi

    (d) Bacteria

    (e) Viruses

    (f) Weeds

### (a) Insects and Mites

The samples should be examined either by naked eye or with magnifying glass, illuminating magnifier or stereoscope binocular. Initially approximately 60 per cent samples are examined by visual inspection, subject to further confirmation by other methods. Symptoms of damage caused by various pests during visual inspection are the major criterion for detection. The insects and mites of high quarantine significance intercepted in exotic germplasm include: in *Phaseolus vulgaris* (Italy and Nigeria); *Anthonomus grandis* in *Hibiscus* spp., *Vigna unguiculata* (Italy and Zambia) and *Ephestia elutella* in *Vigna* spp. (USA) (Agarwal *et al.*, 1998). Insects like Acanthoscelides obtectus in *Cajanus cajan* and *Phaseolus vulgaris* from Brazil, Colombia, Italy and Nigeria.

Some practical issues are being mentioned below:

    (I) Types of insects

        1. Mandibulate-Grass hopper

        2. Haustellate-Bugs and Thrips

    (II) Symptoms of insect damage

        **1.** Under ground plant parts-tunneling, girdling (white grub, termites)

        **2.** Shoots, stems and twigs-sucking (mealy bugs, scales); mining (Dipterous fly); silken webs comprising wood excreta (bark eating cater pillar); cancerous growth (mustard saw fly); galls, twig and bud galls, vagrant, witches broom, big bud (mites) *etc.*

        **3.** Leaves- cutting edges to inner sides (hoppers, Laphygma); leaving veins (Bihar hairy cater pillar); silver yellow spots on tips (rice thrips); skeletonized (cabbage butter fly); honey dew or sooty mold (syllids); necrotic patches (leaf hoppers) *etc.*

        **4.** Buds, flowers and fruits-pods with bore holes (bean pod borer); chewing (moth larvae); sucking (bugs); drying of panicles (Rice Gundhi Bug) *etc.*

        **5.** Seeds

          1. External

          **(a)** Primary-saw toothed grain beetle

          **(b)** Secondary-rust red flour beetle

          2. Hidden

          Two types of development

          **(a)** Field-pea *Bruchid* and *Chalcids*

          **(b)** Storage- pulse beetle and lesser grain borer

**(III) Detection methods**

1. Unaided and microscopic examination
2. **Transparency techniques-**Phenol crystals (2 parts in g), lactic acid (2 parts in ml), glycerin (1 part in ml); distilled water (2 parts in ml.), heat the seeds on water bath at 100 °C for 30 minutes to 3 hours according to seed hardiness.
3. **X-ray radiography-**The list of 340 plant genera known to carry hidden infestation should be compulsorily subjected to X-ray radiography. X-ray radiography, a non-destructive technique, should be used to detect seeds infested with phytophagous *Bruchid*, *Chalcids* and certain other insect groups that do not exhibit any external symptoms on seed surface. Upon exposure to X-rays, the seeds clearly show insect infestation which should be hand-picked and removed from the healthy seeds which are then released to the indentor.

   **(a) Developer:** Metol (3.5 g); Sodium sulphate (60 g); Hydroquinone (9 g); Sodium carbonate anhydrate (40 g) dissolved in approximately in 750 ml of water and make up to 1000 ml. solution

   **(b) Fixer:** Sodium thiosulphate (250 g); Potassium metabisulphate (15 g) dissolved in and make up to 1000 ml. solution

   **(c) X-ray:** Dose rate would be 22Kv, 3mA, 10 seconds at a distance of 30 cm.

The seeds of plant genera (associated as vegetables) which have to be subjected to x-ray radiography immediately on import to detect the presence of hidden infestation are as under:

**Table 6.12: Crop Subjected to X-ray Radiography for Detecting the Presence of Hidden Infestation**

| Sl.No. | Plant Genera | Sl.No. | Plant Genera |
|---|---|---|---|
| 1. | Abelmoschus | 11. | Ipomoea |
| 2. | Bambusa | 12. | Lotus |
| 3. | Bauhinia | 13. | Mentha |
| 4. | Cannavalia | 14. | Mucuna |
| 5. | Chenopodium | 15. | Nelumbo |
| 6. | Coriandrum | 16. | Pachyrrizus |
| 7. | Crotalaria | 17. | Phaseolus |
| 8. | Cyamopsis | 18. | Pisum |
| 9. | Daucus | 19. | Vicia |
| 10. | Dolichos | 20. | Vigna |

**(IV) Treatment**

1. Spray and dips

   Calculation

   $$S.S. = \frac{Volume \times Required\ Concentration}{Given\ Concentration}$$

2. Fumigation

   i- EDCT-mixture: Ethylene dichloride (3 parts) and carbon tetrachloride (1 part)

   ii- HCN: $2NaCN + H_2SP_4 = Na_2SO_4\ 2\ HCN$
   NaCN (1 g); $H_2SO_4$ (1.6 ml); water (3.2 ml)

   iii- MB: $mg/l = M*1000/V$

   M=Weight of fumigant; V= Volume

   iv- $CS_2$: $mg/l = V_1*1225*1000/V_2$

   $V_1$=Volume of $CS_2$

   $V_2$=Volume of area

   1225= Weight of 1 ml of $CS_2$

   v- Collection of insect/mites specimens

   vi- Preservation of insects

   vii- Identification of insects and mites: by use of reference collections, published keys, and taxonomists.

**(b) Nematode**

Plant parasitic nematodes are found in and around roots of host plant, but they can be spotted in leaves, stems, buds, flowers and seeds of certain plant species. More than 65 species of plant parasitic nematodes were intercepted. Among the major interceptions, nematodes not reported in India include: *Heterodera schachitii* (from Denmark, Germany and Italy), *H. glycines, Ditylenchus dipsaci, D. destructor* and *Rhadinaphelenchus cocoohilus* (Agarwal *et al.*, 1990) in soil clods and plant debris. Their identification was based by examination test as follows:

☆ Galls or swelling on roots, tubers, or rhizomes.

☆ White, yellow, or brown pinched size bodies adhering to roots

☆ Swollen or malformed leaf, stem or other tissue

☆ Root lesion, or unusual root proliferation (Lal and Mathur, 1988)

☆ Seeds known to/suspected to carry seed-borne nematodes should be soaked in water for 24 hours. Softened seeds should be teased/crushed under a microscope to observe the nematodes, if present.

☆ Plant material/soil should be soaked overnight and sieved through nematological sieves. These should be recovered and examined under the compound microscope for identification of nematodes.

☆ Staining technique should be used for quick detection of nematodes in vegetative propagules. The nematodes, if present, retain the red stain more deeply than the plant tissue and can easily be detected under a stereo-binocular microscope. Examination of accompanying soil should be done to detect the presence of viable nematodes, especially ectoparasites and cyst of cyst-forming nematodes.

### (c) Fungi

Blotter test of seeds should be used for detection of any fungal pathogen capable of producing mycelial growth, fruiting structures and symptoms on the seedlings. Examination of suspension after seed washing method was used for detecting surface borne fungi. The seeds were shaken along with water (having a few drops of detergent) for a fixed period with the help of a shaker or manually). After centrifugation the sediment was diluted in waste and observed under compound microscope. This has been used for detection of downy mildew, powdery mildew and also spores of *Alternaria, Cercospora, Fusarium etc.* Each seed samples were examined visually and under low power of stereo-binocular microscope (fungi like *Protomyces macrospores*), before subjecting to incubation or other specialized tests. Visual examinations were made for the presence of *Sclerotia*, smut balls, discoloration, malformation, hypertrophies, fungal, spores, *Pycnidia, Perithecia etc.* Most commonly used incubation methods for fungal is moist blotter and agar tests wherein seeds are incubated on these media for a specific time. Generally it is for a week at a suitable temperature under alternating cycles of 12 hours light and darkness. These two media reveal a wide range of internal seed borne infection by fungal and bacterial pathogens in a wide variety of crops. Seedling symptom test and grow out test are also very versatile and reveal symptoms produced by any category of plant pathogens. Downy mildew of soybean is not yet reported from India. However, it was intercepted on seed of soybean imported from various countries (Mukewar *et al.*, 1980 and Agarwal *et al.*, 1990a). Agarwal *et al.* (1998) reported that while processing of 10,336 soybean samples during 1978-1995 downy mildew was intercepted in 1514 samples from 15 countries. In soybean *Phomopsis manshurica* was intercepted on seeds imported from Malaysia (Agarwal and Khetarpal, 1985) and Indonesia (Anitha *et al.*, 1993) where it was not reported as well. Latter on *Phomopsis longicola* was intercepted on *Glycine max* from USA. Beet rust caused by *Uromyces betae* was intercepted for the first time was intercepted in 1983 in seeds imported from Belgium, Denmark, Germany, UK, USA, U.S.S.R and Italy (Agarwal *et al.*, 1984), which 1987-96 was not yet reported from India. Similarly *Phoma lingum* on *Brassica* from several countries has been intercepted. Later, a quarantine treatment was developed by Nath *et al.*, 1986. They have found concentrated sulphuric acid

highly effective in eliminating the inoculum from seed. Seed-borne rust spores can cause epidemics in new areas (Emdal and Foldo, 1979).

### (d) Bacteria

Most commonly used incubation methods bacterial is moist blotter and agar tests wherein seeds are incubated on these media for a specific time. Discoloured, deformed and shriveled seeds were also taken as a criterion for the presence of bacterial pathogens. *Xanthomonas campestris* pv. *campestris*, black rot of crucifer was detected in *Brassica* spp. from several countries (Canada, France, Pakistan, Sweden, Taiwan, UK and USA) and *Pseudomonas syringae* pv. *syringae* was detected in *Hibiscus cannabinus* seeds from Bangladesh (Singh *et al.*, 1995). Plant pathogenic bacteria belonging to genera *Erwinia, Pseudomonas, Xanthomonas, Xylella* and *Xylophyllus* are gram-negative while *Clavibacter, Curtobacterium* and *Rhodococcus* are gram positive. Gram reaction should be confirmed using stains or by KOH test. For example Yellow discoloration around with hilum was used as indication of the presence of bacterial blight of French bean (*Xanthomonas campestris* pv. *phaseoli*) and vascular wilt in bean (*Curtobacterium flaccumfaciens* pv. *flaccumfaciens*). Similarly, halo blight of French bean (*Pseudomonas syringae* pv. *phaseolicola*) was detected on the basis of bluish white fluorescence in white seeded beans under ultra-violet light.

### (e) Viruses

Grow out test is the simplest of the tests used quite extensively for the detection of viruses, viroids, mycoplasma like organisms. Viral infections were identified on the basis of symptoms observed on plants/planting material. In case of seeds, mottling of seeds, mottling of seed coat, split or stained seed coat, necrotic spots, shriveled seeds were considered as a positive signal to viral infection. However, it should be used in combination with other tests like indexing, serology because some of these may be carried symptomlessly in the plants. Plant viruses detected in legumes like pea seed-borne mosaic virus-*Broad bean stain virus* on pea (The Netherlands and Syria), in faba bean (Syria and Bulgaria); bean common mosaic virus in soybean (USA); soybean mosaic virus in soybean (Argentina, Philippines and USA) and yellow mosaic virus in French bean (Syria) (Agarwal *et al.*, 1998). Cowpea mottle virus on *V. subterranean* from Ghana and *V. unguiculata* from the Philippines has been intercepted. *Raspberry ring-spot virus* on soybean from World Vegetable Centre (earlier AVRDC), Taiwan, Sri Lanka, Thailand, USA, *etc.* and Cherry leaf roll virus on *Glycine max* from Taiwan, Sri Lanka, Thailand and USA and *Phaseolus vulgaris* from Colombia. Seeds known or suspected to carry seed-transmitted viruses should be grown in insect-proof PEQ facility. Seedlings showing viral symptoms should be uprooted and burn. Produce from only healthy plants should be released to the indentors. Observation of samples from leaf showing viral symptoms and representative healthy-looking samples should be carried out under the transmission electron microscope which reveals the size and shape of the virus particles, giving indication about the group to which the virus belongs. The samples should be further tested by ELISA and if required reverse transcription PCR (RT-PCR).

### (f) Weeds

Several weeds like *Cichorium spinosum* and *Echinochloa crus-pavonis*, were intercepted which are not yet reported from India.

## II. Specilised Tests

Which are employed for the detection of specific pests.

### (a) Insect/Mites

X-Ray radiography technique has been found to be the most accurate method of studying the seed/stem and the identity of insect could also be ascertained without damaging the seed (Milner *et al.,* 1950; Wadhi *et al.,* 1967). Seeds infested with phytophagous chalcidoids, bruchids (hidden infestation) and certain other insects do not produce any external evidence on seed. Seeds were spread in a single layer and were exposed to X-ray @22ky, 3MA for 10 seconds at a distance of 30 cm. Infested and healthy seeds were clearly distinguishable on the x-ray radiograph. The seed transparency test (boiling seeds in lactophenol to make them transparent) can also be used for the detection of hidden infestation by insects in small size seeds and for extraction of insects for identification.

### (b) Nematode

For seed-borne nematodes, seeds are soaked in water for about 24 hours. This helps the nematode to become active which comes out of the soften seeds in the water. Seeds can also be teased out with the help of needles in water and examined under the microscope for the presence of nematodes. Rooted plants, rhizomes, corms *etc.,* soil and plant debris can similarly be soaked in water and the nematodes can be extracted for identification using nematological sieves or the tissue paper.

### (c) Fungi, Bacteria and Viruses

Serological tests are very effective for the detection and identification of viral and bacterial pathogens. Phage-plaque technique is still more sensitive for bacterial pathogens as even strains can be identified. Indicator test plants are also very effective as they may reveal pathogenic races within a species of fungus, bacterium and specific strains within a virus. Modifications of the moist blotter and agar tests can also be effective used for the detection of specific bacterial and fungal pathogens.

☆ Agar plate method should be used for seeds suspected to carry pathogenic fungi and bacteria, wherein seeds are plated on agar plates for detection of fungi/bacteria and identification is carried out based on types of growth and colony characteristics.

☆ Seedlings symptoms test should be undertaken for fungi/bacteria which are capable of infecting seeds, resulting in either rotting or necrosis of seeds or causing symptoms on seedlings.

☆ Embryo count method should be used for detecting embryo-borne fungi capable of causing disease in the plants in the next generation such as loose smut of wheat, downy mildew of pearl millet, *etc.*

☆ Molecular methods such as polymerase chain reaction (PCR) should be used to detect both fungi and bacteria using species-specific primers if required.

### Salvaging of Infested/Infected Germplasm

No residual inoculum of exotic pest (free from all angles) is utmost important before the material is released to the indentors. As tolerance for quarantine purposes are zero. Fumigation at atmospheric pressure or under reduced pressure (depending on the type of infestation) has been found acceptable as a quarantine treatment for insects and mites.

ICAR-National Bureau of Plant Genetic Resources (ICAR-NBPGR), New Delhi has standardized several methods for salvaging the germplasm against various pathogens like-

(i) **Mechanical cleaning:** Any soil clods, plant debris, weed seeds, ergot sclerotia, smut/bunt balls, discoloured, deformed, malformed seeds, larvae, insect bodies or parts thereof should be mechanically cleaned by hand picking. The vegetative propagules should be cleaned by excising the infected portion.

(ii) **Fumigation:** It is one of the most effective methods used in quarantine for eliminating insects and mites. Atmospheric fumigation should be done at normal air pressure in an air tight container using approved fumigants. Fumigants like Methyl bromide, EDCT mixture, HCN and Phosphine are commonly used.

(iii) **X-ray radiography:** X-ray radiography is a unique technique used for both detection and salvaging, and should be used to separate insect infested seeds (which do not have any external symptoms) from healthy ones. Upon exposure to X-rays, the infested seeds can be easily distinguished and should be hand-picked from the seed lot/geometry. In casse of real-time X-ray machine, the process is much faster and salvaging can be done immediately after the image of infested sample appears on the monitor screen.

(iv) **Hot water/air treatment (HW/AT):** Hot water or hot air treatment (different temperature and time durations) are also used for eradication of insects, mites, nematodes, fungi, bacteria and viruses. The temperature has to be very accurately controlled so as to kill the target pest but not the host. The hot water treatment should be carried out in tank fitted with heaters of different capacities, stirrer, thermostat and contact thermometer for controlling the water temperature. In some cases, pre-soaking gives better results.

(v) **Vapour heat treatment (VHT):** Vapour heat treatment (VHT) is also used for making samples free from pathogens but time of vapour enforcement should be standardized as per crop or samples.

(vi) **Thermotherapy:** Meristem tip culture coupled with thermotherapy could be successfully used for salvaging materials infected with systemic pathogens like viruses, viroids, downy mildews *etc.*

(vii) **Chemical (Pesticidal) treatment:** Chemical/pesticidal treatments are most practical method in quarantine for effective control of surface feeding insects, mites, nematodes *etc.* which is used as dry, slurry, dip or spray treatment against various pests. Pesticidal dip/spray for vegetative propagules and pestcidal dressings of suitable systemic and contact pesticides should be used for seeds at various concetrations and for different time duration. Therefore it is essential to standardize the actual doses of the chemical as per the target pest but not as per the host.

(viii) **Isolation growing:** Isolation growing under quarantine greenhouse conditions are used not only for detection but for salvaging the valuable materials.

The following methods are used for salvaging of fungi:

(a) Spirit wash

(b) Acid wash

### (a) Spirit Wash

A modified technique of seed washing through spirit was adopted to eradicate the seed born pathogen like rust spores of *Puccinia carthami* and *P. helianthi*, the safflower and sunflower rusts, respectively. The contaminated seeds were taken in a test tube. Ethyl alcohol and a pinch of river sand were added to it. The contents were stirred on the medicinal stirrer for 30 seconds. The spores adhering to the seed surface got agitated and were separated out in the alcohol (Agarwal *et al.*, 1990).

### (b) Acid wash

Concentrated sulphuric acid ($H_2SO_4$) was used for destroying the rust spores of sugar beet (*Uromyces betae*) adhering to seeds. The contaminated seeds were stirred with a glass rod in the acid for 1 minute. The treated seeds were immediately washed under the running tap water to remove the traces of acid and then sun dried (Nath *et al.*, 1986).

(i) **Dissection techniques:** True seeds infected with nematodes do not show any clear cut symptoms and therefore difficult to separate from healthy ones. Nematodes from such seeds were recovered by soaking the seeds in water at 25°C for about 24 hours and observing under stereoscopic microscope after crushing the seeds or removing the seed coat.

(ii) **Staining techniques:** The most common staining technique used for root samples was to colour tissue with acid fuchsine dissolved in lactophenol or lactoglycerol, and then destain it with clear lactophenol/lactoglycerol (Goodey, 1937).

(iii) **Soil examination:** Cyst forming and ecto-parasitic nematodes were spread with small amounts of soil with seeds or adhered to roots. To detect them, 5-10g of soil was mixed with 20 ml water and directly observed under stereoscopic microscope. Observations on nematodes in introduced material were also taken by growing the material in post entry quarantine nurseries at ICAR-NBPGR, New Delhi (Lal and Mathur, 1988).

### (c) Fungi and Bacteria

(i) **Moist blotter:** Seeds were incubated on moist blotting paper for a period of seven days at 20°C with alternating cycles of 12 hours light and darkness.

(ii) **Growing on test**

(a) **Water agar method**: The method proposed by Srinivasan *et al.* (1973) was used, to detect *Xanthomonas campestris* pv. *campestris* in crucifer seed.

(b) **Seedling grown in green house:** Bacterial pathogen in beans was detected by this method.

(c) **Specialized tests:** Specialized tests were to detect bacteria and fungi in various crops and have been summered in the Table 6.15.

### (d) Viruses

For detection of viruses following tests were used:

(i) Infectivity test,

**Table 6.13: Specialized Tests Used for the Detection of Fungi and Bacteria in Various Crops**

| Pathogen | Test Method Based on | Crop |
|---|---|---|
| Fungal pathogen | Acid wash | Sugar bet |
| Bacterial pathogen | Water-Agar method | *Brassica* species |
| | Washing test | *Brassica* species |
| | Paper towel method | *Cucumis sativus* |

(ii) Staining for detection of inclusion bodies

(iii) Virus indexing by ELISA (in cases where the specific antisera were available)

## Post Entry Quarantine (PEQ)

Post Entry Quarantine (PEQ) is very important aspects under plant quarantine system. It includes:

(a) Insects

(b) Fungi and Bacteria

(c) Viruses

### (a) Insects

It has been observed that by visual or through common testing methods sometimes detection of the insects is not possible. Therefore, suspected material, where no pathogen was intercepted was grown in isolation in glass/net/screen houses and inspected

**Table 6.14: Interception of Insects/Nematodes**

| Insects/Nematodes | In Association with | From |
|---|---|---|
| *Acanthoscelides obtectus* | *Cajanus cajan* | Brazil and Columbia |
| *Acanthoscelides obtectus* | *Phaseolus vulgaris* | Italy and Nigeria |
| *Acanthoscelides obtectus* | *Phaseolus vulgaris* | USA |
| *Callosobruchus maculatus* | *Vigna unguiculata* | Italy |
| *Bruchus sp.* | *Lathyrus sativus* | ICARDA |
| *Anthonomus grandis* | *Gossypium* and *Vigna* species | USA |
| *Ditylenchus dipsaci* (nematode) | *Allium cepa* | UK |
| *Heterodera schachtii* (nematode) | *Beta vulgaris* | Germany, Denmark, Italy |
| *Globodera rostochiensis* (nematode) | *Potato* (Golden nematode) | - |

**Table 6.15: Important Vegetable Diseases Believed to have been Introduced in India**

| Host | Diseases | Pathogen | First record in India | From |
|---|---|---|---|---|
| Beans | *Fuscans* blight | *Xanthomonos phaseoli* var. *fuscans* | Pantnagar, Shimla, Solan, Delhi, 1972 | |
| | Common blight | *Xanthomonos phaseoli* | Pune, 1972 | |
| Cabbage and cauliflower | Black rot | *Xanthomonos campestris* | Delhi, 1968 | |
| Chillies | Leaf spot | *Xanthomonos vasicatoria* | Maharashtra, 1950 | |
| Onion | Downy mildew | *Peronospora destructor* | Jammu and Kashmir, 1977 | |
| Potato | Late blight | *Phytophthora infestans* | U.P., 1983 | |
| | Wart | *Synchytrium endobioticum* | West Bengal, 1953 | |
| Okra | - | *Fusarium solani* | - | Taiwan |
| *Luffa* sp. | - | *Fusarium solani* | - | USA |
| Tomato | | *Fusarium solani* | - | Netherland and USA |
| Brinjal | | *Fusarium solani* | - | Taiwan and Spain |
| Chilli | | *Fusarium oxysporum* | | Taiwan |
| | | *Colletotrichum capsici* | - | Taiwan |
| Cabbage | - | *Alternaria brassicicola* | - | Netherlands |
| | - | *Xanthomonas capmpestris* pv. *campestris* | - | Netherland |
| Cauliflower | - | *Alternaria brassicicola* | - | Netherlands |
| | - | *A. brassicae* | - | Netherlands |
| | - | *Xanthomonas capmpestris* pv. *campestris* | - | Netherlands |
| Muskmelon | - | *Sclerotinia sclerotiorum* | - | USA |
| Bottle gourd | - | *Sclerotinia sclerotiorum* | - | USA |

regularly. As and when the symptoms observed or the identification of the emerged adult stages is completed, the planting material was suitably treated, destroyed or released to the indenter. Due to doubt, in some cases, material was grown for one full cropping season and after satisfaction only healthy produce was released to the indenters.

### (b) Fungi and Bacteria

It is mandatory that infected seeds must be sown in the isolated net house or in isolation nurseries in the field. The seeds were harvest only from the disease free plants and collected carefully. This technique was normally used for chemically treated germplasm of cucurbits against downy mildew infection.

### (c) Viruses

For Post Entry Quarantine (PEQ), seed samples were grown under controlled conditions. The symptoms on growing were observed and accordingly the material was released.

In the recent past several viral diseases were also introduced with the vegetable seeds. Some of them are being listed below:

**Table 6.16: Interception of Bean Viruses**

| Viral Disease Intercepted | Crop | Source |
|---|---|---|
| Bean common mosaic virus | Phaseolus vulgaris | USA |
| Bean common mosaic necrosis virus | P. vulgaris | USA |
| Bean pod mottle virus* | P. vulgaris | USA |
| Grapevine fan leaf virus | V. unguiculata** | Nigeria |
| Red clover vein mosaic virus* | P. vulgaris | USA |
| Soybean mosaic virus | P. vulgaris** | USA |
| Bean common mosaic virus | Phaseolus vulgaris | USA |

\* Not reported from India; ** Virus present in India but not recorded on the host on which intercepted.

### Interception of Weeds

| Weeds intercepted | Crop | Source |
|---|---|---|
| Convolvulus arvensis | P. vulgaris | USA |

During 2005-2010, a total of 7,461 samples of vegetables were supplied by the ICAR-NBPGR, New Delhi under the Material Transfer Agreement (MTA).

### Regional Cooperation

Developing countries in particular have become more vulnerable to the dangers of introduction of exotic pests. Lack of adequate funds and trained manpower is a serious constraint for them to develop effective quarantine systems to take care of the variety and the volume of material required for increasing food production. Regional cooperation in plant quarantine as in other fields could be very helpful as various types of resources could be pooled to develop a modern and effective quarantine system for the entire SAARC region. Such a regional cooperation in plant quarantine has been developed by ASEAN countries with common plant quarantine regulations promulgated by each country to protect agri-horticultural and forestry wealth of the entire region against exotic pests. This concept of co-operation is based on the premise *i.e.* the larger the land mass covered by the same set of quarantine regulations, greater is the protection afforded to the agriculture of the region against exotic pests.

### Gene Protection Technology

Gene protection technology is essential or not essential tool with respect to "Terminator Gene" is much concerned and debating issues among the researchers and policy makers. Numerous reports are appearing on the sensationally entitled this issue. Monsanto would prefer to term this as "Gene Protection Technology". A number of concerns have been raised in which Monsanto

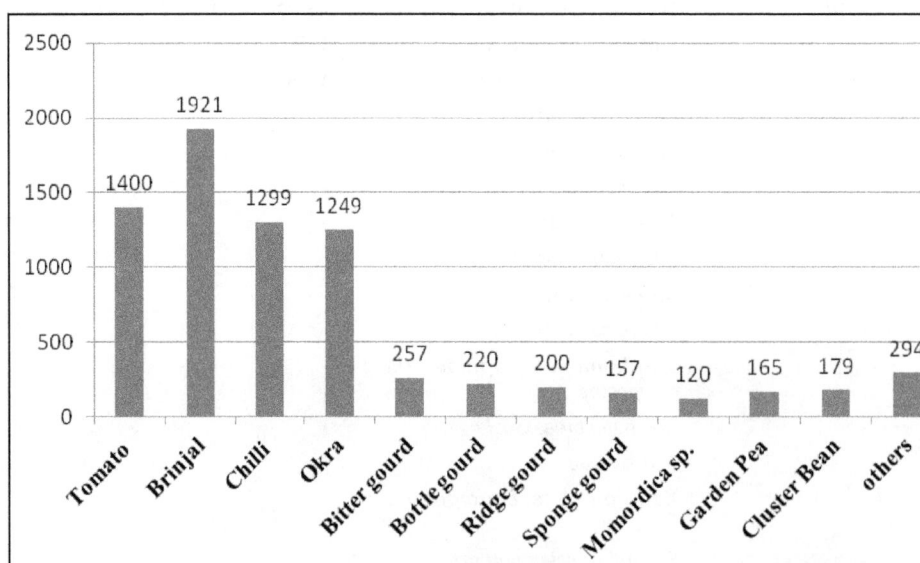

**Figure 6.3: Distribution of Important Vegetable Germplasm during 2005-2010.**

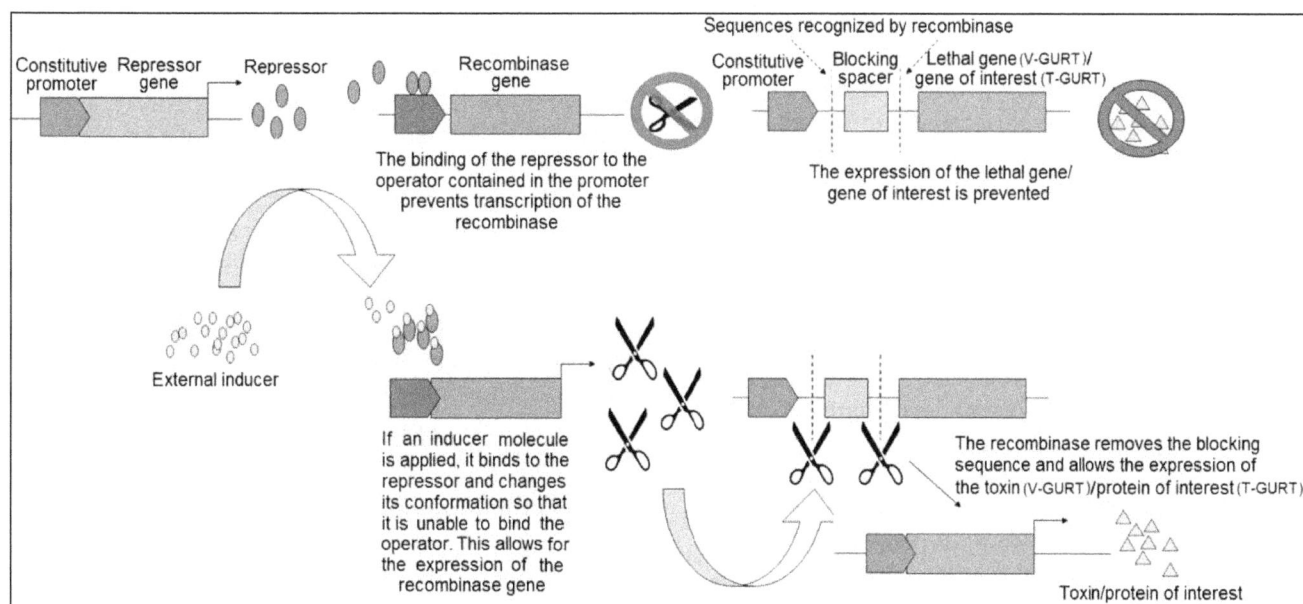

**Figure 6.4: Schematic Representation of the Mechanism of GURTs.**

**In a variant of this scheme, the repressor proteins are normally unable to bind to the pertinent DNA and the inducer is applied to bind the repressor that undergoes a change in its conformation so that it can bind to the DNA and activate transcription.**

*Source*: **Luca Lombardo, 2014.**

has been implicated but fact is that Monsanto does not use Gene protection technology in any of its products. Genetic use restriction technologies (GURTs) developed to secure return on investments through protection of plant varieties, are among the most controversial and opposed genetic engineering biotechnologies as they are perceived as a tool to force farmers to depend on multinational corporations' seed monopolies. In this work, the currently proposed strategies are described and compared with some of the principal techniques implemented for preventing transgene flow and/ or seed saving, with a simultaneous analysis of the future perspectives of GURTs taking into account potential benefits, possible impacts on farmers and local plant genetic resources (PGR), hypothetical negative environmental issues and ethical concerns related to intellectual property that have led to the ban of this technology (Luca Lombardo, 2014).

### Illegal Entry of the "Terminator Seeds" in India

"Terminator Seeds" was bought in picture by Monsanto and Delta and Pine Land, separately. Monsanto claim that "Terminator Gene" is only a theoretical Gene Protection Technology and does not exist in the real sense. Delta and Pine Land and USDA first applied for a prophetic patent, which really means a conceptual patent. In this case, the idea was to develop a process that would provide gene protection in plants and tobacco was used as experimental plant. Illegal entry of the "Terminator Seeds" in India is going to create an adverse situation among farming community.

### Monopoly of Multinational Company in the Indian Market

Monopoly of Multinational Company in the Indian seed market will dominate with the gene protection technology and would limit the Indian farmer's choice. But it is worth nothing that hybrid seeds today provide the farmer with only one productive growing season. Second generation use of hybrids will result in poor yields and low productivity, yet farmers choose to use these highly productive first generation seeds every year.

**Genetic use restriction technology (GURT)**, colloquially known as **terminator technology** or **suicide seeds**, is the name given to proposed methods for restricting the use of genetically modified plants by causing second generation seeds to be sterile. The technology was developed under a cooperative research and development agreement between the Agricultural Research Service of the United States Department of Agriculture and Delta and Pine Land company in the 1990s, but it is not yet commercially available.[1]

The technology was discussed during the 8th Conference of the Parties to the United Nations Convention on Biological Diversity in Curitiba, Brazil, March 20–31, 2006.

There are conceptually two types of GURT (*From Wikipedia, the free encyclopedia; retrieve on 02-01-2017*)

1. Variety-level Genetic Use Restriction Technologies (V-GURTs): This type of GURT

produces sterile seeds, so the seed from this crop could not be used as seeds, but only for sale as food or fodder (www.worldseed.org, *2003*). This would not have an immediate impact on the large number of primarily western farmers who use hybrid seeds, as they do not produce their own planting seeds, and instead buy specialized hybrid seeds from seed production companies. However, currently around 80 per cent of farmers in both Brazil and Pakistan grow crops based on saved seeds from previous harvests (Rizvi, 2006). Consequentially, resistance to the introduction of GURT technology into developing countries is strong (Rizvi, 2006).The technology is restricted at the plant variety level, hence the term V-GURT. Manufacturers of genetically enhanced crops would use this technology to protect their products from unauthorized use.

2. T-GURT: A second type of GURT modifies a crop in such a way that the genetic enhancement engineered into the crop does not function until the crop plant is treated with a chemical that is sold by the biotechnology company (www.worldseed.org, 2003). Farmers can save seeds for use each year. However, they do not get to use the enhanced trait in the crop unless they purchase the activator compound. The technology is restricted at the trait level, hence the term T-GURT.

For V-GURTs, essentially three different restriction mechanisms have been proposed (Visser *et al.*, 2001). The first mechanism of action is that described in the patent (U.S. 5,723,765) by the USDA and Delta and Pine Land (nominally the first V-GURT). This GURT is based on the transfer of a combination of three genes (transgenes), two derived from bacteria and one from another plant, into a plant's cells:

## Opposition

Terminator seeds were initially developed as a concept by the United States Department of Agriculture and multinational seed companies. As of 2006, they had not been commercialized anywhere in the world due to opposition from farmers, indigenous peoples, NGOs, and some governments. In 2000, the United Nations Convention on Biological Diversity recommended a *de facto* moratorium on field-testing and commercial sale of terminator seeds; the moratorium was re-affirmed and the language strengthened in March 2006, at the COP8 meeting of the UNCBD (*Moratorium 2013*). India and Brazil have passed national laws to prohibit the technology (Rizvi, 2006).

## Cross Pollination Leading to Sterility of other Crops

Cross pollination leading to sterility of other crops are common concerns.

## Potential Uses

☆ Where effective intellectual monopoly, specifically biological patents, doesn't exist or is not enforced, GURTs could be an alternative to stimulate plant developing activities by biotech firms (www.worldseed.org, *2003*).

☆ Non-viable seeds produced on V-GURT plants may reduce the propagation of volunteer plants. Volunteer plants can become an economic problem for larger-scale mechanized farming systems that incorporate crop rotation (www.worldseed.org, *2003*).

☆ Under warm, wet harvest conditions non V-GURT grain can sprout, which lowers the quality of grain produced. It is likely that this problem would not occur with the use of V-GURT grain varieties.[1]

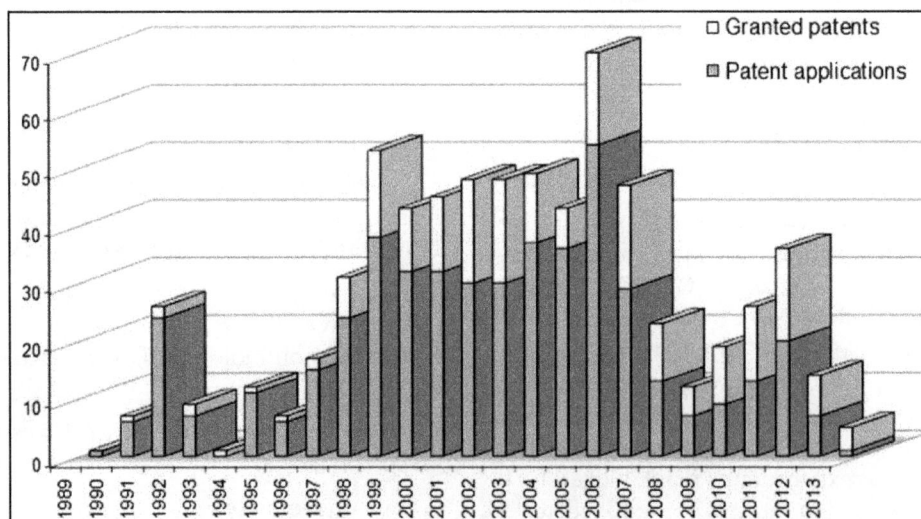

**Figure 6.5: Granted and Applied GURT Patents.**
*Source*: **Luca Lombardo, 2014.**

✩ Use of V-GURT technology could prevent escape of transgenes into wild relatives and prevent any impact on biodiversity. Crops modified to produce non-food products could be armed with GURT technology to prevent accidental transmission of these traits into crops destined for foods (www.worldseed.org, *2003*).

# References

1. Agarwal, P.C. and Khetarpal, R.K. 1985. Interception of *Pernospora manshurica* on soybean seeds imported from Malaysia. *FAO Plant Prot. Bull.*, **33**: 49.

2. Agarwal, P.C., Majumdar, A., Dev, U., Nath, R. and Khetarpal, R.K. 1990a. Seed borne fungi of quarantine importance in exotic germplasm of soybean (*Glycine max*). *Indian J. Agric. Sci.*, **60**: 361-363.

3. Agarwal, P.C., Majumdar, A., Dev, U., Nath, R. and Khetarpal, R.K. 1990b. Quarantine salvaging of safflower seed infected with safflower rust (*Puccinia carthami*). *FAO Plant Prot. Bull.*, **38**: 141-143.

4. Agarwal, P.C., Nath, R., Dev, U. and Majumdar, A. 1990. Quarantine salvaging of sunflower seed infected with safflower rust (*Puccinia carthami*). *FAO Plant Protection Bulletin*, **38**: 141-143.

5. Agarwal, P.C., Singh, B. and Gautam, P.L. 1998. Plant quarantine as an aid to safer exchange, conservation and utilization of Plant Genetic Resources. *Indian J. Plant Genet. Resour.*, **11(2):** 165-171.

6. Anitha, K., Agarwal, P.C. and Nath, R. 1993. Interception of *Pernospora manshurica* in soybean seeds imported from Indonesia. *FAO Newsletter APPPC*, **36:** 1.

7. Brenner, C. 1995. Technology Transfer: Public and Private Sector Roles. In: Kommen, J., Cohen, J.J. and Lee, S.K. (eds.) Turning, Priorities into Feasible Programmes: Proc. of a Regional Seminar on Planning, Priorities and Policies for Agricultural Biotechnologies in Southeast Asia. Intermediary Biotechnology Service/Nayang Technological University. The Hague/Singapore.

8. Burton, G.W. 1985. *Field Crops Research*, **11:** 132-129.

9. Emdal, P.S. and Foldo, N.E.1979. Seed-borne inoculum of *Uromyces betae*. *Seed Sci. and Technol.*, **7:** 93-102.

10. Goodey, T. 1937. Two methods for staining nematodes in plant tissues. *Journal of Helminthology*, **15:** 137-144.

11. Heffer, P. 2001. Biotechnology: a modern tool for food production improvement in Seed Policy and Programmes for the Central and Eastern European Countries, Commonwealth of Independent States and Other Countries in Transition. FAO Plant Production and Protection Paper 168. Food and Agriculture Organization of the United Nations.

12. Jacob, C.T. 2016. Mainstreaming biodiversity into the agricultural sector. 1st International Agrobiodiversity Congress (Abs. Session: IRPs, ABS and Farmers Rights), 6-9 November 2016; held at New Delhi, p. 270.

13. Jefferson, R.A., Byth, D., Correa, C., Otero, G. and Qualset, C. (1999). Genetic use restriction technologies: technical assessment of the set of new technologies which sterilize or reduce the agronomic value of second generation seed, as exemplified by U.S. Patent No. 5,723,765, and WO 94/03619. Convention on Biological Diversity (UNEP/CBD/SBSTTA/4/9/Rev.1).

14. Kenza Le Mentec. 2016. Trade and Biodiversity. 1st International Agrobiodiversity Congress (Abs. Session: III. Quarantine, Biosafety and Biosecurity), 6-9 November 2016; held at New Delhi, p. 257.

15. Khetarpal, R.K. 1985. Interception of *Peronospora monshurica* on soybean seeds imported from Malaysia. *FAO Plant Prot. Bull.* 33: 49.

16. Khoury, C.K. *et al.*, 2016. Origin of food crops connects countries worldwide. Proceedings of the Royal Society of London: Biological Sciences, 283: 20160792.http://dx.doi.org/10.1098/rspb.2016.0792.

17. Klinkowski, M. 1970. Catastrophic plant diseases. *Annual Review of Phytopathology*, **8:** 37-60.

18. Krattiger, A.F. and Lesser, W.H. 1994. Biosafety-an environment impact assessment tool and the role of the Convention on biological diversity. In: Krattiger, A.F., Lesser, W.H., miller, K.R., Hill, St. Y. and Senananayake (eds.) Widening Perspectives on Biodiversity, IUCN, Gland, Switzerland and International Academy of the Environment, Geneva, Switzerland.

19. Lal, A. and Mathur, V.K. 1988. Seed health testing for plant parasitic nematodes. In: Proceedings of Seed Science and Technology (eds. Yadav, T.P. and Chandgi, R.), p. 318-320. HAU Press, Hisar (Haryana).

20. Luca Lombardo, 2014. Genetic use restriction technologies: a review. *Plant Biotechnology Journal*, **12:** 995–1005.

21. Milner, M., Lee, M.R. and Kaiz, R. 1950. Application of X-ray techniques to the detection of internal insect infestation of grain. *Journal of Economic Entomology,* **43:** 933-935.

22. *"Moratorium". Ban Terminator.* Retrieved 12 December 2013.

23. Mukwar, P.M., Nath, R., Lambdat, A.K., Kapoor, U., Khetarpal, R.K. and Rani, I. 1980. Interception of *Pernospora manshurica* imported seed of soybean. *Seed Res.,* **8:** 267-270.

24. Nath, R., Agarwal, P.C., R., Dev and Lambat, A.K. 1986. Quarantine treatment of sugar beet seed infection with rust (*Uromyces betae*) *FAO Plant Protection Bulletin,* **348:** 205-207.

25. Neergaard, P., 1979. Seed Pathology. Vol. 1. The Macmillan Press Ltd., London, UK.

26. Persley, G.J., Giddings, L.V. and Juna, C. 1992. Biosafety: The safe application of Biotechnology in Agriculture and the Environment. International Service for National Agricultural Research. The Hague/Singapore.

27. Rizvi, H. 2006. "BIODIVERSITY: Don't Sell "Suicide Seeds", Activists Warn", Inter Press Service News Agency, 21st March.

28. Roos, E.E. 1984. Genetic shifts in mixed bean populations. II Effects of regeneration. *Crop Sci.,* **24:** 711-715.

29. Singh, B. 1994. Regeneration and multiplication of germplasm resources for gene bank. In: Plant Genetic Resources: Exploration, Evaluation and Maintenance. (eds.) Rana, R.S., Singh, B., Koppar, M.N., Rai, M., Kochhar, S. and Duhoon, S.S., ICAR-NBPGR, New Delhi, p. 103-114.

30. Singh, B. and Williams, J.T. 1984. Maintenance and multiplication of plant genetic resources. In: Conservation and Evaluation. Holden, J.H.W. and Williams, J.T. (eds.). George Allen and Unwin Ltd. London.

31. Singh, B. Anitha, K., Agarwal, P.C. and Nath, R. 1995. Interception of *Pseudomonas syringae* pv. *syringae* on *Hibiscus cannaabinus. Indian Phytopath.,* **48:** 352.

32. Spence, N., Woodhall, J. Bishop, S. and Smith, J.J. 2006. Patterns of new plant disease and pest management-Evidence of growing problem? In: I. Banker, J. Brownlie, C. Peckham, J. Pickett, W. Stewart, J. Wage, P. Wilson and M. Woolhouse. Infection Diseases: Preparing for the Future: A Vision of Future Detection, Identification and Monitoring Systems. London, UK: Office of Science and Technology.

33. Srinivasan, M.C., Neergaard, P. and Mathur, S.B. 1973. A technique for detection of *Xanthomonas campestris* in seed health testing of crucifers. *Seed Sci. and Technol.,* **1:** 853-859.

34. Visser, B., Van der Meer, I., Louwaars, N., Beekwilder, J. and Eaton, D. 2001. The impact of 'terminator' technology. *Biotechnol. Dev. Monit.,* 48: 9–12.

35. Wadhi, S.R., Verma, B.R., Thomas, S. and Lal, R. 1967. Detection of phytophagous chalcidoids in seeds for quarantine purposes. *Indian Journal of Entomology,* **29(2):** 197-199.

36. Walsh, V. 1993. Demand, public markets and innovation in biotechnology. *Science and Public Policy,* **25:** 138-156.

37. *Wikipedia, the free encyclopedia;* Retrieve on 02-01-2017.

38. www.worldseed.org, International Seed Federation. "Genetic Use Restriction Technologies (Bengaluru, June 2003)" (PDF). (Position Paper Supporting V-GURT development).

# 7

# Monitoring and Management System of Plant Genetic Resources

## Introduction

Plant genetic resources occupy an important place within the overall umbrella of biodiversity. India which is one of the 12 mega centres of diversity is rich in plant wealth which includes 356 domesticated species of economic importance and 326 species of their wild form/relatives, native to India. The Indian National Plant Genetic Resource System (IN-PGRS) under the aegis of the Indian Council of Agricultural Research with ICAR-National Bureau of Plant Genetic Resources (ICAR-NBPGR), New Delhi as apex body, is very strong system and holds a prominent place among the global gene bank. ICAR-National Bureau of Plant Genetic Resources (ICAR-NBPGR), New Delhi has the responsibility to plan, conduct, promote, coordinate and take lead in the activities related to plant germplasm collection, introduction, exchange, evaluation, documentation, conservation and sustainable management of germplasm of crop plants and their wild relatives.

At conceptual level, the need for establishment of an agency for "Organized Plant Introduction" in India was expressed by late Dr. B.P. Pal and " Crops and Soil Wing" of the Board of Agriculture and Animal Husbandry in India way back in 1935. Historically, Plant Genetic Resources (PGR) activities were started as early as in 1946 with the initiation of 'Plant Introduction Scheme' in Botany Division of Indian Agricultural Research Institute, New Delhi. This scheme was further strengthened into a full-fledged Plant Introduction and Exploration Organization in 1956 and was upgraded into an independent Division of Plant Introduction

of the ICAR-Indian Agricultural Research Institute, New Delhi in 1961. On the basis of recommendations of a high level committee appointed by government of India in 1970 and through foresight of visionaries like Dr. B.P. Pal, Dr. M.S. Swami Nathan and Dr. A.B. Joshi, the Division of Plant Introduction was upgraded to an independent level and called National Bureau of Plant Introduction (NBPI), New Delhi in August 1976 and rechristened as ICAR-National Bureau of Plant Genetic Resources (ICAR-NBPGR), New Delhi in January, 1977. This is operated under the ICAR system controlled by the Department of Agricultural Research and Education (DARE), Ministry of Agriculture and Cooperation, Govt. of India. Besides main activities of Plant Genetic Resources (PGR) management at the Headquarters at New Delhi, National Bureau of Plant Genetic Resources (NBPGR) has 11 regional stations at Akola, Amaravati, Bhowali, Cuttack, Hyderabad, Jodhpur, Ranchi, Shillong, Shimla, Srinagar and Thrissur. As per the mandate crops of the regional centres, the vegetable crops included are winged bean (Akola), chilli and French-bean (Bhowali), chilli, tomato and brinjal (Hyderabad), cowpea, cluster bean (guar) and month bean (Jodhpur), ginger, colocasia, turmeric and chilli (Shillong), French-bean, amaranth and meetha karela (Shimla), okra (Thrissur) as elaborated by Gautam *et al.* (2000) and Dhillon *et al.* (2001). Further, ICAR- Indian Institute of Vegetable Research, Varanasi (U.P.) has one of the major mandates on vegetable germplasm collection, characterization, evaluation, conservation and utilization. The Indian National Plant Genetic Resources System (IN-PGRS) tends to integrate PGR activities into a system providing for maintenance of active collections of crop genepool at several ecosites, assigned with situation specific responsibility, which are ultimately linked to a single base collection (Rana, 1992).

Plant genetic resources management involves broadly five stages:

(i) Exploration and collection

(ii) Exchange

(iii) Characterization and evaluation

(iv) Conservation and utilization

(v) Documentation

## Definition of Germplasm Monitoring

"Germplasm is the regular checking of the quality (the viability) and the quantity (the existing stocks in number in weight) of seeds or plant materials of germplasm accessions stored in a gene bank".

## Germplasm Advisory Committees (GAC)

Bureau's effort towards organizing a strong National PGR System is noteworthy in constitution of Germplasm Advisory Committees (GAC) whicg habe been set up for specific crops or groups of crops. They play an important role by advising the Bureau regarding the status of its current holding of different crops, shortcomings in storage and management system as well as gaps in exploration and collection of indigenous genetic variability of native and naturalised crops. These committees also suggest promising countries/regions in the world that need to be explored or approached for introduction of new crops/genetic variability to sustain our crop improvement. Internationally, The IPGRI Advisory Committee has been set up to monitor status of germplasm worldwide. For example germplasm stored in the gene bank (-20°C) for long term conservation but how long it will be in the genebank, need time base monitoring. The IPGRI Advisory Committee on seed storage recommends that seeds stored under the preferred standards in base collection should be monitored at least every ten years and for active collection monitoring is recommended every 5 years. These guidelines given by IPGRI hold good for temperate regions (European countries) but for tropical countries (South-east Asian countries) it doesn't hold good due to temperature fluctuations, power failure *etc.* Therefore, suitable monitoring intervals have to be worked out for the species stored. In most species, accession should be regenerated when the seed viability falls below 85 per cent and these accessions should be sent to the concerned scientist or to active germplasm sites for rejuvenation of seeds with the request to send fresh seed for storage in the gene bank.

## National Advisory Board on Management of Genetic Resources

Constituted by president, Indian Council of Agricultural Research (ICAR), vide F. No. 8(2)/2011-Cdn. (Tech.) (Part) dated October 14, 2011 and February 21, 2014.

| | | |
|---|---|---|
| 1. | Dr R.S. Paroda, Chairman, TAAS, New Delhi | Chairman |
| 2. | Secretary, DARE and DG, ICAR, new Delhi | Co- Chairman |
| 3. | Chairperson, PPV&FRA, New Delhi | Member |
| 4. | Chairperson, NBA, Chennai | Member |
| 5. | DDG (Crop Science), ICAR, New Delhi | Member |
| 6. | DDG (Horticultural Science), ICAR, New Delhi | Member |
| 7. | DDG (Animal Science), ICAR, New Delhi | Member |
| 8. | DDG (Fisheries Science), ICAR, New Delhi | Member |
| 9. | Chairman, Research Advisory Committee, ICAR-NBPGR, New Delhi | Member |
| 10. | Chairman, Research Advisory Committee, ICAR-NBFGR, Lucknow (Uttar Pradesh) | Member |
| 11. | Chairman, Research Advisory Committee, ICAR-NBAGR, Karnal (Haryana) | Member |
| 12. | Chairman, Research Advisory Committee, ICAR-NBAIR, Bengaluru (Karnataka) | Member |
| 13. | Chairman, Research Advisory Committee, ICAR-NBAIM, Mau (Uttar Pradesh) | Member |
| 14. | Executive Director, MSSRF, Chennai (Tamil Nadu) | Member |
| 15. | Director, NBRI (CSIR), Lucknow (Uttar Pradesh) | Member |
| 16. | CEO, National Medicinal Plants Board, New Delhi | Member |
| 17. | Regional Coordinator, South and South East Asia, Biodiversity International, New Delhi | Member |
| 18. | Dr P.L. Gautam, Former Chairperson, NBA, Chennai and PPV&FRA, New Delhi | Member |
| 19. | Dr M. Mahadevappa, Former Chairperson, ASRB and Director, JSS Rural Development Foundation, Mysore, Karnataka | Member |
| 20. | Dr Lalji Singh, Vice-Chancellor, B.H.U., Varanasi (Uttar Pradesh) | Member |
| 21. | Dr Suman Sahai, Gene Campaign, New Delhi | Member |
| 22. | Dr Sushma R. Chaphalkar, Director, Vidya Pratishthan's School of Biotechnology, Vidyanagar, Baramati (Maharashtra) | Member |
| 23. | Director, ICAR-NBFGR, Lucknow (Uttar Pradesh) | Member |
| 24. | Director, ICAR-NBAGR, Karnal (Haryana) | Member |
| 25. | Director, ICAR-NBAIR, Bengaluru (Karnataka) | Member |
| 26. | Director, ICAR-NBAIM, Mau (Uttar Pradesh) | Member |
| 27. | ADG (IP and TM), ICAR, New Delhi | Member |
| 28. | ADG (Cdn.), ICAR, New Delhi | Member |
| 29. | Joint Secretary, MoEF and CC, New Delhi | Member |
| 30. | Joint Secretary, (Seeds), DAC, MoA and FW, New Delhi | Member |
| 31. | Joint Secretary, DADF, MoA and FW, New Delhi | Member |
| 32. | Joint Secretary, DBT, MST, New Delhi | Member |
| 33. | Director, DARE, MoA and FW, New Delhi | Member |
| 34 | Director, ICAR-NBPGR, New Delhi | Member-Secretary |

## Data File for Monitoring the Germplasm

The following points are necessary during monitoring of the germplasm kept in the gene bank (Saxena, 1998):

☆ Name of crop/species

☆ Initial weight of an accession (better will be crop base number decided by ICAR-NBPGR or 100 or 1,000 seeds weight as applicable)

☆ Date of storage

☆ Number of accessions stored

☆ Moisture percentage in the seeds

☐ Initial germination percentage

☐ Before performing monitoring test for viability, check each accession number in the data file to make sure that sufficient seeds are available for the test. If few seeds are available, do not test now because the material will have to be regenerated soon.

☐ Count the number of seeds needed for the test from each accession. Work in a dehumidified room, if possible, and close the containers as soon as the seeds are removed, to prevent the seeds absorbing the moisture from the air.

☐ Containers should be opened after conditioning to the room temperature to avoid condensation on the seed and change in moisture content.

## Monitoring of Germination

Periodical germination test of conserved germplasm is very essential. For monitoring of the germplasm for germination, following points should be kept in the mind (Saxena, 1998):

1. Date of monitoring test should be noted.

2. Fixed sample size germination test is done using a minimum of 200 seeds in two replications of 100 seeds each. Sequential germination test is done by forming group of 40 seeds. Calculate the germination from the result of test (Table 7.1).

3. Count number of normal seedlings, and abnormal seedlings.

4. If germination is above 85 per cent accept the test as valid.

5. If germination is less than 85 per cent repeat the test.

6. Calculate the mean percentage viability from the result of two tests and use this as the overall test result; if the percentage decreases to 85 per cent or less, regenerate the accessions. If the percentage is above 85 per cent keep in store.

## Regeneration of Seeds

The most important activity of conservation is to maintain the viability of each accession in the gene bank. Therefore, at a fixed interval monitoring of viability is an essential activity of the gene bank. Some gene bank managers set higher limits since some genes are rare in a population and even slight deterioration of accession viability could mean the loss of some potentially valuable genes, where as a lower rate of germination is acceptable for relatively modern varieties than landraces and wild species. Thus, regeneration standards for each crop species are set so as to balance the risk of valuable gene losses against high cost of regeneration. The IPGRI recommends that if the viability goes down by 10 per cent, the accession should be sent for regeneration.

## Germplasm Management Efforts

### (A) International Level

Importance of plant genetic resources is being recognized worldwide. There are several international organizations which concerned with vegetable genetic resources. The details are as under:

**Table 7.1: Sequential Germination Test Plan for 85 per cent Regeneration Standard for Groups of 40 Seeds**

| No. of Seeds | Regenerate if Number Germinated is Less than Equal to | Repeat Test if Germination is between or Equal to | Store if No. of Seeds Germinated is more than or |
|---|---|---|---|
| 40 | 29 | 30-40 | - |
| 80 | 64 | 65-75 | 76 |
| 120 | 100 | 101-110 | 111 |
| 160 | 135 | 136-145 | 146 |
| 200 | 170 | 171-180 | 191 |
| 240 | 205 | 206-215 | 216 |
| 280 | 240 | 241-250 | 251 |
| 320 | 275 | 276-285 | 286 |
| 360 | 310 | 311-320 | 321 |

*Source*: Ellis, *et al.*, 1980.

**Table 7.2: Prominent Sources of Exotic Genetic Resources**

| Vegetable Crops (Genera) | Source Countries |
|---|---|
| *Solanum* species (Tomato) | Australia, Bulgaria, Canada, Cuba, Denmark, France, Italy, Japan, the Philippines, Poland, Taiwan, USA, USSR |
| *Capsicum* species | Australia, Bulgaria, Czechoslovakia, Hungary, Italy, the Netherlands, Nigeria, the Philippines, Taiwan, USA, USSR, Yemen |
| *Allium* species | Australia, Brazil, Denmark, France, Germany, Hungary, Indonesia, Iran, Italy, Japan, the Netherlands, Nigeria, Poland, Syria, Taiwan, USA, USSR |
| *Brassica* species | Canada, France, Italy, Japan, Korea, the Netherlands, UK, USA, USSR |
| *Cucurbita* species | France, Hungary, Korea, Mexico, the Netherlands, Nigeria, the Philippines, Turkey, UK, USA, USSR |
| Leafy vegetables (Spinach, Amaranths) | Italy, the Netherlands, Poland, USA, USSR |
| *Pisum* species (Garden pea) | Australia, Bangladesh, Bulgaria, Czechoslovakia, Italy, New Zealand, USA, USSR |
| *Phaseolus* species (French bean) | Australia, Bangladesh, Czechoslovakia, Italy, New Zealand, USA, USSR |
| *Daucus* species (Carrot) | Australia, Brazil, Denmark, Egypt, France, Hungary |
| *Solanum* species (Potato) | Australia, Bangladesh, Canada, China, Germany, Hungary, Indonesia, Japan, Kenya, Mexico, Nepal, the Netherlands, New Zealand, Peru, Sri Lanka, Sweden, Switzerland, Taiwan, UK, USA |
| *Solanum* species (Eggplant) | Australia, Bangladesh, Denmark, France, Hungary, Japan, Nigeria, the Philippines, Sri Lanka, USA, USSR |
| *Abelmoschus* species (Okra) | Bangladesh, Brazil, Ivory Coast, Nigeria, the Philippines, Singapore, Sri Lanka, Sudan, Turkey, UK, USSR, Zambia |

## Food and Agriculture Organization (FAO)

An Introduction of newsletter containing genetic resource lists is being published by the Food and Agriculture Organization (FAO) since 1957 at regular intervals.

## Svallbard Global Seed Vault

The Norwegian government has built the Svallbard Global Seed Vault as service to the humanity. This vault became operational in February, 2008. It is located on an island near the North Pole, and is funded by the NGO Global Crop Diversity Trust, Rome. More than 2, 30,000 accessions of germplasm from gene banks of the Consultative Group on International Agricultural Research, NGOs and national programmes have been stored in this vault. The vault popularly termed as "doomsday vault", is aimed to protect the genetic resources from catastrophes like nuclear wars, natural disasters, *etc.* for the future generations of humankind.

## International Board for Plant Genetic Resources (IBPGR)

Coinciding with the global awakening in the field of plant genetic resources International Board for Plant Genetic Resources (IBPGR) was established in the year 1974.

## World Collection

When a germplasm collection is sufficiently large to include entries or accessions from all over the world, it is called world collection.

**Table 7.3: Important Germplasm World Collections (Except international institutes)**

| Germplasm World Collections | Crop |
|---|---|
| Bambey, Senegal | Groundnut |
| Beltsville, U.S.A. | Small grain crops |
| Cambridge, U.K. | Potato |
| Canal Point, Florida | Sugarcane |
| National Centre for Genetic Resources Preservation, Fort, Collins, U.S.A. | 268,000 accessions |
| Plant Introduction Stations, U.S.A. | 197,000 accessions of 4,000 species |
| Ethiopia, Africa | Coffee |
| Institute of Plant Industry, Leningard (USSR) | 1,60,000entries of crop plants |
| Near Tashkent, USSR | Annual New world Cotton |
| New Zealand | Sweet potato |
| Royal Botanic Garden, Kew, England | Over 45,000 entries of crop plants |
| Wisconsin, U.S.A. | Potato |

## World Vegetable Centre (Earlier Asian Vegetable Research and Development Centre (AVRDC) Taiwan)

World Vegetable Centre (Earlier Asian Vegetable Research and Development Centre AVRDC) established in 1971 at Shanhua, Taiwan, Republic of China (ROC) or Taiwan. World Vegetable Centre is a principal international agricultural research centre dedicated to vegetable research and development. The staff includes 21 internationally recruited senior scientists supported by 35 junior researchers and over 300 support personnels. The core budget is about 10 million US dollars. Much of the research work of this centre is done at the Headquarters in South-West, Taiwan. Outreach facilities are in Thailand and Tanzania and project offices are located in Bangladesh, Philippines and Costa Rica. World Vegetable Centre's a not-for-profit institute governed by an International board of Directors. It is funded by a number of donor countries (Australia, ROC, France, Germany, Japan, Republic of Korea, Philippines, Thailand, and USA) and banks and foundations (World Bank, Asian Development Bank, IDRC, USAID).

## World Vegetable Centre's Strategic Programme Direction

☆ Innovative germplasm enhancement for productivity, consumer acceptance, biofortification in tomatoes, peppers, vegetable soybean, mungbean and onion.

☆ Year round supply of safe and nutritious vegetables.

☆ Indigenous vegetables for biodiversity, healthy diet and marketing opportunities.

☆ Interactive user-friendly information management for vegetables in the tropics.

### Innovative Germplasm Enhancement

☆ Important traits transferred from wild species to broaden the genetic base

☆ High yielding, disease resistant, tropically adapted tomatoes, peppers, vegetable legumes and onions

☆ Tomatoes and sweet peppers with improved beta carotene

☆ Mungbean with improved methionine and iron levels

☆ Disease and insect resistant lines that reduce pesticide use

### World Vegetable Centre's Output

☆ Expanded and accessible collection of vegetable germplasm

☆ Enhanced vegetable hybrids and lines for tropical adaption and increased productivity

☆ Improved quality and nutritional contents.

☆ More than 230 improved varieties released to farmers in 90 countries

☆ Use of grafting

☆ Simple hydroponics

☆ Slow release fertilizers

☆ Protective culture

☆ Growth regulators

☆ Improved cropping systems

☆ Management practices for sustainable, safe and economically viable production

## Genetic Resource and Seed Unit at World Vegetable Centre

Genetic Resources work at World Vegetable Centre has expanded in a big way. The key to this qualitative gap is the completed 1344m$^2$ facility with a tripled cold store capacity (AVRDC, 1985). The entire facility-which provides 115 m$^3$ of long term (-20°C, frost free), 193 m$^3$ of medium term (2-5°C, 40-50 per cent RH), 254 m$^3$ of short term (15°C, 40 per cent RH) and 87 m$^3$ of tuber (15°C, 80-95 per cent RH) cold store, 600 m$^3$ of seed conditioning rooms, 49 m$^3$ of seed drying chambers and 106 m$^3$ of germination and preparation room was constructed with a capital quality of just over US$ 500,000; 275,000 for building and 225,000 for equipment.

### Objectives of Genetic Resource and Seed Unit at World Vegetable Centre

☆ To have base and active collection of major vegetable species of the tropics

☆ To undertake characterization, evaluation, documentation and promote utilization of the collected vegetable germplasm

☆ To cooperate within the international network of gene banks

☆ To ensure that incoming and outgoing seed samples are free of pests and diseases

### CIAT (Centro Internacional de Agricultura Tropical = International Centre for Tropical Agriculture)

International Centre for Tropical Agriculture (CIAT) is a not-for-profit organization that conducts socially and environmentally progressive research aimed at reducing hunger and poverty and preserving natural resources in developing countries. CIAT is one of the food and environmental research centres working toward these goals around the world in partnership with farmers, scientists and policy makers; known as the 'Future Harvest Centres'. They are funded mainly by the 58 countries, private foundations and international organizations that make up the Consultative Group on International Agricultural Research (CGIAR). The CIAT is located at Cali, Colombia in South-West Latin America. Although Colombia is in equatorial zone, the Andes produces a varied topography which ranges from rain forest and tropical plains at sea level to paramos and perpetual snow. On account of its geo-astronomical position, Colombia receives constant solar radiation throughout the year which results in day and night being roughly of the same duration. For the region, climate variations are not determined by seasons but rather are due to altitude and to the effects of trade wind, rain and humidity. Temperatures vary as a result of altitude and can thus be called climate zones which diminish

2°F for each 1000 feet of altitude. Cali lies 90 miles north of Popayan and is the state capital in the territory of the Calima Indians. Today the region possesses the most important sugar miles and has one of the largest industrial zones in the country, to the point that Cali in the country is third city in industrial and commercial importance. Situated at 3500 feet above sea level, the city enjoys mean temperatures of 23°C.

### Project Portfolio

- ☆ Agro-biodiversity and genetics
- ☆ Ecology and management of insect pests and diseases
- ☆ Soil ecology and improvement
- ☆ Analysis of spatial information
- ☆ Socioeconomic analysis
- ☆ Rural innovation

### Major Crops

CIAT has a mandate to conduct international research on four commodities that are vital for the poor.

- ☆ Beans
- ☆ Cassava
- ☆ Tropical forages and
- ☆ Rice

Among these, common bean (*Phaseolus vulgaris* L.) was considered to be an important and relevant crop for understanding of PGR management and related research issues since French bean is an important crop in the state. It is consumed both as green pods and dry seeds. French bean is considered to be the most important food legume for more than 300 million people, most of them in Latin America, where the crop was domesticated, and in Africa. Rich in Protein, iron and other dietary necessities, common bean (*Phaseolus vulgaris* L.) has come to the known as the nearly perfect food. In addition to bolstering human nutrition, it is of considerable economic importance, generating income for millions of small farmers. In Africa, the vast majority of bean producers are women (CIAT, 2002).

### (B) Vegetable Crop Introduction and Management Agencies in India

Indian Council of Agricultural Research (ICAR), New Delhi addresses the issue of management of genetic resources through "ICAR-National Bureau of Plant Genetic Resources (ICAR-NBPGR), New Delhi".

### ICAR-NBPGR, New Delhi

As a whole, a centralized plant introduction agency was initiated in 1946 at ICAR-IARI, New Delhi with the name of "Plant Introduction Scheme" in the Division of Botany and was funded by ICAR, New Delhi. In the five year plan, the scheme was expanded as "Plant Introduction and Exploration Organization", and in 1961, it was an independent division "Division of Plant Introduction". Ultimately, the Division was recognized as "National Bureau of Plant Genetic Resources (NBPGR), New Delhi" established in the year 1976 by the Indian Council of Agricultural Research (ICAR), New Delhi as a nodal institute for assembly of global diversity of plant genetic resources (PGR) that are of direct or indirect value to human. The ICAR-NBPGR headquarters is located within the campus of ICAR-IARI, New Delhi.

### Divisions of ICAR-NBPGR

At present, introduction and improvement of agricultural plants (cereals, legumes, fruits, vegetables, medicinal *etc.*) are being looked after by ICAR-NBPGR, New Delhi. The bureau has the responsibility to set up natural gene sanctuaries of plants where genetic resources are endangered. ICAR-NBPGR comprises of following divisions:

- (i) Plant Exploration and Germplasm collection
- (ii) Germplasm Evaluation
- (iii) Germplasm Conservation
- (iv) Plant Quarantine
- (v) Germplasm Exchange Unit
- (vi) Tissue Culture and Cryopreservation Unit
- (vii) Plant Genetic Resources Policy Planning Unit
- (viii) National Research Centre on DNA Fingerprinting

### Mandate of ICAR-NBPGR

The mandate of NBGR is to plan, conduct, promote and coordinate all activities, including collection, conservation, exchange, quarantine, evaluation, documentation and utilization of plant genetic resources. The ICAR-NBPGR is the nodal agency for import, export and national supply of seed and planting material for research purposes, undertaking their quarantine clearance and distribution of pest free germplasm to user scientists in India and abroad. The Bureau realizing the importance of horticultural crops in daily diet of mankind have made efforts in the past for introduction of the trait specific genetic resources from all over the world. Exotic accessions of various crops were introduced as source of desirable traits. Germplasm introductions of various horticultural crops had tremendous impact on Indian agriculture and increasing the yield and production. The well-defined mandate for ICAR-NBPGR, New Delhi is as under:

1. To plan, organize, conduct and coordinate plant explorations for collection of genetic diversity with particular reference to native and naturalized crops and their wild relatives in India and, in specific cases, abroad.

2. To undertake introduction, distribution and exchange of plant germplasm for research.

3. To conduct plant quarantine by examining seed and plant propagules under exchange for the presence of associated pests and pathogens and also salvage healthy material from the infected/infested/contaminated samples.

4. To characterize, evaluate and document germplasm collections of agri-horticultural crops and their wild relatives.

5. To promotes safe *ex situ* conservation of plant genetic resources on a long term basis.

6. To develop and operate National Database for strong retrieval of information on plant genetic resources.

7. To conduct basic research in the area of biosystematics, germplasm conservation and characterization for providing a sound scientific backup to Bureau's services and to develop the National Programme plant genetic resources.

8. To organize suitable training programme at the national, regional and international levels.

9. To assist policy development and offer expert consultancies on plant genetic resources, and also to implement work plants based on memoranda of understanding under bilateral, national and international agreements.

10. To undertake teaching and training on plant genetic resources leading to post-graduate (M.Sc.) and Doctoral (Ph.D) degrees linked with PG School, Indian Agricultural Research Institute (IARI), New Delhi.

11. To actively participate in on-farm and *in situ* conservation of agro-biodiversity in diversity-rich agro-ecosystems in collaboration with other departments/institutions, agencies and NGOs.

12. To act as the sole repository of all notified/released varieties/cultivars along with herbarium specimens and relevant information, as also repository of original/parental lines of patented varieties.

13. To assist in the implementation of new seed policy through introduction of protected varieties, transgenic lines and germplasm under strict quarantine regulation and to assist Plant Variety Protection Authority in identification of material in the event of disputes.

14. To develop national database on farmer's knowledge, innovations, practices and ethnic knowledge related to agro-biodiversity.

## Regional Stations of ICAR-NBPGR

1. **Regional Station, Akola (Maharashtra):** This centre was established in the year 1977. It has 50 acres of land. The major responsibility is exploration and germplasm conservation from Maharashtra, Chhattisgarh, Madhya Pradesh and adjoining regions of Gujarat and Karnataka, and their evaluation, maintenance and conservation. A satellite station is maintained at Amravati for field evaluation of some crops.

2. **Regional Station, Bhowali (Uttar Pradesh):** This centre was established in the year1943 as Imperial Potato Research Station and transferred to ICAR-NBPGR, New Delhi in 1986. The major responsibility is exploration, evaluation, maintenance and multiplication of germplasm of agricultural crop of sub-temperate regions.

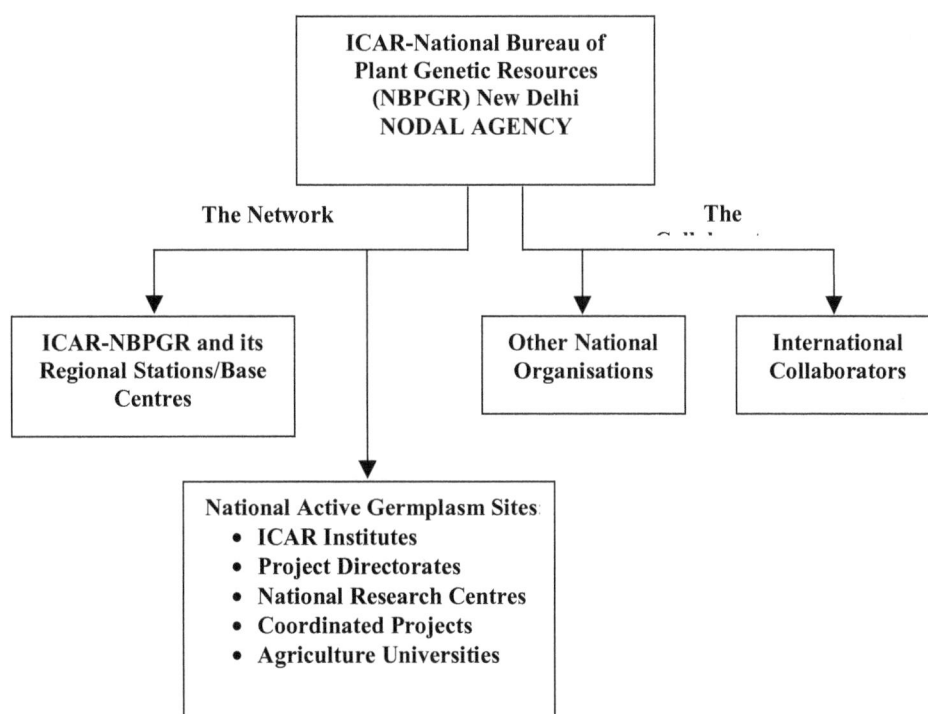

3. **Regional Station, Hydreabad (Telangana):** This centre was established in the year1985. The major responsibility is exploration, in Telangana and adjoining areas and germplasm exchange and quarantine of mandate crops of ICRISAT, Hyderabad.

4. **Regional Station, Jodhpur (Rajasthan):** The major responsibility is exploration, evaluation, maintenance and multiplication of germplasm of agricultural crop of arid regions.

5. **Regional Station, Shilong (Meghalaya):** It carries out exploration and evaluation of agri-horticultural germplasm of north-eastern region, including Sikkim and parts of North Bengal.

6. **Regional Station, Shimla (Himachal Preadesh):** This centre is located at 2.07 m above mean sea level; represent temperate and sub-temperate region. The major responsibility is exploration, evaluation, maintenance and multiplication of germplasm of agricultural crop of sub-temperate regions.

7. **Regional Station, Thrissur (Kerala):** It is involved in collection and evaluation of agri-cultural germplasm of southern peninsular region with special emphasis on spices and plantation crops.

8. **Regional Station, Srinagar (J&K):** This centre was established in the year1989. The major responsibility is exploration, evaluation, characterization, maintenance and multiplication of germplasm of agricultural crops from the region.

## Exploration Base Centre

1. **Exploration Base Centre, Cuttack (Odisha):** This centre was established in the year1986. It is located in the campus of ICAR-C.R.R.I., Cuttack, Odisha. The major responsibility is exploration, evaluation, characterization, maintenance and multiplication of germplasm of agri-cultural crops of Odisha and adjoining regions and conservation of rice germplasm of northern and eastern plains of India.

2. **Exploration Base Centre,, Ranchi (Jharkhand):** The major responsibility is exploration, evaluation, characterization, maintenance and multiplication of germplasm of agri-cultural crops of Bihar and Jharkhand.

## Other Organizations

☆ Central Potato Research Institute, Shimla (H.P.) for potato multiplication and conservation of genetic resources.

☆ National Research Centre for Onion and Garlic, Pune (Maharashtra) for onion and garlic multiplication and conservation of genetic resources.

## ICAR-Central Tuber Crops Research Institute, Trivandrum (Kerala)

During the 5th plan, the Central Government started in a modest way research activity on tuber crops, especially on cassava in Thiruvananthapuram (Kerala). Over the three years, there has been a steady growth of the research infrastructure which has now gained international recognition with the establishment of the ICAR-Central Tuber Crops Research Institute (CTCRI), Trivandrum (Kerala). ICAR-CTCRI is engaged for tuber crops (15 tuber crops) multiplication and conservation of genetic resources including two major crops cassava and sweet potato, three yam species (*Dioscorea alata, D. esculenta* and *D. rotundata*), five aroids species (*Colocasia esculenta, Xanthosoma saggitifolium, Amorphophallus paeniifoilius, Alocasia macrorrhiza* and *Cyrtosperma chamissionis*) and five minor tuber crops (*Solenostemon rotundifolius, Pachyrrhizus erosus, Maranta arundiaceae, Prophocarpus tetragonolobus* and *Canna edulis*). This institute is the only one of its kind in the world carrying out research and development exclusively on tuber crops and has established effective links with various international organizations in furthering its research agenda. The organization includes CIAT, Cali, Columbia: IITA, Ebadon, Nigeria; CIP, Lima, Peru; EMBRAPA, Belem, Brazil; The World Vegetable Centre, Shanhua, Taiwan and Philippines Root Crop Research Institute, Baybag Leyte7122, Philippines. ICAR-CTCRI has a rich diversity of germplasm of all tuber crops consisting of 1578 accessions of cassava, 807 of sweet potato, 452 of aroids, 591 of yams and 151 of other tuber crops.

## International Cooperation and Agreement

Cooperation among the countries and regions is important for effective germplasm conservation and evaluation of accessions. It would greatly facilitate collection, rejuvenation, characterization and evaluation of accessions, and would promote exchange of germplasm and technologies. Since 1950s, several agencies, particularly FAO and UNO have promoted and supported such cooperation. FAO hosted three International Technical conferences during 1967, 1973 and 1981, each of which resulted in the publication of a book summarizing the state of knowledge and technical advances in conservation and utilization of plant genetic resources (PGRs). International Agricultural Research Centres have also promoted and facilitated international technical cooperation.

## International Board for Plant Genetic Resources (IBPGR)

International Board for Plant Genetic Resources (IBPGR) was established in FAO in 1974 with a specific mandate to promote international cooperation on the conservation of plant genetic resources. In 1991, IBPGR was reorganize as International Plant Genetic Resources Institute (IPGRI), Rome as a fully independent institute that works in close cooperation with FAO and other international research centres. IPGRI has been renamed as Bioversity International since December, 2006. The commission on Plant Genetic Resources was established in FAO in 1983 as a forum that would negotiate, develop and monitor international agreements and regulations on genetic resources. Till 1983, 114 countries were members of this commission. Some important agreements on PGRs of agricultural interest are as follows:

(i) The International Undertaking on Plant Genetic Resources (1983)

(ii) International Code of Conduct for Collection and transfer of PGRs

### (i) The International Undertaking on Plant Genetic Resources (1983)

The international agreement was approved by FAO in 1983. It is a nonbinding formal agreement based on the principal that PGRs are common heritage of mankind and should be available without restriction for plant breeding and scientific purposes. This agreement defines the responsibility of countries and provides a framework for international cooperation. Annexures to the undertaking recognize:

(a) The rights of both donors of germplasm and of technology to seek adequate compensation for their contributions, as well as

(b) The farmer's rights

FAO and IPGRI are developing an international network of base collections in gene banks under the auspices of FAO that guarantee unrestricted access to the germplasm contained therein.

### (ii) International Code of Conduct for Collection and transfer of PGRs

International Code of Conduct for Collection and transfer of PGRs has been developed and negotiated through the Commission on PGRs. It aims to promote collection and safe transfer of germplasm; it also provides the procedures to be followed by collectors for making requests for PGR collection mission (=exploration) and by governments for issuing licenses for the same.

### (iii) Global System for the Conservation and utilization

The compressive system was developed by the Commission on PGRs to promote and monitor systematic cooperation and coordination of PGR activities. By 1993, 132 countries have joined the system. FAO, in close collaboration with IPGRI, is developing global information and early working mechanisms,

networks for *in situ* conservation areas and for *ex situ* collections, and periodical reports on the state of world's PGRs.

## Crop Base Germplasm Management

Active collections are basically for sustainable use in various crop improvement programmes. The 40 National Active Germplasm Sites as decided by ICAR-NBPGR, New Delhi, are holding active collections of relevant crop species based at premier crop or crop group-based institutes. Eleven of them have the medium term seed storage facilities in the form of cold storage modules maintained at 4°C and 35-40 per cent relative humidity. The crop-based institutes have multi-disciplinary team of scientists, better equipped for evaluation of germplasm for agronomic potential and various yield reducing factors to identify accessions with desirable traits. This helps to generate information on potential value of various accessions to provide the needed support for their utilization in breeding programmes. ICAR-Indian Institute of Vegetable Research, Varanasi (U.P.) is a National Active Germplasm Site (NAGS) for vegetable crops.

**Table 7.4: Crop Base Germplasm Management Network in India**

| Crops | Associated Organizations |
|---|---|
| Tomato | ICAR-IIVR, Varanasi (UP), ICAR-IIHR, Bengaluru (Karnataka), PAU, Ludhiana (Punjab), ICAR-IARI, New Delhi, TNAU, Tamil Nadu (T. Nadu), NDUA&T, Faizabad (UP), CSAUA&T, Kanpur (UP), ICAR-NBPGR, New Delhi, BCKV, (WB), OUA&T, Bhubaneswar (Odisha), HARP, Ranchi (Jharkhand) |
| Chilli | ICAR-IIVR, Varanasi (UP), ANG Agril. Univ. Hyderabad (Andhra Pradesh), Regional Station Lam, UAS, Dharwad (Karnataka), ICAR-IIHR, Bengaluru (Karnataka), ICAR-IARI, New Delhi |
| Capsicum | ICAR-IARI, Regional Station Katrain (HP), Y.S. Parmar Univ. of Hort. and Forestry, Solan (HP), Palalmpur (HP), |
| Brinjal | ICAR-IIVR, Varanasi (UP), ICAR-IIHR, Bengaluru (Karnataka), OUA&T., Bhubaneswar (Odisha), PAU, Ludhiana (Punjab), ICAR-IARI, New Delhi, TNAU, Tamil Nadu (T. Nadu), ICAR-IIHR, Bengaluru (Karnataka) NDUA&T, Faizabad (UP), CSAUA&T, Kanpur (UP), ICAR-NBPGR, New Delhi, BCKV, (WB), OUA&T, Bhubaneswar (Odisha), HARP, Ranchi (Jharkhand) |
| Amaranth | Tamil Nadu agricultural University, Coimbatore (Tamil Nadu), KAU., Vellanikkara (Kerala), ICAR-NBPGR, Regional Station, Trishur (Kerala), AAU, Jorhat (Asom) |
| Muskmelon | ICAR-IIVR, Varanasi (UP), RARS, Durgapura (Rajsthan), PAU, Ludhiana, ICAR-IARI, New Delhi, ICAR-IIHR, Bengaluru (Karnataka), NDUA&T, Faizabad (UP) |
| Watermelon | RARS, Durgapura (Rajasthan), ICAR-IIHR, Bengaluru, PAU, Ludhiana (Punjab), MPKV, Rahuri (MS) |

| Crops | Associated Organizations |
|---|---|
| Pumpkin | ICAR-IIVR, Varanasi (UP), ICAR-IARI, New Delhi, ICAR-IIHR, Bengaluru (Karnataka), NDUA&T, Faizabad (UP), ICAR-NBPGR, New Delhi, PAU, Ludhiana (Punjab), TNAU, Tamilnadu (T. Nadu), Parbhani (Gujarat), KAU, Vellannikara (Kerala) |
| Bitter gourd | K. A. U., Vellanikkara (Kerala), CSAUA&T, Kanpur (UP), ICAR-IIVR, Varanasi (UP) |
| Bottle gourd | ICAR-IIVR, Varanasi (UP), NDUA&T, Faizabad (UP), ICAR-NBPGR, New Delhi, JNKVV, Jabalpur (M.P.), MPKV, Rahuri (Maharashtra) |
| Ivy gourd | IGKVV, Raipur (Chhattisgarh), ICAR-IIVR, Varanasi, GAU (Gujarat), Navsari (Gujarat), ICAR-IIHR, Bengaluru (Karnataka) |
| Pointed gourd | ICAR-IIVR, Varanasi (UP), BCKV, (WB), NDUA&T, Faizabad (UP), Sabour Agril Univ. Sabour (Bihar), Navsari (Gujarat) |
| Spine gourd | ICAR-IIHR, Central Experiment Station, Bhubaneswar (Odisha), ICAR-IIVR, Varanasi (U.P.) |
| Sweet gourd | ICAR-IIHR, Central Experiment Station, Bhubaneswar (Odisha), ICAR Research Complex, Barapani (Meghalya) |
| Cucumber | ICAR-IIVR, Varanasi (UP), ICAR-IARI, New Delhi, MPKV, Rahuri (Maharashtra), TNAU, Coimbatore (T. Nadu), ICAR-IIHR, Bengaluru (Karnatak), JNKVV, Jabalpur (MP) |
| Squash | ICAR-IIHR, Bengaluru (Karnataka), ICAR-IARI, Regional Station Katrain (HP), ICAR-IIHR, Bengaluru (Karnataka) |
| Roundmelon | PAU, Ludhiana (Punjab), ICAR-IIHR, Bengaluru (Karnataka), ICAR-IIVR, Varanasi (UP) |
| Brinjal | ICAR-IIVR, Varanasi (UP), ICAR-IIHR, Bangalorre (Karnatak), OUA&T, Bhubaneswar (Odisha), RAU, Sabour (Bihar), HARP, Ranchi (Jharkhand) |
| Cabbage | ICAR-IARI, Regional Station Katrain (HP), Y.S. Parmar Univ. of Hort. and Forestry, Solan (HP) |
| Capsicum | ICAR-IARI, Regional Station Katrain (HP), Y.S. Parmar Univ. of Hort. and Forestry, Solan (HP), Sher-e-Kashmir Univ. of Agril. and Tech., Srinagar (Kashmir) |
| Carrot (Tropical) | CCSHAU, Hisar (Haryana), ICAR-IARI, New Delhi, ICAR-IIVR, Varanasi, PAU, Ludhiana (Punjab) |
| Carrot (Temperate) | ICAR-IARI, Regional Station Katrain (HP), Y.S. Parmar Univ. of Hort. and Forestry, Solan (HP) |
| Cauliflower (Early) | ICAR-IIVR, Varanasi, (UP), ICAR-IARI, New Delhi, Sabour Agril Univ., Sabour (Bihar) |
| Cauliflower (Mid) | ICAR-IARI, New Delhi, ICAR-IIVR, Varanasi, ICAR-IARI, Regional Station Katrain (HP), Sabour Agril Univ., Sabour (Bihar) |
| Cauliflower (Late) | ICAR-IARI, Regional Station Katrain (HP), Y.S. Parmar Univ. of Hort. and Forestry, Solan (HP) |
| Lablab bean | ICAR-IIVR, Varanasi (UP), IGKVV, Raipur, (Chhattisgarh), ICAR-IIHR, Bengaluru (Karnataka), ICAR-NBPGR, New Delhi, CSAUA&T, Kanpur (UP), ICAR- IARI, New Delhi, HARP, Ranchi (Jharkhand) |
| French bean | ICAR-IIVR, Varanasi (UP), ICAR-IIHR, Bengaluru (Karnataka), Y.S. Parmar Univ. of Hort. and Forestry, Solan (HP), GB Pant Univ. of Agri. and Tech., Pantnagar (Uttarakhand), MPKV, Rahuri (MS) |
| Peas | ICAR-IIVR, Varanasi (UP), ICAR-IIHR, Bengaluru (Karnataka), CSAUA&T, Kanpur (UP), ICAR-IARI, New Delhi, PAU, Ludhiana (Punjab), VPKAS, Almora (Uttarakhand), JNKVV, Jabalpur |

| Crops | Associated Organizations |
|---|---|
| Garlic | ICAR-NRC Onion and Garlic, Pune (MS), NHRDF, Nasik (MS), NHRDF, Karnal (Haryana), ICAR-IIHR, Bengaluru (Karnataka) |
| Onion | ICAR-NRC Onion and Garlic, Pune (MS), NHRDF, Nasik (MS), NHRDF, Karnal (Haryana), ICAR-IIHR, Bengaluru (Karnataka) |
| Okra | ICAR-IIVR, Varanasi (UP), ICAR-IARI, New Delhi, ICAR-IIHR, Bengaluru (Karnataka), ICAR-NBPGR, New Delhi, PAU, Ludhiana (Punjab), CCSHAU, Hisar (Haryana), Parbhani (MS) |

# National and International Linkages/Collaborations for Germplasm Management

## (A) National Linkages

The ICAR-NBPGR, New Delhi collaborates with the Plant Protection Advisor, Government of India, under the powers delegated to the Director, ICAR-NBPGR, New Delhi for quarantine of incoming and outgoing germplasm for research purpose. The ICAR-NBPGR, New Delhi has active collaboration with Ministry of Environment and Forests also. This organization has developed close linkages with the Department of Biotechnology which supported the National Plant Tissue Culture Repository to carryout work on in-vitro conservation and cryopreservation of germplasm. National coordination is accomplished through associates and linkages with 30 major crop-based ICAR institutes, agricultural universities, research foundations and non-governmental organizations (NGOs) and other stakeholders in the area of biodiversity and plant genetic resources. The ICAR-NBPGR, New Delhi has developed strong linkages, with a large number of crop-based institutes, national research centres, All India Coordinated Crops Improvement Projects and State Agricultural Universities.

## (B) International Linkages

There are several international organizations which are having the viable collaboration for effective plant genetic resource collection, conservation and utilization. The ICAR-NBPGR, New Delhi interacts with the FAO programmes in the region and also collaborates with several International Agricultural Research Centres of CGIAR.

### 1. Collaboration with CGIAR Crop Based Institute

Active collaborations has been established with International Agricultural Research Centres located in India and abroad. International Plant Genetic Resources Institute (IPGRI) has contributed significantly to the nation by offering expertise, training and research support for vegetable crops. ICRISAT, IRRI, CIMMYT, ICARDA, INIBAP, IJO (Bangladesh) *etc.* are active collaborators on joint exploration and multiplication evaluation programme in different crops. International Plant Genetic Resources Institute (IPGRI) office for South

Asia is located in the ICAR-NBPGR, New Delhi campus and there is an active collaboration between them based on biennial work plans. In addition to supporting joint exploration and collection programme in this region, the IPGRI helps in developing training programmes in conservation and management of plant genetic resources primarily for the Asian region.

## 2. Collaboration under Bilateral Programmes

Besides the International Research Centres, many countries have well developed systems for assemblage, enrichment, documentation and conservation of plant genetic resources and also have computerized database network. ICAR has Memorandum of Understanding (MOU) as well as bilateral agreements with several such international organizations and national programmes. ICAR-NBPGR, New Delhi has memoranda of understanding/bilateral agreements with several nations. Prominent among these are USA, UK and other European countries Mexico, Israel, Australia, Russia, New Zealand, Iran, Turkey and Zimbabwe. Presently, exchange of plant genetic resources activities is carried out through ICAR-NBPGR, New Delhi with over 80 countries.

## 3. Collaboration as Global Responsibility

As a global collaboration, the first International Okra workshop was organised at ICAR-NBPGR, New Delhi in 1990 under the sponsorship of IPGRI. Besides the IPGRI experts, representatives from Belgium, France, Brazil, China, India, Indonesia, Ivory Coast, Nigeria, Papua New Guinea, Philippines, Senegal, Sri Lanka, Sudan and USA participated in the deliberations.

**Table 7.5: Some International and National Organization Associated with Germplasm Management**

| International | |
|---|---|
| AVRDC | Asian Vegetable Research and Development Centre, Taipei (Taiwan) |
| CGIAR | Consultative Group for International Agriculture Research, Washington (DC) |
| CIP | Centro Inte~acional de la Papa (International Potato Centre), Peru I (South Amenca) |
| Bioversity International | International Board for Plant Genetic Resources, Rome (Italy) |
| ICGEB | International Centre for Genetic Engineering and Biotechnology, Trieste ~ (Italy) and New Delhi (India) |
| IIB | International Institute of Biotechnology, Canterbury, Kent (U.K.) |
| IIH | International Institute of Horticulture, Brazil |
| ISHS | International Society for Horticulture Science, Wageningen (Netherland) |
| SAVERNET | South Asian Vegetable Research Network |
| **National** | |
| AADF | Associated Agricultural Development Foundation, Nasik (Maharashtra) |
| APEDA | Agricultural and Processed Food Product Export Development Authority, New Delhi |

| CARIANGI | Central Agricultural Research Institute for Andman and Nicobar Groups of Islands, Port Blair (Andman and Nicobar) |
|---|---|
| CAZRI | Central Arid Zone Research Institute, Jodhpur (Rajasthan) |
| CFTRI | Central Food Technological Research Institute, Mysore (Karnataka). CIAH Central Institute for Arid Horticulture, Bikaner (Rajasthan) |
| CIMAP | Central Institute for Medicinal and Aromatic Plants, Lucknow (Uttar Pradesh) |
| CIPHET | ICAR-Central Institute for Post-Harvest Engineering and Technology, Ludhianl (Punjab) |
| CISH | ICAR-Central Institute for Sub-Tropical Horticulture, Lucknow (Uttar Pradesh) |
| CITH | ICAR-Central Institute for Temperate Horticulture, Srinagar (Jammu and Kashmir) |
| CPCRI | ICAR-Central Plantation Crops Research Institute, Kasaragod (Kerala) |
| CPRI | ICAR-Central Potato Research Institute, Shimla (Himachal Pradesh) |
| CTCRI | ICAR-Central Tuber Crops Research Institute, Trivandrum (Kerala) |
| DARE | Department of Agricultural Research and Education, New Delhi |
| HSI | Horticultural Society of India, New Delhi |
| IARI | ICAR-Indian Agricultural Research Institute, New Delhi |
| ICAR | Indian Council of Agricultural Research, New Delhi |
| IIHR | ICAR-Indian Institute of Horticultural Research, Hessarghatta, Bengaluru (Karnataka) |
| IISR | ICAR-Indian Institute of Spice Research, Calicut (Kerala) |
| IIVR | ICAR-Indian Institute of Vegetable Research, Varanasi (Uttar Pradesh) |
| NAARM | ICAR-National Academy of Agricultural Research Management, Hyderabad (Andhra Pradesh) |
| NAFED | ICAR-National Agricultural Co-operative and Marketing Federation of India |
| NBPGR | ICAR-National Bureau of Plant Genetic Resources, New Delhi |
| NBRI | National Botanical Research Institute, Lucknow (Uttar Pradesh) |
| NCDC | National Co-operative Development Corporation, New Delhi |
| NHB | National Horticulture Board, Gurgaon (Haryana) |
| NHRDF | National Horticultural Research and Development Foundation, Nasik (Maharashtra) |
| VPKAS | ICAR-Vivekanand Parvatiya Krishi Anusandhan Shala, Almora (Uttranchal) |

## National Research Centres associated with Germplasm Management (other crops)

a) National Research Centre for Banana, Tiruchirapalli' (Kerala).

b) National Research Centre for Cardamom, Affangald, Mereata (Karnataka).

c) National Research Centre for Onion and Garlic, Rajguru Nagar, Pune (Maharashtra).

**LINKAGES**

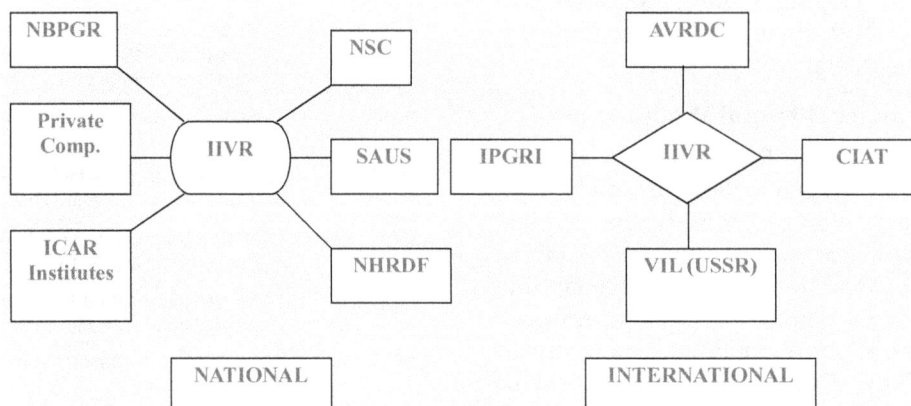

d) National Centre for Mushroom Research and Training (ICAR), Chambaghat, Solan (Himachal Pradesh).

e) National Research Centre for Soybean, Indore (Madhya Pradesh).

f) National Research Centre for Seed Spices, Tabiji, Ajmer (Rajasthan).

## Global Information System for Plant Genetic Resources Management

**Table 7.6: Global Information System for Plant Genetic Resources Management**

| Institute/Sources | Website | Prime Mandate/Information |
|---|---|---|
| World Information and Early Warning System (WIEWS) on plant genetic resources | http://apps3,fao.org/wiews/wiews.jsp | *Ex-situ* collection database of member countries. World list of seed source and crop varieties. |
| CGIAR System-wide Information Network for Genetic Resources (SINGER) | http://singer.grinfo. | Collection of genetic resources held by the CGIAR centres. |
| Commission on Genetic Resources for Food and Agriculture (CGRFA) | http://www.fao.org/WAICENT/FAOINFO/AGRICULT/cgrfa | Conservation and sustainable utilization of genetic resources; specific information on global strategies for animal and plant genetic resources. |
| Global Plan of Action for the Conservation and Sustainable Utilization of Plant Genetic Resources for Food and Agriculture. | http://www.fao.org/WAICENT/FAOINFO/AGRICULT/AGP/AGPS/GpaEN/gpatoc.htm | Four theme areas, namely *In situ* conservation and development, *Ex-situ* conservation and institution and capacity building. |
| International Undertaking on Plant genetic resources for food and agriculture | http://www.fao.org/ag/cgrfa/IU.htm | Complete text of the resolution adopted by FAO in 1983 on international agreement dealing with plant genetic resources for food and agriculture. |
| International Plant Genetic Resources Institute (IPGRI) | http://www.ipgri.cgiar.org | One of the 16 research centres of CGIAR, online information on germplasm and publications of IPGRI. |
| The International Treaty on Plant Genetic Resources for food and Agriculture | http://www.fao.org/ag/cgrfa/itpgr.htm | Deartiled text document about the treaty. |
| The International Union for the Protection of New Varieties of Plants (UPOV) | www.upov.int | All documents related to various meetings and technical committees of UPOV. |
| CIAT Database on Plant Genetic Resources | http://www.ciat.cgiar.org/pgr/ | Contains information on 60,000 accessions mostly related to unimproved landraces of 720 species of bean, cassava and forages. |
| International and Regional Crop related networks supported by FAO | http://www.fao.org/WAICENT/FAOINFO/AGRICULT/AGP/AGPS/cnet.htm | Individual crop related networks under The International Agreement on Genebanks. |
| International Neem Network | http://www.fao.org/forestry/foris/webview/forestry2index.jsp?sitId=2021 and langId=1 | It comprises national institutions from 23 countries, provides information about activities of the network. |
| Seed and Plant Genetic Resources Service:AGPS | http://www.fao.org/waicent/faoinfo/agricult/AGP/AGPS/ | Technical and policy advice on Plant Genetic Resources their conservation and use, seed and planting material. |
| Genetic Resources Action International (GRAIN) | http://www.grain.org | Rich website on issues related to PGR, Biodiversity and Environment |

| Institute/Sources | Website | Prime Mandate/Information |
|---|---|---|
| The Convention on Biological Diversity (CBD) | http://www.biodiv.org | Details about the Convention, The Biosafety Protocol, Programs and issues. |
| Plant Genetic Resources Abstracts (PGRA) | http://www.cabi-publishing.org/AbstractDatabase.asp? SubjectArea= and PID=46 | Derived from the CAB Abstracts database provides access to 10 year backfile of nearly 34,000 records. It is updated weekly. |

## Nodal Database Centre on Plant Genetic Resources

There is certain Nodal Database Centre on Plant Genetic Resources, which provide detail information about germplasm availability.

**Table 7.7: Nodal Database Centre on Plant Genetic Resources**

| Institute/Sources | Website | Prime Mandate/Information |
|---|---|---|
| Federal Information System on Genetic Resources (BIG) | http://www.big-flora.de/index_e.html | Genetic Resources for cultivated and wild flora in Germany. |
| The Institute for Plant Genetic and Crop Plant Research | http://www.ipk-gatersleben.de/englisch | Genebank having 100,000 accessions of cultivated plants. |
| Centre for Genetic Resources (CGN) | http://www.plant.wageningen-ur.nl/about/Biodiversity/Cgn/Default.htm. | Collection of agricultural and horticultural crops. |
| Danida Forest Tree Seed Centre (DESC) | http://www.dfsc.dk/ | Tropical and sub-tropical tree species. |
| EUFORGEN Bibliography | http://www.ipgri.cgiar.org/networks/euforgen/biblio/select.asp. | 1674 refrences on conservation and use of forest genetic resources. |
| Information System on Genetic Resources (GENRES) | http://www.genres.de/genres-e.htm | Cultivated and wild plants, forest plants, domestic animals, microorganism and the policy framework. |
| Nordic Gene Bank | http://www.ngb.sel/nmnmm | Plant species in Nordic agriculture and horticulture including their wild relatives. |
| National Plant Germplasm System of USDA | http://www.ars.grin.gov/npgs/ | Germplasm of plants, animals, microbes and invertibrates. |
| NIAR database | http://www.gene.affrc.go.jp/plant/image/gbsys.html | Plant Genetic Resources in 4 plant groups, namely legume, vegetables, flowers and ornamentals and millets and forage crops |
|  | http://www.gene.affrc.go.jp/plant/image/flower.html | Plant Genetic Resources in 4 plant groups, namely legume, vegetables, flowers and ornamentals and millets and forage crops |
|  | http://www.gene.affrc.go.jp/plant/image/vegetable.html | Plant Genetic Resources in 4 plant groups, namely legume, vegetables, flowers and ornamentals and millets and forage crops |
| N.I. Vavilov All-Russian Scientific Research Institute of Plant Industry | http://www.vir.nw.ru/datadbf.html | Plant Genetic Resources Collection and Conservation |
| Indian Plant Genetic Resources Information System | director@nbpgr.ernet.in | Passport data, evaluation and characterization data of all agricultural and horticultural plants. |

## Community Management Mechanisms in Crop Diversity

Capacity building and partnership with farming communities for Community based conservation and management mechanism and policy processes of crop diversity to bring about improved food security/livelihood in different agro-climatic zones. As example in Maharashtra, partnerships and capacity building of farming communities especially women farmers has share enormous experience and share communities traditional knowledge and integrated with new technique knowledge from database of crop diversity in cultivation and value addition to the researchers. Community based conservation and management process helped us to improve livelihood through value addition and market linkages. Consultative mechanisms have helped in establishing behavior changes and practices *in situ* conservation of crop diversity. Partnership and capacity building procedures helped in generating morphological and nutritional and also knowledge data. Participation of communities has shown interest to develop protocol for value addition and to develop seed banks. More number of community members has shown interest to form network of traditional seed keepers. This network of seed keepers not only document the processes besides crop diversities they are documenting wild vegetables. Regenerating traditional knowledge of food will help to generate food security and nutrition of farming community. Many of experiences of related activities and more importantly will facilitate a coordinated and coherent approach to capacity building, sharing of resources and knowledge will strengthen farming and backward communities to improve livelihood and food sovereignty (Dhakate, 2016).

## References

1. Dhakate, S. 2016. Building partnership with farming communities for community management mechanisms in crop diversity. 1st International Agrobiodiversity Congress (Abs. Session: III. Partnership, Networks and Capacity Building); 6-9 November, held at New Delhi, p. 283.

2. Dhillon, B.S., Varaprasad, K.S., Srinivasan, K., Singh, M., Archak, S., Srivastava, U. and Sharma, G.D. (eds.). 2001. National Bureau of Plant Genetic Resources : A Compendium of Achievments. National Bureau of Plant Genetic Resources (NBPGR), New Delhi. p. 329 (+xvi).

3. Gautam, P.L., Sharma, G.D., Srivastava, U., Singh, B.M., Kumar, A., Saxena, R.K. and Srinivasan, K. (eds.). 2000. Twenty Glorious Years of ICAR-NBPGR (1976-1996). ICAR-National Bureau of Plant Genetic Resources (ICAR-NBPGR), New Delhi, p. 333.

4. Rana, R.S. 1992. Indian National Plant Genetic Resources System. In: Plant Genetic Resources: Documentation and Information Management. Rana *et al.* (eds.), ICAR-NBPGR, New Delhi, p. 1-16.

5. Saxena, R.K. 1998. Monitoring and maintenance of germplasm in genebank. In: Plant Germplasm Colecting: Principles and Procedures (eds. Gautam, P.L., Dabas, B.S., Srivastava, U. and Duhoon, S.S.), ICAR-NBPGR, New Delhi, p. 198-200.

# 8

# Documentation and Registration of Vegetable Genetic Resources

## Introduction

Germplasm or the genetic resources is the continuity of the genetic material responsible for a plant's characteristics, transmitted from one generation to the next. Usefulness of any germplasm (species, genetic material/accessions) depends on progressive collection, evaluation and documentation. The success of the whole genetic resources programme is dependent upon the descriptive information of the conserved material which enables plant breeders to make decisions regarding the material too be used in the breeding programme. Where such information is not available, passport data on characteristics of the natural habitats of species are of great importance. This dependence on information grows exponentially with the size of collections (Sapra, 1991). For example in potato, late blight resistant species are found in Mexican genepool where late blight fungus (*Phytophthora infestans*) has been found to reproduce sexually. Similarly frost resistance is found in species capable of growing at altitudes above 3,500 m. in case of potato the information on passport data, morphological characteristics and reaction to various biotic and abiotic stresses of *tuberosum* and *andigena* germplasm has been published time-to-time in the form of germplasm catalogues. Documentation is, therefore, one of the most critical functions concerned with genetic resources as the collected material *per se*. It helps the PGR workers and breeders in assessing the potential usefulness of the germplasm, the extent and magnitude of diversity and the cultural management practices being followed in particular area/eco-geographic region. It also aids in mapping of collected diversity and provides the basis to address the intellectual property right (IPR) related issues as well as in planning future collection programmes. The information generated in this manner over the years reveals the shift, if any, in cropping pattern.

## Plant Genetic Resource (PGR) Database

A Plant Genetic Resource (PGR) Database is defined as:

'*A centralized repository of information on PGR conserved ex situ in national genebank in various forms (seeds, tissue cultures, cryo-sample, live plants) at a single or many locations*'.

The function of a PGR database is to enhance accessibility to information content associated with genebank collections, conserved *ex situ*.

## Development and Function of Plant Genetic Resource (PGR) Database

i. **For breeders and researchers-** Increase utilization of the genetic resources by making available pre-compiled information on passport, characterization and evaluation.

ii. **For genebank curators-** Systematically manage regeneration and safety-duplications and to rationalize the size of collections by identifying probable duplication between collections.

iii. **For crop managers-** Allow multi-disciplinary activities such as core collections formation and evaluation projects.

iv. **For explorers and collectors-** Identify gaps in the collections to efficiently target collection expeditions and avoid redundant expeditions.

v. **For policy makers-**Make policy decisions related to ownership and sharing of germplasm as well as funding priorities.

Plant genetic resources management involves broadly five stages:

(i) Exploration and collection

(ii) Exchange

(iii) Characterization and evaluation

(iv) Conservation and utilization

(v) Documentation.

In addition, it is also concerned directly or indirectly with the plant quarantine. At each of the various stages in the above process, information about plant material is used for communication and decision making. It is estimated that the scientists and technicians spend at least 30 per cent of their time in handling of data generated at various stages (Rogers *et al.*, 1975).

## Database Management in PGR with Reference to Exploration Characterization and Evaluation

During exploration trips, detailed passport information is collected and some of them are species name, common name, latitude, longitude, source, topography *etc.* In order to keep this detailed information for further use so that this information can be documented, queried, grouped according to various criteria, it becomes mandatory to properly document this information with the use of computers. In other words, the database management activities of PGR should give an important place and role to exploration activities.

### Table 8.1: Characterization and Evaluation of Genetic Resources

| Characterization | Evaluation |
|---|---|
| Documentation | Documentation |
| The characterization data should be documented for collation, analysis and retrieval after through verification. | The recorded data should be standardized to a uniform format and then entered in MS-Excel sheet for integrating into the main database at the earliest to avoid any loss of data. Wherever data loggers are used, these should be downloaded immediately into the computer. |
| The qualitative data may be analyzed for estimating allelic richness and evenness. | The quantitative data should be statistically analyzed following the standard experimental designs. |
| Catalogues may be developed using the qualitative and quantitative data. | - |

## Data Content

All data and information generated throughout the process of germplasm acquisition, establishment, field management, conservation, regeneration, characterization, evaluation and distribution need to be recorded. It is important to note that all these activities may generate independent data and database. An information management system must be created in each genebank. This database must be a user-friendly, searchable one, by the genebank curators and staff for specific information through a range of queries. Identity of the collection is the only attributes that connects information across databases. This identity is called *"Accession Number"* in the genebank parlance and remains unique to every single germplasm accession.

## Introduction to Documentation Systems

Nearly 1.7 million species of plants, animals and microbes have been documented so far while 10-15 times of this number are expected to be waiting for their turn to be recognized and given scientific names (WCMC, 1992; Heywood, 1995). IBPGR has recently published a guide book (Painting *et al.*, 1993) for genetic resources documentation which provides a self-teaching approach to the understanding, analysis and development of genetic resources documentation. It has also developed gene bank management system software (GMS) along with a tutorial and reference guide for use with the genebank management system software (Perry *et al.*, 1993). The self-teaching guide covers in detail, introduction to genebanks and documentation system, information processing in genebanks, analysis of data generation and use in genebank, data recording, organization of different types of data, computer basic, database basic, building the system and implementation and maintenance of system *etc.*

In introduction to documentation systems, following terminology/definition must be cleared:

☆ **Data:** Quantitative and qualitative values derived from observations.

☆ **Information:** When data are recorded, classified, related or interpreted, become information.

☆ **Descriptor:** An identifiable and measurable characters uses to facilitate data classification, storage, retrieval and use. Each descriptor must have a clear definition so as to facilitate the meaningful exchange of information among the cooperating scientists.

☆ **Descriptor list:** A collection of all the individual descriptors used for a particular species or crop.

☆ **Descriptor state:** A clearly definable state which a descriptor can take.

Any way of storing and maintaining data is called documentation system. A documentation system can be maintained as manual methods (such as hand written

## Table 8.2: Data Content

---

**I. Passport Data-**Passport data for all accessions have to be documented using the standard multi-crop passport descriptors (*Annexure IV*)

| | |
|---|---|
| Site of collection | Information on the exact site from where collections are made, is indispensible, to carry out recollection missions in case of loss, to expand the representativeness in case of trait-specific material, to carry out basic population genetic structure studies and to measure location-specific genetic changes (including genetic erosion) information at the exact site from where collection(s) have been made. Since genebanks function as custodians and ownership rights remain with communities, locational information may turn out to be vital in case of rewards, benefit-sharing and dispute resolution. |
| | Modern tools and technologies based on geography information system (GIS) combine site information with data on climate, soil, cultivation *etc.* facilitate decision making for collection, recollection, identification of new areas of introduction (especially in relation to climate change) *etc.* With hand-held tools, one can record latitude and longitude, which later can be translated into actual locational description. |
| | In case data are recorded manually, name of the village (with mandal/taluk/tehsil) is required in order to map the site using GIS tools. Recording the site as a larger entry, such as a district/state, renders data unusable for downstream analytics. |
| Biological status | Consequent upon the implementation of many national and international treaties and conventions including ITPGRFA, National Biodiversity Act, Protection of Plant Varieties and Farmers Rights Act, Seed Act *etc.* individual, community or sovereign ownership of the genetic material, are determined by the description of origin such as landrace, farmers variety, improved variety, hybrid variety, wild and weedy relatives, recent introductions from abroad *etc.* |
| | In case of varieties/improved cultivars/hybrids, pedigree information must be recorded. Categorization and decision making downstream, in terms of ownership-sharing, benefit-sharing, *etc.* depend upon the faithful documentation of the nature of the accession in terms of origin. |
| | Currently, ICAR-NBPGR databases classify the accessions into breeding lines, clonal selections, elite lines, genetic stocks, hybrids, improved cultivars, landraces, parental lines, primitive cultivars, released cultivar, traditional or farmers variety, weedy, wild and others (not fitting into the above.) |
| Date of collecting | Date of collection is essential to connect the incidence of diseases and pests or climate variations available from secondary and compiled sources with characterization and evaluation data of germplasm accessions, to carry out studies on temporal changes in genetic structure, and to compute time-scale attributes of PGR activities. It is recorded in date (dd), month (mm) and year (yyyy). |
| | ICAR-NBPGR databases record, order and search data on the basis of crop name (common name), bionomial name, accession number, collection date, collector number, other ID, cultivar name, pedigree and site information. |

**II. Herbarium Data-** Reference collections (herbarium specimens, photographs and/or description of the original accessions are essential for conducting the true-to-type verification. Information must include source of material, passport data, botanical drawings, digital images, taxonomic description *etc.*)

**III. Seed Genebank Data-**The information system must keep a record of genebank operation data including storage location, stocks, monitoring, health tests and the distribution status. The same system must also manage germplasm indents, shipment related information and content information of curators. Information, in case of orthodox seeds conserved *ex situ* include:

> Passport data (especially including source of material such as organization/individual) for all accessions based on the standard descriptor lists (ICAR-NBPGR/internationally accepted). The use of passport information simplifies data exchange between genebanks.

> Name of the curator

> Accession receipt date, batch number and number of samples in the batch.

> Information should include storage details such as wing, module, rack, *etc.* Old locations, if any, should be retained in record for reference.

> Information on moisture content, germination (per cent), date of last monitoring, monitoring cycle, next monitoring schedule, dormancy treatment (if any), curator's remarks.

> Information should also include user feedback based information on resistance to biotic and abiotic stresses, growth and development features of the accession, quality characteristics of yield *etc.* adding this type of information facilities more focused identification of germplasm to meet prospective indents.

> All data should be kept up to date. They should also be duplicated at regular intervals and stored at a remote site to guard against loss from fire, computer failure *etc.*

> Any other information added post-conservation such as registration, sharing with other genebanks, *etc.*

> barcoding is a useful tool that can complement a genebbank information system.

**IV. Field Genebank Data-** In case of field genebanks, field management processes and cultural practices need to be documented. Details should include:

> Passport data for all accessions using the ICAR-NBPGR/standard multi-crop passport descriptors.

> Inventory, map (hard copy and digital); old maps should be retained for reference.

> Planting and harvesting dates and notes on the identity verification.

> Field management processes and cultural practices.

> Representative images (preferably digital photos).

Regeneration, characterization, and evaluation (including biochemical and molecular if any).

Indents, distribution data and user feedback.

All data should be kept up-to-date. They should also be duplicated at regular intervals and stored at a remote site to guard against loss from fire, computer failure, *etc.*

**V.** *In vitro* **Genebank and Cryo Genebank Data-**Information for *ex situ*, *in vitro* conserved and cryo-preserved germplasm, should include:

Passport data (especially including source of material such as organization/individual) as per ICAR-NBPGR/standard multi-crop passport descriptors.

Name of the curator

Accession receipt date, original identification number, new identity, taxonomical determination.

Information on donor material maintained in field genebank or greenhouse.

Type of storage (*in vitro* and cryo), storage details (culture room, rack, cryo-tank, box, *etc.*); previous location, if any, should be retained in record for reference.

Protocol for establishment and conservation (along with reference) including explants and methods.

Details of subculture (for establishment and maintenance).

Details of safety duplication (duplication date, location/institution/person, corresponding identity numbers.)

All data should be kept up-to-date. They should also be duplicated at regular intervals and stored at a remote site to guard against loss from fire, computer failure, *etc.*

**V. Characterization and Evaluation Data**

**a. Characterization Data-**Data should be documented only in standardized and calibrated measuring formats

Data should be documented as per standardized descriptor lists that are publicly available. Using descriptors publishing by Bioversity International (formerly International Plant Genetic Resource Institute), ICAR-NBPGR, PPV and FRA (DUS), Union for the Protection of New Varieties of Plants (UPOV), USDAs, National plant Germplasm System (NPGS), *etc.* is helpful for characterization.

Unlike breeding experiments, PGR characterization throws rare phenotypes in a wider spectrum. Documentation should facilitate the preservation of rare alleles or for improving access to defined alleles. Documenting how this is achieved is also extremely important.

Molecular marker technologies and genomics are increasingly used for characterization, in combination with phenotypic observations because they have advantages in the estimation of uniqueness of a source of variation within or among accessions. Databases should create provisions to add this additional layer of information.

**b. Evaluation Data-**Generation of evaluation data is time consuming and more expensive than obtaining characterization data. Genebank's documentation systems should allow accumulation of data from multiple sources (*e.g.* users whom seeds have been distributed, crop-based institutes, consortia projects, AICRPs *etc.*) at least after the user has published the evaluation results. Sustainable arrangements in this regard should be worked out between the genebanks and the recipients/users of the material.

Data should be documented as per standardized descriptor lists that are publicly available and calibrated measuring formats. Evaluation data to be documented as per traits and multiple observations within each trait (relevant international standards in respective experimentation to be employed). For evaluation data:

i. First traits are defined

ii. Descriptors for each trait defined

iii. Descriptor states with units, if any, defined

In case of molecular data (may include any of genomics, proteomics, metabolomics and bioinformatics data), descriptors and descriptor states are defined (qualitative, nominal continuous, categorical *etc.*)

Documentation of evaluation trials is not complete before obtaining data from at least three environmentally diverse locations collected over at least three growing seasons. Documentation of data of check is important.

The data need to be formally validated by curator before being uploaded into the database.

Evaluation data should also be include digital images of representative expressions, and can include videos and specific remarks.

Documentation of evaluation data should allow addition of traits on continuous basis (unlike characterization data where descriptors and states are fixed).

**VI. Other Data and Connecting Databases**

Inland distribution data and user feedback should be documented for every accession (*Annexure V, VI, VII*)

Any other information including but not limited to DNA fingerprinting, Germplasm registration, variety protection, voucher images (photographs or drawings), information on utilization, international use based on alternative IDs, map *etc.* should be documented in appropriate databases providing sealess connection between multiple data fields with genebank accessions number as primary key.

*Source*: ICAR-NBPGR, Guidelines for Management of Plant Genetic Resources in India, 2016.

records) and/or completely computerized methods for data storage and maintenance. The system is also designed for information retrieval.

## Desirable Features of a Documentation System

### Data Integrity (Quality and Authenticity of the Data)

Information retrieved from a documentation system must be accurate, reliable and up-to-date for it to be value. For this individual can decide and design for maintenance of accurate data and information. Before recording the data, the code dictionary should be prepared giving the detailed information for interpreting the coded data. Automatic data validation during the inputs phase essentially helps in ensuring the validity of the data in terms of permissible limits for each data item and whether they are of alphabetical or numerical types.

#### Table 8.3: Data Integrity (Quality and Authenticity)

| Data Quality | Data Authenticity |
|---|---|
| Quality is the state of completeness, validity, consistency, timeliness and accuracy that makes PGR data appropriate for intended use in operations, decision making and planning. As data volume increases, quality in terms of internal consistency within data becomes paramount. | Data authenticity is the genuineness of data. This means that the data received at the documentation server are original and are received exactly as they were recorded. Data integrity is the surety that the data records are real and are not faked or modified. |
| International data standards to be adopted to make PGR documentation easier and amenable to use and exchange of India across systems. | Spellings of locations (from villages to countries) to be authentic and consistent across entries. Spellings of botanical names to be authenticated only by experts and not informatics personnel. |
| Blanks in one or more fields to be avoided; matching the common fields like identity number and botanical name across multiple databases (whenever database is being constructed from the pre-existing information) to be ensured or accept only those entries that possess information in all the essential fields. | Content of the data base to be verified by the subject matter specialists and not by the data entry operators or database managers. |

*Source*: NBPGR, Guidelines for Management of Plant Genetic Resources in India, 2016

### Fast Information Retrieval

If system is well designed, retrieving the data and information will be simple and straight forward. If it is not well designed, anyone can spend hours, or worse still, not be able to supply the requisite information at all. Remember the documentation system is working for you, not the other way round.

### Flexible Operation

The documentation system should not be rigid in its operation. It should be able to cope with different requests for information and accommodate changes in gene bank procedures. Take a note, if a new curator was appointed, would the curator have the same information needs?, Probably not. If there were a change in gene bank objectives or a new procedure, would this affect the documentation system? Yes it would! Therefore, you should try to anticipate information requests and changes in gene bank procedures as for as possible.

### Organization of Data

Data not stored in documentation system at random, if they were, maintenance of the system would be tedious and information retrieval would be possible. Instead, the data are organized into groups which are practical to use-practical for data recording, storage and maintenance and practical for information retrieval. In a gene bank documentation system, user's needs must be taken into account when organizing data into practical group.

### Stages in Construction of a Documentation System

The construction of a documentation system requires detailed analysis and planning before the design of any manual forms and/or database. In documentation system six stages can be identified as:

(1) **Stage 1:** Obtaining background information about the genebank
(2) **Stage 2:** Define documentation objectives
(3) **Stage 3:** Analysis of gene bank procedures
(4) **Stage 4:** Identify meaningful sets of descriptors
(5) **Stage 5:** Develop data formats and manual forms and/or computer screen entry forms
(6) **Stage 6:** Develop the documentation procedures and implement the new system

### Relating Data Together from different Activities

The data that are generated and used in a wide variety of gene bank activities will fall into two main categories:

(i) Accession specific data
(ii) Group data

### (i) Accession Specific Data

Accession specific data concerned individual gene bank accessions. Most of the descriptors you use will be accession specific data such as seed moisture content, per cent viability, weight of seed, 1000 seed weight, plant height, seed colour and so on. Accession specific data are very important for gene bank management. How can you relate together and co-ordinate all the different accessions specific data in your documentation system? You need to use the accession numbers, their batch reference (where appropriate) and their scientific name as the basis for building up the documentation system.

**Table 8.4: (1) Stage1: Obtaining Background Information about the Genebank**

| Stage | Purpose | Activities |
|---|---|---|
| **Stage 1:** Obtaining background information about the gene bank | To gather essential background information on the set-up of the gene bank which will help define documentation objectives and facilitate resource management | In order to develop a documentation system tailored to the gene bank's needs, you need to analyse the genebank's and resources with the full cooperation of other staff member. This will give you essential information about the gene bank from which you can later develop documentation objectives. It will also help you to make decisions on how best to use available resources. |
| **Stage 2:** Define documentation objectives | To define priority areas for the documentation effort | You need to have a clear idea of the areas of gene bank work that need documenting and the priorities for documentation. You should write these down as documentation objectives and list them in order of importance. These objectives could include documentation of passport data, inventory data, seed handling procedures, distribution data, characterization and evaluation trials. |
| **Stage 3:** Analysis of gene bank procedures | To identify documentation requirements of each procedures and the relation between different procedures. | Having defined your documentation objectives, you can start the resource requirements of each procedure, and the different type of data which are generated or used in each procedure. This will help you later on in deciding how best to handle the data *e.g.*, the suitability of computers and/or manual forms. The analysis will show how the procedures relate to one another. This information can be used to build up a flow chart showing the relationship between gene bank procedures and information flow. |
| **Stage 4:** Identify meaningful sets of descriptors | To identify meaningful sets of descriptors from the analysis of procedures. | Much of data you will be recording will be concerned with individual gene bank accessions. To facilitate the operation and maintenance of the documentation system, you need to organize the descriptors into practical sets. You can think of these sets as separate books, folders or forms in a manual system or separate files in a computer system. |
| **Stage 5:** Develop data formats and manual forms and/or computer screen entry forms | To develop formats for data which facilitate data recording and flexible information retrieval and to develop manual forms and/or computer database formats and screen entry forms that will be used at each stage of the documentation process. | An important task of the documentation specialist is the design of manual forms and computer screen forms for straight forward data entry which minimize the risk of errors creeping into the system. If these forms are designed well, the accuracy of the data entry is ensured. |
| **Stage 6:** Develop the documentation procedures and implement the new system | To develop the documentation procedures to facilitate operation of the system and to implement the system. | A documentation system can be well thought out and well-constructed but not used unless there are clearly defined documentation procedures and training is given in the use of the system. |

## (ii) Group Data

Group data concern groups of accessions. For instance, you will find data referring to particular species (*e.g.* equilibrium moisture contents, viability test methods, regeneration procedures, methods for determination of moisture contents, *etc.*)

These categories of data will be handled quite differently in your documentation system.

**Data base maintenance and data updating** (As per ICAR-NBPGR Guidelines for Management of Plant Genetic Resources in India, 2016)

A. Software, hardware, tools and technologies

B. Data updating

C. Maintenance of database

D. Data Access

## A. Software, Hardware, Tools and Technologies

☆ The software, hardware, tools and technologies used for management of a PGR database are determined by:

(i) Scale of the database (number of independent entries and diversity of data types) and

(ii) Intensity of use (number and location of users)

☆ In case open source software is employed, it is essential to ensure experts are available on call. It is best to use licensed and enterprise versions of relevant software or a bouquet of software that take of all operations from data entry to database development to online access.

☆ All necessary software and hardware tool for data entry, preparation of documents and reports to be available in enterprise/professional versions wherever available and necessary. It is also important to implement unified threat management (virus, malware, worms, hacking, stealing *etc.*).

☆ Local and remote administrative procedures and, necessary hardware and software should be available. Adequate software and hardware essential for storage, periodic back-up and data safety should be available. Computer hardware gets obsolete fairly quickly (@5 years) and, therefore, sufficient funds need to be earmarked for replacements.

## B. Data Updating

☆ Botanical names and their spellings should be periodically checked for changes/updates. Spellings and geo-political affiliations of locations also should be checked for changes/ updates. For example botanical name of tomato was *Lycopersicon esculentum* but now it is *Solanum lycopersicum*.

☆ Written records of the main passport data, field data books and hard copies of the field maps should be maintained systematically.

☆ Addition of new information, corrections, and deletions should be incorporated on real time basis and all changes to be published on line only after authentication by experts.

## C. Maintenance of Database

☆ Databases are not static and changes are constantly being made as information is added, removed, and moved around. Furthermore, parameters, indexing systems are changed over a period of time. This can cause malfunction in the database. A database that is not maintained can become sluggish, and people may start to experience problems when trying to access records. Database maintenance is an activity designed to keep the database clean and well organized so that it does not lose functionality.

☆ One important aspect of maintaining a database is simply periodic backing up of the data. Database maintenance includes checking for signs of corruption in the database, looking for problem areas, rebuilding indexes, removing duplicate records and checking for any abnormalities in the database that might signal a problem.

☆ Database management being a specialized job, independent functional unit and personnel should be engaged.

## D. Data Access

☆ Access to and use of information should be governed by relevant clauses of the national intellectual property instruments as well as principles of biodiversity conservation and use, as applicable.

☆ As a provider of information, the institute hosting the database should retain all rights to maintain multiple and regulated access to information depending upon the user type. consequently, software and hardware should be prepared for such tired access.

☆ Adequate security (physical and electronic) and regulatory measures have to be implemented to pre-empt unauthorized and un-intended use of the information.

## Use of Information System

With tremendous pile-up of information by various global and national agencies, *viz.*, CGIAR institutes, USDA, ICAR-NBPGR *etc.*, on the collection, characterization and evaluation, storage, *in situ* conservation and usage of the germplasm of various crops and their related wild species during the past few decades, scientists now-a-days spend at least 30 per cent of their time in taking care of the data information associated with collection (Rogers *et al.*, 1975). In most cases, the stage of documentation is manually operated data registers or partially computerized data banks. During recent years, the term documentation is appropriately known as 'Information System'. An information system is much more than simple documenting information. The information system has to be dynamic, vital and flexible and ensuring the reliable and integrity of the data and providing effective methods for its handling. In this chapter details about information systems, database management systems (DBMS) and models, databases, descriptors, descriptor states, standards for data preparation and the national efforts in this endeavor. Hersh and Rogers (1975) have discussed information requirements for genetic resources application from the point of view of systems approach in which there is a dynamic dialogue between the system's design under consideration and the user. They have indicated how to establish such an approach for documentation for global network of genetic resources centres. There is a clearcut advancement towards switching over to computerized database management. But since no standard data retrieval system exists at different plant genetic resources organization, except a few, *e.g.*, GRIN of USDA, IRRIGEN of IRRI *etc.*, the inter-institutional interaction among users becomes quite a difficult task. In the modern era, electronic computers are extensively used in different organisations for multiple purposes.

### Taxonomic Information Retrieval (TAXIR)

Some degree of progress was made towards this aim during the 1970s, when a system called Taxonomic Information Retrieval (TAXIR), a general purpose and computer assisted information system was developed at the Taximetrics Laboratory of the University of Colorado, USA.

### Executive Information Retrieval (EXIR)

Executive Information Retrieval (EXIR) system evolved at the same university to meet specific needs of scientists involves in the data management of genebanks.

These systems were used only in few genebanks in various places (Izmir, Turkey, Bari, Italy) because of difficulties in portability (especially the big size of computers required to run the programme) and have since been overtaken by fully supported commercial Database Management System (DBMS) available for a wide range of computers. The fast development of small sized and low priced personal computers (PCs) has supported or even enabled this development and revolutionized the entire management of information in genebanks. Because of the compatibility of the software and hardware it is not anymore required that everybody is forced to use the same database management system to handle the genebank data as was the case some 35 years ago. Presently, there are 980 genebanks operating in 120 countries and maintaining over 3.6 million of samples. Genebanks differ from one to another in their activities and how the activities are organized and performed. Some of them have developed their own information system fitting to their requirements and based on the availability of the computer system and tailored database management system (DBMS) software used for other purpose. One of the front runners in the management of genetic resources data is certainly the Nordic Genebank at Weibullsholm Plant Breeding Institute in Sweden.

## Geographic Information System (GIS)

Geographic Information Systems (GIS) as developed by Global Resources Information Database (GRID) are powerful tools for integrating and analyzing diverse spatial data (Burrill *et al.*, 1991).

## Germplasm Resources Information Network (GRIN)

Germplasm Resources Information Network (GRIN) system developed in USA, is quite capable of monitoring information on world's largest collection at the National Seed Storage Laboratory (NSSL), Fort Collins and the cooperating Institutions within the USDA research system in USA. Similarly, IRRI, Philippine, CIMMYT, Mexico and several other International Agricultural Research Institute have developed their own information systems of their respective mandate crops for handling the germplasm. IRRIGEN software which is PC based, has been developed at IRRI. For main computers, they have also developed ORACLE based documentation system. Similar is the case with the GRIN system. PCGRIN is a PC based system and GRIN3 is being developed in ORACLE on the main frame system. The system developed at Kew Seed Bank (Royal Botanic Gardens, Wake Hurst Peace, UK) aims to eliminate the duplicate efforts spent in handling the basic data as well as monitoring the progress of accessions into the genebank and the distribution of seeds from it (Smith *et al.*, 1985). Genebank of National Vegetable Research Station (NVRS), Welles Bourne, UK, uses INFO, a pseudo relational database management package which uses a conversational language allowing the user to

create and maintain information in an easily accessible format (Astley, 1985). Using DATATRIEVE-11, Polish Gene Bank at Radzikow has also implemented a documentation system machine independent, globally accepted, easily maintained by data during hardware failure or abuse by unauthorized personnel.

## Databases, Descriptors and Descriptor States

Passport data are important in studying the phenology of the material and variation in distributional pattern with respect to ecological and socio-economic factors. The more information gathered by the explorer on the collected germplasm, the more valuable it is to the users. The recorded information covers wide range of factors relating both to the collection site and to the population from which the sample has been collected. It is advisable to record information on both the essential and optional fields in the passport data sheet at the site of the collection itself by the explorer. However, in any circumstance, the explorer should not leave the information blank on essential fields.

### Database

A database is a combination of related records and a record is a combination of fields or what we call descriptors in genetic resources environment.

### Descriptors

A descriptor is now a widely accepted as a computer term for the character of a plant, as well as for other units of information such as country of origin or the date of collecting. Basic unit in plant genetic resource management is the descriptor which is simply referred as 'field' in computer terminology. Data needs to be standardized in terms of descriptor or descriptor states for making an information system meaningful and more generally applicable. The minimum list of descriptors for the management of gene bank has been suggested as well (Konopka and Hanson, 1985).

### Descriptor State

The descriptor state is then the quantity or the quality of the plant character, or any country name or abbreviations or the actual day, month, and year of collecting. These descriptors can be further distinguished from the point of view of use and content.

## Importance of the Accession Numbering System

In order to facilitate description of accessions, their evaluation and also to improve information and communication between individual scientists in different localities, there had been a continuing effort in several countries especially Germany to provide documentation standards acceptable to all instead of individuals. Each gene bank accession will have its own unique accession number which distinguishes it from all other accessions in the gene bank. If an inappropriate numbering system is used or if none is used at all, the consequences are serious. It will cause confusion, create much unnecessary work through mistakes being made and ultimately lead to loss of valuable germplasm. In order to provide a set of terms a list of descriptors and their terminology for information sets which are associated with accessions of various crops was developed in Germany. The thesaurus of terms provided with exact (or nearly exact) equivalent for each term in some of the major languages, *i.e.* German, French, Spanish and Russian. By use of these terms, workers in various crops and various parts of the world can accurately describe their collections and the results of their observations and tests and be certain that the terminology will be understood by the others.

**Table 8.5: List of Essential Collecting Data Descriptors**

| Sl.No. | Class | Descriptors |
|--------|-------|-------------|
| 1. | Sample labelling | Expedition identifier (or collecting organization) |
|  |  | Name(s) of collector(s) |
|  |  | Collecting number (or collector's number) |
|  |  | Type of material |
| 2. | Sample identification | Genus, species, subspecies, botanical variety |
|  |  | Vernacular name (and language) |
|  |  | Herbarium voucher number |
|  |  | Identification numbers of other associated specimens |
|  |  | Status of sample |
| 3. | Sample information | Number of plants sampled (size estimates of population and of sample) |
|  |  | Sampling method |
|  |  | Collecting source |
| 4. | Collecting site localization | Country |
|  |  | Primary administrative unit |
|  |  | Precise locality |
|  |  | Latitude of collecting site |
|  |  | Altitude of collecting site |

*Source*: Moss and Guarino, 1995.

In accessioning, system should be simple and practical to use:

1. **Use a strictly numeric system which is sequential in operation**: Number should start at "1" and then increase as each new accession is received *i.e.* "2", "3", "4", and so on. You should also add your gene bank acronym to the number. For example, if the acronym of IIVR gene bank is "IIVR" the accession will be labeled IIVR1, IIVR2, IIVR3, and IIVR4.

2. **Do not incorporate additional information in the accession number**: Do not try to add additional information such as the year of deposit or a crop code into the number.

3. **Use only one numbering system**: Operate a single numbering system which is used for all accessions.

4. **Do not use depositor's number**: If you only have a small gene bank, you have may considered using the depositor's designation as basis for your accession numbering system rather than devising your own system. The approach can cause serious operational difficulties in your gene bank and should not be accepted.

5. Do not use the germplasm collector's number.

6. **Do not use a different numbering system for each crop**: Gene bank which contain very large collections of germplasm sometimes operate separate but sequential accession numbering system for each vegetable crop. If you maintain many different vegetable crops or only work in a small to medium sized gene bank, this approach is not recommended.

7. **Do not use a "reserved" numbering system**: You may consider using a single accession numbering system but "reserve" numbers for particular crops, is assigning then it can cause a number of operation difficulties in the long run. So, it will be better that this system should not be attempted.

## Approaches for Recording Quantitative Data

The International board for Plant Genetic Resources (IBPGR) has recognized the need for determining standards for data recording. By avoiding ambiguity in the data standardization one can make the communication of information easier and more effective. With this objective in view, crop advisory committees have been developing minimum descriptors lists accompanied by standards for measurement techniques, units of measurement and data recording and encoding methods.

The different approaches for recording quantitative data are as:

**Table 8.6: Approach and Notes of Quantitative Data**

| Approach | Notes |
|----------|-------|
| Continues scale | Descriptor is scored using standard (SI) using (example meters, grams) |
| Ordinal scale | Descriptor is scored using a series of pre-defined descriptor states. In this example, a plant height of 0.9 meter is scored as either 'short' or '3'. |
| Binary scale | Each descriptor has two descriptor states: + (present) and 0 (absent). In this example, the single descriptor for plant height has been replaced with 9 separate descriptors for the individual ranges, each scored as+ (present) and 0 (absent). |
| Nominal scale | Descriptor is scored using a series of pre-defined descriptor states. In this example, a plant producing orange flowers is scored as 'orange' or '4'. |

## Standards for Data Preparation

In order to make an information system meaningful and more generally applicable, the data needs to be standardized in terms of technology and measurement. All those involved in plant genetic resources work have recognized the need for an internationally accepted system to record, classify, communicate, correct or update information about accessions. The procedures for data preparation need careful consideration, as the future universal use of an existing system for data management will depend upon them.

## Efforts for Germplasm Documentation

### (i) International Efforts towards Documentation

The IBPGR has developed uniform standards for scoring, coding and data recording for a number of crops. Database management activities have been strengthened with the aid of INDO-USAID under PGR database management programme. All the 30 cooperating centres/data generating sites are equipped with microcomputers and necessary software. Database has been developed in ORACLE RDBMS, and centrally located on a Pentium server.

### (ii) National Efforts towards Documentation/ NBPGR's Efforts towards Documentation

The ICAR-National Bureau of Plant Genetic Resources (NBPGR), New Delhi, since its inception in 1976, is catering to the needs of the plant genetic resources community in the country as a central service for all works related to plant exploration and collection, evaluation and its proper conservation for present and the future use. In addition to this the Bureau has an important function to collect all available information regarding the genetic diversity and the same properly documented and disseminated to all concerned breeders/curators in the country as well as outside the country. As a part of its documentation activities, the institute has generated a large amount of information and developed a number of crop catalogues, plant introduction reports, crop inventories, newsletters, information bulletins and monographic volumes for the benefit of PGR community. In ICAR-NBPGR Data model, a separate area for exploration has been identified which can be related to other using Accession Number and the Taxonomic information can also be gathered using the Taxonomic area. The ICAR-NBPGR has plans for the development of National Information System on Plant Genetic Resources in the country. Since the National Information System has to cater to the needs of crop based institutes, coordinated crop projects, agricultural universities and other users agencies in India as well as abroad, a through system analysis has been done in this respect in terms of identification of inputs and outputs of information and their generation and standardization, information flow, security/protection against misuse, hardware and software, man power and expertise and communication facilities *etc.* Efforts for development of on line system, availability of the system on the 'internet' are also under active consideration.

Since 1980, a considerable progress has been made as a follow up action of the recommendations of the First National Workshop on Documentation of Plant Genetic Resources convened in that year. The manual system of data recording/processing is being gradually replaced by a computer based system. Long back ICAR-NBPGR, New Delhi inititated a project on Genetic Resources Information Programme (GRIP) in 1986. This has aimed at the creation and development of a computer based information system for the efficient management of the national plant genetic wealth.

## Documentation Activities

   (i) Plant introduction reporters and crop inventories

  (ii) Quarantine information and check lists

 (iii) Passport information

 (iv) Field evaluation and cataloguing

  (v) Gene bank information

### (i) Plant Introduction Reporters and Crop Inventories

The introduction of germplasm from exotic sources by the breeders in India is one of the regular activities of the Bureau. Since 1940, when the first accession was registered in the National Accession Register, ICAR-NBPGR, New Delhi has introduced germplasm (including trial material) over 9,00,000 samples. Each accession is given EC (Exotic Collection) number at the time of its entry and the other details of information accompanying the accessions, *viz.* botanical name, original identification number/names, source country and address, recipient name and address, number of samples and special notes, if any, of the accessions are recorded in the National Register. The entire information is compiled and a quarterly Plant Introduction Report (PIR) is issued to Indian scientists and cooperating agencies abroad. The information on the germplasm of different crops introduced into India is compiled from time to time and published as inventories.

### (ii) Quarantine Information and Check Lists

All plant introductions, when received, passport through plant quarantine and are assigned Import Quarantine (IQ) number. Country name, type of material, case number, consigned quarantine history *etc.* is also recorded in the import quarantine register along with the information as clearance/rejection, interceptions, salvaging techniques and post quarantine treatment *etc.* these records are maintained manually in the import register. Similar records, as stated for imports, are also maintained for material under export. With a view to know beforehand the risks involved in introducing new pests and pathogens into the country while importing germplasm from exotic sources, check lists are prepared with the help of available literature.

A number of such lists have been compiled by the Division of Plant Quarantine mainly for the purpose of internal use.

### (iii) Passport Information

Plant exploration and collection work has been taken up by the scientists of the Bureau in different agro-climatic zones of the country as well as several regions of Nepal, USSR, Indonesia, USA, Mali, Nigeria, Malawi, Zambia, Kenya, Ethiopia, Bangladesh and Sri Lanka. During the period 1983-1990 the Bureau coordinated and conducted 298 collections missions and assembled over 59,700 germplasm accessions. Site data sheets are used for recording information on a set of passport descriptors *viz.*, collector name and number, Latin name of crop, common/cultivar name, provenance data including longitude, latitude, and altitude, status of the sample, frequency of occurrence, population structure, habit information on pests and pathogens, soil colour and texture and other special attributes, if any. Each accession is assigned an IC (Indigenous Collection) number before it is passed on to the Evaluation Division. In previous years the passport data information was documented manually in the form of Plant Collection Reporter and disseminated to user agencies. All these germplasm passes through the routine process of characterization/evaluation, multiplication and seed testing and finally added to the ICAR-NBPGR's base collection.

### (iv) Field Evaluation and Cataloguing

The Germplasm Evaluation Division is entrusted with the responsibilities of preliminary evaluation and seed increase, characterization, documentation and preparation of catalogues, *etc.* As regards further evaluation, a limited number of characters related to agronomic and production traits, resistance to disease and pests were selected by the different plant breeders in the past. However in the recent past due emphasis has been given to follow the IBPGR list of descriptors as far as possible for recording the data. Further evaluation data are also available with various crop based institutes/coordinated crop projects, agricultural universities, besides such data collected within the Bureau. But so far no standard set of descriptors lists have been prepared which could be uniformly adopted and followed strictly within the country. ICAR-NBPGR, New Delhi has generated considerable evaluation data on different crops and the information has been compiled and documented in the form of 50 crop catalogues. Some of these catalogues also give in detail the estimates of variability, correlation and regression parameters, frequency distribution in respect of quantitative as well as qualitative traits parameters and answers to certain simple and complex queries regarding the database on useful traits or combination of traits. From 1986 onward, the Bureau started computer processing of the evaluation data. A programme called Germplasm Evaluation Information System (GEIS) was developed for handing the information on such data. For better management of the data files, 8 major groups of crops *viz.*, grain, legumes, cereals and *Pseudocereals*, oil seed, millets and minor millets, vegetables, horticultural crops/plants, medicinal and aromatic plants and miscellaneous have been formed. Programmes have been developed in dBASE III PLUS.

### (v) Genebank Information

Earlier it was reported that ICAR-NBPGR, New Delhi holding over 1.5 lakhs of accessions of various crops which have been stored in the National Repository for long-term conservation (Rana, 1992) but today its number increases to more than 3.0 lakhs accessions. Data are maintained on some of the important descriptors, *viz.* crop name, genus, and species, identification number, germination percentage, moisture content and month and year of storage. The data base is being developed for these management descriptors and the information is monitored periodically. Genebanks labels are also printed using the database. Information on samples stored under cryopreservation is also maintained and monitored.

## Training on Documentation and Information Management

Under the Genetic Resources Information Programme (GRIP) at ICAR-NBPGR, New Delhi the institute has organized 10 computer appreciation courses in relation to database management of plant genetic resources since 1986 to train the scientists and technical staffs of the institute as well as of the cooperating centres identified under the INDO-USAID PGR project.

## PGR Database Management Plan

Database management activities are being further strengthened at the ICAR-NBPGR, New Delhi under the INDO-USAID PGR database management. The project has been given high priority and sufficient funds have been allocated for this important area. It has been planned to equip all the 30 cooperating centres/data generating sites with micro-computers and necessary software. A mini computer system with capacity of supporting more than 100 terminals will be located at the ICAR-NBPGR, New Delhi. The development of hardware and software facilities at the ICAR-NBPGR, New Delhi (Headquarter) and at the cooperating centres has begun and 15 of these centres have been equipped with Intel 80386 based micro-computer systems. At the ICAR-NBPGR (Headquarter) recently a prototype computer system (486 EISA) supporting UNIXO.S. and ORACLE RDBMS has been procured to develop the required database design. Data input, information and exchange through magnetic floppies, magnetic types compact disks, external extended memories card, pen drives *etc.* becomes easier through systematic use of computers.

## Cataloguing and Data Processing

**Cataloguing:** A simple and common method for exchange of various information are through crop catalogues which are the printed versions of systematically pre-arranged/analysed data based on a particular set of descriptors and descriptor states decided for a crop or its related species. Crop catalogues are valuable source of information on passport data, various agrobotanical and economic characters and other traits of interest to research workers involved in crop improvement programmes. In the present context, a crop catalogue may be considered primarily a systematic listing of different accessions of a crop, their identification source, characterization and evaluation data along with additional/optional pre-analysed information using different statistical tools. Cataloguing involves frequent use of many softwares and statistical packages, *viz.*, d BASE III plus, Fox Pro, Reflex, Lotus 1-2-3, MSTAT-C, SPSS, BMDP, MICROSTAT, Harward Graphics, Word Star, Word Perfect *etc.* among these, the most frequently used package is any DBMS (Data Base Management System) package. A DBMS package is a computer program that organizes large collection of interrelated data according to a well-defined scheme and produces reports by extracting and reformatting data. Reformatting refers to indexing and/or sorting on a particular field(s). It further range from rearrangement or simply totaling to complex statistical analysis. The DBMS is a system for information storage, retrieval and analysis with emphasis on formatted data. In all the standard softwares on DBMS containing two tracks:

☆ **Field:** A field can be considered as any descriptor [a widely accepted computer term for the character of plant, as well as for other units of information such as country of origin or the date of collection; Sapra and Singh, 1992] from the list of descriptors and

☆ **Record:** A record is an accession with all descriptors. A collection of related records/ information constitutes a file in computer environment.

## Recording of Data

The information about passport data (accession identifiers and information recorded by collectors) and characterization (recording highly heritable characters capable of expressing themselves in all environments), preliminary and further evaluation is recorded on paper or on data loggers, a handy device, from which data can be subsequently transferred to a computer for further storage and analysis. To facilitate description of accessions, results of tests of accessions and to improve communication between scientists in different institutions, internationally accepted ICAR-IBPGR, New Delhi descriptors lists decided for each crop species should be followed. These norms are well defined for qualitative as well as quantitative characters (Rogers

*et al.,* 1975). For scoring in heterogenous populations Hintum (1989) approach should be followed. In modern days, various statistical designs are available in literature which can be extensively used depending upon the population size, availability of seed and resources. in plant genetic resources, the accessions for different crops vary invariably from 100 to 1000. Limited seed quantity in each accession and meager experimental land makes it difficult to replicate the accessions and thus the basic norms of statistical designs are violated. In such situations, augmented randomized design may be adopted (Federer, 1956). The design consists of one set of treatments (the check varieties) replicated a number of times and a second set of treatments (the accessions) appearing only once in the entire layout.

## Data Processing

Extensive data processing is involved in bringing out a catalogue. The computer data processing is a series of operations carried out in order to convert the data into useful information. The voluminous data recorded over a period of time, its tabulation *etc.* and lengthy calculations done manually has been a tedious task for huge pile-up of germplasm information and thus switching over to computer data processing is a good preposition. A computer program for statistical analysis of augmented randomized complete block design was further described in simplified form (Sapra and Agrawal,1992).

## Vegetable Germplasm Registration

A lot of unique and useful genotypes are being collected and developed by researchers. Some are proving world records with their shape, size, quality, colour, texture *etc.*

## Vegetables Registered in Guinness Book of World Records

So if you exclude massive squash and pumpkins from the vegetable class, what is the world's largest vegetable? This title should be limited to vegetables in a typical human diet—because to Australian koalas, the largest vegetable would most certainly be an enormous eucalyptus tree. The 1985 "Guinness Book Of World Records" (UK Edition) lists some of the record-breaking vegetables, including a 35 pound (16.0 kg) turnip, a 45 pound (20.0 kg) red cabbage, a 28 pound (13.0 kg) broccoli, a 52 pound (24 kg) cauliflower, a 25 pound (11.0 kg) lettuce, and a remarkable 124 pound (56.0 kg) cabbage six feet (1.8 m) in diameter. Although this giant cabbage cited in the Guinness Book seems unbeatable for the title of "World's Largest Vegetable," there are tropical yams belonging to the genus *Dioscorea* that may be 6 to 9 feet long (2.0-3.0 m) and weigh 150 pounds (68.0 kg) or more, although they are usually harvested at about 2-6 pounds. These yams are not to be confused with fleshy storage roots of red sweet potatoes (*Ipomoea batatas*) of the Morning Glory Family (Convolvulaceae) which are also called yams. True yams belong to an entirely different and unrelated plant family, the Dioscoreaceae.

With the advent of the regime of ownership and Intellectual Property Right (IPR) in relation to plant genetic resource, it has become imperative to develop mechanism to ensure the developers of plant get due credit. Unlike the developers of released varieties, scientists associated with the development of improved germplasm and genetic stocks (new sources of resistance, male sterility, varied types of mutants, cytogenetic stocks *etc.*) had no mechanism for recognition. Lack of formal recognition of such useful materials and the role of scientist in development of these materials, discourages them from sharing the valuable materials with other researchers. Consequently, most of such valuable material remains underutilized or is lost. This is a mode for protecting landraces, traditional cultivars and other germplasm having unique traits including tolerance/resistance to disease(s), pests and other biotic and abiotic stresses but do not qualify for release and notification because of poor agronomic performance. These materials are novel, unique, and distinct with academic, scientific or applied value but do not have a direct commercial value. With the objective of recognizing the contribution of scientists who have developed or identified such promising experimental materials or promising germplasm, and to facilitate flow of valuable germplasm among the scientists working in the crop improvement programme. ICAR has developed the mechanism to register such germplasm and ICAR-NBPGR, New Delhi is a nodal agency, entrusted with the responsibility for registration of germplasm to assign the registration number through Plant Germplasm Registration Committee (PGRC). The registration of germplasm is an initiative that provides soft protection to the germplasm having unique trait(s). The validity of registered germplasm is fixed in term of duration *i.e.* 15 and 18 years for clonally and seed multiplied vegetable crops, respectively.

## Stepwise Plant Registration Process

### 1. Plant Germplasm Registration Committee (PGRC)

The Plant Germplasm Registration Committee (PGRC) is constituted with the following dignitaries:

(i) Chairman-DDG (CS) Indian Council of Agricultural Research, New Delhi; for maximum of 03 years;

(ii) Permanent member- Director, ICAR-NBPGR;

(iii) Member Secretary Senior Level Scientist, ICAR-NBPGR identified by the Chairman, PGRC;

(iv) Other Members-Co-opted as per the advice of the Chairman.

(v) Need based crop specialist with reference to the material under consideration and with the approval of the Chairman, PGRC.

### 2. Nodal Agency

(i) ICAR-NBPGR, New Delhi, is the nodal agency for registration of germplasm. The application should be addressed to the Director, ICAR-NBPGR, New Delhi-110012, along with seed sample or a certificate of submission of propagules with respective crop/plant based National Active Germplasm Site (NAGS) for establishment/conservation.

(ii) Member Secretary, PGRC will duly acknowledge with date, the receipt of the application and the seed material. In case of vegetatively propagated crops. After ensuring deposition and establishment of genetic material at the relevant NAGS, the acknowledgement would be issued communicating application number and the national identity.

(iii) ICAR-NBPGR, New Delhi maintains a permanent record (register) and data base listing the

**Table 8.7: Vegetable Related National Active Germplasm Site (NAGS)**

| Crop/ Crop Group | Institute | Address | Phone | Fax |
|---|---|---|---|---|
| Onion and garlic | ICAR-DOGR, Pune | ICAR-Drectorate of Onion and Garlic Research (DOGR), Rajgurunagar, Pune-410505, Maharashtra | 02135-222026 | 02135-222026 |
| Potato | ICAR-CPRI, Shimla | ICAR-Central Potato Research Institute (CPRI), Shimla-171001, Himachal Pradesh | 0177-2625182, 2625073 | 0177-2624460 |
| Vegetables | ICAR-IIVR, Varanasi | ICAR-Indian Institute of Vegetable Research (IIVR), Post Bag No. 01; P.O. Jakhini (Shahanshahpur), Varanasi-221305 (Uttar Pradesh) | 0542-2635247; 2635236 | 05443-229007 |
| Temperate vegetables | ICAR-IARI RS, Katrain | ICAR-Indian Agricultural Research Institute (IARI) Regional Station Katrain, Kullu-175129 (Himachal Pradesh) | 01902-240124, 241280 | 01902-240124 |
| Other than ICAR | GBPUA&T, Pantnagar | G.B. Pant University of Agriculture and Technology (GBPUA&T), Pantnagar-263145, Distt., Udham Singh Nagar (Uttarakhand) | 05944-233320, 233350 | 05944-233473 |

germplasm materials approved by PGRC with details on unique trait(s) and other related information.

### 3. Application Form

(i) Application is to be submitted on the prescribed proforma (Form A, Annexure II), which is attached. The PGRC shall meet quarterly (preferably during last month of the quarter) in a year with the concurrence of the chairman, for consideration of applications and related matters following the guidelines (Annexure II).

### 4. Eligibility Criteria for Registration

(i) Germplasm or genetic stock of vegetable plants contains unique, uniform, stable and has potential attributes of academic, scientific or commercial value shall be registered.

(ii) Exotic germplasm (imported) can be registered for a trait other than those published or registered. Similarly, selections made from exotic germplasm can also be registered.

(iii) Selection for unique traits from landraces (other than landrace is known for) may be considered for registration.

### 5. Proof

All claims concerning the germplasm material submitted for registration should accompany scientific evidence for uniqueness, reproducibility and value in the form of:

(i) Performance (yield contributing traits, adaptation traits, quality traits) data for at least four environments (location and year combination under All India Co-ordinated Research Project (AICRP) trial/nursery test supported with relevant extracts of the documents (*e.g.* comparative data of all entries tested) or verification by concerned Project Director/Project Coordinator (PD/PC) under any other relevant system verified by Competent Authority. For qualitative traits (*e.g.* flower colour, leaf venation, seed colour) data of two environments duly supported by documents.

(ii) For resistant/tolerance to biotic and abiotic stresses, data should be obtained for at least four environments under established hot spot locations/under artificial screening (epiphytotic) conditions. All the proposers of the germplasm/genetic stock should sign declaration that standard procedures were followed for testing/screening.

(iii) Supporting biochemical evaluation data should be obtained from at least four environments.

(iv) The proposed genetic stock/germplasm should also be evaluated along with already registered genetic stock(s)/germplasm(s), if available

(v) Supporting documentary evidence on (i), (ii) and (iii) either in Institute Annual report/AICRP Report/peer reviewed journals.

(vi) Recommendation of Institute's Germplasm Identification Committee (IGIC) regarding the novelty and uniqueness of germplasm for trait(s) claimed.

### 6. Germplasm Ineligibility for Registration

(i) Germplasm or genetic stock without accompanying documentary evidence for the claim made in the application.

(ii) Germplasm or genetic stock that does not contain complete passport data (Annexure VI), including authenticated taxonomic identity, parentage, institutional or national identity, geographic location of origin, and all such information relating to the development and contribution, if any, to the uniqueness of the germplasm.

(iii) Exotic material *per se*, with no evidence of human intervention in its improvement.

(iv) Varieties/hybrids of common knowledge or selection from traditional or farmer's varieties without prior approval from the concerned personnel.

(v) Variety prior art, with no evidence of human intervention.

(vi) Varieties and hybrids (including parents) released in the country, zone or state. However, parental lines of non-released hybrids may be submitted for consideration.

(vii) Germplasm of any genera or species, which involves any technology, which is injurious to the life or health of human being, animals or plants.

(viii) Material for which any form of protection has been sought elsewhere.

### 6. Screening of Application(s) and their Consideration by PGRC

(i) The Member Secretary, PGRC screen the proposal(s) on prescribed proforma, as per the guidelines of the checklist (Annexure II)

(ii) Proposal forwarded to the relevant Director, Project Director or Project Coordinator for validation of information, particularly on uniqueness and novelty of the proposed germplasm. When the proposals are received from PC/DC or Directors of the crop based institutes, then these will be sent to concerned experts in the area.

(iii) After initial screening, the incomplete applications would be advised for appropriate revision, if required.

(iv) The application, in which the validation of the data is considerednecessary, the applicant would be required to produce a validation report from an appropriate institute, advised by the Member Secretary. The revised application should accompany such report duly endorsed by the competent authority (CA) of the institute, as advised for the validation.

(v) The proposals complete in all respects along with the comments of concerned Director, PD or PC or expert will be put up to the PGRC for consideration.

(vi) The PGRC consider the proposal as early as possible as and not later than one year.

(vii) The decision of the PGRC will be final.

## 7. Validity of Registration

The period for validity of registration is 18 years for trees and vines; 15 years for other plant species, after which the registered germplasm will be national sovereign property.

## 8. Publication of Registered Germplasm

All germplasm approved for registration will be officially communicated to the applicants along with registration number (INGER). A Certificate will be issued to the applicant. A brief description not more than one page (Annexure II) is to be published in the ensuing issue of Indian Journal of Plant Genetic Resources, New Delhi and update on ICAR-NBPGR Internet Website http://www.nbpgr.ernet.in.

Or appropriate periodicals, *i.e.*

(i) *Indian Journal of Genetics and Plant Breeding*, Published by the ISG and PB, New Delhi-12

(ii) NBPGR, News Letter, ICAR-NBPGR, New Delhi-12

(iii) NBPGR, Internet website http://nbpgr.delhi.in

(iv) In addition (a) Concerned crop newsletter (b) ICAR Annual Report(s)

## 9. Conservation, Maintenance and Sustainable Utilization of Registered Germplasm

(i) Registered germplasm will be conserved either in National Genebank (http://www.nbpgr.ernet. in:8080/ircg/index.htm) or at designated crop/ plant based NAGS (as mentioned above).

(ii) All the material registered with PGRC will also be sent by the developer to the relevant Director, PD/PC or NAGS with the request for sowing/planting of registered germplasm in demonstration plots for annual field days and multiplication.

(iii) The institution associated with the development of the germplasm is to be mandated with the maintenance of working stock of germplasm for supply to *bona fide* users.

## 10. De-registration

A registration may be repealed by the PGRC in case of false claim(s). Appeal for counter claim, if any, should reach the member Secretary, PGRC within a period of three months of the publication of brief note in the Indian Journal of Plant Genetic Resources, New Delhi.

## 11. Procedure for Submission of Proposal/ Germplasm Material

### 1. Submission of Application and Material

### *Novel Genetic Stocks Registration Pro-forma*

(i) All plant germplasm proposed to be registered should be submitted to the following address:

The Director

ICAR-National Bureau of Plant Genetic Resources

Pusa Campus, New Delhi -110 012

Phone: 011-2584 3697; EPABX: 2584 9208, 2584 9211 Extn. 209, 210

FAX: 011-2584 2495

Email: director@nbpgr.ernet.in.

(ii) The material must be accompanied with properly filled Form-A (Annexure-I) duly signed by the applicant and the Head of the institution with official rubber seal (10-15 copies, each with attached documentary evidences to be submitted).

(iii) The Form-A must be accompanied by complete description of the germplasm material using standard descriptors (as per concerned crop AICRP or ICAR-NBPGR descriptors requirements). It may include photograph(s) of plant/plant parts/crop and/or fingerprints (DNA or biochemical profile), if available.

(iv) A declaration to the effect that working-stock for supply to users would be maintained by the institution associated with the development of the material.

(v) Another declaration that such germplasm does not contain any gene or gene sequence involving terminator technology would also be mandatory.

### 2. Guidelines for Submitting the Orthodox Seed Material

These are the seed material that can be dried to low moisture level without loss of seed viability.

(i) A minimum number of 4000 seeds in case of cross-pollinated crop species, 2500 in self-pollinated and 500-1000 in difficult species, such as some vegetables, medicinal and aromatic plants, wild relatives *etc.* should be submitted.

(ii) The seed should be supplied from a fresh harvest and should not be more than 90 days old.

(iii) The seeds supplied should be sound, healthy, physiologically mature and collected from healthy plants.

(iv) For providing good quality healthy seeds, it is advised to dry the seed material in shade immediately after the harvest.

(v) The potential viability of seeds should be more than 85 per cent in most crop species except in special cases, such as cotton, some vegetable crops *etc.*

(vi) Seeds should not be treated with chemicals.

(vii) Seeds should be packed in good quality paper, muslin cloth or plastic packet(s) with proper identity. If required, the packets should be packed in card-board boxes to minimise damage and moisture absorption.

### 3. Guidelines for Submission of Recalcitrant/Intermediate Seed Material

These are generally characterised by large size and high moisture contents (20-80 per cent) at the time of shedding. These can be supplied to ICAR-NBPGR, New Delhi only in cases, where established protocols are available for their conservation using cryogenic technology (see annexure III) following guidelines given below:

(i) Preferably, more than 100 seeds should be supplied. However, recognising the importance of material, even small quantity may be acceptable. Supply of additional seeds may help develop DNA profiles.

(ii) They should be sent as complete fruit. To avoid any injury to the fruit surface they should be sent in aerated polythene bags/cardboard boxes.

(iii) If the fruits are bulky and difficult to transport, the seeds may be extracted without causing any injury and transported within 48hrs, packed in saw dust/charcoal/peat moss *etc.*

(iv) Avoid transporting at high temperature (above 30°C). Store and transport preferably in moist conditions between 15-20°C temperature conditions.

(v) Extracted seeds may be treated with suitable fungicide (0.1 per cent Captan or Thiram powder).

(vi) Avoid air-drying and washing of seeds.

In remaining cases the genetic material (vegetatively propagated) should be supplied to relevant NAGS in the form of propagules for establishment in the field gene bank following the guidelines given below. **An acknowledgement** for deposition and establishment of genetic material has to be obtained from the concerned NAGS and submitted along with application.

### 4. Guidelines for Submission of Propagules

In case of vegetatively propagated crop species, the germplasm material/propagules (tubers, bulbs, rhizomes, cuttings *etc.*) has to be supplied to the concerned crop-based designated as NAGS (National Active Germplasm Site), (Annexure VII) for initial establishment and conservation*. An acknowledgement to this effect has to be obtained from concerned NAGS to accompany the proposal. Additionally, following guidelines need to be followed for safe supply and conservation of germplasm:

(i) At least 10-25 propagules (depending on crops) should be supplied to the concerned NAGS for their maintenance in field repository or *in vitro* repository (if available) with a request for an acknowledgement.

(ii) The concerned NAGS should be informed in advance about the supply of material to facilitate processing and establishment of germplasm.

(iii) The genetic material, stocks, propagules of non-orthodox seed producing crops are generally being maintained in the form of grafts, crafts, slips, propagules, seedlings and plants. While supplying these genetic materials following steps and precautions should be followed depending on the crop:

(a) The slips, grafts, crafts, propagules or plants supplied to the NAGS should be free from insects, weeds and diseases as far as possible. The material should be well labelled and packed properly in aerated polythene bags. During the dry summer the grafts or crafts should be wrapped in moist moss grass to retain the moisture.

(b) In case of crops like coconut, the material should be sent either as embryos or seedlings. If the embryos need to be transferred from the field, the embryos embedded in the endosperm should be packed in the sterile plastic bag with sterile moist cotton. These should be kept in the refrigerator overnight and transferred in the same box with proper labels on it.

In case of seedlings the embryos should be grown using the river sand in plastic bags/boxes. Once the seedlings are established these should be transferred to bigger pots. The healthy, vigorous seedlings should be supplied.

---

* The NAGS at the later stage may supply these materials to the ICAR-NBPGR, New Delhi for *in vitro* maintenance or cryopreservation as *base collections*. Vegetatively propagated germplasm material preferably should be supplied as *in-vitro* cultures (wherever possible). The NAGS will ensure establishment and supply of *in vitro* generated material to ICAR-NBPGR, New Delhi at least of those crops for which protocols are available at ICAR-NBPGR, New Delhi (see Annexure VI).

**Table 8.8: Details of Registered Novel Genetic Stocks of Vegetable Crops**

| Crop Name | Botanical Name | National identity | Donor identity | INGR No. | Year | Pedigree | Developer | Developing Institute | Novel Unique Feature |
|---|---|---|---|---|---|---|---|---|---|
| Tomato | *Lycopersicon esculentum* | IC296468 | Hisar Lalit (NT-8) | INGR No. 3036 | 2003 | Resistant Bengaluru x HS-101 | R.K. Jain, R.D. Bhutani, G. Kalloo, D.S. Bhatti and Kirti Singh | Dept of Vegetable Crops, CCSHAU, Hisar | Resistant to root knot nematodes (*Meloidogyne Javanica* and *M. incognita*) |
| Tomato | *Lycopersicon esculentum* | IC395328 | TLBR-1 | INGR No. 3075 | 2003 | 15 SBSB x H-24 | A.T. Sadashiva, K.M. Reddy, M.K. Reddy, T.H. Singh, M.V. Balaram, B.C. Narasimha Prasad, K.M. Prasanna, L.R. Naveen and S.G. Joshi | IIHR, Vegetable Crop Division, Bengaluru | Resistant to tomato leaf curl virus (TLCV) and bacterial wilt |
| Tomato | *Lycopersicon esculentum* | IC395457 | IIHR-2195 | INGR No. 3076 | 2003 | CLN 2114 Dc 1F1-50-2-16-8-2-17-0 | A.T. Sadashiva, K.M. Reddy, M.K. Reddy, T.H. Singh, M.V. Balaram, B.C. Narasimha Prasad, K.M. Prasanna, L.R. Naveen and S.G. Joshi | IIHR, Vegetable Crop Division, Bengaluru | Resistant to TLCV and bacterial wilt in different genetic background |
| Tomato | *Lycopersicon esculentum* | IC528034 | F-6050 | INGR No. 6036 | 2006 | Selection from segregating population of Swarna hybrid | M. Rai, H.C. Prasanna and Rajesh Kumar | IIVR, Varanasi | High carotene |
| Tomato | *Lycopersicon esculentum* | IC526807 | F-7028-1 | INGR No. 6037 | 2006 | Selection from segregation generation of Maitree | M. Rai, J. Singh, Rajesh Kumar and H.C. Prasanna | IIVR, Varanasi | High lycopene (7.86mg/100g) and carotenoides |
| Tomato | *Lycopersicon peruvianum* | IC565013 | VC-3117A | INGR NO. 8094 | 2008 | EC251790 | K.S. Varaprasad, J.S. Prasad, E.S. Rao and T.Rama Srinivas, N.Sunil, B. Sarath Babu and T. Kiran Babu | NBPGR, RS, Hyderabad | Source of resistance to root knot nematodes (*Meloidogyne javanica*) |
| Tomato | *Lycopersicon peruvianum* | IC565014 | LO-1761 | INGR No. 8096 | 2008 | EC251706 | K.S. Varaprasad, J.S. Prasad, E.S. Rao and T.Rama Srinivas, N.Sunil, B. Sarath Babu and T. Kiran Babu | NBPGR, RS, Hyderabad | Resistant to root knot nematodes (*Meloidogyne javanica*) |
| Tomato | *Lycopersicon esculentum* | IC564448 | NBPGR Tomato-1 | INGR No. 9065 | 2009 | EC 13904 | S.K. Yadav, K.K. Gangopadhyay, S.K. Mishra, Gunjeet Kumar, Chitra Pandey, B.L. Meena, R.K. Mahajan, Mathura Rai, SK Sharma | NBPGR, New Delhi | High TSS (6.00B°) |
| Brinjal | *Solanum melongena* | IC296759 | Hisar Jamuni (H-9) | INGR No. 99037 | 1999 | Aushey x R-34 | G. Kalloo, N.K. Sharma and K.S. Baswana | Deptt. of Vegetable Crops, CCSHAU, Hisar | Intermediate, semi-spreading, medium tall, thornless, better ratooning, long fruit colour retention, less seed content, serves the purpose of long and round types |
| Brinjal | *Solanum melongena* | IC395333 | IIHR-3(96-2-1) | INGR No. 3074 | 2003 | IIHR-124x Arka Keshav | A.T. Sadashiva, T.H. Singh, K.M. Reddy, M.K. Reddy, M.V. Balaram, B.C. Narasimha Prasad, K.M. Prasanna, L.R. Naveen, S.G. Joshi | Vegetable Crop Division, IIHR, Bengaluru | Resistance to bacterial wilt |
| Brinjal | *Solanum melongena* | IC526796 | BWBH-3 | INGR No. 5025 | 2005 | IIHR-3 x IIHR-322 | A.T. Sadashiva and T.H. Singh | Indian Institute of Horticultural Research (IIHR), Bengaluru | Combined resistance to bacterial wilt |

| Crop Name | Botanical Name | National identity | Donor identity | INGR No. | Year | Pedigree | Developer | Developing Institute | Novel Unique Feature |
|---|---|---|---|---|---|---|---|---|---|
| Brinjal | Solanum melongena | IC249349 | P-61 | INGR No. 9122 | 2009 | Collection | K.K. Gangopadhyay, P.Sadhan Kumar, S.K. Mishra, Gunjeet Kumar, S.K. Yadav, Chitra Pandey, B.L Meena, R.K. Mahajan, Mathura Rai, and S.K. Sharma, | NBPGR, New Delhi | Resistance to bacterial wilt |
| Brinjal | Solanum melongena | IC090982 | Dholi-5 | INGR No. 9123 | 2009 | Collection | K.K. Gangopadhyay, P Sadhan Kumar, S.K. Mishra,S.K. Yadav, Gunjeet Kumar, Chitra Pandey, B.L. Meena, R.K. Mahajan, Mathura Rai, SK Sharma, | NBPGR, New Delhi | Resistance to bacterial wilt |
| Brinjal | Solanum melongena | IC0585684 | CARI Brinjal-1 | INGR No. 12015 | 2012 | Local germplasm | Krishna Kumar, P.K. Singh, Ajanta Birah, Shrawan Singh, Naresh Kumar, Awnindra K. Singh, D.R. Singh and R.K. Gautam | Central Agricultural Research Institute (CARI), Port Blair, Andaman and Nicobar Island | Highly resistant to bacterial wilt (Ralstonia solanacearum) disease. Fruit large, oblong and green. Good agronomic performance under Andaman Islands conditions. |
| Chilli/Sweet pepper | Capsicum annuum var. grossum | IC296760 | Capsicum Selection 2 (Nishat1) | INGR No. 1 | 2000 | Oskash x World Beater pedigree 8625-2-8-5-2 | Nazeer Ahmed, M.I. Tanki | Sher -E- Kashmir University of Agricultural Sciences and Technology (SKUAST), Srinagar | Early, high yielding (207.95q/ha), export quality, superior shelf life with high nutritive value |
| Chilli | Capsicum annuum | IC395318 (IC395319) | MS-2A and 2B | INGR No. 3077 | 2003 | Germplasm | K. Madhavi Reddy and A.A. Deshpande | Vegetable Crop Division, IIHR Bengaluru | Cytoplasmic male sterility (CMS) |
| Chilli | Capsicum annuum var. grossum | IC399066 | DRLT-1107 (Naga Jalokia) | INGR No. 3092 | 2003 | Germplasm | Subash Chandra Das | Defense Research Laboratory (DRL), Defense Research and Development Organisation (DRDO), Tezpur | High pungency |
| Chilli | Capsicum annuum | IC296662 (IC296663) | MS1A and 1B | INGR No. 4052 | 2004 | Arka Lohit | K. Madhavi Reddy | IIHR, Bengaluru | CGMS line with good general combining ability (GCA) |
| Chilli | Capsicum annuum | IC296664 (IC296665) | MS3A and 3B | INGR No. 4053 | 2004 | PMR64 (Advanced pedogree of plant Cix IHR 517A) | K. Madhavi Reddy | IIHR, Bengaluru | CGMS line with good GCA in different genetic background |
| Chilli | Capsicum annuum | IC296667 | IHR 14 (PMR -14) | INGR No. 4054 | 2004 | Pant Cl xIHR 517B {IHR 517(a line derived from C. microcarpon)} | K. Madhavi Reddy | IIHR, Bengaluru | Fertility restorer line of MSI and MS 2 |
| Chilli | Capsicum annuum | IC296667 | IHR11-2 (IHR 11#2) | INGR No. 4055 | 2004 | IHR 12-2, pure line selection | K. Madhavi Reddy | IIHR, Bengaluru | Fertility restorer line |
| Chilli | Capsicum annuum | IC526794 | IHR 3315 | INGR NO. 5024 | 2005 | Pure line selectin from ICPN 11#7 | K. Madhavi Reddy | IIHR, Bengaluru | Fertility restorer |

| Crop Name | Botanical Name | National identity | Donor identity | INGR No. | Year | Pedigree | Developer | Developing Institute | Novel Unique Feature |
|---|---|---|---|---|---|---|---|---|---|
| Chilli | *Capsicum frutescens* | IC553284 | BS 35 | INGR No. 7039 | 2007 | Selection from landrace | Sanjeet Kumar, B. Singh, M. Singh, S.K. Rai, S. Kumar and M. Rai | Indian Institute of Vegetable Research (IIVR), Varanasi | Resistant to leaf curl virus disease |
| Chilli | *Capsicum annuum* | IC565015 | PBC-534 | INGR No. 8095 | 2008 | EC 391082 | Someswara Rao Pandravada, B. Sarath Babu, N. Sivaraj, V. Kamala, N. Sunil and K.S. Varaprasad | NBPGR, RS Hyderabad | Resistant to thrips and mites |
| Chilli | *Capsicum annuum* | IC505489 | SDS-4493 | INGR No. 8097 | 2008 | NIC 23906 | Someswara Rao Pandravada, B. Sarath Babu, K. Anitha, S.K. Chakrabarty. N. Sivaraj, V. Kamala, N Sunil, R.D.V.J. Prasada Rao and K.S. Varaprasad | NBPGR, RS Hyderabad | Resistant to thrips and powdery mildew |
| Chilli | *Capsicum annuum* | IC569194 | PANDAV | INGR No. 9129 | 2009 | PANDAV | K. Ramesh | Navaneeta Evergreens, Anandapuram Mandal, Tarulawada, District-Visakhapatnam, Andhra Pradesh | Erect cluster bearing de-stalking nature of fruits, low pungency and bright colour |
| Chilli | *Capsicum chinense* | IC0553688; | IC 553688 | INGR NO. 13068 | 2013 | Germplasm selection | L.K. Bharathi and HS Singh | Central Horticultural Experiment Station, Aiginia, Bhubaneswar, Odisha | Unique material for high capsaicin adapted to the tropical humid climate. |
| Chilli | *Capsicum annuum* | IC0436231 | PSRKK-11287 | INGR No. 14040 | 2014 | Germplasm selection | S.R. Panadravada, N. Sivaraj, V. Kamla, N. Sunil and S.K. Chakrabarty | NBPGR Regional Station, Rajendranagar, Hyderabad, Andhra Pradesh | Purple phenotype as a morphological marker. |
| Chilli | *Capsicum annuum* | IC0570408 | SBT-12549 | INGR NO. 14041 | 2014 | Germplasm selection | K. Anitha, K. Narendra Varma, S.R. Panadravada, G. Suresh Kumar and S.K. Chakrabarty | NBPGR Regional Station, Rajendranagar, Hyderabad, Andhra Pradesh | Immune to Anthracnose caused by *Colletotrichum capsici.* |
| Sweet pepper | *Capsicum annum var. grossum sendtn* | IC0583131 | YCMS-12A | INGR No. 10066 | 2010 | Female: ms-12, Male: Yellow Capsicum | Priya Ranjan Kumar, S.R. Sharma, Chadner Parkash, Satish Kumar Yadav, Reeta Bhatia | IARI, RS, Katrain, Kullu Valley, Himachal Pradesh | Genetic male sterility, bell shaped fruits in non-pungent background |
| Bottle gourd | *Lagenaria siceraria* | IC296733 | ANDRO-MON-6 | INGR No. 99009 | 1999 | Mutant from NDBG-6 | Sheo Pujan Singh, N.K. Singh, I.B. Maurya | Dept of Vegetable Science, NDUA&T, Faizabad | Andromonoecious sex |
| Bottle gourd | *Lagenaria siceraria* | IC571819 | NS/2009/042 | INGR No. 10064 | 2010 | Germplasm collection | N. Sivaraj, Someswara Rao Pandravada, V. Kamala, N. Sunil, K.S. Varaprasad | NBPGR, RS, Rajendranagar, Hyderabad, Andhra Pradesh | Spindle shaped fruit with hard durable rind |
| Bitter gourd | *Momordica charantia* | IC296539 | GY-63 | INGR No. 3037 | 2003 | GY-63 (VRBT 63) | D. Ram, G. Kalloo and Billu Singh | Indian Institute of Vegetable Research (IIVR), Varanasi | Gynoecious line with high yield and attractive fruits |
| Bitter gourd | *Momordica charntia* | IC0591254 | PreGy-1 | INGR No. 12014 | 2012 | Developed by hybridization followed by selection from DBGY-201 (Gynoecious line) and Pusa Do Mausami | T.K. Behera, Anand Pal and A.D. Munshi | Indian Agricultural Research Institute (IARI), New Delhi | Predominately gynoecious habit |

| Crop Name | Botanical Name | National identity | Donor identity | INGR No. | Year | Pedigree | Developer | Developing Institute | Novel Unique Feature |
|---|---|---|---|---|---|---|---|---|---|
| Cucumber | Cucumis spp. | IC296699 | AHC-2 | INGR No. 98017 | 1998 | Local Germplasm | O.P. Pareekh, D.K. Samadia | NRC for Arid Horticulture, Bikaner, Rajasthan | High yielding and long fruit |
| Cucumber | Cucumis sativus | IC296700 | AHC-13 | INGR No. 98018 | 1998 | Local Germplasm | O.P. Pareekh, D.K. Samadia | NRC for Arid Horticulture, Bikaner, Rajasthan | High yielding, small fruit, drought hardy and high temperature insensitive |
| Muskmelon | Cucumis melo | IC557706 | DMDR-1 | INGR No. 8044 | 2008 | Phoot selection-1xM-4 | T.A. More, A.D. Munshi, Anupam Verma, J.P. Misra, V.K. Verma | IARI, New Delhi | Source of resistance to Cucumber Green Mottle Mosaic Virus (CGMMV) and Downy mildew |
| Muskmelon | Cucumis melo | IC557426 | Panjab Anmol | INGR No. 8043 | 2008 | MS-1 x Punjab Sunehari | V.K. Vashisht and Tarsem Lal | PAU, Ludhiana | Fruit wall round, light brown |
| Muskmelon | Cucumis melo L. | IC0599709 | AHMM/ BR-8 | INGR No. 14043 | 2014 | Landrace from Dhrubana, Shivganj, Rajasthan | B.R. Choudhury, S.M. Haldhar, S.K. Maheshwari, R. Bhargava and S.K. Sharma | Central Institute for Arid Horticulture, Beechwal, Bikaner, Rajasthan | Monoecious sex form |
| Kachri | Cucumis callosus | IC296695 | Kachri AHK 119 | INGR No. 98013 | 1998 | Local Genetic Diversity | O.P. Pareek, D.K. Samadia | NRC for Arid Horticulture, Bikaner, Rajasthan | High yielding drought hardy with large fruits, suited for salad |
| Kachri | Cucumis callosus | IC299696 | Kachri AHK 200 | INGR No. 98014 | 1998 | Local Genetic Diversity | O.P. Pareek and D.K. Samadia | NRC for Arid Horticulture, Bikaner, Rajasthan | High yielding drought hardy with large fruits, suited for salad |
| Snapmelon | Cucumis melo var momordica | IC296697 | AHS 10 | INGR No. 98015 | 1998 | Local genetic diversity | O.P. Pareek and D.K. Samadia | NRC for Arid Horticulture, Bikaner, Rajasthan | High yielding and drought hardy with different genetic background |
| Snapmelon | Cucumis melo var momordica | IC296698 | AHS 82 | INGR No. 98016 | 1998 | Local genetic diversity | O.P. Pareek and D.K. Samadia | NRC for Arid Horticulture, Bikaner, Rajasthan | High yielding and drought hardy with different genetic background |
| Snapmelon | Cucumis melo var. mormordica | IC396388, IC553288 | B-159 | INGR No. 7044 | 2007 | Local Collection | S Pandey, B, Singh M Rai and KK Pandey | IIVR, Varanasi | Downy mildew resistance |
| Watermelon | Citrullus lanatus | IC296694 | Mateera-AHW19 | INGR No. 98012 | 1998 | Local genetic diversity | O.P. Pareek and D.K. Sharma | NRC for Arid Horticulture, Bikaner, Rajasthan | High yielding, drought hardy, sweet, juicy with longer shelf life |
| Watermelon | Citrullus lanatus | IC296816 | RW 187-2 | INGR No. 1037 | 2001 | Local collection from Rajasthan | J.P. Luthra and V.S. Yadav | ARS, Durgapura, Jaipur, Rajasthan | Yellow colour flesh |
| Watermelon | Citrullus lanatus | IC296817 | RW 177-3 | INGR No. 1038 | 2001 | Cross between Sugarbaby and K- 3566, a Russian culture | J.P. Luthra and V.S. Yadav | ARS, Durgapura, Jaipur, Rajasthan | Simple unlobed leaf |
| Watermelon | Citrullus lanatus | IC0584139 | IIHR-81-1-3-4-6 | INGR No. 10158 | 2010 | IIHR-81 Selection | M. Pitchaimuthu and K.R.M. Swamy | Indian Institute of Horticultural Research (IIHR), Hassaraghatta Lake Post, Bengaluru, Karnataka | Bushy plant type |

| Crop Name | Botanical Name | National identity | Donor identity | INGR No. | Year | Pedigree | Developer | Developing Institute | Novel Unique Feature |
|---|---|---|---|---|---|---|---|---|---|
| Round-melon | Praecitrullus fistulosus | IC296758 | HT-10 | INGR No. 99036 | 1999 | Hisar Selection x 84# 6 | M.S. Dahiya, K.S. Baswana and B.K. Nehra | Deptt. of Vegetable Crops, CCSHAU, Hisar (Haryana) | Intermediate, semi-spreading vine, green foliage, round bright fruits, tender and sparsely pubescent, tolerant to downy mildew and root rot wilt complex |
| Pumpkin | Cucurbita moschata | IC526803 | SA 90 | INGR No. 5027 | 2005 | Local collection | S Pandey, M Rai and G Kalloo | ICAR-IIVR, Varanasi (U.P.) | High carotenoid content |
| Sponge Gourd | Luffa cylindrica | IC0584054 | VRSG-52-1 | INGR No. 10159 | 2010 | Collection | P.K. Singh, B.R. Choudhary, Ramesh Singh, D.R. Bhardwaj and M. Rai | ICAR-IIVR, Varanasi (U.P.) | Cluster bearing habit of fruiting |
| Sponge gourd | Luffa cylindrica | IC0588956 | DSG -6 | INGR No. 12013 | 2012 | Selection from local material collected from village Bulandi, Hoogly district of West Bengal | A.D. Munshi, Sabina Islam, Ravinder Kumar, Bikash Mandal, T.K. Behera and Amish K. Sureja | Indian Agricultural Research Institute (IARI), New Delhi | Highly resistant to Tomato Leaf Curl New Delhi Virus |
| Pointed gourd | Trichosanthes dioica | IC296492 | IIVRPG-105 | INGR No. 3035 | 2003 | IIVRG-105 | D.Ram, G. Kalloo, M.K. Banerjee and Billu Singh | ICAR-IIVR, Varanasi (U.P.) | Seedless fruit, obligate parthenocarpic with long duration fruiting |
| Ivy gourd | Coccinia grandis | IC553244 | CHIG-15 | INGR No. 9126 | 2009 | Selection | L.K. Bharathi, Vishalnath, G. Naik, S. Mondal, H.S. Singh | CHES (IIHR), Bhubaneswar, Odisha | Fruit length (8.5-9 cm), uniform cylindrical shape |
| Meetha Karela | Cyclanthera pedeata | IC415397 | KP/AK/77 | INGR No. 6020 | 2006 | Selection | J.C. Rana, K. Pradheep and V.D. Verma | NBPGR RS, Shimla | Spineless large fruit (Dia 2.56 cm) |
| Carrot | Daucus carota | IC057007 and IC0570261 | CMS A-line and B-line | INGR No. 10110 | 2010 | Landrace | Praveen Kumar Singh, Ramavtar Sharma, Govind Singh, Mathura Rai, D. Ram | ICAR-IIVR, Varanasi (U.P.) | Petaloid type stable male sterile line of Asiatic carrot, suitable for development of hybrid |
| Soybean | Glycine max | IC296814 | P-1366 | INGR No. 1035 | 2001 | Collection from Pithoragarh, Uttaranchal | K.C. Pant and V.D. Verma | NBPGR, RS, Bhowali | Vegetable type |
| French bean | Phaseolus vulgaris | IC0280837 | JJK-2000-220 | INGR No. 10026 | 2010 | Cultivar | K.C. Muneem, K.S. Negi; Joseph John K, Z Abraham | NBPGR RS, Bhowali, Uttarakhand | Pole type habit, long pod (23-25 cm) and dual purpose type |
| Soybean | Glycine max | 10024; IC0512375; | NRC 105 | INGR No. 10056 | 2010 | Selection from ACCGC 99009-25-9-1-3GC 95024(P)-2-1 XR75 | Anita Rani, Vineet Kumar, R.K. Singh, S.M. Husain, S.K. Pandey | Directorate of Soybean Research, Khandwa Road, Indore, Madhya Pradesh | Vegetable type based on high degree of sweetness and bold seed (60.2 g/100 seeds) at R6 stage. |

| Crop Name | Botanical Name | National identity | Donor identity | INGR No. | Year | Pedigree | Developer | Developing Institute | Novel Unique Feature |
|---|---|---|---|---|---|---|---|---|---|
| Onion | *Allium cepa* L. | IC0598327 | DOGR-1203-DR | INGR No. 14057 | 2014 | Selection from Sukhnagar, Mursidabad, (Landrace of WB) | Amar Jeet Gupta, Vijay Mahajan and Jai Gopal | Directorate of Onion and Garlic Research, (ICAR), Pune, Rajgurunagar, Maharashtra | Very early in maturity (harvested within 90 days after transplanting during rabi). 100 per cent uniform neck-fall. Unique genotype for earliness and uniform neck-fall. |
| Onion | *Allium cepa* L. | IC0598327 | DOGR-1203-DR | INGR No. 14057 | 2014 | Selection from Sukhnagar, Mursidabad, (Landrace of WB) | Amar Jeet Gupta, Vijay Mahajan and Jai Gopal | Directorate of Onion and Garlic Research, (ICAR), Pune, Rajgurunagar, Maharashtra | Very early in maturity (harvested within 90 days after transplanting during rabi). 100 per cent uniform neck-fall. Unique genotype for earliness and uniform neck-fall. |
| Garlic | *Alium sativum* | IC296711 | Agrifound White | INGR No. 98030 | 1998 | Local selection from Biharsharif (Nalanda) | U.B. Pandey, Lallan Singh and S.R. Bhonde | National Horticultural Research and Development Foundation (NHRDF) | Compact bulbs with silvery white skin, cream coloured flesh and big cloves |
| Garlic | *Alium sativum* | IC296712 | Yamuna Safed-3 | INGR No. 98031 | 1998 | Local Selection from Dindgul (TN) | U.B. Pandey, Lallan Singh and R.P. Gupta | NHRDF, Nasik | Compact bulbs, creamy white skin with cream coloured flesh |
| Garlic | *Alium sativum* | IC0596521 | G-389 | INGR No. 14009 | 2014 | Selection from "G-389" genetic stock | R.K. Singh and R.P. Gupta, B.K. Dubey | National Horticultural Research and Development Foundation, Niphad, Nashik, Maharashtra | For earliness, suitable for cultivation in Kharif season, ready for harvest within 72-77 days, in Rabi mature in only 85-95 days. |
| Sowa | *Anethum graveolens* | IC563951 | Sowa (CSSI) | INGR No. 8106 | 2008 | Local collection from Bachhranwa | R.K. Gautam, A.K. Nayak, D.K. Sharma, Ali Qadar and Gurbachan Singh | CSSRI, Karnal | Tolerance to high sodic conditions (pH 9.2) |
| Spinach | *Beta palonga* | IC565527 | Terminal flower mutant phenotype | INGR No. 9124 | 2009 | Mutated line originating after X-Ray treatment | Kalyan Kumar Mukherjee | Bose Institute, Kolkata, West Bengal | Terminal flower, thick leaf, big seed mutant of palak |
| Chenopod | *Chenopdium album* | IC258253 | JCR/TRS-510 | INGR No. 4093 | 2004 | Germplasm | J.C. Rana, T.R. Sharma, V.D. Verma, K. Pradheep | NBPGR, RS, Shimla | Brown seeded chenopod |
| Fenugreek | *Trigonella foenum-graecum* | IC296791 | Hisar Methi 346 | INGR No. 1012 | 2001 | Mutant of IL-355-1 | P.S. Partap, M.K. Rana, Rakesh Mehra | CCSHAU, Hisar | Light green narrow leaves, downy mildew resistant, quick germination, fast initial growth, long pod, bold seed with green tan seed coat colour |
| Sword bean | *Canavalia gladiata* | IC427811 | AHSB-1 | INGR No. 4056 | 2004 | Local germplasm | D.K. Samadia and D.G. Dhandar | CIAH, Bikaner | Drought tolerant, alternative source of vegetables for arid zone |

| Crop Name | Botanical Name | National identity | Donor identity | INGR No. | Year | Pedigree | Developer | Developing Institute | Novel Unique Feature |
|---|---|---|---|---|---|---|---|---|---|
| Cowpea | *Vigna unguiculata* | IC299972 | CAZC-B | INGR No. 4038 | 2004 | Charodi-1 | A. Henry | CAZRI, Jodhpur | Black seed coat colour |
| Cowpea | *Vigna unguiculata* | IC202803 | KM-5501 | INGR No. 8083 | 2008 | Germplasm collection | Saroj Sardana, N.K. Gautam, Sangita Yadav, S.K. Mishra and S.K. Sharma | NBPGR, New Delhi | Bold seed |
| Cowpea | *Vigna unguiculata* | IC0519745 | KDRS-205 | INGR No. 8084 | 2008 | NA | Kamla Venkateswaran, R.D.V.J. Prasada Rao, K.S. Varaprasad, Someswara Rao Pandravada, N. Sivaraj and Sunil Neelam | NBPGR, RS, Hyderabad | Resistance to black eye cowpea mosaic virus |
| Garden pea | *Pisum sativum* L. | IC0598281 | VRP-147 | INGR No. 15028 | 2016 | Selection | Satish Kumar Sanwal, M. Loganathan and B. Singh | Indian Institute of Vegetable Research (IIVR), Post Bag-01, Po-Jakhini, Shahanshahpur, Varanasi-221 305, Uttar Pradesh | Resistance to Downy mildew and Rust |
| Garden pea | *Pisum sativum* L. | IC0598280 | VRP-343 | INGR No. 15029 | 2016 | Selection | Satish Kumar Sanwal, M. Loganathan and B. Singh | Indian Institute of Vegetable Research (IIVR), Post Bag-01, Po-Jakhini, Shahanshahpur, Varanasi-221 305, Uttar Pradesh | Resistance to powdery mildew |
| Garden pea | *Pisum sativum* L. | IC0610501 | VRP-500 | INGR No. 15009 | 2015 | VRP-5 x PC-531 | Satish Kumar Sanwal, Rajesh Kumar and B. Singh | Indian Institute of Vegetable Research (IIVR), Post Bag-01, Po-Jakhini, Shahanshahpur, Varanasi-221 305, Uttar Pradesh | Triple podded at every node. |
| Pea | *Pisum sativum* | IC397028 | DPP-62 | INGR No. 3079 | 2003 | Bonneville x P388 | Pritam Kalia | Division of Vegetable Crops, IARI, New Delhi | Yellow wrinkled seeds with resistance to powdery mildew |
| Dolichos bean | *Lablab purpureus sweet var. typicus* | IC0383192 | BSBS-151 | INGR No. 11031 | 2011 | IC383192 | S.R. Pandravada, B. Sarath Babu, K. Anitha, S.K. Chakrabarty, N. Sivaraj, V. Kamala, N. Sunil and K.S. Varaprasad | National Bureau of Plant Genetic Resources, (NBPGR) Regional Station, Rajendranagar, Hyderabad, Andhra Pradesh | Resistant to aphids (*Aphis craccivora*) and Anthracnose (*Colletotrichum lindemuthianum*) |
| Dolichos bean | *Lablab purpureus sweet var. typicus* | IC0566943 | Ankur Goldy | INGR No. 11032 | 2011 | ARDL-12 x KONKAN BHUSAN | Sonali Suresh Alkari and L.P. Aurangabadkar | Ankur Seeds Pvt. Ltd., 27, New Cotton Market Layout, Opposite MSRTC Bus Station, Nagpur, Maharashtra | For bushiness |
| Mustard | *Brassica juncea* | IC296501 | Heera | INGR No. 3033 | 2002 | ZYR4 x BJ -1058 | A.S. Khalarkar and S.S. Bhadauria | Deptt. of Botany, Nagpur University and Dhara Vegetable Oil and Foods Co. Ltd Vadodara | Low glucosinolate content 16.96 micro moles/g of seed) and low erucic acid in oil (0.12 per cent) |

| Crop Name | Botanical Name | National identity | Donor identity | INGR No. | Year | Pedigree | Developer | Developing Institute | Novel Unique Feature |
|---|---|---|---|---|---|---|---|---|---|
| Mustard | Brassica juncea | IC296507 | NUDH-YJ-5 | INGR No. 3034 | 2002 | NV6 (Pusa bold × Heera) × EH -1 (Early Mutant of Heera) | Y.Y. Barve, S.S. Bhadauria and R.K. Gupta | Deptt. of Botany, Nagpur University and Dhara Vegetable oil and Foods Co. Ltd. Vadodara | Low glucosinolate content (9.3 micro moles/g of seed and low erucic acid in oil (0.14 per cent) |
| Makhana | Euryale ferox Salisb. | IC0610822 | Sel-14 | INGR No. 15013 | 2016 | IC0610822 | Lokendra Kumar | ICAR-RCER Research Centre for Makhana (RCM), Darbhanga (Bihar) | Irregular seed shape (mutant), large fruit size (fruit diameter: 8.1 cm). Highest number of seeds/fruit (139) |

*Source: nbpgr.ernet.in, Singh (2006); Kak et al. (2009); Kak and Tyagi (2010), Indian Horticulture (May-June, 2010), Indian Horticulture (2016).*

**Table 8.9: Germplasm Registered under Tuber Crops**

| Crop Name | Botanical Name | National identity | Donor identity | INGR No. | Year | Pedigree | Developer | Developing Institute | Novel Unique Feature |
|---|---|---|---|---|---|---|---|---|---|
| Potato | Solanum tuberosum ssp. andigena | IC296790 | EX/A680-16 | 1011 | 2001 | Clone of (No 502-14 × Bulk Pollen) | P.C. Gaur, Jai Gopal, S.K. Pandey and M.S. Rana | Central Potato Research Institute (CPRI), Shimla | Short day adapted potato with resistance to early and late blight |
| Potato | Solanum tuberosum | IC296650 | QB/A 9-120 | 4057 | 2004 | JENA 680-16× Kufri Jyoti | Jai Gopal, P.C. Gaur and M.S. Rana | CPRI, Shimla | High resistance to late blight, and GCA for agronomic traits |
| Potato | Solanum tuberosum | IC296651 | QB/B92-4 | 4058 | 2004 | Kufri Red × CP 1755 | Jai Gopal, P.C. Gaur, S.K. Pandey and M.S. Rana | CPRI, Shimla | High dry matter and low reducing sugars, and good GCA |
| Potato | Solanum tuberosum | IC296661 | PS/F-220 | 4059 | 2004 | B-8695 × Kufri Jyoti (cross was made in 1974) | S.K. Pandey, R.B. Singh and S.M. Paul Khurana | CPRI, Shimla | Resistant to Potato stem necrosis |
| Potato | Solanum tuberosum | IC445068 | M/99/322 | 4109 | 2004 | MCP-1× 48-1 | S.K. Pandey, S.V. Singh, Dinesh Kumar and P. Manivel | CPRI, Shimla | High starch, dry matter, low amylose and resistance to late blight |
| Potato | Solanum tuberosum | IC522200 | E/79-42 | 5022 | 2005 | D 42/9 (PBI, Harpenden, UK) × Solanum vernei clone vtn2 62-33-3 (Netherlands) | KS Krishna Prasad, T.A. Joseph, S.K. Pandey, I.A. Khan and D.B. Singh | Central Potato Research Institute (CPRI), RS, Muthorai | Combined resistance to cyst nematodes and late blight disease |
| Potato | Solanum tuberosum | IC524019 | JW 96 | 5023 | 2005 | Kufri Jyoti × CP 1362 | Raj Kumar, G.S. Kang. Jai Gopal and S.K. Pandey | CPRI, RS, Jalandhar | Early maturing (60 days) |
| Potato | Solanum tuberosum | IC547013 | JX-123 | 6021 | 2006 | JE812 × CP 2144 | Raj Kumar, G.S. Kang and S.K. Pandey | CPRI, Shimla | Resistance to early blight |
| Potato | Solanum tuberosum | IC553285 | JN-189 | 7040 | 2007 | Kufri Jawahar × S. andigena | R.B. Singh, S.K. Pandey, S.M. Paul Khurana and P. Manivel | ARS, Ummed Ganj, Kota | Resistance to leaf hopper burn and potato stem necrosis tospi virus |

| Crop Name | Botanical Name | National identity | Donor identity | INGR No. | Year | Pedigree | Developer | Developing Institute | Novel Unique Feature |
|---|---|---|---|---|---|---|---|---|---|
| Potato | *Solanum tuberosum* | IC567213 | D4 | 9067 | 2009 | Androgenic dihaploid of TPS parental line JTH/C-107 | Sushruti Sharma, Debabrata Sarkar, S.K. Pandey | CPRI, Shimla, Himachal Pradesh | Flowering, male fertile androgenic dihaploid of tetraploid potato |
| Potato | *Solanum tuberosum* | IC567214 | C-13 | 9068 | 2009 | Androgenic dihaploid of cv. Kufri Chipsona-II | Sushruti Sharma, Debabrata Sarkar, S.K. Pandey | CPRI, Shimla, Himachal Pradesh | Flowering, male fertile androgenic dihaploid of tetraploid potato, highly resistant to late blight |
| Potato | *Solanum tuberosum* | IC585711 | JX 90 | 9069 | 2009 | CP 1346(Krirrinee)X MS/78-62 | Raj Kumar, G.S. Kang, S.K. Pandey, Jai Gopal | CPRI, Shimla, Himachal Pradesh | Horizontal resistant to late blight and early blight, high yield under early (75 days) and medium (90 days) crop duration |
| Potato | *Solanum tuberosum subsp. andigenum* | EC460686 | SS-2040 (Tetraploid cultivated species) | 9120 | 2009 | Clonal selection from segregating progeny of the accession SS 2040 of cultivated tetraploid species Solanum tuberosum ssp. andigena | S.K. Luthra, J. Gopal, P.Manivel, Vinod Kumar, B.P. Singh, S.K. Pandey | CPRI, RS, Meerut, Uttar Pradesh | Higher frost tolerance introduced through ssp. andigena a cultivated clone |
| Potato | *Solanum spegazzinii* | EC412923 | SS-1725-22-a wild species clone | 9121 | 2009 | Clonal selection from segregating progeny of the accession SS 1725 of diploid wild species Solanum spegazzinii | S.K. Luthra, J. Gopal, P.Manivel, Vinod Kumar, B.P. Singh, S.K. Pandey | CPRI, RS Meerut, Uttar Pradesh | Higher frost tolerance introduced through a wild species a wild species clone (SS-1725-22) of *Solanum spegazzinii* |
| Potato | *Solanum tuberosum* | IC0586781 | YY 6/3 C11 (Tetraploid cultivated species) | 10143 | 2010 | A clonal selection from the segregating progeny of the cross YY-6 x YY-3 using marker assisted selection. | S.K. Kaushik, I.D. Garg, B.P. Singh, S.K. Chakrabarti, Vinay Bhardwaj and SK Pandey. | Central Potato Research Institute (CPRI), Shimla, Himachal Pradesh | An elite parental line possessing Potato virus Y (PVY) extreme resistance gene Ryadg in Triplex (YYYy) condition |
| Potato | *Solanum tuberosum (+) S.etubersoum* | IC0590089 | E 1-3 | 11050 | 2011 | *Solanum tuberosum dihaploid 'C-13' (+) Wild species Solanum etuberosum* | Jagesh Kumar Tiwari, Poonam, D. Sarkar, S.K. Pandey and Jai Gopal | Central Potato Research Institute (CPRI), Shimla, Himachal Pradesh | Tetraploid Somatic male fertile hybrid carrying resistance to potato virus introgressed from *S. etuberosum* |
| Potato | *Solanum tuberosum (+) S. pinnatisectum* | IC0590090 | P-7 | 11051 | 2011 | *Solanum tuberosum dihaploid 'C-13' (+) Wild species Solanum pinnatisectum* | D. Sarkar, Jagesh Kumar Tiwari, Sushruti Sharma, Poonam, Sanjeev Sharma, Jai Gopal, B.P. Singh, S.K. Luthra, S.K. Pandey and D. Pattanayak | Central Potato Research Institute (CPRI), Shimla, Himachal Pradesh | Tetraploid, Somatic male fertile hybrid carrying resistance to Potato late blight introgressed from S. pinnatisectum |

| Crop Name | Botanical Name | National identity | Donor identity | INGR No. | Year | Pedigree | Developer | Developing Institute | Novel Unique Feature |
|---|---|---|---|---|---|---|---|---|---|
| Potato | *Solanum demissum- wild species* | IC0594469 | SS 1735-02 (hexaploid wild potato species) | 13048 | 2013 | Clonal selection From the TPS (True Potato Seed) population GLKS-269? | Vinay Bhardwaj, S.K. Luthra, Dalamu, B.P. Singh, Vinod Kumar, Dinesh Kumar and Sanjeev Sharma | Central Potato Research Institute, Shimla, Himachal Pradesh | An elite wild potato clone (*Solanum demissum*; 2n=6x:4 EBN) possessing very high resistance against late blight and low cold induced sweetening even after 6 months of cold storage (2-40C). |
| Potato | *Solanum tuberosum* | IC0594470 | MP/97-921 | 13049 | 2013 | MP/92-154 × MP/91-65 (Cross was made in 1996). | S.V. Singh, Vinay Bhardwaj, P.Manivel, Dinesh Kumar, B.P. Singh, V.K. Gupta and S.K. Kaushik | Central Potato Research Institute, Shimla, Himachal Pradesh | High dry matter content (>24 per cent in hills and >22 per cent in plains); superior chip colour (even after 6 months cold storage); low reducing sugar and sucrose content; highly resistant to late blight; ideal free amino acid and phenol content; extreme resistant to potato virus PVY. |
| Sweet potato | *Ipomoea batatas* | IC0593650 | ST-14 | 13020 | 2013 | JP-14 | S.K. Naskar and Archana Mukherjee | Central Tuber Research Institute, Trivandrum, Kerela | High carotene (13-14.5 mg/100g) and salinity tolerance. |
| Sweet potato | *Ipomoea batatas* | IC0593651 | ST-13 | 13021 | 2013 | JP-13 | S.K. Naskar and Archana Mukherjee | Central Tuber Research Institute, Trivandrum, Kerela | High anthocyanin (85-90 mg/100g) and salinity tolerance. |
| Sweet potato | *Ipomoea batatas* | IC0593652 | ST-10 | 13022 | 2013 | JP-10 | S.K. Naskar and Archana Mukherjee | Central Tuber Research Institute, Trivandrum, Kerela | High extractable starch (20-21 per cent). |
| Yam | *Dioscorea pubera* | IC202382 | IC202382 | 8061 | 2008 | NA | Asha, K.I. | NBPGR, RS, Thrissur | High diosgenin |
| Yam | *Dioscorea spicata* | IC202383 | IC202383 | 8062 | 2008 | NA | Asha, K.I. | NBPGR, RS, Thrissur | High diosgenin yield |
| Yam | *Dioscorea hispida* | IC202370 | IC202370 | 8063 | 2008 | NA | Asha, K.I. | NBPGR, RS, Thrissur | High diosgenin yield |
| Yam | *Dioscorea hatonian* | IC202328 | AV-222 | 8064 | 2008 | NA | Asha, K.I. | NBPGR, RS, Thrissur | High diosgenin |

(iv) The material should be packed in small wooden/ card-board boxes with proper aeration. Also these boxes should be well marked with labels at 3 or 4 places "To be handled carefully: seedlings" in order to avoid any damage during transit.

(v) The material should be sent to the NAGS immediately after harvest. To avoid any delay in transaction, use speed post or courier services or airfreight.

**Note:** The sample size of propagules and seeds to be submitted may be revised in consultation with the Director, or Head, Division of Germplasm Conservation, ICAR-NBPGR in exceptional cases.

## Status of Registered Vegetable Germplasm in India

Since the inception of germplasm registration mechanism, a large number of elite material/genetic stocks have been registered at ICAR-NBPGR, New Delhi. A total of 61 vegetable germplasm covering 23 crops with unique traits are registered (Table 8.8). Similarly in potato 11 improved germplasm/breeding lines have been registered as elite genetic stock.

In tuber crops several improved germplasm/breeding lines have been registered as elite genetic stock at Table 8.9.

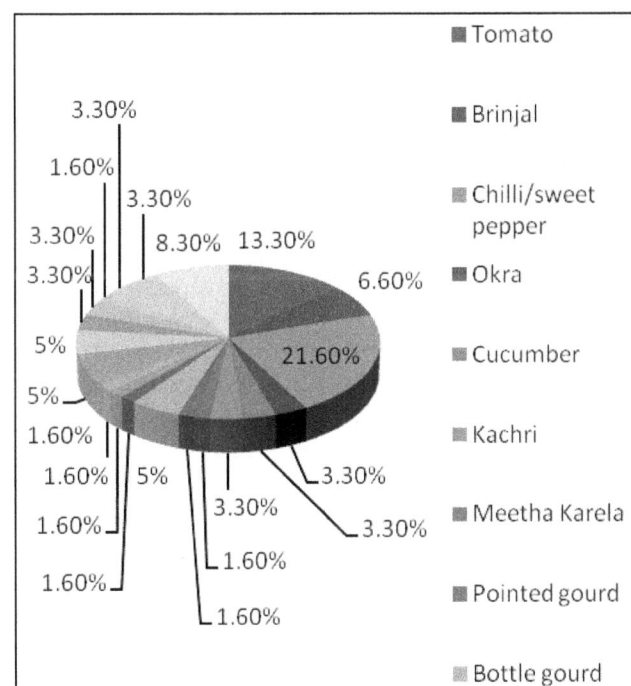

**Figure 8.1: Registration of Trait Specific Vegetable Crop Germplasm during 1996-2010**

## Registration of *Mucuna* Genotypes

1. **COWHAGE/ATMAGUPTA, IIHR-19855 (INGR No. 03045; IC296847)** *Mucuna pruriens* **Baker:** COWHAGE/ATMAGUPTA, IIHR-19855 is a *Mucuna pruriens* accession with high

L-DOPA content (9.5 per cent) in seeds. This high yielding accessions with high L-DOPA content was identified at ICAR-IIHR Bengaluru (Karnataka). L-DOPA contents analyzed using HPLC method at the Section of Medicinal Crops of the ICAR-IIHR. The results were at par with the results obtained at Al Ameen College of Pharmacy. The L-DOPA content recorded in the present accession is much higher than the values reported earlier (4.5 per cent). The genotype possesses intermediate growth habit with extended duration of flowering (200 days) and pod maturity period of 150-180 days. The pods are more or less 'S' shaped and possess 3-4 seeds. However, this species is not grower friendly as the pods have hirsute surface with high itchy hairs, which is a unique feature of this species. Plant growth is highly vigorous and needs strong support. Seeds have dormancy.

2. **COWHAGE/ATMAGUPTA, IIHR-19854 (INGR No. 03046; IC296846)** *Mucuna utilis* **Wall. ex Wright.:** COWHAGE/ATMAGUPTA, IIHR-19854 is a *Mucuna utilis* accession with high L-DOPA content (9.5 per cent) in seeds. This is a cultivated species of *Mucuna* and this accession has been identified with next highest L-DOPA content, which was analyzed using HPLC method at the Section of Medicinal Crops of the IIHR. The results were at par with the results obtained at Al Ameen College of Pharmacy. Plants are determinate type, taking 120-150 days to reach maturity and last for 120 days in field cultivation. The pod surface is smooth and non-itchy. The pod colour is moth greenish when immature, but becomes black on maturity. Seeds can be harvested from the fruits easily. Seeds are cream coloured, bold and non-dormant.

Distinctness, uniformity and stability test (DUS) are essential for registration of varieties under "Protection of Plant Variety and Farmers Right Act 2001 (PPV and FRA) and 114 crops/species have been notified for registration including onion and garlic in India.

**Table 8.10: List of Vegetable Varieties Registered by ICAR under PPV and FRA, 2001**

| Crop and Species | Variety | Registration Number and Date of Issue |
|---|---|---|
| French bean (*Phaseolus vulgaris* L.) | Kashi Param (IVFB-1) | 71 of 2009; 20.07.09 |
| Garden pea (*Pisum sativum* L.) | Kashi Udai (VR-6) | 124 of 2009; 21.12.09 |
| Maize (*Zea mays* L.) | VL Baby Corn-1 (VL-78) | 19 of 2010; 20.10.2010 |
| French bean (*Phaseolus vulgaris* L.) | Arka Suvidha (IIHR-909) | 39 of 2010; 20.10.2010 |
| Garden pea (*Pisum sativum* L.) | Arka Ajit (FC-1) | 40 of 2010; 20.10.2010 |

After India became signatory to the Trade Related Aspects of Intellectual Property Rights Agreement (TRIPs) in 1994, a legislation was required to be formulated. Article 27.3 (b) of this agreement requires the member countries to provide for protection of plant varieties either by a patent or by an effective sui generis system or by any combination thereof. Thus, the member countries had the choice to frame legislations that suit their own system and India exercised this option. The existing Indian Patent Act, 1970 excluded agriculture and horticultural methods of production from patentability. The Protection of Plant Varieties and Farmers' Rights Act (PPV and FR Act) was passed by the Indian Government in 2001. The *sui generis system* for protection of plant varieties was developed integrating the rights of breeders, farmers and village communities, and taking care of the concerns for equitable sharing of benefits.

The objectives of the Act are:

(i) To provide for the establishment of an effective system for protection of plant varieties.

(ii) To provide for the rights of farmers and plant breeders.

(iii) To stimulate investment for research and development and to facilitate growth of the seed industry.

(iv) To ensure availability of high quality seeds and planting materials of improved varieties to farmers.

The Act has come into force after establishment of the Authority known as the "Protection of Plant Varieties and Farmers' Rights Authority" (herein after being called as PPV and FR Authority) in October 2006. It consists of a chairperson and fifteen members as representatives of different concerned ministries and departments, seed industry, farmers' organizations, tribal communities and state level women's organization *etc.* (Sec.3, see full text of the Act at Annexure -VI) The genera and species eligible for protection under the Act are notified from time to time by the Authority at its website: "www. plantauthority.gov.in" and the Plant Variety Journal published by the PPV and FR Authority. In the initial phase 12 crops namely rice, wheat, maize, sorghum, pearl millet, chickpea, pigeon pea, green gram, black gram, lentil, pea and kidney bean have been notified for registration. The applications are being accepted since May 2007. Cotton and jute have also been notified and applications in these crops are being accepted from April 2008 (Table 8.11).

The procedure for filing applications of varieties bred by the public sector including Indian Council of Agricultural Research (ICAR) and State Agricultural Universities (SAUs) is explained in details in the present guidelines. ICAR Guidelines for Intellectual Property Management and Technology Transfer/ Commercialization, which became effective from

**Table 8.11: Genera and Species Notified by the PPV and FR Authority till September 2008**

| Crop Name | Hindi Name | Genus and species |
|---|---|---|
| Rice | Dhan | *Oryza sativa* L. |
| Wheat | Gehun | *Triticum aestivum* L. |
| Maize | Makka | *Zea mays* L. |
| Sorghum | Jowar | *Sorghum bicolor* (L.) Moench |
| Pearl millet | Bajra | *Pennisetum glaucum* (L.) R.Br. |
| Chickpea | Chana | *Cicer arietinum* L. |
| Pigeon pea | Arhar | *Cajanus cajan* (L.) Wilczek |
| Green gram | Mungbean | *Vigna radiata* (L.) Wilczek |
| Black gram | Urdbean | *Vigna mungo* (L.)Hepper |
| Lentil | Masur | *Lens culinaris* Medik |
| Pea | Matar | *Pisum sativum* L. |
| Kidney | Rajmash | *Phaseolus vulgaris* L. |
| Cotton | Kapas | *Gossypium* (*G. barbadensis* L., *G. hirsutum* L., *G. arboreum* L., *G. herbaceum* L.) |
| Jute | Patsun | *Corchorus* (*C. olitorious* L., *C. capsularis* L.) |

2ⁿᵈOctober 2006, require intellectual property protection (IP)/forms the field of agriculture including in agro-biotechnological through patents, plant variety protection and other forms of Intellectual Property Rights. IP ownership generated in ICAR shall vest in the ICAR either solely or jointly. Individual scientists/ staff of ICAR responsible for creation of its IP shall be recognized as the true and First Inventors/Innovators/ Breeders (in case of plant varieties bred by them). ICAR constituted a task force under the Chairmanship of Deputy Director General (Crop Sciences) to streamline the process of filing application of extant- notified varieties through ICAR-National Bureau of Plant Genetic Resources (NBPGR), New Delhi with Project Directorates (PDs)/Project Co-ordinators (PCs)/ or Directors of crop based institutes as Authorized Signatories. These guidelines can be followed by breeders of SAUs also. During the Vice Chancellors conference held at ICAR in January 2008, it was emphasized that all public sector bred varieties released during the last 15 years should be protected under the PPV and FR Act 2001 and SAUs can file applications of plant varieties released at state level by the system developed by ICAR through ICAR-NBPGR, with PD/ PCs or Directors of crop based institutes as authorized signatories.

## Salient Features of the PPV and FR Act 2001

### Period of Protection

The certificate of registration issued under section 24 or sub-section (8) of section 23 shall be valid for 9 years in the case of trees and vines and 6 years in the case of other crops and may be reviewed and renewed for the remaining period on payment of such fees as may be fixed by the rules made, subject to the conditions that the total period of validity shall not exceed-

(i) In the case of trees and vines, 18 years from the date of registration of the variety;

(ii) In the case of extant varieties, 15 years from the date of the notification of that variety by the Central Government under section 5 of the Seed Act, 1996; and

(iii) In the other case, 15 years from the date of registration of the variety.

## Payment of Annual Fee

The Authority may, with the prior approval of the Central Government, by notification in the Official Gazette, impose a fee to be paid annually, by every breeder of a variety, agent and licensee thereof registered under this Act determined on the basis of benefit or royalty gained by such breeder, agent or licensee, as the case may be, in respect of the variety, for the retention of their registration under this Act [Section 35(1)].

## Breeders Rights

The certificate of registration for a variety issued under this Act shall confer an exclusive right on the breeder or his successor or his agent or licensee, to produce, sell, market, distribute, import or export of the variety [Section 28 (1)].

## Researchers' Rights

The researchers have been provided access to protected varieties for bonafide research purposes [Section 30]. This Section states, "Nothing contained in this Act shall prevent (a) the use of any variety registered under this Act by any person using such variety for conducting experiments or research; and (b) the use of a variety by any person as an initial source of a variety for the purpose of creating other varieties provided that the authorisation of the breeder of a registered variety is required where the repeated use of such variety as a parental line is necessary for commercial production of such other newly developed variety".

## Farmers' Rights

The farmers' rights of the Act define the privilege of farmers and their right to protect varieties developed or conserved by them [Chapter VI]. A farmers can save, use, sow, resow, exchange, share and sell farm produce of a protected variety except sale under a commercial marketing arrangement (branded seeds) [Section 39 (1), (i) - (iv)]. Further, the farmers have also been provided protection of innocent infringement when at the time of infringement a farmer is not aware of existence of breeders rights [Section 42 (1)].

A farmer who is engaged in the conservation of genetic resources of landraces and wild relatives of economic plants and their improvement through selection and preservation shall be entitled in the prescribed manner for recognition and reward from the Gene Fund, provided the material so selected and preserved has been used as donors of genes in varieties registerable under the Act (Plant Genome Saviour Award has been instituted).

The expected performance of a variety is to be disclosed to the farmers at the time of sale of seed/ propagating material. A farmer or a group of farmers or an organisation of farmers can claim compensation as per the Act, if a variety or the propagating material fails to give the expected performance under given conditions, as claimed by the breeder of the variety.

## Communities' Rights

The Rights of the communities as defined, provide for compensation for the contribution of communities in the evolution of new varieties in quantums to be determined by the PPV&FR Authority [Section 41 (1)].

## Registration of Essentially Derived Varieties

The breeder of the essentially derived variety shall have the same rights as the plant breeder of other new varieties, which include production, selling, marketing, and distribution including export and import of the variety. The other eligibility criteria for award of registration are also the same as for new variety registration under the Act [Section23(1), 6(1)].

## Compulsory License

The authority can grant compulsory license, in case of any complaints about the availability of the seeds of any registered variety to public at a reasonable price. The license can be granted to any person interested to take up such activities after the expiry of a period of three years from the date of issue of certificate of registration to undertake production, distribution and sale of the seed or other propagating material of the variety [Section 47(1)].

## Benefit Sharing

Sharing of benefits accruing to a breeder from a variety developed from indigenously derived plant genetic resources has also been provided [Section 26(1)]. The authority may invite claims of benefit sharing of any variety registered under the Act, and shall determine the quantum of such award after ascertaining the extent and nature of the benefit claim, after providing an opportunity to be heard, to both the plant breeder and the claimer.

## National Gene Fund

The National Gene Fund to be constituted under the Act shall be creditedthere to:

(a) The benefit sharing from the breeders

(b) The annual fee payable to the authority by way of royalties

(c) By the compensation provided the communities as defined under Section 41(1)

(d) Contribution from any national and international organisation and other sources

The fund will be applied for disbursing shares to benefit claimers either individuals or organisation, and for compensation to village communities. The fund will also be used for supporting conservation and sustainable use of genetic resources including *in situ* and *ex situ* collection and for strengthening the capabilities of the panchayat in carrying out such conservation and sustainable use [Section (45)].

### Application Form

Every application for registration will have to be accompanied with thefollowing information [Section 18 (b-h)]

(a) denomination assigned to such variety by the applicant;

(b) an affidavit sworn by the applicant that such variety does not contain any gene or gene sequence involving terminator technology;

(c) the application should be in such form as may be specified by regulations;

(d) a complete passport data of the parental lines from which the variety has been derived along with the geographical location in India from where the genetic material has been taken.

## Guidelines for Registration of Plant Germplasm

Plant Germplasm Registration Committee (PGRC) is constituted under the Chairmanship of Deputy Director General (Crop Sciences), ICAR for a maximum of three years. The PGRC meets at least twice a year, with the concurrence of the Chairman, for the consideration of applications and related matters.

### Nodal Agency

ICAR-NBPGR, New Delhi is the nodal agency for the registration of germplasm. The application should be addressed to the Director, ICAR-NBPGR, New Delhi along with seed sample or a certificate of submission of propagules with respective crop/plant-based NAGS for establishment/conservation. ICAR-NBPGR, New Delhi is also maintaining permanent register and database listing the germplasm material approved by PGRC with details on unique traits and other related information.

### Eligibility Criteria for Registration

Germplasm or genetic stock of agricultural, horticultural and other economic crops, including agroforestry species, spices, medicinal and aromatic plants, ornamental plants, which is unique, uniform and stable and has potential attributes of academic, scientific or commercial value can be registered.

All claims concerning the material submitted for registration should accompany scientific evidence for uniqueness, reproducibility and value in the form of-

Publication in standard peer reviewedjournal (a copy of reprint to be submitted).

AND/OR

Evaluation data for at least three years under AICRP trial/nursery tests supported with relevant extracts of the document or verification by concerned Project Director/Project Coordinator or three location/year data under any other relevant system.

AND/OR

Publication of information on potential value of proposed germplasm in the institute's annual report or any other such reports.

AND/OR

Certificate of the validation test of the claimed attribute by any institution as per the advice of Member Secretary.

AND

Recommendation of institute's Germplasm Identification Committee regarding the novelty and uniqueness of germplasm for trait(s) claimed.

### Validity of Registration

The period for validity of registration shall be 18 years for trees and vines and 15 years for other plants' species, after which the registered germplasm would be national sovereign property.

### De-registration

A registration may be repealed by the PGRC in case of false claims (s). Appeal for counter claim if any, can be made to PGRC within a period of three months of the publication of brief notes in the Indian Journal of Plant Genetic Resources.

### Important Points for Filing Application by ICAR/SAUs for Extant/New Varieties

The following checklist needs to be kept in mind for submission of applications to PPV and FR Authority

1. Correctly filled copy of Form 1 (Form II is applicable only to essentially derived varieties including the Transgenic varieties).

2. Correctly filled copy of Technical Questionnaire (TQ).

3. Each page of the application form and technical questionnaire to be signed by the authorized signatory who could be Project Director/Project Coordinator/Director of the crop based institute

of the concerned crop in case of ICAR or SAU's.

4. Copies of PV 2 (Authorization by breeders to the applicant *i.e.* ICAR or SAUs) and or PV 1 (Authorization by applicant (ICAR or SAUs) to the authorized signatory for filing application on behalf of the breeders). ICAR has issued a letter authorizing PD/PCs/Directors to file applications of plant variety protection.This letter can be used asPV1 (Annexure 1).

5. Signature and official seal of the authorized signatory as mentioned in 3 above, on last page of Form 1, Form 2 and TQ.

6. A witness has to sign with full name, address/seal on last page of TQ. Witness can be any scientist/breeder from the institute/Project Directorate/Project Coordinator Unit or the University.

7. All applications should be in triplicate (one original with 2 photocopies). The photocopies of all enclosures including demand draft, affidavits *etc.* to be included on second and third copy.

8. Print on single side (one side) of the paper. Photocopy also on single side.

9. Ensure availability of pure seed as per guidelines (follow breeding standards required for maintenance of pure seeds, and also submit the quantity required for each crop as and when requested by the PPV and FR Authority as given in the crop specific guidelines published by the Authority and appearing on the PPV and FR Authority website from time to time).

10. Denomination: The denomination (name) given to the variety being filed for protection shall be such that it is capable of identifying the variety. It should not solely numbers or figures, and should not mislead regarding characteristics, value of the identity of the breeder. It should be different from every other denomination which designates a variety of the same botanical species or of a closely related species registered under the Act. The name should not hurt religious sentiments of any class or section of citizen of India and should not use any emblem (mentioned in the Emblems and Names (Prevention of improper use) Act 1950. It should not solely or partly use geographical names.

It is to be noted that the denomination filled in the application form should be same in all parts of the application form including TQ and affidavits. Use of white space between two parts of the name, use of hyphen ("-") or a comma (",") should be uniform wherever the denomination is to be reproduced *e.g.* Pusa Bold, Pusa-Bold, PusaBold, Pusa/Bold are considered different denominations and the application shall be liable for revision if such difference in denominations are observed by the PPV and FR Authority.

11. The affidavit for certifying that the variety is not using any terminator gene technology should be attested by Notary on a Non-judicial stamp paper of Rs. 100/- denomination. The court fee stamps of same denomination are not accepted by Authority.

12. All applications should be sent to Director, National Bureau of Plant Genetic Resources, Pusa, New Delhi - 110 012, Tel. no. 011 - 2584 3697, Fax no. 011- 25842495, E-mail:dfrector@ nbpgr.ernet.in.

Please note that a soft copy of each application should also accompany with 3 set of application.

13. A copy of release proposal is to be enclosed with the applications of extant-notified varieties. Photocopy with original and duplicate is also required.

14. All columns should be filled with relevant information and no column should be left blank. In case information is not applicable in some columns, "NA" should be written against that column.

15. Demand Draft of Rs. 200/- in the name of "Registrar, PPV and FR Authority", payable at New Delhi should be attached with the application. The draft should have a validity of at least 3 months at the time of filing of application.

16. Please see for any other information required on PPV and FR Authority website www. plantauthority.gov.in

17. Before filing applications with the PPV&FRA a National Accession number (IC No.) to the variety is to be obtained by depositing a sample at the National Gene Bank at ICAR-NBPGR. This is also a pre requisite for getting the variety released through the Central Sub Committee on release and notification, and also for state release of the variety.

18. After acceptance and publications of application in the Plant Varieties Journal (PVJ) published and also available on the website mentioned in 16 above, go through the contents of the applications for any discrepancy, and also check applications of your mandate crops submitted by others. An opposition can be filed for information published for any misuse of public sector material, under intimation to ICAR. These needs to be done within 90 days of publication of the application in the PVJ.

# Annexure I

## Performa for obtaining IC Number for Release Variety

It is mandatory to all breeders to get the IC number of their release varieties from ICAR-NBPGR, New Delhi as per given "Passport Information Data Sheet".

## PASSPORT INFORMATION DATA

**Add:** Indian Institute of Vegetable Research, Post Office-Jakhini (Shahanshahpur) Varanasi-221305 (U.P.)

| National Identity | Alternate Identity, if any | Variety | English name | Botanical Name | Pedigree | Breeding Method | Release at National/ State level | Year of Release | Year of Identification through AICRP (VC) |
|---|---|---|---|---|---|---|---|---|---|
| | | | | | | | | | |

| Developer (s) | Developing Institute Name | Area of Adoption | | | DUS Feature | | | |
|---|---|---|---|---|---|---|---|---|
| | | Latitude | Longitude | Altitude | Botanical | Agronomic | Commercial | Uniformity |
| | | | | | | | | |

| Sl.No. | Collector No. | IC No. | Crop Name | Botanical Name | Vernacular Name | Cultivar/ Hybrid/Wild | Types of Material | Date of Collection/ Develoment |
|---|---|---|---|---|---|---|---|---|
| | | | | | | | | |

| Sl.No. | Source | Frequency | Sample type | Habitat | Village | District | State |
|---|---|---|---|---|---|---|---|
| | | | | | | | |

## Check-List for Screening of Applications

The Member Secretary, Germplasm Registration Committee at ICAR-NBPGR shall screen all applications and make recommendations to the Germplasm Registration Committee for *inter alia* the following points:

(i) Whether this is a new application or a revised one? (Yes/No)

(ii) Whether same or similar material has been registered earlier? (Yes/No)

(iii) Whether unique or distinguishing characteristics of potential value merit consideration for registration? (Yes/No)

(iv) Whether documentary evidence or data is provided in support of the claim on potential value of germplasm? (Yes/No)

(v) State any other economic potential value of germplasm, if possible.

(vi) ICAR-NBPGR viewpoint about the candidate germplasm.

(vii) Whether applicant, institution, university, or centre has given a commitment for maintenance and supply of germplasm for use? (Yes/No)

(viii) Whether appropriate size of germplasm sample for long-term storage at National Gene bank or for conservation and maintenance of active collections at the concerned NAGS has been sent? (Yes/No)

(ix) Whether the applicant has sent maintainer line of the CMS line to the National Genebank, (Yes/No)

(x) Whether acknowledgement receipt of germplasm from concerned NAGS for deposition and establishment is attached, wherever required? (Yes/No)

(xi) Whether detailed address of the corresponding person is given? (Yes/No)

(xii) Whether appropriate institutional authority has duly endorsed the application? (Yes/No)

## Annexure II

### Form A

### Application for Registration of Plant Germplasm

*(To be submitted to the Director, ICAR-NBPGR, New Delhi-110012)*
*Please refer to guidelines for filling the application form appended and Codes*

1. Application status (Code)     N     R

2. Crop Name

3. Botanical name

4. Crop group (Code)

5. Biological status of the material to be registered

       RE     MU     GP     GS

6. Unique identity (Proposer/National)

7. Criteria for registration [Unique feature(s)]

   i.

   ii.

   iii.

8. Nature of genetic material (Code)

9. Quantity deposited (Actual)

10. Value referred to (Code)    SC  CM    AC

11. Basis of eligibility (Code)  PR  CT    OT

12. Has it been registered/protected anywhere  Y  N

13. Manuscript (Ome-page note on proposed Germplasm enclosed)

14. Particulars of the scientist(s)/person(s) who developed germplasm/genetic stock*

**For Use of ICAR-NBPGR**

(i) Application number

(ii) Date of application

(iii) Whether new or revised?   N     R

(iv) If revised, Date of 1st Application

(v) If validation test suggested,
     whether report attached

(vi) Action taken

        a) Forwarded for registration

        b) Sent for validation

        c) Incomplete, sent for revision

(vii) Whether registered or rejected

(viii) Date if registered or rejected

(ix) Registration Number

(x) Notified : on: (xi) Remarks    *INGR No.*

Name (Dr./Ms/Mr.)

Designation:

Address:

Tel:        Fax:        Email:

*(Please attach separate sheet for additional name(s) and address of co-authors (persons responsible).*

15. Name and address of the corresponding person (Developer/Depositor)

Name (Dr./Ms/Mr.)

| | | | | | | | | | | | | | | | | | | | | | | | |
|---|---|---|---|---|---|---|---|---|---|---|---|---|---|---|---|---|---|---|---|---|---|---|---|

Designation:

| | | | | | | | | | | | | | | | | | | | | | | | |
|---|---|---|---|---|---|---|---|---|---|---|---|---|---|---|---|---|---|---|---|---|---|---|---|

Address:

| | | | | | | | | | | | | | | | | | | | | | | | |
|---|---|---|---|---|---|---|---|---|---|---|---|---|---|---|---|---|---|---|---|---|---|---|---|

| | | | | | | | | | | | | | | | | | | | | | | | |
|---|---|---|---|---|---|---|---|---|---|---|---|---|---|---|---|---|---|---|---|---|---|---|---|

Tel: [            ]    Fax: [            ]    Email: [            ]

*(Please attach separate sheet for additional name(s) and address of co-authors (persons responsible).*

16. (A) Passport information of germplasm

| IC/EC No. | Other Identity | Source | Place of origin Tehsil/Distt./Province | Genes (if any) identified |
|---|---|---|---|---|
| [      ] | [      ] | [      ] | [      ] | [      ] |

Remarks [            ]

(B) Pedigree of the Genetic Stock

Pedigree [                                        ]    Breeding method (code) [        ]

17. Salient Characteristics/Chief botanical and morpho-agronomic description** (attach details)

*Sl.No.*                    *Trait Description*

   1.

   2.

   3.

Provide salient description of the material as per All India Co-ordinated Research Programmes (AICRP) performawith comparative data over best two checks for unique features

18. **Usefulness of identified Germplasm/genetic stock**

19. **Year of seed production**    20. **Location of seed production**    21. **Quantity of seed available**

[            ]    [            ]    [            ]

22. Seed viability

23. Additional Information/remarks (if any)

24. Recommendation of institute's Germplasm Identification Committee

**UNDERTAKING**

1. I/We undertake to ensure long term conservation of the aforesaid germplasm/genetic stock at the National Gene bank, ICAR-NBPGR and also its sustainable use by maintaining appropriate quantity of Active/Working Collection and providing access as appropriate on prior informed consent and on mutually agreed terms. I/We also agree to provide any further information or data pertaining to the description and unique characteristics to the ICAR/NBPGR in a transparent manner.

2. That such germplasm does not contain any gene or gene sequence involving terminator technology.

*COUNTERSIGNED*                                                    *SIGNATURE OF THE DEPOSITOR*

*Signature*                                                                          *Name*

*Full Name*                                                      *Designation and Address*

                                                                        *Designation and Address*

COUNTERSIGNED BY HEAD OF THE INSTITUTE

Signature (SEAL)

Full Name

Designation and Address

## Guidelines for Filling Form A and Description of Codes

1. Use capital letters or write legibly. All items are self-explanatory. Give minimal explanation for a particular item in "Remarks" (*Item 19*), wherever needed.

2. Be to the point for *Item 6*, give only the most salient features, traits or alleles, considered suitable for consideration of registration.

3. On the other hand, give detailed description of traits and characteristics of the material in *Item 15*. Follow the format of variety release application or AICRP data sheets for respective crops. Use separate sheet if needed.

   Give main botanical and morpho-agronomic characteristics in description. Include isozyme or DNA profile or other chemical/biochemical characteristics, if available.

4. Use codes for filling in *Item 1,4,5,7,8* [actual], *9,10*, and *14* (B) [Breeding Method]. In case of the code "Other" fill in specific details in *item 19*.

5. For filling crop name (*Item 2*) give English or Hindi name, if known. In case a local name is given then also specify in parenthesis the language or dialect in which this name is used.

6. Give name(s) of all persons associated with development of the material in *Item 12*. Use separate sheet and fill in additional names along with designation, address and phone/fax/email, *etc.* beginning with S. No. 2 on new sheet, in the same format.

7. *Item 14* has two alternate parts, (A) and (B) to fill in:

   (a) In case Nature of the material to be registered as given in *Item 14* is "Germplasm" then you must give its basic passport information that should include National Identity (IC/EC) given by ICAR-NBPGR or other Identity Number allotted and maintained locally. In letter case, ICAR-NBPGR will provide a unique National Identifier number (IC), based on passport data provided.

   (b) In case the nature of the material to be registered is Genetic Stock, then clearly give its Pedigree, including parentage, year of crossing or selection. Also give breeding method used in codes described below.

8. Give particulars of developers in *Item 12* over and that of corresponding person in *Item 13* as the applicant and developer may not be always the same as the first person responsible for development of the material.

9. Undertaking to the effect ensuring long term conservation and maintenance of active material for facilitating access and sustainable use has been given, which may be read and implied before putting signatures.

Codes for filling information in Col. 1,4,5,7,8 [actual], 9,10, and 14 (B) [Breeding Method]

| Item 1: Application Status | | Item 8: Quantity deposited with application | |
|---|---|---|---|
| N | New | | Actual |
| R | Revised | | |
| **Item 2: Crop Groups** | | **Item 9: Value referred to by applicant** | |
| CL | Cereals | SC | Scientific |
| PC | Pseudo Cereals | CM | Commercial |
| MT | Millets | AC | Academic |
| MM | Minor Millets | | |
| GL | Gram Legumes | | |
| OS | Oilseeds | **Item 10: Basis of eligibility for registration** | |
| FC | Fiber Crops | PR | Published with Peer review |
| FG | Forage | CT | All India Co-ordinated Trials Data |
| FT | Fruits | OR | Institute annual report |
| VG | Vegetable Crops | OT | Any other report |
| SP | Spices | **Item 11: Validation Test suggested** | |
| MP | Medicinal and aromatic plants | Y | Yes |
| NC | Narcotics | N | No |
| OR | Ornamentals | | |
| FR | Forestry | **Item 14 B: Breeding method used** | |
| CC | Commercial crops | IN | Introduction and Selection |
| OT | Other (Specify in Col. 19) | MS | Mass selection |
| **Item 5: Biological status of material to be registered** | | PS | Pedigree selection |
| GP | Germplasm collection | PL | Pure line selection |
| GS | Genetic Stock | RS | Recurrent selection |
| RE | Recombinant | BC | Backcross method |
| MU | Mutant | OT | Other (Specify in Item 19) |
| **Item 7: Nature of genetic material** | | | |
| SD | Seed | | |
| TR | Tubers/Roots/Bulbs | | |
| VP | Vegetative Cuttings | | |
| WP | Whole Plant | | |
| OT | Other (Specify in Col. 20) | | |

Furnish the form complete in all respect with requisite quantity of seeds or propagules (propagules to the *concerned NAGS) to the Director, ICAR-NBPGR, Pusa Campus, New Delhi-110012*

# Annexure III

## Instructions to Authors for

## Organization of Manuscript of One Page Note on Registered Germplasm

1. The manuscript should be typed in double-space leaving a margin of 3-4 cm on all sides

2. The total length of the manuscript should be around **one** to **one and a half page**.

3. The title of the paper should be in all CAPS. The author's name with detailed address should appear at the end but before the references.

4. Oxford English spelling should be followed and the consistency of spelling should be maintained throughout the manuscript.

5. The contents of the manuscript should be in the form of a short communication that is structured as follows:

   **(a) Introduction-** It should deal with the unique characteristics for which the germplasm has been registered. It should include information about the parental material, breeding methodologies followed, and the place/institute where the material was developed.

   **(b) Morpho-agronomic Characteristics -** It should provide in brief morpho-agronomic description of the material along with information on multi-locational performance on important traits like yield and yield components.

   **(c) Associated Characters and Cultivated Practices-** It should mention about the other relevant traits of potential value, particularly reaction to major biotic/abiotic stresses. It may include the general recommendations on cultivation practices, and the area of adaptation.

   **(d) References–** References in the text should be cited by author, year of publication (*e.g.,* Joshi, 1995) and multiple citations should be in chronological order. Nothing should be underlined and Latin names should be italicised. These should be listed in alphabetical order under the first author's name citing all authors. The names of journals should be abbreviated according to the latest edition of the "World List of Scientific Periodicals) (eds. P. Brown and G.B. Stratton), Butterworths, London. The following examples may be used for citations:

      i. Bisht IS, RK Mahajan, TR Loknathan and RC Agarwal (1998) Diversity in Indian sesame collection and stratification of germplasm accessions in different diversity groups. *Genet.Resour.Crop Evol.***45**: 325-335.

      ii. Withers LA and F Engelmann (1998) *In vitro* conservation of plant genetic resources. In: A Altman (ed.) *Agricultural Biotechnology.* Marcell Dekker Inc., New York, pp 57-58.

      iii. Wealth of India (1985) (Revised), The Wealth of India-Raw Materials. A Dictionary of IndianRaw Materials and Industrial Products, Raw Material Series, Vol 1: Publications andInformation Directorate (now NISCOM), Council of Scientific and Industrial Research, NewDelhi, 513 p.

      iv. FAO (1998) The State of the World's Plant Genetic Resources for Food and Agriculture. Food and Agriculture Organization of the United Nations, Rome, Italy, 510 p.

      v. Engels, JMM and V RamanathaRao (eds) (1988) Regeneration of Seed Crop and their Wild Relatives. Proceedings of a Consultation Meeting, 4-7 December 1995. ICRISAT, Hyderabad, India and IPGRI, Rome, Italy, 167 p.

6. The manuscript can have a maximum of one table, preferably in continuation of the text. The numerical data along with minimum statistical analysis should be provided after thorough checking.

7. A maximum of one good quality figure or line diagram displaying either the characteristic features or providing clear understanding of the uniqueness of the germplasm is acceptable.

8. Recognized S.I. units should be used in conventionally accepted notation.

# Annexure IV

## Proforma for Passport (descriptors) Information

| Sl.No. | National Identity | Collector No. | Donor Institute | Donor/ other identity | Crop Name | Common Name | Taxonomic Code | Pedigree | Source | Biological Status | Country of Origin | Location | Latitude | Longitude | Altitude | Remarks |
|---|---|---|---|---|---|---|---|---|---|---|---|---|---|---|---|---|
| | | | | | | | | | | | | | | | | |

Source: In=Institute/NRC/IARCs; F= Farmer, M=Market; NGO=Non-Government Organizations; OT=Others

Biological status: W= Wild; RC=Released Cultivar; LR=Landraces; BL=breeding Line; Mu=Mutant; GS=Genetic Stock; OT=Others

Country of origin: Please provide ISO codes

# Annexure V

## Form B

*(See regulation 13 of MOEF and CC Notification C S R No. 827 dated 21ˢᵗ November, 2014)*

Conducting of non-commercial research or Research for emergency purpose outside India by Indian researchers/Government institution using the biological resources

1. Name of the Applicant (Indian research/Government institution)

2. Complete Address*
   a. Permanent
   b. Present

3. Name and address of Institution in India

4. Name of the Supervisor or Head of Institution at the place of work in India

5. Name and contact details of the Institution or Organization who shall guide the proposed research/receiving the biological resources

6. Details of the Supervisor or Head of Institution or Organization who guide the proposed research or recipient of the biological resources

7. Name of the funding agency supporting the proposed research

8. Brief description of the research

9. Details of biological resources proposed to be carried along or sent for the research
   i. Name of the biological resources (scientific/common name)
   ii. Location of collection (Village/Taluk/State)
   ii. Quantity required
   iv. Duration of the research

10. If it is for emergency purpose, specify details

## Undertaking

I, _____ Son/Daughter/Wife/Husband of _____ aged _____ residing at _____ holding a permanent I.D. No. _____ (PAN Card/Aadhaar Card/Passport, *etc.*) hereby declare that all the information provided above is correct and true. I hereby, affirm that the biological resources shall be used only for the purpose as stated in the application. I shall not share/provide/part/leave behind any the biological resource at my collaborator's facility/laboratory without approval of the NBA. I, along with my supervisor and collaborator, individually and severally declare that we shall not put to commercial utilization, nor shall seek any IPR claim on the biological resources and associated traditional knowledge in this research/collaboration. In case such a situation arises we shall apply to National Biodiversity Authority to seek prior approval. Results, process(es), products or other outcomes arising out of this activity shall be shared with the NBA during and/or upon completion of research intended along with the copy of relevant documents and publications.

Date: _____

Place: _____

*Signature*

**Declaration by the Supervisor/Head of Institution**

I, _____ working as_____in (Name of institution) confirm that details provided by Mr./Dr/Mrs/Ms_____are true and correct

Date: _____

Place: _____

*Signature*

*Designation*

*Official Seal*

**Declaration by the Recipient/Collaborator**

I, _____ working as _____ in (Name of institution/Organization) hereby affirm that I or my institution/organization shall use the Biological Resources for the purpose as stated in the application and which were sent by _____ (Name of institution) or being brought by Mr./Dr/Mrs/Ms _____. The said Biological Resources shall be destroyed in full after the completion of the studies/ partnership or upon completion of the studies, the Biological Resources shall be sent back to the institution from where the Biological Resources were received as the case may be. I or the institution I am associated with shall not claim any ownership under instant application nor shall claim any IP Rights over the Biological Resources, derivatives or other such components without prior approval of the applicant, institution affiliated and the National Biodiversity Authority.

*Signature*

*Designation*

*Official Seal*

## Annexure VI

### Requisition for Supply of Seed/Planting Material for Research from/ through ICAR-NBPGR, Pusa Campus, New Delhi-110012

*The Director*

*ICAR-National Bureau Plant Genetic Resources*

*Pusa Campus, New Delhi-110012*

*Ph:011-25843697, 25841129*

*Fax: 011-258442495*

1. Details of seed/planting material required for research

| Sl.No. | Botanical Name | Crop Name | No. of Accessions (IC/EC) | Seed Quantity (per accession/ per sample) | Purpose (Screening/breeding/evaluation/ augmentation/multiplication) |
|--------|----------------|-----------|---------------------------|-------------------------------------------|---------------------------------------------------------------------|
|        |                |           |                           |                                           |                                                                     |

2. Thesis/Project title for which request is made (name of the funding agency _____

3. Objective/Activity for utilization of indented material: (Please attach sheet if required):_____

4. Material Transfer Agreement (enclosed): Yes/No

5. Feedback report submitted on germplasm received earlier (if applicable): Yes/No

6. Have you or your Institute developed any variety based on Germplasm supplied by ICAR-NBPGR? Yes/ No (If yes, please let us know the details)

7. If required by ICAR-NBPGR, will you be able to send viable multiplied seed material of the above seed in sufficient quantity for conservation: Yes/No

8. Signature of the indentor:

   Name:_____ Designation_____

   Address of the Institute_____Phone (with STD code_____

   M_____Fax_____E-mail _____)

9. Signature of the Competent Authority with Seal: _____
   (PI/Head of the Department/Director of the Institute)

# Annexure VII

## Material Transfer Agreement
## Indian Council of Agricultural Research
## Krishi Bhavan, New Delhi – 110001. INDIA

Agreed between

**National Bureau of Plant Genetic Resources (NBPGR), New Delhi-110012**

of the Indian Council of Agricultural Research, Krishi Bhawan, New Delhi – 110001, the apex agricultural research organization of India, being the first Party (Provider of the Material) and

1. Being the Second Party (Recipient of the Material)
2. For the Supply/Exchange/Transfer of Genetic Resources for Food and Agriculture/Germplasm/Genetic Material/Genetic Components for Research[2]

   ❏ Within India, not covering persons as described in Section 3(2) of the Biological Diversity Act, 2002 (18 of 2003) (BDA).

   ❏ Within India, wholly or partly covering persons as described in Sec. 3(2) of BDA.

   ❏ Outside India, with Members of the International Treaty for Food and Agriculture (ITPGRFA), and wholly or partly covering persons as described in Sec. 3(2) of BDA.

   ❏ Outside India, with Non-Members of ITPGRFA, and wholly or partly covering persons as described in Sec. 3(2) of BDA.

AS follows:

Recipient Name

Recipient Institution/Organization/Agency/Centre

Recipient Full Address with PIN Code

Phone number

Fax

E-mail

Nature of activities

| | |
|---|---|
| Germplasm material (specify)[3] | Crop and varieties |
| Supply made through | NBPGR |
| For Official Use of Supplier | 1. Germplasm Identity (Species name, common name, *etc.*) |
| | 2. Accession Number |
| | 3. Short Description of the Material |
| | 4. Passport Data |

*I/We agree to abide by the following terms of the MTA and certify that:*

---

1 Mention Name and address of the Second Party

2 Tick mark the appropriate box

3 Specify the type of material involved for supply/transfer e.g seed, tissue culture, DNA *etc.*

# References

1. Arora, R.K. 1995. Genetic resources of vegetable crops in India: Their diversity and conservation, In: Genetic Resources of Vegetable Crops (Eds., Paroda, R.S., Gupta, P.N., Rai, M. and Kochhar, S.), ICAR-NBPGR, New Delhi, p. 29-39.

2. Bothmer, R. von and Seberg, O. 1995. Strategies for the collecting of wild species. In: Guarino, L., Ramanatha Rao, V. and Reid, R. (Eds.) Collecting Plant Genetic Diversity. Technical Guidelines. CABI International, UK, p. 93-111.

3. Astley, D. 1985. Management Systems at the National Vegetable Research Station Gene Bank. In: Documentation for genetic resources centres: Information handling systems for gene bank management. Konopka, J. and Hanson, J. (eds.). IBPGR, p. 48-55.

4. Arora, R.K. and Nayar, E.R. 1984. Wild relatives of crop plants in India, ICAR-NBPGR Sci. Monograph No.7 ICAR-NBPGR, New Delhi.

5. Burrill, A., Cronze, H. and Simonett, O. 1991. The potential use of the global resources information database (GRID) in plant genetic resources activities. In: Crop Genetic Resources of Africa. Attere, F., Zedan, H., Ng, N.Q., Perrino, P. (eds.), Vol. I. Proceedings of an international symposium, Nairobi, Kenya, 26-30 September, 1988, Rome, Italy, IBPGR: 125-132.

6. Federer, T. Walter. 1956. Augmented (or Hoonuiaku) designs. *The Hawaiian Planter Records*. Vol.IV, 2nd Issue, 1956, 191-208.

7. Harlan, J. 1992. Crops and Man. 2nd ed. American Society of Agronomy, Madison, WI, p. 284.

8. Hersh, G.N. and Rogers, D.J. 1975. Documentation and information requirements for genetic resources application. In: Crop Genetic Resources for today and tomorrow.

9. Hintum, Th. Van. 1989. Scoring heterogenous populations. In: Report of an International Workshop on '*Beta* Genetic Resources'. IBPGR publication. *International Crop Network, Series,* **3**: 106.

10. Heywood, V. 1995. Global Biodiversity Assessment (Ed.). Cambridge University Press, Cambridge.

11. Konopka, J. and Hanson, J. 1985. Management of data in genebank. In: Documentation of genetic resources information handling system for genebank management, IBPGR, Rome. p. 21-28.

12. Moss, H. and Guarino, L. 1995. Gathering and recording data in the field. In: Collecting Plant Genetic Diversity. Technical Guidelines. CABI International, UK, p. 367-417.

13. NBPGR. 2016. Guidelines for Management of Plant Genetic resources in India, p. 6169.

14. Painting, K.A., Perry, M.C., Denning, R.A. and Ayad, W.G. 1993. Guidebook for genetic resources documentation. IBPGR, Rome publication. p. 296.

15. Perry, M.C., Painting, K.A. and Ayad, W.G. 1993. Genebank management system software user's guide. IBPGR, Rome publication. p. 102.

16. Rana, R.S. 1992. Indian national plant genetic resources system. In: Plant Genetic Resources: Documentation and information management. Rana, R.S. *et al.* (eds.). ICAR-NBPGR, New Delhi.

17. Rogers, D.J., Snoad, B. and Seidewitz, L. 1975. Documentation for genetic resources centres. In: Crop genetic resources for today and tomorrow. Frankel, O.H. and J.G. Hawkes (eds.). IBP-2, Cambridge Univ. Press, Cambridge. p. 399-405.

18. Sapra, R.L. 1991. Documentation of plant genetic resources. In: Plant Genetic Resources: Conservation and Management by R.S. Paroda and R.K. Arora (eds.) IBPGR Regional Office for South East Asia, New Delhi. p. 315-325.

19. Sapra, R.L. and Agrawal, R.C. 1992. Germplasm evaluation: Augmented Design. In: Plant Genetic Resources: I. Documentation and Information Management. Rana, R.S., Sapra, R.L., Agrawal, R.C. and Gambhir, R. (eds.), ICAR-NBPGR, Delhi, p. 37-43.

20. Sapra, R.L. and Singh, B. 1992. Database management of plant genetic resources.. In: Plant Genetic Resources: Documentation and Information Management. Rana, R.S., Sapra, R.L., Agrawal, R.C. and Gambhir, R. (eds.), ICAR-NBPGR, New Delhi, p. 17-35.

21. Smith, R.D., Linington, S.H. and Fox, D.J. 1988. The role of the computer in the day to day running of the Kew Seed Bank. In: Documentation of genetic resources: Information handling systems for gene bank management. Konopka, J. and Hanson, J. (eds.). IBPGR, p. 29-42.

22. Singh, H.P. 2010. Managing genetic resources of horti. crops. *Indian Horticulture* (May-June). p. 3-16

23. Verma, S.K., Negi, K.S., Muneem, K.C. and Arya, R.R. (2008a). "*In situ* Conservation Approach for Germplasm of Vegetable Crops" In winter school on conservation and utilization of indigenous germplasm in improvement of vegetable crops. Compiled by Singh, D.K., Department of Vegetable Science, College of Agriculture, GBPUA&T., Pantnagar, Uttarakhand. p. 183-191.

24. Verma, S.K., Negi, K.S., Muneem, K.C. and Arya, R.R. (2008b). "Genetic Diversity of Indigenous Underutilzed cucurbits." In winter school on conservation and utilization of indigenous germplasm in improvement of vegetable crops. Compiled by Singh, D.K., Department of Vegetable Science, College of Agriculture, GBPUA&T., Pantnagar, Uttarakhand. p. 200-213.

25. WCM. 1992. Global Biodiversity: Status of the Earth's Living Resources. Chapman and Hall, London.

26. Yndgaard, Fleming 1982. A documentation system for Nordic genebank, *IBPGR Plant Genetic Resources Newsletter*, **49:** 34-36.

# 9

# Source Availability and Utilization of Vegetable Genetic Resources

## Introduction

Nearly 42 per cent of the world population depends on agriculture directly and indirectly. Conservation and sustainable use of plant biodiversity is absolutely essential for ensuring food, feed, nutrition and healthcare security to the ever-increasing human and domestic animal populations. Among agricultural commodities, vegetables are rich sources of many essential micronutrients, including vitamin C, vitamin K, folate, thiamine, carotenes, minerals and dietary fibre. In addition, vegetables are rich in health related phytochemicals, such as antioxidants. Phytochemicals play an important role in reducing the risk of many chronic diseases. Therefore, it is essential to produce more vegetables to feed our ever growing population. In this context, use of genetically improved, disease resistant and high yielding varieties in precision farming may play a significant role in increasing vegetable productivity and production. Sustainable utilization of economically important components of biological diversity requires the understanding of three important elements:

(i) How much resources are available

(ii) What resources are available

(iii) Where resources are available

Historically, the science of genetics began with the rediscovery of Gregor John Mendal paper in 1900 which was originally published in 1866. Mendel's laws of inheritance provided the foundation of crop improvement. Mendelian variation (generated mainly by gene mutation), site specific hybridization and polyploidy are the three major ways to bring genetic variability in various crop species. Plant genomes are highly dynamic structure offering endless opportunities of reshuffling the gene contents. Realization that the chromosomes are the carriers of genes has led to the development of specialized plant breeding methods of chromosome engineering. Mutations and their recombination upon hybridization usher huge amount of variability yet it needs keenness to timely identify and put it to economic use. It is a fact that modern plant breeding methods have their bases in the genetic and cytogenetic principles. The desired changes in genotypes of crop species and consequent benefits to farmers are brought about by a series of interrelated and largely interdependent activities. These activities include *viz.* creation of variation, selection, evaluation, multiplication and distribution. However, selection is effective in altering a species only when genetic variability exists in the populations of the species. In this context a noteworthy development resulted from the studies of G.H. Shull on maize, who found that inbreeding produced a considerable loss of vigour, and put forth the idea of hybridization and coined the term heterosis. Analysis of data focused on selection of desirable plants, especially during $F_2$ and subsequent generation is largely dependent on the skill and judgment of the breeder. However, the goal of all plant breeding activities has been always to develop an improved crop variety which is utilized by farmers for commercial cultivation. The other developments in the science of crop breeding are totipotency and recombinant DNA technology which have enabled the transfer of desirable genes from any source into plant system. Diseases, insect pests and high temperature are threats to vegetable growers. This, in turn, reduces the availability of vegetables for consumers. Natural genetic resistance is the most effective and environment friendly solution. Genes from wild plants are introduced into crops through crossing techniques. Sometimes, these hybrids are multiplied via tissue culture. Plants with resistance genes are identified through molecular markers or other disease screening tests. Resistance is combined with other desirable horticultural traits such as high yield. The whole process takes about 5-10 years depending upon the traits to be incorporated. The prized genetic materials along with their wild and close relatives are used by plant breeders for development of resistant varieties/hybrids. Wild relatives also constitute a priceless reservoir that contains genes conferring better adaptation to stress environments and resistance to diseases and pests.

Fortunately, there is a universal awareness to explore, collect, maintain, evaluate and utilize the variability before it is lost. Several exotic germplasm of vegetable crops have been utilized directly as promising varieties in India. The assembled germplasm were evaluated and promising ones identified and further tried with the best local checks available and the outstanding ones were released for direct cultivation. In national and international genebanks, total number of germplasm holdings goes on increasing as more collections are made. Similarly, individual breeders are often not able to grow the entire germplasm of target species to select desired parents. Therefore, there is a growing intrest to prepare a set of representative collections for use by plant breeders. In tomato several wild species have been used as donors, for example, genus *Solanum hirsutum* and *S. pimpinellifoium* for fungus resistance, *S. chilense and S. peruvianum* for virus resistance, *S. chmielewskii* for fruit quality and *S. cheesmanii* for tolerance to adverse environments. And, further, use of quality genes such as 'Rin' in tomato resulted into better table and nutritional quality of the crop. The exploitation of resistance genes of *Cucumis hardwickii* helped in saving billions of INR worth of cucumber. However, there is still scope to use these species against different traits and some more species not only for stress resistance, but also for incorporation of quality and yield traits, because they offer rich variability for different desired characteristics. The exploitation of genes from wild relatives has often faced limitations in terms of difficulty of making viable crosses between wild and domesticated species. The biotechnology revolution, including recent developments in recombinant DNA technology (genetic engineering), however,

raised the prospects of transference of useful traits to cultivated species, which was not possible earlier by the conventional means. The vegetable production scenario in this country is undergoing several changes and the demand for fresh vegetables are increasing mainly from middle income segment of the population chiefly from urban and semi-urban areas of the country. Such market demands to adopt off-season production of vegetables, even under unfavorable conditions with lower yield but higher market price for better return. Seasonality in vegetable production, availability and consumption is fast disappearing. Hence genetic resources in vegetables have to be sought not only for increased production, disease and pest tolerance but also for tackling abiotic factors like heat or cold tolerance, rain fed and semiarid conditions *etc*. Besides, there is growing awareness of the advantages of healthy food habits, like increased vegetable consumption and the intake of important nutrients like proteins, vitamins and minerals is facilitated by enhanced vegetable consumption.

## Utilization of Vegetable Genetic Resources

The ultimate goal of plant introduction to any region of the world is the use of germplasm to its fullest potential. Both indigenous and introduced diversity has been utilized to a great extent. To promote the use of promising germplasm/genetic stocks identified after evaluation, germplasm field days should be organized besides publishing research articles, crop catalogues (printed and electronic), *etc*. The development of core set of collection particularly in the crops having large germplasm collection can be a powerful tool for promoting utilization of germplasm. It is also important to accord recognition to those associated with the development of improved/unique germplasm and genetic stocks, such as, plant breeders, farmer-breeders or other developers/innovators. From exotic introductions, several selections/varieties could be directly released after their acclimatization in different locations and initial performance or evaluation. Also the exotic materials possessed several promising traits, which could be incorporated in to indigenous varieties through breeding.

## Ways of Germplasm Utilization

Developing a sustainable utilization plan demands data beyond just numbers. It is not enough to just know how much of what resources we have, but also where these resources are located. Spatial maps of distribution of biological resources help in assessing the gaps between their demand and natural supply and in alerting about possible local extinctions of species due to unsustainable exploitation levels of harvesting. Such data sets are also important to arrive at local specific estimates of sustainable levels of harvest for different species.

**Table 9.1: Important Breeding Methods for Cross-Pollinated Crops**

| Crop | Breeding Methods/Procedures |
|---|---|
| Cauliflower | Backcross method, mass selection, family selection, heterosis breeding, synthetic varieties |
| Cabbage | Backcross method, mass selection, heterosis breeding |
| Radish | Backcross method, mass selection, heterosis breeding |
| Carrot | Mass selection, pedigree selection, synthetic varieties, backcross method |
| Turnip | Backcross method, heterosis breeding |
| Onion | Mass selection, family selection, heterosis breeding, backcross method |
| Cucumber | Backcross method, pedigree method, recurrent selection, heterosis breeding |
| Muskmelon | Selfing and selection, pedigree method, backcross method, heterosis breeding |
| Watermelon | Backcross method, pedigree method, recurrent selection, heterosis breeding, polyploidy breeding |
| Bottle gourd | Selfing and selection, pedigree method, backcross method, bulk method, heterosis breeding |
| Squash | Selfing and selection, heterosis breeding, backcross method |
| Pumpkin | Selfing and selection, heterosis breeding, backcross method |
| Ridge gourd | Selfing and selection, pedigree method, backcross method, single seed descent method heterosis breeding |
| Sponge gourd | Selfing and selection, pedigree method, backcross method, single seed descent method heterosis breeding |
| Potato | Selection, polyploidy breeding, mutation, biotechnological approaches |
| Cassava | Selection, intervarietal hybridization, heterosis breeding, polyploidy breeding, mutation, biotechnological approaches |

Availability of useful variability is a pre-requisite for the improvement of any organism, including plants for the desired traits. Germplasm of wild progenitors and related non-progenitor species are rich reservoir for useful variability for majority of the traits which can be transferred into otherwise elite cultivars and commercially exploited. The access to collection of rich crop diversity, maintained in genebanks, is the only hope to make available to the plant breeders the useful germplasm possessing specific traits required to develop resilient and climate-adapted new crop varieties.

## (i) Use of Collections as Varieties

Top ranking local material can be straightway used as a variety for instance 'Sel.-1' in okra, 'Hara Madhu' in muskmelon, and 'Pusa Purple Long' in brinjal. Similarly, a lot of local materials are available in bottle gourd, sponge gourd, ridge gourd, pointed gourd, spine gourd, sweet gourd, and several other leafy vegetables.

It has been observed that primary introductions are successful only when species are shifted to climate almost similar to its home environment. Many a time such conditions are not fulfilled and performance of a species or a variety remains below expectation. Secondary introduction constitutes "The introduced variety may be subjected to selection or used in a hybridization programme to develop a superior variety". In general, secondary introduction is much more common than primary introduction.

### Use of Indigenous Stocks in Improvement and Release of Varieties

A large number of vegetable varieties have been developed using indigenous stocks either directly or using indigenous stocks in crossing programme. Thus it is obvious that indigenous plant genetic resources

**Table 9.2: Germplasm Introduced and Released Directly as Varieties**

| Crop | Introduction/Selection | Crop | Introduction/Selection |
|---|---|---|---|
| Tomato | Sioux, Marglobe, Best of All, Fire Ball, La Bonita, Balkan, Roma | Peas | Early Badger, Bonneville, Sylvia, Asauji, Arkel, Harbhajan |
| Brinjal | Black Beauty, Pusa Purple Long, Pusa Purple Round | Cowpea | Pusa Barsati, Pusa Phalguni |
| Hot chilli | NP 46A, Pusa Red | Garlic | IC-49373, IC-49382, T86/Sel-1 |
| Guar | Pusa Sadabahar, Pusa Mansuri, Sharadbahar | Onion | Early Grano (America), Red Grano (America), Bermuda Yellow (Philippines) Pusa Red, Pusa Ratnar, Ratnar Selection, Brown Selection |
| Cauliflower | Snowball, Pusa Katki, Improved Japanese (Israel), Snow Ball-16 (Holland) | Carrot | Pusa Keasr, Jeno (Germany) |
| Cabbage | Golden Acre, Drum Head, September (Germany) | Radish | Japanese White |
| Lablab bean | IC-16862, Pusa Bunch | Amaranth | Bari Chauli |
| Sponge gourd | Pusa Chikni | Bottle gourd | Pusa Summer Prolific Long, Pusa Summer Prolific Round |
| Ridge gourd | Pusa Nasdar | Fenugreek | Kasuri Methi, Pusa Early Bunch |
| Okra | Velvet Green, Pusa Makhmali, Ghana Red (Africa) | Watermelon | New Hampshire Midget (US), Sugar Baby (US), Asahi Yamato (Japan), Shipper (US) |
| Cucumber | Poinsette (US) | Summer squash | Patty Pan (US) |

**Table 9.3: Exotic Materials Used as Secondary Introduction**

| Vegetable Crop | Variety Developed | Exotic line(s) Used | Breeding Method Used |
|---|---|---|---|
| Tomato | Pusa Ruby | Sioux x Improved Meeruti | Hybridization |
|  |  | Improved Meeruti x Red Cloud | Hybridization |
| French bean | Pusa Parvati | EC001906 | Mutation breeding |
| Cowpea | Pusa Dofaslai | Pusa Phalguni x Philippines Variety | Hybridization |
|  | Pusa Phalguni | Selection from Dolique du Ton Kin from Canada | Selection |
| Watermelon | Arka Manik | IIHR 21 x Crimson Sweet | Hybridization |
| Cauliflower | Pusa Snowball 1 | EC012013 x EC012012 | Hybridization |
| Cabbage | Pusa Mukta | EC024855 x EC 010109 | Hybridization |
| Turnip | Pusa Chandrima | Japanese White x Snowball | Hybridization |
|  | Pusa Kanchan | Local Red Round x Golden Red | Hybridization |
| Okra | P-7 | Pusa Sawani x Ghana Red | Hybridization |

have played a vital role in the development of superior varieties and hybrids of vegetable crops. Keeping in view their significance in agriculture and food and nutritional security of the country, the plant genetic resource evaluation and their utilization have emerged as the significant areas of plant genetic resources management system. Locally important traditional crops/varieties will be vital to the long term health of regional economies and to the survival of particular traditional cultures. No doubt, some of these crops may become the future (new) crops of international commerce.

**Table 9.4: Useful Indigenous Collections and Release as Cultivars after Desired Improvement**

| | |
|---|---|
| Tomato | HS-101, HS-102, HS-110, S-12, Punjab Kesari, Kuber, Kalyanpur T-1, KS-1, KS-2, Angoorlata, CO-1, CO-2, CO-3. |
| Brinjal | Pusa Purple Long, Pusa Purple Round, Pusa Purple Cluster, Pusa Bhairav, Arka Sheel, Arka Kusumkar, Arka Navneet, CO-1, MDU-1, Junagarh Long. |
| Hot pepper | NB-34 to NB-41, NP-46A to NP-51, Jawala, Pusa Red, Kalyanpur Red, Yellow Kalyanpur, Chaman, Pant C-1, Pant C-2, G-1, G-2, G-3, G-4, G-5. |

A few examples are as given in Table 9.5 (Hazara and Som, 1989 and other sources).

**Table 9.5: Improved Varieties of Vegetables Developed through Indigenous Stocks**

| Crop | Variety | Selection/Parentage |
|---|---|---|
| Tomato | Improved Meeruti | Selection |
| Brinjal | Pusa Purple Long | Selection |
|  | Pusa Purple Cluster | Selection |
|  | Pusa Purple Round | Selection |
|  | Pant Samrat | Selection |
|  | Arka Shirish | Selection |
|  | Arka Kushmakar | Selection |
|  | Arka Sheel | Selection |
|  | Punjab Chamkila | Selection |
|  | T-3 | Selection |
|  | Krishanagar Long Green | Selection |
|  | Punjab Neelam | Selection |
|  | Punjab Bahar | Selection |
|  | Punjab Barsati | Pusa Purple Cluster x H-4 |
|  | Pusa Kranti | (Pusa Purple Long x Hyderpur) x Wynad Giant |
|  | Pusa Hybrid-4 | Pusa Purple Long x Hyderpur |
|  | Pusa Anupam | Pusa Kranti x Pusa Purple Cluster |
|  | Pusa Bindu | GR x Pant Rituraj |
|  | Pusa Uttam | GR x Pant Rituraj |
|  | Hisar Shyamal (H-8) | Aushey x BR-112 |
|  | Pant Rituraj | T-3 x Pusa Purple Cluster |
|  | Pant Brinjal-4 | PB129 x PB-7 |
| Chilli | G-2 | Selection |
|  | G-3 | Selection |
|  | KT-1 | Selection |
|  | CO-1 | Selection |
|  | CO-2 | Selection |
|  | Kalyanpur Yellow | Selection |
|  | Kalyanpur Red | Selection |
|  | Kalyanpur | Selection |
|  | Sabour Arun | Selection |
|  | Pusa Jwala | NP-46A x Puri Red |
|  | NP-46A | Local x Puri Red |
|  | Pant C-1 | NP-46A x Kandhari Local (natural cross) |
| Cowpea | Cowpea-263 | Selection |
|  | Narendra Lobia-1 | Pusa Komal x Varanasi Local |
| Hyacinth bean | Pusa Early Prolific | Selection |
|  | JDL-79 | Selection |
|  | JDL-53 | Selection |
|  | K-6802 | Selection |
|  | HB-18 | Selection |
|  | HB-60 | Selection |

| Crop | Variety | Selection/Parentage |
|------|---------|---------------------|
| | Ragini | Selection |
| | Hebbal Avare-1 | Local Avare x Red Typicus |
| Cluster bean | Pusa Sadabahar | Selection |
| | Pusa Mausmi | Selection |
| | PLG-850 | Selection |
| | Pusa Navbahar | Pusa Sadabahar x Pusa Mausmi |
| Okra | Pusa Sawani | Pusa Makhmali x IC1542 |
| Bottle gourd | Kalyanpur Long Green | Selection |
| | Punjab Komal | Selection |
| | Pant Sankar Lauki-1 | PBOG-22 x PBOG-40 |
| | Pant Sankar Lauki-2 | PBOG-22 x PBOG-61 |
| | Pusa Summer Prolific Long | Selection |
| | Pusa Summer Prolific Round | Selection |
| | Arka Bahar | Selection |
| | Pusa Naveen | Selection |
| | CO-1 | Selection |
| | Pusa Sandesh | Selection |
| | NDBG-1 | Selection |
| | KLG | Selection |
| | KBG-16 | Selection |
| | Singapuri Long | Selection |
| | Phule BTG-1 | Selection |
| Bitter gourd | Kalyanpur Barahmasi | Selection |
| | Kalyanpur Sona | Selection |
| | Pusa Vishesh | Selection |
| | Pusa Do Mausmi | Selection |
| | Pant Karela-1 | Selection |
| | Priya (VK-1) | Selection |
| | Arka Harit | Selection |
| | Coimbatore Long | Selection |
| | Phule BG-6 | Selection |
| | CO-1 | Selection |
| | Punjab BG-14 | Selection |
| | MDU-1 | Selection |
| | MC-23 | Selection |
| | NDB-IMC-84 | Selection |
| | Kashmiri Long | Selection |
| | Pant Karela-1 | Selection |
| Sponge gourd | Pusa Chikni | Selection |
| | Pusa Supriya | Selection |
| | Pusa Sneha | Selection |
| | KLC | Selection |
| | KSG-7 | Selection |
| | Pusa Sadabahar | Selection |
| | DL-1 | Selection |

| Crop | Variety | Selection/Parentage |
|------|---------|---------------------|
| | Kashi Divya | Selection |
| | Kalyanpur Chikani | Selection |
| Ridge gourd | Pusa Nasdar | Selection |
| | CO-1 | Selection |
| | CO-2 | Selection |
| | Arka Sumeet | Selection |
| | Arka Sujata | Selection |
| | PKM-1 | Selection |
| | Pant Tori-1 | Selection |
| Cucumber | Pant Khira-1 | Selection |
| | Sheetal | Selection |
| | Solan Green | Selection |
| | Khira-75 | Selection |
| | Khira-90 | Selection |
| Muskmelon | Hara Madhu | Selection |
| | Pusa Sharbati | Kutana x PMR-6 |
| | Pusa Madhuras | Selection |
| | Pusa Rasraj | MS-3 x Durgapur Madhu |
| | Arka Rajhans | Selection |
| | Arka Jeet | Selection |
| | Punjab Hybrid | MS-1 x Hara Madhu |
| | Punjab Sunheri | Selection |
| | Durgapur Madhu | Selection |
| | Hisar Madhu | Selection |
| | Lucknow Safeda | Selection |
| | Kutana Selection | Selection |
| | GMM-1 | Selection |
| | GMM-2 | Selection |
| | NDM-1 | Selection |
| | NDM-2 | Selection |
| | NDM-3 | Selection |
| | NDM-4 | Selection |
| Watermelon | Durgapura Meetha | Selection |
| | Durgapura Kesar | Selection |
| | Arka Manik | |
| | Pusa Rasal | |
| | Special-1 | Selection |
| Summer Squash | Punjab Chappan Kaddu-1 | Selection |
| | Hisar Sel.-1 | |
| Pumpkin | Arka Chandan | Selection |
| | CM-4 | Selection |
| | Pusa Vishwas | Selection |
| | CO-1 | Selection |
| | CO-2 | Selection |
| | Solan Badami | Selection |
| | Ambili | Selection |
| | CM-346 | Selection |
| | NDPK-224 | Selection |

| Crop | Variety | Selection/Parentage |
|------|---------|---------------------|
| Wax gourd | CO-1 | Selection |
| | CO-2 | Selection |
| | KAU LOcal | Selection |
| Snake gourd | CO-1 | Selection |
| | CO-2 | Selection |
| | MDU-1 | Selection |
| | PKM-1 | Selection |
| | TA-19 | Selection |
| Longmelon | Arka Sheetal | Selection |
| | Karnal sel. | Selection |
| | IIHR-4 | Selection |
| | Pant Kakri-1 | Selection |
| Indian Squash | Arka Tinda | Selection |
| | Punjab Tinda | Selection |
| | S-48 | Selection |
| | KPT-1 | Selection |
| | KPT-2 | Selection |
| Pointed gourd | Kashi Alankar (VRPG-1) | Selection |
| | Kashi Suphal (VRPG-2) | Selection |
| | Faizabad Parwal-1 | Selection |
| | Faizabad Parwal-2 | Selection |
| | Faizabad Parwal-3 | Selection |
| | Faizabad Parwal-4 | Selection |
| | Rajendra Parwal-1 | Selection |

## Intervarietal Mating

The intervarietal crossing, both the varieties belong to the same species whereas inter specific hybridization includes crossing between different species of the same genus.

## Hybridization in Asexually Propagated Crops

In asexually propagated crops, selection of parents should be based upon their general and specific combining ability performance. Since varieties of asexually propagated vegetable crops are heterozygous in nature, segregation of characters will occur in $F_1$ generation following hybridization. Hence, clonal selection should be made in $F_1$ generation and rarely in $F_2$ generation. The important vegetable varieties developed through hybridization in asexually propagated vegetable crops are as follows in Table 9.7.

## Cultivars as Source of Germplasm

In a survey of 25 leading companies engaged in Agri-seed business reveals that 95.7 per cent commercial cultivars were source germplasm for breeders.

## (ii) As a donor parent

The three main germplasm resources basically used for genetic diversity in disease resistance are commercial varities, land races and wild ancestral species. Cultivar mixtures and wild land races have traditionally been used by subsistence farmers to keep disease epidemics at low levels. These subsistence farmers live mainly in developing countries and produce up to 20 per cent of the world's food. Land races have been a valuable source of disease resistance to them because of their already high adaptation in appropriate agronomic backgrounds. Mixtures with improved varieties are now being advocated as an alternative strategy for disease control in many crops. Wild germplasm are also being used to transfer new and valuable genes of disease resisance to cultivated crops. Continuous

### Table 9.6: Vegetable Varieties Developed through Hybridization

| Crop | Variety | Pedigree | Organization |
|------|---------|----------|--------------|
| Carrot | Pusa Kesar | Local Red x Nantes | ICAR-IARI, New Delhi |
| | Pusa Meghali | Pusa Kesar x Nantes | ICAR-IARI, New Delhi |
| | Imperator | Nantes x Chanteny | - |
| | Sel.-233 | Nantes x No. 29 | - |
| Turnip | Pusa Chandrima | Japanese White (Asiatic type) x Snow Ball (European type) | ICAR-IARI, Regional Station Katrain (H.P.) |
| | Pusa Swarnima | Japanese White (Asiatic type) x Golden Ball (European type) | ICAR-IARI, Regional Station Katrain (H.P.) |
| | Pusa Kanchan | Local Red Round x Golden Ball (European type) | ICAR-IARI, New Delhi |
| Radish | Punjab Safed | White-5 x Japanese White | PAU, Ludhiana (Punjab) |
| | Pusa Himani | Radish Black x Japanese White | ICAR-IARI, Regional Station Katrain (H.P.) |
| Cauliflower | Pusa Snow Ball-1 | EC12013 x EC12012 | ICAR-IARI, Regional Station Katrain (H.P.) |
| | Pusa Shubra | A selection between triple cross (MGS23, 15-1 and D-96) | ICAR-IARI, New Delhi |
| Cabbage | Pusa Mukta (Sel.-8) | EC24855 x EC10109 | ICAR-IARI, Regional Station Katrain (H.P.) |
| Muskmelon | Pusa Sharbati | Kutana x PMR-6 (of USA) | ICAR-IARI, New Delhi |
| | Punjab Sunehri | Hara Madhu x Edisto | PAU, Ludhiana (Punjab) |
| Watermelon | Arka Manik | IIHR-21 x Crimson Sweet | ICAR-IIHR, Bengaluru (Karnataka) |

**Table 9.7: Release of Varieties/Hybrids in Asexually Propagated Vegetable Crops through Hybridization**

| Crop | Vaeriety | Pedigree |
|---|---|---|
| Cassava (*Manihot esculenta*) | H-97 | Manjavella x Accession No. 3000 |
| | H-165 | Chadoyaman Galamvella x Kalikalon (both indigenous) |
| | H-226 | Maloyan x Ethakkakaruppan (local) |
| | Sree Vasakham | Accession No. 1051 x S.-2312 |
| Sweet potato (*Ipomoea batatas*) | H-42 | Vella Dampth (Indigenous) x Triumph (America) |
| | Rajendra Sakarkand-5 | Cross 4 x H-5 |
| Potato (Early varieties) | Kufri Alankar | Kennebec x O.N.-2090 |
| | Kufri Bahar | Kufri Red x Gineke |
| | Kufri Chandramukhi | Sd-4485 x Kufri Kuber |
| | Kufri Jyoti | 3069d x 2814a |
| | Kufri Laukar | Serkor x Adina |
| Potato (Medium varieties) | Kufri Badshah | Kufri Jyoti x Kufri Alankar |
| | Kufri Chamatkar | Ekishirazu x Phulwa |
| | Kufri Khasigaro | Taborkyt x Sd. 698D |
| | Kufri Lalima | Kufri Red x AG-14 |
| | Kufri Muthu | (3046 (1) x M-109-330) |
| | Kufri Sheetman | Craigs Defiance x Phulwa |
| | Kufri Sherpa | Utimus x Adina |
| Potato (Late varieties) | Kufri Sindhuri | Kufri Red x Kufri Kundan |
| | Kufri Dewa | Craing Defiance x Phulwa |
| | Kufri Jeeva | M109-3 x 698-D. |

**Table 9.8: Source of Germplasm Used by Breeders (Survey of 25 leading seed companies)**

| Source Germplasm | All Crops (per cent) | Potato (per cent) | Vegetables (per cent) |
|---|---|---|---|
| Commercial cultivars | 81.5 | 50.0 | 95.7 |
| Related minor crops | 1.4 | 8.0 | 0.3 |
| Wild species (from gene bank) | 2.5 | 19.0 | 1.4 |
| Wild species (*in situ*) | 1.0 | 0.0 | 0.1 |
| Landraces (from gene bank) | 1.6 | 1.7 | 1.7 |
| Landraces (*in situ*) | 1.4 | 0 | 0.4 |
| Induced mutations | 2.2 | 3.3 | 0.3 |
| Biotechnology | 4.5 | 1.77 | 0.1 |

*Source*: Timothy Swanson (1977).

efferts are being directed at broadening the genetic base of crops by a search for sources of disease resistance which remain the most practical and environmentally sound means for the control of most major diseases in cucurbitaceous vegetable crops. New collections may not be promising as such but are desirable in one or few traits. Such material can be used as a donor to improve the existing popular varieties lacking in one or few good characters. Orton (1908 and 1911) developed Fusarium wilt (*Fusarium oxysporum f. nivuem*) resistant cultivar "Conqueror". The resistance was transferred from Citron, a semi-wild, non-edible, stock of melon to cultivar 'Eden'.

Tolerant plant species (halophytes) represent about 4 per cent of the flora in the Mediterranean region and offer largely untapped resources for vegetable production under saline conditions. Some of the wild species proved to be potential donors for salt tolerance genes that can be transferred to their respective domesticated species by classical breeding or biotechnology. Exotic halophyte constitutes untapped genetic resources that can be used for developing crops for saline conditions. Salt tolerance in indigenous halophytes and glycophytes is essential.

### (iii) Recombination Breeding

The new collections are hybridized to produce new recombinants better than parental material and after selection new varieties and hybrids are released for commercial cultivation.

**Table 9.9: Intervarietal and Inter Specific Mating**

| Crop | Variety | Pedigree | Organization |
|---|---|---|---|
| **Tomato** | Pusa Ruby | Sioux x Improved Meeruthi | ICAR-IARI, New Delhi |
| | Pusa Early Dwarf | Improved Meeruthi x Red Local | ICAR-IARI, New Delhi (Haryana) |
| | HS-1 | Sel 2-3 x An Exotic Culture | CCSHAU, Hisar (Haryana) |
| | Hisar Anmol (H-24) | Hisar Arun x *S. hirsutum f. glabratum* | CCSHAU, Hisar (Haryana) |
| | Hisar Arun | Pusa Early Dwarf x K-1 | CCSHAU, Hisar (Haryana) |
| | Hisar Lalima | Pusa Early Dwarf x HS-101 | CCSHAU, Hisar (Haryana) |
| | Hisar Lalit | HS-101 x Resistant Bengaluru | CCSHAU, Hisar (Haryana) |
| | HS-102 | S-12 x Pusa Early Dwarf | CCSHAU, Hisar (Haryana) |
| | Punjab Chhuhara | EC55005 x Punjab Tropic | PAU, Ludhiana (Punjab) |
| | Arka Meghali | Arka Vikash x IHR-584 | ICAR-IIHR, Bengaluru (Karnataka) |
| **Brinjal** | Pusa Kranti | (Pusa Purple Long x Hyderpur) x Wynad Giant | ICAR-IARI, New Delhi |
| | PH-4 | Hyderpur x Pusa Purple Long | CCSHAU, Hisar and PAU, Ludhiana |
| | Arka Navneet | IIHR-21 x Supreme | ICAR-IIHR, Bengaluru (Karnataka) |
| | Hisar Shymal | Aushey x BR-112 | CCSHAU, Hisar (Haryana) |
| | Pant Rituraj | T-3 x Pusa Purple Cluster | GBPUA&T, Pantnagar (Uttarakhand) |
| | Azad Kranti | Pusa Purple Long x BGL | CSAUA&T, Kanpur |
| **Chilli** | Andhra Jyoti (G5) | G2 x Bihar Variety | Agri. Res. Stn. Lam, Guntur (A.P.) |
| | X-235 | G4 x Anther Mutant | Agri. Res. Stn. Lam, Guntur (A.P.) |
| | Pusa Jwala | NP-46 x Puri Red | ICAR-IARI, New Delhi |
| | Pusa Sadabahar | Pusa Jwala x IC31339 (*Capsicum frutescens*) | ICAR-IARI, New Delhi |
| | Pant C-1 | NP-46 A x Kandhari | GBPUA&T, Pantnagar (Uttarakhand) |
| | Pant C-2 | NP-46 A x Kandhari | GBPUA&T, Pantnagar (Uttarakhand) |
| | Jawahar-218 | Kalipeeth x Pusa Jwala | JNKVV, Jabalpur (M.P.) |
| | Punjab Lal | Perennial x Long Red | PAU, Ludhiana (Punjab) |
| | K-2 | K-2 x Satoor Lamba | Vilpati (Tamil Nadu) |
| | J-218 | Kalipeeth x Pusa Jwala | JNKVV, Jabalpur (M.P.) |
| | Bhaskar | G4 x Yellow Anther Mutant | Agri. Res. Stn. Lam, Guntur (A.P.) |
| **Okra** | Pusa Sawani | IC1542 x Pusa Makhmali | ICAR-IARI, New Delhi |
| | Pusa Padmini | (Reshmi x Ghana) x (Pusa Sawani x Ghana) | PAU, Ludhiana (Punjab) |
| | Selection-2 | (Pusa Sawani x Best-1) x (Pusa Sawani x IC7194) | ICAR-IARI, New Delhi |
| | Parbhani Kranti | (Pusa Sawani x *A. manihot*) x Pusa Sawani | MAU, Prabhani (Maharashtra) |
| | P-7 | Pusa Sawani x *A. manihot ssp. manihot* | PAU, Ludhiana (Punjab) |
| | Arka Anamika | *A. esculentum* x *A. manihot* | ICAR-IIHR, Bengaluru (Karnataka) |
| | Varsha Uphar | Lam Sel-1 x Parbhani Kranti | CCSHAU, Hisar (Haryana) |
| **Pea** | Jawahar Matar-3 | Type-9 x Early Badger | JNKVV, Jabalpur (M.P.) |
| | Jawahar Matar-4 | Type-19 x Little Marble | JNKVV, Jabalpur (M.P.) |
| | Jawahar Matar-1 | Type-9 x Greater Progress | JNKVV, Jabalpur (M.P.) |
| | Jawahar Matar-2 | Russian 2 x Greater Progress | JNKVV, Jabalpur (M.P.) |
| | Jawahar Matar-83 | (Arkel x JP-829) x (46C x JP-501) | JNKVV, Jabalpur (M.P.) |
| | Jawahar Matar-71 | (Arkel x JP-829) x (46C x JP-501) | JNKVV, Jabalpur (M.P.) |
| | P-88 | Pusa-2 x Morasis-55 | PAU, Ludhiana (Punjab) |
| | VL-3 | Old Sugar x Early Wrinkled Dwarf | VPKAS, Almora (Uttarakhand) |
| | VL Ageti Matar-7 | Pant Uphar (IP-3) x Arkel | ICAR-VPKAS, Almora (Uttarakhand) |
| | PM-2 | Early Badger x IP-3 | GBPUA&T, Pantnagar (Uttarakhand) |
| | Pant Sabji Matar-3 | Arkel x GC-141 | GBPUA&T, Pantnagar (Uttarakhand) |
| | Delwiche Commando | Admiral x Pride | Univ. of Wisconsin |
| **French bean** | Pant Bean-2 (UPF626) | Turkish Brown x Contender | GBPUA&T, Pantnagar (Uttarakhand) |
| | SVM-1 | Contender (*P. vulgaris*) x P.B. L.257 (*P. multiflorus*) | ICAR-VPKAS, Almora (Uttarakhand) |
| | Laxmi | Contender x Local Variety (pole type) | Dr. Y.S. Parmar UH&F, Solan (H.P.) |

| Crop | Variety | Pedigree | Organization |
|------|---------|----------|--------------|
| **Cowpea** | Pusa Do Fasali | Pusa Phalguni x Philippines Introduction | ICAR-IARI, New Delhi |
| | Narendra Lobia-1 | Pusa Komal x Varanasi Local | NDUA&T, Faizabad (Uttar Pradesh) |
| **Dolichos bean** | CO-2 | CO-8 x CO-1 | TNAU, Coimbatore (Tamil Nadu) |
| | Hebbal Aware-3 | Hebbal Aware-1 x US-67-31 | UAS, Hebbal (Karnataka) |
| | Wal Konkan-1 | Wal-2 x Wal 125-36 | Maharashtra |
| | Konkan bhusan | Hebbal Aware-3 x Wal-1 | Maharashtra |
| **Cluster bean** | Pusa Navbahar | Pusa Mausami x Pusa Sadabahar | ICAR-IARI, New Delhi |
| | P-28-1-1 | Pusa Navbahar x IC11521 | ICAR-NBPGR, New Delhi |

## (iv) Genetic Resources Used as Bridge Species

During crossing, in some species desirable traits cannot be transferred easily due to certain barriers. In this situation bridge species are used to transfer the traits. Local land races or wild species may facilitate gene transfer across the species or sub-species by acting as bridge species. In cucurbits *Cucurbita lundelliana* Baily was used as a bridge to transfer certain desirable characters of one cultivated species to another, including tolerance to powdery mildew found in this species long back, and to form an interbreeding population or gene pool. *Cucurbita moschata* was also used as a bridge to transfer diseases resistance (powdery mildew and cucumber mosaic virus), good fruit quality and insect resistance from *C. martinezii* to *C. pepo*. In *Cucumis*, African species carry many desirable characteristics, as disease resistance, not found in Asiatic species. However, strong barriers were found in crosses between annual and perennial *Cucumis* species and no viable seeds were obtained from any cross between African and Asiatic groups. Based on pollen tube behavior, *C. africanus* L. and *C. melo* appear to be the most promising male parents for crossing with *C. sativus*, but special pollination techniques and advanced embryo culture methods are necessary to overcome interspecific barriers in several crosses. The first successful and repeatable cross between African and Asiatic group was made by crossing *C. hystrix* with *C. sativus*. The parental species have different chromosome numbers resulting in a $F_1$ hybrid 2n = 19 (n= 7 from *C. sativus* and n = 12 from *C. hystrix*). The chromosome number of the hybrid was doubled (2n = 4x = 38) and may be useful as a new crop as well as a bridging species for transferring genes between African and Asiatic groups

## (v) Species Used in Wide/Distant Hybridization/ Interspecific Crossing

The genetic base of the cultivated germplasm of most of vegetable crop plants is very narrow. There is, however, still high level of useful variability for various traits in the wild germplasm of primary, secondary and tertiary gene pools. Due to strong genetic barriers to crossability and gene exchange between cultivated species and their wild relatives/weedy taxa remained the main barrier in many cases. Despite the fact, several genes of economic interest are available in wild species their transfers are hindered because of inadequate understanding of genetic mechanism favoring modifications of gene complexes in distant hybrids. Therefore multidisciplinary team efforts are required to exploit such variability. Incompatibility of parental species is the major problem in distant hybrids which can create problem of hybrid sterility, limited genetic recombination and hybrid breakdown. Rescue of hybrid embryos, development of amphidiploids, alien addition, substitution and translocation lines, induced homologous chromosome pairing, chromosome banding, FISH and GISH and use of anchored molecular markers are required to transfer desired traits without any associated undesirable variability for successful exploitation of such variability from wild germplasm. A number of genes for resistance against disease and insects have been transferred and exploited in potato and tomato.

## (a) Interspecific Hybridization

It is possible to transfer the resistant genes from wild and semi-cultivated species into cultivated types but the unequal ploidy levels, incompatibility, sterility and genetic load makes the task difficult and time-consuming. These factors have restricted the exploitation of many valuable species and have also limited the choice to a few species. Interspecific hybridization has been successfully employed at several instances and useful genes stands transferred for example, crossing of *Abelmoschus esculentus* x *A. manihot* for transfer of yellow vein mosaic resistance, *Solanum esculentum* x *S. peruvianum* for transfer of resistance to late blight, *Capsicum annuum* x *C. frutescense* for resistance to leaf curl virus and *S. melongena* x *S. incanum* for transfer of resistance to little leaf MLO. In potato 13 species have been used so far in the varietal improvement programmes in the world. The breeders all over the world have mainly exploited the variability of *S. tuberosum* subsp. *tuberosum*. In addition, wild species such as *S. demissum*, *S. acaule*, *S. chacoense*, *S. spegazzinii*, *S. stoloniferum* and *S. vernei* have also been utilized for the development of cultivars. Genes of *S. demissum* have been incorporated into more than 50 per cent of the world's cultivars for resistance to late blight and PLRV.

## (b) Intergeneric Hybridization

The objective of intergeneric hybridization is to incorporate desirable genes (s) into cultivated species from a distant genus. Such crosses are complicated to attempt and handle. Crossability is low and hybrid sterility further puts barriers in advancing to homozygous state. Now-a-days, innovative approaches like embryo rescue and protoplast fusion technology are being utilized to increase success rate in intergeneric hybridization. Even then there are many species utilized to increase success rate in intergeneric hybridization. Even then there are many instances where intergeneric hybridization has been used either to synthesize new species or to transfer rare genes to cultivated genus. Important examples are synthesis of *Raphanobrassica*. It is all tetraploid (2n = 36) with genomic constitution RRCC and obtained from cross of *Raphanus sativus* x *Brassica oleracea*. Successful intergeneric cross has been attempted between *Solanum tuberosum* (potato) x *Solanum esculentum* (tomato).

### Table 9.10: Species Used in Interspecific Hybridization

| Vegetables | Species Used | Interspecific Cross | Result |
|---|---|---|---|
| Okra | *Abelmoschus* species | *A. esculentus* (2n=130) x *A. tuberculatus* (2n=58) | Sucessful |
| | | *A. esculentus* (2n=72) x *A. manihot* (2n=60) | Sucessful |
| | | *A. esculentus* (2n=130) x *A. manihot* (2n=66) | Sucessful |
| | | *A. esculentus* (2n=130) x *A. tetraphyllus* (2n=138) | Sucessful |
| | | *A. esculentus* (2n=72) x *A. tetraphyllus* (2n=130) | Sucessful |
| | | *A. esculentus* (2n=130) x *A. manihot* var. *ghana* (2n=194) | Sucessful |
| | | *A. esculentus* (2n=130) x *A. ficulneus* (2n=72) | Fail |
| | | *A. esculentus* (2n=130) x *A. moschatus* (2n=72) | Fail |
| Tomato | *Solanum* species | *S. esculentum* x *S. pimpinellifolium* | Sucessful |
| | | *S. esculentum* x *S. mminutum* | Sucessful |
| | | *S. esculentum* x *S. hirsutum* | Sucessful |
| | | *S. esculentum* x *S. peruvianum* | Sucessful |
| | | *S. pimpinellifolium* x *S. minutum* | Sucessful |
| | | *S. pimpinellifolium* x *S. hirsutum* | Sucessful |
| | | *S. minutum* x *S. esculentum* | Sucessful |
| | | *S. peruvianum* x *S. esculentum* | Sucessful |
| Potato | *Solanum* species | *S. tuberosum* x *S. andigenum* | Sucessful |
| | | *S. tuberosum* x *S. stentomum* | Sucessful |
| | | *S. tuberosum* x *S. sparsipilum* | Sucessful |
| | | *S. tuberosum* x *S. demissum* (*via* bridge species) | Sucessful |
| | | *S. curtilobum* x *S. tuberosum* | Sucessful |
| | | *S. acaule* x *S. simplicifolium* | Sucessful |
| | | *S. demissum* x *S. rybinii* | Sucessful |
| | | *S. phureja* x *S. bulbocastanum* | Sucessful |
| Brinajl | *Solanum* species | *S. incanum* x *S. melongena* | Sucessful |
| | | *S. melongena* x *S. aethiopicum* | Sucessful |
| | | *S. melongena* x *S. cumingi* | Sucessful |
| | | *S. melongena* x *S. insanum* | Sucessful |
| | | *S. melongena* x *S. sisymbrifolium* | Fail |
| | | *S. melongena* x *S. incanum* | Sucessful |
| | | *S. melongena* x *S. indicum* | Sucessful |
| | | *S. melongena* x *S. xanthocarpum* | Sucessful |
| | | *S. melongena* x *S. trilobatum* | Sucessful |
| | | *S. melongena* x *S. torvum* | Sucessful |
| | | *S. gelo* x *S. indicum* | Sucessful |
| | | *S. melongena* x *S. grandiflorum* | Sucessful |
| | | *S. indicum* x *S. melongena* | Sucessful |
| | | *S. indicum* x *S. integrifoilum* | Sucessful |
| | | *S. xanthocarpum* x *S. melongena* | Sucessful |

| Vegetables | Species Used | Interspecific Cross | Result |
|---|---|---|---|
| | | *S. integrifolium x S.incanum* | Sucessful |
| | | *S. integrifolium x S. melongena* | Sucessful |
| | | *S. incanum x S. indicum* | Sucessful |
| Chilli | *Capsicum* species | *C. annuum x C. sinense* | Sucessful |
| | | *C. annuum x C. frutescens* | Sucessful |
| | | *C. frutescens x C. sinense* | Sucessful |
| | | *C. sinense x C.pendulum* | Sucessful |
| | | *C.pubescens x C. sinense* | Sucessful |
| | | *C. frutescens x C. annuum* | 2 per cent sucess |
| Cucumber | *Cucumis* species | *C. sativus x C. hardwickii* | Sucessful |
| | | *C. sativus x C. trigonus (C. callosus)* | Sucessful |
| Melon | *Cucumis* species | *C. metuliferus x C. melo* | Sucessful |
| | | *C. Africans x C. dipsaceus* | Sucessful |
| | | *C. Africans x C. anguria* | Sucessful |
| | | *C. anguria x C. prophetarum* | Sucessful |
| | | *C. anguria x C. zeyheri* | Sucessful |
| | | *C. dipsaceus x C. figarei* | Sucessful |
| Pumpkin | *Cucurbita* species | *C. maxima x C. moschata* | Sucessful |
| | | *C. moschata x C. pepo* (through tissue culture) | Sucessful |
| | | *C. lundelian x C. moschata* | Sucessful |
| | | *C. lundelian x C. pepo* | Sucessful |
| | | *C. lundelian x C. mixta* | Sucessful |
| | | *C. lundelian x C. maxima* | Sucessful |
| | | *C. lundelian x C. ficifolia* | Sucessful |
| Onion | *Allium* species | *A. Cepa* var. *cepa x A. fistulosum* | Sucessful |
| | | *A. Cepa* var. *cepa x A. porrum* | Sucessful |
| | | *A. porrum x A. Cepa* var. *cepa* | Sucessful |
| | | *A. porrum x A.ampeloprasum* | Sucessful |
| | | *A. Cepa* var. *cepa x. A. Cepa* var. *aggregatum* | Sucessful |
| | | *A. Cepa x A. altaicum* | Sucessful |
| | | *A. Cepa x A. vavilovii* | Sucessful |
| | | *A. Cepa x A. pskemense* | Sucessful |
| | | *A. Cepa x A. galanthem* | Sucessful |
| | | *A. Cepa x A. drobovii* | Sucessful |
| French bean | *Phaseolus* species | *P. vulgaris x P. coccineus* | Sucessful |
| | | *P. vulgaris x P. dumosus* | Sucessful |
| | | *P. vulgaris x P. flavescens* | Sucessful |
| | | *P. vulgaris x P. acutifolius* | Sucessful |
| | | *P. vulgaris x P. ritensis* | Sucessful |
| | | *P. vulgaris x P. lunatus* | Sucessful |
| | | *P. vulgaris* var. subsp. *compressus x P. multiflorus* Willd. var. *albus* | Sucessful |
| Winged bean | *Psophocarpus* species | *P. tetragonolobus x P. palustris* | Sucessful |
| | | *P. tetragonolobus x P. longepeddunulatus* | Sucessful |
| Bitter gourd | *Momordica* species | *M. dioica x M. charantia* | Fail |
| | | *M. dioica x M. cochinchinensis* | Sucessful |
| | | *M. cochinchinensis x M. charantia* | Sucessful |
| | | *M. cochinchinensis x M. dioica* | Sucessful |
| | | *M. charantia x M. dioica* | Sucessful |
| Luffa | *Luffa* species | *L. acutangula x L. cylindrica* | Sucessful |
| | | *L. acutangula x L. graveolens* | Sucessful |
| | | *L. echinata x L. graveolens* | Sucessful |

| Vegetables | Species Used | Interspecific Cross | Result |
|---|---|---|---|
| Watermelon | *Citrullus* species | *C. lanatus x C. colocynthes* | Sucessful |
| Radish | *Raphanus* species | *R. indicus x R. sativus* | Sucessful |
| | | *R. indicus x R. sensustricto* | Sucessful |
| | | *R. indicus x R. raphanistroides* | Sucessful |
| Cowpea | *Vigna* species | *V. unguiculata* (L.) Walp, subsp. *unguiculata* (syn. *V. unguiculata* var. *radiata*) x V. *unguiculata* subsp. *cylindrica* | Sucessful |
| | | *V. unguiculata* (L.) Walp, subsp. *unguiculata* (syn. *V. unguiculata* var. *radiata*) x V. *unguiculata* subsp. sequipedalis | Sucessful |
| | | *V. unguiculata* (L.) Walp, subsp. *unguiculata* (syn. *V. unguiculata* var. *radiata*) x V. *unguiculata* subsp. *dekindtiana* | Sucessful |
| | | *V. unguiculata* (L.) Walp, subsp. *unguiculata* (syn. *V. unguiculata* var. *radiata*) x V. *unguiculata* subsp. *mensensis* | Sucessful |
| Carrot | *Daucus* species | *D. carota* ssp. *sativum* x *D. carota* ssp. *sativum* | Self-incompatible |
| | | *D. carota* ssp. *sativum* x *D. gingidium* | Cross compatible |
| | | *D. carota* ssp. *sativum* x *D. capillifolium* | Cross compatible |
| | | *D. gingidium* x *D. carota* ssp. *sativum* | Cross compatible |
| | | *D. gingidium* x *D. gingidium* | Self-incompatible |
| | | *D. gingidium* x *D. capillifolium* | Not known |
| | | *D. capillifolium* x *D. carota* ssp. *sativum* | Cross compatible |
| | | *D. capillifolium* x *D. gingidium* | Not known |
| | | *D. capillifolium* x *D. capillifolium* | Self-incompatible |

## Unilateral Incompatibility in Vegetable Crops

In distant crosses some species showed unilateral incompatibility, which are given in Table 9.11.

**Table 9.11: Distant Crosses in some Vegetabe Crops**

| Sl.No. | Cross Combinations |
|---|---|
| 1 | *Solanum indicum* 2n = 24 x *S. melongena* 2n = 24 |
| 2 | *Solanum demissum* 2n = 72 x *S. tuberosum* 2n = 48 |
| 3 | *Solanum verrucosum* 2n = 24 x *S. bulbocastanum* 2n = 48 |
| 4 | *Solanum esculentum* 2n = 24 x *S. peruvianum* 2n = 24 |
| 5 | *Solanum esculentum* 2n = 24 x *S. pennellii* 2n = 24 |
| 6 | *Abelmoschus tuberculatus* 2n = 58 x *A. manihot* 2n = 68 |
| 7 | *Abelmoschus esculentus* 2n = 130 x *A. manihot* var. *ghana* 2n = 194 |
| 8 | *Cucumis metuliferus* x *C. melo* 2n = 24 |
| 9 | *Capscicum frutescens* 2n = 24 x *C. annuum* 2n = 24 |
| 10 | *Raphanus sativus* 2n = 18 x *B. oleracea* var. *botrytis* 2n = 18 |
| 11 | *Phaseolus coccineus* 2n = 14 x *P. vulgaris* 2n = 14 |
| 12 | *Pisum sativum* 2n = 14 x *P. fulvum* 2n = 14 |

## (vi) Utilization of Crop Wild Relatives (CWR)

The Crops Wild Relatives (CWR) includes wild forms, progenitors and those species that are closely related to cultivated crops. The Crops Wild the concept of CWR is relative in the sense that all the related species may not have equal potential as gene donors to crops (Maxted *et al.*, 2006); their relative status is very often inferred rather than based on direct evidence (Heywood *et al.*, 2007). The genepool system devised for understanding genetic relationship between crops and related species by Harlan and de Wet (1971) faciliotated beneficial gene and traits transfer from wild to cultivated. Historically, the conservation of plant genetic resources (PGR) has focused almost explicity on cultivated plants (Maxted *et al.*, 2008), however, in the recent past the importance of CWR have been recognized globally for breeding high yielding varieties to feed increasing population and to tolerate variety of stress arising due to climate change. Thus, future crops species will need to be able to thrive in a drier, warmer, and more variable climate and in an environment increasingly populated by new pathogen organisms. To meet these challenges we need broader crop gene pools and CWR has greater role to play in it (Sandhu and Rana, 2016).

## Crop Wild Relatives (CWR) in India

Indian gene centre is rich in domesticated crops diversity having 168 species out of 2489 species distributed in 12 regions of diversity of cultivated plants (Zeven and De Wet, 1982). Among crop wild relatives, 326 documented species have originated and/or developed diversity in different phyto-geographical regions of India (Pandey and Arora, 2004). However, Pradheep *et al.* (2014) have made a pragmatic exercise to further shortlist and update the CWR of 168 native crops, which resulted into 817 taxa belonging to 730 species, including wild/weedy form(s) or populations of 142 crop species, occurring in India (Table 9.12).

## Importance of CWR

The increasing genetic uniformity of crop varieties combined with climate change effects makes crops

## Table 9.12: Summary of Crop-group-wise Native CWR Occurring in India

| Crop-group (Crops**) | No. of CWR Species* | Taxa | Crop-group (Crops**) | No. of CWR Species* | Taxa |
|---|---|---|---|---|---|
| Cereals and millets (13) | 72 (2) | 83 | Vegetables (25) | 76(11) | 87 |
| Pseudocereals (3) | 13(1) | 13 | Spices and condiments (12) | 50(7) | 54 |
| Grain legumes (10) | 49(4) | 57 | Ornamentals (13) | 141(61) | 152 |
| Oilseeds (4) | 9(1) | 10 | Medicinal and aromatic plants (20) | 70(19) | 81 |
| Fibers (5) | 18(3) | 20 | Plantation crops (3) | 12 | 14 |
| Forages (16) | 58(14) | 63 | Others (8) | 35(7) | 39 |
| Fruits and nuts (36) | 127(12) | 144 | Crops: (168) | 730(142) | 817 |

\* Figures in parenthesis is crop species with wild/weedy form(s) or population occurring in India, which are also included for counting as CWR

\*\* One crop may involve more than one species.

*Source*: Sandhu and Rana, 2016.

more vulnerable to various biotic and abiotic stresses. There has been examples of large scale devastations of crops due to genetic uniformity, for instance potato famine of the 1840s due to late blight epidemic across Ireland, Europe and North America; 1970s southern corn blight outbreak in the US maize and rice losses due to blast in the Philippines, Indonesia and India leading to the great Bengal famine. Intensive modern breeding efforts have contributed to a narrowing of the gene pool by concentrating more on favorable alleles already present in early domesticates (Debouck, 1991). Crops wild relatives are therefore important for maintaining genetic diversity for and preventing such losses, which may have serious consequences for food, nutrition and environmental security. Advancement in the molecular biology in the study and utilization of species as gene source has made CWR a priority in PGR management and crop improvement. At the same time, post-CBD concerns of habitat loss, genetic erosion, policy issues related to access and benefit sharing and sovereign rights have also come in the forefront (Sandhu and Rana, 2016).

### Utilization in Breeding Programmes

Despite having valuable genes with immense value for crops improvement and adaption to changing environmental conditions, utilization of CWR has enjoyed a great success only in few crops, while disaapointing for numerous others. Many genes are still lies untapped in these genetic resources, presumably due to the lack of useful genetic information and genetic bottlenecks as well. Hodgkin and Hajjar (2007) while reviewing the utilization of CWR in 20 years that over 60 wild species were used in 13 major crops and >100 beneficial traits derived from them.

Although, utilization of CWR in crop improvement is limited still many wild species have successfully utilized. Some of example include; use in potato, *Solanum demissum* provided resistance to late blight (National Potato Council, 2003) while in tomato >40 resistance genes have been derived from *S. peruvianum*, *S. pennellii* Corell var. *pennellii*, *Lycoperscon cheemanii*, *L. pimpenellifolium* for traits such as increased soluble solid content, fruit colour, and adaptation to harvesting (Rick and Chetelat, 1995). The species of *Solanum* complex such as *S. incanum*, *S. viarum*, *S. melongena* var. *insanum*, *S. khasianum* have provided gene for resistance to *Fusarium* wilt, bacterial wilt, frost tolerance and fruit and shoot borer. Broccoli varieties producing high levels of anti-cancer compounds have been developed using genes obtained from wild Italian *Brassica oleracia*. The wild species of okra *Abelmoschus tuberculatus* to YVMV and wild cucumbers *Cucumis hardwickii* and *C. callosus* have resistance to downy mildew and fruit fly, *Cucumis melo* var. *chito* for *Fusarium* wilt resistance.

Among pulses, *Vigna trilobata*, *V. mungo* var. *sylvestris*, *V. radiata* var. *sublobata* have provided resistance to yellow mosaic virus, *V. vexillata* has high protein and resistance to cowpea pod sucking bug and bruchid and is crossable with *V. unguiculata* and *V. radiata*. *Phaseolus coccineus* is a source of resistance to anthracnose as well as root rots, white mold, and BYMV in common bean (Sharma and Rana, 2012).

### (vii) Biotechnological/Molecular Approaches for Germplasm Utilization

Recent advances of cellular and molecular biology including micropropagation of plants through tissue culture, cryopreservation, *in vitro* production of secondary metabolites, anchored molecular markers, DNA fingerprinting, recombinant DNA technology, comparative mapping and genomics and gene discovery, and bioinformatics can be extremely useful for conservation and sustainable use of plant biodiversity. Biotechnology offers the germplasm curators more sensitive, robust and reliable methods for characterization of germplasm which can potentially address the limitations associated with morphological markers. In recent years, several modifications of the basic DNA fingerprinting techniques have been developed with common goal of detecting DNA polymorphism. These techniques have application in identification, genetic resources and germplasm using these techniques got a boost with the establishment of National Research Centre on DNA Fingerprinting at

ICAR-NBPGR. Beside characterization, it is important to develop programmes on molecular-aided selection. The biotechnology assisted plant breeding empowers plant breeders to modify the uncultivated crops to make them cultivable or tailor existing crops to produce new products.

The following approaches of biotechnology can be used in combination with the conventional plant breeding methods for directed, rapid and precise improvement of plants.

### (a) Somaclonal Variation and *in vitro* Selection

This approach is especially beneficial for the improvement of vegetatively propagated plants, some of which are highly heterozygous, complex polyploids and do not flower and set seeds regularly. A number of genetic and heritable changes could take place during dedifferentiated callus phase from the explants of a given variety due to chromosomal aberrations, point mutation, mobilization of transposable elements, *etc.* Pre-existing or induced somaclonal variation for a specific trait can be further enriched and screened *in vitro* in the presence of appropriate selection agent/conditions facilitating testing of millions of potential plants in a single petri plate. The regenerated plants from resistant *Abelmoschus caillci* can be further tested for the trait and its transmission. Epigenetic or heritable somaclonal variant cell lines in suspension cultures of some of the endangered species for high level of production of the secondary metabolites under *in vitro* conditions have been isolated and used.

### (b) Protoplast Culture and Somatic Hybridization

Plant cells without cell walls called protoplast can be isolated by using cell wall degrading enzymes such as cellulose, macerozyme and pectinase to give fertile plants. Any two related species, which could not be hybridized using reproductive organs, can now be hybridized using protoplast fusion combining nuclei and cytoplasmic organelles of both the species and the resultant plant is called somatic hybrid. Nucleus of one species can be selectively fused with cytoplasm of another as a single event following acetamide and irradiation treatments, respectively to develop nuclear-cytoplasm alloplasmic lines capable of giving cytoplasmic male sterility for hybrid seed production. Somatic hybrids have been produced in *Brassica*, *Solanum*, *Lycopersicon* genera for transfer of useful variability from related species which could not have been possible through conventional hybrids. Protoplast can also take up isolated DNA for genetic transformation using PEG or electroporation.

### (c) Plant Genome Mapping, Gene Cloning and Sequencing

All the traits of an organism governing its fate from zygote to death are coded by DNA fragments called genes comprising various premutations and combinations of four nucleotides, which are linearly arranged in one to several linkage groups equivalent to the basic or gametic chromosome number of a species. The importance of assigning various traits of an organism to different linkage groups for their efficient manipulation and improvement has been realized ever since the discovery of Mendel's work in the beginning of the 20th century on pea. Linkage maps based on a few morphological traits and biochemical markers were constructed in tomato and other cultivated plants. The use of the classical genetic linkage maps in plant improvement remained inconsequential due to limited markers and failure to tag the genes of economic importance effectively. With the availability of series of molecular markers with high level of polymorphism and maneuverability, high density molecular linkage maps comprising thousands of markers have been constructed and utilized for genome-wide analyses.

Molecular markers tightly linked to different genes of economic importance including quality traits, male sterility and restoration and resistance against biotic and abiotic stresses have been identified in several crop plants which are being used for backcross breeding and gene pyramiding using marker assisted selection. Some of these genes have been tagged after their transfer from wild donor species into cultivars. Gene-for-gene relationship between the gene for resistance in the host plants and corresponding gene for avirulence (*avr*) in the pathogen have been demonstrated for a number of host pathogen combinations in plants. Similarly, a number of genes for resistance against diseases and insects have been cloned using appropriate populations and map based cloning strategy involving molecular maps, linked markers, YAC, BAC and cosmid clones, cDNA libraries and genetic transformation. Martin *et al.* (1993) cloned the *Pto* gene in tomato which confers resistance to races of *P. syringae pv. tomato* carrying the avirulence gene *avrPto* through positional cloning and found that it had amino acid sequence similar to serine threonine protein kinase. RNAi technology can be employed to develop novel Nutraceuticals and Functional Foods. High lysine mustard has been produced by RNAi mediated downregulation of lysine catabolizing genes. In another study, fruit specific RNAi mediated suppression of a photomorphogenesis regulatory gene (DET1) enhanced carotenoid and flavonoid content in tomatoes. Scientists have developed "tear-less onion" by shutting down its lachrymatory factor synthase gene and re-directed the valuable sulpher towards making more nutritive and flavoring compounds.

### (d) Resolution of Quantitative Trait Loci (QTLs) into Mendelian Factors

Most of the economic traits such as yield, quality and resistance to abiotic and biotic stresses are controlled by a number of genes polygene with small effects which are highly influenced by genotype x environment interaction. Molecular maps are being used to resolve

QTLs to discrete Mendelian factors and map and tag to precise location in the map. In one classical attempt Paterson *et al.* (1988) used RFLP map of tomato to resolve six QTLs for fruit mass, four for concentration of soluble solids and five for fruit pH in a population of *Lycopersicon esculentum* and small and green fruited wild relative *L. chmielewskii*. Wild relatives of crop plants are rich source of novel genes and alleles for different QTLs. These QTLs from related wild species are being introgressed into elite cultivars and monitored simultaneously by anchored molecular markers through advanced back cross- QTL (AB-QTL) approach, with the resolution and dissecting of QTLs and their introgression and pyramiding it should be possible to achieve desired tailor crop improvement. In future, the resistance potential of wild species can be exploited through biotechnological approaches.

## Screening of Genetic Resources for Desired Traits

### Resistance to Biotic and Abiotic

Important sources of resistance to major diseases and pests, and to abiotic stresses are essential to vegetable breeders in development of desired cultivars. There is two types of resistance *i.e.* Horizontal Resistance and Vertical Resiatance. The use of either vertical or horizontal resistance has its own advantages and disadvantages.

(a) Horizontal Resistance

(b) Vertical Resiatance

### (a) Horizontal Resistance

The manifestation of horizontal resistance is that, it is very simple and direct. The resistant plants grown under field conditions, show reduced number of infection points or fewer lesions as compared to susceptible ones. The pedigree method is mostly followed to breed horizontal resistance, since this type of resistance is generally inherited polygenically.

### (b) Vertical Resiatance

In general the emphasis should be given on horizontal type of resistance, since it is more stable. The stability of vertical resistance depends upon the management of crops. For example, when a *Meloidogyne javanica* resistant variety of tomato was grown continuously and exclusively in a plot for 5 years, a virulent race of nematode came out and increased. But, when the same resistant variety was grown repeatedly with a susceptibilitle variety in commercial field conditions, the resistance was maintained (Kehr, 1966). Hence, the crop rotation is very important. A comparision of the various features of vertical and horizontal disease resistance is given in Table 9.13.

## Screening Techniques

Use of right screening techniques to screen resistant/tolerant genotypes is very important. A good screening technique is that which clearly discriminates between resistant and susceptible genotypes. Some time screening under natural conditions may not be very rliable, as disease escape and even absence of pathogen are very common. Hence, artificial screening techniques

**Table 9.13: A Comparision of the Various Features of Vertical and Horizontal Disease Resistance**

| Feature | Vertical Resistance | Horizontal Resistance |
|---|---|---|
| Pathotype-specificity | Specific | Nonspecific |
| Nature of gene action | Oligogenic | Polygenic; rarely oligogenic |
| Response to pathogen | Usually, hypersensitive | Resistant response |
| Phenotypic expression | Qualitative | Quantitative |
| Stage of expression | Seedling to maturity | Expression increases as plant matures |
| Selection and evaluation | Relatively easy | Relatively difficult |
| Risk of 'boom and bust' | Present (rarely durable) | Absent (durable) |
| Suitale for: | | |
| 1. Host | (1) Annuals, but not perennials | Both annals and perennials |
| 2. Pathogen | (2) Immobile pathogens, *e.g.*, soil pathogens, but not for mobile air-borne, pathogens | All pathogens |
| Need for specific deployment of resistant varieties | Critical for success with mobile pathogenes | None |
| Need for other control measures | Likely | Much less likely |
| Host-pathogen interaction* | Present | Absent |
| Commonly used, but not strict synonyms | Major gene, race-specific, seedling, monogenic, differential, specific, pathotype-specific resistance | Polygenic, race-nonspecific, pathotype-nonspecific, mature plant, adult plant, field, uniform resistance |
| Efficacy | Highly efficient against specific races | Variable, but operate against all races |

* Detectable by analysis of variance of a suitable experiment.

*Source*: Singh, 2001.

must be adopted. The inoculation of pathogen races may be made by different methods *viz.,*

1. Spraying
2. Dipping
3. Injecting and
4. Rubbing in soil

The seeds, young seedling, adult plants, plant parts, soils, *etc.* may be used for inoculation. The screening can be done under field or laboratory or glass house conditions.

## Techniques of Inoculation

The inoculation techniques, which can be adopted for various types of diseases are as follows:

1. Soil borne diseases
2. Air borne diseases
3. Seed borne diseases
4. Insect transmitted diseases

### 1. Soil Borne Diseases

The soil borne diseases like damping off, seed and root rot, wilts, root knot nematodes are causing heavy losses in vegetable crops. The fungi responsible for these diseases remain in the soil. For screening against these diseases, sick plots are created. In sick plots the inoculation of pathogen is built up for infection of the susceptible plants of the host. The sick plot can be developed by adding the contaminated soil, diseased plants/seeds or inoculum produced in the laboratory. The varieties may be screened under field or laboratory conditions.

### 2. Air Borne Diseases

Most of diseases of vegetable crops belong to this group. The serious air borne diseases are early and late blight, downy mildew and powdery mildew, leaf spots and rusts. The simplest way to produce inoculum is to plant rows of susceptible variety in between the rows of those varieties which are to be screened. In another technique, the inoculum of pathogen is applied on the plant surface by dusting or spraying or injecting.

### 3. Seed Borne Diseases

In Seed borne diseases, spores of pathogen are dusted into seeds of varieties before sowing. In another technique, seeds may be soaked in pathogen suspension before sowing in the field.

### 4. Insect Transmitted Diseases

All the viral diseases are spred through vectors like aphids, whitefly and hoppers. The vectors can be used to transmit viral diseases. In another technique, the juice of diseased plant may be applied to healthy plant through rubbing or injecting.

The main objective of any resistance screening programme, is to distinguish between susceptible and resistat plants. When the resistance is controlled by mono or oligo genes, the susceptible host would develop severe symptoms ad resistant plants show no symptoms. The distribution of the resistance response in segregating population and in this case would be discontinuous and easy to identify. When the resistance is controlled by poly and minor genes, the distribution of resistance response in segregating population would be continuous. Hence, in this case the estimation of disease intensity through a reliable technique is very important.

## Assessment of Resistance

The assessment of disease resistance is done in there ways:

1. **Disease incidence:** It is defined as the proportion of plant units infected, that is, percentage of diseased plants.
2. **Disease severity:** It is defined as the proportion of the total area of plant tissue affected by disease.
3. **Measurement** of number and size of successful infections.

## Use of Score in Screening

### (A) Field Screening and Selection

The screening techniques for different diseases of vegetables have been discussed by various scientists.

### Tomato

Fageria (1994) has done scoring against fruit rot in tomato successfully under field conditions. The system used for scoring is illustrated in following table. Earlier this method was also used by Dadan (1991).

**Table 9.14: Disease Incidence (per cent) and Reaction**

| Grade | | Reaction |
|---|---|---|
| 1 | Nil | Resistant (R) |
| 2 | 0.1-10.0 | Moderately Resistant (MR) |
| 3 | 10.1-25.0 | Moderately Suceptible (MS) |
| 4 | 25.1-45.0 | Suceptible (S) |
| 5 | 45.1 and above | Highly Suceptible (HS) |

The incidences of fruit rot (*Phytophthora nicotianae* var. *parasitica*) in tomato can be calculated on per plant basis by using the following formula.

Disease incidence (per cent)

$$= \frac{\text{Diseased fruits (No.)}}{\text{Total fruits (No.)}} \times 100$$

While working on early blight in tomato, Fageria (1994) screened a large number of tomato genotypes against early blight by using scoring scale of Mckinney

(1923) with some modifications. The genotypes were screened against *Alternaria alternata* and *Alternaria solani*. The scoring scale has been illustrated as below:

**Table 9.15: The Score against Genus *Alternaria***

| Grade | Plant Area Infected (per cent) | Reaction |
|-------|-------------------------------|----------|
| 0 | Nil | Immune (I) |
| 1 | 0.1-10.0 | Resistant (R) |
| 2 | 10.1-25.0 | Moderately Resistant (MR) |
| 3 | 25.1-40.0 | Moderately Suceptoble (MS) |
| 4 | 40.1-50.0 | Suceptoble (S) |
| 5 | >50.0 | Highly Suceptoble (HS) |

Kohli *et al.*, 1996 recorded buckeye rot incidence in tomato on the basis of percentage of fruits infected in each line and computed as per following formula:

Buckeye rot incidence (per cent)

$$= \frac{\text{No. of infected fruits/plant}}{\text{Total no. of fruits/plant}} \times 100$$

The buckeye rot incidence was recorded at weekly interval and classified into seven categories as under, which is a modification of the method earlier used by Fageria (1994).

**Table 9.16: Buckeye Rot Incidence and Scoring Reactions**

| Grade | Buckeye Rot Incidence | Disease Reaction |
|-------|----------------------|------------------|
| 1 | 0 | Immune (I) |
| 2 | 1-10 | Highly Resistant (HR) |
| 3 | 11-20 | Resistant (R) |
| 4 | 21-30 | Moderately Resistant (MR) |
| 5 | 31-50 | Moderately Suceptoble (MS) |
| 6 | 51-70 | Suceptoble (S) |
| 7 | 71-100 | Highly Suceptoble (HS) |

## Chilli

Muthulakshmi and Seetharraman (1994) recorded disease intensity against fruit rot (*Alternaria tenuis*) of chilli 11 days after inoculation by adopting the following ratings as suggested by Ravinder Reddy (1982).

**Table 9.17: Category Value and per cent Fruit Area Diseased**

| Category Value | Per cent Fruit Area Diseased |
|----------------|------------------------------|
| 0 | Healthy |
| 1 | 1-5 |
| 2 | 6-10 |
| 3 | 11-25 |
| 4 | 26-50 |
| 5 | 51 and above |

The per cent disease index (PDI) was calculated using the following formula as suggested by Mckinney (1923).

Per cent disease index (PDI)

$$= \frac{\dfrac{\text{Total numerical ratings}}{\text{Total no. of fruits observed}} \times 100}{\text{Maximum category value}}$$

## Cole Crops

### Downy Mildew

Screening against downy mildew (*Pernospora parasitica*) of cole crops may be done at seedling as well as at adult statges (Kapoor, 1999). For screening, plants are sprayed with 1000-100000 spores per ml of distilled water obtained from the senescent material. Infector rows of susceptible arieties along the rows of best material should be maintained under the mist chamber (21-25°C ≤ 16 hr.light). Visual rating for disease reaction at 5-8 days aftyer inoculation on 0-4 scale is recommended. The per cent disease index (PDI) can be calculated by using the following formula.

Per cent disease index (PDI)

$$= \frac{\text{Sum of all the ratings}}{\text{Total no. of observation}} \times \frac{100}{\text{Maximum grade}}$$

Adult stage screening can be followed in a similar way. However, humidity has to be maintained by regular sparaying.

**Table 9.18: Description and Rating**

| Grade | Description | Rating |
|-------|-------------|--------|
| 0 | No apparent symptoms | Highly Resistant (HR) |
| I | 10 per cent leaf area infected | Resistant (R) |
| II | 11-25 per cent leaf area infected | Moderately Resistant (MR) |
| III | 26-50 per cent leaf area infected | Moderately Suceptoble (MS) |
| IV | >50 per cent leaf area infected | Highly Suceptoble (HS) |

### Alternaria

Screening at seedling and adult stage against against *Alternaria brassicola* in cole crops was made by Kapoor (1999). According to him, plants should be transferred to a mist chamber (25-28°C ≥ 16 hr. light) after spraying with 7-10 days old culture suspension (50,000 spores per ml). Characteristic symptoms start appearing after 5th days of inoculation. Plants are rated visually for disease reaction after eight day of inoculation on a 1-9 scale:

1= No infection

9-89= 100 per cent infection

Plants rated and classified as:

1-3= Resistant

4-6= Moderate

7-9 Susceptible

Entries are classified as:

<25 per cent plants are resistant

25-49.9 per cent moderate

50-74.9 per cent resistant

$\geq$ 75 per cent highly resistant

## Cabbage

Walkey and Neely (1970) have given the scoring system that has been used under field conditions to assess the reaction of cabbage to mosaic virus infection.

**Table 9.19: Symptome Grade and Description of Symptomes**

| Symptome Grade | Description of Symptomes |
|---|---|
| 0 | No necrosis |
| 1 | Necrotic lesions viable, only a few leaves affected with >10 per cent of an individual leaf surface affected |
| 2 | Necrotic lesions on several leaves with 10-25 per cent of the surface affected |
| 3 | Necrotic lesions on many leaves wlth 25-50 per cent of the laef surface affected |
| 4 | Necrotic lesions on most leaves with 50-75 per cent of the laef surface affected |
| 5 | Severe necrotic lesions on all leaves with >75 per cent of the laef surface affected, often accompanied by premature fall. |

## Cauliflower

### Sclerotinia Rot

Sharma *et al.* (1997) recorded disease resistance score against *Sclerotinia sclerotiorum* of cauliflower on individual plant basis after 7th day of inoculation on the following:

**Table 9.20: Disease Resistance Score against *Sclerotinia sclerotiorum***

| Numerical Scale Used | Per cent Leaf Area Affected | Disease Reaction |
|---|---|---|
| 0 | No disease | Immune (I) |
| 1 | Upto 25 per cent leaf area colonised | Highly Resistant (HR) |
| 2 | 26 to 50 per cent leaf area colonised | Resistant (R) |
| 3 | 51 to 75 per cent leaf area colonised | Suceptoble (S) |
| 4 | 76 per cent and more leaf area colonised | Highly Suceptoble (HS) |

## Muskmelon

Somkuwar and More (1996) screened muskmelon hybrids against downy mildew resistance by using 0-4 scale as suggested by Natti (1958). The scale used was as follows:

**Table 9.21: Rating of Downy Mildew Resistance**

| Numerical Scale Used | Per cent Leaf Area Affected | Disease Reaction |
|---|---|---|
| 0 | 0 | Immune (I) |
| 1 | 1-10 | Highly Resistant (HR) |
| 2 | 11-25 | Moderately Resistant (MR) |
| 3 | 26-50 | Moderately Suceptoble (MS) |
| 4 | 51 and above | Highly Suceptoble (HS) |

### Screening against Nematode

The root-knot nematode caused by *Meloidogyne* spp. is serious problem in vegetables and causes enormous losses. The species *M. arenaria*, *M. incognita* and *M. javanica* are very common and attack tomato, chillies, brinjal, potato, cucurbits, okra, carrot and radish. For screening of genotypes against root-knot nematode, a broad based inoculum should be used. It is better to screen the cultivars at seedling stage. After inoculation, the temperature, moisture and aeration should be maintained at optimum level for hatching and growth of larvae. The optimum temperature for growth of nematode is 13°C and the wet soils are found congenial for gall development. In case of pot study about 5000 *Meloidogyne* eggs should be added to near the root zone of each plant. The egg-mass index (EI) gives an appropriate estimation of nematode reproduction based on number of egg masses per plant. The measurement according to EI, value in the degree of resistance is given in following table:

**Table 9.22: Measurement According to EI, Value in the Degree of Resistanc**

| Egg-mass Index (EI) | Host Reaction |
|---|---|
| 0.0-1.0 | Highly Resistant (HR) |
| 1.1-3.0 | Very Resistant (VR) |
| 3.1-3.5 | Moderately Resistant (MR) |
| 3.6-4.0 | Slightly Resistant (SR) |
| 4.1-5.0 | Suceptoble (S) |

The gall index (GI) can also be used for measurement of root-knot nematode reproduction. The gall development can be seen easily in case of susceptible variety but not be in highly resistant or immune plants. The scale is given as below:

**Table 9.23: Gall Index and Host Reacion**

| Gall Index (GI) | Host Reaction |
|---|---|
| 1= No gall formation | Highly Resistant (HR) |
| 2= Trace of galling | Resistant (R) |
| 3= Light galling | Moderately Resistant (MR) |
| 4= Moderate galling | Suceptoble (S) |
| 5= Abundant galling | Highly Suceptoble (HS) |

## (B) *In vitro* Selection for Disease Resistant

### Cole Crops

Kapoor (1999) used this method of selection of disease resistance. Phytotoxins may be useful for indirect selection of disease resistance provided there is a correlation between the reactions to the pathogen stage and may save a significant amount of time and space. Destruction b is the main phytotoxin produced by fungal pathogen *Alternaria brassicae* which causes black spot of crucifers. Endo-PG, Protease and toxic metabolite have been positively correlated with virulent isolates of *Sclerotinia* spp.

### (i) Available Resistance Sources for Biotic Stresses in Vegetable Crops

### (a) Disease Resistance

The vegetables are susceptible to more than 500 diseases and infecting in different stateges of crop growth and development. Some serious diseases causing heavy losses in yield are as shown in Table 9.24.

### Table 9.24: Important Diseases of Vegetables

| Crop | Diseases | Causal Organism |
|---|---|---|
| Tomato | Bacterial wilt | *Pseudomonas solanacearum* |
| | Fruit rot | *Phytophthora nicotianae* var. *parasitica* |
| | Fusarium wilt | *Fusarium oxysporum* f. *lycopersici* |
| | Early blight | *Alternaria solani* |
| | Early blight | *Alternaria alternata* |
| | Virus | Leaf culrl virus [tobacco leaf culrl virus (Gemini viruses group)] |
| | | Mosaic (Tomato Mosaic Virus or TMV, Cucumber Mosaic Virus or CMV, Potato virus X, Potato virus Y) |
| | Damping off | Species of *Pythium, Phytophthora, Sclerotinia, Fusarium, Botrytis, Phoma, Glomerella,* etc. |
| | Late blight | *Phytophthora infestans* |
| Brinjal | Phomopsis blight | *Phomopsis vexans* |
| | Little leaf | Mycoplasma |
| | Bacterial wilt | *Pseudomonas solanacearum* var. *asiaticum* |
| Chilli | Leaf culrl | Tobacco leaf culrl virus |
| | Mosaic | About 14 viruses are responsible |
| | Fruit rot | *Phytophthora nicotianae* var. *parasitica* |
| | Anthracnose | *Colletotrichum capsici* |
| | Fruit ripe rot | *Colletotrichum capsici, Glomerella* spp. |
| Potato | Early blight | *Alternaria solani* |
| | Late blight | *Phytophthora infestans* |
| | Mosaic | Potato virus X |
| | Leaf roll | Potato leaf roll virus of Luteovirus group |
| Cruciferous | Black spot | *Alternaria* spp. |
| | White rust | *Albugo candida* |
| | Black leg of cagbbage | *Phoma lingum* |
| | Club rot | *Plasmodiophora brassicae* |
| | Cabbage Yellows | *Fusarium oxysporum* f. *conglutinans* |
| | Sclerotinia stalk rot | *Sclerotinia sclerotiorum* |
| | Black rot | *Xanthomonas campestris* pv. *campestris* |
| Cowpea | Bacterial blight | *Xanthomonas axonopodis* pv. *vignicola* (Xav.) |
| Okra | Virus | Yellow Vein Mosaic Virus (YVMV) |
| Onion | Purple blotch | *Alternaria porri* |
| | Downy mildew | *Peronospora destructor* |
| | Onion smudge | *Colletotrichum circinans* |
| Cucurbits | Fruits rot | *Pythium aphanidermatum, Pythium butleri* |
| | Downy mildew | *Peronospora cubensis* |
| | Powdery mildew | *Erysiphe cichoracearum* and *Sphaerothea fuliginea* |
| | Anthracnose | *Colletotrichum lagenarium* |
| | Leaf spot | *Cercospora* spp. |

| Crop | Diseases | Causal Organism |
|------|----------|-----------------|
| | Angular leaf spot | *Pseudomonas syringae* pv. *lachrymans* |
| | CMV | Cucumber Mosaic Virus-*Cucumoviruses* group |
| | WMV | Watermelon Mosaic Virus-*Polyviruses* group |
| Pea | Powdery mildew | *Erysiphe polygoni* DC |
| | Seed and root rot | *Pythium* spp. |
| Bean | Bean common mosaic | *Polyviruses* group |
| | Web blight | *Rhizoctonia solani* |
| | Anthracnose | *Colletotrichum lindemuthianum* and *Glomerella lindemuthianum* |
| Beet | Seedling disease | *Pythium, Rhizoctonia, Phoma* and *Sclerotium* |
| | Cercospora leaf spot | *Cercospora beticola, Cercospora bertrandii* |
| | Sclerotium root rot | *Sclerotium rolfsii* |
| | Beet Yellows | Beta Virus 4 or *Corium betae* |
| Spinach | Downy mildew | *Peronospora spinaciae* |
| | Cercospora leaf spot | *Cercospora beticola, Cercospora bertrandii* |
| | Sclerotium root rot | *Sclerotium rolfsii* |
| | Beet Yellows | Beta Virus 4 or *Corium betae* |
| Coriander | Stem Galls | *Protomyces macroporus* |
| Carrot | Cercospora leaf blight | *Cercospora carotae* |
| | Alternaria blight | *Alternaria dauci* |
| | Bacterial soft rot | *Erwinia carotovora* and *Erwinia atrospectica* |
| | Carrot Yellows | Aster yellow virus or *Chlorogenus callistephi* |
| Lettuce | Lettuce Yellows | Aster yellow virus or *Chlorogenus callistephi* |
| Ginger | Rhizome rot or Soft rot | *Pythium* spp. |

### Breeding for Multiple Disease Resistance

As a definition:

*"A variety having resistance against two or more than two diseases is known as multiple disease resistant variety".*

Multiple disease resistance in vegetable crops so far has not been given serious attention in the country. Most of the vegetables are attacked by more than one disease at a time. The linkages between few diseases have been also seen. Hence, development of multiple disease resistant varieties/hybrids is only solution to minimize or avoid spray of pesticides. Multiple resistance breeding largely depends upon availability of resistant sources. In tomato, a large number of wild species and cultivars are available as sources of resistance (Peter and Joseph, 1986). Cheema *et al.* (1992) also suggested that for developing multiple disease resistant lines/genotypes it is essential to simultaneously screen for more than one disease. Work on multiple disease resistance, where phenonmenol success was achieved has been reviewed. 'Florida 1011' tomato breeding

**Table 9.25: Disease Resistance Sources (Wild and line/strain/varieties) in Vegetables**

| Crop | Disease | Resistance Source | |
|------|---------|-------------------|---|
| | | Wild Taxa | Line/Strain/Varieties |
| Tomato | Leaf mold, leaf spot, fruit anthracnose, Verticillium wilt resistance | *Solanum esculentum* var. *cerasiforme* | Pant Bahar |
| | Verticillium wilt, late blight, bacterial canker, bacterial wilt, spotted wilt virus | *S. pimpinellifolium* | |
| | Curly top virus, Tomato mosaic, Verticillium wilt, Powdery mildew and Leaf curl | *S. chilense* | |
| | *Fusarium* wilt, Spotted wilt, Septoria leaf spot, bacterial canker, and Early blight | *S. peruvianum* var. *dentatum* and *S. pimpnellifolium* | |
| | Tomato leaf curl virus, Verticillium wilt, Powdery mildew and Leaf curl | *S. pimpnellifolium, S. hirsutum, S. hirsutum* var. *glabratum, S. peruvianum* | Hisar Anmol, Hisar Gaurav, H-24, H-36, H-86 and H-88 |
| | Fusarium wilt | *S. pimpinellifolium, S. hirsutum* | Pant Bahar, Marglobe, Pan America, Mangla, HS110, Roma, Rosa, EC179883 (USA) |

| Crop | Disease | Resistance Source | |
|---|---|---|---|
| | | *Wild Taxa* | *Line/Strain/Varieties* |
| | Tomato mosaic | *S. chilense, S. pennellii* | Ohio M29 |
| | Anthracnose | | PI272636 |
| | Verticillium wilt, leaf curl virus, tobacco mosaic virus | *S. peruvianum* | |
| | Curly top virus | *S. peruvianum* var. *dentatum, S. peruvianum* var. *humifusum* | |
| | *Fusarium* wilt, Bacterial canker, Mosaic | *S. hirsutum* | |
| | *Fusarium* wilt, *Verticillium* wilt | *S. cheesmani* | |
| | Bacterial wilt, Fusarium wilt, *Verticillium* wilt, Grey leaf spot | - | Neptune |
| | Leaf curl | *S. hirsutum* var. *glabratum* | |
| | Fusarium wilt and Fruit rot | - | EC160195, EC122527, EC021626, EC103598, EC116874, EC117008, EC115947, EC117671, EC117163, EC128769, Flora-Dade Walker for rae 1 and 2 |
| | Late blight | *S. pimpinellifolium, S. esculentum* var. *cerasiforme, S. peruvianum* | TRB-1, TRB-2, Ottawa-30, Ottawa-31, PI112835, PI207407, West Virginia-63 |
| | Early blight | *S. hirsutum* f. *glabratum, L. pimpinellifolium* | EC174076 (*S. pimpnellifolium*)-Fageria *et al.,* 1998. |
| | Bacterial wilt | - | BWR-1, BWR-5, LE-79, BT-10, BT-1 Sonali, Arka Alok, Arka Abha, Utkal Pallavi, Pant Bahar, Sonali, Saturn, Venus, PI127805A, |
| | Leaf spot | *S. glandulosum* | |
| | Leaf spot, Bacterial canker and Powdery mildew | *S. hirsutum* | |
| | Buckeye fruit rot (*Phytophthora* spp.) | | PI207407, EC174023 and EC174041 (Kohli *et al.,* 1996), SGP8M2 (Fageria, 1994) |
| | Verticillium wilt | | Pant Bahar, Mangla, Rajni |
| Brinjal | Little leaf | *Solanum incanum, S. auriculata, S. viarum, S. esticulatum* | Pusa Purple cluster, H-8, S-252-1-1 and S-252-2-1 (Chakrabarty, 1970) |
| | *Fusarium* wilt | *S. incanum, S. indicum* | |
| | Phomopsis blight | *S. nigrum, S. indicum, S. khasianum, S. xanthocarpum, S. gilo, S. sisymbrifolium, S. testiculatum, S. torvum* (Kalda *et al.,* 1976) | Florida Beauty, Florida Market (Decker, 1951), Pusa Bhairav, Pant Samrat |
| | Phomopsis rot | *S. gilo, S. integrifolium* | |
| | Bacterial wilt, Phomopsis blight | - | Pant Samrat |
| | Bacterial wilt | *S. melongena* var. *insanum, S. nigrum, S. xanthocarpum, S. sisymbrifolium* and *S. toxicariu* | Arka Nidhi, Arka Alok, Arka Keshav, Arka Neelkanth, Annamali (Pandey and Pandey, 1998), Pant Rituraj, Pusa Anupam, BB-7, SM-116 (Anonymous, 1999) SM-6-6, ARU-2C, JC-1, JC-2, Gulla, Long Purple, Udipi, Improved muktakeshi, Pusa Purple cluster, Pusa Purple Round, Vijay hybrid, Banaras Giant Green (Nazir, 1974; Sitaramaiah *et al.,* 1981), Dingaras Multiple Purple Sinampiro (from Philippines), Pusa Purple cluster (Java) and Duma Guete long Purple from Philippines (Rao *et al.,* 1976), BB-11, BB-7 (Sharma and Kumar, 1995), BWR-12, Pusa Purple cluster, SM-6-7, BB-44, and Pant Rituraj. |
| Chillies and Capsicum | Damping off, *Cercospora* leaf spot, *Sclerotinia* rot, leaf curl virus, *Pseudomonas* wilt resistance | *Capsicum frutesemce* | - |
| | *Cercospora* leaf spot | | California Wonder, Hungarian Wax |
| | Verticillium wilt | *C. fasiculatum, C. angulosus* | |
| | Bacterial wilt | *C. chacoense* | AAUM-1, AAUM-2, Utkal Rashmi |
| | Wilt and Dieback | - | Punjab Surkh. LCA-235, BC-21-2, LCA-305 |

| Crop | Disease | Resistance Source | |
|---|---|---|---|
| | | Wild Taxa | Line/Strain/Varieties |
| | Fruit rot | *C. chinense* | Chinese Giant, Yolo Y, Hungarian Yellow Wax, Spartan Lorai, S-27, S-41-1 |
| | Root rot, damping off, leaf curl | - | Pant C-1, Pusa Jwala |
| | Phytophthora rot and cucumber mosaic virus | *C. baccatum* | |
| | Leaf curl virus, Tomato Mosaic Virus and Verticillium wilt | *C. chinense* | |
| | Leaf curl virus | *C. frutescens* | Pusa Jwala, Pusa Sadabahar, Pant C-1, Punjab Lal, CCH-01 |
| | CMV and TMV | - | Punjab Lal, Pusa Sadabahar, Pusa Jwala, Puri Red. |
| | Anthracnose | | Punjab Lal |
| Capsicum | Potat virus Y | *C. angulosum, C. pubescens* and *C. microcarpum* | |
| | Fruit rot | *C. chinense* | |
| | Bacterial leaf spot, TMV and Powdery mildew | *C. annuum* and *C. baccatum* | |
| | Fruit rot, Root knot nematode and Powdery mildew | *C. frutescens* | |
| Potato | Late blight | | Kufri Muthu, Kufri Jyoti, Kufri Megha, Kufri Badsah |
| | Late blight and Potato virus | *Solanum demissum, S. stoniferum, S.acaule* and *S. chaecoense* | |
| | Early blight, Late blight, Wart and viruses | | Kufri Jyoti, Kufri Badsah |
| | Wart | | Kufri Sherpa |
| | Early blight and viruses | | Kufri Lalima |
| | Stem necrosis | | Kufri Sinduri, Hybrids PS/F220, JN-189, MS/82-717 and True seed population HPS 25/13 (Singh *et al.*, 1998) |
| | Golden nematode and Late blight and Wart | | Kufri Swarna |
| Okra | Fusarium wilt (*Fusarium oxysporum*) | *A. manihot* | |
| | Yellow Vein Mosaic virus | *A. tetraphyllus, A. manihot, A. manihot* subspecies *manihot, A. tuberculatus, A. pungens, A. crinitus, A. panduraceformis, A. vitifolius* | Arka Anamika (Sel-10 now susceptible), Arka Abhay (Sel-4, IIHR-4, Prabhani Kranti, Sel-2, Punjab Padmini, Punjab-7, Vijay (Rajmony, 1995), Varsha Uphar, Hissar Unnat, Utkal Gaurav, 91-4, Parbhani Kranti (now susceptible), Barkha Bahar, EC305616 (Bangladesh) |
| | Yellow Vein Mosaic virus and Powdery mildew | *Abelmoschus manihot, A. manihot* spp. *manihot* | |
| | Powdery mildew | | Nigeria, EC32598 |
| | Damping off, *Rhizoctonia solani* | | Red Ghana, Sel.7-1. |
| Bottle gourd | CMV, SMV and WMV | | PI271353 |
| | Fusarium wilt (*Fusarium oxysporum*) | | Taiwan variety Renshi |
| Bitter gourd | Distortion Mosaic Virus | | IC68296, IC68335, IC68263, IC68275, IC68250, IC85620, IC68285, IC68312, IC68272 |
| Ash gourd | Mosaic | | AG-1 (Indu), AG-22, AG-50 and AG-53 from KAU, Kerala |
| | Fusarium wilt (*Fusarium oxysporum* f. sp. *melonis* race-1 and race-2) | | Kogannenaski Golden Pear |
| | Watermelon Mosaic Virus Race-1 | | PI391544 and PI391545 |
| Cucumber | Anthracnose and Downy mildew | - | PI197087 |
| | Gummy stem blight | - | M-17 |
| | Anthracnose | | PI175111, PI175120, PI179676, PI182445, Wise-2757(USA), EC398968 to EC398970, Poinsette (Donor), Hybrid 517 |

| Crop | Disease | Resistance Source | |
|------|---------|-------------------|---|
| | | Wild Taxa | Line/Strain/Varieties |
| | Downy mildew | C. hardwickii | B-184, B-159, Wise-2757(USA), EC398974 to EC39890, Palmetto, Ashley, Chinese Long, Stono |
| | Powdery mildew | C. hardwickii | Poinsette, PI200815, PI200818, Wise-2757(USA), Polaris, Ambra, Yamaki |
| | Downy mildew, Powdery mildew, Anthracnose and Angular leaf spot | | Poinsette |
| | Cucumber Green mottle mosaic virus (CGMMV) | C. hardwickii, C. anguria, C. africanus, C. ficifolius, C. asper, C. dinteri, C. sagittatus | - |
| | Powdery mildew and CGMMV | | Yomaki, Saparton Salad, Brasil |
| | Powdery mildew, downy mildew, Anthracnose, Cucumber mosaic | | SC2B4-1 |
| | Cucumber mosaic Virus, Scab and Powdery mildew | | EC320556 |
| | Cucumber mosaic Virus and Powdery mildew | Cucumis species | EC399914 to EC399937 |
| | Downy mildew and Powdery mildew | Cucumis species | PI197087 (Donor) |
| | Cucumber mosaic Virus | | Poinsette |
| | Multiple diseases with gynoecism | | EC329300 |
| | Angular leaf spot | | EC398966 to EC398967, Poinsette, Dixie Gemin |
| | Fruit rot | | EC398971 to EC398973 |
| | Scab | | Wisconsin SR-12, Beleanto, Ashe Fletcher |
| | Cucumber scab and Angular leaf spot | Cucumis species | Poinsette, Yomaki, PI179376 |
| | Fusarium wilt (*Fusarium oxysporum*) | | Hybrids |
| | Bacterial wilt | | Wisconsin SMR-9, Poinsette, MSU 9402, PI169400, PI200815, PI200816, PI200817, PI200818, PI196477 |
| Muskmelon | Green mottle mosaic virus | C. figeri | - |
| | *Fusarium* wilt | C. zeheri, C. meusi, C. melo var. chito, C. inodorus, C. flexuosus, C. reticulates (Donor) | MR1, CM17187, EC382734 |
| | Powdery mildew | | PMR-45, PMR-5, PMR-6, PMR-45,, RM-43 PI124111, EC399934, Campo, Edisto, Georgia-47, FM-5, Jacumba, Arka Rajhans, Punjab Hybrid (Moderately resistant) |
| | Downy mildew | C. callosus | Punjab Rasila, Punjab Hybrid, DMDR-1 DMDR-2, MR-1, PI414723, EC163888, EC399917, EC399924 to EC399931, Seminole, Edisto, Harvest Queen, Phoontee, Annamalai, Budama Types-1, 2 and 3, Goomuk. Georgia-47, Home Garden, Gulf Stream, Planters Jumbo, Snapmelon (*Phut*) collections like SP-1, SP-2, SP-3, KP-2, KP-7 and KP-9 (Singh *et al.*, 1996) |
| | Downy mildew and Powdery mildew | | Punjab Rasila, IIHR-353, Home Garden |
| | Powdery mildew, downy mildew, Muskmelon mosaic | - | MR-12 |
| | Cucumber Green Mottle Virus (CGMV) | C. africanus | DVRM-1, DVRM-2, Hisar Madhur |
| | Cucumber Mosaic virus (CMV) | | PI161375, PI399914, EC399932 to EC399933 (tolerance), DVRM-2, Punjab Hybrid, Durgapur Madhu, Kabul Melon, Ludhiana Col.2 |
| | Watermelon Mosaic Virus | | PI414723, EC 350603 |
| | Watermelon Mosaic Virus and Powdery mildew | | EC350604 |

| Crop | Disease | Resistance Source | |
|------|---------|-------------------|---|
| | | Wild Taxa | Line/Strain/Varieties |
| | Powdery mildew, Watermelon Mosaic Virus and Blight | | EC 178496 (Donor) |
| | Watermelon Mosaic Virus and gummy stem blight | *C. metuliferus* | PI414723, Gulf Coast (Donor) |
| | Watermelon Virus-1 | | B665 |
| | Muskmelon mosaic | | Oriental Pickling Melon |
| | Zuchini Yellow Mosaic Virus | | PI161375 (Pitrat and Lecoq, 1985) |
| | Muskmelon Yellow Virus | | PI161375 and Nagorata Kin Makua (Esteva and Nuea, 1992) |
| Watermelon | Powdery mildew, Downy mildew and Anthracnose | | Arka Manik |
| | *Fusarium* wilt | | Summitte, Conqueror, Charleston Gray, Dixilec, Crimson sweet, EC333659 to EC333669, EC383809Citron, Shipper, Smokylee, Shipper, White Hope, Verons, EC383809 (USA), 378523-24 (USA) |
| | Wilt | | EC333659-69 (USA) |
| | Anthracnose | | Fair Charleston Gray, Congo, PI189225, African Citron-8, Charleston Gray, Congo, Crimson sweet, Jubilee, Rodas Lov, Black Stone, Sheltran Grey Kango |
| | *Fusarium* and Anthracnose | | EC382753, EC 402549 |
| | Watermelon bud necrosis virus (WBNV) | *Citrullus colocynthis* | |
| | *Alternaria cucumerina* | | Fairfax |
| | WMB | | EC350603 (USA) |
| Pumpkin | Yellow mosaic virus | *C. moschata* | |
| | Powdery mildew | *C. lundelliana* | |
| Summer squash | Powdery mildew | *C. pepo* | |
| Wild pumpkin | Powdery mildew tolerance | *C. lundeliana* | |
| Wild pumpkin | Virus resistance | *C. ecudorensis, C. foetidisima, C. martinezi* | |
| Onion | Purple blotch | | Arka nihar, Arka Kalyan, IHR-56-1, VL-67, Pusa Ratnar, Pusa Red, N-2-4-1 |
| | *Alternaria* blight | | N-2-4-1. |
| | Basal rot | | Poona Red, Patna Red, Belari Red, IHR-141 |
| | Smut, Pink rot and neck rot | *A. fistulosum* | |
| Garlic | Pink rot | | EC378476-77 (UK) |
| Pea | Root rot | | CPS-05-06, AP-1 |
| | Powdery mildew | | FP-186, CPS-05-02, Azad Pea-1, Palam Priya, PMR-21, DPP-62, DMR-7, HFP-4,HFP-12, JP-4, JP-71, JP-83, JP-885, P-388, T-6587, T-6588, PRS-4, T-10, T-56, P-185, P-7588 |
| | *Fusarium* wilt | | Sylvia, Sel.-1, Alaska, Early giant, JM-1, JM-2, Sel.-123-3-2, GC-468, JP-501/2, Grey Badger |
| | Pea mosaic | | American Wonder, Little Marvel |
| | Pea rust | | JP-4 |
| | *Ascochyta* blight | | Austrian Winter, Kinnauri |
| Cowpea | Root rot | | V-265 |
| | Bacterial blight | | Pusa Komal, P-426 |
| | Cowpea mosaic | | Goit, Dixies-Cream, Alabunch, Lobiya-263, Arka Garima |
| | | | VRCP-3 and VRCP-4 |
| | *Fusarium* wilt | | Grant |
| | | | EC 496737 |

| Crop | Disease | Resistance Source | |
|------|---------|-------------------|---|
| | | Wild Taxa | Line/Strain/Varieties |
| French bean | *Phytophthora* stem rot | | Chinese Red |
| | Root rot, Helo blight | *Phaseolus coccineus* | |
| | Bacterial blight, wilt | *P. acutifolius* | |
| | Rust | *P. flavescens* | |
| | Multiple disease resistance | *P. ritensis, P. dumosus* | |
| | *Sclerotinia* rot | | MR-1 |
| | Downy mildew | | MR-1 |
| | Powdery mildew | | Pusa Parwati, Long Lkni, Ved, Top Cross, Contender, Long Kidney, Hungarian Yelow, Lady Washington |
| | Bean common mosaic virus | | Pusa Parwati, Pant Anupama, RH-13, ARS6B65, ARS5BP-7, Viva, Robust, Sea Farer |
| | Anthracnose | | RH-13, Sea Farer, Wells Red Kidney, Cornell-49-242, Sea, Way |
| | Angular leaf spot | | Pant Anupama, SVM-1 |
| | *Fusarium* Yellows | | Tenderette, Pintado, Early Gallatin |
| | *Rhizoctonia* wilt | | Cornell-2114-12, PI-165426, Manoa, Wonder |
| Cauliflower | Black rot | - | Pusa Shubhra, Pusa Snowball K-1, Sel.-8 |
| | Black rot and Ricyness | | Pusa Shubhra |
| | Curd/inflorescence blight | - | Pusa Snowball K-1 |
| | Downy mildew | | Pusa Hybris-2, BR-2, MGS-2-3, 3-5-1-1, CC-12C, EWAWH |
| | *Sclerotinia* rot | | EC103576, EC131592, EC162587, Sel.12, RBS-1 Puakea, Janavo, EWAWH, Sn-445, MGS-2-3, BR-2, |

line was found to be resistant to Verticillium wilt, Fusarium wilt, grey leaf spot and leaf mould (Volin *et al.*, 1977). "Rotam 4" a multiple disease resistant fresh market tomato, was resitant to nematodes, bacterial wilt, bacterial canker, Fusarium wilt (race 1 and race 2 of pathogen), Verticillium wilt and bacterial speck (Bosch *et al.*, 1990). It has been observed that 1-7-1 x Patriot, 7-9-6-10 and EC177401 x Sylvestra are resistant to fruit rot and early blight disease (Cheema *et al.*, 1992). *Cucumis hardwickii*, a progenitor of cucumber possess genes for multiple disease resistance especially to downy mildew. Similarly, *Allium fistulosum* has resistance to *Stemphyllium* blight.

Fageria *et al.*, 1996 studied inheritance of resistance to *Alternaria* blight and fruit rot (*Phytophthora nicotianae* var. *parasitica*) of tomato. He reported two breeding lines, which were resistant against both the diseases. Further he developed multiple disease resistant cross (SGP 8M x EC174076) but it was very poor in yield and quality. Peterson *et al.* (1982) developed a breeding line in cucumber which was resistant against as many as nine diseases. The list of some important multiple disease resistant varieties of various vegetables isgiven in Table 9.26.

**Table 9.26: Multiple Disease Resistance Varieties/Sources in Vegetable Crops**

| Crop | Resistant Sources | Resistant aginst |
|------|-------------------|------------------|
| Muskmelon | MR-12 | Downy mildew, powdery mildew, muskmelon mosaic |
| Watermelon | Arka Manik | Anthracnose, Powdery mildew, Downy mildew |
| | EC2170733-74 (USA) | Fusarium and Gummy stem blight |
| | EC330956 (USA) | Multiple disease |
| Tomato | Neptune | Bacterial wilt, Fusarium wilt, Verticillium wilt, Grey-leaf spot |
| | EC399828-38 (USA) | Multiple diseases |
| | EC347396-90 (Taiwan) | TMV and BW |
| | EC321890-928 (Taiwan) | Viral diseases |
| | Floramerica (Gill *et al.*, 1977) | *Fusarium* wilt, grey leaf spot, leaf mould, *Alternaria* leaf spot, Verticillium wilt, TMV, blossom end rot |
| | Inderminate tomato hybrid (Gillbert *et al.*, 1961) | *Fusarium* wilt, grey leaf spot, TMV, root-knot nematode, spotted wilt, leaf mould, early blight, radial cracking, vascular browning. |
| | Rodade (Bosch *et al.*, 1985) | Bacterial wilt, |

| Crop | Resistant Sources | Resistant aginst |
|------|------------------|------------------|
| | 69-069-24 (TVFN) | Tomato mosaic, *Verticillium* and *Fusarium* wilt, root-knot nematode |
| | K4053 (Krustaleva and Shcherbakov, 1987) | Late blight and mosaic disease |
| Pea | FP-186 | Root rot, Powdery mildew, *Verticillium* and *Fusarium* wilt |
| | JP-179 | Powdery mildew, *Fusarium* wilt, bruchus, leaf minor, rust. |
| | JP-9 | Powdery mildew and bruchus |
| | JP-501/A$_2$ | Powdery mildew and *Fusarium* wilt |
| Hot pepper | Pant C-1 | Root rot, damping off, leaf curl |
| Chilli | EC323333 (USA) | Multiple disease |
| | CH-1 (F$_1$ hybrid) | Fruit rot, dieback, CMV and leaf curl virus. This hybrid is developed by PAU, Ludhiana (Punjab) utilizing GMS system |
| Cucumber | SC2B4-1 | Powdery mildew, Downy mildew, Anthracnose, Cucumber mosaic |
| | EC320526 (USA) | Multiple disease |
| | Poinsette | Downy mildew, Powdery mildew, Anthracnose and Angular leaf spot |
| Sweet Potato | Jewel | Internal cork, Fusarium wilt, Souther root rot |
| Cauliflower | EC175800-06 (USA) | Multiple disease |
| Cabbage | Sanihel | Fusarium wilt, Powdery mildew, Rhizoctonia rot,Cabbage mosaic |
| | EC168041-42 (Taiwan) | Club root and black rot |
| Chinese cabbage | EC345978 (Taiwan) | Multiple disease |
| Brinjal | Pant Samrat | Bacterial wilt, Phomopsis blight |
| | Neelima (F$_1$ hybrid) | Bacterial wilt (resistant), Phomopsis blight (tolerat)- Anonymous, 1999. |
| Onion | *Allium fistulosum* | Purple blotch, Pink root, Smut |

The *S. tuberosum* L. is most suited cultivated species of potato in the world owing to its yield potential and wide adaptability. Resistant genes are mostly found scattered in wild and semi-cultivated species available in its centre of origin and diversity in South America. Sources of resistance to biotic and abiotic stresses are listed in Table 9.27.

**Table 9.27: Sources of Resistance to Various Diseases in Wild Potato Species**

| Sl.No. | Diseases | Sources |
|--------|----------|---------|
| 1. | Viruses-PVX | *Solanum acaule, S. berthaultii, S. tuberosum* subsp. *andigena* |
| 2. | PVY | *S. phureja, S. demissum, S. stoloniferum* |
| 3. | PLRV | *S. acaule, S. demissum, S. tuberosum* subsp. *andigena* |
| 4. | Late blight (Vertical resistance) | *S. demissum, S. verrucosum, S. stoloniferum* |
| 5. | Late blight (Horizontal resistance) | *S. berthaultii, S. chacoense, S. microdontum, S. vernei* |
| 6. | Wart | *S. acaule, S. berthaultii* |
| 7. | Common scab | *S. chacoense, S. tuberosum* subsp. *andigena* |
| 8. | Bacterial wilt | *S. chacoense, S. microdontum* |

*Source*: Pandey and Luthra, 2010.

### (b) Available Insect Resistance Sources and its Utilization

Control of insect pests is an important problem in vegetable production. It is estimated in the United States, insects causes annual loss of over 185 million dollar to vegetables in addition to 100 million or more spent for control (Stoner, 1970). In India various kind of cucurbits are grown, and they are damaged by the red pumpkin beetle (*Aulacophora* sp. L.) at the seedling stage, by the fruit fly (*Dacus* sp.) at the fruiting stage, and by aphids (*Aphis* sp.) at any stage of growth (Nath, 1971). Following varieties/lines have been reported to be resistant to different insects in cucurbits. Wild relatives of okra possess resistance, has low susceptibility to fruit and stem borer, a major pest of cultivated okra.

### Mechanisms of Insect Resistance

(i) Non-preference

(ii) Antibiosis

(iii) Tolerance

(iv) Avoidance or Escape

The above three (i, ii, iii) were described by Painter (1951) and the fourth mechanism, the escape or avoidance was added subsequently. A resistance variety may have one, two or more of three mechanisms.

### (i) Non-preference

The term non-preference has subsequently been replaced by antixenosis (Kogan and Ortman, 1978) because non-preference refers to the insect and this is incongruous with the notion of resistance property of the plant. In the host plant some specific features make it unattractive or undesirable to the insects for food, shelter or reproduction. The plant features may be hairiness, leaf angle, colour, light penetration, taste and odour which create non-preference environment (Table 9.28). For example:

#### Table 9.28: Undesirable Features for Non-preference of Insects

| Crop | Insects | Undesirable Features for Non-Preference |
|------|---------|------------------------------------------|
| Pea | Aphid | Yellow green leaves of pea are less desired by aphids than blue green leaves |
| Potato | Colorado potato beetle | The undesirable odour |
| Cabbage | Cabbage Aphid | Cabbage leaves with lesser light reflection |
| Brassica | Cabbage Aphid | Low sinigrin content is less preferred by Cabbage Aphid than high sinigrin content |

### (ii) Antibiosis

As a definition it refers *"the adverse effect of host plant to the biology (such as development, survival and reproroduction of insects feeding on it"*.

Antibiosis is associated with morphological, physiological and biochemical features of the host plant. The strength of antibiosis varies from species to species and even cultivar to cultivar. In some cases, it leads even death of insects. Some successful examples related to antibiosis mechanism of insect resistance in vegetables are as follows:

#### Table 9.29: Insect-Pest and Mechanism of Antibiosis

| Host Crop | Insect-pest | Mechanism of Antibiosis |
|-----------|-------------|--------------------------|
| Tomato | Fruit worm of tomato | Highly active ethanol soluble compound (lethal) in leaves of *Solanum hirsutum* var. *glabratum* |
| Potato | Aphid (*Acrythosiphon pisum*) | Gummy trichome exudates |
| Pea | Pea aphid | Low amno acid content |
| Bean | Mexican bean beetle (*Epilachna varivestis* Mulsant.) | Low amount of carbohydrates and reducing sugars |
| Cole crops | Cabbage aphid | Low sinigrin content in the leaves |
| Celery | Black shallow tail buterfly | Higher sinigrin content |
| Carrot | *Dacus dorsalis* | Lack of thiamine, nicotinic acid, folic acid |

| Host Crop | Insect-pest | Mechanism of Antibiosis |
|-----------|-------------|--------------------------|
| Cucumber | Spotted cucumber Beetle | Low or free cucurbitacin content provide resistant |
| | Spotted spider mite | High cucurbitacin content provide resistant |

### (iii) Tolerance

As a definition it refers *"the extent to which a plant can support an insect infestation without loss of vigour and reduction of crop yield"*. In other words tolerance refers to *"the ability of cultivar to produce higher yield than susceptible cultivar at the same level of insect attack"*.

Panda and Khush (1995) following advantages of tolerance:

☆ Tolerant varieties have a higher economic threshold level than the susceptible ones and thus require less insecticide application and promote bio-control.

☆ Tolerant varieties do not depress insect population nor do they provide any selection pressure on the insect and thus are useful in preventing the development of new insect biotypes.

☆ In varieties with combination of three mechanisms, tolerance increases yield stability by providing at least a moderate level of rsistance, whereas vertical genes providing a high level of resistance through antixenosis and antibiosis succumb to the new biotype.

### (iv) Avoidance or Escape

Escape of variety from insect results due to earliness or its cultivation in the insect free or low insect populated area/season. It is also an effective means of protecting from damage by insect-pests. Production of virus free potato seed, through seed plot technique is an example of avoidance.

The resistant to insect pest can be transferred from five different sources:

### (i) Cultivated Varieties

Cultivated varieties are most potential source for any kind of resistance including insect resistance because cultivated varieties possessed good horticultural traits.

### (ii) Germplasm Collection

Plant breeders always search desirable genes from the nature which is reservoir of all aspects *i.e.* biotic and abiotic. So priority should be given for germplasm collection and evaluation on regular basis.

### (iii) Wild Species

Some times it is not possible to locate desirable gene for a particular trait among the cultivated or collected genepool. In this situation wild or related species must

**Table 9.30: Inheritance of Resistance to Major Insect-Pests in some Vegetable Crops**

| Crops | Insect-pest | Inheritance of Resistance or Causes of Resistance | References |
|---|---|---|---|
| Muskmelon | Melon fruit fly (*Dacus cucurbitae*) | Dominant gene | Nath *et al.*, 1976, Patil *et al.*, 1983 |
| | Melon fruit fly (*Bactrocera cucurbitae*) | Dominant gene | Teuatial Phantar, 1996 |
| | Melon aphid | Single dominant gene | |
| Cucumber | Red Pumpkin Beetle | Single dominant gene | |
| | Stripped cucumber beetle | Polygenic | |
| Squash | Squash bug | Additive | |
| | Two spotted spider mite | Polygenic | |
| Summer squash | Squash bug | Polygenic | |
| Pumpkin | Fruit fly | Monogenic | |
| Watermelon | *Dacus cucurbitae* | Single dominant gene (fwr) | Khandelwal and Nath, 1978 |
| | Red Pumpkin Beetle | Monogenically dominant over susceptible | Vashishta and Choudhary, 1972 |
| Tomato | Arthropod pests | Resistance differ with each pest | |
| Cowpea | Weevil | Additive dominance | |
| Cole (*Brassica*) | Cabbage aphid | Low sinigrin content | |
| Lettuce | Lettuce root aphid | Resistance controlled by the extranuclear factor, modifying gene might also be involved | |
| Potato | Potato aphid | Sticky trichome exudate | |
| Sugarbeet | Aphid | Low free sugar | |

*Source*: Fageria *et al.*, 2001.

**Table 9.31: Cultivars/Genotypes as a Source of Insect Resistance**

| Crop | Insect-pest | Source of Insect Resistance |
|---|---|---|
| Brinjal | *Leucinodes orbonalis* Guen. | Punjab Bahar, Gulabi Doria, Pusa Purple Cluster, Pusa Purple Round, PPI, Nurki (Lale *et al.*, 1986), Aushey, H-128, H-129, SM-17-4, Baramasi, Jalgaon Local, Arka Shirish, Selection-1, Punjab Barsati and BB-17 were resistant and Pant Samrat, KT-4, BB-26, PB-5, PB-29, PB-30, P-31, PB-33, PB-36, DBL-24, BB-46, Composite-2, NDBH-7, NDBH-25, NDBG-8, NDBH-11, Pusa Hybrid-5, PB-40, PB-41, PB-42, PB-43, PB-44, ARBH-527, ARBH-541, ARBH-258 were tolerant (Kumar and Ram, 1998). |
| | Jassids | A-61, B-14, American Black Beauty, Manjari, Pusa Purple Round, Pusa Anmol, Punjab Chamkila, S-5, S-16, Pant Brinjal (Subbaratnam *et al.*, 1983), Manjari Gota, Dorli Jymbli Malayaum |
| | Mites | Sel-1, Long Green, Pusa Purple long, R 34, (Dhooria and Bindra, 1977), SM-227, S-529, S-600 and S-601 (Palaniswamy and Chelliah, 1986) |
| Tamato | *Helicoverpa armigera* | Angurlata (Misra *et al.*, 1996), HT-64, and Hybrid No. 37 were tolerant (Rath and Nath, 1997). Sabour Prabha, Accessions No. 128, 133 and 145 (Kashyap and Kalloo 1983), Atkinsoni, Angurlata, Punjab chhuhara (Gill *et al.*, 1980) were partial resistance. |
| Chillies | Aphids | Kalyanpur Red x 1068, LEC-28, 30, 34. |
| | Mites | Punjab Lal, LEC-1, LEC-1, Kalyanpur Red |
| | Thrips | Chamatkar, NP-46A x 1068, Caleapin Red, Chamatkar, NP46A, BG-4 |
| Bell pepper | Aphid | California Wonder, Koral, Yolo Wonder, Kalyanpur Red x 1068 |
| Potato | *Empoasca fabae* | Sequioa |
| Pea | Aphid and Weevil | LMR-10, LMR-4, LMR-20, PI196027, PI244241 |
| | Stemfly | Sophia-135, Sutton Phinonima (Sandhu *et al.*, 1975), Bonneville, Asaugi, Boach Sel. IP-3 |
| | Leaf minor | LMR-10, LMR-4, LMR-20, JP-179, JP-747, JP-169-1, Early Wonder and Arkel (Das *et al.*, 1988), PS-40, EC16704 |
| | Pod borer | bonneville |
| Beans | Pod weevil | Bat-1235, Vat-1230, Porrillo-1 |
| | Aphid | M.S. 370, TVu2740, P912, P475 (Jayappa and Lingappa, 1988) |
| | Leaf minor | FB-7, IC14913, IC16541, EC313616, EC313617, PI302537, and BDJ-149 were tolerant (Rai and Ram, 1997) |

| Crop | Insect-pest | Source of Insect Resistance |
|---|---|---|
| Vegetable Soybean | Mexican bean beetle (*Epilachna varivestis*) | PI417061, PI417288, PI417310, EC31732 were resistant (Kraemer *et al.*, 1994). |
| Cowpea | Weevil | Land race Tvu2027 (Shade *et al.*, 1996) |
| Onion | Thrips | N-2-4-1, Pusa Ratnar, Kalyanpur Red Round, Sel. 104, Sel. 171, Udaipur-103, Bombay White, Country White, Spanish White, Ludhiana Selection, Arka Suryamukhy, Hisar-2, Red Round |
| Garlic | Thrips | G-2, G-6, G-19, G-20, G-31, G-43, G-84 (Singh, 1987) |
| Bitter gourd | *Dacus cucurbitae* | Short Green Kerala (Lal and Sinha, 1974), Phule BG-4 (Anonymous, 1990), Faizabad Coll.-17, Kerala Coll.-1 (Tewatia, 1994). ACC-23, ACC-5, ACC-25, ACC-28, ACC-33 and C-96 (Thakur *et al.*, 1992) |
| Roundmelon/Tinda | *Dacus cucurbitae* | Arka Tinda (Nath, 1971) |
| Muskmelon | Mites | Durgapura Sel.-1, Early Gold Hara, Hara Madhu (Dhooria and Sandhu, 1974) |
| | Pickle worm (*Diaphania nitidalis*) | Edisto-47 and Honey Dew Melon (Sullivan, 1970), PI140471, PI183311, and PI296354 (Day *et al.*, 1978). |
| | Red pumpkin beetle | Kasawa (Vashishtha and Choudhary, 1971) |
| Cucumber | Cucumber beetle (*Diaborotica undercimpuictata* Howardi) | Edisto-47 (Sullivan and Brett, 1971), Heart of Gold and PMR-4-50 (Wiseman *et al.*, 1961) |
| Pumpkin | Red pumpkin beetle | LC-1-2, LC-7-1, LC-11-2, LC-13-4, LC-17-2, LC-17-6 (Srinivasan *et al.*, 1979) |
| | *Dacus cucurbitae* | IHR-79-2(Nath *et al.*, 1976a). J-18-1, J56-1 (Khandelwal and Nath, 1979) |
| Okra | Leaf minor | U29A-144, U29A-144-3 and Nath Shobha-111 were tolerant (Tripathi *et al.*, 1996) |
| | Shoot and fruit borer | AE-3, AE-22, AE-57, AE-67, AE-79, Wonderful Pink (Mote, 1978). Perkins Long Green, Climpson Spineless, White Show (Kashyap and Verma, 1983), Red-I, Red-II, Red Wonderl and Red Wonder–II. |
| | Jassid | Climpson Spineless, Early Long Green (Teli and Dalaya, 1981), IIHR-21, Crimson Smooth Long, *A. manihot* sp. *manihot* |
| Radish | Aphid | S-6, Pusa Chetki (Singh *et al.*, 1987) |
| Cauliflower | Diamond back moth (*Plutella xylostella*) | EC234599,,Nasika-1 (Khare *et al.*, 1986) |
| | Aphid | Improved Japanese |
| Cabbage | Aphid | Early Quin Glori (Lal *et al.*, 1987). All Season, KK Cross, NY-1381G, NY101181 |
| | Cabbage butterfly and Caterpillar (*Pieris brassicae*) | Red Pickling and Red Rock Mamoth |
| | Diamond back moth (*Plutella xylostella*) | Green Acre, PI3243 |

### Table 9.32: Available Insect-Pests Resistance in Wild Species

| Vegetables | Insect-Pest | Source of Resistance/Tolerance |
|---|---|---|
| Tomato | Fruit borer, Green house fruit fly, Mite and *Bemisia tabaci* | *Solanum hirsutum*, *S. hirsutum* f. *glabratum* |
| Egg plant | *Leucinodes orbonalis* | *Solanum xanthocarpum*, *S. sisymbrifolium*, *S. khasianum*, *S. nigrum*, *S. integrifolium* |
| | Aphid (*Aphis gossypii*) | *S. mammosum* |
| Potato | *Myzus persicae* | *Solanum brachistotrichum* |
| | Colorado potato beetle | *Solanum demissum*, *S. chacoense* |
| Okra | Spotted ball worm, Jassids, leaf hopper, Shoot and fruit borer | *Abelmoschus manihot* var. *ghana* |
| Muskmelon | Fruit fly | *Cucumis callosus* (Chelliah and Sambandhan, 1974) |
| Onion | Thrips, Maggot | *Allium fistulosum* |

*Source*: Fageria *et al.*, 2001.

be identified to transfer the resistance against insect pests.

### (iv) *Mutations*

It has been observed that both spontaneous and induced mutations can also be used as a source of insect resistance. Generally in these phenomenon genes having lethal action against certain pests are isolated from micro-organisms and transferred into the system of crop plants.

### (v) Microorganism

For developing insect resistant generally three methods are applied:

(a) Introduction

(b) Selection

(c) Hybridization.

### Screening Techniques for Insect Resistance

The existence of vertical resistance against insect pest is very limited therefore horizontal resistance must be given priority. It is apparent that vertical resistance prevents the identification of individual plants carrying horizontal resistance in a crop. Breeding for horizontal resistance is easier in cross pollinated crops than self pollinated crops.

### (i) Field Screening

Field condition explores the possibility of screening of large plant population than laboratory condition. In this environment, the plant material is exposed to other prevalent insect pests of the region. Under field condition following steps can be adopted:

(a) Planting of a row of susceptible variety after every two rows of the test material.

(b) Screening the material under insect prone fields

(c) Screening should be made in the season when heavy infestation of the pest occurs.

### (ii) Glass House Screening

In laboratory/glass house screening, it is possible to ensure uniform initial infestation of all plants in a population than in field condition as varied by several uncontrolled environment. In laboratory/glass house, the desired environments can be maintained which is essential for study of behavior of insect pests. The reliability of glass house screening is completely more than field screening.

It is essential to have a reliable and better results, screening should be done in laboratory as well as in field condiotion. It should be mandatory that the material found resistant in the laboratory/glass house should also be evaluated under field condition.

### Screening Criteria

The screening criteria against insect pests proposed by Dahms (1972) are as under:

(i) Visual observation of varieties on growth, lodging, cutting and discolouration.

(ii) Under heavy infestation, counting the surviving plants of varieties/populations at different growth stages.

(iii) Counting population of insect adults or larvae attracted to a genotype under insect favourable conditions.

(iv) Evaluating field performance of variety grown under infested and non-infested plots.

(v) Determination of the number of surviving insects and progeny produced.

(vi) Counting number of eggs laid.

(vii) Recording observations of the comparative effects of forced insect feeding on plants by determining length of insect life cycle, mortality, reproductive rates, or molting.

(viii) Correlation of morphological traits with injury.

(ix) Measuring root damage by amount of force required to pull a plant out of the ground.

(x) Correlation of chemical factors in plants with insect response.

(xi) Measuring of the amount of food utilized by the insects.

Bellotti and Kawano (1980) also proposed usefull screening procedures for evaluating large amount of genetic materials of vegetanle crops. They suggested that as per the basic principal two types of damage key can be used:

(a) The first damage scale may be used to evaluate a large number of varieties when the major objective is to eliminate susceptible material. This scale is usually on a 0 to5 basis.

(b) A second scale can be used for those varieties identified for further evaluation. In this case, a scale is needed with more classes for example, 1 to 10.

During the experimentation they used cassava against mites, thrips, white fly, scales, mealy bugs, stem borer and shoot fly. In this situation this scale is more accurate and reliable while dealing with low levels of resistance or trying to increase resistance by combing low or intermediate levels through crossing.

### (1) Screening against White Flies in Cassava

The cassava genotypes should be screened in an area having heavy natural infestation. After every two lines of varieties/lines, the rows of susceptible variety should be planted to create uniform infestation. The resistance can be evaluated by using three 0 to 5 scales. The large white fly population with few damage symptoms would indicate the tolerance mechanism.

(i) The number of pupae per leaf

0=No pupae

1= Less than 5 pupae per leaf

2= 5 to 10 pupae per leaf

3= 10 to 25 pupae per leaf

4= 25 to 50 pupae per leaf

5= More than 50 pupae per leaf

**Table 9.33: Procedure of Counting the Insect Populations**

| Insect-Pests | Crops | Procedure |
|---|---|---|
| Aphids | Cabbage, cauliflower | Population of nymphs and adults/plant |
| Aphids | Potato, okra, brinjal, tomato, beans, cucurbits | Population of nymphs and adults on three leaves per plant, one each from upper, middle and lower parts of the plant. |
| Aphids | Cauliflower, cabbage, radish at flowering stage | Population of nymphs and adults per 30 cm inflorescence |
| Whitefly | Vegetable crops | Population of pupae is counted on three leaves |
| Leaf hopper or jassids | Brinjal, okra, tomato *etc.* | Population of pupae is counted on three leaves |
| *Epilachna* beetles | Vegetable crops | Population of beetles and grubs per plant is counted. |
| *Epicauta* sp. | Brinjal, French bean, capsicum *etc.* | Population of only adult per plants counted |
| Leaf caterpillars (*Pieris brassicae*), *Plutella xylostella, Plusia fricha* | Cole crops | Total population of larvae per plant is counted |
| Borer | *Leucinodes orbnalis* Brinjal | Damaged fruits at each picking observed and per cent damage on fruit number and fruit weight is counted |
| | *Helicoverpa armigera* tomato | Damaged fruits at each picking observed and per cent damage on fruit number and fruit weight is counted |
| Thrips | Onion, Garlic | Total population of per plant is counted. Whereas, in other crops three leaves per plant may be oberved |
| Mites | Vegetable crops | Population on eggs and mites per square cm from three leaves counted under laboratory. |

(ii) The percentage of leaves infested with pupae

0=No infestation

1= Less than 20 per cent infestation

2= 20 to 40 per cent infestation

3= 40 to 60 per cent infestation

4= 60 to 80 per cent infestation

5= 80 to 100 per cent infestation

(iii) Damage symptoms caused by white fly feeding

0=N damage

1= Single specking of lower leaves

2= Heavy specking of lower leaves

3= Mosaic like symptoms on leaves but little wrinkling, sooty mould on lower and central leaves.

4= Wrinkling and yellowish mottling of lower and apical leaves, some leaf necrosis, considerable sooty mould.

5= Severe wrinkling of apical leaves, leaf necrosis and death of plant.

## (2) Screening against Aphid (Aphis craccivora Koch) in Cowpea

In cowpea during initial growth generally aphid population are small in numbers, hence the absolute count can be done. When population of aphid is in higher number, absolute count of aphid is not possible. In this situation aphid population can be grouped into infestation grades as suggested by Swirski (1954).

**Table 9.34: Infestation Grades for Aphid Population**

| Sl.No. | Grade | Aphid Population |
|---|---|---|
| 1. | 0 | 0 |
| 2. | 1 | 1-10 |
| 3. | 2 | 11-2 |
| 4. | 3 | 21-30 |
| 5. | 4 | 31-51 |
| 6. | 5 | 51-100 |

## (3) Screening against Shoot and Fruit Vorer in Brinjal

Kumar and Ram (1998) screened varieties of brinjal against shoot and fruit borer and proposed the grades as:

**Table 9.35. Reaction against shoot and fruit borer**

| | |
|---|---|
| Non-damaged fruits 90 per cent or less number | Resistant |
| Non-damaged fruits 80-90 per cent or less number | Tolerant |
| Rest as | Susceptible |

Chelliah and Sambandam (1971) compared the resistant species *Cucumis callosus* with the susceptible *C. melo* variety 'Delta Gold' and 'Smith Perfect'. *Cucumis callosus* is resistant to fruitfly. *Dacus cucurbitae* was associated with high silica content of the fruit. Chelliah and Sambandam (1973) reported that high content of amino acids in varieties of *C. melo* was found to be associated with susceptibility to *Dacus cucurbitae*. Pal *et al.* (1983) found that the resistance in *Cucumis callosus* to melon fruitfly (*Dacus cucurbitae*) was associated with low T.S.S. content of fruits. Tewatia (1994) reported that

**Table 9.36: Infestation Grades for Various Insect-Pests**

| Grade | Reaction | Borer infestation (per cent) | Aphids/Plant | Cabbage Caterpillar | | Grey Blister Beetle/15 Plants |
|---|---|---|---|---|---|---|
| | | | | Seed Crop | Vegetable | |
| 0 | Immune | 0 | 0 | 0 | 0 | 0 |
| I | Highly resistant | 1-10 | 1-15 | 1-10 | 1-5 | 1-20 |
| II | Resistant | 11-2 | 51-100 | 11-25 | 6-10 | 21-40 |
| III | Moderately resistant | 21-30 | 101-200 | 26-50 | 11-15 | 41-60 |
| IV | Moderately susceptible | 31-40 | 201-500 | 51-100 | 16-25 | 61-80 |
| V | Susceptible | 41-50 | 501-1000 | 101-150 | 26-50 | 81-200 |
| VI | Highly susceptible | Above 50 | Over 1000 | Over 150 | Over 50 | Over 200 |

higher concentration of ascorbic acid, total ash, silica, tannin and flavanol in the fruits of resistant cultivars (Faizabad Collection-17 and Kerala Collection-1) of bitter gourd was found to be associated with resistance against melon fruit fly.

**Table 9.37: Cucurbits Lines Resistant to Insects in India**

| Cucurbitaceous Crop | Resistant Varieties | Insect Pests |
|---|---|---|
| Sponge gourd | Pusa Chikni | Red pumpkin beetle |
| Ridge gourd | Pusa Nasdar | Red pumpkin beetle |
| Bottle gourd | No. 28, No. 29 | Red pumpkin beetle and fruit fly |
| Watermelon | J-14, J-70 | Red pumpkin beetle |
| Muskmelon | J-C, 20-A, J-103-2, Persian Small, J-6-A | Red pumpkin beetle and fruit fly |
| Pumpkin (*C. moschata*) | J-18, J-32, J-24, J-46 | Red pumpkin beetle |
| Pumpkin (*C. maxima*) | Arka Suryamukhi | Fruit fly |
| Squash (*C. pepo*) | Early Golden Bush Scallop, Royal Acron | Fruit fly |
| Roundmelon | Arka Tinda | Red pumpkin beetle |

**Table 9.38: Insect Resistance Sources in Vegetables**

| Crop | Insect | Resistance Source | |
|---|---|---|---|
| | | Wild Taxa | Germplasm/Lines/Varieties |
| Tomato | Fruit worm | *Solanum hirsutum* | PI 126449 |
| | Root knot nematode | *S. peruvianum, S. cheesmani* | |
| | Fruit fly | *S. hirsutum* | |
| | Fruit borer | | EC110264, EC101552, EC124406, EC129599, EC129597, EC129698, EC129791, EC121451, EC160191, EC161254, EC119121 |
| Potato | Root knot nematode | *S. andigena, S. bulbocastanum, S. vernei, S. sparsipipulum* | |
| | Aphid | *Solanum bulbocastanum, S. berthaultii* | |
| Brinjal | Shoot and fruit borer | *S. khasianum* | |
| | Shoot and fruit borer, aphids, root knot nematodes | *S. sisymbriifolium, S. microcarpon* | |
| Okra | Fruit and shoot borer | *Abelmoschus tuberculatus* var. Red I and Red II, *A. manihot* | |
| | Jassids resistance | *Abelmoschus manihot* subspecies *manihot* | |
| | Root knot nematode | *A. esculentus* var. long green smooth | |
| Bottle gourd | Fruit fly (*Dacus cucurbitae*) | | NB-29 |
| | Red Pumpkin Beetle | | NB-28 |
| Bitter gourd | Pumpkin Beetle | | BG-5, BG-98, BG-102 |
| Cucumber | Fruit fly (*Dacus cucurbitae*) | *Cucumis callosus, C. hardwickii* | |

| Crop | Insect | Resistance Source | |
|------|--------|-------------------|--|
| | | Wild Taxa | Germplasm/Lines/Varieties |
| | White fly | C. anguria, C. africanus, C. ficifolius, C. asper, C. dinteri, C. sagitatus | |
| | Nematode (*Meloidogyne incognita*) | | West Indian Gharkin |
| Muskmelon | Root knot nematode | C. metuliferous | |
| | Fruit fly | C. callosus | |
| | White fly | C. asper, C. dentery, C. dipsaceeus, C. sagitatus | |
| | Aphid | - | EC350602, EC178496 (Donar), LJ90234(An inbred of PI371395) |
| | *Bemisia tabacci* | | EC399923 |
| | Mites | | Durgapur Madhu, Sel;.1, Punjab Sunehri, Pusa Sharbati, Kocha 4 |
| | Red pumpkin beetle | | Casaba, PI170683 |
| Watermelon | Aphid | | EC350602 (USA) |
| Summer squash | Red pumpkin beetle | Cucurbita pepo | |
| *Cucurbita* Spp. | Pumpkin beetle | | BG-5, BG-9 and BG-102 |
| | Cucumber Mosaic Virus | C. pepo cv. inderella, | Gold Rush |
| | Pumpkin Mosaic Virus | | CM214 (Nigerian Local, Immune) |
| | Watermelon Mosaic Virus | C. ecuadorensis, C. joetidissima | |
| | Bacterial wilt | C. ficifolia, C. lundelliana | |
| | Powdery mildew | C. lundelliana, C. okeechobeensid | |
| | Powdery mildew, Squash and Watermelon Virus | C. martinezii | |
| Pumpkin | Fruit fly | | Arka Suryamukhi (*C. maxima*) |
| Hot pepper | Root knot nematode | | Pant C-1 |
| Sweet potato | Flea beetle | | Jewel |
| Cowpea | Aphids | | Tvu 408 P₂, Tvu 46, Tvu 3417, Tvu 2740 and Tvu 3509, PI 473, and PI 476, Mandya Local, MS 370, Tvu 2740, P 912 and PI 475, Tvu 1989, Vs 350, Vs 438 and Vs 452 |
| | *Heliothis armigera* | | TVX 7 |
| | (*Helicoverpa armigera*) | | |
| | *Maruca testulalis* (*Maruca vtrata*) | | GC 82-7 |
| | *Psidia tikora* | | DLPC 16 |
| | (*Cydia tychora*) | | |
| | Bruchids | | C 152, C 190 and GC 82-7 |
| | Root-knot nematode | | Mississipi 57-1, Iron |
| | Root-knot nematode | | IITA 3, Habana-82, Incarita-1, IITA 7, P 902, IT 86D364, IT 87D14638, Vinales 144, |

There are several varieties which have tolerant to several insect pests (Table 9.39).

## Resistance Sources for Nematodes

See the Tables 9.40 and 9.41.

# Resistance Sources for Abiotic Stress in Vegetables

## (I) Heat Tolerance

High temperature for fruit set in tomato is a general problem in subtropical and tropical countries. A large number of literatures were available on the adverse effect of high temperature on fruit set in tomato as well as probable causes of poor fruit set. It was observed that ovule abortion, browning and drying of stigma, poor gamete viability, poor fertilization, embryo sac degeneration and endosperm degeneration at 40°C, disruption in meiosis and prevention of pollen formation are because of high temperature that has been reviewed by Kalloo (1988). Further, splitting of the antheridial cone, induction of pollen sterility, bud and flower drop have been frequently observed at high temperature, due to lack of endothecium formation, style elongation, poor

**Table 9.39: Source against Insect Pests**

| Sl.No. | Crop | Pest | Scientific Name | Varieties/Genotypes |
|---|---|---|---|---|
| 1. | Brinjal | Shoot and Fruit borer | *Leucinodes orbonalis* | SM-17-4, PBR 129-5, Punjab Barsati, ARV 2-C, Pusa Purple Round, Punjab Neelam, Kalyanpur-2, Punjab Chamkila, Gote-2, PBR-91, GB-1, GB-6 |
| | | Aphid | *Aphis gossypii* | SM-17-4, PBR 129-5, Punjab Barsati, ARV 2-C, Pusa Purple Round, Punjab Neelam, Kalyanpur-2, Punjab Chamkila, Gote-2, PBR-91, GB-1, GB-6 |
| | | Jassid | *Amarasca biguttula biguttula* | SM-17-4, PBR 129-5, Punjab Barsati, ARV 2-C, Pusa Purple Round, Punjab Neelam, Kalyanpur-2, Punjab Chamkila, Gote-2, PBR-91, GB-1, GB-6 |
| | | Thrips | *Thrips* sp. | SM-17-4, PBR 129-5, Punjab Barsati, ARV 2-C, Pusa Purple Round, Punjab Neelam, Kalyanpur-2, Punjab Chamkila, Gote-2, PBR-91, GB-1, GB-6 |
| | | Whitfly | *Bemisia tabaci* | SM-17-4, PBR 129-5, Punjab Barsati, ARV 2-C, Pusa Purple Round, Punjab Neelam, Kalyanpur-2, Punjab Chamkila, Gote-2, PBR-91, GB-1, GB-6 |
| 2. | Cabbage | Aphid | *Brevicoryne brassicae* | All Season, Red Drum head, Sure Head, Express Mail |
| 3. | Cauliflower | Stem borer | *Hellula undalis* | Early Patna, EMS-3, KW-5, KW-8, Kathmandu Local |
| 4. | Okra | Jassid | *Amarasca biguttula biguttula* | IC007194, IC013999, New Selection, Punjab Padmini |
| 5. | Onion | Thrips | *Thrips* sp. | PBR-2, PBR-6, Arka Niketan, Pusa Ratnar, PBR-4, PBR-5, PBR-6 |
| 6. | Round gourd | Fruit fly | *Bactrocera cucurbitae* | Arka Tinda |
| 7. | Pumpkin | Fruit fly | *Bactrocera cucurbitae* | Arka Suryamukhi |
| 8. | Bitter gourd | Fruit fly | *Bactrocera cucurbitae* | Hissar-11 |

**Table 9.40: Important Nematode of Vegetables**

| Crop | Diseases | Causal Organism | Resistance Sources |
|---|---|---|---|
| Tomato | Root-knot nematode | *Meloidogyne incognita* | Pusa-120, PNR-7, Hisar Lalit, NRT-3, NRT-8, NRT-12, Mangla |
| | | *Meloidogyne javanica* | |
| | | *Meloidogyne hapla* | |
| Brinjal | Root-knot nematode | *Meloidogyne incognita* | |
| | | *Meloidogyne javanica* | |

**Table 9.41: Resistance Sources for Nematodes**

| Sl.No. | Name of Crop | Nematodes | Sources |
|---|---|---|---|
| 1. | Potato | Cyst nematodes | Solanum tuberosum subsp. andigena, S. berthaultii, S. vernei |
| 2. | Potato | Root knot nematode | *Solanum spegazzinii, S. vernei, S. curtizianum* and *S. multidissiectum* |
| 3. | Tomato | Root knot nematode | *Solanum hirsutum* var. *typicum* |
| 4. | Brinjal | Root knot nematode | *Solanum sisymbrifolium* and *S. elaegnifolium* |

pollen germination are probable causes of poor fruit set at high temperature. High temperatures induced poor pollen development, poor pollination, disintegration of embryonic pistil cells, and hormonal imbalance is reported in literature. Fruit set of 356 cultivated and wild *Lycopersicon* genotypes was evaluated at high temperatures and genotypes having >50 per cent fruit set has been screened. LHT 24 is primarily a source of heat tolerant germplasm, selected from $F_{10}$ of the cross Fresh Market 9 x Saladette. Fruit set, pollen fertility and consuming ability of selected tomato genotypes under high temperature conditions has been studied. The genetics of fruit set at a high temperature in *Lycopersicon* is not clearly understood. Pollen fertility and fruit set under high temperatures were primarily under preponderance of additive gene action for fruit

set, flowers drop and under developed ovaries at high temperature or the involvement of quantitative factors. Heat tolerance exhibits low heritability and highly influenced by environmental factors, thus indicate less scope for selection. The indehescence of the anther or poor anther dehiscence and low pollen viability or pollen sterility are the cause of abscission of flowers at a high temperature and is very common in *Vigna unguiculata*. Super optimal temperature (34/29°C day/ night) stress increases bud and flower production and dramatically increased abscission of buds, flowers and pods compared with controls at 24-19°C in *P. vulgaris*. A rapid method was also reported in the literature for identifying the heat tolerant varieties of Chinese cabbage by sodium dodecylsulfate polyacrylamide gel electrophoresis (SDS-PAGE).

## (II) Cold Tolerance

Tropical and subtropical vegetable crops, *viz.*, tomato, eggplant, pepper, okra and beans, suffer from exposure to low temperature. An extensive review (Kalloo, 1988) is available in the literature. Low temperature seed germination is a heritable characteristic. In *Lycopersicon*, it is mostly controlled by recessive factors and in muskmelon by dominant factors. A diallel analysis to study the genetics of germination was controlled by additive gene action. In *Cucumis sativus*, low temperature seed germination has low heritability. Further genetic studies indicated the equal involvement of both additive and dominant variance. A parthenocarpic variety may have better scope for fruit set under low temperature conditions. Under environmental temperature stress, parthenocarpic lines in tomato can be used for increased fruit set.

## (III) Drought Tolerance

Drought resistance in tomato was assessed on the basis of percentage seed germination and rate of radicle growth in sucrose solution at different concentrations on the basis of plant growth rate and leaf area and on the basis of root system (Kalloo, 1988). To detect the stress tolerance in soybean, many workers follow temperature-mediated leaf injury test. This procedure is based on the observation that when leaf tissue is injured by high temperature, electrolytes diffuse out of the injured cells. The diffused electrolyte is measured by washing the treated tissue in demonized water and then measuring the electrical conductivity of the water. Again leaf specific weight (leaf weight per unit surface area) has been used for drought stress resistance. Resistant varieties cope to drought stress by a decrease in leaf size and then a decrease in leaf number. In *Vigna unguiculata*, reduced early vegetative growth associated with early flowering and early partitioning of dry matter to reproductive parts could improve the drought resistance. Water loss is associated with drought resistance; tolerant *V. unguiculata* lines exhibited less water loss. Determination of proline accumulation capacity under stress may be a useful index while selecting a stress tolerant variety in *P. vulgaris*. A positive correlation was found between water stress tolerance and proline accumulation. Low proline accumulation was associated with drought resistance. Drought stress resistance in watermelon has been associated with a reduced transpiration surface, high water retention and closed stomata at a high temperature. Increase in temperature over the ambient temperature of crop canopies under stress has been commonly used as a criterion of drought resistance (Blum, 1988). Plant breeders have always looked for selection criterions for drought resistance, which could be used, in a large segregating population. The phases of research on identification of single traits for selection for drought resistance are as follows: germination (using mannitol or PEG as osmotic treatments), proline accumulation during germination or growth; response of nitrate reductase; betaine accumulation, and hormonal changes - ABA, cytokinins *etc.* Arnon (1980) stated "Breeding for drought resistance has been a consistent theme for as long as one remembers and probably the greatest source of wasted breeding efforts in the whole field of plant breeding".

## (IV) Excessive Moisture Tolerance

Under excess moisture conditions the production of vegetable crops is very limited. Anaerobiosis develops within a few hours after flooding takes place, due to displacement of oxygen from the soil pore space, which results in a reduction reaction in soil system. Reduction of a submerged soil proceeds roughly in the sequence predicted by thermodynamics (Ponnamperuma, 1972). In tomato, 'Nagcarlan', a primitive cultivar from Philippine LA 1421, a feral member of the *cerasiforme* complex from northeastern Ecuador has been found tolerant to excessive moisture (Rebigan *et al.*, 1977) and can be used in excessive moisture tolerance breeding program of *Lycopersicon*. Opena *et al.* (1983) have designed a detailed breeding program of tomato, chinese cabbage and potato with special reference to the lowland tropics. Plants growing in submerged soils have two adaptations that enable the roots to ward off toxic reduction products, accumulate nutrients, and grow in an oxygen-free medium : oxygen transport from the aerial parts and anaerobic respiration.

## (V) Salt Tolerance

Increasing salinity of the soil has become a serious problem in arid and semiarid regions and thus limits the production of vegetable crops. The interaction of salts with plant physiological process is obviously complex. There are many salt species, many mechanisms and many organs, tissues and cells involved. A number of reasons are given in literature for suppressed growth under salt and water stress which includes: injury to cell membranes, $Ca^{+2}$, $Na^+$ interaction, $Na^+$ - $K^+$ selectivity, transport and leakage, osmotic adjustment through solute accumulation in the symplast, salt accumulation in the apoplast, hormonal balance of CK and ABA and nutrient deficiencies especially of $Na^+$ and K. According to (Rush and Epstein, 1976), the biotypes of *S. cheesmanii* can survive in full strength seawater in a hydroponics medium, whereas *S. esculentum* cannot sustain even half strength seawater. Tolerance was associated with the ability of the cells to cope with increased sodium levels. At higher sodium level. *S. cheesmanii* can absorb and accumulate a greater quantity of salt than *S. esculentum*. (Rush and Epstein, 1981), *Abelmoschus esculentus* is sensitive to salinity. Six varieties were screened at 4.2, 7.8, 11.6 and 16.3 mmhos/cm. 'Pusa Sawani' was found the most tolerant to salinity among other varieties for germination and vegetative growth. This variety can be grown in soil having salinity levels upto 6 mmhos/cm with sodium chloride and 8 mmhos/cm sodium sulfate without significant losses in yield and quality

(Kalloo, 1988). Pea and tomato varieties were screened at 4, 6, 8, and 12 mmhos/cm. In pea, varieties 'New Line Perfection', 'Market Prize' and 'Duke of Albany' in tomato 'P23', 'Pusa Ruby', and 'Best of All', were found tolerant to salinity and can be grown under moderate saline conditions (Kalloo, 1988). Salt tolerance can be induced by continuous growing of the varieties under salinized conditions. Irrigation with a salt solution or saline water also induced salt tolerance. It was reported that by continuous growing of varieties of *Lycopersicon* in salanized conditions, plants become tolerant.

Inheritance of salt tolerance is required to transfer salt tolerance characteristics in the salt susceptible commercial variety. Salt tolerance in *Lycopersicon* is a heritable characteristic (Rush and Epstein 1978, 1981a). The segregating population of *Solanum cheesmanii*, a tolerant and *S. esculentum,* a susceptible genotype has indicated that tolerance is genetically controlled. Thus, there is the possibility to transfer the salt tolerance gene from the resistant/tolerant species.

## Desirable Quality Traits in Vegetable Crops

See the Table 9.43.

## Source for Agronomic Traits in Vegetable Crops

See the Table 9.44.

## Introduction of Male Sterile/Restorer Lines and Hybrids (Source for breeding mechanism)

### Male Sterility

Male sterility is characterized by nonfunctional pollen grains, while female gametes function normally. Male sterility refers to the absence of functional pollen grains in other wise hermaphrodite flowers. Male sterility is of great value in the production of hybrid seed.

### Types of Male Sterility

Male sterility is of two types:

**Table 9.42: Resistance Sources for Abiotic Stress in Vegetables**

| Crop | Abiotic Stress | Resistance Source | |
|------|----------------|-------------------|---|
| | | Wild Taxa | Germplasm/Lines/Varieties |
| Tomato | Low temperature fruit setting | Solanum pimpinellifolium | |
| | Long shelf life | | EC391024 (Australia) |
| | Slow ripening gene | | EC310299 (USA) |
| | Heat tolerant | | EC399828-38 (USA) |
| | Heat tolerant | | EC347359 (Taiwan) |
| | Heat tolerant | | EC32142-26 (Taiwan) |
| | Low irrigation requirement | | EC378682-83 (Australia) |
| Brinjal | Drought | Solanum macrosperma | |
| Capsicum | Water submergence | | EC334206 (Taiwan) |
| Potato | Wider adaptation | Solanum tuberosum | |
| | Frost resistance | S. acaule, S. megistacrolobum, S azanhuiri | |
| | Heat tolerance | S. chacoense, S. commersonii | |
| | Lack of tuber dormancy | S. phureja | |
| | Heat tolerance | S. bulbocastanum | |
| | Frost resistance | S. cartilobum, S. azanhuri | |
| | Frost resistance, day length neutrality | Solanum commersoni | |
| Cucumber | Heat resistance, wider adaptation | Cucumis hardwickii | |
| Wild Pumpkin | Wider cross compatibility | Cucurbita lundeliana | |
| Muskmelon | Sulpher | | EC350603 (USA) |
| | Heat tolerant | | EC350918-28 (Japan) |
| Onion | High temperature tolerance | Allium cepa var. aggregatum | |
| | Cold and drought tolerance | Allium porrum, A. schoenoprasum | |
| Chinese cabbage | Heat tolerant | | EC345978 (Taiwan) |
| Cabbage | Heat tolerant | | EC350918-28 (USA) |
| | Heat tolerant | | EC168041-42 (USA) |
| Cauliflower | Heat tolerant | | EC164414-16 (USA) |

**Table 9.43: Source for Desirable Quality Traits in Vegetable Crops**

| Crop | Quality Traits | Source for Quality Traits | |
|------|---------------|---------------------------|---|
| | | *Wild Taxa* | *Germplasm/Lines/Varieties* |
| Tomato | Ascorbic acid, β carotene | *Solanum pimpinellifolium* | |
| | Ascorbic acid | | AC238, CO3 |
| | Total soluble solid | *Solanum minutam (syn. S. parvifloum), Solanum chmielewskii* | EC031767, EC081841 |
| | β carotene | *Solanum hirsutum* var. *glabratum, Solanum hirsutum* | |
| | Distance marketing and processing | | Arka Saurabh, Arka Vikash, Pusa Uphar, Pb. Chuhara, Roma, La-Bonita, EC391024, EC310299 |
| | Paste making | | EC129604, EC129675, EC129122, Se.4, HS101, HS110, Indo-processII, Indo-process III, Rupali, Sel.11, IIHR674, SB-1, EC321425, EC321426, EC054628, EC052062, EC321425-26 (Taiwan) |
| | Processing type | | EC126676, EC004708, EC129600, EC001154, EC002790, EC159966, EC093739, EC155955, EC024602, EC008820, EC093739, EC164677, EC246028, EC089258, EC160183 |
| | Suitable for fresh market | | EC260636, EC232429, EC024153, EC161245, EC007288, EC027251, EC267729 |
| | Suitable for transportation | | EC161252, EC007352, EC100845, EC110268 |
| | Large fruit type | | EC163708, EC041340, EC267726, EC267728, EC251751, EC168070, EC267725, EC367857, EC361674, EC362945, EC339060, EC367856 |
| | Pear shape fruits | | EC367857, EC035332, EC006192, EC326144 |
| Potato | High starch | *S. vernei* | |
| | Lack of tuber blackening | *S. hjertingi* | |
| | Chip making | | Kufri Chipsona-1, Kufri Chipsona-2 |
| | High protein content | *S. phureja* | |
| Muskmelon | Thick rind, good keeping quality | *Cucumis melo* var. *cantalupensis* | |
| | Sweet and juiciness | | Durgapur Madhu, Pusa Madhuras, Hara Madhu |
| | Good keeping quality | | Arka Rajhans |
| | Vit.C | | Arka Jeet |
| Watermelon | Yellow fleshed | | EC380989-92 (USA) |
| | Red fleshed/round/small | | EC393240-43 (USA) |
| Pumpkin | Good keeping quality, hard rind, strong flavor | *Cucurbita moschata* | |
| | High carotene and keeping quality | | Arka Chandan, Jammu Local |
| | Good transport quality | | Arka Suryamukhi, Chanani Green |
| Chillies and Capsicum | Low pungency, fruit colour | *Capsicum annuum* | |
| | Pungent type | | EC391073 (Taiwan) |
| | High pungent type | | EC347263 (Taiwan) |
| | High capsaicin, more pungency | *Capsicum frutesemce* | |
| | High oleoresin, more colour | *Capsicum chinense* | |
| Onion | Large bulb size | *Allium ampeloprasum* | |
| | More leaf flavor | *Allium kurat, Allium fistulosum, Allium tuberosum* | |
| | Good keeping and storage quality | | Pusa Red, N2-4-1, Pb. Selection, Arka Niketan, Agrifound Light Red, Baswant 780 |
| | Dehydration | | Pb. 48, Pusa White Flat, Pusa White Round, Udaipur 102, N257-9-1 |
| | High TSS and long shelf life | | EC378476-77 (UK) |

| Crop | Quality Traits | Source for Quality Traits | |
|------|----------------|---------------------------|---|
| | | Wild Taxa | Germplasm/Lines/Varieties |
| Garlic | Export and transport quality | | G282, G313 (Agrifound Parvati) |
| | Good storage | | Agrifound white (G41) |
| Carrot | High β carotene | | Pusa Kesar, Pusa Meghali, Pusa Yamdagini |
| | Pickle type | Cucumis sativus | EC565832 (Netherlands) |

**Table 9.44: Source for Agronomic Traits in Vegetable Crops**

| Crop | Traits | Wild Taxa | Accessions |
|------|--------|-----------|------------|
| Tomato | Fruit cracking | Solanum cheesmani | |
| | Early maturing type | | EC357833, EC357827, EC362956, EC362940, EC362944, EC362947, EC362948 |
| | Late maturing type | | EC357829, EC357830, EC362941, EC362942, EC362951, EC241147, EC163594, EC212469, EC241129 |
| Chillies | Early maturity, bunch type | | EC391089-92 (Taiwan) |
| Potato | Short maturity | S. phureja | |
| Cucumber | Early and determinate type | | EC398030 (China) |
| Summer squash | Short maturity | Cucurbita pepo | |
| Watermelon | Early | | EC380989-92 (USA) |
| | Round and small fruit | | EC393240-43 (USA) |
| Onion | High bulb yield | | EC30802-11 (USA) |
| Garlic | High yielding and bold clove | | EC367655-58 (Israel) |

(I) Genetic male sterility

(II) Cytoplasmic male sterility

(III) Cytoplasmic-genetic male sterility

Cytoplasmic male sterility is termed cytoplasmic-genetic when restorer genes are known.

### Utilization of Male Sterility in Plant Breeding

There are several natural mechanisms that control the mode of pollination in crop plants. Two of these mechanisms, self-incompatibility and male sterility are of special significance because of their utilization in hybrid seed production

### (I) Genetic Male Sterility

Genetic male sterility is ordinarily governed by a single recessive gene, *ms*, but dominant genes governing male sterility are also known, *e.g.*, in safflower. Male sterility alleles arise spontaneously or may be artificially induced. A male sterile line may be maintained by crossing it with heterozygous male fertile plants. Such a mating produces 1:1 male sterile and male fertile plants. Genetic male sterility may be used in hybrid seed production. The progeny from **ms ms x Ms ms** crosses are used as female, and are inter planted with a homozygous male fertile (**Ms Ms**) pollinator. The genotypes of the **ms ms** and **Ms ms** lines are identical except for the **ms** locus, *i.e.*, they are isogenic and are known as male sterile (**A**) and maintainer (**B**) lines, respectively. The female line would, therefore, contain

both male sterile and male fertile plants; the latter must be identified and removed before fertile plants; the latter must be identified and removed before pollen shedding. Pollen dispersal from the male (pollinator) line should be good for a satisfactory seed set in the female line. In India, it is being used for hybrid seed production of arhar (*C. caian*) by some private seed companies. It is use in several other crops, *e.g.*, cotton, barley, tomato, sunflower, cucurbits *etc.*, but it is not yet practically feasible.

### (II) Cytoplasmic Male Sterility

This type of male sterility is determined by the cytoplasm. Since the cytoylasm of a zygote comes primarily from egg cell, the progeny of such male sterile plants would always be male sterile. The male sterile line is maintained by crossing it with the pollinator strain used as the recurrent parent in the backcross programme since it s nuclear genotype is identical with that of the male sterile line. Such a male fertile line is known as the maintainer line or B line as it is used to maintain the male sterile line. The male sterile line is also known as A line. It is observe in some important plant crops *viz*. Maize, (*Zea mays*), *Nicotiama tabacum*, *Triticum aestivum*, *Gossypium hrisutum*, *Sorghum bicolor*, *Helianthus annus*, *Oryza sativa etc*. Cytoplasmic male sterility may be utilized for producing hybrid seed in certain ornamental species, or in species where a vegetative part is of economic value. But in those crop plants where seed is the economic part, it is of no use because the hybrid progeny would be male sterile.

### (III) Cytoplasmic-Genetic Male Sterility

This is a case of cytoplasmic male sterility where a nuclear gene for restoring fertility in the male sterile line is known. The fertility restorer gene R, is dominant and is found in certain strains of the species, or may be transferred from a related species *e.g.*, in wheat. This gene restores male fertility in the male sterile line, hence it is known as restorer gene. The cases of cytoplasmic male sterility would be included in the cytoplasmic genetic system as and when restorer genes for them would be discovered. This system is known in maize, jowar, bajra, sunflower, rice and wheat. The plants would be male sterile in the presence of male sterile cytoplasm if the nuclear genotype were rr, but would be male fertile if the nucleus were **Rr** or **RR**. The development of new restorer strains is somewhat indirect. First a restorer strain (**R**) is crossed with a male sterile line (**A**). the resulting male fertile plants are used as the female parent in repeated backcrosses with the strain (**C**), used as the recurrent parent, to which the transfer of restorer gene is desired. In each generation, male sterile plants are discarded, and the male fertile plants are used as females for backcrossing to the strain **C**. This acts as a selection device for the restorer gene **R** during the backcross programme. At the end of the backcross programme, a restorer line isogenic to the strain **C** would be recovered. The cytoplasmic-genetic male sterility is used commercially to produce hybrid seed in maize, bajra and jowar. A triple cross may be produced by crossing single cross with a fertility restoring inbred so that all the plants in the triple cross would be male fertile.

### Origin of Male Sterile Cytoplasm

#### (I) Spontaneous Mutation

Mutant male sterile cytoplasms arise spontaneously in low frequencies. Mutant cytoplasms have been isolated in maize, bajra and sunflower.

#### (II) Interspecific Hybridization

Transfer of the full somatic chromo-some complement of a crop species, through repeated backcrossing into the cytoplasm of a related wild species often leads to cytoplasmic male sterility. In cross-pollinated crop species, the male sterile cytoplasms have generally originated through mutation, while in self-pollinated crops they have been transferred from related species.

#### (III) Induction through Ethidium Bromide

Ethidium bromide is a potent mutagen for cytoplasmic genes or plasmagenes. Male sterile cytoplasm may be induced by seed treatment with ethidium bromide *e.g.*, Petunia.

### Limitation of Cytoplasmic-Genetic Male Sterility

1. **Undesirable-effects of the Cytoplasm**: for *e.g.*, the Texas cytoplasm in maize, by far the most successful cytoplasm commercially, slightly retards growth, yield (2.4 per cent) plant height and leaf number; induces earlier silking and delayed pollen shedding; and makes the plants highly susceptible to **Helminthosporium** leaf blight. Restorer genes only restore male fertility; they are unable to remove the side effects of the male sterile cytoplasms.

2. **Unsatisfactory Fertility Restoration:** In many cases, restoration of fertility is not satisfactory and cannot be used in the production of hybrid seed.

3. **Unsatisfactory Pollination:** Natural pollination is often not satisfactory, except in wind-pollinated crops like maize. This reduces the production of hybrid seed, and thereby increases its cost. In some species *e.g.*, *Capsicum*, this has prevented the use of male sterility in hybrid seed production. Poor pollination would always be a major problem in self-pollinators *e.g.* wheat.

4. Modifier genes may reduce the effectiveness of cytoplasmic male sterility.

5. Sometimes, cytoplasm may also be contributed by the sperm which, in the long run, may lead to a breakdown of the male sterility mechanism.

6. Male sterility mechanisms may break down partially under certain environmental conditions resulting in some pollen production by the male sterile lines.

In crops like wheat, polyploid nature of the crop and undesirable linkages with the restorer gene make it very difficult to develop a suitable restorer (**R**) line.

The $F_1$ hybrids introduced from different countries can be exploited gainfully as source of breeding mechanism. The male sterile/fertile and restorer lines of various crops can be used for hybrid production. The details of some such lines are as show in Table 9.45 (Gautam *et al.*, 1999.)

## Utilization of Genetic Resources for Vegetable Crop Improvement

If one analyzes the vegetable crop improvement since beginning through various means, it appears that since beginning from domestication recombinant technology are the various ways involved in vegetable crop improvement. To gain the benefit of genetic modification to resource poor families, it is important that pre-breeding centres should be established. The novel genetioc combinations produced as such pre-breeding centres should be used for developing location specific varieties through participatory breeding with farming families. This will help to combine genetic

**Table 9.45: Source for Breeding Mechanism**

| Crop | Breeding Mechanism | Resistance Source | |
|------|-------------------|-------------------|---|
| | | *Wild Taxa* | *Germplasm/Line/Varieties* |
| Tomato | Male sterility | *Solanum minutam* | |
| | Hybrids ($F_1$) | | EC3473359-68 (Taiwan) |
| | Hybrids ($F_1$)-cherry tomato | | EC343391-95 (Taiwan) |
| | Stamenless line ($F_2$) | | EC346011 (USA) |
| | Corolla and Stamenless line ($F_2$) | | EC346013 (USA) |
| | Green pistillate type | | EC346014 (USA) |
| Summer squash | More female flower | *Cucurbita pepo* | |
| Cucumber | Gynoecious line | | EC382737-39 (USA) |
| | Gynoecious ($G_4$) line | | EC329300 (USA) |
| Muskmelon | $F_1$ Hybrid (early and sweet) | | EC348140-43 (USA) |
| | Male sterile gene (MS2 to M6) | | EC382726-36 (France) |
| Onion | Production of bulb in clusters | *Allium cepa* var. *aggregatum* | |
| | Viviparous habit | *Allium cepa* var. *viviparum* | |
| | Easy crossability with *A. cepa* | *Allium vavilovi* | |
| Cabbage | Cytoplasmic male sterile and restorer lines | | EC304718-25 (USA) |
| | Male fertile lines | | EC187228-30 (Canada0 |
| Bitter gourd | Gynoecious line | | GX63 |

efficiency with genetic diversity at field levels. Also it brings wisdom of past and innovations of the present into synergetic blend (Singh, 2010).

## (1) Tomato

Tomato is a unique vegetable crop which is broadly used for genetic, cytogenetic, plant breeding and molecular research. Many useful genetic stocks have been generated by tomato breeders and conversely studies in basic genetics like linkage maps. Later, interspecific hybridization especially with the closely related and genetically conspecific *S. pimpinellifolium* and molecular genetics have received more attention and wild sources have been found to be vastly richer than the cultivated tomatoes. Among the allied species *S. pimpinellifolium* belonging to Eulycopersicon subgenus, has contributed genes governing resistance to several diseases like *Fusarium* wilt (*F. oxysporum* f. *lycopersici*), leaf mould (*Cladosporium fulvum*), bacterial wilt (*Pseudomonas solanacearum*), bacterial canker (*Corynebacterium michiganense*), buck-eye-rot (*Phytophthora parasitica nicotianae* var. *parasitica*), bacterial speck (*Phytophthora syringe pv. tomato*), gray leaf spot (*Stemphylium solani*), early blight (*Alternaria solani*), septoria leaf spot (*Septoria lycopersici*), tomato leaf curl and tomato yellow leaf curl virus. Besides, *S. pimpinellifolium* has special attributes like high TSS improved fruit colour, high ascorbic acid and high acidity. *S. peruvianum*, a member of Eriopersicon subgenus with a large variability, is another species which has been identified as a rich source of resistant genes numerous diseases like TMV, Tomato Yellow Leaf Curl Virus, Corky root, root knot nematode *etc.* and because of interspecific barriers, this species

could not be utilzed much in breeding, although its cross with *S. esculentum* is successful by embryo culture. From the same *S. peruvianum* complex is another species *S. chilense* which is more compatible with *S. esculentum* and which is a source for resistant to Tmato Yellow Leaf Curl Virus and Corky root (*Pyrienocheta lycopersici*). It has also special attributes like tolerance to cool temperature in germination and water economy. There are also species like *S. esculentum* complex like *S. esculentum* var. *cerasiforme* carrying resistance to *Verticillium* wilt, early blight (*Alternaria solani*) and anthracnose (*Colletotrichum coccoides*). *S. hirsutum* is an important source of resistance to *Septoria* leaf spot, corky root, *Botrytis* mould, tomato leaf curl and Tmato Yellow Leaf Curl Viruses, and more importantly to tomato fruit borer (*Heliothis zea*). Besides, it has cold and frost tolerance. There are also some interspecific cross derivatives from *S. esculentum* x *S. pimpinellifolium* which have the special attributes of parthenocarpy useful in glasshouse tomatoes, *S. esculentum* x *S. hirsutum* which is the rich source of β-carotene and rich red colour and *S. esculentum* x *S. peruvianum* which is known for high vitamin C content. High sugar content has been recorded in *S. chimielewskii* and drought and salt tolerance in *S. pennellii*.

Donor germplasm lines were identified with definite source of resistance, *i.e.* ability to tolerate temperature stress and drought conditions. These can be used to develop lines possessing resistance one or more stresses. There was a wide variation among genotypes for fruit set under high temperature conditions, and a review of the evaluation studies (at Delhi with temperature of 38-40°C) reveals that 43 accessions set fruit at high temperature. Varieties HS-101 and HS-102 had the ability to set fruit in April, when temperature was 35-39°C in May. EC 130042, *S. cheesmanii* and EC-162935 set fruits at high temperatures due to the stigma exertion of less than 1 mm, whereas other sensitive genotypes produced more than 1 mm of stigma exertion. The most serious problem of high temperature is the reduced size of fruits. Generally there is fruit set in heat tolerant lines but development of such fruit is very slow and poor, with the result that fruits remain smaller. Under Delhi conditions, heat tolerant accessions EC-168070, 130053-1, 165393 were record to have large fruit (Chandra and Gupta, 1994). Wild species show remarkable variation in their inherent adaptation to drought stress. *S. pennellii* and *S. esculentum* var. *cerasiforme* are drought tolerant genotypes. *L. hirsutum* species is tolerant to cold and drought. EC-130042, Sel-28, EC-65992 and *S. pimpinellifoloium* (PI-205009) require more days to express wilting, thus showing tolerance to drought conditions. Sel-28, *S. cheesmanii*, K-14, EC-104395, *S. pimpinellifolium* (EC-65992) and *L. pimpinellifollium* (Pan American) are the best potential source for drought tolerance.

In this crop cold set (Pusa Sheetal), hot set (Pusa hybrid-1) and bacterial wilt resistant varieties have enabled to grow tomato in many non-traditional areas. 'Pusa-120' of tomato has made it possible to achieve higher yields and quality produce in root knot infected soils. The hybrid of tomato specially 'Naveen', 'Rupali', 'vaishali', 'Mangla', 'Avinash-1' and 'Avinash-2' has revolutionalised the production of tomato throughout the country. Hybrid 'Naveen' has now become very popular in off season in Himachal Pradesh. Earlier, 'Solan Gola' was grown in the state.

## (2) Chilli and *Capsicum*

*Capsicum* species are broadly classified into two groups *viz.* (i) purple flowered which include *C. cardenasii, C. eximum, C. pubescens* and *C. itovari* and (ii) white flowered *C. annuum* (wild and domesticated), *C. baccatum* (wild and domesticated), *C. chinense* (wild and domesticated), *C. frutescens* (wild and domesticated), *C. chacoense, C. coccineum, C. galapagoense* and *C. praetermissum*. In improvement programme resistance to pests and diseases has been identified in several species like *C. baccatum, C. chinense, C. annuum* and *C. frutescens*. Sources of resistance to important diseases like *Phytophthora* rot and cucumber mosaic virus have been found in *C. baccatum*, to tobacco mosaic virus, leaf curl virus and *Verticillium* wilt in *C. chinense* and to potato virus Y in *C. barbatum*. *C. frutescens* is reported to have resistant genes to diseases like *Phytophthora* rot and *Verticillium* wilt. There are resistant cultivars developed by the breeders like Ssantanka (bacterial leaf spot), College No. 9 (*Fusarium* wilt), Pant C-1, KAUcluster and white Khandhari (bacterial wilt), World Beater (*Sclerotium*) and Roma (CMV).

**Table 9.46: Utilization of Selected Genotypes in *Capsicum* spp. Improvement**

| Genotypes | Species | Remarks, Especially on Origin/Pedigree, Degree of Pungency, Resistance to Leaf Curl |
|---|---|---|
| Bhut Jholokia | *Capsicum frutescens* x *Capsicum chinense* | Interspecific derivative, landrace of north east India, world hottest known so far, highly resistant to leaf curl (HR), partial restorer |
| BS-35 | *Capsicum frutescens* x *Capsicum chinense* | Interspecific derivative, landrace of north east India, very hot, leaf curl (symptomless) partial restorer |
| CO-O309 | *Capsicum frutescens* | Introduced from AVRDC, originally collected from Taiwan, highly pungent, highly resistant, partial restorer, resistant to tomato etch virus |
| Lankamura Tripura collection | *Capsicum frutescens* x *Capsicum chinense* | Interspecific derivative, landrace of north east India, highly pungent, leaf curl resistant (R), partial restorer |
| CO-5635 | *Capsicum baccatum* | Introduced from AVRDC, originally collected from Brazil, non-pungent, susceptible to leaf curl (S), non-crossable |
| COO-304 | *Capsicum chinense* | Introduced from AVRDC, originally collected from the United States, mild pungent, highly resistant to leaf curl (HR), resistant to tomato etch virus, non-crossable |
| IC-383072 | *Capsicum frutescens* | Collected by ICAR-NBPGR, New Delhi (India), highly pungent, highly resistant to leaf curl (HR), partial restorer |
| GKC-29 | *Capsicum annuum* x *Capsicum chinense* | Interspecific derivative, landrace of north east India, highly pungent, SL, non-crossable |
| Punjab Lal | *Capsicum annuum* | Improved variety developed by PAU, pungent, leaf curl resistant (R), restorer |
| NMCA-40008 | *Capsicum frutescens* | Introduced from NMSU, USA, highly pungent, Tabasco, leaf curl resistant (R), partial restorer |
| NMCA-50003 | *Capsicum annuum* | Introduced from NMSU, USA, non-pungent, Yellow Wonder, moderately susceptible to leaf curl (MS), maintainer |
| AMK-11 | *Capsicum annuum* | Collection from Dharwad (Karnatak), pungent, partial restorer |
| ISC-9 | *Capsicum annuum* | Introduced from Israel, pungent, moderately susceptible to leaf curl (MS), partial restorer |
| EC-619636 | *Capsicum annuum* | Introduced from AVRDC, mild pungent, susceptible, restorer |
| Sel-4 | *Capsicum annuum* | Developed at Solan (Himachal Pradesh), improved sweet pepper line (non-pungent), S, maintainer |
| Solan Local | *Capsicum annuum* | Developed at Solan (Himachal Pradesh), improved sweet pepper line (non-pungent), S, maintainer |

| Genotypes | Species | Remarks, Especially on Origin/Pedigree, Degree of Pungency, Resistance to Leaf Curl |
|---|---|---|
| AKC-89/38 | *Capsicum annuum* | Developed at PDKV, Akola, round fruited, upright bearing, pungent, MS, restorer |
| DSL-2 | *Capsicum annuum* | Collected from north eastern region, pungent, MS,, restorer |
| VR-339 | *Capsicum annuum* | Developed by IIVR, Varanasi (UP), pungent and MS, restorer |
| 0337-7545 | *Capsicum annuum* | AVRDC, Taiwan, pungent, MS, restorer |
| IVPBC-535 | *Capsicum annuum* | Introduced from AVRDC, purified and identified for oleoresin extraction at IIVR, non-pungent, S, Resistant to PVY, BW, partial restorer |
| VR-338 | *Capsicum annuum* | Developed by IIVR, Varanasi (UP), mild pungent and S, restorer |
| Japani Longi | *Capsicum annuum* | Improved population from IIVR, pungent, moderately susceptible to leaf curl |
| Taiwan-2 | *Capsicum annuum* | Introduced from AVRDC, highly pungent, MR, restorer |
| KTPL-19 | *Capsicum annuum* | Introduced from Spain at IARI, New Delhi, non-pungent, suitable for oleoresin extraction, MR, partial restorer |
| NuMex Twilight | *Capsicum annuum* | Introduced from NMSU, USA, ornamental type, mild pungent, S, restorer |
| NuMex Senteniales | *Capsicum annuum* | Introduced from NMSU, USA, ornamental type, mild pungent, S, restorer |
| DC-16 | *Capsicum annuum* | Collected from north east India, purple fruit, pungent, MS, restorer |
| CM-334 | *Capsicum annuum* | Mexican landrace, pungent, resistant to PVY, PeMV, *Phytophthora* and nematodes, MS, restorer |
| Pant C-1 | *Capsicum annuum* | Developed at Pantnagar (Uttarakhand) India, pungent, MS, strong, fertility restorer. |

In chilli, not much progress has been made on the exploitation desirable gene(s) from wild taxa or cultivated species other than *annuum*. There is ample scope of diversification of germplasm and utilization of resistance sources available in the elsewhere-cultivated species like *chinense*, *frutescens* and *baccatum*. All these species are crossable, although with little difficulty. The derivatives of the crosses *annuum* x *chinense* and *annuum* x *baccatum* are most potential sources of development of CMS line and resistant lines against several biotic stresses. There is need to identify more potential source of thrips, mites and dieback, which could be utilized to develop improved resistant lines. Available sweet pepper varieties are still facing problem of adoptability in North India plains. Furthermore, occurrence of diseases and pests are the major for handle and its popularity among the farmers. Considering these views, the target should be to available PGR for the development of parental lines for the hybrids, which are having generally more adaptability and incorporation of desirable resistant gene in comparatively easier in hybrid variety. Since most of the sweet pepper lines lack restorer gene, at this stage more emphasis should be given on the development and utilization of nuclear male sterile (genetic male sterile) lines for economic hybrid seed production. The local cultivars of chillies including 'Mathania Long' of Rajasthan becoming very popular in the state.

## (3) Brinjal

There are several species allied to *S. melongena* like *S. incanum* and *S. integrifolium* resistant to Fusarium wilt. *S. integrifolium*, *S. sisymbrifolium*, *S. torvum*, *S. melongena var. insanum* and *S. xanthocarpum* (now called *S. viarum*) which were found resistant to Verticillium wilt. The species S. *nigrum*, *S. sisymbrifolium*, *S. texanum* and *S. hispisdum* have been found to be resistant to bacterial wilt S. *gilo* and *S. integrifolium* resistant to Phomopsis rot

and *S. macrocarpon* resistant to *Cercospora* spot. Root knot nematode resistant has been found in *S. sisymbrifolium*, *S. torvum* and *S. aethiopicum*. For viral diseases like Big Bud, resistant has been located in *S. integrifolium*. Insect resistance has been recorded in *S. sisymbrifolium*, *S. integrifolium* and *S. xanthocarpum*, for shoot and fruit borer and aphid resiatance has been noted in *S. sisymbrifolium*. Interspecific barriers in brinjal can be got over by newer techniques like protoplast fusion or other biotechnological methods. For abiotic stresses like drought, *S. macrocarpon* has been found to be tolerant, while cv. R-34 has tolerance to high temperature. In this crop several cultivars which have been identified as resistant to different diseases and pests. Cv. "Florida Market" has been reported to be resistant to *Verticillium* wilt, *Phomopsis* fruit rot and root knot nematode. The Japanese cultivar Nihon Nassau has resistant genes for bacterial wilt and anthracnose. Brazilian cultivars have been found to have resistance to leaf spot (*Alternaria*) and anthracnose. Indian cultivar like Gulka has been recorded to be resistant to *Cercospora* and root knot nematode and "Improved Muktakeshi" to Cercospora and jassids. Pusa Purple Cluster genes for tolerance to bacterial wilt, little leaf diseases (mycoplasmal) and jassids and Pusa Bhairav has been evolved with resitance to *Phomopsis* rot. Likewise, several *S. melongena* cultivars have been evolved/identified as disease resistant source.

In brinjal, *S. macrosperma* and *S. integrifolium* are resistant and tolerant against phomopsis rot and drought, respectively. Both these traits can be transferred to cultivated *melongena* very easily through conventional method, since there is no crossing barrier between these species. Similarly, *Solanum khasianum*, *S. sisymbriifolium* and *S. torvum* have been identified for basic as well as applied studies. *S. khasianum* is a potential source of resistance to shoot and fruit borer. *S. sisymbriifolium* is source of resistance to root knot nematode and *S. torvum* is a perennial species having resistance to a number of root diseases. There is need and emphasis from ICAR to break crossability barrier between these potential wild spp. and eggplant with the aid of protoplast fusion or embryo rescue.

## (4) Cabbage

It is well established that *Brassica oleracea* is a polymorphic species of family Brassicaceae encompassing several vegetable crops like cabbage (var. *capitata*), cauliflower (var. *botrytis*), broccoli (var. *italica*), brussels sprout (var. *gemmifera*), kale (var. *acephala*) and knolkohl (var. *gongylodes*). The diversification in this species, has taken place through single gene mutants which are the primary sources of genetic variability. The genetic variability in these crops has been promoted by cross pollination and more specifically by the self-incompatibility system. Within *B. oleracea* itself, several important sources of resistance to diseases and pests and special attributes like heat tolerance, have been located. There are now several cultivars in cabbage having been bred for resistance to different diseases. 'Globelle', 'Hybelle' and 'Sanibel' have been reported for resistance to powdery mildew and tip burn. Wisconsin varieties, 'Jersey Wakefield', 'All Head Early', 'Glory of Enkhuizen' and 'Copenghagen Market' carry resistance to Fusarium yellows. Jones and Gilman (1915) developed cabbage yellows (*Fusarium oxysporum* conglutinans (W.r.) Synder and Hansen) resistant cultivar "Wisconsin Hollander" through selection. They demonstrated that two types of resistance to yellows occur. Type A is monogenic, and Type B is multigenic. Some cultivars have both types of resistance. Type A is the most potent form and has been incorporated into many cultivars. Yellows are now successfully controlled by the use of resistant cultivars. 'Golden Acre' and 'Pride of India' varieties of cabbage have spread rapidly throughout the country.

Savoy Cabbage is known to be resistant to downy mildew. The cultivars Badger Shipper, Acadie, Richesse, Richelai, Oregon 100 and Oregon 123 have been developed for resistance to clubroot. Resistance to black rot has been bred in varieties like Fuji, Early Fuji, Badger inbreds and Hugehot, Sel.8, and in Pusa Mukta. Similarly, against viruses, the resistant cultivars 'Ace High', 'De Brunswick' and 'Winter High III' have been evolved. 'Suprettee' and hybrid 'RI' carry resistance to cauliflower mosaic virus. Cultivars Lares and Winter High III are resistant to pepper spot (Black Speck, Spotted Necrosis, Grey Speck) and physiological disorders. Insect resistance breeding has made some headway. Bugner, Geneva Cabbage and few others carry resistance to cabbage looper; Green Acre and Geneva Cabbage have been bred for resistance to Diamond back moth. Early Jersey Wakefield, Danish Ballhead, Falcon and Geneva Cabbage have been found to resistant to Thrips attack. Against aphids, N.Y. cultivars of cabbage are found to be resistant. Japanese breeders were 1st to develop heat tolerant cabbages and KK cross and KY

cross were the early F[1] hybrids found to successful in SE Asia. The original source for heat tolerance was from the Japanese cultivars 'Yosin' and 'Kawasaki'. Now cabbage is no longer considered as a purely temperate vegetable crop and heat tolerant cultivars and hybrids are being developed by several countries.

## (5) Cauliflower

The tropical cauliflower varieties grown in India are distinct types, acclimatized to high temperature and high rainfall conditions during vegetative phase. These are essentially developed from an European biennial group-Cornish and later recombined with many other cauliflowers both annuals and biennials. Most of the Indian cauliflower varieties now belonging to the various maturity groups "Kunwari", "Katki", "Aghani", 'Poosi' and " Maghi" are highly heterozygous for vegetative, maturity and curd characters. The genetic diversity in these groups has been enlarged by the strong self-incompatability mechanism, not prevalent in the groups like late maturing Snowball, which is the annual summer cauliflower of Western Europe. Genetic resources present in Indian (tropical) cauliflower have also been utilized by breeders in Japan, Hawaii and South America. The indigenous types developed through involuntary selection during the 19th and 20th century, exhibit a high heterosis better than the Snowball types. Several self-incompatability alleles have been identified in Pusa Katki, an improved variety of Indian cauliflower. Hence for exploitation of heterosis, Indian cauliflower varieties are more suitable and hybrid seed production is facilitated by strong self-incompatability. Indian cauliflowers were found to be a very good source for black rot, curd and inflorescense blight caused by *Alternaria* species and downy mildew at seed production stage. The cauliflower variety 'Pusa Early Synthetic' has adapted to the warm climate of Tamil Nadu whereas varieties belonging to snowball group are becoming very popular due to their better quality and grown in winter season throughout the country. Hybrid seed production has become easier through the development of self-incompatible lines in cauliflower.

## (6) Onion

In *Allium* species sterility exhibited by interspecific hybrids indicates that their use in the genetic improvement of the bulb onion will be difficult. In India, short day onions are grown in large areas and the varieties are pungent, mainy used for cooking. In developed countries long day onions which are sweet, yellow and used for salad, are more popular and considerable work has been done in this crop. Within *A. cepa* improved inbreds have been selected with the resistance to diseases like pink root and *Fusarium* basal rot. The cultivar White Persian (glossy variety) has been reported to be resistant to thrips and USDA has developed four other glossy inbreds. Among the allied species, *A. fistulosum* is resistant to *Botrytis squamosa*, pink root, smut and onion maggot. *A. ampeloprasum* carries resistance to white rot and *A. royleii* to downy mildew. The hybrid of the cross between *A. cepa* x *A. fistulosum* is sterile. However, *A. royleii* may be a good source for incorporation of resistance to downy mildew in to *A. cepa* as their hybrid is found to be fertile. Onion cultivars Granex and Excel L35 are resistance to pink root disease. The Indian cultivars have long storage life, such as Bombay Red, Pusa Red, Arka Niketan and Agrifound Dark Red. Jones and Clarke (1943 and 1963) were the first who used cytoplasmic male sterility for producing hybrid seeds in onion at a cheaper rate. The hybrid onion programme originated in 1925 with the discovery by Dr. H.A. jones of a bulb 13-53 in material of 'Italian Red'. This bulb (13-53) was male sterile and could be propagated vegetatively as it produced bulbils in the umbels instead of seed. Male sterility carried by 13-53 was found to be stable under a wide range of environmental conditions. Identification of onion 'N-53' and development of technology for *kharif* onion has enabled to grow two crops annually. Now scientists are engaged to develop specific varieties of onion as per demand of different countries to strengthen its export. The promising varieties/hybrids of onion are 'Arka Kirtiman', 'Arka Lalima', 'Arka Pitamber', 'Arka Niketan', 'Arka Kalyan', 'Pusa Safed', 'Pusa Red', 'Pusa Madhavi', 'Pusa Ratnar', 'Pusa White Round', 'Pusa White Flat', 'Red Round, 'Punjab Selection', 'Punjab Red Round' and CO4.

## (7) Cucumber

In cucumber (*Cucumis sativus*), large variability is met with, within the species *C. sativus* itself. There is only one allied species known *viz. C. hardwickii* (distributed in lower Himalayas) which is crosses readily with *C. sativus* and has a good attribute of absence of fruit set inhibition. It has resistance to green mottle virus and some nematodes. As sources resistance to fungal diseases like anthracnose (PI lines), leaf spot (Hybrid 72502), scab (Davis Perfect), downy mildew (Poinsett), Powdery mildew (Natsufushinari) and *Fusarium* wilt (Wisconsin 28); to bacterial diseases like angular leaf spot (MS 9402, GY14A), bacterial wilt PI lines) and to virus diseases like CMV (Wisconsin and SMR lines), WMV1 (Surinam), WMV2 (Kyoto 3) and ZYMV (TMG-1). Hybrid seed production has become easier through the development of gynoecious lines in cucumber.

## (8) Muskmelon

Muskmelon is desert and non-dessert kinds, some of which are used like cucumber (Vellarikkai of T. Nadu; Nakkadosa Kaya of Andhra Pradesh and Kakri or longmelon of North India). There is snapmelon or Phoot in north India (*C. melo* var. *momordica*) which has been found to be resistant to virus diseases like CGMMV, ZYMV and WMV. Snapmelon has been used for breeding resistance to downy mildew in muskmelon Punjab Agricultural University, Ludhiana (Punjab). The

resistance to powdery mildew in muskmelon originally derived from the local Indian varieties were used to evolve resistant varieties in USA. In Africa there is a large distribution of allied species of *C. melo* consisting of annuals and perennials, monoecious and dioecious and polyploids. These do not cross readily with *C. melo* and several disease resistant genes located in these allied species could not be yet transferred to *C.melo*. However, among the annuals resistant sources identified are *C. africanus* (Green Mottle Virus and bean spider mite), *C. anguria* var. *longipes* (Green Mottle Virus, bean spider mite and nematodes), *C. sagittatus* (white fly). Among the perennial species, resistant sources are *C. ficifolius* (Green Mottle Virus and nematodes), *C. zeyheri* (Green Mottle Virus and nematodes), *C. asper* (Green Mottle Virus and bean spider mite) and *C. heptadactylus* (Green Mottle Virus and nematodes). Intercrossability was however successful in case of hybridization of *C. metuliferus* with *C. melo*.

A notable milestone in plant breeding history occurred in 1973 when Jaggar and Scott (1937) reported research that led to the release of Powdery Mildew (*Spaerotheca fulginea* Schecht ex fr Poll.) resistant Cantaloupe-45. In 1925, Professor J.T. Rosa of the University of California, Davis, received a shipment of muskmelon seed from India through one of his farmer students. Dr. Rosa in collaboration with I.C. Jagger of the USDA found a few plants in this collection that showed field resistance to powdery mildew. These Indian muskmelons had all sorts of objectionable characters. This was used as a source of resistance in several backcrosses to the commercial type. In 1936 a superior cultivar 'Powdery Mildew Resistant Cantaloupe-45' was released to seedsmen. The release of PMR-45 (resistance transferred from Indian muskmelons) caused a major revolution in the cantaloupe industry. Even today PMR-45 is one of the most widely grown cultivars of cantaloupe for shipping to distant markets due to thick, well developed colour, firm good flavor and sugar content. 'Homegarden' a cantaloupe line resistant to aphid, downy mildew and powdery mildew was developed by Ivanoff (1957). Mc Creight *et al.* (1984) made cross between PMR-45 x 90234 (resistant) and derived four aphid resistant lines *viz.* 'Hales Best', 'Jumbo', 'AR5' and ARTOP Mark. Downy mildew and Powdery mildew resistant line 'Edisto' was developed by Copeland in 1957. An another important cantaloupe line 'Seminole', resistant to powdery mildew, downy mildew along with high yield was developed by Whitner in 1960. Thomas and Webb (1981) developed multiple disease resistant lines 'W1', 'W3', 'W4' and 'W6' against powdery mildew, *Alternaria* blight and watermelon mosaic virus-1. Powdery mildew race-2 resistant lines, 'PMR-45' and 'PMR-4-50' were developed by Bohn and Whitaker in 1964. Webb (1967) developed watermelon mosaic virus-1 resistant cultivar, 'B 66-5'. Norton and Cosper (1985) developed gummy stem blight resistant lines 'Chilton' and 'Gulf Coast'. The local muskmelon (of netted type) grown in Banas river bed of Tonk district, Rajasthan is very famous in the state. Hybrid seed production has become easier through the development of gynoecious lines in muskmelon. Two downy mildew resistant varieties 'DMDR-1' and 'DMDR-2' and two viral resistant varieties 'DVRM-1' and 'DVRM-2' of muskmelon have also been developed.

### (9) Watermelon

The African citron- an intermediate domesticate between the wild species and cultivated varieties, is the main source of resistant genes for diseases like *Fusarium* wilt and anthracnose. In fact watermelon breeding started with the hybridization with *C. citron* to transfer resistant genes. There are now resistant cultivars like Charleston Gray and a few others which are resistant to both *Fusarium* wilt and anthracnose. Allied species like *C. naudinianus* carries immunity to both these diseases. CMV and ZYMV resistant genes can be used from *C. colocynthis* and WMV2 tolerant gene is also available in this species. The development of triploid seedless watermelon by Dr. Kihara in 1939 and subsequent success in producing commercial triploids in 1951 to 1952 is other major significant achievement. Triploid watermelon is a cross of tetraploid (4x) x diploid (2x).

### (10) Pumpkin

In *Cucurbita* species, *C. lundelliana*, *C. martinezii* and *C. okeechobeansis* are resistance to powdery mildew. *C. martinezii* is also resistant to squash and watermelon virus. *C. foetidissima* is highly resistant to drought. The species *C. foetidissima* when crossed with *C. moschata* produces sterile hybrids, while *C. lundelliana* gives partially fertile hybrids with *C. moschata, C. maxima* and *C. ficifolia* and its crosses with *C. pepo* and *C. mixta* produce fruits having seeds with small embryos. The species *C. martinezii* and *C. okeechobeansis* crossed with *C. moschata* gave partially fertile hybrids.

### (11) Okra

In *Abelmoschus* species is highly polymorphic in nature. In India *A. esculentus* and in Africa *A. manihot* are more widely cultivated. There are other species like *A. tuberculatus* found in India which has been identified as one of the genome ancestor of the cultivated allopolyploid okra. This species has resistance to fruit and shoot borer. Two major viral diseases, YVMV and OLCV infest the okra crop regularly and cause severe loss. The first YVMV tolerant variety "Pusa Sawani" was developed from the resistant source within *A. esculentus*. However, because of the breakdown of tolerance, efforts were made to transfer resistance genes from other species like *A. manihot* introduced from Ghana. Parbhani Kranti, Punjab Padmini and Punjab-7 were developed from this source, while Arka Abhay and Arka Anamika were evolved from *A. manihot* var. *tetraphyllus*. Complete immunity to YVMV has been reported in *A. pungens*, but it is not crossable with *A. esculentus*.

Through inter-specific hybridization, it was possible to transfer symptomless carrier type of tolerance to YVMV from *A. callei*. The other diseases such as *Cercospora* leaf spot and powdery mildew appear sporadically, but have not attracted the attention of breeders for breeding. The transfer of resistance from wild relatives has been hampered by sterility problems. *A. callei* and *A. tetraphyllus* have been utilized in the improvement of okra, largely.

### (12) Radish

Many Japanese varieties of radish basically of the Minowase and Nerima groups as well as monosomic addition lines with one kale chromosome, are resistant to mosaic virus diseases (TuMV, CMV, CaMV, REMV). The American variety Red Prince, Otable from Japan and China Rose and Kinmon-Aka from China have been found to be resistant to yellows disease (*Fusarium oxysporum* f sp. *congulutinnans*). Similarly the Japanese and Chinese radish cultivars were resistant to soft rot (*Erwinia carotovora* subsp. *carotovora*) and club rot (*Plasmodiophora brassicae*). The development of 'Pusa Chetaki', 'Pusa himani' and 'Punjab Ageti' varieties of radish, made it possible to grow radish throughout the year.

### (13) Carrot

The carrot (*Daucus carota*) variety "Brasilia" is resistant to nematode *Meloidogyne incognita* while the species *Daucus carota* subsp. *hispanicus* has resistance to the other nematode.

### (14) Vegetable Cowpea

Cowpea golden mosaic virus (CGMV) is a major disease that infests vegetable cowpea. Genotypes, CNC-043, TVu-612, CE-315, BR-1-Poty, Arka Samrudhi and Arka Garima (IIHR, Bengaluru), Cowpea-263 (PAU, Ludhiana), BC-244002 (BCKV, Kalyani), NDCP-3(NDUA&T, Faizabad) are resistant to CGMV. Resistant varieties have been developed at ICAR-IIVR, Varanasi (Kashi Gauri, Kashi Unnati, Kashi Kanchan and Kashi Sudha) by utilizing these varieties (Ferry and Dukes, 1995).

### (15) French Bean

'Arka Anoop' is resistant to both rust and bacterial blight. Lines, 'SP6c', 'SP178', 'SP61', 'SP180D' and 'SP377C' have been developed, which are resistant to BCM, BYM, CYV and BBW. Scully *et al.* (1995) developed 'Bellersnap' and 'Tender Cream' of cowpea, which are resistant to root knot nematode and *Cerospora* leaf spot.

### (16) Vegetable Pea

In vegetable pea, powdery mildew is a major disease causing heavy yield loss. Early varieties grown for table purpose escape disease incidence. In addition, rust also infests pea crop and cause significant yield loss. Varieties, 'PRS-4', 'JP-4', 'JP-83', 'NDVP-4' and 'FC-1' are resistant source for powdery mildew. The varieties 'Kashi Mukti', 'Kashi Shakti', 'Arka Ajit', and 'DPP9411' are resistant to powdery mildew. The varieties, 'Arka Karthik' and 'Arka Sampoorna' are resistant to both rust and powdery mildew.

## (17) Lettuce

Thompson and Ryder (1961) developed an amphidiploid "Vanguard" through inter specific hybridization between *L. virosa* and *L. sativa* using backcross breeding.

## (18) Potato

In potato a large number of improved varieties have been developed which is found to be suitable for growing under different agro-climatic conditions. Most of them also have resistance/tolerance to late blight, wart, cyst nematode and viruses.

**Table 9.47: Details of Area of Adaptation and Tuber Characters in Potato**

| Variety | Parantage | Year | Tuber Characters | Area of Adaptation |
|---|---|---|---|---|
| Kufri Kisan | Up-to-date x Sd.16 | 1958 | Large, round, white, deep eyes with prominent eye brows flesh white. Tubers tend to crack under high doses of nitrogen | Released for cultivation in North Indian Plains |
| Kufri Kuber | (*S. curtilobum* x *S. tuberosum*) x *S. andigenum* | 1958 | Medium, oval, tapering towards crownend, white, medium-deep eyes, flesh white. Poor keeping quality | An early variety released for cultivation in North Indian Plains and Plateau Region (Punjab, Bihar, and Maharashtra) |
| Kufri Kumar | Lumbri x Katahdin | 1958 | Medium, oval, tapering towards heel end, white, fleet eyes, flesh white | Released for cultivation in North Indian hills |
| Kufri Kundan | Ekishirazu x Katahdin | 1958 | Medium, round-oval, flattened, white, medium, deep eyes, flesh yellow | Released for cultivation in Himachal Pradesh and hills of Uttrakhand |
| Kufri Neela | Katahdin x Shamrock | 1963 | Medium, round, white, medium-deep eyes, flesh white | Released for cultivation in South Indian Hills |
| Kufri Sindhuri | Kufri Red x Kufri Kundan | 1967 | It is medium late variety and takes 110-120 days to mature. The tubers are medium, round, red skin, medium deep eyes, and dull white flesh. The tubers become hollow under high fertility condition. It is moderately resistant to early blight. It possesses very good keeping quality. Average yield is 30.0 t/ha. | Released for cultivation in North Indian Plains and plateau region |
| Kufri Jyoti | | 1968 | It is suitable for planting in the spring season. Plants are tall, erect, compact and vigorous with green foliage and white flowers. It matures in about 100 days. Tubers are large, oval, and white with fleet eyes and white flesh. It is resistant to wart disease. Average yield is 30.0 t/ha. | This variety was released for northern and southern hills |
| Kufri Alankar | Kennebec x ON 2090 | 1968 | Large, oblong, white, fleet eyes, flesh dull white, tubers turn purple on exposure to light | Early variety released for cultivation in North Indian hills and plateau region of peninsular India |
| Kufri Chamatkar | Ekishirazu x Phulwa | 1968 | Medium, round, white, medium-deep eyes, flesh yellow | Released for cultivation in North Indian Plains |
| Kufri Jeevan | M 109-3 x Seedling 698-D | 1968 | Medium, oval, white, fleet eyes, flesh pale yellow | Released for cultivation in North Indian Hills |
| Kufri Chandramukhi | Seedling 4485 x Kufri Kuber | 1968 | This is suitable for cultivation in North India and Plateau Region of peninsular India. It matures in 80-90 days. Its plants are medium tall and spreading with gray green foliage. It has good keeping quality and attractive white, large and smooth tubers with fleet eyes. Average yield is 25.0t/ha. | Released for cultivation in North Indian Plains and plateau region |
| Kufri Khasigaro | Taborky x Seedling 698-D | 1968 | Medium, round, round-oval, white, deep eyes, flesh pale yellow | Released for cultivation in North Eastern Hills |
| Kufri Naveen | 3070-D (4) x Seddling 692-D | 1968 | Medium, oval, white, fleet eyes, flesh pale yellow | Released for cultivation in North Indian Hills |
| Kufri Neelamani | Kufri Kundan x 134-D | 1968 | Medium, oval, flattened, white, fleet eyes, tubers turn purple on exposure to light, flesh yellow | Released for cultivation in South Indian Hills |
| Kufri Sheetman | Craigs Defiance x Phulwa | 1968 | Medium, oval, white, fleet eyes, flesh dull white | Released for cultivation in North Western Plains |
| Kufri Muthu | 3046 (1) x M 109-3 | 1971 | Large, round-oval, white, medium deep eyes, flesh white | Released for cultivation in South Indian Hills |
| Kufri Deva | Craigs Defiance x Phulwa | 1973 | Medium, round, white, with purple splashes, deep and picked purple eyes, flesh white | Released for cultivation in North Indian Plains |

| Variety | Parantage | Year | Tuber Characters | Area of Adaptation |
|---|---|---|---|---|
| Kufri Badashah | Kufri Jyoti x Kufri Alankar | 1979 | It matures in about 90-100 days. Plants are medium compact, vigorous with grey green foliage and white flowers. Tubers are large oval, white, fleet eyes; tubers turn purple on exposure to light, and having flesh dull white. Average yield is 31.5 t/ha. | Released for cultivation in North Indian Plains and plateau region |
| Kufri Bahar | Kufri Red x Gineke | 1980 | Large, round-oval, white, medium deep eyes, flesh white | Released for cultivation in North Indian plains |
| Kufri Lalima | Kufri Red x AG 14 (Wis. X 37) | 1982 | It matures in about 90-100 days. Plants are tall, erect, medium compact, vigorous with green foliage with purplish flowers. Large to medium, round red, medium deep eyes with white flesh.Average yield is 22.0 t/ha. It has good keeping quality and floury texture. | Released for cultivation in North Indian plains |
| Kufri Sherpa | Ultimus x Adina | 1983 | Tubers are medium, round, flattened, white, medium deep eyes, and having flesh pale yellow | Released for cultivation in North Bengal Hills |
| Kufri Swarna | Kufri Jyoti x (VTn)² 62.33.3 | 1985 | The plants are tall, erect, medium compact and vigorous with white flowers. It matures in about 120 days and produces large to medium, oval, white, flesh white with fleet eyes tubers. The variety is highly resistant to late blight and cyst nematodes. Average yield is 27.0-28.0 t/ha. | Released for cultivation in South Indian Hills |
| Kufri Megha | SLB/K-31 x SLB/Z-73 | 1989 | Medium, round, oval, white, fleet eyes, flesh white | Released for cultivation in North Eastern Hills |
| Kufri Giriraj | | 1992 | The plants are medium tall, semi erect, medium compact and vigorous with light purple flowers. It matures in 100 days and produces tubers that are medium to large, white and oval with fleet eyes. The variety is resistant to late blight. Average yield is 20.0-25.0 t/ha. | Released for cultivation in hilly regions of Himachal Pradesh. |
| Kufri Jawahar | Kufri Neelamani x Kufri Jyoti | 1996 | It takes 80 days to reach maturity. Its plants are short, erect, compact and vigorous with light green foliage and white flowers. Its tubers are oval, medium sized with fleet eyes and good keeping quality. This is moderately resistant to late blight.it is suitable for inter-cropping and mechanized cultivation. Average yield is 35.0 t/ha. | Released for cultivation in North Indian plains (Punjab, Haryana) and plateau region |
| Kufri Ashoka | EM/C-1020 x Allerfuheste Gelbe | 1996 | It takes 75-80 days to reach maturity. Its plants are medium tall and spreading with glossy-green foliage. The tubers are oval to oblong, large sized with fleet eyes. Its yield is 27.5t/ha. | Released for cultivation in Indo-Gangetic plains of central and eastern Uttar Pradesh Bihar and west Bengal. |
| Kufri Sutlej | Kufri Bahar x Kufri Alankar | 1996 | Tubers are medium-large, smooth, round-oval, creamy white, fleet eyesand flesh pale yellow. It matures in about 90-95 days. It is moderately resistant to late blight. Average yield is about 35.0t/ha | Released for cultivation in North Indian plains |
| Kufri Chipsona-1 | Mex.750825 x MS/78-79 | 1998 | It takes 100 days to reach maturity. Plants are medium tall, semi-erect and vigorous. Tubers are medium to large, oval, white, fleet eyes, with dull white flesh. It is low in reducing sugars, low phenols and high dry matter. It is resistant to late blight. Average yield is 25.0-30.0 t/ha. | Released for cultivation in North Indian plains |
| Kufri Chipsona-2 | F-6 x QB/B-92-4 | 1998 | It takes 100 days to reach maturity. Tubers are medium, round, white, fleet eyes, with light yellow flesh. The tubers are low in reducing sugars and phenols but rich in dry matter. It is tolerant to late blight and frost. Average yield is 25.0-30.0 t/ha. | Released for cultivation in North Indian plains |
| Kufri Giriraj | SLB/J-132 x EX/A 680-16 | 1998 | Medium to large, oval, white, fleet eyes, flesh white | Released for cultivation in North Indian Hills |
| Kufri Pukhraj | Craigs Defiance x JEX/B-687 | 1998 | It takes 75 days to reach maturity. Its plants are tall, erect, medium, compact and vigorous with grey green foliage and white flowers. Its tubers are medium sized, oval, slightly tapered with fleet eyes. The variety is resistance to early blight, moderately resistant to late blight and has an average yield of 32.5 t/ha. | Released for cultivation in North Indian plains (Punjab, Uttar Pradesh, Madhya Pradesh, Karnataka, Gujarat and Maharashtra) and plateau region |
| Kufri Anand | Kufri Ashoka x PH/F-1430 | 1999 | Large to medium, oval-long flattened white, fleet eyes, flesh white | North Indian Plains |
| Kufri Kanchan | SLB/Z-405 (a) x Pimpernel | 1999 | Medium, oval, red, eyes medium deep, flesh white | North Bengal Hills |
| Kufri Arun | Kufri Lalima x MS/82-797 | 2005 | Medium to large, oval, red, shallow eyes, flesh light yellow | Released for cultivation in North Indian Plains |

| Variety | Parantage | Year | Tuber Characters | Area of Adaptation |
|---|---|---|---|---|
| Kufri Pushkar | QB/A 9-120 x Spatz | 2005 | Medium, oval, white, shallow eyes, flesh light yellow | Released for cultivation in North Indian Plains |
| Kufri Shalija | Kufri jyoti x EX/A 680-16 | 2005 | Medium, oval, white, eyes fleet, flesh dull white | Released for cultivation in North Indian Plains |
| Kufri Himalini | I-1062 x Bulk pollen | 2006 | Medium, oval, white, shallow eyes, flesh pale yellow | Released for cultivation in North Indian Hills |
| Kufri Chipsona-3 | MP/91-86 x Kufri Chipsona-2 | 2006 | Medium, oval, white, fleet eyes, flesh cream | Released for cultivation in North Indian Plains |
| Kufri Surya | Kufri Lauvkar x LT-1 | 2006 | Medium, oval, light yellow shallow, flesh dull white | Released for cultivation in North Indian plains and plateau region |
| Kufri Sadabahar | MS/81-145 x PH/F-1545 | 2008 | Medium, oval, white, cream, eyes shallow, flesh white | Released for cultivation in North Central Plains |
| Kufri Himsona | MP/92-35 x Kufri Chipsona-2 | 2008 | Medium, round, white cream, shallow eyes, flesh cream | Released for cultivation in North Indian Hills |
| Kufri Girdhari | Kufri Megha x Bulk Pollen (10 genotypes) | 2008 | Medium, oval, white cream, shallow eyes, flesh white | Released for cultivation in Indian Hills |
| Kufri Khyati | MS/82-638 x Kufri Pukhraj | 2008 | Medium, oval, white cream, eyes medium, flesh cream | Released for cultivation in North Indian plains and plateau region |
| Kufri Frysona | MP/92-30 x MP/90-94 | 2009 | Medium to large, long-oblong, eyes shallow, flesh white | Released for cultivation in North Indian Plains |

## Use of Wild Genetic Resources

Global demand for biological resources from the wild has been continuously on the rise.

## Use of Genetic Resources for Grafting

Plant grafting is a propagation technique where two portions of plant which have similar organic texture are joined in such a manner so as to continue their development as a single plant. When grafting is performed, it is important to increase chances for vascular bundles of the scion and root stock to come into contact by maximizing the area of the cut surface that are spliced together and by pressing the spliced cut surface together. Care should be taken during grafting; the cut surface should not be allowed to dry out. In

### Table 9.48: Crop-wise Objectives of Grafting

| Vegetable Crops | Objectives of Grafting and Used Rootstocks |
|---|---|
| Cucumber | Tolerance to *Fusarium* wilt (*Cucurbita ficifolia* and *C. moschata* Shin-Tosa F₁), *Phytophthora melonis* (Squash cv. Kurondane and Shin-tosa), Cold hardiness, Favourable sex ratio, Bloom-less fruit, nematode (KJ-100, pumpkin (cv. Yunnan Black seeded)) |
| Brinjal | Tolerance to *Pseudomonas solanacearum* (*Solanum nigrum*), *Verticillium alborattum*, *Fusarium oxysporum*, low temperature, nematodes (*Solanum nigrum*), induced vigour and enhanced yield. |
| Tomato | Tolerance to *Pyrenochaeta lycopersici* (KNVF (r)), *Fusarium oxysporum* f.sp. *radicis* lycopersici, better colour and lycopene content, tolerance to nematodes (Bruisma (r), Nematex (r) and VNF (s)) |
| Melons | Tolerance to *Fusarium oxysporum* (*Lagenaria siciraria*, *Cucurbita moschata*, *Benincasa hispida*), wilting due to physiological disorders, *Phytophthora* diseases, cold hardiness and enhanced growth |
| Watermelon | Tolerance to *Fusarium oxysporum* (*Luffa aegyptica* (r), Summer squash (s)), wilting due to physiological disorders, cold hardiness and drought tolerance |
| Bitter gourd | Tolerance to *Fusarium oxysporum* f.sp. *momordicae* |

r: Resistant; s: Susceptible.

### Table 9.49. Rootstocks for grafting of vegetables

| Scion | Rootstocks |
|---|---|
| Cucumber | *Cucurbita moschata, C. ficifolia, C. maxima, Sicyos angulatus* |
| Melons | *Cucurbita sp., C. moschata X C. maxima, Cucumis melo, Benincasa hispida* |
| Watermeoln | *Citrulus lanatus, Cucurbita moschata, C. moschata X C. maxima, C. maxima,* |
| Bitter gourd | *Cucurbita ficifolia, C. moschata, Lagenaria siceraria, Luffa aegyptica* |
| Tomato | *Solanum pimpinellifolium, S. lycopersicon, S. nigrum, S. melongena* |
| Brinjal | *Solanum torvum, S. integrifolium, S. melongena, S. nigrum* |

*Source*: Heywood *et al.*, 2007.

the present era, grafting is very important area which is being utilized for crop improvement particularly in horticultural crops considering different biotic and abiotic stresses. The crop-wise major objectives being fulfilled through grafting along with the used rootstocks are given in Tables 9.48 and 9.49.

# References

1. Anonymous, 1990. Annual Report, Project Directorate on Vegetables, ICAR, New Delhi.

2. Anonymous, 1999. Annual Report, Directorate on Vegetables, Varanasi (U.P.).

3. Arnon, I. 1980. Breeding for higher coordinator's report on the 1st session. In: Physiological Aspects of Crop Productivity. Proceeding of the 15th Colloquium of the International Potash Institute, Wageningen, p. 77-81.

4. Bellotti, A. and Kawano, K. 1980. Breeding approaches in cassava. In: F.G. Maxwell and P.R. Jennings (eds.). Nreeding Plant Resistant to Insects. p. 314-369, John Willey and Sons, New Delhi.

5. Blumm, A. 1988. Plant breeding for stress resistance. CRC Press, FL.

6. Bohn, G.W. and Whitaker, T.W. 1964. Genetics of resistance to powdery mildew race in muskmelon. *Phytopathol.*, **54:** 587.

7. Bosch, S.L., Boelema, B.H., Serfontein, J.J. and Wanepoel, A.E. 1990. "Rotam 4" a multiple disease resistant fresh market tomato. *HortSci.*, **25(5):** 508-509

8. Chandra, U. and Gupta, P.N. 1994. Evaluation of tomato germplasm adaptable to abiotic stress condition of Northern India. *Indian J. Plant Genet. Resour.* 7(2): 165-172.

9. Cheema, D.S. 1985. Studies on improvement in chillies for resitance to anthracnose and Cercospora leaf spot disease. *Mysore J. Agric. Sci.*, **17(4):** 388-389.

10. Cheema, D.S., Dhiman, J.S. and Singh, S.1992. Multiple disease rsistance in hotset tomato genotypes in relation to their economic performance. Symposium on Integrated Disease Management and plant Health, Sept., 25-26, 1992. Held at Dr. Y.S. Parmar, UHF, Solan (H.P.).

11. Chelliah, S. and Sambandam, C.N. 1971. Distribution of amino acid in *Cucumis callosus* (Rottl.) Cogn. and *Cucumis melo* L. in relation to their resistance and susceptibility to *Dacus cucurbitae* (Diptera: Tephtritidae). *Lab Deve J. Sci. Tech. Bull.*, **11:** 41-43.

12. Chelliah, S. and Sambandam, C.N. 1973. Role of certain mechanical factors in Cucumis callosus (Rottl.) Cogn. in imparting resistance to *Dacus cucurbitae*. *AUARA.* **3:** 48-53.

13. Copeland, J.1957. New 'Edisto' cantaloupe. *Seed World.* **81(9): 8.**

14. Dadan, D.S. 1991. Variability in *Phytophthora nicotianae* var. *parasitica* (Dast.) Water house and management of buckeye rot of tomato in Himachal Pradesh. Ph.D. Thesis, Department of Plant Pathology, Dr. Y.S. Parmar UHF, Solan, (H.P.), India.

15. Dahms, R.G. 1972. Techniques in the evaluation and development of host plant resistance. *J. Environ. Qual.*, **1:** 254-259.

16. Debouck, D.G. 1999. Diversity in *Phaseolus* species in relation to the common bean, in: Singh, S.R. (ed.) *Common bean Improvement in the 21st Century.* Kluwer Academic Publisher, Dordrecht, p. 25-52.

17. Decker, P. 1951. Phomopsis blight resistant eggplants. *Phytopathology*, **41:** 9.

18. Esteva, J. and Nuea, F. 1992. Tolerance to white fly transmitted virus causing nuskmelon yellow disease in Spain. *Theor. Appl. Genet.*, **5-6:** 693-697.

19. Fageria, M.S. 1994. Studies on developing tomato hybrids with multiple diseases resistance. Ph. D. Thesis, Dr. Y.S. Parmar Univ. of Hort. and Forestry, Solan (H.P.).

20. Fageria, M.S., Arya, P.S. and Choudhary, A.K. 2001. In: Breeding for resistance. Kalyani Publishers, p. 177.

21. Fageria, M.S., Dhaka, R.S. and Jat, R.G.1998. Field evaluation of tomato genotypes for resistance to buckeye rot and *Alternaria* blight. *J. Mycol. Pl. Pathol.*, **28(1):** 52-53.

22. Gautam, P.L., Ashok Kumar and Singh, B.M. 1999. Germplasm collection, maintenance and utilization in improvement of vegetable crops. In: Narendra Singh (eds.) Summer School. In: Advanced Technology in Improvement of Vegetables, ICAR-IARI, New Delhi.

23. Gillbert, J.C., McGuire, D.C. and Tanaka, J. 1961. Indeterminate tomato hybrid with resistance to eight diseases. *Hawaii Farm Science*, 9(3): 1-3.

24. Hajjar, R. and Hudgkin, T. 2007. The use of wild relatives in crop improvement: A survey of developments over the last 20 year. *Euphytica*, 156: 1-130.

25. Harlan, J.R. and de Wet, J.M.J. 1971. Towards a rational classification of cultivated plants. *Taxon.*, **20:** 509–517

26. Hazara, P. and Som, M.G. 1989. Technology for vegetable production and improvement. Naya Prokash, Kolkata. p. 395.

27. Heywood, V., Casas, A. and Ford-Llyod, B. 2007. Conservation and sustainable use of crop wild relatives. *Agr. Ecosyst. Environ.* 121: 245-255.

28. Hodgkin, T., and Hajjar, R. 2007. Using crop wild relatives for crop improvement: trends and perspectives.

29. Ivanoff, S.S. 1957. The 'Homegarden' cantaloupe, a variety with combined resistant to downy mildew, powdery mildew and aphids. *Phytopathol.*, **47:** 552.

30. Jagger, I.C. and Scott, G.W. 1937. Development of powdery mildew resistant cantaloupe No.45. U.S. Deptt. Cir. Agr. p.141.

31. Jones, H.A. and Clarke, A.E. 1943. Inheritance of male sterility in the onion and production of hybrid seed. *Proc. Amer. Soc. Hort. Sci.*, **43:** 189-194.

32. Jones, H.A. and Clarke, A.E. 1963. Onions and their allies, Leonard Hill ltd. London, p. 286.

33. Kalda, T.S., Swarup, V. and Choudhary, B. 1976. Studies on resistance to Phomopsis blight in eggplant. *Veg. Sci.*, **3:** 65-70.

34. Kapoor, K.S. 1999. Screening methodology for effective breeding of cole crops against biotic stresses. In: Narendra Singh (eds.), Summer School on Advanced Technologies in Improvement of Vegetable Crops. ICAR-IARI, New Delhi.

35. Kashyap, R.K. and Kalloo, G. 1983. An appraisal of insect resistance in vegetable crops: A review. *Haryana J. Hort. Sci.*, **12:** 101-118.

36. Kogan, M. and Ortman, E.F. 1978. Antixenosis: A new term proposed to define Painter's "non-preference" modality to resistance. *Bull. Entomol. Soc. Am.*, 24: 175-176.

37. Kohli, U.K., Dev, H. and Thakur, M.C. 1996. Classification of tomato germplasm for resistance to buckeye fruit rot (*Phytophthora nicotianae* var. *parasitica*) and fruit size. *Ann. Agric. Res.*, **17(3):** 244-248.

38. Kraemer. M.E., Mebrahtu, T. and Rangappa, M. 1994. Evaluation of vegetable soybean genotypes for resistance to Mexican bean beetle (Coleoptera cooinellidae). *J. Econ. Entomol.* **87(1):** 252-257.

39. Krustaleva, V.V. and Shcherbakov, V.E. 1987. Semi-cultivated type mutants from a *Phytophthora* resistant sample of current tomato. Mutation Breeding *Newsletter.*, **29:** 9-12.

40. Kumar, M. and Ram, H.H. 1998. Path analysis for shoot and fruit borer resistance in brinjal (*Soloanum melongena* L.). *Ann. Agric. Res.*, **19(3):** 269-272.

41. Lal, B.S. and Sinha, R.P. 1974. Reaction of different cucurbit varieties to invasion by melon flies. Coq. Proc. *Bihar Acad. Agri. Sci.*, **22/23:** 10-103.

42. Lal, O.P. 1999. Recent advances in screening techniques in vegetable crops against insect pests. In: Narendra Singh (eds.). Summer School on Advanced Technologies in Improvement of Vegetables. ICAR-IARI, New Delhi.

43. Martin, G.B., Brommonchenkel, S.H., Chun Wongse, J., Frary, A., Ganal, M.W., Spivey, R., Wu, T., Earle, E.D. and Tanksley, S.D. 1993. Map-based cloning of a protein kinase gene conferring disease resistance in tomato. *Science,* 262: 1432-1436.

44. Maxted, N., Ford, B.V., Kell, S.P. and Turok, J. 2008. Crop wild relatives conservation and use. Wallingford, UK, CABI Publishing, p. 23-28.

45. Maxted, N., Ford-Lloyd, B.V., Jury, S.L., Kell, S.P. an Scholten, M.A. 2006. Towards a definition of a crop wild relative. *Biodiversity and Conservation,* 15(8): 2673-2685.

46. Mc Creight, J.D., Kishaba, A.N. and Bohn, G.W. 1984. 'Hale's Best', 'Jumbo', 'AB-5', 'AR-Topmark' melon aphid resistant muskmelon breeding lines. *HortSci.*, **19:** 309.

47. Mckinney, H.H. 1923. Influence of soil, temperature and moisture on infection of wheat seedlings by *Heliminthosporium sativum. J. Agric. Res.*, **26:** 195-917.

48. Misra, P.N, Singh, M.P. and Nautiya, M.C. 1996. Varietal resistance in tomato against fruit borer,(Heliothis armigera). *Indian J. Entomol.*, **58(3):** 222-225.

49. Muthulakshmi, P. and Seetharraman, K. 1994. Screening chilli genotypes for resistance to fruit rot disease caused by *Alternaria tenuis* Nees. In: K. Sivaprakasam and K. Seetharaman (eds.). Crop Diseases-Innovative Techniques and Management. Kalyani Publishers, Ludhiana.

50. Nath, P. 1971. Breeding Cucurbitaceous crops for resistance to insect. *SABRAO Newsletter,* **3:** 127-134.

51. Natti, J.J. 1958. Resistance of broccoli and other crucifers to downy mildew. *Pl. Dis. Reptr.* 42: 656.

52. Nazir, A.K., 1974. Studies on *Pseudomonas solanacearum* causing wilt of brinjal and tomato in Mysore state. *Mysore J. Agric. Sci.*, **8:** 477-478.

53. Norton, J.D. and Cosper, R.D. 1985. 'A.C. 70-154' a gummy stem blight resistant muskmelon breeding line with desirable horticultural characters. *Rep. Cucurbit Genet. Coop.*, **8:** 46.

54. Orton, W.A. 1908. The development of farm crops resistant to disease. U.S. Dept. Agr. *Year Book of Agri.*, **1908:** 453-504.

55. Orton, W.A. 1911. The development of disease resistant varieties of plants. IV. Cont. Intern. Genetics, Paris, *Compt. Rend. Et. Rapp.*, 247-265.

56. Painter, R.H. (1951. Insect resistance in crop plants. Mc Millan Co., New York.

57. Pal, A.B., Srinivasan, K. and Vani, A. 1983. Development of breeding line of muskmelon for resistance to fruit fly.*Prog. Hort.*, **15:** 100-104.

58. Pandey, S.K. and Luthra, S.K. 2010. Genetic resources for further potato revolution. *Indian Horticulture*, **55(35):** 63-68.

59. Paterson, A.H., Lander, E.S., Hewitt, J.D., Paterson, S.E., Lincoln, S.E. and Tanksley, S.D. 1988. Resolution of quantitative traits into Mendelian factors by using a complete linkage map of restriction fragment length polymorphism. *Nature*, **335:** 721-726.

60. Peter, K.V. and Joseph, M. 1986. Disease resistance in tomato, chilli and brinjal- A review. *Veg. Sci.,* **13(1):** 250-283.

61. Peterson, C.F., Williams, P.H. Palmer, M. and Louward, P. 1982. Wisconsin 2757 Cucumber. *Hort. Sci.,* 17: 268.

62. Pitrat, M. and Lecoq, H. 1985. Inheritance of resistance to cucumber mosaic virus transmission by Aphis gossypii in *Cucumis melo* L. *Phyt.,* **70:** 958-961.

63. Ponnamperuma, F.N. 1972. In: Soil Chemistry (J. Bremner and G. Chesters, eds.). Dekker, New York.

64. Rai, S. and Ram, D. 1997. Screening of French bean genotypes against leaf minor (*Phytomyza horticola* Meigen). *Veg. Sci.,* **24(1):** 58-60.

65. Rao, M.V.B., Sohi, H.S. and vijay, O.P. 1976. Reaction of some varieties of brinjal to *Pseudomonas solanacearum. Veg. Sci.,* **3:** 61-64.

66. Ravinder, Reddy, M. 1982. Evaluation of fungicides against major diseases of chili. M.Sc. Thesis, TNAU, Coimbatore, p. 64.

67. Rick, C.M. and Chetelat, R.T. 1995. Utilization of related wild species for tomato improvement. *Acta Horticulture*, **4125:** 21-28.

68. Sandhu, J.S. and Rana, J.C. 2016. Utilization of Crop Wild Relatives in the breeding programmes: Progress, Impact and Challenges. *Indian J. Plant Genet. Resour.*, **29(3):** 448-451.

69. Shade, R.E., Kitch, L.W., Mentzer, P. and Murdock, L.L. 1996. Selection of cowpea weevil (Colcoltera: Bruchidae) biotype virulent to cowpea weevil resistant land race TVu2927. *J. Econ. Entomol.* **87(5):** 1325-1331.

70. Sharma, S.K. and Rana, J.C. 2012. Strategies for the Conservation of Crops Wild Relatives-Indian Context, In: Sharma, A.K., Ray, D., Ghosh, S.N. (eds.) Biological Diversity-Origin, Evolution and Conservation. Viva Books Private ltd. New Delhi, p. 433-468.

71. Sharma, S.R., Gill, H.S. and Kapoor, K.S. 1997. Inheritance of resistance to white rot in cauliflower. *Indian J. Hort.*, **54(1):** 86-90.

72. Singh, B.D. 2001. Plant Breeding: Principles and Methods. 9th Revised Edition, Kalyani Publishers, p. 475.

73. Singh, D.P. 1987. Breeding for resistance to diseases and insect pest. Springer-Verlag, New York.

74. Singh, K.P. 2010. Exploitation of genetic resources of vegetable crops for precision farming. In: *Precision Farming in Horticulture*, p. 185-207.

75. Singh, P.P., Thinrd, T.S. and Lal, T. 1996. Reaction of some muskmelon genotypes against *Pseudomonas cubensis* under field and artificial epiphytotic conditions. *Indian Phytopathology*, **49(2):** 188-190.

76. Singh, R.B., Paul Khurana, S.M., Pandey, S.K. and Srivastava, K.K. 1998. Screening of potato germplasm for resistance to stem necrosis. *Indian Phytopathology*, **51(3):** 222-224.

77. Sitaramaiah, K., Singh, R.S., Vishwakarma, S.N. and Dubey, G.S. 1981. Brinjal cultivars resistant to *Pseudomonas* wilt. *Indian Phytopathology*, **34:** 113.

78. Somkuwar, R.G. and More, T.A. 1996. Inheritance of downy mildew resistance in muskmelon (*Cucumis melo* L.). *Indian J. Hort.*, **53(1):** 50-52.

79. Stoner, A.K. 1970. Breeding for insect resistance in vegetables. *Hort. Sci.,* 5: 76-79.

80. Sullivan, M.J. and Brett, C.H. 1971. Resistance of cucurbit varieties to the spotted cucumber beetle in the coastal plains of South Carolina. *J. Econ. Entomol.,* **64:** 1205.

81. Swanson, T. 1997. Global Action for Biodiversity. IUCN. Earth Scan Publication, London.

82. Tewatia, A.S. 1994. Resistance studies in bitter gourd against fruit fly. Ph.D. Thesis, CCSHAU, Hisar, Haryana.

83. Thakur, J.C., Khattra, A.S. and Brar, K.S. 1992. Comparative resistance to fruit fly in itter gourd. *Haryana J. Hort. Sci.,* **21:** 285-288.

84. Thomas, C.E. and Webb, R.E. 1981. W1, W2, W3, W4, W5 and W6 multi-disease resistant muskmelon breeding lines. *HortSci.,* **16:** 196.

85. Thompson, R.C. and Ryder, F.J. 1961. Description and pedigree of nine varieties of lettuce. U.S., Dept. Agr. Tech. Bull., 1244.

86. Tripathi, L.K., Jakhmela, S.S. and Bhaduaria, N.S. 1996. Comparative resistance to fruit fly in bitter gourd. *Haryana J. Hort. Sci.,* 21: 285-288.

87. Volin, R.B., Augustine, T.Y., Bryan, H.H., Burgis, D.S., Crill, R., Strobel, J.W. and John, C.A. 1977. "Florida 1011 tomato breeding line". *HortSci.,* **12(5):** 508-509.

88. Walkey, D.G.A. 1985. Applied Plant Virology. William Heinemann Ltd., Londan.

89. Walkey, D.G.A. and Neely, H.A. 1970. Resistance in white cabbage to necrosis caused by turnip and cauliflower mosaic virus and pepper spot. *Journal of Agricultural Sciences,* Cambridge, 95: 803-713.

90. Webb, R.E. 1967. Cantaloupe breeding line B66-5 highly resistant to watermelon mosaic virus-1. *HortSci.*, **2:** 58.

91. Whitner, B.F.J. 1960. 'Seminole' a high yielding good quality downy and powdery mildew resistance cantaloupe. *Univ. Agric. Exp. Stn.*, **5:** 122.

92. Wiseman, B.R., Hall, C.V. and Painter, R.H. 1961. Interaction among cucurbit varieties and breeding responses of the striped and spotted cucumber beetles. Proc. *Amer. Soc. Horti. Sci.*, **78:** 379.

93. Zeven, A.C. and DeWet, J.M.J. 1982. Dictionary of Cultivated Plants and their Regions of Diversity. Centre for Agricultural Publishing and Documentation, Wageningen.

# 10

# Genetic Erosion and Conservation of Vegetable Genetic Resources

## Introduction

The diversity of biological resources is the foundation of the maintenance of the vitality of crops, further development and improvement of crop varieties and ability to react to changing circumstance, such as pest and disease prevalence or climate change. Since the beginning of agriculture, the pioneer farmers of the world used 10, 000 plant species for food and fodder. While today only 150 crops feed the world's population, just 12 crops provide 80.0 per cent dietary energy from plants with rice, wheat, maize, and potato alone is responsible for 60.0 per cent. Rapid and large scale destruction of naturally distributed wild and weedy relatives of certain crops and replacement of landraces by high yielding varieties at an alarming rate is true picture of nature. Indiscriminate extraction of biological material from the wild threatens their survival, even to the extent of pushing some species into extinction, without ever being known. Some scientists believe that we will lose 60,000 of the 2,40,000 known plant species over next 30 years as a result of deforestation alone. Once they are gone, there is no alternative source for acquiring the unique qualities they provide. It is hence important to develop sustainable harvesting regimes that do not undermine the capability of a plant species to survive in the wild. The biologists/scientists throughout the world have shown great concern over the conservation and preservation of environment, biodiversity and ecosystem.

## Components Affecting Biodiversity

The persistence and in some cases intensification of the live principal pressures on biodiversity provide more evidence that the rate of biodiversity loss is not being significantly reduced. The overwhelming majority of the governments reporting to the CBD cite these pressures or direct drivers as affecting biodiversity in their countries. They are show in Table 10.1.

## Decline of Genetic Diversity and Genetic Erosion

The decline in species populations, combined with the fragmentation of landscapes, inland water bodies and marine habitats, have necessarily led to an overall significant decline in the genetic diversity of life on earth. While this decline is of concern for many reasons, there is particular anxiety about the loss of diversity in the varieties and breeds of plants and animals used to sustain human livelihood. A general homogenization of landscapes and agricultural varieties can make rural populations vulnerable to future changes, if genetic traits kept over thousands of years are allowed to disappear. The continued loss of genetic diversity has major implications for current and future human well-being. The loss of genetic diversity in agricultural systems is of particular concern as rural communities face ever-greater challenges in adapting to future climate conditions. In dry lands, where production is often operating at the limit of heat and drought tolerance, this challenge is particularly stark. Genetic resources

## Table 10.1: Components Affecting Biodiversity

| Sl.No. | Components | Descriptions |
|---|---|---|
| 1. | Habitat loss and degradation | Habitat loss and degradation create the biggest single source of pressure on biodiversity worldwide. For terrestrial ecosystems, habitat loss is largely accounted for by conversion of wild lands to agriculture, which now accounts for some 30 per cent of land globally. In some areas, it has recently been partly driven by the demand for biofuels. The IUCN Red List assessments show habitat loss driven by agriculture and unsustainable forest management to be the greatest cause of species moving closer towards extinction. Infrastructure developments, such as housing, industrial developments, mines and transport networks, are also an important contributor to conversion of terrestrial habitats, as is afforestation of non-forested lands. With more than half of the world's population now living in urban areas, urban sprawl has also led to the disappearance of many habitats. |
| 2. | Climate change | Climate change is already having an impact on biodiversity, and is projected to become a progressively more significant threat in the coming decades. Loss of Arctic sea ice threatens biodiversity across an entire biome and beyond. The related pressure of ocean acidification, resulting from higher concentrations of carbon dioxide in the atmosphere, is also already being observed. In addition to warming temperatures. More frequent extreme weather events and changing patterns of rainfall and drought can be expected to have significant impacts on biodiversity. The linked challenges of biodiversity loss and climate change must be addressed by policymakers with equal priority and in close coordination. Already, changes to the timing of flowering and migration patterns as well as to the distribution of species have been observed worldwide. These types of changes can alter food chains and create mismatches within ecosystems where different species have evolved synchronized inter-dependence, for example between nesting and food availability, pollinators and fertilization. Climate change is also projected to shift the ranges of disease-carrying organisms, bringing them into contact with potential hosts that have not developed immunity. Fresh water habitats and wetlands, mangroves, coral reefs, Arctic and alpine ecosystems dry and sub-humid lands and cloud forests are particularly vulnerable to the impacts of climate change. No doubt some species will benefit from climate change. |
| | | Climate change is projected to cause species to migrate to higher latitudes (*i.e.* towards the poles) and to the higher altitudes, as average temperatures rise. In high altitude habitats where species are already at the extreme of their range, local or global extinction becomes more likely as there are no suitable habitats to which they can migrate. The specific impacts of climate change on biodiversity will largely depend on the ability of species to migrate and cope with more extreme climatic conditions. Ecosystems have adjusted to relatively stable climate conditions and when those conditions are disrupted, the only options for species are to adapt, move or die. It is expected that many species will be unable to keep up with the pace and scale of projected climate change, and as a result will be at an increased risk of extinction, both locally and globally. |
| | | In general climate change will test the resilience of ecosystems, and their capacity for adaptation will be greatly affected by the intensity of other pressures that continue to be imposed. Those ecosystems that are already at, or close to, the extremes of temperature and precipitation tolerance are at particularly high risk. Over the past 200 years, the oceans have absorbed approximately a quarter of the carbon dioxide produced from human activities, which would otherwise have accumulated in the atmosphere. |
| 3. | Excessive nutrient load and other forms of pollution | Excessive nutrient load and other forms of pollutions like coal base industries, chemical industries, brick industries, cement industries *etc.* creating problem for natural survival of genotypes. Without completing their life cycle; plants are dying and restricting to future generation. |
| 4. | Over-exploitation and unsustainable use | Over-exploitation and unsustainable use are also important factors and responsible in loss of genetic resources. |
| 5. | Invasive alien species | Invasive alien species to be a major threat to all types of ecosystems and species. There are no signs of a significant reduction of this pressure on biodiversity, and some indications that it is increasing. Intervention to control alien invasive species has been successful in particular cases, but it is outweighed by the threat to biodiversity from new invasions. |

are critically important for the development of farming system that capture more carbon and emit lower quantities of greenhouse gases, and for underpinning the breeding of new varieties.

The databases, holding mainly passport data, can be analyzed for the identification of duplications and gaps among collections. International projects on the characterization and evaluation of vegetables germplasm, including molecular tools, are generating new data and making them increasingly available. Thus genebank material is becoming more attractive to breeders. At the same time, the management of collections can be based on better knowledge of the diversity in stock. The enhancement of the links between germplasm conservation and use will continue to depend, inter alia, on easy access to the genetic material.

The recently approved International Treaty on Plant Genetic Resources for Food and Agriculture established a Multilateral System for facilitated access to germplasm of a number of crops. This includes vegetables such as asparagus, beet, the Brassica complex, carrot and eggplant, but excludes tomato, pepper, cucurbits, Alliums, *etc.*, with possible implications on the use of these crops' diversity in the near future. The rich heritage and ethnic culture has favored to preserve the richest diversity including rare landraces/primitive types of useful vegetables like eggplant, cucumber, ridge and sponge gourd and a number of root and tuber crop species. But due course of time several important land races of 'cultivated and weedy forms' were wiped-out form the vicinity, not only from India but from other countries too. The conservation of biodiversity has now emerged as a priority across many levels of the society in virtually all countries because destiny of agro-biodiversity and people are in separately inter-connected. Gaining knowledge and growing awareness of the implication and consequences of this basic fact during the last six decades brought the issues related to conservation and sustainable utilization of natural bio-resources to the centre stage paving the way for the 1992 convention on Biological Diversity. Genetic diversity is being lost in natural ecosystems and in systems of crop production. Important progress is being made to conserve plant genetic diversity, especially using *ex situ* seed banks. The conservation of biodiversity makes a critical contribution to moderating the scale of climate change and reducing its negative impacts by making ecosystems- and therefore human societies-more resilient. It is therefore essential that the challenges related to biodiversity and climate change are tackled in a coordinated manner and given equal priority. In several important areas, national and international action to support biodiversity is moving in a positive direction. More land and sea areas are being protected, more countries are fighting the serious threat of invasive alien species, and more money is being set aside for implementing the Convention on Biological Diversity (CBD). Frankel (1975) first time reported in a survey of genetic resources that many collections had suffered genetic erosion due to hybridization, selection, genetic drift, unsuitable growing conditions, or human error during propagation. Quantification of genetic erosion requires time series analysis and using morphometric traits has always proven to be problematic (Thormann and Engels, 2015). These changes were detected both by SNP and SNP haplotype analysis during regeneration in maize and significant differences in allelic frequencies were also reported in barley, rye and Brassica (Parzies *et al.*, 2000; Soengas *et al.*, 2009), although in using limited number of accessions.

## Definition of Genetic Erosion

The genetic erosion can be defined as *"Gradual loss of variability from cultivated species and their wild forms and wild relatives"*.

Or

Genetic erosion refers to loss of genetic diversity between and within populations of the same species over a period of time.

Or

Gradual reduction in genetic diversity in the populations of a species, due to elimination of various genotypes/genetic base.

## Undesirable Consequences of Genetic Erosion

Singh (2012) pointed out following causes and undesirable consequences for genetic erosion:

1. **Replacements of Land Races (Heterogenous) by Improve Varieties (Homogenous) or Hybrids**

Th available improved varieties are commonly pure lines (self-pollinated) or hybrids (cross pollinated). The main features of modern cultivars are high yield, uniform (homogenous), narrow genetic base and narrow adaptability. On the other hand land races and primitive cultivars have more genetic diversity, broad genetic base, wider adaptability and low yield potential. It results in disappearance of land races, open pollinated varieties *etc.* which were reservoirs of genetic variability. A small number of improved varieties of each crop become predominant and rapidly replace the heterogeneous local varieties leading to a rapid depletion of genetic variability (genetic erosion).

2. **Narrow Genetic Base**

Many of the improved varieties of a given crop have one or more parents (immediate or somewhat removed in the ancestry) in common with each other. Thus, resulting improved varieties of a crop species are becoming increasingly similar to each other due to the chromosome of one or more parents in their ancestry. This has led to the narrowing own of the genetic base of the developed varieties. Genetic base refers to: *"The genetic variability present among the cultivated varieties of a crop species"*. Narrow genetic base has created genetic vulnerability, which refers to the susceptibility of most of the cultivated varieties of a crop species to a disease, insect pest or some other stress to similar in their genotypes (Singh, 2012). The narrow genetic base has created genetic vulnerability, which refers to: *"The susceptibility of cultivars of a crop species to a disease, insect pest or some other stress due to a similarity in their genotypes"*. Genetic vulnerability can be avoided by using diverse and unrelated parents in breeding programmes, and by using unrelated sources of male sterility, semi-dwarfness, *etc."*.

3. **Increased Susceptibility to Minor Disease**

Most of the resistance related research programmes are being initiated for major diseases and insect pests. This has often increased the level of susceptibility to various minor diseases. As a result, day by day minor diseases have gained importance and in some cases, produced severe epidemics.

4. **Adoption of Improved Crop Management Practices**

Improved management practices has virtually eliminated the wild forms of many crops. In general these forms existed as the so called crop-weed complexes, in which gene introgression occurred from weed to the crop and *vice-versa*. Thus these complexes were sources of considerable genetic diversity.

5. **Human Interventions**

Increasing human needs have extended farming

and grazing into forests, the habitats of most wild species. This has resulted to the extinction of many wild relatives of crops. Deforestation is also playing major role in erosion of genetic resources.

## 6. Developmental Activities

Developmental activities like hydro-electric projects, roads, industrial areas, railways, buildings, *etc.* have also disturbed the wild habitat. Often wild relatives of crops are destroyed due to these activities. The biodiversity is being eroded at an ever alarming rate of nearly 100 species a day due to habitat destruction and fragmentation.

## 7. Introduction of Weedy Species

Sometimes introduction (deliberate or accidental) of weedy species may result in the invasion of wild habitats by this species and lead to the elimination of native wild relatives of crop plants. Even the cultivated forms derived from such introduced species may contribute to genetic erosion.

## 8. Yield Plateau

It is true that if in a crop species variability exhausted and no further yield increases are obtained through breeding *i.e.*, the yield reaches a plateau. Therefore it is essential that old variability must be safeguard and new variability has to be introduced in the breeding population in order to break the plateau.

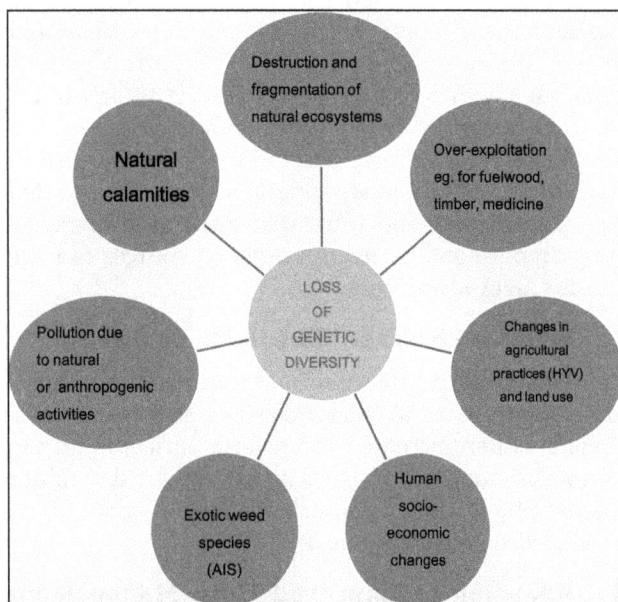

**Figure 10.1: Causes of Genetic Erosion.**

# Variability, Distribution Status, Genetic Erosion and General Crop Priorities

Concentration of genetic diversity comprising native species and landraces are mostly dominating in Western Ghats and North-Eastern Himalayas. The richness of plant diversity is largely due to ecological diversity superimposed with tribal and

**Table 10.2: Variability Distribution Status and Conservation Priorities of Vegetable Crops**

| Crop | Genera | CS | DS | GVS | GES | GCP |
|---|---|---|---|---|---|---|
| Brinjal | *Solanum* species | C | W | W | M | H |
| Chillies | *Capsicum* species | C | W | H | M | H |
| Tomato | *Solanum* species | C | W | H | M | H |
| Cluster bean | *Cyamopsis* species | C | R | M | M | M |
| Cowpea | *Vigna* species | C | W | H | M | H |
| French bean | *Phaseolus* species | C | W | M | M | M |
| Lablab bean | *Lablab* species | C | W | H | M | M |
| Velvet bean | *Mucuna* species | C | R | H | M | M |
| Pea | *Pisum* species | C | R | M | M | H |
| Carrot | *Daucus* species | C | R | M | M | H |
| Elephant Foot Yam | *Amorphophallus* species | C | R | H | M | M |
| Giant taro | *Alocasia* species | W | R | H | M | M |
| Radish | *Raphanus* species | C | W | H | M | H |
| Taro | *Colocasia* species | W | W | H | M | M |
| Yam | *Dioscoria* species | W | R | H | M | M |
| Tannia | *Xanthosoma* | C | R | H | M | M |
| Garlic | *Allium* species | C | W | M | M | H |
| Onion | *Allium* species | C | R | M | M | H |
| Okra | *Abelmoschus* species | C | W | H | M | M |
| Amaranth | *Amaranthus* species | C | W | M | M | M |
| Climbing spinach | *Basella* species | C | W | M | M | M |
| Fenugreek | *Trigonella* species | W | R | M | M | M |
| Palak | *Beata* species | C | R | M | M | H |
| Bitter gourd | *Momordica* species | C | W | H | M | H |
| Bottle gourd | *Lagenaria* species | C | W | H | M | N |
| Cucumber | *Cucumis* species | C | W | H | M | M |
| Muskmelon | *Cucumis* species | C | L | H | H | H |
| Snapmelon | *Cucumis* species | C | W | H | M | H |
| Pointed gourd | *Trichosanthes dioica* | C | L | H | H | M |
| Pumpkin | *Cucurbita* species | C | W | H | M | H |
| Ridge gourd | *Luffa* species | C | W | H | H | H |
| Sponge gourd | *Luffa* species | C | W | H | M | H |
| Tinda | *Prae-citrullus fistulosus* | C | L | M | M | M |
| Watermelon | *Citrullus* species | C | R | H | M | M |
| Roundmelon | *Citrullus* species | C | L | M | M | M |
| Ash gourd | *Benincasa hispida* | C | W | H | M | M |
| Snake gourd | *Trichosanthes anguina* | C | W | H | M | H |
| Coriander | *Coriandrum* species | C | R | H | M | H |
| Turmeric | *Curcuma* species | C | R | H | M | H |
| Turmeric | *Curcuma* species | W | R | H | M | M |
| Ginger | *Gingeber* species | W | L | H | M | M |

*Source*: Genetic Resources of Vegetable crops (Gupta *et al.*, 1995).

**CS:** Crop status (C-cultivated, W-Wild), **DS:** Distribution status (**W**-wide spread distribution, **R:** regional distribution, **L:** Localized distribution), **GVS:** Germplasm variability status (H-high, M- medium, L-low), **GES:** Genetic erosion status (H-high, M-medium, L-low) and **GCP:** General crop priorities (H-high, M- medium).

**Table 10.3: Variability and Distribution Status of Tree Vegetable**

| Crop | Genera | NIS | CS | DS | GVS | GES | GCP |
|------|--------|-----|----|----|-----|-----|-----|
| Jackfruit | *Artocarpus heterophyllus* | 19 | C | W | H | M | M |
|  |  |  | W | W | H | M | M |
| Khejri | *Prosopis cineraria* | 1 | W | R | H | M | M |
| Lasora (Indian cherry) | *Cordia myxa* | 18 | C | R | H | M | M |
|  |  |  | W | R | M | M | M |

**NIS:** Number of Indigenous species, **CS:** Crop status (C-cultivated, W-Wild), **DS:** Distribution status (**W**-wide spread distribution, **R**-regional distribution, **L**- localized distribution), **GVS:** Germplasm variability status (H-high, M- medium, L-low), **GES:** Genetic erosion status (H-high, M-medium, L-low) and **GCP:** General crop priorities (H-high, M- medium).

ethnic diversification, plant usages and religious rituals. Diversity and variability both are very high in wild and cultivated species in crops of Indian origin. At the same time genetic erosion status is also low to medium but in some cases it is high.

Similarly, wide arrays of variability have been observed in several tree vegetables, which should be exploited.

## Extinction of Genetic Resources

There are several species which are being extinct day by day in Indian conditions with the changing scenario in cultivation especially monoculture, development of hybrids, urbanization, fragmentation of habitats, deforestation, overexploitation, rapid changes in the hydrological regime and land use patterns, soil degradation, air and water pollution, the adverse impact of development and increase in the population. Incidence of biotic and abiotic stress and many natural and biological factors, there is threat to the existing genetic wealth. Some endangered cucurbitaceous species and its biogeographic zones are given in Table 10.4. As a result, a significant number of species are considered vulnerable and endangered.

**Table 10.4: Rare and Endangered Cucurbitaceous Species in India**

| Sl.No. | Cucurbitaceous Species | Biogeographic Zones |
|--------|------------------------|---------------------|
| 1. | *Corallocarpus gracillipes* (Naud.) Cogn. | WG |
| 2. | *Gomphogyne macrocarpa* Cogn. | EH |
| 3. | *Indogevillea khasiana* Chatterjee | NE |
| 4. | *Luffa umbellata* (Kleir) Roem. | WG |
| 5. | *Melothria amplexicaulis* Cogn. | DP |
| 6. | *Momordica sub angulata* Bl. | DP, WG |
| 7. | *Trichosanthes lepiniana* (Naud.) Cogn. | DP, WG |
| 8. | *Trichosanthes perrotteliana* Cogn. | WG |
| 9. | *Trichosanthes villosula* Cogn. | DP, WG |

WG: Western Ghats; EH: East Himalaya; NE: North East India; DP: Deccan Peninsula.

*Source*: R.P. Rao. Biodiversity in India (Floristic Aspects) pp. 281.

## Genetic Erosion in Genebank

Genetic erosion occurs in all gene banks due to a lack/deficiency of one or more of the followings: refrigerated storerooms, field space, trained personnel, and administrative and financial support. Repeated regeneration in small plots also leads to undesirable effects in terms of genetic identity and population structure. Genetic erosion is more frequent in gene banks located in the tropics.

## Factors Causing Genetic Erosion in Gene Bank

Loss of genetic resources (reduction of the genetic base of a species) is due to several factors encompassing human activities as well as natural extremities and their combination as well. The important factors that lead to genetic erosion are as follows:

(i) Discontinuity in programme and/or personnel

(ii) Human neglect

(iii) Shifts in programme direction or methodology of maintenance

(iv) Poor storage facilities

(v) Disappearance of some records

(vi) Lack of periodic monitoring of seed viability.

Some of the factors causing genetic erosion in *ex situ* collections are physiological changes in seed, inappropriate storage conditions and management procedures, accidental errors/mixing of seed samples before regeneration, lack of adequate financial resources for maintaining collections, human behaviors (accidental destruction, fire), damage of field collection (animals, meteorological anomalies including natural disasters), armed conflicts, war, regeneration backlogs, economic instability, pest and disease outbreaks, abiotic stresses (heat, drought), lack of resources and skills, and loss of samples during regeneration *etc*. Genetic erosion in genebank collections depends on the quality and quantity of the original material stored, and on the conditions under which the germplasm is maintained, multiplied and regenerated.

### Population Size at the Time of Collecting and at the Time of Regeneration

Number of accessions and their isolation during collection and the time of regeneration are also indicators for genetic erosion. Small populations are at risk of loss of alleles, increased inbreeding and extinction due to random environmental events. Secondly, sample size analyzed-small sample size may result in missing the allele detection, especially for rare alleles. Direct comparison of sample collected at different times in the *ex situ* collections are warranted to assess the genetic erosion.

### Avoiding Genetic Drift

Genetic drift describes random fluctuations in the numbers of gene variants in a population. Genetic drift takes place when the occurrence of variant forms of a gene, called alleles, increases and decreases by chance over time. These variations in the presence of alleles are measured as change in alleles frequencies. Hence, the genetic purity of the conserved sample by avoiding genetic drift and inbreeding; the population size/population genetics theory of sample at the time of regeneration to maintain the genetic integrity of the accessions in the genebank. Typically, genetic drift occurs in small populations, where infrequently occurring alleles face a greater chance of being lost. Genetic drift can result in the loss of rare alleles and decrease the genepool. Genetic drift can cause a new population to be genetically distinct from its original population, which has led to the hypothesis that genetic drift plays a vital role in the evolution of new species.

## Constraints of Gene Bank

There are several constraints of gene bank:

(i) The labour-intensive and costly seed regeneration step is a serious constraint.

(ii) In humid tropics, seed drying is the major problems.

(iii) For gene banks with limited financial support, cut of power is a major problem.

(iv) Availability of technically capable and dedicated staff is a problem for many gene banks.

(v) Conservation of some types of germplasm, *e.g.* un-adapted to adverse environments, disease and pest susceptible accessions, those having relatively poor seed longevity, *etc.*, may be difficult.

(vi) Adequate financial support is a major problem. The estimated requirement in 1990 was US $ 500 million per annum, while the funding for 1982 was merely US $ 55 million.

## Methods to Overcome Genetic Erosion in the Genebank

It is true that in many gene banks particularly in the tropics face rampant genetic erosion; even genebanks with a constant power supply and regularly serviced machinery have some difficulties in maintaining samples because of:

1. Availability of staff,
2. Lack of storage space,
3. Limited facilities for regenerating materials,
4. Lack of trained personnel, and
5. Administrative and financial support

It has also been observed that popular accessions deplete rapidly, and the rest can seriously decline in quality viability and may be in urgent need of regeneration. For example, at the National Seed Storage Laboratory in Fort Collins, Colorado (USA), the viability of some collections has fallen as low as 60 per cent due to problems mentioned above. Therefore, duplicate collections held at different locations help circumvent problems of germplasm access and avoid a total loss of germplasm in the event of power failures, fire or natural hazards. The supply of electricity is often erratic. A short interruption of power to a gene bank is unlikely to jeopardize collections. Widely fluctuating voltage is fairly common and sometimes causes failure of refrigeration equipment. Servicing of broken-down equipment is difficult. For example Iran has lost valuable stored germplasm because of the failure of refrigeration equipment. Another example, improper wiring ruined germplasm; at the University of Vicosa in Brazil, by a short-circuit fire largely destroyed lima bean collections.

To overcome all these problems, following steps should be ensured:

1. Genebank manager must be very vigilant about the continuous working of gene bank

2. To protect the gene bank against losses due to fire, fire extinguishers are to be installed in the gene bank.

3. A stand-by diesel power generator adequate to maintain the refrigeration unit should be provided so that the power failure may not affect the stored seed.

4. Careful and continuous supervision by the trained person is essential for maintaining the efficiency of the genebank.

5. Monitoring of the germplasm at proper interval is a pre-requisite for minimizing genetic erosion in the genebank.

## Genetic Resource Conservation

Genetic diversity present in many species and crops is not adequately conserved (Padulosi *et al.*, 2002; Ford Lloyd *et al.*, 2011), that the continuing contributions by farmers growing traditional varieties of crops are not recognized and understood (Jarvis *et al.*, 2016; Zimmerer, 2003), that the links between *ex situ* and *in situ* conservation programmes are poor (Brush, 1995) and that, overall, the resources for conservation remain inadequate (Gepts, 2006).

Germplasm conservation refers:

*"To maintain the collected germplasm in such a state that there is minimum risk for its loss and that either it can be planted directly in the field or it can be prepare for planting with relative ease whenever necessary"*.

or

*"Germplasm collections aim at minimizing the detrimental effects of genetic erosion by collecting and preserving the variability in crops and their related species"*

Conservation of germplasm is one of the most important activities in PGR management. The status of conservation of vegetable crops germplasm has always received less attention than that of the major staple crops such as cereals and legumes. Information on vegetable germplasm can, however, increasingly be obtained from online international databases. Maintenance and updating of this information requires a high level of international collaboration. This can be exemplified by the activity of the Working Groups of the European Cooperative Programme for Crop Genetic Resources Networks (ECP/GR) on: *Allium; Brassicea; Solanaceae* and *Umbelliferae* crops.

## Conservation of Vegetable Genetic Resources

Ever since humans planted their first crop, they have saved seeds for use in future. Understanding the conservation dynamics of smallholder managed landrace populations requires understanding their human ecology and the selective pressures that farmers exert when they manipulate diversity. Ongoing evolution of crop genetic resources in their centre of origin and diversity, where they are exposed to a dynamic state of management, environmental stress, proximity to crop wild relatives among other forces of natural and human selection, is commonly considered an essential contribution of farmer-driven conservation.

## Conservation Contributing Pathways

At least four pathways contribute to conservation dynamics:

1. Geneflow among landraces and/or compatible wild relatives and the eventual incorporation of new genotypes into farmer landrace stocks (Bonnave *et al.*, 2015).

2. The collection of collected semi-wild genotypes that are brought into cultivation.

3. Mutation leading to intra-clonal variation.

4. Darwinist selection based on exposure of landrace pools to stressors and other selection pressures resulting in a 'survival of the fittest' of best-adapted genotypes, change in phenology, and/or shifts in agro-ecologies (Vigouroux *et al.*, 2011).

The climax of plant genetic resources (PGR) handling systems is two folds:

(i) Conservation management and

(ii) Utilization management.

Conservation is done for posterity in order to save the precious genetic resources of crop plants, from genetic vulnerability and genetic erosion. The magnitude of genetic erosion due to habitat loss, climatic changes and intensive agricultural practices is unimaginable. Biodiversity of vegetatively propagated plants in general and especially those propagated through rhizomes, roots, tubers, bulbs and corms, which are used as economic products in particular is highly vulnerable to conservation due to over exploitation. Their productivity and multiplication rate in nature is further reduced due to series of systemic viral, mycoplasmal, fungal and bacterial diseases. Tropical forests are the enriched regions of biodiversity where everyday witness a loss of 3 species (Sharma, 2009). Gene elimination is not only confined to cultivated species but now wild and weedy species are also under severe threat due to various human activities. Therefore, germplasm conservation works call for sustained support from governments and vigilance and loving care from all concerned workers. De Boef and Thijssen illustrated agro-biodiversity as a dynamic and constantly changing patchwork of relations between people, plants, animals, other organisms and the environment, always coping with new problems and always finding new ways. Conservation and sustainable use of biodiversity are subjects that have been high on the policy agenda since the first Earth Summit in Rio in 1992. Agro-biodiversity is often approached as an ecological service rather than as a resource, owing to its broader and ecological association and to its demarcation within the three levels of biodiversity *i.e.* system, species and genetic diversity.

## Base for Biological Diversity Conservation

### Threat to Biological Diversity

It is true that human activities have contributed significantly towards enhancing the rate of "extinction" of species and rapidly degrading/destroying their natural habitats. This is happening largely owing to unplanned and over-exploitation of natural plants and large scale alteration in land use patterns due to ever growing population pressure and developmental needs.

Speciation and extinction are continuous, contemporary and natural components of the evolutionary process. Extinction of a new species that disappear to make way for others having ability to cope with demands of the sudden change in environment should not, therefore, cause any worry unless it happens abruptly on a large scale accompanied by loss of habitats. Fossil records reveal that there have been at least five occasions when over 50 per cent of the then existing animal species became extinct abruptly. Even on averaging in these periods of mass extinction over nearly four billion years ("Deep time") species per million years (Raup, 1988). Impact of human activities on biodiversity in recent years may be gazed from the current global extinction rate that is estimated to be at the level of 5-10 per cent per decade. Projected into the middle of the 21$^{st}$ century, these activities may lead to a mass extinction on the scale of any in the fossil records with the possibility of losing 25-50 per cent of all existing species.

### Conversion of Natural Resources

Since the beginning of agriculture about 10,000 years ago, humans are continuously engaged in converting natural assets into their selected and more productive forms (for example, conversion of forests into farmlands, factories, real estate *etc.*). This conversion in the guise of development is the root cause of the endangerment of biological resources.

### Non-renewable Nature of Biodiversity

Genes carry encoded information that determines the capabilities of all organisms besides controlling their characteristics. Greater the diversity in a gene pool, greater is the variety of organism that exists or will appear in near future. Genetic variation arises through mutation and also from reshuffling of genes and their alleles (alternate forms of a gene) into new combinations. Selection pressure acts upon alleles (via their function) altering their frequencies in a population. At level of organisms, their reproductive abilities and adaption to sudden changes in environments act as sieves for selection of better forms.

### Dependence of Sustainable Agriculture

It is true that traditional farming is being practiced in stress and fragile environment (like agri-ecosystem in mountains/hills, coastal zones, arid and semi-arid areas) is their diversified nature. It relies more upon diversity of crops and their varieties and also on organic inputs rather than using high cash-inputs. Crops and varieties of this system are adapted to local condition, possess resistance to pathogens and pests and also suit the preferences of farmers and consumers. Because of these attributes, prevailing cultivars have not been replaces to a sizeable extent by high yielding varieties developed for non-limiting environments. Farming under stress environments, through sustainable has remained largely of the subsistence type.

## Need of PGR Conservation

Biodiversity conservation means essentially:

*"The conservation of unique characteristics of the output from rigorous evolutionary process".*

Encroachment of natural habitat by human dwellings and spread of green revolution in many Asian countries as well as transformation of agricultural production systems in many developing countries are posing threat to evolution of variation in wild flora and cultivation of traditional varieties, respectively. Latest threat to genetic resources is now coming from spread of high yielding varieties/hybrids available in several vegetable crops (tomato, brinjal, chillies, cucumber, bottle gourd, bitter gourd, watermelon, muskmelon, capsicum, cauliflower, cabbage, knol-khol *etc.*) because farmers, the keepers of crop diversity are rapidly shifting to hybrid cultivation. Evidently, large number of species and plant communities is becoming dependent for survival on protection offered by human being. The fast shrinking genetic diversity of commercially grown crops renders them vulnerable to wide spread epidemics and pest rampage. In spite of the urgent need and unquestioned value of germplasm collection and conservation, only less than 30 per cent of the countries have formal national germplasm conservation programme. It is not important that a breeder how many collections have made, but important is that how many he has conserved safely. Therefore, germplasm has to be maintained in such a state that minimizes the risk of its loss and allows either it's direct planting in the field or its preparation for planting with relative ease; this called germplasm conservations (Singh, 2012). Now we can say that germplasm conservation has become social responsibility of great concern. It needs to be extended to wild relatives of cultivated plants, wild species used by man and animals besides cultivated plants. However conservational strategies are based on social time scale of concern and are hardly analyzed for eco-biological perspective (Frankel, 1974). The conservation efforts have to be broad based having blend of personalities and technologies from political, social, biological and ecological sciences. Hence, scientific management of these invaluable resources is of prime importance today. Conservation efforts have gained remarkable momentum in recent years since candidate genes for genetic manipulation through new tools of biotechnology are to be obtained from the germplasm collections and also because of challenges/threats caused by patenting of genes/genotypes restricting thereby availability to breeders and other researchers.

## Pre-requisites for Conservation

Before conservation of germplasm is undertaken, the strategy of conservation should be selected judiciously, based on availability of infrastructure, technical expertise and finance besides the nature and requirement of PGR. Germplasm meant for conservation should have been acquired legally, in accordance with prevailing national and international laws, and accompanied with at least passport information/data sheet. Additional information such as characterization, evaluation, biological and taxonomical status *etc.*, are also desirable. Irrespective of the conservation method adopted, it should be ensured that germplasm identity, viability, genetic integrity and associated information are metriculously maintained. The goal is to ensure access to, and availability of, high quality seeds/ propagules of the conserved PGR.

## Agencies/Organizations Working Exclusively for Conservation of Biodiversity

Although there are not many agencies/organizations working exclusively for diversity conservation in the north-east *per se*, the activities taken up by many organizations including non-governmental and traditional institutions, government departments and scientific institutions have direct or indirect implications for diversity conservation.

☆ **State Government Agencies:** Many state agencies are now involved in such diversity conservation activities and establishment of germplasm banks for horticultural crops.

☆ **Research Organizations:** Many state and central government research organizations including universities of the region are engaged in research, inventory and conservation of diversity in the region. Such organizations are Botanical Survey of India, Shillong; GB Pant Institute of Himalayan Environment and Development, North-East Unit, Itanagar; Indian Council of Agricultural Research for North-Eastern Hill Region, Barapani, Shillong with campuses throughout the north-east; State Forest Research Institute, Itanagar, ICAR-NBPGR, Shillong; North-Eastern Hill University, Shillong; Nagaland University, Kohima; Mizoram University, Aizawl; Arunachal University, Itanagar; Tripura University, Agartala; Asom University, Silchar; Tezpur University, Tezpur; Gauhati University, Guwahati; Asom Agricultural University, Jorhat; Regional Research Laboratory, Jorhat; Dibrugarh University, Dibrugarh.

☆ **Non-Governmental Organizations:** Many non-governmental organizations are now working for the conservation of diversity in the north-east.

☆ **International Agencies:** International donor agencies in Meghalaya, Manipur, Asom and Nagaland have been playing crucial role in conserving the diversity through their respective projects.

☆ **International and National Policies and Conventions:** Many of the international treaties and national policies have significant impact on the conservation of diversity in the north-east.

☆ **Shifting Cultivators:** The shifting cultivators and other traditional farming communities of north-east have played a key role in conserving the rich horticultural crops germplasm of the region. In spite of the availability of many hybrid and high yielding varieties, these farmers have been cultivating the traditional varieties for generations.

## Mechanisms/Methods/Strategies of PGR Conservation

The rich heritage and ethnic culture has favored to preserve the richest diversity including rare landraces/ primitive types of useful vegetables like eggplant, cucumber, ridge gourd, sponge gourd and a number of root and tuber crop species. A number of vegetable crops were brought to India from other regions by travelers, invaders like Persian, Turkey, Mughals, Portuguese, Dutch, French and British which acclimatized and developed considerable diversity. Wild relatives/ species of some of the important vegetable crops having commercial importance have been conserved to a great extent. Most of these wild species grown in the natural habitats possess genes resistant to biotic and abiotic stress. However, diversity for valuable genetic resources is threatened in recent times. Therefore, conservation of the vegetable genetic wealth particularly their wild relatives are thus essentially required for future utilization.

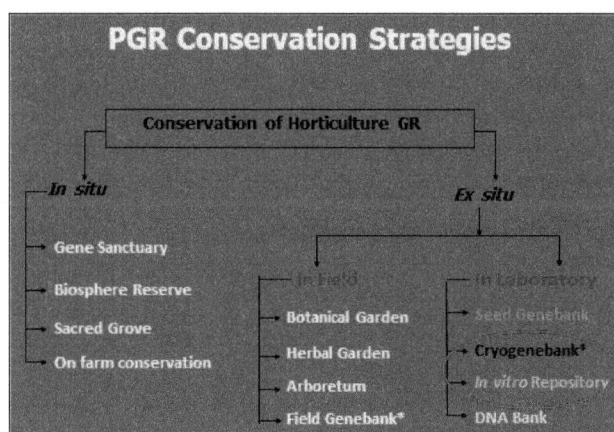

**Figure 10.2: PGR Conservation Strategies.**

To ensure availability of maximum genetic diversity, complementary approaches are needed. The conservation of PGR involves two basic strategies *i.e. in situ* and *ex situ*; as per objective and priorities. Germplasm can be conserved by following mechanisms:

1. *In situ*
2. *Ex-situ*

## 1. *In situ* Conservation

*In situ* conservation sometimes called as "Dynamic conservation" where continuous evolution is taking place. As a definition:

*"Conservation of germplasm in its natural habitat or in the area where it grows naturally is known as In situ conservation".*

Or

*"Conservation of ecosystems and natural habitats, for recovery of viable populations of species in their surroundings".*

In domesticated species, it refers to surroundings where they have developed their distinctive properties. *In situ* conservation (*e.g.*, gene sanctuaries) where preservation of whole ecosystem under minimal human interference in order to conserve plant species in their natural environment is preferred and preservation of threatened species are done in botanical gardens. *In situ* conservation of PGR has to be carried on farm, natural habitats/wild, landraces and locally adapted materials are cultivated, utilized and conserved as part of traditional farming systems by paying due regard to natural ecosystems (protected by farmers for centuries in certain areas). This strategy is ideal for wild relatives of the crops, however, natural and man-made disasters and development (hydropower, mineral extraction, tourism, and land conservation for crops, illicit felling and fires) put species in natural reserves and in unprotected areas under considerable pressure. Of greatest conservation concern is the fate of long-lived species, as their replacement may take decades to centuries. Man-made fragmentation is also an important contributor to loss of genetic diversity. On farm also involves continued cultivation and management by farmers of a diverse set of crop populations in the agro-systems where the crop has either evolved or has secondary centre of diversity. Predicting the consequences of contracting distribution on species survival is the subject of ongoing research. On farm conservation requires a multidisplinary approach.

### Merits of *in situ* Conservation

The main aim of *in situ* conservation is to allow the populations to maintain/perpetuate itself within the natural community environment, to which it is adapted so that it has the potential for continued evolution. Landraces/primitive cultivars developed in traditional agriculture system also need *in situ* conservation in their respective areas. These farming systems are of particular importance in maintaining local genetic diversity and providing food for local consumption and local markets. It will provide an option for further evolutionary changes likely to take place. However, for majority of the situation, *in situ* conservation is ideal method of conserving wild plant genetic resources and perennial vegetables, which either do not set or set recalcitrant seed, do not produce plants true to type. Protection of forests and its biodiversity is an important aspect in the *in situ* conservation of tuber crops. *In situ* conservation of plant genetic resources have a number of advantages as compared to *ex situ* conservation.

1. It usually increases probabilities of conserving a large range of potentially interesting alleles.

2. It is specially adapted to species, which cannot establish or regenerated outside their natural habitats.

3. It allows natural evaluation to continue, which is a valuable option for conservation of diseases', and pest resistant species, which can evolve with their parasites, providing, breeders with dynamic source of resistance.

4. It can serve several factors at once, since gene pool of value to different sectors (crop breeding, production *etc.*) may often overlap and so can be maintained in the same protected area.

5. It facilitates research on species in their natural habitats.

6. It assures protection of associated species of economic importance.

### Demerits of *in situ* Germplasm Conservation

While carrying out *in situ* conservation, there is always a threat of a species becoming extinct or its population declining due to genetic drift, inbreeding, demographic and environmental variations, habitat loss, competition from exotic species, pest incidence or over-exploitation and human disturbances.

1. They are easiest to demarcate, difficult to established and very difficult to maintain. This is particularly so in a country like India, which has an ever increasing human population.

2. Each protected area will cover only very small portion of total diversity of a crop species, hence several areas will have to be conserved for a single species.

3. The management of such areas also poses several problems.

4. This is a costly method of germplasm conservation.

5. These cannot be conserve due to the variability found in crop plants, for which *ex situ* conservations is the only answer.

Conservation of root and tuber crops germplasm in their original habitat hasn't received much attention, even though it is being practiced consciously or otherwise in the case of wild species especially of yams and aroids in the protected forests of Western Ghats. On-farm conservation of germplasm exists in root and tuber crops because these are cultivated by tribal and poor farmers, whose preference for specific local cultivars varies depending upon various factors like size, shape and number of tubers, medicinal value, commercial use, taste *etc*. Tribes play an important role in the *in situ* conservation of wild species of tuber crops. Even though they consume wild tubers, they mostly leave a piece of tuber in the cavity from which the collection is made, to ensure its availability for the next season. Generally, *in situ* conservation can be achieved by following ways, protecting this area from human interference; such an area is called natural park, biosphere reserve or gene sanctuary.

### Role of Biosphere Reserve/Gene Sanctuaries for *in situ* Conservation

The 'National Man and Biosphere Committee' of the department of Environment has already identified sites as potential areas for biosphere reserve (Table 10.5). This covers all the major bio-geographic regions of the Indian subcontinent where flora and fauna can be conserved in natural habitats. These areas have been identified based on their rich genetic diversity; floristic uniqueness, endemic wealth flora/fauna and they are in totality representative of ecosystems occurring in different bio-geographic regions. A total of 18 biosphere reserves have been set up in India to protect representative ecosystem and wild plant species.

A gene sanctuary is best located within the centre of origin of the crop species concerned, preferably covering the micro-centre within the centre of origin. In Ethiopia gene sanctuary for conservation of wild relatives of coffee was setup in 1984. In India, ICAR-NBPGR, New Delhi has established gene sanctuaries in Meghalaya for *Citrus* and in the North-Eastern region for *Musa, Citrus, Oryza, Saccharum* and *Mangifera*.

### Advantages of Gene Sanctuaries

1. It protects the loss of genetic diversity caused by human intervention.
2. It allows natural selection and evolution to operate.
3. The risk associated with *ex situ* conservation is not operative.

### Disadvantages of Gene Sanctuaries

1. Entire variability of a crop species cannot be conserved.
2. Its maintenance and establishment is a difficult task.
3. It is a very good method of *in situ* conservation.

### Kulagar: A Potential *in situ* Conservation System of the Crop Diversity

Goa and Konkan region of Maharashtra are blessed with the diversity of tropical flora and fauna due to the nearness to the Western Ghats. The hot humid climate and the presence of heavy monsoon have made this region a biodiversity hotspot with beautiful landscape. Rural Goa and Konkan region farmers have a conventional homestead system of gardening transmitted from their ancestors called 'Kulagar' to

**Table 10.5: List of 18 Indian Biosphere Reserve, their Area, Location and Demarcation of Area**

| Biosphere Reserve with Area (km²) | State(s) Location |
|---|---|
| Nilgiri (5520) | Part of Wynad, Nagarhole, Bandipur and Madumalai, Nilambur, Silent Valley and Siruvani hills in Tamil Nadu, Kerala and Karnataka |
| Pachmarhi (4981.72) | Part of Betul, Hoshangabad and Chhindwara districts in Madhya Pradesh. |
| Nanda Devi (5860.69) | Part of Chamoli, Pithoragarh and Almora districts in Uttarakhand. |
| Achanakmar- Amarkantak (3,835.51) | Part of Anuppur and Dindori districts of Madhya Pradesh and Bilaspur district of Chattisgarh. |
| Great Nicobar (885) | Southernmost island of Andaman and Nicobar Islands. |
| Gulf of Mannar (10500) | India part of Gulf of Mannar extending from Rameswaram island in the North to Kanyakumari in the South of Tamil Nadu. |
| Seshachalam (4755.997) | Seshachalam hill ranges in Eastern Ghats encompassing part of Chittoor and Kadapa districts in Andhra Pradesh. |
| Sunderban (9630) | Part of delta of Ganges and Brahamaputra river system in West Bengal. |
| Cold Desert (7,770) | Pin Valley National Park and surroundings; Chandratal and Sarchu; and Kibber Wildlife sanctuary in Himachal Pradesh. |
| Manas (2837) | Part of Kokrajhar, Bongaigaon, Barpeta, Nalbari, Kamprup and Darang districts in Asom |
| Panna (2998.98) | Part of Panna and Chhattarpur districts in Madhya Pradesh |
| Nokrek (820) | Part of East, West and South Garo Hill districts in Meghalaya. |
| Kachchh (12,454) | Part of Kachchh, Rajkot, Surendranagar and Patan districts in Gujarat. |
| Similipal (4374) | Part of Mayurbhanj district in Odisha |
| Dibru-Saikhova (765) | Part of Dibrugarh and Tinsukia districts in Asom |
| Dehang-Dibang (5111.5) | Part of Upper Siang, West Siang and Dibang Valley districts in Arunachal Pradesh. |
| Khangchendzonga (2931.12) | Part of North and West districts in Sikkim |
| Agasthyamalai (3500.36) | Part of Thirunelveli and Kanyakumari districts in Tamil Nadu and Thiruvanthapuram, Kollam and Pathanmthitta districts in Kerala |

*Source: Anonymous, 2015* (Ministry of Environment, Forest and Climate change).

cultivate and conserve the local crop plants near their household. It is an integrated system which includes cash crops, plantation crops, spices, fruits, local vegetables, medicinal and aromatic plants and flower crops. Some of the 'Kulagar' includes animal components to make the system economic and complete. The cash crop component in a 'Kulagar' can be arecanutm, coconut, cashew nut, betel vine *etc.* In Goa, mostly areca nut based (rarely coconut based) 'Kulagar' are common. Shade tolerant vegetables and tuber crops like elephant foot yams, *etc.* are common component. Crop diversification, recycling of the resources, value addition and processing and byproduct utilization are important features of a 'Kulagar'. Advanced crop production technologies are being incorporated in 'Kulagar' by the new generation farmers to make it sustainable and economically viable (Maneesha *et al.*, 2016).

## (2) *Ex situ* Conservation

*Ex situ* conservation or "Static Conservation" (a viable alternative over *in situ* to conserve diversity of PGR) requires collection and systematic storage of seeds/propagules outside the natural habitats of species for short, medium and long term. The main objective of the *ex situ* strategy is to conserve accessions without changing their original genetic makeup. The aim is to preserve, as for as possible genetic integrity of an individual or population storing for longer period by way of seed under controlled conditions.

*"Conservation of germplasm away from its natural habitat is called ex situ conservation".*

Sharma (2009) defined *ex situ* conservation as:

*"Ex situ conservation involves removal of plant propagules (seed, stem, root, meristem part, pollen, protoplast etc.) from its natural environment and storing them in gene bank under ambient conditions that maintain propagules viability and vitality for longer".*

### Merits (Importance) of *ex situ* Conservation

*Ex situ* conservation acts as a back-up for certain segments of diversity that might otherwise be lost in nature and in human-dominated ecosystems. It is generally achieved through seed banks (conventional and cryogenic), field gene banks (for clonal crops), and *in vitro* cultures. DNA and pollen storage also contribute indirectly to *ex situ* conservation of plant genetic resources. There are a large number of important tropical and sub-tropical plant species which produce recalcitrant seeds that quickly lose viability and do not survive desiccation, hence conventional seed storage strategies are not possible. For this reason, alternate strategies of conservation are used and it is in this area that biotechnology contributed significantly by providing complementary *in vitro* conservation options through tissue culture techniques. Cost-effective and reliable *ex situ* conservation remains a challenge that can benefit from sharing responsibilities within crop networks. In

these fora, the discussion of common problems (long-term storage, safety-duplication, regeneration) can lead to effective collaborative solutions. *Ex situ* conservation of PGR in the form of seeds in domesticated plant species for use in their genetic improvement was initiated with the establishment of Imperial Agricultural Research Institute (IARI) in 1905 in village Pusa, Darbhanga district, Bihar (earlier in Bengal). Now ICAR-NBPGR, New Delhi, is the nodal agency for *ex situ* conservation of PGR for food and agriculture, is maintaining active and base collections of various crop species and their wild relatives including vegetables in a network of gene banks in the country. *Ex situ* approach is the preferred option for the conservation of potato genetic resources. The traditional method of *ex situ* conservation entails the maintenance and multiplication of tubers in fields and glasshouse. However, this method, based on multiplication in every year is highly labour-intensive, require large space, exposes the germplasm to various biotic and abiotic stresses, and most importantly, results in the progressive accumulation of pathogens in tuber stocks over successive vegetative generation.

1. It is possible to preserve entire genetic diversity of a crop species at one place.
2. Handling of germplasm is also easy.
3. This is a cheap method of germplasm conservation

### Demerits (Importance) of *ex situ* Conservation

*Ex situ* collections are vulnerable to genetic erosion resulting in the loss of amount of genetic diversity and loss in genetic integrity, in terms of presence and absence of genes (alleles).

### Challenges in *ex situ* Conservation

The conservation of the diverse plant genetic resources, as a possible source of wide gene/allelic diversity, under *ex situ* genebanks faces immense challenges and issues some of which are not entirely within managers' control. In *ex situ* crop wild relatives are threatened by changed habitat (soil, water and soil microbial populations *etc.*) agricultural/silvicultural practices, human selection *etc.*, that impact on its fertility and seed production. The challenges are broadly defined by environmental factors for *in situ/on-farm conservations*. The extent of the impact of environmental factors varies from species to species; for example, whereas the cultivated plant species are faced with the challenges of habitat destruction, fragmentation, climate change and restoration efforts while the crop wild relatives of the plant species could be facing the challenges of climate change on a much bigger scale than the habitat destruction and fragmentation *in situ* (Guarino *et al.*, 2011). These challenges precipitate into genetically less diverse populations of the plant species. Jump *et al.* (2009) stated similarly that, genetic erosion can check the resilience, evolutionary potential for adaptation in the short term and survival of any plant species in the long-term, in the face of rapid environmental change. Reducing and managing the loss of genetic integrity and genetic variation of the conserved germplasm during regeneration is an important objective of genetic resource conservation programmes, *i.e.* reduction in genetic drift and genetic shift. Genetic integrity may be lost due to inadvertent selection and reduction in genetic variation due to cumulative bottleneck effects that would have started at the time of collecting through small seed samples used for subsequent regeneration/ multiplication. The management of seed accessions in different genebanks can lead to differential loss of

#### Table 10.6: *Ex situ* Preservation Priority of Vegetable Crops (Annex I, ITPGRFA)

| Crop | Genus | Material for Storage | Example of Preservation* |
|---|---|---|---|
| Breadfruit | *Artocarpus* | Embryo axis | Vitrification/cryopreservation of embryonic axes of *Artocarpus heterophyllus* with 50 per cent developing into plants (Thammasiri, 1999). No reliable method apparent for *Artocarpus altilis* embryonic axis cryopreservation. |
| Asparagus | *Asparagus* | Seed | Survival of *A. officinalis* seed known over c.5 y storage under room or cool conditions (RBG Kew, 2016). |
| Beet | *Beta* | Seed | Seed viability constants: *B. vulgaris*: Ke 8.943; Cw4.723;Ch0.0329' Cq0.000478 (RBG Kew, 2016). |
| Brassica complex | *Brassica* | Seed | [Includes species in: *Brassica, Armoracia, Barbarea, Camelina, Crambe, Diplotaxis, Eruca, Isatis, Lepidium, Raphanobrassica, Raphanus, Rorippa, Sinapis* ]. |
| | | | P50s: *Brassica juncea, B. napus, B. oleracea*=23-59 y; *Crambe abyssinica*=21 y; *Isatis tinctoria*=27 y; *Lepidium sativum*=26 y; *Raphanus sativus*=120 y; *Sinapis alba* P50= 76 y (Walters *et al.*, 2005). |
| | | | Seed viability constants: *Brassica juncea*, Ke 7.768; Cw 4.56; Ch 0.0329; Cq 0.000478. *Brassica napus*, Ke 7.718; Cw 4.54; ch 0.0329; Cq 0.000478 (RBG Kew, 2016). |
| | | | Ultra-dry seeds of many species of Brassicaceae survived after almost 40 years of storage at -5 to -10°C, for example in the genera *Barbarea, Brassica, Sinapis* (Perez-Garcia *et al.*, 2007) |
| Carrot | *Daucus* | Seed | P50: *D. carota*=30 y (Walters *et al.*, 2005). |
| Yams | *Dioscorea* | Shoot tip | Following cryopreservation by a modified droplet vitrification technique, 52 per cent of surviving explants developed further within 1 month (Leunufna and Keller, 2003). |
| Sweet potato | *Ipomoea* | Shoot tip | Shoot tips of 24 sweet potato accessions had shoot formation levels of 2-66 per cent after droplet vitrification cryopreservation (Vollmer *et al.*, 2014). |
| Cassava | *Manihot* | Shoot tip (+ seed) | *In vitro* shoot tips of cassava had 79 per cent average recovery after cryopreservation via droplet vitrification (Dumet *et al.*, 2013). |
| | | | Seed had no loss in viability after 14 y dry, hermetic storage at -20°C (RBG Kew, 2016). |
| Banana/ plantain | *Musa* | Shoot tip | *In vitro* shoot tips of 56 accessions (8 genomic groups of *Musa* spp. and one *Ensete* spp.) had 53 per cent average post-thaw regeneration after droplet vitrification cryopreservation (Panis *et al.*, 2005) |
| Beans | *Phaseolus* | Seed | [Except: *P. polyanthus*]. |
| | | | P50: *P. vulgaris*=31 y (Walters *et al.*, 2005). |
| | | | Seed viaility constants: *P. vulgaris*, Ke 9.09; Cw 4.761; Ch 0.0329; Cq 0.000478 (RBG Kew, 2016). |
| | | | *P. lunatus* seed held under international standard seed bank conditions (RBG Kew, 2016). |
| | | | *P. maculatus* seed had 100 per cent viability after 15 per cent RH and 11 weeks at -20°C (RBG Kew, 2016). |
| | | | *P. macrocarpus* seed had 100 per cent viability after 15 per cent RH and 46 days at -20°C (RBG Kew, 2016). |
| Pea | *Pisum* | Seed | P50: *P. sativum*=97 y (Walters *et al.*, 2005). |
| | | | Seed viability constants: *P. sativum*, Ke 9.858; Cw 5.39; Ch 0.0329; Cq 0.000478 (RBG Kew, 2016). |
| Potato | *Solanum* | Shoot tip | [Except: *S. phureja*]. |
| | | | Full plant recovery after cryopreservation of 1028 accessions of nine *Solanum* species/subspecies ranged from 34 to 59 per cent for PVS2 treated *in vitro* shoot tips (Vollmer *et al.*, 2014). |
| Eggplant | *Solanum* | Seed | P50: *S. melongena* = 46 y (Walters *et al.*, 2005). |
| Faba bean/ Vetch | *Vicia* | Seed | P50: *Vicia* sp.= 71 y (Walters *et al.*, 2005). |
| Cowpea | *Vigna* | Seed | P50: *V. radiata* = 457 y (Walters *et al.*, 2005). |
| | | | Seed viability constants: *V. radiata*, Ke 10.858; Cw 6.27; Ch 0.0329; Cq 0.000478. |
| | | | *V. unguiculata* Ke 9.401; Cw 5.118; Ch 0.0329; Cq 0.000478 (RBG Kew, 2016). |
| Canavalia | *Canavalia* | Seed | [NB six species in the genus are thought to be orthodox in storage response, and one species 'uncertain'] (RBG Kew, 2016). |
| | | | Information needed on *C. ensiformis*. |

genetic integrity. Identification and rationalization of duplicate accessions in genebanks requires information on the genetic integrity of the accessions. In addition, different genebanks may use different methods of identification of duplicate samples and rationalization of collections which can lead to further genetic erosion. To recommend better practices for maintaining panmictic populations of germplasm accessions, studies on genetic integrity during seed multiplication and regeneration using molecular makers from other seed or clonally propagated crops can be useful (Anishetty *et al.*, 2016).

**Ex situ conservation of vegetable biodiversity in Field Genebank (FGB)**

| Sl. No. | Crop group | Main species | No. of accessions | Institutes involved |
|---|---|---|---|---|
| 1 | Bulbous crops | *A. sativum*<br>Wild *Allium* species | 714<br>35 | NBPGR |
| 2 | Tuberous and root crops | Cassava<br>sweet potato<br>yams<br>Taros, etc. | 9605 (cultivated and wild relatives) | CTCRI, Thiruvananthapuram<br>CTCRI, Bhubaneswar<br>NBPGR, Thrissur<br>AICRP(TC) |

*Source of information:* Dhillon et al., 2005

**Figure 10.3: *Ex situ* Conservation of PGR.**

## Approaches for *ex situ* Conservation

The *ex situ* form of conservation includes, in a broad sense, the botanical gardens, fields and clonal repositories, and conservation of seed or vegetative material in genebanks. Biotechnology has generated new opportunities for PGR conservation. Techniques like *in vitro* conservation and cryo-preservation have made it possible to collect and conserve genetic resources, especially of species that are difficult to conserve in form of seeds. DNA and pollen storage also contribute to *ex situ* conservation. Thus, there are a number of options for conserving PGR *ex situ*, and the choice is more often based on its merits in utility, security, complementarity and the advantages over the other available techniques. *Ex situ* conservation is also convenient because it allows plant breeders easier access to genetic resources than is provided by *in situ* conservation.

Options for *ex situ* conservation of PGR, based on storage facility, nature of propagule and the form in which storage may be carried out are given in Table 10.7.

Guidelines for *ex situ* conservation in genebank in India, in the form of seeds, tissue cultures, embryos/embryonic axes, pollen, whole plants, clonal propagules and DNA can be achieved in the following ways:

(i) Plant conservation/field genebanks

(ii) Seed conservations/seed genebanks

(iii) *In vitro* conservation

(iv) Meristem conservation/Meristem genebanks

(v) Cell and organ genebanks

(vi) DNA conservation/DNA genebanks

## (i) Plant Conservation/Field Genebanks

### *(a) Botanical Gardens*

**Table 10.7: Options for *ex situ* Conservation of PGR**

| *Types of Conservation Facility* | *Condition of Storage* | *Category of PGR* | *Form in which Stored* |
|---|---|---|---|
| Seed genebank | Low temperatures, usually-18 to -20°C (long-term storage) or 4-10°C (medium-term storage) | Species producing orthodox seeds (desiccation-tolerant) | Seeds, that are dried to low moisture content without loss of viability |
| *In vitro* genebank | Ambient temperatures, 25±2°C (short-term storage) or low temperatures of 4-15°C (medium-term storage) | Species which do not produce seeds, or if they do, produce few seeds (threatened, wild and/or endemic species); species which are propagated vegetatively or as clones; species that produce non-orthodox seeds (desiccation-sensitive); species that require a long life cycle to generate breeding and/or planting material | Tissue cultures (plantlets, shoot cultures, somatic embryos, root cultures, meristem cultures, embryogenic callus cultures, cell suspensions) which may be actively or slow growing. |
| Cryo genebank | Ultra-low temperatures, ranging from-130 to -196°C (using liquid nitrogen) | As above | Seeds, embryos, embryonic axes, buds, shoot tips, meristems, pollen, cell suspensions, DNA |
| Field genebank (including clonal repository) | Ambient temperature and conditions of an open field or in screen house/net house/green house | As above | Whole plants |
| DNA Bank | Low temperatures of 4°C (short-term storage for 1-2 years) and -20°C (medium-term storage for 3-5 years); Ultra-low temperatures, ranging from -70 to -196°C (long-term storage for > years) | Any species, especially wherever genomic resources are being generated | Genomic, mitochondrial or chloroplast DNA; cloned genes, promoters fused with reporter genes; sub-genomic, cDNA, EST, repeat enriched libraries; cloning vectors, expression vectors, binary vectors, RFLp probes; BAC, YAC, PAC clones. |

*Source*: ICAR-NBPGR, Guidelines for Management of Plant Genetic Resources in India, 2016.

India has more than 100 botanical gardens under different management systems located in different bio-geographic regions. Central and State Government manage 33 botanical gardens that maintain the diversity in the form of plants or plant populations (MoEF, 1998). Few good plant banks exist in India; the 8 botanical gardens managed by "Botanical Survey of India" maintain 4,000 different species of which 200 species are of rare and endangered taxa.

### (b) Arboreta

An arboretum generally refers to a place established for conservation of tree species or vegetatively propagated crops. The Regional Plant Resource Centre at Bhubaneswar (Odisha) has established an arboretum with 1,430 species of tree, a palmeretum of 100 different types of palms, a bombastum with 61 collections of bamboo and an orchidarium housing 220 species of orchids. At the national level the interest in establishment of arboreta is very weak probably because of the cost involved.

### (c) Herbal Garden

This generally refers to the gardens that predominantly conserve herbs, shrubs that are of medicinal and aromatic value. The concept of herbal garden has been picked up by the non-governmental organization (NGOs) in India. Several non-governmental organizations in different parts of the country, particularly in tribal areas in Gujarat, Karnataka, Maharashtra, Madhya Pradesh, Chhattisgarh and Uttarakhand have established herbal gardens with the objective of conservation of local biodiversity in medicinal and aromatic plants and other economically important species. At community level effort for such herbal gardens may increase income of local communities.

### (d) Field Genebank

Most institution dealing with perennial or vegetative propagated (domesticated) crop species (pointed gourd, spine gourd, potato, sweet potato, cassava, yam, taro *etc.*) have field repositories for conservation of Plant Genetic Resources (PGR) with the reason that are sterile or do not easily produce seed, or seed is highly heterozygous. Essentially, a field or plant gene bank is an orchard or a field, in which accessions of fruit trees or vegetatively propagated crops are grown and maintained. Many horticultural species are either difficult or impossible to conserve as seeds because of seeds being recalcitrant or reproduce vegetatively. In ICAR several crop base research institute maintain gene bank of mandated perennial and vegetatively propagated species in the field condition. Hence they are conserved in field genebanks (FGBs). ICAR-NBPGR, New Delhi has established field repositories of perennial/tree species at Akola, Bhowali, Cuttack, Hyderabad, Jodhpur, Ranchi, Shillong, Shimla and Thrissur. Field genebanks (FGBs)

may run a risk of being damaged by natural calamities, infection, neglect or abuse. *Ex situ* conservation of tree species using field genebanks (FGBs) require a substantial number of individual genotypes to be an effective conservation measure. Thus FGBs require more space, especially for large tree species, and they may be relatively expensive to maintain depending upon the location and the complexity of alternative techniques available. However, in species with long juvenile phase as in tree fruits species, FGBs provide easy and ready access to conserved material for research as well as for use. Further for a number of plant species, the alternatives conservation methods are not fully developed yet so that they can be effectively used. Thus it is clear that establishment of FGBs plays a major role in conservation strategy for horticultural genetic resources. Although field gene banks provide easy access to conserved material for use, they run the risk of destruction by natural calamities, pests and diseases. Field collections of horticultural germplasm are located in the NPGR network of regional stations and Experimental Farm, Issapur, New Delhi besides horticultural crop base National Active Germplasm Sites (NAGS).

**Table 10.8: Germplasm Conserved under Field Genebank**

| Method of Conservation | Location | Crop (Number of Accessions) |
|---|---|---|
| Field gene bank | ICAR-NBPGR, New Delhi | Garlic (730) |
| | ICAR-NBPGR, Regional station, Bhowali (MS) | *Alium* spp. (44) |
| | ICAR-NBPGR, Base Centre, Ranchi (Jharkhand0 | Jackfruit (51) |
| | ICAR-NBPGR, PQ Regional station, Hyderabad (A.P.) | Drum stick (25) |
| | ICAR-NBPGR, Regional station, Thrissur (Kerala) | Jackfruit (67, 15) |

*Ex situ* conservation of germplasm of root and tuber crops requires special attention due to the fact they are stored and propagated as vegetative propagules. *Ex situ* conservation of germplasm of various species is generally carried out through field gene banks. The field gene banks have certain advantages, namely, cost effectiveness, particularly in tropical countries and require very less technical expertise as compared to biotechnological methods. However, in field gene banks the germplasm is exposed to natural vagaries. The field gene banks of root and tuber crops are maintained by annual replanting except in the case of sweet potato, which is to be replanted every quarter. The maintenance of field gene bank is extremely difficult due to the possible loss of accessions, chances of mixing up and requirement of heavy inputs (labour intensive) particularly in sweet potato. Conservation of wild species, especially the collections from tropical forests fail to survive in open field because of their specific adaptation and requirements. Shade net houses with 50 per cent to 75 per cent shade has been found effective in growing wild species of *Dioscorea, Amorphophallus, Colocasia, Ipomoea* (sweet potato collections),*Coleus* species and other medicinal tuberous plants (Unnikrishnan *et al.*, 2005). At Central Tuber Crops Research Institute, Thiruvananthapuram (Kerala) 5545 accessions of different crops germplasm are being maintained as field gene bank (Table 10.9).

**Table 10.9: Conservation of Tuber Crops in Field Gene Bank**

| Sl.No. | Crop | Number of Accession |
|---|---|---|
| 1. | Cassava | 1693 |
| 2. | Sweet potato | 927 |
| 3. | Yams | 859 |
| 4. | Aroids | 1184 |
| 5. | Minor Tubers | 284 |
| 6. | Total (Central Tuber Crops Research Institute, Thiruvananthapuram, Kerala) | 5545 |
| 7. | Total (Central Tuber Crops Research Institute, Bhubaneswar, Odisha) | 598 |

In addition to field gene bank at Central Tuber Crops Research Institute, Thiruvananthapuram (Kerala), 184 landraces of cassava are conserved at ICAR-NBPGR, Thrissur (Edison, *et al.*, 2005) and 569 cassava accessions in different Agricultural Universities under the All India Co-ordinated Research Project on Root and Tuber crops (416 accessions at Tamil Nadu Agricultural University, Coimbatore).

### (e) Clonal Repositories or On-Farm Conservation

The alternatives/supplementary method of 'Field Gene Banks' in vegetatively/clonally propagated crops could be maintenance of plants in clonal repository. In outcrossing species, individuals are heterozygous and populations are heterogenous. Some are dioecious, like pointed gourd (parwal), ivy gourd, spine gourd, sweet gourd *etc.* the heterozygocity is increased by the accumulation of point mutations, which are usually recessive and are not expressed under clonal propagation. Being heterozygous, segregation in the progeny is unavoidable. Thus the valuable combination of quality attributes (size, shape and colour, organoleptic quality *etc.*), vigour and resistance to stress in a genotype can be conserved by clonal propagation. The advantages of this method over field gene banks (FGBs) are less space requirement, easy management and cost effectiveness. In addition, clonal collections being dwarf in size can be maintained in protected controlled green houses and the danger of natural calamities is minimal.

As we know that the genetic diversity of traditional varieties of crops is the most economically valuable part of global biodiversity and is of paramount importance

for future world crop production. Increasing genetic diversification, combined with farmer's experimental abilities and underpinned by the formal system, will ensure greater on-farm conservation of more useful genetic resources. On-farm conservation of agro-biodiversity in tuber crops, which is based on traditional/indigenous knowledge, innovations and practices, should focus on local people's needs, constraints and priorities, besides achieving overall objectives of the conservation approach. Key to the success of on-farm conservation is the active community involvement in the diagnosis of local problems, in the formulation of a set of actions, actual implementation of action plans at farm level, and sharing of benefits resulting from the plan of action. This concern should be worked out and materialized in the coming years. Therefore, it is vitally important to identify the pathways along which people and community themselves will behave in making the traditional systems become more efficient and sustainable.

### (ii) National Seed Conservation/Seed Genebank

Among the various *ex situ* conservation methods, seed storage is the most convenient for long-term conservation of plant genetic resources. Seeds are relatively easy to collect, can represent a range of genetic diversity in the species if harvested from a population of individuals and can be stored in a relatively small space. The germplasm of the most seed producing crops is conserved as seeds (seed banks) stored at ambient temperature, low temperature, or ultra-low temperature, for short-, medium-, and long term storage, respectively. Seed banks play an important role in conserving the diversity of plant species and crop varieties for future generations. There is a global emphasis (internationally agreed standards) on conservation of germplasm in seed gene banks. This method is almost universally applied to the orthodox seed species like vegetables, which can maintain high viability when conditioned to low moisture content and stored under low temperature. It is well recognized that seed should be maintained under conditions in which the life process in seeds are minimized so that they can be stored for a longer period and with little loss in genetic diversity, genetic integrity and viability. Due to the extended life span of seeds stored under optimum conditions, it will not be necessary to regenerate the seeds at frequent intervals. Reducing frequency of regeneration, results in cost-effective maintenance of germplasm. More importantly it minimizes the genetic erosion resulting from genetic drift in small populations that may be grown for regeneration; genetic shifts resulting from natural selection when material is grown out in the field.

The Indian National Genebank has been established by the ICAR-NBPGR, New Delhi to conserve national heritage of germplasm collections in the form of seeds, vegetative propagules, tissue/cell cultures, embryos, gametes, *etc.* Based on experience gained from working with a built-in cold storage vault obtained from UK in 1983, four modules (two units of 100 m³ and two 176 m³ capacity) were installed in 1986 for long-term storage of seeds of orthodox species kept in laminated aluminium foils at -20 °C after drying them to 50 per cent moisture content. In seed gene banks germplasm accessions are stored as seeds; virtually all major gene banks are seed banks. Seed conservation is quite easy relatively safe and ordinarily needs minimum space. Under suitable conditions, seeds of many species can be stored for up to 50-100 years. However, seeds of all crops cannot be stored at low temperature in the seed banks. Usually storage of seeds is done in containers of glass, tin, plastic or a combination of these products. Seed banking can also be, and has been, scaled-up in terms of facilities (from chest freezer to walk-in cold stores) and effort (targeting the conservation of tens of thousands of species).

In the seed banks, there are three types of conservation, *viz.*,

(a) **Short term-** Working collections are stored for short term (3-5 years) at 5-10 °C.

(b) **Medium term-**Active collections are stores for medium term (10-15 years) at 4 °C.

(c) **Long term-**Base collections are conserved for long term (50 years or more) at -18 °C or -20 °C.

Work is also in progress on alternative methods of storage of seeds such as the maintenance of seed imbibed storage (recalcitrant seed storage at higher moisture content), cryopreservation of seeds (for extended life span than in long term storage), and storage of ultra-dry seeds (drying to moisture content of 2.5 per cent) in hermetically sealed containers under ambient conditions.

In this regard, the largest seed banks have been devoted to crops and other economically important species. In India the in National Gene Bank, ICAR-NBPGR, New Delhi, holds a total of 3,70,208 accessions of which 23,893 accessions belonging to different vegetable crops. Preset status of vegetable germplasm stored has been presented in Table 10.10.

**Table 10.10: Status of Vegetable Crops Conserved in National Gene Bank**

| Sl.No. | Crop Group | Germplasm Conserved |
|--------|------------|---------------------|
| 1. | Solanaceous vegetables | 9,118 |
| 2. | Cucurbitaceous vegetables | 4,682 |
| 3. | Leguminous vegetables | 4,850 |
| 4. | Malvaceous vegetables | 3,099 |
| 5. | Cole crops | 461 |
| 6. | Bulbous vegetables | 932 |
| 7. | Root vegetables | 391 |
| 8. | Leafy vegetables | 1,012 |
| | **Total** | **24,545** |

At both the NAGS centre ICAR-IIVR Varanasi (6792) and ICAR-IIHR, Bengaluru (2616) accessions, are being maintained. However, some of the collections are likely to be duplicates in various centres. Most of these collections have been characterized and evaluated for agronomic potential and other traits. These collections are available for distribution to plant breeders for their experimental use and crop improvement programms.

**Table 10.11: Germplasm Holdings of Cucurbitaceous Vegetable Crops at Various Places**

| Vegetables | No. of Collections | Vegetables | No. of Collections |
|------------|--------------------|------------|--------------------|
| Cucumber | 2638 | Bitter gourd | 973 |
| Snapmelon | 697 | Spine gourd | 80 |
| Muskmelon | 1480 | Sweet gourd | 65 |
| Kachri | 648 | Sponge gourd | 1852 |
| Pumpkin | 1880 | Ridge gourd | 1628 |
| Bottle gourd | 2313 | Satputia | 19 |
| Snake gourd | 289 | Ash gourd | 1139 |
| Pointed gourd | 324 | Round gourd | 164 |
| Ivy gourd | 154 | Watermelon | 529 |
| *C. hardwickii* | 158 | Melothria | 15 |

*Source*: IIVR ICAR-NBPGR, GBPUA&T, KAU, NDUA&T, IARI, IIHR, HARP, CIAH PAU, ANGRAU, RAU-ARS, Durgapura, 2016.

Although precise information on the current number of vegetable accessions conserved in the world gene bank is not provided by any global documentation systems. Analysis of available data allows us to estimate that more than 40 per cent of the *ex situ* vegetable germplasm is conserved in European gene banks. World Vegetable Centre (AVRDC), Taiwan houses one of the largest collections of vegetable germplasm in the world. Over 50,000 accessions belonging to 434 species and 153 genera collected from 151 countries are held in 'Trust of Global Community'. Of these, more than 40,000 accessions are made up of principal crops of global importance. The National Centre for Genetic Resources Preservation, Fort Collins, USA, established in 1958, holds a collection of half a million seed accessions, plus the largest collection of apple germplasm in the world. The National Centre for Crop Germplasm Preservation at the Chinese Academy of Agricultural Science in Beijing is a nationwide network of seed banks that preserves around 400000 accessions of crop seeds. Partly because of the increasing alarm about climate change and its impact on world food production, the 'Svalbard Global Seed Vault' in Norway in February 2008 as a coordinated effort to consolidate and systematize conservation of the world's crops, particularly those listed on the International Treaty on Plant Genetic Resources for Food and Agriculture (ITPGRFA), which entered into force in 2004.

**Classification of Seed and Conservation**

As proposed by Roberts in 1973, seeds are classified

into two major groups (on the basis of storability/conservation) and one special group "Poiklohydrous and Homoiohydrous" (Bejrak *et al.*, 1990):

(a) Orthodox seeds

(b) Recalcitrant seeds

(c) Poiklohydrous and Homoiohydrous

### (a) Orthodox Seeds

More than 90 per cent of plant species produce orthodox seeds, and most of the agri-horticultural crop species fall into this category. Seeds of this type can be dried to moisture content of 5 per cent or lower without lowering their viability and probably down to -273°C storage temperature. Such seeds can be easily stored for long periods; their longevity increases in response to lower seed moisture and storage temperature. Cooling and re-warming rates used are rather simple. Roberts (1973) proved that long term seed storage of orthodox seeds present no serious difficulty, although, the principles are not yet fully understood. The precise relationship between moisture content, temperatures, storage period and percentage viability conforms to three basic viability equations over a wide range of conditions in all species (Roberts and Ellis, 1977). Oxygen pressure is ignored, since in the sealed containers, it is dropped to zero due to respiration. The equations have been used to develop viability monographs for number of species for which the required experimental evidence is available. These three factors, if manipulated carefully, add to the stability and help stop chromosome breakage in ageing process of seed. In general, longevity is doubled for each 5°C fall in storage temperature and each 2 per cent drop in moisture content. Low temperature storage of orthodox seeds (Seed Ggenebank or National Genebank at ICAR-NBPGR, New Delhi) having12 storage modules with a capacity to hold about one million accessions including vegetable at -20°C for conservation of seed accessions on long-term basis as base collections for posterity. Listing of seed species reported to have survived liquid nitrogen (LN) exposure was prepared extensively by Stanwood (1985) and it enlisted several agri-horticultural and tree species.

### (b) Recalcitrant Seeds

In contrast, recalcitrant seeds cannot be dried below relatively high moisture contents without immediate sub-cellular damage and death. Viability of this group of seeds drops drastically if their moisture content is reduced below 12-30 per cent, *e.g.*, aquatic species, seeds of many forest and fruit trees, and of several tropical plantation crops like citrus, cocoa, coffee, rubber, oil palm, mango, jackfruit, *etc.* Such seeds present considerable difficulties in storage. All these species have short storage life (few months or a few years *i.e.* less than 5 years) even when desiccation is prevented, and many can survive only for a few weeks (King and Roberts, 1979). All these can withstand desiccation up to 10 per cent but cannot withstand storage temperature lower than -40°C. Therefore, germplasm of such plants is conserved by alternative approaches and more research is needed to devise cryopreservation techniques for these species. Although the diagnostic feature of recalcitrant seeds (*i.e.* their inability to withstand desiccation) seeds simple, the interpretation of data is not always straight forward and recent work has shown that a number of species previously thought to be recalcitrant are almost certainly orthodox.

### Problems Associated in Handling of Recalcitrant Seeds

There are several problems associated in handling of recalcitrant seeds:

☆ Precocious germination as seen in jackfruit and cocoa.

☆ Microbial contamination.

☆ Absence of guidelines for standard germination requirements and moisture content determination.

☆ Large seed size and availability of less number of seeds per sample due to bulk weight.

☆ Seed to seed variation in moisture content.

### (c) Poiklohydrous and Homoiohydrous

One special category was proposed by Bejrak *et al.* (1990) and they described them as poiklohydrous and homoiohydrous. They differ from each other in water relations and in whether or not they undergo a period of maturation drying as the final pre-shedding developmental phase. Homoiohydrous seeds undergo no maturation drying and apparently cannot acquire desiccation tolerance. Additionally, such seeds initiate germination on shedding. These seeds become increasingly desiccation sensitive with storage time, even when their water content is maintained at the newly shed level. Cryopreservation has been found to be the only method for storage of these difficult seed species.

Seed conservation includes:

(a) *Low temperature storage of orthodox seeds (seed gene bank):* The seed gene bank is responsible for conservation of seed accessions on long-term basis as base collections for posterity. It has 12 storage modules with a capacity to hold about one million accessions including vegetable at -20°C. The present base collection holding of vegetables in the National Gene Bank (NGB) is 15032.

(b) *Cryopreservation–storage of orthodox seeds and recalcitrant (embryonic axes) seeds:* Cryopreservation is storage of biological samples in viable conditions at ultra-low temperature of liquid nitrogen at -150 to -196 °C. A total of 388 samples have been cryopreserved at moisture content of 5-8 per cent in the vapor phase of liquid nitrogen. Further, investigations are on to develop protocols for cryopreservation of more recalcitrant seed species and establish their base collections. As an alternative complementary method, attempts are also being made to cryopreserve pollen in case of trees or vegetative propagated species. The main issue when exposing biological tissues to low temperatures is the formation of lethal ice crystals during cooling or thawing. Crystals thus formed penetrate membranous cell structures like nuclear and plasma membranes, causing irreversible damage, and a loss of their semi-permeability. The only way to avoid the formation of ice crystal in a solution is vitrification. Vitrification or glass transition refers to the transformation of a glass-forming liquid into a glass. Two requirements must be met for a liquid to vitrify. The solution must be:

i. Rapidly cooled and

ii. "Glass forming" or highly concentrated.

These two requirements form the basis for developing efficient cryopreservation techniques (Panis *et al.*, 2001).

(c) *Low carbon footprint seed conservation technique: The desiccated-ambient storage system using Zeolite beads*: Informal seed system fulfills 70 per cent of total seed requirement in India. Farmers use seed produced in their own field, exchange or sell among themselves when unable to access commercial seeds. India is a tropical country and on-farm conditions do not allow storing seed even for a year. Higher temperature accompanied with high relative humidity deteriorates seed quality rapidly and make them prone to insects and diseases. In genebanks, long or medium term conservation is done under sub-zero temperatures which is totally dependent on electricity and also produce a big carbon footprint. Different studies have reported that seed longevity can be enhanced similar to medium term storage (10-20 years) by storing dried seeds up to 3-5 per cent. Sun-drying cannot bring down the seed moisture content of seeds less than 10 per cent under normal conditions. A novel technique has been proposed where seed moisture content is brought down by using Zeolite beads and subsequently the reduced moisture is maintained by storing the dried seed in air-tight containers. Zeolite beads adsorb moisture from the surroundings and lower down the relative humidity (RH) of the air to >1 per cent, in-turn reducing the seed moisture content of the seeds being kept adjacent hermetically. These beads can adsorb moisture about 20-25 per cent of its own weight. The beads

are rechargeable and can be reused thousand times by heating them in an oven at 200-250°C for 3 hours. The technique is easy to use, farmer friendly and doesn't require a strong expertise with seed storage being possible under ambient temperatures (Gupta *et al.*, 2016).

### (iii) *In vitro* Conservation

*In vitro* conservation includes:

(i) Meristem conservation/Meristem genebanks

(ii) Cell and organ genebanks

(iii) DNA conservation/DNA genebanks

Crops such as potato, yam, cassava and sweet potato have either sterile genotypes or produce orthodox seeds which are highly heterozygous. Therefore, making seed storage of limited interest for the conservation of particular gene combinations. These species are mainly propagated vegetatively to maintain clonal genotypes. At present the most common method to conserve the genetic resources of these difficult crops is as whole plants in the field. There are, however, several serious problems with field gene banks. Distribution and exchange from field gene bank is difficult because of the vegetative nature of the material and the greater risk of disease transfer. The development of biotechnology has led to the production of a new category of germplasm, including clones obtained from elite genotypes, cell lines with special attributes and genetically transformed material (Engelmann, 1994) which is difficult to multiply. The conservation of rare and endangered plant species has also become an issue of concern.

The realization of the potential of *in vitro* conservation came about in the early 1970s, at a time when the storage of microbial cultures was a routine procedure. Since then, tissue culture techniques have been applied to more than 1,000 plant species. Subsequently, the technique has progressed from mere speculation to development and today it is routinely being used for conservation of vegetatively propagated crops and perennial species (Tyagi and Yusuf, 2003).

### *Merits of in vitro Conservation*

☆ Suitable for vegetatively propagated species or species which do not produce true to type seed.

☆ Cultures are maintained under slow growth conditions

☆ Low temperature incubation – a widely used method for slow growth

☆ 37,600 accessions were reportedly conserved *in vitro* worldwide (FAO, 1994)

### *Limitations of in vitro Conservation*

☆ Maintenance of cultures at low temperatures expensive and uncertain

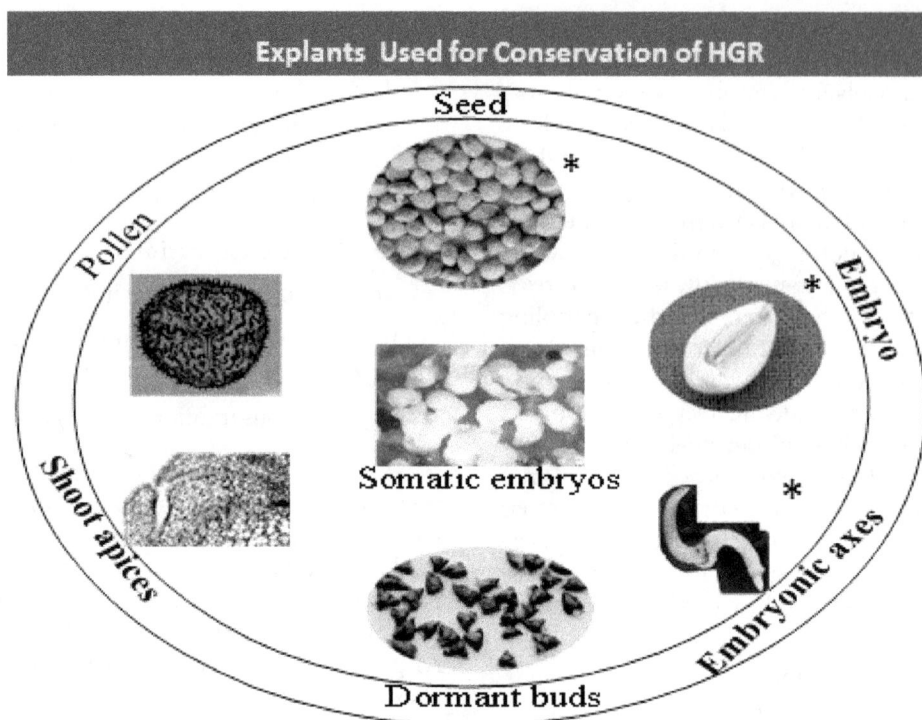

**Figure 10.4: Explants Used for Conservation.**

☆ Risk of culture contamination and loss over long periods of conservation

☆ Slow growth conditions can create stress situations for plans leading to abnormal growth

*In vitro* method can be used in two ways:

Plant tissue culture repository and cryopreservation techniques are employed to conserve the above described genetic resources.

**(a) Plant Tissue Culture Repository:** Conservation of cells, tissues, organs in glass or plastic containers under aseptic conditions through slow growth of cultures

**(b) Cryopreservation:** Basic cryo-principles is to store endangered exceptional species that produce few seeds, tropical and subtropical oilseeds especially orchids, species with inherently short-lived seeds (Type I seeds), despite being orthodox in nature, neglected and underutilized plant species (NUS), with focus on species of biodiversity hot-spots in the moist tropics as well as biotechnolofgically important lines and transgenics. Cryopreservation of cultures (tissues, organs, pollen, somatic/zygotic embryos or embryogenic cell cultures in liquid nitrogen at -150 to -196 °C). At this temperature, all biological processes virtually come to a stop. It may be called cell and organ bank. Both *in vitro* conservation and cryopreservation techniques use tissue culture principles for conservation (Rosa *et al.*, 1989; Reed, 1993; Mandal, 2003). The priority for cryobanking depends upon the

designed infrastructure, scope of expansion, networking with other institutes with common goals and policy statement of the organization or the country.

***Expanding Applications of Cryopreservation***

Panis *et al.* (2016) reported that the future cryobanking activities need a fresh look based on experience gained by laboratories that are adopting large scale banking. Regular brainstorming with international and national experts at ICAR-NBPGR, New Delhi in recent times contributed to refine and fine tune the concepts, practical approaches to match them with the expectations from the current researches. Laboratories are striving to put in correct perspectives the role of cryopreservation to handle the challenges faced by biodiversioty conservation at national and international levels.

The issues needing focus to:

☆ Role of conventional storage and trends in *ex situ* conservation using cryopreservation techniques;

☆ Networking approaches- nationally and internationally;

☆ Best practice, knowledge transfer and infrastructure;

☆ Need to initiate banking even before rechniques are perfected for high survival;

☆ Competitive basic research to maximize storage success:

   a. Systems biology research to identify suitable markers;

b. Identifying the model species for basic studies on desiccation tolerance and for freezing tolerance.

## (C) Vapour Phase Storage

The use of vapour phase temperature (-150°C) rather utilizing -196 °C as a storage temperature found to reduce the cost per accession storage at genebanks along with the thriving seed recovery. This was conceptualized during 2014-2015 in the Department of Plant Genetic Resources, Tamil Nadu Agricultural University, Coimbatore (T. Nadu) through the study of few major vegetable seeds with the morphometric methods as a parameters for investigation and found its relative amenability under sub-zero vapour phase storage condition ((-150°C) exposing it fitness towards extra-long term conservation (Vidhya and Ram, 2016).

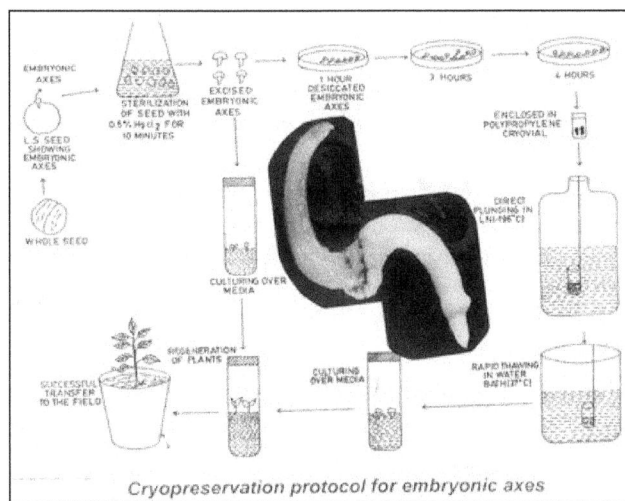

*Cryopreservation protocol for embryonic axes*

**Figure 10.5: Cryopreservation Protocol for Embryonic Axes.**

Most of the *Allium* species, selectively or exclusively vegetatively propagated, are maintained in the field genebank which is not only labour intensive but also suffer from disease outbreak and climatic vagaries. *In vitro* conservation though promising, requires periodic subculture, constant monitoring and is also challenged with accumulation of endophytes particularly in case of garlic (*A. sativum*). Cryopreservation is the preferred choice for safe and secures long-term conservation of *Allium* germplasm resources owing to aforementioned problems. In recent years, vitrification-based cryopreservation techniques have been successfully applied to many vegetatively propagated crops including Alliums. These comprise species which are not only used as a vegetable or a spice but to some extent, as a medicinal plant also. Though garlic is more popular and widely cultivated, species like *A. chinense*, *A. hookeri* and *A. tuberosum* hold potential as under-utilized, commercial crops. Several exotic species (*A. fistulosum*, *A. ramosum* and *A. scorodoprasum*) along with above mentioned *Allium* species are being maintained

in the field genebank and *in vitro* genebank at ICAR-NBPGR, New Delhi. Shoot tips, excised from cloves (in garlic) or from *in vitro* plantlets (in other *Allium* species) served as the explants for cryopreservation. Three cryopreservation techniques, namely: (1)Vitrification (V); (2) Droplet vitrification (DV) and (3) Encapsulation-dehydration (ED) were tested. Depending on the species/genotype, varying degree of success (10-60 per cent) was achieved, with shoots regenerating without intermediary callus. Cryobanking has been initiated in some *Allium* species. Present findings will focus on achievements and challenges with respect to *Allium* germplasm cryopreservation at ICAR-NBPGR, New Delhi (Pandey *et al.*, 2016).

## Plant Tissue Culture Repository

In general, tissue culture-based techniques fall under two categories:

(i) Slow-growth procedures (germplasm accessions are kept as sterile plant tissues or plantlets on nutrient media, and

(ii) Cryopreservation (plant material is stored in liquid nitrogen).

Tissue culture techniques are of great interest for collecting, multiplication and storage of plant germplasm. Tissue culture systems allow propagation of plant material with high multiplication rates in an aseptic condition. Virus free plants can be obtained through meristem culture in combination with thermotherapy, thus ensuring disease free plants and simplifying quarantine procedures. The minimizations of explants allow reduction in space requirements and reduce labour costs. However there are two technical problems associated with this method and need improvement. Firstly, the genetic instability of the material conserved as tissue culture due to somaclonal variation at the time of regeneration of the tissue in to seedlings, and secondly, the length of storage the tissue has been limited. The art and science of plant tissue culture is based on devising media for each genotype that would elicit the optimal response in terms of growth rate of the explants. Efficient conservation of genetic resources in case of vegetatively propagated plants and recalcitrant seed species has been hampered due to problems faced during application of their conventional method of *ex situ* conservation in field gene banks. To tackle these challenges, *in vitro* techniques have been increasingly used for conservation and its related activities like collecting and exchange of germplasm of these problem species. Cryopreservation refers to the non-lethal storage of biological tissues at ultra-low temperature, usually that of liquid nitrogen (LN) at -196 °C, is the only option currently available for the long term conservation of these PGR (vegetatively propagated plants and recalcitrant seed) avoiding exogenous contamination, requiring small space and minimum maintenance. At this very low temperature, all metabolic activities of cell ceases, and theoretically the cell or tissue can be stored for an indefinitely period. In brief:

Cryopreservation is the non-lethal storage of tissues at ultra-low temperature (-150 °C to -196°C) usually using liquid nitrogen

☆ Seeds, embryonic axes, buds, cells

☆ Various methods are available

 ❐ **Slow freezing:** conventional method using programmable freezer.

 ❐ **Fast freezing:** desiccation, pre-growth, pre-growth-desiccation, encapsulation-dehydration, vitrification, encapsulation-vitrification, droplet freezing.

### *Horticultural Germplasm Conservation: Logical Considerations for Cryobanking of Selected Genera*

☆ Cryopreservation has been attempted using seed, embryo and embryonic axes in those species that are exclusively seed propagated (and have intermediate/recalcitrant seed storage behaviour) and are semi wild/wild and those where no known cultivars are available.

☆ Cryostorage attempted to safe guard the existing/threatened diversity especially species of Indian origin or where India is secondary centre of origin. Also seeds of those species are conserved till their specific elite cultivars are identified which may be attempted later for conservation as true-to-type using vegetative tissues.

☆ For vegetables only wild relatives, varieties (modern and obsolete) and other prioritized germplasm like genetic stocks have been cryoconserved as seeds.

☆ For temperate fruits species cryopreservation protocols are being developed using dormant buds.

☆ In some of the highly recalcitrant seed species where no success has been achieved in developing protocol for cryopreservation using seeds, vegetative tissues *etc.*, pollen cryopreservation is being undertaken.

### *Development of Cryopreservation Protocols*

☆ Seed storage behaviour

 Sensitivity to desiccation

 Sensitivity to low temperature

 Longevity of explant

☆ Morphology and structure of explant

☆ Technique for cryopreservation

### *Cryopreservation of Non-Orthodox Seeds*

☆ **Explants:** Whole seeds, zygotic embryos, embryonic axes

☆ **Method used:** Air desiccation (preferably fast drying) followed by fast freezing

### *Cryopreservation Protocols for Hydrated Tissues*

It is true that over a span of 28 years, ICAR-NBPGR New Delhi has developed useful cryoprotocols, in most cases on species-specific basis, using explants like seeds, embryos, embryonic axes, pollen and dormant buds of the difficult-to-store types. Cryopreservation protocols have been standardized primarily for tropical species of horticultural, plantation, agro-forestry and industrial importance.

Roughly, three main cryopreservation protocols can be distinguished for hydarated tissues:

i. The "classical" slow freezing protocol (currently mainly applied to non-organized tissues such as cell suspensions and calluses) (Withers and King, 1980).

ii. The encapsulation/dehydration method that relies on synthetic seeds (Fabre and Dereuddre, 1990).

iii. The methods relying on the application of highly concentrated vitrification solutions such as PVS2 (Plant Vitrification Solution 2) (Sakai *et al.*, 1990).

The latter two methods are more suitable for organised plant tissues such as shoot cultures.

### Shoot Tip Genebank through *In vitro* Slow/Normal Growth Techniques

*In vitro* slow/normal growth techniques offer short- to medium term storage option (Engelmann and Drew, 1998; Sarkar and Naik, 1998), avoiding risk of losses of germplasm on field gene bank due to insects, nematodes, disease attacks and natural disasters. In such gene banks, germplasm is conserved as slow growth cultures of shoot-tips and nodal segments. It is commonly used for vegetatively propagated species, nonorthodox seeded species and wild species which produce little or no seeds. Their regeneration consists of sub-culturing the cultures, which may be done every 6 months to 3 years. The largest cryopreserved collection of *in vitro* shoot-tips comprises currently about 1,454 potato accessions and is stored at the Leibniz Institute of Plant Genetics and Crop Plant Research (IPK), Gatersleben (Germany) (Keller *et al.*, 2016). Other large collections of *in vitro* shoot-tips are Bioversity International's International Transit Centre (ITC), in Leuven, Belgium (950 banana accessions) and the Centre for International Potato (CIP)'s genebank, in Lima, Peru (1,450 potato accessions) (CIP, 2016).

A methodical procedure for conservation of *Capparis decidua* (commonly known as karira, is a useful plant in its marginal habitat and its spicy fruits are used for preparing vegetables, curry and fine pickles) explant was established employing *in vitro* generated shoot tips and nodal segments for encapsulation and complexation in 3 per cent in alginate solution and

**Table 10.12: Retesting of Viability of Vegetable Seeds after Cryostorage for various Periods**

| Sl.No. | Species | Accessions/ Varieties | Per cent Viability before Cryostorage | Per cent Viability after Cryostorage for (years) | | |
|--------|---------|----------------------|-----------------------------------|-----------|-----------|-----------|
| | | | | *I Testing* | *II Testing* | *III Testing* |
| 1 | *Amaranthus hypocondriacus* | IC - 38655 | 86 | 85 (6) | 78.00 (14) | - |
| | | VHC- 7502 | 99 | 86.94 (14) | - | - |
| 2 | *Amaranthus acruentus* | EC - 150200 | 85 | 78.33 (14) | - | - |
| 3 | *Vigna radiata* | PLM - 818 | 95 | 96.00 (2) | 94.5 (10) | - |
| | | IC – 8592-3 | 90 | 92.86 (2) | - | - |
| 4 | *Solanum melongena* | NIC - 6875 | 84 | 66.00 (6) | 80 (12) | - |
| 5 | *Solanum insanum* | NIC - 4237 | 82 | 87.5 (6) | - | - |
| 6 | *Allium cepa* | Pusa Ratnar | 96 | 100.00 (1) | 92.00 (3) | 100.00 (18) |
| | | Ratnar Selection | 95 | 95.00 (1) | 94.00 (3) | 100.00 (18) |
| 7 | *Raphanus sativus* | IC – 76305 | 82 | 89.50 (13) | - | - |
| | | Pusa Reshmi | 100 | 100.00 (13) | - | - |
| | | Pusa Hemani | 98 | 100.00 (13) | - | - |

100nM calcium chloride. Shoot tips nodal segments were encapsulated and stored at 4°C and 8°C temperature whereas immediate cultured synthetic beads kept as control. Seed longevity was investigated by storing synthetic seed at 4°C and 8°C for 20, 30 and 40 days, respectively. The viability/germination was depicted 80 per cent (maximum) in the seeds stored at 4°C and 8°C, respectively for 30 days on MS Media MS1, MS 2, MSA and MSB. The time taken for germination was comparatively swift on MSA containing 1 mg/l BAP+1 mg/lNAA+0.2 per cent charcoal, media MSB containing 1/2MS+0.1 mg/l BAP+0.7 mg/l IBA+2.5 gm phytagel, responded good in case of nodal segments. Based on the results of the moisture content percentage, synthetic seeds stored for 30 days at 4°C and 8°C, respectively manifested moisture content 69.7 per cent and 50.9 per cent, which shows highest germination (90 per cent on MSA and 80 per cent on MSB). The encapsulated shoot tips of *Capparis decidua* responded 100 per cent conversion to plantlets as compared to nodal segment. Intact plant regeneration was attained from the synthetic seeds of shoot tips. This finding could effectively be used for establishing a certain system for conservation, short term storage and production of synthetic seeds to be released for commercial purposes (Ahlawat and Sehrawat, 2016).

### *Merits for the Conservation of Vegetatively Propagated Crops/Germplasm*

The chief merits for the conservation of germplasm of vegetatively propagated crops and trees species.

☆ Genotypes of the accessions can be conserved in definitely free from disease and pests.

☆ They can be used for such crops, which either do into produce seeds or produce recalcitrant seeds.

☆ Subculture becomes necessary only after relatively long periods (every 6-36 months).

☆ Regeneration *i.e.*, sub-culturing, requires a comparatively very short time.

In addition, cuttings, bulbs and tubers can be maintained under controlled humidity and temperature conditions; however, this approach is practical for the short and medium term storage, and it should be used in conjunction with a field gene bank.

### **Appropriate Methods**

The various methods used to achieve this include the following:

☆ Use of growth retardants,

☆ Use of minimal growth media,

☆ Use of somatic regulators,

☆ Reduction in oxygen concentration

☆ Size and type of culture vessels

☆ Type of enclosures

☆ Maintenance under reduced temperature and

☆ For reduced light intensity

Here more than one treatment can be used under *in vitro*/slow growth mechanisms. Explants used for *in vitro* conservation must be of right type as well as physiological stage.

### *Meristem Conservation/Meristem Genebanks*

*In vitro* meristem culture technique offers the possibility of eliminating viruses and thus, exchange of virus-free germplasm is possible. Germplasm of asexually propagated species can be conserved in the form of meristems. This method is widely used for conservation and propagation of horticultural species. The apical and auxiliary meristems of very small size are the preferred explants for *in vitro* storage. In fact, organized explants have proved better than unorganized tissues, in terms genetic stability of the germplasm.

### (iv) Cell and Organ Genebanks

A germplasm collection based on cryopreservation (at -196°C in liquid nitrogen) embryogenic cell cultures, shoot-tips and or somatic/zygotic embryos may be called cell and organ banks. The techniques for cryopreservation of plant cells and tissues are being rapidly refined, and some such banks have been established, *e.g.*, for potato in Germany.

### *Droplet Freezing*

The droplet vitrification protocol was established because it combines the application of highly concentrated cryoprotective vitrification solutions (often PVS2) with ultra-fast freezing and thawing. Such high freezing and thawing rates are obtained by transferring the biological material (often meristems) to small strips of aluminium foil and exposing them directly to liquid nitrogen (Panis *et al.*, 2005)

☆ Apices are pretreated with liquid cryoprotective medium

☆ Placed on aluminium foil in minute droplets

☆ Frozen rapidly (potato)

☆ Frozen slowly (apple)

### *Pollen Storage*

Pollen grains are a rich source for diverse alleles in any genepool, holding large genetic diversity in small sample, size offering an effective propagule for germplasm conservastion. Its inherent ability of hardiness and stability in harsh conditions enables its long-term cryogeneic storage. The usefulness are as under:

☆ Cryogenic storage preferred for germplasm conservation since it is expected that pollen should remain viable for many years

  ❒ Sp. producing trinucleate pollen *e.g.* Gramineae, Cruciferae, Compositae *etc.*

  ❒ Sp. producing recalcitrant seeds

☆ Hybridization between plants cultivated/grown in different regions or showing nonsynchronous flowering

☆ Elimination of the need to continuously grow male lines used in plant breeding

☆ To facilitate supplementary pollinations for improving the yields particularly in Orchard sp.

As example, cryopreservation was used to conserve pollens of six potato varieties Kufri Girdhari, Kufri Himalini, Kufri Himsona, Kufri Jyoti, Kufri Kanchan and Kufri Shailja at ICAR-Central Potato Research Institute, Shimla (Himachal Preadesh). Pollen collected were packed in gelusil capsules, kept in cryo-vials, immersed in liquid nitrogen cryocan. A single lot was tested after 24-hrs of cryopreservation using pollen germination and double staining with Alexanders stain, containing malachite green for pollen cell wall staining (indigo-blue colour) and acid fuscine for pollen protoplasm (purple). Under similar conditions, the germination response for fresh and cryopreserved varieties was observed to be statistically at par, with heat-killed pollen as negative control showing nil germination. The germination response among varieties varied from 4.6 per cent (Kufri Jyoti) to 24.6 per cent (Kufri Himalini). Low germination response was mainly attributable to pollen grains showing no protoplasm and having only indigo-blue walls, this may reportedly be due to abnormal meiotic segregation producing inviable pollen in potato. The pollen tube lengths varied from 10 μm to 500 μm after 3 hrs of incubation in medium, indicative of differences in pollen vigour. The positive germination response after 24-hrs cryogenic treatment is a direct indicative of its prognostic long-term storage, for potato germplasm conservation, directly channeled into breeding programmes, also offering ease of transport and germplasm exchange (Kaur *et al.*, 2016).

#### Table 10.13: Pollen Cryostorage for Vegetable Species to Aid in Plant Breeding

| Sl.No. | Plant Species |
|--------|---------------|
| 1 | *Cucumis sativus* |
| 2 | *C. hardwickii* |
| 3 | *Abelmoschus callei* |
| 4 | *A. manihot* |
| 5 | *A. pungens* |

### (4) DNA Conservation/DNA Genebanks

In these banks, DNA segments from the genomes of germplasm accessions are maintained as cosmid clones, phage lysates or pure DNA (that last one being for relatively short periods). These DNA segments can be evaluated and the desired ones may be used to produce transgenic plants. This approach is applicable to the conservation of genetic materials of already extinct species since DNA extracted from well preserved herbarium specimens can often be cloned. However, it is very expensive and highly sophisticated. A world-wide network of DNA banks for threatened/endangered species has been established.

### Factors Responsible for Cryopreservation of Seeds

While attempting to cryopreserve these seeds, several factors have been found to be important:

1. Drying rate: In a number of species, the more rapidly water loss is achieved, the lower the moisture content before viability is affected. The process of flash drying, whereby excised axes are extremely rapidly dehydrated in a air stream, has exploited this property.

2. Seeds at various developmental stages are not tolerant to desiccation. During development they acquire desiccation tolerance at a particular

**Table 10.14: Cryopreservation Methods Used in World for Storing Vegetatively Propagated Germplasm**

| Institute | Country | Crop | Cryopreservation Method |
|---|---|---|---|
| Bioversity International, Leuven | Belgium | Banana | Droplet vitrification |
| Crop Research Institute, Prague | Czech Republic | Potato, garlic, hops | Droplet vitrification |
| International Centre for Tropical Agriculture (CIAT), Cali | Colombia | Cassava | Droplet vitrification Encapsulation/ dehydration |
| International Institute of Tropical Agriculture (IITA), Ibadan | Nigeria | Yam, banana, cassava | Droplet vitrification |
| International Potato Centre (CIP), Lima, Peru | Peru | Potato | Straw vitrification |
| | | | Droplet vitrification |
| Julius Kühn-Institut (JKI), Institut für Züchtungsforschung an Obst, Dresden, Germany | Germany | Strawberry/Fruit trees | Vitrification |
| | | | Dormant bud freezing |
| Leibniz Institut of Plant genetics and Crop Plant Research (IPK), genebank Department, Gatersleben, Germany | Germany | Potato, garlic, mint | Droplet freezing |
| | | | Droplet vitrification |
| National Agrobiodiversity Centre (NAAS), RDA, Suwon | South Korea | Garlic | Droplet vitrification |
| Tissue Culture and Cryopreservation Unit, ICAR-NBPGR, New Delhi | India | Banana, chives, medicinal plants, berries, fruit trees | Vitrification |
| | | | Droplet vitrification |
| | | | Slow freezing |
| | | | Dormant bud freezing |
| National Institute of Agrobiological Sciences (NIAS), Tsukuba, Japan | Japan | Mulberry | Dormant bud freezing |
| USDA-ARS, Fort Collins and Corvallis | USA | Citrus species, grape, garlic, mint, fruit trees | Vitrification |
| | | | Droplet vitrification |
| | | | Slow freezing |
| | | | Dormant bud freezing |

stage which appears to be variable in different species. In some cases it has been found that immature embryos might be more adaptable to manipulation than mature embryos or axes.

### Safety Back-up of Cryopreserved collection

Cryoconserved samples randomly tested for various periods upto 25 years (Choudary *et al.*, 2013; Choudhury and Malik, 2014) and after 28 years storage (pers. comn.) Until now each institute that stores cryopreserved crop germplasm has applied its own conservation standards with respect to 'what is a good regeneration rate to consider the germplasm safely stored'. Some of them consider a crop accession as safely cryopreserved for the long term provided the regeneration rate of the representative sample is at least 30 per cent. Others use 40 per cent, 60 per cent or even 80 per cent as cut-off.

### Status of Cryobanking

### 1. International Level

Several countries of the world have adopted large scale banking for long-term conservation of germplasm as zygotic seeds and its components, pollen, dormant buds, *in vitro* raised explants with each institute/ organization focusing on each type separately and in few cases all the forms in the same bank.

☆ National Centre for Genetic Resources Preservation (NCGRP), USA, besides being a

seed bank, is also a repository for animal genetic resources in the form of semen and plant genetic resources as graftable buds or *in vitro* explants.

☆ Bioversity International accepts a banana accession as safely stored provided:

i. Three successful independent repetitions are executed, and

ii. For each successful repetition stored in the cryobank

The ITC [International Transit Centre (previously called INIBAP)] is about 95 per cent confident that at least one plant can be regenerated (Dussert *et al.*, 2003). As an additional "security" measure, one replicate set of all accessions has been transferred under frozen condition to the Institut de Recherche pour le Development (IRD), Montpellier, France using a dry shipper. This 'black-box' back-up at IRD is located 1000 km from Leuven thus reducing the risk of germplasm loss due to political and/or environmental hazards.

Bioversity International proposes creating a 'Global Cryo Vault' for a range of vegetatively propagated crops. The proposed location will be in Leuven, Belgium, although other locations are being considered. This 'Global Cryo Vault', will ensure that a copy of cryopreserved crop samples conserved elsewhere in the world will be safely stored. It will act as a complementary facility to the 'Svalbard Global Seed Vault', in the Arctic Tundra, which conserve crops that reproduce through 'storable' seeds. With these two facilities, the majority of existing crop diversity-tens of thousands of species and varieties of all food crops and wild relatives will be preserved for present and future generations.

### 2. National Level (Cryobanking in India)

ICAR-National Bureau of Plant Genetic Resources New Delhi is leading the plant cryobanking efforst since 1987 equipped with large capacity cryobanks have successfully banked non-orthodox (intermediate and recalcitrant species) seeded species, prioritized orthodox seeds, pollen, plant origin explants, dormant buds and genomic resources at ultra-low temperatures with retesting data generated for 28 years of cryostorage. The cryobanking facility with six cryobanks of extra-large capacity (1000 liters Liquid Nitrogen) can hold quarter million samples. Presently it holds about 42,000 containers (cryovials of various capacities, polyolefin tubings of different diameter) storing more than 11,700 accessions belonging to about 794 plant species (Anon., 2015).

### International Status of PGR in Genebanks

Over 1750 genebanks and 2,500 botanical gardens conserve a total of 7.4 million germplasm accessions and 80,000 species, respectively around the world. Currently, 7,74,601 samples are deposited at 'Svalbard Global Seed Vault', Norway by 53 genebanks. It is estimated that more than one third of the globally distinct accessions of 156 crop genera stored in genebanks as orthodox seeds are conserved in the global seed vault. Global holdings of commodity groups *viz.*, cereals (31,57,578); food legumes (10,69,897); roots and tubers (2,04,408); vegetables (5,02,889); nuts, fruits and berries (4,23,401); oil crops (181,752); forages (6,51,024); sugar crops (63,474), fiber crops (1,69,969); medicinal, aromatic and spice crops (1,60,050); industrial and ornamental plants (1,52,325) and others (2,62,993) as reported by FAO (2014).

### National Status of PGR Conservation under *in vitro* Genebank (ICAR-NBPGR, New Delhi)

Lenka *et al.* (2016) reported that flow of unique germplasm during the last 15 years for conservation under medium- or long-term storage condition has revealed that out of a total 2,15,737 accessions processed by Germplasm Handling Unit (GHU), 1,99,294 accessions (92.5 per cent) have been found suitable for long-term storage based on their germination and viability. During the seed health testing, 5.43 per cent of the accessions were not found suitable as per genebank standards. Remaining accessions with less seed quantity or poor quality of germplasm could noty qualify for conservation as per genebank standards.

### *In vitro* Conservation of Root and Tuber Crops

In general, clonally propagated crops (produce very few seeds, are vegetatively propagated for breeding reasonsand/or species that require a long life cycle

**Table 10.15: Status of Crops at ICAR-NBPGR, New Delhi**

| | |
|---|---|
| Total collections | 2038 |
| Total Species | 131 (48 genera) |
| Total no. of Cultures | 40,000 |
| Fruit crops (Banana, temperate fruits) | 728 |
| Tuber crops (Taros, yams, sweet potato *etc.*) | 590 |
| Bulbous crops (garlic) | 171 |
| MAP (more than 20 spp.) | 169 |
| Spices and industrial crops (Turmeric, *Piper* spp., *Zingiber etc.*) | 380 |

to generate breeding and/or planting materials) such as root and tuber crops, essentially, are not conserved as orthodox seeds. The germplasm is either conserved as live plants in fields, as potted plants in enclosed structures or different plants parts are conserved into *in vitro*, either in slow growth (medium term) or cryopreservation (long term). Traditionally, the conservation and maintenance of root and tuber crops is done in field conditions. Many guidelines and training manuals exist for the management of root and tuber crops germplasm collections held in field genebanks (Reed *et al.*, 2004a; Geburek and Turok, 2005). However, major challenges of field conservation of root and tuber crops are long reproductive cycles, associated with low multiplication rate, high cost, pest and diseases, mislabeling and duplication leading to germplasm losses. These disadvantages are magnified by climatic factors. *In vitro* biotechnological approaches are reliable, complementary and/or alternative system to support multiplication, safer and longer conservation of root and tuber crops genetic diversity, and their sustainable utilization. They have the potential to address future, technical, scientific, economical and environmental demands on root and tuber crops (Pilatti *et al.*, 2011).

### Advantages of *In vitro* Conservation of Root and Tuber Crops

The major advantage of *in vitro* propagation of root and tuber crops material is the potential of large multiplication for the seed system, *e.g.* as an efficient and cost-effective propagation system (Asiedu *et al.*, 198; Quin, 1998; Thro *et al.*, 1999). The latter authors report two projects in Latin America that used cassava *in vitro* culture to address priorities of small-scale cassava farmers. Cassava propagation is generally done using lignified stem cuttings (Thro *et al.*, 1999). The multiplication rate is as low as 1:10 compared to at least 1:100 in some cereals, creating a bottleneck for transfer and adoption ov new varieties.

### Status of *In vitro* Slow Growth Conservation of Root and Tuber Crops (RTBs)

Slow growth conservation leads to the reduction of loss risks associated with the field banks, and constitute a viable alternative to complement and reduce the large size required for field banks. RTBs genebanks around the world have *in vitro* tissue culture facilities as a complementary conservation system, giving the possibility to clean the germplasm from diseases and pest *via* meristem culture and/or coupled with other cleaning methods. This conservation method requires technical expertise, facilities and operating budget. Though, they are generally more economical and less risky in a long-term perspective; as compared to field collections. Slow growth storage is suitable for short to medium term conservation, after which the plantlets are subcultured when signs of deterioration/necrosis are visible (Balogum, 2009). The conservation of RTBs needs small quantity of material and allows longer duration between two regenerations or subcultures, using slow growth storage. The principle is to place the *in vitro* plantlets under slow growth conditions, through adaptation to physical factors (light, temperature, culture medium, growth retardants). According to Ng and Ng (1997), 47 countries were holding cassava collections but only 12 maintained *in vitro* facilities for conservation. *In vitro* conservation of cassava is still far less common than field conservation. The largest national *in vitro* collections are hold in Brazil and Argentina. There appear to be very few RTBs *in vitro* genebanks in Africa. International collections are held at CIAT, CIP, Bioversity International and IITA, while all other *in vitro* genebanks have a national or regional focus.

**Table 10.16: *In vitro* Conservation of Vegetable Tuber Crops Genetic Diversity at Genebank of ICAR-NBPGR**

| Species | Accessions (No.) | Storage Period (Months) |
|---|---|---|
| Alocasia indica | 3 | 7 |
| Colocasia esculenta | 183 | 7 |
| Dioscorea spp. | 140 | 6 |
| Ipomoea batatas | 255 | 6 |
| Xanthosoma sagittifolia | 9 | 6 |

Cryopreservation, almost systematically associated with *in vitro* conservation, is another conservation method for ICAR-NBPGR germplasm. It allows maintenance of plant material at ultra-low temperature (in liquid nitrogen at -196°C) using cryogenic rechniques. At such low temperature, plant cell biological activities and metabolism are stopped, eliminating the need to regular rejuvenate or regenerate the plant.

Similarly, at what temperature crops are being stored under *in vitro* conditions are given below:

**Table 10.17: Status of *In vitro* Conservation at the Tissue Culture Repository at ICAR-NBPGR, New Delhi**

| Crop | Accessions (No.) | Storage Temp. (°C) | Storage Period (Months) |
|---|---|---|---|
| Allium sativum | 64 | 10 | 16 |
| Allium spp. | 15 | 10 | 12 |
| Ipomoea batatas | 230 | 25 | 12 |
| Dioscorea spp. | 31 | 25 | 12 |
| Zingiber officinale | 120 | 25 | 12 |
| Curcuma spp. | 24 | 25 | 8 |
| Musa spp. | 246 | 25 | 12 |
| Others | 35 | 25 | 10 |

## Current Status of Germplasm of Tuber Crops in India

At the ICAR-Central Tuber Crops Research Institute, Kerala, 4840 accessions of different tuber crops germplasm are maintained as field gene bank besides 598 accessions at the Regional Centre, Bhubaneswar (Odisha).

## Genetic Stability of the Conserved Germplasm

Now a day, various biochemical and molecular techniques are available to assess the stability of conserved material. Isozyme analysis of *in vitro* conserved germplasm was studied and some biochemical markers were identified. For organized structures, plants obtained from cryopreserved meristem of *Manihot*

**Table 10.18: Status of Horticultural Germplasm at National Cryogenebank (as on 31ˢᵗ march, 2011)**

| Categories | No. of Genera and Species | Accessions (No.) |
|---|---|---|
| *Recalcitrant and Intermediate* | | |
| Fruits and Nuts (mainly species of *Carissa, Citrus, Cordia, Diospyros, Grewia, Manilkara, Poncirus, Prunus, Ribes* etc.) | Genera – 54, Species – 176 | 2585 |
| Spices and Condiments (mainly *Piper, Amomum, Anethum, Illicium* etc.) | Genera – 9, Species – 10 | 148 |
| Plantation Crops - (*Camellia, Coffea* spp.) | Genera – 2, Species – 2 | 22 |
| *Orthodox* | | |
| Vegetables (Mainly *Abelmoschus, Allium, Cucumis* and *Solanum* species | Genera – 23, Species – 65 | 437 |
| Medicinal and Aromatic Plants | Genera – 190, Species – 313 | 852 |
| (mainly spp *Andrographis, Catharanthus, Chlorophytum, Datura, Digitalis, Mucuna, Ocimum, Plantago, Rauvolfia, Terminalia* etc) | | |
| Dormant buds | Genera – 4, Species – 16 | 337 |
| Pollen grains | Genera – 8, Species – 9 | 345 |
| **Total** | **4726** | |

**Table 10.19: Current Status of Germplasm of Tuber Crops in India**

| Crop | ICAR-CTCRI, Kerala | ICAR-CTCRI, Regional Centre | ICAR-NBPGR, Thrissur | ICAR-NBPGR, New Delhi | Centres of AICRP (TC) | Total |
|---|---|---|---|---|---|---|
| Cassava | 1693 | 33 | 184 | - | 569 | 2479 |
| *Manihot* species | 27 | - | - | - | - | 27 |
| Sweet potato | 927 | 243 | - | 260 | 1858 | 3288 |
| *Ipomoea* species | 6 | 78 | - | - | - | 84 |
| Greater Yam | 318 | 44 | 195 | - | 211 | 768 |
| White Yam | 258 | - | - | - | - | 258 |
| Lesser Yam | 125 | - | 68 | - | - | 193 |
| *Dioscorea* species | 82 | - | 184 | 44 | - | 310 |
| Taro | 1043 | 120 | 487 | 49 | 834 | 2533 |
| *Colocasia* (wild) | 7 | - | - | - | - | 7 |
| Tannia | 49 | - | - | 3 | 58 | 110 |
| Elephant foot yam | 85 | 32 | 56 | - | 120 | 293 |
| *Amorphophallus* species | 5 | - | 95 | - | - | 100 |
| Chinese potato | 88 | 1 | 50 | 1 | - | 140 |
| *Coleus* species | 8 | - | - | - | - | 8 |
| Costus | 2 | - | - | - | - | 2 |
| *Curcuma* species | 11 | - | - | - | - | 11 |
| Canna | 5 | 1 | 7 | - | - | 13 |
| Arrowroot | 7 | 1 | 5 | - | - | 13 |
| Yam bean | 63 | 45 | - | - | - | 108 |
| *Alocasia* | 1 | - | 3 | 1 | - | 5 |
| Other tuberizing species | 27 | - | - | - | - | 27 |
| *Typhonium* species | 3 | - | - | - | - | 3 |
| **Total** | **4840** | 598 | 1334 | 358 | 3650 | **10780** |

*Source*: Edison *et al.*, 2007.

*esculenta, Solanum tuberosum* and somatic embryos of oil palm were phenotypically normal. RFLP analysis of cryopreserved shoot tips of potato showed no changes. Flow cytometry revealed that cryopreservation did not appear to induce ploidy changes in sensitive dihaploids of potato.

Tissue culture techniques together with cryopreservation are very useful for conservation of vegetatively propagated crop germplasm. A number of techniques have been discussed for *in vitro* conservation. Based on the system and its related problems, a suitable amalgamation of the available techniques has to be standardized for a successful storage system which should have the ability to (Tyagi, 1998):

(1) Minimizes growth and development of the plant to enable the increased intervals between subculture.

(2) Maintain the viability of the stored material at the highest possible level, together with the minimum risk of genetic stability.

(3) Maintain the full developmental and functional potential of the conserved germplasm when it is retrieved for utilization.

(4) Make significant savings in inputs.

## Classification/Types of Seed Collections (on the basis of storage)

The conditions for seed storage depend mainly on the duration of storage. In general, seedbank collections are classified into three groups:

(i) Base collections

(ii) Active collections

(iii) Working collections

### (i) Base Collections

Base collections consist of all the accessions present in the germplasm of a crop, which are stored at about -20°C with 5 per cent moisture content, considering long term storage. They are disturbed only for regeneration, and germination tests are done every 10-15 years. Therefore, these are not used as a routine source for distribution of germplasm to user agencies but as a security against loss. Seeds are removed from base collections at infrequent intervals to monitor viability and for regeneration, when seed viability drops or when stocks of an accessions are no longer available. When the germination of an accession falls below, usually, 95 per cent of its germination at the start of storage, the accession is regenerated. For reasons of safety, duplicates of base collections should be conserved in other germplasm banks as well. High quality orthodox seeds can maintain good viability up to 100 years, while for crops like peas this period may be as short as 20-30 years only.

### (ii) Active Collections

The accessions in an active collection are stored at temperatures below 15 °C (often near 0°C), and the seed moisture is kept at 5-8 per cent. The storage is for medium term duration, *i.e.*, 10-15 years. These collections are used for evaluation, multiplication/regeneration and distribution. Active collections are usually maintained by multiplying the seeds of their own accessions. But from time to time, base collection material should be used for regeneration of active collections to prevent any appreciable shift in the genetic make-up of the accessions. Storage condition are usually less demanding than those of base collections because seeds are stored shorter period. These collections are generally large, since these are meant for various purposes. Base and active collections are closely linked. All the accessions stored in base collections should also be available in active collections. Generally, the genetic resources centres are responsible for conserving the base collections and active collections. This is essential to prevent any appreciable shift in these genetic make-ups of the collections.

### (iii) Working Collections

The accessions being actively used in crop improvement programmes constitute the working collection. Seeds of these accessions are stored for 3-5 years at <15°C, and usually, about 10 per cent moisture. These collections are maintained by the breeders who will be using them at medium or long term storage for better upkeep. The running cost of the long term storage facilities are to be maintained low as far as possible.

## Duplicate Base Collection

Duplicate base collections are again for long term storage which is housed for security reasons at different locations, from the corresponding base collections.

## Design of Seed Storage for PGR Conservation

Design of seed storage technique for plant genetic resource conservation is an important component which reflects the quality and quantity of stored seeds. The IPGRI, Rome has specified standards and has recommended that orthodox seed should be stored at -20°C at a moisture content of 5-7 per cent and that in general, regeneration should be done, when viability has fallen by 5 per cent. In fact -20°C has been found to

be most suitable temperature since most of equipment (compressor) operates economically at-20°C and this practice is being followed in many international gene banks.

## Guidelines for Sending Seeds to Genebank

In India, national facility has been created to conserve the genetic resources (seed materials of cultivated plants, and their wild relatives) with the name of "National Genebank" for longer time. Not only this, Indian gene bank has also accepted global responsibility for providing safety duplicate conservation of germplasm holdings of several crops on behalf of IARCs under terms set out in memoranda of understanding.

Before sending the seed for storage, following requirements should be met:

1. Seed should be fully developed and physiologically matured.
2. Undersized, shriveled and immature seeds should be discarded.
3. Off types and discoloured seeds should be discarded.
4. Seed should be free from insect and pests.
5. As for as possible, harvesting in rain should be avoided because seeds will absorb more moisture from the atmosphere and during storage, such seed will develop diseases and become infested with insect pests.
6. Seeds should be dried soon after the harvest (under shade only). If possible, dry the seeds in an air-conditioned or dehumidified room; it will increase the life span of seed.
7. Samples should contain at least 3,000 seeds for self-pollinated crops 6,000 seeds for cross-pollinated crops so as to fully represent variability of the original samples and also allow sufficient seeds for monitoring of viability during storage and subsequent regeneration. Sample size should be reduced in case of wild relatives where seeds are available in limited quantity.
8. Seeds should be packed in good quality paper pouches or cloth bags and wrapped in polythene bags. Entire seed lot after packing should be kept either in tin (metallic) boxes or cardboard boxes to minimize damage during transit. While packing, place label inside and also write the accession number outside the pouches for double ensuring the authenticity of the accessions number. Packing in gunny bags should be avoided.
9. Only untreated seeds are accepted in the gene bank.
10. Seed lots must be accompanied by minimum passport (*e.g.*, name of the crop, location

of collection, accession number *etc.* and characterization data). Other specific attributes like stress tolerance, resistance to diseases and pests may be recorded for each accession.

11. After harvest and cleaning, seed should be stored under ideal conditions *i.e.*, a cool place with low humidity to avoid spoilage due to pathogen and pests. Minimum time should be taken for dispatch to the gene bank.

12. Typed list of germplasm should be sent along with the material.

13. Prior information regarding the supply of seed material for storage should be sent to Germplasm Conservation Division, ICAR-NBPGR, New Delhi so that all arrangements can be made to receive the material.

## Steps before Dispatch of Germplasm

Ensure the following points before dispatch of germplasm:

1. Month and year of harvest on the bag/envelop/container.

2. Ensure healthy seeds in the lots and free from treatment.

3. Seeds should be in sufficient quantity (number).

4. Seeds should be properly dried and packed in dry bag/envelop/container.

5. Seeds should be accompanied with minimum passport and evaluation data.

6. Proper labeling of accessions is essential.

## Processing Procedures of Seeds for Conservation

The ICAR-NBPGR, New Delhi, receives regularly exotic and indigenous germplasm of agri-horticultural crops (orthodox seeds) from different sources for long term storage in the National Gene Bank. Before it is finally stored/conserved, it passes through following channels:

1. Registration
2. Seed cleaning and purity analysis
3. Seed germination test
4. Seed drying
5. Moisture testing
6. Seed packing and storage
7. Documentation

### 1. Registration

Germplasm seed received from different sources are checked and duplicate samples if found, must be rejected. Only new arrivals (samples) are registered when they first enter into the genebank and based on passport information, the accession is assigned

IC (Indigenous collection) or EC (Exotic collection) numbers. Accession number assigned is not re-assigned to any other accession even if seed viability is lost during storage.

### 2. Seed Cleaning and Purity Analysis

Seed cleaning and purity analysis following points should be kept in mind:

☆ Seeds are inspected and undersized, shriveled, immature seeds are screened

☆ Broken and damaged seeds are rejected

☆ Diseased seeds, insect attacked and seeds affected by nematodes are rejected

☆ Grass and cereal florets, empty glumes, lemmas, sterile florets of grass and cereals are rejected and

☆ Weed seeds are separated out.

### 3. Seed Health Status for Pest-Free Conservation in the National Genebank

Conservation of plant genetic resources in National Genebank (NGB) for long term storage is one of the mandates of ICAR-NBPGR, New Delhi for their utilization towards food security and sustainability. Seed health testing (SHT) has been major application in conservation of PGR for long term storage in the National Genebank by detecting and identifying seed-borne pests. Hence, seed health testing with specialized tests is essential undershot for long-term conservation of PGR free from pests. During 2011 to 2015, a total of 67,279 samples of indigenous PGR samples have been processed for seed health testing and observations of incubated seeds resulted in detection of a large number of economically important fungal pathogens from 5,922 samples (8.8 per cent), X-ray radiography revealed hidden infestation of insects in 4,673 samples (6.9 per cent) and nematode infestation in 175 samples out of 1,874 paddy samples (9.3 per cent) with varying infection/infestation level. Based on morphological characteristics, some of the important pathogens identified were *Alternaria brassicola, A. brassicae, Botrytis cinerea, Fusarium oxysporum, F. solani, Lasiodiplodia theobromae, Verticillium albo-atrum, etc.* Apart from immature stages of bruchid, some insect species such as *Callosobruchus chinensis, Pectinophora gossypiella, Tribolium castaneum, Rhizopertha dominica, Sitotroga cerealella, Sitophilus oryzae, etc.* were identified. (Akhtar *et al.,* 2016).

### 4. Seed Germination Test

The seed sample must conform to a standard of 85 per cent normal germination to be acceptable for storage.

For seed germination test following substrates are used:

a) Rolled towel test

b) Filter paper

c) Agar media

d) Sand media

## a) Rolled Towel Test

Rolled towel test method is generally used to test germination in medium and large seeds like cereals, maize, grain legumes, oilseed crop (sunflower, safflower *etc.*) and vegetables (okra, cucurbits *etc.*). Towel paper are spread on top of each other and then soaked in water. A wax paper is kept below the towel paper. On rough side of waxy paper, date of experiment and accession number is written. Normally 25 or 50 seeds are places on top of the uppermost sheet in fairly regular pattern so that they are approximately equidistant. Another towel is then placed on top of the seeds. The basal 2-30 mm portion of the longer edge of all towels is turned up to form a lip to prevent the seeds falling out. The towels can then be gently and closely rolled to form a tube of about 50-60 mm and tied with rubber band kept in germinator at desirable temperature and period. The towels are rolled loosely, (tight rolling will prevent air circulating between them and results in either no germination or germination or abnormal germination) and are placed in the germinator. Where germinator maintains high humidity, the rolled towels can be held with rubber band and kept upright in cages within the germination. If the germination test is to be conducted incubators where relative humidity is not controlled, water is sprayed over the towels every 2 to 3 days.

## b) Filter Paper

For filter paper testing, normally 9 cm diameter filter papers are used. In some cases, larger and smaller filter papers and petri dishes are used. The small seeded crops like tomato, brinjal, chillies, *Brassica* species *etc.*, are subjected to germination test in petri dishes. The filter papers are soaked in water and placed in the petri dish lid by glass pencil. Petri dishes are kept in the germinator for germination.

## c) Agar Media

Agar media is an alternative to routine germination test. Agar is a polysaccharide complex, which though insoluble in cold water, dissolves slowly in hot water to a viscid solution. New Zealand agar or Japan agar is suitable for germination test. In general1 per cent agar solution forms a stiff jelly on cooling, on to which the seeds can be placed for the germination test. The advantage of agar over paper as a germination test substrate is that the control of moisture availability is better, for example, in a germination test at 20°C, the agar remains moist for about a month, while in a germination test at 30°C, the agar remains moist for 2 weeks or so, the initial seed imbibitions environment may be more favourable than paper substrates and potential problems from ambition injury can be reduced.

## d) Sand Media

The sand consists of uniform particles and have a pH between 6.0- 7.5. Before use following points should be considered:

☆ Sand should be sieved with 0.8 mm diameter holes, which passes through the sieve holes and

☆ Sand must be sterilized at for 2 hours.

During the heating, the sand must be turned regularly in order to dry uniformly. Sufficient quantity of water is added to aluminum or plastic dishes (containers) filled with the sterilized sand and seeds are sown in rows on top of the sand and then covered with 10-20 mm loose sand. The dish is covered with flat plate, and placed in the germinator. The dish is labeled with proper accession number. Those accessions having more than 85 per cent germination are processed for drying.

## 5. Seed Drying

Each accession is placed either in porous paper bags or muslin cloth bags. Seeds are dried in dryer maintaining low relative humidity and at low temperature *i.e.* 15 per cent R.H. (relative humidity) and 15°C temperature. Under these conditions, seeds dry within 10-15 days.

## 6. Moisture Testing

Moisture is tested by oven drying method and when seed attains 4-6 per cent moisture (base collection), the seed samples are considered ready for storage and then it is hermetically sealed. For active collections, seed should be dried to 7 per cent moisture content or less is recommended.

## 7. Seed Packing and Storage

The sealed packets are ready for storage in gene bank. When sample size is enough for storage, a part of it goes for long term storage and another part is passed on for multiplication and storage in the active collection.

## 8. Documentation

All the information pertaining to the germplasm stored (long term), is properly documented and the print out of computerized database is available with the following indication:

☆ The module number

☆ Rack number

☆ Self-number and

☆ Basket number

The IPGRI has defined several categories of items of data about samples, passport, characterization and preliminary evaluation.

(i) **Accession data:** It includes accessions number, date of collection/acquisition date, principal attributes and distribution status.

**(ii) Sample data:** It includes serial number, initial data of storage, location of seed within the store, data of initial viability testing initial seed viability, seed moisture content, number/weight of seed in store, regeneration *etc.*. On receipt, the sample is registered and records maintained in a ledger and card index and these are kept in duplicate.

## 9. Choice of Containers

Generally accepted containers used in gene bank are:

(i) Glass bottles with rubber caps with hermetic sealing

(ii) Glass jars with screw caps-those are rather more convenient but the sealing is not perfect

(iii) Aluminium metal cans, are in use at many places and

(iv) Laminated aluminium foil packets consists of an inner layer of polythene (250 gauges) middle layer of aluminium foil (12µ) and an outer layer of polyester (12µ). It is convenient to use aluminium foil packets. Different sizes can be used to meet the needs of gene bank. These are inexpensive, occupying very little space and can be easily spaced and are total barrier to moisture. The packets are sealed by heat and pressure.

## 10. Sizes of Accessions and Volume of Cold Room

As per the latest recommendation of International Plant Genetic Resources Institute (IPGRI) experts committee, the following minimum number of seeds per population sample for storage is accepted:

**Table 10.20: Number of Seeds Required per Population Sample for Storage**

| Population Type | Base Collection | Duplicate of Base Collection | Active Collection |
|---|---|---|---|
| Fairly uniform | 1,500 | 1,000 | 3,000 |
| Highly variable | 3,000 | 3,000 | 5,000 |

These are arbitrary number and gene bank can certainly adopt the principle and modify as per the requirements. These weight and volume occupied by seeds very greatly between species. The Germplasm Conservation Division at the ICAR-NBGR, New Delhi has worked out the weight, volume, number of seeds in a litre space and volume of seed per kg for 60 of our important crops. The space requirements are to be worked out as per holdings of priority and other crops.

## 11. Shelving

It is not advisable to fill the internal volume of cold store with containers. Sufficient air circulation is to be allowed between the inner surfaces of the cold room. There should be a gap of 20 cm between the cold room wall and the mobile shelving: 10 cm between the floor and the lower shelf and 50 cm gap at the top of the shelves to allow air circulation.

## Monitoring of Seed Viability Stored under LTS and MTS

The most important activity of conservation is to maintain the viability of each accession in the genebank. The viability of stored seed is monitored by periodic germination tests when stored seed has to be replaced because seed stocks are nearly exhausted or viability has dropped, then regeneration, *i.e.*, rejuvenation of accessions becomes necessary at a fixed interval.

✰ The longevity of seed stored under LTS and MTS would vary among species and batches, and warrants monitoring of viability by the genebank curator.

✰ As per FAO 'Genebank Standards', viability monitoring test intervals should be set at one-third of the predicted for viability to fall to 85 per cent of initial value or lower depending on the species or specific accessions, but no longer than 40 years.

✰ If this deterioration period cannot be estimated, for accessions held in LTS at -18°C in hermetically sealed containers, the interval should be 10 years for species expected to be long-lived and 5 years or less for species expected to be short-lived. Monitoring of seed viability of accessions conserved in MTS should be carried out after 5 years.

✰ Viability monitoring should be ideally done by sampling 200 seeds in two replicates of 100 each. However, in exceptional cases, depending upon the availability/limitation of conserved seed quantity, 50-100 seeds may be tested in two replicates of 25-50 seeds each.

✰ The viability threshold for regeneration or recollection should be 85 per cent or lower depending on the species or initial viability of specific accessions.

## Regeneration of Seed Stored under LTS and MTS

The stage at which drop in viability, makes regeneration necessary is subject to judgment and argument. Some gene bank managers set higher limits since some genes are rare in a population and even slight deterioration of accession viability could mean the loss of some potentially valuable genes, where as a lower rate of germination is acceptable for relatively modern varieties than landraces and wild species.

✰ During monitoring of seeds conserved in genebanks, if seed viability/or number is found below the acceptability standards, regeneration should be done.

☆ Regeneration involves multiplication activities leading to an increase of the seeds conserved in the genebank (in case of insufficient material as per genebank standards) and/or to an increase of the viability of the seeds equal to or above an agreed minimum level (regeneration threshold).

☆ The regenerated seed material should have same characteristics as the original population.

☆ The FAO panel recommends regeneration when germination has dropped below the initial value by 5 per cent (preferred standard) or by 10 per cent (acceptable standard). However in some species or races, the threshold can be lowered especially if the initial viability is appreciable lower than that obtained in other species, as elaborated in table.

**Table 10.21: Minimum Standards of Seed Viability and Quantity in some Vegetable Species for Long Term Conservation**

| Botanical Name | Minimum Germination (per cent) | Minimum Seed Quantity (No.) |
|---|---|---|
| *Abelmoschus moschatus* var. *betulifolius* (Mast.) Hocher. | 50 | 500 |
| *A. anguuosus* Wight and Arn. | 50 | 500 |
| *A. bitiliolius* L. | 50 | 500 |
| *A. caillei* (A. Chev.) Stevels | 50 | 500 |
| *A. crinitus* Wall. | 50 | 500 |
| *A. ficulneus* (L.) W. and A. ex Wight | 50 | 500 |
| *A. manihot* (L.) Moench | 50 | 500 |
| *A. manihot* spp. *manihot* (L.) Medik | 30 | 1000 |
| *A. manihot* var. *pungens* (L.) Medik | 50 | 500 |
| *A. manihot* var. *tetraphyllus* (Hornem.) Borss. Waalk. | 50 | 500 |
| *A. moschuts* Medik | 50 | 500 |
| *A. moschuts* spp. *tuberosus* (Span.) Borss. | 50 | 500 |
| *A. pungens* Wall. | 50 | 500 |
| *A. tetraphyllus* Wall. | 30 | 1000 |
| *A. tuberculatus* Pal and Singh | 30 | 1000 |
| *Allium* species | 30 | 500 |
| *A. tuberosum* Rottl. Ex Spreng. | 50 | 500 |
| *Asparagus* spp. | 90 | 500 |
| *Brassica tournfortii* Gouan | ≥75 | 2000 |
| *Canavalia ensiformis* (L.) DC. | 30 | 100 |
| *Canna Indica* L. | 30 | 500 |
| *Cichorium intybus* L. | 30 | 500 |
| *Citrullus vulgaris* var. *citroid* L.H. Bailey | 50 | 500 |
| *Hibiscus pungens* Roxb. | 30 | 500 |
| *Lepidium sativum* L. | ≥75 | 500-1000 |
| *Luffa echinata* Roxb. | 50 | 500 |
| *L. pentandra* Roxb. | 50 | 500 |

| Botanical Name | Minimum Germination (per cent) | Minimum Seed Quantity (No.) |
|---|---|---|
| *Momordica dioica* Roxb. ex Willd | 60 | 500 |
| *M. sahyadrica* Kattuk. and V.T. Antony | 30 | 500 |
| *M. subangulata* ssp. *renigera* | 30 | 500 |
| *M. tuberosa* (Roxb.) Cogn. | 30 | 500 |
| *Solanum* spp. | 30 | 500 |
| *Trichosanthes bracteata* (L.) Voigt | 30 | 500 |
| *T. cucumerina* L. | 30 | 500 |
| *T. cuspidate* Lam. | 30 | 500 |
| *T. lobata* Wall. | 30 | 500 |
| *T. nervifolia* L. | 30 | 500 |
| *T. palmata* L. | 30 | 500 |
| *T. tricuspidata* Lour. | 50 | 500 |

Thus, regeneration standards for each crop species are set so as to balance the risk of valuable genes losses against high cost regeneration. The IPGRI recommends that if the viability goes down by 10 per cent the accession should be sent for regeneration.

**Table 10.22: Threshold Germination Percentages for Regeneration of Accessions**

| Initial Germination (per cent) | Final Germination (per cent) |
|---|---|
| 100 | 85 |
| 99 | 84 |
| 98 | 83 |
| 97 | 82 |
| 96 | 82 |
| 95 | 81 |
| 94 | 80 |
| 93 | 79 |
| 92 | 78 |
| 91 | 77 |
| 90 | 77 |
| 89 | 76 |
| 88 | 75 |
| 87 | 74 |
| 86 | 73 |
| 85 | 72 |

☆ As far as possible, the original sample should be used to regenerate accessions.

☆ The size of population for regeneration depends on the mode of pollination. For self-pollinated, genetically homogenous crops, a minimum of 100 randomly selected seeds should be used for regeneration, based on viability and availability of the seeds, so as to ensure a minimum plant stand of 75 or more. For cross-pollinated genetically heterogenous crops, a minimum of

200 randomly selected seeds should be used for regeneration, so as to ensure a minimum plant stand of 150 or more.

☆ Regeneration should be carried out in such a manner that the genetic integrity of the accession is maintained. Species-specific regeneration measures should be taken to prevent admixtures or genetic contamination arising from pollen flow from other accessions of the same species or from other species around the regeneration fields.

☆ After regeneration, the seed material should be processed as per the genebank standards mentioned in Section 5.2. Regenerated samples should be conserved along with the original material in base collection.

## Storage Periods of Seeds

Storage periods can be addressed as:

1. Long term storage methods
2. Medium term storage methods

## 1. Long Term Storage Methods

The original sample, near-original sample and/ or safety duplicate sample of an accession should be conserved under long term storage (LTS) conditions of temperature (-18±3°C) and RH (15±3 per cent).

Such samples are termed as 'base collections'. Seeds are dried to 5 per cent moisture content and can be packed in hermetically sealed containers. Therefore, immediately on receipt of material, the seeds are dried to 5± 2°C moisture (wb). To determine that, if the drying is necessary, the oven drying method, prescribed by ISTA is adopted. The techniques for drying the seeds to 5 per cent moisture should be very simple. Drying room/seed dryer provided at the gene bank operate at 15°C and 10-15 per cent relative humidity (RH) with good air circulation. This can also be done by using air dehumidifier and providing air conditioning. A computerized data processing unit is established, for an efficient management of germplasm holdings under long term conservation. In order to maintain viability of accessions in the long term the seed moisture content must be maintained after seed drying, in hermetic containers. At the ICAR-National Bureau of Plant Genetic Resources New Delhi, tri-layered laminated alumium foil pouches/packets are used for keeping the seed for long term storage of vegetable seeds. Seeds sample for LTS should have ≤ 85 per cent initial viability, 3-7 per cent moisture content and these should not have been treated with any chemical (pesticides/fumigants, *etc.*).

In case of vegetables varieties like cole crops, root crops onion, okra, leguminous vegetables tomato, brinjal and chillies, cucurbits leafy vegetable *etc.* are being conserved effectively.

**Table 10.23: Released Varieties of Vegetable Crops Stored for Long-Term Conservation**

| Crop | Botanical name | Source | Variety | IC No. |
|---|---|---|---|---|
| Spinach | *Spinacia oleracea* L. | ICAR-IARI, Regional Station Katrain, H. P. | Virginia Savoy | 75028 |
| Methi | *Trigonella foenumgraecum* L. | SKN College of Agriculture, Jobner, Rajasthan | Co1 | 78380 |
| Lettuce | *Lactuca sativa* L. | ICAR-IARI, Regional Station Katrain, H. P. | Chinese Yellow | 76300 |
| Longmelon | *Cucumis melo* L. var. *utilissimus* Dutch | ICAR-Indian Institute of Horticultural Research, Bengaluru (Karnataka) | Arka Sheetal | 76262 |
| Muskmelon | *Cucumis melo* L. | ICAR-Indian Institute of Horticultural Research, Bengaluru (Karnataka) | Arka Jeet | 76259 |
| Muskmelon | *Cucumis melo* L. | Agricultural Research Station, Durgapura, Rajasthan | Durgapura Meetha | 113138 |
| Watermelon | *Citrullus vulgaris* Schrad. | Agricultural Research Station, Durgapura Rajasthan | Durgapura Keshar | 113139 |
| Pumpkin | *Cucurbita moschata* Poir. | ICAR-Indian Institute of Horticultural Research, Bengaluru (Karnataka) | Arka Chandan | 76256 |
| Summer squash | *Cucurbita pepo* L. | ICAR-IARI, Regional Station Katrain, H. P. | Australlian Green | 76308 |
| Summer squash | *Cucurbita pepo* L. | ICAR-IARI, Regional Station Katrain, H. P. | Pusa Alankar | 76309 |
| Winter squash | *Cucurbita maxima* L. | ICAR-Indian Institute of Horticultural Research, Bengaluru (Karnataka) | Arka Suryamukhi | 75257 |
| Bitter gourd | *Momordica charantia* L. | ICAR-Indian Institute of Horticultural Research, Bengaluru (Karnataka) | Arka Harit | 75263 |
| Cucumber | *Cucumis sativus* L. | ICAR-IARI, Regional Station Katrain, H. P. | Pusa Sanyog | 76296 |
| Cucumber | *Cucumis sativus* L. | ICAR-IARI, Regional Station Katrain, H. P. | Japanese Long Green | 76297 |
| Brinjal | *Solanum melongena* L. | ICAR-IARI, Regional Station Katrain, H. P. | Pusa Purple Long | 76286 |
| Brinjal | *Solanum melongena* L. | ICAR-IARI, Regional Station Katrain, H. P. | Pusa Purple Cluster | 76287 |
| Brinjal | *Solanum melongena* L. | ICAR-IARI, Regional Station Katrain, H. P. | Pusa Anupam | 118806 |
| Brinjal | *Solanum melongena* L. | Tamil Nadu Agril. University, Coimbatore, T. Nadu | PLR-1 | 113889 |

| Crop | Botanical name | Source | Variety | IC No. |
|------|----------------|--------|---------|--------|
| Okra | *Abelmoschus esculentus* L. Monech | ICAR-IARI, Regional Station Katrain, H. P. | Pusa Makhmali | 76282 |
| Okra | *Abelmoschus esculentus* L. Monech | ICAR-IARI, Regional Station Katrain, H. P. | Perkin's Long Green | 76283 |
| Sweet pepper | *Capsicum annuum* L. | ICAR-IARI, Regional Station Katrain, H. P. | California Wonder 1 | |
| Sweet pepper | *Capsicum annuum* L. | ICAR-IARI, Regional Station Katrain, H. P. | Yolo Wonder | |
| Hot pepper | *Capsicum frutescens* L. | Regional Research Station, Lam, A. P. | Sindhur | 73158 |
| Hot pepper | *Capsicum frutescens* L. | Gujarat Agriculural University, Anand, Gujarat | Jwala | 77004 |
| Tomato | *Solanum lycopersicum* L. | ICAR-IARI, Regional Station Katrain, H. P. | Roma | 76310 |
| Tomato | *Solanum lycopersicum* L. | ICAR-IARI, Regional Station Katrain, H. P. | Best of All | 76311 |
| Tomato | *Solanum lycopersicum* L. | ICAR-IARI, Regional Station Katrain, H. P. | Marglobe | 76312 |
| Tomato | *Solanum lycopersicum* L. | ICAR-IARI, Regional Station Katrain, H. P. | Sious | 76313 |
| Tomato | *Solanum lycopersicum* L. | Gujarat Agriculural University, Anand, Gujarat | S-120 | 76997 |
| Tomato | *Solanum lycopersicum* L. | Tamil Nadu Agril. University, Regional Station, Paiyur, T. Nadu | Paiyur1 | 13891 |
| Cabbage | *Brassica oleracea* L. var. *capitata* L. | ICAR-IARI, Regional Station Katrain, H. P. | Golden Acre | 76238 |
| Cabbage | *Brassica oleracea* L. var. *capitata* L. | ICAR-IARI, Regional Station Katrain, H. P. | Pusa Drumhead | 76289 |
| Knol khol | *Brassica oleracea* L. var. *gongyloides* L. | ICAR-IARI, Regional Station Katrain, H. P. | White Voenna | 76318 |
| Cauliflower | *Brassica oleracea* L. var. *botrytis* L. | ICAR-IARI, Regional Station Katrain, H. P. | Pusa Snow Ball-1 | 76293 |
| Garden beet | *Beta vulgaris* L. | ICAR-IARI, Regional Station Katrain, H. P. | Detroit Dark Red | 76285 |
| Garden beet | *Beta vulgaris* L. | ICAR-IARI, Regional Station Katrain, H. P. | Crimson Globe | 76284 |
| Turnip | *Brassica rapa* L. | ICAR-IARI, Regional Station Katrain, H. P. | Pusa Chandrima | 76314 |
| Turnip | *Brassica rapa* L. | ICAR-IARI, Regional Station Katrain, H. P. | Purple Top White Globe | 76316 |
| Radish | *Raphanus sativus* L. | ICAR-IARI, Regional Station Katrain, H. P. | Pusa Himani | 76305 |
| Radish | *Raphanus sativus* L. | ICAR-IARI, Regional Station Katrain, H. P. | Japanese White | 76306 |

*Source*: Singh and Jain, 1993.

## 2. Medium term Storage Methods

Many active collections are stored at temperature between 0°C and are classified as 'medium term stores'. Sample of an accession which have been repeatedly multiplied and/or regenerated should be conserved under medium term storage (MTS) conditions of temperature (5-10°C) and RH (±10 per cent). The sample size of accessions under MTS should be larger than that of LTS. The recommended seed moisture content can be controlled either by using hermetic containers or if the samples are withdrawn more frequently, it may be considered more convenient to use unsealed containers (and control RH, but it is expensive). In some genebanks, open containers are used and in that case dehydrant with colour indicators are used. Silica gel with cobalt chloride is the usual indicator. In these stores, material is generally kept for 5 to 10 years. Seed samples in MTS can be additionally be stored in other types of containers such as cloth bags, metal cans or glass jars. Accessions in MTS have to be regenerated periodically to maintain the stocks for distribution. Seeds sample for MTS should have ≤ 85 per cent initial viability, 3-7 per cent moisture content and these should not have been treated with any chemical (pesticides/fumigants, *etc.*).

## Indian National Genebank (NGB)

The ICAR-National Bureau of Plant Genetic Resources (NBPGR), New Delhi is the nodal organization in India for planning, conducting, promoting and coordinating all activities concerning management of PGR including conservation and exchange of diverse germplasm for use of breeders and other researchers. ICAR-NBPGR houses the Indian National Gene Bank (NGB), which was commissioned in 1997 and maintaining germplasm in seed banks, field banks, as slow-growth cultures and as cryopreserved accessions. Indian National Genebank (NGB) was commissioned in 1997, and has 13 modules, each of which can store 50,000-76,000 samples. The seed bank accessions are maintained as base collections at ICAR-NBPGR headquarters, while medium-term storage of active germplasm collections is done at the ICAR-NBPGR headquarters and its regional stations. One of these modules is used for medium-term storage, while the remaining 12 are used for long-term storage at -18°C. Presently National Genebank houses ~0.43 million accessions of over 1800 species in seed genebank, 11,000 accessions in cryo genebank and 1,900 accessions in *in vitro* genebank (Mohapatra, 2016).

### Advantages of Seed Genebank

Advantages of genebank are (Sharma, 2009):

☆ Many collections can be accommodated in minimum space.

☆ Easy to maintain and handling.

☆ Help endangered flora to tide over natural adversaries.

☆ Easy accessibility for utilization

☆ Diverse genetic resources can be stored at one place

☆ Germplasm is conserved under pathogen and insect free environment.

### Disadvantages of Seed Genebanks

☆ Seeds of recalcitrant species cannot be stored in seed banks.

☆ Failure of power supply may lead to loss of variability and thereby loss of germplasm

☆ It requires periodical evaluation of seed viability. After some time multiplication is essential to get new or fresh seeds for storage.

**Table 10.24: Accessions in Module (-20°C), Cryobank (-196°C) and *In vitro* Cultures (4°C-27°C)**

| Crop/Group | Accessions in | | In vitro |
|---|---|---|---|
| | Module (-20°C) | Cryobank (-196°C) | Cultures (4°C-27°C) |
| Cereals and Pseudocreals | 90456 | 60 | - |
| Millets and Minor Millets | 16935 | 150 | - |
| Pulses | 31856 | 117 | - |
| Oilseeds | 25015 | 208 | - |
| fiber crops | 6755 | - | - |
| Vegetable, spices and fruit plants | 10990 | 369 | 513 |
| Medicinal and aromatic plants | 689 | 367 | 69 |
| Others (including Registered germplasm) | 1029 | - | 428 |
| Released cultivars | 1320 | - | - |
| Safety duplicate samples | 10146 | - | - |
| Recalcitrant 7 intermediate seed crops | - | 298 | - |
| Pollen species | - | 65 | - |

## Community Genebank/Community Seed Banks

The 'Community Seed Banks' (CSB) will help farmers to conserve existing indigenous crop diversity in a region, maintaining the availability which is normally unavailable in the formal seed systems. With this the seed supply within a community can be streamlined and a big diversity of seeds of various crops can be made available to the farmers. The 'community seed

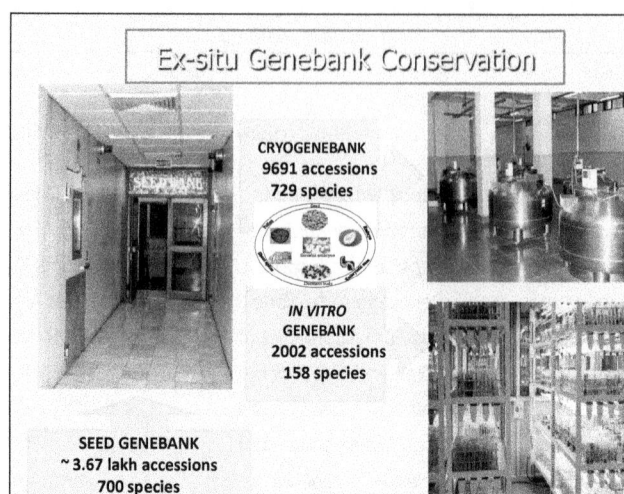

**Figure 10.6: *Ex-situ* Conservation of Genetic Resources.**

banks' can conserve as the centre of activity where women can be engaged in its maintenance and operation in conservation work. So a self-sustaining cycle can be established which consists of selection of different crops/varieties, their seed production, conservation in the community seed banks and cultivation (Nivedhitha *et al.*, 2016).

Community genebank is very useful in conserving the diversity. Community genebanks developed by MS Swaminathan Research Foundation (MSSRF) are worthy of replication:

1. **Genetic garden of halophytes:** This is to preserve genepools for breeding crop varieties tolerant to sea water in coastal areas, as a part of anticipatory research to face the challenge of sea level rise.

2. **Genetic garden of biofortified plants:** Species like Moringa, sweet potato, *etc.* This is for helping to find agricultural remedies for prevailing nutritional maladies, in particular micronutrient deficiencies like vitamin A, Vitamin $B_{12}$, zinc, iron, *etc.*

3. **Farm genebank for the *in situ* conservation:** These will help agrobiodiversity relevant to promoting a climate smart agriculture.

The type of three community genebanks may serve to illustrate the need and opportunities for conserving agrobiodiversity for launching an era of biohappiness. At the same time, there is need for greater emphasis on developing sustainable water security systems (Swaminathan, 2016).

## Conservation of Wild Relatives/Species

Wild relatives/species of some of the important vegetable crops having commercial importance have been conserved to a great extent. Most of these wild species grown in the natural habitats possess genes resistant to biotic and abiotic stress. However, diversity for valuable genetic resources is threatened in recent

**Table 10.25: Status of Cultivated and Wild Germplasm of Vegetables Conserved at ICAR-NBPGR, ICAR-IIVR and ICAR-IIHR**

| Crop Name | Botanical Name | No. of Accessions Conserved | | | |
|---|---|---|---|---|---|
| | | Centre of Origin | NBPGR | IIVR | IIHR |
| Tomato | *Solanum lycopersicum* | Mexico | 2337 | 1250 | 520 |
| | *S. esculentum* var. *cerasiforme, S. hirsutum, S. peruvianum, S. pimpinellifolium, S. Chilense* | Coastal Peru | 162 | 25 | 36 |
| Brinjal | *Solanum melongena* | India | 3723 | 295 | 222 |
| | *S. aculeatissimum, S. aethiopicum, S. albicans, S. viarum, S. albicaule, S. anguivi, S. aviculare, S. giganteum, S. gilo, S. khasianum, S. incanum, S. indicum, S. insanum, S. nigrum, S. macrocarpon, S. pubescens, S. verbascifolium, S.torvum, S. seaforthianum, S. setosissimum, S. torvum, S. lassiocarpum* | Indo-Burma | 763 | | |
| Chilli | *Capsicum annuum, C. baccatum, C. chinense, C. frutescens* | American tropics | 2568 | 425 | 665 |
| Okra | *Abelmoschus esculentus* | Ethiopia | 2416 | 245 | 129 |
| | *Abelmoschus angulosus, A. crinitus, A. ficulneus, A. manihot* subsp. *manihot, A. pungens, A. manihot* subsp. *tetraphyllus, A. moschatus, A. tuberculatus, A. caillei* | India, Africa | 716 | | |
| Onion | *Allium cepa* | Central Asia | 834 | - | - |
| | *Allium albidum, A.altaicum, A.auriculatum, A. stracheyi, A. fistulosum, A. griffithianum, A. tuberosum, A. senescens, A. ledebouranum, A. oreoprasum, A. oschaninii, A. pskemense, A. ramosum* | India, Russia, Austria, Switzerland | 117 | - | - |
| Ash gourd | *Benincasa hispida* | Indo-China | 222 | 293 | - |
| Bottle gourd | *Lagenaria siceraria* | Egypt | 722 | 183 | - |
| Bitter gourd | *Momordica charantia* | Asia | 369 | 230 | |
| | *M. dioica, M. cochinchinensis, M. balsamina, M. subangulata* | | 159 | 160 | - |
| Ridge gourd | *Luffa acutangula, L. hermaphrodita, L. pentandra, L. echinata, L. tuberosa* | Asia, Africa | 294 | 68 | - |
| Sponge gourd | *Luffa cylindrica* | Asia | 383 | 93 | - |
| Round gourd | *Praecitrullus fistulosus* | India, Asia | 108 | | |
| Ivy gourd | *Coccinia grandis, C. indica, C. cordifolia* | India | 23 | 26 | |
| Snake gourd | *Trichosanthes anguina, T. bracteata, T. cucumerina, T. cuspidata, T. lobata* | Indo- Malayan | 218 | | |
| Cucumber | *Cucumis sativus* | India | 343 | 104 | - |
| | *C. melo* var. *utilissimus, C. prophetarum, C. argrestis, C. callosus, C. hardwickii, C. trigonus* | South East Himalayas | 544 | 34 | - |
| *Cucurbita* spp. | *Cucurbita maxima, C. pepo, C. moschata* | America, Africa | 353 | 423 | - |
| Muskmelon | *Cucumis melo* | Himalayas | 793 | 619 | 98 |
| Watermelon | *Citrullus lanatus* | Africa | 257 | | |
| Pea | *Pisum sativum* | Ethiopia, Asia | 260 | 425 | - |
| Leafy vegetable | *Trigonella corniculata, Spinacia oleracea, Amaranthus cruentus* | India, Europe | 723 | 45 | 56 |
| Cabbage | *B. oleracea* var. *capitata* | Europe, India | 86 | - | - |
| Cauliflower | *B. oleracea* var. *botrytis* | Syria | 149 | 48 | 56 |
| Beet root | *Beta vulgaris* | Indo China | 46 | | |
| Carrot | *Daucus carota* | Europe, Asia | 74 | 25 | 80 |
| Radish | *Raphanus sativus* | East Mediterranean | 253 | 45 | - |
| Lablab bean | *Lablab purpureus* | Near East | 293 | 129 | 175 |
| Cowpea | *Vigna unguiculata* | West Africa | 225 | 217 | 235 |
| French bean | *Phaseolus vulgaris* | Mexico | 118 | 136 | 80 |

*Source*: National Genebank, ICAR-NBPGR, New Delhi.

times. Therefore, conservation of the vegetable genetic wealth particularly their wild relatives are thus essentially required for future utilization.

## National Active Germplasm Sites (NAGS) for Vegetables Conservation

The germplasm is supplied by ICAR-NBPGR, New Delhi from its active germplasm collections at Headquarters and Regional Stations or through the National Active Germplasm Sites for different crops located all over the country. The 40 National Active Germplasm Sites are holding active collections of relevant crop species. Eleven of them have the medium term seed storage facilities in the form of cold storage modules maintained at 4°C and 35-40 per cent relative humidity. Active collections are basically for sustainable use in various crop improvement programmes. The NAGS are based at premier crop or crop group-based institutes. The crop-based institutes have multi-disciplinary team of scientists, better equipped for evaluation of germplasm for agronomic potential and various yield reducing factors to identify accessions with desirable traits. This helps to generate information on potential value of various accessions to provide the needed support for their utilization in breeding programmes.

There are two designated major institutes designated as National Active Germplasm Sites to maintained active collections of vegetable germplasm:

(i) ICAR-Indian Institute of Vegetable Research (IIVR), Varanasi and

(ii) ICAR-IIHR, Indian Institute Horticultural Research (IIHR) Bengaluru

**Table 10.26: National Active Germplasm Sites (NAGS) for Vegetable Crops and Number of Holdings**

| Institution | Total Germplasm/ Accessions |
|---|---|
| ICAR-Indian Institute of Vegetable Research, Varanasi (U.P.) | 4007 |
| ICAR-Central Potato Research Institute, Shimla (Himachal Pradesh) | 2,500 |
| ICAR-Central Tuber Crops Research Institute, Thiruvananthapuram (Kerala) | 3,871 |
| ICAR-Indian Institute of Horticultural Research, Bengaluru (Karnataka) | 2, 900 |

## Development of Ecological Niche Modeling (ENM) for Plant Conservation

Over the years the plant conservation strategies have focused on simple measures of ensuring population build-up of endangered species to more holistic approaches by integrating new technologies and their applications in the conservation programme. To this end, computer based modeling techniques have become an important component of conservation planning in recent years (Guisan and Thuiller, 200). The ecological niche based modeling (ENM) technique is fast gaining wide popularity among conservation practitioners. It is a simple and user friendly technique which generates robust amounts of information. It uses computer algorithms to generate predictive maps of species distribution in a geographical space by correlating the point distribution data with a set of environmental raster data. The techniques have transformed the approaches of conservation by their ability to recreate prehistoric distribution of target species (Waltari *et al.*, 2007), predict existence of unknown populations and species (giriral *et al.*, 2008), identify sites for translocation and reintroductions (Wilson *et al.*, 2011), plan area selection for conservation (Papers *et al.*, 2006) and also forecast future distribution under changing climatic scenarios (Barik and Adhikari, 2011).

### Generating Data for ENM

☆ **Distributional data-** Occurrence data of target species can be obtained from primary and secondary. While the former comprises of geo referenced points collected mainly through field survey, the latter is obtained by geo-referencing (providing coordinates) species locality data from published sources (monographs and papers), museum records, herbaria.

☆ **Environmental data-** Environmental data used in ENM are common variables relating to climate (*e.g.* temperature, precipitation), topography (*i.e.* elevation, aspect), soil type and land cover type. climatic data where measurements are done on some continuous scale are referred as continuous while discreet data sets such as land use, forest types, vegetation types *etc.* are referred categorical data. Compiled data sets for both types of environmental data is freely available at various web URLs. For example, climatic variables could be sourced from http://www.worldclim.org; vegetation type from http://www.edcdeec.usgs.bov/glcc/glccversion1.html#global; land use from http://www.glcf.umiacs.umd.edu.data/, *etc.*

### Generating Niche Model

A number of both freely available and paid algorithms are available for ecological niche modeling such as BIOMAPPER, BIOCLIM, MAXENT, GARP, DIVA-GIS and many others. Each model carries with them their own data requirements, advantages and disadvantages. For instance certain software requires presence data only, others, functions with both presence and absence data along with abundance data. Using distributional data and environmental data, the software is able to prepare species distribution maps with high level of statistical confidence, predict areas of potential existence based on the niche of the species, and also identify suitable areas for reintroduction.

## Prediction Models

Ecological niche for *Gymnocladus Asomicus* and *Coptis teeta* was developed using MAXENT software viewed on DIVA platform. MAXENT is a niche modeling programme that uses BIOCLIM (Bio climatic analysis and prediction system) to produces output having probability distribution from the present occurrence data and the environment variables (Phillips and Dudik, 2008). Some of the advantages that MAXENT offers include use of occurrence data and does not model what is unknown. It makes use of continuous as well as categorical variables and gives a continuous prediction. It can work with small sample size. The model is primarily looking for new sites where the species may be found and also for identifying areas for species introduction.

## ICAR-National Bureau of Plant Genetic Resources (NBPGR), New Delhi

ICAR-National Bureau of Plant Genetic Resources (NBPGR), New Delhi is the nodal agency for the Indian National Plant Genetic Resources System that holds the primary responsibility for acquisition, augmentation, characterization and evaluation, documentation and maintenance of germplasm and is effectively assisted by 30 National Active Germplasm Sites (NAGS) that are responsible for the maintenance of situation specific subsets of collections for the crops assigned to them as a part of the system. A careful advance planning and an effective, systematic follow-up can obviously help achieve this enormous task. The Bureau operates its activities at its Headquarters through five administrative Divisions related to different fields of plant genetic resource activities *viz.* Exploration and collection, Introduction and Exchange, Plant Quarantine, Germplasm Evaluation and Germplasm Conservation besides a National Facility on Plant Tissue Culture Repository. Further, ICAR-NBPGR, New Delhi has 10 well established regional stations/satellite Stations/base stations/quarantine station located in different agro-ecological zones of the country that hold crop specific responsibilities for evaluation, region specific responsibility for exploration and collection or post entry quarantine responsibility. India's contribution to plant genetic resources has been impressive with repositories of over 50, 000 varieties of rice, 5,000 of sorghum, and 1000 of mango with a large number of horticultural, medicinal, aromatic and other crops. ICAR-NBPGR, New Delhi is having following regional and base centres with modern facilities to safe guard the precious genetic resources of their area.

- ☆ Regional Station, Shimla
- ☆ Regional Station, Bhowali
- ☆ Regional Station, Shilong
- ☆ Regional Station, Jodhpur
- ☆ Regional Station, Akola

- ☆ Satellite Station, Amravati
- ☆ Regional Plant Quarantine Station, Hyderabad
- ☆ Regional Station, Thrissur
- ☆ Base Centre, Cuttack
- ☆ Base Centre, Ranchi

All the above centres has the responsibility for characterization, maintenance, preliminary evaluation and multiplication of all the cereals, legumes, oil seed crops, forage grasses, fodder crops, fruits, vegetables (tomato, brinjal, chillies, cucurbits, okra, onion, garlic *etc.*), fiber crops, medicinal plants *etc.* The details about the centres are given below:

### (1) Regional Station, Shimla

This Regional Station, established in 1961, is located at Phagli, Shimla (HP) at 31°06′ N/77°E latitude/longitude and at an altitude of 2076 m above msl. It has the responsibility for exploration and collection of crop diversity in Himachal Pradesh and adjoining areas of Uttar Pradesh, Uttarakhand and Jammu and Kashmir. It has the responsibility for characterization, evaluation and maintenance of amaranths (Hills types), Buckwheat, chenopods, grain legumes (adzuki bean, lima bean, multiflorus bean), temperate fruits and nuts, ornamental plants and hops in its 7 ha farm area.

### (2) Regional Station, Bhowali

This Regional Station, established in 1986, is located about 15 Km from Nainital at 29°20′ N/79°E latitude/longitude at an altitude of 1460 m above msl and is responsible for characterization, preliminary evaluation and maintenance of hill collection of maize, rice, minor millets, lentil, cole crops, temperate forage grasses and legumes, medicinal and aromatic plants and wild relatives of temperate crops. It also undertakes explorations for collection of crop diversity in the central Himalayas. The station has a farm area of 10 ha.

### (3) Regional Station, Shilong

The regional established in August, 1978 is located at the ICAR Research Complex for NEH Region campus, Barapani (Meghalaya) at 25°34′ N latitude/91°56′E longitude at an altitude of 1000 m above msl. The farm area is about 3.0 ha. The major responsibility is characterization, preliminary evaluation and maintenance of paddy, maize, rice bean, citrus, and other minor economic plants, and undertakes explorations for the collection of crop diversity in all the seven north-eastern states and adjoining areas of north Bengal and Sikkim.

### (4) Regional Station, Jodhpur

This regional established in 1965, is located in the campus of Central Arid Zone Research Institute, Jodhpur (Rajasthan) at latitude/longitude of 26°20′ N/73°04 E at an altitude of 224 m above msl. Deep Tubewell facilities were created at Jodhpur centre in May, 1994. It has a farm area of 6 ha for characterization, preliminary evaluation and maintenance of crops suited to arid climate *viz.* Pearl millet (grain), Castor, Moth bean, Under-utilized plants (jojoba, euphorbia, tumba *etc.*). It also undertakes plant exploration and collection programmes in the entire Rajasthan, Saurashtra region of Gujarat and adjoining areas of Madhya Pradesh.

### (5) Regional Station, Akola

This regional station established in November, 1977, is located in the Punjabrao Krishi Vidyapeeth Campus at Akola (Maharashtra) and is situated at 24°43′ N/77° 64 E latitude/longitude at an altitude of 281 m above msl. The station has a 20 ha farm with black cotton soil. It has the responsibility for the characterization, preliminary evaluation and maintenance of germplasm of sorghum (grain), minor millets, safflower, sesame, groundnut, niger, linseed, castor, soybean, chickpea, pigeon pea, kulthi, winged bean, okra and grain amaranths (for plains). The station is also responsible for exploration and collection of crop diversity in Maharashtra, eastern Gujarat, northern Karnataka and southern parts of Madhya Pradesh.

### (6) Satellite Station, Amravati

The satellite station at Amravati, having farm area of 5.2 ha, is located about 90 km east of Akola (Maharashtra), at an altitude of 365 m above msl. It is mainly responsible for preliminary evaluation, multiplication and maintenance of mung bean, Indian bean (sem), chillies, sweet potato, certain horticultural and economic plants suited to soil and climate conditions of the region.

### (7) Regional Plant Quarantine Station, Hyderabad

This regional established in April, 1986 is located in the campus of APAU, Hyderabad at 17°20′ N/78°30 E latitude/longitude at an altitude of 542 m above msl. It has 6.4 ha farm area and is responsible for quarantine processing and clearance of pearl millet, sorghum, chickpea, pigeon pea, groundnut, small millets, paddy and other crops meant for the research institutes/universities located in southern parts of India. it also serves as an exploration base centre for the collection of germplasm of various crops in Andhra Pradesh, Telangana and adjoining areas of other states.

### (8) Regional Station, Thrissur

This regional established in 1977, is located in the Vellanikkara Campus of Kerala Agricultural University, at 10° 50′ N/76° 20 E latitude/longitude at an altitude of 50 m above msl. The station has 10.5 ha farm and is responsible for characterization, preliminary evaluation and maintenance of germplasm of paddy (plains), bitter gourd, cassava, *Dioscorea, Colocasia, Xanthosoma, Amorphophallus*, Coleus, ginger, turmeric, pepper, banana, jack fruit, and wild relatives of tropical crops and collection of crop diversity in Tamil Nadu, Kerala and southern parts of Karnataka.

### (9) Base Centre, Cuttack

The Base Centre, established in July, 1986, is located in the campus of ICAR-Central Rice Research Institute, Cuttack (Odisha) at latitude/longitude of 20°40′ N/85° 52 E at an altitude of 132 m above msl. It has a farm area of 2.0 ha and has the responsibility for exploration and collection of germplasm in the state of Odisha and adjoining parts of West Bengal and preliminary evaluation and maintenance of rice germplasm.

### (10) Base Centre, Ranchi

ICAR-NBPGR, New Delhi, Base centre, Ranchi, Jharkhand established in 1987 is located at 23°23′ N/85° 23 E latitude/longitude at an altitude of 625 m above msl in the tribal belt of Chhotanagpur. The region is known for its immensely rich primitive crop plant diversity including wild relative of many crop plants. The centre has farm area of 20 ha and is primarily responsible for exploration and collection of crop diversity in Bihar and adjoining area.

## ICAR-Indian Institute of Vegetable Research, Varanasi (U.P.)

In view of the importance of vegetables, systemic research in network mode on vegetables was initiated in the country with the establishment of All India Coordinated Research Project on Vegetable Crops-AICRP (VC) in the year 1971. AICRP (VC) was later upgraded as Project Directorate of Vegetable Research (PDVR) and subsequently shifted to Varanasi in the year 1992. Seven years later, PDVR was elevated to a national level institute as ICAR-Indian Institute of Vegetable Research (ICAR-IIVR). Since the establishment in 1999, ICAR-IIVR has been the leading institute on vegetable research in India. The main campus of the institute is located near Adalpura village on south of Varanasi. Geographically it is at 83.530E longitude and 18.520N latitude. It is spread in an area of 60 ha. The campus is 25 km from Varanasi rail-way station and 45km from Lal Bahadur Shastri International Airport, Babatpur, Varanasi. It receives an annual rainfall of 1000 mm. The minimum temperature is recorded during January (<5°C) and maximum during May-June (45°C).

### Mandate

☆ To contribute significantly to the nutritional security of India through the development of production technologies of vegetable crops, which are resources sustainable, economically viable and environmentally safe.

☆ To undertake basic, strategic and applied research for developing technologies to enhance productivity of vegetable crops.

☆ To provide scientific leadership in coordinated network research for solving location specific problems of production and to monitor breeder's seed production of released/notified varieties.

☆ To act as a national repository of scientific information relevant to vegetable crops and as a centre of training for up-gradation of scientific manpower in vegetable crops.

☆ To disseminate the vegetable production technologies to the farmers and to provide consultancy in vegetable research and development

☆ To collaborate with relevant national and international agencies for achieving the above mandate.

## ICAR-IIVR as a National Active Germplasm Site (NAGS) for Vegetables

Medium storage gene bank facility available at the institute helps in maintaining and conserving the valuable genetic resources of vegetables. ICAR-IIVR facilitates multiplication, conservation and evaluation of vegetable germplasm in collaboration with ICAR-NBPGR, New Delhi. IIVR holds 5320 germplasm accessions of 23 vegetables.

☆ Vibrant pre-breeding programs of ICAR-IIVR continuously augment the germplasm repository of vegetables.

☆ Germplasm holdings of ICAR-IIVR are enhanced through regular introductions from international genetic resource centres and indigenous collections.

☆ Germplasm resources available at ICAR-IIVR, Varanasi (U.P.) are exchanged with the members of national agriculture research system (NARS) on request

### Table 10.27: Status of Vegetable Germplasm at ICAR-IIVR, Varanasi

| Group of Vegetbles | Crop | No. of Accessions |
| --- | --- | --- |
| Solanaceous vegetables | Tomato | 1250 |
| | Chillies | 295 |
| | Brinjal | 295 |
| Malvaceous vegetable | Okra | 245 |
| Leguminous vegetables | Peas | 425 |
| | Cowpea | 217 |
| | French bean | 136 |
| | Lablab bean | 129 |
| Cucurbitaceous vegetables | Bitter gourd | 230 |
| | Bottle gourd | 145 |
| | Ridge gourd | 68 |
| | Sponge gourd | 93 |
| | Sathputia | 09 |

| Group of Vegetbles | Crop | No. of Accessions |
| --- | --- | --- |
| | Pointed gourd | 160 |
| | Muskmelon | 619 |
| | Pumpkin | 423 |
| | Ash gourd | 293 |
| | Cucumber | 104 |
| | Ivy gourd | 26 |
| Cruciferous vegetable | Cauliflower | 48 |
| Leafy vegetables | Amaranth | 40 |
| | Basella | 15 |
| | Chenopod | 8 |
| Root vegetables | Radish | 45 |
| | Carrot | 25 |

IIVR, Varanasi has developed several elite lines and registered with ICAR-NBPGR, New Delhi like Gynoecious bitter gourd (INGR03037), seedless pointed gourd (INGR03035), Leaf Curl resistant chilly (INGR07039), Downy mildew resistant Snapmelon (INGR07044), High carotenoids Pumpkin (INGR05027), Joint less tomato mutant (INGR06036), High carotenoids tomato (INGR06037), Dwarf Okra (INGR05026), Thin and long fruited Okra (INGR09125), Male sterile Carrot line flowers (INGR10110) *etc.* which are being utilized in breeding programme.

## Other National Organizations

In addition to ICAR-NBPGR, New Delhi several institutes under the umbrella of Indian Council of Agricultural Research (ICAR) like Indian Agricultural Research Institute (IARI), New Delhi; ICAR-Indian Institute of Vegetable Research (IIVR), Varanasi (U.P.); ICAR-Indian Institute of Horticulture Research (IIHR), Bengaluru (Karnataka); ICAR-Vivekananda Parvatiya Krishi Anusandhan Sansthan (VPKAS), Almora (Uttarakhand); ICAR-IARI Regional Station, Katrain (H.P.), Horticulture and Agro-forestry Research Programme (HARP), Ranchi (Jharkhand) and several State Agricultural Universities (SAUs) under All India Coordinated Research Project on Vegetable Crops are regularly collecting the germplasm of different vegetable crops from various parts of the country. These base collections are being maintained by ICAR-NBPGR in the National Gene Genk (NGB) in form of seeds in seed gene bank (where seeds dehydrated to ~5 per cent moisture content and sealed in tri-layered laminated aluminum foil packets and conserved at -18°C), in form of tissue culture in the *in vitro* genebank and in cryobank (-196°C), and in form of live plants in field genebanks. In addition, a large number of active collections have been maintained in Medium Term Storage (MTS) at ICAR-NBPGR, New Delhi (5081), and its regional stations at Thrissur (6634), Hyderabad (4310), Akola (2049), Shimla (392), Cuttack (36), Shillong (102) and Srinagar (23).

## International Organization Involved in Genetic Resources Conservation

List of the CGIAR Ceners associated with Natural Resources Management and conservation

1. Africa Rice Centre
2. Biversity International
3. CIAT: Centro Internacional de Agricultura Tropical
4. CIFOR; Centre for International Forestyry Research
5. CIMMYT: Centro Internacional de Mejoramiento de Maiy Trigo
6. CIP: Centro Internacional de la Papa
7. ICARDA: International Centre for Agricultural Research in the Dry Areas
8. ICRISAT: International Crops Research Institute for the Semi-Arid Tropics, Hyderabad, India
9. IFPRI: International Food Policy Research Institute
10. IITA: International Institute for Tropical Agriculture
11. International Livestock Research Institute
12. IRRI: International Rice Research Institute
13. IWMI: International Water Management Institute
14. World Agroforestry Centre (ICARAF)
15. World Fish Centre

Several international organizations are actively involved in germplasm collection, evaluation, exchange and conservation.

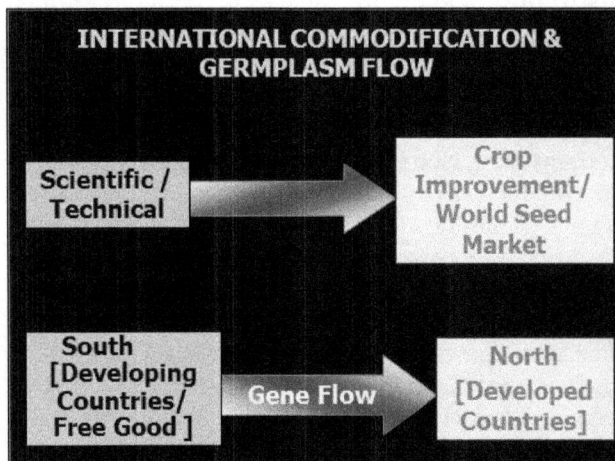

**Figure 10.7: International Commodification and Germplasm Flow.**

There are several examples like Germplasm Resource Information Network (GRIN) USDA, BLDG.003, BARC-West 10300 Baltimore Ave., Beltsville, MD 20705-2350, USA, contributing a lot in sharing the germplasm and related information to the users.

## (A) Consultative Group on International Agricultural Research (CGIAR)

The CGIAR arose in response to the widespread concern in the mid-20th century that rapid increases in human populations would soon lead to widespread famine. Starting in 1943, the Rockefeller Foundation and the Mexican government laid the seeds for the Green Revolution when they established the Office of Special Studies, which became the International Maize and Wheat Improvement Centre (CIMMYT) in 1963. CIMMYT and the International Rice Research Institute (IRRI), established in 1960 with support from the Rockefeller Foundation and Ford Foundation, developed high-yielding, disease-resistant varieties that dramatically increased production of these staple cereals, and turned India, for example, from a country regularly facing starvation in the 1960s to a net exporter of cereals by the late 1970s. But it was clear that these foundations alone could not fund all the agricultural research and development efforts needed to feed the world's population. In 1969, the Pearson Commission on International Development urged the international community to undertake "intensive international effort" to support "research specializing in food supplies and tropical agriculture". In 1970, the Rockefeller Foundation proposed a worldwide network of agricultural research centres under a permanent secretariat. This was further supported and developed by the World Bank, FAO and UNDP, and the Consultative Group on International Agricultural Research (CGIAR) was established in May, 1971which is well supported by (members of the CGIAR Consortium) a network if 15 (earlier number was 18) international agricultural research centres in close collaboration with hundreds of partners, including national and regional research institutes, civil society organizations, academia, development organizations and the private sector play greater role in global partnership that unites organizations engaged in research for a food secure future. The membership of CGIAR includes country governments, institutions, and philanthropic foundations including the USA, Canada, the UK, Germany, Switzerland, and Japan, the Ford Foundation, the Food and Agriculture Organization of the United Nations (FAO), the International Fund for Agricultural Development (IFAD), the United Nations Development Programme (UNDP), the World Bank, the European Commission, the Asian Development Bank, the African Development Bank, and the Fund of the Organization of the Petroleum Exporting Countries (OPEC Fund). This international organization is dedicated to reducing rural poverty, increasing food security, improving human health and nutrition, and ensuring sustainable management of natural resources. The 11 CGIAR genebanks with 730,000 accessions in 35 collections manage the world's largest and genetically most diverse collections, and have distributed ~380,000 samples to over 120 countries between 2012 and 2014 for their own research and also to meet the demands of national crop improvemen programmes (CGIAR, 2014; Galluzzi *et al.*, 2015). The 11 CGIAR genebanks were also credited with the highest annual number of International Germplasm Exchanges, especially under the International Treaty on Plant Genetic Resources for Food and Agriculture (ITPGRFA).

### The Vision of CGIAR's

The CGIAR's vision is supported by four strategic objectives:

1. Reducing rural poverty
2. Improving food security
3. Improving nutrition and health
4. Sustainably managing natural resources

### Reform in CGIAR

Seeking to increase its efficiency and build on its previous successes, CGIAR embarked on a program of reform in 2001. Key among the changes implemented was the adoption of Challenge Program as a means of harnessing the strengths of the diverse centres to address major global or regional issues. Three Challenge Programs were established within the supported research centres and a fourth to FARA, a research forum in Africa:

- ☆ Water and Food, aimed at producing more food using less water;https://en.wikipedia.org/wiki/CGIAR - cite_note-13

- ☆ Harvest Plus, to improve the micronutrient content of staple foods; and

- ☆ Generation, aimed at increasing the use of crop genetic resources to create a new generation of plants that meet farmers and consumer's needs.

**Table 10.28. Active members of the CGIAR Consortium of International Agricultural Research Centre**

| Sl.No. | Active CGIAR Centres | Headquarters Location |
|---|---|---|
| 1. | Africa Rice Centre (West Africa Rice Development Association, WARDA) | Bouaké, Côte, d'Ivoire/ Cotonou, Benin |
| 2. | Bioversity International | Maccarese, Rome, Italy |
| 3. | Centre for International Forestry Research (CIFOR) | Bogor, Indonesia |
| 4. | International Centre for Tropical Agriculture (CIAT) | Cali, Colombia |
| 5. | International Centre for Agricultural Research in the Dry Areas (ICARDA) | Beirut, Lebanon |
| 6. | International Crops Research Institute for the Semi-Arid Tropics (ICRISAT) | Hyderabad (Patan- cheru), India |
| 7. | International Food Policy Research Institute (IFPRI) | Washington, D.C., United States |
| 8. | International Institute of Tropical Agriculture (IITA) | Ibadan, Nigeria |

| Sl.No. | Active CGIAR Centres | Headquarters Location |
|---|---|---|
| 9. | International Livestock Research Institute (ILRI) | Nairobi, Kenya |
| 10. | International Maize and Wheat Improvement Centre (CIMMYT) | El Batán, Mexico State, Mexico |
| 11. | International Potato Centre (CIP) | Lima, Peru |
| 12. | International Rice Research Institute (IRRI) | Los Baños, Laguna, Philippines |
| 13. | International Water Management Institute (IWMI) | Battaramulla, Sri Lanka |
| 14. | World Agroforestry Centre (International Centre for Research in Agroforestry, ICRAF) | Nairobi, Kenya |
| 15. | World Fish Centre (International Centre for Living Aquatic Resources Management, ICLARM) | Penang, Malaysia |

### Change/Deleted Members of CGIAR

**Table 10.29: Members of CGIAR**

| Sl.No. | Inactive CGIAR Centres | Headquarters | Change |
|---|---|---|---|
| 1. | International Laboratory for Research on Animal Diseases (ILRAD) | Nairobi, Kenya | 1994: merged with ILCA to become ILRI |
| 2. | International Livestock Centre for Africa (ILCA) | Addis Ababa, Ethiopia | 1994: merged with ILRAD to become ILRI |
| 3. | International Network for the Improvement of Banana and Plantain (INIBAP) | Montpellier, France | 1994: became a programme of Bioversity International |
| 4. | International Service for National Agricultural Research (ISNAR) | The Hague, Netherlands | 2004: dissolved, main programmes moved to IFPRI |

## International Plant Genetic Resources (IPGRI)

International plant Genetic Resources (IPGRI) is the legal successor to the International Board for Plant Genetic Resources (IBPGR) and became operational in 1993. IPGRI has four main elements/objectives in its work on plant genetic resources:

1. Strengthening national programmes
2. Controlling to international collaboration
3. Improving strategies and technologies for conservation
4. Providing an international information service.

### The Mandate of IPGRI

The mandate of IPGRI is to advance the conservation and use of plant genetic resources for the benefit of present and future generations. IPGRI's mission is to encourage, support and engage in activities to strengthen the conservation and use of plan genetic resources worldwide with special emphasis on the needs of developing countries. IPGRI considers national programmes to be the basic elements of any global effort in plant genetic resources. IPGRI will undertake research and training and provide scientific and technical advice and information. The strategy of the IPGRI "Diversity for Development" elaborately deals with IPGRI's projections. Its programme is built on the basis of multidisciplinary projects, comprising a series of activities.

## IPGRI Offices

To facilitate overall collaboration in PGR activities the institute has adopted and implemented a structure of eight programme groups; 5 regional Groups and 3 thematic Groups, the Headquarters is established in Italy. The regional structure comprises 5 Regional groups which are responsible for the institute's works are:

(i) Sub-Saharan Africa

(ii) West Asia and North Africa

(iii) Asia, the Pacific and Oceania (APO) Regional Office is located at Singapore and has two more offices under it:

(a) The Office for East Asia at Beijing and

(b) The Office for South Asia at New Delhi

(iv) The Americas and

(v) Europe

These regional groups are responsible for developing, coordinating research and reviewing regional strategies in PGR activities. In addition information work of inter-regional or global relevance in their respective subject areas are also considered. The groups are also responsible for providing scientific and technical support to the regions. The three groups comprise:

(a) Genetic Diversity

(b) Germplasm Maintenance and Use

(c) Documentation, Information and Training

## International Plant Genetic Resources (IPGRI) for South Asia

International Plant Genetic Resources (IPGRI) for South Asia located in the premises of the ICAR-National Bureau of Plant Genetic Resources, New Delhi. The establishment of this office (initially Regional Office for South and Southeast Asia) was formalized with Government of India under a Memorandum of Understanding with Indian Council of Agricultural Research (ICAR), New Delhi in November, 1987 and the Office became operational in July, 1988. The South Asia region covered by this office includes Bangladesh, Bhutan, India, Maldives, Nepal and Sri Lanka. It is very diverse agro-climatically, topographically, culturally and a seat of diversification for several crop plants; rice, grain legumes, minor millets, spices, oilseeds, several tropical fruits and vegetables *etc*. With India, its programme was jointly developed under the biennial work plans with ICAR/NBPGR, for plant genetic resources activities as per national priorities.

## Objectives

The major mandate of International Plant Genetic Resources (IPGRI) for South Asia is to promote and strengthen plant genetic resources activities in the region with emphasis on their conservation and use. The following activities are trusted:

(i) Assist the programmes in the region in strengthening their PGR activities through advice, small projects, *etc.*, as per request.

(ii) Coordinate and support PGR activities on a regional level wherever desired by the countries and whenever meaningful (South Asian PGR national coordinators meeting, workshops on crop networks, training courses *etc.*)

(iii) Create awareness on national and regional level and promote the conservation and utilization of PGR.

(iv) Coordinate/undertake information synthesis and documentation of data on plant genetic resources held by national programmes.

(v) Liaise with IARCs and other regional, global organizations/FAO and the NGOs.

(vi) Disseminate PGR information/literature *etc.*

**Table 10.30: Exploration under by International Plant Genetic Resources (IPGRI) for South Asia**

| Sl.No. | Activities | Organizations involved | Crop |
|--------|-----------|------------------------|------|
| (i) | Facilitate joint explorations in the region and provide guidelines/logistics. | Bangladesh (NBPGR-BARI), Nepal (NBPGR-NARC) and Sri Lanka (NBPGR-PGRC) | Eggplant, okra |
| (ii) | Support within country explorations on priority | NBPGR, New Delhi | Okra, eggplant, maize, sesame |
| (iii) | Assess gaps in collecting, including wild relatives, genetic erosion, national priorities | - | - |
| (iv) | Assist in preparing Project proposals, country reports to prioritize germplasm collecting activities | - | - |

## The World Vegetable Centre (formerly as AVRDC), Taiwan

Founded in 1971 as the Asian Vegetable Research and Development Centre (AVRDC) with a focus on tropical Asia, today the work of AVRDC–The World Vegetable Centre spans the globe. Headquarter in Taiwan, with regional bases in West and Central Africa, Eastern and Southern Africa, East and Southeast Asia, South Asia, West and Central Asia and Oceania, the Centre has 44 international scientists and 300 national scientists and support staff dedicated to the mission of alleviating poverty and malnutrition through the increased production and consumption of nutritious, health-promoting vegetables. For 40 years AVRDC–The World Vegetable Centre has been the world's leading international centre focused on vegetable research and development. We maintain the world's largest public sector vegetable genebank, with a focus on hardy traditional vegetables important as food for the poor as well as wild relatives of common vegetables. Our improved varieties are planted on millions of hectares around the world and our production and postharvest technologies have made major improvements in smallholder incomes. Through the dissemination of good agricultural practices and effective postharvest value-addition and marketing mechanisms, the Centre fosters opportunities for increasing employment and incomes for small-scale farmers, landless laborers and communities.

### Strategy of World Vegetable Centre (AVRDC) Genebank

The AVRDC Genebank maintains the world's largest public vegetable germplasm collection with more than 61,000 accessions from 155 countries, including about 12,000 accessions of indigenous vegetables. Collecting and conservation work is done in collaboration with national partners who maintain duplicate collections. Molecular characterization and genetic diversity analysis of selected germplasm collections is done to identify markers and map genes linked to important agronomic traits such as disease resistance, stress tolerance, or high nutritional value. This significantly enhances the efficiency of breeding programs as key genes can be identified for introgression into improved lines.

☆ Conserve and distribute vegetable germplasm to improve crops

☆ Identify superior sources of genes for important horticultural traits

☆ Characterize the Centre's germplasm to make better use of its diversity

☆ Develop DNA markers for improved traits for marker-assisted selection

☆ Use molecular technologies to isolate and validate genes affecting important traits

☆ Share the benefits of the Centre's germplasm collections

☆ Train partners in germplasm conservation, use, and gene discovery

Since its founding, the centre has distributed more than 600,000 seed samples to researchers in the public and private sectors in at least 180 countries. This has led to the release of hundreds of varieties throughout the world, especially in developing countries. The AVRDC Genebank is part of **Genesys**-the global information and germplasm exchange network.

## The Vegetable Genetic Improvement Network (VeGIN), Wellesbourne

The Vegetable Genetic Improvement Network (VeGIN) falls under Warwick Crop Centre, The University of Warwick, Wellesbourne, Warwick CV35 9EF. VeGIN combines the expertise of plant geneticists, breeders, physiologists, entomologists, pathologists, virologists and bioinformaticians. Using traditional breeding methods in union with advanced modern breeding tools, aim is to improve vegetable crops, specifically, Brassicas, lettuce, onion and carrot for a diverse range of traits of economic and social importance. The output from this work contributes towards meeting the challenges of future vegetable crop production

## Centre for Genetic Resources, the Netherlands

CGN focuses on conservation and use of vegetable crops, farm animal breeds and autochthonous forest species. The Centre for Genetic Resources, the Netherlands (CGN) conducts, on behalf of the Dutch government, statutory research tasks associated with the genetic diversity and identity of species that are important for agriculture and forestry. CGN is an independent research unit within DLO Foundation that assists the government in its statutory tasks. Its activities are aimed at the *ex situ* conservation, support for *in situ* conservation, and promotion of the use of genetic propagation material in support of breeding and research, and as part of our bio-cultural heritage. Policy support of the Dutch government and international organizations is provided as a complementarily activity. The programme concerns crops and forest species as well as domestic animals.

## The Svalbard Global Seed Vault (SGSV)

Deep inside a mountain on a remote island in the Svalbard archipelago, halfway between the northern-most tip of mainland Norway and the North Pole, lies the Svalbard Global Seed Vault. This is fail-safe, last chance backup facility for the world's crop diversity. It currently holds more than 860,000 samples of crop diversity from more than 60 genebanks, and nearly every country

**Table 10.31: Genebanks Supported by Crop Trust through Long-Term Grants from the Endowment Fund and the Genebanks CRP**

| CGIAR Centre | Country | Crops | # of Crop Accessions |
|---|---|---|---|
| Africa Rice | Benin | Rice | 19,983 |
| International Institute for Tropical Agriculture | Nigeria | Cowpea | 30,388 |
| Biversity International | Belgium | Banana | 1,455 |
| International Maize and Wheat Improvement Centre | Mexico | Maize and wheat | 175,526 |
| International Centre for Agricultural Research in the Dry Areas | Syria | Barley, chickpea, faba bean, forages, lentil and wheat | 136,350 |
| International Centre for Tropical Agriculture | Colombia | Bean, cassava and forages | 67,574 |
| International Crops Research Institute for the Semi-Arid Tropics | India | Chickpea, groundnut, minor millet, pearl millet and sorghum | 129,081 |
| International Livestock Research Institute | Kenya | Forages | 17,716 |
| International Potato Centre | Peru | Andean roots and tubers, potato and sweet potato | 15,756 |
| International Rice Research Institute | Philippines | Rice | 121,595 |
| World Agroforestry Centre | Kenya | Agroforestry | 5,490 |
| Centre for PCIFIC Crops and trees | Fiji | Aroids and yams | 1,467 |

in the world. The 'Crop Trust' maintains the Vault in partnership with the Norwegian government and the Nordic Genetic Resources Centre, which is responsible for its management and operation. In October, 2015, seeds were withdrawn from the Global Seed Vault for the first time when ICARDA, the International Agricultural Research Centre formerly based in Aleppo, Syria withdrew 40,000 of its crop accessions in order to re-establish collections in facilities in Lebanon and Morocco. The Crop Trust's work is not restricted to long-term funding for plant genebanks. It also implements short-term, strategies projects that underpin the global system and strengthen crop diversity conservation worldwide. This includes efforts to collect, conserve and uses the wild relatives of 29 crops of global importance to food security, adding them to the pool of resources available under the Plant Treaty.

## The Crop Trust

The Crop Trust is also building an information system to ensure ready access not just to the diversity itself, but also to any information that exists about it: Genesys, online portal bringing together information from genebanks worldwide. The Crop Trust has also supported, in close partnership with the United States Department of Agriculture, the development of the GRIN-Global Genebank Data Management Software.

Crop Trust also provides assistance to specific crops, are as under:

**Table 10.32: Long-Term Grants Provided by the Crop Trust**

| | |
|---|---|
| Edible Aroids | SPC |
| Banana and Plantain | Bioversity International |
| Barley | ICARDA |
| Bean | CIAT |
| Cassava | CIAT |
| Cassava | IITA |
| Chickpea | ICRISAT |
| Faba bean | ICARDA |
| Forages | ICARDA |
| Forages | ILRI |
| Grass pea | ICARDA |
| Lentil | ICARDA |
| Maize | CIMMYT |
| Pearl millet | ICRISAT |
| Rice | IRRI |
| Sorghum | ICRISAT |
| Sweet Potato | CIP |
| Wheat | CIMMYT |
| Yam | SPC |
| Yam | IITA |

## International Institute of Tropical Agriculture (IITA)

International Institute of Tropical Agriculture (IITA) in Nigeria is one of the CGIAR ceners with an international genebank conserving germplasm of its six mandate crops: banana and plantain (*Musa* spp.), cassava, cowpea and wild *Vigna* species, maize, soybean, and yam. Since its establishment in 1967 the centre has collected and conserved over 33,000 accessions of landraces and wild relatives representing global collections (cowpea and yam) as well as regional resources (cassava, maize, *Musa* and soybean). The centre also has a vibrant international breeding programme developing improved varieties of its six mandate crops. Both the genebank and breeding programmes frequently exchange germplasm for research and other uses. IITA conserve and distributes cowpea and other Vigna spp., maize, and soybean as botanical seeds, germplasm of cassava, *Musa* and yam as *in vitro* tissue culture plants or plant propagules such as stems, corms, or tubers.

## Conservation Strategy in Permafrost Conditions at Extreme Altitude of Trans Himalaya in Leh-Ladakh

The ladakh located in Trans Himalaya, is one such geographical entity having unique culture and the customary practice of indigenous knowledge to sustain man and animal life in harsh cold climatic condition. The vegetation of the cold desert supports temperate, sub-alpine and alpine vegetation comprise of herbs, shrubs and few species of trees which are of economically importance mainly in the form of food, fodder, medicine, fuel and ornamental purpose. To collect the germplasm of wild edibles and to record the uses of traditional vegetables and healthcare remedies practiced by the local inhabitants and Amchis (local herbal doctor), the survey was carried out in the three valeys of Ladakh *viz.*, Nubra, Indus and Suru. Since the dawn of history, in the barren land of Ladakh, where otherwise the existence of life has been questanable; the indigenous people have been using wild plants as only source of their food, fodder and medicines. Of the known wild edibles plant species and varieties including *Atriplex hortensis, Capparis spinosa, Nepta floccose, Allium carolinianum, Rheum spiciforme, Rhodiola tibetica, Rhodiola imbricate, Arnebia euchroma, Carum carvi, Teraxicum officinale, Lepidium latifolium, Potentill anserine, Hippophae rhamnoides* etc., constitute a small number of the imperative and nutritious un-trapped plants consumed in different areas across Ladakh. Collection and conservation strategies for PGR including rare endangered wild relative plants species at Permafrost facility situated at extreme high altitude, Changla (17, 600 ft amsl) in Leh-Ladakh (J&K, India) (Singh *et al.*, 2016).

## Global Conservation Status of Indian Germplasm and its Implications

Every nation has a unique cropping system that involves both ethnic and exotic food crops, with the latter forming a significant part of the staple diet, in many countries. These exotic crops have been introduced in those regions over a period of time through political and cultural interventions and their crop production has been sustained by an unrestricted global flow of plant genetic resources. This global flow of germplasm has always been channelized under an umbrella of policy networks that have been framed through international consensus. Through India has ratified all major global policies/treaties pertaining to germplasm access, repeated concerns have been raised globally, about the restrictions placed on flow and access of Indian germplasm from within the country. Global conservation status of Indian-origin PGR accessions

were analysed using Genesys Global Portal on Plant Genetic Resources. According to Genesys, there are a total of 2,802,770 accessions in the global germplasm pool, accounted by 446 organizations. Amongst these, 100,607 are flow and access of Indian germplasm from within the country. Global conservation status of Indian-origin accessions, with 62,920 being conserved by CG genebanks. Similarly, out of 824,625 accessions of PGR conserved by 60 genebanks in Svalbard Global Seed Vault (SGSV), 66,339 accessions are of Indian-origin (Jacob *et al.*, 2016).

## Gaps in Diversity Conservation

The depletion of diversity and inadequacy in actions to conserve the diversity of the region may be attributed to several factors, which range from inadequate knowledge about diversity and its components to adoption of wrong and inappropriate policies by the concerned stakeholders.

### 1. Gaps in Knowledge and Information

- ☆ Information on urban diversity is scanty
- ☆ Species inventory in inaccessible areas of Arunachal Pradesh, Nagaland, Karbi Anglong and North Cachar hills of Asom, and parts of Mizoram and Manipur is yet to be made.
- ☆ Information on genetic diversity is extremely poor

### 2. Gaps in Vision

Most of the programmes and activities being undertaken by the state governments are shortsighted. Long-term planning based on sustainable development strategies and integration of diversity conservation issues with development planning are the needs of the hour.

- ☆ **Monoculture plantations:** In order to increase the revenue generation, the State Horticulture Departments pursue the policy of raising plantations of commercially important species by clearing and burning the natural diversity areas.
- ☆ **Introduction of high yielding varieties/hybrids of crops:** The horticulture departments are introducing various high yielding varieties/hybrids of cucurbitaceous crops. This is associated with increasing use of inorganic fertilizers and chemicals for plant protection. Such policies not only ignore the indigenous species and varieties but also have adverse effects on existing flora and fauna.
- ☆ **Gaps in crop wild relatives at Indian National genebank:** Crop wild relatives are important resource but despite their recognized value their conservation has been largely neglected. To achieve the objective of effective conservation

there is a need to priorities the crops for conservation. The total number of species of crop wild relatives in the genebank is 1,247, including both indigenous as well as exotic species. Among the 1,073 indigenous species the maximum number of species conserved belongs to medicinal and aromatic plants followed by fodder crops, agroforestry, ornamental and vegetables, reflecting the similar trend as per their occurrence in Indian sub-continent, albeit a very small number of species conserved in National Genebank. In the exotic crop wild relatives species the maximum number of species conserved belongs to medicinal and aromatic group followed by cereals and vegetables. Taking into consideration only native status, economic value, national distribution and *ex situ* conservation status of indigenous crop wild relatives germplasm in National Genebank, 238 species were identified for gap analysis. Initial results of gap analysis highlight that out of 238 crop wild relatives species, 210 were ranked as high priority crop wild relatives species as they are either not present in the genebank or are represented by less than 20 accessions, 14 species each were ranked as medium priority (with 20-40 accessions are conserved in the genebank) and low priority (>40 accessions are conserved in genebank). Further analysis of various agro-biodiversity zones and occurrence of these species Western Himalayan region, north-eastern hill region, Deccan plateau, Konkan and Malabar regions are the targeted areas for future collections and subsequently for long term conservation (Kak *et al.,* 2016).

### Gaps in Policies and Legal Structure

☆ The wrong conservation policies with focus on economically important species have been harmful to diversity. Such policies as adopted in Tripura, Mizoram, Nagaland and Meghalaya have not only decreased the species diversity in natural/rehabilitated forests but have also resulted in accelerated soil erosion and loss of soil moisture.

☆ The policy of rehabilitation of *Jhumias* through rubber plantation as has been done in Tripura may prove to be a disaster for other floral species in such areas.

☆ The policy of promoting high yielding varieties and assessment of progress and success on the basis of consumption of fertilizer and plant protection chemicals has led to ignoring the indigenous varieties. The government subsidy and credit policy is instrumental in adopting these schemes.

☆ Through the Public Distribution System only HYV are distributed. There is a need to include distribution of indigenous varieties too.

☆ The planners have not considered the role and value of diversity in preparing developmental plans.

☆ Most of the problems are related to increase in population. The rate of population growth in the northeast is unusually high. This causes tremendous strain on the natural resources and adoption of certain policies that are not very friendly to conserve diversity. No population policy has been adopted for future planning.

☆ Education policy does not include teaching on diversity conservation. The school curriculum should be able to mould the young minds in favour of diversity conservation.

☆ No policy as such is operational to create awareness among masses for diversity.

### 4. Gaps in Institutional and Human Capacity

☆ The number of trained taxonomists in the region is grossly inadequate. This is one of the most important bottlenecks for completing the inventorization of diversity.

☆ Not all persons concerned with management of genetic resources understand the concept of diversity in proper perspective. Many of them suffer from biased attitudes. So it is imperative that those who plan, decide and implement the developmental programmes are adequately trained and educated in favour of diversity conservation.

☆ There are a number of institutions, departments, colleges, universities, NGOs, local community groups that follow certain programmes having bearing on genetic resources. While framing their programmes, these agencies are motivated to pursue their own goals in watertight compartments without considering their impact on other programmes or existing resources. There is no institution, which can make them sit together and discuss the programmes in a holistic manner.

### 5. Gaps in Diversity Related Research and Development

☆ Regeneration and cultural practices for many species need to be researched and standardized for their cultivation. Threatened species need immediate action for ensuring their continued existence.

☆ Identification and classification of threatened species need to be done.

☆ Richness of diversity of horticultural crop species is yet to be fully inventorized and documented.

☆ There is a serious gap between research and field needs. The established formal institutions like university departments, departmental research stations and others rarely consult the farmers and local communities about their problems while pursuing research. Need-based research needs to be encouraged.

# References

1. Ahlawat, J. and Sehrawat, J.R. 2016. Germplasm conservation *via* encapsulating *in vitro* generated shoot tips and nodal segments of *Capparis decidua*. 1st International Agrobiodiversity Congress (Session: I. *In vitro*, Cryo and DNA Banking); held at New Delhi, 6-9 November; p. 91.

2. Ahuja, S., Mandal, B.B. and Srivastava, P.S. 2000. Vitrification: A potential cryopreservation technique for long-term conservation of yam germplasm. Abs. In: National symposium on Prospects and Potentials of Plant Biotechnology in India in the 21st Century. Jodhpur, Oct. 18th-21st, 2000. Pp. 159.

3. Akhtar, J., Singh, B., Bhalla, S., Parakh, D.B., Chalam, V.C., Gupta, K., Singh, M.C., Khan, Z., Singh, S.P., Kandan, A., Gawade, B.H., Kumar, P., Gupta, V., Pandey, S., Chand, D., Maurya, A.K., Meena, D.S., Jain, S.L. and Dubey, S.C. 2016. 1st International Agrobiodiversity Congress (Session: I. Seed Genebanks Resources); held at New Delhi, p. 68.

4. Anishetty, N.M., Rao, V.R., Sivaraj, N., Babu, B.S., Sunil, N. and Varaprasad, K.S. 2016. Genebanks: Management of genetic erosion in *ex situ* collections. *Indian J. Plant Genet. Resour.*, **29(3):** 268-271.

5. Anonymous, 2015. List of Biosphere Reserves in India. Ministry of Environment and Forests, New Delhi.

6. Asiedu, R., Ng, S.Y.C., Bai, K.V., Ekanayake, I.J. and Wanyera, N.M.W. 1988. Genetic Improvement. In: Orkwor, G.C., Asiedu, R. and K.V., Ekanayake (eds.). Food Yams. Advances in Research. IITA/NRCRI, p. 63-104.

7. Bonnave, M., Bleeckx, T., Terrazas, T. and Bertin, P. 2015. Effect of the management of seed flows and mode of propagation on the genetic diversity in an Andean farming system: The case of oca (*Oxalis tuberosa* Mol.) *Agric. Hum. Values*, **33(673):** doi:10.1007/s10460-015-9646-3.

8. Brush, S. 1995. *In situ* conservation of landraces in centres of crop diversity. *Crop Sci.*, **35:** 346-354.

9. CGIAR, 2014. Report from CGIAR Consortium to the FAO Commission on Genetic Resources for Food and Agriculture. CGRFA+15/15/Inf.32, p. 15.

10. Chandel, K.P.S., Chaudhury, R., Radhamani, J. and Malik, S.K. 1995. Desiccation and freezing sensitivity in recalcitrant seeds of tea, cocoa and jackfruit. *Annals of Botany*, **76:** 44-45.

11. Choudary, R., Chaudhury, R., Malik, S.K., Kumar, S. and Pal, D. 2013. Genetic stability of mulberry germplasm after cryopreservation by two-step freezing technique. *African J. Biotech.*, **12:** 5983-5993

12. Chaudhury, M.K.U. and Vasil, I.K. 1993. Molecular analysis of plant regenerated from embryogenic cultures of apple. *Genet.*, **86:** 181-188.

13. Chaudhury, R. and Malik, S.K. 2014. Implementing cryotechniques for plant germplasm: storing seeds, embryo axes, pollen and dormant buds. *Acta Horticultuae*, **1039:** 273-280.

14. CIP. 2016. Retrieved from http://research.cip.cgiar.org/confluence/display/GEN/Cryopreservation on 25 Sept. 2016.

15. Dumet, D., Diebiru, E., Adeyemi, A., Akinyemi, O., Gueye, B. and Franco, J. 2013. Cryopreservation for the 'in perpetuity' conservation of yam and cassava genetic resources. *Cryo Letters*, **34:** 107-118.

16. Dussert, S., Engelmann, F. and Noirot, M. 2003. Development of probabilistic tools to assist in the establishment and management of cryopreserved plant germplasm collections. *Cryo Letters*, **24(3):** 149-160.

17. Ellis, R.H., Roberts, E.H. and Whitehead, J. 1980. A new more economic and accurate approach to monitoring the viability of accessions during storage in seed banks. *Plant Genetic Resources News Letter*, **41:** 3-18.

18. Engelmann, F. and Drew, R.A. 1998. *In vitro* germplasm conservation, *Acta Hort.*, **461:** 41-47

19. Engelmann, F. 1994. Cryopreservation for the long term conservation of tropical crops of commercial importance. In: Proceedings of the International Symposium on the 'Application of Plant *in vitro* technology', Univ. Pertanian, Salangor, Malaysia, 16-18 Nov. 1993. p. 64-77.

20. Fabre, J. and Dereuddre, J. 1990. Encapsulation-dehydration: A new approach to cryopreservation of *Solanum*) shoot-tips. *Cryo Letters*, **11:** 413-126.

21. Ford Lloyd, B.V., Schmidt, M., Armstrong, S.J., Barazani, O., Engels, J., Hadas, R. and Li, Y. 2011. Crop wild relatives-undervalued, underutilized and under threat? *Bio Science*, **61(7):** 559-565.

22. Frankel, O.H. 1975. Genetic resources survey as a basis for exploration. In: Frankel, O.H., Hawkes, J.G. (eds.) Crop Genetic Resources for Today and Tomorrow. p. 99-109. *International Biological Programme*, 2. Cambridge University Press.

23. Galluzzi, G., Halewood, M., Noriega, I.L. and Vernooy, R. 2015. Twenty-five years of international exchanges of plant genetic resources facilitated by the CGIAR genebanks: a case study in international interdependence. ITPGREA Research Study, No. 9.http://www.planttreaty.org/sites/default/files/Research per cent 20Paper per cent 209_20150528.pdf. *Biodiversity and Conservation*, 25(8): 1421-1446.

24. Geburek, Th. and Turok, J. 2005. (eds.) Conservation and management of forest genetic resources in Europe. Arbora Publishers, Zvolen, 2005, xix-xxii, ISBN80-967088-1-3.

25. Gepts, P. 2006. Plant genetic resources conservation and utilization. *Crop Sci.*, **46(5):** 2278-2292.

26. Guarino, L., Rao, V.R., and editors, *E.G.* 2011. Collecting Plant Genetic Diversity: Technical Guidelines-2011 Update. Bioversity International, Rome, Italy. ISBN 978-92-9043-922-6. Available online: http://cropgenebank.sgrp.cgiar.org/index.php?option=com_content and view=article and id=390 and Itemid=557.

27. Gupta, A., Dadalani, M. and Sharma, A. 2016. Low carbon footprint seed conservation technique: The desiccated-ambient storage system using Zeolite beads. 1st International Agrobiodiversity Congress (Session: I. Seed Genebanks Resources); held at New Delhi, p. 63.

28. Gupta, P.N., Rai, M. and Rana, R.S. 1995. Centres of origin and genetic variability of vegetable crops. In: Genetic Resources of Vegetable Crops (eds. Rana, R.S., Gupta, P.N., Rai, M. and Kochhar, S.), ICAR-NBPGR, New Delhi, p.52-62.

29. Jacob, S.R., Tyagi, V., Agrawal, A., Chakrabarty, S.K. and Tyagi, R.K. 2016. The global conservation status of Indian germplasm and its implications. 1st International Agrobiodiversity Congress (Abs. Session: III. IPRs, ABS and Farmers' Rights); 6-9 November 2016, held at New Delhi, p. 269.

30. Jarvis, A., Upadhyaya, H., Gowda, C.L.L., Aggarwal, P.K., Fujisaka, S. and Anderson, B. 2016. Plant genetic resources for food and agriculture and climate change. In: *Coping with Climate Change-the Roles of Genetic Resources for Food and Agriculture*, FAO, Rome.

31. Jump, A.S., Marchant, R. and Penuelas, J. 2009. Environmental change and the option value of genetic diversity. *Trends in Plant Science*, **14(1):** 51-58.

32. Kak, A, Gupta, V. and Tyagi, R.K. 2016. Conservation gaps in crop wild relatives at Indian National Genebank. 1st International Agrobiodiversity Congress (Session:I. Seed Genebanks Resources); held at New Delhi, p. 68.

33. Kaur, R.P., Girimalla, V., Kumar, V. and Bhardwaj, V. 2016. Pollen cryopreservation in aid to conservation of potato genetic resources. 1st International Agrobiodiversity Congress. (Session: *In vitro*, Cryo and DNA Banking); held at New Delhi, p. 87.

34. Keller, E.R.J., Grübe, M., Hajirezaei, M.R., Melzer, M., Mock, H.P., Rolletschek, H., Senula, A. and Subbarayan, K. 2016. Experience in large-scale cryopreservation and links to applied research for safe storage of plant germplasm. *Acta Hort.*, **1113:** 239-250.

35. King, M.V. and E.H., Roberts. 1979. The storage of recalcitrant seeds achievement and possible. IBPGR, Rome.

36. King, M.V. and Roberts, E.H. 1979. The storage of recalcitrant seeds achievement and possible. IBPGR, Rome.

37. Lenka, S., Gupta, V. and Tyagi, R.K. 2016. Germplasm handling for long-and medium-term conservation of plant Germplasm at national genebank in India. 1st International Agrobiodiversity Congress. (Abst. Session:I. Seed Bank Resources), held at New Delhi, 6-9 November 2016, p. 72.

38. Leunufna, S. and Keller, E.R.J. 2003. Investigating a new cryopreservation protocol for yams (*Dioscorea* spp.). *Plant Cell Rep.*, **21:** 1159-1166.

39. Malaurie, B., Trouslot, M. Engelmann, F. and Chavlvillange, N. 1998. Effect of pretreatment conditions on the cryopreservation of *in vitro* cultured yam (*D. alata* 'Brazofuerte' and *D. bulbifera* 'Noumeaimboro') shoot apices by encapsulation-dehydration. *Cryo-Lett.*, **19:** 15-26.

40. Mandal, B.B.1999. Conservation biotechnology of endemic and other economically important plant species of India. In: Benson, E.E. (ed.) *Plant conservation Biotechnology*, Taylor and Francis, London. p. 211-225.

41. Mandal, B.B. 2003. Cryopreservation techniques for plant germplasm conservation. In: B.B., Mandal; R. Chaudhury, F. Engelmann, B. Mal; K.L. Tao; B.S. Dhillon (Eds.). Conservation Biotechnology of Plant Germplasm. ICAR-NBPGR, New Delhi/FAO, Rome. p. 177-193.

42. Mandal, B.B. and Chandel, K.P.S. 1996. Conservation of genetic diversity in sweet potato and yams using *in vitro* strategies. In: Kurup, G.T., Palaniswami, M.S., Potty, V.P., Padmaja, G., Kabeerathumma, S. and Pillai, S.V. (Eds.) Tropical Tuber Crops: Problems, Prospects and Future Strategies; Oxford and IBH, Publishing Co. Pvt. Ltd., New Delhi, p. 49-59.

43. Mandal, B.B., Chandel, K.P.S. and Dwivedi, S. 1996. Cryopreservation of yams (*Dioscorea* spp.) shoots apices by encapsulation-dehydration. *Cry-Lett.*, **17:** 165-174.

44. Mandal, B.B. and Dixit, S. 2000. Cryopreservation of shoot-tips of *Dioscorea deltoidea* Wall. - An endangered medicinal yam, for long-term conservation. *IPGRI News Letter for Asia, the Pacific and Oceania*. No. 33. Sep.-Dec. 2000. p. 23.

45. Maneesha, S.R., Priya, S. and Singh, N.P. 2016. 'Kulagar': A potential system to conserve the crop diversity. 1st International Agrobiodiversity Congress (Session: III. *In-situ* and On-farm Conservation); held at New Delhi, p.79.

46. Mantell, S.H., Haque, S.Q. and Whitehall, A.P. 1978. Clonal multiplication of *Dioscorea alata* L. and *D. rotundata* Poir yams by tissue culture. *J. Hort. Sci.*, **53:** 95-98.

47. Mitchell, S.A., Asemota, H.N. and Ahmad, M.H. 1995. Effects of explants sources, culture medium strength anf growth regulators on the *in vitro* propagation of three Jamaican Yams (*Dioscoreacayenensis, D. trifida* and *D. rotundata*). *Journal of the Science of Food and Agriculture*, **67(2):** 173-180.

48. Mohapatra, T. 2016. Indian Agrobiodiversity System. *Indian J. Plant Genet. Resour.*, **29(3):** 230-233.

49. Morishita, M. and Yamada, K. 1978. Studies on tissue culture of taro (*Colocasia esculenta* Schott.) (1) Inducing plantlets from apical meristem. *Bull. Osaka Agri. Res. Cent.*, **15:** 9-12.

50. Murashige, T. and Skoog, F. 1962. A revised medium for rapid growth and bio-assays with tobacco tissue cultures. *Physiol. Plant.*, **15:** 473-497.

51. Nair, N.G. and Chandrababu, S. 1994. A slow growth medium for *in vitro* conservation of edible yams. *J. Root Crops*, **20(1):** 68-69.

52. Nair, N.G. and Chandrababu, S. 1996. *In vitro* production and micro-propagation of three species of edible yams. In: Kurup, G.T., Palaniswami, M.S., Potty, V.P., Padmaja, G., Kabeerathumma, S. and Pillai, S.V. (Eds.) Tropical Tuber Crops: Problems, Prospects and Future Strategies; Oxford and IBH, Publishing Co. Pvt. Ltd., New Delhi, p. 55-60.

53. Ng, S.Y.C. and Hahn, S.K. 1985. Application of tissue culture to tuber crops at IIT. In: Biotechnology in International Agricultural Research, International Rice Research Institute, Manila, the Philippines, p. 27-40.

54. NBPGR, 2016. Guidelines for Management of Plant Genetic Resources in India, p. 27.

55. Nivedhitha, S., Gupta, A., Sharma, N. and Mathur, P. 2016. Community Seed Banks to conserve local crop diversity and ensure resilience against climate change. 1st International Agrobiodiversity Congress (Session: *In-situ* and On-farm Conservation); held at New Delhi, p. 79.

56. Padulosi, S., Hodgkin, T., Williams, J.T. and Haq, N. 2002. Underutilized Crops: Tends, Challenges and Opportunities in the 21st Century. J.M.M. Engels, Rao, V.R., Brown, A.H.D. and Jackson, M.T. (eds.). New York, USA: CABI, p. 323-338.

57. Pandey, R., Pandey, A., Sharma, N., Negi, K.S. and Tyagi, R.K. 2016. Cryopreservation of *Allium* germplasm resources at ICAR-NBPGR: Achievments and Challenges. 1st International Agrobiodiversity Congress. (Session: *In vitro*, Cryo and DNA Banking); held at New Delhi, p. 88.

58. Panis, B., Piette, B. and Andre, E., Van den houwe, I. and Swennen, R. 2011. Droplet vitrification: the first genetic cryopreservation protocol for organized plant tissues? *Acta Hort.*, **908:** 157-164.

59. Panis, B., Piette, B. and Swennen, R. 2005. Droplet vitrification of apical meristems: A cryopreservation protocol applicable to all Musaceae. *Plant Science*, **168:** 45-55.

60. Panis, B., Swennen, R. and Engelmann, F. 2001. Cryopreservation of plant germplasm. *Acta Hort.*, **560:** 79-86.

61. Panis, B., Van den houwe, I., Swennen, R., Rhee, J. and Roux, N. 2016. Securing Plant Genetic Resources for perpetuity through cryopreservation. *Indian, J. Plant Genet. Resour.*, **29(3):** 300-302.

62. Parzies, H.K., Spoor, W. and Ennos, R.A. 2000. Genetic diversity of barley landrace accessions (*Hordeum vulgare* ssp. *vulgare*) conserved for different lengths of time in *ex situ* genebanks. *Heredity*, **84(4):** 476-486.

63. Peacock, W.J. 1989. Molecular biology and genetic resources In: A.H.D. Brown, D. Marshall, R.H. Frankel and O.T. Williams (Eds.). The Use of Plant Genetic Resources, Cambridge Univ. Press Cambridge, p. 363-376.

64. Pence, V.C. 1993. Cryopreservation of embryo axes for germplasm storage of large seeded temperate seeds. Presented at SLTB 93, U.K.

65. Perez-Garcia, F., Gonzalez-Benito, M.E. and Gomez-Campo, C. 2007. High viability recorded in ultra-dry seeds of 37 species of Brassicaceae after almost 40 years of storage. *Seed Sci. and Technol.*, **35:** 143-153.

66. Pilatti, F.K., Aguiar, T., Simoes, T., Benson, E.E. and Viana, A.M. 2011. *In vitro* and cryogenic preservation of plant biodiversity in Brazil. *In vitro Cellular and Development Biology. Plant*, **47(1):** 82-98.

67. Quin, F.M. 1998. An overview of Yam Research. In: Orkwor, G.C., Asiedu, R. and Ekanayake, I.J. IITA/ NECRI Food Yams; Advances in research, p. 215-229.

68. Raup, D. 1988. Diversity crisis in the geological past. In: Wilson, E.O. (ed.) Biodiversity. National Academy Press, Washington, D.C.

69. RBG (Royal Botanic garden) Kew. 2016. Seed Information Database, http://data.kew.org/sid/.

70. Reed, B.M. 1993. Improved survival of *in vitro* stored *Rubus* germplasm. *J. Am. Soc. Hort. Sci.*, **118**: 890-895.

71. Reed, B.M., Engelmann, F., Dullo, M.E., Engels, J.M.M. 2004a. Technical guidelines for the management of field and *in vitro* germplasm collections. IPGRI Handbooks for Genebank, No.7. International Plant Genetic Resources Institute, Rome.

72. Roberts, E.H. and Ellis, R.H. 1977. Prediction of seed longevity and sub-zero temperatures. *Nature,* **268**: 431-433. (04 August 1977); doi:10.1038/268431a.

73. Roca, W.M., Reyes, R. and Beltran, J. 1984. Effects of various factors on minimal growth in tissue culture storage of cassava germplasm. In: Proc. 6th Symp. *Intl. Soc. Tropical Crops.* p. 441-446.

74. Sakai, A., Kobayashi, S. and Oiyama, I. 1990. Cryopreservation of nucellar cells of navel orange (*Citrus sinensis* Osb. var. *Brasiliensis* Tanaka) by vitrification. *Plant Cell Rep.*, **9**: 30-33.

75. Sarkar, D. and Naik, P.S. 1998. Factors effecting minimal growth conservation of potato micro plant *in vitro*. Plant Genetic Resources Institute, Rome.

76. Sharma, J.P.2009. *Principles of Vegetable Breeding*. Kalyani Publishers. p. 32.

77. Singh, B.B. and Jain, S.K. 1993. Long term conservation of released vegetable varieties. *Indian J. Plant Genet. Resour.*, **6(1)**: 79-84.

78. Singh, N., Rinchen, T. and Maurya, S.B. 2016. Phyto-diversity and its conservation strategies in Permafrost conditions at extreme altitude of Trans Himalaya in Leh-Ladakh (India). 1st International Agrobiodiversity Congress. (Session: Seed Genebanks Resources); held at New Delhi, p. 64.

79. Soengas, P., Cartea, M.E., Lema, M. and Velasco, P. 2009. Effect of regeneration procedures on the genetic integrity of *Brassica oleracea* accessions. *Molecular Breeding*, **23(3)**: 389-395.

80. Stanwood, P.C. 1995. Cryopreservation of seed germplasm for genetic conservation. In: Cryopreservation of Plant Cells and Organs, Katha, K.K. (ed.). CRC Press Boca Raton, Florida. p. 199-226.

81. Staritsky, G. Dekkers, A.J., Lowwaars, N.P. and Zandvoort, E.A. 1986. *In vitro* conservation of aroid germplasm at reduced temperature and under osmotic stress. In: Withers, L.A. and Anderson, P.G. (Eds.) *Plant Tissue Culture and its Agriculture Applications*. Butterworth, London, p. 277-283.

82. Swaminathan, M.S. 2016. Agrobiodiversity and achieving the zero hunger challenge. *Indian J. Plant Genet. Resour.*, **29(3)**: 249-250.

83. Takagi, H., Tien, N., Islam, O.M., Senboku, T. and Sakai, A. 1997. Cryopreservation of *in vitro* grown shoot tips of taro (*Colocasia esculenta* (L.) Schott) by vitrification. 1. Investigation on basic conditions of the vitrification procedure. *Pl. Cell Rep.*, **16**: 594-599.

84. Thammasiri, K. 1999. Cryopreservation of embryonic axes of jackfruit by vitrification. In: Marzalina, M *et al.* (eds), *Recalcitrant Seeds-IUFRO Seed Symposium 1998*. FRIM, Malaysia, p. 153-158.

85. Thinh, T.N., Takagi, H. and Sakai, A. 2000. Cryopreservation of *in vitro* grown shoot tips of five vegetatively propagated monocots by vitrification. In: Cryopreservation of tropical plant germplasm. Current Research Progress and application (Engelmann, F. and Takagi, H. (Eds.) *JIRCAS/IPGRI*, p. 227-232.

86. Thormann, I. and Engels, J.M.M. 2015. Genetic diversity and erosion: A global perspective. p. 263-294. In: *Genetic Diversity and Erosion in Plants-Indicators and Prevention*, Ahuja, M.R.

87. Thro, A.M., Roca, W.M., Restrepo, J., Caballero, H., Poats, S., Escobar, R., Maflaand, G. and Hernandez, C. 1999. Can *In vitro* biology have farm-level impact for small-scale cassava farmers in Latin America? *In vitro Cellular and Developmental Biology. Plant* **35(5)**: 382-387.

88. Tyagi, R.K. 1998. Techniques of micro-propagation and *in vitro* conservation. In: Plant germplasm collecting: Principles and Procedures (eds. Gautam, P.L., Dabas, B.B., Srivastava, U. and Duhoon, S.S., 1998), ICAR-NBPGR, New Delhi, p. 201-211.

89. Tyagi, R.K. and Yusuf, A. 2003. *In vitro* medium term storage of germplasm. In: Mandal, B.B., Chaudhury, Engelmann, R. F., Mal, B., Tao, K.L. Dhillon, B.S. (Eds.). Conservation Biotechnology of Plant Germplasm. ICAR-NBPGR, New Delhi/ FAO, Rome. p. 115-121, 157-161.

90. Unnikrishnan, M., Mukherjee, A. and Nair, N.G. 1990. *In vitro* conservation of tuber crops through slow growth cultures. *J. Root Crops*: ISRC Nat. Sym. 1990. Special p. 296-301.

91. Unnikrishnan, M., Edison, S. and Sheela, M.N. 2005. Tuber Crops and their wild relatives. In: Tamil Nadu Biodiversity Strategy and Action Plan. (Ed.) Annamalai, R., Tamil Nadu Biodiversity strategy and action plan, Published by Tamil Nadu Forests Department Chennai, p. 159-168.

92. Unnikrishnan, M.and Sheela, M.N. 2000. Studies on media explants and incubation conditions for *in vitro* conservation of cassava germplasm. In: Cassava Biotechnology IV International Scientific meeting-CBN. Luiz. J.C.B. Carvalho, Ann Marie Thro and Alberto Duarte Vilarinhos (eds.) Published by Brazilian Agricultural Research Corporation-EMBRAPA, Genetic Resources and Biotechnology-CENARGEN and Cassava Biotechnology Network-CBN. p. 425-430.

93. Unnikrishnan, M. and Sheela, M.N. 2000. Biotechnology in conservation and improvement of tuber crops. In: Biotechnology in Horticulture and Plantation Crops; Chadha, K.L., Ravindran, P.N. and Leela, S. (Eds.), Malhotra Publishing House, New Delhi, India.

94. Unnikrishnan, M., Sheela, M.N., Gayathri, V. and Nair, J.R. 2002. Genotype response variation to micro propagation and effect of retardants on *in vitro* growth in cassava. CBN-V. *Fifth International Scientific meeting of the Cassava Biotechnology Network*, Nov. 4-9, 2001. Donald Danforth Plant Science Centre, St. Louis-Missouri-USA. Abstract Book eds. Taylor, N.J., Ogbe, F. and Fauquet, C.M.S., 7-33.

95. Vidhya, V. and Ram, G. 2016. Vapour phase storaghe-potential technique for extra-long term seeds conservation. 1st International Agrobiodiversity Congress. (Session: *In vitro*, Cryo and DNA Banking)-Abstract; held at New Delhi, 6-9 November 2016; p. 90.

96. Vigouroux, Y., Mariac, C., Mita, S.D., Pham, J.L., Gerard, B., Kapran, I., Sagnard, F., Deu, M., Chantereau, J., Ali, A., Ndjeunga, J., Luong, V., Thuillet, A.C., Saidou, A.A. and Bezancon, G. 2011. Selection for earlier flowering crop associated with climatic variations in the Sahel. *Plos One*, **6(5):** e19563.doi:10.1371/journal.pone.0019563.

97. Vollmer, R., Panta, A., Tay, D., Roca, W. and Ellis, D. 2014. Effect of sucrose preculture and PVS2 exposure on the cryopreservation of sweet potato [*Ipomoea batatas* (I.) Lam.] shoot tips using the PVS2 droplet vitrification. *Acta Hort.*, **1039:** doi:10.17660/ActaHortic.2014.1039.33.

98. Vollmer, R., Villagaray, R. Egusquiza, V., Espirilla, J., Garcia, M., Torres, A., Rojas, E., Panta, A., Barkley, N.A. and Ellis, D. 2014. The potato cryobank at the International Potato Centre (CIP): A model for long-term conservation of clonal plant genetic resources collections of the future. *CryoLetters*, **37(5):** 318-319.

99. Walters, C., Wheeler, L.M. and Grotenhuis, J.M. 2005. Longevity of seeds stored in a genebank: species characteristics. *Seed Sci. Res.*, **15:** 1-20.

100. Withers, L.A. 1984.Germplasm conservation *in vitro*. Present status of research and its application. In: *Crop Genetic Resources: Conservation and Evaluation*. Holden, J.H.W. and Williams, C.J. (eds.) George Allan and Unwinn pub. London, U.K., p. 138-157.

101. Withers, L.A. 1991. *In vitro* conservation. *Biol. J. Linnean Soc.*, 31-42.

102. Withers, L.A. 1993.new technologies for the conservation of plant genetic resources. International Crop Science 1. Crop Science Society of America, Madison, USA.

103. Withers, L.A. and King, P.J. 1980. A simple freezing unit and routine cryopreservation method for plant cell cultures. *Cryo Letter*, **1:** 213-220.

104. Zimmerer, K.S. 2003. Geographics of seed networks for food plants (Potato, Ulluco) and approaches to agro-biodiversity conservation in the Andean Countries, *Society and Natural Resour., An Intern. J.,* **16(7):** 583-601.

# 11

# Biotechnological Approaches in Vegetable Genetic Resource Management and Utilization

Deforestation, urbanization, pollution, habitat destruction, fragmentation and degradation, spread of invasive alien species, climate change, changing life styles, globalization, market economics, over-grazing and changes in land-use pattern are contributing indirectly to the loss of biodiversity. No doubt, biotechnological approaches have made it easy to collect and conserve plant genetic resources amicably in the form of traditional varieties, modern cultivars, wild relatives and other wild species, especially of those species which are rare and difficult to conserve through seeds. Advances in biotechnology, especially in the area of *in vitro* culture techniques and molecular biology provide some important tools for improved conservation and management of plant genetic resources.

Vegetables are becoming increasingly popular due to their indispensable role in human nutrition and agricultural diversification. It's also playing an important role in Indian agriculture. Vegetables are considered protective foods and also play an important role in providing a balanced diet to the human beings. To get maximum benefits of their nutrients, vegetables should be consumed fresh as far as possible. According to World Health Organization recommendation, daily consumption should be of 400g of vegetables excluding potato. According to this recommendation, per capita vegetable requirement calculated as 146 kg per year. But the average world consumption of vegetables is around 85 kg per person per year. In developed countries, it is around 120 kg per person per year. However, India still lags behind many countries in terms of vegetable

production and productivity. During past four decades, India has achieved substantial growth in vegetable production and ranks next to China. Major global emphasis during the past three decades has been to conserve germplasm diversity *ex situ* in medium-and long-term gene banks. The different approaches adapted for this is botanical garden, arboreta, herbal garden, field gene banks, seed gene banks, cryo-gene banks, *in vitro* gene banks and DNA banks. There are several economically important species which are sexually propagated and can be conserved in conventional seed banks and cryo-banks. There are others which are asexually propagated or lack of natural mechanism of seed production or possess high heterozygosity and can not be propagated through seeds *e.g.* potato, sweet potato, yams, cassava, several bulbous crops, tuberous crops, spices, *etc.* hence these need to be propagated vegetatively in field gene bank or *in vitro* gene banks. The word biotechnology was coined early in the 20th century by an agricultural engineer from Hungary, named Karl Freky. According to him "*biotechnology is the technology which includes all such work by which products are produced from raw material with the aid of living organism*". A grand challenge for the 21st century scientists is to harness knowledge of earth's biological diversity on which life depends. This knowledge is critical to science and society for formulating rational policy for managing natural systems, sustaining human health, maintaining economic stability and improving the quality of human life. Biodiversity research is inherently a global enterprise. Systematists and ecologists in the world for over a century have collaborated across political borders to document the distribution of species and of biological diversity. In recent decades much work has been focused on understanding the dynamics of ecological processes responsible for the creation and maintenance of diverse and sustainable systems. Biotech crops are also known as genetically modified (GM) or genetically engineered (GE) crops. Phenotypically they look just like their traditional counterparts. With the rapid progress in advanced biology, biotech crops have been developed with the help of genetic engineering tools to possess special characteristics (traits) that make them better. The most common traits developed in biotech crops include insect resistance, herbicide tolerance, virus resistance and improved product quality. The "staking' (use of more than one trait in a single crop) of these traits is an important feature that has been used increasingly to tackle multiple constraints in agriculture. It is expected that development of crops with tolerance to drought and salinity, improved nitrogen use efficiency, enhanced yield, quality and nutritional status coupled with existing traits will make food better and safer. At the national level, it will make agriculture more efficient and competitive to meet the challenges of hunger, poverty, malnutrition and food security in tomorrow's world (Global Knowledge Centre on Crop Biotechnology, 2008). Over the years, gene exchange between two plants have been attempted to produce offspring that inherited desired traits. This was done by transferring pollens from male flower of one plant to the female flower of another. Such traditional cross-breeding methods have some limitations:

1. The gene exchange is only feasible between the same or very closely related species.

2. It is usually takes a long time to achieve desired results and characteristics of interest may often not exist, or not be available at the required level in the species or in any related species.

3. There is little or no guarantee of obtaining specific gene combination from millions of crosses generated.

4. Undesired genes also get transferred along with desired genes or while one desired gene is gained, another may be lost because genes of both parents are mixed together and re-assorted more or less randomly in the offspring.

An *Agrobacterium*-mediated transformation system using transient transformation assays was used to evaluate conditions influencing transformation for Broad leaf (*Typha latifolia* Roxb.). These studies were aimed at the long-term objective of evaluating candidate genes for phytoremediation. The binary plasmid vector pCAMBIA1301/EHA105, containing the beta-glucuronidase coding sequence was used in combination with factors known to affect transformation. These

included callus age at the time of co-cultivation with *Agrobacterium tumefaciens* type and concentration of auxin for explant growth, light or dark culture environment, the presence or absence of acetosyringone (AS), explant type, explant wounding and the number of days used for co-cultivation. The number of days needed for the first detection of transient expression of the beta-glucuronidase gene was also examined. Three days of *Agrobacterium* co-cultivation of 50 days old seedlings derived calluses, grown on 20.7 micro M (5 mg l$^{-1}$) picloram supplemented medium, in the dark, resulted in higher levels of transient beta-glucuronidase expression than were seen in calluses cultured on 4.5 or 22.6 micro M (1 or 5 mg l$^{-1}$), 2, 4-dichlorophenoxyacetic acid containing media. The addition of 100 micro M acetosyringone significantly enhanced transient beta-glucuronidase activity. Wounding of explants by cutting into two or three pieces, 3 days before co-cultivation, increased expression of beta-glucuronidase only in calluses cultured under light conditions. Transient beta-glucuronidase expression was observed as early as 24 h after co-cultivation and increased as the days post cultivationincreased. The developed transient system can be used for stable transformation of *Typha* species (Nandakumar *et al.*, 2004.)

Dealing the potentiality of biotechnology and its impact on the developments that have taken place in developing biosafety measures, including the finalization of the 'International Technical Guidelines for safety in Biotechnology' (UNEP, 1995).

## Impacts of Biotechnology

Biotechnology provides a range of tools and methods for assessment, monitoring and managing biological resources, such as clarifying taxonomic and evolutionary relationships among groups of organisms and assessing the effects of ecosystem disturbance on components of biological diversity and biological processes. The tools of biotechnology could also be utilized for *in situ* conservation *i.e.* assessment of optimal or minimal population size and *ex situ* conservation *i.e.* enhancing the quality of characteristics and efficiency, through compact storage of DNA libraries and sequence database (Apples *et al.*, 1995). The tools current employed for the sustainable use of genetic resources are in the area of breeding, genetic engineering, the development of novel genes and gene products, and environmental remediation. Some area of application of biotechnology in utilization of biological diversity as identified by Montagu *et al.* (1995), are given in Table 11.1.

### I. Direct Impacts

The introduction of any Living Modified Organisms (LMOs) in a biological community can have various undesirable impacts (Tzotzos *et al.*, 1995)

**Table 11.1: Areas of Applications of Biotechnology in the Management of Biological Resources**

| Areas of Biotechnological Applications | Specific Area |
|---|---|
| Using technologies as sources of: | Proteins and peptide |
| | Lipids and fatty acids |
| | Carbohydrates |
| | Secondary metabolites for pharmaceuticals |
| Genetic engineering, breeding and in vitro culture systems can be used to enhance agronomic performance | - |
| Improved environmental conditions through | Identification of soil microorganisms and determination of best combinations for soil rehabilitation |
| | Use of plants to mitigate heavy metal pollution |
| | Engineering key genes in bacteria for pollutant degradation |
| | Improving plant microbe symbiotic systems for waste water treatments |
| | Production of biosurfactants |
| Enhancing the efficiency of microorganisms in industrial processes such as: | Microbial-enhanced secondary recovery of oil from reservoirs |
| | Bioleaching microbiological extraction of metals from low grade ores |
| | Production of industrial enzymes |
| | Production of endogenous products *e.g.* antibiotics |

*Source*: Montagu *et al.*, 1995.

☆ Displacement or destruction of indigenous/endangered or endemic species.

☆ Exposure of species to new pathogenic or toxic agents;

☆ Pollution of the gene pool;

☆ Loss of species diversity and disruption of energy and nutrient cycling.

These impacts are largely ecologically and evolutionary and can be scientifically assessed and tested through simulated conditions. As part of an overall risk assessment strategy, the main problems arising from direct impact could be dealt with including the processes of introgression, weediness, pathogenicity, altered nutrient cycling *etc.* however, there is an inadequate understanding of the possible direct impact of LMOs on soil micro flora and fauna (Angels, 1994 and Morra, 1994) and the potential of virus resistant plants on the host range of some viruses (Rissler and Mellon, 1993).

## II. Indirect Impacts

The possible number and types of indirect effects of biotechnology could be immense. These effects are mostly socio-economic in nature and can be of major importance particularly to middle to low income developing countries where people are dependent on biological resources for subsistence. Indirect impacts may be secondary or tertiary effects. Tzotzos *et al.* (1995) has listed some indirect impacts:

☆ Pressure on natural habitats because of the increasing value of genetic resources

☆ Lack of immediately perceivable incentives for conservation

☆ Moral/ethical problems of ownership of genetic resources and benefit-sharing

☆ Increase in agricultural productivity

☆ Replacement of traditional landraces and

☆ Decline or opportunities loss for disadvantaged groups in areas of marginal production.

In addition, various sociological and socio-economic impacts could also be visualized with these impacts depending on the type of LMOs, kind of release, and the precaution taken for any harmful effect.

## Public Understanding of Biotechnological Need for Biosafety Mechanism

The public debate on the application of biotechnology has been marked by apprehensions of two kinds.

1. The first is that there may be adverse impacts on the environment and human health.

2. The second is the fear that control of the new technology could give some nations or groups the power to use this in unfair ways.

The public perception of biotechnology as unpredictable and dangerous was highlighted by Perlas (1993) through examples such as the creation of a 'Sper Aids Virus', 'Super pigs and cows' reporting of serious ailments using a genetically altered version of L-tryptophan and insulin; the use of biotechnologically developed bovine growth harmone (BGH); stunting in corn and other crops due to the presence of *Clavabacter xyli*, a vector used to transfer Bt endotoxin gene; the requirement of six times more pesticides for Unliver cloned oil palms; novel mutant carps, catfish, trout and salmon polluting native species; and illegal trials on pseudo-rabies in Argentina. This short list of unpredictable and negative impacts of biotechnologies clearly shows the possibility of the adverse impact of biotechnologically developed products on human health and the environment. The views of Holmes (1993), Williamson (1991) and Ellstrand and Hoffman (1990) further support these perceptions.

## Legislative and Regulatory Mechanisms for Biosafety

Virgin *et al.* (1995) extensively analysed biosafety regulations and reported that 24 countries with high to high-middle income economies have laws and regulations in place taking into account the specific concerns arising from new recombinant techniques. The European countries have instituted new laws, which are similar in scope requirements and impacts. In developing countries, the situation is significantly different. In Latin America, Argentina, Brazil, Mexico, Chile, Costa Rica and Cuba have regulatory mechanisms in place. The African continent is represented only by South Africa and Egypt. Kenya, Zimbabwe and Nigeria are at various stages of drafting regulations and are likely to finalise these very soon. In Eastern Europe, Hungary has an ad hoc review process, and Russia has submitted a biosafety law for official approval. Of the developing countries in Asia, only India, China, Thailand and the Philippines have guidelines. Malaysia and Indonesia is preparing new legislation and drafts.

### Table 11.2: Status of Adoption of Biosafety Regulations in different Countries

| | |
|---|---|
| Industrialized countries | Australia, Austria, Belgium, Canada, Denmark, Finland, France, Germany, Greece, Ireland, Israel, Italy, Japan, Luxembourg, New Zealand, Norway, Portugal, South Africa, Spain, Sweden, Switzerland, The Netherlands, United Kingdom, United States |
| Developing countries | Argentina, Brazil, Chile*, China*, Costa Rica*, Cuba*, Egypt*, Hungary**, India*, Indonesia*, **, Kenya*, **, Malaysia*, **, Mexico, Nigeria*, Philippines*, Russia**, Thailand*. |

*Lower-Middle to Low Income Economy.

** Currently drafting regulations.

*Source*: Virgin *et al.*, 1995.

It may be stated that the implementations of these biosafety regulations varies from developed countries to developing countries, from being very rigidly effective to non-effective because of the lack of a well-defined institutional structure. It is also pertinent that the guidelines evolved by most developing countries are very similar in scope and requirements, and have been adopted as a part of their National Environment Acts. However, these provisions are inadequate in respect of modalities/protocols for access to and transfer of biotechnology on 'Mutually Agreed Terms', procedures for 'Adavanced Informed Agreements' and procedures for risk assessment and management (Chauhan, 1996).

The majority of countries believed that Application of Advance Informed Agreements (AIA) procedures constituted a very important part of a protocol dealing with Trans boundary Movement of LMOs, taking into account the provisions of the Basal Convention and the operational guidelines and principles developed by the Forest Stewardship Council.

### Table 11.3: Major Articles of the Proposed 'Protocols on Biosafety'

| Articles | Applications |
|---|---|
| **A-Some Major Articles** | |
| Article 3 | Application of Advance Informed Agreements (AIA) Procedure |
| Article 4 | Notification Procedure for Application of Advance Informed Agreements (AIA) |
| Article 5 | Response to Application of Advance Informed Agreements (AIA) Notification |
| Article 6 | Decision Procedure for Application of Advance Informed Agreements (AIA) |
| Article 7 | Review of Decision Under Application of Advance Informed Agreements (AIA) |
| Article 8 | Notification of Transit |
| Article 9 | Simplified Procedures |
| Article 10 | Subsequent Imports |
| Article 11 | Bilateral and Regional Agreements |
| Article 12 | Risk Assessment |
| Article 13 | Risk Management |
| Article 14 | Minimum National Standards |
| Article 15 | Unintentional Trans boundary Movement |
| Article 16 | Emergencies Measures |
| Article 17 | Handling |
| Article 18 | Competent Authority Focal Point |
| Article 19 | Information Sharing/Biosafety Clearing House |
| Article 20 | Confidential Information |
| Article 21 | Capacity Building |
| Article 22 | Public Awareness |
| Article 23 | Public Awareness/Public Participation |
| Article 24 | Non-Parties |
| Article 25 | Non-Discrimination |
| Article 26 | Socio-economic Considerations |
| Article 35 | Monitoring and Compliance |

| Articles | Applications |
|---|---|
| **B-Other items (text to be consolidated and negotiated)** | |
| Preamble, Objectives and Jurisdictional Scope | Accession |
| Use of terms/Definition | Depository |
| Relationship with other international agreements | Reservation and declaration |
| Entry into Force | Review and adoption |
| Settlement of disputes | Authentic Texts |
| Financial Issues | Annexes |
| Right to Vote | Withdrawal and Signature |

*Source*: Working Group Report, 1998.

## Biodiversity Conservation through Biotechnological Approach

Biotechnology has vast applications in characterization/evaluation, conservation (Choudhury and Malik, 2008) and utilization of plant genetic resources (PGR). The plant multiplication is done under optimized conditions by producing adventitious or non-adventitious organs or adventitious embryos. Biotechnological approaches are being utilized largely to enhance the productivity through quality planting materials. Quality planting material is of prime importance to get good quality products apart from the total yield. Planting material comprised of seeds, tubers, stems, cuttings, leaf, *etc.* Sometimes these planting materials possess dormant seeds, poor germination, weak seedlings, less availability *etc.* In nature, several vegetable crops are being propagated through vegetative means like pointed gourd, ivy gourd, sweet gourd, spine gourd, ginger, turmeric, elephant foot yam *etc.* For commercial cultivation/propagation of these crops, maximum amount of planting materials are required, which is becoming limiting factor among the growers. In dioecious vegetable crops lack of quality and disease free planting materials not only decreasing total yield but also decreasing net area under these crops. The tissue culture technique is now widely employed as an aid, not only to get sufficient quantity of better and disease free planting materials but also in crop improvement. Biotechnology also plays an important role in international plant conservation programs and in preservation of the world's genetic resources (Bajaj, 1995; Benson, 1999). Advances in biotechnology provide new methods for plant genetic resources and evaluation (Paunesca, 2009). Berg *et al.* (1974) voiced their concern about the potential biological hazards with respect to r-DNA experimentations. This led to the finalization of 'The Guidelines for Research Involving r-DNA' by the US Government. By the mid-1980s, however, the context of biotechnology had shifted from research to commerce. An intense debate ensued between molecular biologists and ecologists, on controversial issues regarding risk assessment related to the release of new biotechnology products into the environment (Krimsky, 1991). The debate was also joined by prominent environmental groups, politicians, corporations and trade unions. This led to the formulation of a 'Coordinated Framework for Regulation of Biotechnology' by the US Government. Subsequently, biosafety guidelines were developed by the OECD and European Union (EU) and its member countries. Several surveys conducted to analyse public perceptions revealed that they were very similar irrespective of the geographical situations. However, there were also considerable differences between European Union (EU) countries (OTA, 1987), Hoban and Kendall (1993) and Mailier (1992).

# Type of PGR Conservation through Biotechnological Means

## *Ex situ* Conservation

*Ex situ* conservation of plant germplasm by storing seeds is arguably the most effective and efficient method. Low storage costs, combined with ease of seed distribution and regeneration of whole plants from genetically diverse material, offers distinct advantages. Cryopreservation is one of the *ex situ* conservation methods which works based on the removal of all freezable water from tissues by physical or osmotic dehydration, followed by ultra-rapid freezing.

## *In vitro* Conservation

*In vitro* conservation is an alternative method for multiplication and genetic conservation of crop species where re-generable cells or cultures are maintained *in vitro* for long time. It is sometimes called as micropropagation which denotes plant propagation using small explants. However this term is used frequently used interchangeably as clonal propagation and multiplication. Explants are mostly shoot, leaf, flower pieces, immature embryos, hypocotyls fragments or cotyledons (Paunescu, 2009). Generally, younger and more rapidly growing tissues are suitable. The criteria for a proper quality explants are normal, true-to-type donor plant, vigorous and disease free (Fay, 1992). As a rule, fragile tissues including meristems, immature embryos, cotyledons and hypocotyls requires less exposure to sterilizing agents than seeds or lignified organs (Paunescu, 2009). Explants may be obtained from seedlings grown from sterilized seeds. *In vitro* conservation techniques, using slow growth storage, have been developed for a wide range of species, including temperate woody plants, fruit trees, horticultural and numerous tropical species (Shikhamany, 2006). *In vitro* storage based on slow growth techniques has been pointed out as alternative strategies for conservation of genetic resources of plants. In particular, it is useful where the seed banking is not possible, such as vegetatively propagated plants, recalcitrant seed species, and plants with unavailable or non-viable seeds due to damage of grazing or diseases, and large and fleshy seeds. Some vegetable species conserved at *in vitro* conditions are *Allium* spp., *Solanum* spp., *Musa* spp., *Colocasia esculentum, Manihot* spp., and *Ipomaea batatas* (Henshaw, 1975; Zapartan and Deliu, 1994; Withers, 1995; Ashmore, 1997; Withers and Engelmann, 1997; Engelmann and Engels, 2002; Gonzalez-Benito *et al.,* 2004; Paunescu and Hololobiuc, 2005; Sarasan *et al.,* 2006).

## Merits of *in vitro* Conservation

Cryopreservation, developed during the last 25 years, is an important and the most valuable method for long-term conservation of biological materials. The main advantages in cryopreservation are simplicity and the applicability to a wide range of genotypes (Engelmann, 2004). True-to-type and disease-free plants can be obtained through *In vitro* culture in large numbers throughout the year from appropriate explants of a given species to restore the area under endangered species and productivity of the old varieties to the original level.

*In vitro* conservation technology offers several advantages like:

(i) *In vitro* conservation enables conservation of plants not producing fertile seeds such as male sterile plants.

(ii) Maintains heterozygosity indefinitely. The same is required in conservation of $F_1$ hybrid varieties.

(iii) A wide range of genotypes can be maintained under similar simulated conditions within a limited space.

(iv) Storage is in pathogen free state (aseptic conditions).

(v) The collection remains unexposed to natural and climatic vagaries

(vi) It is easier to get rid of the fungal and bacterial pathogens but special techniques such as meristem culture coupled with chemo-or thermo-therapy are required to eliminate viral or mycoplasmal pathogens.

(vii) The maintenance is independent of seasons.

(viii) Long term storage of crops having short seed viability or plantlets under various stages are possible.

(ix) Multiplication rate is very high.

(x) Genetic erosion is decreased to zero under optimal conditions

(xi) Decrease the labour cost.

(xii) Ready availability of material for distribution.

(xiii) A constant flow of disease free plantlets/clones can be realized

(xiv) Easy transportation within the country for research purpose

(xv) The micro-propagated plants (*in vitro* material) being free from diseases and pests and multiplied under aseptic conditions can be exported/exchanged internationally with limited quarantine restrictions.

### Demerits of *in vitro* Conservation

Some species do not respond desirably in tissue culture, which limits the use of these techniques.

The problems or disadvantages with *in vitro* conservation could be:

(i) Tissue culture is time consuming for initial preparation and maintenance.

(ii) An adequate number of representative species/genotypes for conservation are not possible.

(iii) *In vitro* culture conditions have to be worked out for each species (sometimes genotype) that has to be conserved. So, it a matter of trial and error.

(iv) In general, there may be loss of morphogenic potential with the passage of time

(v) Genetic variability in cultures due to chromosomal changes both structural as well as numerical.

(vi) Somaclonal variations that can distort genetic uniformity of *in vitro* cultures (Larkin and Scowcroft, 1981).

(vii) Maximum period for which the *in vitro* techniques can be applied, is not ascertained and

(viii) The choice of explants for micro-propagation differs from species to species.

Whereas clonal propagation is essentially production of true-to-type plants through tissue culture on a suitable medium and its multiplication. Realization of *in vitro* conservation came about in the early 1970's at a time when the storage of microbial cultures was a routine procedure. Since then tissue culture techniques have been applied to more than 1,000 plant species. Subsequently *in vitro* conservation has progressed from mere speculation to development and implementation. As per the cell Theory put forward by Schleiden and Schwann in 1839 and as a historical remark, the potential for regenerating a whole plant from a single cell was first hypothesized by the German botanist Haberlandt in 1902. The ultimate aim of tissue culture system employed for the purpose of genetic conservation is that of the preservation of specific and unique individual genomes which is governed by several factors. *Ex situ* conservation techniques which are recently being used are (Pawar and Kale, 2011):

(i) Tissue culture methods (*In vitro* conservation) and

(ii) Cryopreservation (in liquid nitrogen).

Recently with the advancement in biotechnology, the triploid progenies from immature fruits/seeds of 4x (tetraploid) x 2x (diploid) crosses of watermelon could be salvaged through embryo culture. Micropropagation of multiple shoots from seedling tips has been achieved. *In vitro* propagation of tetraploid and diploid parental lines of watermelon to facilitate large scale production of 3x watermelon seed is being carried out. The possibility for largest periods of seed storage is advocated using crypreservation.

### (i) Tissue Culture Techniques (*In vitro* conservation)

Tissue culture techniques have been adopted to conserve germplasm of wild, endangered, threatened species as well as selected elite varieties. The tissue cultures derived from rooted plants are appropriately, acclimatized and hardened for their higher survival under field conditions. Many plant species have a tendency of somatic embryogenesis directly from the explant or following a short intervening callus phase under appropriate auxin and cytokinin ratio. Thiaduron is very effective in inducing somatic embryogenesis in many plants. It aimed to provide a brief account of large-scale *in vitro* propagation techniques employed in vegetable crops with their commercial utility. The tissue culture techniques in general have been reviewed by Biondy (1986) and Boxus (1990) in several crops. Reports on commercial multiplication have been presented by McCown and McCown (1999), Zimmermann (1996) and Giles (1990). Commercial micro propagation with special reference to vegetables has been reviewed by Walkey (1987) and Joshi *et al.* (1995). We have tried to compile the most successful reports of the past 15-20 years which could be utilized commercially.

Sixty three countries have a tissue culture facility and the total number of accessions stored *in vitro* worldwide is approximately 38, 000 (FAO, 1996). In India, different aspects of plant tissue culture research have been carried out at institute of government departments. A number of research groups are working in about 100 different universities and research institutions all over the country. Prominent among these are National Chemical Laboratory (NCL), Poona, Botany Department, Delhi University, Tata Energy Research Institute (TERI), Delhi, Central Institute for Medical and Aromatic Plants (CIMAP), Lucknow (U.P.), Sugarcane Breeding Institute (SBI), Coimbatore (Tamil Nadu), ICAR-Indian Agricultural Research Institute (IARI), New Delhi, Indian Institute of Spices Research (IISR), Calicut, ICAR-CTCRI, Trivendrum and ICAR-NBPGR, New Delhi. *In vitro* repository at ICAR-NBPGR maintains large collection of various tropical and temperate crops of more than 52 genera of 158 species (Mandal and Chaudhury, 2004; Anonymous, 2010).

### Where Tissue Culture Conservation is Applicable?

Agriculture is the backbone of Indian economy, the success of which depends on the availability of good quality planting materials. In case of clonally propagated crops, the availability of large scale true to type planting material is always a limitation faced by Indian farmer. The tissue culture technology has revolutionized the industry by supplying good quality, disease free and true to type planting material. Science and art of plant tissue culture have developed through the distinguished contributions of numerous dedicated scientists since the beginning of the 20th century. In 1960 Cocking was the first to isolate the plant protoplast enzymatically. Protoplast have been isolated from many plant tissues and calli, manipulated and differentiated into fertile plants. The totipotency of both somatic and gametic cells of plants has been achieved beyond doubts. Tissue culture as conservation method can be applied to a plant species that are:

(1) Exclusively or predominantly, vegetatively propagated crops

(2) Have a long life-cycle and seeds are produced after long periods

(3) With recalcitrant seeds

(4) Sterile individuals that possess important characteristics, and

(5) Species where disease elimination is desired for successful conservation and subsequent propagation.

### Tissue Culture in Crop Improvement

☆ Hybridization
  ❏ *In vitro* fertilization
  ❏ Embryo rescue
  ❏ Protoplast fusion

☆ Tissue and Organ culture
  ❏ Micropropagation
  ❏ Anther and microspore culture
  ❏ Ovule and ovary culture
  ❏ Artificial seeds

☆ Mutation breeding
  ❏ Somaclonal variations
  ❏ *In vitro* selection

☆ Recombinant DNA technology
  ❏ Somatic embryogenesis
  ❏ Protoplast culture

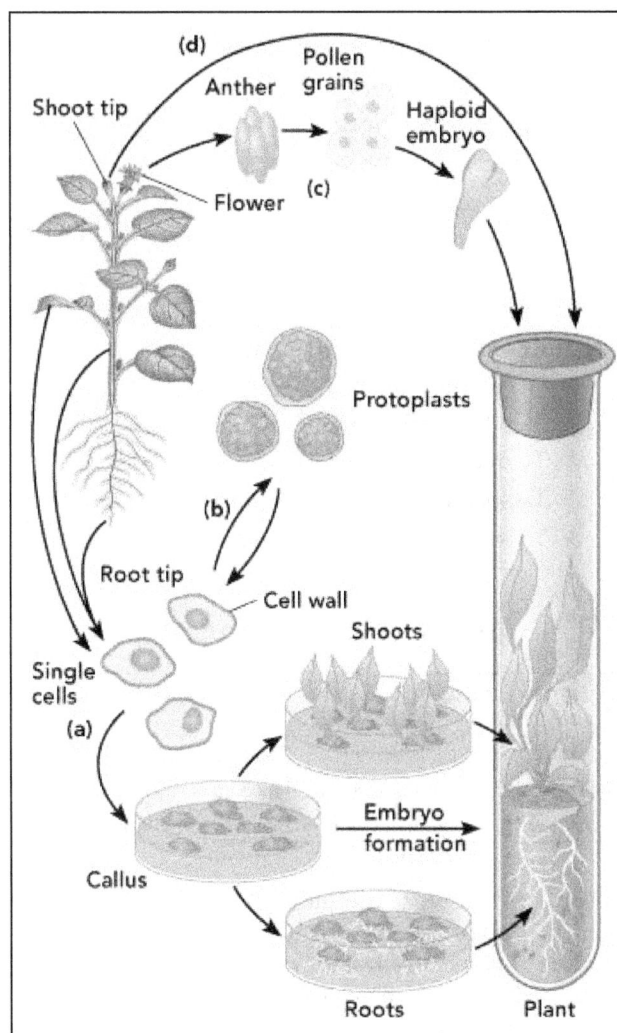

**Figure 11.1: Tissue Culture Technique.**

Mass multiplication of plants, using tissue culture technique, is more commonly used in cauliflower and other cole crops, especially in case of self-incompatible and male sterile lines. Explants like leaf, peduncle, pedicel, anther, meristem, tip and segments of root, stump and stem can be used for *in vitro* multiplication. For seed production, industrial production of *in vitro* hybrid parent plants has been performed by curd explants. Thousands plants can be obtained by this way (Kieffer *et al.*, 1994). Production of disease resistant plants *in vitro* have been discussed in detail by Ross (1980).

### Somaclonal Variation

Decapitation of cauliflower does not follow development of axillary branches as axillary buds are not commonly produced/developed in this crop. However, sometimes branches may develop from leaf scars near the base of stem or exposed roots. These shoots shows abnormality for morphological as well as reproductive characters, which may be analogous to somaclonal variation as recovered from cell culture or callus culture in many crops (Crisp and Tapsell, 1993).

### In vitro Pollination/Fertilization

☆ Adopted in various cultures
  ❐ *In vitro* flowering
  ❐ Flower Culture
  ❐ Ovary Culture
  ❐ Ovule Culture
☆ Overcomes incompatibility barriers

### Uses of Embryo Culture

☆ **Rescue F₁ hybrid from a wide cross**
☆ Overcome seed dormancy, usually with addition of hormone to media (GA)
☆ To overcome immaturity in seed
☆ To speed generations in a breeding program
☆ To rescue a cross or self (valuable genotype) from dead or dying plant

#### Table 11.4: Hybrid Embryo Rescue

| | |
|---|---|
| *Solanum lycoperiscum* x *S. peruvianum* | Thomas and Pratt, 1981 |
| *P. vulgaris* x *P. angustissimus* | Belvanis and Dore, 1986 |
| *Vigna pubescens* x *V. unguiculata* | Fatokum and Singh, 1987 |
| *S. melongena* x *S. torvum* | Blestos *et al.*, 1998 |

In cauliflower, the embryo culture technique (embryo rescue) *in vitro* has been used to rescue non-variable interspecific hybrids in *Brassica* species (Ayotte *et al.*, 1987). Embryos of the cross of *Brassica oleracea* and *Brassica napus* were rescued by them between 11-17 days after pollination. Later, they transferred triazine-resistance characters from *Brassica napus* into cauliflower, cabbage, broccoli, and kale. Usually it is difficult to obtain F₁ hybrid seeds by crossing *Brassica oleracea* with *B. campestris* but by using embryo rescue techniques, Inomata (1977) got success. Somatic embryos were induced from leaf tissues of *Dioscorea rotundata* in culture medium containing 2,4-D and incubated in darkness. However, low induction frequencies (<30 per cent) were recorded and protocols need to be optimized. In *D. alata* and *D. opposita*, embryogenic cell masses were induced from root explants in liquid MS supplemented with 2,4-D and cultured in light (Twyford and Mantell, 1996; Nagasawa and Finner, 1989). Germination of somatic embryos of *D. alata* increased in the presence of GA3 (Deng and Cornu, 1992; Twyford and Mantell, 1996). Plantlet recovery from somatic embryos of *D. rotundata* was enhanced at 4.5 per cent sucrose but not affected by benzyl aminopurine (Okezie *et al.*, 1994; Pandro *et al.*, 2011). These reports pointed out probable genotype-dependent protocol for yam embryogenesis.

### Advantages of Somatic Hybridization

☆ Combine two complete genomes
  ❐ Another way to create allopolyploids
  ❐ Novel interspecific and intergeneric hybrid
  ❐ Fertile diploids and polyploids from sexually sterile haploids, triploids and aneuploids
☆ Partial genome transfer
  ❐ Exchange single or few traits between species
  ❐ May or may not require ionizing radiation
☆ Genetic engineering
  ❐ Micro-injection, Electroporation, *Agrobacterium*
☆ Transfer of organelles
  ❐ Unique hybrids of nucleus and cytoplasm "Cybrids"
  ❐ Transfer of mitochondria and/or chloroplasts between species

### Wide Crosses via Protoplast Fusion and Tissue Culture Regeneration

*Brassica juncea* (Pbᵗ) x *Thlaspi caerulescens* (Znᵗ and Niᵗ) has been performed.

#### Table 11.5: Some Distant Somatic Gybrids

| Symmetric or Near-Symmetric Hybrids | Asymmetric Hybrids |
|---|---|
| *Solanum tuberosum* + *Solanum lycoperiscum* | *Daucus carota* + *Aegopodium podagraria* |
| *Arabidopsis thaliana* + *Brassica campestris* | *Daucus carota* + *Petroselinum hortense* |
| *Atropa belladonna* + *Nicotiana chinensis* | *Nicotiana tabacum* + *Daucus carota* |

## Table 11.6: Comparison of different Explants available for Germplasm Conservation

| Explants | Advantages | Disadvantages |
| --- | --- | --- |
| Apical and auxiliary meristem | Lowest frequency of somaclonal variation | Rooting problem in certain species |
| | Easy to propagate | - |
| | True-to-type of parent material | - |
| | Genotype independent | |
| | Full genome present | |
| | Little work required | |
| | Potential for early maturation | |
| Adventitious meristem (organogenesis) | Full genome present | Highest frequency of somaclonal variation |
| | True-to-type of parent material | Rooting problem in certain species |
| | | Genotype dependent |
| | | Labour intensive |
| Somatic embryos (adventitious embryogenesis) | Low frequency of somaclonal variation in most species | Genotype dependent |
| | | Labour intensive |
| | Root and aerial system well developed | - |
| | True-to-type from starting tissue | - |
| Gametogenic embryos (adventitious embryogenesis) | Bypass fertilization process | Haploid genome |
| | | Genotype dependent |
| | Need only male or female organs | Not True-to-type; results of meiosis but true to type in the same line |
| | | Only from gametes |
| | | Labour intensive |
| | | Only a few species of trees and shrubs worked on |
| Zygotic embryos (excised from seeds, not tissue cultured per se) | Genotype independent | Seeds must be available |
| | No early maturation | Not true-to-type, result of sexual process |
| | Little work required | - |
| Isolated DNA | Requires minute quantity of plant material | No possibility of whole plant regeneration |
| | Exact copy of parental DNA | Only conserves genetic information of parent tree |
| | Can be amplified artificially | A last resort |
| | Little work required | Only a few genes could be transferred at a time into different host genome |

Plant tissue culture techniques, in combination with recombinant DNA technology, are the essential requirements for development of transgenic plants. The advantages and disadvantages of using various explants for *in vitro* conservation are given in Table 11.6.

Cytogenetically variation can be minimized either by appropriate choice of the explants or by selective growth medium (Chopra and Narasimhulu, 1986). The juvenile shoots or buds are usually free from infections, more responsive and re-generable.

### Techniques of In vitro Mmicro-propagation

The term *In vitro* micro-propagation covers a wide range of techniques which can be broadly categorized as following:

1. Organ culture
2. Callus culture and embryogenesis
3. Suspension culture
4. Isolation and culture of single cell and protoplast and
5. Anther and pollen culture.

### Explants Used for in vitro Conservation

A tissue culture technique owes its success to the inherent feature of totipotency of plant cells *i.e.* "*Each cell in a plant has the potential to give rise to a new progeny*".

Multiplication in cultures is achieved through subculture of:

☆ Organized tissues (*e.g.* meristems, somatic/ zygotic embryos) or

☆ Re-differentiated tissues (cells, protoplast *etc.*).

### Steps of In vitro Micro-propagation

Micro-propagation is prerequisite for *in vitro* conservation, briefly the various steps that are followed for developing successful micro-propagation protocols are enumerated below:

1. Selection of stock/mother plant
2. Establishment of aseptic culture
3. Multiplication of propagules
4. Hardening of plantlets
5. Transfer to green house/fields

### Types of In vitro Culture

Micro-propagation could be accomplished through one of the following processes:

1. Meristem culture
2. Proliferation of axillary buds
3. Nucellar embryos
4. Somatic embryos
5. *In vitro* tuberization and cormlet production

For *in vitro* conservation programme, the organized explants have proved better than the unorganized tissue. Suitable explants are processed, surface sterilized and cultured on solidified or liquid medium for establishment aseptic cultures. The aseptic cultures are then transferred to proliferation medium with hormonal combination avoiding any intervening callus phase. The proliferated shoots are then rooted for transplanting from *in vitro* conditions. With the advent of Habertlandt's hypothesis of totipotency of plant cells, there has been an enormous upsurge of interest in possible uses and applications of *in vitro* techniques. Nowadays, this technique is being commercially exploited for large-scale rapid multiplication of elite stock, creation of variability and selection of resistant lines. *In vitro* clonal multiplication aims at the production of larger numbers of disease-free and true to type plants in quicker time. This is generally accomplished by *in vitro* methods involving meristematic explants, shoot tip culture, adventive buds induction or by stimulation of axillary or lateral meristems into their growth and development, or through culture of non-meristematic explants (leaf, petiole, root, *etc.*). The Murashige and Skoog's medium is most common in use. Stages in plantlet formation include shoot bud initiation, development of buds into shoots, rooting of shoots and preparation of plantlets for transplantation from culture vessels to soil. Such *in vitro* techniques are very powerful tool but caution is required towards the conformity of daughter plants.

### Embryo Culture

Embryo culture consists of isolation (excision of embryo from mature seed) and culture of embryo under aseptic conditions on nutrient medium. Embryos are more totipotent and easily grow into seedlings or proliferate into callus on *in vitro* culture medium. *In vitro* culture of immature embryos is followed mostly in case of F$_1$ hybrids to protect it from degeneration. Embryo culture have application in following area of vegetable crop improvement:

i. Distant hybridization
ii. Propagation of seeds having short viability
iii. Shortening of breeding cycle
iv. Overcoming dormancy

### Meristem Tip Culture

The apical dome (*i.e.*, growing point) of shoots could be cultured resulting to continued and organized growth. This method is known as meristem tip culture, which gives enhanced axillary proliferation ensuring clean and variably stable culture with high purity and reliability. These shoot apices finally give rise to small shoots which could be rooted. These are of practical significance and can be utilized to produce disease free plants. As we know that shoot primordial are closed structure and generally they are free from bacteria, viruses, *etc.* Thus, the extreme tip of the shoot (0.5-2.0 mm in length) can be cultured. The technique involves manipulation of culture condition so that the meristem could be organized to form shoot/root, leading to plantlet formation (George, 1993a). By far, the most versatile method of ensuring uniformity of the crop is through culture of shoot apices in a majority of plant species but rather restricted in commercial propagation, unless multiple shoot induction through axillary meristems can be secured.

**Table 11.7: Meristem Tip Culture in Vegetable Crops**

| Vegetables | Target | Reference |
| --- | --- | --- |
| *P. sativum* | Cryopreservation | Kartha *et al.*, 1979 |
| *Pisum sativum* | Seed born mosaic virus | Kartha and Gamborg, 1978 |
| *L. esculentum* | Cryopreservation | Grout *et al.*, 1978 |
| *Allium sativum* | Yellow dwarf virus | Walkey *et al.*, 1987 |
| *A. ascalonicum* | Yellow dwarf virus | Walkey *et al.*, 1987 |
| *S. melongena* | EMCV | Raj *et al.*, 1991 |

Getting an organized structure such as bud into aseptic culture is a time consuming process and initially the contamination becomes a hurdle. The multiplication rate is also slow at first, but later on in better culture conditions higher multiplication rate could be achieved. Axillary buds of shoot tips are induced to produce numerous shoots by adjusting concentrations and combinations of growth regulators, preferably cytokinins. The rapid proliferation of shoot results in masses of shoots being produced from a single shoot tip. This multiplication process is of commercial value, where one can multiply the elite stock, hybrid lines, seedless lines, male sterile line *etc.* Several hundred shoots could be produced within 12 weeks (4-6 subcultures). The multiplication rate greatly varies with genotypes and species. Shoots are removed at regular interval and a portion of mass is replaced on fresh medium to continue proliferation. The small shoots are removed from the cultures and then rooted on rooting medium. These plantlets could be generally grown in the greenhouse with good survival rate.

### Stem Node Culture

The term referred to culture of explants bearing intact shoot meristem/nodal portions. It is widely used for clonal propagation. The main purpose of node culture is to produce large number of shoots with repeated formation of axillary branches. Thus the newly developed shoots serve as stock culture and used as explants for repeated proliferation. The usual size of explants is 10-20 mm in length as compared to meristem culture. Commercially this technique is very much preferred to produce virus-free stocks. The most favourable feature of this technique is high multiplication frequency with uniformity. The growth is generally promoted in addition of growth regulators (cytokinins) into the growth medium. The primary explant is derived from apical or nodal segment having meristematic stem/few leaf initials. Developing meristem, shoot tip or nodal portion will result in development of small shoots. After the appropriate treatment in stage I, II, III and IV these original cultures give rise root to produce plantlets. At the same time these axillary buds can be induced (at stage II) to develop multiple shoot or shoot clusters.

### Cell/Callus Culture

Cell/Callus culture can also be used to produce multiple embryos which can give rise to new plants. Theoretically unlimited number of plants can be propagated from a single superior plant. Though many of the callus derived plants are genetically identical to the original plant but fractions of them are variants which is called somaclonal variations. Occurrence of somaclonal variations limits its use at commercial scale where the main aim is to maintain uniformity (Bhojwani and Rajdan, 1996). It is a big problem, however, the commercial laboratories has utilized this technique to create variability, *e.g.*, carrot and pepper varieties with better taste has been selected from callus culture. In some cases commercial ventures has used these techniques with extreme caution.

### Anther Culture

Using immature anthers taking from young inflorescence or floral buds and culturing on nutrient medium for androgenesis is called anther culture. Immature anthers possessing several microspores at the right stage for *in vitro* culturing after surface sterilization. In general 4-6 anthers per culture tube are adequate in case of Solanaceous plants such as tomato, potato,

brinjal and capsicum. The cell wall tissue turns brown in responsive anthers and within 3-6 weeks, they burst due to pressure exerted by growing callus or microspore plantlets. Anther culture is simple, quick and efficient technique of haploid production but has disadvantage of producing plantlets of some other ploidy levels. This problem is taken care by opting for microspore culture. It involves the oozing out of microspore from anthers by squeezing. Microspores in anthers produce haploid plantlets in two ways:

i. Direct androgenesis-Embryos are directly developing from responding microspores without forming the callus.

ii. Indirect androgenesis-Microspores follow the organogenic pathway that is they undergo cell division to form callus, which is then induced to diffenetiate into plants. Plantlets originating from the callus may exhibit different ploidy levels.

### Protoplast Culture

Removal of cell walls by mechanical and/or enzymatic methods and culturing of protoplast (naked plant cells) is called protoplast culture. The protoplasts are produced by subjecting the plasmolyzed cells to treatment of mixture of enzyme (macerozyme and cellulose). The protoplasts are stabilized by addition of an osmoticum (mannitol or sorbitol). Optimum plating density is about $5 \times 10^4$ protoplast/ml. culture medium of protoplast is similar to plant tissue culture but devoid of ammonium and increased concentration of calcium. Glucose and sucrose are preferred carbon source. Cell wall regeneration starts just after a few hours of isolation. First cell division is completed within 2-7 days. Small callus colonies develop within 3 weeks and attaining size of 1 mm diameter by $6^{th}$ week. Protoplast regeneration is prerequisite to its application in crop improvement and it has been achieved in potato, tomato, carrot, radish and cauliflower *etc*.

### Approaches of In vitro Conservation

Many national and international organizations have used *in vitro* techniques and applied it to more than 1000 species, as a complementary conservation method to field genebanks. Approaches of *In vitro* conservation, includes:

(a) Conservation of cells, tissues, organs in glass or plastic containers under aseptic conditions through slow growth of cultures

(b) Cryopreservation of cultures (tissues, organs, pollen or cultures in liquid nitrogen at −150 to −196°C). At ICAR-NBPGR, New Delhi, 85 accessions of garlic and 14 accessions of *Allium* spp. are being maintained in the *in vitro* Bank.

*In vitro* conservation is practiced by two approaches:

**(1)** Cryopreservation (long term conservation)

**(2)** Slow growth method (short and medium term conservation)

### (1) Cryopreservation

Cryopreservation can be defined as:

"Cryopreservation is storage of biological samples in viable conditions at ultra-low temperature such as on solid carbon dioxide (-79°C), in low temperature deep freezers ((-80°C), in vapor phase of nitrogen (-150°C) or in liquid (−150 to −196 °C)".

or

Preservation of cells, tissues, and organs in liquid nitrogen at -196 0C is called cryopreservation.

The science pertaining to cryopreservation is known as cryobiology. Since the discovery of glycerol as a cryoprotectant 60 years back, cryoprotectant have become accepted as antifreezes needed to protect life in the frozen state. Theoretically, the cells and tissues can be cryopreserved in live state indefinitely, and cells/tissues recovered after thawing should be unchanged and live. This reduces the problems associated with seed regeneration schemes especially the risk of loss of genes/analyses caused by environmental problems. Cryopreservation can be applied for short and medium term conservation. Even if the intervals between sub-cultures are greatly increased, conventional medium term conservation techniques poses considerable problems in management of large collections. Thus, cryopreservation is preferred for long term conservation of germplasm. The basis of cryopreservation is the changes of water present in the cells to the solid state when subjected to sub-zero temperature. The cells initially undergo super-cooling but regain thermodynamic equilibrium subsequently either by losing water or by intracellular freezing. The metabolic processes are considerably slowed down and longevity of the material enhances. Seeds of desiccation tolerant plants can be easily stored under low temperature (-20°C) and low humidity for several years. For desiccation sensitive and recalcitrant seeds and vegetatively propagated plants, certain tissue culture methods have to be used for medium and long term storage of germplasm without affecting its genetic stability. The utilization of liquid nitrogen (LN) as a storage medium is predicted on the capability of seeds to survive liquid nitrogen exposure without significant damage to variability. Medium term storage can be done by meristem culture and prolonging the period between subculture through medium modification with growth retardants and osmotic and incubation at low temperature (4°C). Long term storage can be done through cryopreservation of appropriate plant tissue cultures at liquid nitrogen (-196°C) following slow preliminary cooling to some level (Withers and Engelmann, 1977).

### Cryopreservation Techniques

Cryopreservation technique has been successfully applied to callus, protoplast, pollen, meristems, zygotic and somatic embryos and suspension cultures of around 42 crop species including various horticultural crops. Under cryopreservation technique includes two ways:

**(1)** Classical cryopreservation technique (slow freezing).

**(2)** Recent/new cryopreservation technique

### (1) Classical Cryopreservation Technique (Slow freezing)

These techniques were in usage during all initial studies using plant cell and tissue cultures. It involves chemical cryoprotection and freeze-induced dehydration. Basically 2 methods have been used

    **i. Two step freezing:** This includes first slow controlled cooling of sample to a defined pre-freezing temperature using a programme freezer followed by rapid immersion in liquid nitrogen. Freezing rates between 0.5 °C and 2 °C min$^{-1}$ down to freezing temperature around -40 °C; generally give satisfactory results in several cases. In order to avoid use of sophisticated programmable freezer, domestic or laboratory deep freezers have been successfully employed to perform the slow cooling step.

    **ii. Use of cryoprotectants:** In this procedure samples are submitted to a cryoprotective treatment for a defined period before freezing. Various cryoprotectants like dimethysulphoxide, glycerol, sorbitol, mannitol, sucrose or polyethylene glycol, are used either alone or in binary or tertiary mixtures. Mixtures of cryoprotectants have generally proved to be more effective than using single cryoprotectant.

These techniques have been successfully applied to undifferentiated culture systems such as cell suspensions and calluses (Withers and Engelmann, 1997). In the case of differentiated structures, these techniques have been used for freezing apices of cold-tolerant species (Reed and Chang, 1997) with no success so far in tropical species.

### Steps of Cryopreservation

Cryopreservation process involves following successive steps which have to be defined for each species.

☆ Selection

☆ Excision of tissues or organs

☆ Culture of source material

☆ Select healthy cultures

☆ Apply cryoprotectants

☆ Pregrowth treatments

☆ Cooling/freezing

☆ Storage

☆ Warming and thawing

☆ Recovery growth

☆ Viability testing

☆ Post-thawing

Some steps are given as under:

☆ **Selection of material:** Generally rapidly growing meristems which are small, relatively resistant to freezing, possessing fewer or smaller victualed cells and dense cytoplasm are preferred for cryopreservation (Englemann, 1991). The material can be sampled preferably from *in vitro* cultures to avoid contamination. Physiological stage of explants is also very important. Material should be used from fresh active cultures (Harding *et al.*, 1991).

☆ **Pre-growth:** It refers to the period in culture before cryopreservation begins. It offers to induce physiological changes in tissues to increase freeze-tolerance. The plant tissues can be cold-hardened using effective pre-growth additives such as mannitol, sorbitol, proline or dimethyl sulphoxide (DMSO) on pre-growth culture medium.

☆ **Cryo-protection:** Just before freezing, the effective compounds as cryoprotectants are applied which includes Dimethyl Sulphoxide (DMSO), glycerol, proline, sucrose, sorbitol and polyethylene glycol (PEG). These compounds can be used individually or in combination.

☆ **Freezing, storage and thawing:** Steps like freezing, storage and thawing are followed very carefully; otherwise damage during freezing may occur. Slow freezing is carried out by using controlled freezing apparatus and may cause extra-cellular ice formation. In contrast, during rapid freezing intra- cellular ice crystals do not

have time to form to a damaging size before all the intra- cellular water has frozen. Depending on the species, the freezing rate can be very precise as in strawberry or fall within a much broader range as in somatic embryos of oil palm (Englemann and Dereuddre, 1988).

☆ **Recovery growth:** It is important that viability level achieved on thawing should be high. Usually recovery growth is carried out on a defined medium supplemented with activated charcoal to adsorb toxins released by damage cells and growth substances to stimulate the desired responses. It is sometimes necessary to avoid the osmotic shock caused by an intermediate transfer onto a medium with low potential by successive transfer of material to progressively less concentrated medium (Englemann *et al.*, 1985). To avoid photo-oxidation which may be harmful, recovery can be made in dark (Benson *et al.*, 1989)

## *(2) Recent Cryopreservation Technique*

These are based on the vitrification phenomenon where cell desiccation is performed either by exposure of samples to concentrated cryoprotective solutions or by air desiccation. Vitrification based procedures are superior to classical freezing techniques and are more appropriate for complex organs like shoot-tips and embryos which contain a variety of cell types. Vitrification procedures are operationally less complex than classical ones (*e.g.* they do not require the use of controlled freezers) and have greater applicability (Sakai, 2000). This can be achieved using different procedures, such as pre-growth, desiccation, pregrowth-desiccation, ED, vitrification, encapsulation-vitrification and droplet-freezing (Engelmann, 2004).

☆ **Desiccation:** Zygotic embryos are extracted from the seeds, dehydrated in laminar flow or an air-tight container with silica gel. Freezing is rapidly done by placing the embryos in liquid

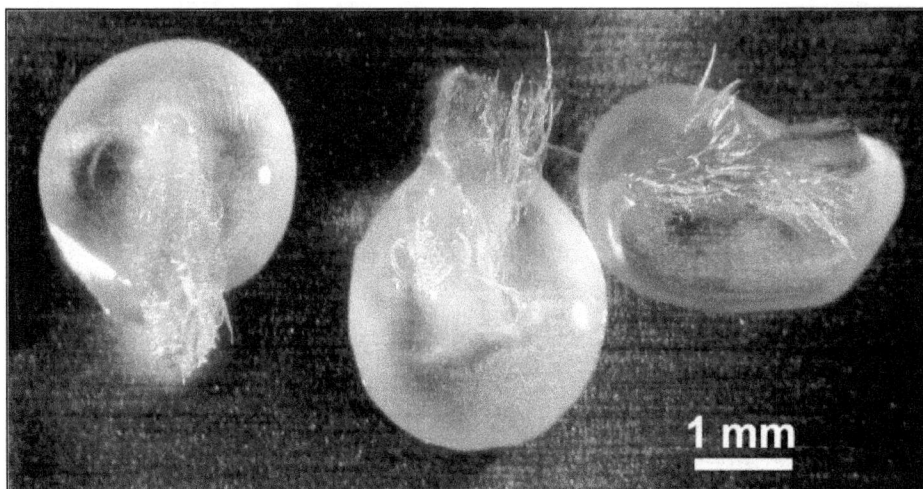

**Figure 11.2: Potato Apices Embedded in 10 per cent DMSO-Droplets on Aluminium Foil.**

nitrogen and re-growth of embryos after freezing is direct on a suitable medium. Dehydration must be just sufficient and should not be of very high degree which can induce the injury. By using this technique, the excised embryonic axes of tea, jack fruit, trifoliate orange, almond, cocoa, walnut and embryos of coconut and coffee have been preserved successfully (Chaudhury and Chandel, 1995; Chandel *et al.*, 1995).

☆ **Pre-growth Desiccation:** This technique consists of growing/initiating the growth of the plant material for different duration (hours to weeks depending on the material) on media supplemented with cryoprotectants, partially desiccating the material before rapid freezing in liquid nitrogen. It has been successfully applied to somatic embryos of oil palm, coffee, pea and melon and zygotic embryos of coconut.

☆ **Encapsulation-dehydration:** This technique is based on the development of 'synthetic seeds' *i.e.*, somatic embryos, shoot/axillary buds and shoot apices are encapsulated in a gel matrix of calcium alginate beads. For cryopreservation, explants are dissected under microscope and cultured for a definite period on a definite medium. These are subsequently encapsulated in alginate beads and allowed to grow in liquid medium with high concentration of sucrose (0.5-1.25M). Encapsulated explants are dehydrated using slow or rapid desiccation techniques under a laminar flow to decrease the moisture of beads to calcium 20 per cent. Freezing is done by placing the beads in liquid nitrogen. Recovery can be

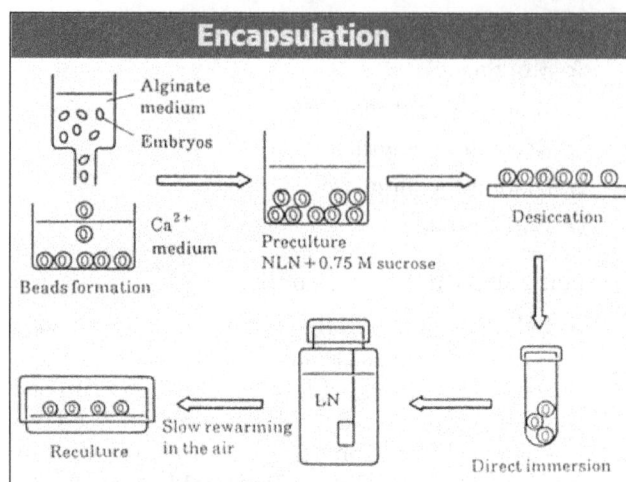

**Figure 11.3: Encapsulation.**

made after slow thawing and using a standard protocol for a particular species. Encapsulation-dehydration has been reported successful for potato, sweet potato, yams *etc.*

☆ **Vitrification:** Vitrification is phase transition of water from a liquid directional a non-crystalline or amorphous glass. It consists of treating the plant material with extremely concentrated solutions of cryoprotectant (PVS2, DMSO) and freezing them ultra-rapidly. Consequently, intracellular water vitrifies *i.e.* forms an amorphous glassy appearance. However, it cannot be used for cryoprotectant-sensitive plant material. Vitrification procedures have been developed for more than 20 different species using protoplasts,

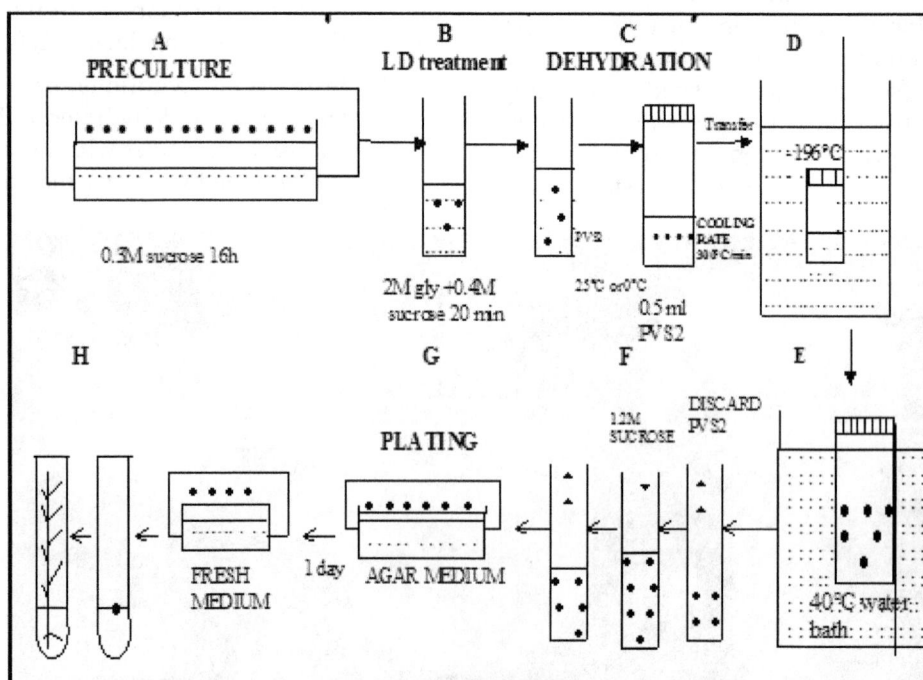

**Figure 11.4: Vitrification Procedure for Embryonic Axes.**

cell suspension, apices and somatic embryos. Vitrification has been successful with shoot tips/meristems of various species such as garlic, yams, cassava and *Solanum* species (Sakai, 2000).

☆ **Encapsulation-Vitrification:** This technique is a combination of encapsulation-dehydration and vitrification procedures. The alginate beads containing the explants are subjected to freezing by vitrification. It has been observed that recovery rate of apices frozen using this technique was 30 per cent higher than with the encapsulation-dehydration technique.

### Desiccation Tolerance

Optimal survival through freezing was obtained when the axes were dried to a moisture level just above that producing desiccation damage. Detailed investigations regarding the nature of water present in the recalcitrant seeds using differential scanning calorimetry have revealed that degree of desiccation tolerance of any seed or axes is determined by amount of freezable water present at any stage and the ability to withstand the loss of substantial proportion of non-freezable water. In studies on tea embryonic axes cooling rates have been shown to affect the properties of water in axes. Increasing the cooling rates and drying the axes to moisture content between 1.6 and $1.10g^{-1}$ could increase the survival of axes in culture. Thus in different recalcitrant seed species, various physiological factors play role in imparting different degrees of desiccation tolerance. Thus envisages development of separate freezing protocols for each of them.

### Application of Cryotechniques

In most of the reports of successful cryopreservation, the technique involved has been the excision of embryonic axes aseptically, desiccation to around 11-16 per cent moisture level using air drying or by addition of cryoprotectants and exposing to temperatures of liquid nitrogen (LN) either directly or controlled rate of freezing. The axes were later thawed rapidly in a water bath maintained at +38°C and cultured on the defined media to obtain complete plants. In some cases, the plants have also been successfully transferred to the field. More recently pretreatment of zygotic/somatic embryos by culturing on medium with high sugar content and partial desiccation has proved successful. The results obtained with embryo of coffee, oil palm and coconut (Englemann *et al.*, 1983) have demonstrated that the critical of the various cryopreservation processes developed is the pretreatment phase, during which desiccation tolerance of embryos is induced using high levels of sugar and/or desiccation.

### What Materials can be Cryopreserved?

Cryopreservation has proved to be extremely useful in long term conservation of seeds, vegetative propagules, meristematic tissues and organs, and *in vitro* cultures. Zygotic embryos and embryonic axes of recalcitrant seeds can be successfully used for cryopreservation (Chandel *et al.*, 1993). The embryonic axes are preferred because of its organized small structure, independent identity and the presence of appreciable protection of meristematic tissues. In fact, cryopreservation is the only technique available presently, for long term storage of vegetatively propagated species and non-orthodox sedd species (Bajaj, 1995).

☆ **Seed Cryostorage:** Based on their storage characteristics, seeds have been classified as orthodox or non-orthodox (intermediate and recalcitrant), and different protocols are used for their cryopreservation (Chaudhury *et al.*, 2003).

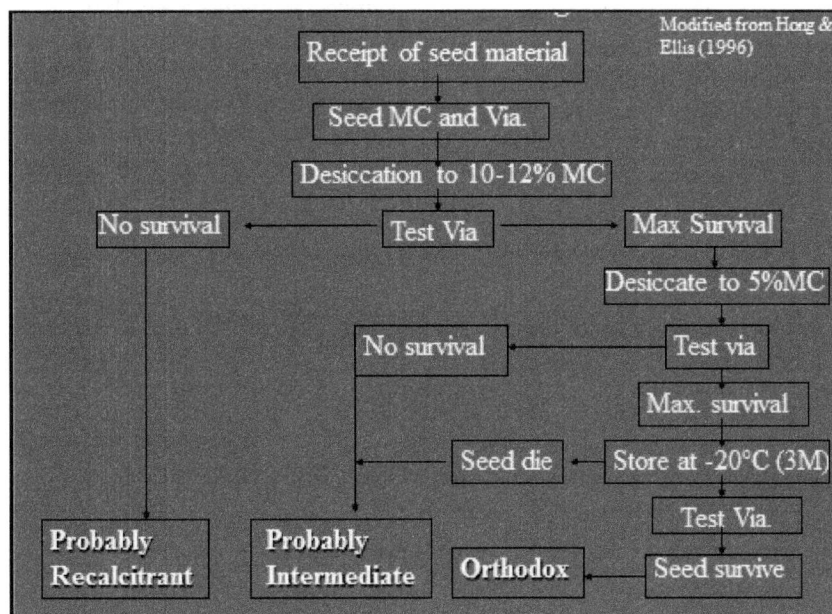

**Figure 11.5: Protocol to Determine Seed Storage Behaviour.**

☐ **Orthodox seeds** are capable of retaining viability even after being dried to less than about 5 per cent moisture content and are tolerant to liquid nitrogen. Seeds of most common agricultural species belong to this category and can be safely cryopreserved, thus providing a means of greatly extending their viability than is possible through conventional conservation at -20 °C. Pre-desiccation followed by direct immersion in liquid nitrogen or after slow cooling is the commonly applied method of cryopreservation for orthodox seeds. At ICAR-NBPGR, New Delhi, long term conservation of a large number of orthodox seed collections comprising cereals, pulses, oilseeds, vegetable crops and other economically important species especially their varieties, genetic stocks, wild species has been successfully achieved using liquid nitrogen based cryopreservation (Chaudhury *et al.*, 1989).

☐ **Non-orthodox** seeds are shed with high moisture levels and are desiccation and freeze-intolerant. They lose viability after being dried below a critical limit, usually between 12-30 per cent moisture content. At these moisture levels, the seeds cannot be subjected to subzero temperature since they undergo freeze injury. In large-seeded species of several tropical and temperate non-orthodox species, the excised embryonic axis is the preferred explant because of its organized small structure, independent identity and the presence of appreciable proportion of mersitmatic tissues.

☆ **Excised embryonic axes:** The aseptically excised embryonic axes are either desiccated to around 11-16 per cent moisture level using air-drying or treated with cryoprotectants before immersing in liquid nitrogen. The axes are later thawed rapidly in a water bath maintained at 38 °C and cultured on defined media to obtain complete plants. Embryonic axes of tea (Chaudhury *et al.*, 1991), jackfruit (Chandel *et al.*, 1995), and neem, black pepper, almond, litchi, and whole seeds of cardamom can be cryopreserved (Chaudhury and Chandel, 1991, 1994, 1995a, 1995b). The use of embryonic axes for cryostorage of temperate tree species with short lived large seeds has been done for *Juglans, Aesculus, Castanea* and *Quercus* species (Pence, 1992).

☆ **Preserving pollen:** Preserving pollen of wild species and cultivated crops assume importance for safeguarding the pollen parents in any crop improvement programme and dispense away with the need for growing the same frequently. Long term pollen preservation forms a part of integrated gene resources conservation (Roberts, 1975). Pollen can be made available in abundant quantities and forms an ideal compact material for conservation recombined genetic information present in pollen at the gametophytic stage. Pollen viability and longevity varies with different crop species, methods are available for preserving vegetable pollen for extended durations (Alexander and Ganeshan, 1989a) under cryogenic conditions. For long term conservation potential wild species of vegetable crop can be considered. Successfully, wild species of tomato (*Solanum pimpinellifolium* Mill.), wild eggplant (*Solanum indicum* L.) and cultivar of brinjal (*S. melongena* L.) 'Arka Kusumakar' and 'Arka Shirish' are conserved through liquid nitrogen.

After storage results obtained showing in Table 11.9.

Pollen of *Cucumis sativus*, wild species of *C. hardwickii* and *Abelmoschus* species namely *A. callei, A. manihot, A. pungens* etc. have been cryostored for use by breeders in crossing programmes (Chaudhury and Malik, 2009). *In vitro* cultures of tuber crops mainly of sweet potato, yams and taro conserved for last more than 20 years have been conserved in the *in vitro* repository. In onion cryostored pollens gave significant results in

**Table 11.8: Pollen Viability in Brinjal and Tomato under Fresh and Liquid Nitrogen Storage**

| | Crop/Cultivars/Species | | | |
| --- | --- | --- | --- | --- |
| | Brinjal | | | Tomato |
| | Arka Kusumakar (9 Years cryostored in Liquid Nitrogen) | Arka Shirish (3.5 Years cryostored in Liquid Nitrogen) | Solanum indicum (9 Months cryostored in Liquid Nitrogen) | S. pimpinellifolium (2 Years cryostored in Liquid Nitrogen) |
| Fresh pollen viability | 45.96 (51.71) | 48.32 (55.72) | 50.50 (59.55) | 45.55 (51.00) |
| Stored pollen viability | 20.70(12.74) | 54.24 (65.88) | 46.73 (53.05) | 44.43 (49.04) |
| SeM | 1.622 | 2.366 | 1.333 | 0.356 |
| LSD (p=0.05) | 6.369 | NS | NS | NS |
| CV per cent | 8.430 | 7.992 | 4.749 | 1.373 |

*Source*: Rajasekharan *et al.*, 1998.

**Table 11.9: Controlled Pollinations with Cryostored Pollens**

| Crop | Species/Cultivar | Parents | | No. Flowers Pollinated | No. of Fruits Set | No. of Seeds/Fruit Recovered |
| | | Female | Male | | | |
|---|---|---|---|---|---|---|
| Brinjal | | Arka Kusumaker | Arka Kusumaker (control) | 46 | 41 | 429.60 |
| | Arka Kusumaker | Arka Kusumaker | Arka Kusumaker (9 Years cryostored in Liquid Nitrogen) | 37 | 34 | 569.50 |
| | Arka Shirish | Arka Shirish | Arka Shirish (control) | 20 | 13 | 237.75 |
| | | Arka Shirish | Arka Shirish (3.5 Years cryostored in Liquid Nitrogen) | 17 | 10 | 405.00 |
| | Solanum indicum | Solanum indicum | Solanum indicum (control) | 10 | 6 | 46.33 |
| | | Solanum indicum | (9 Months cryo stored in Liquid Nitrogen) | 7 | 7 | 25.33 |
| Tomato | S. pimpinellifolium | Arka Shirish | S. pimpinellifolium (2 Years cryostored in Liquid Nitrogen) | 24 | 23 | 60.00 |

Source: Rajasekharan et al., 1998.

fruit and seed set (Ganeshan, 1986). Lotus pollen can be successfully stored at -20±1 °C for 14 days with silica gel as a desiccant. Seed formation is possible with stored pollen for about 1 year but the success rate greatly diminishes (Kasumi et al., 2000).

### Current Status of Cryopreserved Materials

Various desiccation techniques for seeds and embryonic axes, namely air desiccation, pre-growth-desiccation, vitrification and encapsulation-dehydration are currently been used at cryolab of ICAR-NBPGR, New Delhi (Chaudhury and Malik, 1999; Chaudhury, et al., 2000). Cryoproptocols of vitrification and encapsulation have been successfully attempted in embryonic axes of Artocarpus heterophyllus, Litchi chinensis, Poncirus trifoliate and Citrus species (Chaudhury, 2000; Malik and Chaudhury, 2006) and in Dioscorea species using encapsulation-dehydration (Mandal et al., 1996). Various cultivated and wild species of Allium are being maintained as in vitro cultures and cryopreservation to achieve short-medium and long term conservation and more than 150 accessions are presently being maintained in the in vitro repository (Anonymous, 2010).

Using the cryoprotocols a total of 433 accessions of 23 genera representing wide genetic diversity in vegetables, including varieties and wild species, has been cryopreserved in the form of seeds at moisture content of 5-8 per cent in the vapour phase of liquid nitrogen at ICAR-NBPGR's National Cryogenebank in the form of seeds. Further, investigations are on to develop protocols for cryopreservation of more recalcitrant seed species and establish their base collections. As an alternative complementary method, attempts are also being made to cryopreserve pollen in case of trees or vegetatively propagated species. Attempts of cryopreservation at the ICAR-NBPGR, New Delhi include successful cryopreservation of embryonic axes of tea (Chaudhury et al., 1991), jackfruit (Chandel et al., 1995), and neem, black pepper, almond, litchi, and whole seeds of cardamom (Chaudhury and Chandel, 1991, 1994, 1995a, 1995b). At ICAR-NBPGR, New Delhi,

254 accessions of okra, brinjal, onion, tomato and Cole crops are being maintained in the Cryo-Bank.

**Table 11.10: Conservation Status of Germplasm of Vegetable Crops**

| Method of Conservation | Location | Crop (Number of accessions) |
|---|---|---|
| Cryo bank (-196°C) | ICAR-NBPGR, New Delhi | Okra, eggplant, onion, tomato and Cole crops (254) |
| In vitro bank | ICAR-NBPGR, New Delhi | Garlic (85), Allium species (44) |

In addition of legume vegetables comprising 34 accessions have so far been cryostored. Efforts are oyrsued to conserve sizable variability of additional species and to cryostore these accessions as meristems and shoot tips. This will complement the conservation in the field genebank where they are difficult to maintain due to various biotic and abiotic stresses.

Genetic diversity of highly nutritious and are rare underutilized fruits commonly used as vegetables in several parts of India such as Capparis decidua (Ker), Carissa species (Karonda), Cordia species (Lasora), Emblica officinalis (Aonla) and Moringa oleifera (drumstick) have also been conserved in the Cryogenebank at ICAR-NBPGR, New Delhi (Malik et al., 2010).

### Cryogenebanks Centres

Important world cryobanks besides ICAR-NBPGR, other cryogenebanks include those of National Centre for Genetic Resources Preservation (NCGRP), Fort Collins, Colorado, USA; Association Foret Cellulose (AFOCEL), France; National Institute of Agrobiological Resources (NIAR), Japan' and Institute of Plant Genetic Resources and Crop Plant Research (IPk), Germany.

### (2) Slow-Growth Method

Slow-growth Method is also known as growth limitation approach. Withers (1991) defined Slow-growth Method as:

*"Slow-growth Methods refer to any tissue culture procedure aimed at decreasing the growth rate of plant tissues in vitro".*

This involves reduction in the growth rate of cultures with the aim of increasing the duration of sub-culturing events. The goal of this method is to obtain the longest period of subculture without detrimental effects on the plant tissues. Cells are cultures and stored at non-freezing conditions but their growth rate is curtailed to minimum by manipulation of temperature, media composition and culture environment. The sub-culturing period is extended by 100 per cent. Lowering of temperature, use of osmotic retardants like manitol and sorbitol, partial desiccation, lowering of atmospheric pressure and use of CCC (Cycocel) are the basis of slow growth method. In this method there is no requirement of additional equipment but demands careful imposition of stress so that growth occurs at a little pace and degenerative processes do not in. For example potato meristem, tomato callus and somatic embryos of carrot have survived storage through growth limitation approach. The approach has great commercial application as during lean or off-period propagules can be safely stored and active planting material regenerated on seed basis by removing the growth limitation factor(s).

### Techniques for Achieving Slow-growth Cultures

Slow-growth of cultures can be achieved by using one or a combination for the following techniques:

(i) **Minimal growth at reduced temperature and/ or light intensity:** The most commonly used and single method for restricting the growth of tissues is by decreasing the temperature in combination with decreased light intensity at which cultures are maintained. Cultures that are normally kept at temperature around 25°C exhibit a reduction in growth when transferred to 10-15°C. A basic storage technique can be developed by decreasing the temperature for prolonging the subculture period without significant injuries (Withers, 1985). Tropical plant species are generally cold-sensitive. The physiological damages induced by low temperature are referred to as chilling injury. Therefore, storage temperature depends on the cold-sensitivity of the species.

(ii) **Use of growth retardants in the medium:** Inhibitory level of growth retardants such as Meleic Hydrazide (MH), Abscicic Acid (ABA), N-Dimethyl Succinamic acid (DSA), Cycocel (CCC) and Phosphone D may also be added to the culture medium in a few cases with fairly good success. Although the optimal level of growth retardants will be determined by experimentation on a particular system. It has been generally found that ABA and DSA (5-50 mg/l) and CCA (2-20 mg/l) can be used safely.

(iii) **Decreasing nutrient contents in the culture medium:** Lowering the nutrient contents and/ or sucrose have proved a successful and simple technique to restrict the growth. Kartha *et al.* (1981) could conserve *Coffea arabica* plantlets for two years on a medium devoid of sugar and only half of the mineral solution of the standard medium.

(iv) **Use of osmotic regulators:** Use of osmotic like Manitol, Sorbitol and Sucrose is recommended as they are relatively metabolic inert where the intention is to minimize the growth by imposing a level of osmotic stress on the cultures. Selected osmotic inhibitor is incorporated into fresh culture medium generally in the range of 3-6 per cent w/w.

(v) **Type of enclosures:** It has been observed that type of enclosures of the culture vessel influences the rate of evaporation of water content of the medium. Therefore, by using the propylene caps instead of cotton plugs as enclosure, storage period of the cultures can be increased (Balachandran, *et. al.*, 1990).

(vi) **Size and type of the culture vessels:** Size and type of the culture vessels play very important role in slow growth culture. Roca *et al.* (1984) indicated that when storing cassava plantlets in 50 x 50 mm bottles instead 25 x 15 mm test tubes, the rate of shoot elongation in larger vessel was almost twice, however, leaf fall, diminished and culture viability increased. They also observed that leaves and roots remained healthier in large vessels than in small vessels. Reed (1991) has reported the use of sterile, heat sealable polypropylene bags with the advantage of reduction of occupied surface in culture rooms.

(vii) **Physiological stage of explants:** The type and physiological stage of explants are very important factor in slow growth culture. Minimum size of explants should be selected for conservation. The duration between the last transfer and the moment when cultures are placed in storage condition can be of great importance. It is sometimes better to culture immediately after the transfer, which avoids the necrosis of the tissues and production of phenolic compounds.

(viii) **Reduction in oxygen concentration:** Minimal growth can be achieved by lowering the available oxygen level to tissues of culture containers. The simplest method is to cover the tissues with mineral oil layer. However, some associated problems with decreased oxygen concentration encountered vitrification and partial or complete necrosis. To simplify the technique, cultures are covered with mineral oil, paraffin oil or liquid medium.

**(ix) Modification of gaseous environment:** Some gases like $CO_2$ and ethylene, inside culture vessel also influences the growth rate of the tissues. However, it is complicated and expensive method to control and maintain the gas atmosphere of a large number of individual culture vessels. The development of gas permeable culture vessels used for strawberry (Reed, 1991) opens up the possibility of restricting growth by controlling the gaseous environment of entire culture room which may be more feasible than controlling in individual culture vessel. Although the technique is simple with great potential but still viewed as an experimental technique by virtue of this limitations.

**(x) Combined treatments:** It is true that very little systematic work has been done on judicious use of combination of treatments outlined above may be more useful for *in vitro* conservation, *e.g.*, reduced temperature with a growth retardant has been used successfully for *Solanum tuberosum*. Investigations on the potential of combined treatments rests with the individual worker, and the safest general approach is to evaluate the effects of each of the treatments before attempting to use them in combination.

### In vitro Gene Bank

### (a) Cell and Organ/Shoot-tip Gene Vank

The possible genetic diversity of the particular genetic stock can be maintained using *in vitro* gene bank technique. Several international organizations are engaged potentially to conserve disease tree, clean and elite class of genetic stock and using this process mainly following slow growth, *in vitro* techniques, cryopreservation of several stocks together with volume analysis of genetic diversity. The international plant genetic resources as well as consultative group on International Agricultural Research is heavily involved in the conservation of rare and endangered plant species *in vitro* gene bank (Reed *et al.*, 2004; ICARD, 2014.)

A germplasm collection based on cryopreserved (stored at -196°C in liquid nitrogen) embryogenic cell cultures, shoot-tips and/or somatic/zygotic embryos may be called cell and organ bank. The techniques for cryopreservation of plant cells and tissues are being rapidly refined and used for germplasm conservation, *e.g.*, for potato in Germany. In 2006, ICAR-NBPGR, New Delhi had 7,000 germplasm accessions belonging to 710 species cryopreserved in the form of seeds, embryos, embryonic axes, pollen grains, and dormant buds.

In such gene bank, germplasm is conserved as slow-growth cultures of shoot-tips and nodal segments. Their regeneration consists of sub-culturing the cultures, which may be done every 6 months to 3 years. This approach offers the following chief merits for the conservation of germplasm of vegetatively propagated crops and tree species:

(i) Genotypes of the accessions can be conserved indefinitely free from diseases and pests.

(ii) They can be used for crops, which produce no seed or recalcitrant seeds

(iii) Sub-culture becomes necessary only after relatively long periods (every 6-36 months).

(iv) Regeneration, *i.e.* sub-culturing requires a comparatively very short time.

In addition, cuttings, bulbs and tubers can be maintained under controlled humidity and temperature conditions; however, this approach is practical for short and medium term storage, and it should be used in conjunction with a field gene bank.

### (1) DNA Banks/Genome Resources Conservation DNA Bank

The DNA sequences in the genomes of germplasm accessions are the sources of the genes required in breeding programs. When identified and isolated by cloning, these genes may be used to prepare transgenic organisms that express them. The basic objective in conservation of PGR is conservation of genetic diversity existing in the form of a functional unit called 'gene'. Cloned genes, genomic libraries, and sequence data bases have significant potential uses in germplasm conservation and management.

☆ Article 2 of the CBD defines the term "genetic resources' to mean "any material of actual or potential value of plant, animal, microbial or other origin containing functional units of heredity". In the same logic, genome resources are whole or parts of the genome (DNA) or it functional units (RNA) of actual or potential value. Whereas genetic resources can actually give rise to whole organisms, genomic resources can at best recover traits (directly or indirectly) of these organisms.

☆ With the advancement in recombinant DNA, molecular marker and other technologies, DNAs from the nucleus, mitochondria and chloroplast are now routinely extracted, purified and used in PCR and non-PCR based assays. This advancement has led to the need for conservation of the genetic resources in the form of nucleic acid (DNA or RNA sequences) popularly referred as "DNA Banking".

☆ Current research (both cloning experiments and genome sequence projects) generates a lot of genome resources. These genomic resources are indispensable tools for post-genomic research, be it physiological and morphological characterization of a species or functional analysis of genes or comparative genomics or plant breeding. Therefore, it is essential to maintain an efficient system for conservation and management of spin-off DNA material.

☆ Due to the availability of techniques that help characterize and utilize DNA sequences (without the requirement of whole organism), value added products of gene banks can attract new clients involved in *allele-mining* and *cisgenesis*, such as molecular biologists and geneticists alongside the traditional plant breeders.

☆ DNA storage should be regarded as an insurance policy rather than a replacement for conventional modes for germplasm storage.

☆ ICAR-NBPGR has established a 'National Genomic Resources Repository (NGRR)' for conservation of genomic resources.

*Source: ICAR-NBPGR, Guidelines for Management of Plant Genetic Resources, 2016*

The discovery of DNA endonucleases, automated synthesis of oligonucleotides and DNA sequencing and PCR amplification of DNA have made available a series of molecular markers for unequivocal classification, cataloguing, clustering, DNA fingerprinting of diverse germplasm collections. The whole genome in the form of genomic library or a sequence of DNA in the form of DNA library may be conserved following the appropriate DNA conservation method. Each technology is selected based on its merits in terms of utility, security, complementary and the advantages over the others. The feasibility of storing the genomic information in the form of DNA libraries has been established (Peacock, 1989). In DNA banks, DNA segments from the genomes of germplasm accessions are maintained as cosmid clones, phage lysates or pure DNA (the last one being for relatively short periods). These DNA segments can be evaluated and the desired ones may be used to produce transgenic plants. This approach is applicable to the conservation of genetic materials easily and indefinitely of already extinct species. Since DNA extracted from well preserved herbarium specimens can often be cloned. However, it is very expansive and highly sophisticated. A major demerit of this method is that the existing engineering techniques will not allow for the recovery of the whole genome and also it is not able to identify many genes of interest which are to be subsequently cloned and transferred. By conserving the DNA of a species, adequate genetic representation of that particular species is difficult. However a positive point can be made that the horticultural crops are

### Table 11.11: Comparison of the Three Approaches for *in vitro* Germplasm Conservation

| Sl.No. | Features | Cryopreservation | Slow-growth | DNA Bank |
|---|---|---|---|---|
| 1. | Tissue/organ preserved | Shoot-tip, zygotic and somatic embryo, cells, protoplast | Shoot-tip, slow-growing culture | DNA pieces or phage clones |
| 2. | Storage temperature | -196 °C | 4-9 °C or 15-20 °C | 4 °C |
| 3. | Storage in | Liquid nitrogen | Refrigerators | Deep freezers, Lyophilized state |
| 4. | Metabolic activity | Nil | Slow | Nil |
| 5. | Monitoring during storage | Replenishing liquid nitrogen | Sub-culture at every 6-36 months | Nil |
| 6. | Sophistication | Sophisticated | Less sophisticated | Highly sophisticated |
| 7. | Applicable to | All species amenable to tissue culture | All species amenable to tissue culture | All species |
| 8. | Costs involved | Costly equipment required | Relatively cheap | Very costly |
| 9. | Risks | Loss of viability with time | Contamination | Nil |

### Table 11.12: Essential Observations Related to *in vitro* Genomics

| **Deposition of genomic resources** | |
|---|---|
| Types of genomic resources that can be deposited | Cloning vectors, expression vectors, binary vectors, RFLP probes. |
| | Cloned genes, promoters fused to reporter genes |
| | Sub-genomic, cDNA, EST, repeat enriched libraries |
| | BAC, YAC, PAC clone set from sequencing projects |
| | Genomic, mitochondrial or chloroplast DNA |
| | Cloned DNA from wild and weedy species produced exclusively for the repository. |
| Who can deposit | Only the principal investigator (PI) of a project can deposit the material (*i.e.* co-investigator and research associates need PI's consent. A research student needs the consent of the supervisor guide). |
| | The repository shall not be responsible for any disputes once the material is deposited. Disputed material shall be removed from the distribution pipeline/repository as shall be found appropriate. |
| Storage methodologies | Genomic resources are conserved for one to two years at +4°C; four to seven years at -20°C and greater than five years when stored at -70°C. |

ESTs, full-length cDNAs, BACs, PACs and YACs, should be maintained in 96-well or 384 well microplates at -80°C whereas cDNA clones as plasmid DNA should be maintained at -20°C.

For long-term un-undisturbed storage, lyophilized DNA should be preserved.

All tools and methods should be followed/developed to adopt storage at ambient temperature in order to save energy and environment.

**Quality and quantity of material**

| | |
|---|---|
| Genomic DNA | Every sample of genomic DNA meant for deposition should contain a minimum of 100 µg, which is to be used only for repository and no for distribution. In case distribution is expected replicates should be deposited. |
| | The quality of the DNA should be as follows: A260/A280=1.7-2.0; A262/A230>1.5. |
| | One agarose gel electrophoresis photo with at least one size marker should be submitted. |
| Library | The library should have less than 5 per cent empty vectors and should be free from any contamination. |
| | Libraries should preferably be deposited as amplified libraries @$10^{10}$pfu/ml. No library shall be accepted below $10^6$pfu/ml. |
| | Libraries can be deposited as 384 well plate (BACs) or 96 well plate (cDNA, Shotgun, EST) |

**Accompanying data/voucher**

| | |
|---|---|
| Genomic DNA | Name of the plant, genus, species, common name, family, TaxID, Genebank Accn. No. be accepted below $10^6$pfu/ml. |
| | Extraction procedure, DNA dissolved in, concentration (ng per µL) |
| | Importance of the plant |
| | Reasons for the choice of the variety/landrace/accession |
| | Source of biological material |
| | Clear from MTA for plant material |
| | Information on ITK, if any |
| | Publication (copies to be attached) |
| | IP rights/patents *etc.* (copies to be attached) |
| Cloned DNA/library | Name of the plant, genus, species, common name, family, TaxID, Genebank Accn. No. |
| | Vector, antibiotic marker, cloning sites, fragment size (or size range) |
| | Bacterial strain, transformation method, culture conditions, storage conditions |
| | Importance of the plant |
| | Reasons for the choice of the variety/landrace/accession |
| | Source of biological material |
| | Clear from MTA for plant material |
| | Information on ITK, if any |
| | Publication (copies to be attached) |
| | IP rights/patents *etc.* (copies to be attached) |
| | Supplying seeds/other planting material/herbarium specimen along with the genomic resources |
| | If so, ICAR-NBPGR genebank accession number (if known) |

*Source*: ICAR-NBPGR, Guidelines for Management of Plant Genetic Resources, 2016.

generally conserved in either field gene bank or *in vitro* gene bank where the genetic representation problem is just the same. Hence, it is worth considering the establishment of DNA libraries in view of molecular technological advancement which may enable us to use DNA in genetic resources activities in future (Ford-Lloyd and Jackson, 1991).

Some essential observations related to *in vitro* genomics are as shown in Table 11.12.

In 2006, ICAR-NBPGR, New Delhi had a total of 1,904 accessions of germplasm conserved as slow-growth cultures. All the above mentioned techniques are being utilized at ICAR-NBPGR, New Delhi for conservation of a number of vegetatively propagated species, their wild relatives and elite types for further utilization through optimization of conservation and retrieval protocols.

# Micro-propagation Protocols in Vegetable Crops

Plant tissue culture has been widely used as an experimental system not only to study the basic problem in the physiology of cell growth and differentiation but also in applied area like micro-propagation and genetic engineering for production of transgenic in number of vegetable crops.

## *In vitro* Culture and Plant Regeneration of Solanaceous Crops

### Chilli (*Capsicum annum*)

Efficient micro propagation system in *Capsicum* spp. serves for propagation of elite plants like male sterile plant, plant displaying heterosis *etc.* Capsicum members are recalcitrant to differentiation and plant

**Table 11.13: Micropropagation in Chilli (*Capsicum annum*)**

| Explants | Medium | Response | Reference |
|---|---|---|---|
| Apical/axillaries buds | 3 mg l⁻¹ BA + 0.9 mg l⁻¹ IAA/5 mg l⁻¹ BA + 1.5 mg l⁻¹ IAA | High frequency of plantlet regeneration | Mohammed *et al.*, 1998 |
| Apical/axillaries buds | 3 mg l⁻¹ BA + 1.5 mg l⁻¹ IAA | Rooted plantlets | Gupta *et al.*, 1998 |
| Shoot/shoot tips | 5 mg l⁻¹ BA, 0.5 mg l⁻¹ IAA | Rooted plantlets | Mirza and Narkhede, 1996 |
| Hypocotyls | MS medium | Shoot proliferation | Remirez and Ochoa, 1996 |
| Immature zygotic embryo | 9 μM 2,4-D +10 per cent CW + 8 per cent sucrose | Normal plantlet regeneration | Binzel *et al.*, 1996 |
| Cotyledon | PD+MS inorganic salt+ ZH organic + 5 BA+0.2 mg l⁻¹ IAA+ 2AgNO₃ | High frequency of plantlet regeneration | Zhou *et al.*, 1994 |
| Shoot/shoot tips | B 5+300 mg l⁻¹ chloramquet or B5 + 50μM thiophnate methyl | 4.78 million plants/year | Ma *et al.*, 1991 |
| Node/axillary bud | Half strength MS medium | Rooted plantlets | Yamamoto, 1994 |
| Shoot/shoot tips | MS basal medium | Normal plantlet regeneration | Ezura *et al.*, 1993 |
| Cotyledon | 2.5 mg l⁻¹ TDZ + 0.1 mg l⁻¹ IAA + 2 mg l⁻¹ GA₃ | Plantlet regeneration | Arce and Lopez, 1994 |
| Node | 5 mg l⁻¹ TDZ | Increased productivity | EL-Zeiny-OAH, 2002 |
| shoot meristem explants | 0.22 mM l⁻¹ BA + 0.48 mM l⁻¹ IAA | *in vitro* plant propagation | Peddaboina P. *et al.*,2006 |

regeneration. However, in recent past numerous studies have been done on micro propagation *via* shoot organogenesis induced on explants prepared from seedling at different stages. In few cases shoot regeneration on a medium devoid of any growth regulator has been observed (Ezura *et al.*, 1993; Binzel *et al.*, 1996). Usually, in all papers cited thereafter, micro propagation has been induced and accomplished by exogenous growth regulators supplied to the medium. The maximum utilized explants consist of half seed explants (Rogozinska and Orozdowska, 1996), cotyledon and hypocotyls (Ramirez and Ochoa, 1996). Commonly explants were placed on to a semisolid shoot induction medium supplemented with cytokinin (BA or TDZ) and often an auxin (IAA, IBA) (Arec and Lopez, 1994). Many a times shoot regeneration was achieved by transferring the bud cluster to an elongation medium in *vitro* because, shoot elongation is a problem in chilli cultures (Steinitz *et al.*, 1999). Gibberelic acid is usually added to the elongation medium. Zhou *et al.* (1994) used PD medium with MS inorganic salts and silver nitrate to achieve rapid regeneration of *Capsicum* cultures. In most of the studies the number of plants regenerated per explant was 1⁻¹⁰. However, Steinitz *et al.* (1999)

reported several problems with micro propagation of chilli plants. Out of the 40 genotypes studied by Steinitz and his group only four genotypes has consistent regeneration capability. Many times buds develop either from leaf structure become stunted and forms aberrant shoot. A very successful micropropagation system has been reported by Ma *et al.* (1991), where they produced 4.78 million plants by micropropagation based on multiplication rate 9: 1. In an interesting study of *in vitro* propagation of chilli hybrids, EL-Zeiny (2002) found TDZ best in stimulation of lateral bud break and increasing shoot branching. Plants derived by this method produce plantlets that were shorter, thicker, had more chlorophyll, bigger leaf area, and gave higher yield than those derived by seed sowing.

## Tomato (*Solanum lycopersicum* syn. *Lycopersicon esculentum*)

Tomato is a major vegetable crop that has achieved tremendous popularity over last century. It is grown in almost every country of the world, and ranks first amongst the important vegetables crops in many countries. It is grown in both open field and greenhouse condition for commercial and home use. The *in vitro*

**Table 11.14: Micropropagation in Tomato (*Solanum lycopersicon*)**

| Explant | Medium | Response | Reference |
|---|---|---|---|
| Cotyledon | 2.5 mg l⁻¹ NAA + 1.5 mg l⁻¹ Kin | Normal plantlet regeneration | Jawahar *et al.*, 1998 |
| Seedling leaf | 20 μM BA + 1 mg l⁻¹ NAA | True to type plantlets | Geetha *et al.*, 1998 |
| Node/shoot/shoot tips | MS major salts + B5 | Flowering plantlets | Gamburg *et al.*, 1997 |
| Hypocotyl | BA | Shoot proliferation | Newman *et al.*, 1996 |
| Apical shoot | Various regulators | True to type plantlets | Mirghis *et al.*, 1995 |
| Cotyledon | 0.1 mg l⁻¹ Zeatin | Shoot proliferation | Ichimura and oda, 1995 |
| Hypocotyl/Cotyledon | 1 mg l⁻¹ NAA + 1 mg l⁻¹ Zeatin | Normal plantlet regeneration | Ye *et al.*, 1994 |
| Cotyledon | 6 mg l⁻¹ IAA + 5 mg l⁻¹ Kin | High yielding rapid multiplication | Farash *et al.*, 1993 |
| Shoot/shoot tips | 4 mg l⁻¹ BA + 4 mg l⁻¹ IAA | Normal plantlet regeneration | DeVilliers *et al.*, 1993 |

culture of tomato tissues and organ is highly advanced now. It is evident from many reports on tomato transformation. However, limited reports are available on shoot tip culture or axillary branching method of micropropagation in tomato. Tomato can produce a number of shoots without exogenous supply of any plant growth regulators after removal of apical dominance *via* decapitation. Decapitation has been repeated several times to achieve multiplication (Fari *et al.*, 1992). Many auxin and cytokinin combinations could induce shoot proliferation in tomato, including BA and NAA, or zeatin with NAA or IAA resulted in shoot development. In last few years many reports on successful field transfer of micropropagated tomato plants are available, Geetha *et al.* (1998) used seedling explants and the multiplication was achieved on MS medium with 20 μM BA. Hypocotyl and cotyledonary explants are the most preferred explant in tomato micro propagation (Table 11.14). A shoot regeneration protocol has been designed by Gamburg *et al.* (1997), where nodal segments and shoot tips were used and gave rise true to type tomato plants bearing normal flowers. The variation occur where callus culture takes place but Mirghis *et al.* (1995) produced true to type, normal plantlets using callus culture with apical and axillary buds.

## Brinjal (*Solanum melongena* L.)

Eggplant along with tomato is one of the most studied systems in vegetable tissue culture. It produces very good material for tissue culture studies as it can be cultured in semisolid as well as liquid medium without loss of regeneration ability. Like other solanaceous vegetables, in brinjal also hypocotyl and cotyledon are the most utilized explants. Sharma *et al.* (1995) succeeded in regenerating eggplant on MS medium with 11.1 μM BA and 2.9 μM IAA (Table 11.15). Bhansali and Ramavat (1993) obtained shoot proliferation of *S. melongena* and by placing axillary buds on Woody Plant medium with 0.1 mg l$^{-1}$ BA.

## *In vitro* Culture and Plant Regeneration of Cucurbitaceous Crops

The fruits of various cucurbits are popular throughout the world and are eaten as raw (salad, pickles or fruits) or cooked (vegetable). The *in vitro* multiplication techniques are extremely useful for cucurbit family, as the family contains many dioecious species and gynoecious lines. Gynoecious lines are being used in commercial cultivation and hybrid seed production. In order to develop gynoecious inbred lines for use in hybrid breeding programme, individually

selected plants must be propagated in sufficient numbers for crossing. The practice of use of cuttings is slow, inconvenient, labour intensive, requires lot of vines, and also not feasible for all Cucurbit species. Therefore, the micropropagation technique holds promise for large-scale production of such precious material, and is now being used to multiply the quality materials (Kumar *et al.*, 2001, 2002). In most of the studies with cucurbit species MS medium has been widely used. A number of reports have been described for different type of regeneration in various species (Table 11.16). Kathal (1988) described adventitious shoot formulation from cotyledon explant on a medium containing 1 μM BA and 2ip each. For this study seeds were germinated in *vitro* and cotyledons of 0.3- 1.3 cm from seedling were excised on different interval. Rooting was done on MS medium containing 0.2 mgl$^{-1}$ BA or 2.0 mg l$^{-1}$ IAA. Bud explant was used from triploid watermelon. Seedlings with their cotyledon were removed. 93.33 per cent plantlet regeneration was achieved on MS medium with 3 mg l$^{-1}$ BA. In this way 80,000 plants were produced in 6 month involving six generations of sub culturing. As mentioned earlier many researchers are using micro propagation for two dioecious species and reports are available in *Coccinia* and pointed gourd. Micro propagation using apical and axillary bud explants (vines) has been achieved in both the genotypes and studies in several other species are in progress (Kumar *et al.*, 2002).

## Cucumber

*In vitro* multiplication of cucumber (*Cucumis sativus*) is well advanced. Reports are available on micro propagation through leaf cotyledon, apical meristem and nodal portions (Table 11.16). A number of cucumber lines were tested for micro propagation conditions by Navratilova (1987). Apical meristem was excised from 9-daya old plants and cultured on to MS medium with 0.01 mg l$^{-1}$ BA and 0.01 mg l$^{-1}$ Ascorbic acid. Szymanska (1995) reported direct regeneration of plantlets from leaf explants on MS medium supplemented with 0.5 mg l$^{-1}$ IAA, 2 mg l$^{-1}$ BA and 1 mg l$^{-1}$ AlCl$_3$. Burza and Malepszy (1995) used young leaves to develop rapid and efficient regeneration in cucumber. Completely normal plants were produced with normal flowers after four weeks of culture.

## Watermelon

Micropropagation has also been considered as an alternative method for propagation of extensive hybrids *e.g.*, triploid (seedless) watermelons are now commercially produced though micropropagation.

### Table 11.15: Micropropagation Reports in Brinjal (*Solanum melongena*)

| Explant | Medium | Response | Reference |
|---|---|---|---|
| Leaf explant | 2.9 μM Kin | High frequency of plantlet regeneration | Raghuveer *et al.*, 1994 |
| Cotyledon | 3.5 mg l-1 IAA + 0.7 mg l-1 Zeatin | Shoot formation | Brancha *et al.*, 1990 |
| Cotyledon, Hypocotyls, leaf | 11.1 μM BA + 2.9 μM IAA | Plantlet regeneration | Sharma *et al.*, 1995 |

## Table 11.16: Micropropagation Reports in Cucurbit Vegetables

| Explant | Medium | Response | Reference |
|---------|--------|----------|-----------|
| **Cucumber (*Cucumis sativus*)** | | | |
| Leaf explant | 0.5 mg l⁻¹ IAA + 2.0 mg l⁻¹ BA + 1.0 mg l⁻¹ AlCl₃ | Direct organogenesis | Szymanska and Molas, 1995 |
| Cotyledon/leaf/node | 3.0 mg l⁻¹ BA | Field transfer of plants | Mishra and Bhatnagar, 1995 |
| Cotyledon/leaf/node | 9.0 per cent Sucrose | Somatic embryo | Lou and Kako, 1994 |
| Apical meristem | 0.01 mg l⁻¹ IBA + 0.1 mg l⁻¹ BA + 10.0 Ascorbic acid | Regenerated plants | Kathal *et al.*, 1994 |
| **Pumpkin (*Cucurbita moschata*)** | | | |
| Shoot tip/axillary bud | 1 mg l⁻¹ BA + 0.5 mg l⁻¹ IAA | Direct organogenesis | Zhao *et al.*, 1998 |
| Shoot tip/node | 1 mg l⁻¹ BA + 0.1 mg l⁻¹ NAA | Field transfer of plants | Islam *et al.*, 1992 |
| Shoot tip/node | 1 mg l⁻¹ BA | Field transfer of plants | Seo *et al.*, 1991 |
| **Ivy gourd (*Coccinia cordifolia*)** | | | |
| Leaf/Shoot tip/node | 1.5 mg l⁻¹ BA + 0.5 mg l⁻¹ Kin + 0.01 mg l⁻¹ IBA | Easy rooting field transfer of plants | Joskutty *et al.*, 1993 |
| Shoot tip/node | | Field transfer of plants | Kumar *et al.*, 2002 |
| **Pointed gourd (*Trichosanthes dioica*)** | | | |
| Leaf/Shoot tip/node | 1.5 mg l⁻¹ BA + 0.5 mg l⁻¹ Kin + 0.01 mg l⁻¹ IBA | Easy rooting field transfer of plants | Hoque *et al.*, 1997 |
| Shoot tip node | 2 mg l⁻¹ BA | Field transfer of plants | Mythili and Thomas, 1998 |
| Shoot tip/nodal portion | 8.88 µM BA | Field transfer of plants | Kumar *et al.*, 2001 |
| **Watermelon (*Citrullus lanatus*)** | | | |
| Axillary buds | 2.0 mg l⁻¹ BA + 0.1 mg l⁻¹ IAA | Multiple shoot | Spestidis 1996 |
| Seedling buds | 3.0 mg l⁻¹ BA | Field transfer of plants | Tang *et al.*, 1994 |
| Root explants | 1.0 mg l⁻¹ BA | Field transfer of plants | Kathal *et al.*, 1994 |
| Cotyledon | 2.0 mg l⁻¹ BA | Plants with quality fruits | Song *et al.*, 1998 |
| **Bitter gourd (*Momordica charantia*)** | | | |
| Nodal segments | 22.0 mg l⁻¹ BA + 0.2 mg l⁻¹ NAA | Plant regeneration, multiplication | Sultana and Miah, 2003 |
| Meristem | 1.0 mg l⁻¹ BA + 0.1 mg l⁻¹ IBA + 0.3 mg l⁻¹ GA3 | Apical Meristem culture | Huda and B. Sikdar, 2006 |
| **Spine gourd (*Momordica dioica*)** | | | |
| Immature embryo explant (female x female cross) | 10.8 mg l⁻¹ BA, 1.08 mg l⁻¹ NAA and 0.54 mg l⁻¹ GA | rapid and prolific shoot regeneration | Hoque *et al.*, 2007 |
| Node, shoot tip, leaf and the cotyledon, | 1.0 mg l⁻¹ BAP and 0.1 Mg l⁻¹ NAA | Oranogenesis and propagation | Nabi *et al.*, 2002 |

### Squash

The micropropagation in squash has been reported by Chee (1991) through shoot apices. Seedling derived shoot apices were cultured on MS medium with 0.8 mg l⁻¹ BAP and 0.1 mg l⁻¹ kinetin and the regenerated plantlets were successfully transferred to soil.

### Pumpkin

An efficient micropropagation protocol has been developed in pumpkin, which could by exploited commercially. Sheo *et al.* (1991) and Zhao *et. al.* (1998) reported micropropagation using shoot tip/axillary buds, the multiple shoot were induced on the MS medium with 1.0 mg l⁻¹ BA. In both the cases large number of plantlets were transferred to soil. Shoots of cucurbitaceous species regenerate directly or indirectly from explant, rooted easily when cultured *in vitro* on MS basal or half-strength MS basal medium, with 3 per cent sucrose. In most of the species there is no need of growth regulators but in some cases, a low level of auxin

(0.01 NAA/IAA) was used to induce rooting in painted gourd (Mythili and Thomas, 1999).

### Bitter Gourd

In bitter gourd two year old calluses derived from cotyledons were subjected to NAA (0-1mg/litre) + adenine (13.5-67.5 mg/litre) and IBA+BA (both 0.25-10.0 mg/litre) treatments. The NAA (1.0 mg/litre) + adenine (33.75 mg/litre) induced buds in this crop. Fresh hypocotyls callus cultures did not induce into buds and plantlets, but 23 month old cultures could be induced to buds and plantlets by the addition of NAA (1.0 mg/litre) and adenine (33.75 mg/litre) (Halder and Gadgil, 1982).

### Bottle Gourd

In bottle gourd [*Lagenaria siceraria* (Mol.) Standl.] morphological observations during several plantings of the collected seed, as well as results from two genetic analyses (random amplified polymorphic DNA and chloroplast sequencing), indicate that the Zimbabwe collection is part of a genetically distinct and wild lineage of *Lagenaria siceraria* (Mol.) Standl. Similarly, Walters *et al.* (2001) assessed diversity in landraces and cultivars of bottle gourd using RAPD. Total 64 RAPD markers representing 30 primers were examined in 31 landraces and 43 cultivars accessions of *Lagenaria siceraria* and a wild relative *Lagenaria sphaerica*. An *in vitro* study was carried out by Ahmad *et al.* (1997) with bottle gourd (*Lagenaria siceraria* (CV. Pusa Naveen) to regenerate plantlets directly by organogenesis and indirectly by callogenesis with the aim of standardizing a protocol which could be used further for screening of somaclonal variants. Cotyledon, hypocotyls, petiole, leaf and shoot tip explants collected from 10 day old *in vitro* grown seedlings were cultured on Murashige and Skoog medium supplemented with benzyladenine (BA, 0.1-5.0 mg/litre), NAA (0.01-1.0 mg/litre) and 2,4-D (0.1-5.0 mg/litre). Callus development was best in hypocotyls explants (followed by cotyledon and petiole explants) cultured on medium supplemented with 0.4 mg/litre, 2, 4-D mg/litre. Leaf explants showed very poor callusing. Shoot bud differentiation occurred on calluses subcultured on medium containing NAA (0.1 mg/litre) and BA (2.5 mg/litre). Shoots induced from the basal end of the cotyledon rooted on medium containing NAA (0.1 mg/litre) and complete plants were obtained. In shoot tip explants there was no callus formation and shoot regeneration occurred in 10-15 days on medium supplemented with NAA (0.05 mg/litre) and BA (0.1 mg/liter). These shoots rooted after 15-20 days on the same medium and complete plantlets were obtained.

## Pointed Gourd

Rapid *in vitro* multiplication of Swarna Rekha and Swarna Alaukik was achieved at ICAR-IIHR, Bengaluru by culturing shoot tip and nodal explants on Murashiage and Skoog medium containing indol-3-acetic acid (1.0µM) and indol -3- butyric acid (0.2µ M). Results revealed that Swarna Alaukik showed highest rate (6.6 cuttings per plant) followed by Swarna Rekha (4.9). the basal rooted stumps and plantlets from one week of culture were acclimatized and established *ex vitro* with 75-95 per cent efficiency in different genotypes (Krishna Prasad, 2006). Female sex associated RAPD marker in pointed gourd (*Trichosanthes dioica*) was studied by Singh *et al.* (2002). RAPD marker associated with females and absent in males of pointed gourd was identified. The importance of this finding is the early detection of gender as well as possible implication in understanding the molecular basis of sex determination in this dioecious species. Hoque *et al.* (1998) conducted an experiment to study the *in vitro* culture of pointed gourd (*Trichosanthes dioica*) and ivy gourd (*Coccinia indica*). Cotyledons rescued from physiologically mature seeds (PMS) immature seeds (IMS) of pointed gourd were used as explants. Cotyledons excised from PMS responded very well in all culture conditions. Plant regeneration from these Cotyledons ranged from 62 to 92 per cent in different combinations of media. It was highest in MS +

1 mg/litre BA and lowest in hormone free MS medium. The establishment rate of regenerated plantlets ranged from 61 to 94 per cent in vermiculture and 70 to 90 per cent in soil medium. Plantlets obtained from MS + 1 mg/litre NAA performed best in both media. Mythili and Thomas (1999) works with rapid *in vitro* multiplication of 2 female cultivars (Swarna Alaukik and Swarna Rekha) and one male line of pointed gourd (*Trichosanthes dioica*) was achieved by culturing shoot tip and nodal explants on Murashige and Skoog medium containing IAA (1.0 µM) and IBA (0.2µM). Single shoot with short internodes (1.0-1.05 cm) accompanied by rooting were obtained in all genotypes within 4 weeks. Shoots were subcultured at 4-week intervals excluding the basal 3 nodes. Genotypes differed significantly with respect to multiplication rate. Swarna Alauk showed the highest rate (6.6 cuttings per explant) followed by the male line (5.3) and Swarna Rekha (4.90). The basal rooted stumps and planlets from 1 week of culture were acclimatized and established *ex vitro* with 75-95 per cent efficiency in different genotypes. In Kakrol (*Momordica dioica* Roxb.), an *in vitro* plant regeneration study was conducted. The combination of 1 mg/liter BAP (Benzylaminopurine [benzyladenine]) and 0.1 mg/liter NAA in ms media was found most suitable in callus induction followed by 0.2 mg/liter BAP. The combination of 1 mg/liter BAP and.01 mg/liter NAA produced the highest number of multiple shoots from the callus. Half strength ms medium supplemented with IBA showed better performance than with IAA for root number but the latter produced tallest shoot. Four types of explants, node, shoot tip, leaf and cotyledon were used but only the cotyledon exhibited the best performance. Plantlets were successfully exhibited in normal soils (Nabl, 2002).

## *In vitro* Regeneration of Okra

Okra is recalcitrant to *in vitro* regeneration. For testing, surface sterilized seed of cv. Kashi Kranti grown *in vitro* on half strength of Murashige and Skoog medium. For regeneration purpose hypocotyl, cotyledon and meristamatic tissues were selected as explants from 5-15 days old seedlings. More than 112 combination of media supplemented with different concentration of plant growth regulators. Plant growth regulators BA, 2iP, 4-CPPU, and TDZ in combination with IAA ranging 0.1-5.0 mg/l. In all the tested combination only callus and root were induced and only few (1-2) shoot induction was reported in any combination on hypocotyls and cotyledons explant. Callus induced in all combination IAA, BA, and 2iP were light green, globular and friable while in NAA, TDZ and CPPU were dark green, compact and larger in size in comparison to both BA and 2iP. On all the combination of medium evaluated for regeneration, on hypocotyl and cotyledon explants induce either callus or root or both no any combination induced shoot. The IAA in combination with BA and 2iP induced light green, globular and friable callus while on the combination of IAA with

TDZ and 4-CPPU induced dark green, nodular and compact callus. The callus induction frequencies were higher than BA and 2iP in TDZ and 4-CPPU. The roots induction frequencies were higher in the combination supplemented 0.1-0.2 mg/l IAA and 1.0-2.0 BA or 2iP, the rooting were prominent in 2iP supplemented medium than BA. Higher concentration of plant growth regulators than the above induces callus of different characteristics (Ranjan *et al.*, 2016).

### *In vitro* Culture and Plant Regeneration of Cole Crops

The genus *Brassica* includes several field crops and a great diversity of vegetables. The successful reports on micropropagation have been given in (Table 11.16). Vegetative propagation of *Brassica* is limited to its use in plant breeding but because the several species are self-fertile and some self-incompatible its application vary for this purpose. The growth and manipulation of *Brassica* are slightly easy *in vitro* and now the methods are available to propagate many genotypes. The rapid micropropagation method for selected plants is already providing advantages in breeding programmes. $F_1$ varieties of *Brassica* vegetables are popular for micropropagation. The production of hybrid seeds is much costlier, because the *Brassica* are generally self-incompatible and the maintenance of parental lines is done by tissue culture.

In recent past many workers have achieved direct regeneration from meristematic tissues and curd pieces. Mainly the micro propagation work has been limited to maintenance of male sterile lines, self-incompatible lines *etc.* In case of cauliflower, the most responsive explant reported is curd pieces. Vandermoortele *et al.* (1991) achieved multiple shoots from curd pieces on MS medium containing 2.5 µM BA, 5 µM NAA and 168 µM sucrose. Another very successful report is by Borovec (1988) where he achieved 95 per cent survival of cauliflower plants after transfer to soil. There are efficient regeneration systems available for Indian cauliflower to multiply plants using different explants like hypocotyl,

cotyledon (Narsimhulu and Chopra (1983), Singh *et al.* (2002). The cabbage has been mostly utilized for genetic transformation but not for large-scale multiplication, which is evident from few reports on micro propagation (Table 11.17). Souza *et al.* (1998) successfully regenerated cabbage plants from axiallary buds on MS medium supplemented with 28.5 µM IAA, 118.58 µM kinetin and 0.87 µM TDZ, and transferred the plantlets to soil with high survival rate.

### *In vitro* Culture and Plant Regeneration of Bulb Crops

The bulb crops Onion (*Allium cepa*) and Garlic (*Allium sativum*) are most common vegetable used as food, salad or spices. As the onion and garlic both are vegetatively propagated crops, the main purpose is to generate variability initially. There was not much attention on multiplication of garlic through biotechnological means. Now in recent past many studies have been reported on micro propagation aspect that too for producing virus free stocks. The most utilized explants in garlic are cloves and immature bulbs. It has been reported that up to 60,000 plantlets could be obtained from single bulb in one year (Yaseen *et al.*, 1994).

### *In vitro* Culture and Plant Regeneration of Rhizomatous Crops

In ginger MS medium supplemented with different combinations and concentrations of hormone were studied to obtain a suitable protocol of plantlet regeneration. Three explants of two varieties of ginger *viz.*, Suruchi and BARI Ada-1 were cultured on MS medium supplemented with 0.5 mg/l Dicamba, 0.75 mg/l Dicamba and 1 mg/l, 2,4-D. Assessments on callus induction were studied through five quantitative traits such as days required for callus initiation, size of callus, color of callus, nature of callus and percentage of callus induction. Suruchi showed 62.64 per cent callus induction, 63.98 per cent shoot induction and 68.76 per cent root induction. Leaf explant gave best

**Table 11.17: Micropropagation Reports in Cole Crops**

| Explant | Medium | Response | Reference |
|---------|--------|----------|-----------|
| **Cauliflower (*B. oleracea* var. *botrytis*)** | | | |
| Hypocotyl/Curd pices/ branches | 2.44 µM BA + 5.0 µM NAA + 168 µM Sucrose | Multiple shoots | Vandermoortele *et al.*, 1999 |
| Curd meristem | 2.0 Kin Liquid | Field transfer of plants | Keiffer *et al.*, 1995 |
| Head explant | 1.0 mg l$^{-1}$ NAA + 0.1 IAA mg l$^{-1}$ | Plants with high survival rate | Borovec, 1988 |
| Internode and inflorescence axis | 5.0 mg l$^{-1}$ BA | Regenerated Plants | Singh,1988 |
| Cotyledon | Various combinations | Shoot bud formation | Narsimhulu and Chopra, 1988 |
| Leaf and Stem | 8.88 µM BA | Regenerated plants | Singh *et al.*, 2002 |
| Cabbage (*B. oleracea* var. *capitata*) | | | |
| Axillary buds | 28.5 µM IAA + 18.58 µM Kin + 0.87 µM TDZ | Regenerated plants, field transfer | Souza *et al.*, 1998 |
| Shoots | 0.5 mg l$^{-1}$ BA + 0.1 mg l$^{-1}$ IBA | Regenerated plants, field transfer | Sretenovic *et al.*, 1997 |

**Table 11.18: Micropropagation in Bulb Crops**

| Explant | Medium | Response | Reference |
|---|---|---|---|
| **Onion (*Allium cepa*)** | | | |
| Immature flower/ovary | 1.0 mg l⁻¹ NAA + 2.0 mg l-1 BA + 0.02 mg l⁻¹ TDZ | Large no. of plantlets through direct organogenesis | Gems and Martinovitch, 1998 |
| Immature flower bud | BDS + 2.0 mg l⁻¹ BA + 2.0 2,4-D + 100 mg l⁻¹ myoinositol + 10 per cent sucrose | High frequency of plantlets | Jeong *et al.*, 1998 |
| Shoot primordial | 1.0 mg l⁻¹ 2, 4-D + 0.5mg l⁻¹ picloram | High frequency of plantlets | Jeong *et al.*, 1997 |
| Sliced basal plate | 4.44 µM BA | Shoot proliferation with no variation in regenerated plants | Barringer *et al.*, 1996 |
| **Garlic (*Allium sativum*)** | | | |
| Shoot tip | 0.3 mg l⁻¹ IAA + 2.0 mg l⁻¹ 2ip | Plantlet formation | Yun *et al.*, 1998 |
| Leaf explants | 2 Kin + 0.5 IAA mg l⁻¹ | High frequency of plantlets through globular bodies | Zheng *et al.*, 1998 |
| Meristems | 2 mg l⁻¹ ip + 5 mg l⁻¹ jasmonic acid | High frequency of plantlet regeneration | Uchman *et al.*, 1998 |
| Root and leaves | 5 mg l⁻¹ IAA | Somatic embryogenesis | Backish *et al.*, 1997 |
| Root explants | 1 mg l⁻¹ NAA + 6 mg l⁻¹ BA | High frequency of plantlet regeneration | Haque *et al.*, 1997 |
| Bud and basal plate | B5 + pH7.5 + 15°C | Bulblet formation | Liu *et al.*, 1996 |
| Meristem | MS + B5 basal | Microbulbils | Hernandez *et al.*, 1994 |
| Shoot tips | MS + 0.5 per cent ascorbic acid + 12 per cent sucrose | Commercial low cost bulbils | Yaseen *et al.*, 1994 |
| Clove shoot | 0.05 mg l⁻¹ NAA | Virus free plantlets | Masuda *et al.*, 1994 |
| Scale tip | 0.05 mg l⁻¹ NAA + 0.05 mg l⁻¹ Kin | High frequency of Virus free plantlets | Ma *et al.*, 1994 |

result over shoot tip and root explant. Leaf explant produced 67.07 per cent callus, 67.77 per cent shoot and 66.93 per cent root. MS medium supplemented with 0.5 mg/1 Dicamba was the best (70.20 per cent) for callusing, MS +1.0 mg/1 Kn + 1.0 mg/1 BAP best (72.03 per cent) for shooting and Ms + 1 mg/1 IBA gave best (66.43 per cent) result for rooting over other treatments. The highest (73.60 per cent) callus induction was found from the leaf x Suruchi interaction. Leaf x Suruchi gave 74 per cent shooting and 74.13 per cent rooting. Per cent callus induction was maximum (76.47 per cent) with Suruchi x 0.5 mg/1 Dicamba interaction and it was significantly different from all other values. Per cent shoot induction were highest (76.33 per cent) by the interaction of Suruchi x 1.0 mg/1 Kn +1.0 mg/1 BAP. Highest (76.87 per cent) percentage of root was produced by 1 mg/1 IBA x Suruchi interaction. Highest percentage of callus induction (87.60 per cent) was obtained from leaf explants of Suruchi and 0.5 Dicamba interactions. The interaction of Suruchi x leaf x 1.0 mg/1 Kn + 1.0 mg/1 BAP produced highest (87.60 per cent) percentage of shoot. Root induction was best (85.40 per cent) from Suruchi x leaf x MS + 1 mg/1 IBA interaction. The regenerated plantlets were successfully established into pot after proper hardening (Azra *et al.*, 2009).

India has been known as the 'land of spices' from time immemorial. Among the spices, ginger (*Zingiber officinale* Rosc.) contributes greatly towards human health and is regarded as a food medicine for several

ailments. The major limitation in increasing production and productivity of ginger is lack of adequate disease.-free planting materials of high yielding varieties. As the major diseases are spread through contaminated seed-rhizomes, the possibility of producing pathogen-free planting materials using tissue culture is attractive. Therefore, the present investigation was undertaken to standardise a rapid, efficient, and reliable regeneration protocol for *in vitro* propagation of a high yielding ginger, cv. Suravi, collected from the high altitude research station at Pottangi (Koraput, a tribal district of Odisha), India. The axillary bud (0.2–0.5 mm size) from the sprouted rhizome was taken as the explant. The most ideal surface sterilant was found to be 0.1 $HgCl_2$ for 13 min, which reduced the total infection (fungal + bacterial) significantly to 3.3 per cent and took shortest time for bud emergence (9.3 days) in standard Murashige and Skoog (MS) medium. The extent of survival (96.7 per cent) and production of buds per explant (2.7) were maximum with this sterilant. MS medium supplemented with 3.0 mg/L benzyl amino purine (BAP) and 0.4 mg/L naphthalene acetic acid (NAA) was ideal for shoot proliferation and resulted in maximum number of total shoots from a single explant (36.0), maximum shoot length (6.1 cm) with 4.7 leaves after a second sub-culturing. For rooting, MS supplemented with NAA (0.5 mg/L) was found to be more effective and produced the maximum number of roots per shoot (13.3) and the maximum root length (2.0 cm) plus taking the least time for root initiation (10.3 days). The *in vitro* plantlets were prehardened in ½ MS liquid medium. The hardening and acclimatization media mixture of soil: sand: farm yard manure (1:1:1) was found to be best for survival of the plantlets in ginger (Jagadev *et al.*, 2006).

*In vitro* microrhizome production was obtained in turmeric (*Curcuma longa* Linn.). Freshly sprouted buds with small rhizome portions excised from stored mature rhizomes were cultured on semi-solid culture initiation medium — MS basal medium + 0.88μ M BAP (6-benzylaminopurine) + 0.92 μ M kinetin + 5 per cent coconut water + 2 per cent sucrose + 0.5 per cent agar — resulting in bud elongation. Multiple shoots were produced from these elongated buds by culturing in liquid shoot multiplication medium –- MS basal medium + 2.2μ M BAP + 0.92μ M kinetin + 5 per cent coconut water + 2 per cent sucrose — at 25±1°C and 16-h light (at 11.7 μ mol m$^{-2}$ s$^{-1}$)/8-h dark cycles. Clumps of four to five multiple shoots/single shoots were used in various experiments. Cultures were incubated in the dark at 25±1°C. Half strength MS basal medium supplemented with 80 g l$^{-1}$ sucrose was found to be optimal for microrhizome production. Cytokinin BAP had an inhibitory effect on microrhizome production. At the highest concentration of BAP tried (35.2 μ M) microrhizome production was totally inhibited. Microrhizome production depended on the size of the multiple shoots used. Microrhizomes produced were of a wide range in size (0.1–2.0 g) and, readily regenerated when isolated and cultured *in vitro* on culture initiation medium or shoot multiplication medium. Under *in vivo* conditions, small (0.1–0.4 g), medium (0.41–0.8 g) and big (>0.81 g) microrhizomes regenerated. Plantlets developed from big microrhizomes grew faster (Mrudul *et al.*, 2001). Plant regeneration from cultured immature inflorescence of *Curcuma longa* was obtained by direct shoot development on Murashige and Skoog's basal medium supplemented with BA (5 or 10 mg l$^{-1}$) in combination with 2, 4-D (0.2 mg l$^{-1}$) or NAA (0.1 mg l$^{-1}$) and TDZ (1 or 2 mg l$^{-1}$) in combination with IAA (0.1 mg l$^{-1}$). Regenerated shoots were grown on MS medium for further development and later transferred to medium supplemented with 0.1 mg l$^{-1}$ NAA for induction of roots. Complete plants thus obtained were transferred to sterilized soil in paper cups for 3–4 weeks and then to field, where 95 per cent of the plants survived to maturity.

A simple procedure has been outlined for plant regeneration of *Amorphophallus albus* Liu and Wei, a native medicinal plant in China, from petiole-derived callus. Calli were induced at a high frequency of 76.4±3.2 per cent from petiole explants excised from two-month-old plants on Murashige and Skoog (MS) medium supplemented with 5.37 μM α-naphthaleneacetic acid (NAA) and 4.44 μM 6-benzyladenine (BA). Of the different types of callus induced, type III callus was selected for morphogenesis induction. Culture of the callus on MS medium containing proper NAA and BA or KT combinations resulted in formation of corm-like structure (CLS) that produced shoots and roots during further culture. The optimal morphogenetic response was observed on the media with a cytokinin/auxin ratio of about 4:1, which resulted in more than 70 per cent CLS formation and 6~8 CLSs per callus. Complete plantlets with well-developed root systems were obtained from these CLSs by sub-culturing them on the original media from which they had been derived without a separate rooting culture. Transfer of the plantlets with roots to soil resulted in a more than 90 per cent survival rate. Analysis of 20 regenerated plants by two molecular markers, randomly amplified polymorphic DNA (RAPD) and inter simple sequence repeat (ISSR), revealed somaclonal variation in the regenerated plants. The percentage of polymorphic bands in RAPD and ISSR analysis were respectively 20.8 per cent and 39.0 per cent for the 20 plants. Cluster analysis indicated that the genetic similarity values calculated on the basis of RAPD and ISSR data among the 21 plants (20 regenerated and one donor plant) were, respectively, 0.973 and 0.917, which allowed classification of the plants into distinct groups. A high-frequency somaclonal variation induced in *A. albus* tissue culture may help in the selection of useful variants that may be induced to improve this important crop (Neeta *et al.*, 2004). A protocol was developed for Agrobacterium-mediated transformation of embryogenic suspension cultures of cassava. The bacterial strain ABI containing the binary vector pMON977 with the nptII gene as selectable marker and an intron-interrupted uidA gene (encoding β-glucuronidase) as visible marker was used for the experiments. Selection of transformed tissue with paromomycin resulted in the establishment of antibiotic-resistant, β-glucuronidase-expressing lines of friable embryogenic callus, from which embryos and subsequently plants were regenerated. Southern blot analysis demonstrated stable integration of the uidA gene into the cassava genome in five lines of transformed embryogenic suspension cultures and in two plant lines (Jianbin *et al.*, 2008).

## Photo Autotrophic Propagation (PAP)

Photo autotrophic propagation (PAP) is another system that should be used for root and tuber crops. In this system, explants are directed towards autotrophy while in culture by reducing or completely substituting sucrose with carbon dioxide. Plantlet growth is enhanced in photoautotrophic more than heterotrophic conditions if environmental control is adequate (Hazarika *et al.*, 2003; Xiao, 2004; Afreen, 2005). However, after about four weeks, plants may stop responding to carbon dioxide and growth is reduced due to slight closing of stomata on the bottom of the leaves as plants sense high $CO_2$. However, use of $CO_2$ enrichment for two weeks, then a week off intermittently will ensure higher yields as the plant continuously seeks as much $CO_2$ as possible (Andrew, 2002). In potato, much research work has been done on photo autotrophic propagation and photo-mixotrophic propagation (Mohamed and Alasdon, 2010; Santana *et al.*, 2012).

## *In vitro* Culture and Plant Regeneration of Bamboo

Micro-propagation is a method using tissue culture technology to produce large quantities of plants in a very short time. Plants are grown in growth chambers under controlled environmental conditions *viz.*, on 1m² shelf area in a conditioned growth room, 2000 plants can be grown at one time. Stated otherwise 1 m² can hold enough plants to plant 4 hectares of tropical bamboos in plantations.

Starting from only a few nodal pieces, within a period of 4-8 months thousands of new plants can be produced. These plants are multiplied every 3-5 weeks and yield 3-6 new plants each. This is an excellent method to propagate new introductions or new selections very rapidly. Moreover, micro-propagation plants are small but vigorous growers, free from diseases and pests. Tissue culture plants have certain advantages; the most important are outlined here for large-scale plantations:

- ☆ Starting from elite selected material the overall yield will be higher than in seedling populations of bamboo.
- ☆ Vigorous growers.
- ☆ Pre-formation of roots and rhizomes for a more rapid establishment.
- ☆ Healthy and disease free plants and
- ☆ Small and compact plants, easy for packing and transportation.

### Elimination of Viruses

The viruses travel through the vascular system of the plants. The actively growing apical meristem tissues without the differentiated vascular bundles are usually free from viruses. The vegetatively propagated plants with viral diseases can be freed from the virus by excising and *in vitro* culture of 0.2 to 0.4 mm meristem tip.

### *In vitro* Culture of Bamboos

See the Table 11.19.

## *In vitro* Culture and Plant Regeneration of Root and Tuber Crops

Healthy propagules of convenient size, form and a good storage (shelf) are essential for conservation of germplasm for prolonged period and for their safe exchange. Such as requirements have led to the utilization of *in vitro* conservation and micro-tuberization in many root and tuber crops that are being stored in major gene banks all over the world. During the past two decades *in vitro* techniques have been extensively used for conservation of vegetatively propagated plants involving all the above components (Withers, 1993; Ashmore, 1997). *In vitro* technology can be utilized at various stages in conservation and utilization of root and tuber crops. These stages include efficient collection, disease elimination, fast multiplication, exchange and storage. Establishment and exchange of disease-free material is of paramount importance in genetic resources conservation programs. *In vitro* culture methods are used for both medium- and long- term *ex situ* conservation of plants (Engelmann, 1991). In the slow-growth method meant for medium term conservation (up to 5 years), cultures are maintained under conditions that can reduce the rate of growth, thus leading to extension of subculture interval. Maintenance of shoot cultures tub in less than 4 months. This is achieved by modifying the culture medium and culture maintenance conditions. Addition of growth retardants and osmotic regulators and reduction of nutrient content are some of the ways of modifying the growth medium. However, in order to reduce the frequency of somaclonal variations, the protocols involving addition of osmotic and growth retardants in the media are usually avoided. Since all cell metabolic process stop at these temperatures, the live material can be stored without variation for unlimited periods. Since this is a laborious and time consuming process and there is the increased risk of microbial contamination, culture maintenance in continued growth at normal rate is considered to be an unsatisfactory approach to germplasm conservation (Withers, 1984). The application of controlled osmotic stress using manitol or sorbitol in combination with other treatment such as reduced temperature, low light intensity and varied photoperiod has been used with success. Cryopreservation, for long term conservation, involves the storage of propagules at ultra-low temperature (-15°C to -196°C) using liquid nitrogen (LN).

**Table 11.19: Studies to Exploit the Modern Techniques for Bamboo (*D. asper*) Multiplication**

| Source | Shoot Multiplication | | | | Survival (per cent) of Plants after | |
| --- | --- | --- | --- | --- | --- | --- |
| | I Subculture | III Subculture | X Subculture | Rooting per cent | Hardening | Field Transfer |
| Single axillary bud (TFRI plant) | 1-2 | 1-3 fold | 15 fold | 100 | 95 | 70-80 |
| Seed (Ten lines) | 1-20 | 5-7 fold | 10-12 fold | 100 | 95 | 80-90 |

*Source*: Arya, and Arya, 1996.

## *In vitro* Culture and Plant Regeneration of Cassava

Cassava germplasm are being conserved under *in vitro* condition through slow growth cultures at various genetic resource centres in Africa, South America and Asia. Being vegetatively propagated crop, various conservation methods like low temperature incubation, use of low light intensity, use of osmotically active agents or growth retardants, manipulation in the concentration of the growth hormones, macro and micro nutrients and sucrose have been used individually or in combination for the medium term storage are being commonly used. Unnikrishnan and Sheela (2000) conducted the experiments with minimal and mannitol enriched media indicated that combination of both media was usually effective to induce slow growth. In their experiments, osmotic retardants like mannitol (3 per cent) or sorbitol (3 per cent) along with 3 per cent sucrose (as carbon source) in the medium was found effective in extending the subculture period from 3 to 12 months. Sorbitol was found better in imparting slow and healthy growth than mannitol. Silver nitrate (1mM) and activated charcoal (1g/l) in MS medium with sucrose (3 per cent) and growth regulators (0.05) μM NAA, 0.1μM BA and 0.3μ GA₃) was found to help in maintaining the cultures up to 18 months (Table 11.20). Addenda like silver nitrate (1mM) and activated charcoal (1.0 g/l) was found effective in reducing leaf chlorosis and preventing leaf shedding, thereby contributing to culture life and regeneration capacity. Throughout the storage period, the cultures continued to produce axillary buds, indicating the viability of the cultures.

Lower level of (0.1 μM) of BA, GA₃ and NAA was found more effective in development of meristem cultures of a wide number of genotypes. Murashige and Skoog (1962) medium with sucrose (3 per cent) and vitamins (Meso inositol and Thiamine HCL) was found effective in micro-propagation of cassava. In slow growth cassava experiments, Mannitol media indicated that combination of both media was effective to induce slow growth. Osmotic retardants like Mannitol or sorbitol (3 per cent) along with sucrose (3 per cent) in the medium was found effective in extending the subculture period from 10-12 months. Sorbitol was found better in imparting slow and healthy growth than Mannitol (Unnikrishnan *et al.,* 2000 and 2002). Bajaj (1987) reported cryopreservation technique for long term conservation of cassava using shoot tips. He found that isolated meristem pre-cultured on a medium containing DMSO and treated with a mixture of DMSO, sucrose and glycerol, could be frozen in liquid nitrogen using rapid freezing as well as slow freezing techniques. By adopting the method of procedure followed by slow freezing using the programmable freezer, the protocol for cryopreservation, using shoot tips has been developed and tested for 15 genotypes of cassava at CIAT and in some cases; regeneration up to 70 per cent has been achieved.

## *In vitro* Culture and Plant Regeneration of Sweet Potato

Sweet potato gemplasm were earlier being conserved through *in vitro* methods at several international centres like Centro International de la Papa (CIP), Peru, the International Institute of Tropical Agriculture (IITA), Nigeria, and World Vegetable Centre (AVRDC), Taiwan. In USA, Costa Rica and India have been maintaining a large number of collections through *in vitro* techniques over the last 15 years. Meristem cultures in conjunction with thermotherapy along with the use of antiviral drugs have been found effective in virus elimination in sweet potato (Ashmore, 1997). Meristem culture techniques coupled with indexing by ELISA and grafting on indicator plants was used to established healthy cultures, at ICAR-NBPGR, New Delhi Mandal and Chandel (1996)., has refined the *in vitro* techniques so as to develop a single medium capable of eliciting desirable response from all varieties and related species. MS medium (Murashige and Skoog, 1962) supplemented with 4 per cent sucrose, 0.2 mg l⁻¹ kinetin and 0.1 mg l⁻¹ IAA has been successfully used to establish cultures of 260 sweet potato accessions maintained at ICAR-NBPGR. For sweet potato conservation various others methods like incubation at low temperature (16-20°C) as proposed by Ng and Hahn (1985), addition of Mannitol to the medium coupled with reduction in incubation temperature and use of sucrose (3 per cent) with 3 per cent Mannitol (Dodds and Roberts, 1985) have been used for medium term storage. Though low temperature incubation (18°C) is effective in reducing growth rate, maintenance of large collections at this temperature during the summer months when the ambient temperature exceeds 40°C, is expensive and risky in view of frequent power breakdowns or sometimes an equipment failure. Mandal and Chandel (1991) after modification observed that the medium with 1-2 per cent Mannitol was very effective and using this method,

**Table 11.20: Effect of Silver Nitrate (AgNO₃) in Slow Growth of Cassava (Mean of 36 accessions)**

| Type of Accessions | Shoot Length (cms) | | | Number of Leaves | | | Number of Nodes | | |
|---|---|---|---|---|---|---|---|---|---|
| | Mannitol | Sorbitol | Silver Nitrate | Mannitol | Sorbitol | Silver Nitrate | Mannitol | Sorbitol | Silver Nitrate |
| CI | 4.61 | 2.08 | 5.40 | 1.50 | 3.500 | 11.40 | 2.50 | 3.54 | 16.2 |
| CE | 2.50 | 1.75 | 6.20 | 1.00 | 3.00 | 15.30 | 2.00 | 3.00 | 18.0 |
| Hybrids | 3.86 | 1.66 | 4.00 | 2.00 | 3.00 | 13.10 | 2.40 | 2.86 | 14.00 |

260 accessions have been maintained at ICAR-CTCRI, Thiruvananthapuram (Kerala) culture room temperature of 25±2°C for ten years with an annual subculture. While genotypic variation in response to Mannitol does exist the concentration of 1-2 per cent was suitable for most of the accessions. Application of cryopreservation technology for sweet potato conservation has not been satisfactory. The use of somatic embryos or embryogenic tissues which seems to hold more promise since recent studies (Bhatti *et al.*, 1997; Blackesley *et al.*, 1997), have demonstrated the recovery of complete plantlets from Liquid Nitrogen exposed embryogenic tissues.

**Table 11.21: Media Composition for *In vitro* Conservation of Sweet Potato**

| Crop | Media Code | Media Composition |
|------|-----------|-------------------|
| Sweet potato | Medium A | MS + 30.0 g sucrose/l + 30.0 gms manitol/l |
| | Medium B | MS + 20.0 g sucrose/l + 20.0 gms manitol/l |
| | Medium C | MS + 20.0 g sucrose/l |
| | Medium D | MS + 20.0 g sucrose/l + 1mM AgNO$_3$ |
| | Medium E (control) | Basal MS + 30.0 gms sucrose/l |

*Source*: Unnikrishnan *et al.*, 1990.

**Table 11.22: Slow Growth of Sweet Potato after Ten Months of Storage (mean of 35 accessions)**

| Media | Shoot Length (cm) | Number of Leaves | Number of Nodes | Remarks |
|-------|-------|-------|-------|---------|
| Medium A | 1.48 | 5.71 | 6.03 | All the four media combinations were found to be effective for storage |
| Medium B | 1.34 | 5.34 | 6.06 | |
| Medium C | 2.80 | 5.17 | 7.34 | |
| Medium D | 1.07 | 5.08 | 4.83 | |
| Control | 3.28 | 5.77 | 9.17 | |

Studies conducted with lower level of sucrose (20.0 g/l), sucrose (30.0 g/l) in combination with mannitol (30.0 g/l), lower levels of sucrose and Mannitol (20.0 g/l) were found equally effective in extending the subculture

period up to 10 months. Silver nitrate (1mM) along with sucrose (20.0 g/l) was also found to impart slow growth up to 10 months and beyond in studies conducted at ICAR-CTCRI, Thiruvananthapuram (Kerala).

## Yams

Yams are primarily propagated through vegetative means and thereby seriously affected by various pathogens. To overcome the problems of field maintenance, IITA, and IPGRI have recommended the use of *in vitro* techniques for yam conservation (Hanson, 1986a&b). In yams, two kinds of *in vitro* germplasm preservation were considered:

(i) Slow growth condition for medium term conservation and

(ii) Cryopreservation using the encapsulation/ dehydration technique for long term conservation.

*In vitro* conditions, axillary buds and nodal cultures have been found to be used most frequently. Mitchell *et al.* (1995) reported the addition of BAP@0.5 mg/l for the *in vitro* establishment of *D. rutundata*.

The use of MS medium with 3.0 per cent Mannitol was found to support healthy slow growth of the shoots of white yam upto 12-15 months (Nair and Chandrababu, 1994). Nair *et al.* (1996) found that best regeneration for white yam could be obtained in the presence of NAA (1.0 mM) and BAP@1.0 or 2.0 mM. In India, micropropagation techniques for yam conservation have been established by ICAR-CTCRI, Thiruvananthapuram (Kerala) and ICAR-NBPGR (Mandal and Chandel, 1996; Nair and Chandrababu, 1996). Using nodal cuttings as explants, *in vitro* tuber development has been recorded in several species (Ng and Ng, 1997; Mandal, 1999). They advocated that nodal segments with one or two axillary buds have been cultured on MS medium alone or supplemented with kinetin or NAA @ 0.25 mg l$^{-1}$ each. However, for routine sub culturing, a single medium *i.e.* MS supplemented with 0.15 mg l$^{-1}$ NAA, is applicable to all the accessions maintained in the repository. Incubation temperatures play a significant role in the

**Table 11.23: Different Media Evaluated for *in vitro* Conservation of Yams**

| Code | Media | Sucrose (per cent) | Mannitol (per cent) | Sorbitol (per cent) | NAA (μM) | BAP (μM) | GA$_3$ (μM) | Kinetin (μM) |
|------|-------|---------|----------|----------|------|------|------|---------|
| A | MS | 3 | - | - | 1.00 | - | - | 5.00 |
| B | MS | 3 | - | - | 1.00 | 5.00 | - | - |
| C | MS | 3 | - | - | 1.00 | 2.00 | 0.10 | - |
| D | MS | 3 | - | - | 0.01 | 0.10 | 0.30 | - |
| E | MS | 3 | - | - | 0.10 | 0.10 | 0.10 | - |
| F | MS | 6 | - | - | 0.05 | 0.10 | 0.30 | - |
| G | MS | 6 | 0.05 | - | 0.01 | 0.01 | 0.30 | - |
| H | | 6 | - | 0.05 | 0.10 | 0.10 | 0.10 | - |
| I | MS | 3 | - | - | - | - | - | - |
| J | MS | 4 | - | - | 0.10 | 0.50 | 0.1 | - |

growth rate of yam plantlets. Plantlets storage period could be extended to 1-2 years by incubating them at 18-20°C. Use of Mannitol and high level of sucrose, either alone or in combination, has significant effect on plantlet growth rate of the culture at room temperature (26±2°C). Addition of Mannitol to the medium can prolong storage of culture at normal room temperature of 25-30°C for a period up to 12 months with high recovery rates (80 per cent). Use of BA (2.0 mg l⁻¹) or kinetin (2.0 mg l⁻¹) helped in maintaining cultures for two years. Induction of micro tubers increased the shelf life by an additional 3-6 months. At ICAR-CTCRI, Thiruvananthapuram (Kerala), *D. alata, D. rotundata* and *D. esculenta* nodal cultures could be conserved up to 12 months without sub culture on MS basal medium containing 3 per cent Mannitol and without growth regulators.

Cryopreservation of yam shoot tips of two cultivated edible species (*Dioscorea alata* and *Dioscorea bulbifera*), a medicinal yam (*D. floribunda*) and one wild edible type (*D. wallichii*) were successfully carried out using the technique of encapsulation-dehydration (Mandal and Chandel, 1996). Though survival and growth initiation from cryopreserved shoot-tips was recorded at a frequency of 28-71 per cent, regeneration of complete plantlets were recorded at a frequency of 21-37 per cent in *D. alata* and *D. wallichii* (Mandal *et al.* 1996). In *D. alata* sorbitol @ 0.05 per cent in MS media with NAA (0.10μM), BA (0.10μM) and GA₃ (0.10μM) and sucrose (6 per cent) enhanced the storage period up to 12 months. In earlier reports nodal segments of *D. alata* and *D. rotundata* grown under long days (16 hours light) developed into plantlets on MS basal medium without any growth regulators (Mantell *et al.*, 1978). Malaurie *et al.* (1998) reported successful cryopreservation of shoot tips of *D. alata* and *D. bulbifera* using the technique of encapsulation-dehydration. In cryopreservation, success has been obtained by using shoot-tips/somatic embryos of several edible and medicinal yams adapting vitrification and encapsulation-dehydration. These include, vitrification of shoot-tip *D. rotundata* (Ahuja *et al.*, 2000), *D. floribunda* (Mandal, 2000), and *D. pentaphylla* (Ahuja *et al.*, 2000), *D. deltoidea* (Mandal and Dixit, 2000) and somatic embryos of *D. bulbifera* (Dixit *et al.* 2000). The generated plants obtained from cryopreserved shoot-tips/somatic embryos were transferred to field with 90-100 per cent establishment. The field grown plants obtained from cryopreserved explants were morphologically similar to those of their mother stocks maintained *in vitro* conditions and transferred to the field. In the three edible *Dioscorea* species studied *viz.*: *D. esculenta, D. alata* and *D. rotundata*, sucrose at 40.0

g/l were found better than lower level (30.0 g/l) for imparting healthy growth in stored cultures. *D. alata* and *D. rotundata* cultures could be stored upto 20 months and *D. esculenta* upto 10 months in MS basal medium + 40.0 g sucrose/l + 1.0 g activated charcoal/l + 0.1 μM NAA, 0.5 BA, 0.1 μM GA₃ (Table 11.24).

Some of the wild species like *D. oppositifolia* and *D. bulbifera* also could be conserved using MS basal medium + 40.0 g sucrose/l + 1.0 g activated charcoal/l +0.1 μM NAA, 0.5 μM BA, 0.1 μM GA₃). Medium term storage of the cultures up to 33 months in *D. oppositifolia* and 25 months in *D. bulbifera* was achieved in the same medium. Lower light intensity (1500 Lux for 8 hours) and 25-28°C temperature were found sufficient for *in vitro* storage of the edible *Dioscorea* species.

### *In vitro* Culture and Plant Regeneration of Aroids

*In vitro* technology in aroids is well developed and is being routinely used in national and international centres for conservation of these crops (Staritsky *et al.* 1986; Bessembinder *et al.* 1993). Morishita and Yamada (1978) taken shoot tips (meristem) dome plus one or two-leaf primordia, 0.2 to 0.3 mm in size and cultured on MS medium with 0.1 mg l⁻¹ NAA at 25°C under 12 hours day length. These meristems regenerated to shoots and produced roots after four months of culture. Successful cormel tip culture of *Colocasia esculenta* leading to the elimination of Dasheen mosaic virus has been done at ICAR-CTCRI, Trivendrum (Kerala). For medium term storage, the cultures of taro (*Colocasia esculenta*) are maintained in murashige and Skoog's (MS) medium, with 6 per cent sucrose supplemented with 2 mgL⁻¹ BAP, 0.1 mgL⁻¹ NAA and 0.8 per cent agar at 25±2°C temperature. The cultures maintained in these conditions produces only single shoot along with a micro-corm and the growth is restricted to extend the subculture till 12-15 months. This optimized protocol is cost effective and can be utilized further for standardized long term cryopreservation protocol for taro in *in vitro* gene bank. In *Xanthosoma sagittifolium*, addition of spermine, arginine and ornithine was found to enhance plant regeneration from shoot tip cultures (Sabapathy and Nair, 1995). In India work on medium term *in vitro* conservation has been carried out at CTCRI, Trivendrum (Kerala) and ICAR-NBPGR. Taro cultures could be stored upto 10 months under light (3000 lux) on MS basal medium containing sucrose and mannitol (3 per cent) without vitamins or growth regulators. Half strength basal medium was also found to induce slow growth (Unnikrishnan and Sheela, 2000).

**Table 11.24: *In vitro* Conservation of *Dioscorea* Species**

| Species | Shoot Length (cm) | Number of Leaves | Number of Nodes | Subculture Interval (Months) |
|---|---|---|---|---|
| D. rotundata | 7.15 | 11.03 | 10.08 | 25 |
| D. alata | 6.34 | 12.11 | 14.88 | 21 |
| D. esculenta | 6.20 | 10.30 | 4.16 | 10 |

*Colocasia* and *Alocasia* cultures could be conserved at normal culture room temperature of 25°C±2°C upto 10 months without sub-culturing (Mandal, 1999). For long-term conservation, slow freezing of embryogenic callus prior to cryopreservation resulted in 70 per cent survival, without loss of regenerability (Takagi *et al.* 1997). Vitrification of axillary buds, which initially met with only partial success, was modified for cryopreservation of *in vitro* shoot tips resulting in almost 80 per cent shoot recovery (Takagi *et al.* 1997; Thinh *et al.*, 2000). In Chinese potato (*Solenostermon rotundifolius*), standardized protocols for *in vitro* conservation and micro-propagation are available. Direct regeneration was obtained from lamina explants and intermodal cuttings at higher levels of BA (1 to 10 μM). Higher levels of NAA (1 to 10 μM) produced callus and indirect regeneration.

### *In vitro* Culture and Plant Regeneration of *Sesbania grandiflora*

Studies were carried out with the objective to evaluate the effects of growth regulators like auxin and cytokinin on *Sesbania grandiflora* regeneration potentials through callus induction. Explants were cultured on to solid MS medium enriched with different concentrations of auxin (IAA and NAA) and cytokinins (kinetin and BAP). The MS medium containing 0.5 mg/L BAP and 0.2 mg/L IAA showed up to 76 per cent regeneration activity. The shoot formation from callus was observed 60 per cent in MS media with 1.5 mh/L BAP and 0.5 mg/L NAA, while for combination of MS media 0.5 mg/L Kinetin and 0.1 mg/L IAA, the shooting percentage was recorded up to 80 per cent for the cotyledons explants and MS media, 1.5 mg/L Kinetin and 0.2 mg/L NAA revealed the shoot regeneration of 68 per cent grown on solid medium. The highest root induction of cotyledons callus was observed in MS medium containing 0.1 mg/L IBA and MS medium having 0.1 mg/L IAA which was 70 per cent and 100 per cent, respectively. The auxins (IAA, NAA) and cytokinins (Kinetin, BAP) could possibly be used for the root and shoot regeneration in *Sesbania grandiflora*.

### *In vitro* Culture and Plant Regeneration of Aquatic Vegetables

See the Table 11.25.

## Guidelines for Germplasm Acquisition

☆ Germplasm meant for *in vitro* conservation should be sourced from *ex vitro* plants (preferably growing in field genebank/green house/glass house/net house) or from *in vitro* cultures. The source material should have authentic taxonomic identity, be disease-free (if possible, virus free) and have passport data. Data on evaluation/characterization is desirable.

☆ The type and number of explants required for culture initiation should be determined based on crop species, season, propagation rate and health status. Since *in vitro* conservation is generally applied to clonally propagated germplasm, the

### Table 11.25: Application of Tissue Culture in Vegetable Crops

| Crop | | Descriptions | References |
|---|---|---|---|
| Water Chestnut | *Trapa bispinosa* Roxb. | Stem segments + axillary buds were cultured an half strength MS (Murshige and Skoog) medium containing IAA or GA at 0.5 mg +BA@2 mg/l and 3 per cent sucrose in light (1500-2000 lx) at 25 ± 1 °C. when 1 cm of a young stem segments was cultured for 2 weeks, buds formed profusely and when these were separated and sub-cultured on the original medium, new buds again formed profusely. When the buds were transferred to half strength MS medium containing only IAA at 0.5 mg/l, roots were produced after 2 weeks and the plants grew vigorously. When the plantlets were transplanted to half strength MS medium without growth regulators, growth was encouraged by soaking the roots in Nanhu lake water. | Zhou *et al.*, 1983 |
| Broadleaf Cattail | *Typha latifolia* L. | In three Typha species, plantlet production *via* organogenesis was optimum when the explants were first placed on Linsmaier and Skoog (LS) medium + 5.0 mg/l 2,4-D to initiate callus production and after 9.5 weeks the callus was re-cultured on LS+1.0 mg/l BA. Of the species tested, *T glauca* produced callus and new shoots more rapidly than *T. latifolia* or *T. angustifolia*. As inflorescences matured, an increased level of an auxin-like plant growth regulator, picloram (2.5 mg/l) was necessary for callus induction. Excision and separate culture of green spot or clumps that formed on the calli enhanced shoot regeneration from the callus. | Zimmermann and Read, 1986 |
| Broadleaf Cattail | *Typha latifolia* L. | Callus of broadleaf cattail was initiated from leaf sections as well as from pistillate and staminate spikes. Two basal media in combination with three growth regulator regimes were tested for their capacity to induce callus from the explants. Pistillate spikes maintained in the dark on $B_5$ medium supplemented with 5 mg.$L^{-1}$ dicamba and 1 mg.$^{-L}$ BA produced the fastest growing cell line compared to other explants and media combinations. A growth curve in suspension culture was generated for this cell line on $B_5$ medium. The mass of the callus increased by 150 per cent by the end of the growth curve. Upon transfer of the callus to MS media without growth regulators but with 3 per cent sucrose and 3 per cent phytagel, plants could be regenerated from 22 per cent of the cultures. | Estime *et al.*, 2002 |

| Crop | | Descriptions | References |
|------|------|------|------|
| Wasabi | *Wasabia japonica* (Miq.) Matsum. | A protocol for the propagation of Wasabi from axillary bud explants is presented. Of the 3 cytokinins tested benzyladenine was the most effective at promising shoot proliferation. Shoots were rooted in growth regulator free medium. | Hosokawa *et al.*, 1999 |
| | *Wasabia japonica* (Miq.) Matsum. | Media supplemented with cytokinins alone were the most effective for multiple shoot formation without callus induction. Benzyladenine (BA) and kinetin were more efficient than zeatin for shoot multiplication. The best results were obtained using 0.2-1.0 mg BA or kinetin/l. After apical meristems were cultured on medium with 1.0 mg BA/l for 60 days in primary culture, 8.2 shoots/explant were obtained. Plantlets divided into small parts from multiple shoots developed 3-4 well developed shoots with 4-5 true leaves every 2 months. Rooting and root growth was most effective on MS medium with 0.01 mg IBA/l. after rooting, regenerated plants were washed and transferred to pots containing sterilized soil. | Eun *et al.*, 1998 |
| | *Wasabia japonica* (Miq.) Matsum. | Immature zygotic embryos of Wasabi were isolated from immature seeds and cultured on modified MS media supplemented with auxins and cytokinins. The early cotyledonary zygotic embryos were more productive than the torpedo stage or late cotyledonary ones in somatic embryogenesis, and they also produced greater numbers of somatic embryos. At the early cotyledonary stage, the highest overall embryogenesis was achieved in basal media containing 2.0 mg IAA/l. Regeneration of somatic embryos was best in media with 2.0 mg of kinetin/l, and plantlets generated from somatic embryos had phenotypically normal leaves and roots. Shoots lacking roots were separated individually and transferred to medium for multiple shoot and root formation. Rooting and root growth were best in media with 0.01 mg IBA/l, and multiple shoots were produced in media containing BA [benzyladenine]. Combinations of BA and 0.01 mg IBA/l showed little or no effect on rooting. | Eun *et al.*, 1999 |
| | *Wasabia japonica* (Miq.) Matsum. | The proliferation of Wasabi on BA and KT, NAA and IBA were higher compared to ZT and IAA, respectively. The optimum medium for plantlets propagation was MS+0.5 per cent BA (or KT) mg/l+0.05 NAA mg/l. CPPU enhanced wasabi propagation and the suitable medium was MS+0.4 per cent CPPU mg/l. the combination of BA and KT had a propagation coefficient which reached 5.1 on MS+0.2BA mg/l+0.5KT mg/l+0.05NAA mg/l. the culture medium containing 0.02 mg/l GA3 could promote growth of wasabi plantlet in vitro. | Wu *et al.*, 2002 |
| | *Wasabia japonica* (Miq.) Matsum. | Best rooting (97 per cent) with 0.1 per cent NAA and 0.05 per cent IBA + 0.3 per cent AC. These media also resulted in strong viability of the root system in wasabi. After the rooted plantlets were hardened and transferred to sterile soil, a high survival percentage (>94 per cent) was obtained. | Wang and Long, 2002 |
| Water dropwort | *Oenanthe javanica* (Blume) DC. | The effects of medium composition and cultural environment on the uniformity and maturity of embryos of Water dropwort were investigated. Regenerated plantlets had the greatest potential for embryogenic callus induction, 2, 4-D at 1.0 mg/l$^{-1}$ was the most effective for induction. Liquid MS medium containing 2, 4-D at 0.5 mg/l$^{-1}$ was effective for the maintenance and proliferation of embryogenic callus. The induction and proliferation of the callus occurred to the same extent in the dark or light (16-h light). Somatic embryogenesis was strongly induced in a growth regulator free liquid medium. Continuous sub culture with fresh medium at 5 days was further promoted the induction of somatic embryos. Treatment of somatic embryos at the globular stage with ABA at 0.05 mg/l$^{-1}$ was effective for increasing uniformity and inhibiting abnormality. The maturity somatic embryos were enhanced on MS medium supplemented with 6 per cent sucrose. Growth regulators had no positive effect on embryo maturity. | Kim and Lee, 1995 |
| | *javanica* (Blume) DC. | For callus induction and/or proliferation, dark condition was more effective than that of light illumination. Among the auxins, 1.0 mg/l$^{-1}$, 2,4-D and 0.5 mg/l$^{-1}$, 2,4-D levels were found to be suitable for the solid and liquid media, respectively. Air-lift bioreactor was incorporated for cell growth and the optimum air flow rate was determined at 0.5 vvm. During the bioreactor, cultivation, the final cell volume was adjusted to 20 per cent by feeding fresh medium. Medium feeding was conducted when final volume reached 30 per cent. Total cell growth using bioreactor revealed as two times higher than that of flask. For somatic embryogenesis, plastic capping revealed at the best until 3 weeks later among the caps tested. The optimum cell inoculation density of somatic embryogenesis was 600 mg/200 ml with 5 cm rolling bottle. Homogenization of the cultured cells at 16,000 rpm for 10 second gave most effective result for synchronization of induced somatic embryos. | Kim *et al.*, 2002 |
| Water pepper | *Polygonum hydropiper* L. | The callus and suspension-cultured cells induced from hypocotyls of *Polygonum hydropiper* L. Seedlings. Both the callus and suspension-cultured cells produced mainly (+)-catechin accompanied by (-)-epicatechin and (-)-epicatechin-3-O-gallate. The (+)-catechin production of suspension-cultured cells increased with cell growth and reached the maximal value (29.0 mg g$^{-1}$ wt) after 6 days from the start of subculture. This was the highest value of (+)-catechin. The amount of (-)-epicatechin was in the range of 1.1-7.7 mg g$^{-1}$ dry weight (wt) and that of (-)-epicatechin-3-O- | Ono *et al.*, 1998 |

| Crop | | Descriptions | References |
|---|---|---|---|
| | | gal-late was in the range of 2.6-6.4 mg g$^{-1}$ dry weight (wt) for a culture period of 15 days, respectively. Comparing with plant parts in regard to (+)-catechin, the amount of suspension-cultured cells was about 1.5 times as much as callus cells, about 9 times that of leaves (3.2 mg g$^{-1}$ dry weight (wt) and about 7 times that of stems (4.0 mg g$^{-1}$ dry weight (wt). the maximal yield of total catechins in suspension-cultured cells was 4.3 per cent dry weight (wt). | |
| Paracress | *Spilanthes acmella* L. A.H. Moore | *Spilanthes acmella* L. Was successfully micro-propagated using axillary buds asexplants. The aseptic axillary buds formed multiple shoots within five weeks when cultured on MS medium supplemented with 2.0 mg/l N6-Benzyl adenine (BA). The addition of Indol-3-butyric acid (IBA) had no significant effect on the multiple shoots formation of this plant. MS medium supplemented with 0.5 mg/l N6-Benzyl adenine (BA) was sufficient for the proliferation of rooted multiple shoots of *S. acmella* L. First subculturing of the in vitro individual shoot in the same proliferation medium could double the formation of multiple shoots. | Haw and Keng, 2003 |
| Watercress | [*Rorippa nasturtium aquaticum* (L.) Hayek] | Nodal explants of watercress taken from 4 uniforms, pot grown clones were culture *in vitro* on growth regulator free MS medium. The plantlets produced were grown on 25, 50 and 100 per cent strengthened MS media supplemented with 1, 2 or 4 per cent sucrose. The highest number of nodes (40.9) occured on 25 per cent MS with 2 per cent sucrose, but some shoots (14 per cent) become vitrified. Two node propagules produced more nodes than shoot tip or single node propaule because both auxillary buds developed shoots, although the number of nodes/shoot was lower than for other types of propagule and there was not difference in final dry weights.propagules derived from the distal part of the shoot produced new shoot with significantly more nodes (37.5) than those derived from the middle or proximal (20.5) sections. In vivo, 99 per cent of the two node propagules were successfully established by rooting directly in a 50:50 mixture of vermiculite and compost, compared with 79 per cent in 100 per cent vermiculite. | Wainwright and Marsh, 1986 |
| | [*Rorippa nasturtium aquaticum* (L.) Hayek] | Callus was initiated on MS medium supplemented with auxin and cytokinin in the dark at 25 °C. The regeneration media contained no auxin and the callus was cultured 25 °C with a 16 h photoperiod. The cytokinin thidiazuron was better than benzyladenine for shoot regeneration and auxin was regulated in the initiation medium. Naa stimulated shoot production from calus on initiation medium while with 2, 4-D shoot production was at its highest frequency at 24-32 weeks. | Gilby and Wainwright, 1989. |

number of explants does not affect the genetic make-up of the conserved germplasm. The disease status (including virus) may, however, differ in individual explants.

☆ Shoot tip/nodal cutting/meristem should be used as explants in most herbaceous plants. For tree and woody species, shoot tip/bud/ meristem explants should be derived from bud woods. In vegetatively propagated species with per-ennating organs, shoot tip/bud/meristem explants can also be derived from underground storage organs such as tubers, bulbs, suckers, corms *etc.* Use of floral buds should be avoided, except in case of triploid/seed-sterile plants.

☆ The material for *in vitro* conservation should be transported from the site of collection to gene bank in the shortest possible time and under the best possible conditions.

## Haploid Development

Protocol standardization of haploid development in vegetable crops is the need of the day. India, second largest producers of vegetables in the world but less productivity probably due to the cultivation of open pollinated varieties only. Productivity can be increased by growing hybrids which is possible through inbred parents. Development of inbreds takes long time to establish particular traits for stability. For example, inbred line development in onion takes more than 12 years owing to its two years per generation cycle. Hence haploids and subsequently doubled haploids (DHs) production through either *in vitro* gynogenesis or androgenesis is needed. Mathapati *et al.* (2016) used onion variety Pusa Ridhi to know the different media (MS and B5) on haploid induction efficiency through *in-vitro* gynogenesis. It was observed that use of PPM (1mg/l) in culture medium was efficient to prohibit contamination from field grown flower buds, in B5 medium, 2,4-D and BA (2 mg/l each) and NAA (0.5 mg/l) + BA (3 mg/l) alongwith 7.5 per cent sucrose were successful in plantlet induction. In MS medium, again 2, 4-D and BAP (2 mg/l each), NAA 2 mg/l + 2iP (1 and 2 mg/l) and NAA 0.5 mg/l + 1.0 mg/l were efficient. Best medium combination was 2,4-D+BAP (2 mg/l each) with 10 per cent sucrose. Kinetin, meta-topolin, Thidiazuron, sucrose did not show any positive influence on embryo induction in B5 but in MS only sucrose (10 per cent) led to increase in induction frequency upto 5.0 per cent. Days taken for plantlet formation was 76-122 days with an average of 98.9 days in B5 medium and 73.95 days with an average of 83.1 days in MS medium. In B5 medium, cultured 3552 flower buds induced 18 plantlets with an induction rate of 0.51 per cent, average induction days

98.9 days with a survival mean of 27.8 per cent whereas in MS medium, cultured 1328 flower buds leading to 10 plantlet, 0.75 per cent induction frequency, average induction days 83.1 and overall plant survival of 30 per cent. Plants were tested cytologically and 3 haploid, 1 mixoploid and 1 diploid was observed in B5 medium whereas in MS medium, 2 haploid and one diploid was obtained. Flow cytometry confirmed the haploid status of HAP02 and three SSR markers successfully identified the status of mother plant and gynogenic haploids.

# References

1. Afreen, F. 2005. Physiological and anatomical characteristics of *in vitro* photoautotrophic plants. In: Kozai, T., Afreen, F. and Zobayed, SMA (eds.). *Photoautotrophic (Sugar-Free Medium) Micropropagation as a New Propagation and Transplant Production System.* 1st edn. Dordrecht, Springer, p. 61-90

2. Alexander, M.P. and Ganeshan, S. 1989a. Preserving viability and fertility of tomato and eggplant pollen in liquid nitrogen. *Indial J. Plant Genet. Resour.*, **2:** 140-144.

3. Alexander, M.P. and Ganeshan, S. 1989b. An improved cellophane method for *in vitro* germination of recalcitrant pollen. *Stain Technology*, **64:** 225-227.

4. Andrew, S. 2002. Carbon dioxide enrichment.www. hydrocentre.com.au

5. Angels, J.S. 1994. Release of transgenic plants: biodiversity and population level considerations. *Molecular Ecology*, **3:** 45-50.

6. Arce, T.P. and Lopez, C.B. 1994. *Capsicum Eggplant Newsletter*, **13:** 68-71.

7. Arya, I.D. and Arya, S. 1996. Introduction, mass multiplication and establishment of edible bamboo *Dendrocalamus asper* in India. *Indian J. Pl. Genet. Resour.*, **9(1):** 115-121.

8. Ashmore, S.E. 1997. Status report on the development and application of *in vitro* techniques for the conservation and use of plant genetic resources. IPGRI. Rome.

9. Ayotte, R., Harney, P.M. and Souza-Machado, V.1987. The transfer of triazine resistance from *Brassica napus* L to *B oleracea* L. Production of $F_1$ hybrids through embryo rescue. *Euphytica*, **38:** 615.

10. Azra S., Lutful H., Syed, D., Ahmad, A.H., Shah F., Batool, M.A., Islam, R.R. and Moonmoon, S. 2009. *In vitro* regeneration of ginger using leaf, shoot tip and root explants. *Pak. J. Bot.*, **41(4):** 1667-1676.

11. Bajaj, Y.P.S. 1995. Biotechnology in Agriculture and Forestry, Vol. 32. Cryopreservation of Plant Germplasm I. Springer, Berlin, Germany.

12. Balachandran, S.M., Bhat, S.R. and Chandel, K.P.S. 1990. *In vitro* clonal multiplication of turmeric (*Curcuma* species) and ginger (*Zingiber officinale*). *Pl. Cell Rep.*, **8:** 521-524.

13. Barringer, S.A. Yasseen, M.Y. Schloupt, R.M. and Splittstoesser, W.E. 1996. *J. Veg. Crop Prod.*, **2:** 27-33.

14. Benson, E.E., Harding, K. and Smith, H. 1989. The effects of pre and post freeze light on the recovery of cryopreserved shoot tips of *Solanum tuberosum*. *Cryo-Lett.*, **10:** 323-344.

15. Berg, P., Baltimore, D., Boyer, H.W., Cohen, S.N., Davis, R.W., Hogness, D., Nathans, D., Roblin, R., Watson, J.D., Weissman, S. and Zinder, N.D. 1974. Potential biohazards of recombinant DNA molecules. *Science*, **185:** 303.

16. Bessembinder, J.J.E., Stritsky, G. and Zandvoort, E.A. 1993. Long term *in vitro* storage of *Colocasia esculenta* (L.) Schott under minimal growth conditions. *Pl. Tiss. Org. Cult.*, **33:** 121-127.

17. Bhansali, R.R. and Ramawat, K.G. 1993. Micropropagation of little leaf diseased eggplants infected with mycoplasma-like organisms. *J. Hort. Sci.*, **68:** 25-30.

18. Bhatti, M.H., Percival, T., Davey, C.D.M., Henshaw, G.G. and Blakesley, D. 1997. Cryopreservation of embryogenic tissue of a range of genotypes of sweet potato (*Ipomoea batatas* L.) using an encapsulation protocol. *Pl. Cell Rep.*, **16:** 802-806.

19. Bhojwani, S.S. and Rajdan, M.K. 1996. Plant Tissue Culture Theory and Practices; A review edition. Elsvier Science BV, The Netherlands.

20. Binzel, M.L., Sankhla, N., Joshi, S. and Sankhla, D. 1996. Induction of direct somatic embryogenesis and plant regeneration in pepper (*Capsicum annuum* L.). *Plant Cell Rep.*, **15:** 536-540.

21. Biondy, S. 1986. *Giornale Bot. Italiana.*, **120:**1-6.

22. Blackesley, D., Percival, T., Bhatti, M.H. and Henshaw, G.G. 1997. A simplified protocol for utilizing sucrose pre-culture only. *Cryo-Lett.*, **18:** 77-80.

23. Bockish, L., Saranga, Y. Altman, A. and Ziv, M. 1997. *Acta Hort.*, **447:** 241-242.

24. Borovec, V. 1988. *Bull. Vyzkmny Slech. Univ.*, **15:** 31-45.

25. Boxus, P. 1990. *Annales Gembloux*, **96:** 33-42.

26. Branca, C., Torelli, A. and Bassi, M. 1990. *Plant Cell, Tissue and Organ Culture*, **21:**17-19.

27. Burza, W. and Malepszy, S. 1995. Direct plant regeneration from leaf explants in cucumber is free of stable genetic variation. *Plant Breed.*, **114:** 341-345.

28. Chandel, K.P.S., Chaudhury, R., Radhamani, J. and Malik, S.K. 1995. Desiccation and freezing sensitivity in recalcitrant seed of tea, cocoa and jackfruit. *Ann. Bot.,* 76: 443-450.

29. Chaudhury, R. and Chandel, K.P.S. 1985. Cryopreservation of embryonic axes of almond (*Prunus amygdalus*) seeds. *Cryo-Lett.,* **16:** 51-56.

30. Chaudhury, R., Radhamani, J. and Chandel, K.P.S.1991. Preliminary observations on the cryopreservation of desiccated embryonic axes of tea (*Camellia sinensis* (L.) O. Kuntze) seeds for genetic conservation. *Cryo-Lett.,* **12:** 31-36.

31. Chauhan, K.P.S. 1996. Legal requirements for fulfilling obligations under the Convention on Biological Diversity. *Indian J. Plant Genet. Resour.,* **9(2):** 183-191.

32. Chee, P.P. 1991. Somatic embryogenesis and plant regeneration of squash *Cucurbita pepo* L.cr. Yc-60. *Plant Cell Report,* 9(11): 620-622.

33. Chopra, V.L. and Narasimhulu, S.B. 1986. Modern techniques for genetic conservation. In: Conservation for productive agriculture (Chopra, V.L. and Khoshoo, T.N. eds.), PID, ICAR, New Delhi. p. 161-173.

34. Choudhury, R. and Malik, S.K. 2008. Germplasm storage. In: Plant Biotechnology: Methods in Tissue Culture and Gene Transfer. R. Keshavachandran and K.V. Peter (Eds.), Universities Press (India) Private limited, India. p. 190-202.

35. Crisp, P. and Tapsell, C.R. 1993. Cauliflower (*Brassica oleracea* L.). In: Genetic Improvement of Vegetable Crops. Eds. G. Kalloo and B.O. Bergh, Pergamon Press. Oxford, p.157-178.

36. Deng, M. and Cornu, D. 1992. Maturation and germination of walnut somatic embryos. *Plant Cell, Tissue and Organ Culture,* **28(2):** 195-202.

37. DeVilliers, R.P., Van, R.J., Janse, V.R. and Van, S.J. 1993. *J. South African Soc. HortSci.,* **3:** 24-27.

38. Dixit, S., Mandal, B.B. and Srivastava, P.S. 2000. Cryopreservation of somatic embryos/ embryogenic tissues: Application in Conservation and Biotechnology. Abstract, In: National symposium on Prospects and Potentials of Plant Biotechnology in India in the 21st Century. Jodhpur, Oct. 18th-21st, 2000. p. 161.

39. Dixon, A. and Gonzalez, R.A. 1994. In: Plant Cell Culture: A Practical Manual; (eds Dixon and Gonzalez) IRS Press, Oxford, U. K.

40. Dodds, J.H. and Roberts, L.W. 1985. Experiments in Plant Tissue Culture, 2nd Edn. Cambridge, Easwari Amma, New York: Cambridge University Press.

41. Ellstrand, N.C. and Hoffman, C.A. 1990. Hybridization as an avenue of escape for engineered genes. *Bioscience,* **40:** 438-442.

42. EL-Zeiny, O.A.H. 2002. Using tissue culture as a tool for increasing the productivity of seedlings and total yield of some pepper hybrids. *Journal of Agricultural Siences,* **10 (1):** 273-285.

43. Engelmann, F and Engels, J.M.M. 2002. Technologies and strategies for *ex situ* conservation. In: Managing Plant Genetic Diversity. Ramanatha Rao, V., Brown, A.H.D. and Jackson, M.T. (eds.). Walling Ford, Rome, CAB International IPGRI, p. 89-104.

44. Engelmann, F. 2004. Plant cryopreservation: Progress and Prospects. *In vitro Cell Dev. Biol. Plant,* 40: 427-433.

45. Englemann, F. 1991. *In vitro* conservation of horticultural species. *Acta Hort.,* **298:** 327-334.

46. Englemann, F. 1991. *In vitro* conservation of tropical germplasm- a review. *Euphytica,* 57: 227-243.

47. Englemann, F. and Dereuddre, J. 1988. Cryopreservation for oil palm somatic embryos: importance of freezing process. *Cryo-Lett.,* **7:** 220-235.

48. Englemann, F., Duval, Y. and Dereuddre, J. 1985. Survival and proliferation of oil palm (*Elaeis guineesis* Jacq.) somatic embryos after freezing to liquid nitrogen. *CR Acad. Sci.,* **301:** 111-116.

49. Estime, L., O'-Shea, M., Borst, M., Gerrity, J. and Liao, S.L. 2002. Suspension culture and plant regeneration of *Typha latifolia*. *HortSci.,* **37(2):** 406-408.

50. Eun, J.S., Ko, J.A. and Kim, Y.S. 1998. Mass propagation of *Wasabia japonica* Matsum. by apical meristem culture. *Journal of the Korean Society for Horticultural Science,* **39(3):** 278-282.

51. Eun, J.S., Lee, J.M., Gross, K.C., Watada, A.E. and Lee, S.K. 1999. Production of high quality seedlings by immature zygotic embryo culture of *Wasabia japonica* Matsum. Proceedings of the International Symposium on "Quality of Fresh and Fermented Vegetables", Seoul, Korea Republic, 27-30 October, 1997. *Acta Horticulture,* **483:** 357-366.

52. Ezura, H., Nishimiya, S. and Kasumi, M. 1993. Efficient regeneration of plants independent of exogenous growth regulators in bell pepper (*Capsicum annuum* L.). *Plant Cell Rep.,* **12:** 676-680.

53. Farash, E.M., Abdalla, H.L, Taghian, A. S. and Ahmad, M. H. 1993. *Australian J. Agril. Sci.,* **24:** 3-14.

54. Farham, M.W. and Nelson, B.V. 1993. Utility of *in Vitro* Propagation for Field-grown Broccoli: Effect of Genotype and Growing Season. *HortSci.,* **28:**655-656.

55. Fari, M., Szasz, A. Mityko, J., Nagy, I., Csanyi, M. and Andrasfalvy, A. 1992. Induced organogenesis *via* the seedling decapitation method (SDM) in three Solanaceous vegetable species. *Capsicum Newslette.,* **2:** 243-248.

56. Fay, M.F. 1992. Conservation of rare and endangered plants using *in vitro* methods. *In vitro Cell Dev. Biol. Plant*, 28: 1-4

57. Ford-Lloyd and Jackson, M.J. 1991. Biotechnological methods of conservation of plant genetic resources. *J. Biotechnol.*, **17:** 247-256.

58. Gamburg, K.Z., Semenova, L.A. and Ziv, M. 1997. *Acta Hort.*, **447:** 147-148.

59. Ganeshan, S. 1986. Viability and fertilizing capacity of onion (*Allium cepa* L.) pollen stored in liquid nitrogen (-196 °C). *Tropical Agriculture* (Trinidad). **63:** 45-48.

60. Geetha, N., Venkatachalam, P., Reddy, P.S. and Rajaseger, G. 1998. *In vitro* plant regeneration from leaf callus cultures of tomato (*Lycopersicon esculentum* Mill.). *Adv. Plant Sci.*, **11:** 253-257.

61. Gemes, J.A. and Martinovich, L. 1998. *Zoldsegtermesztesi-Kutato-Intezet-Bulletinje.*, **28:** 39-45.

62. George, E.F. 1993a. Plant propagation by tissue culture: Part I (George, E. F. ed.) Exegetics Limited, Edington, UK.

63. George, E.F. 1993b. Plant propagation by tissue culture: Part II (George, E.F. ed.) Exegetics Limited, Edington, UK.

64. Gilby, A.C. and Wainwright, H. 1989. Use of tissue culture in improvement of watercress (*Rorippa nasturtium aquaticum* L. Hayek). *Acta Horticulture*, **244:** 105-113.

65. Giles K.L. and Pierik, R.L.M. 1990. Current plant science and biotechnology 'in agriculture, Vol. 12 (Prakash, J. ed.) Kluwer Academic Publishers; Dordrecht; Netherlands p.155-160.

66. Global Knowledge Centre on Crop Biotechnology, 2008. Biotech Crop Background and Pocket Ks, 1-33, ISAAA.

67. Gonzalez-Benito, M.E., Clavera-Ramirez, I. and Lopez-Aranda, J.M. 2004. The use of cryopreservation for germplasm conservation of vegetatively propagated crops. *Span J. Agric. Res.*, 2(3): 341-351.

68. Gupta, C.G., Lakshmi, N. and Srivalli, T. 1998. *Capsicum Eggplant Newslett.*, **17:** 42-45.

69. Hanson, J. 1986. Methods of storing tropical root and germplasm with special reference to yam. FAO/IBPGR. *Plant Genetic Resources Newsletter*, 64: 24-32.

70. Haque, M.S., Wada. T. and Hattori, K. 1997. High frequency shoot regeneration and plantlet formation from root tip of garlic. *Plant Cell, Tissue and Organ Culture*, **50:** 83-89.

71. Harding, K., Benson, E.E. and Smith, H. 1991. The effects of *In vitro* culture period on the recovery of cryopreserved shoot tips of *Solanum tuberosum*. *Cryo-lett.*, **12:** 17-22.

72. Haw, A.B. and Keng, C.L. 2003. Micropropagation of *Spilanthes acmella* L., a bio-insecticide plant, through proliferation of multiple shoots. *J. Appl. Hort.*, **5(2):** 65-68.

73. Hazarika, V., Parthasarathy, V.A. and Nagaraju, V. 2003. *Photoautotrophic Micropropagation*-a review. *Agric. Rev.*, **24(2):** 152-156.

74. Henshaw, G.G. 1975. Technical aspects of tissue culture storage for genetic conservation. In: Frankel, O.H., Hawkes, J.G. (eds.) Crop genetic resources for today and tomorrow. p. 349-357. Cambridge University Press.

75. Hernandez, P.R., Prado, O., Gil, D.V., Perez, P.J. and Ruiz, M. 1994. *Centro- Agricola*, **21:** 76-84.

76. Hoban, T.J. and Kendall, P.A. 1993. Consumer attitudes about the use of biotechnology in agriculture and food production, US Dept. of Agriculture, Raleigh, North Carolina.

77. Holmes, B. 1993. The Perils of Planting Pesticides. New Scientist, 28 August, p. 34.

78. Hoque, A., Hossain, M., Alam, S., Arima, S. and Islam, R. 2007. Adventitious Shoot Regeneration from Immature Embryo Explant Obtained from Female × Female *Momordica dioica* Roxb. *Plant Tissue Cult. and Biotech.*, **17(1):** 29-36.

79. Hoque, M.E., Bhomik, A. and Khalequzzaman, M. 1998. *In vitro* culture of pointed gourd. *Thai J. Agric. Sci.*, **31:** 369-374.

80. Hosokawa, K., Oikawa, Y. And Yamamura, S. 1999. Clonal propagation of *Wasabia japonica* by shoot tip culture. *Planta Medica*, **65(7):** 676.

81. Huda, A.K.M.N. and Sikdar, B. 2006. *In vitro* Plant Production through Apical Meristem culture of bitter gourd (*Momordica charantia* L.). *Plant Tissue Cult. and Biotech.*, **16(1):** 31-36.

82. ICARDA. 2014. Genebank. International Centre for Agricultural Research in the Dry Areas. Available on http://www.icarda.cgiar.org/research_sub/biodiversity-and-its utilization access on May, 2014.

83. Ichimura, K. and Oda, M. 1995. Stimulation of shoot regeneration from cotyledon segments of tomato (*Lycopersicum esculentum* Mill.) by agar and its extract. *J. Japanse Soc. Hort. Sci.*, **64:** 135-141.

84. Inomata, N. 1977. Production of interspecific hybrids between *Brassica campestris* and *B. oleracea* by culture *in vitro* of excised ovaries. I. Effects of yeast extracts and casein hydrolysable on the development of excised ovaries. *Japanese J Breed.*, **27:** 295.

85. Islam, A. K., M.R., Joarder, O.L., Rahman, S. M. and Hossain, M. 1992. *Ind. J. Hort.* **49:** 249-252.

86. Jagadev, P.N., K.N. Panda, and S. Beura 2006. A fast protocol for *in vitro* propagation of ginger (*Zingiber officinale* Rosc.) of a tribal district of India. ISHS *Acta Horticulturae*, 765: XXVII International Horticultural Congress-IHC2006: International Symposium on "Plants as Food and Medicine: The Utilization and Development of Horticultural Plants for Human Health".

87. Jawahar, M., Mohamed, S.V. and Jayabalan, N. 1998. *Adv. Plant Sci.*, 11: 199-204.

88. Jeong, H.B. and Park, H.G Jeong, H.B., Cho, M.A., Ha, S.H. and Kang, K.Y. 1998. *RDA-J. Hort. Sci.*, 40:78-82.

89. Jeong, H.B. and Prak, H.G. 1997. Plant re-differentiation and *in vitro* multiplication of onion by shoot primodium culture. *J. Kor. Soc. Hor. Sci.*, 38: 123-128.

90. Jianbin H.U.I., Xiaoxi, G., Jun, L.I.U., Conghua X.I.E., and Jianwu L.I., 2008. Plant regeneration from petiole callus of *Amorphophallus albus* and analysis of somaclonal variation of regenerated plants by RAPD and ISSR markers. *Botanical Studies*, 49: 189-197.

91. Josekutty, P.C., Swati, S. and Prathapasenan, G. 1993. *J. HortSci.*, 68: 31-35.

92. Joshi, A.K., Kanwar, J.S. and Saimbhi, M.S. 1995. *Punjab Vegetable Grower*, 30: 14-22.

93. Kartha, K.K., Leung, N.L. and Pahi, K. 1980. Cryopreservation of strawberry meristem and mass propagation of plantlets. *Jm. Soc. HortSci.* 105: 481-484.

94. Kasumi, M., Yashira, K. and Hayashi, M. 2000. Effectiveness of cold stored pollen on seed formation in East Indian lotus (*Nelumbo nucifera* Gaertn.). *Journal of the Japanese Society of Horticultural Science*, 65(6): 732-735.

95. Kathal, R., Bhatnagar, S.P. and Bhajwani, S.S. 1994. Plant regeneration from the callus derived from root explants of *Cucumis* melo L. cv. Pusa Sharbati. *Plant Sci.*, 96: 137-142.

96. Kieffer, M., Fuller, M.P and Jellings, A.J. 1995. Rapid mass production of cauliflower propagate from fractionated and graded curd. *Plant Sci. Limerick.*, 107: 229-235.

97. Kieffer, M., Fuller, M.P. and Jellings, A.J. 1994. The rapid mass production of cauliflower propagules from fractionated and graded curd. ISHS Symposium on *Brassicas*, Ninth Crucifer Genetics Workshop. 15-19 November 1994. Lisbon-Portugal.

98. Kim, H.S. and Lee, B.Y. 1995. *In vitro* production of somatic embryo in *Oenanthe stolonifera* DC. *Journal of the Korean Society for Horticultural Science*, 36(1): 38-45.

99. Kim, J.A., Lee, B.Y., Lee, Y.H. and Son, S.H. 2002. Scale up and stable production of somatic embryos bioreactor in *Oenanthe stolonifera* DC. *Journal of the Korean Society for Horticultural Science*, 43(2): 173-177.

100. Krimsky, S. 1991. Biotechnics and Society-The Rise of Industrial Genetics. Praeger Publisher, New York.

101. Krishna Prasad, V.S.R. 2006. Improvement of pointed gourd (*Trichosanthes dioica* Roxb.), *Cucurbits Breeding and Production Technology* (Edit. Gautam, P. L., Ram, H. and Singh, H. P.) Pub. G.B. Pant University of Agriculture and Technology, Pantnagar (Uttrakhand). p.165-168.

102. Kumar, P.A. 1999. Genetic engineering for insect resistance in vegetable crop. In: Summer school on advanced technologies in improvement of vegetable crops including Cole crops. ICAR-IARI, New Delhi.

103. Kumar, S, Singh, M., Kalloo, G. and Banerjee, M.K. 2001. Large scale multiplication of *Coccinia grandisli* (L.) Voige. using *in vitro* techniques. National Conference on Role of Biotechnology in Indian Agriculture. Vasant Rao Naik Mahavidyalaya, Aurangabad (MS), 7-8 Oct. 2001 (Abstract).

104. Kumar, S., Singh, M., Ram, D. and Banerjee M.K. 2002. In: International Conference on Vegetables, PNASF, Bengaluru (India), 11-14 Nov, 2002, p. 135.

105. Kyte, L. and Kleyn. J. 1996. Plants from test tubes: an introduction to micro propagation. 240 pp Timber Press, Inc. Oregon; USA.

106. Larkin, P.J. and Scowcroft, W.R. 1981. Somaclonal variation: a novel source of variability from cell cultures for plant improvement. *Theor. Applied Genet.*, 60: 197-214.

107. Liu, G.Q., Li, S.J. and Zhang, X.P. 1996. *J. Nanjing Agric. Univer.*, 19: 31-36.

108. Lou, H. and Kako, S. 1994. *HortSci.*, 29: 906-909.

109. Ma, W.J., Mi, S.E and Li, B. 1991. *Gansu-Nongye-Daxue-Xuebao.*, 26: 409-416.

110. Ma, Y., Wang, H.L. and Zhang, C.L. Kang, Y. Q. 1994. *Plant Cell Rep.*, 14: 65-68.

111. Mailier, E. 1992. Eurobarometer 35.1: Opinions of Europeans on Biotechnology in 1991. In: J. Durant (ed.) Biotechnology in Public: A Review of Recent Research. Science Museum for the European Federation, Biotechnology, London.

112. Mandal, B.B. and Chandel, K.P.S. 1996. Conservation of genetic diversity in sweet potato and yams using *in vitro* strategies, Kurup *et al.* (eds.). Oxford & IBH Publication, p. 49-54.

113. Mandal, B.B. and Chaudhury, R. 2004. Recent techniques for *ex situ* conservation of plant genetic resources. IN: Dhillon, B.S., Tyagi, R.K., Lal, A. and Saxena, S. (Eds.)-Plant Genetic Resource Management. Narosa Publishing House, New Delhi. p. 194-204.

114. Masuda, K., Hatakeyama, E., Ita, A., Takahashi, S. and Inoue, M. 1994. *Bull. Akita Prefectural College Agri.*, **20**: 43-48.

115. Mathapati, G., Khar, A., Kalia, P., Islam, S. and Saini, N. 2016. Protocol standardization of haploid development in short day onion (Abs.: 14; Session III: Biotechnological Interventions in Horticulture). 7th Indian Horticulture Congress, held at ICAR-IARI, New Delhi, 15-18th November 2016, p. 372-373.

116. McCown, B.H. and McCown, D.D. 1999. *In* Vitro *Cell Develop. Biol. Plant*, **35**: 276·277.

117. Mirghis, E., Mirghis, R., Lacatus, V., Fernandez, R., Cuartero, J. and Gomez-Guillamen, M.L. 1995. Analysis of tomato cultivars and hybrids for *in vitro* callus formation and regeneration. *Acta Horticulturae*, **412**: 111-116.

118. Mirza, M.N. and Narkhede, M.N. 1996. Shoot tip culture in chilli (*Capsicum annum* L.) *Annals Plant Physio.*, **10**: 148-152.

119. Misra, A. K. and Bhatnagar, S.P. 1995. Direct shoot regeneration from the leaf explant of cucumber (*Cucumis sativus* L.). *Phytomorphology*, **45**: 47-55.

120. Mitchell, S.A., Asemota, H.N. and Ahmad, M.H. 1995a. Effect of explant source, culture medium strength and growth regulators on the *in vitro* propagation of three Jamaican yams (*Dioscorea cayenensis*), *D. trifida* and *D. rotundata*. *J. Sci. Food Agric.*, **67**: 173-180.

121. Mitchell, S.A., Asemota, H.N. and Ahmad, M.H. 1995b. Factors effecting the *in vitro* establishment of Jamaican yams (*Dioscorea* spp.) from nodal pieces. *J. Sci. Food Agric.*, **67**: 541-550.

122. Mohamed, M.A.H. and Alasdon, A.A. 2010. Influence of ventilation and sucrose on growth and leaf anatomy of micropropagated potato plantlets. *Scientia Horticulturae*, **123**: 295-300.

123. Mohamed, M.E. 1998. *Acta Hort.*, **457**: 243-253.

124. Montagu, van M., Tiedje, J.M., Powell, D., Simoens, C., Tzotzos, G.T. and Barlow, B.A. 1995. Application of biotechnology for the utilization of biodiversity. In: Heywood, V.H. (ed.) Global Biodiversity Assessment. Cambridge University Press, Cambridge.

125. Morra, M.J. 1994. Assessing the impact of plant products on soil organisms. *Molecular Ecology*, **3**: 53-55.

126. Mrudul, V., Shirgurkar1, C.K., John L. and Rajani S. N. 2001. Factors affecting *in vitro* microrhizome production in turmeric. *Plant Cell, Tissue and Organ Culture*, **64 (1):**5-11.

127. Murashige, T. 1974. *Ann. Rev. Plant Physiol.*, **25:** 135-166.

128. Murashige, T. and Skoog, E. 1962. A revised medium for rapid growth and bioassays with tobacco tissue culture. *Physiol. Plant.*, **15:** 473-497.

129. Mythili, J.B. and Thomas, P. 1999. *In vitro* multiplication of pointed gourd (*Trichosanthes dioica* Roxb.) *Sci. Hort.*, **79:** 87·90.

130. Nabi, S.A., Rashid, M.M., Al-Amin, M. and Rasul, M.G. 2002. Organogenesis in Teasel gourd (*Momordica dioica* Roxb.). *Plant Tissue Cult.*, **12(2):** 173-180.

131. Nagasawa, A. and Finner, J. 1989. Plant regeneration from embryogenic suspension cultures of Chinese yam (*Dioscorea opposita* Thunb.). *Pl. Sci.*, **60:** 263-271.

132. Nair, N.G. and Chandrababu, S. 1994. A slow growth medium for *in vitro* conservation of edible yams. *J. Root Crops*, 20: 68-69.

133. Nandkumar, R., Chen, L. and Rogers, S.M.D. 2004. Factors affecting the Agrobacterium-mediated transient transformation of the wetland monocot, *Typha latifolia*. *Plant Cell, Tissue and Organ Culture*, **79(1):** 31-38.

134. Navratilova, B. 1987. *Bull. Vyzkmny Slech. Univ.*, **31:** 35-44.

135. NBPGR. 2016. Guidelines for Management of Plant Genetic Resources, ICAR-NBPGR, New Delhi.

136. Neeta D.S., Leela, G. and Susan, E. 2004. Direct regeneration of shoots from immature inflorescence cultures of turmeric. *Plant Cell, Tissue and Organ Culturre*, **62(3):** 235-238.

137. Newman, P.O., Krishnaraj, S. and Saxena, P.K. 1996.Regeneration of tomato (*Lycopersicon esculentum* L.) somatic embryogenesis and shoot organogenesis from hypocoty explants induced with 6-Benzyladenine. *Int. J. Plant Sci.*, **157:** 554-560.

138. Office of Technology Assessment (OTA). 1987. New Developments in Biotechnology-Background Paper; Public Perceptions of biotechnology, OTA-BP-45, United States Govt. Printing Office, Washington, DC.

139. Okezie, C., Okonkwo, S. and Nwoke, F. 1994. Carbon source requirement for the culture of white yam (*Diocorea rotundata*) embryos *in vitro*. *Acta Hortic.*, **380:** 329-334.

140. Ono, K., Nakao, m., Toyota, M., Terashi, Y., Masashi Yamada, M., Kohno, T. and Asakawa, Y. 1998. Catechin production in cultured *Polygonum hydropiper* cells. *Phytochemistry*, **49(7):** 1935-1939.

141. Pandron, *I.E.S.*, Torres-Arizal, L.A. and Litz, R. 2011. Somatic embryogenesis in Yam (*Dioscorea rotundata*). *Rev. Fac. Natl. Agril. Mendellin*, **64(2)**.

142. Paunescu, A. 2009. Biotechnology for endangered plant: A critical overview. *Romanian Biotechnological Letters*. 14: 4095-4103.

143. Paunescu, A. and Hololobiuc, I. 2005. Preliminary researches concerning micropropagation of some endemic plants from Romanion Flora. *Acta Hort. Bot.*, Buccurest, 32: 103-108.

144. Pawar, B.D. and Kale, P.B. 2011. Molecular Plant Breeding for Sustainable Agriculture: An overview. *Agrobios NewsLetter*, **10(7):** 66-68.

145. Peddaboina, V., Thamidala, C., Karampuri, S. 2006. *In vitro* shoot multiplication and plant regeneration in four Capsicum species using thidiazuron. *Scientia Horticulturae*, **107:** 117–122.

146. Pence, V.C. 1992. Desiccation and survival of Aesculus, castanea and quercus embryo axes through cryopreservation. *Caryobiology*, 29: 391-399.

147. Perlas, N. 193. When what could go wrong did go wrong. Third World Resurgence, Penang, Malaysia, p.8.

148. Raghuveer, P., Khetarpal, S. and Patil, P. 1994. *Seed Res.*, **22:** 112-118.

149. Rajasekharan, P.E., Alexander, M.P. and Ganeshan, S. 1998. Long term pollen preservation of wild tomato, brinjal species and cultivars in liquid nitrogen. *Indian J. Plant Genet. Resour.*, **11(1):** 117-120.

150. Ramirez, M.R. and Ochoa, A.N. 1996. An improved and reliable chili pepper (*Capsicum annuum* L.) plant regeneration method. *Plant Cell Rep.*, **16:** 226-231.

151. Ranjan, J.K., Krishna, R., Singh, A.K., Singh, M., Singh, P.M. and Singh, B. 2016. *In vitro* regeneration response in okra (*Abelmoschus esculentus* L. Moench). (Abs.: 15; Session III: Biotechnological Interventions in Horticulture).7th Indian Horticulture Congress, held at ICAR-IARI, New Delhi, 15-18th November 2016, p. 373-374.

152. Reed, B.M. and Chang, Y., 1997, Medium- and long-term storage of *in vitro* cultures of temper-ate fruit and nut crops, In: Conservation of Plant Genetic Resources *in vitro*, Vol. 1, M.K. Razdan and E.C. Cocking (Eds.), p. 67-105, Enfield. NH, USA: Science Publishers. Inc.

153. Reed, B.M., Engelmann, F., Dullo, M.E. and Engels, J.N.M. 2004. Technical guidelines for the management of feed and *in vitro* germplasm collection. *Handbook for Genebanks* No. 7, IPGRI/SGRP, Rome.

154. Rissler, J. and Mellon, M. 1993. Perils Amidst the Promise: Ecological risks of transgenic crops in a global market. Union of concerned Scientists, Cambridge, Mass.

155. Roberts, E.H. 1975. Problems of long term storage of seeds and pollen for genetic resources conservation in O.H. Frankel and H.G. Hawkes (Eds.). *Crop Genetic Resources for Today and Tomorrow*, Cambridge University Press, Cambridge, p. 269-296.

156. Roca, W.M., Rodriguez, J.A., Mafla, G. and Roa, J. Procedure for recovering cassava clones distributed *in vitro*. CIAT Colombia, p. 8.

157. Rogozinska, J. and Drozdowska, L. 1996. *J. Appl. Genet.*, **37:** 357-366.

158. Ross, C.L. 1980. Embryo culture in production of disease resistant Brassica. In: Tissue Culture Methods for Plant Pathologist, Ingram, D.S. and Helgeson, J.P. Eds., Blackwell Scientific Publication, Oxford, UK. 225.

159. Sakai, A. 2000. Development of cryopreservation techniques. In: F. Engelmann and H. Takagi (Eds.), cryopreservation of Tropical Plant Germplasm. International Plant Genetic Resources Institute, Roam.

160. Santana, A., Jesus, S., Larrayoz, M.A. and Filho, R.M. 2012. "Supercritical carbon dioxide extraction of algal lipids for the biodiesel production," Procedia *Engineering*, **42:** 1755-1761.

161. Sarasan, V., Cripps, R., Ramsay, M.M., Atherion, C., McMichen, M., Prenderfast, G. and Rowntree, J.K. 2006. Conservation *in vitro* of threatened plants: Progress in the past decade. *In vitro Cell Dev. Biol. Plant*, 42: 206-214.

162. Seo, Y.K., Paek, K.Y., Hwang, J.K. and Cho, Y. H. 1991. *J. Korean Soc. Hort. Sci.*, **32:** 137-145.

163. Shah, G.A., Deping, G. and Guanhgwan, Z. 1993. *HortSci.*, **28:** 677.

164. Sharma, P. and Rajam, M.V. 1995. *J. Exp. Bot.*, **46:** 135-141.

165. Shashank, Bansal, N., Sharma, K. and Gahlawat, S.K. 2011. Biodiversity informatics. *AgroBios NewsLetter*, **10(7):** 68-69.

166. Shikhamany, S.D. 2006. Horticultural genetic resources. Role of *ex situ* conservation. In: ICAR, Short course on *in vitro* conservation and cryopreservation new options to conserve horticultural genetic resources. ICAR-IIHR, p 6-15.

167. Singh, M., Kumar, S., Singh, B. and Banerjee, M.K. 2002. In: International Conference on Vegetables, PNASF, Bengaluru (India), 11-14 November.

168. Singh, S. 1988. *Curr. Sci.*, **13:** 730-732.

169. Song, P.L., Pen, W.Z. and Yang, Y.H. 1988. *Acta. Sci. Nature Univer. Hunan.*, **11:** 340-345.

170. Souza, C.M., Pinto, J.E.B.P, Rodrigues, B.M., Furtado, D.E, Morais, A. R., Arrigoni, B. M. and Morais, A.R 1998. *Ciencia-e-Agrotecnologia.*, **22:** 52-56.

171. Spetsidis, N., Sapountzakis, G. and Tsaftaris, A. S. 1996. *Cucurbit Genet Cooper. Rep.,* **19:** 63-65.

172. Sretenovic, R.T., Vinterhalter, D., Vinterhalter, B. and Lazic, B. 1997. *Acta Hort.,* **462:** 589-593.

173. Stenitz, B., Wolf, D., Josef, T.M., and Zelcer, A. 1999. *Capsicum Eggplant Newslett.,* **18:** 9-15.

174. Sultana, R.S. and Miah Bari, M.A. 2003. *In vitro* propagation of karalla (*Momordica charantia* L.) from nodal segment and shoot tip. *Journal of Biological Sciences,* **3(12):** 1134-1139.

175. Szymanska, M. and Malas, J. 1995. *Acta Agrobotanica,* **48:** 75-82.

176. Thomus, B.R. and Pratt, D. 1981. Efficient hybridization between *Lycopersicon esculentum* and *L. peruvianum via* embryo callus. *Theoretical and Applied Genetics,* 59(4): 215-219.

177. Twyford, C. and Mantell, S. 1996. Production of somatic embryos and plantlets from root cells of the Greater Yam. *Plant Cell, Tissue and Organ Culture,* **46(1):** 17-26.

178. Tzotzos, G.T., Lesser, W.H., Powell, P.J. and Dale, P.J. 1995. Impacts of biotechnology on Biodiversity. In: Heywood, V.H. (ed.) Global Biodiversity Assessment, Cambridge University Press, Cambridge.

179. Ucman, R., Zel, J. and Ravnikar, M. 1998. *Scientia Horti.,* **73:** 193-202.

180. Vandemoortele, J.L., Billard, J.P., Baucaud, J. and Gaspar, T. 1999. *In Vitro Cell. Develop. Biol. Plant,* **35:** 13-17.

181. Virgin, I., Fredrick, R.J. and Ramachandran, S.C. 1995. Impact of International Harmonisation on Biosafety Regulations in Biotechnology. Proc. on Biosafety Regulations in Biotechnology, Bangkok, Thailand, p. 115-125.

182. Wainwright, H. and Marsh, J. 1986. The micro-propagation of watercress (*Rorippa nasturtium aquaticum* L.). *Journal of Horticultural Science,* **61(2):** 251-256.

183. Walkey, D.G.A. 1987. Micro propagation of vegetables. In: Institute of Horticulture Symposium, University of Nottingham School of Agriculture, p. 97-112.

184. Wang, L. and Long, C.L. 2002. Technique of *in vitro* propagation of *Eutrema wasabia. China Vegetables,* **2:** 16.

185. Williamson, M. 1991. Biocontrol Risk. *Nature,* 353: 394.

186. Withers, L.A. 1985. Cryopreservation and storage of germplasm. In: Dixon, R.A. (ed.) Plant Cell Culture-A practical Approach. IRL Press Oxford. p. 169-192.

187. Withers, L.A. and Engelmann, F. 1997. *In vitro* conservation of plant genetic resources. In: Altman A (ed) Biotechnology in agriculture, Marcel Dekker, NY, p. 57-88.

188. Wu, Z., Wang, G.D., Liu, Q. and Li, S. 2002. Effect of different hormones with various concentrations on apical buds proliferation of wasabi (*Wasabia japonica* Marsum.). *Southwest China Journal of Agricultural Sciences,* **15(3):** 66-69.

189. Xiao, Y. and Kozai, T. 2004. Commercial application of a photoautotrophic Micropropagation system using vessels with forced ventilation: plantlet growth and production cost. *HortSci.,* **39(6):** 1387-1391

190. Yamamoto, T. 1994. In: 22nd annual meeting of The-International-Plant Propagators'-Society, Wellington, New Zealand, p. 361-365.

191. Yasseen, M.Y., Splittstoesser, W.E. and Litz, R.E. 1994. *Plant Cell, Tissue and Organ Culture,* **36:** 243-247.

192. Ye, Z.S., Li, H.X. and Zhou, G.L. 1994. *J. Huazhang Agricultural University,* **13:** 291-295.

193. Yun, J.S., Hwang S.G., Song, I.G., Lee, C.H., Yun, T. Jeong, I.M., Paek, K.Y. 1998. *RDA-J. Hort. Sci.,* **40:** 14-19.

194. Zapartan, M. and Deliu, C. 1994. Conservation of endemic rare and endangered species in the Romania flora using *in vitro* methods. Proc. 8th Nat. Symp. *Ind. Microbiol. Biot. Chrol.* Bucharest, p. 432-436.

195. Zhang, L.Y., Li, G.G., Li, K.L. and Chen, R.Z. 1989. *Acta Bot. Sinica,* **4:** 177—181.

196. Zhao, J.P. Bai, K.H., Jiang, X.M. and Bai, X.E. 1998. *Plant Physiol. Commun.,* **29:** 435-437.

197. Zheng, H.R., Shen, M.J, Zhong, W.J., Zhang, Z.Q. and Zhou, Y. 1998. *Acta Agric. Shanghai,* **14:** 33-38.

198. Zhou, J.Q., Zhang, Z.P., Zhou, G.Y., Liu, S.Z. and Chen, B.Q. 1983. Culture of axiliary buds of Nanhu Singhara nut. Plant Physiology Communication. *Zhiwu Shenglixue, Tongxun,* **1:** 31.

199. Zhou, Z.X., Zhang, Z.J., Jiu, Y.J., Jiang, Q.Y. and Mi, J.J. 1994. *Acta Agri. Boreali Sinica,* **9:** 59-63.

200. Zimmerman, R.H. 1996. *J. Korean Society Hort. Sci.,* **37:** 486-490.

201. Zimmermann, E.S. and Read, P.E. 1986. Micro-propagation of *Typha* species. *HortSci.,* **21(5):** 1214-1216.

# 12

# Novel Genomic Tools for Vegetable PGR Management and Crop Improvement

## Introduction

Recent advances in molecular technology have provided powerful genetic tools, which can provide rapid and detailed genetic resolution. Genetic markers, deficiencies differences at the genotype (DNA) level, can be used to answer and explain questions (Paterson *et al.,* 1991a,b). To be useful as a genetic markers, the marker locus has to show experimentally detectable variation among individuals in the test population (Liu, 1998.) In this direction National Bioresource Project was started in 2002 by the support of Ministry of Education, Culture, Science, Sports and Technology (MEXT) in Japan. The important purpose of this project is to collect, preserve, and distribute bioresource (such as experimental animals, plants and microbes *etc.*) that are essential experimental materials for life sciences research. Developing fundamental technologies to improve the value of each bioresources, and enriching genome information are also important target of this project. At the same period, the Experimental Plant division of RIKEN BioResource Cwenter (RIKEN BRC:http://epd.brc.riken.jp/en/) was established in the RIKEN Tsukuba Institute in 2001 to promote resource activities of *Arabidopsis* plant (Kobayashi, 2011). Today, stability and identify of crop variety has assumed great importance for predicting plant breeder's right/farmer's right because of use of several modern tools with high cost. The unambiguous pattern of crop varieties can be obtained using DNA markers and it has been termed as "DNA Finger Printing". The technique was developed by Alec Jeffery in 1985 in human and 1st time it was used

in rice crop by Dallas (1988) for cultivar identification. Molecular markers poses several distinct advantages over morphological or other conventional markers:

☆ They can be established at cellular or tissue level,

☆ They are not influenced by the environmental fluctuation

☆ They usually show high degree of polymorphism *etc.*

Different application of molecular technique in finding solutions for problems of developing world includes: study of plant genetic resources and their utilization for vegetable crop improvement, to develop disease resistance in crops using natural resources and improve nutritional value in food and feed supplements. Transgenic technology is useful in creating crops with desired phenotype or 'designer crops". Traditionally, evaluation and conservation of biodiversity/genetic variability is based on comparative anatomy, morphology, embryology, physiology *etc.*, which provide information data but of low genetic resolution. Earlier electrophoretic banding homology is considered to be powerful approach in estimating genetic relationship between two or more species (Johnson, 1968; Johnson and Thein, 1970), as well as in the identification and classification of cultivars (Cooke and Draper, 1983). For example the evolutionary relationships between cultivated and wild species of Asiatic *Vigna* species were investigated through biochemical parameters utilizing isozyme variation patterns of peroxidase, catalase and seed protein polymorphism studied through polyacrylamide gel electrophoresis. Only through this study "Mung" and "Urid" beans are two distinct species and they have independently evolved from wild progenitors *V. radiata* var. *sublobata* and *V. mungo* var. *silvestris*, respectively, which show remarkable similarties in isozyme profiles particularly in number of bands, position and Rf values (Sehgal and Chandel, 1992). In vegetable crops, molecular markers are now increasingly being used for the germplasm management, gene or QTL tagging, marker assisted selection, identification of heterotic parents, hybrid seed purity test, comparative mapping *etc*. Indigenously developed insect resistant *Bt*-brinjal is under consideration for approval of its commercial cultivation is a valid example. Genomics may greatly enhance utilization of plant genetic resources (Brozynska *et al.*, 2016). This is especially critical with the prospects of major climate change (Abberton *et al.*, 2016).

## Biotech crops

Since the first demonstration of the introduction and expression of foreign genes in tobacco in 1984, more than 150 plant species in at least 50 plant families have been experimentally transformed and transgenic events reported. Thirteen transgenic crops are grown commercially in 25 different countries, including 15 developing countries (James, 2008). Regulatory approvals for 24 transgenic crops, for the importation for food and feed use and for release into the environment have been granted in another 30 countries (James, 2008). Novel genomic tools have great potential for alleviating some of the production constraints such as insect pests, pathogens, salinity and drought. Biotech crops viable options to counter these limitations. Today a number of transgenic plants with different toxin genes have been developed which are resistant to different insect pests.

**Table 12.1: List of some Developed Important *Bt* Transgenic Plants**

| Crop | Toxin Gene | Target Insect |
|------|------------|---------------|
| Tomato | Synthetic Cry 1Ab, Cry 1AC | *Manduca sexta* |
| Potato | Synthetic Cry IIIA | *Leptinotarsa decimineata* |
| Brinjal | Synthetic Cry 1Ab | *Leucidous* |

Genetic transformation for resistant transgenic in Indian cauliflower and cabbage was carried out by Kalia *et al.* (2016) using promising cultivar against Diamond back moth (*Plutella xylostella* L.). Reproducible and highly efficient protocol for Agrobacterium mediated genetic transformation and regeneration was developed and established in Indian cauliflower varieties 'Pusa Meghana', 'Pusa Snowball-K1', 'Pusa Snowball-K25' and 'Golden Acre' cabbage. Cotyledon and hypocotyl from aseptically grown seedlings were employed as explants for transformation, using *A. tumefaciens* strains EHA105, GV3101 and LBA4404 harbouring a binary vector pPIPRA *Bt* of 16.787 Kb size. The TDNA of plasmid pPIPRA *Bt* contains a Bar/NPTII gene for plant selection, under 34SFMV promoter *Bt* genes Cry 1b and Cry 1c provides resistance against Diamond back moth. *Cry* toxins produced by *Bacillus thuringiensis* (*Bt*) are effective biological insecticides against borer insect pests. Successful *Agrobacterium* mediated transformation, shoot and root regeneration protocol for the popular varieties of cauliflower and cabbage was developed. Confirmation of transgenic integration will be carried out through molecular analysis, ELISA and Southern blotting whereas for efficiency of transgene bioassay will be conducted. The *Bt* genes which have been used to produce transgenic in different vegetables are given:

**Table 12.2: Insect Pests and *Bt* Toxins**

| Crop | Insect Pest | Bt Toxins |
|------|-------------|-----------|
| Brinjal | *Leucinodes orbonalis* | GY 1, Ab, Gy 1B |
| Tomato | *Helicoverpa armigera* | GY 1, Ac, Gy2Aa |
| Cabbage | *Plutella xylostella* | GY 1, Ab, Gy 1B |
| Cauliflower | *Plutella xylostella* | GY 1, ab, Gy 1B |
| Potato | *Phthormaea operculata* | GY 1, ab, Gy 1B |
| Okra | *Earias vitella* | Gy 1 AC |
| Red pepper | *Helicoverpa armigera* | GY 1, Ac, Gy2Aa |

*Source*: Kumar, 1999.

Among the vegetable crops, potato (*Solanum tuberosum*) which belongs to the same genus as brinjal (*Solanum melongena*), was one of the first biotech crops developed and commercialized in the USA. Other Solanaceaeous vegetables like tomato, and pepper were also developed and commercialized in China.

**Table 12.3: Transgenic Vegetable Crops Engineered for Enhanced Resistance against different Insect Pests**

| Transgenic Crop | Gene | Insect | Reference |
|-----------------|------|--------|-----------|
| Potato | CpTi | *Lacanobia oleraceae* (Linnaeus) | Gatehouse *et al.*, 1999 |
| Potato | cry3a | *Leptinotarsa decemlineata* (Say) | Perlak *et al.*, 1993 |

As early as 1994, and onwards a series of vegetables with extended shelf-life (known as *flavr Savr* tomato), tomato with superior quality and attractive deep red colour, and '*Endless Summer*' tomato and squash with novel virus resistance. Presently, a number of biotech vegetables are being either field tested or commercialized in different parts of the world.

In India several genetically modified (GM) related biotechnological research works are being initiated to cope up the biotic and abiotic stresses along with quality traits. Related genes must be known and conserved for future.

**Table 12.4: Transgenic Vegetable Crops Engineered for Enhanced Resistance against Fungal Pathogen**

| Transgenic Crop | Gene/gene product | Donor | Target pathogen | References |
|---|---|---|---|---|
| Carrot | Acidic wheat class IV chitinase + acidic wheat 1, 3- glucanase + rice cationic peroxidase (*POC1*) | Wheat, Rice | *Verticillium dahliae, Fusarium oxysporum* f. sp. *vasinfectum, Rhizoctonia solani* and *Alternaria alternata* | Wally *et al.,* 2009 |
| Carrot | Microbial factor 3 (*MF3*) | Pseudomonas fluorescence | *Alternaria dauci, Alternaria radicina* and Botrytis cinerea | Baranski *et al.,* 2007 |
| Carrot | Lipid transfer protein gene and chitinase | Wheat, Barley | Foliar fungal pathogen | Jayaraj and Pubja, 2007 |
| Carrot | Human lysozyme | Human | *Erysiphe heraclei, Alternaria dauci* | Takaichi and Oeda, 2000 |
| Carrot | Chitinase | Tobacco | *Alternaria dauci, A. radicina; Colletrichum carotae* | Melchers and Stuvier, 2000 |
| Tomato | *NPR1* | Arabidopsis | Resistance to fungal and bacterial disease | Lin *et al.,* 2004 |
| Tomato | *Pn-AMPs* (hevein like protein) | *Pharbitis nil* | *Phytophthora* spp. *Fusarium* spp. | Lee *et al.,* 2003 |
| Tomato | *PR2* | Tobacco | *Fusarium oxysporum* | Foolad *et al.,* 2002 |
| Tomato | *Oxalate decarboxylase* | *Collybia velutipes* | *Sclerotinia sclerotoorum* | Kesarwani *et al.,* 2000 |
| Tomato | *cercosporin-melittin cationic peptide* | Synthetic gene | *Multiple pathogens* | Osusky *et al.,* 2000 |
| Tomato | Defensins | Raddish | *Alternaria solani* | Parashina *et al.,* 2000 |
| Potato | Defensins (alfAFP) | *Alfalfa* | *Verticillium dahliae* | Gao *et al.,* 2000 |
| Potato | Gene 1-2 | *Tomato* | *Fusarium* spp. | Mes *et al.,* 2000 |
| Potato | Lactoferrin | Human | Not tested | Chong and Langridge, 2000 |
| Potato | Osmotin gene | Tobacco | *Phytophthora infestans* | Li *et al.,* 1999 |
| Potato | Endochitinase | *Trichoderma harzianum* | Foliar and soil borne fungal pathogen | Lorito *et al.,* 1998 |
| Potato | 34-aa chimaeric peptide MsrA1+melittin | Bee venom | *E. carotovora* ssp. *atroseptica* | Osusky *et al.,* 2000 |
| Potato | *sarco* gene coding for sarcotoxin IA | Sarcophaga peregrina | *E. carotovora, P. syringae* pv. *lachrymans* and *R. solanacearum* | Galun *et al.,* 1996 |

**Table 12.5: Transgenic Vegetable Crops Engineered for Enhanced Resistance against Viral Pathogens**

| Transgenic Crop | Mechanism/Strategies Employed | Source/Gene Product | Virus | Reference |
|---|---|---|---|---|
| Common bean | RNA interference | Replication initiator protein (*rep; AC1*) | Bean golden mosaic virus (BGMV) | Aragao and Faira, 2009. |
| Tomato | Coat protein mediated resistance | *N* gene | Tomato spotted wilt virus (TSWV) | Goldbach *et al.,* 2003 |
| Tomato | | Coat protein (CP) | Cucumber mosaic virus | Xue *et al.,* 1994 |
| Tomato | | Coat protein (CP) | Cucumber mosaic virus | Gielen *et al.,* 1996 |
| Tomato | | Coat protein (CP) | Cucumber mosaic virus | Fuchs and Provvidenti, 1996 |
| Tomato | | Coat protein (CP) | Cucumber mosaic virus | Tomassoli *et al.,* 1999 |
| Tomato | | Coat protein (CP) | Cucumber mosaic virus | Kaniewski *et al.,* 1999 |
| Pepper | | Coat protein (CP) | Cucumber mosaic virus | Shin *et al.,* 2002a |
| Potato | RIP | PAP (*Phytolacca americana*) | Cucumber mosaic virus | Lodge *et al.,* 1993 |
| Pepper | *RNA satellites* | Sat-117N | Cucumber mosaic virus | Kim *et al.,* 1997 |
| Cucumber | | Coat protein (CP) | Cucumber mosaic virus | Gonsalves *et al.,* 1992 |
| Melon | | Coat protein (CP) | Cucumber mosaic virus | Gonsalves *et al.,* 1994 |
| Squash | | Coat protein (CP) | | Tricoli *et al.,* 1995 |
| Tomato | | Sat-S | Cucumber mosaic virus | Stommel *et al.,* 1998 |
| Pepper | Enhancement of *HR/SAR* | *Tsil* (Tobacco) | Cucumber mosaic virus | Shin *et al.,* 2002b |
| Tomato | *Plantibodies* | *ScFv* antibodies | Cucumber mosaic virus | Villani *et al.,* 2005 |
| Potato | | *ScFv* antibodies | PVY | Gargouri-Bouzid *et al.,* 2006 |

**Table 12.6: Transgenic Vegetable Crops Engineered for Enhanced Resistance against Nematode**

| Transgenic Crop | Gene/Gene Product | Donor | Target Nematode | Reference |
|---|---|---|---|---|
| Tomato | Cysteine proteinase inhibitor (*CeCPI*) | *Colocasia esculenta* | *Meloidogyne incognita* | Chan *et al.*, 2010 |
| Tomato | Cry6A | *Bacillus thuringiensis* (Berliner) | Root knot nematode | Li *et al.*, 2007 |
| Tomato | CaMi | Hot pepper (*Capsicum annuum* L.) | *Meloidogyne* spp. | Chen *et al.*, 2007 |
| Brinjal | cry1Ab | *Bacillus thuringiensis* (Berliner) | *Meloidogyne incognita* | Phap *et al.* 2010 |

**Table 12.7: Research and Development of Biotech/Genetically Modified (GM) Vegetables in the World**

| Sl.No. | Vegetable Crop | Botanical Name | Trait | Source Country |
|---|---|---|---|---|
| 1. | Brinjal | *Solanum melongena* L. | IR, DST | India, Bangladesh, the Philippines, Italy |
| 2. | Broccoli | *Brassica oleracea* var. *italica* | IR, HT | New Zealand, Japan |
| 3. | Cabbage | *Brassica oleracea* var. *capitata* | IR | India, Australia, New Zealand |
| 4. | Cassava | *Manihot esculenta* | PQ, MG, VR | India, USA |
| 5. | Carrot | *Daucus carota* | NR, PQ, HT | USA, New Zealand |
| 6. | Cauliflower | *Brassica oleracea* var. *botrytis* | IR, HT | India, Japan, Australia, New Zealand |
| 7. | Cucumber | *Cucumis sativus* | AP, VR, HT, PQ, IR | USA, Poland, Japan |
| 8. | Garlic | *Allium sativum* | AP, PQ | New Zealand |
| 9. | Lettuce | *Lactuca sativa* | VR, HT, FR, PQ | USA, Japan |
| 10. | Okra | *Abelmoschus esculentus* L. | IR | India |
| 11. | Onion | *Allium cepa* L. | HT, FR, DR, AP | India, New Zealand, USA |
| 12. | Pea | *Pisum sativum* | OO, HT, VR, PMP, DR | USA, Germany, United Kingdom |
| 13. | Potato | *Solanum tuberosum* L. | FR, VR, OO, PQ, IR, AP, BR, HT | India, Canada, New Zealand, USA, Germany, Spain, UK, Netherlands, Czech Republic, France, Poland, Ireland, Sweden, Finland, Japan |
| 14. | Squash | *Cucurbita* spp. | VR | USA, Canada |
| 15. | Sweet potato | *Ipomoea batatas* | HT, VR | USA |
| 16. | Tomato | *Solanum lycopersicum* L. | PQ, FR, IR, VR, AP, BR, OO, HT, NR | India, Canada, USA, Italy, Japan, China |
| 17. | Watermelon | *Citrullus lanatus* | AP, OO, VR | USA |

AP: Agronomic Properties; BR: Bacterial Resistance; DR: Disease Resistance; DST: Drought and Salt Tolerance; FR: Fungal Resistance; IR: Insect Resistance; Mg: Selectable Marker; NR: Nematode Resistance; OO: Cold/drought Resistance; VR: Virus Resistance; HT: Herbicide tolerance; PMP: Plant manufacturing pharmaceuticals; PQ: Product Quality

*Source*: ERMA, New Zealand; OGTR, Australia; CFIA, Canada; and IGMORIS, 2008, India; US Regulatory Agencies Unified biotechnology; ISB, USA; JRC, European Commission; FAO, 2008; Katie Hagen, 2006.

**Table 12.8: Status of Laboratory and Field Trials of Biotech/Genetically Modified (GM) Vegetables in India**

| Sl.No. | Crop | Botanical name | Trait | Gene/Event | Developer |
|---|---|---|---|---|---|
| 1. | Brinjal | *Solanum melongena* L. | IR | EE-1 | Mahyco Seeds Co. |
| | | | IR | EE-1 | Sungro Seeds Co. |
| | | | IR | EE-1 | Tamil Nadu Agril. Univesity, T. Nadu |
| | | | IR | *EE-1* | Univ. of Agril. Sciences, Dharwad (Karnataka) |
| | | | IR | *cry1Fa1 gene* | Bejo Sheetal Seeds Co. |
| | | | IR | *cry1Fa1 gene* | Krishidhan Seeds Co. |
| | | | IR | *cry1Fa1 gene* | Nath Seeds Co. |
| | | | IR | *cry1Fa1 gene* | Vibha Agrotech |
| | | | IR | *cry1abc gene* | NRCPB/ICAR |
| | | | DST | *otsB-A gene* | NRCPB/ICAR |
| | | | IR | *cry1Ab gene* | ICAR-IIHR |
| | | | IR | *cry1Ac gene* | ICAR-IIVR |
| | | | IR | *cry1ia5 gene* | Nirmal Seeds Co. |
| | | | IR | *vip gene* | Nirmal Seeds Co |

| Sl.No. | Crop | Botanical name | Trait | Gene/Event | Developer |
|--------|------|----------------|-------|------------|-----------|
| 2. | Cabbage | *Brassica oleracea* var. *capitata* | IR | *cry1Ba; cri1 Ca gene* | Nunhems Seeds Co. |
| | | | IR | *cry1Ac gene* | Mahyco Seeds Co. |
| | | | IR | *cry1Ac gene* | Sungro Seeds Co. |
| 3. | Cassava | *Manihot esculenta* | VR | *Rep antisense gene* | ICAR-CTCRI, Kerala |
| | | | DR | *Replicase gene* | Tamil Nadu Agril. Univesity, T. Nadu |
| 4. | Cauliflower | *Brassica oleracea* var. *botrytis* | IR | *cry1Ac gene* | Sungro Seeds Co. |
| | | | IR | *cry1Ba; cry1 Ca gene* | Nunhems Seeds Co. |
| | | | IR | *cry1Ac gene* | Mahyco Seeds Co. |
| | | | IR | *cry1Ac/vip gene* | Bejo Sheetal Seeds Co. |
| 5. | Okra | *Abelmoschus esculentus* L. | IR | *cry1Ac gene* | Mahyco Seeds Co. |
| | | | IR | *cry1Ac gene* | Sungro Seeds Co. |
| | | | IR | *cry1Ac gene* | Bejo Sheetal Seeds Co. |
| | | | IR | *CP-AV1 gene* | Arya Seeds Co. |
| 6. | Onion | *Allium cepa* L. | DR | *n/a* | NRCOG/ICAR |
| | | | DR | *n/a* | ICAR-IIHR |
| 7. | Potato | *Solanum tuberosum* L. | NE | *Ama1 gene* | ICAR-NIPGR/CPRI |
| | | | LBR | *RB gene* | ICAR-CPRI |
| | | | LBR | *cry1Ab gene* | ICAR-CPRI |
| | | | LCV | *Cp sense gene* | ICAR-CPRI |
| | | | DST | *Osmotin gene* | ICAR-NRCPB/CPRI |
| 8. | Tomato | *Solanum lycopersicum* L. | IR | *cry1Ac gene* | Mahyco Seeds Co. |
| | | | LCV | *Rep antisense gene* | ICAR-NRCPB |
| | | | DST | *Osmotin gene* | ICAR-NRCPB |
| | | | AP | *ACS gene* | ICAR-NRCPB |
| | | | AP | *Expansin gene* | ICAR-NRCPB |
| | | | DST | *DREB 1a gene* | ICAR-IIVR/NRCPB |
| | | | DST | *DREB 1a gene* | ICAR-IIHR/NRCPB |
| | | | LCV | *Truncated Rep gene* | ICAR-NRCPB |
| | | | LCV | *Truncated Rep gene* | ICAR-IIHR |
| | | | IR | *cry1Aa gene* | NRCPB/IIHR/ICAR |
| | | | AP | *n/a* | CSIR-CFTRI |
| | | | AP and NE | *unedited NAD9 gene* | Avesthagen |
| | | | VR and FR | *n/a* | Indo-American Hybrid Seeds Co. |
| | | | IR | *cry1Ac gene* | Bejo Sheetal Seeds Co. |
| 9. | Watermelon | *Citrullus lanatus* | VR | *n/a* | ICAR-IIHR |
| | | | VR | *N gene* | UAS, Dharwad/Indian Inst. of Sci. |

AP: Agronomic Properties; CFTRI: Central Food Technological Research Institute; BR: Bacterial Resistance; CPRI: Central Potato Research Institute; DR: Disease Resistance; CSIR: Council of Scientific and Industrial Research; DST: Drought and salt Tolerance; CTCRI: Central Tuber Crop Research Institute; FR: Fungal Resistance; IARI: Indian Agricultural Research Institute; IR: Insect Resistance; ICAR: Indian Council of Agricultural Research; LBR: Late Blight Resistance; LCV Leaf Curl Virus; IIVR: Indian Institute of Vegetable Research; NE: Nutritional Enhancement; NRCPB: National Research Centre on Plant Biotechnology; ST: Submergence Tolerance; UAS: University of Agricultural Sciences; VR: Virus Resistance; NRCOG: National Research Centre for Onion and Garlic.

## Molecular Markers

Molecular markers directly reveal the polymorphism at the level of DNA. These are tags that can be used to identify specific genes and locate them in relation to other genes. There are three different types of markers, *viz.*, morphological, biochemical and molecular.

## Role of Molecular Markers in Germplasm Management

Molecular markers are used for development of saturated genetic maps; DNA fingerprinting; phylogenetic and evolutionary studies; heterotic breeding; gene tagging and marker assisted selection (MAS). Molecular markers provide greater discrimination

**Table 12.9: Features of Molecular Markers**

| Sl.No. | Character | Descriptions |
|---|---|---|
| 1. | Abundance | Genetic markers should be in abundance covering the entire genomics for the development of high density linkage maps or genome wide DNA fingerprinting. |
| 2. | Level of polymorphism | A marker must be polymorphic *i.e.* it must exist in different forms. The appropriate genetic marker technique having high level of polymorphism should be employed in genome mapping/DNA fingerprinting. The level of polymorphism among the genetic markers depends on the type of markers methods used for detection. |
| 3. | Number of alleles | There are two possible types of markers: markers with a single alternative allele (biallelic) and markers with several alternative alleles (polyallelic). |
| 4. | Locus specificity | Markers are classified in two groups: as single locus markers (unique location on the genome) and multilocus (several locations on the genome) markers. Single locus markers are preferred for genome mapping while the markers of multilocus nature are employed for DNA profiling. |
| 5. | Nature of alleles | Markers of biallelic nature are considered as co-dominant when both the alleles are observed in the hybrid. Co-dominant markers are more informative than the dominant markers. It can distinguish heterozygotes from homozygotes. This allows the determination of genotypes and allele frequencies at loci more precisely. |
| 6. | Reproducibility | Reproducibility is more important. |
| 7. | Ease of operation | There is no single molecular marker which meets all these requirements. A wide range of molecular techniques are available which detect polymorphism at the DNA level. |

**Table 12.10: Chronological Evolution of DNA Markers**

| Year | Acronym | Nomenclature | Reference |
|---|---|---|---|
| | | **First generation DNA Markers** | |
| 1974 | RFLP | Restriction Fragment Length Polymorphism | Grodziker *et al.*, 1974 |
| 1985 | VNTR | Variable number Tandem Repeats | Jeffreys *et al.*, 1985 |
| 1986 | ASO | Allele Specific Oligonucleotides | Saiki *et al.*, 1986 |
| 1988 | AS-PCR | Allele Specific Polymerase Chain Reaction | Landegren *et al.*, 1988 |
| 1988 | OP | Oligonucleotides Polymorphism | Beckman, 1988 |
| 1989 | SSCP | Single Stranded Conformational Polymorphism | Orita *et al.*, 1989 |
| 1989 | STS | Sequence Tagged Site | Olsen *et al.*, 1989 |
| | | **Second generation DNA Markers** | |
| 1990 | RAPD | Randomly Amplified Polymorphic DNA | Williams *et al.*, 1990 |
| 1990 | AP-PCR | Arbitrarily Primed Polymerase Chain Reaction | Welsh and McClelland, 1990 |
| 1990 | STMS | Sequence Tagged micro Satellite Sites | Beckmann and Soller, 1990 |
| 1991 | RLGS | Restriction Landmark Genome Scanning | Hadda *et al.*, 1991 |
| 1992 | CAPS | Cleaved Amplified Polymorphic Sequence | Akopyanz *et al.*, 1992 |
| 1992 | DOP-PCR | Degenerate Oligonucleotide Primer- Polymerase Chain Reaction | Telenius, 1992 |
| 1992 | SSR | Simple Sequence Repeat | Akkaya *et al.*, 1992 |
| 1992 | MAAP | Multiple Arbitrary Amplicon Profiling | Caetano Anoles *et al.*, 1993 |
| 1993 | SCAR | Sequence Characterized Amplified Region | Paran and Michelmore, 1993 |
| | | **New generation DNA Markers** | |
| 1994 | ISSR | Inter Simple Sequence Repeat | Zietiewicz, *et al.*, 1994 |
| 1994 | SNP | Single Nucleotide Polymorphisms | Jorden and Humphines, 1994 |
| 1995 | AFLP (SRFA) | Amplified Fragment Length Polymorphism (Selective Restriction Fragment Amplification) | Vos *et al.*, 1995 |
| 1995 | ASFLP | Allele Specific Fragment Length Polymorphism | Gu *et al.*, 1995 |
| 1996 | CFLPCleavge | Fragment Length Polymorphism | Brow, 1996 |
| 1996 | ISTR | Inverse Sequence Tagged Repeats | Rhode, 1996 |
| 1997 | DAMD-PCR | Direct Amplification of Mini Satellite DNA-Polymerase Chain Reaction | Babeli *et al.*, 1997 |
| 1997 | S-SAP | Sequence Specific Amplified Polymorphism | Waugh *et al.*, 1997 |
| 1998 | RBIP | Retrotransposon Based Insertional Polymorphism | Flavell *et al.*, 1998 |
| 1999 | IRAP | Inter Retrotransposon Amplified Polymorphism | Kalender *et al.*, 1999 |
| 1999 | REMAP | Retrotransposon microsatellite Amplified Polymorphism | Kalender *et al.*, 1999 |
| 2000 | MITE | Miniature Inverted Repeat Transposable | Casa *et al.*, 2000 |

| Year | Acronym | Nomenclature | Reference |
|------|---------|--------------|-----------|
| 2000 | TE-AFLP | Three Endonuclease- Amplified Fragment Length Polymorphism | Vander Wurff *et al.*, 2000 |
| 2001 | IMP | Inter Miniature Inverted Repeat Transposable Polymorphism | Chang *et al.*, 2001 |
| 2001 | SRAP | Sequence Related Amplified Polymorphism | Li and Quiros, 2001 |

*Source*: Maheswaran, 2004.

and sample throughout efficiency are emerging rapidly. Using molecular markers, it is possible to differentiate individuals, identify varieties and specific quality attributes in mixtures of grain or flour. Molecular markers and genomics are increasingly being used as they are less influenced by environments, and ideally should be used in combination with morphological and biochemical markers. Markers are capable of detecting single nucleotide differences (Nickerson *et al.*, 1990; Baranyi, 1991). Markers are heritable characteristics associated with and useful for the identification and characterization of specific genotypes and are DNA sequences which are readily detected and whose inheritance can easily be monitored.

The chronological evolution of DNA markers are given in Table 12.10

## Advantages of Molecular Marker for Genetic Purity Testing

The advancement of molecular biotechnology opens new vistas to fastened the breeding programme of vegetable crops by using a various molecular marker in the different methods of breeding and their steps for enhancing of the improvement programme.

☆ Accurate, rapid and reproducible cultivar identification

☆ A random scattering across the genome and established variation between the cultivar at the level of nucleotide sequence

☆ Great informative power

☆ Simple co-dominant inheritance

☆ Unaffected with respect to developmental stage, season, location and agronomic practices

☆ Potential for automation

☆ Less labour-consumption

☆ Conventional methods may fail to differentiate many vegetable cultivars because of their narrow genetic variation.

Now various technique of identification for vegetable varieties by molecular markers in tomato (Noli *et al.*, 1999), potato (Ashkenazi *et al.*, 2001), onion, garlic and related species (Fischer and Bachmann, 2000) has been developed.

## Morphological Makers

There are several references on application of molecular markers in vegetable crops. Morphological makers in hybrid development is important

## Implications of DNA Analysis Technologies

DNA analysis technologies developed in the 1990's such as microarrays dramatically increased the throughput and reduced the costs of identifying and scoring genetic variants. For the first time it became practical to analyses the whole genome rather than a small set of genetic markers. Single Nucleotide Polymorphism chips assaying in parallel 9,000 to 90,000 markers are now widely used. Improvement in cost and throughput of DNA sequencing is now making it one method of choice to identify and score genetic

### Table 12.11: Utilization of Morphological Markers in Hybrid Development

| Vegetable | Marker | Mechanism | Advantage |
|-----------|--------|-----------|-----------|
| Tomato | Brown seed: bs, bs2, bs3 and bs4 genes | No emasculation + HP | Reduced cost on hybrid seed production; only 10 per cent brown seeds (non-hybrid=selfed) which can be eliminated easily |
| Tomato | Anthocynin less (aa) linked with ms gene | Male sterile line + HP | Planting of 100 per cent male sterile plants in hybrid seed production field |
| Tomato | Enzyme marker (prx-2) linked with ms-10 gene | - | Proposed indirect use of marker in male sterile line development |
| Tomato | Anthocynin less (aa) and exerted stigma | No emasculation + HP | Reduction in hybrid seed production cost and identification of selfed seeds at seedling stage |
| Watermelon | Linkage of ms gene with delayed-green (dg) seedling marker gene | Male sterile line + HP | Planting of 100 per cent male sterile plants in hybrid seed production field |
| Onion | Brown seed | cms line + NP | Eliminate necessity of specific planting arrangement of female (cms) and male parents |
| Chilli* | The ms-10 gene is linked with taller plant height, erect growth and dark purple anther | gms line + NP | Identification of male sterile plants at comparative early stage in the hybrid seed production. |

*Exploited at commercial scale, where: HP: Hand Pollination; NP: Natural Pollination; gms: Genetic male sterility; cms: Cytoplasmic male sterility.

*Source*: Kumar and Singh, 2005.

variants. For species with large genomes, genotyping by sequencing. Where, a reproducible fraction of the genome-1to10 per cent -is sequenced, allows to determine marker profiles with a density of thousands to tens of thousands markers. The process costs about 5,000 USD for 96 samples. Prompted by technological development, genetic data production services have become available commercially. For most breeding programmes, it is now more cost-effective to outsource high density data production to commercial facilities, than to produce marker data in house. Data quality from commercial facilities is also higher.

## DNA Base Makers

DNA molecular technologies, are based on sequence variation of specific genomic regions, provide powerful tools for cultivar identification (Liu *et al.*, 2007). Molecular marker is an identifier (sometimes called a 'tag') of a particular aspect of phenotype and/ or genotype; its inheritance can easily be followed from generation to generation. Germplasm characterization and estimation of genetic diversity in plants on the basis of morphological characteristics are not much reliable. But, with DNA-based markers it is more desirable due to precise identification and quantitative estimation of genetic diversity (Glaszmann *et al.*, 2010). Molecular techniques can be utilized to develop genetic diversity data, molecular characterization, trait mapping, and elite parent selection. Molecular plant breeding utilizes molecular markers for allele-selection to investigate natural variation. Today, mutation breeding, transgenic approach and reverse genetic technologies are utilized to expand the natural variation. Estimation of genetic diversity or variability is important in vegetable crop improvement; for this purpose, molecular markers have been found superior to morphological, and biochemical markers. Molecular marker techniques give consistent information of genetic variability; it may develop screening for traits by showing linkage with quantitative and qualitative traits. This will result in the understanding of first hand architecture of traits at the level of DNA. Molecular markers are useful for early generation testing in traditional plant breeding programs. This molecular approach not only decreases the time required for cultivar development but also decrease the material for further tests. Selection of contrasting parents for crossing or breeding program is an equally important application of molecular marker. Using this tool it is possible to survey in wild or unusual germplasm for new genetic assortment. Tightly linked markers must be identified for a gene(s) controlling trait of interest; this can be used in MAS with conventional breeding programmes. The desirable properties of molecular markers for use in marker assisted selection includes: Markers should be tightly linked to target loci, should have high reliability to predict phenotype, should be highly polymorphic, should be able to discriminate between different genotypes and technique should be simple, cost effective and rapid. It is necessary about markers to get validated their associated with a diverse genetic backgrounds. In case where, cost ratio of phenotypic selection to the score marker loci is less than one and heritability is greater than 0.3, MAS is not useful. Hence, hard work needed to find out the genetic distance between molecular markers and the genes responsible for traits of interest. Sadashiva (2014) able to introgress *Ty-2* gene from *S. habrochaites* into bacterial wilt and early blight resistance lines through marker assisted selection and was successful in developing two tomato hybrids 'Arka Rakshak' and 'Arka Samrat' with triple disease resistance to *TyLCKav*, bacterial wilt and early blight.

The success of a marker-assisted breeding depends on following factors (Pawar and Kale, 2011):

(i) Availability of genetic map with an adequate number of uniformly spaced polymorphic markers to accurately locate desired QTLs or major gene(s);

(ii) Close linkage between the QTL or a major gene of interest and

(iii) Adjacent markers; adequate recombination between the markers and rest of the genome.

Among the molecular markers, isozyme markers have already proven their usefulness in genetic studies and breeding programmes of various crops. Nevertheless, isozyme loci suffer several limitations which are not inherent to the use of DNA markers.

In past two decades, however, attention has been paid on the construction of a detailed molecular map in several important vegetables including tomato, chilli, eggplant, cucumber *etc.*

## Molecular Markers in Quantitative Trait Locus (QTL) Mapping

Molecular markers can be used to analyze quantitative traits by quantitative trait locus (QTL) mapping so as to target individual loci in marker assisted selection (MAS). Usually QTL mapping is essential for MAS which can be used to pyramid several different QTLs (Pawar and Kale, 2011).

Molecular markers are classified in to four groups:

i. Probe (Southern hybridization) based- Restriction Fragment Length Polymorphism (RFLP)

ii. Amplification of PCR based-Random Amplified Polymorphic DNA (RAPD)' Simple Sequence Repeats (SSR), Inter Simple Sequence Repeats (ISSR) and Sequence Characterized Amplified Region Markers (SCAR)

iii. Combination of probe based and PCR base markers- Amplification Fragment Length Polymorphism (AFLP)

iv. New generation markers-Single Nucleotide polymorphism (SNP)

**Table 12.12: Properties of Systems for Generating Genetic Markers**

|  | RFLP | RFLP | RFLP | RFLP |
|---|---|---|---|---|
| Principle | Endonuclease restriction, Southern blotting Hybridization | DNA amplification with random primers | PCR of simple sequence repeats | Endonuclease restriction and PCR |
| Type of polymorphism | Single base changes Insertions Deletions | Single base changes Insertions Deletions | Changes in length of repeats | Single base changes Insertions Deletions |
| Genomic advance | High | Very high |  |  |
| Level of polymorphism | Medium | Medium | High | High |
| Dominance | Co-dominant | Dominant | Co-dominant | Dominant |
| Amount of DNA required | 2-10µg | 10-25 ng | 50-100 ng | 2-10 µg |
| Sequence information required | No | No | Yes | No |
| Radio detection required | Yes/No | No | No | Yes |
| Start-up costs | Medium/High | Low | High | Medium |

**Table 12.13: Details of PCR-based Diagnostics Developed in Vegetable Crops at ICAR-NBPGR, New Delhi**

| Transgenic Crops | Event/Trait | Transgene/Marker Genes | Species-Specific Gene |
|---|---|---|---|
| Brinjal | Insect resistance | cry1Ac, nptII, aadA, 35S promoter, nos terminator | exon 7 of β-fructosidase |
|  | Insect resistance | cry1Ab, 35S promoter, nos terminator | exon 7 of β-fructosidase |
| Tomato | Drought and salt tolerance | osmotin, 35S promoter | Lat52 |
|  | Drought and salt tolerance | Avp1, nptII, 35S promoter, nos terminator | exon 7 of β-fructosidase |
| Cauliflower | Insect resistance | cry1Ac, 35S promoter | SRK |
| Mustard | Male sterility | barnase, barstar, 35S promoter | HMG1 |
| Soybean | Roundup Ready soybean herbicide tolerance | cp4epsps, 35S promoter, nos terminator | Lectin |
| Maize | Roundup Ready maize | cp4epsps, 35S promoter, nos terminator | Zein |
|  | MON810 | cry1Ab, 35S promoter, nos terminator | Zein |
| Okra | Insect resistance | cry1Ac, nptII, 35S promoter, nos terminator | Chloroplast-tRNA |
| Potato | Insect resistance | cry1Ab, nptII,35S promoter, nos terminator | exon 7 of β-fructosidase |
|  | Better nutritional quality | Ama1, nptII,35S promoter, nos terminator | exon 7 of β-fructosidase |
|  | Late blight resistance | RB, nptII,35S promoter, nos terminator | exon 7 of β-fructosidase |

**Table 12.14: Total Number of Amplified Fragments and Number of Polymorphic Fragments Generated by PCR using Seven Random Decamer Oligonucleotide Primers in Cabbage**

| Sl.No. | Primer Name | Sequence (5´-3´) | Total Number of Amplified Bands | Total Number of Polymorphic Amplified Bands | Total Number of Monomorphic Amplified Bands | Genotype Specific Bands | Polymorphism Ratio, Percentage (per cent) |
|---|---|---|---|---|---|---|---|
| 1. | OPB-01 | GTTTCGCTCC | 2 | 2 | 0 | 0 | 100 |
| 2. | OPB-2 | TGATCCCTGG | 4 | 4 | 0 | 0 | 100 |
| 3. | OPB-3 | CATCCCCCTG | 2 | 1 | 0 | 1 | 50 |
| 4. | OPB-4 | GGACTGGAGT | 4 | 4 | 0 | 0 | 100 |
| 5. | OPB-5 | TGCGCCCTTC | 3 | 3 | 0 | 0 | 100 |
| 6. | OPB-7 | GGTGACGCAG | 4 | 3 | 0 | 1 | 50 |
| 7. | OPB-09 | TGGGGGACTC | 3 | 3 | 0 | 0 | 100 |
|  | Total |  | 22 | 20 | 0 | 2 | - |

*Source*: Thakur and Srivastava (2016).

## Classification of Molecular Markers

Molecular markers generally classified in the following three groups:

### (i) Southern Hybridization based DNA Markers

Which includes restriction fragment length polymorphisms (RFLP) and oligonucleotide fingerprinting. The Southern Hybridization based approach involves restriction digestion of genomic DNA, size fractionation of digested DNA on gel electrophoresis, transfer of fractionated DNA fragments on nylon membranes and finally hybridization with labeled probe to visualize DNA polymorphisms. Restriction Fragment Length polymorphisms (RFLP) is the most widely used techniques under this approach.

### (ii) Polymerase Chain Reaction (PCR) Based DNA Markers

Such as random amplified polymorphic DNAs (RAPD), sequence characterized amplified regions (SCAR), simple sequence repeats (SSR), sequence tagged sites (STS), amplified fragment length polymorphism (AFLP), inter simple sequence repeat (ISSR), cleaved amplified polymorphic sequences (CAPS) and amplicon length polymorphisms (ALPs) are widely used techniques. Polymerase Chain Reaction (PCR) based approaches involves, amplification of specific/random regions of genomic DNA using primers of different specificities, through polymerase chain reaction followed by visualization of amplified products on gel after standing with Ethidium bromide or fluorescence. In several vegetable crops linkages map has been constructed containing loci for morphological markers, resistant genes and several mutants affecting physiological functions.

Thakur and Srivastava (2016) studied the genetic relationship among 16 genotypes (Golden Acre, Best of All, BC-79, Green Emperor, Green Europium, KGAT-3, Pusa Drum Head, DARL Cabbage, EC030191, AC204, NO-29, NO-4, Giddeon, Green Kid, Cabbage Mangla and General Cabbage) of cabbage (*Brassica oleracea* L. var. *capitata*) using RAPD markers. A total of 17 primers were used to generate RAPD profiles, out of these 7 primers gave reproducible banding patterns. A total of 22 bands were obtained and all were polymorphic with cultivar specific bands. The percentage of primer polymorphism was 100 per cent. Similarity index was on Jaccard's similarity coefficient and used for cluster analysis based on UPGMA. At 62 per cent similarity level sixteen genotypes were grouped into two clusters. Results indicates that RAPD technology could be useful for identification of different genotypes as-well-as accessing the genetic similarity among different accessions of cabbage.

### Sequence Characterized Amplified Region Markers (SCAR)

Common bean variety Coyne (PI 655574) has the *Ur-3* and *Ur-6* genes for resistance to rust and carries the single dominant hypersensitive *I* gene that provides resistance to all non-necrotic strains of Bean Common Mosaic virus (Urreaa *et al.*,2009). SCAR markers identified in French bean for rust disease are listed in Table 12.15.

### (iii) DNA Chip and Sequencing Based DNA Markers

Such as single nucleotide polymorphisms (SNPs). A large number of DNA markers have been developed and many of them have been found to be linked with the gene of interest especially in tomato, one of the model

crops for the molecular biologist. Although this map has served well in genetic studies, its application in plant breeding is rather limited. Therefore, now attention is being shift to construct detailed molecular map based on protein and DNA markers in many vegetables and use them in the breeding programmes. Among all PCR-based markers are preferable because they are efficient, cost effective and require only a small quantity of genome DNA for genotyping and are enabling genotyping even at early seedling stages. Initially RAPD, AFLP, and RFLP are most commonly employed for marker trait association and diversity analysis, but there use for MAS is not preferred by conventional plant breeders because of the poor reproducibility, difficulty in handling, requirement of highly skilled person and use of radioactive elements for generating these markers. Polymerase chain reaction (PCR)-based SSR and SNP markers are likely to be preferred by breeders because these markers can be easily be employed in genotyping of large-segregating populations. SSR markers are highly reproducible, co-dominant in inheritance, relatively simple and cheap to use and highly polymorphic. Requirement of polyacrylamide gel electrophoresis system was the only disadvantages of SSRs and generally it gives information only about a single locus, although advances in electrophoresis and multiplexing approaches are possible to overcome the problems (Pawar and Kale, 2011). The available information in tomato have served many purpose *e.g.*, verifying the purity of hybrid seed, tagging of genes of economic importance, detecting genes and chromosomes from wild species following introgression, identification of somatic hybrids and mapping genes underlying quantitative variations. The genetic relationships of cultivated Chinese water chestnut in Australia were investigated by Li and Midmore (1999) using random amplified polymorphic DNA (RAPD) analysis. Ninety six RAPD markers generated by 14 primers separated the samples from Taiwan (cv. Shu-Lin), Hangzhou of mainland China (cv. Da Hong Pao), New South Wales of Australia and the USA from the remainder of the samples from Australia. These remaining samples were too closely related to be differentiated. The dissimilarity observed between these remaining samples (0.78-4.4 per cent) may be due to more scoring errors of undetectable bands and sampling error rather than to real genetic variation. It is therefore suggested that the observed morphological and physiological variations in Chinese water chestnut produced in Australia (*e.g.* corn sweetness) are phenotypic and reflect the difference of environment and cultivation rather than genetic diversity.

### Single Nucleotide Polymorphisms (SNPs)

Single nucleotide polymorphisms (SNPs) are the most common form of sequence variation between individuals within the same species. In SNPs, sequences of DDNA from individuals or lines are conflicting by

**Table 12.15: SCAR Markers Linked to Rust (*Uromyces phaseoli*) Resistance**

| SCAR Name | Size (bp)/Orientation | Sequences of SCARS | Tagged Locus | Reference |
|---|---|---|---|---|
| SK14 | 620 cis | CCC GCT ACA CAC CAA TAC CTG | Ur-3 | Haley *et al.*, 1994 |
|  |  | CCC GCT ACA CTT GAT AAA ATG TTA G |  | Nemchin. and Stavely, 1998 |
|  |  |  |  | *Miklas *et al.*, 2002 |
| SA14 | 1079/800 Codominant | CTA TCT GCC ATT ATC AAC TCA AAC | Ur-4 | Miklas *et al.* 1993 |
|  |  | GTG CTG GGA AAC ATT ACC TAT T |  | Miene *et al.* 2004 |
|  |  |  |  | *Miklas *et al.*, 2002 |
| SI19 | 460 cis | AAT GCG GGA GAT ATT AAA AGG AAA G | Ur-5 | Haley *et al.*, 1993; |
|  |  | AAT GCG GGA GTT CAA TAG AAA AAC C |  | Melotto *et al.*, 1998 |
|  |  |  |  | *Miklas *et al.*, 2000c |
| SBC6 | 308 cis | GAA GGC GAG AAG AAA AAG AAA AAT | Ur-6 | Park *et al.*, 2003b, 2004b |
|  |  | GAA GGC GAG AGC ACC TAG CTG AAG |  | *Miklas *et al.* 2002 |
| SAD12 | 537 cis | AAG AGG GCG TGA GAT CGT CG | Ur-7 | Park *et al.*, 2003a, 2004a, 2008 |
|  |  | AAG AGG GCG TCT TGA AGG TT |  |  |
| SAE19 | 890 trans | CAG TCC CTG ACA ACA TAA CAC C | Ur-11 | Johnson *et al.*, 1995 |
|  |  | CAG TCC CTA AAG TAG TTT GTC CCT A |  | Queiroz *et al.*, 2004c |
|  |  |  |  | *Miklas *et al.*, 2002 |
| UR11-GT2 | 450 cis | CGC ACT TAG GAG CAC AAA | Ur-11 | Boone *et al.*, 1999, |
|  |  | TGG TGG GTC CCA TAT TTT G |  | Miklas *et al.*, 2002 |
| KB126 | 405/430 Codominant | GAA TTC AAC CTC GGC CAC TAC C | Ur-13 | Mienie *et al.*, 2005 |
|  |  | TTA AAC CTT CCG GAG GAT TC |  |  |
| SF10 | 1072 cis | GGA AGC TTG GTG AGC AAG GA | Ouro Negro | Correa *et al.*, 2000; |
|  |  | GGA AGC TTG GCT ATGATG GT |  | *Miklas *et al.*,2002 |
| SBA8 | 530 cis | CCA CAG CCG ACG GAG GAG | Ouro Negro | Correa *et al.*, 2000; |
|  |  | GCC ATG TTT TTT GTC CCC |  | *Miklas *et al.*, 2002 |

single base. They can act as high-throughput molecular markers and can be examined and exploited. The high frequency of SNPs exists in many plant species. Increase in available expressed sequence tags (ESTs) has recommended that SNP-based marker systems are likely to become the marker of choice for plant breeders. Several techniques such as SNP microarray hybridization-based methods, pyro-sequencing, polymerase chain reaction and enzyme-based methods as have been considered for SNP detection. Second generation sequencing can be applied for the detection of markers which can be useful for MAS. With the development of technologies the cost of genotyping SNP in large populations continues to decline (Pawar and Kale, 2011).

## Simple Sequence Repeats (SSR) Markers

Genetic diversity in any crop species is critical for sustaining and thus continuing our efforts towards successful development of desirable varieties. Among the several methods available to assess diversity among germplasm, the SSR markers based DNA profiling is one of the most reliable approaches to assess differences across accessions or varieties unambiguously. SSR markers are ideal due to their co-dominant inheritance, high hyper variability, wide genomic distribution, high reproducibility, multi-allelic nature, chromosome specific locations and ease to score (Varshney *et al.*, 2005). In vegetable *Brassica*, the genic-SSR markers showed polymorphism across different species and showed close association with distant species like *B. juncea* and *B. napus* (Kumar and Radhamani, 2016).

## Sequence Related Amplified Polymorphism (SRAP)

In tomato, the diversity analysis was carried out with respect to nine morphological traits of 16 genotypes along with one check. For diversity analysis at DNA level, five forward and reverse sequences related amplified polymorphism (SRAP) primers were used. The 18 combinations of SRAP primers produced 177 amplified products among which 106 were polymorphic (59.92 per cent polymorphism). Comparison of UPGMA dendrograms of morphological and SRAP markers using the Mantel's test indicated a non-significant correlation of r=0.019 (P=0.56). Among all tomato genotypes, SRAP similarities ranged from 72 to 98 per cent, indicating the available genotypes as an important source of genetic diversity that can be exploited in future breeding programmes (Mane *et al.*, 2013).

## Diversity Array Technology

Diversity array technology is a high throughput hybridization based technique which allows real-time typing of thousands of polymorphic loci spread over a genome without any previous sequence information about these loci. In this technology, polymorphic DNAs of genome are identified based on differential hybridization using a diversity 'genotyping array'. These markers are biallelic and dominant or co-dominant. Generally genomic DNA, 50-100 ng is used for genotyping nearly 5000-8000 genomic loci in parallel in a single-reaction assay to discover polymorphic markers. With the help of this technology discovery and scoring markers can be done simultaneously.

## Use of Modern Genetic Tools in Improvement of Vegetable Crops

Retrotransposons are mobile elements that move through RNA intermediary, thereby creating new permanent loci upon insertion. The retrotransposon insertional sites have been effectively used for developing markers in many plant species. There are many types of markers developed based on insertional site variation like REMAP, IEAP and SSAP. Among the methods,

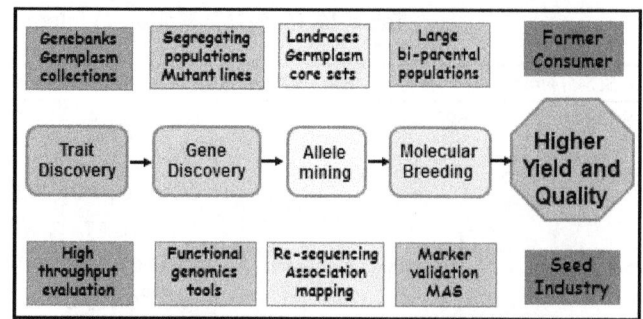

**Figure 12.1: Stages of Allele Mining and Utilization.**

sequence specific amplification polymorphism (SSAP) is more useful in revealing polymorphism. The method is similar to AFLP and involves DNA digestion, ligation of adapter and amplification of flanking region. The application is more useful in species where there is limited knowledge on genome data. But unlike AFLP, SSAP requires information on retrotransposon sequence.

Crop wise applications of some modern tools are given below:

## Tomato

Rai *et al.* (2013) conducted a study to identify PCR-based markers linked to gene(s) conferring resistance to *Tomato Leaf Curl New Delhi Virus* (ToLCNDV) by generating a $BC_1F_1$ population derived from a reciprocal cross between a ToLCNDV tolerant accession of *Solanum habrochaites* LA1777 and susceptyible 15SBSB (*Solanum lycopersicum* L.). Genetic analysis of 135 plants $BC_1F_1$ population indicated that three dominant genes confer resistance to ToLCNDV in the accession *S. habrochaites* LA1777. Using bulk-Segregant-Analysis, an ISSR marker was identified, which produced a 564 bp fragment in the tolerant wild accession and also in the tolerant bulk sample. The identified marker has been validated in a set of 18 diverse tomato accessions and can be used as a diagnostic marker to assist marker-assisted-breeding for ToLCNDV tolerance in tomato.

Hussain *et al.* (2016) screened 15 genotypes carrying various combinations of *Ty* genes; varieties including 'Pusa Rohini', 'Pusa-120', 'Pusa Ruby', 'PVB-2' along with 15 $F_1$s for reaction to ToLCV. Polymerase Chain Reaction (PCR) detection using degenerate and *Tomato Leaf Curl New Delhi Virus* (ToLCNDV) specific primers of begomo virus and *Tomato Leaf Curl New Delhi Virus* confirmed the presence of ToLCNDV in the plant genomic DNA sample extracted during field screening. All the varieties including 'Pusa Rohini', 'Pusa-120', 'Pusa Ruby' and 'Pusa Sadabahar' were 100 per cent susceptible to ToLCV. However, *Ty* genes carrying donors with PDI less than 50 per cent were EC814916 (20) and EC814917 (25). The $F_1$s Punjab Varkha Bahar-II x EC814915 (26), EC814913 (30), Pusa Rohini x EC687094 (49), Pusa Rohini x EC814916 (45) were found promising for resistant against ToLCV. Characteristic SCAR, SSR and CAPS markers, specific for different *Ty-2* and *Ty-3*

**Table 12.16: List of some Markers Linked to Disease Resistance Gene in Vegetable Crops**

| Host | Disease/Pathogen | Mapping Population* | Type of Marker | Resistance | Locus/Distance (cM)** |
|---|---|---|---|---|---|
| **A. Bulked segregant analysis (BSA)** | | | | | |
| Common bean (*Phaseolus vulgaris* L.) | Common bacterial blight (*Xanthomonos campestris* pv. *phaseoli*) | $F_2$ (77) | RAPD | *Xcp*-2 | $OPL7_{750}$ (2.4) |
| | | | | *Xcp*-3 | $OPU10_{400}$ (2.4) |
| Lettuce (*Lactuca sativa* L.) | Downy mildew (*Bremia lactucae*) | $F_2$ (66) | RAPD | *Dm5/8* | $OPF12_{1400}$ (6): 1[+] |
| | | $F_2$ (80) | RFLP | *DM16* | $CL922(0)$ : 2[+] |
| | | $F_2$ (133) | RAPD | *R17* | $OPX04_{1300}$(3.5): 4[+] |
| Soybean (*Glycine max* L.) | Frog eye leaf spot (*Cercospora sojina*) | $F_2$ (219) | RAPD | *Rcs3* | $CsoPAZ_{1250c}$(3.5): 1[+] |
| Tomato (*Solanum lycopersicon* L.) | Tomato spotted wilt virus | $F_2$ (100) | RAPD | *Sw-5* | $R1$ (1.1) : 3[+] |
| | Powdery mildew (Leveillula) | $BC_1$ (167) | RFLP | *Lv* | *CT211/CT219* (5.5) : 2[+] |
| **B. Recombinant inbred lines** | | | | | |
| Pea (*Pisum sativum* L.) | Fusarium wilt (*Fusarium oxysporum* f.sp. *pisi*) | F7 (53), BSA | RAPD | *Fwf* | *U693a* (5.6) |
| **C. Near-isogenic lines** | | | | | |
| Lettuce (*Lactuca sativa* L.) | Downy mildew (*Bremia lactucae*) | $F_{2:3}$ (66) | RAPD | *Dm4/11* | $OPA01_{860}$(1.8) : 2[+] |
| | | $F_2$ (159) | | | |
| | | $F_{2:3}$ (80) | RFLP | *Dm15/16* | *CL922* (0): 1[+] |
| Tomato (*Solanum lycopersicon* L.) | *Pseudomonas syringae* pv. *tomato* | $BC_6F_2$ (12), (35), (80) | RFLP | *Pto* | *TG96* (6.8) |
| | Verticillium wilt (*Verticillium dahliae*) | $BC_4F_2$ (49) | RAPD | *Ve* | *UBC458* (3.5) |

**Table 12.17: Infectivity Analysis of Tomato Leaf Curl New Delhi Virus in Tomato Accessions**

| Accessions | Source$ | Origin@ | Plant Infected/ Inoculated | First Symptom Appearance | Symptom Severity# | Overall Grade |
|---|---|---|---|---|---|---|
| LA1777 | *S. habrochaites* | IIHR, Bengaluru | 9/75 (12 per cent) | 12 | + | HT |
| CLN2116B | *S. hirsutum* | AVRDC, Taiwan | 28/75 (37.3 per cent) | 10 | ++ | MT |
| TLBRH-5 | *S. habrochaites* | IIHR, Bengaluru | 29/75 (38.6 per cent) | 10 | ++ | MT |
| TLBRH-6 | *S. habrochaites* | IIHR, Bengaluru | 29/75 (38.6 per cent) | 10 | ++ | MT |
| EC520071 | *S. peruvianum* | IIHR, Bengaluru | 27/75 (36 per cent) | 10 | ++ | MT |
| IIHR2202 | *S. habrochaites* | IIHR, Bengaluru | 45/75 (60 per cent) | 8 | +++ | S |
| H-24 | *S. habrochaites* | IIHR, Bengaluru | 39/75 (52 per cent) | 9 | +++ | S |
| FLA496-11-61-0 | *S. chilense* | AVRDC, Taiwan | 41/75 (54.6 per cent) | 8 | +++ | S |
| FLA478-6-3-1-11 | *S. chilense* | AVRDC, Taiwan | 42/75 (56 per cent) | 9 | +++ | S |
| FLA653-3-1-0 | *S. chilense* | AVRDC, Taiwan | 44/75 (58.6 per cent) | 8 | +++ | S |
| FLA456-4 | *S. chilense* | AVRDC, Taiwan | 41/75 (54.6 per cent) | 8 | +++ | S |
| CLN2026 | *S. hirsutum* | AVRDC, Taiwan | 38/75 (50.6 per cent) | 9 | +++ | S |
| CLN2498D | *S. hirsutum* | AVRDC, Taiwan | 39/75 (52 per cent) | 9 | +++ | S |
| 99S-C-39-20-11-24-17-0 | NA | AVRDC, Taiwan | 43/75 (57.3 per cent) | 8 | +++ | S |
| BL1172 | NA | IIHR, Bengaluru | 45/75 (60 per cent) | 8 | +++ | S |
| EC20060 | *S. hirsutum* | IIHR, Bengaluru | 68/75 (90.6 per cent) | 6 | ++++ | HS |
| 15SBSB | *S. lycopersicon* | IIHR, Bengaluru | 71/75 (94.6 per cent) | 6 | ++++ | HS |
| EC520070 | *S. peruvianum* | IIHR, Bengaluru | 67/75 (89.3 per cent) | 6 | ++++ | HS |
| TY55 | *S. chilense* | AVRDC, Taiwan | 70/75 (93.3 per cent) | 6 | ++++ | HS |
| TY52 | *S. chilense* | AVRDC, Taiwan | 71/75 (94.6 per cent) | 5 | ++++ | HS |

#+: Least severe; ++: Moderately severe; +++: Severe; ++++: Highly severe.

* T: Tolerant (1-20 per cent); MT: Moderately tolerant (20.1-40 per cent); S: Susceptible (41.1-60 per cent); HS: Highly susceptible (60.1-100 per cent)

@: ICAR-IIHR, Indian Institute of Horticultural Research; AVRDC: Asian Vegetable Research and Development Centre

$: NA: Not available.

genes validated on resistant lines, recurrent parents and F$_1$s. *Ty-2* genes with resistant fragment and susceptible allele was validated by marker TG302 as well as by CAPS marker TG36. The *Ty-3* gene with resistant allele, susceptible allele and *Ty-3a* was validated by a gene based SCAR marker P6-25. These *Ty* genes were validated in breeding lines EC814915, EC814916, EC814917 and EC814918. Based on field screening for reaction to ToLCV, the breeding lines carrying *Ty-2* and *Ty-3* were resistant to *Tomato Leaf Curl New Delhi Virus*. The breeding lines and F$_1$s carrying *Ty-1, Ty-2, Ty-5* and *Ty-6*, alone as well as in combinations, were not effective against ToLCNDV. Molecular markers associated with *Ph-2* and *Ph-3* gene conferring late blight (LB) for resistance will be extremely useful in the screening and selection of tomato. Three co-dominant markers (TG 422, TG 328 and TG 591) were found to be associated with *Ph-2* and *Ph-3*, all of which were sequence characterized amplified region (SCAR) type. The tomato line PAU 2372 and LBR-12 were crossed to produce a single cross hybrid (F$_1$). A total of 120 plants were screened for the target loci and tagged, from 120 F$_2$ plants, 17 were resistance for *Ph-2* and 26 for *Ph-3* gene 51 plants were susceptible and 15 plants were found to be positive for both *Ph-2* and *Ph-3*gene (Dubey *et al.,* 2016).

One approach to improving the colour and lycopene content has been the use of genes that affect the colour and biosynthesis of carotenoids. The most important genes are *at*, **B** and its allele *agc, Del, dps, gf, gh, Gr, hp-1, hp-2, Ip* and its allele *dg, Mo$_B$, ry, t, r, sucr,* etc. (Wann *et al.,* 1985; Chalukova and Manueyan, 1991; Chetelat *et al.,* 1993, 1995). Many of these genes have been used in improving fruit quality, sometimes by employing marker-assisted selection. RAPD and AFLP (Zhang and Stommel, 2000) and SCAR and CAPS linked to the *B* and *Mo$_B$* genes have been identified, thus facilitating their introgression in breeding lines. Delayed fruit ripening (antisense RNA-technology) cultivar 'Flavr Savr' (trademark) was commercialized in the year 1994 by Calgene. Delayed ripening (antisense RNA-technology) cultivar 'Endless Summer' (trademark) was commercialized in the year 1995 by DNA Plant Technology. In the year 1995 delayed ripening was given more attention Monsanto. Remarkable work on important traits like thicker skin and altered pectin content was started in the year 1995 by Zeneca/Petaseed,

Molecular markers are linked to major disease resistance in tomato such as *Meloidogyne incognita* (Williamson *et al.,* 1994) and tomato mosaic virus (Sabir *et al.,* 2000).

## Brinjal

The SSR markers were utilized successfully for development of commercial improved varieties of brinjal for higher yield. Genetic diversity was estimated by Mahalanobis D$^2$ statistic of multivariate analysis, which was recognized as a powerful tool in quantifying the degree of genetic divergence among the population. To understand, the actual genetic diversity among genotypes at molecular level SSR markers with D$^2$ statistic can give more precise results and information on genetic diversity of brinjal with a final goal to improve molecular breeding and identification of high yielding elite genotypes. Moreover, the work also evaluates the actual genetic and morphological diversity performance of genotypes at molecular level. The nucleotide sequence and phylogeography analysis of 16Sr RNA region of phytoplasma infecting brinjal was worked out by Kodandaram *et al.* (2016) and reported that it belongs to Clover proliferation group (16Sr-IV). Screening of 55 brinjal varieties, lines and 17 wild *Solanum* species under natural conditions showed, cv. Uttara and all wild *Solanum* species were found to be immune without any diseases incidence and cv. Pusa Ankur as resistant with 6.5 per cent disease incidence, remaining 12 lines were moderately resistant, 37 lines were susceptible and 4 lines were highly susceptible. PCR analysis using the universal primer pair (P1/P7) corresponding to 16Sr RNA gene also showed no amplification in cv. 'Uttara' and 17 wild *Solanum* species.

## *Capsicum* spp.

In chilli out of 60 polymorphic SSR markers screened, only four differentiated the individual constituents of resistant and susceptible bulks against anthracnose (*Colletotrichum* spp.). Of these four, only one (HpmsE081) was found associated with genomic regions controlling anthracnose resistance. However, the association was weak as suggested by low contribution of the marker towards the variance of response to anthracnose disease in terms of lesion size (Nanda *et al.,* 2016). Verma *et al.* (2016) characterized 106 germplasm of chilli including cultivars (hot and sweet pepper) and popular land races like 'King-chilli', 'Cherry chilli', 'Dalle chilli' and 'Bird Eye chilli' for yield contributing traits and also differentiate on the basis of microsatellite markers. Wide range of variation observed for all yield contributing traits. A total of 205 alleles were generated with an average of 4.36 alleles per locus using 47 microsatellite markers. The number of alleles ranged from two (in HPMSE 7) to eight (in HPMSE 72 and CAMS 91). Based on the value of Shannon's information index CAMS 91 (1.96) was identified as the most informative marker in this study. Out of 205 alleles only 10 alleles were common in all different *Capsicum* spp. under study. Nine alleles were found specific to Dalle chilli (*Capsicum chinense*) and five to Bird eye chilli (*C. frutescens*). Two alleles were also found specific to King chilli (*C. chinense*) and 21 alleles were specific to chilli (*C. annum*). The results of cluster analysis revealed that the King chilli is closer to Bird eye chilli than the Dalle chilli. The popular land races were found more diverse over commercial cultivars of both chilli and capsicum. High level of genetic diversity assessed in this study at both morphological and molecular level emphasized the importance of this region for conserving *Capsicum* spp. germplasm. The pattern of genetic diversity identified in this region may be used to develop *in situ* conservation strategies for *Capsicum* spp.

## French Bean

DNA markers tightly linked to agronomically important traits are used as molecular tools for marker assisted selection (MAS) in plant breeding (Ribaut and et al., 1997). MAS involve identifying the molecular markers tightly linked to target gene and using such markers for phenotypic selection at the early stage. Thus, molecular markers are efficient, effective and reliable in selection of desirable plant types. Besides, it saves cost and time compared to conventional breeding procedures. This new approach of using of DNA markers as breeding tools in MAS is called 'Molecular Breeding' (Rafalski and Tingey, 1993). Torres *et al.* (2008) used MAS in the genetic improvement of several legume crops like soybean, common bean and peas. MAS for genetic improvement of faba bean are currently under progress. He even reported that, recently, markers linked to a gene controlling growth habit or to select against traits affecting the nutritional value of seeds (tannins, vicine and convicine content) have also been reported, which may facilitate a more efficient selection of new cultivars free of anti-nutritional compounds.

Ender *et al.* (2008) studied on marker-assisted selection (MAS) of two recombinant inbred line (RIL) populations of common bean (*Phaseolus vulgaris* L.) to test the effectiveness of MAS for resistance to white mold (*Sclerotinia sclerotiorum*). Markers for quantitative trait loci (QTL) on linkage groups $B_2$ and $B_7$ that were previously associated with resistance and plant architectural avoidance traits in the resistant parent Bunsi were chosen. In the 'Bunsi' or Midland population 10 RILs included in the MAS selected group developed significantly less disease than the control group based on two years of field evaluation under white mold pressure. Growth habit had no significant effect on disease severity or incidence. In the Bunsi or Raven RIL population, disease scores in the MAS selected group were significantly lower than scores in the control group over two years. Additional progress in enhancing resistance to white mold was detected when yield and plant architecture were included in the selection process. Lower disease scores among RILs were observed when comparisons were made to RILs selected using MAS alone. Yield is an important trait that should be considered while selecting for resistance to white mold. Finally, the potential of Bunsi as a genetic donor of QTL for white mold resistance was confirmed in both populations studied. This study supported the effectiveness of MAS to enhance selection for a complexly inherited trait such as resistance to white mold in common bean.

## DNA Based Markers in French Bean

The term 'marker' implies that the detected loci are phenotypically neutral. These DNA based makers differentiate the organisms at DNA level and are inherited in simple Mendelian fashion (Waltson, 1993). DNA markers can detect differences in genetic information carried by two or more individuals. Markers are applied in a number of studies *viz.*, forensic studies, identifying genes responsible for disease resistance, evolutionary linkage mapping, paternity testing; map based cloning and genetic diversity studies. They are important for management of genetic resources and for effective use of genetic resources in selection programs. An ideal genetic marker should be polymorphic, multi-allelic, co-dominant, non-epistatic and not affected by environment. Gene tagging and marker assisted selection for disease resistance has made rapid progress in recent years and has provided effective tool for indirect selection for resistance to a number of major diseases.

## Use of Molecular Markers in French Bean

Markers independent from the environment are necessary for reliable identification and discrimination of genotypes and cultivars. Different types of marker systems have been used for genetic analysis and genotyping, including morphological, cytological, biochemical and DNA markers. The values of markers depend on their heritability and on the level of polymorphism they can reveal (Porter and Smith, 1982). DNA markers are independent from environmental interactions and show high level of polymorphism. Therefore, they are considered valuable tools for determining genetic relationships or diversity. The molecular markers are based on different principles, are obtained by using procedures of varying complexity and generate different amounts of polymorphic data. One of the main uses of DNA markers in agricultural research has been in the construction of linkage maps for diverse crop species and analysis of quantitative trait loci (QTL's). In *Phaseolus*, integration of the genes and QTL mappings and use of larger segregating populations enable better resolution of minor effect QTLs and characterization of gene clusters. Once an important QTL is found, the region is saturated with markers using phenotype and map-based bulk-segregant analysis in an effort to obtain tightly linked flanking markers. Further fine mapping, using BACs for development of contigs, may lead to identification and cloning of the gene responsible for the QTL. As more resistance-linked markers are characterized and mapped, the utility of MAS for developing durable and multiple disease resistant cultivars will increase in bean breeding programme around the world (Miklas *et al.*, 2002). Rivkin *et al.* (1999) studied primers based on a conserved nucleotide binding site (NBS) found in several cloned plant for disease resistant genes which were used to amplify DNA fragments from the genome of common bean (*Phaseolus vulgaris* L.). Cloning and sequence analysis of these fragments uncovered eight unique classes of disease-resistance related sequences. All eight classes contained the conserved kinase 2 motif and five classes contained the kinase 3a motif. Gene expression was noted for five of the eight classes of sequences. A clone from the SB3 class mapped 17.8 cM from the *UR-6* gene that confers resistance to several races of the bean rust pathogen *Uromyces appendiculatus* L. Linkage mapping identified micro-clusters of disease resistance related sequence in common bean and sequences mapped to four linkage groups in one population. Comparison with similar sequences from soybean (*Glycine max*) revealed that any one class of common bean disease-resistance related sequences was more identical to a soybean NBS containing sequence than to the sequence of another common bean class. Miklas *et al.* (1992) found that the $Up_2$ gene of common bean (*Phaseolus vulgaris* L.) is an important source of dominant genetic resistance to the bean rust pathogen [*Uromyces appendiculatus* (Pers. ex Pers.) Unger. var. *appendiculatus* [syn *U. phaseoli* (Reben) Wint.]. A strategy that employed bulked DNA samples formed separately from the DNA of three $BC_6F_2$ individuals with $Up_2$ and three without $Up_2$ as contrasting near isogenic lines (NILs) was used to identify random amplified polymorphic DNA fragments (RAPDs) tightly linked to the $Up_2$ locus. Only 1 of 931 fragments amplified by 167 10-mer primers of arbitrary sequence in the polymerase chain reaction (PCR) was polymorphic. The RAPD marker ($OA14_{1100}$) amplified by the 5'-TCTGTGCTGG-3' primer was repeatable and its presence and absence was easy to score. No recombination was observed between $OA14_{1100}$ and the dominant $Up_2$ allele within a segregating $BC_6F_2$ population of 84 individuals. This result suggests that $OA14_{1100}$ and $Up_2$ were tightly linked. Andean and Mesoamerican bean germplasm, with and without the $Up_2$ allele, were assayed for the presence of $OA14_{1100}$. The marker possibly is of Andean origin. Because all Andean lines, with or without the $Up_2$ allele, had the marker and this marker was absent in all Mesoamerican germplasm except the lines to which $Up_2$ had been transferred. These results suggest that $OA14_{1100}$

will be most useful for pyramiding $Up_2$ with other rust resistance genes into germplasm of Meso-American origin where the marker does not traditionally exist. The use of bulked DNA samples may have concentrated resources toward the identification of RAPDs that were tightly linked to the target locus.

Park *et al.* (2004b) carried out bulk segregant analysis to identify random amplified polymorphic DNA (RAPD) markers linked to genes for specific resistance to a rust pathotype and indeterminate growth habit in an $F_2$ population from the common bean cross PC-50 (resistant to rust and determinate growth habit) X Chichara 83-109 (susceptible to rust and indeterminate growth habit). Six RAPD markers were mapped in a coupling phase linkage with the gene (*Ur-9*) for specific rust resistance. The linkage group spanned a distance of 41 cM. A RAPD marker OA4.1050 was the most closely linked to the *Ur-9* gene at a distance of 8.6 cM. Twenty-eight RAPD markers were mapped in a coupling phase linkage with the gene (*Fin*) for indeterminate growth habit. The linkage group spanned a distance of 77 cM. RAPD markers OQ3.450 and OA17.600 were linked to the *Fin* allele as flanking markers at a distance of 1.2 cM and 3.8 cM, respectively. The RAPD markers linked to the gene for specific rust resistance of Andean origin were detected along with other independent rust resistance genes from other germplasm could be utilized to pyramid the different genes into a bean cultivar for durable rust resistance.

## SCAR Markers Related to Rust Resistance in French Bean

SCAR (Sequence Characterized Amplification Polymorphism) was developed by sequencing single RAPD band and designing primers (usually 18 to 24 nucleotides) to amplify band of a specific size. These SCAR markers overcome the problem of low reproducibility of RAPD and ISSR markers. Correa *et al.* (2000) attempted to identify molecular markers linked to rust resistance gene block present in bean cultivar *Ouro Negro*. They grouped the $BC_3F_{2:3}$ into resistant and susceptible bulk. The bulks were amplified with

**Table 12.18: Some Selected Examples of RAPD Markers Linked to QTLs in Common Bean**

| QTL/Pathogen | Per cent Variance | Location on Chromosome | Marker |
|---|---|---|---|
| White mold disease (*Sclerotinia sclerotiorum*) severity index | 11.6 | Linkage group B2 | $BC20_{1800}$ |
| Days to maturity | 13.5 | B2 | $O12_{1600}$ |
| Branching pattern | 9.1 | B7 | $I07_{1200}$ |
| Lodging | 9.1 | B7 | $I07_{1200}$ |
| Seed size (g 100 seed$^{-1}$) | 19.5 | B7 | $I07_{1200}$ |
| Yield (kg ha$^{-1}$) | 36.5 | B7 | $I07_{1200}$ |
| Yield (kg ha$^{-1}$) | 27.2 | B7 | $G17_{820}$ |
| Web blight (*Thanatephorus cucumeris*) | 10 | Linkage group3 | $O16_{1200}$ |
| | 7 | Linkage group1 | $H19_{1050}$ |
| | 5 | Linkage group2 | $D13_{1300}$ |

605 random primers. Two bands were polymorphic between the bulks and those bands were transformed into SCAR namely SCARBA08 and SCARF10. Corrales *et al.* (2008) identified SSR markers linked to the rust resistance gene or genes present in PI 260418. Liebenberg *et al.* (2006a and b) studied and confirmed that the *Ur-13* gene in Kranskop and Redlands Pioneer gives protection to many races of *Uromyces appendiculatus* L. In this study, three co-dominant SCAR markers (SEAACMACC430/405, SEACAMCTT310/288 and SEAGGMCGT436) were used to trace the origin of *Ur-13*. Liebenberg *et al.* (2006b) found that the SCAR marker SAE19890 derived from RAPD marker OAE19890 is linked in repulsion to *Ur-11* (from PI 181996) and it can be used as a coupling marker for *Ur-*(3+11). Mahuku *et al.* (2004) reported that three AFLP markers (E-ACA/MCTT330, E-AAC/M-CAG310 and E-AAC/M-CAT285) segregated in coupling phase with the resistance gene in G 10474 line and even the E-ACA/M-CTT330 marker was successfully converted to a co-dominant sequence characterized amplified region (SCAR) marker at 5 cM from the resistant gene. Further, validation of the SCAR marker outside the mapping population showed that the utility of this marker for marker-assisted selection (MAS) was limited to the Andean gene pool of *P. vulgaris*. Mienie *et al.* (2005) used $F_2$ population derived from a cross between Kranskop and a susceptible South African cultivar Bonus. They used bulk segregant analysis in combination with the amplified fragment length polymorphism (AFLP) technique. In their study, they found that the seven AFLP fragments linked significantly to the rust resistant gene *Ur-13*. Out of which, five were successfully converted to SCAR markers. KB1 Taq 1, KB4 Hha1, KB7, KB85, KB126 were successfully developed SCAR markers from AFLPs. Great Northern common bean germplasm line ABC-Weihing has *UR-3* and *UR-6* genes for resistance to common bean rust and carries the single dominant hypersensitive *I* gene that provides resistance to all non-necrotic strains of the *Bean common mosaic virus* (BCMV) (Mutlu *et al.*,2008). Park *et al.* (2004a) identified random amplified polymorphic DNA (RAPD) markers linked to *UR-6* controlling SR (Specific resistance) to race 51 using bulk segregant analysis in an $F_2$ population from the Meso-American (MA) common bean cross Olathe (resistant) x Nebr1 sel. 27 (susceptible) and determined the presence or absence of these identified markers in 70 Mesoamerican and Andean bean cultivars and breeding lines. A single dominant gene controlling SR to race 51 was found in the $F_2$ and confirmed in the $F_3$. The gene was flanked by two coupling-phase markers OBC06.300 and OAG15.300 at 1.3 and 2.0 cM. Sequence characterized amplified region (SCAR) marker SOBC06.308 was developed based on the specific primer pair designed from the sequence of the RAPD marker OBC06.300.

A coupling-phase random amplified polymorphic DNA (RAPD) marker OAD12.550 previously was identified to be linked (no recombination) to *Ur-7* of Middle American (MA) origin for specific rust resistance in the common bean cross of Great Northern (GN) 1140 × GN Nebr 1. They found that RAPD marker OAD12.550 most tightly linked to *Ur-7* to a SCAR marker SOAD12.537 for use as a marker assisted selection tool and survey the presence or absence of the SCAR marker SOAD12.537 in 90 MA and Andean bean genotypes for determining the genetic relationship of *Ur-7* with *UR-6* (Park *et al.*, 2008)

Common bean variety Coyne (PI 655574) has the *Ur-3* and *Ur-6* genes for resistance to rust and carries the single dominant hypersensitive *I* gene that provides resistance to all non-necrotic strains of Bean Common Mosaic Virus (Urreaa *et al.*,2009). SCAR markers identified in French bean for rust disease are listed in Table 12.19.

## *Vigna* Species

Improved cultivars of asparagus bean in China generally had a narrow genetic basis compared with landraces. This suggested that breeding programs of asparagus bean need to utilize landrace Germplasm to enhance genetic variability, ensure long-term gains from selection, and reduce genetic vulnerability to pathogen or pest epidemics. Tantasawat *et al.* (2010) estimated genetic diversity and relatedness of 23 yardlong bean (*Vigna unguiculata* ssp. ssp. *sesquipedalis*) accessions and 7 accessions of a hybrid between cowpea (*V. unguiculata* ssp. *unguiculata*) and dwarf yard long bean in Thailand by morphological characters, SSR and ISSR markers. Five morphological characters were diverse among most accessions. However, five groups of 2-3 accessions could not be distinguished from one another based on these morphological characters alone. The comparison of average marker index of the multilocus marker and mantel test indicated higher efficiency of IISR for estimating the levels of genetic diversity and relationships among yard long bean and dwarf yard long beans in the study. Xu *et al.* (2010) assessed the genetic diversity of asparagus bean cultivars from different geographical origins in China by EST-derived and GSS-derived SSR markers. PCA (principal coordinate analysis) and phylogenetic clustering based on 62 alleles detected by 14 polymorphic SSR markers distinguished ssp. *unguiculata* and *sesquipedalis* into separate groups. Xu *et al.* (2011) reported the first genetic map of asparagus bean based on SNP and SSR markers. The current map consists of 375 loci mapped onto 11 linkage groups, with 191 loci detected by SNP markers and 184 loci by SSR markers. The overall map length is 745 cm, with an average marker distance of 1.98 cm.

## Cluster Bean

Kumar *et al.* (2016) assessed the genetic diversity among 30 genotypes of cluster bean belonging to Rajasthan and Punjab were analyzed by using 21 ISSR markers. Out of the 263 total fragments generated across

**Table 12.19: SCAR Markers Linked to Common Mosaic Rust (*Uromyces phaseoli*) Resistance**

| SCAR Name | Size (bp)/Orientation | Sequences of SCARS | Tagged Locus | Reference |
|---|---|---|---|---|
| SK14 | 620 cis | CCC GCT ACA CAC CAA TAC CTG<br>CCC GCT ACA CTT GAT AAA ATG TTA G | Ur-3 | Haley *et al.*, 1994<br>Nemchin. and Stavely, 1998<br>Miklas *et al.*, 2002 |
| SA14 | 1079/800<br>Codominant | CTA TCT GCC ATT ATC AAC TCA AAC<br>GTG CTG GGA AAC ATT ACC TAT T | Ur-4 | Miklas *et al.* 1993<br>Miene *et al.* 2004<br>Miklas *et al.*, 2002 |
| SI19 | 460 cis | AAT GCG GGA GAT ATT AAA AGG AAA G<br>AAT GCG GGA GTT CAA TAG AAA AAC C | Ur-5 | Haley *et al.*, 1993;<br>Melotto *et al.*, 1998<br>Miklas *et al.*, 2000c |
| SBC6 | 308 cis | GAA GGC GAG AAG AAA AAG AAA AAT<br>GAA GGC GAG AGC ACC TAG CTG AAG | Ur-6 | Park *et al.*, 2003b, 2004b<br>Miklas *et al.* 2002 |
| SAD12 | 537 cis | AAG AGG GCG TGA GAT CGT CG<br>AAG AGG GCG TCT TGA AGG TT | Ur-7 | Park *et al.*, 2003a, 2004a, 2008 |
| SAE19 | 890 trans | CAG TCC CTG ACA ACA TAA CAC C<br>CAG TCC CTA AAG TAG TTT GTC CCT A | Ur-11 | Johnson *et al.*, 1995<br>Queiroz *et al.*, 2004c<br>Miklas *et al.*, 2002 |
| UR11-GT2 | 450 cis | CGC ACT TAG GAG CAC AAA<br>TGG TGG GTC CCA TAT TTT G | Ur-11 | Boone *et al.*, 1999,<br>Miklas *et al.*, 2002 |
| KB126 | 405/430<br>Codominant | GAA TTC AAC CTC GGC CAC TAC C<br>TTA AAC CTT CCG GAG GAT TC | Ur-13 | Mienie *et al.*, 2005 |
| SF10 | 1072 cis | GGA AGC TTG GTG AGC AAG GA<br>GGA AGC TTG GCT ATGATG GT | Ouro Negro | Correa *et al.*, 2000;<br>Miklas *et al.*,2002 |
| SBA8 | 530 cis | CCA CAG CCG ACG GAG GAG<br>GCC ATG TTT TTT GTC CCC | Ouro Negro | Correa *et al.*, 2000;<br>Miklas *et al.*, 2002 |

*Source*: Urreaa *et al.* (2009).

all germplasm, 204 (77.57 per cent) were polymorphic and 59 (22.43 per cent 0 were monomorphic. The number of polymorphic fragments per primer ranged from four (markers UBC-854) to fifteen (ISSR-5 and IS-7 marker), with an average of 9.71. The percentage polymorphism ranged from 58.33 per cent (UBC-856) to 100 per cent (UBC-820 and ISSR-8). The average percentage polymorphism for ISSR markers was 77.06. the observed number of alleles, effective number of alleles, Nei's genetic diversity and Shannon's information index for cluster bean genotypes using 21 ISSR markers were found to be 1.78120 ±0.4167, 1.4627 ±0.3844, 0.2671 ±0.1939 and 0.3988 ±0.268, respectively. The value of total genotype diversity among population (Ht) was

0.2639 ±0.0378 whereas diversity within population (Hs) was found to be 0.253 ±0.035. Mean coefficient of gene differentiation (Gst) value was 0.041 and the estimated gene flow in the population was 11.549. ANOVA revealed that molecular variances were 8 per cent of the total variance and was among the sub-populations, while 92 per cent was among individuals within the populations. The estimated variance based on ISSR marker data was 0.608 (among the population) and 6.783 (within the population). The average estimated variance was 3.965. Coefficient of variability, estimates of heritability and genetic advance as per cent of mean, as well as correlation coefficient among yield attributes of cluster bean observed in the present study substantially reveal the great scope for improvement in many of the desirable economic traits of the crop. Most desirable genotypes on the basis of morphological characterization were IC421834, IC421855, IC421828, IC421855, HG-365, IC421809 and IC421806. The promising germplasm identified will serve as useful resources for functional and comparative genomics, mapping and cloning genes, and in applied breeding for enhancing the genetic potential of the cluster bean. Most desirable genotypes investigate on the basis of molecular characterization in this study are IC258087, IC258092, IC282272, IC311440, IC311441, IC369789 and IC369868.

## Field Bean

An efficient, reproducible and genotype independent *Agrobacterium*-mediated in-planta transformation protocol has been standardized for field bean varieties using half seed as explants by Sivabalan *et al.* (2016). The field bean seeds were infected with three *A. tumefaciens* strains harbouring pCAMBIA1304-plasmid, and the transformed were selected against *hptII*. Various parameters influencing the field bean genetic transformation system including, *Agrobacterium tumefaciens* strains, pre-culture duration, acetosyring one concentration, co-cultivation, sonication and vacuum infiltration duration, have been evaluated. The putatively transformed (T0) plants were screened by GUS and GFP visual assay. Among the various parameters and concentration tested, 12-h pre-cultured seeds were sonicated for 6 min. and 3 min. vacuum infiltered in 150μM acetosyringone containing *A. tumefaciens* EHA 105 suspension and co-cultivated for 72-h showed transformation efficiency of 29.6 per cent (cv. Co13). The amenability of the standardized protocol was evaluated on five different varieties. Among the five field bean varieties evaluated using the standardized protocol, with var. Co (Gb) 14 showed the highest transformation efficiency at 38.6 per cent. The transgene was successfully transmitted to progeny plants (T1) which confirmed the GUS. Gene integration by GUS, GFP assay and expression whereas polymerase chain reaction (PCR) and southern blot hybridization confirmed the gene integration and copy number in the transformed field bean genome. The transformed field

bean plants (T0 and T1) were fertile and morphologically normal. The developed protocol is cost-effective, efficient, short duration, genotype independent and applicable to transform the plants with insect-resistant traits, into different varieties of field bean and the transformed plants can be generated within 60 days.

## Muskmelon (*Cucumis melo* L.)

Garcia Rodrigez *et al.* (1996) used RAPD to determine genetic relationships between nine genotypes of *Cucumis melo* belonging to different varietal types. Only 39 amplification products, among the 107 generated using 19 polymorphic primers, revealed polymorphism. Dani-Poleg *et al.* (1996) studied microsatellite and RAPD marker in 8 varieties of melon. Microsatellite markers detected a high level of polymorphism (78 per cent), while RAPD was able to detect only 38 per cent polymorphism. According to these results, microsatellite markers are a very promising tool for the detection of polymorphism in the Cucurbitaceous family. Silberstein *et al.* (1996) assessed DNA polymorphism among *Cucumis melo* accessions using RAPD and RFLPs. Thirteen varieties that represent diverse melon types were surveyed using 18 RAPD primers. According to the phylogenetic tree, the longest branch separates *C. metuliferus* from all the melon accessions. The largest divergence among melon types occurred between *C. melo* var. *momordica* from India and the North American and European cultivars. Diversity among 17 melon landraces and inbred line was assessed using RAPD (Staub *et al.*, 2004). Total 19 RAPD primers indicate that all but one of the accessions from Greece showed genetic affinities among themselves and with 23 RA of various origins.

## Cucumber (*Cucumis sativus* L.)

Horejsi *et al.* (1999) studied efficiency of RAPD to SCAR in cucumber (*C. sativus*). The data suggest that although SCAR markers may demonstrate enhanced performance over the RAPD markers from which they are derived, careful consideration must be given to both the costs and potential benefits of SCAR marker development in cucumber. Development and characterization of PCR markers in cucumber was studied by Fazio *et al.* (2005) who developed 110 SSR, 4 SCAR and 2SNP markers. The identified markers will enhance the use of genetic breeding, diversity analysis, variety identification and the protection of cucumber germplasm. Xie *et al.* (2002) studied PCR based single strand conformation polymorphism (SSCP) analysis to clone nine aquaporin genes in cucumber. Nine sequences from the cucumber genome were successfully identified and cloned that encoded two well conserved asparagines-proline-alanine (NPA) homologus to aquaporin genes. Jat *et al.* (2016) screened two parental lines with 485SSR primers pairs resulted in the detection of 54 polymorphic markers between PPC-2 and Pusa Uday. The 11 SSR markers were found polymorphic for the chromosome 5 where the gene for uniform immature fruit colour (u) and dull fruit skin (D) are present. In the linkage analysis for chromosome 6, opne co-dominant marker SSR 13251 was found to be tightly linked to the gynoecy (F locus) at 1.0cM. Genetic mapping information from this study opens the way for marker-assisted selection of these horticulturally important traits in cucumber. Punetha *et al.* (2016) access genetic diversity of 13 genotypes using 08 ISSR primers. The 06 ISSR primers generated 52 alleles. A total of 52 loci were amplified that exhibited 92.30 per cent polymorphism. A maximum of 11 loci were detected, by the primer UBC-808, UBC-840 and UBC-855, the similarity value ranged from 22 per cent to 80 per cent with the average value of 46 per cent. *Pgyn*-1 was found most diverse than others with 33 per cent similarity. The 13 genotypes were grouped into two major clusters (A and B) based on ISSR markers. Cluster B contained the maximum number of genotypes and further divided in two major sub clusters IB and IIB. Sub cluster IB includes small sub clusters of seven genotypes in which *Pgyn*-1 was found separately whereas, sub clusters A having only one genotype *Pgyn*-1 which was found most diverse from the other.

## Cucurbita Species

Brown *et al.* (2002) used RAPD markers to construct a partial map of the *Cucurbita* genome. The map contains 148 RAPD markers in 28 linkage groups. Loci controlling five morphological traits were placed on the map. QTL associated with fruit shape and the depths of indentations between primary leaf veins were identified. In summer squash transgenic variety 'Freedom' was released commercially by Asgrow in the year 1995.

## Pointed Gourd

Female sex associated RAPD marker in pointed gourd (*Trichosanthes dioica* Roxb.) was studied by Singh *et al.* (2001). RAPD marker associated with females, and is absent in males of pointed gourd was identified. The importance of this finding is the early detection of gender in *T. dioica* Roxb., as well as possible implication in understanding the molecular basis of sex determination in this dioecious species.

## Bottle Gourd

Xu *et al.* (2011) reported a total of 150,253 sequence reads, which were assembled into 3,994 contigs and 82, 522 singletons were generated. The total length of the non-redundant singleton/assemblies is 32 Mb, theoretically covering ~10 per cent of the bottle gourd genome. Functional annotation of the sequence revealed a broad range of functional types, covering all the three top-level ontologies. Comparison of the gene sequences between bottle gourd and the model cucurbit cucumber (*Cucumis sativus*) a 90 per cent sequence similarity on average. Using the sequence information, 4395 microsatellite-containing sequences were identified and 400 SSR markers were developed, of which 94 per cent amplified bands of anticipated sizes. Transferability of

these markers to four other cucurbit species showed obvious decline with increased phylogenetic distance. From analyzing polymorphisms of a subset of 14 SSR markers assayed on 44 representative China bottle gourd varieties/landraces, a principal coordinates (PCO) analysis output and a UPGMA-based dendrogram were constructed. Bottle gourd accessions tended to group by fruit shape rather than geographic origin; although in certain subclades the lines from the same or close origin did tend to cluster. Yang *et al.* (2016) developed a cDNA library from root RNA sampled from bottle gourd seedling at 0 days after inoculation (DAI) and 10 DAI were sequenced using Illumina highseq 2000 platform. more than 41,831,589 high quality reads were generated and was assembled into 56,905 unigenes with an average swequence length of 735.00 base pairs. A total of 38,515 (67.7 per cent) unigenes were annotated and of which 8,171 were assigned to specific metabolic pathways by the Kyoto Encyclopaedia of Genes and Genomes (KEGG). Of all unigenes, 2,469 genes were differently expressed, including 1,294 up-regulated genes and 1,175 down-regulated genes. With clusters of Orthologous Groups of Proteins (COG) and Gene Ontology (GO) functional classifications, all differently expressed-transcripts were grouped into 23 and 55 categories, respectively, which includes many functional categories, such as energy metabolism, signaling, transcripton regulation and defence reactions. Genes involving the defence biosynthesis were analysed. 37 gene sequences encoding the defence metabolites were discovered in the transcriptome. It concludes that the dynmics of the root transcriptome at the early development stage. The transcriptome data also provides an important resource for understanding the formation of defence constituents in the root extract from bottle gourd.

## Bitter Gourd

DNA based molecular markers are more useful for genetic characterization and diversity assessments. Several dominant DNA markers like RAPD, ISSR and AFLP have been reported for genetic analysis of bitter gourd. Microsatellite or simple sequence repeat (SSR) markers have gained considerable importance in plant genetics and breeding owing to many desirable attributes like their multi-allelic nature, co-dominant transmission, extensive genome coverage, small amount of starting DNA, andease of detection by polymerase chain reaction. SSRs are known to have high heterozygous values and are more informative than dominant DNA markers. Although in various genetic analyses, microsatellite markers provide accurate results with minimum number of loci/alleles employed in the study. It is established that greater number of markers are necessary for the development of genetic map and marker-assisted selection. The utility of these markers in analysis of intra-specific genetic diversity, as well as cross-species transferability to six species of Momordica is also demonstrated (Adarsh *et al.*, 2016). Behera *et al.*

(2016) studied genetic diversity and genetic relatedness among 51 bitter gourd lines using 61 SSR markers. Among 61 SSR primers screened, 30 were polymorphic and highly informative to differentiate these genotypes. Based on genotyping, a moderate to high level of genetic diversity was observed with a total of 101 alleles. The PIC values were ranged from 0.038 for BG_SSR-8 to 0.723 for SSR marker S-24 with an average of 0.429. The number of alleles was ranged from 2 to 5 with an average of 3.3 alleles per locus and heterozygosity was ranged from 0.02 to 0.24 suggesting a wide variation among 51 bitter gourd accessions. The cluster-I comprises of 4 small fruited type genotypes which are commercially cultivated in central and eastern India. Cluster-II comprises of 35 genotypes which were abundant and diverse group consisting of medium to long size fruited genotypes. Cluster-III comprises of 12 genotypes with long and extra-long fruited genotypes. The STRUCTURE analysis result was also in accordance with clustering method of UPGMA tree and principal coordinate analysis (PC0A). Therefore these polymorphic SSR markers will have potential applications in assessment of genetic diversity, molecular fingerprinting and genetic mapping in *Momordica* species for economic traits. Total raw reads in FASTQ file size 14.62 GB for Gyno and 15.06 GB for Mono were obtained. Total numbers of reads were 32946510 (32.95 millions) for Gyno and,33912199 (33.91 millions) for Mono whereas Total Number of HQ Bases in Mb were 2202.59542 Mb for Gyno and 2355.78336 Mb for Mono. Percentage of HQ Bases was -96 per cent for both genotypes. Total 65540 transcript contigs for Gyno and 61490 transcript contigs for Mono were assembled using *de-novo* assembly tool Velvet_1.1.07 fallowed by Oases_O.2.0l. Total filtered transcript contigs having >20Qbp (-54667 for gyno and -51324 for mono) were deposited in TSA (Transcriptome Shotgun Assembly) submission portal of NeB! database. The primary accession numbers for gyno GANFOOOOOOO and mono GANGOOOOOOO. Based on available sequences 13491 SSR for Gyno and to biotic and abiotic stresses, and in development. 13526 SSR for Mono were identified. SNP study in this research, we identified 80 putative WRKY revealed that 21065 SNP for gyno and 19871 SNP genes from the bitter gourd genome, 56 ARF genes for Mono were identified successfully. Identification which play a pivotal role in auxin-regulated gene of the candidate gene for disease management and expression of primary response genes. The nutritional quality improvement play improvement availability of transcriptome data gives us role in any crop system. So based on availability of opportunities for identification of candidate genes transcriptome sequences we have identified some and their functional characterization using *In silico*, important transcription factor associated genes *In vivo* and *In vitro* experiment. WRKY proteins play significant roles in responses. (Singh *et al.*, 2014).

## Teasel Gourd

Teasel gourd (*Momordica subangulata* subsp. *renigera*) is a semi-domesticated vegetable crop of eastern and north-eastern India. It is second most important vegetable of genus *Momordica* grown for its immature fruits. Although sufficient diversity exists in teasel gourd, it has never been estimated using molecular markers. In an experiment one hundred anchored ISSR primers were screened for identification of primers useful for amplification in teasel gourd. Fifty primers were identified and these were used for diversity analysis among seventy two accessions collected from the states of Odisha (16), Tripura (30), Mizoram (19) and Asom (7). A total of 504 markers were amplified of which 464 markers (92.10 per cent) were polymorphic. An average of 10.10 markers were amplified per primer with the primer UBC869 and UBC826 amplifying five (minimum) and sixteen (maximum) bands, respectively. The primer UBC888 had the maximum resolving power while the primer UBC844 had the lowest resolving power. The average heterozygosity detected was 0.28. As observed by the Nei's gene diversity index, the accessions from Tripura were most diverse followed by those from Mizoram. Maximum number of markers detected polymorphism among the accessions from Tripura (83 per cent) followed by Mizoram (76.80 per cent), Odisha (66.20 per cent) and Asom (60 per cent). The present study indicates that ISSR markers detect moderate genetic diversity in the accessions of teasel gourd from Tripura, Mizoram, Odisha and Asom (Solanki *et al.*, 2016).

## *Luffa* Species

The genus *Luffa* (Miller) is one of the most important and widely adapted vegetable of the humid tropics with several health benefits. There are nine species of *Luffa*, of which seven are found in India. Among them the edible species are: *Luffa acutangula*, *Luffa cylindrica* and *Luffa hermaphrodita*. These species are cross compatible which makes them amenable to traits transferability for genetic enhancement of *Luffa* species utilizing the prevailing diversity. *Luffa hermaphrodita* (locally known as Satputia) is characterized by short viny growth, earliness and most importantly, bears small smooth fruits in clusters. These traits can be utilized to enhance the genetic architecture of *Luffa acutangula* and *Luffa cylindrica* as they are commercially cultivated. To assess the genetic diversity in *Luffa* species, both phenotypic and genotypic approaches have been employed. Morphological characteristics of 40 *Luffa* accessions were recorded for both qualitative and quantitative traits. For genotypic analysis, PCR based markers *i.e.* ISSR and EST-SSR (developed using NGS approach from the sponge gourd leaf transcriptome) were used for molecular diversity study. Of the 191-based SSR markers of sponge gourd tested across the accessions, 132 EST showed band amplification indicating likelihood of markers transferability across the species. The 20 ISSR primers tested produced clear and polymorphic bands in most of the accessions (Bhutia *et al.*, 2016.).

## Ash Gourd/Wax Gourd

Two Chinese wax gourd (*Benincasa hispida*) cultivars (one with and one without wax) and one chieh-qua (*Benincasa hispida* var. *chieh-qua*) cultivar were examined for random amplified polymorphic DNA (RAPD) genetic markers with twenty 10-mer random primers, and 12 gave reproducible polymorphic DNA amplification patterns. The DNA band patterns generated were primer- and genotype dependent. Nine primers produced unique banding patterns for each of the 2 Chinese wax gourd cultivars, 6 primers for the Chinese wax gourd cv. 'Without Wax' and 'Chieh-Qua' and 12 primers for the Chinese wax gourd cv. with wax and 'Chieh-Qua'. Only 3 primers gave unique banding patterns for each of the 3 cultivars. The three cultivars were very similar when comparisons were made on the basis of RAPD patterns (Meng *et al.*, 1996).

## Allium

Development of reliable molecular markers is a challenging task in genus *Allium* due to its huge nuclear genome among monocots as onion genome (16,415 Mbp per 1C), which is nearly 36 times larger than rice and 6 times bigger than maize (Arumuganathan and Earle, 1991). Although, in *Allium* RAPDs were developed and used for diversity analysis (Bradeen and Havey, 1995; Ennequin *et al.*, 1997; Tanikawa *et al.*, 2002) but counter repeatability problem (Eicht *et al.*, 1992) has made them less attractive to the researchers. Use of RFLPs and AFLPs has also been reported (Bark and Havey, 1995; Lampasona *et al.*, 2010), but RFLPs cannot be employed for studies in routine due to their high cost, labour intensive procedure and low polymorphism, while dominant inheritance of AFLPs makes it less attractive to plant biologists (McCallum *et al.*, 2008; Lampasona *et al.*, 2010). Now-a-days due to the advances in next generation sequencing technologies a wealth of genomic and transchriptome data was developed in various crop plants including genus Allium which were explored using different bioinformaic tools (table 1) to find out SSR markers (Varshney *et al.*, 2005; Baldwin *et al.*, 2012; Sun *et al.*, 2012; Duagnjit *et al.*, 2013). As result, in genus *Allium*, SSR markers have been developed in bulb onion (Fischer and Bachmann, 2000; Jakse *et al.*, 2005; McCallum *et al.*, 2008; Baldwin *et al.*, 2012), bunching onion (Wako *et al.*, 2002; Song *et al.*, 2004) and garlic (Ma *et al.*, 2009; Khar, 2012) which are utilized in different studies for realizing the worth of *Allium* germplasm.

It is asexually propagated *Allium*, which originated in central Asia is cultivated throughout the world (Brewester, 2008). The development of garlic cultivars has been limited to selection which hugely depends on the extent of variability available (Lampasona *et al.*, 201). SSR markers were also developed in garlic to study genetic diversity for planning efficient strategies for germplasm conservation (Ma *et al.*, 2009; Cunha *et al.*, 2012). These studies indicated that SSR markers act as useful tool for characterization of garlic germplasm. In this context, Acharya and Simon (2010) used 20 SSR markers on 48 accessions collected from different countries and got sufficient divergence in the germplasm. The Tunisian and French germplasm was characterized using IISR markers where factor analysis of distances table (AFTD) did not classify accessions on the basis of geographical origin or morpho-physiological characters, particularly bolting ability, but confirmed the appurtenance of analyzed accessions to *sativum* botanical subspecies (Jabbes *et al.*, 2011). A core collection was developed, consisting of 95 accessions using heuristic approach which were the representative sample of whole germplasm. The degree of variability within accession (84.4 per cent) was more than between the accessions (15.6 per cent). There were four groups with $-F_{ST}$ value of 0.1560, indicating a moderate differentiation among the groups (Zhao *et al.*, 2011). In China, considerable variability was observed in the garlic collections as 40 accessions were clustered into three groups (Xia *et al.*, 2012). The garlic accessions collected from four countries were grouped into four clusters according to their geographical origin with SSR markers. It suggests that local selection pressure and differences in adaptability of garlic germplasm in particular regions is important cause of variability (Jo *et al.*, 2012). Average number of alleles, observed heterozygosity, expected heterozygosity, Hardy Weinberg equilibrium, Shannon index and polymorphic information content values were 4.4, 0.468, 0.576, 1.073 and 0.518, respectively. Recently, EST-SSR markers were also developed in garlic and used for genetic diversity analysis (Ipek *et al.*, 2012; Khar, 2012). Indian garlic germplasm has considerable amount of variability and their clustering was independent of the geographical origin but based on the basis of vernalization requirement for bulb formation in garlic (Khar, 2012).

**Table 12.20: Different Bioinformatics Softwares available to Discover SSR Markers**

| Sl.No. | Software Names | Reference |
|--------|----------------|-----------|
| 1. | Tandem Repeat Finder (TRF) | Benson *et al.* (1999) |
| 2. | Tandem Repeat Occurrence Locator (TROLL) | Castelo *et al.* (2002) |
| 3. | SSR Identification Tool (SSRIT) | Kantety *et al.* (2002) |
| 4. | SSR Finder | Gao *et al.* (2003) |
| 5. | MIcroSAtellite (MISA) | Theil *et al.* (2003) |
| 6. | Build SSR | Rungis *et al.* (2004) |
| 7. | SAT | Dereeper *et al.* (2007) |
| 8. | SSR Locator | Maia *et al.* (2008) |
| 9. | Rapid Identification of SSRs and Analysis (RISA) | Kim *et al.* (2012) |
| 10. | GMATo | Wang *et al.* (2013) |

# Onion

Genetic diversity studies in 20 *Allium* species (Bhima Super, *A. alticum* (CGN-14770), *A. fistulosum* (Taiwan), PRAN-1, *A. ampeloprasum* (EC609483), *A. alticum* (CGN-14769), *A. alticum* (ALL-233), *A. alticum* (CGN-16417), *A. ampeloprasum* (ALL-697), *A. cepa* (Uzbekistan), *A. cepa* var. *aggrigattum* (-5), *A. tuberosum* (CGN-20779), *A. griffithianum* (IC250414), *A. sativum*, *A. tuberosum*, *A. aschaninii* (EC328495), *A. schenoprasum*, *A. fistulosum* (ALL-646), *A. guttatum* (CGN-16418), *A. cepa* (Eshing Tilong)) were carried out by SRAP markers. A total of 11 SRAP primers were used, out of which 8 primers amplified. All the 8 amplified primers showed polymorphism.

## SSR Transferability in *Allium*

Availability of only limited number of SSR markers in *Allium* species is one of the major concerns for *Allium* research community. However, SSR transferability to related taxa is very useful approach as it was employed in various crops, which lacks sufficient sequence information (Varshney *et al.*, 2005). Initially genomic SSR markers were screened for transferability studies but they exhibited low transferability in different Allium species (Fischer and Bachmann, 2000; Tsukazaki *et al.*, 2008; Araki *et al.*, 2010). It implies that markers developed from conserved coding regions would be more informative in transferability studies (Varshney *et al.*, 2005). In this context, bulb onion derived EST-SSR markers were employed for cross amplification in bunching onion and show high (75.10 per cent) degree of transferability than genomic SSR (43.30 per cent) markers (Tsukazaki *et al.*, 2008). Latter, bulb onion derived SSR markers were used to estimate degree of transferability in wide range of *Allium* species and high transferability (100 per cent, 87.17 per cent, 25.0-91.70 per cent) was observed in different studies (Khar *et al.*, 2011; Khosa *et al.*, 2013; Mallor *et al.*, 2014). The garlic derived SSR markers were tested for transferability in five related Allium species where highest transferability was observed in *A. porum* (73.00 per cent) followed by *A. fistulosum* (48.00 per cent) and *A. altaicum* (47.60 per cent) which implies genome conservation among alliums (Lee *et al.*, 2011). Overall these findings suggested that in near future, SSR markers developed in bulb onion, bunching onion and garlic can be used in other *Allium* species for better understanding of *Allium* germplasm and their utilization (Khosa *et al.*, 2015; Pandey *et al.*, 2008).

## (i) Bulb Onion

*Allium cepa* L. is a highly valuable crop which is reported to have originated in central Asian region. Most cultivars developed through selection and precise estimation of genetic variability using improved techniques is prerequisite for efficient breeding (Brewester, 2008). In this context, sequence tagged microsatellites were developed from a genomic library enriched for microsatellite markers to determine genetic relationships between onion accessions. The eighty three bulb onion accessions were analyzed for diversity analysis where germplasm was partly grouped according to their geographical origins (Fischer and Bachmann, 2000). But, these markers had not proven reliable in various genetic studies due to their complex requirements and low transferability (McCallum *et al.*, 2008). To overcome these barriers, Expressed Sequence Tags (ESTs) were generated and microsatellite markers were found in EST libraries during sequence analyses (Khul *et al.*, 2004). To test their utility in germplasm analysis, 35 elite populations from specific companies or breeding programme were analyzed using EST-SSR markers and it was observed that different populations could be distinguished by them. Germplasm was found to be closely related with each other and grouping was according to their geographical origins (Jakse *et al.*, 2005). Latter, more diverse germplasm was surveyed to estimate variation using 56 EST-SSR markers and four genomic SSR markers (McCallum *et al.*, 2008). Population grown in long day, short day and Indian region clustered apart from each other. They also suggested that resequencing of EST markers can readily provide SNP markers for purity testing of inbreds and other application in *Allium* genetics. Different diversity studies have reported uniqueness of Indian bulb onion germplasm than exotic germplasm but very few accessing/extant varieties were utilized (Mahajan *et al.*, 2009). Later, greater numbers of accessions were used to assess the diversity of tropical Indian onion (Khar *et al.*, 2011). It was reported that indigenous Indian short day onion formed separate cluster from the exotic short day and long day onions. Also, Indian and North American germplasm are quite different from each other and central Asian revealed close relationship with Indian material (McCallum *et al.*, 2008). It suggests that Indian cultivars and land races might provide novel germplasm sources to broaden bulb onion breeding base. There is one concern with EST-SSR markers as they detect less number of amplicons while analyzing bulb onion germplasm. Also, onion has large genome size and low gene diversity because of which only a small region of onion genome has been explored. A new set of genome SSR markers were developed that exhibited high allelic diversity during germplasm analysis. Most of the germplasm clustered according to their geographical origins but Algerian sample that s appeared to cluster across three groups and this may indicate multiple introductions into the country or independent selections (Baldwin *et al.*, 2012). In future, SSR markers will be useful to determine the levels of inbreeding and population structure for associations mapping. The sequence specific amplification polymorphism (SSAP) for onion was developed based on *Copia* like elements in onion is available in data base (Anandhan *et al.*, 2016). The primers were designed based on LTR sequences. The DNA of different varieties of onion was digested with Taq1 restriction enzyme and adapter ligated to digested DNA. The DNA was amplified with retrotransposon specific primer and adapter specific primer. The amplicons were re-amplified with selective primer from adapter based sequence and labeled retrotransposon primer. The SSAP could able to detect around 300 polymorphic loci per primer set. Similarity between varieties varied from 0.08 to 0.63. Highest similarity was found between Pusa White Flat and ALR and lowest being Pusa White Flat and Cadillac. The marker could able to detect high level of polymorphism between varieties. Cluster analysis reveals two major clusters with Cluster I with exotic varieties and Cluster II with Indian varieties. The Indian cluster was further subdivided with four minor clusters. The detection of polymorphism is higher than any earlier reported TRAP marker. The newly developed marker will be great use for diversity analysis genetic and mapping studies.

### (ii) Bunching Onion

It is believed that bunching onion originated in North West China and is also being grown on a large scale in Japan and Kore (Brewester, 2008). In bunching onion, first set of SSR markers were developed from a genomic library and used for linkage map construction (Wako *et al.*, 2002; Song *et al.*, 2004; Tsukazaki *et al.*, 2007; 2008). SSR markers were utilized to determine genetic uniformity in bunching onion hybrids and very low were detected at different polymorphic loci suggesting high level of heterozygosity (Tsukazaki *et al.*, 2006). According to these results, SSR tagged breeding scheme was proposed in bunching onion and utilized to realize the utility of this scheme (Tsukazaki *et al.*, 2009). Thirty bunching onion cultivars were classified using SSR markers and it was consistent with the previously reported results based up on morphological characters. Average number of amplicons per primer in this study was high (10.6) in comparison to previous studies (Song *et al.*, 2004) which indicates that high degree of genetic variability occurs in bunching onion germplasm (Tsukazaki *et al.*, 2010). In bunching onion germplasm, SSR markers were more polymorphic than Sequence Related Amplified Polymorphism (SRAP) on the basis of an average number of polymorphic alleles (Li *et al.*, 2008). But, the information given by SRAP was more consistent with morphological variability than that of SSR markers (Li *et al.*, 2008). Bunching onion accessions collected from different regions of India found to be distinctive from exotic accessions (Khosa *et al.*, 2013). In future, estimation of genetic variability in bunching onion germplasm of wide geographical origins must be investigated to detect some novel alleles. The genetic homogeneity eight bunching onion varieties, including six $F_1$ hybrids were evaluated using 14 SSR markers. Two or more polymorphic alleles were detected at all of the SSR loci examined in each variety. The number of alleles detected in the eight varieties ranged from 3 to 7 among the 14 SSR loci and the polymorphism in formation content from O.41 to O.76. All the varieties examined displayed very low degree of uniformity at all of these polymorphic loci, Based on these results, it may be impossible to determine an appropriate genotype identity or any of the existing bunching onion varieties. To facilitate enhance the accuracy of variety identification, proposed here an "SSR-tagged breeding "scheme in which the plant homozygous at a few SSR loci would be selected out of a foundations need field. This scheme may enable to achieve efficient variety identification on purity determination of Fi seeds not only in bunching onion but also in any allogamous crops exhibiting severe inbreeding depression (Tsukazak *et al.*, 2006).

## Garlic

Phenotypic traits and dominant molecular markers are predominantly used to evaluate the genetic diversity of garlic clones. However, 24 SSR markers (codominant) specific for garlic are available in the literature, fostering germplasm researches. In this study, we genotyped 130 garlic accessions from Brazil and abroad using 17 polymorphic SSR markers to assess the genetic diversity and structure. This is the first attempt to evaluate a large set of accessions maintained by Brazilian institutions. A high level of redundancy was detected in the collection (50 per cent of the accessions represented eight haplotypes). However, non-redundant accessions presented high genetic diversity. We detected on average five alleles per locus, Shannon index of 1.2, HO of 0.5, and HE of 0.6. A core collection was set with 17 accessions, covering 100 per cent of the alleles with minimum redundancy. Overall FST and D values indicate a strong genetic structure within accessions. Two major groups identified by both model-based (Bayesian approach) and hierarchical clustering (UPGMA dendrogram) techniques were coherent with the classification of accessions according to maturity time (growth cycle): early-late and midseason accessions. Assessing genetic diversity and structure of garlic collections is the first step towards an efficient management and conservation of accessions in genebanks, as well as to advance future genetic studies and improvement of garlic worldwide (da Cunha *et al.*, 2014). Seven selected simple sequence repeats (SSRs) revealed a total of 37 alleles across 120 garlic accessions, with an average of seven alleles per locus. The values for observed heterozygosity ranged from 0 to 0.99 (mean = 0.71). The average genetic diversity and polymorphic information content values were 0.586 and 0.518, respectively. Based on the 37 alleles obtained from the seven SSRs, a phylogram was constructed to understand the relationships among the 120 accessions. The garlic accessions were clustered into four main groups (G1-G4) in the phylogram. Group1 consisted of accessions of 'Aomori', Group2 consisted of 64 accessions, Group3 consisted of25 accessions, and Group 4 consisted of 20 accessions. Our results indicate that genetic diversity is correlated with geographical region. There may have been local selection

**Table 12.21: Total Number of Alleles and Genetic Diversity Index for Seven Simple Sequence Repeat (SSR) Loci in the 120 Garlic Accessions**

| Locus Name | GeneBank Accession No. | Size Range | Specific Alleles[a] | Rare Alleles[b] | NA[c] | MAF[d] | Ho[e] | GD[f] | PIC[g] |
|---|---|---|---|---|---|---|---|---|---|
| GB-ASM-035* | EU909132 | 286-313 | 0 | 1 | 4 | 0.495 | 0.991 | 0.574 | 0.484 |
| GB-ASM-040* | EU909133 | 290-310 | 0 | 4 | 8 | 0.364 | 0.830 | 0.735 | 0.696 |
| GB-ASM-053* | EU909134 | 228-302 | 1 | 2 | 5 | 0.648 | 0.567 | 0.509 | 0.450 |
| GB-ASM-072* | EU909136 | 237-410 | 2 | 5 | 7 | 0.457 | 0.990 | 0.625 | 0.553 |
| GB-ASM-078* | EU909137 | 206-232 | 0 | 2 | 4 | 0.538 | 0.866 | 0.565 | 0.480 |
| GB-ASM-080* | EU909138 | 171-223 | 1 | 3 | 5 | 0.612 | 0.765 | 0.520 | 0.441 |
| GB-ASM-109* | EU909139 | 216-223 | 0 | 0 | 4 | 0.596 | 0 | 0.574 | 0.522 |
| **Total** | | | **4** | **17** | **37** | | | | |
| Mean | | | | | 5.285 | 0.530 | 0.716 | 0.586 | **0518** |

a: Allele found only in one accession, b: Number of alleles that <5 allele frequency, c: Number of alleles, d: Major allele frequency, e: Observed Heterozygosity, f: Gene diversity, g: Polymorphic information content.

\* Locus which deviate from Hardy-Weinberg Equilibrium.

pressure and differences in adaptability of the garlic to different geographical conditions. All of the tested loci deviated significantly (P < 0.01) from Hardy-Weinberg equilibrium. Thus, a number of disturbances occurred in the garlic population tested, including natural selection (Jo *et al.*, 2012).

## Potato

Chandel *et al.* (2016) developed interspecific potato somatic hybrids between cultivated *S. tuberosum* dihaploid C-13 and wild species *S. cardiophyllum* via protoplast fusion. Out of 26 regenerants, only 4 were confirmed as true somatic hybrids containing both parental genomes based on molecular markers and phenotypes. Somatic hybrids were identified by RAPD, ISSR, SSR, AFLP and cytoplasm (chloroplast and mitochondrial genomes) type molecular markers. Intermediate phenotypes of somatic hybrids were confirmed by leaf, flower and tuber traits. Late blight resistance of the hybrids was assessed by challenge inoculation of *P. infestans* under controlled conditions. Somatic hybrids were found tetraploid by flow cytometry, exhibited high pollen stainability by acetocarmine staining and formed berries and viable seeds after crossing with a common potato variety. Somatic hybrids possessed diverse cytoplasm types (W/α, W/ã and T/β) as assessed by chloroplast and mitochondrial genome-specific markers. Further, cluster analysis based on the Jaccard's coefficient of molecular profiles generated by all above markers showed genetic distinctness in somatic hybrids and parents. In potato *cry* gene incorporated (resistant to Colorado potato beetle) cultivar New Leaf (trademark) was released for commercial cultivation in 1995. Similarly, insect resistance *cry* gene incorporated variety was commercialized in 1995 by Monsanto.

## Greater Yam

Jyothy *et al.* (2016) characterized the landraces of greater yam on the basis of molecular markers. The

genetic diversity of 40 accessions representing different districts of Kerala was analyzed on the molecular basis using 9 simple sequence repeat (SSR) and 15 inter simple sequence repeats (ISSR). All the SSR primers and ISSR primers studied showed high polymorphism.

## Turmeric

Jain *et al.* (2016) conducted experiment to identify the level of variations among 42 accessions (wild species 13, cultivars 14 and 15 other germplasm) of turmeric (*Curcuma longa* L.) collected from 10 states using sequence-related amplified polymorphism (SRAP) markers. The study comprised of *C. aeruginosa*, *C. amada*, *C. amarissima*, *C. aromatic*, *C. brog*, *C. caesia*, *C. ferrugenia*, *C. latifolia*, *Curcuma* spp. (unidentified), *C. xantorrhiza*, *C. zeodaria* etc., *C. longa* cultivars *viz.*, CLI-329, Lakadong, NDH-9, NDH-79, NDH-98, Palam Lalima, Palam Pitamber, PTS-38, PTS-59, PTS-62, Pratibha, Punjab Haldi-1, Salem and Suvarna was taken for genetic diversity comparison. High genetic similarity (81.1 per cent) observed between two *Curcuma* cultivars collected from Himachal Pradesh followed by 78.0 per cent genetic similarity noted from *C. longa* collected from Kerala and Himachal Pradesh. Most of the *Curcuma* wild species was grouped in single cluster, whereas *Curcuma* cultivars collected from different regions of India grouped as another major cluster. Wild species *Curcuma zeodaria* collected from Kerala not grouped in any of the two major clusters. The grouping of wild species and cultivars are not based on the geographical regions collected for this study.

## Carrot

A total of 24 carrot carotenoid biosynthesis pathway genes were investigated by Ghemeray *et al.* (2016). The sequence were retrieved from gene bank data base and STS primers are designed, synthesized, PCR amplified, PCR product eluted, sequenced, sequences are aligned and variations studied in black, red and orange carrots. Allelic variations are observed like deletions. The three main PSY 1 variations differed from each other by 1-2 amino acids positions at $168^{th}$ – $175^{th}$ positions. PSY 1 allelic variations observed at 7 genomic regions, whereas in PSY 1 gene; deletions were observed at three gene genomic regions *i.e.* 137-139, 504-509, 677, 688 nucleotide regions. Phytoene desaturase enzyme (PDS) deletions was observed from 308-325 gene genomic area in both red and orange carrot, similarly in æ-Carotene desaturase 2 (ZDS2) more numbers of SNP and deletions are observed in black and red carrots compared to orange carrot at 710, $714^{th}$, $756^{th}$-$759^{th}$. Carotenoid isomerase enzyme (CRTISO) few variations observed in all three colour carrots Lycopene β-cyclase (LCYB) a long stretch of deletions from 365-424 observed in black as compared to red and orange carrot. In case of Lycopene ε-cyclase (LCYE) more number of structural variations are observed in all carrot colour lines. Carotenoid cleavage dioxygenase 3 (CCD 3) was also most similar with black, red and orange with respect to sequence. In 9-cis-Epoxycarotenoid dioxygenase 1 (NCED 1), more variations observed in black and orange compared to red carrot. Similarly (9-cis-Epoxycarotenoid dioxygenase 3) NCED 3 sequence retrieval very narrow and variation identified. Selvakumar and Kalia (2016) working on carrot for identification of marker(s) linked with anthocyanin loci, $F_2$ mapping population was developed by crossing White Pale (yellow colour epidermal layer, phloem and xylem) with IPC-126 (purple coloured epidermal layer, phloem and xylem). Out of 420 SSR markers used in this study, 41 showed the polymorphism between the parental inbred lines of carrot. However, purple and yellow colour could be differentiated by eight markers GSSR-134, GSSR-128, GSSR-107, GSSR-14, GSSR-3, GSSR-9 and BSSR-43 which were used for genotyping of $F_2$ mapping population. These markers had particular location on chromosome 3 which are linked with anthocyanin gene (*P*), according to order up to 10.2cM distance. The markers of BSSR-43 and GSSR-14 were found to be tightly linked with anthocyanin locus (*P*) at 1.4cM and 4.0cM distance, respectively. The other markers, namely GSSR-14, GSSR-128, GSSR-107, GSSR-3 and GSSR-9 were also located on the same chromosome 3 at a distance of 6.3cM, 5.6cM, 8.3cM, 8.7cM and 10.2 cM, respectively. GSSR and BSSR markers will, therefore, play an important role in hastening the process of selection of anthocyanin rich lines enhancing the breeding efficiency thereby.

## Interspecific Relationship in *Sesbania* spp.

Six species of *Sesbania* were fingerprinted through RAPD and ISSR primers. Both types of markers produced 249 bands out of which 243 were polymorphic in nature indicating high degrees of genetic diversity in the genus. Cluster analysis using the combined data revealed segregation of the lone species *Sesbania grandiflora* from rest of the species. High boot strap values in the dendograms show the accuracy and authenticity of the result. All other five species got separated to a distinct cluster. This supports the taxonomic division of the genus *Sesbania* into *Agati* comprising the only species *Sesbania grandiflora* and the sub-genus *usesbania* containing other five species.

## Cauliflower

It is now possible to select desirable genotypes from a segregating population with very high degree of precision and predictability by using biotechnological tools. Restriction Fragment Length Polymorphism

(RFLP) and Random Amplified Polymorphic DNA (RAPD) are the most fundamental tools, which have raised high hopes among the breeders. A detailed analysis of Brassica species has been carried out by many researchers using RFLP markers. A series of monosomic *Brassica oleracea* chromosome addition lines were established in genetic background of B. campestris, using RFLP markers, plant morphology and isozymes and these gene markers were mapped on some *Brassica oleracea* synteny groups (McGrath *et al.*, 1990). A detailed RFLP based genetic maps constructed in *Brassica oleracea* (Landry *et al.*, 1992), have also been used to study the origin of *Brassica oleracea* and its relationship with other Brassica species (Song *et al.*, 1988, 1990) and among *Brassica oleracea* varieties (Osborn *et al.*, 1989). RAPD has been used to identify club root resistant plants in segregating population (Grandclément *et al.*, 1996). Molecular Assisted Selection (MAS) is still in infancy stage in India in cole crops. However, some work is in progress to develop tropical cabbage lines having resistance/tolerance to a serious pest diamond back moth (Unpublished) using transformation technique. The purple (*Pr*) gene in the cauliflower (*Brassica oleracea* L. var. *botrytis*) represents a unique spontaneous mutant in conferring ectopic anthocyanin biosynthesis, which is caused by the tissue-specific activation of *BoMYB2* transcription factor, an ortholog of *Arabidopsis PAP2* or *MYB113*. The successful isolation and cloning of *Pr* gene in cauliflower offers a genetic resource for development of new varieties and for breeding vegetable crops with enhanced health-promoting properties and visual appeal. Among the molecular breeding techniques, the marker-assisted backcrossing (MABC) at the present is the most widely and successfully used method in practical molecular breeding. MABC aims to transfer one or a few genes/QTLs of interest from one genetic source into a superior cultivar or elite breeding line (serving as the recurrent parent) to improve the targeted trait. Unlike traditional backcrossing, MABC is based on the alleles of markers associated with or linked to gene(s) QTL(s) of interest instead of phenotypic performance of largest trait. Marker assisted backcross breeding has been used successfully to incorporate genes or QTLs for both qualitative and quantitative traits in a number of vegetables specially tomato, cucumber, potato, but in cauliflower there is no work reported regarding to introgression of *Pr* gene. Therefore, as the closely linked molecular markers are available for the *Pr* locus, marker assisted backcross breeding can be efficiently used to transform the Indian cauliflower using *Pr* gene for the enhancement of anthocyanin accumulation and providing anthocyanin rich food to consumers and mitigating malnutrition problem (Singh *et al.* 2016). Soi *et al.* (2016) while screening 584 markers including 486 SSR and 98 dCAPs; 24 SSR and 6 dCAP found to be polymorphic among identified black rot resistant (BR161 and BR207) and susceptible (Pusa Himjyoti and Pusa Sharad) cauliflower genotypes. Three markers will be used for mapping black rot resistance gene(s) in $F_2$ generation of cauliflower for employing them in marker assisted breeding. Screened including 166 SSR and 98 dCAP, 24 SSR and 62 CAPs were found to be polymorphic among identified black rot resistant (BR161 and BR207) susceptible cauliflower genotypes (Pusa Himjyoti and Pusa Sharad). These markers will be illustrating the validation of resistant lines of cauliflowers (BR161 and BR207) to black rot using gene-specific SSR markers. For the introgression black rot disease resistance in susceptible lines of cauliflower (Pusa Himjyoti and Pusa Sharad). These markers will be used for mapping black rot resistance gene(s) in $F_2$ generation of cauliflower for employing them in marker assisted breeding. A number of mitochondria specific PCR based markers were screened to identify cytoplasmic male sterile plants of cauliflower by Bansal *et al.* (2016). These markers targeted *atpa*, *atp6*, *atp9* and *coxl*, *coxll*, *orf224*, *orf38* and *orfB* loci in the mitochondrial genomes of cauliflower. Among 14 genotypes of cauliflower employed for the study, DC8498 and DC8441-5 were sterile and DC98-10 and DC41-5 were fertile in the early maturity group. Whereas DC8409, DC8410, DC8410-22, DCDB15. DC8401 were sterile and DC309. DC310, DC310-22, DC15, DC401 were fertile in mid maturity group. Out of 20 primer pairs used for discriminating CMS and fertile plants, four were promising. Two of these primers, which were developed from *orf224* and *atp6-orf224* loci, respectively gave alternate alleles in CMS and fertile plants; whereas the remaining two primers developed from *orf138* locus showed dominant pattern with band in CMS plants only. An experiment was carried out by Saha *et al.* (2016) to find out molecular markers linked to downy mildew resistance gene using 92 recombinant inbred lines (RILs) developed from a cross of Pusa Himjyoti (susceptible) and BR-2 (resistant). Segregation analysis in progenies of RILs indicated single dominant gene governed resistance to downy mildew in BR-2. Out of 74 simple sequence repeat (SSR) markers, 18 (24 per cent) were found to be polymorphic between the parents. Bulk Segregant analysis (BSA) in resistant and susceptible bulks showed four differentiating polymorphic markers out of 18 polymorphic. Three markers BoGMS0030$_{260/250}$' BoGMS1330$_{193/183}$' and BoGMS742$_{249/220}$' were mapped at 4.2cM, 4.3 cM and 6.4 cM distance, respectively on upper arm of the resistance gene. The SSR marker BoGMS1322$_{126/116}$' was found to be linked at 8.6 cM distance on lower arm of the resistance

gene. Ther closely linked SSR markers BoGMS0030$_{260/250}$' (4.2 cM) and BoGMS1330$_{193/183}$'(4.3 cM) can be employed for marker assisted selection in developing of downy mildew resistant varieties of cauliflower in future.

## Okra

Molecular characterization of identified lines ($F_{12}$-1, $F_{12}$-2, $F_{12}$-6 and $F_{12}$-7,) and their parental varieties were carried out using RAPD and ISSR markers by Arunkumar *et al.* (2016). The selected ten RAPD primers produced a total of 71 amplicons of which 58 were polymorphic giving a polymorphism of 81.69 per cent. RAPD assay showed clear and distinct variations between YVMV resistant and susceptible lines. Among the selected RAPD primers, the unique bands produced by OPC-2 (400 bp and 450 bp), OPT 02 (1200 bp) and OPA 02 (450 bp) were specific to only resistant lines. The ISSR primers produced a total of 92 amplicons of which 68 were polymorphic giving a polymorphism of 73.91 per cent. Among the selected ISSR primers, three primers *viz.*, UBC 834 (700 bp, 750 bp and 800 bp), UBC 818 (1150 bp) and UBC 830 (1815 bp and 1500 bp) produced unique amplicons which were present only in the resistant lines. Molecular marker analysis could assess the variability among the advanced generation selections and their parents. The study could locate some markers in the resistant genotypes which on further in-depth study will aid in marker assisted selection for YVMV resistance.

## Search of Genetic Resistance

An antimicrobial protein *i.e.* WjAMP-1, has been purified from leaves of wasabi possessing antimicrobial activity against both fungi and bacteria. The deducted amino acid sequence of cDNA of WjAMP-1 showed 60 per cent and 70 per cent identity with a hevein from *Hevea brasiliansis* and a hevein-like protein from *Arabidopsis thaliana*, respectively. However, matured WjAMP-1 lacked the hevein domain and may correspond to the C-terminal domain of hevein. Southern blot analysis showed that one or two copies of the WjAMP-1 gene were presented in the genome of wasabi. Expression of WjAMP-1 was detected in all organs tested, and was especially strong in petioles. Expression of WjAMP-1 was induced by the inoculation with fungal pathogens and treatment with methyl jasmonate. Recombinant WjAMP-1 expressed in *Nicotiana benthamiana* using potato virus X vector also inhibited not only growth of fungi but also bacteria. These results suggest that WjAMP-1 may be the C-terminal domain of hevein and one of the defense genes in Wasabi. WjAMP-1 gene may be useful genes to generate resistant plant against fungal and bacterial pathogens (Kiba *et al.*, 2003).

## Genomics and Post-Genomics Activities

While harnessing the benefits of genetically modified (GM) crops, for better yield, nutritional quality and resistance to biotic and abiotic stresses, it is important to safeguard the genetic diversity. Inadvertent and accidental occurrence of GM events may lead to adventitious presence of transgenes in the *ex situ* collections conserved in the genebanks. The genebanks collect, conserve and provide the genetic resources to the breeders, hence maintaining the purity of genetic identity of the germplasm is of critical importance. Therefore, all possible efforts should be made to prevent the unintentional introgression of transgenes into the conserved samples.

## Genomics

1. Structural: Deciphering the complete genetic code of the organism
2. Functional: Systematic assignment of functions to the predicted genes

## Post-Genomics

1. Allele mining for the identified genes by way of correlating gene sequence polymorphism with altered functions (traits).
2. Use of markers and genes in crop improvement programs.

## Sequencing of Plant

Sequencing of 10 plants completed and 6 are in progress. These are- *Arabidopsis*, Rice, Poplar, *Medicago*, Sorghum, Papaya, Cassava, Cucumber, Tomato, Potato, Maize, Soybean, Citrus, Grape, Banana, Wheat are in Progress.

## International Project on Vegetable Crops

Some international Projects on vegetable crops are:

☆ The International Solanaceae Genome Project

☆ November 2003 an international meeting was held in Washington DC, attended by 69 scientists from 11 countries.

☆ Shaping of a 10 years project for on *Solanaceae* and allied species.

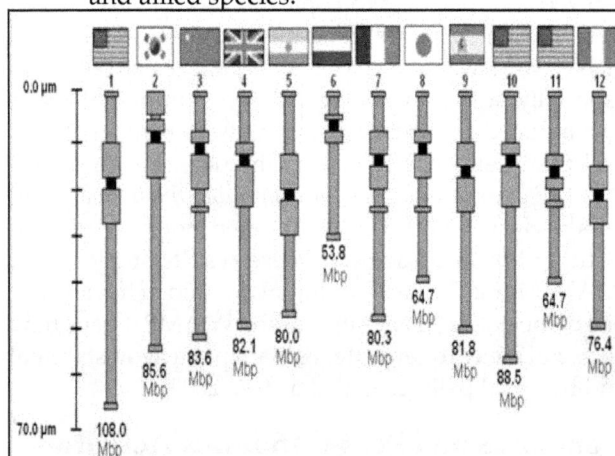

**Figure 12.2: International Tomato Genome Sequencing Consortium**

**(Based on Song-Bin Chang Ph.D. Thesis 2004).**

## Main Questions

1. How can a common set of genes give rise to a wide range of morphologically and ecologically distinct organisms?
2. How a deeper understanding of genetic basis of plant diversity can be harnessed to better meet the needs of society?

☆ A cornerstone was the tomato genome project to a golden standard.

## Issues and Gaps in GM Detection

There are several issues pertaining to the development of methods for GM testing which need to be addresses:

(i) Self-certification/disclosure from source countries or personnel

(ii) Obtaining gene sequence for synthesizing probes and primers from the developer of GM crop in case of indigenously developed GM crop and from the importer in case of imported GM crop.

(iii) Classifying areas and crops as per the possibility of contamination: certain hot-spots, where the possibility of accidental contamination by transgenic seeds is possible have to be identified.

(iv) Cost of detection: less for protein detection in the field, through developmental cost for the kits is high; moderate, if it is for one gene and one sample in case the primer/probe(s) of the transgenes are already available and high if several samples are to be analyzed and it is not known as to which transgene should be analyzed and fresh primer/probe(s) needs to be designed.

## Initiative at National Level

For GM-free conservation, efficiency strategy is to check the unintentional introgression/adventitious presence of transgene in *ex situ* collections employing DNA-based GM diagnostics (Tiwari and Randhawa, 2010). The expertise and capacity fr DNA based GM detection has been strengthened and upgraded significantly at ICAR-NBPGR, New Delhi during the last decade. Cost effective strategies based on GMO matrix, polymerase chain reaction (PCR), real-time PCR, loop-mediated isothermal amplification (LAMP) and multi-target real-time PCR system, which facilitate testing of GM events have been reported by GM detection laboratory. PCR and real-time PCR based diagnostics for *Bt* crops, commercialized (*Bt* cotton) or under field trials (*Bt* okra, *Bt* eggplant, *Bt* rice) in the country, have been developed, which could be employed for monitoring adventitious presence of transgenes in respective crops (Randhawa *et al.*, 2010). To confirm the GM status of a sample irrespective of specific crop and GM trait, a hexaplex PCR-based screening assay targeting marker genes (*aadA*, *bar*, *hpt*, *nptII*, *pat*, *uidA*) commonly employed in the GM events, was developed (Randhawa *et al.,* 2009). GMO matrix of 141 GM events of 21 crops with 106 genetic elements was developed as a decision support system to check for authorized GM events (Randhawa *et al.,* 2014a). a TaqMan real-time PCR based multitarget system simultaneously detecting 47 targets for six GM crops was developed (Randhawa *et al.,* 2014b).

To ensure GM free conservation preliminary studies for checking adventitious presence of transgenes in *ex situ* collections of brinjal (150 accessions) and okra (50 accessions) employing PCR and Real Time PCR based markers were undertaken. The strategies for selecting these accessions were:

**(1)** Collecting sites in proximity with the regions where fields trials of specific GM events of a particular crop were conducted

**(2)** The year of collection either after commercialization of the particular crop or the year after which field trials of GM have been conducted. None of the tested accessions ofthese crops showed adventitious presence of transgenes; Bairwa *et al.,* 2016; Randhawa *et al.,* 2015; Parimalan *et al.,* 2015.

Case studies for monitoring of adventitious presence of transgenes in brinjal and okra, employing DNA based markers could be used as models for monitoring other major crops with rich diversity or having the Centre of Origin in a particular country, where field trials of GM crops are being conducted in close proximity of those areas rich in their biodiversity.

# References

1. Abberton, M., Abbott, A., Batley, J., Bentley, A., Blakeney, M. Bryant, J., Cai, H., Cockram, J., Costa de Oliveira, A., Cseke, L.J., Dempewolf, H., De Pace, C., Edwards, D., Gepts, P., Greenland, A., Hall, A.E., Henry, R., Hori, K., Howe, G.T., Hughes, S., Humphreys, M., Ismail, A.M., Lightfoot, D., Marshall, A., Mayes, S., Nguyen, H.T., Ogbonnaya, F.C., Ortiz, R., Paterson, A.H., Simon, P.W., Tohme, J., Tuberosa, R., Valliyodan, B., Varshney, R., Wullschleger, S.D. and Yano, M. 2016. Global agricultural intensification during climate change: a role for genomics. *Plant Biotechnol. J.,* **14:** 1095-1098.

2. Acharya, L. and Simon, P.W. 2010. Diversity of garlic (*Allium sativum* L.) using SSR, EST and AFLP markers. *Plant Animal Genome Conf.* San Diego, CA, USA: 124.

3. Adarsh, A., Bhardwaj, A., Kumar, R. and Kumar, A. 2016. Molecular marker in bitter gourd for genetic characterization and fidelity assessment. 7$^{th}$ Indian Horticulture Congress (Abs: 12, Session III: Biotechnological Interventions in Horticulture), held at ICAR-IARI, New Delhi, 15-18 November, p. 371.

4. Anandhan, S., Nair, A. and Mahajan, V. 2016. Development of SSAP marker for onion (*Allium cepa* L.). 7$^{th}$ Indian Horticulture Congress (Abs: 25, Session III: Biotechnological Interventions in Horticulture), held at ICAR-IARI, New Delhi, 15-18 November, p. 381.

5. Araki, N., Masuzaki, S.I., Tsukazaki, H., Yaguchi, S., Wako, T., Tashiro, Y., Yamuchi, N. and Shigyo, M. 2010. Development of microsatellite markers in cultivated and wild species of section *cepa* and *phyllodolon* in *Allium. Euphytica,* **173:** 321-328.

6. Arumuganathan, K. and Earle, E. 1991. Nuclear DNA content of some important plant species. *Plant Mol. Biol. Rep.,* **9:** 208-218.

7. Arunkumar, B., Sureshbabu, K.V., Shylaja, M.R. and Pradeepkumar, T. 2016. Molecular characterizationof interspecifically derived YVMV resistant lines of okra (*Abelmoschus esculentus* (L.) Moench). 7$^{th}$ Indian Horticulture Congress (Abs: 27, Session III: Biotechnological Interventions in Horticulture), held at ICAR-IARI, New Delhi, 15-18 November, p. 382-383.

8. Ashmore, S.E. 1997. Status report on the development and application of *in vitro* techniques for the conservation and use of plant genetic resources. IPGRI. Rome.

9. Bairwa, R.K., Singh, M., Bhoge, R.K., and Devi, C. 2016. Randhawa, G.J. 2016. Monitoring adeventitious presence of transgenes in *ex situ* okra (*Abelmoschus esculentus*) collections conserved in genebank: A case study. *Genet. Resour. Crop Evol.,* **63(2):** 175-184.

10. Balachandran, S.M., Bhat, S.R. and Chandel, K.P.S.1990. *In vitro* clonal multiplication of turmeric (*Curcuma* species) and ginger (*Zingiber officinale*). *Pl. Cell Rep.,* **8:** 521-524.

11. Baldwin, S., Joyce, M.P., Wright, K., Chen, L. and McCallum, J. 2012. Development of robust genomic simple sequence repeats markers for estimation of genetic diversity within and among bulb onion (*Allium cepa* L.) populations. *Mol. Breed.,* **30:** 1401-141.

12. Bansal, V.P., Soi, S., Mangal. M., Singh, S. and Kalia, P. 2016. PCR-based marker development for identification of CMS system in cauliflower. (Abs:24, Session III: Biotechnological Interventions in Horticulture). 7$^{th}$ Indian Horticulture Congress, held at ICAR-IARI, New Delhi, 15-18$^{th}$ November 2016, p. 380-381.

13. Bark, O.H. and Havey, M.J. 1995. Similarities and relationships among populations of bulb onion (*Allium cepa* L.) as estimated by nuclear RFLPs. *Theor. Appl. Genet.,* **90:** 407-414.

14. Behera, T.K., Jat, G.S., Devi, M.B., Munshi, A.D. and Kumar, S. 2016.Analysis of population structure and assessment of genetic diversity in *Momordica* accessions using microsatellite (SSR) markers. 1st International Agrobiodiversity congress (Abs. PGR and Genomics), 6-9 November, held at New Delhi, p. 218.

15. Benson, E.E., Harding, K. and Smith, H. 1989. The effects of pre and post freeze light on the recovery of cryopreserved shoot tips of *Solanum tuberosum*. *Cryo-Lett.*, **10:** 323-344.

16. Bessembinder, J.J.E., Stritsky, G. and Zandvoort, E.A. 1993. Long term *in vitro* storage of *Colocasia esculenta* (L.) Schott under minimal growth conditions. *Pl. Tiss. Org. Cult.*, **33:** 121-127.

17. Bhatti, M.H., Percival, T., Davey, C.D.M., Henshaw, G.G. and Blakesley, D. 1997. Cryopreservation of embryogenic tissue of a range of genotypes of sweet potato (*Ipomoea batatas* (L.) Lam.) using an encapsulation protocol. *Pl. Cell Rep.*, **16:** 802-806.

18. Bhutia, N.D., Sureja, A.K., Munshi, A.D., Arya, L. and Verma, M. Diversity of Luffa species in India: Prospects for future utilization. Abs: 1st International Agrobiodiversity Congress, held at New Delhi, 6-9 November, 2016, p. 212.

19. Blackesley, D., Percival, T., Bhatti, M.H. and Henshaw, G.G. 1997. A simplified protocol for utilizing sucrose preculture only. *Cryo-lett.*, **18:** 77-80.

20. Bradeen, J.M. and Havey, M.J. 1995. Randomly Amplified Polymorphic DNA in bulb onion (*Allium cepa* L.) and its use to assess inbred integrity. *J. Am. Soc. HortSci.*, **120:** 752-758.

21. Brewester, J.L. 2008. Onions and other vegetable Alliums 2nd Edition, CABI, Wallingford, p. 1-22.

22. Brozynska, M., Agnelo Furtado, A. and Henry, R.J. 2016. Genomics of crop wild relatives: Expanding the genepool for crop improvement. *Plant Biotechnol. J.*, **14:** 1070-1085.

23. Chalukova, M. and Manueyan, H. 1991. Breeding for carotenoids pigments in tomato. In: G. Kalloo (ed.). Genetic Improvement of Tomato (Monographs on *Theor. Appl. Genet.* 14). Springer-Verlag, Berlin.

24. Chandel, P., Tiwari, J.K., Ali, N., Devi, S., Sharma, S., Sharma, S., Luthra, S.K. and Singh, B.P. 2016. Interspecific potato somatic hybrids between *Solanum tuberosum* and *S. cardiophyllum*: A potential source of late blight resistance breeding. 7th Indian Horticulture Congress (Abs: 94, Session: Breeding of Horticultural Crops), held at ICAR-IARI, New Delhi, 15-18 November, p. 72-73.

25. Chaudhury, R. and Chandel, K.P.S. 1985. Cryopreservation of embryonic axes of almond (*Prunus amygdalus*) seeds. *Cryo-Lett.*, **16:** 51-56.

26. Chaudhury, R., Radhamani, J. and Chandel, K.P.S.1991. Preliminary observations on the cryopreservation of desiccated embryonic axes of tea (*Camellia sinensis* (L.) O. Kuntze) seeds for genetic conservation. *Cryo-Lett.*, **12:** 31-36.

27. Chetelat, R.T., DeVerna, J.W. and Bennet, A.B. 1995. Effect of the *Lycopersicon chmielewskii* sucrose accumulator gene (*sucr*) on fruit yield and quality parameters following introgression into tomato. *Theor. Pl. Genet.*, **91:** 334-339.

28. Chetelat, R.T., DeVerna, J.W., Kalnn, E. and Bennett, A.B. 1993. Sucrose accumulator gene (*sucr*), a gene controlling sugar composition in fruit of *Lycopersicon chmielewskii* and *L. hirsutum*. *TGR Report*, **43:** 14-15

29. Chopra, V.L. and Narasimhulu, S.B. 1986. Modern techniques for genetic conservation. In: Conservation for productive agriculture (Chopra, V.L. and Khoshoo, T.N. eds.), PID, ICAR, New Delhi. p. 161-173.

30. Cook, R.J. and Draper, S.R. 1983. Potential application of ultrathin layer isoelectric focusing for characterization of cultivars of crop species. *J. Nat. Inst. Agri. Bot.*, **16:** 173-181.

31. da Cunha, C.P., Hoogerheide, E.S., Zucchi, M.I., Monteiro, M. and Pinheiro, J.B. 2012. New microsatellite markers for garlic, *Allium sativum* (Alliaceae). *Am. J. Bot.*, 99: http://dx.Doi: 10.3732/ajb.1100278.

32. da Cunha, C.P., Resende, F.V., Zucchi, M.I. and Pinheiro, J.B.2014. SSR-based genetic diversity and structure of garlic accessions from Brazil. *Genetica*, **142(5):** 419-31. doi: 10.1007/s10709-014-9786-1.

33. Dallas, J.F. 1988. Detection of DNA fingerprints of cultivated rice by hybridization with a human minisatellite DNA probe. *Proc. Acad. Sci.*, USA, 85(18): 6831-6835.

34. Damon, S., Hewitt, J.D., Zamin, D., Rabinowitch, H.D. Lincoln, S.E., Lander, E.S. and Tanksley, S.D., 1991a. Mendelian factors underlying quantitative traits in tomato. Comparison across species, generations and environments. *Genetics*, 127: 181-197.

35. Dixit, S., Mandal, B.B. and Srivastava, P.S. 2000. Cryopreservation of somatic embryos/embryogenic tissues: Application in Conservation and Biotechnology. Abs. In: National symposium on Prospects and Potentials of Plant Biotechnology in India in the 21st Century. Jodhpur, Oct. 18th-21st 2000. p. 161.

36. Dodds, J.h. and Roberts, L.W. 1985. Experiments in Plant Tissue culture, 2nd edn. Cambridge, Easwari Amma, N.Yark: Cambridge University Press.

37. Duangjit, J., Bohanec, B. Chan, A.P., Town, C.D. and Havey, M.J. 2013. Transcriptome sequencing to produce SNP-based genetic maps of onion. *Theor. Appl. Genet.*, **126**: 2093-2101.

38. Dubey, M., Dhaliwal, M.S., Jindal, S.K., Sharma, A. and Kaur, S. 2016. Marker assisted screening of $F_2$ population for late blight resistant in indeterminate tomato in net house. 7th Indian Horticulture Congress (Abs: 28, Session III: Biotechnological Interventions in Horticulture), held at ICAR-IARI, New Delhi, 15-18 November, p. 383.

39. Eicht, C.S., Erdahl, L.A. and McCoy, T.J. 1992. Genetic segregation of random amplified polymorphic DNA in diploid cultivated alfalfa. *Genome,* **35**: 84-86.

40. Englemann, F. 1991. *In vitro* conservation of horticultural species. *Acta Hort.*, **298**: 327-334.

41. Englemann, F. 1991. *In vitro* conservation of tropical germplasm- A review. *Euphytica,* **57**: 227-243.

42. Englemann, F. and Dereuddre, J. 1988. Cryopreservation for oil palm somatic embryos: importance of freezing process. *Cryo-Lett.,* **7**: 220-235.

43. Englemann, F., Duval, Y. and Dereuddre, J. 1985. Survival and proliferation of oil palm (*Elaeis guineesis* Jacq.) somatic embryos after freezing to liquid nitrogen. *CR Acad. Sci.,* **301**: 111-116.

44. Ennequin, M.L.T.D., Olivier, P., Thierry, R. and Agnes, R. 1997. Assessment of genetic relationship among sexual and asexual forms of *Allium cepa* L. using morphological traits and RAPD markers. *Heredity,* **78**: 403-409.

45. Fischer, D. and Bachmann, K. 2000. Onion microsatellites for germplasm analysis and their use in assessing intra and interspecific relationship with in subgenus *Rhizirideum*. *Theor. Appl. Genet.*, **101**: 153-164.

46. Ford-Lloyd and Jackson, M.J. 1991. Biotechnological methods of conservation of plant genetic resources. *J. Biotechnol.*, **17**: 247-256.

47. Friesen, N., Fritsch, R.M. and Blattner, F.R. 2006. Phylogeny and intrageneric classification of *Allium* L. (Alliaceae) based on nuclear ribosomal DNA ITS sequence. *Aliso,* **22**: 372-395.

48. Ghemeray, H., Gowda, R.V., Reddy, L.D.C. and Pattnaik, A. 2016. Identification of allelic variation in structural genes coding for carotenoids biosynthesis in carrot (Abs: 66). 7th Indian Horticulture Congress held at ICAR-IARI, New Delhi, 15-18th November 2016, p. 52-53.

49. Glaszmann, J.C., Kilian, B., Upadhyaya, H.D. and Varshney, R.K. 2010. Accessing genetic diversity for crop improvement. *Curr. Opini. Plant Biol.*, **13**: 1-7.

50. Harding, K., Benson, E.E. and Smith, H. 1991. The effects of *In vitro* culture period on the recovery of cryopreserved shoot tips of *Solanum tuberosum*. *Cryo-lett.,* **12**: 17-22.

51. Hussain, Z., Mangal, M., Saritha, R.K., Gosavi, G.U., Kumar, P., Monika, Nimisha and Yadav, S.K. 2016. Molecular breeding for improvement of tolerance to tomato leaf curl virus in tomato. (Abs:18, Session III: Biotechnological Interventions in Horticulture).7th Indian Horticulture Congress, held at ICAR-IARI, New Delhi, 15-18th November 2016, p. 376.

52. Ipek, M., Ipek, A., Cansev, A., Seniz, V. and Simon, P.W. 2012. Development of EST based SSR markers for garlic genome VI International Symposium on Edible Alliaceae ISHS *Acta Hort.*, 969.

53. Jabbes, N., Geoffriau, E., Clerc, V.L., Dridi, B. and Hannechi, C. 2011. Inter simple sequence repeat fingerprints for assess genetic diversity of Tunisian garlic population. *J. Agri. Sci.,* **3**: doi:5539/jas.v3n4p77.

54. Jain, A., Kandan, A., Yadav, R.P. and Rana, N. 2016. Genetic diversity analysis of *Curcuma* germplasm using sequence related amplified polymorphic (SRAP) marker. (Abs.: 16; Session III: Biotechnological Interventions in Horticulture).7th Indian Horticulture Congress, held at ICAR-IARI, New Delhi, 15-18th November 2016, p. 374-75.

55. Jakse, J., Martin, W., McCallum, J. and Havey, M.J. 2005. Single nucleotide polymorphism, indels and simple sequence repeats for onion (*Allium cepa* L.) cultivar identification. *J. Amer. Soc. Hort. Sci.,* **130**: 912-917.

56. James, C. 2008. Global status of commercialized biotech/GM crops: 2008.ISAAA Brief No. 39.

57. Jat, G.S., Munshi, A.D., Behera, T.K., Choudhary, H. and Das, P. 2016. Genetic study and molecular mapping of Horticulturally important traits in Indian cucumber (*Cucumis sativus* L.). (Abs.: 13; Session III: Biotechnological Interventions in Horticulture). 7th Indian Horticulture Congress, held at ICAR-IARI, New Delhi, 15-18th November 2016, p. 372.

58. Jo, M.H., Ham, I.K., Moe, K.T., Kwon, S.W., Lu, F.H., Park, Y.I., Kim, W.S., Won, M.K., Kim, T.I. and Lee, E.M. 2012. Classification of genetic variation in garlic (*Allium sativum* L.) using SSR markers. *Australian J. Crop Sci.,* **6**: 625-631.

59. Johnson, B.L. 1968. The protein electrophoretic approach to species relationship in wheat. In: Genetic lectures, Vol. I, Bogait, R. (ed.), Oregon state University Press, Cornwallis, p. 18-44.

60. Johnson, B.L. and Thein, M.M. 1970. Assessment of evolutionary affinities in *Gossypium* by protein electrophoresis. *Amer. J. Bot.,* **57**: 1081-1092.

61. Joothy, A., Sheela, M.N., Anwar, I., Radhika, N.K., Krishnan, B.S.P. and Abhilash, P.V. 2016. Genetic diversity analysis of greater yam (*Dioscorea alata* L.). 1st International Agrobiodiversity congress (Abs. Session: Trait Discovery and Enhanced Use of PGR), 6-9 November, held at New Delhi, p. 181.

62. Kalia, P., Singh, A.K., Rawat, S., Chaudhary, P. and Behera, T.K. 2016. Development of DBM resistant *Bt* transgenic in Indian cauliflower and cabbage. (Abs.: 22; Session III: Biotechnological Interventions in Horticulture).7th Indian Horticulture Congress, held at ICAR-IARI, New Delhi, 15-18th November 2016, p. 379.

63. Kartha, K.K., Leung, N.L. and Pahi, K. 1980. Cryopreservation of strawberry meristem and mass propagation of plantlets. *Jm. Soc. Hortic. Sci.*, **105**: 481-484.

64. Khar, A. 2012. Cross amplification of onion derived microsatellites and mining of garlic EST database for assessment of genetic diversity in garlic. VI International Symposium on Edible Alliaceae ISHS *Acta Hort.*, 969.

65. Khar, A., Lawande, K.E. and Negi, K.S. 2011. Microsatellite marker based analysis of genetic diversity in short day tropical Indian onion and cross amplification in related *Allium* species. *Genet. Resour. Crop Evol.*, **58**: 741-752.

66. Khosa, J.S., Dhatt, A.S., Negi, K.S. and Khar, A. 2015. Utility of Simple Sequence Repeat (SSR) markers to realize worth of germplasm in genus *Allium*.

67. Khosa, J.S., Dhatt, A.S., Negi, K.S. and Singh, K. 2013. Characterization of *Allium* germplasm using bulb onion derived SSR markers. *Crop Improv.*, **40**: 69-73.

68. Khosa, J.S., Dhatt, and Negi, A.S. 2014. Morphological characterization of *Allium* species using multivariate analysis. *Indian J. Pl. Genet. Resour.*, **27**: 24-27.

69. Khul, J., cheung, F., Yuan, Q., Martin, W., Zewdie, Y., McCallum, J., Catanach, A., Rutherford, P., Sink, K.C., Jenderek, M., Prince, J.P., Town, C.D. and Havey, M.J. 2004. A unique set of 11, 008 expressed sequence tags (EST) reveals expressed sequence and genomic differences between monocot order asparagales and poales. *Plant Cell.*, **16**: 114-125.

70. Kiba, A., Saitoh, H., Nishihara, M., Omiya, K. and Yamamura, S. 2003. C-terminal domain of a hevein-like protein from Wasabia japonica has potent antimicrobial activity. *Plant and Cell Physiology*, **44(3)**: 296-303.

71. Kim, J., Choi, J.P., Ahmad, R., Oh, S.K., Kwon, S.Y. and Hur, C.G. 2012. RISA: A new web-tool for Rapid Identification of SSRs and Analysis of primers. *Genes Genomics*, **34**: 583-590.

72. Kobayashi, M. 2011. Overview of *Arabidopsis* Resource Project in Japan. *Interdisciplinary Bio Central*, **3**: 1-5.

73. Kodandaram, M.H., Venkataravanappa, V., Nagendran, K., Chauhan, N.S., Rai, A.B. Tiwari, S.K. and Singh, B. 2016. Probing for resistant source in different brinjal varieties, lines and wild *Solanum* species to little leaf disease (Abs.: 64; Session I: Breeding of Horticultural Crops). 7th Indian Horticulture Congress, held at ICAR-IARI, New Delhi, 15-18th November 2016, p. 50-51.

74. Kumar, R. and Radhamani, J. 2016. Identification of genic SSR markers for diversity analysis in vegetable brassica *viz.-a-viz.* related species. *Indian J. Hort.*, **73(1)**: 133-136.

75. Kumar, S. and Singh, P.K. 2005. Mechanism for hybrid development in vegetables. Journal of new seeds, DOI: 10.1300/J153v06n04_05.

76. Kumar, V., Ram, R.B. and Rajvanshi, S.K. 2016. Studies on genetic diversity and variability in cluster bean [*Cyamopsis tetragonoloba* (L.) Taub.]. (Abs.:17, Session III: Biotechnological Interventions in Horticulture).7th Indian Horticulture Congress, held at ICAR-IARI, New Delhi, 15-18th November 2016, p. 375-376.

77. Lampasona, S.G., Burba, J.L. and Simon, P.W. 2010. Molecular Markers: Are they really useful to detect variability in local garlic collections? *Am. J. Plant Sci. Biotechnol.*, **4**: 104-112.

78. Larkin, P.J. and Scowcroft, W.R. 1981. Somaclonal variation: a novel source of variability from cell cultures for plant improvement. *Theor. Applied Genet.*, **60**: 197-214.

79. Li, H.Z., Yin, Y.P., Zhang, M. and Li, J.M. 2008. Comparison of characteristics of SRAP and SSR markers in genetic diversity analysis of cultivars of bunching onion (*Allium fistulosum* L.). *Seed Sci. Technol.* **36**: 423-434.

80. Li, M. and Midmore D.J. 1999. Estimating the genetic relationships of Chinese water chestnut (*Eleocharis dulcis* (Burm.f.)Hensch) cultivated in Australia, using random amplified polymorphic DNAs (RAPDs). *Journal of Horticultural Science and Biotechnology*, **74(2)**: 224-231.

81. Ma, K.H., Kwag, J.G., Zhao, W., Dixit, A., Lee, G.A., Kim, H.H., Chung, I.M., Kim, N.S., Lee, J.S., Ji, J.J., Kim, T.S. and Park, Y.J. 2009. Isolation and characteristics of eight novel polymorphic microsatellite loci from the genome of garlic (*Allium sativum* L.) *Sci. Hortic.*, **122**: 355-361.

82. Mahajan, V., Jakse, J., Havey, M.J. and Lawande, K.E. 2009. Genetic fingerprinting of onion cultivars using SSR markers. *Indian J. Hort.*, **66**: 62-68.

83. Maheswaran, M. 2004. Molecular markers: History, features and applications. *Advance Biotech.*, 17-24.

84. Maia, L.C.D., Palmieri, D.A., Souza, V.Q.D., Kopp, M.M., Carvalho, F.I.F.D. and Oliveira, AC. 2008. SSR Locator: Tool for Simple Sequence Repeat discovery integrated with Primer Design and PCR Simulation. *Int. J. Plant Genom.*: http.//dx.doi.org/10.1155/2008/412696.

85. Mallor, C., Andres, M.S.A., and Claver, A.G. 2014. Assessing the genetic diversity of Spanish *Allium cepa* landraces for onion breeding using microsatellite markers. *Sci. Hortic.*, **170**: 24-31.

86. Mane, R., Sridevi, O., Nishani, S. and Salimath, P.M. 2013. Evaluation of genetic diversity and relationships among tomato genotypes using morphological parameters and SRAP markers. *Indian J. Hort.*, **70(3)**: 357-363.

87. McCallum, J., Thomson, S., Pitcher-Joyce, M., Kenel, F, Clarke, A. and Havey, M.J. 2008. Genetic diversity analysis and single nucleotide polymorphism marker development in cultivated bulb onion on Expressed Sequence Tag-Simple-Sequence Repeats Markers. *J. Am. Soc. Hortic. Sci.*, **113**: 810-818.

88. Nanda, C., Prathibha, V.H., Mohan Rao, A., Ramesh, S., Shailaja, H. and Pai, S. 2016. Tagging SSR markers associated with genomic regions controlling anthracnose resistance in chilli. *Indian J. Hort.*, **73(3)**: 350-355.

89. Negi, K.S. 2006. *Allium* species in Himalayas and their uses with special reference to Uttaranchal. *Ethnobotany*, **18**: 53-66.

90. Noli, E., Conti, S. Maccaferri, M. and Sanguineti, M.C. 1999. Molecular characterization of tomato cultivars. *Seed Sci. Technol.* 27: 1-10.

91. Pandey, A., Pandey, R., Negi, K.S. and Radhamani, J. 2008. Realizing value of genetic resources of Allium in India. *Genet. Resour. Crop Evol.*, **55**: 985-994.

92. Parimalan, R., Bhoge, R.K., Randhawa, G.J. and Pandey, C.D. 2015. Assessment of transgene flow in eggplant germplasm conserved at National Genebanks. *Indian J. Biotechnol.*, **14**: 357-363.

93. Paterson, A.H., Damon, S., Hewitt, J.D., Zamin, D., Rabinowitch, H.D. Lincoln, S.E., Lander, E.S. and Tanksley, S.D., 1991a. Mendelian factors underlying quantitative traits in tomato: Comparison across species, generations and environments. *Genetics*, 127: 181-197.

94. Paterson, A.H., Tanksley, S.D. and Sorvels, M.E. 1991b. DNA markers in plant-improvement. *Advances in Agronomy*, 44: 39-90.

95. Pawar, B.D. and Kale, P.B. 2011. Molecular Plant Breeding for Sustainable Agriculture: An overview. *Agrobios News Letter*, **10(7)**: 66-68.

96. Punetha, S., Singh, D.K., Singh, N.K. and Basavaraj, H.M. 2016. Fingerprinting in cucumber (Cucumis sativus L.) using ISSR markers. 7th Indian Horticulture Congress (Abs: 29, Session III: Biotechnological Interventions in Horticulture), held at ICAR-IARI, New Delhi, 15-18 November, p. 384.

97. Rai, N.K., Sahu, P.K., Gupta, S., Reddy, M.K., Ravishankar, K.V., Singh, M., Sadashiva, A.T. and Prasad, M. 2013. Identification and validation of an ISSR marker linked to tomato leaf curl New Delhi Virus resistant gene in a core set of tomato accessions. *Veg Sci.*, **40(1)**: 1-6.

98. Randhawa, G.J., Chhabra, R. and Singh, M. 2009. Multiplex PCR-based simultaneous amplification of selectable marker and reporter genes for screening of GM crops. *J. Agric. Food Chem.*, **57**: 5167-5172.

99. Randhawa, G.J., Chhabra, R., Bhoge, R.K. and Singh, M. 2015. Visual and real-time event-specific loop-mediated isothermal amplification based detection assays for Bt cotton events MON531 and MON15985. *J. AOAC Intern.*, **98(5)**: 1207-1214.

100. Randhawa, G.J., Morisset, D., Singh, M. and Zel, J. 2014a. GMO matrix: A cost-effective approach for screening for unauthorized genetically modified events in India. *Food Control*, **38**: 124-129.

101. Randhawa, G.J., Singh, M., Chhabra, R. and Sharma, R. 2010. Qualitative and quantitative molecular testing methodologies and traceability systems for *Bt* crops commercialized or under fields trials in India. *Food Anal Meth.* **3(4)**: 295-303.

102. Randhawa, G.J., Singh, M., Sood, P. and Bhoge, R.K. 2014b. Multitarget real-time PCR-based system: Monityoring for unauthorized genetically modified events in India. *J. Agric. Food. Chem.*, **62(29)**: 7118-7130.

103. Saha, P., Shrivastava, M. and Kalia, P. 2016. Molecular mapping of downy mildew resistance gene in Indian cauliflower (*Brassica oleracea* var. *botrytis* L.). 7th Indian Horticulture Congress (Abs: 26, Session III: Biotechnological Interventions in Horticulture), held at ICAR-IARI, New Delhi, 15-18 November, p. 382.

104. Sehgal, A. and Chandel, K.P.S.1992.Phylogeny of *Vigna* species as indicated by seed protein and isozyme electrophoresis. *Indian J. Plant Genet. Resour.*, **5(1)**: 39-50.

105. Selvakumar, R. and Kalia, P. 2016. Inheritance and molecular mapping of anthocyanin locus (*P*) on chromosome 3 in carrot (*Daucus carota* L.). (Abs.: 20; Session III: Biotechnological Interventions in Horticulture). 7th Indian Horticulture Congress, held at ICAR-IARI, New Delhi, 15-18th November 2016, p. 377-378.

106. Shashank, Bansal, N., Sharma, K. and Gahlawat, S.K. 2011. Biodiversity informatics. *AgroBios New Letter,* **10(7):** 68-69.

107. Singh, M., Bhardwaj, D.R., Shukla, A., Naik, P.S. Kumar, R. and Singh V.K.. 2014. Bitter gourd transcriptorne for disease management and agricultural economics. *AGROBIOS NEWSLETTER,* **13(1):** 62-63.

108. Singh, M., Kumar, S., Singh, A.K., Ram, D. and G. Kalloo. 2002. Female sex-associated RAPD marker in pointed gourd (*Trichosanthes dioica* Roxb.). *Current Science,* **82:** 131-132.

109. Singh, S., Kumar, R. and Dey, S.S. 2016. Molecular breeding for enhancing anthocyanin accumulation in Indian cauliflower (*Brassica oleracea* var. *botrytis* L.). (Abs. 42, Session II: Breeding of Horticultural Crops). 7th Indian Horticulture Congress, held at ICAR-IARI, New Delhi, 15-18th November 2016, p. 117-118.

110. Sivbalan, K. and Manickavasagam, M. 2016. Factors influencing the in-planta genetic transformationefficiency of field bean varieties (*Lablab purpureus* L.). 7th Indian Horticulture Congress (Abs: 30, Session III: Biotechnological Interventions in Horticulture), held at ICAR-IARI, New Delhi, 15-18 November, p. 384-385..

111. Soi, S., Bansal, V.P., Mangal, M., Singh, S. and Kalia, P. 2016. Searching polymorphic markers for tagging black rot resistance in cauliflower (*Brassica oleracea* var. *botrytis* L.). (Abs.: 23; Session III: Biotechnological Interventions in Horticulture).7th Indian Horticulture Congress, held at ICAR-IARI, New Delhi, 15-18th November 2016, p. 380.

112. Solanki, S., Gautam, D., Bharathi, L.K., Archak, S. and Gaikwad, A.B. 2016. Genetic diversity analysis of *Momordica subangulata* subsp. *renigera* (Teasel gourd) collections from Eastern India using ISSR markers. Abs: 1st International Agrobiodiversity Congress, held at New Delhi, 6-9 November, 2016, p. 212.

113. Song, Y.S., Suwabe, K., Wako, T., Ohara, T., Nunome, T. and Kojima, A. 2004. Development of microsatellite markers in bunching onion (*Allium fistulosum* L.). *Breeding Sci.,* **54:** 361-365.

114. Sun, X., Zhou, S., meng, F. and Liu, S. 2012. De novo assembly and characterization of the garlic bud transchriptome by Illumina sequencing. *Plant Cell Rep.,* **31:** 1823-1828.

115. Tanikawa, T., Takagi, M. and Ichii, M. 2002. Cultivar identification and genetic diversity in onion (*Allium cepa* L) as evaluated by random amplified polymorphic DNA (RAPD) analysis. *J. Jpn. Soc. Hortic. Sci.,* **71:** 249-251.

116. Tantasawat, P., Trongchuen, J. Prajongjai, T., Seehalak, W. and Jittayasothorn, Y. 2010. Variety identification and comparative analysis of genetic diversity in yardlong bean (*Vigna unguiculata* ssp. *sesquipedalis*) using morphological characters, SSR and ISSR analysis. *Scientia Horticulturae,* 124: 204-216.

117. Thakur, N. and Srivastava, D.K. 2016. Molecular characterization of cabbage (*Brassica oleracea* L. var. *capitata*) genotypes using RAPD Markers. *Indian J. Plant Genet. Resour.,* **29(1):** 79-82.

118. Tiwari, S.P. and Randhawa, G.J. 2010. Strategy to monitor the adventitious presence of transgenes in *ex situ* collections. *Indian J. Agric. Sci.,* **80(5):** 351-356.

119. Tsukazaki, H., Fukuoka, H., Yeon-Sang Songi, Yamashita, K. and Wakoi, T. 2006. Considerable heterogeneity in commercial $F_1$ varieties of bunching onion (*Allium fistulosum*) and proposal of breeding scheme for conferring variety traceability using SSR Markers. *Breeding Science,* **56:** 321-326.

120. Tsukazaki, H., Fukuoko, H., Song, Y.S., Yamashita, K., Wako, T. and Kojima, A. 2006. Considerable heterogeneity in commercial $F_1$ varieties of bunching onion (*Allium fistulosum* L.) and proposal of breeding for conferring variety traceability using SSR markers. *Breeding Sci.,* **56:** 321-326.

121. Tsukazaki, H., Honjo, M., Yamashita, K., Ohara, T., Kojima, A., Ohsawa and Wako, T. 2010. Classification and identification of bunching onion (*Allium fistulosum* L.) varieties based on SSR markers. *Breeding Sci.,* **60:** 139-152.

122. Tsukazaki, H., Nunome, T., Fukuoko, H., Kanamori, H., Kono, I., Yamashita, K., Wako, T. and Kojima, A. 2007. Isolation of 1,796 SSR clones from SSR enriched DNA libraries of bunching onion (*Allium fistulosum* L.). *Euphytica,* **157:** 83-94.

123. Tsukazaki, H., Yamashita, K., Yaguchi, S., Masuzaki, S., Fukuoka, H., Yonemaru, J., Kanamori, H., Kono, I., Hang, TTM and Shigyo, M. 2008. Construction of SSR based chromosome map in bunching onion (*Allium fistulosum* L.). *Theor. Appl. Genet.,* **117:** 1213-1223.

124. Tsukazaki, H., Yamashita, K.I, Kojima, A. and Wako, T. 2009. SSR-tagged breeding scheme for allogamous crops: A trial in bunching onion (*Allium fistulosum* L.) *Euphytica,* **169:** 327-334.

125. Varshney, R.K., Graner, A. and Sorrells, M.E. 2005. Genic microsatellite markers: features and applications. *Trends Biotechnol.,* **23:** 48-55.

126. Verma, V.K., Pandey, A., Ryntathiang, I. and Jha, A.K. 2016. Genetic diversity in popular land races over cultivars of Capasicum spp. grown in North Eastern Region of India based on morphological and molecular analysis. 1^st International Agrobiodiversity congress (Abs. Session Trait Discovery and Enhanced Use of PGR), 6-9 November, held at New Delhi, p. 176.

127. Wako, T., Ohara, T., Song, Y.S. and Kojima, A. 2002. Development of SSR markers in bunching onion. *Breeding Res.*, **4(1):** 83.

128. Wann, E.V., Jourdain, E.L., Pressey, R. and Lyyon, B.G. 1985. Effect of mutant genotypes *"hp og"* and *"dg og"* on tomato fruit quality. *J. Am. Soc. Hort. Sci*, **110:** 212-215.

129. Williamson, V.M., Ho, J.Y., Wu, F.F., Miller, N. and Kaloshian, I.A. 1994. PCR based markers tightly linked to the nematode resistance gene, *Mi* in tomato. *Theor. Appl. Genetc.*, 87: 757-763.

130. Withers, L.A. 1985. Cryopreservation and storage of germplasm. In: Dixon, R.A. (ed.) Plant Cell Culture: A practical Approach. IRL Press Oxford. p. 169-192.

131. Xia, C.S., Yan, X.C., Zhou, J., Jun, N.D., Zhi, H.C. and Huan, W.M. 2012. Genetic diversity of garlic (*Allium sativum* L.) germplasm by simple sequence repeats. *J. Agric. Biotechnol.*, **20:** 372-381.

132. Xu, P., Wu, X., luo, J, Wang, B., Liu, Y., Ehlers, J.D., Wang, S., Lu, Z. and Li, G. 2011. Partial sequencing of the bottle gourd genome reveals markers useful for phylogenetic analysis and breeding. BMC Genomics, **12:**467, doi:10.1186/1471-2164-1 2-4673.

133. Xu, P.,Wu, X.H., Wang, B. G., Liu, Y.H., Qin, D. H. and Ehlers, J.D. 2011. A SNP and SSR based genetic map of asparagus bean (*Vigna unguiculata* ssp. *sesquipedalis*) and comparison with the broader species, *PLoS ONE*, **6(1):** 1-8.

134. Xu, P.,Wu, X.H., Wang, B. G., Liu, Y.H., Qin, D. H. and Ehlers, J.D. 2010. Development and polymorphism of *Vigna unguiculata* ssp. *unguiculata* microsatellite markers used for phylogenetic analysis in asparagus bean (*Vigna unguiculata* ssp. *sesquipedalis*) (L.) Verdc.). *Molecular Breeding*, **25(4):** 675-684.

135. Yang, X.P., Zhang, M., Xu, J.H., Liu, G., Yao, X.F., Li, P.F. and Zhu, L.L. 2016. Transcriptome analysis of root development in bottle gourd (*Lagenaria siceraria*). *Acta Hortic.*, 41-47, DOI 10.17660/1110.7.

136. Zhang, Y. and Stommel, J.R. 2000. RAPD and AFLP tagging and mapping of *Beta* (*B*) and *Beta* modifier (*Mo*$_B$), two genes which influence beta-carotene accumulation in fruit of tomato (*Lycopersicon esculentum* Mill.). *Theor. Appl. Genet.*, **100:** 368-375.

137. Zhao, W.G., Chung, J.W., Lee, G.A., Ma, K.H., Kim, H.H., Kim, K.T., Chung, I.M., Lee, J.K., Kim, N.S., Kim, S.M. and Park, Y.J. 2011. Molecular genetic diversity and population structure of a selected core set in garlic and its relatives using novel SSR markers. *Plant Breeding*, **130:** 46-54.

# 13

# National and International Legal Frameworks to Access Biological Resources

## Introduction

Vegetables are our most important source of the micronutrients, fiber, vitamins and minerals which are essential for a balanced and healthy diet. They are also considered a major source of cash income for smallholder farmers. But in most countries of the world, production is too low to provide their populations with even the minimum intake required for good health. Diets in many developing countries are commonly overloaded with more accessible carbohydrates and fats resulting in increasing global rates of obesity. Contamination from microbial sources and pesticides also reduces the safety of many vegetables in developing countries and high post harvest losses further reduce the availability of the relatively little that is grown. Vegetables provide smallholder farmers with much higher income and more jobs per hectare than staple crops (AVRDC, 2006). Further, there are strong scientific indicators that the altered growing conditions caused by climate change such as heat, drought, pests, and diseases are going to adversely affect food production in the country. Unfortunately, adapting crops to the changing edapho-climatic conditions will not be a straightforward process because there is no such thing in existence as a "climate change gene". This makes it immensely difficult to convince the planners and policy/decision makers to invest in projects/programmes like maintaining gene banks and specific crop breeding focused on climate change adaptation and mitigation. Plant genetic resources are the important components of human being and animals as food, fiber, medicine,

feed *etc.* and it is worldwide recognized for crop improvement. But due course of time, a lot of changes occurred in access and utilization which lead a legal binding of plant genetic resources management. India is known as vegetable bowl of the world; ranks second in their production after China with annual production of vegetables 162.0 million tones covering an area of 9.0 million hectare). More than 40 kinds of vegetables belonging to different groups, namely, solanaceous, cucurbitaceous, leguminous, cruciferous, root crops and leafy vegetables are grown in India in various tropical, subtropical and temperate regions of India. Among vegetables; potato accounts for 4.2 million tonnes annual production followed by Tomato (1.6 million tonnes) and Onion (1.5 million tonnes) (Indian Horticulture Database 2011: National Horticulture Board). The worldwide production of vegetables has doubled over the past quarter century and the credit mainly goes to high yielding varieties. Now days, the breeder's variety have reached close to the top of the curve for yields but, will continue to make significant strides with quality and innovation that will improve growers' returns and increase demand through the introduction of new vegetable products; and these new vegetables varieties will expand the chances of winning the war against malnutrition. Creation of vegetable hybrids is a key means towards the development of varieties for modern vegetable production. Hybrid seed production is high technology and a cost intensive venture but require specific genotypes. Only well-organized seed companies with good scientific manpower and well equipped research facilities can afford seed production. The increased complexity of markets and the higher demands force modern plant breeding to reduce the time for new varieties development thus, further escalating the cost. The short span of life of a variety in the environment is resulting in shorter earn back period. This double impact is bound to put more pressure on investors to recover their investment through protection of intellectual property, and consequently royalties. Protection of intellectual property in plant breeding is not the primary driver to develop new, innovative varieties but, it is an adequate tool to protect the new varieties in the market against illegal reproduction and sales (Kumar *et al.*, 2008). Promotion of vegetable research and development has huge potential to contribute towards all the nine Millennium Development Goals (MDGs) of United Nations in developing and under-developed countries (Anonymous, 2005). In the recent past, there has been a paradigm shift of genetic resources being treated as "common heritage of mankind" to the "sovereign rights of the nations". Sovereign rights over these resources were highlighted in the context of managing and accessing genetic resources by different countries. Thus, access to genetic resources post-CBD is a high deliberated issue amongst the researchers/breeders/community dealing with their utilization and conservation. This significant change has led to a whole lot of new international developments with a crucial debate on access and utilization plant genetic resources (PGR). In the last 30 year legal development at national and international levels have completely reshaped the ways in which plant genetic resources are used in global agriculture.

## Why Conserve Biodiversity?

As we all know by now, biodiversity is essential for maintaining the ecological functions, including stabilizing of the water cycle, maintenance and replenishment of soil fertility, pollination and cross-fertilization of crops and other vegetation, protection against soil erosion and stability of food producing and other ecosystems. Conservation of biological diversity leads to conservation of essential ecological diversity to preserve the continuity of food chains. Biodiversity provides the base for the livelihood, cultures and economies of several hundred millions of people, including farmers, fisher folk, forest dwellers and artisans. It provides raw material for a diverse medicinal and health care systems. It also provides the genetic base for continuous up-gradation of agriculture, fisheries and for critical discoveries in scientific industrial and other sectors. The rapid erosion of biodiversity in the last few decades has impacted on the health of the land, water-bodies and people. Biodiversity is a wealth to which no value can be put. In the final analysis, the very survival of the human race is dependent on conservation of biodiversity. It is evident that this invaluable heritage is being destroyed at an alarming rate due to several reasons. Measures are being taken up at national and international levels to address this issue.

## Historical Aspects of Biodiversity Issues

As a history, the basic principles that nations have sovereign rights over their genetic resources was started earlier by the UN conference on Human Environment in 1972 at Stockholm (politically and legally recognized at international level) as sharp focused by International World Conservation Union (IUCN). These concerns were brought into the public domain by the acclaimed report *Our Common Future* (WCED, 1987). Through its decision 14/26 of June, 1987 established an ad hoc working Group of Experts on Biological Diversity. Based on inputs provided by the Working Group during 1988-90, the Governing Council made this group the inter-governmental negotiating Committee for the Convention on biological Diversity. The actual negotiations took place during 1991-92. India contributed significantly to the draft of the Convention. After the UNEP (United Nations Environment Programme) Conference at Stockholm, and the Earth Summit (UNCED); the Convention on Biological Diversity (CBD) held at Rio, Brazil in June, 1992 clearly recognized for the first time the Sovereign right of States (countries) over their natural resources including plant genetic resources and came into force in December, 1993. As per the negotiations progressed through five normal sessions, the elements of the proposed convention went through a sea change, becoming more complex. Sanchez (1994), who was the Chairperson of the Negotiating Committee, listed the following issues, which became the core of the negotiations:

1. The cost of taking measures to conserve biological diversity versus the cost of not taking any measures.

2. Access to genetic resources and different possibilities of regulating it;

3. Whether the focus should be on wild species or should include both wild and domesticated species

4. Access to fund transfer of technology, including biotechnology, which must be considered for conservation and rational use of the components of biological diversity.

5. The eventual source and method of funding the costs of the measures that would be agreed upon; and

6. The consequences and impact of biodiversity on trade and development.

The FAO commission's non-binding International Undertaking on Plant Genetic Resources (IUPGR) advocates that these resources should be freely available being "heritage of mankind" while stating the same time, in a revised "agreed interpretation" that the "free access" does not mean free of charge. The realization of the Breeder's Right (BR), the Farmer's Right (FR) or the issue of access on mutually agreed terms to the plant genetic resources, including *ex situ* base collections of CGIAR centres, provide a strong interface between the two institutions/mechanisms. The instrument of "mutually agreed terms" and "prior information consent (PIC)" has been brought in by the CBD as well as the International code of conduct for Plant Germplasm Collecting and Transfer. However, exchange of plant genetic resources (PGR) is essential for all crop improvement programme. Apprehension related to access to biological resources raise several question such as ownership and accessibility (Evenson, 1999; Hamilton *et al.*, 2005) such as who owns the biological resources conserved at national level or at International Research Centres.

As per the definition provided for 'Biological Resources' under BDA, 2002, "biological resources include plants, animals and microorganism or parts thereof, their genetic material and by products (excluding value added products) with actual or potential use or value, but does not include human genetic material". Therefore, PGRFA were included under the definition of biological resources. BDA, 2002, hence does not differentiate between biological resources and PGRFA, which are of utmost importance for food and nutritional security, and sustainable agriculture. Most of the cultivated crops are part of PGRFA but not all of them are of Indian origin. The term research is also defined under the Act as "study or systematic investigation of any biological resource or technological application that uses biological systems, living organisms or derivatives thereof to make or modify products or process for any use". Similarly, the value added products defined under the Act are "products which may contain portions or extracts of plants and animals in unrecognizable and physically inseparable form". This definition does not specifically mention vegetable oils, proteins, starch, *etc.* which are used directly as food products. Another definition in the Act on commercial utilization says, the end uses of biological resources such as drugs, industrial enzymes, food flavours, fragrances, cosmetics, emulsions, oleoresins, colours, extracts, and genes used for improving crops and livestock through genetic interventions, but does not include conventional breeding or traditional practices in use in any agriculture, horticulture, poultry, dairy farming, animal husbandry or bee keeping. The definition and use of term conventional breeding or traditional practices is being interpreted differently by different stakeholders. Do conventional practices include hybrid seed production, or use of artificial insemination in animals to produce desirable progeny was highly debated.

## Responsibility and Holding of Plant Genetic Resources

Responsibilities for safeguarding natural diversity become prime duties at:

1. National level responsibility for Plant Genetic Resources
2. International level responsibility for Plant Genetic Resources

### (1) National Responsibility for Plant Genetic Resources

From the national point of view, the Convention (Art. 3) establishes the most important principle of States having sovereign rights to exploit their own resources, while ensuring that their activities do not cause cross-boundary damage to the environment of other States.

#### General Measures

Art.6 establishes a set of norms for developing national strategies, plans and programmes for the conservation and sustainable use of biological diversity and their integration into sectorial or cross-sectorial policies, plans, which is the National Focal point/Nodal Agencies for this purpose, had (prior to UNCED) already finalized the National Conservation Strategy and Policy Statement on Environment and Development (GOI, 1992). This document incorporates a broad strategy for promoting conservation of biological diversity in the country, which is supported by earlier policy instruments including the National Forest Policy (GOI, 1988) and the National Wildlife Action Plan. Ministry of Environment and Forest (MoEF) has also undertaken a Country Status Report. Also involved are the universities and various non-government organizations (NGOs) such as the World Wide Fund for Nature (WWF), the M.S. Swaminathan Research Foundation (MMSRF), the Bombay Natural History Society (BNHS), Centre for Science and Environment (CSE) *etc.* The gaps identified would help to strengthen the activities (Chauhan, 1996).

#### Identification, Monitoring and Article-wise Justification in India

Conservation and sustainable use of biological diversity requires actions at several levels. In order to:

a. Identify critical components *i.e.* ecosystems, species and genomes (Art. 7a);
b. Monitor such components to determine priorities (Art. 7b);
c. Identify activities and monitor their adverse impacts (Art. 7c);
d. Improve data acquisition and management (Art. 7a).

India has established a system of protected areas including special conservation areas like wetlands, mangroves and biosphere reserves. The multifarious tasks are shared by a network of official organizations, including the Botanical, Zoological and Forest Surveys of India (BSI, ZSI and FSI), the Indian council of Agricultural Research (ICAR) and the Council of Scientific and Industrial Research (CSIR) systems, the Wildlife Institute of India (WII), the Indian Council of Forest Research and Education (ICFRE), and the Central and State Pollution control Boards (Chauhan, 1996).

### *In-situ* Conservation

Article 8 of the convention relates to *in situ* conservation; it calls for establishing a national system of protected areas (PAs); development of guidelines for selection, establishment and management of Pas; regulating access to biological resources within or outside the Pas; and promoting protection in natural surroundings and conservation areas adjacent to Pas. India already has 80 national parks and 441 Sanctuaries covering 1, 48, 700 sq. km area which is about 4.5 per cent of the total geographical area (GOI, 1996). Further expansion of the Pas is envisaged on the basis of a systematic study carried out by Rodgers and Panwar (1988). This would reduce the gaps in various biogeographic units and biomes. In addition, special conservation areas including 16 wetlands, 15 mangroves, 4 coral reefs and 8 biosphere reserves have been particularly designated by MoEF for scientific management and conservation. The National Afforestation and Ecodevelopment Board (NAEB) has been given specific mandate for conserving areas adjacent to Pas and overall ecodevelopment activities for restoring degraded ecosystems. Recovery of endangered and threatened species (Art. 8f) is covered through *ex-situ* preservation areas (GOI, 1994).

Of particularly concern is to control the use of living organisms modified through biotechnologies and to guard against the associated risk (Art. 8g). In India, in 1989 MoEF in consultation with the Department of Biotechnology (DBT) had already framed rules to govern the manufacture, use, import, export and storage of hazardous microorganisms/cells. These rules regulate aspects of safety in research and production and provide for granting approval to proposals for use of hazardous microorganisms and recombinants in research and industrial production, including their release into the environment. MoEF has proposed to expand the scope of these rules through further guidelines and introducing risk-assessment procedures for the release of living modified organisms in contained and open environments and also for granting environmental impact clearance (Chauhan, 1996).

The obligation to control threatening alien or exotic species (Art. 8h) necessitates extensive review (in consultation with the concerned government agencies) of the scope of existing quarantine measures for screening biomaterials entering the country, with reference to fungal, bacterial and viral diseases and pests. Similarly,

the aspects relating to the need of maintain compatibility of physical components especially land and water, with the conservation of biological diversity shall have to be deliberated upon.

Article 8 (j) stresses the need to respect, preserve and maintain the knowledge innovations, skills and practices of indigenous/local communities, and to promote their wider application. The National Forest Policy and the National Conservation Strategy endorse this. MoEF, the Indian Council of Medical Research (ICMR) and the Anthropological Survey of India (ASI) are engaged in collecting information towards this end. In the light of the convention, the existing arrangements need further review, for which MoEF has already initiated appropriate action. MoEF has already done a comprehensive review of existing legislations and is in the process of formulating specific legal proposals for the proposed Biological Diversity (Conservation) Act (Chauhan, 1996).

Both the financial and other kinds of support for *in situ* conservation (Art. 8m) is being provided under existing national policy, legislation and institutional mechanisms. However, it is desirable that the existing mechanisms be further strengthened through internal and external support.

### *Ex-situ* Conservation

To complement the *in-situ* measures (Art. 8), the Convention (Art. 9) also covers measures to be adopted for *ex-situ* conservation. These include establishment facilities for research, recovery and rehabilitation of threatened species; regulating and managing the collection of biological resources from natural habitats for *ex-situ* conservation purposes; and adequate financial and other support. India has already made concerted efforts in this regard. For example, under the aegis of ICAR, the five National Bureaus have been established for *ex-situ* conservation and for undertaking research on domesticated biological diversity:

1. ICAR-National Bureau of Plant Genetic Resources (NBPGR), New Delhi.

2. ICAR-National Bureau of Animal Genetic Resources (NBAGR), NDRI, Karnal (Haryana)

3. ICAR-National Bureau of Fish Genetic Resources (NBFGR), Lucknow (Uttar Pradesh)

4. ICAR-National Bureau of Agricultural Insects Resources (NBAIR), Bengaluru (Karnataka)

5. ICAR-National Bureau of Agriculturally Important Microorganisms (NBAIM), Mau (Uttar Pradesh)

Wild biological diversity has been conserved through 275 *ex-situ* areas and 66 botanic gardens. The research component is pursued through universities and the institutions of ICAR, CSIR, BSI, ZSI, WII, ICFRE, the Salim Ali Centre of Ornithology (SACON) and the G.B. Pant Institute of Himalayan Environment and Development *etc.* the research on biotechnological aspects is supported by DBT through the national facility for Microbial Type Culture, Blue-green Algae, Marine cyno-bacteria, Plant Tissue Culture Repository, Animal Tissue and Cell Cultures, Animal House, Seed Banks, medicinal Plants, genetic Engineering Units, Biotechnological Engineering, Immunology and Biosafety *etc.* However, the financial and technical/scientific inputs need to be augmented for which the cooperation of other Contracting Parties to the Convention is essential (GOI, 1994).

## Sustainable Use of Components of Biological Diversity

Article 10 emphasizes that sustainable usage must be integrated into national decision making processes, through measures to control adverse impacts, protecting customary and traditional use, providing support to local populations in remedial action in degraded areas and encouraging cooperation between government and private sectors.

Through the various policy instruments relating to wildlife and forests, India has tried to integrate sustainable usage into decision making. However, the lack of linkages between various agencies both at the national and state levels has resulted in a rather unsatisfactory situation. Conservation of marine biological diversity and characterization of micro-organisms have not been adequately addressed. On-farm conservation of agri-biological diversity is also rather poor. As a result, a significant number of traditional varieties of important crops have been lost forever. For the purpose of sustainable use of ecosystems, criteria for determining ecological fragility have been developed and are being implemented (Chauhan, 1996).

## Incentive Measures

The Convention states that States should adopt economically and socially sound incentives (Art. 11) for conservation and sustainable use of biological diversity. In this direction, the government of India has launched several programmes. In view of the provisions of the Convention, further incentives need to be evolved for on and off farm conservation of agri-biological diversity. Similarly, incentives will have to be given to local communities for conserving and sustainability using rich freshwater and marine biological diversity.

## Research and Training

Education, training and research concerning certain aspects of biological diversity (Art. 12) are currently being promoted through:

a. The universities;

b. Government agencies such as MoEF, Min. of Sci. and Technol. (MST), DBT, CSIR, ICAR, ICFRE, WII, SACON, G.B. Pant Institute of Himalayan Environment and Development and other associated institutes/organisations and

c. NGOs such as WWF (India), BNHS, Centre of Environmental Education (CEE) *etc.*

## Public Education and Awareness

The National Environmental Awareness Campaign has had 'Biological Diversity' as its theme. Communication media including television, radio, the national and regional Press (print media) have also been used to generate awareness about the Convention. NGOs are being actively involved throughout the country to spread information and understanding of the importance of conservation and sustainable use of biological diversity, and the measures required for this (Art. 13a). The University grants Commission (UGC) and the National Council of Educational Research and Training (NCERT) also has initiated undergraduate and postgraduate programme/courses. The cooperation of other Contracting Parties has also been sought including nations in the South Asian region. Collaboration with international agencies such as Bioversity International; FAO; UNDP. UNEP; IUCN, World Vegetable Centre (AVRDC) and the World Resources Institute are being maintained (Chauhan, 1996).

## Impact Assessment and Minimizing Adverse Impacts

Article 14 of the Convention deals with:

a. Introducing procedures requiring Environmental Impact Assessment (EIA);

b. Arrangements to account for the environmental consequences of programmes and policies;

c. To account for the adverse impact beyond national jurisdiction

d. Notification of adverse effects upon other States as well as initiation of action to prevent damage and;

e. Promoting national arrangements for emergency responses to activities threatening biological diversity and establishing joint contingency plans with other Contracting Parties.

India's existing arrangements, under the Environment (Protection) Act, provide for impact assessment of development projects in the public sector alone, such as projects for river valleys and irrigation, industry and mining, thermal power, ports and harbours, railways, highways, airports, townships, *etc.* It is proposed to extend the scope of EIA to private sector development projects as well and to provide statuary backing for this.

## Access to Genetic Resources

Access to Genetic Resources is covered by Art. 15 which provides;

a. Recognition of the rights of sovereign States;

b. Creating conditions to facilitates access to genetic resources by other countries;

c. Add;

d. Access on mutually agreed terms with prior informed consent; and;

e. Developing and carrying out research with full participation of the Contracting Parties.

Access to genetic resources is facilitated through bilateral arrangements under FAO which regulates the import/export of seeds and the country's export-import policy under the Foreign (Trade) Act. However, arrangements to govern access to genetic resources in the future are yet to be worked out in accordance with the provisions of Articles 15, 16 and 19 of the Convention.

### Access to and Transfer of Technology

The Convention recognizes the importance of access to and transfer of technology and advocates that countries take facilitating, legislative, administrative and policy measures, under fair and most favourable terms to both government and private sectors. Granting access to and transfer of technology must take into account the issues relating to patents and intellectual property rights.

Since this is closely linked to access to genetic resources and will be crucial for successfully implementing the provisions of the convention, India is having care full assess and review the existing protocols/treaties such as provisions of GATT and TRIPPS alone with other own patent law and other related legal regimes. Cooperation of other Contracting Parties and TRIPS along with our own patent law and other related legal regimes should be in place. The Cooperation of other Contracting Parties may be sought, as and when it is necessary (Chauhan, 1996).

### Exchange of Information (Art. 17) and Technical and Scientific Cooperation

Regarding the exchange of information (Art. 17) and technical and scientific cooperation (Art. 18), in the light of these two Articles, ongoing programmes could emphasize and integrate aspects related to biological diversity.

### Handling of Biotechnology and Distribution of its Benefits

India already has rules for the manufacture, use, import, export and storage of genetically engineered organisms/cells which were adopted in 1889 under the Environment (Protection) Act. Besides, DBT has been actively promoting research activities related to biotechnologies through bilateral joint ventures in the frontier areas. India is also participating in developing the modalities for biosafety protocol and other related matters (Art. 19).

### Financial Resources

For effective implementation of the provisions of the Convention, India is already providing financial support for national activities in accordance with its capacity. A National Action Plan is setup to identify the areas for formulation of specific project proposals for bilateral and multilateral funding (Chauhan, 1996).

### National and International Treaties on Biological Resources

The rapid degradation of major ecosystems and their biological components has become a major international concern. Developing and establishing adequate conservation measures and mechanisms for sustainable utilization of biological resources pose a multidimensional challenge, involving scientific, socio-economic, administrative, legal and political issues. Scientists, policy makers, administrators, legal experts and the political leadership have attempted to formalize a common understanding of the scientific, socio-economic and legal issues and to develop viable mechanisms at the national and international levels. The process of negotiating ecologically sound and politically equitable agreements is still incomplete. However during the last 38 years, significant developments have taken place at the national and international levels (Chauhan, 1998).

#### i. Historical Perspectives

The significant changes in the land-use pattern, intensive agriculture, industrial development and rapidly unscientific exploitation of natural resources have seriously altered the dynamics and the functioning of major ecosystems; consequently depleted their wide range of biological resources. This had led to the concern for conserving species and their natural habitats. At the same time, the potential value of genetic resources has been enhanced by the fast-emerging biotechnologies and possibilities of genetically tailoring the biological resources to meet growing human requirements for food, medicines, and industrial products. Both at national and international levels, there have been extensive debates in understanding the intricacies of the elements essential for an effective international legal regime for the conservation of biological resources and their habitats.

#### ii. Scientific Perspectives

Since the 1960s, international scientific community contributed to sharpening the focus on the conservation of biological resources. This is re-elected in the International Biological Programme, IUCN-the World Conservation Union and MAB Programme of the UNESCO. The World Conservation Strategies (1980) published jointly by IUCN, UNEP and WWF laid down three main conservation

☆ The maintenance of essential ecological process

☆ The preservation of genetic diversity

☆ The sustainable use of species and ecosystems

This document had considerable influence on various nations in recognizing the principles for conserving biological diversity and helped them in developing their national policies and legislations, as well as treaties like the 'ASEAN Treaty' finalized in 1985. The follow up document "Caring for the Earth- a Strategies for Sustainable living" highlighted concrete action points such as enhancing *ex situ* and *in situ* conservation, sustainable harvesting of wildlife resources, and generating incentives to conserve biological diversity.

### iii. Political Perspectives

The United Nations Conferences on the Human Environment, 1972, brought political leaders together for the first time, who adopted the "Stockholm Declaration" and approved the creation of UNEP. The declaration laid the foundation for internationally accepted conservation objectives. Moreover, it started a process of interaction among governments to make national and international legal instruments compatible, while upholding the principles of national sovereignty over natural resources.

### iv. Leagal Perspectives

With increasing scientific and political awareness, it became clear that more specific legal regimes were necessary to effectively conserve biological resources, to implement and monitor programmes of action, and several international agreements were adopted. These agreements are not legally binding but are a common declaration of principles, charters or resolutions. These "soft law instruments" (de Klemm and Shine, 1993), which express a broad consensus of the world community, include the following:

☆ Principle 2 and 4 of the Stockholm Declaration of 1972 which emphasis the need to protect both species and their habitats and need to be integrated in planning for economic development.

☆ The World Charter for Nature adopted by the United Nations general Assembly in 1982 which proclaims principles of conservation "by which all human conduct affecting nature is to be guided and judged" and incorporates the three objectives outlined in the World Conservation Strategies:

❑ The Brundtland Report of the World Commission on Environment and Development, adopted by the United Nations General Assembly in 1987 as a framework for sustainable yield in the use of natural animals and plant resources, and for future cooperation in the field of environment and development.

❑ Precautionary Principle 2 of the Rio Declaration adopted at UNCED, 1992 in conjunction with the Convention on biological Diversity and

❑ Agenda 21-an action plan drawn up by UNCED, 1992, wherein Chapter 15 outlines action points relating to the Conservation on biological Diversity.

### International Treaties

International treaties relating to biological resources are the result of a long process of negotiation and hard-won consensus. Broadly these treaties:

☆ Establish uniform conservation rules

☆ Express the commitment of the Contracting Parties to conserve certain species

☆ Organise effective international cooperation.

The treaties define the principles, objective and methods for both conservation and utilization of biological resources at the species (including migratory species) level. The conservation of habitats is envisaged through joint environmental impact assessment in respect of activities undertaken by one country which may have an adverse cross-boundary environmental impact on neighboring states. Besides the international agreements regional treaties and other sectorial instruments have led an important role in reporting the objectives of conservation (Chauhan, 1998). Some treaties are as under:

☆ The only treaty in the Asian region is the Conservation of Nature and Natural Resources in Asia, considered to be one of the most comprehensive conservation treaties. The 1985 agreement was concluded by the ASEAN countries, Brunei, Indonesia, Malaysia, the Philippines, Singapore and Thailand to ensure global environmental protection to the region. It is based on the World Conservation Strategy and takes into account air, water, soil, forests, fauna, flora and also ecological processes.

☆ In the Pacific region too there is only one treaty entitled "The Convention on the Conservation of Nature in the South Pacific" which came into force in 1990. It provides for the creation of protected areas and the protection of both plant and animal species.

☆ The African Convention on Conservation of Nature and Natural Resources (1968) was signed by 30 African States. Article 2 of the Convention sets out the principle that "The Contracting States shall undertake to adopt measures to ensure conservation, utilization, development of soil, flora and faunal resources in accordance with scientific principles and with due regard to the best interest of the people". The text of the Convention deals in detail with each element mentioned, and lays down general obligations.

☆ There are two major treaties in the American region-the "Western Hemisphere Convention (1940)" and the more recent "Convention for the Conservation of Biodiversity and the Protection of Priority Wild areas in Central America (1929)". Specific provisions of the 1992 Convention include development of national legislation for the conservation and use of biodiversity; promotion of species recovery plans; control of illegal traffic and collection in wild flora and fauna; and regulation of domestic trade in such resources.

☆ In Europe, besides European Community Legislation, various other conventions are in operation, such as the Berne Convention, 1979; the Benelux Convention of 1970; and the Alpine Convention (Chauhan, 1998).

    a. The Directive 79/409 on the Conservation of Wild Birds (1979) was the first legal instrument of the European Community. It implements the Berne Convention's provisions and is limited to the bird community only. The other species are covered by the Directive (97/43) on the Conservation of Natural Habitats and of Wild Fauna and Flora (1992). Nearly 160 plant species and 22 animal species are identified as "Priority species" for conservation including their habitats. The Directive aims to establish "a coherent European ecological network" called 'Natura 2000' which will comprise the special conservation areas as well as special protection areas created by member States.

    b. The European Community is also a party to the Bonn Convention, the Regional Seas for East Africa and the Caribbean, and the Convention on the Conservation of Antarctic Marine Living Resources of 1980.

    c. The convention on the Conservation of European Wildlife and Natural Habitats (Berne, 1979) deals with the protection of plant (517 species) and animal species (including most of the European mammal, bird, reptile, amphibian, and some fresh water fish and invertebrate).

    d. The 1970 Benelux Convention on Hunting and the Protection of the Birds is mainly intended to harmonise the legislation of Belgium, Luxemburg and the Netherlands. The three countries also signed a Convention on Nature Conservation and Landscape Protection in 1982.

    e. The convention on the Protection of the Alps is the first international treaty which covers a complete terrestrial ecological unit. It was signed in 1991 by all Alpine countries *viz.* France, Italy, Switzerland, Liechtenstein, Germany and Austria, and Slovenia was also expected to join. But it has not yet come into force. The obligations are of a general kind and cover a wide range of environmental problems of the Alpine region.

### Regional Treaties

UNEP has established a marine conservation programme which currently covers 11 seas or coastal regions to preserve the natural marine and coastal habitats of the world. So for, Protocols relating to the conservation of natural areas in respect of four regional seas have been concluded. These are the Mediterranean (Geneva, 1982), East Africa (Nairobi, 1985), the South-East Pacific (Colombia, 1989) and the Caribbean region (Jamaica, 1990).

Besides there are 5 international instruments which relate to the protection and preservation of the Antarctic region. These are the:

1. Antarctic Treaty (1959) and its
2. Protocol (1991);
3. The Convention on Antarctic Marine Living Resources (1980);
4. The convention on the Regulation of Antarctic mineral Resources Activities (1986), which has since been suspended by the
5. Madrid Protocol (1991).

Antarctic treaty, meant for only the Antarctic region, is a political treaty which was to freeze all claims to national sovereignty in the region, to ban nuclear tests, to authorize peaceful activities and to reinforce cooperation between the Parties in the field of scientific research and indirectly addresses the issues relating to the conservation of marine living resources (Chauhan, 1998).

### Sectorial Treaties

Sectorial conservation treaties are those which deal with the components of biological diversity such as species, natural habitats or protected areas and are either regional or global.

### Species-Based

National conservation measures may be invalidated by differences in the legal provisions of neighbours, affecting the states of species, and even leading to their extinction. Therefore, effective protection measures for their protection have to be through international agreements for joint conservation and management of stocks, controlled withdrawal and trade and the preservation of the regional habitats of the concerned species. For example, an agreement on the conservation of polar bear was concluded by the five circumpolar States in 1973.

Migratory species for both terrestrial and marine categories are usually considered international biological resources. There are numerous bilateral and multilateral agreements and conventions which have dealt with protection of species inhabiting international waters or those which migrate from one country to another. Some important agreements are (Chauhan, 1998):

☆ The Law of Sea Convention, 1982 dealing with certain migratory marine species.

☆ Convention on Migratory Species of Wild Animals, 1983. This is not very effective as the Russian Federation, USA, Canada, most Latin American countries, china, Japan and some South East Asian countries are still outside the Convention; and

☆ Bilateral agreements concluded for North America, the Pacific and parts of Asia in the 1970 and 1980 for the protection of migratory birds.

The most important treaty responsible for the regulation of trade in endangered wild species is the Convention on International Trade in Endangered Wild Fauna and Flora (CITES) which came into force in 1975. Since most countries are parties to this convention, it is felt that the aspect of trade may not have to be dealt with in future.

Certain species-based treaties are not really conservation oriented but deal mainly with extraction from natural habitats. These provide joint regulatory mechanisms to be adopted and implemented and to share the results of scientific research in population and management of concerned wild species. Major agreements are:

☆ International Convention for the regulation of Whaling 1946

☆ Convention on the Conservation of Antarctic Marine Living Resources (CCAMLR), 1980; and

☆ Resolution 44(225) of the U.N. General Assembly, 1989 for high seas as a whole.

## Area-Based

The international community has also two significant global treaties on the conservation of specific ecosystems:

☆ Convention on wetlands of International Importance especially as waterfowl Habitat (commonly known as Ramsar Convention), 1971.

☆ Convention concerning the Protection of the World Cultural and Natural Heritage (also known as World Heritage Convention)

The Ramsar Convention focuses primarily on the conservation and management of wetlands identified as 'Ramsar Sites' of International Importance. Parties have to promote the "Wise Use" of wetlands and take measures for their conservation. The Contracting Parties have initiated both short and long term programmes to achieve the objective of this convention, including setting up the Wetland Fund in 1990.

The World Heritage Convention has two main legal provisions:

1. States are to conserve their natural and cultural heritage.

2. It calls for Parties to contribute financial to the conservation of Natural heritage as most sites are located in low income countries. Special financial mechanism has been established to assist Parties to undertake their obligation more effectively.

## Law of the Sea

In order to protect and preserve the marine environment, the Law of Sea was signed in 1982 and has yet to come in force. The scope of this convention is very broad-defining boundaries of each sea, legal regime for exploitation of mineral resources on the deep sea-bed and ocean flow beyond national jurisdiction, control of pollution of marine environments, and conservation and exploitation of marine species. Under the convention, states are fully sovereign over their "internal waters", "territorial seas" and : "Exclusive Economiz Zone" (EEZ) for the purpose of exploring, exploiting, conserving and managing the natural resources found there. These areas also include the continental shelf and its natural resources. The convention is largely confined to the high sea beyond the 200-mile outer limit of the EEZ and both living and non-living resources found there are declared "common heritage of mankind". Once the convention comes into force, the International Sea-bed Authority shall be responsible for organizing, exploring and exploiting the area for the benefit of mankind, through equitable sharing of economic benefits derived from the activities (de Klemm and Shine, 1993).

## Regulation on Access to Biological Diversity (ABS)

### I. Access to Biological Diversity (ABS) at the National and International Levels

Global regulation of access to biological resources and sharing of benefits arising from their commercial utilization is one of the three most discussed topics today, the other two being the multilateral trade and climate change. Negotiations under various for a in this context require a clear understanding of complex interplay of international laws and national legislation so as to maintain a harmonious and synergistic relationship among them. An international law, or the law of nation, is primarily a system governing the relationship among nations who have become contracting parties to legally binding international treaties and are required to adapt their national legal framework to meet their national obligations under the provisions of those treaties. However, it happens sometimes that the national obligations under some international treaties on specific topics, like access to biological resources and benefit sharing, may appear to be in conflict as seen from the relevant provisions under the Convention on Biological Diversity (CBD) and WTO-Trade related Intellectual Property rights (TRIPS) or even those of CBD and the International Treaty on Plant Genetic Resources for Food and Agriculture (ITPGRFA or the Plant Treaty). To go a step further, perceptions on the relationship between international law and national law also differ regarding supremacy of one over the other. There are two contending concepts as proposed by (Park and Yanos, 2006; Muller, 2013):

(i) **Monism:** The idea of monism assumes that international law and national law are simply two components of a single legal system and regards 'law' as one entity. In other words, both are interrelated parts of one single legal structure and form a unity, through international law may have supremacy over the national law in cases of conflict. Monism in practice envisages that the legal institutions of a country, such as its judiciary, legislature and executive, should ensure that national rights and obligations in this context conform to international law.

(ii) **Dualism:** On the other hand dualism assumes that international law and national law of States are two separate and district legal systems. Being different legal orders, international law would not as such form part of the domestic law of a state. Where rules of international law apply within a State, they do so as a result of their adoption under the national law and not under the international law. Dualism refrains from any controversy as to supremacy of one legal system over the other recognizing that each one is considered supreme in one's own sphere and operates on a different level but recognizes that, ultimately, national interests can override the international interests in some situations.

There are, however, some distinctions between national law and international law.

(a) Firstly, the subjects of national law are individuals, while the subjects of international law are Nation States.

(b) Secondly, juridical origins of the two legal systems are different, *i.e.* the source of national law is the will of the representatives of public expressed through the State while the source of international law is the common will of contracting parties (Nation States).

(c) Thirdly, national law is a law of the Sovereign (national government) over individuals (citizen) whereas international law is a law, not above, but among sovereign States.

## II. Access to Biological Resources and Benefit Sharing

Governance systems for access to genetic resources and sharing of benefits (ABS) arising from their commercial utilization may be seen from several positions such as perspectives of the primary stakeholders, provisions of the national regulatory framework and the country's legally binding obligations under international treaties to which it is a Contracting Party (Halewood *et al.*, 2014). Regulation of genetic resources has 3 distinct dimensions which are legally binding treaties andIndia is a contracting party to them (Rana, 2010):

☆ **Perspective of their developers and users:** The first dimension comprises local farming communities (developers, conservers and end-users of their genetic resources who are the primary stakeholders), public sector research institutions and genebanks (trustee custodians of germplasm collections and users for public good), and seed companies/corporations (users for commercial utilization meant for private benefits). They together represent the main stakeholders and key beneficiaries. This treaty highlights the conservation of bioresources, their sustainable use, regulated access and fair and equitable benefit sharing.

☆ **Governance at the state and national levels:** The second dimension involves policy makers. Legislators, managers and administrators regulating authorized access while promoting conservation with sustainable use. This treaty also highlights the conservation of bioresources, their sustainable use, regulated access and fair and equitable benefit sharing.

☆ **National obligations under international treaties/agreements:** The third dimension related to national obligations under multilateral environment and trade agreements. Under this category, 3 major international agreements, namely CBD, ITPGRFA and WTO-TRIPS have impacted the access to genetic resources globally and also at the national level, more so in bio-diverse developing countries. This treaty focuses mainly on patenting crop variety protection laws that grant monopolistic/exclusive rights to IPR holders/breeders to the exclusion of the rights of farmers and other primary beneficiaries.

Considering that agro-biodiversity constitutes a subset of the total biological diversity, and a very important one, it is imperative that all these international agreements need to be implemented in harmony with each other, more particularly in countries like India whose national economy is primary based on agriculture (Tvedt, 2014).

## III. Global Development on Access to Biological Diversity

In order to understand the landscape of international governance of genetic resources, it is important to appreciate the on-going governance efforts and identify the problematic areas where more attention is required for moving forward is mentioned below:

### 1. International Laws and the National Legal and Policy Framework

National policies describe the objectives and missions of a government indicating how it proposes to achieve those objectives by issuing relevant guidelines. Laws, on the other hand, are the standard rules and regulations that are compulsory to be followed by all the people of that country and there are provisions in those laws for punishment for those who contravene them. In other words, laws help a government in setting up legal and institutional framework to achieve the aims out in its policy statements. National laws are enacted by the parliament and enforced by the national government within its national boundaries. International laws, in contrast, arise from customary laws, judiciary pronouncement, and more often, from legally binding national obligations under international agreements, treaties, conferences and conventions. In a way, they expand the jurisdiction of national laws beyond their national boundaries but they also impact them.

### 2. Impact of the UN Conference on the Human Environment, 1972

UN Conference on the Human Environment, held in Stockholm in 1972, took some important decisions concerning environment and sustainable development and had a significant impact in India. impact of these decisions in the Indian context may be clearly seen in the 42nd Amendment to the Indian Constitution, enacted in 1976, adopting Article 48A (One among the Directive Principles which cannot be challenged in any court of law) stating that the State shall endeavor to protect and improve the environment and to safeguard the forests and wildlife of the country. The subject of wildlife and forests was thus transferred from the state list to the concurrent list of the constitution through this decree, providing enormous powers to the Central Government in this area. In addition, Article 51A (g) (Fundamental Duties) was also introduced to protect and improve the natural environment including forests, lakes, rivers, wildlife and to have compassion for living creatures. The wildlife (Protection) Act 1972 (as amended in 1991), the Water (Prevention and control of pollution) Act, 1974, the Air (Prevention and control of pollution) Act, 1981 and the Environment Protection Act, 1986 were enacted to fulfill the commitments made by India during the Stockholm Conference. In addition, a separate Department of Environment was created in 1980 and a separate Union Ministry of Environment and Forests (with Climate Change added subsequently) was established in 1985.

### 3. The International Undertaking on Plant Genetic Resources, 1983: A New Initiative

In 1983, the FAO established a Commission on Plant Genetic resources (later renamed the Commission on Genetic Resources for Food and Agriculture), the first permanent intergovernmental forum devoted to conservation and development of genetic resources. Commission's first major action was to adopt a non-binding resolution in 1983 for setting up the International Undertaking (IU) on Plant Genetic resources (PGR). It worked on the basic principle that PGRR are common heritage of humankind and, hence should be made available to researchers without restriction. Many commercial seed companies disliked the IU because it required that elite genetic stocks (including improved and current breeder's lines) should also be made available without restriction. Under their influence, the United States and many other developed countries declined to sign the IU and adhere to it. Efforts to conciliate the concerns of developed and developing countries resulted in two 1989 amendments to the undertaking paving the way for the United States and Canada signing the IU but they still declined to adhere to its binding obligations. In 1993, FAO adopted the Resolution 7/93, calling for intergovernmental negotiations for revision of the IU to harmonise its contents with those of the CBD, acknowledging thereby the sovereignty of States over their natural resources. Accordingly, provisions of the IU were suitably revised to bring them in harmony with those of the CBD and the revised version was adopted in 2001 as the legally-binding International Treaty on Plant Genetic Resources for Food and Agriculture.

### 4. The Convention on Biological Diversity (CBD): the Turning Point

In 1992, the United Nations hosted an Earth Summit in Rio de Janeiro Brazil in June, 1992 and it gave birth to the legally binding Convention on Biological Diversity (CBD) under UNEP besides several other treaties. The Convention on Biological Diversity came into force on 29th December, 1993 under an international treaty. The CBD has 193 Contracting Parties making it almost globally accepted treaty though the USA is the only major country that has not yet ratified it. Through its 42 Articles, the Convention recognizes national sovereignty over biological resources. Its objectives include the conservation of biological diversity, sustainable use of its components and fair equitable sharing of benefits arising out of the utilization of genetic resources. It is a conscious of intrinsic value of biological diversity in terms of ecological, social, economic, scientific, educational, cultural, recreational and aesthetic importance. In this agreement 168 signatory countries including India, are members. To recapitulate, the CBD marked the end of the 'common heritage' concept of genetic resources and it asserted that nations have sovereign rights over natural resources within their boundaries, and that the authority to determine access to genetic resources rests with the national governments and it is subject to national legislation. The provisions of the CBD facilitate the access to genetic resources on 'mutually agreed terms' and with 'prior informed consent' of the country providing the resources, with the recipient country being committed to share accruing benefits.

## Elements of Convention on Biological Diversity

After the enactment of BDA, 2002 and dealing with exchange of Plant Genetic Resources (PGR) in the country at ICAR-NBPGR, it was noted that the flow of Plant Genetic Resources for Food and Agriculture (PGRFA) was much restricted after the CBD came into force, December 1993 when compared to before its implementation by different countries passed legislation to claim sovereign rights over their bio-resources and to implement CBD's provisions. For example, the Philippines established a system for access to biological resources by an executive order issued in 1995 and the Andean Community, in its Decision No. 391 taken in 1996, adopted a Common Regime on Access to Genetic Resources. India enacted the Biological Diversity Act in 2002 and framed Rules under it in 2004. Chapter II of the BDA, 2002 defined the Regulation for access to biological diversity and the Section 3 mentions that certain persons cannot obtain any biological resource occurring in India for research or commercial utilization, without the approval of the NBA (www.nbaindia.org). These persons include those who are not citizens of India, or are non-resident Indians, or companies not registered or incorporated in India. These also include Indian companies registered in India with foreigners managing the company or provide fund to these. Section 5 of the Act provides for exemptions from Section 3 and Section 4 for access to Indian genetic resources by nationals of other countries, covered under the collaborative research projects/Memoranda of Understanding (MoU)/Work Plans. Such collaborative research projects must conform to the policy guidelines issued by the Ministry of Environment, Forests and Climate Change (MoEF and CC), S.O 1911 (E). however, if any research/institutions was intending to send any biological material for deposition in international repositories for the purpose of patents or for new species recognition or taxonomic review were not able to do so under Section 5. A cost of paying requisite fees was also involved in transfer of material. For all other requests not covered under any collaborative research projects the non-Indian applicant as described above, was required to obtain prior approval of NBA. The above provisions were subjected to individual interpretations with regard to access under BDA, 2002. After detailed deliberations in

various meetings/for a, notifications/regulations were put in place by NBA and MoEF and CC to bring more clarity on access of biological resources by different types of users of biological resources for example, the term 'occurring in India' is ambiguous as all the PGRFA occurring in the country are not 'Indian'. Thus, these questions are being raised on the interpretation of the definitions from time to time.

However, collections from International Research Centres were not affected by CBD (Tyagi *et al.*, 2006). On the other hand countries that were party to International Treaty on Plant Genetic Resources for Food and Agriculture (ITPGRFA) including India were obliged to provide facilitated access to PGRFA to all member countries for the crops listed under ITPGRFA treaty (Table 13.1). The BDA, 2002, therefore, needs to be implemented in synergy with ITPGRFA obligations. Specific provisions in the BDA, 2002 were needed to be included/revised/amended for accessing PGRFA which includes PGRFA beyond their origin/sovereignty [*e.g. In trust* material-held by consultative Group on International Agricultural Research (CGIAR) Centres collected before 1993 and referred to as CG institute hereafter]. For such policy deliberations, Indian Council of Agricultural Research (ICAR) had established a high level national Advisory Board on Management of genetic Resources (NABMGR) to discuss and advise the concerned departments for harmonization of access and benefit sharing issues. During its tenure of 3 years, 5 meetings were organized and several recommendations were made to streamline the process of access of germplasm. It was one of the recommendations of the Board that the implementation of the ITPGRFA in the national system needed to be adopted urgently and the proposal put forth by India for designation of material for the Multilateral System (MLS) of the ITPGRFA. In response, the guidelines for 'Implementation of the ITPGRFA were developed and Central Government, in consultation with National Biodiversity Authority (NBA) notified exemption for access to crops of the treaty/ITPGRFA, and declared that Department of Agriculture and Co-operation (DAC), Ministry of Agriculture (MoA) as the nodal department. DAC may from time to time notify the listed crops of the treaty to NBA for exemption. This exemption came after a series of detailed discussions in the board meetings and its sub-committees.

## Recent Notification

The notification issued recently has further stressed some of the provisions, as indicated below:

G.S.R. 827 dated November 21, 2014 on access to genetic resources and the fair and equitable sharing of benefits arising from their utilization (Crops of the Treaty; http://nbaindia.org/uploadded/pdf/Gazette_notification_of_ABS_Guidlines.pdf).

## Table 13.1: Central Legal Instruments Relevant to Biological Diversity

| Sl.No. | Acts | Key Points |
|---|---|---|
| 1. | Agricultural and Processed Food Products Export Development Authority Act, 1985/1986 | ☆ Promotion and regulation of export of agricultural products specified in schedules.<br>☆ Includes medicinal plants |
| 2. | Agricultural Produce (Grading and Marking) Act, 1937 | ☆ Fixing grade designations to indicate quality of any specified agricultural produce (3a, b)<br>☆ Prohibition or restriction on trade in wrongly marked/graded produce (3g)<br>☆ Extension of such provisions to any other article (includes non-agricultural articles) [6] |
| 3. | Cardamom Act, 1965 | ☆ Provisions as in Rubber Act (see below), includes seeds.<br>☆ Provisions for prohibiting/restricting export/import of cardamom (21). Applicable to *Elettaria cardamomum* Maton, but extendable to any other plant notified by Cardamom Board [3] |
| 4. | Coconut Development board Act, 1979 | ☆ As in Rubber Act, Tea Act, Cardamom Act, *etc.* |
| 5. | Customs Act, 1962 | ☆ Regulation of import-export specifically for:<br>(a) The protection of human, animal or plant life or health (11) [t]<br>(b) The conservation of exhaustible natural resources (11) [m]<br>(c) Regulation of transportation and storage of notified items (11) [j, k, l, m] |
| 6. | Destructive Insects and Pests Act, 1914 | ☆ Prohibition or regulation of import of any "articles" which may cause infection to any plant (1) [3]<br>☆ Prohibition or regulation or movement, between states within India, of articles likely to cause infection to any plant [4A]<br>**Note:** 'Articles' includes insects and plants |
| 7. | Environment (Protection)Act, 1983 | ☆ General measures to protect environment (i) [3]<br>☆ Restriction of industrial and other processes/activities in specified areas (v) [3] (2)<br>☆ Prevention and control of hazardous substances, including their manufacture, use, release and movement [3] (2), 7, 8 |
| 8. | Fisheries Act, 1987 | ☆ Prohibition on us of explosive for fishing (1) [4]<br>☆ Prohibition on use of poisons for fishing [5]<br>☆ Regulation on fishing in private waters, with consent of owners/right holders [6] (2) and (3).<br>☆ Prohibition of al fishing in specified waters for maximum 2 years [6] (4) |
| 9. | Forest Act, 1927 | ☆ Setting up and managing reserved forests [Chapter II]<br>☆ Setting up and managing village forests [Chapter III]<br>☆ Setting up and managing protected forests [Chapter IV]<br>☆ Protection on non-government forests and lands [Chapter V]<br>☆ Control of movement of forests produce [Chapter VII]<br>☆ Control of grazing/trespass by cattle in forest land [Chapter X] |
| 10. | Forest (Conservation) Act, 1980 | ☆ Prohibiting or regulating non-forest use of forest lands [2] |
| 11. | Import and Export (Control) Act, 1947 | ☆ Prohibition or restriction on import and export of specified items [3]<br>☆ Regulation on transportation of specified items [4e] |
| 12. | Marine Products Export Development Authority, 1972 | ☆ Establishment of an Authority for developing and controlling marine products [4, 9] (1)<br>☆ Developing and regulating off-shore and deep-sea fishing; taking measures for conservation; fixing standards for export; regulating exports [9] (2a, c, f)<br>☆ Prohibition or restriction on export and import of marine products [2] (1) |
| 13. | Maritime Zones of India (Regulation of Fishing by foreign vessels) Act, 1980 | ☆ Regulation of fishing in India's EEZ by people using foreign vessels [3]<br>☆ Permits only to be granted within definition of public interest and for scientific research, experiments, *etc.* [5 (3) 8] |
| 14. | National Dairy Development Board Act, 1987 | ☆ Establishment of Board which promotes dairy development and other agricultural based industries [4, 16 (1a)]<br>☆ Financing and facilitating animal husbandry, agriculture, high yielding cattle (including import of semen), import-export of milch animals and bulls and general enhancement of cattle wealth [16(1)] |

| 15. | National Oilseeds and Vegetable Oils Development Board, 1983 | ☆ As in Rubber Act, *etc.* special focus on providing farmers, especially seeds, and certified seeds of high quality and for improved methods of cultivation |
|-----|---|---|
| 16. | New Seed Development Policy, 1988 | - |
| 17. | Prevention of Cruelty to Animals Act, 1960 | ☆ Restriction on cruel treatment of animals, including use, transpotation, and trade [Chapter II, and Rules under Section 38] |
| | | ☆ Restriction on use of animals for purpose of experimentation and performances [Chapter IV and V] |
| 18. | Rubber (Production and Marketing) Act, 1947 | ☆ Establishment of Indian Rubber board, with function of developing/encouraging improved rubber cultivation and marketing, advising or import/export [B(1) and (2)] |
| | | ☆ Restriction on right of rubber planters-license required to plant or replant where to plant, *etc.* [17] |
| | | **Note:** Applicable to 4 species of rubber initially, more if board so notifies [Definitions] |
| 19. | Seeds Act, 1966 | ☆ Regulation on quality of seeds of notified food crops, cotton and fodder, to be sold for agricultural purpose [5,6] |
| | | ☆ Restriction on export/import of notified seeds [12] |
| | | ☆ Exemption to persons selling/delivering on own premises, seeds grown by them [24] |
| 20. | Spices Board Act, 1986 | ☆ As in Rubber Act *etc.*, for Cardamom; for other spices, restricted to export-import development and regulation. |
| 21. | Tea Act, 1935 | ☆ As in Rubber Act *etc.*, includes restrictions on export of tea seeds [17]. Applicable to one species, *Camellia sinensis*, presumably to all its varieties |
| 22. | Territorial Waters, Continental Shelf, Exclusive Economic Zone and other Maritime Zones Act, 1976 | ☆ Establishment of sovereign rights over waters and seabed within the continental shelf and exclusive economic zone (200 nautical miles from nearest appropriate point on Indian territory) [3 (1), 5 (1). 6 (2), 7(4)] |
| | | ☆ Sovereign rights to explore, exploit, conserve and manage resources of continental and conservation of marine environment [6 (5), 7 (6)] |
| 23. | Tobacco Board Act, 1975 | ☆ As in Rubber Act, Tea Act, for Cardamom *etc.* |
| 24. | Wildlife (Protection) Act, 1972 and Wildlife (Protection) Amendment Act, 1991 | ☆ Restriction or prohibition on hunting of animals [Chapter III] |
| | | ☆ Protection of specified plants [Chapter IIIA] |
| | | ☆ Setting up and managing sanctuaries and national parks [Chapter IV] |
| | | ☆ Setting up of zoo authority, control of zoos, and captive breeding [Chapter IV A] |
| | | ☆ Control of trade and commerce in wild animals, animal articles and trophies [Chapter V and Chapter VA] |

*Source*: Kothari and Singh, 1992.

## (A) Regulation 13

Conducting of non-commercial research or research for emergency purposes outside India by Indian researchers/government institutions-

(1) Any government institution which intends to send biological resources to carry out certain urgent studies to avert emergency like epidemics, *etc.* shall apply to NBA.

(2) The NBA shall, on being satisfied with the application accord its approval within a period of 45 days from the date of receipt of the application. The applicant shall be required to deposit voucher specimens in the designated national repositories before carrying/sending the biological resources outside India and a copy of proof of such deposits shall be endorsed to NBA.

### Implications

This provision implies that any Indian researchers going abroad for any approved basic research activity could carry biological resource with the approval of NBA which shall be granted within 45 days, as opposed to 6 months as prescribed for other cases of access and sending material outside India.

Secondly, any government institution could send material in the interest of the nation in the same manner for testing/screening or taxonomic studies to a foreign institution in cases of emergent threats to biological resources and complete taxonomic studies, without the need to formulate any collaborative research project. Such approval shall be granted by NBA within 45 days. Fees for access shall not be required in such cases.

## (B) Regulation 17

Certain activities or persons are exempted from approval of NBA or State Biodiversity Board (SBB), namely:

(a) Indian citizens or entities accessing biological resources and/or associated knowledge, occurring in or obtained from India, for the purposes of research or bio-survey and bio-utilization for research in India;

(b) collaborative research projects, involving the transfer or exchange of biological resources or related information, if such collaborative research projects have been approved by the concerned Ministry/Department of the State or Central Government and conform to the policy guidelines issued by the Central Government for such collaborative research projects;

(c) Local people and communities of the area, including growers and cultivators of biological resources. *vaids* and *hakims*, practicing indigenous medicine, except for obtaining intellectual property rights;

(d) Accessing biological resources for conventional breeding or traditional practices in use in any agriculture, horticulture, poultry, dairy farming, animal husbandry or bee keeping, in India;

(e) Publication of research papers or dissemination of knowledge, in any seminar or workshop, if such publication is in conformity with the guidelines issued by the Central Government from time to time;

(f) Accessing value added products, which are products containing portions or extracts of plants and animals in unrecognizable and physically inseparable from and

(g) Biological resources, normally traded as commodities notified by the Central Government under Section 40 of the Act.

### Implications

This implies that all Indians could access biological resources for research use within India. Secondly, local communities and traditional medical practitioners could use biological resources occurring in India without the need for prior intimation to the SBB except when seeking a patent.

Also, any Indian or non-Indian entity could access biological resource for traditional plant breeding and for other traditional practices of plant and animal improvement if the research was being carried out within India. Therefore, any seed company or private research organization doing conventional (traditional) plant breeding in India could access plant varieties and germplasm from public or private resources. Such access shall however be based on the mutual agreements/terms and conditions as laid down by the provider(s)/institution(s).

## 5. The Nagoya Protocol to CBD on ABS, 2010: A New Beginning

The Nagoya Protocol (Rcherzhagen, 2014) to CBD on ABS is a new international treaty on ABS, adopted in October, 2010 to support implementation of the third objectives of CBD, namely, the fair and equitable sharing of benefits arising from the utilization of genetic resources. It is based on the twin principles of prior informed consent (PIC) and mutually agreed terms (MAT) enshrined in the CBD. This Protocol on ABS entered into force on 9 October, 2014 prompting the Parties to CBD to prepare for its implementation by taking appropriate policy, legislative and administrative measures. Sixty nations have already ratified this legally binding Protocol and many more are in the process of doing so. India signed the Protocol on 11 May, 2011 and ratified it on 9 October, 2012. This Protocol requires that Provider Parties adopt measures that need to:

☆ Create legal certainty, clarity and transparency for access to genetic resources

☆ Provide fair and non-arbitrary rules and procedures

☆ Establish clear rules and procedures for prior informed consent and mutually agreed terms

☆ Provide for issuance of an internationally recognized certificate of compliance when access is granted.

The Nagoya Protocol establishes clear rules for accessing, trading, sharing and monitoring the use of the world's genetic resources that can be used for pharmaceutical, agricultural, industrial, cosmetic, and other purposes. By establishing this framework, it seeks to ensure that genetic resources are not used without prior consent of the countries that provide them, and that the communities, that possess the traditional knowledge associated with the use of these resources, also share the benefits arising from its commercial utilization. The Protocol seeks to increase transparency in transfer of genetic resources through its Access Benefit-Sharing Clearing House (ABS-CH), which is an online platform for exchanging relevant information (Morgera *et al.*, 2014). Its goal is to enhance clarity on procedures in provider countries for access to genetic resources and also to monitor their commercial utilization in user countries.

## 6. International Treaty on Plant Genetic Resources for Food and Agriculture (ITPGRFA): New Approach to Promote Global Food Security

Recognizing the interdependence among countries regarding crop genetic resources, representative of 135 member-nations of FAO approved in Rome on 3 November, 2001 a new International Treaty on Plant Genetic Resources for Food and Agriculture (PGRFA) to promote global food security. This treaty relates to Plant Genetic Resources for Food and Agriculture (PGRFA) and defines PGRFA as any genetic material of plant origin of actual or potential value for food and agriculture (http://www.planttreaty.org). The FAO had revised the text of the IU on PGR to bring its provisions in harmony with those of the CBD and then adopted it as the legally binding International Treaty.

Farmers' Rights were recognized under this Treaty but its realization was left to the national governments in their jurisdiction. India's legislation on Protection of Plant Varieties and Farmers' Rights, 2001 has led the way in this direction but there is need to evaluate its effectiveness. It is now widely recognized that a reliable way to realize Farmers' Rights is to enable them to save, exchange and sell the seeds of improved and IPR protected varieties grown by them, and also to assist them in improving their locally adapted crop varieties through participatory breeding while paying greater attention to the needs and circumstances of resource-poor farmers who are the real guardians of much of the agricultural biodiversity (FAO, 2010). Genetic material is defined as any material of plant origin, including reproductive and vegetative propagating material, containing functional units of heredity. Thus the scope of BDA, 2002 being all biological resources, it does not differentiate PGRFA from other resources but these are of utmost importance for food and nutritional security and sustainable agriculture.

Of the nations participating in that FAO conference, that adopted the ITPGRFA, only the United States and Japan abstained, citing concerns about a lack of clarity regarding the effect of the Treaty on intellectual property rights (IPR). The Plant Treaty, which entered into force on 29 June, 2004, provides a Multilateral Systems (MLS) of Access and Benefit-Sharing to facilitate exchange of PGRFA. The MLS presently applies to an initial annex of 35 food crops and 29 genera of forages.

**Table 13.2: Crops Listed under ITPGRFA (of the Treaty)**

| Sl.No. | Crop | Genus | Observations |
|---|---|---|---|
| | **Food crops (35)** | | |
| 1. | Breadfruit | Artocarpus | Breadfruit only |
| 2. | Asparagus | Asparagus | |
| 3. | Oat | Avena | |
| 4. | Beet | Beta | |
| 5. | Brassica complex | Brassica | Genera included are: Brassica, Armoracia, barbarea, Camelina, Crambe, Diplotaxis, Eruca, Isatis, Lepidium, Raphanobrassica, Raphanus, Rorippa, and Sinapis. This comprises oilseed and vegetable crops such as cabbage, rapeseed, mustard, cress, rocket, radish, and turnip. |
| 6. | Pigeonpea | Cajanus | |
| 7. | Chickpea | Cicer | |
| 8. | Citrus | Citrus | genera Poncirus and Fortunella are included as root stock |
| 9. | Coconut | Cocos | |

| Sl.No. | Crop | Genus | Observations |
|---|---|---|---|
| 10. | Colocasia Xanthosoma | Colocasia, Xanthosoma | Major aroids include taro, cocoyam, dasheen and tannia |
| 11. | Carrot | Daucus | |
| 12. | Yams | Dioscorea | |
| 13. | Finger Millet | Eleusine | |
| 14. | Strawberry | fragaria | |
| 15. | Sunflower | Heliathus | |
| 16. | Barley | Hordeum | |
| 17. | Sweet potato | Ipooea | |
| 18. | Grass pea | Lathyrus | |
| 19. | Lentil | Lens | |
| 20. | Apple | Malus | |
| 21. | Cassava | Manihot | Manihot esculenta only |
| 22. | Banana/Plantain | Musa | Except Musa textilis |
| 23. | Rice | Oryza | |
| 24. | Pearl Millet | Pennisetum | |
| 25. | Beans | Phaseolus | Except Phaseolus polyanthus |
| 26. | Pea | Pisum | |
| 27. | Rye | Secale | |
| 28. | Potato | Solanum | Section tuberosa included, except Solanum phureja |
| 29. | Eggplant | Solanum | Section melongena included |
| 30. | Sorghum | Sorghum | |
| 31. | Triticale | Triticosecale | |
| 32. | Wheat | Triticum | Including Agropyron, Elymus, and Secale |
| 33. | Faba bean | vicia | |
| 34. | Cowpea | Vigna | |
| 35. | Maize | Zea | Excluding Zea peremis, Z. diploperennis, and Z. luxurians |

| Sl.No. | Genera | Species |
|---|---|---|
| | **Forages (29)** | |
| 1. | Astragalus | chinensis, cicer, arenarius |
| 2. | Canavalia | ensiformis |
| 3. | Coronilla | varia |
| 4. | Hedysarum | coronarium |
| 5. | Lathyrus | cicero, ciliolatus, hirsutus, ochrus, odoratus, sativus |
| 6. | Lespedeza | cuneata, striata, stipulacea |
| 7. | Lotus | corniculatus, subbiflorus, uliginosus |
| 8. | Lupinus | albus, angustifolius, luteus |
| 9. | Medicago | arborea, falcate, sativa, scutellata, rigidula, truncatula |
| 10. | Melilotus | albus, officinalis |
| 11. | Onobrychis | vicifolia |
| 12. | Ornithopus | sativus |
| 13. | Prosopis | offinis, alba, chilensis, nigra, pallida |
| 14. | Pueraria | Phaseoloides |

| Sl.No. | Genera | Species |
|--------|--------|---------|
| 15. | *Trifolium* | *alexandrinum, alpestre, ambiguum, angustifolium, arvense, agrocicerum, hybridum, incarnatum, pretense, repens, resupinatum, rueppellianum, semipilosum, subterraneum, vesiculosum* |
| **Gross Forages** | | |
| 16. | *Andropogan* | *gayanus* |
| 17. | *Agropyron* | *cristatum, desertorum* |
| 18. | *Agrostis* | *stolonifera, tenuis* |
| 19. | *Alopecurus* | *pratensis* |
| 20. | *Arrhenatherum* | *elatius* |
| 21. | *Dactylis* | *glomerata* |
| 22. | *Festuca* | *arundinacea, gigantea, heterophylla, ovina, pratensis, rubra* |
| 23. | *Lolium* | *hybridum, multiflorum, perenne, rigidum, temulentum* |
| 24. | *Phalaris* | *aquatic, arundinacea* |
| 25. | *Phleum* | *pratense* |
| 26. | *Poa* | *alpine, annua, pratensis* |
| 27. | *Tripsacum* | *laxum* |
| **Other Forages** | | |
| 28. | *Atriplex* | *halimus, nummularia* |
| 29. | *Salsola* | *vermiculata* |

The PGRFA covered under the MLS established under the Treaty is covered under Article 11 which includes all PGRFA which are in the public domain, *ex situ* collections of the CG institutes, and in other international institutions, and encourages contracting parties to put their collections in the multilateral system.

The PGRFA covered under the MLS and referred to as Treaty Crop according to criteria of food security and interdependence. Because this list is a result of political compromises, some crops that might have been expected to be covered, such as soybean, groundnuts, and sugar cane are conspicuously missing. It is notable that the MLS covers only those PGRFA which are 'in the public domain" and those which are held in trust, in *ex situ* collections by IARCs. The Standard Material Transfer Agreement (SMTA) provides a mechanism for overcoming potential difficulties of enforcement by empowering FAOs as the entity chosen by the Governing Body, to represent its interests as a third party beneficiary under the SMTA, and to initiate action where necessary to resolve disputes. The Treaty forbids recipients of PGRFA through the MLS to claim any IPR on them in the form received from the MLS as that may limit access to them or their genetics parts or components. There are, however, some hazy areas on this aspect that need to be addressed (Andersen *et al.*, 2010).

The ITPGRFA differs substantially from the CBD, as this treaty as whole applies to one specific group of organisms, *i.e.*, plant genetic resources for food and agriculture (PGRFA). The MLS for ABS has become the legal instrument for the already ongoing exchange of accessions of PGR stored in the international collections, while adding a number of national; collections to the MLS (see Box 1). It still remains to be validated whether the MLS has led to more exchange of, and better access to, PGRFA. Further, the issue of genetic resources collected from countries of their origin, prior to CBD, still hangs on though the designated accessions stored

---

**Box 1 Benefit-Sharing under the Multilateral System of ITPGRFA**

Facilitated access to genetic resources that are in public domain and are included in the Multilateral System, is itself recognized as a major benefit for researchers and plant breeders. Other benefits arising from the use of PGRFA that are to be shared on a 'fair and equitable' basis include:

(1) **Exchange of information:** This includes catalogues and inventories, information on technologies and results of technical, scientific and socio-economic research on PGRFA including data on characterization and evaluation.

(2) **Access to and transfer of technology:** Contracting Parties agree to provide or facilitate access to technologies for the conservation, characterization, evaluation and use of PGRFA. The Treaty points out various means by which transfer of technology is to be carried out, including participation in crop-based or thematic networks and partnerships, commercial joint ventures, human resource development and through making research facilities available. Access to technology, including that protected by IPR, is to be provided and/or facilitated under fair and most favourable terms, including on concessional and preferential terms where mutually agreed. Access to these technologies is provided while respecting applicable property rights and access laws.

(3) **Capacity building:** This Treaty assigns priority to programmes for scientific education and training in the conservation and use of PGRFA, to the development of facilities for conserving and using PGRFA and to the carrying out of joint scientific research.

(4) **Sharing of monetary and other benefits arising from commercialization:** Monetary benefits include payment into a special Benefit-Sharing fund of the MLS of a share of the revenues arising from the sale of PGRFA products that incorporate material accessed from the MLS. Such payment is mandatory where the product is not available for further research and breeding, for example, as a result of certain types of patent protection. In the SMTA, adopted by the Governing Body at its First Session in 2006, the payment is set at 1.1 per cent of the gross sales generated by the product less 30 per cent (*i.e.* 0.77 per cent)

*Source*: Rana, 2015

in CGIAR's International Genebanks have been brought under the jurisdiction of FAO.

The Multilateral System for ABS under the treaty, as mentioned above, applies only to PGRFA under specific circumstances, *i.e.* when certain accessions of PGRFA are in the public domain, are accessed for specific uses, and under the condition that no IPRs hinder the further exchange and access of the material received from the MLS. These limitations in the scope of the MLS need to be better understood if we are to clarify the legal relationship between the two instruments.

## Regulations of Access to Plant Genetic Resources for Food and Agriculture (PGRFA)

The provisions for facilitated access to PGRFA are laid down in Article 12 of the Treaty. The access to be provided solely for the purpose of utilization and conservation for research, breeding and training for food and agriculture, provided these are not for chemical, pharmaceutical and/or other non-food/feed industrial uses. Access is provided with all available passport data. However, Intellectual Property (IP) or other rights which limit the facilitated access cannot be claimed. There is a provision for the material which is under development and access to such material is at the discretion of the developer.

The issue of approval and involvement of holders of traditional knowledge for sustainable use of biological diversity is considered and addressed by the CBD. However, CBD does not specifically outlined as to how the domestic PIC should be obtained. Conferences of parties to the CBD, during its 3rd meeting held at Buenos Aires in September 1996, established a multiyear programme of work on agricultural biodiversity stating elements of a joint work programme with FAO. These elements fall into five broad categories of work as listed below:

(i) Sectoral assessment of plant, animal, forest and fisheries genetic resources related ecosystem diversity of relevance to food and agriculture;

(ii) System level work on the sustainable management of natural resources and development of technologies and practices that contribute to enhancing and sustaining biological diversity for food and agriculture;

(iii) Cross-sectoral work on the gender and other socio-economic issues, including local and indigenous knowledge, of relevance to the conservation and sustainable use of agricultural biodiversity, and so analysis of the economics of valuation and of impacts of biotechnology of commodity trade;

(iv) Development of policies, standards and codes of conduct to promote sustainable use of biotechnology on commodity trade;

(v) Development of methodologies and indicators for agricultural biodiversity as well as information and monitoring systems for diagnostic purposes and as tools for policy decision making and country level reporting.

The CBD agreement comprises 42 articles. Important among them, related directly to biological diversity are Articles 8, 9, 10, 11, 12, 15, 16 and 19 (Table 13.3). Article 12.4 describes the provisions of providing access to PGRFA pursuant to a standard material transfer agreement (SMTA), adopted by the Governing Body, which embodies the benefit-sharing provisions set forth in Article 13.2d.

Article 10 defines the establishment of a multilateral system (MLS), which is efficient, effective, and transparent, both to facilitate access to PGRFA, and to share the benefits arising from the utilization of these resources in a fair and equitable way. Treaty also recognizes the sovereign rights of State over their own PGRFA and access should be consistent with national legislation.

Article 15 states that PGRFA listed as in table of this Treaty and held by the CG institutes and those collected before its entry into force and held by the CG institutes shall be made available in accordance with the provisions of the Material Transfer Agreement (MTA) pursuant to agreements between the CG institutes and the Food and Agriculture Organization (FAO).

More appropriately the agreement imposes several but flexible obligations on member nations to formulate legislations ensuring conservation and sustainable use of biological diversity.it recognizes sovereign rights to the nations and establishes a tenable participatory relationship among the country, provider and user of genetic resources.

A total of 4, 87,793 accessions (as on February 2015) are included in the MLS by 33 contracting parties (countries) *viz.* Armenia, Austria, Belgium, Brazil, Canada, Cyprus, Czech Republic, Egypt, Estonia, France, Germany, Italy, Jordan, Kenya, Lebanon, Madagascar, Malawi, Morocco, Namibia, The Netherlands, Poland, Portugal, Romania, Rwanda, Senegal, Spain, Sudan, Sweden, Switzerland, Tanzania, U.K., Uruguay and Zambia. India being signatory to ITPGRFA has an obligation as a Contracting Party and needs to provide facilitated access to the Contracting Parties of Treaty for the crops listed. Notification for exemption of listed crops is issued by the Government of India through DAC, MoA on February 16, 2014 including the approved Guidelines for the implementation of the ITPGRFA (Office Memorandum No. 13-5/2013 SD-V). MoEF and Cc has also issued a notification for exemption of listed crops of the Treaty. (http://nbindia.org/upload/pdf/Gazette_Notification_on_exemption_of_crops_listed_in the_Annex._of _the_ITPGRFA.pdf) (Annexure)

**Table 13.3: Subjective Articles of Convention on Biological Diversity (CBD)**

| Articles | Elements | Remarks |
|---|---|---|
| Article 1 | Objectives | The objective of this convention to be pursued in accordance with its relevant provisions, are the conservation of biological diversity, the sustainable use of its components and the fair and equitable sharing of benefits arising out of the utilization of genetic resources, including by appropriate access to genetic resources and by appropriate transfer of relevant technology, taking into account all rights over those resources and to technologies, and by appropriate funding. The Convention on Biological Diversity (CBD) held at Rio, Brazil in June, 1992. The Convention on Biological Diversity, which came into force on 29th December, 1993 under an international treaty with the following objective:<br><br>(i) The conservation of biological diversity<br><br>(ii) The sustainable use genetic resources<br><br>(iii) Fair and equitable sharing of benefits arising from *inter alia* use of plant genetic resources<br><br>Further CBD considers two other factors to be important *i.e.*:<br><br>(a) The sovereign rights must be exercised in accordance with the charter of the United Nations and Principles of International Law and,<br><br>(b) The activities of nations selected to exploration of resources within their own beyond the limits of their national jurisdiction. |
| Article 8 | *In situ* Conservation | Each Contracting Party shall, as for as possible and as appropriate:<br><br>(a) Establish a system of protected areas or areas where special measures need to be taken to conserve biological diversity.<br><br>(b) Develop, where necessary, guidelines for the selection, establishment and management of protected areas or areas where special measures need to be taken to conserve biological diversity.<br><br>(c) Regulate or manage biological resources important for the conservation of biological diversity whether within or outside protected areas, with a view to ensuring their conservation and sustainable use.<br><br>(d) Promote the protection of ecosystem, natural habitats and the maintenance of viable populations of species in natural surroundings.<br><br>(e) Promote environmentally sound and sustainable development in areas adjacent to protected areas with a view to furthering protection of these areas.<br><br>(f) Rehabilitate and restore degraded ecosystem and promote the recovery of threatened species, inter alia, through the development and implementation of plants or other management strategies.<br><br>(g) Establishment or maintain means to regulate, manage or control the risk associated with the use and release of living modified organism resulting from biotechnology which are likely to have adverse environmental impacts that could affect the conservation and sustainable use of biological diversity.<br><br>(h) Prevent the introduction of, control or eradicate those alien species which threaten ecosystems, habitats or species.<br><br>(i) Endeavor to provide the conditions needed for compatibility between present uses and the conservation of biological diversity and the sustainable use of its components.<br><br>(j) Subject to its national legislation, respect, preserve and maintain knowledge, innovations and practices of indigenous and local communities embodying traditional lifestyles relevant for the conservation and sustainable use of biological diversity and promote their wider application with the approval and involvement of the holders of such knowledge, innovations and practices and encourage the equitable sharing of the benefits arising from the utilization of such knowledge, innovations and practices. |
| Article 9 | *Ex situ* Conservation | Each Contracting Party shall, as far as possible and as appropriate, and predominantly for the purpose of complementing in situ measures:<br><br>(a) Adopt measures for the *ex situ* conservation of components of biological diversity, preferably in the country of origin of such components<br><br>(b) Establish and maintain facilities for ex situ conservation of and research on plants, animals and micro-organisms, preferably in the country of origin of genetic resources<br><br>(c) Adopt measure for recovery and rehabilitation of threatened species and for their reintroduction into their natural habitats under appropriate conditions<br><br>(d) Regulate and manage collection of biological resources from natural habitats for *ex situ* conservation purposes so as not to threaten ecosystems and *in situ* population of species, except where special temporary *ex situ* measures are required under subparagraph (c) above and<br><br>(e) Cooperate in providing financial and other support for *ex situ* conservation outlined in subparagraphs (a) to (d) above and in the establishment and maintenance of *ex situ* conservation facilities in developing countries. |
| Article 10 | Sustainable use of components of biological diversity | Each Contracting Party shall, as far as possible and as appropriate:<br><br>(a) Integrate consideration of the conservation and sustainable use of biological resources into national decision-making<br><br>(b) Adopt measures relating to the use of biological resources to avoid or minimize adverse impacts on biological diversity |

| Articles | Elements | Remarks |
|---|---|---|
| | | (c) Protect and encourage customary use of biological resources in accordance with traditional cultural practices that are compatible with conservation or sustainable use requirements |
| | | (d) Support local populations to develop and implement remedial action in degraded areas where biological diversity has been reduced and |
| | | (e) Encourage cooperation between its governmental authorities and its private sector in developing methods for sustainable use of biological resources. |
| Article 12 | Research and Training | The Contracting Parties, taking into account the special needs of developing countries, shall: |
| | | (a) Establish and maintain programmes for scientific and technical education and training in measures for the identification, conservation and sustainable use of biological diversity and its components and provide support for such education and training for the specific needs of developing countries |
| | | (b) Promote and encourage research which contributes to the conservation and sustainable use of biological diversity, particularly in developing countries, *inter alia* in accordance with decisions of the Conference of the Parties taken in consequence of recommendations of the Subsidiary Body on Scientific, Technical and Technological Advice; and |
| | | (c) In keeping with the provisions of Articles 16, 18 and 20, promote and cooperate in the use of scientific advances in biological diversity research in developing methods for conservation and sustainable use of biological resources. |
| Article 15 | Access to Genetic Resources | (1) Recognizing the sovereign rights of States over their natural resources. The authority to determine access to genetic resources rests with the national governments and is subject to national legislation. |
| | | (2) Each Contracting Party shall endeavor to create conditions to facilitate access to genetic resources for environmentally sound uses by either Contracting Parties and not to impose restriction that run counter to the objectives of this Convention. |
| | | (3) For the purpose of this Convention, the genetic resources being provided by a Contracting Party, as referred to in this Articles and Articles 16 and 19, are only those that are provided by Contracting Parties that are countries of origin of such resources or by the Parties that have acquired the genetic resources in accordance with this Convention. |
| | | (4) Access, where granted, shall be on mutually agreed terms and subject to the provisions of this Article. |
| | | (5) Access to genetic resources shall be subject to prior informed consent of the Contracting Party providing such resources, unless otherwise determined by that party. |
| | | (6) Each Contracting Party shall endeavor to develop and carry out scientific research based on genetic resources provided by other Contracting Parties with the full participation of, and where possible in, such Contracting Parties. |
| | | (7) Each Contracting Party shall take legislative, administrative or policy measures, as appropriate, and in accordance with Articles 16 and 19 and, where necessary, through the financial mechanism established by Article 20 and 21 with the aim of sharing in a fair and equitable way the results of research and development and benefits arising from the commercial and other utilization of genetic resources with the Contracting Party providing such resources. Such sharing shall be upon mutually agreed terms. |
| Article 16 | Access to and Transfer of Technology | (1) Each Contracting Party, recognizing that technology includes biotechnology, and that both access to and transfer of technology among Contracting Parties essential elements for the attainment of the objectives of this Convention, Undertakes subject to the provisions of this Article to provide and/or facilitate access for and transfer to other Contracting Parties of technologies that are relevant to the conservation and sustainable use of biological diversity or make use of genetic resources and do not cause significant damage to the environment. |
| | | (2) Access to and transfer to technology referred to in paragraph I above to developing countries shall be provided and/or facilitated under fair and most favorable terms, including on confessional and preferential terms where mutually agreed, and, where necessary, in accordance with the financial mechanism established by Article 20 and 21. In the case of technology subject to patents and other intellectual property rights, such access and transfer shall be provided on terms which recognize and are consistent with the adequate and effective protection of intellectual property rights. The application of this paragraph shall be consistent with paragraphs 3, 4 and 5 below. |
| | | (3) Each Contracting Party shall take legislative, administrative or policy measures, as appropriate, with the aim that Contracting Parties, in particular those that are developing countries, which provide genetic resources are provided access to and transfer of technology which makes use of those resources, on mutually agreed terms, including technology protected by patents and other intellectual property rights, where necessary, through the provisions of Article 20 and 21 and in accordance with international law and consistent with paragraphs 4 and 5 below. |
| | | (4) Each Contracting Party shall take legislative, administrative or policy measures, as appropriate, with the aim that private sector facilitates access to joint development and transfer of technology referred to in paragraph 1 above for the benefit of both governmental institutions and the private sector of developing countries and in this regard shall abide by the obligations included in paragraphs 1, 2 and 3 above. |
| | | (5) The Contracting Parties, recognizing that patents and other intellectual property rights may have an influence n the implementation of this Convention, shall cooperate in this regard subject to national legislation and international law in order to ensure that such rights are supportive of and do not run counter to its objectives. |

| Articles | Elements | Remarks | |
|---|---|---|---|
| Article 19 | Handling of biotechnology and Distribution of its benefits | (1) | Each Contracting Party shall take legislative, administrative or policy measures, as appropriate, to provide for the effective participation in biotechnological research activities by those Contracting Parties, especially developing countries, which provide genetic resources for such research, and where feasible in such Contracting Parties. |
| | | (2) | Each Contracting Party shall take all practical measures to promote and advance priority access on fair and equitable basis by Contracting Parties, especially developing countries, to the results and benefits arising from biotechnologies based upon genetic resources provided by those Contracting Parties. Such access shall be mutually agreed terms. |
| | | (3) | The parties shall consider the need for and modalities of a protocol setting out appropriate procedures, including, in particular, advance informal agreement in the field of the safe transfer, handling and use of any living modified organism resulting from biotechnology that may have adverse effect on the conservation and sustainable use of biological diversity. |
| | | (4) | Each Contracting Party shall, directly or by requiring any natural or legal person under its jurisdiction providing the organisms referred to in paragraph 3 above, provide any available information about the use and safety regulations required by that Contracting Party in handling such organisms, as well as any available information on the potential adverse impact of the specific organisms concerned to the Contracting Party into which those organisms are to be introduced. |

## International Cooperation and outlook for Vegetable Germplasm

International Plant Genetic Resources Institute (IPGRI; now Bioversity International) has been supporting national programmes on conservation, characterization and documentation of genetic resources. Under the agreement between IPGRI and ICAR, okra is one among the three South Asian crops (Sesame and maize being the others) on which the focus was given. The okra germplasm that was formerly housed in the Ivory Coast was transferred to ICAR-NBPGR through these efforts during late 1980's. In another project, eggplant was included for joint exploration missions with other South Asian countries. This mission has collected 1835 accessions of okra and its wild relatives and 3070 of eggplant (Karihaloo *et al.*, 2000). Through the Asian Development Bank funding the South Asian Vegetable Research Network (SAVERNET) was established in 1992 with the SAARC countries and AVRDC, Taiwan was the executing agencies. In collaboration with the SAARC countries, AVRDC assembled a total of 5622 accessions of 23 different vegetable crops. In return the AVRDC, Taiwan has distributed 39,859 seed packets of priority crops. Access to the results of germplasm enhancement research was facilitated and following the network master plan, genetic diversity in pathogens such as bacterial wilt and yellow leaf curl of tomato and anthracnose and viruses on chilli were surveyed. India has integrated its economy with the World Trade Organization (WTO) and dealing Intellectual Property Rights (IPRs) with international and national laws is one of the commitments, which is going to affect section of the society *i.e.* researchers, research managers, policy makers, political leaders and the farmers.

## 7. Trade Related Intellectual Property Rights (TRIPS) Agreement under WTO

World Trade Organization (WTO) emphasizes a foolproof system of protection of intellectual property rights by member nations. It intends to establish a legitimate reward system to the inventors for their creative inputs under the framework of Trade Related Intellectual Property Rights (TRIPS). So, it is one of the Agreements adopted in WTO framework and is in force since 1995. Each member country has the option to frame its own patent laws within the broad framework defined in the GATT agreement (Ganguli, 1998). TRIPS Agreements seeks by members to provide protection to different intellectual property forms namely patents, trademarks, copyrights, designs *etc*. Articles 27 of TRIPS in paragraph 1 provides in part the patents shall be available for any invention whether products or process in all fields of technology. Articles 27.3 (b) of TRIPS provides that members may exclude from patentability of plants and animals other than microorganisms, and essentially biological process for the production of plants or animals other than biological and microbial process. However, members shall ensure for the protection of plant varieties either by patents by an effective *sui generis* system or by any combination thereof. Under Article 65, (Ganguli, 1998) developing economics and least developed economies are allowed a grace period of 4 years and 10 years, respectively to implement these provisions (Gautam and Singh, 1998). TRIPS Articles 27.3 (b) that is related to life patents and plant variety protection provides that its provisions shall be reviewed 4 years after the date of entry into force of WTO agreement. The TRIPS Agreement also provides for a minimum level of protection of commercial marks such as trademarks and geographical indications (GI). Trade secrets have been also accorded the status of IPRs. The various aspects of agriculture encompass:

## World Trade Organization (WTO)

The 8th round of GATT held at Uruguay in 1986 in which 'Agriculture' was included first time as a new element. WTO was successfully shaped into agreeable document by December 15, 1993. The negotiations were signed on April 15, 1994 at Marrakash, Morocco by

**Table 13.4: Various Aspects of Agriculture**

| Sl.No. | Aspects | Details |
|---|---|---|
| 1 | Agreement on Agriculture | (1) To establish a fair and market led agricultural trading system which is to be achieved through a reform process with negotiations of commitments on support and protection. |
| | | (2) To provide for substantial progressive reductions in agriculture support and protection. |
| | | (3) The provisions of agreement cover market access, domestic support and export competition |
| 2 | Agreement on sanitary and phytosanitry measures | The agreement recognizes that members have the right to adopt or enforce measures that are necessary to protect human, animal or plant life or health. This right is subject to the condition that such measures should not act as a means of arbitrary or unjustifiable discrimination between the members as a disguised restriction on international trade. |
| 3 | Agreement on technical barriers | The agreement is for the establishment of international standards and conformity assessment system in packaging, marketing and labeling so as to ensure that technical regulations and standards and procedures for assessment of conformity technical regulations and standards do not create unnecessary obstacles to international trade. |
| 4 | Agreement on trade related aspects of Intellectual Property Rights (IPR) | (1) The agreement is to promote effective and adequate protection of related aspects of Intellectual Property Rights (IPR) and also to ensure that the measures taken in this direction do not become an impediment to legitimate trade. |
| | | (2) It is through Article 27.3 (b) of the agreement that the subject of agriculture is brought under this agreement. The said article requires the members to provide for the protection of plant varieties either by patent or by an effective *sui generis* system or by any combination thereof. |

*Source*: Sahai, 2003.

about 124 countries. This event leads to formation of World Trade Organization (WTO), on 1ˢᵗ January, 1995. It requires Member countries to make patents available for any inventions, whether products or processes, in all fields of technology without discrimination, subject to the normal tests of novelty, inventiveness and industrial applicability. The WTO now has 153 member nations and 29 others with observer status.

## Objectives of World Trade Organization (WTO)

World Trade Organization (WTO), the final verdict of Uruguay round was established with the following Objectives:

1. To develop a fair and marketed oriented agricultural trading system

2. To provide platform for reform process through negotiations of commodities on support and protection.

3. To reduce the trade barriers between member nations through mutually advantageous agreements.

4. To establish operationally more effective GATT rules and disciplines.

5. To settle all trade related disputes amicably among the member nations.

There are three permissible exceptions to the basic rule on patentability:

☆ One is for inventions contrary to order public or morality including inventions dangerous to human, animal or plant life or health or seriously prejudicial to the environment.

☆ The second exception is that Members may exclude from patentability the diagnostic, therapeutic and surgical methods for the treatment of humans or animals.

☆ The third exception is that Members may exclude plants and animals other than micro-organisms and essentially biological processes for the production of plants and animals other than non-biological and microbiological processes.

However, any country excluding plant varieties from patent protection must provide for an effective *sui generis* system of protection. The term of protection available shall not end before the expiry of a period of 20 years counted from the filing date. Members shall require that an application for a patent shall disclose the invention in a manner sufficiently clear and complete for the invention to be carried out by a person skilled in the art. Compulsory licensing and the government use, without the authorization of the right holder, are allowed. It is noteworthy that the agriculture plants sector is currently the only one where access is granted under two ABS systems, operated by the CBD and the FAO. In addition, two other systems are available for securing IPRs over them, namely, patents and plant breeders' rights (Box 2).

## General Agreement on Trade and Tariffs (GATT)

The second world war which shook the world trade and economy, it was decided to streamline world trade through the General Agreement on Trade and Tariffs (GATT). GATT dealt with the quotas and duties of tradable commodities between the member nations. Its framework provided a temporary arrangement to amicably settle disputes between the trading countries. Subsequently GATT became a code of rules

---

**Box 2 Some Notable points**

☆ Animal genetic resources are not covered under the ITPGRFA and WTO-TRIPS whereas CBD and also its Protocol on Abs cover all genetic resources holistically. CGRFA under FAO is now engaged in developing an international treaty on animal genetic resources on the patter of ITPGRFA.

☆ Accessions of PGR, that are stored in the National Genebank, need to be carefully scrutinized to identify designated accessions for placing them under public domain and making them available for exchange under ITPGRFA. For the PGR, not covered under the ITPGRFA, there is need to continue regulating their access and ensure benefit sharing as provided under the Biological Diversity Act, 2002. There is also an urgent need to prepare inventory of high value traits of our genetic resources and to monitor their commercial utilization abroad.

*Source*: Rana, 2015

---

for international trade and forum to discuss and find solutions to trade related problems amongst the member nations (Sharma, 2009).

## India and TRIPS

Pending modification of patent laws as per TRIPS provision, the developing countries are expected to provide 'mail box' protection [Article 70.8(a)] and under this provision country will accept patent application for products related to pharmaceuticals and agricultural chemicals from January1, 1995. In India, U.S.A. complained to WTO that India has failed to meet the basic commitments to TRIPS. The Dispute Settlement Body of WTO observed that India has failed to provide 'mail box' system of protection to the concerned products. Ultimately, India has been given time till April, 1999 to make the above provisions, failing which U.S.A. could call for appropriate sanctions.

## IV. Overlapping Provisions on ABS under CBD, ITPGRFA and TRIPS

Unlike the CBD, which provides for bilateral negotiations to establish the terms of access and benefit sharing for each specific exchange of materials, all multilateral germplasm exchanges under the MLS would be subject to SMTA. Monetary benefits would be paid to the Global Crop Diversity Trust fund to be used primarily to support farmers who conserve and sustainably use PGRFA. However, the financing of germplasm conservation activities has been addressed only in general terms, making this aspect of the treaty potentially difficult to implement (Rana, 2015).

Concerns on the relationship between ABS provisions under CBD and other international legal regimes bearing on genetic resources led to inclusion of Article 4 in the Nagoya Protocol stating that the Nagoya Protocol does not apply to Parties to the specialized instrument in respect of specific genetic Resources covered by and for the purpose of that specialized instrument. The scope of other existing regimes would therefore be crucial to define which genetic resources are covered by the Nagoya Protocol. The International Treaty on Plant Genetic Resources for Food and Agriculture (ITPGRFA), for example, has been in force since 2004. It is a global instrument

designed to promote conservation of PGRFA, and to help protect farmers' rights, and also to ensure fair and equitable sharing of benefits arising from the use of PGRFA. The Plant Treaty has established a Multilateral System (MLS) under which genetic resources of listed crops can be exchanged without individual regulation, subject to a standard material transfer agreement (SMT). One challenge concerning this instrument is that not all parties to the CBD are members of the Plant Treaty. Another concern is that ABS in the Plant Treaty differs from the Abs regime of the CBD.

Another alarming development is that the FAO; CGRFA is now discussing ABS mechanisms for six more groups of genetic resources, namely animals; aquatic; invertebrates; plants forests and microbial genetic resources (FAO, 2015). Any agreement in the Commission on need for specialized regimes for ABS holds potential to exclude commercially valuable groups of ABS governed by the CBD and the Nagoya Protocol. Another international platform for regulating access and benefit sharing has also reached agreement with the World Health Organization in 2011 giving green signal to two SMTAs concerning exchange and use of viral genetic resources with pandemic potential for humans. The question of access and benefit sharing from genetic resources in the area beyond national jurisdiction has also been on the agenda of the UN Convention on the Law of the Seas and this may include, for example, genetic resources taken from the seabed and/or the high seas, taking them out of purview of CBD. In addition, discussion under the auspices of the Antarctic Treaty is also progressing on how to regulate genetic resource material from one of the world's most remote, yet biologically unique areas.

## International Union for the Protection of New Varieties of Plants (UPOV)

International Union for the Protection of New Varieties of Plants (UPOV) is an international organization, established in the year 1961 in Paris which provides 111 protections of crop varieties. The protection manifests in the form of plant variety protection or plant breeders right. The purpose is that member countries acknowledge the achievements of breeders of new plant varieties and provide the exclusive rights on certain set principles or norms. The exclusive rights serve as an incentive to breeders for further genetic improvement and development of new varieties and hybrids in agriculture, horticulture and forestry. By 1997, UPOV has membership of 32 countries mostly from the developed world. Some salient points of UPOV are:

☆ The intent of rights means to provide an opportunity to breeders to a return from the release of cultivars.

☆ Breeder is required to maintain the variety and actively participate in its seed distribution.

☆ Signatory countries must align national policy and legislation. UPOV encourages the adoption of sui generis laws for protecting new plant varieties.

☆ A variety should be new distinct, uniform and stable to qualify for protection.

☆ Plant breeders' rights do not restrict anyone from using protected cultivar as an initial sopurce of variation for the purpose of developing other cultivars.

☆ Plant breeders and developing nations whether or not they are members of UPOV may use any protected cultivars for research and evaluation to develop improved cultivars.

## International Undertaking on Plant Genetic Resources (IUPGR)

International Undertaking on Plant Genetic Resources (IUPGR) non-legal binding agreement and so far adopted by 150 countries and 40 organizations (Gautam and Singh, 1998). It was established in 1983 at the FAO conference to deal with handling of plant genetic resources, so, it was called as 'FAO Commission of genetic Resources for Food and Agriculture'. The commission envisions a global system that comprises following areas of prior activities on plant genetic resources:

(i) Safe conservation

(ii) Sustainable utilization of plant genetic resources for food and agriculture

(iii) Flexible framework for sharing benefits as well as responsibilities

Global plan of action (GPA) for the conservation and sustainable utilization of plant genetic resources for food and agriculture is the major element of International Undertaking on Plant Genetic Resources (IUPGR). GPA has identified 20 priority areas which are placed in four groups *viz.*,

☆ *In situ* conservation and development

☆ *Ex situ* conservation

☆ Utilization of plant genetic resources

☆ Institution and capacity building

## Regulating Access and Benefit Sharing in India: Procedures

Under CBD, the sovereign authority to determine access to genetic resources rests with the national governments and it is subject to national legislation. To fulfill national obligations under the CBD, India enacted the Biological Diversity Act in 2002 through a systematic consultation process and also framed the Biological Diversity Rules under it in 2004. In addition to promoting conservation and sustainable use of all categories of bio-resources, this umbrella legislation regulates access to them while determining mode/ quantum of fair and equitable benefit sharing and signing agreements with the users based on mutually agreed terms (Box 3).

---

**Box 3 Salient Features of the Biological Diversity Act, 2002**

☆ Regulates access to biological resources of the country with the purpose of securing equitable and fair sharing of benefits arising out of the use of biological resources; and associated traditional knowledge (TK) relating to biological resources;

☆ Promotes conservation and sustainable use of all components of biological diversity;

☆ Aims at respecting and protecting traditional knowledge of local communities related to biodiversity;

☆ Provides for sharing of benefits with local people as developers and conservers of biological resources and holders of knowledge and information associated with their use;

☆ Promotes conservation and development of areas of importance from the standpoint of biological diversity by declaring them as biological diversity heritage sites;

☆ Lends support to on-going programmes on protection and rehabilitation of rare, endangered and threatened species;

☆ Encourages increasing involvement of institutions and state governments in the broad scheme of implementing the Biological Diversity Act, through constitution of appropriate committees;

☆ Recognizes for broad categories of users who are required to apply in different kinds of specified forms along with payment of prescribed fees. These categories are based on stated objectives of the applicants and include accessing biological resources for research/bio-survey and bio-utilization/commercial utilization, transferring results of research on bio-resources, seeking IPR over Products/innovations based on use of bio-resources an third party transfer of already accessed bio-resources.

*Source*: Rana, 2015

This Act also provides further support to other relevant national laws in force, namely, the Wildlife (Protection) Act, 1992 as amended in 1991, and the Protection of Plant Varieties and Farmers' Rights (PPVFR) Act, 2001. It also provides suitable linkage to the provision for patenting of products and processes/technologies, based on the use of bio-resources and associated indigenous traditional knowledge (ITK), under Section 10 (4) of the Patents (Amendment) Act, 2002. The stage is thus set for developing a national movement for implementing these combined provisions for access and benefit sharing to ensure food and livelihood security based on conservation, inclusive development and suitable use of bio-resources.

The act provides for its implementation through a 3-tier system comprising the:

(i) National Biodiversity Authority (NBA)

(ii) The State Biodiversity Boards (SBBs) and

(iii) The Biodiversity Management committees (BMCs) at the local communities level.

## Functions

Functions of this system at all the three levels have been clearly defined and all the Union States have constituted SBBs. In exercise of the powers conferred by Sub-Section (1) (4) of Section 8 of the biological Diversity Act, 2002, the Central Government established the National Biodiversity Authority (NBA) on 1st October, 2003. For major functions of NBA see Box 4.

Recognizing that the Indian citizen owe allegiance to the Indian Constutution and can be called upon by the courts in person to ensure compliance to this Act's provisions, a differentiating way has been adopted under which the following categories of persons/body corporate/associations/organizations are required to obtain prior approval of the NBA for seeking access to India's buio-resources (and associated TK) for research and commercial use or engaging in bio-survey and bio-utilization activities [Section3(2) and Section 19]:

☆ A person who is not a citizen of India

☆ A of citizen of India, who is non-resident

☆ A body corporate, association or organization-not incorporated or registered in India; or incorporated or registered in India but has any non-Indian participation in its share capital or management, See Box-5.

Access to bio-resources for research by Indian citizens, and companies registered in India and not having any foreign share in their management, is unrestricted and free. However, Section 7 states that no person, who is a citizen of India or a body corporate, association or organization which is registered in India, shall obtain any biological resource for commercial utilization, or bio-survey and bio-utilization for commercial use except after giving prior intimation to the concerned State Biodiversity Board which grant the required approval based on relevant rules and procedures for this purpose (Section 23 and 24) and imposes benefit sharing terms based on the guidelines notified for this purpose in November, 2014.

## Restriction Imposed on Granting Access to Bio-resources

See Boxes 6 and 7 for some restrictions imposed on granting access to bioresources and also some exemptions from the provisions of this Act.

---

**Box 4 Main Functions of the National Biodiversity**

1. To lay down procedures and guidelines to govern the activities provided under section 3, 4, and 6: Permission to foreigners/non-resident Indians/foreign entities.

2. To regulate activities and advice the government of India on research/commercial use of bio-resources, bio-survey and bio-utilization.

3. To grant approval under section 3, 4, and 6 based on the following considerations:

   (i) Certain persons not to undertake Biodiversity related activities without approval of National Biodiversity Authority (Section 3)

   (ii) Results of research not to be transferred to certain persons without approval of National Biodiversity Authority (Section 4) (transfer research results)

   (iii) Application for seeking IPR rights not to be made without prior approval of the NBA (Section 6)

4. To grant approval to certain persons seeking transfer of already accessed biological resources/associated traditional knowledge (Third Party Transfer) (Section 20)

5. To determine and impose terms of equitable benefit sharing arising out of the use of accessed biological resources/associated traditional knowledge (Section 21)

6. To establish and operate the National Biodiversity Fund

7. To provide the State Governments in the selection of areas of biodiversity importance to be notified under Section 37 (1) as heritage sites and measures for their management.

8. To take any measure, on behalf of the Central Governments, necessary to oppose the grant of IPR in any country outside India on any bio-resources obtained from India or knowledge associated with it which is derived from India.

*Source*: Rana, 2015

---

**Box 5 The Principal of Common but Differentiated Responsibilities**

1. The biologiocal Diversity Act differentiates between Indian citizens/Indian entities and foreign citizens/foreign entities (including Persons of Indian Origin) requiring the latter category to obtain prior approval of NBA for accessing India's bioresources for research/commercial utilization or for undertaking biosurvey and bio-utilization. Whereas the Indian citizens are required to approach the concerned State Biodiversity Board, the foreign citizens/ foreign entities need to apply to NBNA

2. The Act also requires Indian citizens to take prior approval of NBA for transferring the results of research, conducted on bioresources, to foreign citizens/foreign entities.

*Source*: Rana, 2015

---

**Box 6 Restriction Imposed on Granting Access to Bio-resources**

Certain restrictions have been imposed under Rule 16 on NBAs, and also SBBs approvals for activities related to access to bio-resources, requiring the Authority to take steps to restrict or prohibit requests for such access on considering the following reasons:

☆ The request for access is for any endangered taxa;

☆ The request for access is for any endemic and rare species;

☆ The request for access may result in advance effect on the livelihoods of the local people;

☆ The request for access may result in adverse environmental impact which may be difficult to control and mitigate;

☆ The request for access may cause genetic erosion or adverse effect ecosystem functioning;

☆ When the use of resources is for purposes contrary to national interest and other related international agreements entered into by India.

*Source*: Rana, 2015

---

Similarly, exemptions have been provided under the Biological Diversity Act to use the bio-resources for research and non-commercial purposes:

---

**Box 7 Exemptions Provided under the Biological Diversity Act**

The following exemptions have been provided under this Act to promote bona fide use of bio-resources for research and non-commercial use:

☆ Indian citizens/entities accessing bio-resources for research/bio-survey and bio-utilization for research in India are exempted from provisions of this Act.

☆ Provision of Section 3 (access to bio-resouce and Section 4 (transfer of research results) shall not apply to the approved collaborative research projects, conforming to the extant policy and guidelines issued by the Ministry of Environment and Forests such as the notification dared 8 November, 2006.

☆ Provision of Section 6 shall not apply to any person making an application for any right under the Protection of Plant Varieties and Farmers' Rights Act, 2001. Where any right is granted under this law, the concerned authority granting such right shall endorse a copy of such document (granting the right) to the NBA.

☆ Accessing biological resources for conventional breeding or traditional practice in use in any agriculture, horticulture, poultry, dairy farming, animal husbandry, bee keeping, *etc.* in India is exempted from the provisions of this Act. However, "End Uses" of biological resources for "Commercial Utilization" (such as drugs, industrial enzymes, food flavours, fragrance, cosmetics, emulsifiers, oleoresins, colours, extracts and genes) used for improving crops and livestock through genetic interventions, covered u/s 2(f), are not exempted.

☆ Publication of research papers or dissemination of knowledge, in any workshop exempted from provisions of Section 4 of the Act if it is in country with the guidelines issued by the Central Government for this purpose.

☆ "Value added products", which may contain portions or extracts of plants and animals in unrecognizable and physically inseparable form as defined u/s 2(P).

☆ Provision of Section 7 (prior intimation to SBB for commercial use) shall not apply to the local people and communities including village healers/*vaids*, farmers and other traditional growers and also to Indian users of these bio-resources for research (not when seeking intellectual property rights).

☆ Items such as normally traded commodities, as notified by the Central government u/s. 40 would be exempt from purview of this Act.

☆ Exchange of designated accessions of genetic resources of listed crops of the ITPGRFA have been exempted but cannot apply for any IPR without prior approval of the NBA.

*Source*: Rana, 2015

However, all the users, Indian citizens as well as foreigners, are required to seek prior approval of NBA for transferring results of their research on bio-resources to foreign persons/entities (Section 4), for applying for IPR on products/processes based on bio-resources (Section 6) and also for third party transfer of the already accessed bio-resources (Section 20), by submitting applications in specified formats along with payment of prescribed fee for each of the above mentioned purpose.

The exemption thus ensures facilitated access to PGRFA as per the provisions of the treaty and use of SMTA for such transfer of PGRFA. The following are the major categories of accessing PGRFA

☆ Access to crops listed under the Treaty only

☆ Access to and from contracting parties only

☆ Access to and from CG Centres

☆ Access for the purpose of utilization and conservation for research, breeding and training for food and agriculture only

☆ Access to PGRFA held by CG Centres including those collected before 1993 referred to as 'in trust" or "FAO designated material"

☆ Access with the terms and conditions of the SMTA

☆ Use of SMTA required for material accessed through SMTA to any third or subsequent user

☆ No Intellectual Property Rights (IPR) is to be granted on the material "in the form received"

☆ Transfer of any material accessed from MLS and received under SMTA to any third party after signing the SMTA

## Authorised Access to Biological Resources Required Prior to Seeking IPR

Any person seeking any kind of IPR in or outside of India for any invention/technology/product or process based on any biological resource (or associated information) obtained from India, is required to obtain prior permission of the NBA [Section 6]. In addition, the Patent (Amendment) Act, 2002, requires the patent applicant to disclose the source and geographical origin of the used biological material in the patent application, when used in an invention [Section 10 (4)].

When the CBD was finalized, negotiating parties recognized that some important issues were left without satisfactory solutions in international law as reflected in section 4 of Resolution 3 adopted by the Nairobi conference, where the text of the CBD was agreed, which reads: 'Further recognizes the need to seek solutions to outstanding matters concerning plant genetic resources within the global System for the Conservation and Sustainable Use of Plant Genetic Resources for Food and Sustainable Agriculture, in Particular:

(a) Access to *ex situ* collections not acquired in accordance with this Convention; and

(b) The question of 'farmers rights'.

These issues were referred to the FAO in the context of suitable revising the contents of the International Understanding on PGRFA. The issues of genetic resources collected from countries of their origin, prior to CBD, still hangs on but the designated accessions stored in CGIAR's International Genebanks have been brought under the jurisdiction of FAO. Thus as per the notification issued the access under the MLS is exempted from Section 3 and 4 of the BDA, 2002 (18 of 2003) and is on par with Section 5 under which the collaborative research projects are exempted from Section 3 and 4. The MLS is highly relevant for ABS because it is the first sectoral approach to ABS, and could provide useful lessons for the implementation of ABS, including whether and if so, how, sectoral ABS can be dealt with to meet the objectives of the CBD (under NP Art. 4 and Art. 19). Sixth Session of the Governing Body of the ITPGRFA, held in Rome on 5-9 October, 2015 (Brahmi and Tyagi, 2015).

---

**Ministry of Environment, Forests and Climate Change (National Biodiversity authority); Notification New Delhi, the 21st November, 2014**

G.S.R.827.- In exercise of the powers conferred by section 64 read with sub-section (1) of section 18 and sub-section (4) of section 21 of the Biological Diversity Act, 2002 (18 of 2003), hereinafter referred to as the Act, and in pursuance of the Nagoya Protocol on access to genetic resources and in the fair and equitable sharing of benefits arising from their utilization to the Convention on biological Diversity dated the 29th October, 2010, the National Biodiversity authority hereby makes the following regulations, namely -

13. **Conducting of non-commercial research or research for emergency purposes outside India by Indian researchers/ Governmental institutions.-**

   (1) Any Indian researchers/Governmental Institution who intends to carry/send the biological resources outside India to undertake basic research other than collaborative research referred to in section 5 of the Act shall apply to the NBA in Form 'B' annexed to these regulations.

   (2) Any Governmental Institution which intends to send biological resources to carry out certain urgent studies to avert emergencies like epidemics, *etc.*, shall apply in Form 'B' annexed to these regulations.

   (3) The NBA shall, on being satisfied with the application under sub-regulation (1) or sub-regulation (2), accord its approval within a period of 45 days from the date of receipt of the application.

(4) On receipt of approval of the NBA under sub-regulation (3), the applicant shall deposit voucher specimens in the designated national repositories before carrying/sending the biological resources outside India and a copy of proof of such deposits shall be endorsed to NBA.

**17. Certain activities or persons are exempted from approval of NBA or State biodiversity Board (SBB), namely:**

(a) Indian citizens or entities accessing biological resources and/or associated knowledge, occurring in or obtained from India, for the purposes of research or bio-survey and bio-utilization for research in India;

(b) collaborative research projects, involving the transfer or exchange of biological resources or related information, if such collaborative research projects have been approved by the concerned Ministry/Department of the State or Central Government and conform to the policy guidelines issued by the Central Government for such collaborative research projects;

(c) Local people and communities of the area, including growers and cultivators of biological resources. *vaids* and *hakims*, practicing indigenous medicine, except for obtaining intellectual property rights;

(d) Accessing biological resources for conventional breeding or traditional practices in use in any agriculture, horticulture, poultry, dairy farming, animal husbandry or bee keeping, in India;

(e) Publication of research papers or dissemination of knowledge, in any seminar or workshop, if such publication is in conformity with the guidelines issued by the Central Government from time to time;

(f) Accessing value added products, which are products containing portions or extracts of plants and animals in unrecognizable and physically inseparable from and

(g) Biological resources, normally traded as commodities notified by the Central Government under Section 40 of the Act.

*Source*: Brahmi and Tyagi, 2015

Similarly other guidelines put forth by ministry are as:

**Ministry of Environment, Forests and Climate Change (National Biodiversity authority); Notification New Delhi, the 17ᵗʰ December, 2014**

**S.O.3232 (E.-)** Whereas, India is party to the International Treaty on Plant Genetic Resources for Food and Agriculture (ITPGRFA) having signed and ratified the said treaty on 10ᵗʰ June, 2002;

☆ and Whereas, the objectives of the ITPGRFA are conservation and sustainable use of plant genetic resources for food and agriculture and fair and equitable sharing of the benefits arising out of their use, in harmony with the Conservation of Biological Diversity, for sustainable agriculture and food security;

☆ and Whereas, Article 12 of the ITPGRFA provides for facilitated access to plant genetic resources for food and agriculture under the Multilateral System by the contracting parties; and

☆ and Whereas, the Nagoya Protocol on Access to Genetic Resources and the Fair and Equitable sharing of Benefits arising from their Utilization to the Convention on Biological Diversity dated 29ᵗʰ October, 2010 is the instrument for implementation of access for benefit sharing provisions of the Convention on Biological Diversity;

☆ and Whereas, Article 4 of the said Nagoya Protocol provides that the protocol does not apply for the party or parties to the specialized instrument in respect of the specific genetic resource covered by and for the purpose of the specialized instrument;

☆ and Whereas, Section 40 of the Biological Diversity Act, 2002 (18 of 2003) empowers the Central Government to exempt certain biological resources from the provisions of the said Act.

Now, therefore, in exercise of the powers conferred by section 40 of the Biological Diversity Act, 2002 (hereinafter referred to as the said Act), and in fulfillment of the obligation of the Government of India to the ITPGRFA for providing facilitated access to the plant genetic resources for food and agriculture, the Central Government, in consultation with the National Biodiversity Authority, hereby declares that the Department of Agriculture and Cooperation may, from time to time specify such crops as it considers necessary from amongst the crops listed in the treaty of the ITPGRFA, being food crops and forage covered under the Multilateral System thereof, and accordingly exempts them from Section 3 and 4 of the said Act, for the purpose of utilization and conservation for research, breeding and training for food and agriculture: Provided that such purposes shall not include chemical, pharmaceutical and/orother non-food orfeed industrial uses.2 The Department of Agriculture and Cooperation shall keep the National Biodiversity Authority informed of all crops as may be specified by it from time to time, for providing access to plant genetic resources for food and agriculture under the ITPGRFA for the purposes aforesaid.

*Source*: Brahmi and Tyagi, 2015

## Implications

As per the guidelines developed which are notified by DAC, MoA, Government of India (GOI), the supply of PGFRA of listed crops and held in 'trust' by the CG institutes shall also be made available in accordance with the provisions of SMTA with the concurrence of the DAC, MoA, the National Focal Point (Figure 13.1)

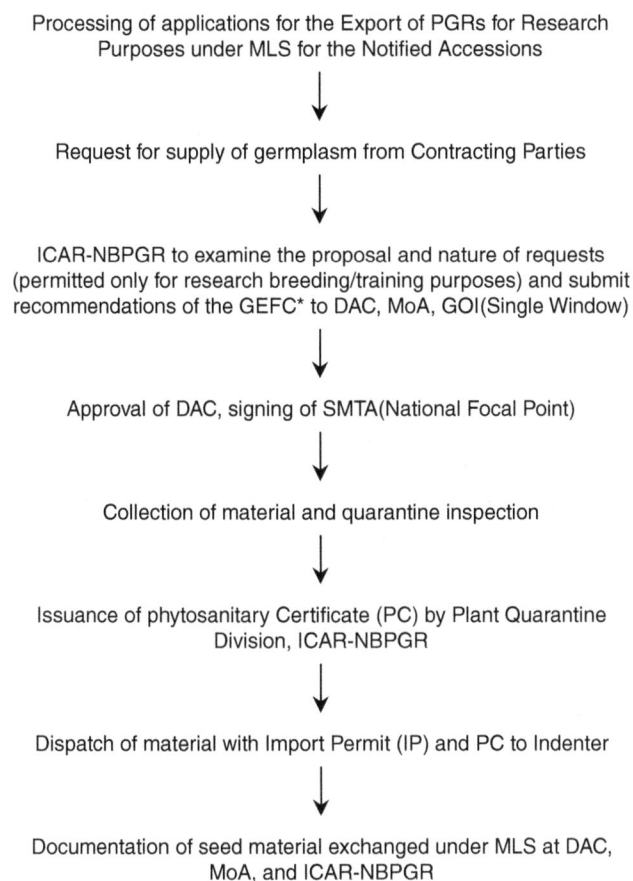

Processing of applications for the Export of PGRs for Research Purposes under MLS for the Notified Accessions

↓

Request for supply of germplasm from Contracting Parties

↓

ICAR-NBPGR to examine the proposal and nature of requests (permitted only for research breeding/training purposes) and submit recommendations of the GEFC* to DAC, MoA, GOI(Single Window)

↓

Approval of DAC, signing of SMTA(National Focal Point)

↓

Collection of material and quarantine inspection

↓

Issuance of phytosanitary Certificate (PC) by Plant Quarantine Division, ICAR-NBPGR

↓

Dispatch of material with Import Permit (IP) and PC to Indenter

↓

Documentation of seed material exchanged under MLS at DAC, MoA, and ICAR-NBPGR

### Figure 13.1

*Source*: Brahmi and Tyagi, 2015.

*Genetic Resources Export Facilitation committee (GEFC) comprising: Director, ICAR-NBPGR; Director/Dy. Secretary (DARE); Two representative from DAC, Assistant Director General (concerned crop); Assistant Director General (IP TM and PME), ICAR; Director/PD/PC (concerned crop); Head of Plant Quarantine Division, ICAR-NBPGR; Officer In charge of Germplasm Exchange Unit, and Policy Planning Unit, ICAR-NBPGR

## Notification on Guidelines on Access and Benefit Sharing

Regulation of Access to biological Resources (and associated TK) and benefit Sharing: Notified under Biological Diversity Act on 21 November, 2014. These guidelines provide:

☆ Legal certainty,

☆ Clarity and transparency,

☆ Simplified procedure for the Indian researchers/ government institutes to carry out basic research outside India. Options of benefit sharing for different users.

☆ Graded benefit sharing system,

☆ Establishing supply chain from source to manufacture.

☆ Upfront payment on high economic valued bio-resources (Red sanders, Sandalwood *etc.*)

☆ Apportioning accrued benefits to the local communities/BMCs.

## Facilitating Non-commercial Research Abroad by Indian Researchers/Government Institutions

Through this guideline, NBA introduced a special Form for the Indian research/scientists or Government Institutes to carry/send the biological resources outside India for doing research (like CSIR, ICAR, ZSI, BSI, Government universities) Government institutes may send the biological resources outside to carry out studies to avert emergencies like epidemics *etc.* (Rana, 2015).

### Determination of Benefit Sharing

Determination of benefit sharing; Monetary and/or non-monetary modes, as agreed upon by the applicant and the NBA/SBB concerned in consultation with the BMC/Benefit claimer, *etc.*

### Determination of the Amount of Benefit Sharing

a. **Benefit sharing for Commercial Utilization of Bio-resources**

| Annual Gross Ex-factory Sale of the Product (minus Govt. taxes) | Benefit Sharing Component |
|---|---|
| Upto to Rs. 1,00,00,000 | 0.1 per cent |
| Rs. 1,00,00,000 to Rs. 3,00,00,000 | 0.2 per cent |
| Above Rs. Rs. 3,00,00,000 | 0.5 per cent |

### b. Transfer of Results of Research

The benefit sharing obligation shall be 3.0 to 5.0 per cent of the monetary consideration received.

### c. Intellectual Property Rights

| Relevance | Terms of Benefit Sharing |
|---|---|
| When applicant himself commercialises the process/ product/innovation | 0.2 to 1.0 per cent of the annual ex-factory gross sale (minus Govt. taxes) |
| When applicant assigns/ licenses the process/product/ innovation to a third party for commercialisations | 3.0 to 5.0 per cent of the received in any form and 2.0 to 5.0 per cent of the royalty received. |

*Source*: Rana, 2015.

**d. Alternative Option for procurement of Bio-resources through a Supply Chain**

When the traders sells the biological resource purchased by him to another trader or manufacturer, the buyer,

☆ If he is a trader, he is to pay @ 1.0 to 3.0 per cent of his purchase price.

☆ If he is a manufacturer, he is to pay 3.0 to 5.0 per cent of his purchase price.

☆ If the buyer submits proof of benefit sharing paid by the immediate seller in the supply chain, then the buyer shall pay benefit sharing on that portion of the purchase price for which the benefit has not been paid along the supply chain.

In cases of biological resources having high economic value, such as sandalwood and red sanders, the benefit sharing may include an upfront payment of not less than 5 per cent, on the proceeds of the auction or sale amount, as decided by the NBA or SBB, as the case may be. If the sale is through auction, the successful bidder or the purchaser shall pay the amount to the designated fund, before accessing the biological resources. Information on penalties for contravention of this Act or abetting such as contraventions is provided in Box 8.

## Present Status of Implementing the Biological Diversity Act

India's 5th National Report to CBD, submitted in 2014, provides an overview of the status of implementing the provisions of CBD along with the progress made in implementing the Biological Diversity Act and the rules framed under it. Following the establishment of MBA in 2003, SBBs have also been constituted in all the 29 states and over 38,000 BMCs set up. Nearly 1900 Peoples' Biodiversity Registers (PBRs), documenting local bio-resources and associated traditional knowledge, have also been developed and validated. NBA has provided financial assistance of over rupees 100 million towards strengthening SBBs and BMCs and developing PBRs. NBA has also set up expert committees on 'access per cent benefit sharing', 'agro-biodiversity' and 'normally traded commodities. Fifteen National Designated Repositories have been recognized for safekeeping of voucher specimen/reference samples

### Core Expert Group

☆ A core expert group has also been constituted to address concerns of these repositories and develop guidelines.

☆ Another core expert group has been constituted to develop a unified model agreement deed to replace the existing four kinds of agreement forms.

☆ Effort is also on to develop on-line processing of all applications

Guidelines on collaborative research projects were notified in 2006 for claiming exemption from the Act's provisions under Section 5. Much awaited guidelines on ABS have also been notified in November, 2014 to speed up implementation of Abs provisions under the Act. NBA has already approved 633 application out of 985 applications received by it under different categories and entered into 171 benefit sharing agreements with the users of bio-resources [See Table 13.5].

Guidelines, exempting designated accessions of listed crops under ITPGRFA from Sections 3 and 4 of the Act, have also been notified for smooth implementation of this treaty. The expert committee on 'Normally Traded Commodities' has now prepared a list of over 420 species for exemption under Section 40 of the Act so long as these are traded as commodities. Over 14 States have developed and notified lists of threatened species' under their jurisdiction. More than 160 million rupees have

---

**Box 8 Penalties for Contravention of Biological Diversity Act**

☆ Whoever, contravenes or attempts to contravene or abets the contravention of the provisions of section 3 or section 4 or section 6 shall be punishable with imprisonment for a term which may extend to 5 years, or with fine which may extend to Rs. 10 lakh rupees and where the damage caused exceeds Rs. 10 lakh rupees fine may commensurate with the damage caused, or with both [Section 55 (1)].

☆ Whoever, contravenes or attempts to contravene or abets the contravention of the provisions of section 7 or any other order made under sub-section 2 of section 24 shall be punishable with imprisonment for a term which may extend to 3 years, or with fine which may extend to Rs. 5 lakh rupees or both. [Section 55 (1)].

☆ If any person contravenes any direction given or order made by the Cenral Government, the NBA or the SBB for which no punishment has been separately provided under this Act, he shall be punished with a fine which may extent to 1 lakh rupees and in case of a second or subsequent offence, with fine which may extend to 2 lakh rupees and in case of continuous contravention with the additional fine which any extend to 2 lakh rupees every day during which the default continues. [Section 56].

☆ The offences under this Act shall be cognizable and non-bailable. [Section 58].

☆ The provisions of this Act shall be in addition to, and not in derogation of, the provisions in any other law, for the time being in force, relating to forests or wildlife. [Section 59].

*Source:* Rana, 2015

**Table 13.5: Statutes of Processing and Approval of Applications as on 30 April, 2015**

| Items | Form I | Form II | Form III | Form IV | Total |
|---|---|---|---|---|---|
| Number of applications received | 186 | 40 | 681 | 78 | 985 |
| Number of applications approved | 85 | 16 | 494 | 38 | 633 |
| Number of applications in process | 66 | 13 | 159 | 24 | 262 |
| Number of application closed | 44 | 14 | 34 | 17 | 109 |
| Number of BS agreements signed | 40 | 12 | 93 | 26 | 171 |
| **Total** | **421** | **95** | **1461** | **183** | **2160** |

*Source*: Rana, 2015.

been deposited in the National Biodiversity Fund for payment to benefit claimers and to promote conservation and sustainable use of bio-resources. Some SBBs, notably those of Gujarat, West Bengal and Uttarakhand, have gone ahead and signed a large number of benefit sharing agreements with the user companies in their states, securing thereby the much needed funds required for their functioning aimed at conservation of bio-resources and their sustainable use at the ground level, besides ensuring fair and equitable sharing of benefits arising from their commercial utilization.

High level consultations have been held with the Council of Scientific and Industrial Research, New Delhi, Indian Council of Agricultural Research, New Delhi and the Indian Patent Office to address concerns and ensure smooth implementation of the Act. National consultations have also been held with major stakeholders, including pharmaceutical and seed sectors, to provide further clarifications on definition of some terms under Section 2 of the Act. Effort is now on to expand the list of Frequently Asked Questions displayed on NBA's website to provide the required guidance. An Expert Committee is also working on developing a unified version of the four types of the benefit sharing agreements, presently in vogue, with a common basic format attached with four kinds of annexes meant for granting access for research and commercial utilization, transfer of the results of research on bio-resources, IPR and third party transfer of bio-resources. In another significant move forward, it has been agreed to notify 427 plant species as Normally Traded commodities which will be exempted from provisions of the BD Act as long as they are traded as commodities (Rana, 2015).

## Implementing Provisions of CBD, ITPGRFA and TRIPS Agreement in India

In India, the Union Ministry of Environment, Forests and Climate change is the nodal ministry for implementing the CBD and also the Nagoya Protocol on ASB. Under the CBD, the Sovereign Authority to determine access to genetic resources rests with the national governments and it is subject to their national legislation. India has been party to several international multilateral agreements and programmes concerned with aspects of PGR, including the CBD (1993), TRIPS

Agreement under the WTO (1994), Biological Diversity Act, 2002, Cartagena Protocol on Biosafety (2003), ITPGRFA (2004) and Nagoya Protocol (2010).

The Biological Diversity Act, 2002, was enacted in India to fulfill this requirement and also to provide further support to other complementary national laws in force, namely, the Wildlife Protection Act, 1972 (as amended in 1991), and the Protection of Plant Varieties and Farmers Rights Act (PPV and FRA), 2001. It also provides suitable linkages to the provision for patenting of products and processes/technologies, based on the use of bio-resources and associated indigenous traditional knowledge (ITK), under Section 10(4) of the Patents (Second Amendment), Act 2002 and Patent (Third Amendment), Act 2005 which provide for exclusion of plants and animals from the purview of patentability, exclusion of an invention which in effects is traditional knowledge from patentability, mandatory disclosure of the source and geographical origin of the biological material in the specification when used in an invention and provision for opposition to grant of patent or revocation of patent in case of non-disclosure or wrongful disclosure of the source of biological material and any associated knowledge. The stage is thus set for developing a national movement for implementing these combined provisions for access and benefit sharing to ensure food and livelihood security based on conservation, inclusive development and sustainable use of bio-resources (Rana, 2012). National biodiversity Authority (NBA) has been designated as the National Competent Authority for this purpose. Format for the Internationally Recognized Certificate of compliance has been approved. Efforts are now on to designate the Check Points to monitor commercial utilization of India's bio-resources or patenting of their derivatives/products in other countries. Other key provisions to make the ABS regime functional include developing user and provider country measures.

Union Ministry of Agriculture and Cooperation is the nodal ministry for implementing the ITPGRFA and Joint Secretary (Seeds) is the National Focal Point, assisted by the DARE and ICAR-NBPGR, New Delhi. A notification has been issued exempting the exchange og designated accessions of genetic resources of listed crops of the ITPGRFA from the provisions of Sections 3 and 4 of the Biological Diversity Act for research, breeding and training purposes. Germplasm exchange would be based on signing the SMTA as approved under this Treaty.

For implementing the WTO-TRIPS agreement, Union Ministry of Commerce is the nodal ministry in India. India amended its Patents Act, 1970 to permit patenting of products and also of micro-organisms, as required under the TRIPS Agreement, and also enacted the Protection of Plant varieties and Farmers Rights Act (PPVFR), 2001.

In a bid to harmonise provisions of the CBD and WTO-TRIPs, the Doha Ministerial Declaration had asked for 'Disclosure of Source and Origin' to be made mandatory in patent applications which were also required to have an international Certificate of Compliance to the CBD confirming PIC and MAT provisions. Doha Round of negotiations is, however, underway since 2001 but the progress made so far is much below the developing countries expectations.

## References

1. Andersen, R., Tvedt, M.W., Fauchald, O.K., Rosendal, T.W.K. and Schei, P.J. 2010. International Agreement and Processes Affecting an International Regime on Access and Benefit-Sharing under the Convention on Biological Diversity: Implications for its Scope and Possibilities of a Sectoral Approach. FNI Report 3/2010. Lysaker, FNI.

2. AVRDC (2006). Vegetables Matter. AVRDC-The World Vegetable Centre. Shanhua, Taiwan.

3. Brahmi, P. and Chaudhary, V. 2011. Protection of plant varieties: systems across countries. *Plant Genetic Resources*, **9(3):** 384-391.

4. Brahmi, P. and Tyagi, V. 2015. Recent developments for access to plant genetic resources (PGR) in India and their implications. *Indian J Plant Genet. Resour.*, **28(2):** 237-246.

5. Chauhan, K.P.S. 1996. Implications of the convention on biological diversity: Indian Approach. *Indian J. Plant Genet. Resour.*, **9(1):** 1-10.

6. Chauhan, K.P.S. 1996. Legal requirements for fulfilling obligations under the Conservation on Biological Diversity in India. *Indian J. Pl. Genet. Resour.*, **9(1):** 1-10.

7. Chauhan, K.P.S. 1998. International treaties on biological resources: Implications for the Asian Region. *Indian J. Pl. Genet. Resour.*, **11(1):** 1-12.

8. Chawla, H.S. 2002. Intellectual property rights. In: Introduction to Plant Biotechnology. Oxford and IBH, Publishing House, New Delhi. p. 467-483.

9. de Klemm, C. and Shine, C. 1993. Biological diversity conservation and the law, 292. IUCN, Gland, Switzerland and Cambridge, U.K.

10. Dias, J.S. 2010. Impact of improved vegetable cultivars in overcoming food insecurity. *Euphytica*, **176:**125-136.

11. Evenson. 1999. *TCABR*, paper.

12. FAO. 2010. Second Report on the State of the World's Plant Genetic Resources for Food and Agriculture. FAO, Rome.

13. FAO. 2015. Summary of the 15th session of the Commission on Genetic Resources for Food and Agriculture: 19-23 January, 2015. Earth Negotiations Bulletin 9; 26 January, 2015.

14. Ganguli, P. 1998. Gearing up for Patents-The Indian Scenario Universities Press, Hyderabad.

15. Ganguli, P. 1998. Patenting innovations: New demands in emerging contexts. *Current Sci.*, **75:** 433-439.

16. Gautam, P.L. and Singh, A.K. 1998. Agro-biodiversity and intellectual property rights related issues. *Indian J. Plant Genet. Resour.*, **1(2):** 129-152.

17. Gautam, P.L; Singh, A. K.; Srivastava, M. and Singh, P.K. 2012. Protection of Plant Varieties and Farmers' rights: A review. *Indian J. Pl. Genet. Res.*, **25(1):** 9-30

18. Ghijsen, H. 1998. PVP essential to crop improvement in Asia Pacific. *Asian Seed and Planting* Material, **5:** 31-33.

19. GOI. 1988. National Forest Policy. Ministry of Environment and Forests, Government of India, New Delhi.

20. GOI. 1992. National conservation Strategy and Policy Statement on Environment and Development. Ministry of Environment and Forests, Government of India, New Delhi.

21. GOI. 1994. Country Paper-India. International Consultation on biological Diversity. MoEF, UNEP and IISC, Bengaluru.

22. GOI.1996. Annual Report. Ministry of Environment and Forests, Government of India, New Delhi.

23. Halewood, M.Noriega, I.L. and Louafi, S. 2014. Global Biodiversity Governance: Genetic Resources, Species, and Ecosystems. In: R.S. Axelrod and S.D. Van Deveer (eds.). *The Global Environment. Institutions, Law and Policy*. Sage, Los Angeles, p. 283-304.

24. Hamilton, S.K., Bunn, S.E., Thoms, M.C. and Marshall, J.C. 2005. Persistence of aquatic refugia between flow pulses in a dryland river system (Cooper Greek, Australia). *Limnology and Oceanography*, 50: 743-754.

25. Kochupillai Mrinalini. 2011. The Indian PPV and FR Act, 2001: Historical and Implementation Perspectives. *J. Intellectual Property Rights*, **16:** 88-101.

26. Kothari, A. and Singh, S. 1992. Convention on biological diversity and the law in India. Indian Institute of Public Administration, Delhi.

27. Kumar, S., Rai, A. and Kuamer, S. 2008. Vegetables. In: Intellectual Property Rights in Horticulture. (Eds. Kannaiyan, S., Parthasarathy, V.A. and Prasath, D.). Associated Publishing Co., New Delhi, p. 201-220.

28. Lacey, H. 2002. Ethics and values in agricultural biotechnology. In: 11th Annual Meeting of the Association for Practical and Professional Ethics, Cincinnati, March 1, 2002. (Available at http://www.swarthmore.edu/Humanities/hlacey/transgenic.doc)

29. Medakkar, A. and Vijayaraghavan,V. 2007. Successful commercialization of insect-resistant eggplant by public private partnership: Reaching and benefit resource-poor farmers. In: Intellectual property management in health and agriculture innovations: A handbook of best practices (Eds. A Krat-tiger. Mahoney, R.T., and Nelsen, L.) p. 1829-1831. MIHR Oxford, U.K. and PIPRA: Davis, U.S.A. Available online at www.ipHandbook.org.

30. Morgera, E., Noriega, I. L. and Louafi, S. 2014. Unraveling the Nagoya Protocol: A commentary on the Nagoya Protocol on Access and benefit-Sharing to the Convention on biological Diversity (Legal Studies on Access and Benefit-Sharing). Matinus Nijhoff Publishers, Leiden.

31. Muller, A. 2013. Relationship between national and international law, Public International Law, University of Oslo, Oslo, Norway, Natl_Internatl_law2013pdf.

32. Nagarajan, S., Yadav, S.P. and Singh, A.K. 2008. Farmers' variety in the context of Protection of Plant Varieties and Farmers' Rights Act, 2001. *Current Science*, **94(6):** 709-713.

33. Park, W.W. and Yanos, A.A. 2006. Treaty obligations and national law: Emerging conflicts in international arbitration. Hastings Law Rev. 58:www.arbitration-icca.org.

34. Prasanna, B.M; Rao, S.K; Kalloo, G. and Singh, R.B. 2008. Varieties of common knowledge' in the context of plant variety protection. *Current Science,* **95 (11):** 1522-1524.

35. Rana, R.S. 1995. Intellectual Property Rights: Protection of Plant Varieties. In: Genetic Resources of Vegetable Crops: Management, Conservation and Utilization (Eds.) Rana, R.S., Gupta, P.N., Rai M. and Kochhar, S. ICAR-NBPGR, New Delhi, p. 421-427.

36. Rana, R.S. 2010. Relating access to genetic resources and promoting benefit sharing in India. *Indian J. Plant Genet. Resour.,* **25:** 31-51.

37. Rana, R.S. 2012. Accessing plant genetic resources and sharing the benefits: Experiances in India. *Indian J. Plant Genet. Resour.,* **23:** 253-264.

38. Rana, R.S. 2015. Interplay of National and International Laws on Access to Biological Resources and Benefit Sharing. *Indian J. Plant Genet. Resour.,* **28(2):** 165-179.

39. Rcherzhagen, C. 2014. The Nagoya Protocol: Fragmentation or Consolidation? *Resources*, **3:** 135-151.

40. Rodgers, W.A. and Panwar, H.S. 1988. Planning a protected area network in India. Wildlife Institute of India, Dehradun.

41. Sahai, S. 2003. India's Plant Variety Protection and Farmers Right Act, 2001. *Current Science*, **84(3):** 407-412.

42. Sanchez, V. 1994. "The Convention on Biological Diversity; Negotiations and Contents". In: Sanchez, V. and Jauna, C. (eds.) Biodiplomacy. African Centre for Technology studies, Nairobi.

43. Saxena, S. and Singh, A.K. 2006. Revisit to definitions and need for inventorization or registration of landrace, folk, farmers' and traditional varieties. *Current Science*, **91(11):** 1451-1454.

44. Sharma, J.P. 2009. Intellectual Property Rights: An Agricultural Perspective. In: Principles of Vegetable Breeding, p. 391-413.

45. Singh, B.D. 2005. Plant Breeding: Principles and Methods. Kalyani Publishers, New Delhi.

46. Soam, S.K. 2005. Analysis of prospective Geographical Indication of India. *Journal of World Intellectual Property*, **8(5):** 679-705.

47. Soam, S.K., Tiwari, S.P. and Sastry, R.K. 2007. Sustainable use and conservation of agro-biodiversity through protecting as goods of geographical indications. In: Agro-biodiversity-Crop Genetic Resources and Conservation (Eds. Kannaiyan, S. and Gopalam, A.), Associated Publishing Co., New Delhi, p. 229-241.

48. Srivastava, M., Bhardwaj, M., Agarwal, R.C. and Singh, B. 2013. Protecting Vegetable crops varieties through Protection of Plant Varieties and Farmers' Rights Act, 2001. In Souvenir: Biotic and Abiotic Stress in Vegetable Crops, held at ICAR-IIVR, Varanasi (U.P.), pp111-117.

49. Srivastava, U. 2006. New Seed Bill 2004. Issues and Analysis. *Indian J. Plant Genet. Resource,* **19(2):** 141-153.

50. The Biological Diversity Act, 2002, http://www.nbaindia.org

51. The Gazette of India (2011a). Notification of bodies to which a copy of certificate of registration issued under the PPV and FR Act, 2001 shall be sent. SO 1912 (E) August 18, 2011.

52. The Protection of Plant Varieties and Farmers' Rights Act, 2001, http://www.plantauthority.gov.in

53. TRIPS Agreement, Article 27.3 (b), http://www.wto.org/english/docs_e/legal_e/27-TRIPS.pdf, www.cbd.int

54. Tvedt, M.W. 2014. Beyond Nagoya: Towards a legally functional system of access and benefit-sharing. In: Oberthur, S. and G.K. Rosendal (eds.) Global Governance of Genetic Resources: Access and Benefit Sharing after the Nagoya Protocol. Routledge, London/New York, p. 158-177.

55. UNEP.1992. Convention on Biological Diversity. United Nations Environment Programme, Nairobi.

56. WCED.1987. Our Common Future. World Commission on Environment and Development. Oxford University Press, Oxford.

## Notes of CGIAR

1. Research for Development > Water and Food Challenge Programme". *DFID. 2008-11-14.* Retrieved 2012-07-18.

2. "HarvestPlus/International/S and T Organisations/Home - Knowledge for Development". *Knowledge.cta.int*. Retrieved 2012-07-18.

3. Jump up^ "The Generation Challenge Programme Platform: Semantic Standards and Workbench for Crop Science". *Hindawi.com. 2007-09-22.* Retrieved 2012-07-18.

# 14

# Intellectual Property Rights, Protection of Plant Varieties and DUS Guidelines

## Introduction

Right on inventors to obtain legal protection to exclusive use of their creative innovations are widely recognized and granted in the form of patents in many developed countries. These are intended to reward the inventors by enabling them to earn exclusive profits from the commercial application of their inventions over a reasonable period of time and also to ensure that the new inventions and discoveries are utilized on a commercial scale for public good, particular *in situ*ations where inventors do not have adequate resources to do so on their own. Licensing is the primary mechanism by which intellectual property rights are transferred from inventors to investors in a way that benefits may reach the users.

India had already made a significant policy shift towards a pro-intellectual property (IP) position in the seed sector two decades ago, when it became a member of the World Trade Organisation (WTO) in 1995. Many existing laws were amended, including three amendments to the Patent Act of 1970, which allow for the patenting of seeds produced by non-biological methods such as found in modern biotechnology. Also, new IP were made that have an impact on agriculture: the Geographic Indications Act, 1999 and more importantly the Protection of Plant Varieties and Farmers Rights (PPV&FR) Act, 2001 to bring India's seed sector in line with the WTO TRIPS' requirements. At the beginning of the 20th century, seeds were still a public good that scientists improved according to the latest discoveries in genetics, especially Mendel's laws of inheritance that saw a 'rediscovery' at that time. Public institutions systematically categorised seeds and made them available to farmers.

## What IPR offers

☆ Provides IPR legal monopolistic rights

☆ Provide protection to research costs in innovative activity

☆ Reward and encourage investment in technical innovation

☆ Assisting incremental development of technology

☆ IPR in agriculture
   ❐ Utility patents
   ❐ Breeders' rights

## Harmonization of International Intellectual Property Rights

In addition to the general strengthening of IPR for living organisms, harmonization of international IPR has been pursued at different forums with varying degree of success. In 1984, the World Intellectual Property Organization (WIPO)-the United Nations' agency that administers most international IPR treaties-initiated its attempt to harmonize certain provisions of national patent laws of several MDCs.

## Property Rights over Plant Genetic Resources

Biodiversity Treaty is important because its

provisions on national sovereignty, prior informed consent and participation in research and benefit sharing-while not new ideas in that they reflect the basic principles of the FAO Undertaking-are significant advances toward recognizing the interests of LDCs.

## Indian Responses to the IPR Regime

Indian culture and traditions emphasize the sacrifice of individual interests for the benefit of the society at large, while IPR regimes seek to reverse this trend. However, in order to survive in an IPR hungry world, systematic effective and continued efforts must be made to bring about the following (Ganguli, 1998) and possibly more.

(1) Scientists and technologists must be trained to read patents, interpret claims and map claims into prior art so that they are able to provide technical support to the following:

(i) Writing 'world class' patents based on their innovations.

(ii) Defending patents.

(iii) Formulating opposition/revocation cases.

(iv) Identifying infringements.

(v) Preparing well-focused R&D programmes.

(vi) Striking collaborations with business houses.

(vii) Technology transfer/liaison officers in the institutions, and

(viii) Evaluating the effectiveness of IPR portfolios.

(2) Access to international databases on patents for their technical content, legal status, *etc.* needs to be ensured.

(3) Information scientists need to be trained to extract relevant information from patents granted worldwide in a timely and cost-effective manner.

(4) Knowledge databases of our national resources such as biodiversity, traditional medicinal and cosmeceutical practices, techniques of our craftsmen, *etc.* have to be created and made accessible to various practitioners. This is important as bringing innovations based on traditional knowledge and practices under the IPR regime are current tissues awaiting explosive growth.

(5) In order to enhance working and commercialization of patents a formal 'patent market' may be set up. This could serve as 'one-stop-shop' for industries and entrepreneurs.

(6) Technical personnel may be trained formally as patent attorneys so that they are able to handle the complex techno-legal issues related to patents.

(7) Active liaison units with technical expertise in IPR may be established; these would serve as resource centres for various institutions in the country.

(8) Up-to-date information on patents filed and granted in India should be made available on-line.

(9) Technology mapping based on patent information and human resource (technical/business) databases should be undertaken to create knowledge directories. This would enable the identification of national priorities and formulation of a national science and technology directory.

(10) Efficient and effective system of IPR portfolio management must be evolved to make IPR acquisition a rewarding activity.

(11) Educational institutions must introduce IPR into their curriculum in order to generate awareness among students about the meaning of opportunities due to IPR and its impact in innovation, trade, industry and the nation.

(12) The Indian Patent Office (IPO) needs to modernize in every respect so that it is able to perform its functions speedily. It is heartening that the government has already initiated this process. For example, e-filing of patent and trademark application is now in place. IPO proposes to achieve USPTO efficiency with its second phase of modernization.

(13) The time taken for grant of patent by IPO needs to be substantially reduced. At present IPO takes 3-4 years to grant a patent as against 2 years by USPTO. Patent Rules (2006) are designated to simplify the procedure and also to ensure time-bound disposal of patent applications. Examinations of patent applications seems to be the critical factor in this regard; the average number of applications examined each year during the last 4 years was less than 10,000 as against ~33,000 new applications filed each year.

(14) The role of IPO should be widened from a mere governance of the patent system to being a proactive partner in national awareness of IPR and technology/business development.

(15) Our institutions will have to evolve a new ethos by setting up frameworks/processes and respond to business driven R&D functions, identify focused projects, learn to negotiate terms and at the same time deliver 'world class' scientists and engineers as a part of their academic commitment.

(16) A drastic change in the attitudes of scientists/technologists is essential. This calls for appropriate and motivating changes in the working atmosphere, reward-punishment

regimes, *etc.* so that the scientists/technologists/academicians are motivated to adopt a serious and committed work culture.

(17) The administrative and civil legal/judicial system needs to be streamlined to be able to handle the huge number of disputes that are likely to arise due to the wider implementation of the various IPR regimes.

(18) Our educational system need to be drastically reformed to produce vibrant, innovative, committed and motivated individuals, especially scientists and technologists.

Intellectual property protection of plant varieties is somewhat different from that of the other forms of technology because a new plant variety does not normally arise out of a single innovative step. It is rather developed in a cumulative manner in discernable stages with addition of a new advantages or genetic important at each level to a basic genotype which often happens to be a popular locally adapted variety. Even though a new plant variety may not involve an inventive step in the strict sense, its economic usefulness to society is unquestionable. It is, hence, important to encourage development of new varieties and spread. General Agreement on Tariffs and Trade (GATT with its new form of World Trade Organization) was finally concluded and signed by over 120 nations in December, 1993. India is among the 64 countries, who had ratified this multilateral agreement by June 1994 and the agreement is legally binding. It includes Trade Related Intellectual Property Rights (TRIPS).

## Patents

*"Patent is governmental granted permission or exclusive right to an inventor to others from making, using or selling the patented product or process".*

Or

*"Patent is personal property that can be sold, leased or licensed to any other person".*

Or

*"A patent is the right granted by government to an inventor to exclude others from imitating, manufacturing, using or selling the invention in question for commercial use during the specified period".*

Patents on plant have always been a 'hot potato'. Similarly, the relationship between patents on plant and plant breeder's rights has been subject on inverse societal and academic debate. Most recently, the grant of patents for plant resulting from essentially biological processes, notably tomatoes with reduced fruit water contents and broccoli with anti-cancer potential, spurred stormy disputes.

Patents are granted for:

(i) An invention (including a product)

(ii) Innovation/improvement in an invention

(iii) The process/product of an invention, and

(iv) A concept

## Types of Patent

Three types of patents are granted:

(i) **Plant patent-**Plant patents are granted to noble types in sexually and asexually propagated plants. The protection is for 20 years

(ii) **Design patent-**A design patent protects the shape, structure and other relief characteristics of a material. The life span of design is usually 14 years.

(iii) **Utility patent-**Utility patent is the regular form used by universities, research organization and companies to protect their research techniques, protocols and products.

Patent information available in the area of plant transformation can be judged by the fact that a search of the US application database alone for "transgenic plant" and "method" returned records on 6th September, 2007.

**Table 14.1a: Selection of US Patents on Transgenic Plants**

| Number | Company/Institution | Subject |
|---|---|---|
| 7,265,280 | Senesco Inc. | Senescence |
| 7,265,278 | Unknown | Flowering |
| 7,265,269 | Bayer Bioscience | *Bt* protein |
| 7,265,267 | Crop Design N.V. | Cyclin-depedent kinase |
| 7,265,266 | University of Arizona | Salt tolerance |
| 7,265,265 | Pioneer Hi-Breed International | Galactomannan |
| 7,265,264 | Washington University | Plant size |
| 7,265,219 | Pioneer Hi-Breed International, Inc. | Dormancy promoter |
| 7,265,207 | Calgene | Tocopherol synthesis |
| 7,264,970 | Univ. California, Oregon, Arizona | Gene silencing |
| 20070209092 | CSIRO, BASF | Vernolic acid |
| 20070209089 | Simplot Co. | Marker free transgenics |
| 20070209088 | Unknown | Starch |
| 20070209087 | BASF | Sugar and lipids |
| 20070209086 | Mendel Biotech | Yield |
| 20070209085 | Monsanto | Enzyme gene promoter |
| 20070209068 | Monsanto | *Bt* protein |
| 20070207525 | Genencor | Phytase enzymes |

## Patent Requirement and Patenting in India

All the contracting parties are bound to provide certain minimum levels of protection to new plant varieties through an effective Sui generis system or patenting or a suitable combination of both. A product or process to qualify for patent to meet three criteria of patentability which are as shown in Table 14.1b.

**Table 14.1b: Criteria for Granting Protection to a New Plant Variety**

| Characters | Remarks |
|---|---|
| Novelty | The variety must be new and should not be already known to the public. If the same thing is already known the claim loses its patentability. |
| Inventiveness | The invention should not be obvious to a person skilled in the art, and should represent an innovation. In other words, the product required highly skillful, technical tough steps to evolve, means it is beyond the thinking or vision of an ordinary mind. |
| Industrial Application and Usefulness | The subject matter of the patent must have an industrial application, either immediate or in the future, that is useful to the society/nation, indicating the complete reproducibility of the product or process. The patent claim should be associated with characters: <br> 1. Through written description <br> 2. Enables others to reproduce the results <br> 3. Should be adequate <br> 4. Have a deposit mechanism |
| Patentability | The subject matter of the patent must be patentable under the existing law and its current interpretation. |
| Disclosure | The patent must be elaborately disclosed to the public through appropriate documents. |

**Table 14.2: Products, Patent and Amended Rules**

| Products | Patent | Amended |
|---|---|---|
| Protection of design | Indian Patents and Design Act (1911) | Amended in 1978 and amended rules in 1985 |
| Trademark | 1940 Act | In force since June 1, 1948 |
| Indian Trade and Merchandise | Indian Trade and Merchandise Marks Act (1958) | In force since November 25, 1959 |
| Copyright laws | The Indian Copyright Act (1957) | Amended in 1999 |

The first law on patent was passed in Venice in 1474, which gave monopoly rights to artisans for their inventions. In 1623, the House of Commons of U.K. passed the Act of Proprietorships. In the human society, things (goods, means *etc.*) were developed as per day to day need but peoples were not aware about novel background for future use and stand. In Indian ethnic groups, innovations and novel technique were retained within the families/small social groups. It is also true that there was no formal system of protecting their rights to the knowledge so generated. In 1856, the then Government of India introduced the Act of Protection of Inventions based on the British Patent Law of 1852. Later, Patents and Design Protection Act was passed in 1872. In 1883, the Protection of Inventions Act was introduced; it was consolidated as Inventions and Design Act in 1888. On August 15, 1947, the Indian patents and designs came under the management of Controller of Patents and Designs. The Indian patents Act (1970) was introduced in the Parliament in 1965, was modified in 1967, and was passed in 1970. It was subsequently amended as Indian Patents (Amended) Act (1999). The Act was amended again in 2002, and 2005, and the patent rules were last amended in 2006 as Patents (Amended) Rules, 2006 made effective from May, 2006.

The Indian Patent Act of 1970 did not allow product patents in pharmaceuticals, foods and agrochemicals. But this Act has now been amended as Indian Patent (Amended) Act (1999):

(i) The new Act allows product patents, except for some specified medicine/drugs

(ii) A provision for grant of 'exclusive marketing rights' (Section 41.5.1) has been made up to December 31, 2004.

(iii) The provision in relation to compulsory license shall, subject to necessary modifications apply to 'exclusive marketing rights' as well.

(iv) Considering security views, the government of India may not disclose any information relating any patentable invention, and take action including the revocation of any patent, provided the intention for the same is modified in the official gazette before taking any action.

The Patent Act, 1970 specifies what are not considered inventions under Section 3. Section 3(b) of the Act earlier recited: "What are not inventions–An invention the primary or intended use of which would be contrary to public health".

## Patenting Procedure

The Indian Patent Office (IPO) has its headquarters at Kolkata and branch office at Chennai, New Delhi and Mumbai. A properly prepared application (on prescribed proforma) should be submitted/filed to the territorial limits of the applicant/first applicant normally resides. The application is screened and verify, if found in order for publication, it is process but in incomplete

**Table 14.3: Limits of Patent**

| Limitations | Remarks |
|---|---|
| Limitation of time | A patent is valid for a specified period, 15-20 years in most countries, of time from the date of its award |
| Limitation of space | A patent is valid only in the country of its award and not in other countries. A group of nations may agree to honour the patents awarded by any member country, *e.g.*, in European Community. WTO has similar provisions in that a patent awarded by WTO will be valid in all member countries. |

condition, time is allowed for their rectification. Complete application in all respect including requisite fee, is processed within 18 months after the application was filed. Anyone who wishes to challenge the award of patent can do so within a specified period of time, *e.g.*, within 4 months from the date of publication in India and again it is examined for their suitability for the award of patent. Report along with the objection (if any) is communicated to the applicant. The inventor may with draw the filed application, modify and resubmit it or submit it with an explanation for the objection raised by the patent office. In India, normally all the objection must be addressed to within 15 months of the first examination of the application. In case a patent application is not challenged the patent is awarded within a specified period, *e.g.* within six months from the date of publication in India, provided the objections have been removed. But if a patent is challenged, the arguments and counter arguments of both the applicant and the person challenging the application are heard by a competent authority of the patent office and a final decision is taken on the award of patent. Thus, if a patent application is rejected due to a contest following its publication, the main features of the invention stand disclosed. In India, generally all the objection must be addressed to within 15 months of the 1$^{st}$ examination of the application.

## Revocation of Patent

Patent is a subjective contract and may be revoked in the light of some discrepancies. There could be various reasons for revocation of a patent (Chawla, 2002), which are enlisted below:

(i) Non-working of patent

(ii) Patent obtained by concealing the factual information

(iii) Inadequate elaboration of process or technique

(iv) Results not reproducible

(v) The techniques or process lacks inventiveness

(vi) Revocation necessary in the public interest

(vii) Patentee not entitled to the patent

(viii) Non-compliance of the authority instructions

(ix) Claims no clearly defined

## The Patent Complication (Imbroglio)

Developments of biotechnological products need several technology and/or raw materials. For example,

production of transgenic plant would require the following:

(i) Outstanding variety of concerned crop

(ii) Suitable promoter and/or enhancer sequences

(iii) Appropriate reporter genes

(iv) The gene specifying the concerned trait

(v) Appropriate genetic transformation method and sometimes

(vi) A specific technology, *e.g.* antisense RNA technology for suppression of endogenous genes.

During development of transgenic plant many of the above, would have been patented, while the ownership of some of others may be uncertain. Patent/license holders generally do not raise questions of infringement during the research and development phase and prefer to wait till the product is ready to be marketed. At that time, the patent/license holders initiate legal proceedings claiming infringement. For example, DNA Plant Technology (U.S.A.) developed 'Endless Summer' transgenic tomato by using the following:

(i) Antisense RNA technology

(ii) *Agrobacterium*-mediated genetic transformation

(iii) CaMV *35S* promoter

(iv) Selective marker *nptII* with its promoter and terminator

(v) The ACC synthase

All these are protected by patents. DNA Plant Technology was to release 'Endless Summer' in 1996, but it faced difficulties due to use of patented subject matter. It is trying to substitute some of the components used to produce 'Endless Summer', *e.g.* use of ALS (acetolactate synthase gene providing resistance against several herbicides) selected marker system in place of *nptII*, *etc.*, and to obtain licenses for the others. ALS gene has been patented by DuPont, and DNA Plant Technology has entered collaboration with DuPont.

## What Options will be if Patented Process/Products are Used?

The various options available to an inventor in cases of use of patented processes/products in an invention are as follows:

(1) License of the process/products may be obtained

(2) The inventor may enter into a collaboration with the patent/license holder

(3) An effort may be made for cross-licensing

(4) Merger/takeover of the company holding the license, and

(5) Replacement of the patented technology/ product with another available technology/ product.

## Patented Crops

Among the higher organisms, plants were first to be protected in U.S.A. in 1985, a maize line overproducing tryptophan was granted a patent, since then patenting of plants became a common practice in U.S.A. and a large number of plant lines have been patented. In vegetable crops, a large number of patents are available worldwide including U.S patents (Table 14.4).

**Table 14.4: Total Number of US Patents Awarded for different Vegetable Crops**

| Vegetable Crops | Number of US Patents | Vegetable Crops | Number of US Patents |
|---|---|---|---|
| Asparagus | 09 | Garlic | 25 |
| Broccoli | 04 | *Momordica* | 01 |
| Cabbage | 06 | *Moringa* | 14 |
| Carrot | 06 | Okra | 07 |
| Cauliflower | 03 | Onion | 13 |
| *Coccinia* | 01 | Pepper | 09 |
| Coriander | 01 | Pea | 11 |
| Cucumber | 15 | Radish | 02 |
| *Cucumis* | 06 | Tomato | 42 |

*Source*: USPTO, 2007.

However, to demonstrate some examples, a comprehensive list of U.S Patents on various aspects of vegetable crops has been synthesized and presented (Table 14.5b). A representative list of the U.S. patents awarded for selected vegetable crops.

**Table 14.5a: Plant Variety Protection Certificates Granted**

| Crop | Number Granted between | |
|---|---|---|
| | 1971-1985 | 1971-1987a |
| Peas | 113(0) | 187(0) |
| Beans | 110(2) | 169(4) |
| Lettuce | 44(0) | 69(0) |
| Tomato | 9(0) | 28(4) |
| Onion | 14(0) | 25(4) |
| Watermelon | 10(1) | 24(6) |
| Cauliflower | - | 19(0) |

Figures in parentheses indicates the number of public varieties.

## Protection of Biotechnological Inventions or IPRs and Transgenic

Transgenic crops are now a reality in India, and many more are to be expected in the market in near future. Transgenic products have 'Social Value', which results in benefits to the society at large (for example, use of less chemicals and pesticides in the case of transgenic disease resistant plant) and 'Commercial Value' from increased yield or improved product quality (for example, high yielding, early maturing and value-added crops) (Lacey, 2002). The biotechnology companies in Europe strongly favour the patenting of inventions based on living materials at par with that in

**Table 14.5b: U.S. Patents Awarded for Selected Vegetable Crops**

| Vegetables | Year | Patent Number | Brief Description about Invention (Product/Process) |
|---|---|---|---|
| Asparagus | 2006 | 6994874 | A skin whitening composition includes an extract of asparagus obtained by sequentially exposing the asparagus to two or more solvents |
| | 2006 | 7025995 | Herbal synergistic formulation comprises plant extracts for treatment of acute and chronic stomach ulcers. |
| | 2006 | 7014872 | Disease preventive neutraceutical herbal formulation(s) for diabetics. Formulation comprises seed powders mixture from selected genes including *Asparagus*. |
| | 2005 | 6914075 | The invention provides cysteine derivatives which may be in a free form, a salt form, a solvate form. |
| | 2005 | 6381425 | Neutraceutical herbal formulation for women and lactating mothers. Formulation comprises roasted seed powders mixture from selected genus including *Asparagus*. |
| | 2002 | 6386778 | A dispensing system for a multi-component product |
| | 1990 | 4963370 | A process for producing a proteinous material. |
| | 1951 | 2559625 | Isolation of an organic sulphur compound from *Asparagus*. |
| | 1936 | 2052219 | Isolation of vitamin concentrates and process of making the same. |
| Bitter gourd | 2006 | 7014872 | Health protective, promotive and disease preventive neutraceutical herbal formulation(s) for diabetics and a process for its preparation. |
| | 2005 | 6964786 | Novel oil extracted from the seeds for topical application to a body of mammal and used as anti-inflammatory, anti-arthritic, vasculo dilatory and wound healing agent. |
| | 2005 | 6852695 | A water soluble extract (MC6). Methods for its preparation and methods for its use in the treatment of hyperglycemic disorders. |

| Vegetables | Year | Patent Number | Brief Description about Invention (Product/Process) |
|---|---|---|---|
| | 2005 | 6911577 | Methods and compositions for modulating development and defence responses and nucleotide sequences encoding defence in proteins. |
| | 2005 | 6960348 | A novel cosmetic preparation can be obtained by incorporating bitter gourd fruit and leaves into genetic paraffin and cosmetic clay. |
| | 2004 | 6831162 | A novel and highly effective hypoglycemic protein called polypeptide-k, extracted, which is useful in the treatment of diabetics mellitus. |
| | 2004 | - | The invention is directed to secreted and trans membrane polypeptides and to nucleic acid molecule encoding those polypeptides. |
| | 2004 | - | Novel nucleic acids, novel cathepsin V-like polypeptides sequences encoded by these nucleic acids and their uses. |
| | 2004 | 6800726 | Isolation of nucleic acids and their encoded polypeptides that are involved in enhancing the essential amino acids. |
| | 2004 | 6673988 | Isolation of nucleic acid fragment encoding a lipase. |
| | 2003 | - | Pharmaceutical products are provided comprising EC progenitors for use in methods for regulating angiogenesis. |
| | 2003 | - | Variant immunoglobulins, particularly humanized humanized antibody polypeptides along with methods for their preparation and use |
| | 2003 | 6593514 | The preparation and use of nucleic acid fragments encoding plant fatty acid modifying enzymes associated with modification of the delta-9 position offatty acids, in particular, formation of conjugated double bonds are disclosed. |
| | 2003 | 6562379 | Methods of inducing weight loss and treating adult-onset diabetes in a mammal by administering to isolated lectin. Lectin pharmaceutical compositions are also disclosed. |
| | 2002 | 6379718 | Novel herbal extracts provide potent efficacy in the treatment of acne and furuncle. The formulated extracts are from either the whole plant or parts of the plant. |
| | 2000 | 6042829 | Cytotoxic bio-therapeutic agents effective for treating cel1ain types of cancer in humans. This comprises the TP-3 murine monoclonal antibody chemically conjugated to pokeweed antiviral protein. |
| | 1999 | 5900240 | An edible composition comprising a mixture of at least two herbs selected from the bitter gourd and eggplant. The herbal mixtures are useful as dietary supplements including humans suffering from diabetes mellitus. |
| | 1999 | 5929047 | An anti-viral agent comprising as the effective component, an alkali extract of mangroves, bitter gourd and *Aspalathus linearis* belonging to Leguminosae family. |
| | 1996 | - | Baits for diabroticine beetles are microspherical particles containing a homogeneous mixture of a toxicant for diabroticine beetles and a feeding stimulant in a binder containing a gelatin and a gum. |
| | 1996 | 5484889 | A protein, in particular MAP 30, obtainable from both the fruit and seeds or produced by recombinant means useful for treating tumors and HIV infections is disclosed. |
| | 1990 | 4958009 | Immunotoxins comprising a cytotoxic moiety and monoclonal antibodies, which bind to human ovarian cancer tissue. Methods of killing human ovarian cancer cells, retarding the growth of human ovarian cancer tumors in mammals or extending the survival of mammals carrying human ovarian cancer tumors are claimed. |
| | 1989 | 4795739 | A method of inhibiting expression of HIV antigens in human blood cells infected with HIV. The infected cells are exposed to a plant protein or glycoprotein, such as trichosanthin or momorcharin. |
| | 1989 | 4869903 | A method of inhibiting HIV replication in and cellular proliferation of HIV-infected cells. |
| | 1983 | 4368149 | A protein hybrid having cytotoxicity obtained by covalently bonding an immunoglobulin or its fragment, which is capable of binding selectively to an antigen possessed by a cell to be destroyed, to a protein. |
| Cabbage | 2004 | 6825321 | Invention of thermogenic genes (*SfUCPa* and *SfUCPb*) derived from skunk cabbage. |
| | 1994 | 5288626 | A method for increasing the proportion of mutants in a first plant species having a recognized and established phenotype. |
| Pepper | 2005 | 6919095 | A method of providing an essential oil extracts (capsaicinoid and terpene). |
| | 2004 | | An anti-inflammatory composition for treatment of joint and muscle pain through transdermnal delivery of a capsacinoid in conjunction with glucosamine. |
| | 2003 | 6517832 | A prophylactic treatment for the human malady clinically (migraine headache) by daily prescribed dosage of a first formulation, which is derivatives from beet root, powder, watercress, celery, dandelion, |
| | 2003 | 663239 | A method and formulation for sterilizing and disinfecting surfaces, and for killing bacteria on contact, particularly adapted for sterilization and disinfecting food and meat stuffs, and for food and meat processing equipment and facilities. |
| | 2003 | 6523298 | A method for exterminating existing infestations of ants, termites, insects or other living organisms in structures, soils and other materials using a Capsicum-containing killing solution in either a liquid or vapor form. |
| | - | 7097867 | A process for obtaining oleoresin of improved color and pungency. |

| Vegetables | Year | Patent Number | Brief Description about Invention (Product/Process) |
|---|---|---|---|
| | 2002 | 7097867 | A cancer cell growth suppressor or a cell differentiation inducer, which contains as an effective component a water-soluble component from Capsicum. |
| | 2001 | - | This invention relates to the novel improvements of pungency factors of Capsicum in particular the capsaicinoids and intermediates of phenyl propanoid pathway leading to capsaicinoid biosynthesis. |
| | 2000 | 6143349 | Distinct and stable cultivars of no-heat Jalapeno peppers are disclosed. Non-pungent Jalapeno cultivars in which substantially all.the fruits produce no capsaicin. |
| | 2000 | 6074687 | Principal components of paprika, red pepper, pungent chili, or other plants of the genus Capsicum containing carotenoid pigments are simultaneously extracted and concentrated with an edible solvent in a series of mixing and high temperature and pressure mechanical pressing steps. |
| | 2000 | 6159474 | A repellant composition for repelling both domesticated and wild animals which comprises between 0.05 per cent and 2 per cent by weight of an essential oil of either black pepper or *Capsicum*. |
| | 2000 | 6069173 | Insecticidal compositions comprising a synthetic surfactant and capsaicin or other capsaicinoid exhibit synergistic effects against numerous insects. |
| | 2000 | 6060060 | Analgesic compositions obtained from the fruit and therapeutic uses of this analgesic composition. |
| | 1999 | 5945580 | Polynucleotides of hemicellulase gene and compositions and its uses in controlling plant development and other characteristics. |
| | 1999 | 5910512 | A water-based topical analgesic and method of application wherein the analgesic contains capsicum, capsicum oleoresin and/or capsaicin. This analgesic is applied to the skin to provide relief for rheumatoid arthritis and osteoarthritis. |
| | 1998 | 5773075 | Principal components of paprika, red pepper, pungent chilli, or other plants of the genus Capsicum containing carotenoid pigments are simultaneously extracted and concentrated with an edible solvent. |
| | 1998 | 5811640 | The novel variety (JZA) of Capsicum chinense is the product of an organized pedigree breeding program. |
| | 1997 | 5599803 | Insecticidal compositions, comprising normally-employed insecticides but comprising also an effective activity-enhancing amount of capsaicin or other capsaicinoid. |
| | 1997 | 5698191 | Insecticidal compositions, comprising normally-employed insecticides but comprising also an effective activity-enhancing amount of capsaicin or other capsaicinoid. |
| | 1990 | 4931277 | Medicaments for treatment of alcoholic toxicomania comprise at least one extract of vegetable origin, particularly those obtained by maceration, decoction and/or infusion in an aqueous alcoholic solvent of Capsicum and/or bark or wood of *Populus poplars*. |
| | 1986 | 4592912 | A composition for the prevention and relief of muscular aches and pains, aches caused by tension such as headaches and backaches and aches and pain caused by inflamed muscles and inflammation surrounding muscles. |
| | 1953 | 2636824 | Spice substance and method of its preparation. |
| Cauliflower | 1998 | 5710364 | A novel cruciferous plant containing a large quantity of carotene even in those parts where a conventional cruciferous plant contains little carotene. |
| | 1997 | 5629175 | Novel constructs are provided for expression of physiologically active mammalian proteins in plant cells, either in culture or under cultivation. |
| | 1971 | 3578466 | Pickled vegetable product. |
| Coccinia | 206 | 7014872 | A health protective and disease preventive nutraceutical herbal formulation(s) for diabetics. The formulation comprises the base product of microwave roasted seed powders mixture from selected genus including *Coccinia*. |
| | 1999 | 5856487 | A process for isolating berberine from plants. |
| | 1995 | 5466455 | Processes for polyphase fluid extraction, active therapeutic components from parts of selected medicinal plants, which have been identified as chemotaxonomically. |
| Coriander | 2003 | 6579543 | A composition for topical application to an animal's skin for relief from a variety of symptoms caused by medical conditions or physical injuries. |
| | 2002 | 6365175 | Edible compositions containing petroselinic acid are used for the preparation offood compositions or food supplements and used as anti-inflammatory compositions. |
| | 2000 | 6017373 | An artificial firelog which contain 2 per cent to about 6 per cent sub. weight coriander seed added to create a crackling sound that mimics the sounds produced during the burning of natural logs. |
| | 1999 | 5959131 | A nutritionally superior fat for food compositions, which comprises triglycerides containing cis-asymmetric monounsaturated fatty acids. |
| | 1995 | 5430134 | A process for producing lipids containing the fatty acid namely, petroselinic acid. |
| | 1993 | 5256405 | A stick deodorant composition that has active antibacterial constituents consisting essentially of natural materials. The active antibacterial constituents include coriander oil. |
| | 1972 | 3637859 | The coriander oil has been used for the fragrance. |
| | 1970 | 3527827 | The monocyclic terpenes were made by this process from the seeds of coriander. |
| | 1963 | 3082095 | Method for dust proving of an edible, dry, finely divided product and the resulting product. |

| Vegetables | Year | Patent Number | Brief Description about Invention (Product/Process) |
|---|---|---|---|
| | 1925 | 1523840 | The oil is used in preparation of tooth pest. |
| | 1879 | 222187 | Coriander seeds are used in preparation of medicated herbal beverages. |
| Cucumber | 2004 | 527165 | Cucumber sandwich. |
| | 2004 | 6765130 | Seeds of inbred line (80-5079) and to methods for producing a cucumber plant, either inbred or hybrid. |
| | 2000 | 6084152 | A transgenic plant that produces high levels of super oxide dismutase (SOD) and to a method for producing the transgenic plant. |
| | 1997 | 5623066 | A DNA fragment which encodes the coat protein of cucumber mosaic virus strain c (CMY -C), the method of preparing it,its use to develop transgenic plants. |
| | 1997 | 5654414 | Chemically regulatable DNA sequences capable of regulating transcription of an associated DNA sequence in plants or plant tissues. |
| | 1989 | 4822949 | Production of F1 hybrid seeds and a method wherein the pollen parent bears only male flowers and thus lacks the capability to bear fruits. |
| | 1909 | 915186 | Pickles from fruits |
| Eggplant | 2006 | 6984725 | A method for the separation of a triglycoalkaloid from roots. |
| | 2006 | 7078063 | A water-soluble extract from Solanum genus consists essentially of at least 60 per cent -90 per cent of solamargine and solasonine. |
| | 2006 | 7012172 | Methods for interfering with expression of the genes in plant cells by using replicating recombinant viral vectors. |
| | 2004 | 6753462 | Transgenic plants over-expressing a transgenic encoding a calcium binding protein or peptide (CaBP). |
| | 2003 | 6639050 | A new approach inthe field of plant gums is described which presents a new solution to the production of hydroxyproline (Hyp)-rich glycoproteins (HRGPs), repetitive proline-rich proteins (RPRPs) and arabinogalactan-proteins (AG Ps). |
| | 2002 | 6369296 | Nucleic acid vectors may be used as expression vectors or for achieving viral induced gene silencing (VIGS) of a target gene. |
| | 2002 | 453238 | Eggplant soap. |
| | 2002 | 6483012 | Use of the promoter region of the DefH9 gene of *Anthirrhinum majus* or of a promoter of a homologous gene displaying the same expression pattern and characteristics for the establishment of parthenocarpy or female sterility in plants. |
| | 2001 | 6207881 | Control of fruit ripening through genetic control of ACC synthase synthesis. |
| | 2001 | 6225528 | Pathogen-resistant transgenic plants and methods of making the plants. |
| | 2000 | 6072105 | Transgenic plants along with improved culture media and methods enabling efficient regeneration of shoots from cultured explants. |
| | 2000 | 6043409 | The cDNA and genomic DNA encoding the ACC oxidase of broccoli are provided along with recombinant materials containing antisense constructs of these DNA sequences to permit control of the level of ACC oxidase. |
| | 2000 | 6133505 | Nucleotide sequences produced by mutation of C I nucleotide sequences present in a Pathogenic Gemini Virus genome in plants with one or more mutations capable of producing a dominant negative phenotype for the replication of the pathogenic virus. |
| | 2000 | 6023013 | Transgenic plants and transformed host cells, which express modified Cly3B genes with enhanced toxicity to Coleopteran insects. |
| | 1999 | 5955652 | Nucleic acid sequences for ethylene insensitive, *EIN* loci and corresponding amino acid sequences. |
| | 1999 | 5900240 | An edible composition comprising a mixture of at least two herbs selected from the group consisting of *Syzygium cumini, Gymnema sylvestre./Momordica charantia* and *Solanum melongena*.The herbal mixtures are useful as dietary supplements, particularly humans suffering from diabetes mellitus. |
| | 2000 | 6156956 | ACC synthase of higher plants are coded by multigene families; only certain members of these families are responsible for various plant development characteristics affected by ethylene. |
| | 1998 | 5723766 | ACC synthases of higher plants are coded by multigene families; only certain members of these families are responsible for various plant development characteristics effected by ethylene. |
| | 1998 | 5744334 | A Blec plant promoter sequence and a method of transforming plants with a Blec promoter-gene. |
| | 1997 | 5614408 | A hybridoma cell line produces and secretes a monoclonal antibody, which selectively binds to the glycoalkaloids of potatoes, tomatoes, and eggplants, as well as their corresponding a glycones. |
| | 1997 | 5627216 | A composition for the treatment of hemorrhoids. The composition effectively reduces swelling and pain in the anal area is formed by mixing powdered eggplant leaves and boiling virgin olive oil in a covered container for approximately 30 minutes |
| | 1997 | 5674701 | A process for identifying a plant having disease tolerance, comprising administering to a plant an inhibitory amount of ethylene and screening for ethylene insensitivity. |

| Vegetables | Year | Patent Number | Brief Description about Invention (Product/Process) |
|---|---|---|---|
| | 1996 | 5589623 | A method for control of ethylene biosynthesis in plants, comprising a vector containing codons for a functional heterologous polypeptide having AdoMetase activity. |
| | 1990 | 4921804 | An enzyme preparation having specific bilirubin degrading activity. Specific plant sources include eggplant, tomato and potato |
| Garlic | 2003 | 6511674 | A composition of a garlic extract solution having a concentration greater than 10 per cent by weight of a garlic extract. |
| | 2003 | 6641836 | A dietary composition and method for enhancing immune response and improving the overall health of canines. |
| | 2002 | 6468571 | A method for processing fresh garlic, which will not leave an unpleasant odor. |
| | 2002 | 12761 | A cultivar (Melany), characterized by early harvesting of the plant, high yield of bulbs, disease-free vegetation. |
| | 2001 | 12272 | A cultivar (Angelique), characterized by the presence of flower escape, bigger-sized bulbs, vigorous foliage. |
| | 2001 | 6270803 | An orally administrable formulation for the controlled release of granulated garlic. |
| | 1999 | 5913729 | Efficient methods for cultivating garlic plants. |
| | 1998 | 574604 | Large quantities of true seeds are obtained from garlic using a process that involves the growing of a garlic parent plant from a virus-free garlic propagule under virus-free conditions. |
| | 1977 | 4022923 | A method of processing garlic to form a composition of matter affected by a stable emulsion for the retention of flavor thereof and to permit freezing of the same. |
| | 1966 | 3258343 | The process of dry and powdered garlic from row garlic. |
| | 1951 | 2554088 | A process by which the cloves of garlic are treated in such a manner that this servers as a very good antibiotic toward both gram positive and gram negative bacteria. |
| | 1949 | 2490424 | The extract of garlic is an ingredient of carminatives. |
| Drumstick | 2006 | 7070817 | An herbal composition comprising extract of plant, for treating or alleviating of vascular headaches, neurological conditions and neuro-generative diseases. |
| | 2005 | 6858588 | A novel nitrile glycoside of Formula I named NIAZIRIDIN and to analogues and derivatives. |
| | 2005 | 6890565 | A process for preparing proteins that can act as effective coagulants in the treatment and purification of contaminated water. |
| | 2004 | 6780441 | A pharmaceutical or medicinal preparation, which comprises a mixture of herbs including *Moringa*. |
| | 2004 | 6750256 | Methods and compositions based upon natural aromatic aldehydes, which may be used as pesticides. |
| | 2003 | 6667047 | Cosmetic compositions with enhanced slip and/or break strength are described, said compositions comprising ultra-stable *Moringa* oils, or its derivatives. |
| | 2003 | 6517861 | An herbal dietary supplement composition for lactating mothers, comprising the required quantum of one or more herbs. |
| | 2002 | 6440437 | A skin health enhancing soft wet wipe or wipe-type product, such as a baby wipe, hand wipe or faces wipe. |
| | 2002 | 6383495 | A novel herbal formulation useful for the treatment of skin disorders and comprising two or more plant extracts including from *Moringa*. |
| | 2001 | 6217874 | A fat composition for cosmetic or pharmaceutical emulsion products. |
| | 2001 | 6271001 | Certain cultured plant cell gums, including those produced in suspension culture. |
| | 2001 | 6287581 | A superior skin barrier enhancing body facing material, such as a body side-liner. |
| | 2000 | 6080401 | The curative action of drugs, including herbal remedies, allopathic remedies, and periodontal remedies, is enhanced and accelerated by administering such drugs. |
| | 1999 | 5994404 | A baby or infant composition, which comprises one or more nutrient materials and, as a supplement, nervonic acid. |
| Okra | 2002 | 6379719 | Use of at least one protein fraction extracted from Hibiscus esculentus seeds and to a cosmetic composition containing such a fraction. |
| | 2000 | 6124248 | Mucilages and their extracts those are effective as water-soluble, non-toxic, biodegradable, environmentally benign lubricants and/or coolants for a variety of industrial and machining purposes. |
| | 1998 | 5851963 | A new organic lubricant for lubricating and cooling tools. |
| | 1980 | 4233075 | A food base preservative solution made from okra, water and a food preservative agent. Useful for material surfaces such as finished furniture surfaces, painted surfaces, glass surfaces, vinyl plastic surfaces. |
| | 1979 | 4154822 | Polysaccharide substances-essentially consisting of rhamnose, galactose and galacturonic acid and preferably derived by extraction and purification of okra plant. |
| | 1960 | 2932610 | Brightening material of plant origin for electroplating. |
| | 1936 | 2060336 | Composition of matter containing mucilaginous extracts from plants. |
| | 1865 | 51251 | Improvement in the manufacture of paper. |

| Vegetables | Year | Patent Number | Brief Description about Invention (Product/Process) |
|---|---|---|---|
|  | 1951 | 8184 | Making hemp from okra |
| Onion | 2002 | 6468565 | A cycloalliin-enriched onion extract. |
|  | 1926 | 1612255 | Hair lotion and its method of preparation. |
|  | 1924 | 1492823 | The ointment made with the onion extract. |
| Pea | 2005 | 6916787 | A method for producing haemin proteins. |
|  | 2002 | 6344600 | A method for producing human hemoglobin proteins. |
|  | 1987 | 4677065 | Seeds of a grain legume of relatively low lipid content are processed under conditions found to yield an improved protein isolate, which is well suited for human consumption. |
| Radish | 2004 | 6686517 | Plants containing increased levels of anthocyanins. |
|  | 2002 | 6428822 | A mixed substance extracted from carrot and whole radish for treating hypertension, constipation, detoxification and boosting immune system. |
|  | 1998 | 5736144 | A mixed substance extracted from carrot and whole radish for treating hypertension, constipation, detoxification and boosting immune system. |
|  | 1997 | 5650559 | Male sterile plants possessing Ogura cytoplasms derived from a Japanese radish with plants having pure nuclei of genus *Brassica*. |
|  | 1994 | 5324707 | A method for the in vitro application of bioregulator compound. |
| Tomato | 2006 | 015277 | Materials and methods for providing genetically engineered resistance in plants to geminivirus, using polynucleotides containing all or a portion of a replication associated protein (Rep) gene of TYLCY. |
|  | 2004 | 6806399 | A genotype independent method for efficiently carrying out pollen mediated transformation of maize, tomato or melon id described. |
|  | 2002 | 6429299 | Nucleotide sequences TDET1 (*HP-2*) gene. The sequences, if modified, result in a light hypersensitive phenotype. Vectors and uses for the production of transgenic plants. |
|  | 2002 | 6340748 | A tomato promoter (*LeExp-1*) which can direct a high level of fruit-specific expression. |
|  | 2001 | 6252141 | An isolated complementary and genomic DNA segment encoding lycopene cyclase of the B locus of tomato. |
|  | 2001 | 6239331 | A method for enhancing the expression ofa tomato phytoene synthese gene in a plant, while avoiding or reducing co-suppression. |
|  | 2001 | 6180854 | Determinate, delayed-ripening hybrids cherry tomato plants, derived from a determinate, non-ripening parental line. |
|  | 2001 | 6207881 | Recombinant materials for the production of tomato ACC synthase. |
|  | 2000 | 6060648 | Seedless tomatoes developed by crossing a tomato plant containing at least one parthenocarpic gene as the male parent with a male sterile tomato plant containing at least one parthenocarpic gene as the female parent. |
|  | 2000 | 069389 | A method for breeding to'lnato plants that produce tomatoes with reduced fruit water content. |
|  | 1999 | 5871574 | A tomato pigment is obtained by centrifuging a treated mass of tomato. |
|  | 1997 | 845539 | A protein expansion (Ex!) from tomato, melon and strawberry that is highly abundant and specifically expressed in ripening fruit. |
|  | 1999 | 6414226 | A new and distinct inbred tomato line (FOR 16-2045). |
|  | 1999 | 5952546 | Tomato plants exhibiting a delayed ripening phenotype, comprise of a T-DNA insert comprising a tomato A ee synthase gene. |
|  | 1998 | 5821398 | An isolated nucleotide sequence encoding an inducible soft fruit promoter, particularly the alcohol dehydrogenase-2 promoter. |
|  | 1998 | 5824873 | A method for modifying fruit ripening characteristics in plants and is particularly suitable for modification of tomato plants. |
|  | 1998 | 5817913 | A method for breeding tomato plants that produce tomatoes having superior taste characteristics including the step. |
|  | 1997 | 5614408 | Hybridoma cell lines, which produces and secretes a monoclonal antibody that selectively binds to the glycoalkaloids of potatoes, tomatoes, and eggplants and their corresponding aglycones. |
|  | 1997 | 5656474 | Two osmotic stress- and ABA-responsive members of the endochitinase gene family has been isolated and identified from leaves of drought-stressed *Lyeopersieon ehilense* plants. |
|  | 1997 | 5700506 | A method by which the shelf life of fresh tomato pieces can be substantially increased. |
|  | 1996 | 5585542 | An isolated DNA sequence encodes at least part of the tomato enzyme endopolygalacturonase PGI beta-subunit, which may be used to produce genetically engineered tomato plants with modified ripening characteristics. |
|  | 1996 | 5495071 | A method for producing genetically transformed plants exhibiting toxicity to Coleopteran insects. |
|  | 1996 | 5489745 | Novel tomato lines having disease resistance, and producing a fruits having a weight of at least 140 g with high pigment and reduced blossom end scar. |

| Vegetables | Year | Patent Number | Brief Description about Invention (Product/Process) |
|---|---|---|---|
| | 1996 | 5536653 | Promotes isolated from potato, which cause expression of a gene of choice in tomato; tomato plant cells and plants containing them. |
| | 1996 | 5569831 | Methods of creating transgenic tomatoes containing a lowered level polygalacturonase (beta-subunit) isoform1. |
| | 1996 | 4835339 | A method for forming haploid male sterile tomato plants and doubled haploid true breeding lines. |
| | 1995 | 479384 | A process for the inhibition of the production of a gene product in a plant cell. |
| | 1995 | 479384 | Fertile hybrid seed or hybrid seed comprising fertile and sterile seed using male-sterile plants created by employing molecular techniques to manipulate anti-sense gene and other gene that are capable of controlling the production of fertile pollen in plants. |
| | 1995 | 5387757 | Fruit especially tomatoes, having lowered expression of fruit-softening enzymes caused by antisense gene expression. |
| | 1995 | - | Chimeric isopotentyl transferase (*ipt*) gene constructs were prepared and introduced into tomato plants via Agrobacterium mediated transformation. |
| | 1993 | 5254800 | DNA constructs comprise a DNA sequence homologous to some of the gene encoded by the clone pTOM36. Fruit from the transformed plants are expected to have modified ripening properties. |
| | 1991 | 5073676 | The antisense mRNA produced delays softening of fruit, in particular tomatoes. |
| | 1990 | 509673 | Anew tomato cultivar with homozygous recessive genetic factor, which confers the ability to bear fruit that accumulate sucrose. |
| | 1990 | 4940839 | A method for producing a hybrid plant of a wild species and a cultivated species of *Lycopersicon* using protoplast fusion of the wild and cultivated species. |
| | 1980 | PP4539 | A new hybrid tomato plant characterized as novel when compared to Walter, the most similar variety to it, by uniform crop characteristics, low grading loss, early maturity and multiple disease resistance and tolerance. |
| | 1974 | 3826851 | Process for enhancing fresh tomato flavor in tomato products. |

U.S.A. and Japan. The European Commission proposed in 1988 a Council Directive on the Legal Protection of biotechnological inventions. The various production process and the products obtained from them are protected by patents in all countries, including India.

*"The processes used for genetic modification of various organisms are patentable".*

Biotechnological inventions involve following life forms:

(i) Various methods/processes of generating useful biotechnological products.

(ii) Various biotechnological products, *e.g.* antibiotics, purified vitamins, *etc.*

(iii) Application of the various processes/products like application of a bio control agent to manage a pest, *etc.*

(iv) Various microorganisms, cell lines, plant lines obtained through biotechnological approaches.

(v) Various DNA sequences and the protein encoded, if any, by them.

(vi) Biotechnological processes/technologies for modification of properties of various organisms.

## Types of Patent Allowed to Biotechnological Inventions

### (i) Genes and DNA Sequences

An artificially synthesized gene is patentable inn almost all countries. Genes isolated from natural organisms was controversial earlier but now patents are allowed on such genes in USA. Under this, *aroA* gene was isolated from a mutant bacterial strain and was transferred into plants to confer glyphosate resistance. Seeing its practical utility means 'useful' trait, patent was granted to Calgene, Inc., USA. Court in U.K. viewed that natural genes are not patentable but European parliament approved provisions for patenting of DNA sequence where a use or technological process is specified.

### (ii) Gene Patent and Genetic Resources

The developing countries are technology poor, but gene rich. In contrast, developed countries are technology rich, but gene poor. For example, not a single crop of significance grown in USA. has originated there but developing countries having limited financial capabilities usually weak infrastructure and often a misplaced sense of social and ethical values. In contrast developed countries are strong financially, well equipped infrastructural setup and in general have a society responsive to the challenges of changing world agriculture/technology situations. The developed countries have made extensive collections of germplasm of all important crops, conserved and characterized them and are now developing them for the development of new improved cultivars. However, some developing countries have made adequate efforts to collect, conserve and characterize germplasm of their own crops, and fewer have made serious efforts for collections from elsewhere. Under these circumstances, many developing countries may virtually depend for their germplasm

**Table 14.6: Patenting Offices and Views**

| Patenting Office | Remarks |
| --- | --- |
| European Patent Office (EPO) | EPO suggested that isolation of substance from nature is merely a 'discovery' and therefore, should not be patentable. However, the process developed for the isolation of this product is patentable. But if the substance is characterized and is found to be 'new' having no previously recognized existence, the substance *per se* should be patentable. |
| U.S. Patents and Trademark Office (USPTO) | USPTO views that natural products were not patentable but in 1970, the US Court of Customs and Appeals clarified that although a natural product *per se* is not patentable, a new 'form' or 'composition' of product can be patented. |

supply on rthe developed countries, the price tag of which is likely to match the need. Germplasm is a common heritage of humanity and every nation must have foresightness for collection and conservation for posterity. Many argue that patents should not be allowed for germplasm, gene sequence and genes, and some suggest banning of patent on life forms. It has been proposed that only 'the technology' to engineer gene sequence and the technical scientific knowledge should be patented. The use for genetic transformed (germplasm collection made by developing countries and patenting of useful genes isolated from developed countries) such a gene by any one may be prohibited or, at the least could carry a suitable fee.

## Achievements

MAHYCO is the first Indian company to have received the rights under license for the use of the Bt *cry1Ac* gene technology for insect pest management from Monsanto company. This licensed *cry*-gene technology was used by MAHYCO to develop and generate eggplant hybrids. The *cry*-gene technology was licensed to several public institutes in south and South Asia that were participating in a public-private consortium created to develop eggplant shoot and fruit borer resistant (EFSB) open pollinated varieties of eggplant that would improve the conditions of resource constrained farmers in developing countries. The Agricultural biotechnology Support Project II (ABSP II) played a pivotal role in this venture by funding all the consortium partners for their R&D roles in developing the EFSB resistant eggplant. The technology was sublicensed by MAHYCO on a royalty free basis to public research institutes in India (IIVR, Varanasi; TNAU, Tamil Nadu; UAS, Dharwad, Karnataka); in Bangladesh (Bangladesh Agricultural Research Institute), and in the Philippines (University of Philippines, Los Banos). MAHYCO also sub-linceed this technology to East West Seeds, a private corporation in Bangladesh, on commercial royalty bearing terms (Medakkar and Vijayaraghavan, 2007).

## International Harmonization of Patent Laws

The first concrete effort by developed country was made to harmonize patent laws of different countries under Paris Convention for the Protection of Industrial Property signed in 1983. The Paris Convention has 100 member states; India has joined the *Paris Convention on December 7, 1998*. The provision of Paris and subsequent conventions on IPR are administered but not enforced by the World Intellectual Property Organization (WIPO), Geneva, asking member states to ratify a convention and to introduce the agreed basic principles into their national laws. The European Patent Convention (EPC) began to operate in 1978 and has 17 member states. EPC was the first to introduce specific provisions for biotechnology inventions, including:

(1) The need for depositing cultures of microorganisms, for which patents are sought, and

(2) Exclusion of plant varieties bred through classical methods from patent coverage.

## Copyright

It is true that certain intellectual properties are not patentable but they can be protected under copyright like books, audio and video cassettes, style or form of expression of ideas, *etc.* The creativity could be arrangement of words, musical notes, colours, *etc.*, also. Creations of writers, authors, artists, composers are given protection under this act. The term of copyright for published work is lifetime of author plus 60 years. In case of plant biotechnology, copyright protection is available for DNA sequences (Chawla, 2002). But one may get around this protection by designing an alternative sequence to encode the same protein taking advantage of degeneracy of genetic code.

## Trademarks

Trademarks are the symbol; numbers, designs, slogan *etc.* are the identification marker of a company and also provide opportunity to consumer to identify the products of his choice. The purpose of trademarks is to differentiate similar business activities from other enterprises. Thus by providing quality service to customers, a company can make a benchmark in its business. Salient features of trademarks are:

(i) They are generally endowed with commercial products

(ii) They are company or institution specific

(iii) They may be product specific

(iv) It is binding on company to use registered trademark.

## Conservation of Vegetable Biodiversity through Geographical Indications (GIs)

The global validation has been provided for the protection of Geographical Indications (GIs) as defined in Art. 22:1 'Agreement on Trade Related Intellectual Property' (TRIPs), aspects. Prior to the TRIPs agreement, there were three main instruments (convention and agreement) regulating indications of geographical origin:

**Table 14.7: Agreement on Trade Related Intellectual Property (TRIP)**

| Convention and Agreement | Remarks |
|---|---|
| Paris Convention | The Paris Convention for the Protection of Industrial Property |
| Madrid Agreement | The Madrid Agreement for the repression of False o Deceptive Indications of Source on goods |
| Lisbon Agreement | The Lisbon Agreement for the Protection of Appellations of Origin and their International Registration |

Since the emergence of the TRIPs Agreement, the concept of protection of geographical indications has been added as multilateral protection mechanism, eventually leading to the development of national protection system in India. The legal provisions are in complete coherence with TRIPs with certain additional features.

Geographical Indications may be defined as:

*"Geographical Indications are, indications which identify a good originating in the territory of a Member, or a region or locality in that territory, where a given quality, reputation or other characteristic of the goods is essentially attributed to its geographic origin".*

In India, the process of protection of geographical indications was initiated with the inception of the legislative Act called "The Geographical Indications of Goods (Registration and Protection) Act 1999" that came into force in September 2003. This Act provides procedural requirements for registration of GIs, provisions of negative protection, and protection of homonymous indications and additional protection. Geographical indications must be seen in the light of provisions of trademark; therefore, in 1999, the new law called "The Trademarks Act 1999" also came into existence (Soam *et al.*, 2007).

## What is GI in Terms of Vegetable Diversity?

Geographical location brings out reputation to certain goods (buiological and non-biological) and also to services such as "Njavarakizhi" massage or therapy in Kerala but services are not object of GI protection system. In case of agro-biodiversity, there are several biological resources as in vegetable which have specific characteristics and thus attain a particular reputation of being associated with a certain geographical location or region. For example: 'Ramnagar Giant' Brinjal' (region-Varanasi-U.P.), 'Cuttack Pindi Brinjal' (region-Cuttack-Odisha.), Bhoot Jholkia/Naga Jholkia/King Chilli (region-NEH) 'Jaunpuri Mooli' radish'(region-Jaunpur-U.P.), expressed geographical location or region. There are certain other GIs have got reputation due to human interventions of development of localized specific human skill. Thus it is seen that associated of Geographical location is always there. In many cases, this association can be attributed to the greater role played by the locally biological resources in the development of a product of high reputation *e.g.* in the production of "Kalamkari" of Machilipatnam and Kalahasti region in Andhra Pradesh where various parts of the locally available plants are used as vegetable dyes in this art; earlier the cloth used for purpose was also locally produced (Soam *et al.*, 2007).

## Implication of Geographical Indications (GIs)

The GI in agro-biodiversity can be used mainly for the following products:

(i) Asexually and sexually propagating material

(ii) Food grains or economically important produce of various types

(iii) Value added products such as flour (potato), juices (bottle gourd, bitter gourd, watermelon *etc.*) or dairy products and other similar products.

(iv) Live plants (various colour of pumpkins and chilies) or flower plants from 'Kullu valley' (Himachal Pradesh)

(v) Live animals available in the farming systems

(vi) Agricultural implements and agro-chemicals

(vii) Rural niche products such as fermented bamboo shoots, vinegar *etc.*

(vii) Finished products from wood *etc.*, such as furniture and sports material.

Vegetable geographical indications covered Nasik for onion and Guntur for Chilli.

## Key Stakeholders in GI Protection System

As per section 2(1) (n) of GI Act association of producer or persons or any organization may be registered in part A of register as the proprietors of a GI. The GI Act has provision for three types of owners:

(i) Proprietor

(ii) Producer and

(iii) Authorized users

The group of producers can be the proprietors, and the non-producers like any NGO or government department can also be the proprietors. It thus, means that all proprietors are not producer. Some of the producers can be authorized users (those who join an association after registration of GI), and a non-producer can never be an authorized user.

## GI Act 1999 and the Biological Diversity Act 2002

There is no direct relationship between these two Acts. The major differences and issues of critical observations are that the basic objectives of both the acts are different, whereas GI Act tries to retain the basic components of GI through limiting its local content; Biological Diversity Act (BD Act) allows the exchange/transfer of biological resources as per provisions. In GI, there is no scope of transfer/exchange of resources (inputs) from other areas, regions or countries, but there is possibility of improving or modernizing a GI through incorporating an external input, but in that case, a GI loses its local content and thus significance also. Biological Diversity Act covers genetic resources or germplasm but not goods as Section 40 mentions clearly about normally traded commodities "Notwithstanding anything contained in this act, the Central Government may, in consultation with the National Biodiversity Authority, by notification in the Official Gazette, declare that the provisions of this act shall not apply to many items, including biological resources normally traded as commodities". Therefore, before registering, some land races or farmers varieties must be characterized, and IPR on genetic resources and associated knowledge must be protected before registering them as GI. There is no because once these registered as GI, these qualify to be normally traded commodities. Therefore, germplasm can be exported without any intervention of National Biodiversity Authority (NBA) as Section 19 of Biological Diversity Act 2002 would not be applicable. A few examples in this category may be 'Ramnagar Giant' Brinjal' (region-Varanasi-U.P.), 'Cuttack Pindi' Brinjal (region-Cuttack-Odisha.), Bhoot Jholkia/Naga Jholkia/King Chilli (region-NEH), 'Jaunpuri Mooli' radish' (region-Jaunpur-U.P.), expressed geographical location or region.

As per class 31 of Fourth Schedule of GI Rules 2002, seeds and natural plants which can be registered as GI, mean a 'Good'. On this issue there is no relationship between GI Act, BD Act and PPV and FR Act. A particular situation may arise where intervention of NBA is essential with respect to export of a GI product which normally traded as commodity *e.g.* Bhoot Jholkia/Naga Jholkia/King Chilli (region-NEH), Ramnagar Giant' Brinjal' (region-Varanasi-U.P.) *etc.*, because these goods may have wonderful genetic material (Soam *et al.*, 2007).

The approaches of rural livelihood and conservation and sustainable utilization of genetic resources are different in GI and BD Act. In BD Act, the benefit sharing is a regulatory mechanism. National Biodiversity

Authority (NBA) determines the benefit sharing (amount) and manner as provided under section 21 (2). In GI Act, there is no such arrangement of benefit sharing but the basic objectives of the registration of a GI are protection of consumers interests, and enhancement of producers profitability through restricting non-qualified product and also an increase in the market revenue. Some amount of benefit so accrued as registered GI can be used for conservation of biological resources, the intention here being "profit sharing" for a particular cause. Therefore, marketing associations, producers association, authorized users and proprietors of a registered GI must ensure the profit sharing for conserving biological resources (Soam *et al.*, 2007).

As per Biological Diversity Act section 41(3), the local biodiversity management committee may levy the charges by way of collection fees for accessing or collecting biological resources for commercial purposes. The fee for the registered GI may be charged on the basis of the present commercial potential of a GI and benefit accrual in future. This would not be possible without a facilitator. Therefore government, non-government or community-based organizations must function as a link between the GI registry and a biodiversity management committee (Soam *et al.*, 2007).

## GI Act and Protection of Plant Varieties and Farmers Right Act 2001

A Plant Varieties and Farmer's Rights Protection Authority has been established which will undertake registration of extant and new plant varieties through the Plant Varieties Registry on the basis of varietal characteristics. Under PPVFR Act, there is no mention of a geographical indication except at one place *i.e.* section 15(4) (viii), which states that a new variety shall not be registered under this Act if the, denomination given to such variety, "is comprised of solely or partly of geographical name: provided that the Registrar may register a variety. The denomination of which comprises solely or partly of geographical name, if he considers that use of such denomination in respect of such variety is an honest use under the circumstances of the case". For verification of circumstances of the case, there is no linkage between Geographical Indications Act and Protection of Plant Varieties and Farmers Right Act. The protections of plant variety and farmers rights authority have to keep a vigil so that no GI registered as a variety. In both the Acts, there is a provision of registration (Soam *et al.*, 2007). Therefore plant variety registry and GI have been constituted. The trademark Act 1999 also touches PPV and FR Act and Geographical Indication Act because there can be a situation where a registered geographical indication is a registered variety also, and it may have a trademark also. The best form of closeness of these two legal mechanisms would occur when farmer's variety/land races/folk varieties are registered variety and these fetch good market due to their reputation as a registered GI. The registration of geographical indication may be of various products from a single variety *e.g.* propagating material *i.e.* seed, economic yield *i.e.* grain or fruits (horticultural crops) and value added product *i.e.* flour, flakes or juice. The patent of a variety is not available in India but it is there in some other countries. Therefore, if option is the patent then registry of GI must be discouraged. In some cases, in India patenting is possible for a variety *e.g.* mushrooms and if any GI is available in mushrooms, it must preferably be registered after the patent of the variety (Soam *et al.*, 2007).

## Geographical Indication Act and Patent Act, 1970 and Amendments up to 2005

It is obvious as patents protect the individual IPRs, while geographical indication protects the community IPRs or a public good. Further, patents envisage monopoly over invention and restrict others to use invented technology without proper authority; geographical indication on the other hand, invites and includes other producers to produce the goods using the registered method of production, maintaining the standards of the geographical indication for particular characteristics which thus contribute to its reputation. But unlike patents, the goods can never be produced and sold by the persons who are not authorized patents, the goods can never be produced and sold by the persons who are not authorized users as per Act, by those registered as proprietors or so authorized cannot authorize anyone to produce a good on their behalf. The goods cannot be produced outside the traditional producing area and also registered for the purpose. The registered geographical indication functions as negative protection for the traditional knowledge and IPR on genetic resources. As a process of preparation of document for geographical indication registry, the special characteristics, contributing factors, history of the product and associated traditional knowledge are function for prior art search, thus protecting the IPR by placing them to public domain (Soam *et al.*, 2007).

## Geographical Indication Act and Trademark Act 1999

In agreement with TRIPs provisions have been made in Geographical Indication Act in some circumstances Geographical Indication holds supremacy over trademark and *vice versa* in other situations, and in some cases both Geographical Indication and Trademark can co-exist. In Indian context, Soam (2005) has given a detailed analysis of relationship of geographical indication with trademark on the given situation of supremacy of geographical indication or trademark or other co-existence. Supremacy to geographical indication over trademark is accorded under two situations:

(1) If goods not originating in a geographical region uses a GI as trademark in such a way as to confuse or mislead the persons as the true place of origin of such goods [Section 25(a) of GI Act];

(2) If goods have been provided additional protection under Section 25(b), trademark such as "Champagne fruit juice" or bottle gourd or bitter gourd juice can be permitted in India.

Supremacy to trademark over geographical indication is accorded if trademark contains or consists of GI and has been applied or registered in good faith, or acquired through use in good faith before commencement of GI Act or before filing of GI registration application ]Section 26(1) of GI Act]. Trademark prevails over GI in certain legal situations pursuant to Section 26(4) of GI Act, which provides that:

*" No action in connection with use or registration of trademark shall be taken after the expiry of 5 years from date on which such use or registration infringes any GI registered under this Act become known to registered proprietor or authorized user registered in respect of such GI under this Act (it means for opposition or infringement action one has to be either proprietor or authorized user) or after the datef registration if such date is earlier than the date on which such infringement became known to such proprietor or authorized user and such GI is not used or registered in bad faith".*

The coexistence of both trademark and geographical indication is possible, when there is no ground of prohibition of registration of GI Section 9 of GI Act; or under provisions of Section 26(1), 26(2) and 26(3) of GI Act in accordance with Art. 24.5, 24.6 and 24.8 of TRIPs, respectively (Soam *et al.*, 2007).

The application of GI can be addressed: Geographical Indications Registry, Intellectual Property Office Building, Industrial Estate, G.S.T. Road, Guindy, Chennai-600 032; Ph: 044-22502091-93 and 98; Fax: 044-22502090; e-mail: gir_ipo@nic.in; website: www. ipindia.gov.in

## Trade Secrets

"When the individual/organization owning an intellectual property does not disclose the property to any one and keeps it a secret to make profit, it is called as a trade secret" (Singh, 2005). It may be a formula, or a process. Trade secrets are private proprietary information and provide a fair edge to a company or institution over other involved in the same business. in vegetable seed industry this is a common practice for protection of intellectual property in the form of trade secret of parental lines of hybrid seeds. Since exact characteristics of the parental lines of a hybrid cross are difficult for others to ascertain the owner of the hybrid may maintain the parental lines as trade secret. The disadvantage with this kind of protection is the maintaining the secrecy, and if the secret is out to public then no more protection is available. In addition, trade secret cannot be challenged in the court of laws.

**Table 14.8a: Registration Details of GI in Vegetable Crops**

| Sl.No. | Application No. | Geographical Indications | Goods [As per Sec. 2(f) of GI Act (1999)] | Year | State |
|--------|-----------------|--------------------------|-------------------------------------------|------|-------|
| 1 | 109 | Naga Mircha | Agricultural | 2008-2009 | Nagaland |
| 2 | 143 | Guntun Sannam Chilli | Agricultural | 2010-2011 | Andhra Pradesh |
| 3 | 129 | Byadagi Chilli | Agricultural | 2010-2011 | Karnataka |
| 4 | 199 | Udupi Mattu Brinjal | Agricultural | 2011-2012 | Karnataka |
| 5 | 212 | Bangalore Rose Onion | Agricultural | – | Karnataka |
| 6 | 374 | Naga Tree Tomato | Agricultural | – | Nagaland |
| 7 | 377 | Mizo Chilli | Agricultural | – | Mizoram |
| 8 | 435 | Assam Karbi Anglong Ginger | Agricultural | – | Asom |
| 9 | 471 | Waigaon Turmeric | Agricultural | – | Maharashtra |
| 10 | 473 | Bhiwapur Chilli | Agricultural | – | Maharashtra |
| 11 | 491 | Lasalgaon Onion | Agricultural | – | Maharashtra |
| 12 | 501 | Jalgaon Bharit Brinjal | Agricultural | – | Maharashtra |

**Table 14.8b: Different Types of Stakeholders and their Importance**

| Stakeholders | Details |
|--------------|---------|
| Stakeholders 'A' | Stakeholders in 'A' are of low influence but of high importance; special initiatives are required to protect their interest (4, 6), to involve them in GI protection system (14, 16). State governments would play a major role in defining the geographical boundaries and conflict resolution especially when the matters are subject of more than one state. |
| Stakeholders 'B' | Stakeholders in 'B' have high influence and are also of high importance in order to get GI registered and to fetch benefits out of it, and construct good working relationship with them. Agricultural Products Export Development Authority (APEDA) and Marine Products Development Authority (MPEDA) are the central government organizations and can function as key players in developing GI systems. |
| Stakeholders 'C' | Stakeholders in 'C' appear to be of low influence and low importance, through unlikely to be the subject of GI project activities and management but required limited monitoring and evaluation from involvement point of view. Unlike patents, GI is not a monopolistic instrument. Therefore, some stakeholders may not be interested in GI registration process but may be interested in fulfilling social commitments (2) profit earning through registered GI (10) or shattering the GI registration (17) if GI product is competitive to their own product. Some stakeholders (5, 18, 19) are not direct parties but would be interested in GI registration for their own benefits; others such as Self-Help Groups (SHGs) and Development of Women and Children in Rural Areas (DWACRA) groups are very efficient community groups in some states. |
| Stakeholders 'D' | Stakeholders in 'D' are of high influence and can therefore; affect project output, but their interests are not the target of the GI registration project. They need careful involvement and management; otherwise there shall be significant risk in GI registration. In this column, there can be various kinds of stakeholders, who are immense importance: <br><br>1. First kind of stakeholders are those who get reward or remuneration through facilitating the registration of a GI or enhancement of trade of a registered GI e.g. attorneys and marketing intelligence consultants. <br><br>2. Second kinds of stakeholders are those who would like to spoil the efforts of registration of a GI. They may be opposition parties whose business is going to be affected after registration of a GI, or those groups who would oppose the registration of a GI on the pretext that the GI is a generic product. <br><br>3. Third type of stakeholders in this category are those who are not directly or indirectly involved with the product or registration but can be of very high influence in enhancing the trade, including export, once the product has the GI registration mark. |

*Source*: Soam *et al.*, 2007.

Trade secrets are mostly protected via an agreement between employer and employee. In case if employees violate the agreement, he can be penalized by court laws.

Some characteristics of trade secrets are:

(i) Trade secrets have unlimited duration of protection or till they become publically known.

(ii) It is not obligatory on the part of company or institution to disclose the inventive or creative ideas.

## Trade Secrets in Agriculture

Trade secrets in agriculture sectors may include:

(i) Parental cross combination of a hybrid variety.

(ii) Development of transgenic cell lines

(iii) *In vitro* regeneration protocol

(iv) Customers list *etc.*

## Designs

A design manifests the shape, size, configuration, pictorial presentation or pattern of arrangement of a material. The conception of a design is an intellectual property and provided protection. In India the design Act 1911 as amended in 1999 is in force. To qualify for registration, a design must be novel, distinct and not seen previously.

## Know-Hows

*"Know-how is a process or procedural protocol leading to development of a product or formulation".*

It could be a database providing predictable results. Production technologies also come under this category. R&D institutions are often benefited out of this type of IPRs.

## Stakeholder Management and Stakeholder's Importance-Influence Matrix

In order of importance to get GI registered and to protect interests and influence the process of GI registration, and derive benefits out of registered GI, various stakeholders can be placed in four various columns (A, B, C and D) of the importance-influence matrix from the strategies management perspectives.

## Rights Empowered under the Act

### 1. Plant Breeder's Rights (PBR)

Plant varieties are generally protected through plant breeders rights (PBRs). Intellectual property protection of plant varieties is somewhat different from that of the other forms of technology because a new plant variety does not normally arise out of a single innovative step. It is rather developed in a cumulative manner in discernible stages with addition of a new advantages or genetic improvement at each level to basic genotypes which often happens to be popular locally adapted variety. Even though a new plant variety may not involve an inventive step in the strict sense, its economic usefulness to society is unquestionable. Therefore it is important to encourage development of new varieties by providing incentives such as Breeder's Rights over exclusive production and marketing of seed of an improved variety.

*"Plant Breeder's Rights are rights granted by a government to a plant breeder, originator or owner of variety to exclude others from producing or commercializing the propagating material of that variety for a minimum period of 15-20 years".*

If a person holding Plant Breeder's Rights (PBR) title to a variety can keep the commercial right with him or authorize other interested persons/organizations to produce, sale in the market, distribute, and import or export the propagating material in form of variety. The object of protection in PBR is the variety, and that genetic component and the breeding procedures are not protected. The approach is based on the notion that if a continuous prosperous commercial plant breeding

is to be encouraged then it must be rewarded through confirming some supremacy to the breeders, further strengthened (Sahai, 2003). PBR systems also contain some form of 'breeder's exemption and 'farmer's privilege' (Section 41.7.5, 41.7.6).

## The Need for Plant Breeder's Rights

Considerable points for PBR systems are:

(i) Plant Breeder's Rights (PBR) system allows breeders to benefit from the developed varieties which encourages plant breeding activities

(ii) Private sector is encouraged to invest in plant breeding and seed industry

(iii) Development of a plant variety is as much of an innovations

## History of Plant Breeder's Rights

An act to provide patents on plants was first introduced in Germany in 1866. Subsequently, during the beginning of 20th century, legislations for patenting plants were introduced in other countries; including U.S.A. The International Organisation for Plant Variety Protection (ASSINSEL) was established in 1938 with the objective of persuading governments of different countries for introducing laws to protect plant varieties. For example Europe, developed their own system of Plant Breeder's Rights (PBR). The most significant event in the development of PBRsystem was the effort to harmonize PBR laws of different countries through UPOV (Union Internationale pour la Protection des Obstentions Vegetales, International Union for Protection of New Plant Varieties). The first UPOV convention was signed in 1961 in Paris; in 1993, UPOV had 24 member states, *e.g.* Australia, Belgium, Canada, France, Germany, Israel, Italy, Japan, South Africa, Switzerland, U.K., U.S.A. *etc.* the member states of UPOV adopt PBR systems conforming to the broad framework agreed upon in the convention. In addition, nationals of one member state have rights in other member countries. In India, mostly agricultural research works are being carried out by public sector institutions (Agriculture Universities and ICAR Institutes/Centres) and private sector is yet to emerge and progress. In other countries, multinational corporations like Monsanto, U.S.A., an up word surge in the activities may be anticipated.

## Advantages from Plant Breeder's Rights

1. The opportunity to breeders for obtaining profits from varieties developed by them will act as an incentive in promoting plant breeding research.

2. A PBR system encourages private companies to invest in plant breeding activities.

3. It will enable access to varieties developed in other countries and protected by IPR laws.

4. It will encourage competition among various

organizations engaged in plant breeding, which is likely to be beneficial to both the farmers and the nation.

## Disadvantages from Plant Breeder's Rights

1. PBR will encourage monopolies in genetic material for specific traits.
2. It suppresses free exchange of genetic material and may encourage unhealthy practices.
3. The holder of PBR-title may produce less seed than the demand in order to increase prices for achieving more profit
4. Farmer's privilege to re-sow the seed produced by him may become gradually diluted/eliminated.
5. PBR may result in increased cost of seed, which will be burdensome to the poor farmers of India, and would confine the benefits from new varieties to a small segment of rich farmers.

## Requirement of Plant Breeder's Rights

To fulfill the requirements under the provisions of UPOV 1991 Act, a plant variety must satisfy the following four criteria for protection

(1) **Novelty:** A variety should be novel and that a variety should not have been commercially exploited for more than one year before the grant of PBR protection.

(2) **Distinctiveness:** The new variety must be distinguishable from other varieties by one or more identifiable morphological, physiological or other characteristics.

(3) **Uniformity:** The new variety must be uniform in appearance under the specified environment of its adaptation.

(4) **Stability:** The new variety must be stable in appearance and its clone characteristics over successive generation under the specified environment to satisfy the criterion of stability.

## The Extent of Protection of Plant Breeder's Rights

The provisions of UPOV 1991 Act offer the following protections to the concerned variety:

(i) Production for commercial purpose, offering for sale and selling all material becomes the exclusive right of the holder of the PBR title.

(ii) A grower may be allowed to reserve a portion of his harvest for use as seed for next generation without the permission of the holder of the PBR title.

(iii) Exchange of propagating material of different cultivars between farmers is not allowed.

(iv) The minimum period of protection is 20 years but many UPOV member states have established a longer period of protection of 20-25 years, with 30 years in France for inbred lines of maize, and for clovers and a few grasses.

(v) The use of propagating material from a protected cultivar for scientific purpose is not dependent on permission of the holder of the PBR title.

(vi) The use of protected cultivar for the creation of genetic variability for plant breeding purposes does not require permission from the holder of the PBR title.

(vii) PBR protection does not cover breeding methods.

(viii) PBR protection covers the new variety, but does not protect the parents of the variety, except in the case of hybrid varieties.

Asian countries should evolve their own system of PBR (Ghijsen, 1998):

(i) That recognizes community interests, *e.g.*, the informal seed system of open pollinated varieties

(ii) That extends, in public interest, the concept of "essentially derived" variety to varieties developed from unprotected varieties

Now India has its own PBR law (PPVFR Act, 2001).

The International Convention of the Protection of New Variety (UPOV) which administers PBR across a number of developed member countries revised its convention to strengthen PBR in 1991.

## Breeder's Exemption

The use of PBR protected variety (*the initial variety*) for the development of new varieties is permitted but PBR for the new varieties will be of the breeder who developed. The holder of the PBR title of the initial variety will have no claim to their PBR titles and this provision is called breeder's exemption. Under the UPOV 1978 Act, all new varieties evolved using protected varieties were exempted from protection under this provision. Previously available breeder's exemption was eliminated in the case of an "essentially derived variety."

## Farmer's Privilege

Plant Breeder's Right system generally allows farmers to use the developed and protected variety on their farm for planting of their new crop without any obligation to the PBR holder but PBR does not allow farmers to exchange seeds of protected varieties produced on their farms. This exemption is usually referred to as 'Farmer's privilege' which was also given under the UPOV 1978 Act but in the UPOV 1991 Act this privilege has been made 'optional' and each UPOV member state can either allow or disallow this privilege.

## Penalties for Infringement

Violation of breeder's rights at any level leads to infringement. It may be established in any of the following means:

(i) Use of false denomination

(ii) Use of false name of originating country or place

(iii) Duplication of packing (Similar looking package)

(iv) Use of registered variety by unauthorized person or company

Penalties vary depending upon the extent of damage caused. It includes both fine and imprisonment. Thus, wrong use, possession or marketing of a registered variety by anyone is punishable with imprisonment ranging from three months to three years and it also attracts fine ranging from Rs. 50,000 to 20.0 lakhs. As per the decision of authority either or both of them can be imposed upon the person or company. Protect holders of PBR from appropriation by other breeders, especially biotechnology companies.

## 2. Researcher's Rights

The researcher's rights have been clearly described. In this provision of plant variety protection are:

(i) The registered varieties can be used for research or developing new varieties

(ii) Breeder's authorization is needed if a registered variety is to be used respectively in the development of new variety.

(iii) As per the Indian Act, breeder's authorization is needed for developing Essentially Derived Varieties (EDVs)

Essentially Derived Varieties (EDVs) encompasses natural selection, mutant selection, somaclonal variants, backcrosses and transformation by genetic engineering, in other words, all methods of creating new varieties will be covered. This reflects complete control of breeders on promising germplasm leaving a little scope for other researchers to undertake genetic improvement of a crop.

## 3. Farmer's Rights

On earth, agriculture started long back (probably 10,000 years ago) and stepping with up and down, natural resources played vital role in survival of human and animals. According to he needs, useful genetic resources were collected, selected, developed and conserved by farmers/farming community knowingly/ unknowingly for posterity. These raw materials become source in evolving the modern high yielding varieties of various crops. Seed sales/trades of these improved varieties earn huge profits for the seed corporations. At various forums, discussions go on that the farmers should be allowed a share in this profit in recognition of their contribution by way of development of the germplasm of crops. FAO recognized the facts (Resolution No. 5/89) and put forth 'Farmer's rights', which arise from the past, present and future contributions of farmers in conserving, improving and making available plant genetic resources, particularly in the centres of origin/ diversity. The resolution stands adopted and endorsed by all the member countries of WTO.

The resolution recognizes the following facts:

(i) Farming communities have conserved plant genetic resources over centuries.

(ii) Their selection for consumer's preferences led to development of several land races.

(iii) Indigenous knowledge systems served as blue print for scientific knowledge systems.

(iv) Farmer's varieties or land races are often endowed with enormous genetic diversity the basic raw material for modern plant breeding.

It has been emphasized that the farmer's rights should be obligatory and should not be relegated as privileges because the contributions made by them are base for all development (Sahai, 2003). The various modalities related to the farmer's right are yet to be developed and their implementation remains a far cry. So far, India is the only country to develop a system to recognize and implement 'Farmer's rights'. "The farmers shall be deemed to be entitled to save, use, sow, exchange, share or sell his farm produce including seed of a variety protected under this act in the same manner as he was entitled before the coming into force of this Act provided that the farmers shall not be entitled to sell branded seed of a variety protected under this Act.

### Important Features of Farmer's Rights

(i) **Right on seed:** A farmer is entitled to grow registered variety, save seed, sow, re-sow, exchange, share it or sale to other farmers but he will not use brand name (commercial marketing arrangement) or denomination of the variety and similar package or containers.

(ii) **Right to register their varieties:** If a farmer has bred or developed a variety (traditional varieties developed or conserved) he is authorized to get it registered in the manners similar to a plant breeder for recognition.

(iii) **Right for reward and recognition:** Reward to the farmers or farming community involved in conservation of genetic resources of crop plant (wild relatives of economic plants and their improvement through selection and preservation) in case their material so selected and preserved has been used as donor of genes in varieties registered under the Act. He shall be entitled in the prescribed manner for recognition and reward from the Gene Fund.

(iv) **Right for benefit sharing:** The Act provides for equitable sharing of the benefit earned from the new variety with farming or tribal communities contributed varieties used as parents.

(v) **Right to receive compensation for undisclosed use of traditional varieties:** If the breeder of a new variety may not disclose the correct identity of parental varieties or knowledge, under such situations, any third party (an Non-Government Organization-NGO, an individual, a government or private institution) who has a reasonable knowledge on the possible identity of the traditional varieties or knowledge used in the breeding of the new variety is eligible to prefer a claim for compensation on behalf of the concerned local or tribal community.

(vi) **Right for protection against innocent infringement:** According to this, a Court is prevented from prosecution of a farmer on charges of infringement of the Act, if the respondent farmer makes an affirmation that she/he was not aware of the legal provision deemed to have been violated by him or her at the time of such commission. This exceptional provision is provided in the Act in view of the low legal literacy of tradition-bound Indian farmers and to discourage petty legal harassment to farmers from seed companies.

(vii) **Right to get compensation for the loss suffered from the registered variety:** A breeder is bound to disclose the potential of the variety to the farmers or farming community interested in purchase of the seed of the variety. If the variety does not perform to its standard, the breeder is supposed to pay compensation if deemed fit by the authority.

(viii) **Exemption from fees:** A farmer or group of farmers is exempted to pay any fee to authority for any service under the Act.

(ix) **Disclose of terminator gene:** Breeder will have to inform that variety does not contain terminator gene.

## 4. Communities Rights

The rights of the communities as defined provide compensation for the contribution of communities in the evolution of new varieties in quantum to be determined by the PPV and FR Authority.

### Plant Variety Protection (PVP) System

Plant varieties are developed through years of painstaking and scientifically planned work. Therefore varieties should be regarded as intellectual properties of the breeders who have developed them. Many countries recognize varieties as an intellectual property and grant a protection to them through a patent or a suitable form of plant breeder's rights. Patent to plant varieties are now being granted under the provisions of this act. In U.S.A., three system of protection are available for IPR related to plants:

**Table 14.9: IPR Protection System**

| System of Protection | Remarks |
|---|---|
| The Plant Patents Act 1930 | Covers varieties of asexually propagated crops. |
| The Plant Variety Protection Act of 1970 | US version of the plant breeders rights system followed by European Union and several other countries, including India. |
| The Utility Patents Act 1985 | Cover man made industrial inventions and processes. |

As a historical background WTO, TRIPs Agreement finalized at the Uruguay round of multi-trade negotiations on April 15, 1994, made it mandatory for member nations to grant patent to all inventive processes or products including micro-organisms, plants and animals. Thus for India it become binding to provide protection to plant variety either by patenting or by a *sui generis* system.

### What is *Sui generis* System?

"*Sui generis*" means system of its own kind.

Ministry of Agriculture Govt. of India considered *sui generis* system as better option for plant variety protection. A rational for *sui generis* system needed on account of following reasons:

(i) Incorporation of equity concerns

(ii) To protect rights of farmers

(iii) To protect rights of researchers

(iv) To provide protection to genera and species

(v) To fix appropriate level and period of protection

(vi) To ensure sustainable development of agro-biodiversity

### *Advantages of Plant Variety Protection (PVP) System*

Plant Variety Protection (PVP) System in general has following merits:

(i) Plant Variety Protection (PVP) confirms ownership rights to breeder and its marketing assures a reasonable return.

(ii) Plant breeders are encouraged to take over and mission oriented research for development of varieties.

(iii) Unauthorized exploitation of varieties is prevented.

(iv) Indian farmers and Indian seed industries can have access to exotic varieties of other countries where IPR laws are in force.

(v) The process will attract investment from private sectors.

(vi) Private sector will be encouraged to establish own varietal development program. Enhancement of varietal spectrum because of better access to more genetic variability.

### Disadvantages of Plant Variety Protection (PVP) System

Plant Variety Protection (PVP) System has some demerits:

(i) Monopoly of a breeder or company on the varieties.

(ii) The holder of plant breeder rights may charge seed cost at a higher price.

(iii) Small and marginal farmers will have reduced access to quality seed of newly evolved varieties.

(iv) Threat to bio-diversity as landraces will be thrown out of cultivation.

(v) It may foster some undesirable practices because of restricted exchange of planting material.

### Objective of Plant Variety Protection (PVP) System

No doubt, Plant Variety Protection (PVP) System has provided reward to the developers of new and novel varieties. Thus, this system will encourage to breeders for more concentrate efforts on evolution of varieties. Few major objectives of this system are as under:

**(1) Encouragement for better varietal development programme-**A continuous breeding of newer and better varieties is necessary to sustain the growing human population on the earth. A legal protection to varieties ensures better return and economizes the varietal development programme of larger magnitude. This encourages both public and private sectors to invest more on variety generation technologies. Moreover, this is necessary for private sectors to make their seed production and distribution programs sustainable.

**(2) Promotion of quality seed industry-**The system promotes competition and only the fittest will survive indicating that only best varieties will dominate. Quality firms and institutions will invest for their survival. This will ensure availability of high quality seeds and plant material to Indian farmers.

**(3) Recognition to farmers on community contribution-**farmers are not only cultivators but they are also conservator and domestication of desirable genotypes. The system affirms role of farmers, rural tribal communities to a country's agro-biodiversity and advocates of benefit sharing with farmers and protects their rights.

## The Protection of Plant Varieties and Farmers Right Act 2001(PPV and FRs Act 2001)

The Protection of Plant Varieties and Farmers Rights Act, 2001 (PPV and FRs Act 2001) was passed on 9th August (53 of 2001) by the Lok Sabha is now a PVP Act under Department of Agriculture and Co-operation of Ministry of Agriculture, Government of India. The Act has come into force from 11th November, 2005 Act vide S.O. 1589 (E) in part and in complete from 19th October 2006. This act "Protection of Plant Variety and Farmer's Rights Act, 2001 was adopted to meet the obligation under Article 27(3) (b) of TRIPs Agreement, to protect rights of farmers and plant breeders and to encourage the development of new plant varieties. The Act is based on a **sui generis** system and is unique in sense that it concurrently recognizes the rights of breeders, farmers, communities and researchers. The Central government has established headquarters at New Delhi (on November 11, 2007), which began registration of plant varieties in 2007. This office publishes the Plant Variety (PV) Journal of India, which is the official gazette of the Authority.

### The Aims of PPV and FRs Act

*"An act to provide for the establishment of an effective system for protection of plant varieties, the rights of farmers and plant breeders and to encourage the development of new varieties of plants".*

The main features of this Act are briefly summarized below:

1. Registration of 'farmer's varieties', 'extant varieties' and 'new varieties' of such genera and species as notified in the Official Gazette by the Central government. "*A farmer variety is a variety that has been traditionally cultivated and evolved by farmers, or is a wild relative or land race in common knowledge of farmers*". Similarly, "*An extant variety is a notified variety or a farmer's variety that is public domain*". Registration of the extant varieties will be done within a specified period and subject to their meeting the criteria of distinctiveness, uniformity and stability.

2. A new variety shall be registered if it is meets the criteria of novelty, distinctiveness, uniformity and stability. The criterion of novelty requires a variety to be in commercial use for less than one year in India, or 4 years (6 years in case of trees and vines) outside India. The variety must be distinguishable for at least one essential characteristic from any other variety whose existence is a common knowledge in any country (distinctiveness). Essential characteristic is a heritable trait that contributes to the 'principal feature, performance or value of the plant variety'. Further, a variety in 'common knowledge' means any variety, for which an application for grant of PBR or for entering the variety in the official register of varieties has been filed in any convention country. The criteria of uniformity and stability are essential comparable to those for UPOV, 1991.

3. Any variety that involves any technology including 'gene use restriction' and 'terminator technologies', which is injurious to life or health of human beings, animals or plants shall not be registered.

4. Any variety that has been 'essentially-derived' from an 'initial variety' can be registered as a new variety. The breeder of such a variety must obtain authorization from the breeder of the initial variety since the essentially derived variety is subject to the PER of the initial variety. The definition of an *"an essentially-derived variety is comparable to that given for UPOV Act, 1991 with an additional clarification that such a variety must be distinguishable from the 'initial variety' and otherwise conform to the latter in the expression of heritable essential characteristics"*.

5. The duration of protection of the varieties will be 15 years for the extant varieties, 18 years for varieties of trees and vines and 15 years for new varieties of other crops.

6. Registration of a variety confers on the breeder of that variety or his successor or agent or licensee an exclusive right 'to produce, sell, market, distribute, import or export the variety'. Apparently, the protection is not limited to seed or propagules and extends to all materials of the protected variety.

7. The provision for researcher's right allows any person to use any registered variety for research and for creation of new varieties, except essentially derived varieties, without paying any royalty to the PBR title holder.

8. The Act recognizes the farmer's rights in the following respects:

   (i) Registration of farmer's varieties.

   (ii) Reward from the 'national gene fund' for those farmers who are "engaged in the conservation of genetic resources of land races and wild relatives of economic plants and their improvement through selection and preservation" provided that the "materials so selected and preserved have been used as donors of genes in varieties registered under this Act.

   (iii) Freedom of farmers "to save, use, sow, re-sow, exchange, share or sell" their "farm produce, including seed (except for 'brand seed') of a variety protected under this Act in the same manner" as they were "entitled before the coming into force of this Act".

   (iv) Requirement of the breeder to disclose to the farmers the expected performance of the variety under given conditions; the farmers can claim compensation if this expectation is not fulfilled.

9. The procedure for making a "claim attributable to the contribution in the evolution of any variety" and seeking reward from the 'gene fund' has been specified.

10. The central Government is to constitute a 'National Gene Fund' from the earnings of benefit sharing of registered varieties, compensations deposited in the fund, and contribution from national and international organizations. The gene fund shall be used for paying compensation to communities for their contributions to the development of a variety, for benefit sharing (as determined under the provisions of this Act) and for 'conservation and sustainable use of genetic resources' and for 'strengthening the capability of Panchayats in carrying out such conservation and sustainable use'.

11. Compulsory license may be granted after 3 years of registration of a variety if seed of the variety is not available to the public either in adequate quantity or at a reasonable price.

12. The central Government shall establish the Protection of Plant Varieties and Farmer's Rights Authority (PPV&FRA). It shall be the duty of the authority to promote the development of new varieties of plants and to protect the rights of the farmers and breeders.

   Government of India have established Protection of Plant Varieties and Farmer's Rights Authority (PPV&FRA), with the headquarters at New Delhi (on November 11, 2007), which began registration of plant varieties in 2007. This office publishes the Plant Variety (PV) Journal of India, which is the official gazette of the Authority.

13. The central Government shall establish a Plant Variety Registry for the registration of plant varieties. The registry shall maintain a 'national register of plant varieties' containing names of all registered varieties, names and addresses of their breeders and other relevant details.

14. The breeder shall be required to deposit specified quantities of seeds/propagules of the registered variety as well as its parental lines in the National Gene Bank as specified by the Protection of Plant Varieties and Farmer's Rights Authority.

15. Citizens of convention countries will have the same rights as citizens of India under the Act. A convention country is a country that is member of such an international convention for protection of plant varieties, to which India is also a member, or a country with which India has agreed to grant PBR titles to citizens of both the countries.

16. Application for registration of a variety may be made in India within 12 months from the date of application for registration of the same plant

variety made in a convention country. If such a variety is registered, the date of registration in India shall be the date of application in the convention country.

17. The rights of PBR holder "shall not be deemed infringed by a farmer who at the time of application in the convention country.

## Constitution of PPV&FRA Committee

PPV&FRA consisted of a chairperson and fifteen members. The details of Protection of Plant Varieties and Farmer's Rights Authority body are as shown in Table 14.10.

## Powers and Duties of the PPV and FR Authority

The Registrar concerned with PPV and FR Authority, shall have all the powers of a civil court for the purpose of receiving evidences, administering oaths, enforcing the attendance of witnesses, compiling the discovery and production of documents and issuing commissions for examination of witness. Duties of the authority encompasses the promotion, by such measures as it thinks fit, the encouragement for the development of new varieties of plants and to protect the rights of farmers and breeders. All this may be achieved through:

(i) The registration of plant varieties that may include:

(a) Such genera and species as notified by central government

(b) New varieties

(c) Extant varieties

(d) Farmer's varieties

(e) Varieties of common knowledge

(ii) Developing characterization and documentation of varieties registered under this Act.

(iii) Compulsory cataloguing facilities for all varieties of plants.

(iv) Ensuring the seeds of varieties registered under this act are available to the farmers and providing for compulsory licensing of such varieties if the breeder of such varieties or any other person entitled to produce such variety does not arrange for production to meet the demand in the manner as may be prescribed.

(v) Documentation, indexing and cataloging of farmers varieties.

(vi) Collecting statistics with regard to plant varieties including the contribution of any person at any time in the evolution or development of any plant variety in India or in any other country for compilation and publication.

## Progress of Plant Variety (Including vegetable crop) Registry Included for Protection

Protection of plant varieties in India was started by the Protection of Plant Varieties and Farmers' Right

### Table 14.10: Constitution of Committee

| Sl.No. | Person and Division Involved | Designation |
|---|---|---|
| 1 | 1. Dr. Nagarajan (First Chairperson)<br>2. Dr. P. L. Gautam (Second Chairperson)<br>3. Dr. R. R. Hanchinal (Third Chairperson)<br>4. Dr. K.V. Prabhu (Forth and Present Chairperson) | Chairperson (The term is 5 years) |
| 2 | The Agricultural Commissioner, Govt. of India, Department of Agriculture and Cooperation, New Delhi | Member *ex-officio* |
| 3 | The Deputy Director General (Crop Science), Indian Council of Agricultural Research, New Delhi | Member *ex-officio* |
| 4 | The Joint Secretary, Incharge of Seeds, Govt. of India, Department of Agriculture and Cooperation, New Delhi | Member *ex-officio* |
| 5 | The Horticulture Commissioner, Govt. of India, Department of Agriculture and Cooperation, New Delhi | Member *ex-officio* |
| 6 | The Director, ICAR-National Bureau of Plant Genetic Resources, New Delhi | Member *ex-officio* |
| 7 | Advisor Grade I, Govt. of India, Department of Biotechnology, New Delhi | Member *ex-officio* |
| 8 | The Joint Secretary (Conservation), Govt. of India, Ministry of Environment and Forests, New Delhi | Member *ex-officio* |
| 9 | The Joint Secretary, Legal Advisor, Govt. of India, Ministry of Law and Justice, New Delhi | Member *ex-officio* |
| 10 | One representative from a National or State Level Farmers, Organization | Member (The term is 3 years) |
| 11 | One representative from a tribal, Organization | Member (The term is 3 years) |
| 12 | One representative from Seed Industry | Member (The term is 3 years) |
| 13 | Director of Research, Punjab Agricultural University, Ludhiana | Member (The term is 3 years) |
| 14 | One representative from a National or State Level Women's Organization associated with Agricultural Universities | Member (The term is 3 years) |
| 15 | Secretary (Agriculture), Govt. of Maharashtra | Member (The term is 3 years) |
| 16 | Secretary (Agriculture), Govt. of West Bengal | Member (The term is 3 years) |

The Registrar General is the *ex-officio* member-secretary of the Authority.

**Table 14.11: Crops Covered under the PPV and FRA**

| Cereals | 8 | Rice, Bread Wheat, Maize, Sorghum, Pearl Millet, Durum wheat *Triticum durum* Desf., Dicoccum wheat *Triticum dicoccum* L, Other *Triticum* species |
|---|---|---|
| Legumes | 6 | Chickpea, Mungbean, Urid bean, Kidney bean, Lentil, Pigeon pea |
| Oil seed | 11 | Indian mustard, Karan rai, Rapeseed, Gobhi sarson, Groundnut, Soyabean, Sunflower, Safflower, Castor, Sesame, Linseed |
| Fibre Crops | 6 | Diploid Cotton (2 Species), Tetraploid Cotton(2 Species), Jute (2 Species) |
| Sugar Crop | 1 | Sugarcane |
| Spices | 4 | Black pepper, Small cardamom, Turmeric, Ginger |
| Vegetables | **9** | Brinjal, Cabbage, Cauliflower, Garlic, Okra, Onion, Potato, Tomato, Field pea |
| Flowers | 5 | Rose (Rosa spp. other than R. damascene), Chrysanthemum, Bamboo Leaf Orchid or Boat Orchid, Spray Orchid, Vanda or Blue Orchid |
| Fruit crop | 2 | Mango, Coconut |
| M and APs | 5 | Isabgol, Menthol Mint, Damask Rose, Periwinkle, Brahmi |

Authority with 12 crop species in 2007 and started receiving applications for registration and protection of eligible varieties of notified genera and species of the crops from 21 May 2007 and presently 57 crop species have been covered under PPV and FRA (Table 14.11) and the Authority is in process of developing and validating guidelines for DUS testing of more than 35 crop species at various institutions of Indian Council of Agricultural Research (ICAR), Indian Council of Forestry and Education Research (ICFRE), State Agricultural Universities (SAUs), *etc.* Some of the prioritized crops includes Apple, Pear, Almond, Walnut, Cherry, Apricot, Citrus species, Banana, Litchi, Guava, Papaya, Indian Gooseberry, Pomegranate, Indian Jujube, Pineapple, Bamboo, Teak, Shisham, Tendu, Sandal wood, Deodar, Chir, bottle gourd, bitter gourd, cucumber, pumpkin, pointed gourd, watermelon and muskmelon. The DUS test guidelines for three species of vegetables namely Bitter gourd (*Momordica charantia* L.), Bottle gourd (*Lagenaria siceraria* (Mol.) Standl.), Cucumber (*Cucumis sativus* L.) and Pumpkin (*Cucurbita moschata* Duch. ex Poir.) has already published in "Plant Variety Journal of India, Vol. 07(3), March 01, 2013".

## Varieties Registerable under the Plant Variety Act

The various categories of varieties eligible for registration under the Act are:

☆ New plant varieties

☆ Essentially derived varieties

☆ Extant varieties notified under the Seeds Act, 1966

☆ Extant varieties about which there is a common knowledge

☆ Farmer's varieties

Plant variety as per the Act is a plant grouping except microorganism within a single variety botanical taxon of the lowest known rank, which can be defined by the expression of the characteristics resulting from

a given genotype of the plant grouping, distinguished from any other plant grouping by expression of at least one of the said characteristics and considered as a unit with regard to its suitability for being propagated which remains unchanged after such propagation. It includes propagating material of a variety which may be a new, extant, transgenic, farmers' and essentially derived variety.

### 1. New Variety

It is defined by the expression of the characteristics resulting from a given genotype of a plant of that plant grouping and distinguished from any other plant grouping by the expression of at least one of the characteristics. Considered a unit with regard to its suitability for being propagated which remains unchanged after repeated propagation. A variety which conforms to the criteria of novelty *i.e.* not been sold or otherwise disposed of in India, earlier than 1 year (before that date of filing) and outside India (in case of trees and vines earlier than six years, or, in any other case, earlier than four years.)

### 2. Essentially Derived Variety (EDV)

A variety which has been predominantly derived from an initial variety or from a variety that itself is predominantly derived from such initial variety/from another variety notified under Seed Act, 1966 (protected or otherwise) and conforms to the initial variety in all aspects except for the differences which result from the act of derivation (while retaining the expression of the essential characteristics of the initial variety), yet is clearly distinguishable from such initial variety.

### 3. Extant Notified Variety

A variety which has been notified under the Seeds Act under Section 5 of the Seeds Act, 1966 (54 of 1966) and have not completed 15 years from the date of release or farmer's Variety or a Variety of Common Knowledge (VCK) a variety which has not been released and notified under the Seeds Act, 1966, have been sold or otherwise disposed off in India for more than a year from the date of filing. It should be under cultivation in a State/region/country, even as "truthfully labeled" variety, finds an entry in official list/register of varieties in any country granting PVP, including filing of an application for PBR, inclusion in a recognized publicly accessible collection, including an accession in a National/International Gene Bank and adequate description of the variety in a publication that may be considered a part of the public technical knowledge. These varieties are protected for the period of 15 year from the date of notification in Gazette.

### Farmers' Variety

A variety which has been traditionally cultivated and evolved by the farmers in their fields. This includes the landraces, folk varieties, wild relative/species or a variety about which the farmers possess the common and useful knowledge. Most of these varieties are usually developed and conserved collectively by community of farmers rather than by individual farmers. Hence, the Act recognizes that varieties conserved by communities of farmers are also eligible for registration.

As a signatory to the Trade Related Aspects of Intellectual Property Rights (TRIPs) Agreement of 1994, India was obliged to enact legislation that brought plant varieties within the general purview of intellectual property. The Protection of Plant Varieties and Farmers' Rights Act, 2001 was enacted in fulfilment of that obligation.

### Example Variety

Illustrates the DUS characteristic and/or provides the basis for ascribing the appropriate state of expression to each variety. All DUS characteristics and their states of expression included in the DUS test guidelines of a crop species are covered by the minimum total number of example varieties. The set of example varieties for a given characteristic provides information on the range of expression of the DUS characteristic.

### Reference Variety

A variety has similar grouping DUS characteristics as compared to that of candidate variety is known as reference variety. Such reference varieties are planted along with the candidate variety for comparison with the candidate variety during the conduct of DUS trial. The reference variety may be either a notified variety or a variety registered under the Act and the number of reference variety included in the DUS trial may be one or more than one.

## Standard Criteria for Protection of Plant Varieties

It is true that all varieties are not registerable means a variety must fulfill certain criteria to qualify for registration or standard protection. The criteria is similar under UPOV style of recommendation which states that

not withstanding anything a variety has to meet criteria of DUS (distinctiveness, uniformity and stability) or DHS (distinctiveness, homogeneity and stability).

## What are Non-Registrable Plant Varieties in India

All plant varieties cannot get legal protection in India. Certain plant varieties are excluded from the protection under PPV&FR Act, 2001. Any variety where prevention of commercial exploitation of such variety is necessary to protect public order or public morality or human, animal and plant life and health or to avoid serious prejudice to the environment or any varieties which has terminator, technology or any variety belonging to the species or genera which is not listed in the notification issued by the Central Government cannot be registered for the protection under the Act.

## What is DUS?

*"The character of a variety which reflects distinctiveness, uniformity and stability in a particular environment is called as DUS".*

In general DUS testing is first done for extant or obsolete varieties or varieties of common knowledge which may be of national or international standard. Subsequently other registered varieties are also used as standard. Thus documentation of all the released varieties at national or international levels assumes significance as these can be used as reference varieties under DUS testing. DUS testing comprises a comparative growing sampling and analysis of data of reference and new varieties at one or more locations for one or two years. Tests are conducted by the Authority themselves or by contracted breeders or any agency on behalf of the Govt. or by the applicant breeders themselves. Crop specific guidelines stand prepared by the PPV and FRA.

## Criteria for Granting Protection to a New Plant Variety

The criterion of 'novelty' replaces that of 'non-obvious' which is the requirement for protection of a new invention. It is used in the sense that the new variety should not have been commercially exploited for more than a year before granting protection. The new variety must possess a trait or traits that make it distinguishable from other known varieties by one or more identifiable morphological, physiological, or other characteristics.

After getting all the information about the variety and receiving DUS test fee and seed; the variety is send to DUS testing centres, located at various locations of India for testing DUS characteristics. DUS testing centre for vegetable crops have been given in Table 14.13.

After DUS testing at one or two locations (depending upon the category and type of crop) results are submitted by the respective DUS Centre to the Authority which are being analyzed for the claimed characters by the

**Table 14.12: Criteria for Granting Protection to a New Plant Variety**

| Character | Guidelines |
|---|---|
| Novelty | The criterion of 'novelty' replaces that of 'non-obvious' which is the requirement for protection of a new invention. It is used in the sense that the new variety should not have been commercially exploited for more than a year before granting protection. A variety is deemed to be novel if at the date of filing of the application for registration for protection, the propagating or harvested material of such variety has not been sold or otherwise disposed of by or with the consent of its breeder or his successor for the purposes of exploitation of such variety:<br><br>(a) In India, earlier than one year or<br><br>(b) Outside India in case of trees or vines earlier than six years or in any other case earlier than four years, before the date of filing application. |
| Distinctness | A variety is considered distinct provided it is clearly distinguishable at least by one essential characteristic from any other variety whose existence is a matter of common knowledge at the time of filing of the application. Distinctness is considered clear if it occurs even at 1 per cent probability of an error. For years variation, the combined over years distinctness (COYD) method is in vogue. Identifiable characteristics for each crop are available but some characters useful in adding identification can be added to expand the table (Article 7 of UPOV 1991). |
| Uniformity | As per Article 7 of UPOV 1991 is homogeneous manifestation of characters and deemed to be uniform if it is expressed in expected range. Variation permissible for self-pollinated and vegetatively propagated crops while relative tolerance limits for variation based standard varieties is used in cross pollinated varieties and synthetics. A variety is not considered homogenous if its variance exceeds 1.6 times the average of the varieties used for comparison. Single cross hybrids are considered at par with self-pollinated crops. Their inbreds are also tested for uniformity. For evaluation of uniformity the combined over years uniformity (COYU) method has been developed. |
| Stability | As per Article 7 of UPOV 1991 Act states that a variety shall be deemed to be stable if its relevant characters remain unchanged even after repeated cultivation or propagation. In other words, after every cycle of reproduction, the variety manifests the same phenotype. The testing for stability needs 2-3 years. Homogeneity of a variety also confers its stability. |

**Table 14.13: DUS Testing Centre for Various Vegetable Crops**

| Vegetable Crop | DUS Centre-1 | DUS Centre-2 | DUS Centre-3 |
|---|---|---|---|
| Tomato | ICAR-IIVR, Varansi (U.P.) | ICAR-IIHR, Hessaraghatta, Bengaluru (Karnataka) | |
| Brinjal | ICAR-IIVR, Varansi (U.P.) | ICAR-IIHR, Hessaraghatta, Bengaluru (Karnataka) | |
| Okra | ICAR-IIVR, Varansi (U.P.) | ICAR-IIHR, Hessaraghatta, Bengaluru (Karnataka) | |
| Cauliflower | ICAR-IIVR, Varansi (U.P.) | ICAR-IARI, Regional Station Katrain (H.P.) | |
| Cabbage | ICAR-IIVR, Varansi (U.P.) | ICAR-IARI, Regional Station Katrain (H.P.) | |
| Potato | ICAR-CPRI, Shimla (H.P.) | | ICAR-IARI, New Delhi |
| Onion | ICAR-DOGR, Pune (Maharashtra) | | |
| Garlic | ICAR-DOGR, Pune (Maharashtra) | | |
| Pea (Vegetable type) | ICAR-IIVR, Varanasi (U.P.) | ICAR-IIHR, Hessaraghatta, Bengaluru (Karnataka) | |

applicant and subjected to issue of certificate only when at least one of the essential character is present in the claimed variety as per DUS guidelines.

## Procedure of Registration of Plant Varieties

☆ Denomination given on such variety is comprised of solely or partly of geographical name.

☆ Where prevention of commercial exploitation of such variety is necessary to protect public order or public morality or human, animal and plant life and health or to avoid serious prejudice to the environment.

☆ Variety which involves any technology which is injurious to the life or health of human beings, animals or plants shall be registered under this Act, including genetic use restriction technology and terminator technology.

☆ Non-notified genera.

## Grouping and Categorization of Proposals

The data obtained with the registration proposal is studied for grouping of variety. Most distinguishable or distinct characteristics are considered for categorization in the beginning which follows comparison with list of table characteristics available crop wise. Maximum numbers of descriptors are used to facilitate grouping of varieties.

(i) In India, varieties are released through All India Coordinated Research Projects (AICRP) based on comparative yields and diseases and pest reactions.

(ii) European Economic Community demands distinctiveness, uniformity and stability (DUS) test as well as value for cultivation and use (VCU).

## Procedure of Registration of Plant Varieties

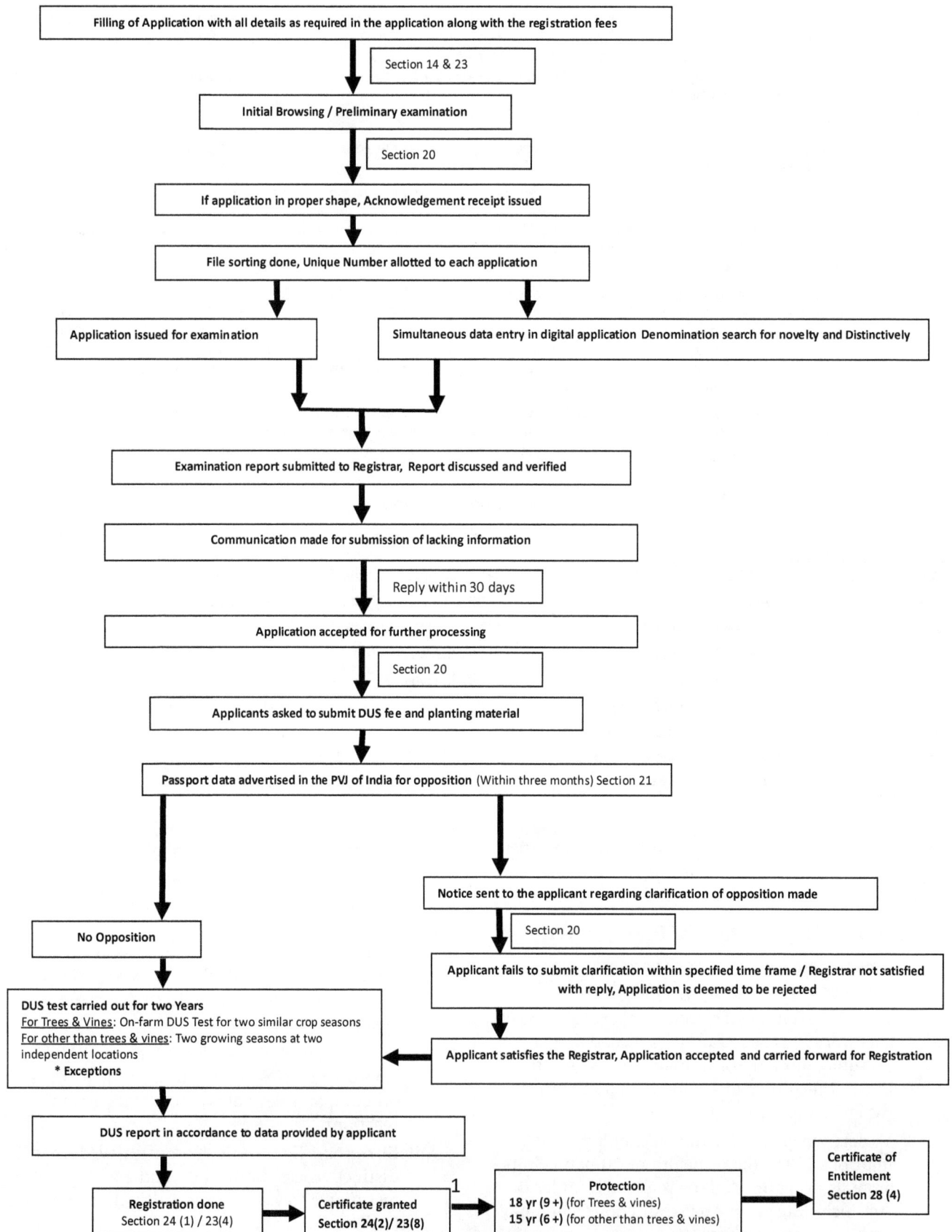

Filling of Application with all details as required in the application along with the registration fees

Section 14 & 23

Initial Browsing / Preliminary examination

Section 20

If application in proper shape, Acknowledgement receipt issued

File sorting done, Unique Number allotted to each application

Application issued for examination

Simultaneous data entry in digital application Denomination search for novelty and Distinctively

Examination report submitted to Registrar, Report discussed and verified

Communication made for submission of lacking information

Reply within 30 days

Application accepted for further processing

Section 20

Applicants asked to submit DUS fee and planting material

Passport data advertised in the PVJ of India for opposition (Within three months) Section 21

Notice sent to the applicant regarding clarification of opposition made

Section 20

No Opposition

Applicant fails to submit clarification within specified time frame / Registrar not satisfied with reply, Application is deemed to be rejected

DUS test carried out for two Years
For Trees & Vines: On-farm DUS Test for two similar crop seasons
For other than trees & vines: Two growing seasons at two independent locations
* Exceptions

Applicant satisfies the Registrar, Application accepted and carried forward for Registration

DUS report in accordance to data provided by applicant

Certificate of Entitlement Section 28 (4)

Registration done
Section 24 (1) / 23(4)

Certificate granted
Section 24(2)/ 23(8)

1

Protection
18 yr (9 +) (for Trees & vines)
15 yr (6 +) (for other than trees & vines)

(iii) In USA, the entire responsibilities of seed material and DUS testing rest with the breeder and records provided by breeder are kept at PPV office.

(iv) In Germany, both DUS and VCU tests are necessary for release of variety and conducted by a single agency Bundessortnamt (BSA).

## Who Can Apply for Variety Protection?

An application for registration of variety can be made by following claimants:

(i) Any person claiming to be the breeder of the variety or

(ii) Any successor of the breeder of the variety or

(iii) Any person being the assignee of the breeder of the variety in respect of the right to make such application or

(iv) Any farmer or group of farmers or community of farmers claiming to be the breeder of the variety or

(v) Any person authorized by a person specified under (i) to (iv) to make application on his behalf or

(vi) Any university or publically funded agricultural institution claiming to be the breeder of the variety or

(vii) An application may be made by any of the person referred as above individually or jointly with any other person.

## Form of Application

An application for registration of a variety must be filled in prescribed format. It must be in respect of a variety, must state the denomination assigned to the variety and must be accompanied by an affidavit stating that such variety does not contain any gene or gene sequence involving terminator technology. It must further contain a complete passport data of the parental lines from which the variety has been derived along with the geographical location in India from where the genetic material was taken. The applicant must also be accompanied by a statement containing a brief description of the variety bringing out its characteristics of novelty, distinctiveness, uniformity and stability. The application must contain a declaration that the genetic material or parental material acquired for breeding, covering or developing the variety has been lawfully acquired. The application for registration of a variety must be accompanied with prescribed fees and enough quantity of seeds of the said variety for the purpose of conducting DUS tests to evaluate whether seeds along with parental material conform to the specified standards. Copies of all the forms and details are available at Protection of Plant Varieties and Farmer's

Rights Authority, India site (www.plantauthority.gov. in).

## Acceptance and Advertisement of Registration Application

The Registrar may accept the application absolutely or with limitations on being satisfied about the particulars contained in such application. If the Registrar is satisfied that the application does not comply with the requirements of this Act, the application may be either rejected by him or the applicant may be required to attend the application to the satisfaction of the Registrar. However, the Registrar cannot reject the application for registration without affording an opportunity to the applicant to defend the case. Once the application for registration of a plant variety is accepted, the Registrar will advertise the application with limitations, if any, and the specifications of the variety including its photographs or drawing in the prescribed manner for calling objections from the persons interested in the matter.

## Opposition to the Acceptance

After the advertisement of acceptance, within three months from the date of the advertisement, any person, on payment of the prescribed fees, may give notice of opposition to Registration, in writing, to the Registrar. The grounds for opposition may be a person opposing the application is entitles to the breeder's right as against the applicant; or the variety is not registerable under this Act; or the grant of certificate of registration may not be in public interest; or the variety may have adverse effect on the environment. The Registrar is empowered to pass an order upholding or rejecting the application by giving reasons for the same after considering all the grounds on which the application has been opposed.

## Duration, Terms and Effect of Registration

As per PPV and FRA norm registration of variety has to be completed within 12 months from the date of application. Once the application for registration of a variety, other than an essentially derived variety, has been accepted with or without opposition, the Registrar shall register the variety; issue a certificate of registration to the applicant sealed with the seal of the Registrar. The breeder is required to deposit such quantity of seeds or propagating material of the registered variety in the National Gene Bank as may be specified for reproduction purpose at breeder's expense. The certificate of registration is valid for 9 years in case of trees and vines and 6 years in case of other crops. It may be reviewed and renewed (on payment of the prescribed fees) subject to the conditions that total period shall not exceed:

(i) 18 years in case of trees and vines from the date of registration of the variety.

(ii) 15 years in case of extant varieties from the date of notification of that variety by the Central Government under section 5 of Seed Act, 1966 (54 of 1966).

(iii) In other cases 15 years from the date of registration of the variety.

## Compulsory Licensing

The PPV and FRA has power to make order for compulsory license in relevant circumstances after 3 years of registration of a variety. Duration of compulsory licensing will vary case to case but will not exceed the limit of registration period. It is recommended only situations such as:

(i) Responsible requirements of public are not made of seeds or planting material.

(ii) Seeds or planting material of the variety is not available to the public at a reasonable price.

## Exceptions

☆ **Extant notified varieties:** Exempted from DUS testing. Seven Extant Variety Recommendation committee (EVRC) members examine the suitability for registration.

☆ **Extant Variety [(Variety of common knowledge (VCK)]** DUS testing is done for 1 growing season at two DUS testing locations with half the quantity of seeds as specified for new variety

☆ **Special tests:** Special Tests are to be conducted only when DUS test fails to establish the requirement of distinctiveness. The DUS testing shall be field and laboratory based, multi-locational for at least two crop seasons. The Authority shall charge separate fees for conducting special test. Broadly, these tests can be classified into five main groups: physical tests, biochemical tests, molecular tests, response tests and organoleptic tests.

## Fee Structure

Application for **registration** of plant varieties should be accompanied with the fee of registration notified by the Authority (new and EDV Individual- Rs 5000/, Institutional- Rs 7000/-, Commercial- Rs 10000/-); Extant variety notified under the seeds Act, 1966- Rs 1000/- and variety about which there is a common knowledge (Individual- Rs 2000/, Institutional- Rs 3000/-, Commercial- Rs 5000/) (The Gazette of India, 2008 and 2009e). No fee is to be paid by the farmer for registration of a farmers' variety. While the annual fee shall be determined on the basis of declaration given by the registered breeder or agent or licensee regarding the sales value of the seeds of the variety registered under the Act during the previous year and royalty, if any, received during the previous year from the sale proceed of seeds of the registered variety and verified by the Authority. The applicant has to submit the DUS test fee (as notified for different crops) for conducting the DUS test of the candidate variety (Table 14.14).

**Table 14.14: DUS Test Fee of Various Crops**

| Crop species | DUS Test Fees (Rs) |
|---|---|
| Wheat, Rice, Maize, Sorghum, Pearl millet, Pigeon pea, Chickpea, Lentil, Mungbean, Urid bean, Field pea, kidney bean and Oilseed crops | 20,000/- |
| Cotton, Jute, Sugarcane | 35,000/- |
| Black pepper, Small cardamom, Ginger, Turmeric, Rose, Chrysanthemum | 45,000/- |
| Tomato, Brinjal, Okra, Cabbage, Cauliflower, Onion, Garlic | 40,000/- |
| Potato | 48,000/- |
| Mango | 30,000/- |

## Issuance of Registration Certificate

The Authority has issued a total of 600 registration certificates (Upto15 March, 2013) of different crop species. Of these, 537 pertain to that of extant varieties, 56 under new, 6 under farmers' variety and 1 under EDV. The applicant wise share shows 488 certificates to Public, 106 to Private and 6 to Farmers. The Authority has also opened a National Register of "Registered Plant Varieties" having all details of the registered plant varieties and is the authentication of the plant varieties registered. As a requirement under the Act, for the purpose of benefit sharing, the authority also send a copy of the certificate of registration to the National Biodiversity authority and Indian Council of agricultural Research (The Gazette of India, 2011a).

## Maintenance of Registered Varieties

As per the Act, it is mandatory to maintain the seed samples/propagating material of the registered plant varieties up to a period of protection provided to the candidate variety and also to address the issues for intellectual property of plant breeders' rights, compulsory license *etc.* Authority has established the National Gene Bank at old campus of ICAR-National Bureau of Plant Genetic Resources (NBPGR), New Delhi for medium term storage of true samples of orthodox seed of all registered varieties for their entire period of protection. After the expiry of protection period, the seed may be submitted to ICAR-NBPGR/ any public repository (Brahmi and Choudhary, 2011). For recalcitrant varieties 3 Field Gene Bank have been developed at Dapoli (Tropical and sub-tropical crops), Ranchi (Eastern Ecosystem) and Mashobra (Temperate crops). Further 2 gene banks have been planned in Tamil Nadu (coastal ecosystem) and Rajasthan (Arid ecosystem).

## DUS in Vegetable Crops

Basic information has been developed in tomato, brinjal, chilli, bitter gourd, bottle gourd *etc.* as per the DUS guide lines. For example tomato and bottle gourd are given below for understanding the concept.

## Tomato

### *Breeding of Candidate Variety*

Parental material (name of the parental material, characteristics of the parental material, distinguishable from the candidate variety). If the variety was developed

### Table 14.15: Characteristics of Female and Male Parents

| DUS Sl.No. | Type of Assessment | Description of Parental Materials (DUS Characteristics) | Parbhani Kranti | | | VRO-3 | | |
|---|---|---|---|---|---|---|---|---|
| | | | R1 | R2 | R3 | R1 | R2 | R3 |
| 1. (*) | VG | Stem: Colour | 1 | 1 | 1 | 1 | 1 | 1 |
| 2 | VG | Stem: Intensity of green colour | 5 | 5 | 5 | 3 | 3 | 3 |
| 3.(*) (+) | VG | Leaf blade: Depth of lobing | 5 | 5 | 5 | 5 | 5 | 5 |
| 4.(*) | MS | Stem: Number of nodes at first flowering (upto and including the first flowering node) | 5 | 5 | 5 | 5 | 5 | 5 |
| | | (Average of 10 plants/replication) | 5.7 | 5.8 | 6.0 | 5.8 | 5.8 | 5.6 |
| 5.(*) | MG | Flowering: Time (50 per cent of the plants with at least one open flower) | 7 | 7 | 7 | 7 | 7 | 7 |
| | | (Average of 10 plants/replication) | 49 | 50 | 50 | 48 | 48 | 49 |
| 6 | MS | Leaf blade: Length | 7 | 7 | 7 | 5 | 5 | 5 |
| | | (Average of 10 plants/replication) | 26.25 | 27.00 | 28.00 | 18.10 | 18.27 | 17.90 |
| 7 | MS | Leaf blade: Width | 7 | 7 | 7 | 7 | 7 | 7 |
| | | (Average of 10 plants/replication) | 28.25 | 29.65 | 28.57 | 24.73 | 24.30 | 24.07 |
| 8 | VS | Leaf blade: Serration of margin | 7 | 7 | 7 | 7 | 7 | 7 |
| 9. (*) | VS | Leaf blade: Colour between veins | 1 | 1 | 1 | 1 | 1 | 1 |
| 10 | VG | Leaf blade: Intensity of colour between veins | 5 | 5 | 5 | 3 | 3 | 3 |
| 11 | VG | Vein: Colour | 2 | 2 | 2 | 2 | 2 | 2 |
| 12 | MS | Petiole: Length | 5 | 5 | 5 | 5 | 5 | 5 |
| | | (Average of 10 plants/replication) | 23.25 | 24.75 | 25.90 | 22.44 | 22.10 | 22.83 |
| 13 | VG | Flower: Petal colour | 2 | 2 | 2 | 2 | 2 | 2 |
| 14 | VG | Flower: Petal base colour (purple) | 2 | 2 | 2 | 2 | 2 | 2 |
| 15 | MS | Flower: Length (cm) | 7 | 7 | 7 | 7 | 7 | 7 |
| | | (Average of 10 plants/replication) | 5.28 | 5.50 | 5.60 | 5.90 | 5.87 | 5.80 |
| 16 | MS | Flower: Diameter(at the top of flower) | 7 | 7 | 7 | 7 | 7 | 7 |
| | | (Average of 10 plants/replication) | 5.80 | 5.80 | 5.90 | 5.57 | 5.80 | 5.70 |
| 17. (*) | VG | Fruit: Colour | 2 | 2 | 2 | 1 | 1 | 1 |
| 18 | MS | Fruit: Length (cm) at marketable stage (four days after anthesis) | 5 | 5 | 5 | 7 | 7 | 7 |
| | | (Average of 10 plants/replication) | 11.75 | 11.25 | 11.90 | 13.00 | 12.25 | 13.45 |
| 19 | MS | Fruit: Diameter (at mid length) (cm) | 7 | 7 | 7 | 7 | 7 | 7 |
| | | (Average of 10 plants/replication) | 1.60 | 1.55 | 1.65 | 1.60 | 1.60 | 1.55 |
| 20. (*)(+) | VG | Fruit: Surface between ridges | 5 | 5 | 5 | 5 | 5 | 5 |
| 21 | VG | Fruit: pubescence | 5 | 5 | 5 | 7 | 7 | 7 |
| 22.(+) | VG | Fruit: constriction of basal part | 1 | 1 | 1 | 3 | 3 | 3 |
| 23.(+) | VG | Fruit: Shape of apex | 3 | 3 | 3 | 2 | 2 | 2 |
| 24.(*) | MS | Fruit: Number of locules | 1 | 1 | 1 | 1 | 1 | 1 |
| | | (Average of 10 plants/replication) | 5.0 | 5.0 | 5.0 | 5.0 | 5.0 | 5.0 |
| 25. (*) | MS | Plant: Number of branches | 7 | 7 | 7 | 7 | 7 | 7 |

| DUS Sl.No. | Type of Assessment | Description of Parental Materials (DUS Characteristics) | Parbhani Kranti | | | VRO-3 | | |
|---|---|---|---|---|---|---|---|---|
| | | | R1 | R2 | R3 | R1 | R2 | R3 |
| | | (Average of 10 plants/replication) | 5.00 | 5.20 | 5.50 | 6.00 | 5.80 | 6.20 |
| 26 | MS | Stem: Diameter(at 10cm above ground level) (cm) | 7 | 7 | 7 | 7 | 7 | 7 |
| | | (Average of 10 plants/replication) | 2.50 | 2.40 | 2.30 | 2.70 | 2.70 | 3.00 |
| 27 | MS | Plant: Height (cm) | 7 | 7 | 7 | 7 | 7 | 7 |
| | | (Average of 10 plants/replication) | 125.25 | 124.50 | 128.00 | 145.00 | 148.00 | 145.75 |
| 28.(*) (+) | MG | Fruit: Length of physiologically mature fruit (cm) | 7 | 7 | 7 | 7 | 7 | 7 |
| | | (Average of 10 plants/replication) | 15.25 | 16.75 | 15.00 | 16.25 | 15.85 | 17.00 |
| 29 | MS | Fruit: Diameter (at mid length) (cm) | 5 | 5 | 5 | 5 | 5 | 5 |
| | | (Average of 10 plants/replication) | 1.60 | 1.70 | 1.80 | 1.63 | 1.70 | 1.70 |
| 30 | VG | Seed: Colour | 1 | 1 | 1 | 2 | 2 | 2 |
| 31 | VG | Seed : Hairiness | 1 | 1 | 1 | 1 | 1 | 1 |

by selection, then the number of selection cycles completed before fixing it.

### Characteristics of the Candidate Variety

Describe characteristics of the variety in the subheadings: Plant, Stem, Leaf, Inflorescence, Flower and flower parts, Fruit and fruit parts, Seed *etc.* Describe characters within subheadings generally in the following order: habit, height, length, width, size, shape, colour (RHS colour chart reference with edition). Refer the specific guideline wherever necessary for clarity of description.

**Table 14.16: Distinguishing Characteristics (Descriptive or elaborate)**

| DUS Sl.No. | Type of Assessment | DUS Characteristics | Kashi Vardan (VRO-25) | | |
|---|---|---|---|---|---|
| | | | R1 | R2 | R3 |
| 1. (*) | VG | Stem: Colour | 2 | 2 | 2 |
| 2 | VG | Stem: Intensity of green colour | 3 | 3 | 3 |
| 3.(*) (+) | VG | Leaf blade: Depth of lobing | 5 | 5 | 5 |
| 4.(*) | MS | Stem: Number of nodes at first flowering (upto and including the first flowering node) | 5 | 5 | 5 |
| | | (Average of 10 plants/replication) | 6.0 | 5.0 | 6.0 |
| 5.(*) | MG | Flowering: Time (50 per cent of the plants with at least one open flower) | 5 | 5 | 5 |
| | | (Average of 10 plants/replication) | 42 | 43 | 42 |
| 6 | MS | Leaf blade: Length | 5 | 5 | 5 |
| | | (Average of 10 plants/replication) | 18.70 | 19.20 | 21.80 |
| 7 | MS | Leaf blade: Width | 7 | 7 | 7 |
| | | (Average of 10 plants/replication) | 27.30 | 25.80 | 27.50 |
| 8 | VS | Leaf blade: Serration of margin | 5 | 5 | 5 |
| 9. (*) | VS | Leaf blade: Colour between veins | 1 | 1 | 1 |
| 10 | VG | Leaf blade: Intensity of colour between veins | 7 | 7 | 7 |
| 11 | VG | Vein: Colour | 1 | 1 | 1 |
| 12 | MS | Petiole: Length | 5 | 5 | 5 |
| | | (Average of 10 plants/replication) | 33.40 | 30.50 | 30.10 |
| 13 | VG | Flower: Petal colour | 1 | 1 | 1 |
| 14 | VG | Flower: Petal base colour (purple) | 2 | 2 | 2 |
| 15 | MS | Flower: Length (cm) | 5 | 5 | 5 |
| | | (Average of 10 plants/replication) | 6.30 | 5.70 | 5.80 |
| 16 | MS | Flower: Diameter(at the top of flower) | 5 | 5 | 5 |
| | | (Average of 10 plants/replication) | 6.70 | 6.50 | 7.00 |
| 17. (*) | VG | Fruit: Colour | 2 | 2 | 2 |

| DUS Sl.No. | Type of Assessment | DUS Characteristics | Kashi Vardan (VRO-25) | | |
|---|---|---|---|---|---|
| | | | R1 | R2 | R3 |
| 18 | MS | Fruit: Length (cm) at marketable stage (four days after anthesis) | 5 | 5 | 5 |
| | | (Average of 10 plants/replication) | 12.00 | 11.8 | 10.5 |
| 19 | MS | Fruit: Diameter (at mid length) (cm) | 7 | 7 | 7 |
| | | (Average of 10 plants/replication) | 1.60 | 1.55 | 1.52 |
| 20. (*)(+) | VG | Fruit: Surface between ridges | 3 | 3 | 3 |
| 21 | VG | Fruit: pubescence | 5 | 5 | 5 |
| 22.(+) | VG | Fruit: constriction of basal part | 3 | 3 | 3 |
| 23.(+) | VG | Fruit: Shape of apex | 2 | 2 | 2 |
| 24.(*) | MS | Fruit: Number of locules | 1 | 1 | 1 |
| | | (Average of 10 plants/replication) | 5.0 | 5.0 | 5.0 |
| 25. (*) | MS | Plant: Number of branches | 3 | 3 | 3 |
| | | (Average of 10 plants/replication) | 3.00 | 2.00 | 1.00 |
| 26 | MS | Stem: Diameter(at 10cm above ground level) (cm) | 7 | 7 | 7 |
| | | (Average of 10 plants/replication) | 2.00 | 2.20 | 2.00 |
| 27 | MS | Plant: Height (cm) | 5 | 5 | 5 |
| | | (Average of 10 plants/replication) | 111.80 | 118.40 | 118.50 |
| 28.(*) (+) | MG | Fruit: Length of physiologically mature fruit (cm) | 5 | 5 | 5 |
| | | (Average of 10 plants/replication) | 15 | 14.6 | 13.8 |
| 29 | MS | Fruit: Diameter (at mid length) (cm) | 5 | 5 | 5 |
| | | (Average of 10 plants/replication) | 1.50 | 1.80 | 1.70 |
| 30 | VG | Seed: Colour | 1 | 1 | 1 |
| 31 | VG | Seed : Hairiness | 1 | 1 | 1 |

## Statement of Distinctness of Candidate Variety

Please give a distinctness statement covering a brief summary of the characteristics that distinguish the candidate variety from all varieties of common knowledge. The distinctness statement should include:

(i) Names of reference variety (ies) that have been observed most similar to the candidate variety, and

(ii) Salient comparison for major distinguishing characteristics between the candidate variety and the similar/reference variety (ies).

## National Gene Fund (NGF)

The National Gene Fund (NGF) was constituted by the DAC, ministry of Agriculture, Government of India under the PPV and FR Act, 2001 and which include contributions from:

i. Benefit sharing received from the breeder of a variety or an essential derived variety registered under the PPV and FR Act, 2001.

ii. Annual fee received by PPV and FR Authority

iii. Compensation deposited and

iv. Contributions by National and International organizations.

As per the Act, the National Gene Fund (NGF) can be utilized for supporting conservation and sustainable use of genetic resources, including *in situ* and *ex situ* collections. Some allocation may be earmarked for *ex situ* conservation of varieties and maintenance of gene banks. The funds are also to be used for recognizing and rewarding the contributions of farmers engaged in the conservation and enhancement of agro-biodiversity.

## Seed Bill 2004

In the year 2004 a new seed bill was placed before the Indian Parliament which aimed at to reduce the direct involvement of government in seed production and marketing and to encourage the private sector participation in agriculture and seed production including development of new varieties, in other words a complementation in functioning of private and public sectors was started. In fact, Seed Act 1966 needed immediate amendment in view of expanding scenario of seed industry in India because it suffered from many shortcomings such as:

(i) Registration of seed variety was not essential

(ii) It did not cover non-notified varieties.

(iii) The Act also did not cover commercial and plantation crops.

**Table 14.17: Distinctness of Candidate Variety**

| DUS Sl.No. | Type of Assessment | DUS Characteristics | Arka Anamika | | | Pusa Sawani | | | Arka Abhay | | | Kashi Vardan | | |
|---|---|---|---|---|---|---|---|---|---|---|---|---|---|---|
| | | | R1 | R2 | R3 | R1 | R2 | R3 | R1 | R2 | R3 | R1 | R2 | R3 |
| 1. (*) | VG | Stem: Colour | 1 | 1 | 1 | 1 | 1 | 1 | 1 | 1 | 1 | 2 | 2 | 2 |
| 2 | VG | Stem: Intensity of green colour | 5 | 5 | 5 | 5 | 5 | 5 | 5 | 5 | 5 | 3 | 3 | 3 |
| 3.(*) (+) | VG | Leaf blade: Depth of lobing | 5 | 5 | 5 | 7 | 7 | 7 | 7 | 7 | 7 | 5 | 5 | 5 |
| 4.(*) | MS | Stem: Number of nodes at first flowering (upto and including the first flowering node) | 7 | 3 | 5 | 5 | 5 | 5 | 5 | 5 | 5 | 5 | 5 | 5 |
| | | (Average of 10 plants/ replication) | 11.0 | 4.0 | 7.0 | 6.8 | 6.6 | 6.4 | 6.4 | 6.6 | 6.53 | 6.0 | 5.0 | 6.0 |
| 5.(*) | MG | Flowering: Time (50 per cent of the plants with at least one open flower) | 5 | 5 | 5 | 7 | 7 | 7 | 5 | 5 | 5 | 5 | 5 | 5 |
| | | (Average of 10 plants/ replication) | 42 | 44 | 38 | 49 | 49 | 48 | 45 | 45 | 44 | 42 | 43 | 42 |
| 6 | MS | Leaf blade: Length | 3 | 3 | 3 | 5 | 5 | 5 | 5 | 5 | 5 | 5 | 5 | 5 |
| | | (Average of 10 plants/ replication) | 9.6 | 9.9 | 9.8 | 18.33 | 18.47 | 18.50 | 17.83 | 18.03 | 17.83 | 18.70 | 19.20 | 21.80 |
| 7 | MS | Leaf blade: Width | 5 | 5 | 5 | 7 | 7 | 7 | 7 | 7 | 7 | 7 | 7 | 7 |
| | | (Average of 10 plants/ replication) | 15.3 | 15.4 | 20.1 | 25.33 | 26.17 | 25.32 | 24.50 | 24.83 | 24.13 | 27.30 | 25.80 | 27.50 |
| 8 | VS | Leaf blade: Serration of margin | 5 | 5 | 5 | 5 | 5 | 5 | 5 | 5 | 5 | 5 | 5 | 5 |
| 9. (*) | VS | Leaf blade: Colour between veins | 1 | 1 | 1 | 1 | 1 | 1 | 1 | 1 | 1 | 1 | 1 | 1 |
| 10 | VG | Leaf blade: Intensity of colour between veins | 7 | 7 | 7 | 5 | 5 | 5 | 5 | 5 | 5 | 7 | 7 | 7 |
| 11 | VG | Vein: Colour | 1 | 1 | 1 | 2 | 2 | 2 | 2 | 2 | 2 | 1 | 1 | 1 |
| 12 | MS | Petiole: Length | 5 | 5 | 5 | 5 | 5 | 5 | 5 | 5 | 5 | 5 | 5 | 5 |
| | | (Average of 10 plants/ replication) | 15.1 | 15.4 | 15.6 | 24.83 | 25.27 | 24.10 | 24.83 | 25.17 | 26.17 | 33.40 | 30.50 | 30.10 |
| 13 | VG | Flower: Petal colour | 2 | 2 | 2 | 2 | 2 | 2 | 2 | 2 | 2 | 1 | 1 | 1 |
| 14 | VG | Flower: Petal base colour (purple) | 2 | 2 | 2 | 2 | 2 | 2 | 2 | 2 | 2 | 2 | 2 | 2 |
| 15 | MS | Flower: Length (cm) | 5 | 5 | 5 | 7 | 7 | 7 | 5 | 5 | 5 | 5 | 5 | 5 |
| | | (Average of 10 plants/ replication) | 4.7 | 4.3 | 4.9 | 5.73 | 5.83 | 5.50 | 4.80 | 4.97 | 5.00 | 6.30 | 5.70 | 5.80 |
| 16 | MS | Flower: Diameter (at the top of flower) | 5 | 5 | 3 | 5 | 5 | 5 | 5 | 5 | 5 | 5 | 5 | 5 |
| | | (Average of 10 plants/ replication) | 3.1 | 3.4 | 2.6 | 4.73 | 5.00 | 4.60 | 4.27 | 4.30 | 4.67 | 6.70 | 6.50 | 7.00 |
| 17. (*) | VG | Fruit: Colour | 2 | 2 | 2 | 2 | 2 | 2 | 1 | 1 | 1 | 2 | 2 | 2 |
| 18 | MS | Fruit: Length (cm) at marketable stage (four days after anthesis) | 7 | 7 | 7 | 5 | 5 | 5 | 5 | 5 | 5 | 5 | 5 | 5 |
| | | (Average of 10 plants/ replication) | 16.9 | 17.2 | 18.3 | 11.67 | 12.00 | 11.50 | 12.00 | 11.75 | 11.25 | 12.00 | 11.8 | 10.5 |
| 19 | MS | Fruit: Diameter (at mid length) (cm) | 5 | 5 | 5 | 7 | 7 | 7 | 5 | 5 | 5 | 7 | 7 | 7 |
| | | (Average of 10 plants/ replication) | 2.4 | 2.2 | 2.7 | 1.58 | 1.70 | 1.60 | 1.50 | 1.45 | 1.48 | 1.60 | 1.55 | 1.52 |
| 20. (*) (+) | VG | Fruit: Surface between ridges | 3 | 3 | 3 | 3 | 3 | 3 | 3 | 3 | 3 | 3 | 3 | 3 |

| DUS Sl.No. | Type of Assess-ment | DUS Characteristics | Arka Anamika | | | Pusa Sawani | | | Arka Abhay | | | Kashi Vardan | | |
|---|---|---|---|---|---|---|---|---|---|---|---|---|---|---|
| | | | R1 | R2 | R3 | R1 | R2 | R3 | R1 | R2 | R3 | R1 | R2 | R3 |
| 21 | VG | Fruit: pubescence | 5 | 5 | 5 | 5 | 5 | 5 | 5 | 5 | 5 | 5 | 5 | 5 |
| 22.(+) | VG | Fruit: constriction of basal part | 3 | 3 | 3 | 7 | 7 | 7 | 3 | 3 | 3 | 3 | 3 | 3 |
| 23.(+) | VG | Fruit: Shape of apex | 1 | 1 | 1 | 2 | 2 | 2 | 2 | 2 | 2 | 2 | 2 | 2 |
| 24.(*) | MS | Fruit: Number of locules | 1 | 2 | 1 | 1 | 1 | 1 | 1 | 1 | 1 | 1 | 1 | 1 |
| | | (Average of 10 plants/ replication) | 5.0 | 7.0 | 4.0 | 5.0 | 5.0 | 5.0 | 5.0 | 5.0 | 5.0 | 5.0 | 5.0 | 5.0 |
| 25. (*) | MS | Plant: Number of branches | 7 | 7 | 5 | 5 | 5 | 5 | 7 | 7 | 7 | 3 | 3 | 3 |
| | | (Average of 10 plants/ replication) | 5 | 5 | 4 | 3.90 | 3.80 | 4.00 | 4.40 | 4.60 | 4.60 | 3.00 | 2.00 | 1.00 |
| 26 | MS | Stem: Diameter (at 10cm above ground level) (cm) | 3 | 3 | 3 | 7 | 7 | 7 | 7 | 7 | 7 | 7 | 7 | 7 |
| | | (Average of 10 plants/ replication) | 0.9 | 0.8 | 0.8 | 2.20 | 2.47 | 2.60 | 2.62 | 2.53 | 2.57 | 2.00 | 2.20 | 2.00 |
| 27 | MS | Plant: Height (cm) | 3 | 3 | 3 | 5 | 5 | 5 | 7 | 7 | 7 | 5 | 5 | 5 |
| | | (Average of 10 plants/ replication) | 65 | 78 | 85 | 116.67 | 118.67 | 116.00 | 128.00 | 128.34 | 130.00 | 111.80 | 118.40 | 118.50 |
| 28.(*) (+) | MG | Fruit: Length of physiologically mature fruit (cm) | 7 | 7 | 7 | 7 | 7 | 7 | 7 | 7 | 7 | 5 | 5 | 5 |
| | | (Average of 10 plants/ replication) | 18.9 | 19.6 | 17.4 | 16.47 | 15.90 | 16.20 | 15.4 | 17.7 | 16.6 | 15 | 14.6 | 13.8 |
| 29 | MS | Fruit: Diameter (at mid length) (cm) | 5 | 5 | 5 | 5 | 5 | 5 | 5 | 5 | 5 | 5 | 5 | 5 |
| | | (Average of 10 plants/ replication) | 1.8 | 1.8 | 1.7 | 1.83 | 1.87 | 1.87 | 1.60 | 1.57 | 1.60 | 1.50 | 1.80 | 1.70 |
| 30 | VG | Seed: Colour | 1 | 1 | 1 | 1 | 1 | 1 | 1 | 1 | 1 | 1 | 1 | 1 |
| 31 | VG | Seed : Hairiness | 1 | 1 | 1 | 1 | 1 | 1 | 1 | 1 | 1 | 1 | 1 | 1 |

(iv) Provision for regulation of transgenic varieties was lacking.

(v) It recommended certification of seed through State Seed Certification Agencies.

## Objective of the Seed Bill 2004

The Seed Bill 2004 provides sound regulation of the quality seed which will be available to the growers on the sale basis. Not only this, import, an export and to facilitate production and supply of quality seeds and for matters connected therewith or incidental thereto, will be major concern. The Seed Bill 2004 was proposed with the following objectives:

(i) To overcome the lacunae of existing Seed Act.

(ii) To enhance seed replacement rates for various crops.

(iii) To create facilitative climate for growth of seed industry.

(iv) To boost export of seeds and encourage import of useful germplasm.

(v) To create conducive atmosphere for application of frontier science in varietal development and enhanced investment in research and development.

## Implementable Aspects

(i) Compulsory registration of all seeds traded in India.

(ii) Compulsory state level registration of seed producer, processing units and traders.

(iii) The bill also introduces compulsory registration of nurseries selling horticultural plants.

(iv) Duration of registration is 15 or 18 years with provision to double this term.

(v) Besides Central and State Seed Testing Laboratories, accriditated individuals and institutions can undertake seed certification.

(vi) Maintenance of National Register of seeds and advocates for centralization of registration.

A comparison between Seed Bill 2004 and Seed Act 1966 are given in Table 14.18.

**Table 14.18: Comparison of Seed Act, 1966 and Seeds Bill, 2004**

| | Seeds Bill, 2004 | Seed Act, 1966 |
|---|---|---|
| Definition | "Agriculture" includes horticulture, forestry, plantation, medicinal and aromatic plants | "Agriculture" includes horticulture |
| | Definition of "Seed" and "Variety" have been changed to make them more specific and technical | |
| | Defines terms such as " Dealer', "Essentially Derived Variety", "Extant Variety', "Farmer", "Horticulture Nursery", "misbranded", "spurious Seed, and "transgenic Variety" | Does not define these terms |
| Registration | All seeds are sale must be registered | Only varieties notified by the government need to be registered |
| Seed committee | Constitutes Central and State Seed Committees. A Registration Sub-Committee would register seeds of all varieties. | Constitutes Central Seed Committees. The Central government, after consulting with the Central Seed Committees (CSC) may notify a seed in order to regulate the quality of seed |
| Transgenic Varieties | Special provisions for registration of transgenic varieties of seed. | No provision for transgenic varieties of seeds. |
| Compensation to Farmers | Provides for compensation to farmers under the Consumer Protection Act ,1986 in the event of underperformance of seeds. | No specific provision for compensation mentioned in the Act. |
| Export and Import | All seed imports are regulated by the Plant Quarantine (Regulation of Import into India) Order,2003 or any corresponding order of the Destructive Insects and Pest Act, 1914; shall conform to minimum limits of germination *etc.* Exports can be restricted if it adversely affects the food security of the country. | A person is restricted from exporting or importing notified variety of seed unless it conforms to minimum limits of germination *etc.* |
| Penalties | Any person who contravenes any provisions of the Act imports, sells or stocks seeds deemed to be misbranded or not registered can be punishable by a fine of Rs. 5,000/- to Rs. 25, 000/-. The penalty for giving false information is a prison term up to six months and/or a fine up to Rs. 50, 000/-. | Any person who contravenes any provisions of the Act prevents a Seed Inspector from taking samples *etc.* shall be punished for the first offence with a fine which may extend to Rs. 500/-. If the offence is repeated he may be imprisoned for a maximum term of six months and or fined up to Rs. 1,000/-. |

*Source*: Srivastava, 2006.

## Salient Features of Seed Bill Act, 2004

Highlights of Seed Act Bill 2004 are enlisted here as described by Srivastava (2006):

(i) The Seed Bill 2004 aims to regulate the quality of seed sold and replace the Seed Act 1966.

(ii) The Bill does not restrict the farmer's right to use or sale his form seeds and planting material, provided he does not sell them under a brand name. All seeds and planting material sold by farmers will have to conform to the minimum standards as applicable to registered seeds.

(iii) If the seed of registered variety fails to perform to expected standards, the farmer can claim compensation from the producer or dealer under the Consumer Protection Act 1986.

(iv) Accreditation of ICAR centres, State Agricultural Universities and Private Organizations to conduct agronomic trials.

(v) Every seed producer and dealer and the horticulture nursery have to be registered *i.e.*, all varieties of seeds for sale have to be registered.

(vi) Maintenance of National registered of varieties.

(vii) To regulate the export and import of seeds.

(viii) Enhancement of penalty for major and minor infringements.

(ix) Inclusion of provisions to regulate genetically modified crops and ban on terminator gene.

Several discrepancies have been pointed out and suggestions given to make the Seed Bill more effective and relevant (Srivastava, 2006). A comparison between Seed Bill 2004 and Protection of Plant Variety and Farmer's Rights Act 2001 are given in Table 14.19.

**Table 14.19: Comparison of Seeds Bill 2004 and PPV and FR Act 2001**

| | Seeds Bill 2004 | PPV and FR Act 2001 |
|---|---|---|
| Definitions | "Farmer" means any person who cultivates crops either by cultivating the land himself or through any other person but does not include any individual, company, trader or dealer who engages in the procurement and sale of seeds on a commercial basis. | "Farmer" means any person who cultivates crops either by cultivating the land himself or cultivates crops by directly supervising the cultivation or land through any other person; or conserves and preserves, severally or jointly with any other person any wild species or traditional varieties or adds value to such wild species or traditional varieties through selection and identification of their useful properties. |
| Registration | Establishes a Registration Sub-Committee which would maintain a National Register of Seeds. | Establishes a Plant Variety Registry, which would maintain a National Register of Plant Varieties. |

| | Seeds Bill 2004 | PPV and FR Act 2001 |
|---|---|---|
| | No specification regarding parentage of variety | Specifies details under which a variety may be registered such as a complete passport data of the parental lines from which a variety has been derived. |
| | Registration is for 15 years for annual/biennial crops and 18 years for long duration perennials. On expiry, registration can be renewed for a similar period. | Registration is for 15 years for annual/biennial crops and 18 years for long duration perennials. registration cannot be renewed |
| Farmer's Rights | A farmer can save, use, exchange, share or sell his farm seeds and planting material. He cannot sell seeds under a brand name. Seeds sold have to conform to the minimum limit of germination, physical purity, genetic purity prescribed by the Act. | A farmer is entitled to save, use, sow, re-sow, exchange, share or sell his farm produce including seed of a variety protected under the Act in the same manner before this Act came into force. He cannot sell branded seed of a variety protected under the Act. |
| Compen-sation | The seed producer, distributor or vender will have to disclose the expected performance of a given conditions. If the seed fails to perform to expected standards, the farmer can claim compensation from the dealer, distributor or vendor under the Consumer Protection Act 1986. | If a breeder of a propagating material of a variety registered under the Act sells his product to a farmer. He has to disclose the expected performance under given conditions. If the propagating material fails to perform, the farmer can claim compensation in the prescribed manner before the Protection of Plant Varieties and Farmer's Rights Authority. |
| Penalties | Any person who contravenes any provisions of the Ac, prevents a Seed Inspector from taking samples *etc.* shall be punished for the first offence with a fine up to Rs. 500/-. If the offence is repeated he may be imprisoned up to six months and/or fined up to Rs. 1,000/-. | Penalty for applying false denomination to a variety is imprisonment up to two years and/or a fine between Rs. 50,000/- and Rs. Five lakhs. Penalty for falsely representing a variety as registered is imprisonment up to three years and/or a fine between Rs. One lakh and Rs. Five lakh or both. Penalty for subsequent offence is imprisonment up to three years and/or a fine between Rs. Two lakh and Rs. Twenty lakhs. |

*Source*: Srivastava, 2006.

# References

1. Brahmi, P. and Chaudhary, V. 2011. Protection of plant varieties: systems across countries. *Plant Genetic Resources*, **9(3)**: 392-403.

2. Chawla, H.S. 2002. Intellectual property rights. In: Introduction to Plant Biotechnology. Oxford and IBH, Publishing House, New Delhi. p. 467-483.

3. Ganguli, P. 1998. Intellectual property rights in transition. World Patent Information, 20: 171-180.

4. Srivastava, U. 2006. New Seed Bill 2004. Issues and Analysis. *Indian J. Plant Genet. Resour.*, **19(2)**: 141-153.

# 15

# Plant Identification and Herbarium

## Taxonomy and Plant Identification

Angiosperms or flowering plants form the largest group of plant kingdom, including about 411 families, 8,000 genera and more than 4,00,000 lakhs; are well known by human being and around 20,000 species of lowering plants have been characterized by scientific body. They are considered to be the highest evolved plants on the surface of the earth. Angiosperms are annuals or perennials herbs, shrubs, trees, climbers, twiners and lianes. If we say for plant, the identity of a plant expressed as a binomial, *viz.*, genus name and species name followed by author(s) who has described and named it, is basic to effective communication of correct gesture of plants and research findings. Furthermore, the ability to identify species correctly and note and record variation in response to differences in environmental factors *viz.* climate, edaphic and biotic, is basic to exploration, collection and characterization of cultivated and wild plants. Before proceedings to group the plant it is also essential to know the terminology of crops.

Besides, information on plant species with respects to the area of availability, variability pattern, flowering /fruiting time, threat status and endemism, other ecological features, economic uses, indigenous traditional knowledge (ITKs), *etc.* gathered from herbarium data, serves as resource for basic and applied research, referral use and for educational programme.

Taxonomic and systematic studies or native crop taxa *viz., Vigna, Abelmoschus, Luffa, Trichosanthes, Allium* and *Moringa* have helped in better understanding the identity/relationship of taxa under the crop genera of PGR relevance (Nayar, 2015). Taxa of dubious identities are studied in detail for further validation using evidences from biochemical and molecular tools under inter-institutional collaborative research works. Many important wild relatives of crop plants in the Indian region have been enumerated (Pradheep *et al.,* 2014). Based on characters of taxa studied, field identification are prepared for *Vigna, Allium, Trichosanthes, Luffa, etc.* (Pandey and Pandey, 2005; Pandey and Nayar, 1994; Pandey and Bhatt, 2008; Pandey *et al.,* 2014a,c).

## Classification of Plants

Taxonomy of cultivated plants has largely been a neglected field (Huaman and Spooner, 2002).

### 1. Bentham and Hooker's System

Benthan and Hooker's system is broadly accepted. Generally, in horticultural crops; plants are broadly classified into two classes:

(1) **Class: Dicotyledons**-Embryo with two cotyledons; stem open bundles; leaves netted (reticulate) venation; flowers usually pentamerous.

**Table 15.1: Class: Dicotyledons**

| Sl.No. | Class | Name | Description |
|--------|-------|------|-------------|
| 1. | Sub-class I | Polypetalae | Flowers usually with two whorls of perianth *i.e.* (calyx and corolla), petals free. |
| 2. | Sub-class II | Gamopetalae | Flowers usually with two whorls of perianth *i.e.* (calyx and corolla), petals united. |
| 3. | Sub-class III | Monochlamydeae | Flowers usually with one whorls of perianth commonly sepaloid or absent |

## Table 15.2: Sub-class I-Polypetalae

**Series (i)**-Petals and stamens hypogonous, disc absent *Thalamiflorae*

| | | | |
|---|---|---|---|
| (1) | a. | Androecium rarely definite, gynoecium free or immersed in torus, rarely united; embryo minute, albuminous | *Ranales* |
| | b. | Receptacle neither hollow nor concave; ovary apocarpous | *Ranunculaceae* |
| (2) | a. | Gynoecium syncarpous, partial placentation | *Parietales* |
| | b. | Flowers actinomorphic, sepals 2 or 3, twice sepals | *Papaveraceae* |
| | c. | Corolla cruciform, androecium tetradynamous | *Cruciferae* |
| | d. | Andro-or gynophore or both present | *Capparidaceae* |
| | e. | Flower zygomorphic, interior petal often spurred | *Violaceae* |
| (3) | a. | Gynoecium syncarpous, free central placentation; herbs, sepals 5 or 4, petals 5 or 4, stamens twice petals, obdiplostemonous | *Caryophyllineae* |
| | b. | Gynoecium 3 or 5, carpellary | *Caryophylliaceae* |
| (4) | a. | Flowers rarely irregular; sepals 5, 2 or 4, free or united; petals as many or (O); stamens indefinite monodelphous; gynoecium 3 to indefinite numbers of carpels, carpels united | *Malvales* |

**Series (ii)**-Stamens hypogynous, disc present, ovary superior *Disciflorae*

| | | | |
|---|---|---|---|
| (5) | a. | Ovary superior or inferior, syncarpous; stamens twice the number of sepals, in two or one whorl | *Geraniales* |
| | b. | Tree or shrubs; leaves gland dotted; sepals 4 or 5, petals 4 or 5, stamens 2, obdiplostemonous; gynoecium 4 or 5 or indefinite capillary, syncarpous | *Rutaceae* |

**Series (iii)**- Flower perygynous or epigynous; ovary sometimes inferior; ovary enclosed by developments of floral axis *Calyciflorae*

| | | | |
|---|---|---|---|
| (6) | a. | Gynoecium one or more carpellary, apocarpous; flower actinomorphic or zygomorphic, perigynous | *Rosales* |
| | b. | Gynoecium monocarpellary, flower zygomorphic (except in Mimosoideae); stamens 5 or 5+5, united | *Leguminosae* |
| | c. | Gynoecium pentacarpellary, apocarpous, flower regular, sepals 4 or 5, petal 4 or 5, stamens 1 to indefinite, gynoecium 1 to indefinite number of carpels | *Rosaceae* |
| (7) | a. | Flower regular, usually bisexual; ovary syncarpous, inferior; styles undivided or very rarely styles are free | *Myrtales* |
| | b. | Evergreen trees or shrubs or creepers, flowers in cymes; flower epigynous, sepals 4 -5 or (4-5), petals 4-5; stamens indefinite, usually free, axile or rarely parietal placentation | *Myrtaceae* |
| (8) | a. | Flowers bisexual or unisexual, partial placentation, styles free or connate, ovary trilocular, flower unisexual | *Passiflorales* |
| (9) | a. | Flowers bisexual, locules in ovary one to indefinite number, inflorescence, umbel | *Umbellales* |
| | b. | Ovary bicarpellary, bilocular | *Umbelliferae* |

## Table 15.3: Sub-class II-Gamopetalae

**Series (i)**- Ovary inferior, stamens usually as many as petals Inferae

| | | | |
|---|---|---|---|
| (10) | a. | Flower regular or irregular, stamens epipetalous, ovary with 2- indefinite number of locules | *Rubiales* |
| | b. | Tree, shrubs herbs; leaves opposite, inter and intra stipules present; cymose inflrorescence, sepals 5-4, petals (5-4), gynoecium with one-indefinite carpels, syncarpous, ovary inferior | *Rubiaceae* |
| (11) | a. | Flower regular or irregular, stamens epipetalous, ovary with one locule and one ovule; stamens, syngenesious | *Asterales* |
| | b. | Stamens syngenesious, basal placentation, inflorescence capitulum, stigma bifid | *Compositae* |

**Series (ii)**- Ovary usually superior; stamens as many as or fewer than corolla lobes, alternipetalous; gynoecium 2 rarely to 1-3-carpellary *Bicarpellatae*

| | | | |
|---|---|---|---|
| (12) | a. | Flower regular, hypogynous; stamens epipetalous; leaves generally opposite | *Gentianales* |
| | b. | Latex axillary present; inflrorescence panicle, dischasial cyme or cincinnus; flower bisexual, regular, sepals 5, petals 5, fused, usually salver or funnel shaped; stamens 5, alternipetalous, epipetalous | *Apocynaceae* |
| | c. | Sepals 5, fused; petals 5, fused; stamens 5, gynoecium bicarpellary, ovary superior; filaments united into a tube, anthers adnate to the stigma; gynostigma and translator, pollinia are present | *Asclepiadaceae* |
| (13) | a. | Flower regular, hypogynous; leaves alternate; stamens epipetalous; ovary 195 loculed | *Polemoniales* |
| | b. | Axillary inflrorescence; ovary bilocular (or tetralocular due to false septum); indefinite number of ovules per loculus; a swellen placenta obliquely placed to the mother axis, sepals 5, fused; petals 5, fused; stamens 5; gynoecium bicarpellary, corolla lobes | *Solanaceae* |
| | c. | Placenta not blique; ovary bilocular with two ovules per loculus but later becomes tetralocular and thus one ovule per loculus, sepals 5 free or fused; petals 5, fused; stamens 5; gynoecium bicarpellary. | |
| (14) | a. | Flowers usually irregular, corolla often bilipped; stamens generally fewer than corolla tubes, usually 4, didynamous or 2; ovary 1-4 locular, ovules usually indefinite | *Personales* |
| | b. | Indefinite number of ovules per locules; sepals 5, fused; petals 3+2, gamopetalous; stamens 4 or 5, gynoecium bicarpellary, syncarpous | *Acanthaceae* |

| (15) | a. | Corolla usually blipped; flower hypogynous, rarely regular; ovary 2-4 loculed, ovules solitary in loculus or rarely more than one; first a drupe or nutlets | *Lamiales* |
|---|---|---|---|
| | b. | Ovary entire to whorled leaves, flowers cymose, umbel, gynoecium usually tetracarpellary, style terminal, form of secon, septa, fruit drupe or schizocarp | *Verbenaceae* |
| | c. | Gynobasic style, ovary becomes tetralocular due to false septum, single ovule in each locule; sepals 5, fused; petals 5, fused, bilabiate; stamens 4-5; gynoecium bicarpellary. | *Labiatae* |

## Table 15.4: Sub-class III- Monochlamydeae
### Flowers usually with one whorl of perianth, commonly sepaloid or perianth absent.

**Series (i)**-Terrestrial plants with usually bisexual flowers; stamens generally equal in number to perianth lobes, ovules usually solitary; embryo curved in floury endosperm *Curvembryae*

| (1) | a. | Ochreate stipules; inflorescence racemose or cymose; flowers bisexual, regular, cycle or acylic; perianth 3+3; stamens 3+3; gynoecium tricarpellary | *Polygonaceae* |
|---|---|---|---|
| | b. | Flowers unisexual; ovary syncarpous or monocarpellary; ovules solitary or two per carpel | *Unisexuales* |
| | c. | Albumin present or absent in seed; perianth sepaloid, much reduced or absent; a inflorescence cyathium. | *Euphorbiaceae* |
| | d. | Inner perianth petloid, ovary inferior, microspermae, rarely axile | *Orchidaceae* |

**Series (ii)**- Perianth partly petaloid; ovary usually inferior; endosperm abundant *Epigynae*

| 1. | a. | Stem a rhizome; flower bisexual or unisexual; inflorescence a spadix | *Musaceae* |
|---|---|---|---|
| | | | *Zingiberaceae* |

**Series (iii)**- Inner perianth petloid, ovary usually free, superior, endosperm abundant, flower bisexual *Coronariae*

| 1. | a. | Flowers regular, bisexual, perianth 3+3; stamens 3+3, epipetalous; gynoecium tricarpellary, syncarpous, ovary superior | *Liliaceae* |
|---|---|---|---|
| | b. | Flowers regular, bisexual, trimerous, blue in colour | *Commelinaceae* |

**Series (iv)**-Perianth sepaloid, herbaceous or membranous ovary usually inferior, Calycinae,

| | a. | flowers usually unisexual | *Palmae* |
|---|---|---|---|

**Series (v)**- *Nudiflorae*

**Series (vi)**- *Apocarpae*

**Series (vii)**- Flower solitary, sessile in the axils of bracts and arranged in heads or spikelets with bracts; perianth of scales or no perianth; ovary usually unilocular with a single ovule; seed endospermic *Glumaceae*

| | | | *Poaceae* |
|---|---|---|---|
| | | | *Cyperaceae* |

## Table 15.5: I-Dicotyledons

| (i) | Corolla consisting of mostly free petals | Archichlamydeae |
|---|---|---|
| (ii) | Corolla consisting of mostly fused petals | Sympetalae |

| (i) Archichlamydeae | | | | |
|---|---|---|---|---|
| A. | Ovary Superior | (1) Flowers unisexual | - | - | *Euphorbiaceae* |
| | | (2) Gynoecium apocarpous, fruit never drupe, calyx and corolla sometimes not distinguished | - | - | *Ranunculaceae* |
| | | (3) Gynoecium either apocarpous or monocarpellary, fruit may be drupe, epicalyx often present | - | - | *Rosaceae* |
| | | (4) Gynoecium monocarpellary | (a) Fruit drupe | - | *Rosaceae* |
| | | | (b) Fruit legume or lomentum | (i) Flowers actinomorphic, stamens mostly indefinite | *Mimosoideae* |
| | | | | (ii) Flowers zygomorphic, aestivation of the corolla ascending imbricate; stamens 10 but never diadelphous | *Caesalpinoideae* |
| | | | | (iii) Flowers zygomorphic, aestivation of the corolla descending imbricate; stamens 10 diadelphous | *Papilionaceae* |

| | | | |
|---|---|---|---|
| (5) Gynoecium syncarpous | (a) Placentation free-central | | *Caryophyllaceae* |
| | (b) Placentation parietal | (i) Gynoecium tricarpellary | *Violaceae* |
| | | (ii) $K_2$ or 3 C 2 + 2 or 3+3 Adeepakddd G (2-deepakddd) | *Papaveraceae* |
| | | (iii) $K_{2+2}$, $C_4$, A6 ordeepakddd, G (2) androphore or gynophore or gynandrophore present | *Capparidaceae* |
| | | (iv) $K_2$+ 2, $C_4$, $A_2$ +4 (tetradynamous), G (2) | *Cruciferae (Brassicaceae)* |
| | (c) Placentation axile | (i) Epicalyx present, stamens monoadelphous | *Malvaceae* |
| | | (ii) Prominent disc present below gynoecium, stamens obdiplostemonous or polyadelphous | *Rutaceae* |
| | (d) Ovary inferior | (i) Leaves exstipulate, inflorescence umbel | *Umbelliferae* |
| | | (ii) Leaves exstipulate, inflorescence mostly cymose or a spike, petals often nearly circular, stamens indefinite | *Myrtaceae* |
| | | (iii) Leaves mostly stipulate, the latter often adnate to the petals, never circular, epicalyx often present | *Rosaceae* |

### Table 15.6: (ii) Sympetalae

| | | | |
|---|---|---|---|
| A. Ovary Superior | (a) Corolla actinomorphic | (i) Stamens often included in the corolla tube, anthers mostly sagitate and connivent round the stigmatic head, stigma dumble shaped | *Apocynaceae* |
| | | (ii) Gynostegium present, gynoecium free below and fused above | *Asclepiadaceae* |
| | | (iii) Ovary bilocular with typically two ovules in each loculus | *Convolvulaceae* |
| | | (iv) Carpels obliquely placed in the flower ovary bilocular, placenta swollen, ovules shining and indefinite. | *Solanaceae* |
| | (b) Corolla zygomorphic | (i) Corolla mostly bilabiate, style gynobasic | *Labiate* |
| | | (ii) Corolla often bilabiate, ovary elongated, style long and terminal | *Acanthaceae* |
| B. Ovary inferior | | (i) Mostly climbing plants with well-developed tendrils, flowers unisexual, androecium complex, anthers twisted | *Cucurbitaceae* |
| | | (ii) Inflorescence capitulum, calyx in the form of pappus, anthers cohering by their edges (syngenesious), placentation basal | *Compositae* |
| | | (iii) Stipules prominent, either inter or intrapetiolar, placentation axile | *Rubiaceae* |

### Table 15.7.II- Monocotyledons

| | |
|---|---|
| (i) Gynoecium monocarpellary with two feathery stigmas, perianth absent, ovary superior | Poaceae |
| (ii) Flowers unisexual, minute and produced in very large numbers, ovary superior | Palmeae |
| (iii) Flowers bisexual, six stamens in two whorls, ovary superior | Liliaceae |
| (iv) Flowers mostly bisexual, stamens 5 in two whorls, ovary inferior | Musaceae |

*Source*: Subrahmanyam, 1999.

### Table 15.8: Traits and their Description

| Sl.No. | Traits | Description |
|---|---|---|
| 1. | **Habitat** | The natural abode or locality of the plant *i.e.*, whether cultivated as an ornamental plant or a food crop or occurs wild |
| 2. | **Habit** | (a) Annual (plant complete their life cycle in one year), biennial (plant complete their life cycle in two year) or perennial (plant complete their life cycle in more than two year) |
| | | (b) Herb (usually low, soft or coarse plant with annual above ground stems), undershrub, shrub (a much branched woody perennial plant, usually without a single trunk) or tree (a tall, woody perennial plant, usually with a single trunk) or vine or liana (an elongate, weak-stemmed, often climbing annual or perennial plant, with herbaceous or woody texture) |
| | | (c) Any other peculiarity such as parasite, epiphyte, xerophyte or hydrophyte |

| Sl.No. | Traits | Description |
|---|---|---|
| 3. | Root | **(a)** Tap (it is found in dicotyledonous plants only. It is developed from the radical of the seed. This is root system with a prominent main root, directed vertically downwards and bearing smaller lateral roots (secondary roots) which in turn produce the tertiary roots) or adventitious (root developing from part of plant other than radical *e.g.* from stem or leaf cutting *etc.* This type is present in monocotyledonous plants only) |
| | | **(b)** Branched or un-branched |
| | | **(c)** Any specialty such as tuberous (in sweet potato-*Ipomoea batatas* and cassava-*Manihot utilisima*, the adventitious roots become tuberous), fasciculate (in *Dahlia, Asparagus, Ruellia etc.* many adventitious root tubers are formed in clusters. Such roots are called fasciculated roots), fleshy, fibrous, nodulose/nodulated (in mango ginger—a ginger which smells like mango, the root becomes suddenly swollen at its tip), moniliform (if the roots show regular bead-like swellings at frequent intervals, they are said to be monilifor roots *e.g.* some grass) aerial, climbing, parasitic, conical (the tuberous root taken up the shape of a cone *i.e.*, the root is broad above and becomes gradually tapering below, *e.g.* carrot-*Daucus carota*), fusiform (the root is broad in the middle and gradually tapers on both ends, *e.g.* radish- *Raphanus sativus*), napiform (the root is broad above and abruptly tapers below, *e.g.* beet root-*Beta vulgaris*) *etc.* |
| 4. | Stem | **(a) Erect** (when the stem grows erect. Erect stems may be herbaceous (Having the characters of a herb; stem containing little woody tissue. A herbaceous stem may be solid or hollow -fistular), suffrutescent (semishrubby), shrubby (woody), arborescent (trees), **caudex** (when the woody stem is tall, unbranched and covered with persistent leaf base), **culm** (when the stem is jointed with swollen and solid nodes and hollow internodes), **caulescent** (when there is an aerial stem on which leaves are borne), **acaulescent** (when the stem is extremely condensed and the leaves appear to arise in a cluster from the ground level, with unbranched aerial inflorescence axis called the scape), **excurrent** (a tree with a pyramidal shape) or deliquescent (a tree with branches spreading in all directions so that the general shape of the tree is like a wide dome), monopodium-*Nerium*; sympodium-*Vitis*), prostrate, twining or climbing. If climbing write the mode of climbing *i.e.* whether by tendrils or spines or hooks or by any other means |
| | | **(b)** Any special modification *i.e.* **rhizome** (prostrate underground stem, bearing buds in axils of reduced scale like leaves; serving as a means of perennation and vegetative propagation *e.g. Canna, Ginger, Mentha*), **bulb** (modified shoot consisting of a stem like a biconvex lens with closely set leaves on the upper surface and fibrous adventitious roots on the under surface *e.g.* onion. garlic), **tuber** (swollen end of underground stem bearing buds in axils of scale-like rudimentary leaves or much thickened, usually short, swollen underground tuberous stem with buds (eye) only *e.g.* potato-*Solanum tuberosum, Helianthus tuberosus*), **corm** (a condensed thickened, underground fleshy stem with buds and scaly leaves growing in a vertical direction *e.g. Colocasia, Amorphophallus*), **phylloclade** (modified stem having appearance and function of leaf. They bear tubercles or areole—tuft of spines at each node *e.g. Euphorbia trigona, Opuntia dilenii*), **Cladode** (modified stems but each representing a single internode only. Cladodes are flattened, non-succulent, green and perform the function of leaves *e.g. Asparagus racemosus*), runner, sucker *etc.*; **root-stock** (Rhizome which remains in vertical position and branches grow horizontally; nodes, internodes, scale leaves and axillary buds are present *e.g. Alocasia indica*); **bulb** (very much shortened vertical, underground stem reduced into a small disc and covered by fleshy scale **(i) Tunicated bulb**-fleshy scale-leaves grow concentrically *e.g.* onion (*Allium cepa*), **(ii) Scaly bulb**-Leaves grow in an imbricating fashion. In garlic, some **leathery scale** develop concentrically and cover the fleshy scale *e.g. Allium sativum*) |
| | | **(c)** Branched or un-branched. If branched, write the mode of branching *i.e.* whether racemose or cymose. If cymose, whether uniparous, biparous or multiparous |
| | | **(d)** Herbaceous or woody-having the characters of a shrub or a tree |
| | | **(e) Cylindrical** (circular stem as seen in transverse section), **angular** (a stem shows many angles in transverse section), compressed or reduced, **flattened** (flat stem in T.S.) |
| | | **(f) Hairy** (stem covered with hair), **glabrous** (smooth stem), **waxy** (stem having wax coating) or spiny (stem having spines) |
| | | **(g) Solid** (interior portion of the stem is filled up with matter) or **fistular** (the stem is without a central pith and is hollow interior) |
| | | **(h) Setose** (covered with bristles), **pilose** (surface covered with soft and elongate hair), **tomentose** (covered with woody hair), **mealy** (stem is covered with a white or bluish powder or bloom), **glaucous** (stem smooth and shining), **hirsute** (stem covered with coarse, stiff hair), **hispid** (stem covered with rigid or stiff bristles), **ribbed** (stem having prominent ridges), **prostrate** (trailing stem lying flat on the ground), **procumbent** (lying flat on the ground but not rooting), **decumbent** (stem lying flat on the ground but with rthe tip ascending), **twiner** (plant having herbaceous stem which twine round the support e. g. *Dolichos, Ipomoea, Clematis*), **runner** (horizontly creeping stem consists of nodes and internods, stolon that roots at tip forming new plant (base of nodes) that eventually is freed from connection with parent by decay of runner or broken parts grow independently *e.g. Oxalis corniculata, Cynodon dactylon*), **sucker** (a creeping stem but growing obliquely upwards directly giving rise to a leafy/aerial shoot *e.g. Mentha, Chrysanthemum coronarium, Artemisia dubia*), **offset** (a horizontal, thickened, prostrate branch producing at the apex a tuft of leaves above and a cluster of small roots beneath, when the connection is cut off two daughter plants are formed *e.g. Pistia stratiotes, Eichhornia crassipes* and terrestrial plants like *Agave Americana, Sansiviera cylindrical* produce offsets), **sobole** (underground runners which may not produce roots and shoots of every node and are very stout and rigid *e.g.* grasses like *Spinifex littoreus, Saccharum spontaneum*) **stolon** (a slender lateral branch originating from the base of the stem and growing horizontal that roots at nodes *e.g. Colocasia* spp.*, Mentha piperita, Fragaria vesca*), |

| Sl.No. | Traits | Description |
|--------|--------|-------------|
| | | **bulbil** (modification of vegetative or floral bud whose tissues (multicellular reproductive body) are fleshy and laden with food material *e.g. Agave americana, Globba bulbifera*), **phylloclade** (when the stem is flattened, green and leaf like both in appearance and function *e.g. Ruscus*), **cladode** (a phylloclade of single internode *e.g. Asparagus, Casuarina*), **stem tendril** (when a place of an axillary bud, there is a tendril, *e.g. Cucurbita, Vitis, Passiflora*), **stem thorn** (when in place of an axillary bud there is a thorn-hard, woody and pointed structures developed from axillary or terminal buds for defense and protection purpose, thorns are endigenous in origin, *e.g. Citrus, Bougainvillea, Duranta repens, Carisa carandas*), **tendril climbers** (tendrils are thin, delicate, wire like/thread like structures with sensitive tips. They coil around the support and help in climbing: tendrils may be present- **i-Modified stipules-***e.g. Smilax,* **ii- Modified leaf-***e.g. Lathyrus aphaca,* **iii- Modified stem** *e.g. Passiflora,* **iv-Modified leaflets** *e.g. Pisum sativum, Bignonia,* **v- Modified leaf apex** *e.g. Gloriosa superba*), |
| 5. | **Leaf** | **(a)** Deciduous or evergreen |
| | | **(b)** Radical, cauline or ramal |
| | | **(c)** Alternate, opposite or verticillate (whorled). If opposite, whether superposed or decussate. |
| | | **(d)** Petiolate or sessile |
| | | **(e)** Stipulate or exstipulate. If stipulate, write the nature of the stipules |
| | | **(f)** Nature of the leaf-base, whether sheathing, connate, perfoliate or ligulate |
| | | **(g)** Simple (i-an apical bud is always present, ii-it develops in the axil, iii-axillary buds are always present,and iv-stipules are present at the base of every leaf; simple leaf: pinnately lobed, *e.g. Brassica*; palmately lobed, *e.g. Passiflora, Gossypium*) or compound (with leaf divided into two or more leaflets, i-no apical bud at the tip, ii-it develops at a node, iii-axillary buds are not present in the axils of leaflets, and iv-stipules, if present are found at the base of the entire compound leaf: **I-pinnate compound** (a) Unipinnate-paripinnate compound, *e.g.* Casia, tamarindus, rose; imparipinnate compound, *e.g.* Tephrosia, neem (b) Bipinnate, *e.g. Mimosa, Delonix, Caesalpinia* (c)Tripinnate, *e.g. Moringa,* Aralia (d) Decompound, *e.g. Parthenium, Foemculum*) **II-Palmate compound** (a) Unifoliate compound, *e.g. Citrus, Berberis* (b) Bifoliate compound, *e.g. Hardwickia, Zornia, Balanites* (c) Trifoliate compound, *e.g. Dolichos, Oxalis, Feronia* (d) Tetrafoliate compound, *e.g. Marsilea* (e) Pentafoliate compound, *e.g. Gynandropsis, Aesculus* (f) Multifoliate compound, *e.g. Bombax, Manihot.* If simple, write about the outline of the lamina, its incisions, nature of the margins, apex, surface and venation (unicostate or multicostate; reticulate or parallel). If compound, whether pinnate or palmate. If pinnate, whether paripinnate or imparipinnate, number and arrangement of leaflets. If palmate, whether uni-, bi-, tri-, tetra-, penta-, or multifoliate. If the leaf is compound, the leaflets should be described in the same way as a simple leaf. |
| | | **(h)** **Blade** (flat, expanded portion of leaf), **leaflet** (a distinct and separate segment of a leaf), **ligule** (an outgrowth or projection from the top of the sheath, as in Poaceae), **midrib** (the central conducting and supporting structure of the blade of a simple leaf), **midvein** (the central conducting and supporting structure of the blade of a leaflet), petiole (leaf stalk), **pulvinus** (the swollen base of a petiole), **rachis** (the main axis of a pinnately compound leaf), **sheath** (any tubular portion of the leaf surrounding stem or culm), **bract** (modified leaf found in the inflorescence), **bracteole** (small leaf, usually on a pedicel), **epicalyx** (group of leaves resembling sepals below of the true calyx), **glume** (bract, usually occurring in pairs, at the base of the grass spiklet), **lemna** (outer leaf subtending grass floret), **palea** (inner scale subtending grass floret), **phyllode** (flattened bladder like petiole or midrib), **scale** (small, non-green leaf on bud or modified stem), **spathe** (an enlarged bract enclosing an inflorescence), **spine** (sharp-pointed petiole, midrib, vein or stipule), **tendril** (usually a coiled rachis or twining leaflet modification), **bipalmate compound** (with two orders of leaflets, each palmately compound *e.g. Prinsepia*), **bipinnate compound** (with two orders of leaflets, each palmately compound *e.g. Delonix, Melia, Azadirachta*), **decompound** (a general term for leaflets in two or more orders *e.g. Coriander*), **imparipinnately compound** (off-pinnately compound, with a terminal leaflets, *e.g. Pisum*), **palmately compound** (with leaflets form one point at end of petiole, *e.g. Crataeva religiosa*), **pinnate compound** (With leaflet arranged oppositely or alternately along a common axis, the rachis *e.g. Azadirachta indica*), **simple** (with leaf not divided into leaflets, *e.g. Hibiscus, Passiflora, Mangifera*), **trifoliate** (with 3 leaflets, pinnately compound with terminal petiolule longer than lateral or palmately compound with petiolules equal in length, *e.g.* Oxalis), **tripinnate** compound (leaflets are arranged on the third order of rachis, *e.g. Moringa*), **petiolate** (with a petiole), **pulvinal** (with a swollen base as in the Fabaceae), **sessile** (without petiole), **winged** (with flattened blade-like margins) |
| | | **Stipules-**The lateral appendages of leaf borne at its base are termed stipules. Stipules are green or of various colours; persistent, deciduous or caduceus; their function is protection of young leaf in bud. (i) **Stipulate-**if the stipules are present the leaves are termed as stipulate (ii) **exstipulate-** if the stipules are absent the leaves are termed as exstipulate (iii) **stipel-**when a small appendage is present at the base of a leaflet it is termed as stipel, *e.g. Clitoria ternatea, Phaseolus mungo.* |
| | | **Kind of stipules-(i) free lateral** *e.g.* China rose, cotton, **(ii) Adnate** *e.g.* Rose, groundnut, strawberry, **(iii) Interpetioler** *e.g. Ixora, Anthocephalus* **(iv) Intrapetioler** *e.g. Gardenia,* **(v) Ochreate** *e.g. Rumex, Polygonum,* **(vi) Foliaceous** *e.g. Lathyrus,* **(vii) Bud scale** *e.g. Magnolia, Mesua* |
| | | **Modification of stipules- (i) tendrillar** *e.g. Smilax,* **(ii) spinuous** *e.g. Acacia, Zizyphus, capparis,* **(iii) Lacerate** *e.g. Passiflora,* **(iv) Adnate** –with stipule attached to petiole *e.g. Rosa,* **(iv) Basal-**with stipules attached near base of petiole **(v) Interpetioler-**with connate stipules from two opposite leaves *e.g. Moringa, Ixora,* **(vi) Free lateral-**with stipules adnate to petiole and free parts of stipules located along the petiole *e.g. Hibiscus* **(vii) Sheathing-**enclosing a bud or flower **(viii) Foliaceous-** Stipules are large green and leafy in appearance, *e.g. Lathyrus,* **(ix) Orchreate-**stipules that form a hollow tube encircling the stem from node to a certain height *e.g. Polygonum* |

| Sl.No. | Traits | Description |
|---|---|---|
| | | **Leaf apex-(i) Acute-** ending in a sharp point forming an acute angle *e.g. Thevetia*, **(ii) Acuminate-**drawn out into long; point *e.g. Dolichos* **(iii) Obtuse-**with blunt or rounded end *e.g. Calotropis* **(iv) Emarginate-**having a notch at apex *e.g. Bauhinia* **(v) Truncate-**terminating abruptly as if tapering and were cut off *e.g. Caryota urens* **(vi) Mucronate-**abrupted terminated by a sharp spine *e.g. Casia auriculata*, **(vii) Cuspidate-**terminating in a point, *e.g. Date palm* (*viii*) *Aristate-*provided with awns or with a well-developed bristle, **(ix) Retuse-**Obtuse with a broad shallow notch in middle, *e.g. Oxalis*, **(x) Cirrhose-**midrib prolongates to form a tendril *e.g. Gloriosa* **(xi) Apiculate-**Forming abruptly to a small tip *e.g. Dalbergia* |
| | | **Veneation of leaf-** (i) Circinnate *e.g. Fern*, (ii) conduplicate *e.g. Abutilon, Thespesia*, (iii) convolute *e.g. Canna, Musa,* (iv) Involute *e.g. Nymphaea, Ottelia* |
| | | **Symmetry of leaf-** (i) Bilateral *e.g. Calotropis* (ii) Oblique-*e.g. Melta, Begonia* |
| | | **Shape of leaf-** (i) **Linear-**long and narrow leaf *e.g. Cynodon*, (ii) **Lanceolate-**lance-shaped leaf *e.g. Polygonum, Nerium*, (iii) **Round-**leaf with a circular leaf blade, (iv) **Elliptical-** an elipse-shaped leaf, *e.g.* Guava, Jack (v) Ovate-leaf with an egg-shaped leaf blade *i.e.* slightly broader at the base than the apex, *e.g. Hibiscus rosa sinensis*, (vi) **Spathulate-** spatula-shaped leaf *e.g.* Calendula, (vii) **Oblique-**leaf with two unequal halves *e.g. Begonia*, (viii) **Oblong-**leaf with wide and long leaf blade *e.g. Calotropis* (ix) Reniform-Kidney shaped leaf *e.g. Centella*, (x) **Cordate-**leaf with heart-shaped leaf blade *e.g. Abutilon, Thespesia*, (xi) **Sagittate-**leaf with an arrow-shaped leaf blade *e.g. Colocasia*, (xii) Hastate-sagittate leaf with its two lobes directed outside *e.g. Ipomoea*, (xiii) **Lyrate-**lyre-shaped leaf *i.e.,* with a large terminal lobe and some smaller lateral lobes *e.g.* Mustared, (xiv) **Acicular-**long, narrow and cylindrical leaf *i.e.* needle-shaped *e.g. Pinus* (xv) **Orbicular-**leaf with circular leaf blade *e.g. Nelumbium*, (xvi) Cuneate-Wedge-shaped leaf *e.g.* water lettuce |
| | | **Leaf arrangement** (i) Radical-proceeding from or near the root *e.g.* Onion, (ii) Cauline-pertaining to the stem *e.g.* Palms (iii) Alternate-a single leaf arising at each node *e.g. Abutilon, Annona, Withania, Delonix*, (iv) Opposite-on different sides of the axis with the bases at the same level *e.g. Barleria, Tecoma, Leucas,* (v) Opposite decussate- in pairs at right angles to one another *e.g. Calotropis* (vi) Opposite superposed-a pair of leaves that stands directly over the lower pair in the same plant *e.g. Quisqualis,* Guava, (vii) Whorled-more than two leaves arranged in a circule round an axis *e.g. Clerodendron, Nerium, Alstonia*, (viii) Sub-sessile- having short petiole (ix) Sessile-without a petiole or stalk *e.g. Dianthus*, (x) Exstipulate-having no stipules *e.g. Ipomoea* |
| | | **Venation of Lamina (i) Reticulate-**when the veins in the lamina branch and anastomoses to form a network or mesh or reticulum *e.g.* most of the dicots **(ii) Parallel-**when the veins in the lamina run almost parallel to one another, *e.g.* most of monocots. |
| | | **Types of Lamina-(a) Unicostate-**having only one principal vein, the midrib in addition to the majority of thin and slender veins (b) **Multicostate-**having many principal veins in addition to the majority of thin and slender veins-**(i) Multicostate divergent-**when the main veins diverge out from the leaf base into the lamina and do not converge towards the apex, (ii) **Multicostate convergent-** when the main veins diverge out from the leaf base but again converge towards the apex; **Pinnate reticulate venation-** *Mangifera, Ficus*; **Palmate reticulate venation, convergent-***Zizyphus, Cinnamomum*; Palmate reticulate venation, divergent- *Gossypium, Ricinus*; **Parallel venation-***Eryngium, Calophyllum*; Pinnate parallel- *Musa, Canna, Curcuma*; **Palmate parallel, convergent-**grasses; **Palmate parallel, divergent-***Brassica, Corypha* |
| 6. | **Inflorescence** | Racemose, cymose, mixed, compound or of any special form. In all case write about the details of the inflorescence |
| | | **Racemose-**Inflorescence with monopodial branching; **Raceme-** Inflorescence having a common axix and stalked flowers in acropetal succession *e.g. Brassica, Crotolaria juncea, Crotolaria verrucosa*; **Compound Raceme or Panicle-** *e.g.* gold mohar (*Caesalpinia*), *Andrographis paniculata*; **Corymb-**a raceme with lower pedicels elongated so that the top is nearly flat *e.g. Iberis amara, Caesalpinia pulcherrima*; **Spike-** Inflorescence with sessile flowers along axis *e.g. Achyranthus aspera, Digera arvensis*; **Catkin-** a spike with unisexual flowers and pendulous rachis *e.g. Acalypha, Morus, Croton*; **Umbel-** an arrangement of flowers springing from a common centre and forming a flat or rounded cluster *e.g. Calotropis gigantean,* Hydrocotyle; **Spadix-**a raceme Inflorescence with elongated axis, sessile flowers and an enveloping spathe *e.g. Colocasia antiquorum, Amorphophallus titanium*; **compound spadix-**Coconut; **Strbile-***Humulus lupulus*; **Spikelet-** Poaceae, Cyperaceae; **Compound umbel-***Foeniculum, Coriandrum, Daucus carota*; Capitulum- an inflorescence of sessile flowers or florets crowded together on a receptacle and usually surrounded by an involucre *e.g. Helianthus annuus, Tridax procumbens.* |
| | | **Cymose-**Sympodially branched; Solitary- *Hibiscus rosa sinensis* **(i) Uniparous (monochasium)-**having a cymose inflorescence with one axis at each branching; it is of tw types-helicoid and scorpioid; **(ii) Biparous (dichasium)-**Dichotomously branched cymose inflorescence *e.g. Stellaria media, Ixora, Nerium odorum, Thevetia nerifolia, Nyctanthes, Clerodendron* **(iii) Multiparous (Polychasial)-**developing several or many lateral axces bearing flowers *e.g. Hamelia patens, Calotropis, Viburnum* **(iv) Helcoid cyme-**an uniparous inflorescence produced by suppression of successive axes on same side, thus causing the sympodium to be spirally twisted *e.g. Begonia, Hamelia patens, Scorpioid cyme, Heliotropium indicum*; **(v) Special cymose inflorescence-** Hypanthodium (*Ficus cunea*), **Coenanthium** (*Dorstenia*), **Cyathium** (*Poinsettia pulcherrima*); Verticillaster (*Leucas aspera*); **Cymose Capitateb** (*Anthocephalus cadamba*); **(vi) Mixed inflorescence-Mixed panicle** (*Ligustrum vulgare*); **mixed spadix** (*Musa paradisica*); **Cymose umbel** (*Allium cepa, Calotropis, Crinum, Lantana*); **Cymose corymb** (*Alstonia scholaris, Oldenlandia corymbosa*); Fascile (*Garcinia*, some members of Caryophyllaceae); **Thyrsus** (*Vitis vinifera*) |
| 7. | **Flower** | **(a)** Pedicellate or sessile |
| | | **(b)** Bracteates orebracteate. If bracteates, the bracts should be described |
| | | **(c)** Complete or incomplete |
| | | **(d)** Bisexual or unisexual |

| Sl.No. | Traits | Description |
|--------|--------|-------------|

**(e)** Actinomorphic or zygomorphic

**(f)** Colour of the flowers

**(g)** Hypogynous, perygynous or epigynous

**Flower-** A typical flower is made up of four sets of flower parts **(i) Sepals-**Collectively called the calyx, are the outermost parts and are commonly leaf-like and green; but they may be coloured like the petals and have a thinner texture, in which case they are being described as being petaloid **(ii) Petal-**Petals, collectively called the corolla, normally occupy a position in the flower between the sepals and the stamens. Petals are often delicate in texture and are often coloured. They are often larger than the sepals, and they may be shed soon after the flower opens.

**Perianth-** When sepals and petals are of one type, then sepals and petals are combinedly called as perianth.

**Complete flower-** When all the characteristics parts-sepals, petals, stamens, and pistils are present, the flower is complete.

**Incomplete flower-**When one or more of these parts may be lacking, in which case the flower is Incomplete *e.g. Achyranthes, Digera, Amaranthus.*

**Naked flower-** When whole of the perianth of sepals and petals is lacking, the flower is naked.

**Apetalous flower-** When the petals are lacking, the flower is apetalous.

**Perfect flower-** Flowers having both stamens and pistils, regardless of the presence or absence of other parts, are perfect flower.

**Unisexual flower-** If a flower has only kind of sexual organ, either stamens or pistils but not both, it is a unisexual flower; male flowers, or those having stamens, are called staminate flowers, and female flowers, or those having pistils, are called pistillate flowers.

**Monoecious-**When the flowers are unisexual and both sexes occur on the same plant, the plants are known as monoecious plants

**Dioecious-** When unisexual flowers occur on separate plants, the plants are dioecious.

**Polygamous-** In some plants, unisexual and bisexual flowers occur on the same plant and such a condition is known as polygamous.

**Regular flower-**It is applied to the perianth as a rule, rarely involve the reproductive parts. Regular flower

**Irregular flower-** (actinomorphic flowers) are those in which the perianth parts of each kind are similar in size and shape, so that the flower may be divided into equal halves by a vertical plane in various directions, the flower being radially symmetrical.

**Irregular flower-** It is sometimes called as zygomorphic flowers, in which the perianth parts of each kind are dissimilar in size and shape, some petals being unlike other petals,or some of the sepals being dissimilar in size or shape to others. Irregular usually involves the petals but it may involve the sepals or the whole perianth. Irregular flowers may be divided into equal halves only by a single vertical plane.

Floral arrangements: **(i) Spiral-**In many primitive flower types; the various floral parts may be inserted on the floral axis in a spiral manner. **(ii) Cyclic-** In more advance flower type, the various parts are inserted in whorls, each whorl at a slightly different level, and this is known as cyclic arrangement *e.g. Solanum.* **(iii) Acyclic-**When flower parts arise spirally *e.g. Ranunculus.*

**Type of flower**

**(i) Bracteate flower-** Those flowers which have bracts, *e.g. Vinca rosea.* **(ii) Bracteolate flower-***Ruellia* **(iii) Ebracteate flower-** Flowers without bract, *e.g. Solanum;* **(iv) Ebracteolate flower-***Cassia.* **(v) Pedicillate flower-** Flowers having pedicel, *e.g. Hibiscus, Dianthus.* **(vi) Sessile flower-** Flowers without pedicel, *e.g.* ray and disc florets of *Helianthus, Adhatoda*

**Actinomorphic flowers-**Those in which floral organs are so arranged that when flowers cut vertically in any plane thorough centre, always two equal parts be obtained or a flower having radial symmetry; regular *e.g. Tribulus terrestris, Hibiscus.*

**Zygomorphic flowers-**Those flowers which can be divided into equal parts only in one vertical plane; bilateral *e.g. Dolichos, Crotolaria, Pisum.*

**Hermaphrodite flower (Bisexual)-** Flowers having both male and female reproductive organs, *e.g. Hibiscus*

**Unisexual flowers-**Those which have either male or female reproductive organs *e.g. Euphorbia, Cucumis, Coccinia*

**Staminate flowers-** Those which have only reproductive organs (stamens)

**Pistillate flower-** Those flowers which have only female reproductive organs (gynoecium)

**Achlamydeous-**Perianth is totally absent *e.g. Euphorbia, Poinsettia*

**Chlamydeous-**Flowers having perianth

**Monochlamydeous-**When flower consists of single whorl of perianth: *Ricinus,* male flower of *Croton.*

| Sl.No. | Traits | Description |
|---|---|---|
| | | **Dichlamydeous-** When the flower consists of two whorl of perianth, *e.g.* homochlamydeous-*Allium, Yucca*, most of the monocot; heterochlamydeous-most of the dicots. |
| | | Polypetalous-When petals are free, *e.g.* Mustard |
| | | Gamopetalous- When petals are fused, *e.g. Datura* |
| | | **Bracts**-Bract is a leaf-like structure at the base of a flower, **(i) Foliaceous** *e.g. Adhatoda* **(ii) Petaloid** *e.g.* Bougainvillea, Poinsettia **(iii) Scaly** *e.g. Casuarina, Dahlia* **(iv) Spathe** *e.g. Musa, Commelina, Rhoeo* **(v) Cymba (tough and woody bract**) *e.g. Borassus, Cocos* **(vi) Involucre-** *e.g. Helianthus, Tagetes, Crinum* **(vii) Involucel-** *e.g.* Coriandrum, Echinops **(viii) Capsule-** *e.g.* Quercus **(ix) Epicalyx-** *e.g. Hibiscus rosa sinensis* **(x) Glume-** *e.g.* Cyperaceae and Poaceae |
| | | Bracteoles- Bracteole is a small bract on the stalk of a flower. |
| 8. | Calyx | **(a)** Number of sepals |
| | | **(b)** Poly- or gamosepalous. If polysepalous, the shape, outline and the apex of the sepals should be described. If gamosepalous, give special forms such as tubular, campanulate, bilabiate, spurred *etc.* |
| | | **(c)** Green or petaloid |
| | | **(d)** Inferior or superior |
| | | **(e)** Aestivation of the calyx |
| | | **Polysepalous-** When sepals are free *e.g. Polyalthia, Tribulus terrestris*, tomato, mustard |
| | | **Gamosepalous-** When sepals are united *e.g. Hibiscus, Datura fastuosa, Thespesia populnea* |
| | | **Deciduous (Caudcous)**-Falling off early, or prematurely *e.g. Argemone mexicana* |
| | | **Persistent, marcescent**-Remaining attached in the fruit also *e.g.* Tomato, brinjal, Chillies, *Ocimum canum* |
| | | **Persistent, acrescent-** A persistent calyx growing in size along with the fruit. *Withania somnifera, e.g. Physalis minima* |
| | | **Petaloid**- *e.g. Mussaenda, Saraca, Tamarindus, Clerodendron* |
| | | **Pappus**- *e.g. Tridax* |
| | | **Bilabiate or Bilipped-** *e.g. Ocimum* |
| | | **Cup-like-** *e.g. Pongamia, Thespesia* |
| | | **Tubular-** *e.g. Datura* |
| | | **Fleshy-** *e.g. Dellenia* |
| 9. | Corolla | **(a)** Number of petals |
| | | **(b)** Poly- or gamosepalous. If polypetalous, the shape, outline and the apex of petals should be described. Give special forms, if any such as caryophyllaceous, rosaceous, papilioanaceous *etc.* If gamosepalous, give special forms such as tubular, campanulate, bilabiate, spurred *etc.* |
| | | **(c)** Colour of the petals |
| | | **(d)** Inferior or superior |
| | | **(e)** Aestivation of the corolla |
| | | **(I) Aestivation of the calyx and corolla**-Aestivation is the way that the sepals and petals overlap one another in the bud. It is often better seen in partially-opened flowers than in fully opened ones. |
| | | **Valvate-** An aestivation when the segments of corolla are so places that their edges touch each other not overlap *e.g. Solanum, Calotropis*, Custard apple. |
| | | **Twisted (contorted) -** One margin of the petal overlaps that of the next one, and the next margin overlaps and third one *e.g.* China rose, Cotton, *Thespesia populnea.* |
| | | **Imbricate-**One member of whorl is outside all the other (*i.e.* its margins are free) and inside all the others (*i.e.* both margins are overlapped) the other overlap by one margin only *e.g. Bauhinia, Cassia auriculata, Tecoma stans* |
| | | **Quincuncial-**An imbricate aestivation with 5 petals or sepals out of which 2 are exterior, 2 interior and the 5$^{th}$ has one margin exterior and one interior, *e.g.* sepals of *Ipomoea cornea, Cassia auriculata* |
| | | **Vexillary-** An imbricate (descending imbricate), in which out of the 5 petals the posterior one is the largest and covers the lateral petals and the lateral petals (wings) overlap the 2 anterior and smallest petals (keels) *e.g.* Papilionaceae |
| | | **(II) Shape of corolla** |
| | | **Cruciform-**There are 4 petals arranged in the form of a cross *e.g. Brassica* |
| | | **Caryophyllaceous-**The corolla with 5 free petals; the petals are with long claws and with limbs placed at right angles to the claws *e.g.* Dianthus, Silene. |
| | | **Polypetalous and deciduous**-*Tribulus terrestris, Gynandropsis pentaphylla* |
| | | Rosaceous-The petals spreading like that of rose *e.g.* rose. |

| Sl.No. | Traits | Description |
|---|---|---|
| | | **Campanulate-**Corolla is bell-shaped *e.g. Campanula* |
| | | **Infundibuliform-**Funnel-shaped corolla, *e.g. Petunia, Datura fastuosa, Ipomoea sepiaria* |
| | | **Hypocrateriform (Salver-shaped)-**A gamopetalous corolla with a long tube horizontal limb *e.g. Clerodendron, Vinca rosea* |
| | | **Rotate-Wheel-shaped gamopetalous corolla-**The gamopetalous corolla with a flat and circular limb at right angles in the short tube *e.g. Solanum nigrum, Solanum xanthocarpum* |
| | | **Papilionaceous butterfly-like-**The corolla has one large posterior standard, 2 lateral wings and 2 innermost and smallest keels *e.g. Pisum sativum, Crotolaria juncea, Sesbania grandiflora.* |
| | | **Bilabiate-**2-lipped, zygomorphic gamopetalous corolla *e.g. Salvia* |
| | | **Personate-** Zygomorphic gamopetalous corolla 2-lips *e.g. Antirrhinum* |
| | | **Ligulate (strap-shaped)-** Zygomorphic gamopetalous corolla forming a short, narrow tube below and ligule-like flat structure above *e.g.* ray floret of *Helianthus annuus, Tridax procumbens* |
| | | **Tubular-**Tube-like gamopetalous corolla *e.g.* disc floret of *Helianthus annuus, Ruellia prostrate, Tecoma stans.* |
| | | **Spur-**A sac-like or tubular projection ofa petal *e.g. Viola* |
| | | **Nectary-**A nectar-secreting gland in petals *e.g. Salvia.* |
| | | **Corona-**The appendages found in between corolla and stamens, or on the corolla, *e.g. Calotropis,* |
| 10. | Perianth | If perianth is present, describe it in a similar way except that the terms poly-and gamophyllous should be used |
| 11. | Androecium | **(a)** Number of stamens. If more than ten, write indefinite |
| | | **(b)** Polyandrous, syngenesious or adelphous. If polyandrous, write special arrangements if any, such as didynamous, tetradynamous *etc.* if adelphous, whether mono-di-polyadelphous. |
| | | **(c)** Whether epipetalous or free from the petals |
| | | **(d)** Nature of the filaments, whether long, short or flattened |
| | | **(e)** Anthers introrse or extrorse |
| | | **(f)** Colour, fixation and dehiscence of the anthers |
| | | **Androecium-**The androecium of a flower is made up of the male reproductive parts-*i.e.* the microsporophyll or stamens. It occupies a position outward from the gynoecium and inward from the corolla, but the stamens are sometimes inserted on the corolla (epipetalsous), and sometimes stamens are attached to the pistil (gyandrous). In all perigynous flowers the stamens are inserted on a hypanthium (the calyx-tube and surrounding tissue which rise to petals and stamens). Each stamen consists of two parts; **(i) Anther-**The sac-like part, which contains pollen; **(ii) Filament-** Filament connects the anther to the floral axis. |
| | | **(a) Inserted-**The stamens do not extend beyond the petals *e.g. Datura*, Bean. |
| | | **(b) Exerted-** The stamens extend beyond the petals *e.g. Acasia, Samanea.* |
| | | **(c) Haplostemonous-**Stamens are arranged in one whorls; alternipetalous *e.g. Solanum* |
| | | **(d) Dilostemonous-**Stamens are arranged in two whorls; outer whorl alternipetalous *e.g.* Cassia |
| | | **(e) Epiphyllous-**The filaments of stamens remain united with perianth leaves and anthers are free *e.g.* Onion. |
| | | **(f) Epipetalous-** The filaments of stamens are united to petals by their filaments *e.g.* Potato, sunflower, *Datura fastuosa, Ipomoea sepiaria*, brinjal. |
| | | **(g) Gynandrous-**The stamens are united to carpel *e.g. Calotropis, Aristolochia bracteata, Cryptostegia.* |
| | | **Monodelphous-** The filaments of all stamens are united to form a single tube like structure and anthers remain free *e.g.* Cotton, *Crotolaria verrucosa, Azadirachta indica.* |
| | | **Diadelphous-** The filaments of stamens are fused to form two bundles *e.g.* pea, *Sesbania grandiflora.* |
| | | **Polydelphous-** The filaments of stamens are united to form a number of bundles and the anther remain free *e.g.* Citrus, *Salmalia malabarica, Melaleuca.* |
| | | **Synandrous-**The anthers of stamens remain fused and filaments remain free *e.g. Cucurbita maxima*, sunflower. |
| | | **Syngenesious-** The stamens are united through their whole length *i.e.* by filaments and anthers *e.g.* in members of family Cucurbitaceae, *Tridax procumbens, Helianthus.* |
| | | **Tetradynamous-**In this condition out of 6 stamens 4 are long and 2 are short *e.g. Brassica.* |
| | | **Didynamous-** In this condition out of 6 stamens 4 are long and 2 are short *e.g. Tecoma stans, Ocimum.* |
| | | **Bisifixed-**When filament is attached to the base of anther *e.g. Brassica, Datura.* |
| | | **Adnate-**When the filament runs up the whole length of anther from the base to apex *e.g. Nicotiana, Magnolia, Cassia auriculata, Nelumbium.* |
| | | **Dorsifixed-** When the filaments attached to the dorsal side of anther *e.g. Sesbania, Passiflora, Annona.* |
| | | Versatile- When the filament is attached to the back of anther (dorsifixed) at one point only, loosely permitting a free motion *e.g. Crinum asiaticum, Delonix*, grasses. |

| Sl.No. | Traits | Description |
|---|---|---|
| | | **Anther parts-(i) Connective-**Filament extension between thecae **(ii) Locule-**Compartment of an anther **(iii) Pollen grain-**Young male gametophyte **(iv) Pollen sac-**Male sporangium |
| | | **Theca-**One half of anther containing two pollen sacs or male sporangia. |
| | | **Anther cross section (i) Monothecous-**When 2 loculi are seen in a transverse section of the anther, *e.g. Phyllanthus, Hibiscus.* **(ii) Dithecous-**When 4 loculi are seen in a transverse section of the anther, *e.g. Michelia.* |
| | | **Orientation of face (i) Extrose-**When in the bud condition, the face of the anther is oriented towards the corolla; anther dehiscing longitudinally outward *e.g. Argemone mexicana, poppy* **(ii) Introse-**When in the bud, the face of the anther is oriented towards the gynoecium; anther dehiscing longitudinally inward *e.g. Helianthus annuus, Dianthes,* Citrus. |
| | | **Dehiscence (i) Longitudinal-**When the anthers dehisce by longitudinal slits *e.g. Datura* **(ii) Porous-** When the pollen is liberated through apical or basal pores *e.g. Cassia, Solanum.* **(iii) Transverse-**Dehiscing at right angles to long axis of theca, *e.g. Hibiscus* **(iv) Valvular-**When the anther dehisces by the separation of part of the anther wall in the form of valves; dehiscence by recurred valves, *e.g. Cassytha, Berberis, Cinnamomum.* |
| 12. | Gynoecium | **(a)** Number of carpels, whether mono-, bi-, tri-, tetra-, penta- or polycarpellary |
| | | **(b)** Syncarpous or apocarpous |
| | | **(c)** Ovary superior or inferior) |
| | | **(d)** Number of loculi |
| | | **(e)** Number of ovules in each loculus or on each placenta |
| | | **(f)** Shape of the ovules |
| | | **(g)** Placentation |
| | | **(h)** Nature and form of the style and the stigma |
| | | **(i)** Nector secreting disc should be described if present |
| | | **Gynoecium-**The gynoecium of a flower is made up of the female reproductive parts-the carpels or pistils. It occupies a central position in the flower. **(i) Trimerous-**In the monocotyledons the flowers usually have a numerical plan of three:3 sepals, 3 petals and usually 3 or a multiple of 3 stamens, generally termed as trimerous flowers **(ii) Tetramerous-** Dicotyledon flowers usually constructed on a numerical plan of four, **(iii) Pentamerous-** Dicotyledon flowers usually constructed on a numerical plan of five. |
| | | (*The parts of a flower are usually arranged in such a manner that the petals alternate with the sepals and the stamens alternate with the petals. The pistil or carpels, however, are often opposite the sepals. This feature of alternation of parts gives us valuable clue to the vestigial or missing flower parts*). |
| | | The gynoecium may consist of a single pistil (lily) or many pistils (buttercup). |
| | | **Parts of pistil** |
| | | **(i) Stigma-**The pollen receptive part at the summit, which may be single or variously lobed or branched **(ii) Style-**The stalk like portion below the stigma; and **(iii) Ovary-**The enlarged portion at the base, which contains one or more ovules or immature seeds. |
| | | **Type of pistil** |
| | | **(i) Simple pistil-**When the pistil is composed of a single megasporophyll, or carpel, it is termed a simple pistil |
| | | **(ii) Compound pistil-**When it is composed of two or more carpel that are more or less united, it is a compound pistil. |
| | | **Placentation-** Placentation is the type of arrangement of placenta in a syncarpous ovary. **(i) Parietal-**Carpels are fused only by their margins, placentation then appearing as internal ridges on ovary wall *e.g. Brassica, Argemone, Carica,* Cucurbitaceae. **(ii) Axile-**Margins of carpel fold inwards, fusing together in centre of ovary to form a single, central placenta; ovary is divided into as many compartments (loculi) as there are carpels *e.g. Hibiscus, Thespesia, Slonum, Citrus, Allium,* tomato. **(iii) Marginal-**The gynoecium in monocarpellary or polycarpellary, apocarpous. **(iv) Basal-**The ovules are few or reduced to one and are borne at the base of ovary, the ovules when solitary often filling the cavity; the ovary is unilocular *e.g.* Compositae **(v) Sperficial (Laminar)-**Carpels numerous, the placenta develop all-round the inner surface of the partition wall; ovary multilocular *e.g. Nymphaea.* **(vi) Free central-**The syncarpous gynoecium is unilocular and the placentae are borne on a central column formed by the extension of base of the gynoecium where the carpels fuse, or the suppression of the septa *e.g. Stellaria, Dianthus, Primula.* |
| | | **Ovule-**Ovules are borne on the placenta. Each ovule is a stalked structure. The stalk is known as the funiculus |
| | | **Parts of ovule (i) Chalaza-**End of ovule opposite micropyle **(ii) Embryo sac-**Female gametophyte **(iii) Integuments-**Outer covering of ovule **(iv) Micropyle-** Hole through integuments(s), **(v) Nucellus-**The body of the ovule is known as nucellus; megasporangium in seed plants **(vi) Raphe-**Longitudinal ridge or outer integument **(vii) Hilum-**The point of attachment of the body of the ovule to the tip of the funiculus |
| | | **Types of ovules- (i) Amphitropous-**With body bent or curved on both sides so that the micropyle is near the medially attached funiculus *e.g.* Centrospermae, Alimaceae, Ranunculus, Butomus, Alisma, **(ii) Anatropous-**When the ovule turns by about 180 degrees just below the chalaza so that the micropyle lies very close to the hilum. The micropyle and chalaza remains in one line but the funiculus is not in the same line *e.g.* Magnoliaceae, Compositae; *Annona, Tridax,* **(iii) Campylotropous-**With body bent or curved on one side so that micropyle is near medially attached to funiculus *e.g.* Fabaceae (old-Papilionaceae), Cruciferae (old-Brassicaceae), Capparis, bean, **(iv) Hemianatropous (Hemitropous)-** When the ovule is attached to the funiculus near about its middle and the line joining the micropyle and the chalaza is at right angle to the funiculus *e.g.* Primula. |

| Sl.No. | Traits | Description |
|---|---|---|

**(v) Orthotropous (Atropous)-**When the micropyle, chalaza and the funiculus lie in one straight line *e.g.* Piperaceae, Vitaceae, *Polygonum, Piper.*

**Carpel-**The basic unit of construction of a pistil is the carpel, which is a single megasporophyll, or modified seed bearing leaf.

**Types-(i) Monocarpellary-***Dolichos*, **(ii) Bicarpellary-***Brassica, Tridax*, **(iii) Tricarpellary-***Cocos*, **(iv) Tetracarpellary-***Gossypium*, **(v) Pentacarpellary-***Hibiscus*, **(vi) Multicarpellary-***Annona, Abutilon.*

**Locules- (i) Unilocular-**Bean, *Tridax, Dianthus*, **(ii) Bi-locular-** *Solanum*, **(iii) Tri-locular-***Allium, Cocos*, **(iv) Tetra-locular-***Gossypium, Datura*, **(v) Penta-locular-***Hibiscus*, **(vi) Multi-locular-** *Abutilon*

**Style-**The elongated apical part of a carpel or ovary bearing the stigma at its tip. **(1) Condupplicate-**Folded with a longitudinal groove **(2) Ecentric-**Off-centre style, **(3) Fimbriate-**Finged, **(4) Flabellate-**Fan-shaped, **(5) Geniculate-**Bent abruptly **(6) Gynobasic-**Attached at base of ovary in central depression *e.g.* Ocimum **(7) Heterostylous-**With styles of different size or lengths or shapes within a species, **(8) Homostylous-**With of same sizes or lengths and shapes, **(9) Stylopodic-**With a stylopodium or discoid base, as in the Apiaceae (Umbelliferae), **(10) Terete-**Cylindrical and elongate, **(11) Tuberculate-**With hard, swollen, persistent base or tubercle, **(12) Umbraculate-**Umbrella-shaped, as in Sarracenia, **(13) Terminal-**Lying in the same straight line with the ovary *e.g.* Datura, *Hibiscus*, **(14) Lateral-**That arises from the side of the ovary, *e.g.* Mango, *Anacardium*, **(15) Capitate-**Head-like, **(16) Clavate-**Club-shaped, **(17) Crested-**With a terminal ridge or tuft, **(18) Decurrent-**Elongate, extending downwards, **(19) Diffuse-**Spread over a wide surface, **(20) Discoid-**Disc-like, **(21) Lineate-**In lines, stigmatic surface linear, **(22) Lobed-**Divided into lobes, **(23) Plumose-** Feather-like.

**Stigma-(i) Capitate or round-***Hibiscus, Citrus*, **(ii) Feathery or brush-like-**grasses **(iii)** Fid or forked-*Tridax*, **(iv) Dump bell shaped-***Cathernthus*, **(v) Persistent and wheel-like-***Poppy*, **(vi) sensitive-***Mimulus, Martynia* **(v) Funnel shaped-***Crocus*, **(vi) Discoid-***Melia*, **(vii) Radiate hood-like-***Poppy*, **(viii) Bifid-***Ixora, Sonchus*, **(ix) Knob-like-***Justica, Cryptostegia, Achyranthus*, **(x)** Drum-shaped-Apocynaceae **(xi) Sticky-***Cleome viscosa.*

A flower should be cut vertically into two half-flowers.

**(A)** If the sepals and petals inserted

**(i) Hypogynous-**Independently on the receptacle below the gynoecium, the flowers are hypogynous. In this case the thalamus is convex and the gynoecium is situated at the apex and other whorls arise below it. This type of flower is said to be hypogynous and ovary as superior *e.g.* Mustard, *Datura, Ranunculus, Annona, Hibiscus.*

**(ii) Perigynous-**On the rim of a saucer-shaped to cylindrical tube around the gynoecium, the flowers are perigynous. In this type the thalamus forms a shallow or deep cup shaped structure around ovary. The gynoecium is situated in the centre and sepals, petals and stamens are arranged outside. This type of flowers said to be perigynous and ovary to be superior, *e.g.* Rose, *Crotolaria, Tephrosia, Delonix.*

**(iii) Epigynous-**At the apex of the gynoecium wholly above it, the flowers are Epigynous. In this type the thalamus enclosed the ovary entirely and fuses. The sepals, petals and stamens arise from the top of ovary. Such flowers known as epigynous and ovary as inferior, *e.g. Pyrus malus, Oenothera, Tridax, Cucurbita.*

**(B)** If the gynoecium inserted above the place of insertion of other floral parts it is superior, *e.g. Argemone mexicana, Datura*; half superior in *Dolichos.*

**(C)** If situated below the place of insertion of the other floral part-gynoecium is inferior, *e.g. Coccinea indica, Citrullus colocynthis, Aristolochia bracteata, Psidium.*

**(D)** Gynoecium is semi-inferior, if with the place of insertion of other floral parts some distance up its side, *e.g. Punica granatum.*

**-Apocarpous-**The gynoecium is apocarpous if the carpels are free, *e.g. Annona, Rosa.* The pistil is simple if it is made up of only one carpel, *e.g. Crotolaria.* The pistil is compound if it is made up of two or more carpels joined together *e.g. Thespesia populnea*, Sub-carpous-carpels partially united *e.g. Dianthus, Calotropis, Hibiscus, Catheranthus.*

**-Syncarpous-**Carpels are united completely, *e.g. Datura*

| 13. | Fruit | (a) Kind of fruit (Seeds enclosed within the mature ovary-so-called fruit). |

**Fruit-**Fruit is the ripened ovary of the flower, enclosing seeds.

**1- Multiple fruits-**Derived from several flowers or an inflorescence (except solitary terminal and solitary axillary). These are two types: **(i) Sorosis –**A multiple fruit developing from a spike or spadix; flowers fuse tgether by their succulent sepals and the axis bearing them grows and becomes fleshy or woody, and the whole inflorescence becomes a compact mass, *e.g.* pineapple (*Ananas sativus*), jack fruit, (ii) Syconus-Develops from a hollow, pear-shaped, fleshy receptacle which enclose a number of minute male and female flowers, the receptacle becomes fleshy and forms the so called fruit; it really enclose a number of true fruits or achenes *e.g.* Fig, Banyan.

**2. Aggregate fruits-**Derived from a single flower with more than one free carpels, *Etaerio follicles e.g. Calotropis; Etaerio of achenes; Clematis; Etaerio of drups; Rubus; Etaerio of berries;* Custard apple-Syncarpous fruit- *Annona squamosa* the carpels are fleshy and are fused together in one mass forming a berry.

**3. Simple fruits-** Derived from a monocarpellary or polycarpellary syncarpous gynoecium. These are two types; Dry and fleshy

**Dry simple fruits-**The fruit wall or the pericarp is dry.

| Sl.No. | Traits | Description |
|---|---|---|

**Type of simple fruits**

**(A) Achenial fruits:** Derived from monocarpellary or polycarpellary syncarpous pistil, with one ovule. These fruits are single seeded and do not open, **(i) Achene**-A small, dry one-chambered and one-seeded fruit developing from a superior monocarpellary ovary, pericarp is free from the seed coat; *e.g. Clematis, Carthamus tinctorius.* (ii) Caryopsis- A very small, dry, one seeded fruit developing from a superior, monocarpellary ovary, with the pericarp fused with the seed *e.g.* Poaceae; *e.g. Oryza.* **(iii) Cypsela**-A dry one-chambered and one seeded fruit developing from an inferior, bicarpellary ovary with the pericarp and the seed coat free *e.g.* sunflower, **(iv) Nut**-A dry one-chambered and one seeded fruit developing from a superior, bi-or polycarpellary ovary with the pericarp hard and woody, *e.g.* Chestnut. **(v) Simple Samara**-A dry indehiscent, 1 or 2 seeded fruit developing from a bi-or trycarpellary ovary with flattened wing-like outgrowths *e.g. Hiptage, Pterolobium indicum, Hardwickia binate, Combretum ovalifolium.*

**(B) Schizocarpic fruits:** Many seeded and indehiscent dry fruit, *e.g. Abutilon, Rcinis, Tribulus,* **(i) Double Samara**-It consists of 2, 3 or 4 samaras *e.g.* Sapindaceae, **(ii) Cremocarp**- A dry indehiscent, 2-chambered fruit developing from an inferior, bicarpellary ovary; when splits into 2 indehiscent, one seeded mericarps which remains attached to the prolonged end (carpophore) of the axis *e.g.* coriander. **(iii) Carcerulus**-Develops from a bicarpellary pistil with a superior ovary which becomes quadrilocular owing to the formulation of 2 false septa; at maturity the 4 mericarps separate from each other towards the middle *e.g. Ocimum.* **(iv) Lomentum**-The legume is constructed or partitioned between the seeds into a number of one-seeded parts *e.g. Acacia.* **(v) Regma-** A dry schizocarpic fruits fruit developing from a tricarpellary, syncarpous, superior ovary and splitting at maturity into 3 cocci *e.g.* Castor.

**(C) Capsular fruits:** Dehiscent many seeded fruits; develop from a simple or compound pistil **(i) Legume or Pod-** A dry monocarpellary fruit developing from a superior, one chambered ovary and dehiscing by both the sutures *e.g. Pisum, Sesbania, Poinciana,* **(ii) Follicle-** A dry monocarpellary superior, one chambered fruit like the legume, but it dehisces by one suture only *e.g. Calotropis gigantean, Sterculia guttata,* **(iii) Capsule-** A many seeded, uni-or multilocular fruit developing from a superior, bi-or polycarpellary ovary, and dehiscing in various ways-**(a)** *Septicidal capsule*-The dehiscence slits appear along the septa or the ventral sutures of the carpels *e.g.* mustard, *Aristolochia* **(b)** *Loculicidal capsule*-The dehiscence slits appear along the dorsal sutures of the fused carpels *e.g. Ruellia, Hibiscus ficulneus* **(c)** *Septifragal capsule*-When the capsule breaks irregularly or by septicidal or loculicidal means, but the seeds remains attached to the central column of placentae *e.g. Datura, Cedrela toona.* **(iv) Siliqua**-A long, narrow many seeded fruit developing from a superior, bicarpellary ovary with two parietal placenta, dehiscing from below upwards by both the sutures *e.g. Brassica.* **(v) Silicula**-When a siliqua is much shorter and flattened being nearly as broad as it is long and contains only a few seed *e.g. Capsella*

**Fleshy simple fruits:** These fruits are derived from simple or compound pistils. The fruit wall or pericarp is thick and fleshy. It is distinguishable epicarp, mesocarp and endocarp

**(i) True fleshy fruits:** The pericarp is fleshy and distinguishable into epicarp, mesocarp and endocarp, these are two types **(1) Drupe**–A fleshy, one or more-chambered and one or more-seeded fruit developing from a monocarpellary or syncarpous pistil *e.g.* mango, peach, plum, coconut, *Calophyllum inophyllum, Thevetia,* **(2) Berry**-A superior indehiscent, many seeded, fleshy or pulpy fruit developing from a single carpel or a syncarpous pistil, with axile or parietal placentation *e.g. Solanum* (tomato, potato *etc.*), grapes, guava, citrus, *Cephalandra.*

**(ii) False fleshy fruits:** The fruit is derived from the ovary is not edible and is hard. The edible portion is juicy or fleshy thalamus or the receptacle.

**Pome**-An inferior, two or more-celled fleshy syncarpous fruit surrounded by the thalamus *e.g.* apple, pear.

| | | |
|---|---|---|
| 14. | Seeds | **(a)** Endospermic or non-endospermic |

**(b)** Position, shape and the size of the embryo

**(c)** Cotyledons, straight or folded

**(d)** Number of cotyledons

**Fleshy seeds-***Punica granatum, Solanum esculentum*

**Winged seeds-***Tecoma, Oroxylum indicum, Dolichandrone, Cedrela toona*

**Compose seeds-***Calotropis gigantean, Alstonia scholaris, Nerium odorum*

**Hairy seeds-***Hibiscus microanthus, Eriodendron anfractusum*

**-Endospermous seeds-** *Hibiscus cannabinus, Phoenix, Coconut, Trigonella foenumgraecum, Delonix regia, Vinca*

**-Ex- Endospermous seeds-***Cicer arietinum, Erythrina indica, Cajanus indicus, Cucurbita, Cucumis,*

**-Perispermic seeds-***Cardamomum, Nymphaea*

**Aril**-This is an extra fleshy or membranous outgrowth covering the seed either partially or wholly and arises from near the micropyle. It is known as caruncle in *Ricinus,* and strophile in *Polygala.* It forms a fleshy covering in *Pithecellobium dulce* and in *Myristica* it appears as coarse net-work (mace)

*Source*: Subrahmanyam, 1999.

**Table 15.9: Floral Formula of Crop Plants**

| *Family* | *Botanical name* | *Floral formula* |
|---|---|---|
| **Dicotyledons** | | |
| Nymphaceae | *Nymphaea nouchali* Burm.f.syn. *N. lotus* | $\oplus$, ⚥, $K_{3\text{-}5}$, $C_{\infty}$, $A_{\infty}$, $G_{\underline{(\infty)}}$ |
| | *Nymphaea stellata* | $\oplus$, ⚥, $K_{4}$, $C_{\infty}$, $A_{\infty}$, $G_{\underline{(\infty)}}$ |
| | *Nelumbo nucifera* Gaertn. | $\oplus$, ⚥, $K_{4}$, $C_{\infty}$, $A_{\infty}$, $G_{\underline{(\infty)}}$ |
| | *Euryale ferox* | $\oplus$, ⚥, $K_{4}$, $C_{\infty}$, $A_{\infty}$, $G\text{-}_{(\infty)}$ |
| | *Cabomba* | $\oplus$, ⚥, $K_{3}$, $C_{3}$, $A_{3+3}$, $G_{\underline{3}}$ |
| Brassicaceae | *Raphanus sativus* L. | Ebr, $\oplus$, ⚥, $K_{4}$, $C_{4}$, $A_{2+4}$, $G_{\underline{(2)}}$ |
| | *Brassica napus* L. | $\oplus$, ⚥, $K_{4}$, $C_{4}$, $A_{4+2}$, $G_{\underline{(2)}}$ |
| | *Brassica juncea* L. | Br, Ebrl, ⚥, $K_{4}$, $C_{4}$, $A_{4+2}$, $G_{\underline{(2)}}$ |
| | *Brassica campestris* Linn. var. *sarson* Prain | $\oplus$, ⚥, $K_{2+2}$, $C_{4}$, $A_{2+4}$, $G_{\underline{(2)}}$ |
| Fabaceae | *Dolichos lablab* L. or *Dolichos purpureus* | Br, Ebrl, %, ⚥, $K_{(5)}$, $C_{5}$, $A_{(9)+1}$, $G_{\underline{1}}$ |
| | *Trigonella foenum-graecum* L. | %, ⚥, $K_{(5)}$, $C_{1+2+(2)}$, $A_{1+(9)}$, $G_{\underline{1}}$ |
| | *Pisum sativum* L. | Br, %, ⚥, $K_{(5)}$, $C_{1+2+(2)}$, $A_{(9)+1}$, $G_{\underline{1}}$ |
| | *Phaseolus vulgaris* L. | Br, %, $_{+}$, $K_{(5)}$, $C_{1+2+(2)}$, $A_{(9)+1}$, $G_{\underline{1}}$ |
| | *Bauhinia variegata* L. | Br, Brl, %, ⚥, $K_{5}$, $C_{(5)}$, $A_{5}$, $G_{\underline{1}}$ |
| | *Crotolaria juncea* l. | Br, %, ⚥, $K_{(2)+(3)}$, $C_{1+2+(2)}$, $A_{(10)}$, $G_{\underline{1}}$ |
| Cactaceae | *Opuntia dillenii* | Asymmetrical, ⚥, $P_{\infty}$, $A_{\infty}$, $G\text{-}_{(\infty)}$ |
| Cucurbitaceae | *Cucurbita maxima* Duch. | Male flower-Br, $\oplus$,♂, $K_{(5)}$, $C_{(5)}$, $A_{1+(2)+(2)}$, $G_{0}$<br>Female flower- Br, $\oplus$,♀, $K_{(5)}$, $C_{(5)}$, $A_{0}$ staminodes, $G\text{-}_{(3)}$ |
| | *Cucurbita moschata* Duch. | Male flower- $\oplus$,♂, $K_{(5)}$, $C_{(5)}$, $A_{(2)+(2)+1}$, $G_{0}$<br>Female flower- $\oplus$,♀, $K_{(5)}$, $C_{(5)}$, $A_{0 \text{ or staminodes}}$, $G\text{-}_{(3)}$ |
| | *Luffa cylindrica* (L.) Roem | Male flower-Br, $\oplus$,♂, $K_{(5)}$, $C_{5}$, $A_{2+2+1}$, $G_{0}$<br>Female flower- Br, $\oplus$,♀, $K_{5}$, $_{C(5)}$, $A_{0}$, $G\text{-}_{(3)}$ |
| | *Coccinia cordifolia* L. | Male flower-Br, $\oplus$,♂, $K_{(5)}$, $C_{(5)}$, $A_{(2)+(2)+1}$, $G_{0}$<br>Female flower- Br, $\oplus$,♀, $K_{(5)}$, $C_{(5)}$, $A_{3}$ staminodes, $G\text{-}_{(3)}$ |
| | *Momordica charantia* L. | Male flower- $\oplus$,♂, $K_{5}$, $C_{5 \text{ or} (5)}$, $A_{(2)+(2)+1}$, $G_{0}$<br>Female flower- $\oplus$,♀, $K_{5}$, $C_{5 \text{ or } (5)}$, $A_{0}$, $G\text{-}_{(3)}$ |
| | *Cucumis sativus* L. | Male flower- $\oplus$,♂, $K_{(5)}$, $C_{(5)}$, $A_{2+2+1}$, G pistillode<br>Female flower- $\oplus$,♀, $K_{5}$, $C_{5 \text{ or } (5)}$, $A_{0}$, $G\text{-}_{(3)}$ |
| | *Lagenaria siceraria* (Mol.) Standl. | Male flower- $\oplus$,♂, $K_{(5)}$, $C_{5}$, $A_{(2)+(2)+1}$, $G_{0}$<br>Female flower- $\oplus$,♀, $K_{(5)}$, $C_{(5)}$, $A_{0}$, $G\text{-}_{(3)}$ |
| | *Citrullus lanatus* (Thunb.)Mansf. | Male flower- $\oplus$,♂, $K_{(5)}$, $C_{(5)}$, $A_{(2)+(2)+1}$, $G_{0}$<br>Female flower- $\oplus$,♀, $K_{(5)}$, $C_{(5)}$, $A_{0}$, $G\text{-}_{(3)}$ |
| Apiaceae | *Coriandrum sativum* L. | Peripheral flower-Br, %, $_{+}$, $K_{5}$, $C_{5}$, $A_{5}$, $G\text{-}_{(2)}$<br>Central flower- Br, $\oplus$, ⚥, $C_{5}$, $A_{5}$, $G\text{-}_{(2)}$ |
| | *Daucus carota* L. | Central flower-Br, $\oplus$, ⚥, $K_{5}$, $C_{5}$, $A_{5}$, $G\text{-}_{(2)}$<br>Peripheral flower-Br, %, ⚥, $K_{5}$, $C_{5}$, $A_{5}$, $G\text{-}_{(2)}$ |
| | *Foeniculum vulgare* Mill. (fennel) | Ebr, $\oplus$, ⚥, $K_{(5)}$, $C_{5}$, $A_{5}$, $G_{(2)}$ |
| | *Anethum graveolens* L. Syn. *Peucedanum graveolens* Benth & Hook. f. (dill or soya) | Ebr, Ebrl, $\oplus$, ⚥, $K_{5\text{scales}}$, $C_{(5)}$, $A_{5}$, $G\text{-}_{(2)}$ |
| | *Trachyspermum ammi* (Linn.) (ajwain) | Br, Brl, $\oplus$, ⚥, $K_{(5)}$, $C_{5}$, $A_{5}$, $G\text{-}_{(2)}$ |
| Convolvulaceae | *Convolvulus arvensis* L. | Br, $\oplus$, ⚥, $K_{(5)}$, $C_{(5)}$, $A_{5}$, $G\text{-}_{(2)}$ |
| | *Convolvulus pluriculis* Choisy. | Br, $\oplus$, ⚥, $K_{(5)}$, $C_{(5)}$, $A_{5}$, $G\text{-}_{(2)}$ |
| | *Ipomoea palmata* | Br, $\oplus$, ⚥, $K_{(5)}$, $C_{(5)}$, $A_{5}$, $G_{\underline{(2)}}$ |
| | *Ipomoea tropica* | Br, Brl,$\oplus$, ⚥, $K_{(5)}$, $C_{(5)}$, $A_{5}$, $G_{\underline{(2)}}$ |
| | *Ipomoea fistulosa* Mart ex Choisy | Br, Brl,$\oplus$, ⚥, $K_{(5)}$, $C_{(5)}$, $A_{5}$, $G_{\underline{(2)}}$ |

| Family | Botanical name | Floral formula |
|---|---|---|
| | *Ipomoea hederacea* | Br, Brl, $\oplus$, $\male\female$, $K_{(5)}$, $C_{(5)}$, $A_{2+3}$, $G_{\underline{(2)}}$ |
| Solanaceae | *Solanum tuberosum* L. | Br, $\oplus$, $\male\female$, $K_{(5)}$, $C_{(5)}$, $A_5$, $G_{\underline{(2)}}$ |
| | *Solanum nigrum* L. | Br, Ebrl, $\oplus$, $\male\female$, $K_{(5)}$, $C_{(5)}$, $A_5$, $G_{\underline{(2)}}$ |
| | *Capsicum frutescens* L. | Br, Ebrl, $\oplus$, $\male\female$, $K_{(5)}$, $C_{(5)}$, $A_5$, $G_{(2)}$ |
| | *Solanum melongena* | Br, Ebrl, $\oplus$, $\male\female$, $K_{(5)}$, $C_{(5)}$, $A_5$, $G_{\underline{(2)}}$ |
| | *Solanum xanthocarpum* | $\oplus$, $\male\female$, $K_{(5)}$, $C_{(5)}$, $A_5$, $G_{\underline{(2)}}$ |
| | *Solanum torvum* Sw. | Ebr, Ebrl, %, $\male\female$, $K_{(5)}$, $C_{(5)}$, $A_5$, $G_{(2)}$ |
| Amaranthaceae | *Amaranthus viridis* L. | Male flower- Br, Brl, $\oplus$, $\male$, $P_3$, $A_3$, $G_0$<br>Female flower- Br, Brl, $\oplus$, $\female$, $P_3$, $A_0$, $G_{(2)}$ |
| | *Amaranthus spinosus* | Male flower- Br, Brl, $\oplus$, $\male$, $P_5$, $A_{4 \text{ or } 5}$, $G_0$<br>Female flower- Br, Brl, $\oplus$, $\female$, $P_5$, $A_0$, $G_{\underline{(2)}}$ |
| Chenopodiaceae | *Chenopodium album* L. | Br, Ebrl, $\oplus$, $\male\female$, $P_{(5)}$, $A_5$, $G_{(2)}$ |
| | *Beta vulgaris* L. | Br, $\oplus$, $\male\female$, $P_5$, $A_5$, $G\text{-}_{\underline{(3)}}$ |
| Polygonaceae | *Polygonum orientale* L. | Br, $\oplus$, $\male\female$, $P_5$, $A_{5+2}$, $G\text{-}_{\underline{(3)}}$ |
| | *Polygonum pleibijum* | $\oplus$, $\male\female$, $P_5$, $A_{5+3}$, $G\text{-}_{\underline{(3)}}$ |
| | *Polygonum barbatum* L. | Br, Ebrl, $\oplus$, $\male\female$, $P_{3+3}$, $A_{4+3+8 \text{ staminodes}}$, $G_{\underline{(2)}}$ |
| | *Polygonum glabrum* Willd. | Br, Ebrl, $\oplus$, $\male\female$, $P_5$, $A_{5+2}$, $G_{(2-3)}$ |
| | *Rumex maritimus* L. | $\oplus$, $\male\female$, $P_{(3+3)}$, $A_{3+3}$, $G_{\underline{(3)}}$ |
| | *Rumex dentatus* L. | Ebrl, $\oplus$, $\male\female$, $P_{(3+3)}$, $A_{3+3}$, $G_{\underline{(3)}}$ |
| Portulaceae | *Portulaca oleracea* L. | $\oplus$, $\male\female$, $K_2$, $C_5$, $A_{8-10}$, $G_{(3-5)}$ |
| Rutaceae | *Murraya paniculata* (L.) Jacq. | Br, $\oplus$, $\male\female$, $K_{(2)}$, $C_5$, $A_{5+5}$, $G_{\underline{(3)}}$ |
| Asteraceae | *Centauraea cyanus* L. (sweet sultan) | Ray-floret-▨, Neuter, $K_0$, $C_{(5)}$, $A_0$, $G_0$ |
| | | Disc-floret-$\oplus$, $\male\female$, $K_{pappus}$, $C_{(5)}$, $A_5$, $G\text{-}_{(2)}$ |
| **Monocotyledons** | | |
| Musaaceae | *Musa paradisiaca* | Ebr, Ebrl, %, $\male\female$, $P_{(3)+3}$, $A_{3+2+1}$, $G\text{-}_{(3)}$ |
| | *Musa sapientum* Linn. | Male flower-%, $\male$, $P_{3+3)}$, $A_{5+1 \text{ staminodes}}$, $G_0$<br>Female flower- %, $\female$, $P_{3+3)}$, $A_0$, $G\text{-}_{(3)}$ |
| Zingiberaceae | *Zingiber officinale* | Br, Ebrl, %, $\male\female$, $K_3$, $C_3$, $A_1$, $G\text{-}_{(3)}$ |
| | *Curcuma longa* | Br, %, $\male\female$, $K_3$, $C_3$, $A_1$, $G\text{-}_{(3)}$ |
| Liliaceae | *Allium cepa* L. | Br, Ebrl, $\oplus$, $\male\female$, $P_{3+3}$, $A_{3+3}$, $G_{(3)}$ |
| | *Asparagus racemosus* Roxb. | Br, Ebrl, $\oplus$, $\male\female$, $P_{3+3}$, $A_{3+3}$, $G_{(3)}$ |
| | *Yucca gloriosa* | Br, $\oplus$, $\male\female$, $P_{3+3}$, $A_{3+3}$, $G_{\underline{(3)}}$ |
| Araceae | *Alocasia indica* Schott. | $\male\female$, or , $P_0$, $A_{3-5}$, $G_{(2-4)}$ |
| | *Colocasia esculenta* (L.) Schott. Syn. *C. antiquorum* | Male flower-$\oplus$, $\male$, $P_0$, $A_{3+3}$, $G_0$ |
| | | Female flower-$\oplus$, $\male\female$ $P_0$, $A_0$, $G_{(3)}$ |
| Poaceae | *Zea mays* | Male flower- Br, %, $\male$, $P_{2(lodicules)}$, $A_3$, $G_0$<br>Female flower- Br, %, , $P_{2(lodicules)}$, $A_0$, $G_1$ |
| | *Bambusa arundinacea* (Retz.) Wild. | %, $\male\female$, $P_{(2lodiculues)}$, $A_{3+3}$, $G_1$ |
| Amaryllidaceae | *Narcissus poeticus* L. | Br, $\oplus$, $\male\female$, $P_{(3+3+corona)}$, $A_{3+3}$, $G_{\underline{(3)}}$ |

(2) **Class: Monocotyledons-** Embryo with a single cotyledon; stem with closed bundles; leaves with parallel venation; flowers usually trimerous.

# 2. Hutchinson's System

This is for identification of the families of Angiosperms (flowering plants)

**I. Dicotledons**

(a Flowers mostly penta, or tetramerous

(b) Calyx and corolla mostly distinct

**II. Monocotledons**

(a) Flowers mostly trimerous

(b) Calyx and corolla mostly not distinguished as separate whorls; perianth present

## Method of Describing a Flowering Plant

The scheme showing in Table 15.8 should be used while describing and identifying a flowering plant.

## Classification of Crop Species on the Basis of their Natural Mode of Pollination

See the Table 15.10.

**Table 15.10: Classification of Vegetable Crop Species**

| Sl.No. | Vegetable Crop | Botanical Name |
|---|---|---|
| **Self-pollinated crops** | | |
| 1. | Pea | *Pisum sativum* L. |
| 2. | Cowpea | *Vigna unguiculata* L. |
| 3. | French bean | *Phaseolus vulgaris* L. |
| 4. | Guar | *Cymopsis tetragonoloba* L. |
| 5. | Tomato | *Solanum lycopersicon* |
| 6. | Okra | *Abelmoschus esculentus* |
| 7. | Lettuce | *Lactuca sativa* |
| 8. | Brinjal | *Solanum melongena* |
| 9. | Chillies | *Capsicum annuum* |
| 10. | Parsnip | *Pastinaca sativa* |
| 11. | Potato | *Solanum tuberosum* |
| **Cross-pollinated crop** | | |
| 12. | Sweet corn | *Zea mays* var. *rugosa* |
| 13. | Cabbage | *Brassica oleracea* |
| 14. | Cauliflower | *Brassica oleracea* |
| 15. | Turnip | *Brassica rapa* |
| 16. | Broccoli | *Brassica oleracea* |
| 17. | Carrot | *Daucus carota* |
| 18. | Onion | *Allium cepa* |
| 19. | Garlic | *Allium sativum* |
| 20. | Radish | *Raphanus sativus* |
| 21. | Cucumber | *Cucumis sativa* |
| 22. | Pumpkin | *Cucurbita moschata* |
| 23. | Muskmelon | *Cucumis melo* |
| 24. | Watermelon | *Citrullus vulgaris* |
| 25. | Sweet potat | *Ipomoea batatas* |
| 26. | Beets | *Beta vulgaris* |
| 27. | Parsley | *Petroselinum hortense* |
| 28. | celery | *Apium graveolens* |
| 29. | Spinach | *Spinacea oleracea* |
| 30. | Asparagus | *Asparagus offinalis* |
| 31. | Coriander | *Coriandrum sativum* |

## Herbarium

A herbarium is defined as:

*A collection of plants that usually have been dried, pressed, preserved/mounted on sheets and arranged in accordance with any accepted system of classification for future reference and study."*

The herbaria present a picture of the plant world through the representative specimens and are the chief basis for future monographic and phytogeographical studies.

## Virtual Herbarium

Digital images are present as virtual herbarium in alphabetical order in family, genus and species folders. These images have been labelled with unique identify numbers (as of herbarium specimens) and linked to database. Digital images for over 4,000 species (~6,000 images) of crop plants and their wild relatives and potentially useful plants are available for use. Digitization process involves scanning of authenticated/identified specimens/taxa, linking with digital images of reference herbarium specimen(s); working out "spot" characters for identification of species (close view, if needed). Digital scans (jpeg images) with good resolution (300 dpi for close up of parts-seed, trichomes and 600-1200 dpi for micro details) not only facilitate fast access of material for identification but drastically reduce chances of damage due to routine handling.

## Importance of Herbarium Specimens

Herbarium specimens of a taxon, represented over time and space, have a collective value, providing a record of the diversity that is significant as germplasm; individually important specimens include vouchers from exploration programmes and experimental studies.

The herbarium specimens are great importance in:

☆ Providing material collected from diverse areas, in different season at one place, for study.

☆ Maintaining authentically identified and correctly labeled specimens and associated records for the known and studied plants.

☆ The individual variation and details of structure and development.

☆ Most comparative studies for taxonomic purposes are made in the herbarium and laboratory.

☆ The newly collected specimens are identified by the comparison of their morphology with morphology of plants collected in herbarium.

☆ Availability of samples representative of variation, expressed phenotypically as plant specimens, photographic records, voucher samples, seed material *etc.*

## Build of Herbarium Specimens

Build-up of material as through specimens/seeds collected during explorations in different agro-ecological zones of India, material introduced from abroad under various research/breeding/selection programme and also vouchers deposited of the systematic studies on crop-groups. Presently there is thrust on target collections towards landraces, CWR and with thrust in collections from unrepresented areas (as neglected regions, across eco-geographical regions, tribal/north-eastern region). To achieve the targeted collection, gaps are revisited from time to time. Some important works have served as baseline for build-up of material in the NHCP: cultivated taxa (and variability within them) of crop/economic species (Ambasta *et al.*, 1986; Nayar *et al.*, 2003), wild relatives of major crop taxa (Arora and Nayar, 1984; Pandey *et al.*, 2005) and wild edible and economic taxa (Arora and Pandey, 1996).

Herbarium specimens representing plant genetic resources introduced from abroad are a distinctive component of the National Herbarium of Cultivated Plants (NHCP). These specimens represent diversity augmented in crops and wild species mostly not native to the Indian region. Herbarium specimens, bearing the unique identity number assigned to germplasm introduced into the Indian region, were screened, checked with primary and secondary data records for identity, source locality/area, and its availability as *ex situ* germplasm (Nayar, 2014). Any herbarium associated with PGR has the distinctive feature of having in its holdings not only native diversity but also of exotic germplasm augmented through exploration or introduction of germplasm in the form of seeds or planting material (Nayar *et al.*, 2011). Thus, these specimens are reference material of taxa not native to the Indian region.

## Systems of Arrangement of Herbarium Specimens

System of arrangement of herbarium specimens differs from that of the other herbaria; specimens are arranged by families, than by genera and than by species; all in an alphabetical order. This was found more convenient for wide use of PGR workers, para-botanists and non-taxonomists and beginners. For efficient access to use the resources, documentation of the herbarium holdings as soft data, images as virtual herbarium, index cards and inventory of digitised taxa (Nayar *et al.*, 2011) can be referred. Facilities such as net-house to grow out for identification and teaching purpose, experimental areas for study on selected taxa and raising material received as vegetative propagules/ seed and are also available. In addition, standardization of methodology for economic/ecofriendly storage, specialized groups such as landraces (variation), difficult groups (succulents, large fruited types, aquatic plants, plants tending to leaf fall on drying) *etc.* (Pandey *et al.*, 2013). Guidelines for effective use of the herbarium for consultation/visit to NHCP and identification/ authentication of species are will in place (Pandey *et al.*, 2015a,b).

## Herbarium Based Material Used for Variability Studies

Herbarium based studies have been done in order to know the area of availability of species, for location of rare and endangered plants, ethnobotanical studies, *i.e.* uses of plants by tribal cultures, diversity patterns in relation to geography, biogeographic studies, phylogenetic studies on crop-weed complexes, *etc.* Material from any study, morphological, anatomical, ecological or genetical, which is correctly identified with clear records and notes, is suitable for deposition at a recognized herbarium for future reference by later workers. Workers on plant genetic resources need to know the procedures for preparation, study and consultation of herbaria and associated literature for its meaningful documentation and use. Traditional herbaria rarely focus on herbarium collections depicting variability within cultivars, primitive forms, landraces and as also obsolete cultivars. The National Herbarium of Cultivated Plants (NHCP) provides reference in material for identification of taxa, baseline information on locality of availability, climatic and habitat preferences of taxa, and ancillary information on source localities.

## Keeping the Plant Records with Preservation

Keeping these records (availability, time and place of collection with other features) for any species identification can be preserved by wet (in jars, bottles *etc.*) or dry (on sheet).

1. **Wet preservation:** Some fleshy plants lose their diagnostic features when dried and pressed and have to be preserved in some liquid medium such as 4 per cent formalin solution of F.A. A. (90 parts 50 or 70 per cent ethyl alcohol, 5 parts glacial acetic acid and 5 parts formalin).

2. **Dry preservation:** Most of the plant can be preserved by drying on sheets but some plant parts such as palm inflorescence, branches and cones of many gymnosperms and several dry fruits which being bulky are not suited for pressing. They are dried as such without pressing and dry stored in special containers.

3. **Digital herbarium:** At ICAR-National Bureau of Plant Genetic Resources (NBPGR), New Delhi, digital scans (jpeg images) of herbarium specimens were created using flat-bed scanner (high resolution HP Scanjet 3500C) at a resolution of 300dpi. These images of herbarium specimens were linked to database for quick access on a family-wise and genus-wise basis, as well as by unique identity numbers assigned to herbarium specimens (HS numbers) and those assigned to the introduced germplasm (EC numbers).

## Locating the Specimens Records

Use of the herbarium requires literature associated with it for use-standard floras-local/regional, for the system of classification and for knowing the synonyms (older and incorrect named used for the same species/genus) under which specimens may be looked up for study. Specimens in the herbarium are arranged in dust-proof cabinets. The general method of arrangement is a hierarchical one, following one of the standard systems of classification:

☆ Bentham and Hooker's System
☆ Engler and Prantl's System

Individual herbaria may make minor changes in family/genus split-up. The group/species of interest for study can be located using the following guidelines:

(i) A family is generally separated from the one above and the one below by marker flaps.

(ii) Collections from different localities/regions are demarcated using different colour folders, corresponding to the different phytogeographic zones of the world, in larger herbaria (over 50, 000 specimens).

(iii) Genera and species under them, are usually arranged alphabetically and sometimes according to a system of classification, as for families.

## Identification of Specimen

The requirements for this are a 'flora' listing species of a region, and basic information on a range of feature of the vegetative and reproductive parts of a plant. A plant species can be easily identified once these basic requirements are fulfilled in the following sequence:

(i) One should know the precise locality and season of collection.

(ii) The material should be complete with leaves, flowers, fruits with seed, notes on plant and habitat features not seen in specimen, *e.g.* height of plant, colour of various parts, hairiness, whether available as a weed of road sides/fields *etc.*

(iii) A 'flora' of the area or similar area provides a list of species available in the area and diagnostic characters and key for identification.

(iv) A description of the plant is worked out in a systematic manner sequentially, starting from the plant as a whole to underground parts, stem, inflorescence, flower, fruit and seed. Use of standardised, technical terms help in developing a precise and brief description.

(v) Following the hierarchical system of classification, and knowing the diagnostic characters, identification is done starting at the highest category-dicot/monocot to family, then genus, species and variety.

## Automated Identification of Specimen

At the present time, automated identification of specimen has become possible. With access to the computer, one can load a database for identification and by feeding in information about the distinctive features; the options available can be narrowed down till the precise species is identified. Such databases are available for identification of families and selected genera *e.g. Cucumis* (Nayar *et al.*, 1994).

## System of Classification in Indian Herbaria and Herbarium Literature

The specimens should be arranged in their cases according to well-known system of classification. In India and most of other countries Bentham and Hooker's system of classification is being used for this purpose. The herbaria in India are mostly regional/local (Botanical Survey of India herbaria at Kolkata and its zonal circles and those associated with universities and major colleges) with maximum representation of species collected from natural habitats and few in cultivation. These herbaria, in common with those in Europe, are mostly arranged following an artificial classification system, that of Bentham and Hooker. All Indian 'floras' also followed the same system. These, irrespective of size and importance, have an identifying acronym, recognized internationally (and published in the Index Herbarium from the New York Botanical Gardens, USA). Almost all species described and renamed since the adoption of the binomial system of naming of plants are published in Index Kewensis (published from the Royal Botanical Gardens, Kew, UK).

## Basic Methods and Procedures of Herbarium Specimens

### (a) Plant Collection

The plant specimens should be collected with maximum representative of from different localities and habitats in every stage of their growth and reproduction. A complete specimen possesses all parts of plant including root system. It is essential to labeled and placed with minimum overlap of parts between blotters. The plants collected should be either pressed on the spot or may be collected in vasculum and may be pressed after some time. The plant specimen should be usually in flowering stage and at least about six specimens should be collected of one plant. The specimens collected must be tagged immediately and record about locality must be noted. Herbarium specimen prepared of a material only at vegetative or reproductive stage is inadequate and therefore of less value for use in systematic study. For a cultivated or wild species one should take as many observations as possible at various growth stages in the field during collection and while processing and grow out in experimental study. In case of *Allium* species, collector needs to go during early growth period (in December-January) for vegetative material and for flowering specimen (in May-June when bulb mature). It demands visiting the same area at least two times or else hunt for early and late sown plants.

Besides herbarium specimens of cultivated plants, some neglected groups: less known domesticated species *viz., Moghania vestita* (Soh-phlong), *Digitaria* (Raishan), *Coix lacryma-jobi* in north-eastern hills, and others such as *Malva verticillata, Imla racemosa, Hodgsonia heteroclita, Brachiaria mutica, Aisandra butyracea* (Cheura), *Adansonia digitata* (Gorakh imli), *Setaria glauca, Momordica dioica, Alliums* spp., rice bean, winged bean, *Vigna vexillata* and taxa of potential/commercial value are represented. Important taxa of crop wild relatives (CWR) maintained in the NHCP include *Vigna cajanus* Atylasia, *Solanum, Abelmoschus, Cucumis, Luffa, Allium, Trichosanthes* and *Amaranthus.* Some specialty collections include wild *Vigna* from north-western Himalaya and wild *Allium* from high altitude areas of western and eastern Himalaya. Type collections of newly distributed taxa by ICAR-NBPGR under genera *viz., Curcuma, Abelmoschus, Vigna, Cucumis, Herpetospermum, Momordica etc.* are some valuable collections maintained by NHCP.

### Observations to be Recorded during Collection for Herbarium Specimens Preparations

Observations to be recorded during collection for herbarium specimens preparations are as per need. For example to discuss the above issues we take the example of *Allium* species (Pandey *et al.*, 2016):

☆ **Habitat:** Notes on habitat conditions for different species of *Allium* (wild, semi-domesticated or cultivation), collected from farmer's field, kitchen garden or experimental conditions; record data on frequency of occurrence (abundant, frequent, rare, occasional, *etc.*), associated species, soil type (rocky, saline, sandy), colour and texture, topography, *etc.*, are essentially needed for wild plants. Data on latitude, longitude and other geographic parameters can be recorded using GPS; this can help in locating the species at similar location elsewhere in the area of explorations.

☆ **Habit:** It is often difficult to collect a complete specimen with fully mature bulb and leaves and floral parts together for preparing into a herbarium specimen; in many cases bulb may be fully mature but leaves. Scape or inflorescence withered away. The freshly dug out specimen with immature bulb does not really represent the actual shape of the particular species. Fresh harvest of Allium sold in the market may only represent part of species (leaves of *A. tuberosum, A. hookeri*) and young bulb with leaves (*A. cepa, A. chinense, A. ampeloprasum*). In such casaes one should try to represent different stages of plant at vegetative growth (bulb, leaf *etc.*) and reproductive stages (scape, inflorescence and flower).

☆ **Bulb:** Tunicated bulb formed from basal leaves is a well demarcating character that helps delimiting *Allium* from many bulbous taxa (*Amaryllis, Drimia* called wild onion in many regions of India). Notes/observations on bulb (single or in cluster, cylindrical, bulb shape, size, level of development, membrane character-outer and inner coat colour, texture, *etc.*) are important in identifying different taxa of *Allium*. Collector should ensure that intact bulb is collected, or if peeling off collect the sample of membrane in paper pouch. Notes on bulb colour at harvest/maturity are to be recorded; in an immature bulb the coat colour often an unstable character and can mislead in the identification of species. Take observations of the inner wall of outer coat making peel mount and examining it for cell shapes under 10X magnification. Presence of underground bulbils is an additional character that delimits taxa. Bulbs in single or in cluster, size and shape of the bulb, and leaf characters can distinguish species even at vegetative stage (*A. chinense, A. cepa* var. *cepa, A. cepa* var. *aggregatum* and *A. proliferum*). *A. chinense* has cylindrical-long necked bulbs, triquetrous leaves and white membrane whereas in others the leaves are fistular (round/elliptical in section, bulb not cylindrical with short neck, membrane white or shades of brown, violet, lilac). Usefulness of bulblets/bulbils to serve as taxonomic identity especially when detached from the mother plant is questionable. Bulblets (cloves) of garlic (*A. sativum*) and *A. ampeloprasum* (great headed garlic) can be distinguished on the basis of size and texture of bulb coat; the former with thin membrane flat leaves but latter has thick, leathery and keeled leaves. Underground/aerial bulbils are always thick walled, smaller than the normal bulblets. Underground bulbils are located below the bulblets or may be with stoliniferous roots. Aerial bulbils are smaller than the underground bulbils, borne in the scape/floral heads. While collecting the position, size and shape (morphological structure) of the bulbils must be carefully noted. In case of doubt grow out tests are essential.

☆ **Rhizome:** Presence of a developed rhizome, its position-horizontal or vertical is characters associated with subgenus and section. Subg. *Amerallium*, rarely produce a notable rhizome; subg. *Anguinum* and *Melano crommyum* consist of small rhizomes and no bulb whereas subg. *Butomissa, Rhizirideum, Allium cepa, Reticulato bulbosa* and *Polyprason* have both rhizome and bulbs (reduced, modified or condensed state). Well-developed rhizome is generally deep-seated and collector should use digging tools to take out carefully. Presence of thick rhizomes in a dried specimen can only be seen if specimen is dug properly and processed (*A. tuberosum*). Condensed rhizome in garlic and onion can be visible through the vertical section of bulb. Fibrous coat on bulb/rhizome needs special observations for colour and texture.

☆ **Root:** Root structure is very important character at infrageneric level identification. Fleshy root character associated with subg. *Bromatorrhiza-A. hookeri, A. fasciculatum* and *A. macranthum* is distinct feature to delineate species. Fleshy roots generally dry or shrink on processing and may be noted or sketched on herbarium label; photo of freshly dug out sample can serve the purpose.

☆ **Leaf:** Number of leaves, shape (triquetrous, round/channeled, flat and keeled) can be examined in fresh material. Longitudinal/transverse section of leaf can delimit *A. cepa, A. cepa* var. *aggregatum, A. fistulosum* and *A. proliferum*. Keeled nature of leaf can distinguish *A. ampeloprasum* at pre-flowering stage from *A. sativum*. Colour of leaf, shape of the leaf tip, leaf orientation with respect to vertical axis, density of leaves, texture membranous or thick, leaf length with respect to scape are important characters to identify taxa. The two species, *A. auriculatum* (subg. *Cyathophora* sect. *Coleoblastus*) and *A. przewalskianum* (subg. *Rhizirideum*) co-occur on dry slopes, ravines and rocky crevices are difficult to distinguish in routinely preserved herbarium specimens as fistular leaf and perianth characters of the latter species, as a distinguishing trait from the former species are lost during processing.

☆ **Inflorescence:** Shape of the inflorescence-globose, compact; spreading pattern, opening of flower (regular, irregular, centrifugal), numbered of flower opened at a time, scape length (longer or shorter than leaves), shape (cylindrical-tapering), hollow or solid (at maturity), shape of flower, can be noted at any stage of collection to drying of herbarium. In pressed condition the inflorescence of *A. cepa* var. *cepa* and *A. cepa* var. *aggregatum* are difficult to distinguish; others like *A. semenovii* and *A. macranthum* can be quickly identified with distinct inflorescence.

☆ **Flower:** Macroscopic characters as colour, streak on the perianth, orientation of perianth in open flower, perianth shape, and stamen length with respect to perianth and orientation; shape of the basal part of inner filament, orientation, presence or absence of teeth and its shape can distinguish closely related taxa. *A. ramosum* and closely related *A. tuberosum* differ in flower characters; the former has pink mid line on the dorsal side of perianth. Orientation of perianth and stamen in

an open flower is lost during drying so need to be recorded in field data book. Notes on flowering time (day hours) *vs.* stage of flower (early-late), pollinators' specificity should be noted. Among the insect pollinators *Aphis dorsata* the common honey bee, and *A. cerana* and *A. mellifera* are the most frequent visitors to *Allium cepa* and *A. schoenoprasum* whereas the common fly prefers to visit *A. ampeloprasum* (Kumar and Gupta, 1993).

☆ **Seed:** Seeds are very small and not readily observed in the field for detailed micro-characters; however an outline and overall shape (angular, round, broader) can be noted with help of hand lens. Seeds shape round (*A. victorialis*), sickle shaped (*A. carolinianum*), tear shaped (*Nothoscodum gracile*), *etc.* can be drawn on the field note book.

☆ **Odour:** Onion or garlic type odour on crushing can be noted when sample is harvested and recorded on label. For example *A. tuberosum* and *A. schoenoprasum* have mild onion flavor whereas *A. ampeloprasum* has garlic odour.

☆ **Photographs and other material:** Plants at different stages of growth, inflorescence at different angles, flower when fully open (close-up view), perianth and anther orientation need to be photographed. Additionally, the dried parts (flowers, stamens, leaves may be collected) and information on important features and ethnobotanical data should be recorded in herbarium specimen (Pandey *et al.*, 2016).

## (b) Poisoning of Specimen

Poisoning of the specimen should be done immediately after collection. Poisoning kills the plants and thereby prevents the formation of abscission layer and decay. For poisoning the specimens mercuric chloride, lauryl pentachlorophenate (LPCP), formalin; fumigants (volatile poisonous liquids) like methyl bromide, carbon disulphide, carbon tetrachloride, paradichlorobenzene (PDB) are suitable.

## (c) Pressing and Drying of Plant

The plants should be pressed in between the sheets of blotting paper/blotters. One plant be arranged on one sheet in a manner that there should be no over-lapping of parts. The larger specimens may be folded in 'V' or 'N' shapes. The blotting papers with plant specimen should be placed in field press for about 24 to 48 hours. The press is then opened, blottings should be changed and rearrange the plants properly. After it press should be again closed and again after 2 or 3 days change the blotting and dry plants in sunlight or in artificial heat (50°C) till the plant specimen ceases to be limp. An ideally dried herbarium with character representation (vegetative characters: roots, tubers, bulbs and rhizome, leaf, stipule, spine, bark, *etc.* and floral character:

inflorescence, flower-spathe, scape, stamen, sepal, petal/tepals and fruit characters: pericarp, placentation, seed) are good resources for taxonomic studies (Lawrence, 1951; Davis and Heywood, 1963; Holmgren and Holmgren, 1998). Herbarium specimen prepared of a material only at vegetative or reproductive stage is inadequate and therefore of less value for use in systematic study. For a cultivated or wild species one should take as many observations as possible at various growth stages in the field during collection and while processing and grow out in experimental study. In case of *Allium* species, collector needs to go during early growth period (in December-January) for vegetative material and for flowering specimen (in May-June when bulb mature). It demands visiting the same area at least two times or else hunt for early and late sown plants. It is recommended that collected material should be pressed fresh (especially flowers) in the field or collected in polythene bag (closed with rubber band) during collection. Standard digging instrument, blotters, plant press, camera, hand lens, bottles (with 10 per cent FAA) were used during initial collection (Jain and Rao, 1977).

## Mounting and Labeling (Information Content) of the Herbarium Specimens

### Herbarium Sheet

The herbarium specimen sheet is records of all studied that have been done using the specimen. This gives basic information on the specimen when collected from its natural habitat.

☆ **Mounting:** After drying, the specimens must be mounted for permanent record on sheets called mounting papers or herbarium sheets. Specimen is mounted using cotton thread on acid free paper, labeled, given an identity number/collection number and stored in a specimen folder. The material is identified and indexed. The mounting paper is of standard size 11½½ x 16½½ and must be heavy and of good quality to support the specimens. Some curators use 100 per cent rag paper, but this is expensive. The paper of lower rag content is more generally used. The specimens are mounted to the sheets with the help of glue or adhesive gummed strips or quick drying liquid paste. After mounting the specimens on herbarium sheets, each sheet should be labeled. A label is pasted or printed on the lower right-hand corner. Herbarium labels are important parts of finished specimens. A label should be of the size 4½½ x 2 ½½ and must give the following information.

| Flora | : |
|---|---|
| Number of plants | : |
| Family | : |
| Genus | : |
| Species | : |

| Locality | : |
|---|---|
| Habit | : |
| Date of collection | : |
| Collector | |

☆ **Determinative slip:** These are small slips on which the re-identification details are given. Any name changes done during subsequent studies are added to the herbarium sheet along with the name of the person making the change. Thus, the record of the studies done with the herbarium sheet.

☆ **Pouches:** Material removed from the specimen for any further studies and detached material, *e.g.* Leaves, flowers, seeds, dissection of plant parts are kept here during mounting of a specimen after study.

☆ **Photography:** Photography of specimen is essential not only for record but also for clarification of doubt/confusion.

☆ **Voucher:** Voucher specimens of experimental studies should be maintained properly.

## Storage of Herbarium

The mounted plant specimens or herbarium sheets may be stored in specially constructed herbarium cases or in 'Steel Almirah'. To protect these herbarium sheets from mold, fungi, insects *etc.* 2 per cent solution of mercuric chloride (which is highly poisonous) should be spread. Moth ball and naphthalene flakes may also be placed in shelves of herbarium.

## Exchange of Herbaria

Herbarium materials are mostly available for exchange and loan for study among recognized herbaria.

## Availability of Herbaria

(1) Important herbaria of the world

(2) Important herbaria of India

### (1) Important Herbaria of the World

See the Table 15.11.

### (2) Important Herbaria of India

The centres showing in Table 15.12 are associated in developing and maintaining the herbarium records.

## Global Herbarium

Global herbarium resources consist of approximately 4,000 recognised herbaria collectively holding over 350,00,00,000 herbarium specimens. Major global herbaria are committed to provide herbarium resources accessible to users with holdings including large representation of vascular plants especially those with major economic value (BM, E, K, P, MO, S, B, UC/JEPS) and yet some other focus on regional flora (F, PE, CAL); only few are rich in representation of cultivated plants (only for cultivated ornamentals). Among the cultivated plant herbaria, The Gatersleben Herbarium (GAT) located in the Department of Gene bank of the Leibniz. Institute of Plant Genetics and Crop Plant Research (IPK) is one of the largest specialized herbaria which serves as a source of reference and working material for the reproduction of accessions maintained in gene bank and other institutional research programme. Over 4.30 lacs specimens of cultivated plants and its wild relatives, seed and fruit collections (about 1 lac samples) and the spike collection (55,000 samples) are well represented. The holding include mainly the vascular plant species of Europe and the Mediterranean and the temperate region, eastern and middle Asia, Mongolia and Cuba.

**Table 15.11: Herbaria of the World**

| Sl.No. | Herbarium/Organizations | Number of Herbaria | Year of Establishment |
|---|---|---|---|
| 1. | The first herbarium of the world was established in University of Padua, Italy | - | 1845 |
| 2. | Herbarium of Royal Botanical Gardens, Kew, Great Britain, (world's greatest herbarium) | 65, 00, 000 | 1853 |
| 3. | Museum National Histoire Naturelle Laboratory de Phanergamie, Paris, France. | - | 1635 |
| 4. | V.I. Komorov Botanical Institute, Leningrad, Russia | 50,00,000 | 1823 |
| 5. | Herbarium of Royal Botanical Gardens, Edinburgh, England | 25,00,000 | 1761 |
| 6. | U.S. National Museum, U.S. Smithsnian Institution, Washington, United States | 30,00,000 | 1868 |
| 7. | Conservatorie et Gordin Botaniques, Geneva, Switzerland | 50,00,000 | 1817 |
| 8. | New York Botanic Garden, New York, United States | 30,00,000 | 1801 |
| 9. | Vienna Botanischer Gaertn, Vienna, United States | 25,00,000 | 1748 |
| 10. | National History Museum, Chicago, Austria | 23,50,000 | 1893 |
| 11. | Missouri Botanic Garden, St Louis, United States | 17,00,000 | 1859 |
| 12. | National Herbarium, Melbourne, Australia | 15, 00, 000 | 1857 |
| 13. | Zurich Botanischer Gaertn, Zurich, Germany | 15, 00, 000 | 1834 |
| 14. | Gray Herbarium, Cambridge, United States | 14,85,000 | 1807 |
| 15. | Harvard University, Philadelphia Academy of Science, Philadelphia, United States | 10,00,000 | 1812 |
| 16. | Arnold Arboretum, Boston, United States | 7,00,000 | 1872 |
| 17. | Garden College, Rawalpindi, Pakistan | 60,000 | 1893 |

**Table 15.12: Centres Associated with Herbarium Specimens**

| Sl.No. | Herbarium/Organizations | Number of Herbaria | Year of Establishment |
|---|---|---|---|
| 1. | Herbarium of the Indian Botanical Garden, Kolkata (Central National Herbarium, Howrah), Kolkata (W.B.) | 25,00,000 | 1793 |
| 2. | Herbarium of Forest Research Institute, Dehradun (Uttarakhand) | 3,00,000 | 1874 |
| 3. | Herbarium of Agriculture College and Research Institute, Coimbatore (Tamil Nadu) | 2,00,000 | 1874 |
| 4. | Blatter Herbarium, St. Xaviers College, Fort, Mumbai (M.S.) | 1,00,000 | 1906 |
| 5. | Botanical Survey of India, Eastern Circle, Shillong (Meghalaya) | 1,00,000 | 1956 |
| 6. | Botanical Survey of India, Western Circle, Poona (M.S.) | 1,35,000 | 1956 |
| 7. | Botanical Survey of India, Northern Circle, Dehradun (Uttarakhand) | 60,000 | 1956 |
| 8. | Botanical Survey of India, Industrial Section, Kolkata (W.B.) | 50,000 | 1887 |
| 9. | Botanical Survey of India, Central Circle, Allahabad (U.P.) | 40,000 | 1955 |
| 10. | National Botanical Garden Herbarium, Lucknow (U.P.) | 1,00,000 | 1948 |

## National Herbarium of Cultivated Plants (NHCP)

The herbarium at ICAR-NBPGR, New Delhi (acronym PI) is established as a National Herbarium of Cultivated Plantings which occupies an important place among 25 major Indian herbaria (Singh, 2010; Nayar *et al.* 2014). The National Herbarium of Cultivated Plants (NHCP) is a specialized herbarium located in the ICAR-National Bureau of Plant Genetic Resources (NBPGR), New Delhi focusing on representation of taxa of use or of potential importance in plant genetic resources (PGR) programmes *viz.* crop species, their wild relatives and wild economic plants. It has the mandate to collect and study diversity with emphasis on crop/cultivated taxa which are traditional less represented in herbaria. NHCP occupies a place among 25 major Indian herbaria. The holding of the NHCP cover a period of nearly seven decades (5000 spcimens); it includes the herbarium specimens of the Plant Introduction Scheme (PI) under the Botany Division of the Indian Agricultural Research Institute (IARI) commencing activities of germplasm augmentation through systematic introductions since 1946 (Setupon 1948) upto 1977. Subsequently, this activity was undertaken by a separate institute for PGR, *i.e.* at the ICAR-NBPGR. Taxa represented in NHCP, categorized according to use, were well represented in the PI as well as the NHCP herbaria for cereals and millets, vegetable crops as well as herbage grasses and legumes. In legumes, grain legumes were the priority for addition in the NHCP, besides species of herbage use belonging to *Crotolaria, Lathyrus, Vicia* and *Vigna*. NHCP presently holds collection of 22,778 herbarium specimens (representing 265 families, 1500 genera and 4,516 species, 23,000 type specimen) taxa mainly of cultivated and wild/weedy relatives, potentially important species, native or introduced species for use in breeding (*Source*: http://sciweb.nybg.org/science2/indexherbariorum. asp). Besides, seed and carpological samples/economic products are represented as complementary collections. It differs in its mandate from other herbaria across the country in having emphasis on collection of variability in crops depicted as cultivars, primitive types landraces, crop wild forms and their wild relatives (CWR/weedy types, wild/domesticated forms and taxa introduced for breeding purpose and potential species from different agro-ecological conditions. It is intended to serve as a reference collection for identification, taxonomic study of plant genetic resources and a resource for teaching on plant taxonomy, ethnobotany and economic botany with Post Graduate School, Indian Agricultural Research Institute, New Delhi. Significant holdings include ~500 crop wild relatives (species number in parenthesis)

under genera: *Oryza* (21), *Sorghum* (6), *Vigna* (25), *Cajanus/Atylosia* (10), *Allium* (29), *Abelmoschus* (14), *Solanum* (30), *Cucumis* (11), *Trichosanthes* (7), *Piper* (21), *Curcuma* (24), *Rosa* (19), *Prunus* (16) *etc.* over 3,500 species of crop genepolls (~6,000 images) are maintained as virtual herbarium. The NHCP provides technical input on identification, validation and taxonomic know-how on taxa of PGR relevance. The NHCP accepts unique and unrepresented genetic resources (as herbarium specimens and seed) and encouraging depositing vouchers for future reference (Pandey *et al.*, 2016). To lay thrust on building-up to infrastructure facilities work was undertaken in project mode in 1985 under institutional project entitled "Establishment, building up and maintenance of herbarium and seed museum of cultivated plants." Specialized training attended by the scientific staff of the herbarium at various national and international herbaria such as the Royal Botanic Gardens at Kew, and New York Botanical Garden and Arnold Arboretum in USA and Wealth of India Herbarium New Delhi and Botanical Survey of India and Forest Research Institute, Dehradun and National Botanical Research Institute Lucknow (U.P.) facilitated in upgradation of infrastructure. The insect-pest and dust-free storage cabinets gradually replaced the traditional pigeon hole cabinets with addition of five new-space saver compactors during 2004-2008, the capacity of herbarium has increased to upto 40,000 specimens.

Voucher herbarium specimens have value as cultivars of crops; selections bred valuable traits and landraces. These are distinct variants with observable traits, which may or may not be recognized as distinct taxonomic entities under the genus or species. However, these are significant from the point of view of PGR study and utilization. These specimens are maintained as representatives of intrspecific variants in the NHCP.

**Table 15.13:Herbarium Specimens of Introduced Germplasm in National Herbarium of Cultivated Plants**

| Sl.No. | Crop Category | Total Species | Specimens of Introduced Accessions | Specimens of Plant Introduction Herbarium (PI) | Germplasm Accessions Conserved (GB/FGB) |
|---|---|---|---|---|---|
| 1. | Cereals/millets (C/M) | 38 | 71 | 26 | 13 |
| 2. | Grain legumes (Gl) | 19 | 43 | 4 | 9 |
| 3. | Fruits (Fr) | 13 | 14 | 3 | 10 |
| 4. | Vegetables (Vg) | 14 | 38 | 11 | 11 |
| 5. | Oilseeds (Os) | 13 | 21 | 5 | 8 |
| 6. | Pseudocereals (Ps) | 3 | 6 | 3 | 3 |
| 7. | Spices/condiments (S/C) | 5 | 5 | 3 | - |
| 8. | Fodder/forage grasses (FoGr) | 35 | 53 | 6 | 2 |
| 9. | Fodder/forage legumes (FoLg) | 124 | 190 | 63 | 7 |
| 10. | Ornamentals | 11 | 11 | 11 | 8 |
| 11. | Medicinal/aromatic plants (M and AP) | 23 | 30 | 24 | 5 |
| | **Total** | **298** | **482** | **167** | **22** |

**FGB:** Field genebank**; GB:** Genebank.

*Source*: Nayar *et al.*, 2014.

**Table 15.14: Significant Herbarium Specimens of Introduced Vegetable Species in NHCP**

| Sl.No. | Botanical Name | EC Number | HS Number | Crop Category | Remarks |
|--------|----------------|-----------|-----------|---------------|---------|
| 1. | *Amaranthus caudatus* L. (=*A. caudatus* subsp. *mantegazzianus* (Pass.) ined.; *A. mantegazzianus* (Pass.)) | EC013953 | HS04651 | Pseudocereals | Local variant |
| 2. | *Amaranthus caudatus* L. (=*A. edulis* L. | EC150188 | HS07061 | Pseudocereals | Genebank |
| 3. | *Amaranthus hypochondriacus* L. | EC169611 | HS07091 | Pseudocereals | Genebank |
| 4. | *Amaranthus hypochondriacus* L. | EC359422 | HS09743 | Pseudocereals | Genebank |
| 5. | *Solanum lycopersicum* L. var. *lycopersicum* (=*Lycopersicon esculentum* Mill.) | EC001129 | HS09095 | Vegetable | cv. Burwood Prize |
| 6. | *Solanum lycopersicum* L. var. *lycopersicum* (=*Lycopersicon esculentum* Mill.) | EC315490 | HS09086 | Vegetable | cv. Mountain Delight |
| 7. | *Solanum lycopersicum* L. var. *lycopersicum* (=*Lycopersicon esculentum* Mill.) | EC130053 | HS09105 | Vegetable | Genebank |
| 8. | *Solanum lycopersicum* L. var. *lycopersicum* (=*Lycopersicon esculentum* Mill.) | EC315478 | HS09109 | Vegetable | Genebank |
| 9. | *Solanum melongena* L. | EC169769 | HS07089 | Vegetable | Genebank |
| 10. | *Solanum melongena* L. (collection from Nepal) | EC316224 | HS18316 | Vegetable | Genebank |
| 11. | *Solanum peruvianum* L. (=*Lycopersicon peruvianum* (L.) Mill. var. *peruvianum*) | EC315457 | HS09101 | Vegetable | Genebank |
| 12. | *S. pimpinellifolium* L. (=*Lycopersicon pimpinellifolium* (L.) Mill.) | EC274046 | HS09099 | Vegetable | Genebank |

*Source*: Nayar *et al.*, 2014.

Furthermore, these local variants or selections from wild species, were often distinguishable by additional features such as colour of leaves, leaf thickness often not observable on the specimens and needing photographic record or additional descriptive data along with specimens for reference. Example of these were, *mantegazzianus* type of *Amaranthus caudatus*, a local cultivar of Argentina bearing condensed and rounded inflorescence spikes in place of long feathery and often drooping inflorescence and earlier considered in distinct sub-species (Zeven and de Wet, 1982; Hanelt, 1986). *Pisum sativum* cv. Dwarf Gray Sugar, an edible snow pea under continuous cultivation for over 200 years (http://casfs.ucsc.edu/documents/for-the-gardener/peas.pdf) and *Solanum esculentum* (old name *Lycopersicon esculentum*) cv. Burwood Prize, an heirloom cultivar from Australia (http://naldc.nal.usda.gov/download/36965/pdf).

## Significant of Achievements of NHCP

### Study on Domestic Trends

Crop taxa important for Indian region–*Luffa, Moringa, Ocimum, Crotalaria, etc.* were studied for available diversity in the region through field and study of herbarium resources. Range of diversity available here facilitates study of trends of domestication thereby facilitated highlighting changes in character during this process (Pradheep *et al.*, 2011; Pradheep *et al.*, 2015; Pandey *et al.*, 2016).

Potential species of PGR value identified for new uses/new records included–*Crotalaria tetragona* (tum thang, Bhatt *et al.*, 2009a), *Bidens pilosa* (Bhatt *et al.*, 2009b), *Plukenetia corniculata* (meetha patta), *Ziziphus nummularia*

(ber), *Hodgsonia heteroclita, Abelmoschus tetraphyllus* (Sukhlai) (Pandey *et al.*, 2010, 2011a,b; Pandey *et al.*, 2014; Pradheep *et al.*, 2015b; Semwal *et al.*, 2014; Pandey *et al.*, 2015; Rathi *et al.*, 2016). Eco-geographic study on taxa of PGR relevance done through resources available in NHCP has supported other research programmes in the institute and has facilitating viewing new dimensions to the existing crop genepool (Pradheep *et al.*, 2011).

## Development of Protocol for Safe Preservation

Protocols are being developed for difficult-to-store groups, pest-free storage, ideal storage conditions for families which are sensitive to pest damage and through standardization of use of low temperature (-20°C) using deep freezer, dusting of naphthalene powder, *etc.* Difficult-to-represent taxa like bulbous group, tuberous/rhizomatous taxa, *Musa, Agave, etc.* are being work out for representation in NHCP. Routine drying techniques using modified methods and microwave drying techniques for difficult material are being further refined based on traditional methods (Jain and Rao 1977; Pandey *et al.*, 2006a; Pandey *et al.*, 2016; Pandey *et al.*, 2013). Keeping in view the material used for advance studies such as biosystematic study, biochemical/phytochemical and molecular study the efforts have been on minimal use of hazardous chemicals for maintaining the specimens. To minimize use of contact poisons/chemicals, and insect repellents (naphthalene balls), deep freezing methods are preferred.

### Significant Documents

Several publications in the form of books, manuals, chapters, and research papers on new geographical distribution have been brought out. Some significant ones include: Wild Relatives of Crop Plants of India (Arora and Nayar, 1984), Wild Edible Plants of India (Arora and Pandey, 1996), Wild Relatives of Crop Plants: Collection and Conservation (Pandey *et al.*, 2008). Genetic Resources of Rosaceae of India (Pandey *et al.*, 2007) and Guidelines for Use of NHC P (Nayar *et al.*, 1999-2005) (krishikosh.egranth.ac.in/.../1/2035781; krishikosh. egranth.ac.in/.../1/.../1/8.pdf) and Importance of Voucher Specimens of Introduced Germplasm (Nayar *et al.*, 2003; 2014).

The work done in project mode on Genetic Resources Study of Economically Important Plant Families- Cucurbitaceae, Malvaceae, Rosaceae and Poaceae' during 1984-1995 served as base line for many taxonomic works undertaken in NHC'P. Study of crop taxa of Indian region (Asiatic *Vigna, Crotalaria, Allium, Prunus* and wild Triticeae); and check-lists of Indian Crop Plants and Crop Wild Relatives pin-pointed gaps in collection and prioritisation for build-up holdings (Arora and Nayar, 1984; Pandey and Nayar, 1994; Nayar *et al.*, 2003; Nayar, 2015).

## Services, Teaching and Linkages

Besides providing the expert consultation service in field of taxonomy, NHCP is actively involved in providing technical input on identification/ authentication, validation of taxa of PGR relevance; it also provides hands on exercise on herbarium procedures to large number of college and school students and researchers especially working in fields of pharmacy, pathology, entomology, breeding, *etc.* It is linked to the other fields of science especially for seeking identification/authentication of material conserved, as host-plant species relationship, introduced germplasm, weed science, agronomy *etc.* for the benefit of different users seeking services provided by the NHCP, special guidelines are in place (Pandey *et al.*, 2015; Annexure 1, 2). Since 1999 this facility is available for teaching courses on taxonomy. ethnobotany and economic botany with PG School. ICAR-Indian Agricultural Research Institute, New Delhi.

The NHCP maintains links with many ICAR institutes, State Agriculture Universities and traditional universities, Botanical Survey of India (BSI), Forest Research Institute (FRI), and Herbarium Cryptogamae Indiae Orientalis (HCIO–a national-fungal herbarium facility at ICAR-IARI, New Delhi), and the Herbarium of Wealth of India, CSIR-NISCAIR, New Delhi.

The NHCP accepts unique and unrepresented genetic resources (as herbarium specimens, seed samples and economic products) and encourages users for depositing vouchers for future reference.

*Solanum/Lycopersicon* (tomato group), native of the south America region, was represented by the wild species, *S. cheesmaniae* (1) and *S. pimpinellifolium* (4), and were the closest relatives of tomato and potential sources of resistance to drought and disease; and *S. chilense* (2), *S. hirsutum* (4) and *S. peruvianum* (5), belonging to its secondary genepool, were sources of disease, pest and drought resistance. *Solanum* (brinjal group), was represented by the wild species *S. sisymbriifolium* (1), native of South America region and a potential source of disease and pest resistance. Some of the introduced species of wild and weedy relatives of crop like *S. sisymbriifolium* extended their distribution as weeds of disturbed areas (Kohli *et al.*, 2012) naturalized in the Indian region.

## References

1. Ambasta, S.P., Ramachandran, K., Kashyapa, K. and Chand, R. 1986. *Useful Plants of India* (eds.). Publication and Information Directorate, Council for Scientific and Industrial Research, New Delhi.

2. Arora, R.K. and Nayar, E.R. 1984. Wild Relatives of Crop Plants in India. *Sci. Mongr. 7*, p. 90.

3. Arora, R.K. and Pandey, A. 1996. *Wild Edible Plants of India: Diversity Conservation and Use.* ICAR-NBPGR, New Delhi.

4. Bhatt, K.C., Pandey, A., Dhariwal, O.P., Panwar, N.S. and Bhandari, D.C. 2009a. "Tum-thang" (*Crotolaria tetragona* Roxb. ex Andr.): A little known wild edible species in the northeastern hill region of India. *Genet. Resour. Crop Evol.* 56: 729-733.

5. Bhatt, K.C., Sharma, N. and Pandey, A. 2009b. "Ladakhi tea" *Bidensa pilosa* L. (Asteraceae): A cultivated species in the cold desert of Ladakh Himalaya, India. *Genet. Resour. Crop Evol.* 56: 879-882.

6. Davis, P.H. and Heywood, V.H. 1963. *Principles of Angiosperm Taxonomy.* University of Edinburgh Press, Great Britain, p. 556.

7. Hanelt, P. 1986. Mansfeld's Encyclopedia of Agricultural and Horticultural Crops (eds.). Institute of Plant Genetics and Crop Plant Research, Springer-Verlag, Berlin, Germany.

8. Holmgren, P.K. and Holemgren, N.H. 1998. *Index Herbariorum: A Global Directory of Public Herbaria and Associated Staff.* New York Botanical Gardens, USA.

9. Huaman, Z. and Spooner, D.M. 2002. Reclassification of landrace population of cultivated potatoes (*Solanum* sect *Petota*). *Am. J. Bot.* 89(6): 947-965.

10. Jain, S.K. and Rao, R.R. 1976. A handbook of field and herbarium methods. Today and Tomorrow Printers and Publishers, New Delhi, p. 157.

11. Jain, S.K. and Rao, R.R. 1977. A handbook of field and herbarium methods. Today and Tomorrow Printers and Publishers, New Delhi.

12. Kohli, R.K., Batish, D.R., Singh, J.S., Singh, H.P. and Bhatt, J.R. 2012. Plant invasive in India: an overview. In: Bhatt, J.R., Singh, J.S., Singh, S.P., Tripathy, R.S. and Kohli, R.K. (eds), *Invasive Alien Plants. An Ecological Appraisal for the Indian Sub-continent.* CAB International, UK.

13. Kumar, J. and Gupta, J.K. 1993. Nector sugar production and honeybee foraging activity in 3 species of onion (*Allium* species). *Apidologie,* **24:** 391-398.

14. Lawrence, G.H.S. 1964. Taxonomy of vascular plants. Oxford and IBH Publishing Co., p. 823.

15. Lawrence, G.H.S. 1951. *Taxonomy of Flowering Plants.* Oxford & IBH Publishing Co., pp. 823.

16. Nayar, E.R., Pandey, A. and Venkateswaran, K. 1994. Plant Identification and Herbarium Procedures. In: Plant Genetic Resources: Exploration, Evaluation and Maintenance, ICAR-NBPGR, New Delhi, p. 320-325.

17. Nayar, E.R., Pandey, A., Pradheep, K. and Gupta, R., 2011. Inventory of Digitized Taxa in the NHCP. ICAR-NBPGR, New Delhi.

18. Nayar, E.R., Pandey, A., Pradheep, K., Gupta, R. and Sharma, S.K. 2014. National Herbarium of Cultivated Plants: Importance of Voucher specimens of introduced germplasm. *Indian, J. Plant Genet. Resour.,* **27(2):** 163-170.

19. Nayar, E.R., Pandey, A., Pradheep, K., Gupta, R. and Sharma, S.K. 2014. National herbarium of cultivated plants (NHCP): importance of voucher specimens of introduced germplasm. *Ind. Polant Genet. Resour.,* 27: 163-170.

20. Nayar, E.R., Pandey, A., Pradheep, K., Gupta, R., Bhandari, D.C. and Bansal, K.C. 2011. National Herbarium of Cultivated Plants. National Bureau of Plant Genetic Resources (NBPGR), New Delhi.

21. Nayar, E.R., Pandey, A., Venkateswaran, K., Gupta, R. and Dhillon, B.S. 2003. Crop Plants India: A Check-list of Scientific Names. Agro-biodiversity (PGR-26). National Agricultural Technologies Project on Sustainable Management of Plant Biodiversity, ICAR-NBPGR, New Delhi.

22. Pandey, A. 2015a. Plant Systematics: Field Inventory, Herbarium Preparation and Management of Important Herbaria and Botanical Gardens of the World and India. Institute of Life Long Learning, Delhi University (http://vle.du.ac.in/mod/resource/view.php?id=13116) ISSN No. 978-93-85611-90-2.

23. Pandey, A. 2015b. Plant Systematics: Documentation: Flora, Monographs, Journals, Online Journals and Keys. Institute of Life Long Learning, Delhi University (http://vle.du.ac.in/mod/resource/view.php?id=13116) ISSN No. 978-93-85611-90-2.

24. Pandey, A. and Bhatt, K.C. 2008. Diversity distribution and collection of genetic resources of cultivated and weedy type in *Perilla frutescens* var. *frutescens* and their utilization in Indian Himalaya. *Genet. Resour. Crop Evol.* 55: 883-892.

25. Pandey, A. and Nayar, E.R. 1994. Some observations of systematics of genus *Crotolaria. Indian J. Plant Genet. Resour.* 7(2): 133-144.

26. Pandey, A. and Pandey, R. 2005. Wild useful species of *Allium* in India: Key to identification. *Indian J. Plant Genet. Resour.* 18: 175-178.

27. Pandey, A., Kuldheep, K., Gupta, R. and Ahlawat S.P. 2017. National herbarium of cultivated plants: A resource for study of crop gene pools. *Indian J. Plant Genet. Resour.,* 30(3): 206-252.

28. Pandey, A., Nayar, E.R., Kuldheep, K. and Gupta, R. 2013. Preparation of herbarium specimens of cultivated plants. In: *Training Manual on Management of Plant Genetic Resource* (eds. Jacob *et al.*). ICAR-NBPGR, New Delhi.

29. Pandey, A., Pradheep, K. and Gupta, 2014c. Chinese chives (*Allium tuberosum*) Rottlen ex Sprengel.: A home garden species and commercial crop in India. *Genet. Resour. Crop Evol.* 61: 1433-1440.

30. Pandey, A., Pradheep, K. and Gupta, R. 2016. Methodology for collecting and preparing herbarium specimen of *Allium. Indian, J. Plant Genet. Resour.,* **29(1):** 32-39.

31. Pandey, A., Pradheep, K. and Semwal, D.P. 2014a. Notes on *Luffa* (Cucurbitaceae) genetic resources in India: Diversity, distribution, germplasm collection, morphology and use. *Indian J. Plant Genet. Resour.* 27: 47-53.

32. Pandey, A., Pradheep, K., Gupta, R. and Ahlawat, S.P. 2016. National herbarium of cultivated plants: A resource for study of crop genotypes. 1$^{st}$ International Agrobiodiversity Congress (Abs. Session: Partnership, Networks and Capacity Building), 6-9 November 2016, held at New Delhi, p. 282.

33. Pandey, A., Semwal, D.P., Bhatt, K.C., Gupta, R. and Ahlawat, S.P. 2016. A new report on cultivation of "Sukhlai" [*Abelmoschus manihot* (L.) Medik. Subsp. *tetraphyllus* (Roxb. Ex. Hornem.) Borss Waalk.]: A species used as organic clearant in jaggery industry in India. *Genet. Resour. Crop Evol.* 63: 1447-1455.

34. Pradheep, K., Pandey, A. and Bhandari, D.C. 2011. Notes on naturalized taxa of plant-genetic resources value in Himachal Pradesh. *Indian J. Plant Genet. Resour.* 24: 293-298.

35. Pradheep, K., Singh, P.K., Pandey, A. and Bhandari, D.C. 2011. Collecting genetic resources of wild *Moringa oleifera* Lam. from Western Himalayas. *Indian J. Plant Genetc. Resour.* 24: 75-81.

36. Semwal, D.P., Bhatt, K.C., Bhandari, D.C. and Panwar, N.S. 2014. A note on distribution, ethnobotany and economic potential of *Hodgsonia heteroclita* (Roxb.) Hook. F. & Thoms. (Cucurbitaceae) in northeastern India. *Indian J. Nat. Product. Resour.,* 5: 88-91.

37. Singh, H.B. 2010. Handbook of Herbaria in India and Neighbouring Countries. National Institute of Science Communication and Information Resources (NISCAIR), New Delhi.

38. Subrahmanyam, N.S. 1999. Laboratory Manual of Plant Taxonomy. Vikas Publishing House Pvt. Ltd. New Delhi, p. 63-68.

39. Zeven, A.C. and de Wet, J.M.J. 1982. *Dictionary of Cultivated Plants and their Regions of Diversity.* Wageningen, the Netherlands.

# 16

# Biodiversity Informatics and Germplasm Management

## Introduction

The term "Biodiversity informatics" was reproduced by Walter Berendsohn and coined in 1992 for the first time by John Whiting. To cover the activities of an entity known as the Canadian Biodiversity Informatics Consortium, a group involved with fusing basic biodiversity information with environmental economics and geospatial information in the form of GPS and GIS. Subsequently, it appears to have lost any obligate connection with the GPS/GIS world and be associated with the computerized management of any aspects of biodiversity information (Bisby, 2000).

*"Biodiversity informatics is the application of informatics to record and yet-to-be discovered information specifically about biodiversity and the linking of this information with genomic, geospatial and other biological and non-biological data sets".*

It can also be defined as *"Biodiversity informatics is the application of information technology (IT) tools and approaches to biodiversity information, principally at the organism level".*

Biodiversity informatics is the application of techniques to biodiversity information for improved management, presentation, discovery, exploration and analysis of primary data. It thus deals with information capture, storage provision, retrieval and analysis focused on individual organism, populations, species and their interactions. It covers information generated by the fields of systematics, evolutionary biology, population biology, ecology as well as more applied fields such as conservation biology and ecological management". Biodiversity informatics is an interdisciplinary research and infrastructure development activity. It thrives on collaborations among environmental biologists, computer scientists and software engineers. This field brings together leading information processing technologies such as semantic frameworks, network communications protocols, standards, software applications and web services, to enable wholly new kinds of analytical and synthetic research in biology. It also attempts to accurately acquire, represent, communicate, integrate, analyze and apply information extracted from natural systems (Shashank *et al.*, 2011).

Progress in genomic technologies and bioinformatics plays a significant role in discovering useful genetic diversity. Gene banks need to embrace the genomics era by developing new strategies and tools to assess the genetic diversity represented in their collections identify traits related to nutritional quality and resistance to pests and drought; support breeding activities.

## Steps of Biodiversity Information

The biodiversity information process comprises of three main steps:

1. **Exploration and discovery:** It includes data collection, documentation of data and management of data

2. **Integration and visualization:** It involves information aggregation and research and analysis of collected information.

3. **Interpretation and use:** As conveyed by the name, this step includes understanding of information, its interpretation and use in decision support mechanisms.

## The Role of Cyber Infrastructure in Biodiversity Research

Continuing advances in computation and communication are transforming the scientific process. Science is increasingly being conducted in virtual laboratory environments; it is increasingly collaborative and increasingly global. Large data sets with hundreds of gigabytes of data (*e.g.* genome sequence) are available online, which means that for a growing number of biologists, "data" is now available on the Web, not in the field (Foster, 2005). These data are being stored in large distributed database complied from many unrelated and independent projects. Biodiversity research, like the disciplines of molecular, structural and proteomic biology, is re-inventing itself with new technology applications, and is evolving into an increasingly predictive and integrative science focused on important research and policy issues.

## Retrieving Useful Databases/Information's about biodiversity

Research in information modeling and standardization of biological databases had been an important issue in the past decade. Some of the databases which can be useful for retrieving information about biodiversity are:

1. **MicrobesOnline:** The MicrobesOnline website is a resource for comparative and functional genomics that serves the scientific community for the analysis of microbial genes and genomics.

2. **Integrated Taxonomic Information System (ITIS):** It provides an automated reference database of scientific and common names for species. As of March 2011, it contains over 638,000 scientific names, synonyms, and common names for terrestrial, marine, and freshwater taxa from all biological kingdoms (animals, plants, fungi, and microbes).

3. **Animal Diversity Web (ADW):** It is online database that collects the natural history, classification, species characteristics, conservation biology, and distribution information of thousands of species of animals.

4. **FishBase:** It is a comprehensive database of information about fish. As of May 2010, it is included descriptions of 31,600 species, 279,100 common names in hundreds of languages, 49,300 pictures, and references to 44,200 works in the scientific literature.

5. **PubMed:** PubMed gives biological data in text format and this service provided by the U.S. National Library of Medicine. It links to more than 17 million resources from different journals within the field of life sciences.

6. **SWISS-PROT:** It is a curated protein sequence database which provides a high level of annotation (such as the description of the function of a protein, its domains structure, post-translational modifications, variants, *etc.*), a minimal level of redundancy and high level of integration with other databases.

7. **PROSITE:** It consists of documentation entries describing protein domains, families and functional sites as well as associated patterns and profiles to identify them.

8. **Bio-molecular Interaction Network Database:** It archives bio-molecular interaction, reaction, complex and pathway information.

9. **Yeast Protein Database:** The Yeast Protein Database (YPD™) in PROTEOME is an on-line resource containing easily searchable, manually-curated details from the *Saccharomyces cerevisiae* PubMed literature, and incorporating valuable tools for gene set analysis and pathway visualization.

Diversity at all levels of biological organization transcends national boundaries, and the study of biodiversity cuts across myriad research disciplines. Biodiversity informatics provides communication and computational infrastructure for this global interdisciplinary research enterprise. Technology continues to be an important driver in the evolution of biodiversity research methods and protocols. Web technologies and network software have become increasingly critical foundations for collaborative interactions and research progress. These services, protocols, applications and standards will emerge and be supported in a global architecture for biodiversity research computing. Analytical and synthetic activities at the tail end of the biodiversity informatics value chain will become more important pieces, applying value and relevance through modeling and environmental predictions. Decision support system will facilitate the conversion of models and analyze into biologically meaningful natural resources policies.

## Global Information Systems

Global information system (GIS) is a system of hardware and software used for storage, retrieval, mapping and analysis of geographic data. Researchers and practitioners also regard the total GIS as including the operating personnel and the data that go into the system. Spatial featured are stored in a coordinate system (latitude/longitude, state plane, UTM *etc.*), which references a particular place on the earth. Spatial data and associated attributes in the same coordinate system can when be layered together for mapping analysis. GIS can be used for scientific investigations, resource management and development planning. GIS differs from CAD and other graphical computer applications in that all spatial data is geographically referenced to a map projection in earth coordinate system into another thus data from various sources can be brought together into a common database and integrated using software. Boundaries of spatial features should "register" or align properly when reprojected into the same coordinate system. Another property of a GIS database is that it has "topology", which defines the spatial, relationship between features. The fundamental components of spatial data in a GIS are points, lines (arcs) and polygons, when topological relationship you can perform analyses, such as modelling the flow through connecting lines in a network combining adjacent polygons that have similar characteristics, and over laying geographic features.

See the Tables 16.1–16.3.

### Table 16.1: Global Information Systems for Plant Genetic Resources

| Institute/Source | Website | Prime Mandate/Information |
|---|---|---|
| World Information and Early Warning System (WIEWS) on plant genetic resources | http://apps3.fao.org/wiews/wiews.jsp | *Ex-situ* collection database of member countries. World list of seed source and crop varieties |
| CGIAR System-wide Information Network for Genetic Resources (SINGER) | http://singer.grinfo.net/ | Collection of genetic resources held by the CGIAR centres |
| Commission on Genetic Resources for Food and Agriculture (CGRFA) | http://www.fao.org/WAICENT/FAOINFO/AGRICULT/cgrfa/ | Conservation and sustainable utilization of genetic resources; Specific information on global strategies for animal and plant genetic resources |
| Global Plan of Action for the Conservation and Sustainable Utilization of Plant Genetic Resources for Food and Agriculture | http://www.fao.org/WAICENT/FAOINFO/AGRICULT/AGP/AGPS/GpaEN/gpatoc.htm | Four theme areas, namely *in situ* conservation and development. *Ex-situ* conservation, use of plant genetic resources and institution and capacity building. |
| International Undertaking on Plant Genetic Resources for Food and Agriculture | http://www.fao.org/ag/cgrfa/IU.htm. | Complete text of the resolution adopted by FAO in 1983 on international agreement dealing with plant genetic resources for food and agriculture |
| International Plant Genetic Resources Institute (IPGRI) | http://www.ipgri.cgiar.org/ | One of the 16 research centres of CGIAR, online information on Germplasm database, training material and publication of IPGRI. |
| The International Treaty on Plant Genetic Resources for Food and Agriculture | http://www.fao.org/ag/cgrfa/itpgr.htm | Detailed text document about the treaty. |
| The International Union for the Protection of New Varieties of Plants (UPOV) | www.upov.int | All documents related to various meetings and technical committees of UPOV |
| CIAT Database on Plant Genetic Resources | http://www.ciat.cgiar.org/pgr/ | Contains information on 60,000 accessions mostly related to unimproved landraces of 720 species of bean, cassava and forages. |
| International and Regional Crop-related Networks Supported by FAO | http://www.fao.org/WAICENT/Faoinfo/Argicult/AGP/AGPS/cnet.htm. | Individual crop-related networks under The International Agreement on Gene banks |
| International Neem Network | http://www.fao.org/forestry/foris/webview/forestry/index.jsp/sitId=2021 and langId= | It comprises national institutions from 23 countries, provides information about activities of the network. |
| Seed and Plant Genetic Resources Service: AGPS | http://www.fao.org/waicent/faoinfo/agriclt/AGP/AGPS | Technical and policy advice on plant genetic resources, their conservation and use, seed and planting material. |
| Genetic Resources Action International (GRAIN) | http://www.grain.org | Rich website on issues related to PGR biodiversity and environment. |
| The Convention on Biological Diversity | http://www.biodiv.org | Details about the Convention: The Biosafety Protocol, Programs and Issues |
| Plant genetics Resources Abstracts (PGRA) | http://www.cabi-publishing.org/Abstract Database.asp/Subject Area= and PID=46 | Derived from the CAB Abstracts database provides access to 10-year backfile of nearly 34,000 records. It is updated weekly. |

**Table 16.2: National Database on Plant Genetic Resources**

| Institute/Source | Website | Prime Mandate/Information |
|---|---|---|
| Federal Information System on Genetic Resources (BIG) | http://www.bigflora.de/index_e.html | Genetic resources for cultivated and wild flora in Germany |
| The Institute for Plant Genetics and Crop Plant Research | http://www.ipkgatersleban.de/englisch | Gene bank having 100,000 accessions of cultivated plants |
| Centre for Genetic Resources (CGN) | http://www.plant.wageningenur.nl/about/biodersity/cgn/default.htm | Collections of agricultural and horticultural crops |
| Danida Forest Tree Seed Centre (DFSC) | http://www.dfsc.dk/ | Tropical and sub-tropical tree species |
| EUFORGEN Bibliography | http://www.ipgri.cgiar.org/ | Total 1674 reference on conservation and use of forest genetic resources |
| Information System on Genetic Resources (GENRES) | http://www.genres.de/genrese.htm | Cultivated and wild plants, forest plants, domestic animals, microorganisms and the policy framework |
| National Gene Bank (NGB) | http://www.ngb.sel/nmnmm | Plant species used in Nordic agriculture and horticulture including their wild relatives |
| National Plant Germplasm System of USDA | http://www.ars.grin.gov/npgs/ | Germplasm of plants, animals, microbes and invertibrates |
| NIAR database | http://www.gene.affrc.go.jp/plant/image/gbsys.html | PGR database in four plant groups, namely legume, vegetables, flowers and ornamentals and millets and forage crops |
| NIAR database | http://www.gene.affrc.go.jp/plant/image/flower.html | PGR database on flower |
| NIAR database | http://www.gene.affrc.go.jp/plant/image/vegetable.html | PGR database on vegetables |

**Table 16.3: Active Members of the CGIAR Consortium of International Agricultural Research Centre**

| Sl.No. | Active CGIAR Centres | Headquarters Location |
|---|---|---|
| 1. | Africa Rice Centre (West Africa Rice Development Association, WARDA) | Bouaké, Côte, d'Ivoire/Cotonou, Benin |
| 2. | Bioversity International | Maccarese, Rome, Italy |
| 3. | Centre for International Forestry Research (CIFOR) | Bogor, Indonesia |
| 4. | International Centre for Tropical Agriculture (CIAT) | Cali, Colombia |
| 5. | International Centre for Agricultural Research in the Dry Areas (ICARDA) | Beirut, Lebanon |
| 6. | International Crops Research Institute for the Semi-Arid Tropics (ICRISAT) | Hyderabad (Patancheru), India |
| 7. | International Food Policy Research Institute (IFPRI) | Washington, D.C., United States |
| 8. | International Institute of Tropical Agriculture (IITA) | Ibadan, Nigeria |
| 9. | International Livestock Research Institute (ILRI) | Nairobi, Kenya |
| 10. | International Maize and Wheat Improvement Centre (CIMMYT) | El Batán, Mexico State, Mexico |
| 11. | International Potato Centre (CIP) | Lima, Peru |
| 12. | International Rice Research Institute (IRRI) | Los Baños, Laguna, Philippines |
| 13. | International Water Management Institute (IWMI) | Battaramulla, Sri Lanka |
| 14. | World Agroforestry Centre (International Centre for Research in Agroforestry, ICRAF) | Nairobi, Kenya |
| 15. | World Fish Centre (International Centre for Living Aquatic Resources Management, ICLARM) | Penang, Malaysia |

**Change/Deleted Members of CGIAR**

| Sl.No. | Inactive CGIAR Centres | Headquarters | Change |
|---|---|---|---|
| 1. | International Laboratory for Research on Animal Diseases (ILRAD) | Nairobi, Kenya | 1994: merged with ILCA to become ILRI |
| 2. | International Livestock Centre for Africa (ILCA) | Addis Ababa, Ethiopia | 1994: merged with ILRAD to become ILRI |
| 3. | International Network for the Improvement of Banana and Plantain (INIBAP) | Montpellier, France | 1994: became a programme of Bioversity International |
| 4. | International Service for National Agricultural Research (ISNAR) | The Hague, Netherlands | 2004: dissolved, main programmes moved to IFPRI |

**Table 16.4: Designated National Active Germplasm Sites (NAGS) for Horticultural Crops**

| Crop(s) | Designated National Active Germplasm Sites |
|---|---|
| Arid fruit | ICAR-Central Institute for Arid Horticulture, Bikaner (Rajasthan) |
| Banana | ICAR-National Research Centre on Banana, Tiruchirapali (Tamil Nadu) |
| Cashew | ICAR-Directorate of Cashew Research, Puttur (Karnataka) |
| Citrus | ICAR-Central Citrus Research Institute, Nagpur (Maharashtra) |
| Grape | ICAR- National Research Centre for Grapes, Pune (Maharashtra) |
| Aonla, Bael and Litchi | ICAR- Research Complex for Eastern Region, Research Centre, Ranchi (Jharkhand) |
| Jack fruit | ICAR-Indian Institute of Horticultural Research, Hessaraghatta Bengaluru (Karnataka) |
| Medicinal and Aromatic plants | ICAR- ICAR-Directorate of Medicinal and Aromatic Plants Research, Anand (Gujarat) |
| Mango | ICAR-Central Institute for Sub-tropical Horticulture, Rehmankhera, Lucknow (Uttar Pradesh) |
| Mulberry | Central Seri-cultural Germplasm Research Institute, Hosur |
| Oil palm | ICAR-Indian Institute of Oil Palm Research, Pedavegi, West Godavari, (Andhra Pradesh) |
| Onion and Garlic | ICAR-Directorate of Onion and Garlic Research, Pune (Maharashtra) |
| Orchids | ICAR- National Research Centre for Orchids, Gangtok (Sikkim) |
| Ornamental crops | CSIR-National Botanical Research Institute, Lucknow (U.P.) |
| Plantation crops | ICAR-Central Plantation Crops Research Institute, Kasaragod (Kerala) |
| Potato | ICAR-Central Potato Research Institute, Shimla (Himachal Pradesh) |
| Spices | ICAR-Indian Institute of Spices Research, Kozhikode (Calicut), Kerala |
| Temperate horticultural crop | ICAR-Central Institute of Temperate Horticulture, Srinagar (J&K) |
| Tropical fruits | ICAR- Indian Institute of Horticultural Research, Hessaraghatta, Bengaluru (Karnataka) |
| Tuber crops | ICAR-Central Tuber Crops Research Institute, Thiruvananthapuram (Kerala) |
| Vegetables | ICAR- Indian Institute of Vegetable Research, Varanasi (Uttar Pradesh) |

# National Information System on Horticultural Crops

In India, information regarding plant collections, introductions, multiplications, conservations *etc.* can be retrieve from ICAR-NBPGR, New Delhi. This premier research organization is maintaining all the information on agriculturally important plants but for maintaining, handling and easy accesses, several National Active Germplasm Sites (NAGS) has been designated.

Designated National Active Germplasm Sites for horticultural crops are mentioned in Table 16.4.

## Reference

Bisby, F.A. 2000. The quiet revolution: Biodiversity informatics and the internet. *Science,* 289: 2309-2312.

# Plant Genetic Resources Related Glossary

**Access:** Acquisition (right to use) biological resources and/or knowledge associated thereof, and their derivatives or, as applicable, intangible components for purpose of research, conservation, biological prospecting and industrial application or commercial use, among others.

**Accession:** A distinct, uniquely identifiable sample of seeds/propagules representing a landrace, cultivar, breeding line or a population, which is maintained in a genebank for conservation and use.

**Accession number:** A unique identifier assigned to an accession, when it is held in a genebank. This number is never assigned again to another accession even after loss of the accession.

**Active collection:** A germplasm accession that is used for regeneration, multiplication, distribution, characterization and evaluation. Active collections are maintained in short-to-medium-term storage and usually duplicated in a base collection maintained in medium to long term storage.

**Adaptation:** The process by which individuals (or parts of individuals), populations, or species change form or function in such away to better survive under given environmental conditions.

**Additional declaration:** A statement that is required by an importing country to be entered on a phytosanitary certificate and which provides specific additional information on a consignment in relation to regulate pests.

**Adventitious regeneration:** Development of shoots or roots from regions (such as callus, leaf segments, intermodal segments) other than usual points of origin (buds, shoot bases) or embryos from sources other than zygotes.

**Agricultural biodiversity:** Also referred as 'agrobiodiversity', encompasses the variety and variability of animals, plants and micro-organisms necessary to sustain key functions of an agro-ecosystem, its structure and processes for, and in support of, food production and food security. It covers, inter alia, crop varieties, including forage and fodder plants and trees, animal breeds, including fish, molluscs, bird species and insects, as well as fungi, yeasts and micro-organisms such as algae and diverse bacteria.

**Agro-ecosystem:** The organisms and environment of an agricultural area considered as an ecosystem.

**Allele or Allelomorph:** One of several alternative forms of a gene occupying the same locus on a particular homologous chromosome. When the alternates exceed two, the alleles form a multiple allelic series. Multiple alleles arise by repeated mutations of a gene, each with different effects.

**Allelic frequency:** A measure of commonness of an allele in a population. It is used to describe the frequencies of polymorphic genes statistically correlated with temporal variation of environmental factors.

**Artificial seed:** *In vitro* micropropagation through somatic embryogenesis of meristem growing tips, shoot buds and protocorn-like bodies of orchids and subsequently incapsulation of somatic embryos in water soluble gel such as Na-alignate, gelrite, poly-oxyethylene or polyacrylamide.

**Asynapsis:** Failure of pairing of homologous chromosome during meiosis.

**Autogamy:** Self-fertilization.

**Avirulent:** Inability of a pathogen to produce a disease on its host.

**Axenic culture:** A culture without foreign or undesired life forms.

**Backcross:** A cross of a hybrid to either of its parent. In genetics, a cross of a hweterozygote to a homozygous recessive.

**Barcode:** A computerized coding system that uses a printed pattern or bars on labels to identify germplasm accessions. Barcodes are read by optically scanning the printed pattern and using a computer program to decode the pattern.

**Base collection:** The widest and most complete collection of germplasm accessions. It is stored for long-term, more for conservation *per se*, and used only for filling gaps in the active collection.

**Benefit sharing:** Means sharing of benefits arising from use of biological resources and associated knowledge based on prior informed consent and mutually agreed terms, with contracting party providing such resources.

**Biased sampling:** Deliberate collecting of desired phenotypes from the observable diversity in a population at a given time.

**Biochemical markers:** Genetic markers that can be classified on the basis of protein and/or isozymes.

**Biological diversity:** The variability among living organisms from all sources including, *inter-alia*, terrestrial, marine and other aquatic ecosystems and the ecological complex of which they are part. It includes diversity within species, between species and of ecosystems.

**Biological material:** Material containing genetic information and capable of reproducing itself or being reproduced in a biological system.

**Biological resources:** Include genetic resources, organisms or its parts, populations, or any other biotic component of ecosystems with actual or potential use or value for humanity.

**Biotechnology:** Any technological application that uses biological systems, living organisms, or derivatives thereof, to make or modify products or processes for specific use. It includes a number of individual techniques, *inter-alia*, recombinant DNA-molecule manipulation, protein engineering, cell fusion, nucleotide synthesis, monoclonal antibody use and production, product recovery, and unique fermentation techniques.

**Bulking:** Combining several samples in a collection and managing them as a single accession on the basis of similarity in phenotypic characters, habitat, geographic locations, *etc.*

**Callus:** A least differentiated, unorganized, proliferative mass of cells usually formed as a wound response.

**Caryopsis:** The fruit which develops from a single carpel with pericarp united to seed is called caryopsis *e.g.* seeds of cereals and grasses.

**Cell suspension:** Cells or aggregate of cells which multiply while suspended in liquid medium.

**Centre of diversity:** A geographical area where a plant species first developed its distinctive properties (in farmer's field or in the wild). A primary centre of diversity is the region of true origin (often referred to as the centre of origin), and secondary centres of diversity are regions of subsequent spread of a crop.

**Characterization:** Recording of highly heritable qualitative traits which are easily observed and expressed in all environments; helpful for distinguishing accessions from each other. Standard descriptors and descriptor states are normally used for description of traits.

**Clonal propagule:** The vegetative plant parts such as cuttings, grafts, or tissue culture that can be collected in plants for raising a new plant.

**Clone:** A group of genetically identical individuals that result from asexual, vegetative multiplication, and are, thereof, a genetic duplicate (true-to-type) of its parent.

**Coarse grid survey:** The preliminary survey and collecting wherein sampling is made at wide intervals (25-50 km, depending upon the terrain), with a few samples taken from each site (*vs.* fine grid survey)

**Collecting:** Includes identifying, locating, acquiring, organizing, cataloguing and post-collection handling the germplasm of importance as PGR.

**Collection:** A group of germplasm accessions maintained under defined conditions for a specific purpose.

**Collector:** A specialist who explore, survey and collects germplasm in the form of plans, seeds, propagules, *etc.* and records related information on diversity distribution, use, environmental features, *etc.*

**Conservation:** The practice that permits the perpetuation of a resource. Seed conservation refers to the practice of protecting the abundance of biological diversity of plant genes and seeds from loss and extinction, and ensuring the preservation of those genetic resources and seeds for the use of future generations.

**Contaminant:** Soil, fungal spores, fruiting bodies, plant debris, live/dead/dormant insects/stages thereof.

**Contamination:** Presence in a commodity, storage place, conveyance or container, of pests or other regulated articles, not constituting an infestation.

**Convention on Biological Diversity:** An international convention adopted in June, 1992 during the United Nations Conference on Environment and Development held in Rio de Janeiro, Brazil. According to Article 1 of the CBD, the Convention aims at *"the conservation of biological diversity the sustainable use of its components and the fair and equitable sharing of the benefits arising out of the utilization of genetic resources, including by appropriate access to genetic resources and by appropriate transfer of relevant technologies, taking into account all right over those resources and to technologies, and by appropriate funding"*.

**Core set:** A subset of accessions (~10 per cent) from the entire collection that captures most of the available diversity in the species which would represent with minimum of repetitiveness, the genetic diversity of a crop and its wild relatives.

**Country of origin (of a consignment of plant):** Country from where the plants were received.

**Critical difference (CD):** The difference between effects of different treatments and used to explain the variation due to treatment effect.

**Crop-specific survey:** Survey to collect maximum observable diversity in a single crop occurring in the region to be explored.

**Cross-pollination :** The process of pollen transfer from the anthers of one flower to the stigma of another flower by different means.

**Cryopreservation or cryostorage:** Storage of cells, tissues, organs or organisms, in a viable condition, at ultra-low temperatures usually below -100°C. in the genebank context, it is used for storage of buds, shoots tips, other meristematic and embryogenic tissue, non-orthodox seeds or their explants, orthodox seeds (in special cases), pollen and somatic embryos, in liquid nitrogen (-196°C) or its vapour phase.

**Cultivar:** A variety defined by a set of common features (morphological, phonological, *etc.*) of a plant that has been created or selected intentionally and maintained through cultivation.

**Database:** An organized set of interrelated data assembled for a specific purpose and held in one or more storage media.

**Descriptor:** An easily identifiable and measurable trait or characteristic of a plant observed in an accession, which is used to facilitate data classification, storage, retrieval and use.

**Descriptor list:** A collection of all individual descriptors of a particular crop or species.

**Description:** The process whereby a substance is released from or through the surface of another substance.

**Distribution:** The process of supplying samples of germplasm accessions to breeders, researchers, curators and other bonafide users.

**Documentation:** Presentation of record data on germplasm accessions in a standard format that describes structure, purpose, operation, maintenance and other requirements.

**Donor:** An institution or individual responsible for donating germplasm0

**Domestication:** The process by which plants, animals or microbes selected from the wild adapt to a special habit created for them by humans, bringing a wild species under human management. In a genetic context, the process in which changes in gene frequencies and performance arise from a new set of selection pressures exerted on a population.

**Dormancy:** The state in which certain live seeds do not germinate, even under optimal suitable conditions.

**Ecosystem:** A dynamic complex of plant, animal and micro-organism communities and their abiotic environment interacting as a function unit.

**Ecotype:** A population of a species that survives as a distinct group through environmental selection and isolation, maintains its genetic identity and usually is linked to one or more environmental factors in which it is favoured (mostly with reference to wild species).

**Endangered:** When used in the context of the IUCN Red List, a taxon is classified as 'Endangered' when there is very high risk of extinction in the wild, in the immediate future.

**Endemic:** Native to, and restricted to, a particular geographical region. Highly endemic species, those with very restricted natural ranges, or especially vulnerable to extinction if their natural habitat is eliminated or significantly disturbed.

**Equilibrium moisture content:** The moisture content at which seed is in equilibrium with the relative humidity of the surrounding air.

**Eradication:** Application of phytosanitary measures to eliminate a pest from an area.

**Ethnobotany:** All studies (containing plants) which describe local people's interaction with the natural environment.

**Evaluation:** The recording of quantitative traits whose expression is often influenced by environmental factors; it provides an assessment of the potential of germplasm for use in breeding/research.

*ex situ* **conservation:** The conservation of biological diversity outside its natural habitat. In case of PGR, it may compare facilities such as botanical gardens, seed genebanks, *in vitro* genebanks, cryo genebanks, field genebanks *etc.*

**Exotic accession:** Accession introduced or imported from a foreign country.

**Exotic species:** A species which is not native to the region in which it occurs.

**Explant:** An excised plant part implanted on a medium for growth or maintenance.

**Extinct:** The taxa (*e.g.* population, subspecies and species) not found even after repeated searches of known and likely areas.

**Fecundity:** Actual reproductive rate of an organism or population.

**Field:** A plot of land with defined boundaries within a place of production on which plants are grown.

**Fine grid survey:** An intensive sampling at sites (10-20 km) for survey and collecting (where earlier coarse grid survey has been done).

**Foreign seeds:** Seeds of weeds and crops other than the variety being tested.

**Fumigation:** Treatment with a chemical agent that reaches the plant or area wholly or primarily in a gaseous state.

**Gene:** The basic unit of heredity transmitted from generation to generation during sexual or asexual reproduction. Genetically, the term is used in relation to the transmission and inheritance of particular identifiable trait.

**Gene flow:** The movement of genes by pollen (dispersal of gametes), seeds (through zygote) and plants from one population to another. The exchange of genes (in one or both directions) at a low rate between two populations, due to the dispersal of gametes or individuals from one population to another, also called migration.

**Genebank:** Is a facility where genetic resources (genetic material) are conserved under suitable conditions to prolong their lives.

**Genepool:** The sum total of all the genes and combinations of the genes that occur in a population of organisms of the same species. The term genepool in reference to wild relatives and related taxa is classified into primary, secondary, tertiary depending upon crossability.

**Genetic conservation:** The collection, maintenance, storage and sustainable management of genetic resources aimed at ensuring their continued existence, evolution and availability for future generations. Also referred as 'gene conservation' and 'genepool conservation'.

**Genetic diversity:** The genetic variability (variety of genetic traits) within a population or a species, arising due to number and relative abundance of alleles. It can be assessed at three levels:(a): Diversity within breeding populations(b): Diversity between breeding populations(c): Diversity within the speciesGenetic diversity occurs at gene (molecular), individual, population, species and ecosystem levels.

**Genetic drift:** Random change in allele frequencies in a population from one generation to the next because of small population size, and not due to selection, migration or mutation. Smaller the population, greater the genetic drift, with the result that some alleles are lost, and genetic diversity is reduced. Minimization genetic drift is an important consideration for conservation of genetic resources.

**Genetic material:** Any material of plant, animal, microbial or other origin containing functional units of heredity.

**Genetic resource:** Genetic material of actual or potential economic, scientific or social value contained within and among species. In a domesticated species, it is the sum of all the genetic combinations produced in the process of evolution. The term includes modern cultivars and breeds, traditional cultivars and breeds, special genetic stocks (breeding lines, mutants, *etc.*), wild relatives of domesticated species and genetic variants of wild species.

**Genetic resource for food and agriculture:** A genetic material of actual or potential value for food and agriculture.

**Genetic shift:** Change in the performance of an accession/genotype if grown over a long period in areas outside its adaption.

**Genetic stocks:** Plants or populations generated and/or selected for genetic studies and represent a unique class of extremely valuable germplasm.

**Genetic variation:** The occurrence of differences among individuals of the same species, arising due to variation in alleles, genes or genotypes. Genetic variation is brought about by a change in genes, as distinct from difference due to environmental factors.

**Genome:** All the genetic material (DNA sequences) in a single (haploid) set of chromosomes of an organism. The genetic material inherited from either parent.

**Genomics:** Study of genomes of organisms, including intensive efforts to determine the entire DNA sequence and genetic mapping.

**Genotype:** The genetic constitution of an individual plant or organism as distinguished from its appearance or phenotype. The genotype interacts with the environment to produce the phenotype.

**Genotyping:** The process of identifying the genetic make-up of an organism by using molecular methods (DNA sequence level).

**Germ:** The embryo of a seed.

**Germination:** The biological process that leads to the emergence and development of a seedling from a seed, to a stage where the aspect of its essential structure indicates whether it is able to develop further into a satisfactory plant under favourable conditions. According to ISTA rules, only seedlings showing normal morphology are considered to have germinated.

**Germplasm:** The genetic material which forms the physical basis of heredity and which is transmitted from one generation to the next by means of germ cells. Often synonymous with genetic material. When applied to PGR, it includes plants in whole or in parts and their propagules including seeds, vegetative parts, tissue cultures, cell cultures, genes and DNA-based sequences that are held in a repository or collected from wild, and are utilized in genetic studies or breeding programmes for crop improvement.

**Germplasm exchange:** Mutual give and take of germplasm or plant genetic resources from all available.

**Glasshouse:** A glass building used for growing plants under controlled environmental conditions.

**Grain:** A commodity class for seeds intended for processing or consumption and not for planting.

**g x e interaction:** Part of phenotypic variation which is the result of interaction between genotype and environment.

**Habitat:** Part of an ecosystem with conditions in which an organism naturally occurs or can establish.

**Hard seed:** Seeds having an impermeable seed coat which hardly absorbs water and requires very long time for germination.

**Herbarium:** Herbarium is a collection of dried plant specimens identified and arranged using a standard system of classification. All parts of the plants (vegetative and reproductive) are represented in a herbarium specimen; information on locality habitat used *etc.* is recorded on the herbarium label.

**Heterozygosity:** Dissimilar allelic combination that provides genetic reserves and potential plasticity and permits a large proportion of the individuals to exhibit combination of phenotypic properties near the optimum.

**Hygroscopic:** Tendency to absorb water in the form of water vapour

**Import permit:** An official document authorizing importation of a commodity/resource in accordance with specified phytosanitary requirements, into the country from outside.

**Inbreeding:** Inbreeding is defined as the production of offspring through mating between related individuals.

**Inbreeding depression:** Loss in vigour due to mating among closely related individuals.

**Indigenous knowledge:** The local knowledge, unique to a given culture or society.

**Indigenous species:** A species that is assumed to be intrinsically part of the ecosystem of ones country, owing to having developed or arrived in the area long before record of such matters was kept, having arrived by natural means (unaided by human action) *etc.*

**Infestation:** Presence of a living pest of the plant or plant product concerned. Infestation includes infection.

***in situ* conservation:** The conservation of ecosystems and natural habitats and the maintenance and recovery of viable populations of species in their natural surroundings and, in the case of domesticated or cultivated species, in the surroundings where they have developed their distinctive properties. Until recenetly, it was narrowly used to describe conservation of genetic resources in their natural surroundings, normally protected from human interference. However, it is increasingly used to designate conservation on the farm, where genetic resources are developed, bred and maintained.

**Intellectual property rights (IPR):** Legal rights that are conferred to the owner of an intellectual creation. The IPR are granted by means of protection through appropriate legislation based on the type of creation that generally include patents, copyright, trademark, industrial design, geographical indications, trade secrets, protection of layout design of integrated circuits and protection of new plant varieties. The IPR protection entitles the owner or his assignee the exclusive right to fully utilize the invention/creation for commercial gain generally for a fixed period of time.

**Interception (of a pest):** The detection of a pest during inspection or testing of an imported consignment.

**intermediate seeds:** A category of non-orthodox seeds that are more tolerant of desiccation than recalcitrant seeds, though that tolerance is much more limited as compared to orthodox seeds, and they generally lose viability more rapidly at low temperature. Seed viability is influenced by initial seed quality, stage of maturation, and initial exposure after collection to conditions favorable for germination. Seeds can be stored in liquid nitrogen for long-term, with specialized handling and processing.

**International Plant Protection Convention (IPPC):** A multilateral treaty overseen by FAO to secure coordinated action to prevent and control the introduction and spread of pest of plants and plant product.

**International Standard for Phytosanitary Measures (ISPM):** Adopted by the members of IPPC, comprise standards, guidelines and recommendations for applying phytosanitary measures by WTO members under the Agreement on application of Sanitary and Phytosanitary Measures.

**International Treaty on plant genetic Resources for Food and Agriculture:** Popularly known as The Treaty, is a comprehensive international agreement in harmony with Convention on Biological Diversity which aims at guaranteeing food security through the conservation exchange and sustainable use of the world's PGRFA as well as the fair and equitable benefit sharing arising from its use. It has a Multilateral System of access and benefit sharing among those countries that ratify the treaty for food and forage crops listed in Annex 1 to The Treaty.

***In vitro* active genebank:** Represents *in vitro* germplasm collections under slow growth (by imposing growth restrictive conditions) to enhance subculture duration. The conservation is thus cost-effective and involves cyclic flow of material following subculture and recovery and facilitating distribution of cultures to users.

***In vitro* base genebank:** Represents *in vitro* germplasm collections under suspended growth (under cryopreservation) for long-term and the germplasm is normally not distributed to users.

***in vitro* collecting:** Plant where seeds or cuttings are not available to a collector or transport is not practical tissue collected using standard *in vitro* methods for obtaining and transporting germplasm. This methodology is generally practiced for collecting wild endangered species for propagation, tissue banking particularly for species with recalcitrant seeds or for species making a few or no seeds.

***In vitro* conservation:** Maintenance of germplasm in a relatively stable form under more or less defined nutrient conditions in an artificial environment.

***In vitro* culture:** The cultivation of plant organs or entire plants on artificial nutrient medium in glass or plastic containers. *In vitro* cultures are used for micropropagation, virus elimination via meristem culture and slow-growth storage. *In vitro* culture is also used as a preparatory phase to cryopreservation as well as for recovery phases after cryopreservation.

***In vitro* genebank:** A genebank involving conservation of vegetatively propagated genetic resources under tissue culture conditions involving normal growth, slow growth or suspended growth (under cryopreservation).

**Laminated aluminum foil packets:** Packets constructed of a laminate consisting of an inner layer of polyethylene, a middle layer of aluminum foil and an inner layer of polyester.

**Landrace:** A crop cultivar or animal breed that evolved with and has been genetically improved by traditional agriculturalists, but has not been influenced by modern breeding practices. It comprises population of individuals that have become adapted to specific local environmental conditions in which it evolved and is usually genetically heterogeneous.

**Liquid nitrogen:** It is the liquid produced industrially by fractional distillation of liquid air. It boils at -196°C and freezes at -210°C. Being a cryogenic fluid it causes rapid frost-bites and when insulated from ambient heat it can be stored and transported in a vacuum flask.

**Long term storage/conservation:** The storage of germplasm for a long period, such as in case collections and duplicate collections. The period of storage before seeds need to be regenerated varies, but is at least several decades and possibly a century or more. Long-term conservation takes place at sub-zero temperatures.

**Material Transfer Agreement (MTA):** A contractual arrangement (can legally binding) in commercial and academic research partnerships involving the transfer of tangible biological materials (such as germplasm, microorganisms and cell cultures) from a provider to a recipient. The MTA sets conditions for access, use, commercialization and sharing of benefits derived from the use of the materials provided.

**Medium-term storage/conservation:** The storage of germplasm in the medium-term such as in active and working collection, it is generally assumed that little loss of viability will occur for approximately ten years. Medium-term storage/conservation takes place at temperatures between 0°C and 10°C.

**Mini core set:** A subset of germplasm accessions comprising 10 per cent of the core set representing the total variability in the core set.

**Minimum seed standards:** Minimum specified limits of seeds of different crops in relation to purity, germination, moisture *etc.* to qualify as quality seed. These standards vary with a crop and class of seed.

**Moisture content (wet-weight basis):** The weight of free moisture divided by the weight of water plus dry matter, expressed as a percentage.

**Molecular markers:** Specific fragments of DNA that can be identified within the whole genome and are used for measuring genetic variation, trait identification, fingerprinting *etc.*

**Monitoring:** The periodic checking of accessions for viability and quantity, to ascertain whether to continue storing the accessions or to produce a fresh stock of seeds by regenerating or replacement from other source.

**Monitoring interval:** The period of storage between monitoring tests.

**Most original sample (MOS):** A sample of seeds that have undergone the lowest number of regeneration since the material was acquired by a genebank, as recommended for storage as a base collection. it may be a sub-sample of the original seed lot or a seed sample from the first regeneration cycle if the original seed lot required regeneration before storage.

**Multilateral system:** Established under the ITPGRFA for efficient, effective and transparent access to PGRFA and to share in a fair and equitable way the benefits arising from utilization of these resources on a complementary and mutually reinforcing basis.

**Multiplication:** The representative sample of an accession grown to multiply the quantitative of conserved material for distribution.

**Mutually agreed terms:** Terms used for access to genetic resources; involves two or more parties at least one party will be supplying the genetic resource and other acquiring them. The terms under which the recipient acquires genetic resources are likely to affect its freedom to supply the same resources or their derivatives to a third party.

**National identity:** The unique national identity number generated in the centralized database of ICAR-NBPGR, prefixed letter IC or EC denotes indigenous collection and exotic collection, respectively. It serves as the identification number in accession table of the centralized database.

**Native species:** A species which is a part of the original flora of the area, occurring naturally in the given area or region.

**Novel and unique trait:** Novel unique distinct and stable trait(s) which has been the basis for registration of germplasm.

**Non-orthodox seeds:** Seeds which are not desiccation-tolerant, do not undergo maturation drying and are shed at a water content in the range of 0.3-4.0 g g$^{-1}$. Any loss of water rapidly results in decreased vigour and viability, and seeds die at relatively high water contents. Non-orthodox seeds comprise two categories-recalcitrant and intermediate. Seeds do not survive drying to any large degree, and are thus not amenable to long-term storage (usually not more than few months), although the critical moisture level for survival varies among species.

**Orthodox seeds:** Seeds that can be dried to low moisture content and stored at low temperatures (without damage) to increase seed longevity.

**Outcrossing:** Controlled or natural mating among unrelated individuals of sexually propagated species.

**Passport data:** Basic information about the origin of an accession, such as details recorded at the collecting site, pedigree or other relevant information that assists in the identification of an accession.

**Pathogen:** A living microorganism such as a virus, bacterium or fungus that causes disease in another organism.

**Pathway:** Any means that allows the entry or spread of a pest.

**Pedigree:** The record of the ancestry of a genetic line or variety.

**Pest:** Any species, strain or biotype of plant, animal or pathogen regarded as injurious or harmful (*e.g.* weeds, insects, nematodes, fungi, bacteria, viruses).

**Pest risk analysis:** The process of evaluating biological or other scientific and economic evidence to determine whether a pest should be regulated and the strength of any phytosanitary measures to be taken against it.

**Phenophases:** An observable stage or phase in the life cycle of a plant that can be defined by start and end point. Phenophases generally have duration of a few days or weeks.

**Phenotype:** The external appearance of an organism, that results from the interaction of its genetic composition (genotype) with the environment.

**Phenotyping:** The method of determining differences in the appearance of an individual at phenotypic level.

**Phytosanitary certificate:** An official paper document or its official electronic equivalent, issued by an authorized officer at the country of origin of consignment or re-export, consistent with the model certificates of the IPPC, attesting that a consignment meets phytosanitary import requirements.

**Phytosanitary measures:** Any legislation, regulation or official procedure having the purpose to prevent the introduction and/or spread of quarantine pests, or to limit the economic impact of regulated non-quarantine pests.

**Phytotron:** An enclosed research greenhouse used for studying interaction between plants and the atmosphere.

**Plant debris:** Dried plant parts or pieces thereof other than seed.

**Plant exploration:** Plant exploration is the survey and search for plants with specific aim and intention.

**Plant genetic resources:** Genetic material of plants of value for present and future generations of people. In crops, it is the sum of all the genetic combinations produced in the process of evolution and breeding. The term includes modern cultivars, traditional cultivars, landraces, special genetic stocks (breeding lines, mutants, *etc.*), wild relatives of domesticated plants and genetic variants of wild species.

**Plantlets:** Rooted shoots derived from plant parts other than seed.

**Plant for planting:** Plants intended to remain planted to be planted or replanted.

**Planting (including replanting):** Any operation for the placing of plants in a growing medium, or by grafting or similar operations, to ensure their subsequent growth, reproduction or propagation.

**Plant quarantine:** All activities designed to prevent the introduction and/or spread of quarantine pests, or to ensure their official control.

**Pollination:** The process in which pollen is transferred from an anther to a respective stigma by pollinating agents such as wind, insects, birds, bats, or the opening of the flower itself.

**Population:** A group of individual plants or animals that share a geographic area or region and have common traits.

**Post-entry quarantine:** Quarantine applied to a consignment after entry.

**Primitive cultivar:** An early cultivated form of a crop species evolved from a wild population.

**Prior informed consent:** Recognizes indigenous people's inherent and prior rights to their lands and resources and respects their legitimate authority to require that third parties enter into an equal and respectful relationship with them based on the principle of informed consent.

**Propagule:** Any structure with the capacity to give rise to a new plant, whether through sexual or asexual (vegetative) reproduction. This includes seeds, spores, and any part of the vegetative body capable of independent growth if detached from the parent.

**Protected areas:** A clearly defined geographical space recognized, dedicated and managed through legal or other effective means to achieve the long-term conservation of nature with associated ecosystem services and cultural values.

**Providers and recipients:** Many include the government sector [*e.g.* government ministers, government agencies (National, regional or local), including those responsible for administration of national parks and government land]; commerce or industry (*e.g.* pharmaceutical, food and agriculture, horticulture, and cosmetics enterprise); research institutions (*e.g.* universities, genebanks, botanic gardens, microbial collections); custodians of genetic resources and traditional knowledge holders (*e.g.* Associates of healers, indigenous peoples or local communities, peoples' organizations, traditional farming communities); and other [*e.g.* private land owners(s), conservation group(s) *etc.*]

**Quarantine:** Official confinement of regulated articles (introduced germplasm) for observation and research or for further inspection, testing or treatment to ensure that it does not carry diseases or pests injurious to the importing country.

**Quarantine pest:** A pest of potential economic importance to the area endangered thereby and not yet present there, or present but not widely distributed and being officially controlled.

**Quarantine inspection:** Official examination of plants, plant products or other regulated articles to determine if pests are present or to determine compliance with phytosanitary regulations.

**Quarantine interception (of a consignment):** The refusal or controlled entry of an imported consignment due to failure to comply with phytosanitary regulations.

**Recalcitrant seeds:** Seeds that are not desiccation-tolerant; which do dry during physiological maturity and are shed at water contents in the range of 0.3-4.0 g g$^{-1}$. The loss of water rapidly results in decreased vigour and viability, and seed death at relatively high water contents, although the critical moisture level for survival varies among species. Hence, they are usually not amenable to long-term storage.

**Rare:** Taxa with few plants/small populations that are not at present endangered but are at risk of loss.

**Reference set:** An ideal set of best promising accessions for a given trait used for allele mining, association genetics mapping and cloning gene(s).

**Regeneration:** Grow-out test of a seed accession to obtain a fresh sample with high viability and numerous seeds. It is renewal of accession such as that it possess the same characteristics as the original sample and is a very critical operation in genebank management.

**Regeneration standard:** The percentage seed viability at or below which the accession must be regenerated to produce fresh seeds.

**Relative humidity:** The ratio of the amount of water vapor in the air at a specific temperature to the maximum amount that the air could hold at the temperature expressed as a per cent. It differs from absolute humidity, which is the amount of water vapour present in a unit volume of air, usually expressed in kilograms per cubic meter.

**Repatriation of germplasm:** Restoration of germplasm to the source countries.

**Rescue collecting:** Germplasm collecting in an area where genetic diversity available is under threats or loss due to natural (floods, earthquakes, tsunami *etc.*) or manmade (construction of dams, barriers *etc.*) factors.

**Safety duplicates:** A duplicate of a base collection stored under similar conditions for long-term conservation, but at a different location to insure against accidental loss of material from the base collection.

**Sample:** A part of a population used to estimates the characteristics of the whole.

**Sampling:** Drawing a part of the whole lot *i.e.* a few individuals out of a population mainly with an assumption that a sample represents the diversity available in the source population.

**Seed dormancy:** Block to the completion of germination of an intact viable seed under favourable conditions.

**Seed-borne:** The pest present on, in or along with a seed.

**Seed deterioration:** An irreversible degenerative change in quality of seeds in respects of viability, vigour and usefulness.

**Seed disinfection:** Ridding the seed surface of organisms which are potentially causing. This is carried either by physical or chemical treatment or it is process of treating the seed with a fungicide or insecticide to kill the pests/pathogens.

**Seed-transmitted:** The pest present in or with a seed and transmitted to the next generation of growing seedlings.

**Seed viability:** The capacity of seeds to germinate under favourable conditions.

**Selection:** Any process, natural or artificial, which permits a change in the genetic structure of population in succeeding generations. Natural selection is the differential reproduction and survival of individuals, such that poorly adapted offspring do not continue in the population. Artificial selection occurs when human manipulation yield specific traits deemed desirable.

**Self-pollination:** The process of pollen transfer from anthers to stigma of the same flower.

**Shoot meristem:** An undifferentiated mass of cells as a shiny dome like structure within the shoot tip, usually measuring less than 0.1 mm in diameter.

**Shoot tip:** Comprises shoot apical meristem along with one to several primordial leaves generally measuring 0.1 to 1.0 mm in diameter.

**Slow growth:** Refers to reducing growth rate of plant tissues without detrimental effects on them.

**Somatic embryogenesis:** Origin of plantlets from somatic cells by a developmental pathway similar to that of zygotic embryogenesis.

**Somatic embryo:** Embryos developed from somatic cells by a developmental pathway similar to that of zygotic embryos.

**Species:** A population of organisms.

**Species diversities:** Species represented as a measure of the diversity of the area.

**Standard Material Transfer:** The FAO has developed and adopted in 2006 a Standard Material Transfer Agreement (SMTA) as required for the implementation of the ITPGRFA. It specifies the conditions for exchange of germplasm between signatories (parties) with members of the ITPGRFA..

**Storage life:** The numbers of years that a seed/propagule can be stored before death occurs.

**Stratification:** It refers to incubation of seeds at a low temperature of 0-5°C over a moist substratum before transferring them to optimum thermal environment for germination.

**Subculture:** A process which involves cutting the tissue or explant into smaller segments (propagules) and transferring them to fresh culture medium.

**Sucker:** An off-shoot that develops from root of lower stem of a plant. It is propagule in ginger, turmeric, banana *etc.*

**Taxon (plural of taxa):** A formal taxonomic unit or category at any level in a classification (family, genus, species etc.)

**Taxonomy:** Classification of plants in hierarchical groups called taxa (singular taxon).

**Threatened species:** Any species which is likely to become endangered within the near future, all throughout or a significant portion of its range.

**Tissue culture medium:** A solid or liquid combination of nutrients and water which meet the requirements of cells tissue or organs grown *in vitro*.

**Trait:** A recognizable quality or attributes resulting from interaction of a gene or a group of genes with the environment.

**True seed:** A true seed is reproductive unit that develops from ovules or it is a fertilized mature ovule.

**Variety:** A recognized division of a species, next in rank below sub-species; it is distinguishable from another variety by a set of heritable characters (such as flower colour, leaf colour etc.). the term is considered synonymous with cultivar.

**Viability:** A test on a sample of seeds from an accession that is designed to estimate the viability of the entire accession.

**Voucher:** Plant material such as herbarium specimen, seed sample, and economic product preserved for future reference.

**Vulnerable:** When used in the context of the IUCN Red List a taxon is classified as 'Vulnerable' when facing a high risk of extinction in the wild in the immediate future.

**Weed:** A plant that is growing where it is not 'wanted' generally used to describe plants which colonize readily and compete for resources with a crop.

**Wild relatives:** Plant species that are taxonomically related to crop species and serve as potential sources for genes in breeding of new variety of those crops.

**Wild species:** Species occurring in relatively undisturbed or less disturbed habitat and areas that have not been subject to breeding to alter them from their native (wild) state.

**Working sample:** A reduced sample taken from submitted sample in the laboratory on which one of the seed quality tests is made.

*Source: ICAR-NBPGR, 2016 (eds.);*
*and Bhardwaj, 2012 (ed.)*

## Reference

1.: ICAR-NBPGR. 2016. Guidelines for Management of Plant Genetic Resources. ICAR-National Bureau of Plant Genetic Resources, New Delhi, India. 142+xxivp.

# Index

# C

Cantaloupe 336, 344

*Capsicum frutescens* 36, 265, 300, 333, 379, 563

Centre of diversity of vegetable crops 16

Centre of origins 2, 8, 9, 10, 11, 12, 13, 14, 17, 18, 20, 22, 37, 38, 86, 205, 318, 351, 355, 366, 584

CGIAR 62, 72, 103, 138, 182, 197, 200, 207, 237, 242, 243, 244, 253, 386, 387, 390, 393, 394, 479, 488, 494, 503, 510, 579, 580

Characterization and documentation of genetic resources 497

Characterization of germplasm 125, 172

Characterization of tuber crops 171

Chemical treatment 591

*Chenopodium album* 3, 36, 40, 563

*C. histella* 3

CIAT 62, 171, 183, 237

*C. lanatus* var. *citroides* 2

Climate change 346, 355, 507

Cluster analysis 173, 426, 464, 466

Clustering of germplasm 172

Coarse and fine grid sampling 90

Cold tolerance 327

Collection of fruit crops 89

Collection of vegetative materials 88

Collection of wild relatives 89

Community genebank 380

Community management 245

Community seed banks 380, 395

Conservation of biological diversity 478

Conservation of tuber crops 354, 396

Copyright 523, 587

Core collection 110, 117, 121, 122, 125, 172-188, 462, 464

Core sets 121, 175, 176, 178

Crop advisory committees 111

Crop diversity 9, 21, 236, 245, 355, 499

Crop genepool 234, 572

Crop networking 181

Crop wild relatives 82, 115, 304, 343, 573

Cryobank 369, 385, 397

Cryopreservation 242, 257, 262, 305, 357, 361-372, 393-397, 403, 410-417, 428-431, 435, 439, 470, 588

*C. sessilifolia* 3

Cultivated genetic resources 27, 29

Cytoplasmic-genetic male sterility 331

Cytoplasmic genetic system 331

# D

Data access 252, 253

Database management 139, 143, 248, 253-257

Data integrity 251

Data processing 258

Data recording 101, 124

Delineate core 177

De-registration 261, 276

Descriptors and descriptor states 119, 120, 250, 258, 584

Desiccation tolerance 362, 365, 368, 413

Development of core collection 175, 176

DIP Act 1914 73

Directional selection 53, 54

Discriminate analysis 179

Disease resistance 27, 32, 33, 50, 61, 63, 88, 110, 115, 117, 122, 298, 299, 301, 307-317, 342, 389, 441, 448, 454-456, 467, 521, 522

Disruptive selection 53, 54

Distant crosses 304

Distribution pattern 27

Diversification 335

Diversity array technology 452

Diversity in agroecological regions 19

DNA based molecular markers 172, 460

DNA chip 450

DNA collection 96

DNA conservation 358, 363, 418

Documentation 374, 375

Domestication syndrome 51, 52

Donor parent 298

Drought 327

Drought tolerance 327

Drying of samples 102

DUS guidelines 511

# E

Ecosystem services 5, 40

Embryo culture 37, 59, 301, 332, 404, 406, 435

Encapsulation 365, 369, 393, 412, 413

Environmental selection 585

Essentially derived varieties 276, 533

Evaluation of germplasm 110, 122, 125, 135, 139, 193, 241, 382

Evaluation of vegetable genetic resources for abiotic stresses 173

Evolution of crop plants 46

Excessive moisture tolerance 327